SCANNING ELECTRON MICROSCOPY

Edited by **Viacheslav Kazmiruk**

Scanning Electron Microscopy
Edited by Viacheslav Kazmiruk

Published by InTech

Janeza Trdine 9, 51000 Rijeka, Croatia

Publishing Process Manager Daria Nahtigal

Technical Editor Teodora Smiljanic

Cover Designer InTech Design Team

Additional hard copies can be obtained from orders@intechopen.com

Scanning Electron Microscopy, Edited by Viacheslav Kazmiruk
p. cm.
ISBN 978-953-51-0092-8

Meet the editor

Dr Viacheslav Kazmiruk is Head of the laboratory of scanning electron microscopy at the Institute of Microelectronics Technology and High Purity Materials, Russian Academy of Sciences (IMT RAS), where he is working since its foundation in 1984. From 1983 to 1990, he was Head of the Joint Department of IMT RAS and Scientific Instrument Factory of RAS, responsible for mini SEM 100™ design and manufacturing. Between 1987 and 1997, he was Chief Designer of SEMs and e-beam Lithography System in the USSR and Russian Federation. From 1995 – 2008, he worked on the development and application of the new SEMs MicroScan MS10™ and MS20™. He is an author and coauthor of 75 papers, most having to do with SEM instrumentation as well as signal formation in a SEM within different operation modes. At the present time, he is working on the development of high resolution low voltage e-beam systems for both defect inspection and lithography.

Contents

Preface

For more than 70 years since its invention, Scanning Electron Microscopes (SEMs) have evolved from relatively simple devices with a resolution of 50 nm to sophisticated computer – controlled systems with wide analytic potentialities and resolutions of 1-2 nm to sub-nm in some particular cases.

At the present time, it is hard to conceive of a science field which would not employ methods and instruments based on the use of fine focused e-beams. Well instrumented and supplemented with advanced methods and techniques, SEMs provide possibilities not only of imaging but quantitative measurement of object topologies, local electrophysical characteristics of semiconductor structures and performing elemental analysis. Moreover, a fine focused e-beam is widely used for the creation of micro and nanostructures by e-beam lithography and Electron Beam Induced Deposition (EBID).

The book's approach covers both theoretical and practical issues related to scanning electron microscopy. The book has 41 chapters divided into six sections: Instrumentation, Methodology, Biology, Medicine, Material Science, Nanostructured materials for Electronic Industry, Thin Films, Membranes, Ceramic, Geoscience, and Mineralogy. The first section considers method strategies and instrumentation aspects of scanning electron microscopy, and the other five sections are devoted to SEM application in different fields of present day science. Each chapter, written by different authors, is a complete work which presupposes that readers have some background knowledge on the subject.

We would like to thank all the authors for their generous contributions. On behalf of the authors, we gratefully acknowledge the efforts of Daria Nahtigal on coordination of this project, as well as of all who made this publication possible. It is our hope that the book provides a deep insight into fields related to Scanning Electron Microscopy, benefitting both beginners, experts, and prospective readers.

Viacheslav Kazmiruk
Head of Laboratory of Scanning Electron Microscopy,
Institute of Microelectronics Technology, Russian Academy of Sciences
Russia

Part 1

Instrumentation, Methodology

1

Gaseous Scanning Electron Microscope (GSEM): Applications and Improvement

Lahcen Khouchaf
Université Lille - Nord de France, Ecole des Mines de Douai, Douai, France

1. Introduction

The imaging and the microanalysis of hydrated and insulating materials using electron beam probe methods (Conventional SEM, Auger Electron Microscopy,...) are very limited by the necessity to keep the sample under high vacuum and the presence of the charge effect. In this case the sample must be coated except when the experiment is performed at very low energy in order to avoid the charge phenomenon. Studies of vegetable and biological samples are almost impossible without degradation. In Conventional SEM (Pressure = 10^{-5} mbar in the specimen chamber) image quality and microanalysis results are strongly related to the size of the electron beam, the accelerating voltage and the nature of the sample. A large description of different aspects on SEM/EDS exists in the literature (Newbury et al, 1986; Goldstein et al, 1992).

In order to overcome the high vacuum in the specimen chamber different types of microscopes with the possibility to introduce different gases inside the sample chamber are now available (Danilatos, 1980, 2009, Carlton, 1997, Wight, 2001). Depending on the pressure value in the specimen chamber different names are given in the literature such as ESEM: Environmental Scanning Electron Microscope, LVSEM: Low Vacuum Scanning Electron Microscope, HPSEM: High Pressure Scanning Electron Microscope, VPSEM: Variable Pressure Scanning Electron Microscope, CPSEM: Controlled pressure Scanning Electron Microscope and depending on the maximum pressure attainable in the specimen chamber (Danilatos 1988, Khouchaf & Vertraete, 2002, 2004; Khouchaf et al., 2006, 2007, 2010, 2011; Kadoun et al, 2003; Gilpin, 1994; Carlton, 1997; Doehne, 1997; Newbury 2002; Gauvin, 1999; Bolon, 1991; Wight, 2001). But all these microscopes differ from CSEM by the capability to introduce the gas as an environment unlike High vacuum in CSEM and the use of gaseous detection system such as Gaseous Secondary Electron Detector (GSED). Indeed, all these microscopes may be called Gaseous Scanning Electron Microscope (GSEM).

Unlike CSEM, with GSEM image quality and microanalysis results are strongly related to the size of the electron beam, the accelerating voltage, the nature of the sample and depend on the pressure in the chamber and the kind of gas used. GSEM showed the enormous use in several fields using materials. The electron beam scattering tends to decrease the resolution. Different correction methods were developed (Bilde-Sorensen et al, 1996; Doehne, 1996-1997; Le Berre et al, 1997; Gauvin et al, 1999) but weren't satisfactory for many reasons such as the time and the difficulty of implementation.

In this chapter different applications using GSEM will be given in the first part. In the second part an introduction of some physical phenomena related to the scattering of the primary electron beam with the gas and their consequence on the image quality and on the microanalysis results is given. The focus here is to present the potential, the limitation and some way to optimize the use of GSEM.

2. Applications of ESEM

GSEM allows imaging and analysis of many types of materials without any preparation with the presence of a gaseous environment. Gas atoms or molecules interact with the primary electron beam and produce positive ions. The presence of the positive ions allows neutralization the negative charge on the surface of the insulating sample. Gas serves also for detection (Danilatos, 1988).

Different types of gases may be introduced such as: N_2, O_2, Ar, He, H_2O...... As an illustration, we give two examples. By introduction of gases, vacuum incompatible materials, and dynamic surface modification may be studied. At pressures in excess of 500 Pa, water can be condensed in situ enabling characterization of hydrated materials.

In this study, the experiments were performed in environmental 'wet' mode using an Environmental Scanning Electron Microscope (ESEM) « ElectroScan 2020 » equipped with EDS Microanalysis system « Oxford Linkisis ». The electron source is a tungsten filament. The energy of the electron beam used was 20 kV with an emission current of 49 μA. The condenser is also fixed at 43 % value and the diameter of the projection aperture is 50 μm. Secondary electrons were detected using a long gaseous secondary electron detector (GESD)at a working distance of 19 mm is used in order to reduce the skirt beam phenomena. The chamber pressure is varied by introducing gas.

The sample is embedded in epoxy resin and polished.

For the estimation of the unscattered fraction and the skirt radius an ESEM electron flight simulator software was used (Electron Flight Simulation Software, version 3.1-E).

2.1 Observation of vegetable

Plant material is insulating and has a fragile structure. Its observation using an electron beam is a delicate operation. Its structure is degraded under the electron beam. Some plants contain Stomata (or epidermis) the structures deposited on the outer leaf skin layer. They consist of two cells, called guard cells that surround a tiny pore called a stoma. Stomata allow communication between the internal and external environments of the plant.

Figure 1a shows an image of a plant after the coating operation in order to neutralize the negative charges. The image is obtained at 20 kV and high vacuum using the Gaseous Secondary Electron Detector (GSED) detector. This specimen is formed by stomata which are tiny openings or pores, found mostly on the underside of a plant leaf and used for gas exchange. The image below shows an elongated and irregular structure.

The same observation is made without coating under a gaseous environment of water vapor (Fig.2b). In this case, the morphological aspect and the structure of the plant are very different compared to the image in Fig. 1a. The sample kept its structure. We can observe the stomata in their natural state. In order to study mechanism of exchange in the plant, it is necessary to keep the sample in its natural state. This is possible by using GSEM. It is interesting to underline that despite the scattering phenomena, the image kept its quality. That will be explained below.

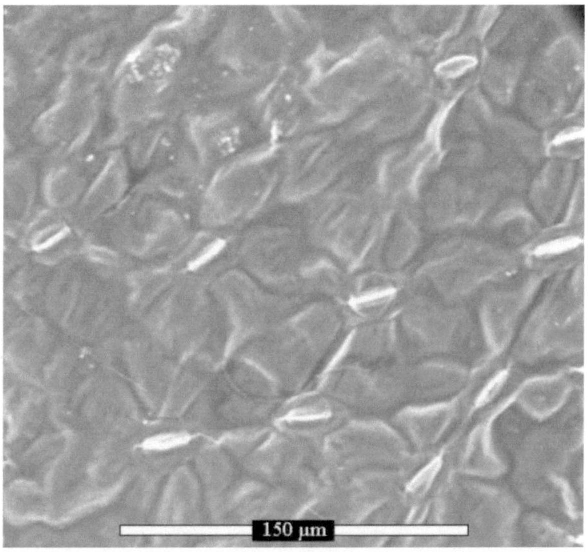

Fig. 1a. ESEM micrographs of a plant before the coating process.

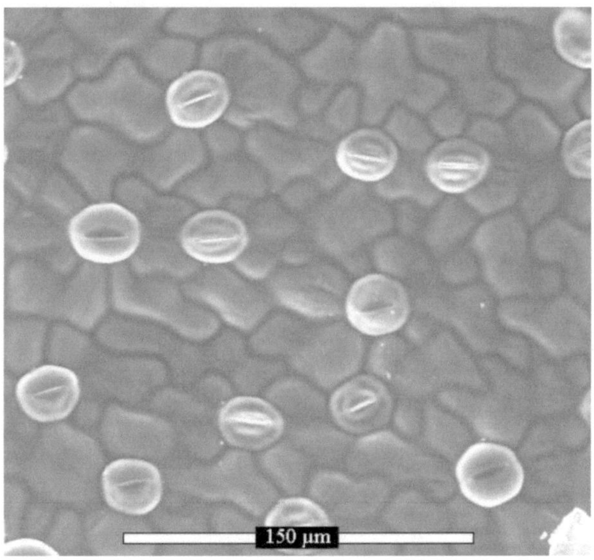

Fig. 1b. ESEM micrographs of a plant before the coating process.

2.2 Detection of calcium potassium inside SiO_2 Framework

Figure 2 shows an example of a flint aggregate subjected to attack by Alkali-Silica Reaction (ASR). ASR is a physicochemical process which takes place during the degradation of concrete. Observation of the ASR effects using Scanning Electron Microscope (SEM) is important because the attack of the aggregate is heterogeneous on the microscopic scale. When using CSEM, sample preparation prior to imaging is required, which may lead to an alteration of the true surface morphology or even the creation of artifacts.

The use of GSEM overcomes these problems and gives direct imaging of the samples in their natural state. The image in Fig. 2 shows the presence of different degraded zones (1 to 8) affected by chemical reaction. These zones have different micronic sizes and under the given conditions, the effect of the beam skirt is not the same. If the volume of the generation of X-rays is lower than the size of the zone then the effect of the skirt may be neglected. If the volume of the generation of X-rays is higher than the size of the zone then the effect of the skirt may be taken into account.

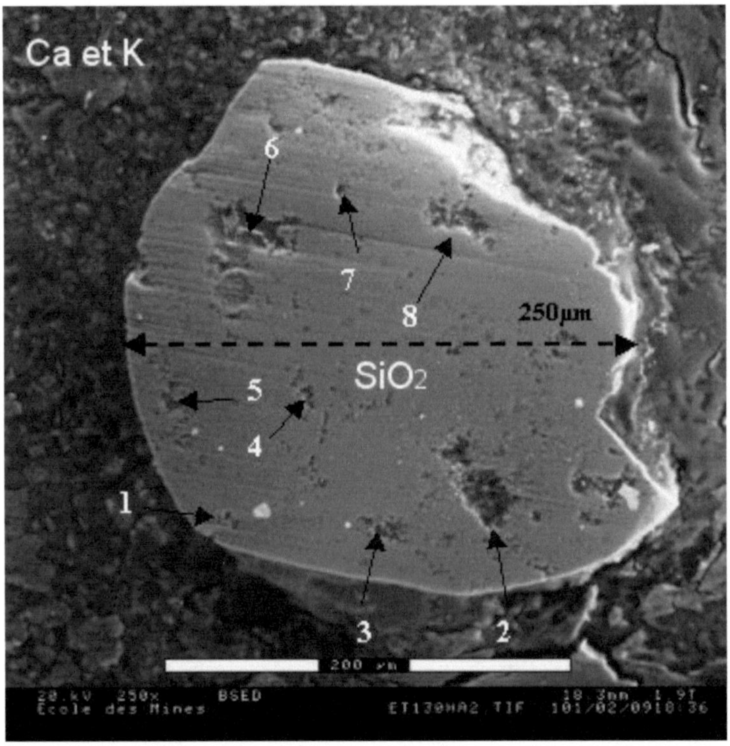

Fig. 2. ESEM micrographs of flint aggregate after reaction at (GSED, P=532 Pa).

The images below (Figures 3) show the effect of calcium cations during the degradation of concrete. Figure 3a presents flint aggregate after 30 hours of reaction with the presence of calcium and potassium and figure 3b without calcium. With the presence of calcium, it is interesting to note that the reaction has not finished particularly in the center of the aggregate (Fig. 3a).

However, when the calcium is removed, the grain is fragmented to small grain clearly separated by showing that the mechanism is different when calcium is present (Fig. 3b). Despite a high pressure (532 Pa) and high accelerating voltage 20 kV, again the resolution of the image is not bad.

Fig. 3. ESEM micrographs of (a) flint aggregate after reaction at (GSED, P=532 Pa) with the presence of calcium, (b) flint aggregate after reaction (GSED, P=545.3 Pa) without calcium.

3. Description of the beam skirt

Experiments above were performed with insulating materials without any preparation and without the coating procedure by introducing a gaseous environment into the specimen chamber. A part of primary electron beam interacts with the atoms or molecules of the gas.

The average collision number with particle gas per electron is given by the equation below:

$$m = \sigma_t \times n \times L \tag{1}$$

Where

σ_t: scattering cross section is specific to each gas molecule
n: gas particle number/volume
L: Working distance (distance between the final aperture PLA1 and the surface of the sample see Figure 4). The Gas Path Length GPL is introduced and corresponds to the distance that electrons have to travel through the gas to reach the sample. m may be expressed as below (Danilatos, 1988):

$$m = \frac{\sigma_T . L . P}{k . T} \tag{2}$$

where σ_T is the total cross section of the gas, L the gas path length, P the gas pressure, k the Boltzmann constant and T the temperature.

Using the number m it is possible to define three different scattering regimes corresponding to:

Minimal scattering: $m < 0.05$
Partial scattering: $0.05 < m < 3$
Plural scattering: $m > 3$

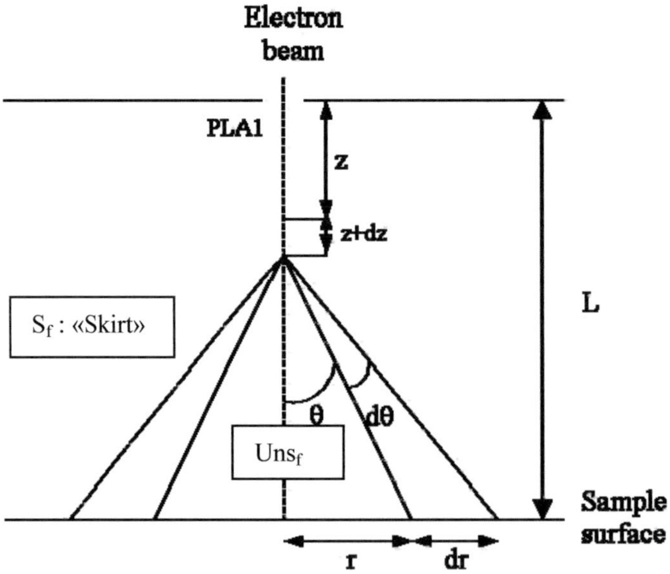

Fig. 4. An electron after PLA1 aperture of the ESEM, moves along the axis of PLA1, undergoes a collision at a distance between z and z + dz in an angle of θ+ dθ; it is then scattered and arrives at the surface of the sample in an annulus between r and r + dr.

After interaction between electron and gas, the primary electron beam is divided into two parts (Fig. 4) called "scattered fraction: S_f" which corresponds to the elastic scattering by the gas atoms or molecules and a second part called "unscattered fraction: Uns_f". The "unscattered fraction" of the electron beam can be written by the equation below when a simple mode of scattering is considered (Danilatos, 1988):

$$Uns_f = \exp(-\frac{P \times L \times \sigma_t}{k \times T}) \qquad (3)$$

P: Pressure in the specimen chamber
L: Working distance (WD, distance between the final aperture PLA1 (Fig. 1) and the surface of the sample). In this study the Gas Path Length GPL is introduced and corresponds to the distance that electrons have to travel through the gas to reach the sample.
σ_t: total scattering cross section is specific to each gas molecule
k: Boltzman constant
T: Temperature in Kelvin.

Elastic scattering leads to the enlargement of the primary electron beam to form a skirt producing the generation of X-rays which are not representative of the zone of interest for X-ray microanalysis. Different correction methods have been developed in order to take into account the contribution of the skirt (Bilde-Sorensen et al, 1996; Doehne, 1996-1997; Le Berre et al, 1997; Gauvin et al, 1999). Up to now these methods have not been successful. Danilatos, 1988) introduced the radius rs which represents the radius containing 90% of the incident beam) as below:

$$r_s = \frac{364.Z}{E}.\left(\frac{P}{T}\right)^{1/2}.\,GPL^{3/2} \qquad (4)$$

where r_s is the skirt radius, Z the gas atomic number, E the incident beam energy, P the pressure, T the temperature and GPL the gas path length.

Considering the condition used in $ 2.2, the simulations of the electron beam scattering were performed using the Electron Flight Simulator software (Figure 5a). In this case the unscattered fraction is about 85.7% with a rs of 26 μm. Worst case X-ray Gen radius means the approximate region where X-ray signals will be generated and given just as indication. Based on our previous sudy using helium gas (Khouchaf et al, 2004, 2007, 2011), the same simulation under helium gas (Fig1.b) leads to Uns_f of 97.6%, rs < 1 μm showing a good improvement of the conditions with a gas having a low average atomic number.

Some authors (John F. Mansfield) have suggested that quantitative analysis is possible with ESEM (with water vapour as the standard gas) only under very restrictive conditions such as : short working distances between 6 mm and 7.2mm, gas path length between 1.2mm and 2.2mm in the 70 to 350Pa range at high accelerating voltage of 30 kV. From these conditions it is easy to notice that the values of working distance, gas path length and pressure must be very low when the accelerating voltage decreases. Using conditions close to that given by Mansfield we perform a simulation by means of Electron Flight Simulator (Fig. 6a and 6b).

At P= 70 Pa and GPL=1mm, Uns_f and r_s are close to 95.5% and <1 μm respectively. When the pressure increases to 350 Pa and GPL to 2mm (Fig. 6b), Uns_f decreases to 93.5% and r_s increases to 2 μm.

Parameters suggested by Mansfield consider a high accelerating voltage of 30 kV which is conform with excitation of heavy or metallic elements and which we can study by using

CSEM. In addition, high voltage leads to large volume of interaction and a degradation of the resolution by generation of unwanted x-rays. More applications in GSEM require a low accelerating voltage which is possible by recent microscope with a FEG gun.

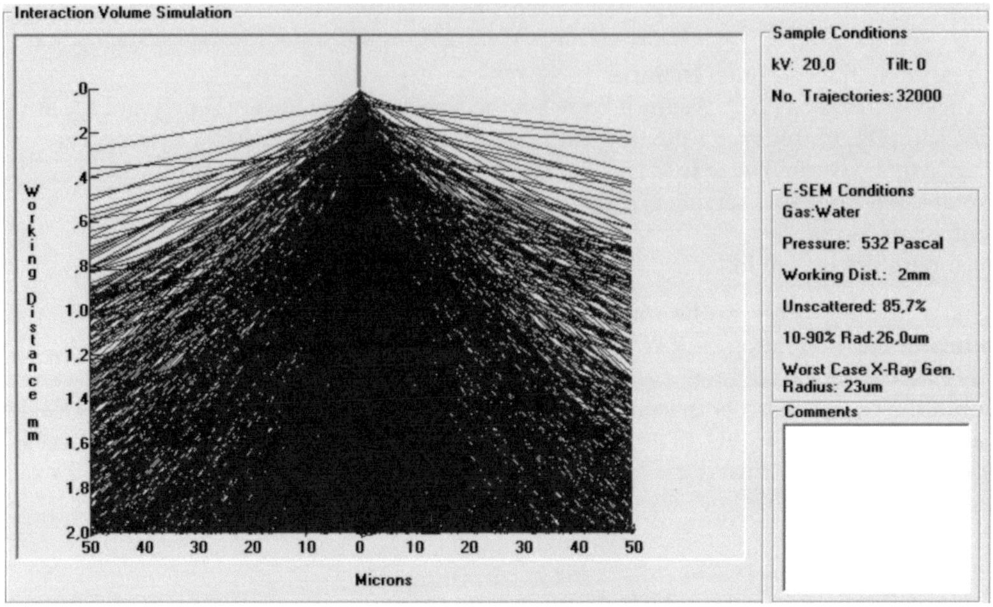

Fig. 5a. Monte Carlo simulation using Electron Flight Simulator of the electron beam scattering under water vapor, V=20kV, P= 532 Pa, GPL = 2mm.

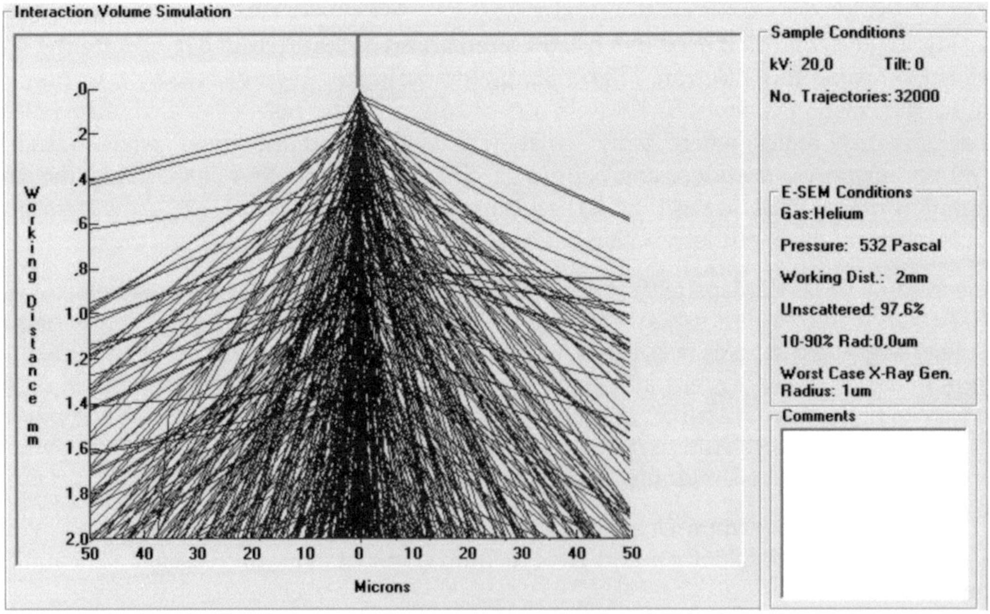

Fig. 5b. Monte Carlo simulation using Electron Flight Simulator of the electron beam scattering under helium, V=20kV, P= 532 Pa, GPL = 2mm.

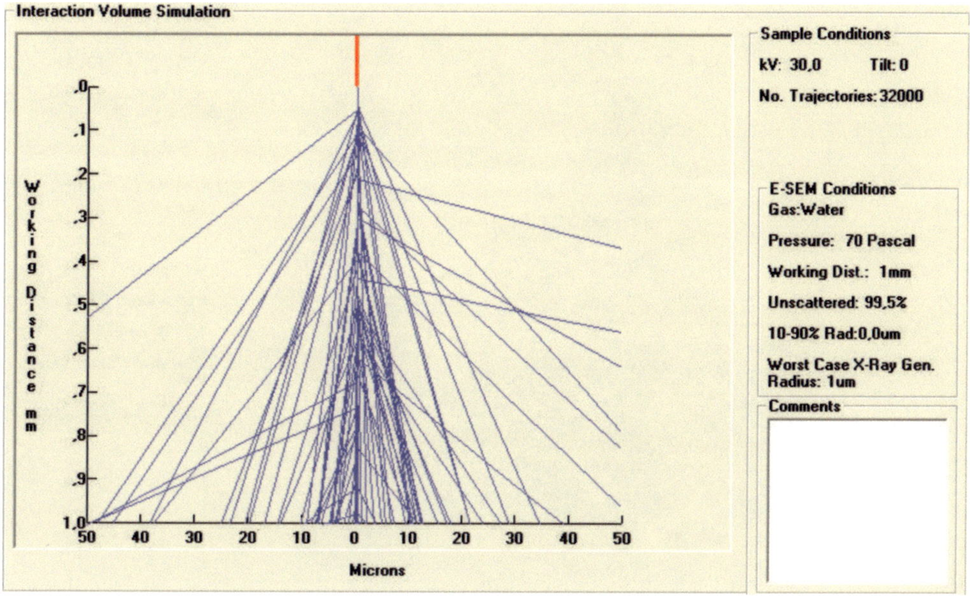

Fig. 6a. Monte Carlo simulation using Electron Flight Simulator of the electron beam scattering under water vapor, V=30kV, P= 70 Pa, GPL = 1mm.

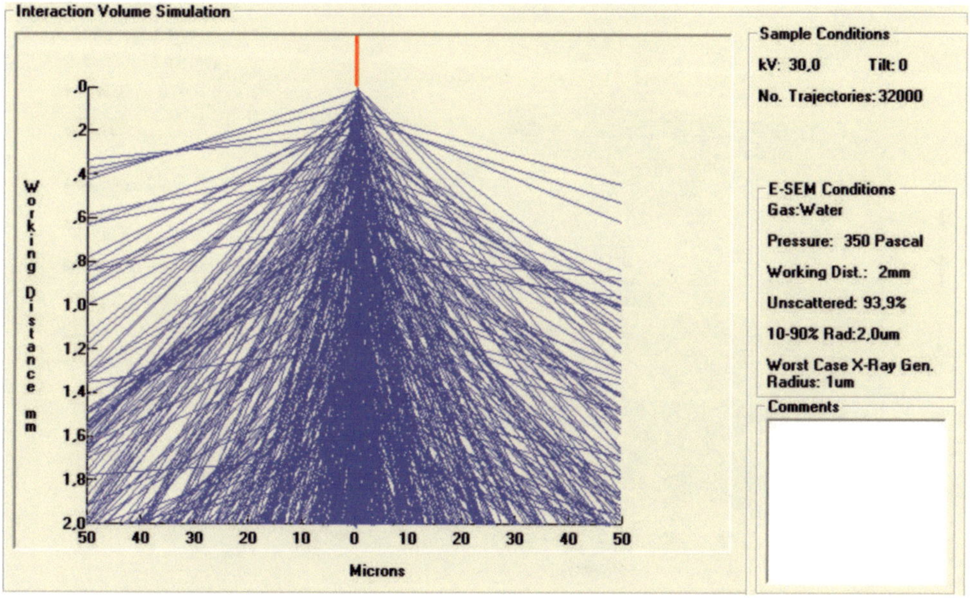

Fig. 6b. Monte Carlo simulation using Electron Flight Simulator of the electron beam scattering under water vapor, V=30kV, P= 350 Pa, GPL = 2mm.

Let us consider the conditions above at low accelerating voltage of 5 kV under water vapor (Fig7a and 7b). At P= 70 Pa and GPL=1mm, Uns_f and r_s are close to 93.2% and 12 μm respectively, but at P= 350 Pa and GPL=2mm, Uns_f and r_s are close to 47% and 109.3 μm respectively . It is easy to conclude that the quality of the results will be affected.

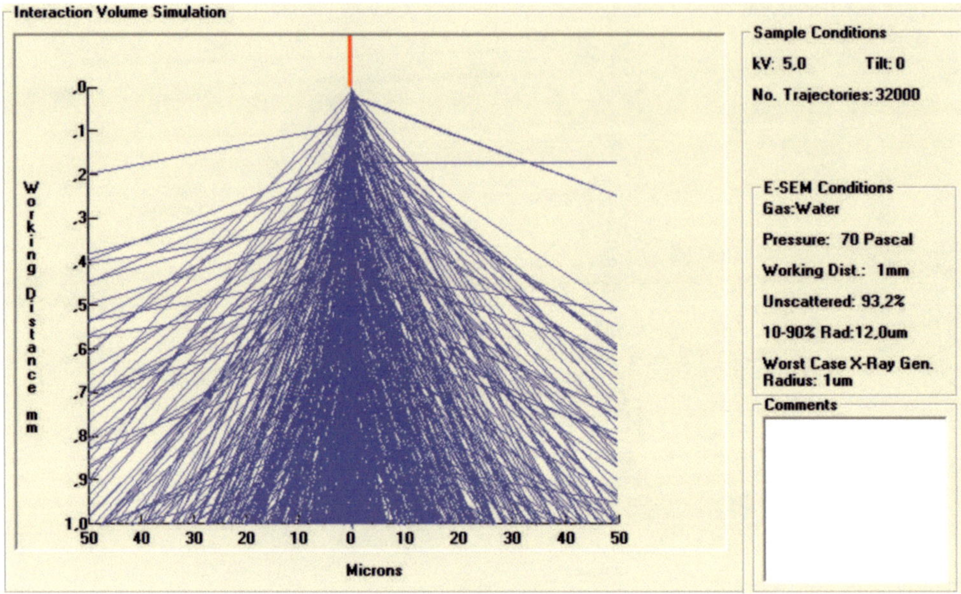

Fig. 7a. Monte Carlo simulation using Electron Flight Simulator of the electron beam scattering under water vapor, V=5kV, P= 70 Pa, GPL = 1mm.

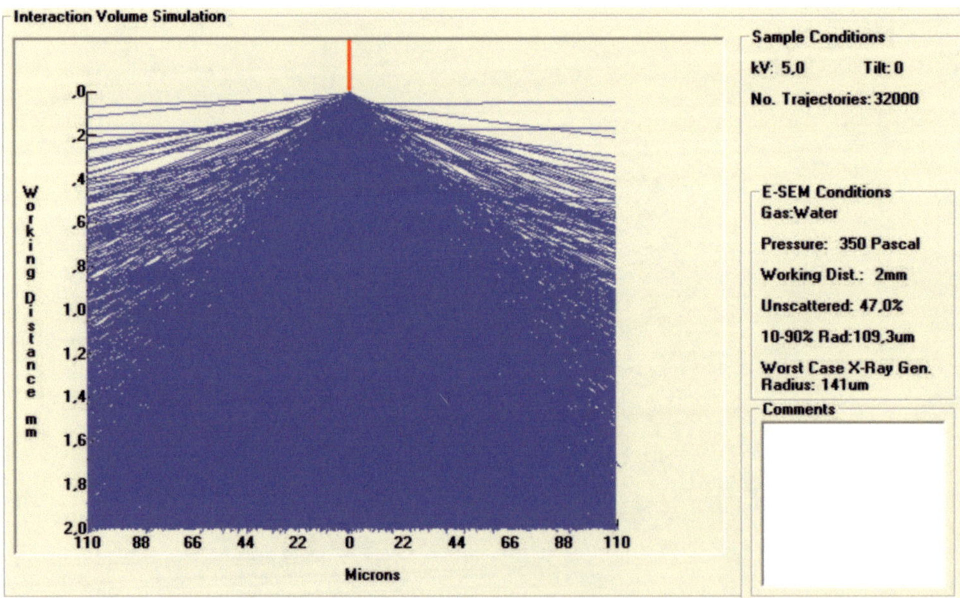

Fig. 7b. Monte Carlo simulation using Electron Flight Simulator of the electron beam scattering under water vapor, V=5kV, P= 350 Pa, GPL = 2mm.

As given by equation 4, the value of rs depends on the gas introduced, the incident energy, the pressure, the temperature and the working distance. Indeed, the good way to minimize the beam skirt phenomena is to optimize different parameters used during X-ray microanalysis. With equation 3 this leads to choose a gas with a low average atomic number, to increase the incident beam energy, to reduce the pressure and the gas path length, to increase the temperature. Unfortunately, these conditions lead to a significant limitation to use GSEM. For example for ESEM, the standard gas is water vapor and the best results are obtained by helium (ref Khouchaf). Increasing the incident beam leads to a minimization of the beam skirt by decreasing the resolution and degrading the fragile materials.

Based on the conditions and parameters such as the bem energy, the pressure, the gas, the GPL it's not sufficient to define the limit of the use of GSEM. In order to obtain the best results it is also necessary to take into account the value oft he average number of scattering events per electron m. The best results will be obtained with a minimal scattering regime corresponding to $m<0.05$. This suggests the use a gas with a low average atomic number (Khouchaf et al, 2011).

Unfortunately most new microscopes use gases with a high average atomic number such as (N_2, air, H_2O vapor). The improvement of the results can also be reached by increasing the temperature. One way is to use a gas with a low average atomic number such as helium. Figures (8a and 8b) below show the electron Flight Simulator spectra obtained under helium environment. The results may be compared to those in figures 6a and 6b.

At P= 70 Pa and GPL=1mm, Uns_f and r_s are close to 99.8% and <1 µm respectively and at P= 350 Pa and GPL=2mm, Uns_f and r_s are close to 99.1% and <1 µm respectively. It is easy to conclude that the quality of the results will be improved.

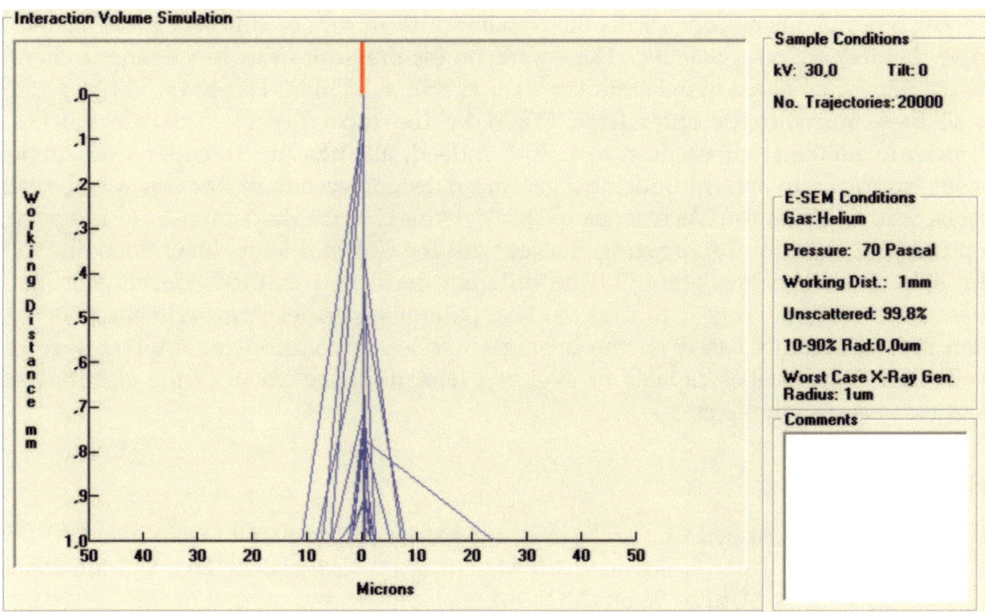

Fig. 8a. Monte Carlo simulation using Electron Flight Simulator of the electron beam scattering under helium, V=5kV, P= 70 Pa, GPL = 1mm.

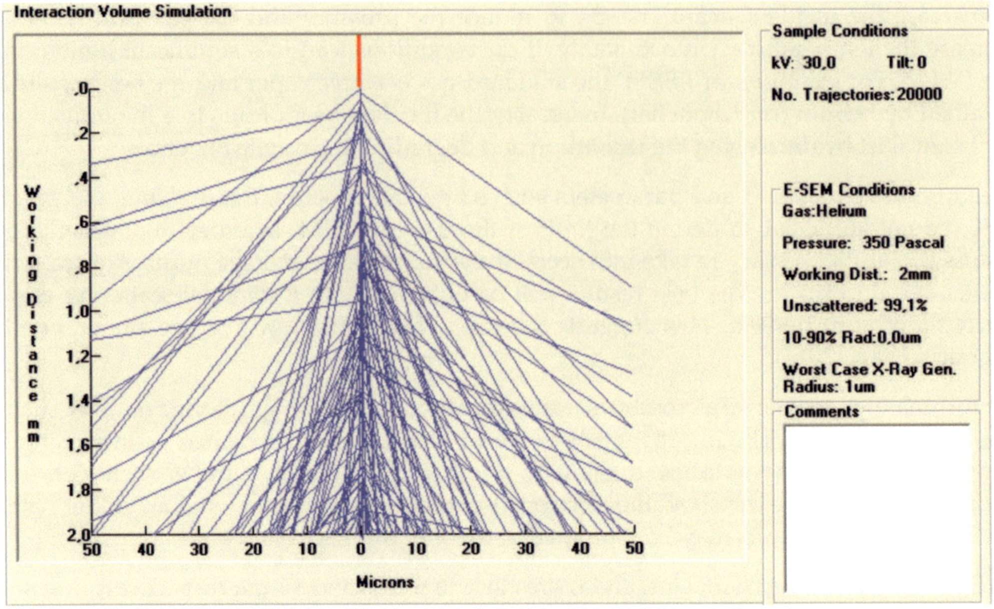

Fig. 8b. Monte Carlo simulation using Electron Flight Simulator of the electron beam scattering under helium, V=5kV, P= 350 Pa, GPL = 2mm.

4. Conclusion

Different types of microscopes with the possibility to introduce different gases inside the sample chamber are now available. Depending on the pressure value in the sample chamber different names are given in the literature such as ESEM, LVSEM, HPSEM, VPSEM, CPSEM. But all these microscopes differ from CSEM by the capability to introduce the gas as environment unlike High vacuum in CSEM. Indeed, all these microscopes work under a gaseous environment and introducing a gaseous detection system in this way may be called gaseous Scanning Electron Microscope (GSEM). In this chapter we demonstrate and confirm the possibility to perform interesting studies with the GSEM if some limitations due to the beam skirt are taken into account. The different correction methods developed are not satisfactory. The good way is to find the best parameters for each experiment in order to obtain the best results based on the average number of collision m. Another way is to develop new microscopes capable of avoiding (isolating) the travel of the electron beam across gaseous environment.

5. References

Bilde-Sorensen, JB & Appel, CC. (1996). Improvements of the spatial resolution of XEDS in the environmental SEM. In *EUREM, 11th Euro Congress on EM*, Dublin, Ireland, Published by EUREM 96 on CD-ROM

Carlton, PA. (1997). The effect of some instrument operating conditions on the x-ray microanalysis of particles in the environmental scanning electron microscope, *Scanning*, 19, pp.(85-91)

Danilatos, G.D. (1980). An atmospheric scanning electron microscope (ASEM). *Scanning,* 3, pp.(215-217)

Danilatos, G.D. (1988). *Foundations of Environmental Scanning Electron Microscopy,* Advances in Electronics and Electron Physics. 71. Academic Press, ISBN 0120146711pp. 109-250

Danilatos, G.D. (2009). Optimum beam transfer in the environmental scanning electron microscope. Journal of Microscopy, 234, 1, pp.(26-37)

Doehne, E. (1996). A new correction method for energy-dispersive spectroscopy analysis under humid conditions. *Scanning,* 18, pp.(164-165)

Doehne, E. (1997). A new correction method for high resolution energy dispersive X-ray analyses in the environmental scanning electron microscope. *Scanning,* 19, pp.(75-78)

Fletcher, A.L.; Thiel, B.L.; & Donald, A.M. (1997). Amplification measurements of potential imaging gases in environmental SEM. *J. Phys,* D 30, pp.(2249-2257)

Gauvin, R. (1999). Some theoretical considerations on X-ray microanalysis in the environmental or variable pressure scanning electron microscope. *Scanning,* 21, pp(388-393)

Goldstein, J.; Newbury, D.; Echlin, P.; Joy, C.; Roming, A.D.; Lyman, E.; & Lifshin, E. (1992). *Scanning Electron Microscopy and X-ray Microanalysis,* 3rd ed. Plenum Press, ISBN 0-306-44175-6, New York , USA

Kadoun, A. ; Belkorissat, R. ; Khelifa, B. ; & Mathieu, C. (2003). Comparative study of electron beam-gas interaction in an SEM operating at pressures up to 300 Pa.*Vacuum,* 69, pp.(537- 543)

Khouchaf, L.; Blondiaux, J.; Hedouin, V.; Gosset, D. ; Dürr, J. & Flipo, R.-M. (2000). La Microscopie Electronique à Balayage Environnementale équipée en Microanalyse X : son utilisation en Pathologie Osseuse Humaine. Perspectives et limites. *Journal. Phys. IV,* 10, pp.(551-559)

Khouchaf, L. & Verstraete J. (2002). X-ray microanalysis in the environmental scanning electron microscope (ESEM): Small size particles analysis limits. *J. Phys. IV,* 12, 6, pp.(341-346)

Khouchaf, L. & Verstraete, J. (2004). J. Electron scattering by gas in the Environmental Scanning Electron Microscope (ESEM): Effects on the image quality and on the X-ray microanalysis. *J. Phys IV,* 118, pp.(237-243)

Khouchaf, L.; Boinski, F. (2007). Environmental Scanning Electron Microscope study of SiO_2 heterogeneous material with helium and water vapor, *Vacuum,* 81, pp.(599-603)

Khouchaf, L.; Mathieu C. & Kadoun, A. (2010). Electron microbeam changes under gaseous environment: CP-SEM case and microanalysis limits, *Microscopy: Science, Technology, Applications and Education,* A. Méndez-Vilas and J. Díaz (Eds.), FORMATEX, ISBN-13: 978-84-614-6190-5, Badajoz, Spain

Khouchaf, L.; Mathieu C. & Kadoun, A. (2011). Microanalysis results with low Z gas inside Environmental SEM. *Vacuum,* 86, pp.(62-65)

Le Berre, J.F; Demopoulos, G.P; Gauvin, G. (2007). Skirting: A Limitation for the Performance of X-ray Microanalysis in the Variable Pressure or Environmental Scanning Electron Microscope. *Scanning,* 29, pp.(114-122)

Mansfield, J.F. (2000). X-ray microanalysis in the environmental SEM. a challenge or a contradiction, *Mikrochim Acta,*132, pp(137-143)

Mansour, O.; Aidaoui, K.; Kadoun, A.; Khouchaf, L. & Mathieu, C. (2010). Monte Carlo simulation of the electron beam scattering under gas mixtures in an HPSEM at low energy. *Vacuum,* 84, pp.(458463)

Newbury, D. (2002). X-Ray Microanalysis in the Variable Pressure (Environmental) Scanning Electron Microscope. *J. Res. Natl. Inst. Stand. Technol.,*107, pp.(567–603)

Newbury, D.; Joy, C.; Echlin, P.; Fiori, C. & Goldstein, J. (1986). *Advanced Scanning Electron Microscopy and X-ray Microanalysis,* Plenum Press, ISBN 0-306-42140-2, New York , USA

Stokes, D. J.; (2008). *Principles and practice of Variable Pressure/Environmental scanning electron microscopy (VP-ESEM),* John Wiley & Sons, Ltd. ISBN: 978-0-470-06540-2, Cornwall, UK

Wight, A. (2001). Experimental data and model simulations of beam spread in the environmental scanning electron microscope. *Scanning,* 23, pp.(320–327)

2

Interactions, Imaging and Spectra in SEM

Rahul Mehta
University of Central Arkansas, USA

1. Introduction

In microscopy the question arises- Why employ electron beams instead of light beams to produce magnified images and the answer has to do with resolution. When doing microscopy to produce magnified image of objects, diffraction (bending of waves around narrow openings and obstacles) limits the resolution and hence the quality of image in terms of fine details one can see. The optical wavelengths from deep UV to IR are in range of hundreds of nanometers while electron beam of energy in keV have wavelengths in fractions of nanometers. The dependence of diffraction on the wavelength of the beam makes electron beam more suitable than beams of wavelengths in the optical region. The diffraction also depends on the size of the objects. A Scanning Electron Microscope (SEM) with electron beams in the keV range allows one to produce image (Fig. 1) of objects in the micro to nanometer range with relatively lower diffraction effects. Using a SEM to produce proper image requires a judicious choice of beam energy, intensity, width and proper preparation of the sample being studied. The electron beam in a SEM is nowadays generated using a field emission filament that uses ideas of quantum tunneling. Other methods are also available. The deflection of electron beam of certain energy E is accomplished by means of electromagnetic lenses. Typical E values for conventional SEM can range from as low as 2-5 keV to 20-40 keV.

A basic SEM consists of an electron gun (field emission type or others) that produces the electron beams, electromagnetic optics guide the beam and focus it. The detectors collect the electrons that come from the sample (either direct scattering or emitted from the sample) and the energy of the detected electron together with their intensity (number density) and location of emission is used to put together image. Present day SEM also offer energy dispersive photon detectors that provide analysis of x-rays that are emitted from the specimen due to the interactions of incident electrons with the atoms of the sample.

2. Interaction

Assume that an electron beam of energy E, with a circular cross-section A and a beam current I is incident on a sample with atomic number Z. We will assume that the energy E is typically much less than 100 keV in the following discussions. As the electron beam enters the sample it interacts with the atoms of the samples. This interaction of the electrons is not confided to the surface layers only but also with the atoms and molecules inside. The electron interaction with the atom consists of coulomb attraction with the nuclear positive charge. The interaction of the electron beam with the electrons from the sample is of repulsive nature as the electrons are deflected by the target electrons. The electrons can undergo change in momentum and/or change in energy or both in these interactions. So an entering electron beam can scatter elastically and/or inelastically.

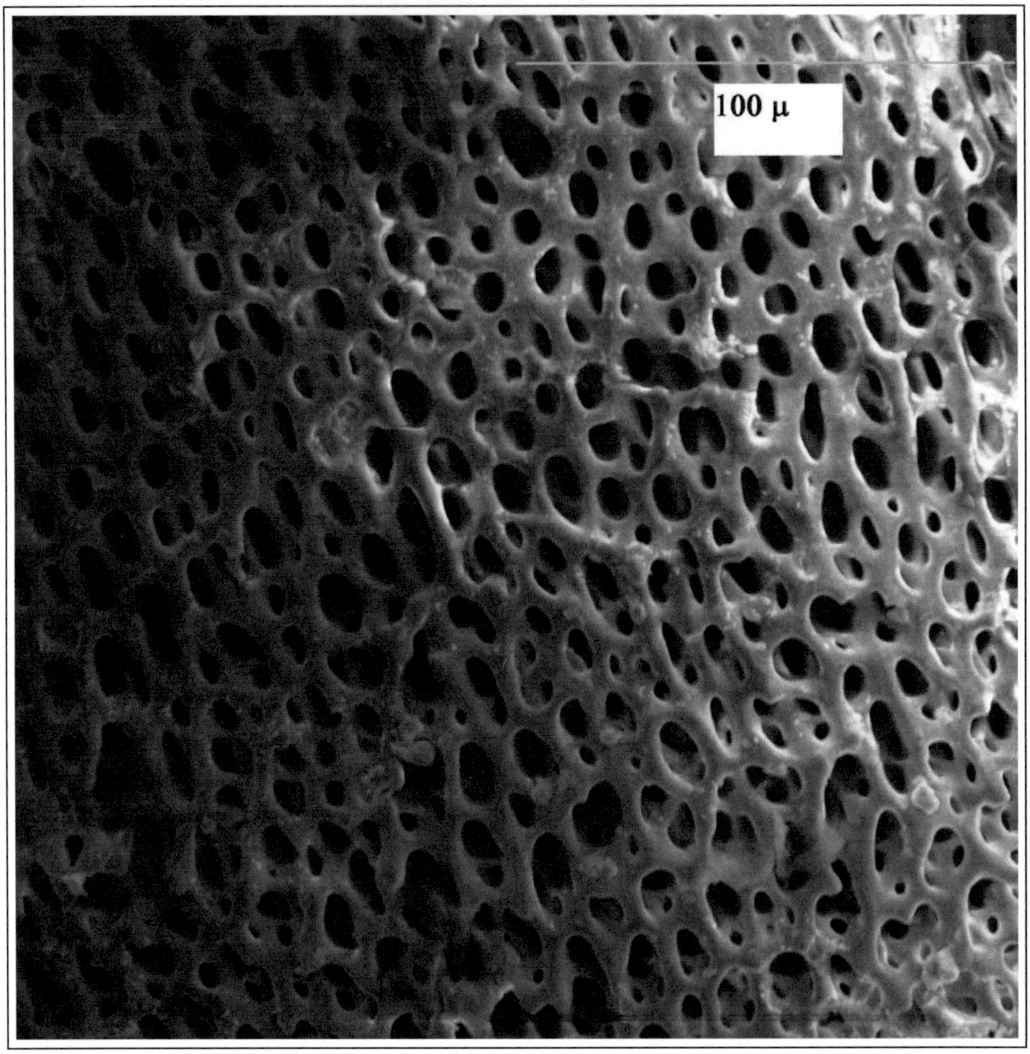

Fig. 1. Biological sample showing kT pores imaged with 20 keV electron beam using a quad backscattered detector. Scale shown by line of 100 μm.

2.1 Elastic scattering

If the scattering involves no loss of energy it is Rutherford scattering (Rutherford, 1911,1914) which is peaked in the forward direction with the probability of scattering decreasing dramatically with increase of angle of scattering and the electron trajectory is modified from some small angle elastic scattering to large angle deviation. Some of the electrons can travel laterally while others can even back scatter. After many of these events it is possible for some of the electrons to leave the sample and these backscattered electrons provide one way of imaging the sample. Probability of elastic scattering depends on inverse square of energy E which means a higher energy beam will start to spread out much later in its path than a smaller energy beam. An electron can transfer energy to the conduction electrons or to a single valence electron - but this will not be important in SEM imaging as the mean free paths for both of these is large, the scattering angles are small and energy loss less than an eV.

2.2 Inelastic scattering

An electron can interact with the solid as a whole generating vibrations (phonon scattering). The energy of the electron goes into overall heating of the solid slightly. The overall energy loss is less than 1 eV and this channel is probably more important near the end of the path of the electron. The scattering results in electron being scattered by larger angle. This effect will be important for image resolution and contrast. The energy loss from inelastic scattering is related inversely with E therefore a higher energy incident electron will keep more of its energy at a depth than a lower energy incident electron at the same depth. If the scattering involves loss of energy then it cannot be described by Rutherford formula. There are many channels by which an electron can lose energy in a sample but here we will look at some that are more pertinent for SEM imaging.

The channels that are useful for imaging are the ones that results in radiative or non-radiative transitions to occur in the sample atom. This is when the electron transfers energy to one of the inner shell electrons and then this result in ionization or electronic rearrangement. The atom that absorbs the energy this way will either give out a photon (radiative process) or eject an electron from same or different shell (Auger process- non-radiative). The radiative photon is generally in the x-ray region of electromagnetic spectrum. The probability of radiative versus non-radiative process taking place defines the fluorescence yield ω. In energy dispersive analysis of a sample using SEM- ω plays an important role in conversion of x-ray intensities (from x-ray spectrum) into absolute numbers. These absolute numbers are related to sample elemental thicknesses and overall compositions.

2.3 Energy loss

The energy loss of the electron in scattering is dependent in a complex way on the atomic number Z of the sample atom, on their mass number A and the density ρ of those atoms. The energy lost by the electron can be transferred to the sample atoms in inelastic scattering. The rate of energy loss with the path length x, dE/dx, was described by Han Bethe (Bethe, 1930) mathematically. Calculations based on this formula suggest that dE/dx increases with Z while increasing E lowers this rate. The dependence on E is much more dramatic than with Z. Monte-Carlo type simulations (Metropolis & Ulam,1949; Newberrry & Myklebust, 1979; Rubinstein & Kroese, 2007) of trajectories of electrons (as they interact with the sample) suggest visualization in terms of an interaction volume. The size and depth of the volume is dependent on energy of the electron beam, their number density and the details about the interacting atoms. volume The probabilities of the electron interactions drops off by a large factor outside this volume.

2.4 Radiative and non-radiative mechanisms

The interaction between the incident electrons and the sample target atoms provides rich information about the chemical environment of the target atoms. This information is in the form of radiative and non-radiative transitions and subsequent emissions that take place in the atoms. The ion-atom collision results in transitions that involve energy transfer through the mechanisms (both radiative and non-radiative type). The radiative transitions in the atoms can lead to emission of photons mainly in the form of x-rays from K, L, M- shells. These x-rays are characteristics of the elements they come from and the x-ray spectra has signature to that effect. Recognizing these x-rays and then measuring them provides relative abundance of elements in the sample. To get an absolute value (e.g. # of atoms of one type as a fraction of all atoms) generally specified as parts per million (ppm)) normalization of the emission yields has to be done. This requires measuring the emission yields from the sample and from a standard sample under identical conditions so that ratios can be formed.. The standard must have been measured independently and sometimes with a different spectroscopic method (e.g. mass spectroscopy or infrared spectroscopy) and for it ppm needs to be available. The non-radiative transitions can result in emission of Auger electrons and Auger spectroscopy can provide information about the intensities there. Normally the standard SEM may not have capability of differentiating and measuring the auger electrons. What is done in that case is to use the value of fluorescence yield ω (which relates the radiative yilds to non-radiative yields) and determine fraction of time an energy transfer to an atom will result in some form of radiative emission. The fluorescence yield then allows one to convert cross section for ionization into cross section for production of x-rays. The fluorescence yield factor F which is related to the ratio of radiative to non-radiative transition has to be carefully used or determined in the normalization procedure and plays a role in correction factors to get the absolute numbers. The correction factors take into account the fact that ratios of intensities are substantially different than the ratio of concentrations of elements in a sample. The atomic number Z and the mass absorption of x-rays in the volume of the sample A are the other two effects that go into the ZAF correction factor and they will be discussed in more detail later on..

2.5 Imaging

In usage the electron beam is incident on a target region from the specimen sample. The energy of the electron E, the mass density of the target, and the atomic number Z of the sample determines the relative intensities of various types of electron scattering. The penetration depth of the electrons, the mean free path and the strengths of different scattering (which are also dependent on both the Z and E) play a role in the information one gets (in the form of images) about the sample. Primarily the back scattering electrons provide an electronic signal that delineates the interaction volume and carries details about the scattering. In addition the information about the specimen is also comes from the production of secondary electrons from the sample.

3. X-ray imaging, analysis and other techniques

3.1 Elemental profile using SEM

Before one can do spectroscopy using a SEM, the sample has to be prepared correctly, mounted on special sample holders and oriented properly. Metallic stubs with sticky carbon surface allows one to present the sample in a particular orientation to the beam. Samples that are placed on a goniometer can even be rotated to image the sample from a different direction. In a typical preparation of samples for SEM analysis: the sample has to be cleaned to remove contamination, dried in most cases and the surface to be analyzed prepared so that the analyzed surface is flat and electrically conducting. The cleaning starts with sample placed in ethanol baths. Part of this fixes the sample and also replaces the water content. For a biological sample -like a bone -first the bone has to be cleaned of most of soft tissues and then the remaining soft tissues are removed by placing the sample with dermestid beetles. The sample is observed under light microscope and if needed other techniques are used to remove any more soft tissue in the area of interest. More ethanol baths for different lengths of time and different concentration of ethanol may have be used. Cleaned samples are sectioned using high speed Dremil and other cutting tools. The surface to be analyzed has to be flat, smooth as possible and without any intruding parts in front of them. The samples are dried using the critical point dryer, if needed, and then sputter coated with Au to make them electrically conductive. For electron beam to be incident on the sample normally, the sample is placed on the mounting stub (with a sticky carbon tape exposed in the normal direction). The prepared flat cross section needs to be positioned correctly on the metal stub. This then ensures the proper orientation of the sample in the beam. The conductive gold layer allows the electrons a path to the local ground - absence of which will result in area of the sample acting as non-conductors (insulator). Electron beam incident, on the non-conductive area, will result in electrical charge getting collected. When seen in the SEM image, the area that is non-conducting will show up as whitish region with very less details to be seen. Over time the whitish area will get brighter losing even more details and also may grow in size (Figure 2). A layer of conducting metal like gold (few atom layer thick) will be sufficient to alleviate this charge clumping and in the SEM image the whitish appearance will disappear. If the image continue to show incomplete charge conduction from an area then a second layer of gold can help to minimize the charge clumping. In extreme cases, one has to use a lower energy and intense electron beam. One of the affect of an extra layer on a sample is to mask some of the features that are being imaged. Other difficulty that arises from a thicker coating of metal is x-ray interference. The metal coating (e.g. gold) emits characteristic x-rays from that metal. These x-rays can overlap partially the x-ray spectra coming from the sample being studied.

Samples that are to be studied in their original conditions have to be handled differently. Some of these are wet samples. Other samples that are not fixed and non- conductive create imaging problems that are tackled differently. These samples generally outgas in vacuum of the SEM chamber and have to be studied in a mode in SEM that allows for differential pumping in different sections of the SEM. For these samples high vacuum (like ~10^{-6} Torr) cannot be achieved and so resolution is not as good and images are not as crystallized as a dry sample will do. But the SEM images will still provide details that are useful for the researcher.

Once the sample is placed in the SEM chamber and the detector is chosen (between secondary electron detector and/or backscattered electron detector) image is generated. The image details including the resolution are dependent on the energy of the electron beam type of sample, its geometry and atomic numbers of the atoms present. When the image shows the proper details and is magnified correctly one can open the energy dispersive system to do x-ray spectroscopy. The Energy Dispersive Analysis (EDS) mode of the SEM provides the x-ray spectra for elemental analysis. In order to quantify the elemental yield one needs standard samples. For example in the study of bones, standards representing Calcium Phosphate, are used. Also to get a good calibration of the detector's response in the energy region being studied, other standard elemental samples are employed. For example a pure copper sample has L-shell x-rays around 1 keV and K-shell x-rays around 9 keV. A pure gold or lead sample will give M-shell x-rays in 2-3 keV range and L-shell x-rays around 10 keV. It is essential that the range of x-ray energies being studied be understood in terms of the response of the detector. This response also needs to be established for the range of electron beam energies to be used. The x-ray spectra from standards and from the samples are analyzed using software that is specially developed for analysis needed with corrections built in for various effects that may be important at some energies and not at others. FLAME (fuzzy logic software for spectral analysis and elemental ratio determination) is one of those software. The software, with statistical capabilities provides identification of the elements, atomic and weight percent of elements, intensities of the x-rays and other parameters that are electron beam and elemental atomic number dependent. The software generates a table showing the elemental ratios (weight and atomic) among the elements detected: e.g. oxygen, phosphorus, and calcium in the bone samples.

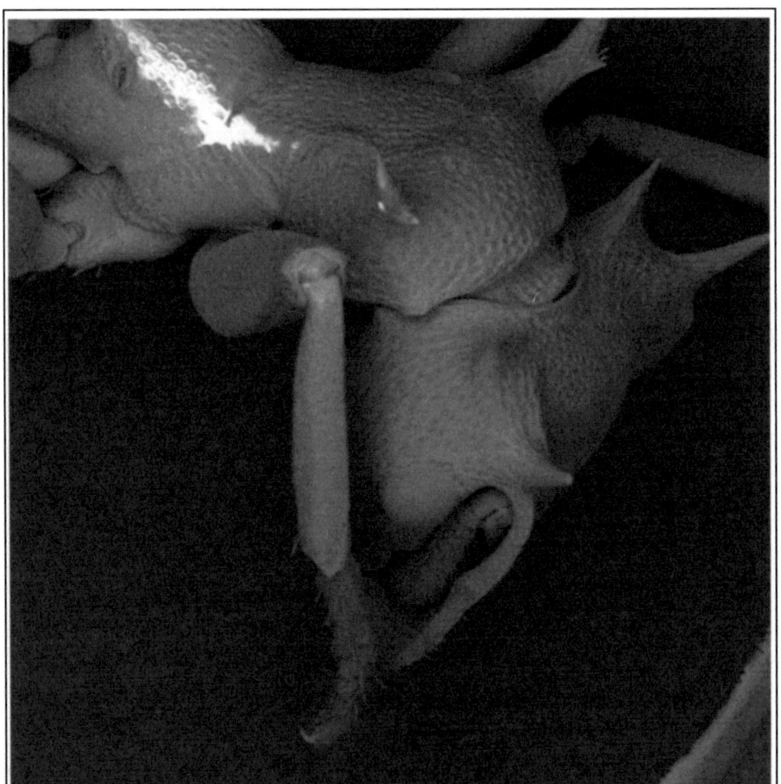

Fig. 2. SEM image of a biological sample(cephalotes) using quad backscattered detector. The sample was not sputter coated resulting in excessive charging(white area) on the sample.

3.2 ZAF correction factors

Castaing (Castaing,1951,1966; Castaing & Henoc, 1966) showed that the k-ratio, which is the ratio of sample x-ray intensity to standard sample x-ray intensities, is proportional to the ratio of the mass fraction of the sample element to that for the standard sample. But experimentation has shown that there are deviation of this k-ratio from the actual concentration ratios. These differences arise from many parameters of the sample but mainly density, electron backscattering, x-ray ionization and production cross section (these are connected by the fluorescene yield) , energy loss of the electron beam and the absorption in the sample matrix. In samples that contain many elements and the mixture is not very homogeneous the measured intensity may vary by a large factor on variation in elastic , inelastic scatterings, and the absorption of the x-rays though the elements of the sample before reaching the detector. In general these various effects coming from the sample matrix on the measured intensity can be lumped into correction due to atomic number (the Z-effect), the absorption of the x-rays in the sample (the A effect) and the F effect due to x-ray fluorescence yield. In total the correction is called ZAF factor and in a simplified equation it is given by eq. (1) as

$$C_i/C_{(i-std)} = \{ZAF\}_i. \left(I_i \middle| I_{(i-std)}\right) \tag{1}$$

where C_i *and* $C_{(i-std)}$ are the fractional sample weight of element i and for the same element in the standard sample. Here $\left(I_i \middle| I_{(i-std)}\right)$ are the intensities as measured for the same element in sample and in the standard sample. In order to understand the Z,A and F factors, one has to visually assimilate the various processes taking place as an electron beam traverses the sample, loses energy by scattering processes and excitation of the host atoms of the sample takes place.

Z-factor: When an electron beam is backscattered, the backscattering mechanism removes part of beam of electrons which then reduces the number of interactions that can lead to ionization and production of x-rays. In samples with many elements the kinematics of scattering results in greater spread of the beam . The scattering results in greater spread in the energy for the scattered electron. Kinematics suggests that a greater number of electrons backscatter when atomic number Z is greater. The higher Z elements then remove a larger fraction of electron energies. The energy loss from inelastic scattering tends to remove electron energy due to thickness (defined as a product of the thickness as measured along the path and the density). The low atomic number remove this energy at a higher rate than higher atomic number. A Monte Carlo simulation of the trajectory of electron suggests that as the electron traverses a sample it is losing energy. The ionization of an atom and subsequent production of x-rays is critically dependent on if the energy available is above the excitation energy for the particular atom. So the energy may be enough to excite L-shell x-rays but not excite higher K-shell x-rays or in the heavy elements like gold the energy may excite M-shell x-rays but not L-shell x-rays and definitely not K-shell x-rays. During elastic scattering, the kinetic energy conservation tends to scatter electrons at larger angles and hence deviate from its path more. These scattered electron would be less likely to produce ionization and x-rays then if it did not interact elastically. Thus the distribution of the electron in the sample, their energies at a point in the sample and the x-ray production depends strongly on the atomic number of sample atoms. This distribution can be defined in terms of a function φ (ρZ). An area under the plot of this function φ (ρZ). versus ρZ allows one to integrate for the intensities that would be generated. The atomic number effect (the Z-factor) for each element is then the ratio of this function φ (ρZ) for the sample versus for the standard sample.

A-Factor: Inner shell ionization followed by x-ray production takes place over a range of thickness in the sample. The volume from which x-rays come from increases with energy of the incident electrons and scattered electrons can come from deeper region and overall a larger volume. Ionization followed by a radiative transfer of energy leads to the production of x-ray. The x-rays on their way to the detector gets absorbed by the matter they have to pass through. This absorption can be defined in terms of an exponential function. This exponential decrease is given as eq. (2)

$$I = I_0\, e^{-\mu\rho t} \qquad (2)$$

where I and I_0 are the intensity of the x-ray at the detector versus intensity when produced, μ is the mass absorption coefficient, ρ is the density of matter the x-ray passes through and t is the path length of this matter layer and ρt gives the thickness in units of mass per unit area. The exponential term representing the fraction by which incident intensity is reduced is calculated for each of the layers the x-rays have to pass through. The direction in which a generated x-ray has to travel to get to the detector defines the path length. This is related to the takeoff angle, the angle between the incident electron beam and the direction of the x-rays. The incident energy of the electron beam and the takeoff angle can affect the fraction absorption by a large factor. X-ray absorption factor A generally is the largest factor in the ZAF factor. Again the plot of ϕ(ρt) versus with ρt is used to determine the A-factor from difference in area under the curves of ϕ for generated x-rays and for emitted one. The emitted x-ray intensity contains the absorption effect using the exponential law.

F-factor: In addition to x-rays being produced following ionization of the atoms by the electron beam, the x-rays themselves can fluoresce more x-rays from the atoms of the sample they pass near. The x-rays fluoresced have energies less than the energy of the x-ray (E_0) that fluoresced them. This has to do with the threshold excitation energy E_c needed for fluorescence. The fluorescing becomes negligible if E_0 is greater than E_c by 5 keV or more.

3.3 Comparative techniques

The x-ray spectra obtained from an SEM is analyzed with special software to determine the yield of x-rays. The spectra is generally shown as intensity versus the energy of the x-rays (Figure 3 and 4). The detector normally ised in a SEM is a (Si(Li) detector with a resolution of about 140 eV at 5.9 keV for $_{54}$Mn x-rays. This resolution is enough to resolve x-rays from adjacent elements and also can differentiate some of the individual transitions within the x-rays from the same element. Si(Li) detectors uses a Silicon crystal which is Lithium doped (has to be cooled below liquid nitrogen temperatures for it to work). The response of the crystal to photons in the 1- 100 keV region is generally depicted with an efficiency curve. This curve shows the percent detection of the photons arriving in the active region of the detector. Other than the geometry of the detection system, a typical efficiency may be 1 out of 10000 (or 1% or less). The physical region between location where x-ray photons are generated and their passage through the in-between matter before reaching the active silicon region of the detector determines the attenuation fraction of the original x-ray signal. In a typical SEM, this attenuation takes place in the layers of air (in the high vacuum chamber), beryllium window layer as the front window of the detector, the gold contact layer and the dead layer of silicon. This absorption and attenuation depends on the energy of the x-ray photon and also the thickness of each layer. For energies above 3-4 keV, the efficiency is smoothly varying (fairly constant in the 5-20 keV range). There are many calibrated photon sources available to measure the efficiency in this region. Experimentally measured efficiencies, together with that predicted and calculated from models are compared. The calculated efficiency includes the attenuation of photon intensities in the layers described above. measured and calculated efficiencies are normalized to each other using the measured energy point (Gallagher & Cipolla,1974; Lennard & Phillips, 1979; Papp, 2005; Maxwell & Campbell, 2005, Mehta etal., 2005)). This procedure results in normalized efficiency curves. The efficiency in the 5-20 keV region can be determined to uncertainties of few percent but for energies of x-rays in the 1-3 keV efficiency is lot more uncertain especially below 1 keV and there lies the problem.

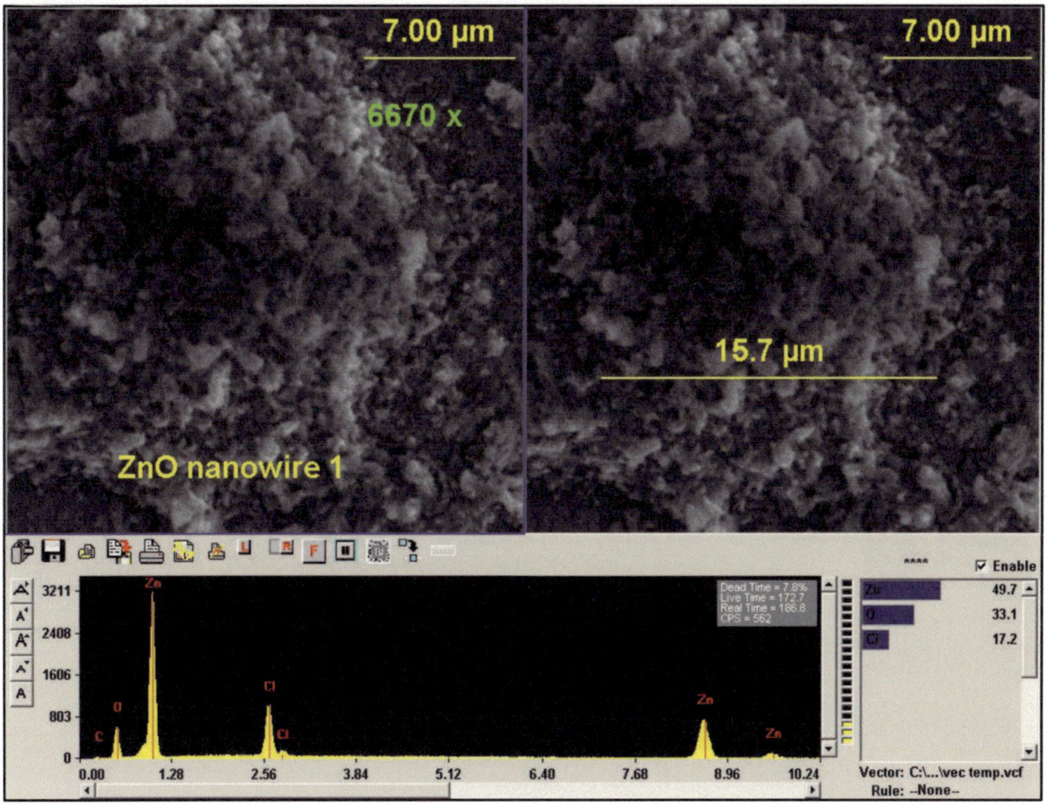

Fig. 3. SEM Image (magnification x6670 and scale as shown) and x-ray spectrum showing L-shell (~1 keV) and K-shell (~ 9 keV) x-rays from zinc in a zinc oxide Nanowire. Also chlorine K_α and K_β can be seen as just resolved. The K-shell x-rays of zinc clearly show separated K_α and K_β peak with a peak intensity ratio of 4:1. Right side table show relative percentages of the elements in the sample (not corrected with k-ratio).

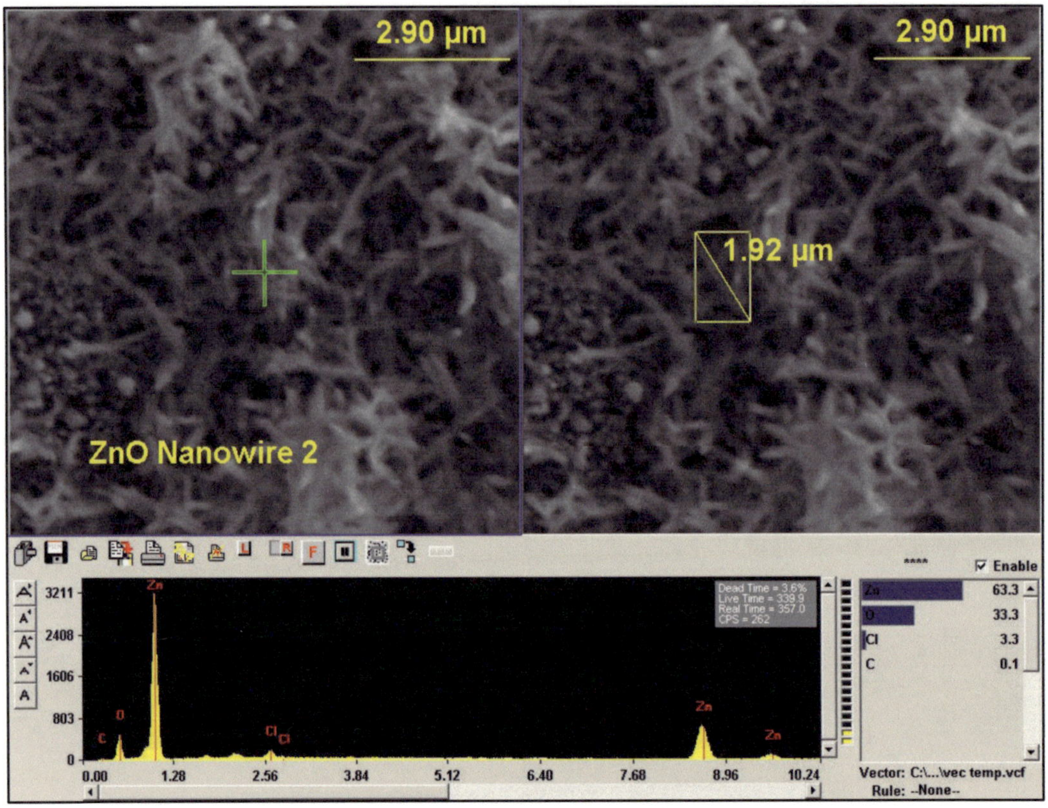

Fig. 4. SEM Image(from a box < 2 µm on the side) and x-ray spectrum showing L-shell (~1 keV) and K-shell (~ 9 keV) x-rays from zinc in a zinc oxide nanowire. The k-shell x-rays clearly show separated K_α and K_β peak with a ratio of 4:1. The image clearly shows the wires of ZnO.

The x-rays from K-shell of carbon, oxygen, up to sodium are all ~1 keV or less. L-shell x-rays below 1 keV come from elements Calcium(Z=20) through Zinc (Z=30) while M_shell x-rays are all less than 3.5 keV (highest M-shell x-rays for Uranium Z=92). For lanthanum (Z=57) the M-shell x-rays are less than 1 keV. The x-rays generated in an SEM are limited by the maximum energy the electrons can have. For a typical SEM that has a maximum voltage available for accelerating of say 20 kV – the electron beam has maximum possible energy of 20 keV. The x-rays generated from samples by such beams can then only be up to 20 keV. So depending upon the elements present in the sample, the x-ray data can give yields that are uncertain by above uncertainties. Yields can be converted to absolute numbers if the number of electrons involved in the generation of x-rays can be determined and standard samples for the elements are available. This leads to the realization that any absolute numbers have to be checked against absolute numbers from other comparable technique. Any normalization procedure among the techniques have to find a unique common point.

3.3.1 X_ray fluorescence (XRF)

For large Z elements (Z> 45) XRF (Bundle et al., 1992) can provide information about x-rays greater than 20 keV that the SEM cannot. XRF is used in that situation and again normalize K-shell x-ray production using XRF with L- or M-shell x-ray production by the electron beam of an SEM. Some of the analyzed samples are fluoresced using radioactive sources of Fe-55, Cm-244 and Am-241 in the XRF. EDS analysis from SEM is energy limited by the electron beam energy used, while XRF is not. XRF spectra is measured to provide x-ray measurements that are outside of the energy range of the SEM measurements. In addition, the lower energy L and M-shell x-rays are measured to provide another set of elemental ratio data. This allows for comparison between elemental ratios determined using SEM and XRF.

3.3.2 Neutron activation analysis

A standard neutron source (Pu-Be in a Howitzer or a neutron generator) can be used to do neutron activation work. The energy of the neutron beam and the flux coming from the source may determine if this technique can allow one to analyze a sample also analyzed with SEM. The incident energy of the neutrons from the source will determine if neutron-atom interaction can lead to compound nucleus formation. In order to see any particular decay mode from this compound nucleus, there has to be appropriate isotopes formed with half lives of transitions in that isotope suitable for decay measurement. Also the yield of these newly made isotopes will depend upon the cross section for absorption of the neutrons in the sample. In order to do neutron activation analysis (NAA), the table of isotopes is used to determine the isotopes that can be produced in activation of the samples. The suitability of the radiation these isotopes produce for analysis has to be established too. Once this is established the uncoated samples are prepared for neutron activation and activated for an optimum length of time. The activated samples are analyzed using gamma ray spectroscopy using a combination of Geiger counter , Sodium iodide detector and/or germanium type high resolution gamma detectors. Intensities of photo peaks can be used to form ratios in a particular photon energy range. This divides out any effect due to efficiency variation. Next taking into account other parameters (like neutron cross section, atomic number, branching ratio etcetera) and comparing the ratios of intensities from a standard sample and from the measured sample, a normalized absolute intensities can be determined. For example standard samples can be used to provide a baseline for radioactivity measurements and dose dependent measurement of other standards to be used. This baseline can provide a scheme for normalizing the intensities from different samples. Comparison among elemental ratios determined using SEM, XRF and NAA is possible then.

3.3.3 Other comparative methods

Another technique that provides absolute weight and atomic percent of the elements in the samples is Particle Induced X-ray Emission (PIXE) (Flewitt & Wild, 2003). This is performed at an accelerator lab facility. PIXE analysis at an accelerator lab can be used to study biological samples using microprobe beam. The samples and standards are mounted as targets on special sample holders. Proton or alpha particle beams interacting with the targets provides an absolute value for weight percent and atomic percent of the elements in the samples. Again an elemental ratio from this technique can be compared to ratios from other techniques described earlier. The goal is to determine a normalization procedure that can be used to efficiently determine a normalized absolute weight or atomic percent of the elements in the sample. The reliability of the results and efficiency of the technique allows researchers to choose one of these techniques to produce reliable results using the normalization procedure established. The goal of any normalization technique is to decrease the uncertainties in the measurements including those done with SEM.

3.3.4 Statistical analysis

A crucial factor in coming to any conclusion in all these techniques is appropriate application of Statistical analysis. It is imperative to the researcher that they analyze the data using statistical packge (e.g. student t-test or ANOVA) after establishing normal distribution of data and homogeneity of variances.

4. Conclusions

SEM is suitable to look at micro- and nano- structural characteristics of solid objects. Visual images obtained from electron detectors combined with characteristic x-rays mapping allow for detailed micro- and nano-compositional analysis. SEM combined with XRF,NAA and PIXE provide a platform to quantify and produce absolute numbers related to compositional elemental and molecular structures.

The sample that is to be investigated has to be specially prepared so as to provide images and spectral information meaningful to the investigation. Many factors play a role here: the type of sample (say biological sample versus a sample for material science study has to be prepared differently at some stage of preparation), the appropriate energy of the beam, angle of incidence, beam intensity (resolution will be affected greatly from this), the counting time and statistics and others. SEM imaging is done differently for a wet cell sample than a critically dried and sputter-coated solar cell slides.

The other crucial factor is the methodology or methodologies adopted for data analysis and the subsequent results determination. Once the images and the spectra have been collected, the data has to be sorted, analyzed and mathematical functionality recognized and established. Statistical analysis then provides the basis for the eventual conclusions and their validity. For example topographical and/or compositional images can be used to generate structural patterns leading to understanding of type of crystalline lattice underlying a bone. This can then provide the basis for determining the strength of a bone or its elasticity or the reason a bone under microgravity conditions leads to Osteoporosis. SEM spectra that can be analyzed to determine the elemental composition of a certain bone have inherent uncertainties. When studying changes in bone composition these uncertainties will affect the determination of the conclusion.

5. Acknowledgment

The author would like to acknowledge the support for his research from the Arkansas Space Grant Symposium, College of Natural Sciences and Mathematics and the department of Physics and Astronomy at University of Central Arkansas. Also the author would like to acknowledge Dr. Jingbio Cui and Alan Thomas, Physics department, University of Arkansas at Little Rock and Devika Mehta (Highschool Senior at Arkansas School for mathematics, Sciences and Arts), for use of the SEM images (Fig 3 and 4) of their ZnO nanowires.

6. References

Bethe H.; *Ann. Phys.* (Leipzig) Vol 5, (1930) 325.

Bundle C. Richard, Evans C.A. Jr, & Wilson S; *Encyclopedia of Material Characterization*, Butterworth-Heinemann, Boston (1992)

Castaing R, *Ph.D. Thesis*, University of Paris (1951)

Castaing R, *Advances in Electronics and Electron Physics*, 13 (1960) 317

Castaing R & Henoc J, *Proc 4th Int. Conf. On X-ray Optics and Microanalysis* (1966) 120

EDS, SEM from RJLee Instruments Limited, Trafford, Pa

Flame, ASPEX corporation of Delmont, Pa. WWW.ASPEXCORP.com

Flewitt P.E.J. and Wild R. K. *Physical methods for Materials Characterization*, Institute of Physics Publishing, Bristol, UK (2003)

Gallagher W.J. and Cipolla S.J., *Nucl. Inst & Meth.* Vol 122 (1974) p.405.

Krane K.S.; *Introductory Nuclear Physics*, John Wiley and Sons (1988) pp.396-405

Lennard W.N. and Phillips D., *Nucl. Inst. & Meth.* vol 166 (1979)p.521

Mehta R.,Puri N.K., Kumar Ajay, Kumar A., Mohanty B.P., Balouria P., Govil I.M., Garg M.L., Nandi T., Ahamad A., and G. Lapicki, *Nucl. Inst. & Meth.* vol B241 (2005)pp.63-68

Metropolis, N. & Ulam S.; "The Monte Carlo Method" *Journal of the American Statistical Association* Vol 44 (1949) pp. 335–341.

Maxwell J.A. and Campbell J.L., X-ray Spctrometry Vol 34 (2005) p. 320

Newberrry D.E. & Myklebust R.L.; *Ultramicrosopy*, vol 3, (1979) 391.

Rubinstein, R. Y. & Kroese, D. P. ; *Simulation and the Mont Carlo Method* (2nd ed.). New York: John Wiley & Sons. (2007)

Rutherford E.;"The scattering of alpha and beta particles by matter and the structure of the atom", *Philosophical Magazine*, , Vol 21 (1911), pp. 669-688.

Rutherford E.,"The structure of the Atom", *Philosophical Magazine,* vol 27 (1914), pp. 488-498.

Papp T., *X-ray Spectrometry* Vol 34 (2005) p.320

3

In Situ Experiments in the Scanning Electron Microscope Chamber

Renaud Podor, Johann Ravaux and Henri-Pierre Brau
Institut de Chimie Séparative de Marcoule, UMR 5257 CEA-CNRS-UM2-ENSCM
Site de Marcoule, Bagnols sur Cèze cedex,
France

1. Introduction

Since the first scanning electron microscope by Knoll (1935) and theoretical developments by von Ardenne (1938a, b) in the 30's, this imaging technique has been widely used by generations of searchers from all the scientific domains to characterize the inner structure of matter. Even if the obtained information is essential for matter description or comprehension of matter transformation, the main constraints associated with classical electron microscopy, i.e. the necessity to work under vacuum and the necessity to prepare the sample before imaging, have always limited the possibilities to "post mortem" characterisation of samples and avoided observation of biological samples.

Electron microscopists early identified the necessity to undergo these limits. The development of a SEM chamber that is capable of maintaining a relatively high pressure and that allows imaging uncoated insulating samples began in the 70's and has been "achieved" in the late 90's – early 00's (Stokes, 2008) with the commercialisation of the low-vacuum and environmental SEM. The availability of new generations of electron guns (and more particularly the field effect electron gun characterized by a very intense brightness), as well as the new generation of electronic columns that are now commonly associated with the environmental scanning electron microscopes opens new possibilities for material characterisation up to the nanometer scale. The development of this generation of microscopes have opened the door for performing real time experiments, using the electron microscope chamber as a microlab allowing direct observation of reactions at the micrometer scale. Many SEM providers or researchers have developed specific stages that can be used for the *in situ* experimentation in the scanning electron microscope chamber. This field is one of the most interesting uses of the ESEM that offers fantastic opportunities for matter properties characterisation. Even if numerous recent articles and reviews are dedicated to *in situ* experimentation in the VP/ESEM (Donald, 2003 ; Mendez-Vilas et al., 2008 ; Stokes, 2008 ; Stabentheiner et al., 2010 ; Gianola et al., 2011 ; Torres & Ramirez, 2011), no one describes all the possibilities of this technique. The present chapter will provide a large – and as exhaustive as possible – overview of the possibilities offered by the new SEM and ESEM generation in terms of "*in situ* experiments" focussing specifically on the more recent results (2000-2011).

This chapter will be split into five parts. We will first discuss the goals of *in situ* experimentation. Then, specific parts will be devoted to *in situ* mechanical tests, experiments under wet conditions, and a forth part dedicated to high temperature experiments in the SEM. Last, a specific part will be devoted to the "future" of in-SEM experiments. In each part, the main limits of the technique as well as the detection modes will be reported. Each part will be focussed on examples of the use of the technique for performing *in situ* experiments.

2. Goals and implementation requirements of *in situ* experimentation

The main goal of *in situ* experimentation in the SEM (or ESEM) chamber is to determine properties of matter through the study of its behaviour under constraint. This requires the combination of data collection over a given duration (on a unique sample) and image treatment for information extraction. The studied properties are generally related to microscopic phenomena and hardly assessable by other techniques. *In situ* experiment in the SEM chamber corresponds to both imaging systems in evolution under a constraint and imaging systems stabilized under controlled conditions.

To achieve this goal, several requirements are necessary:

- The duration of the phenomenon to be observed must be suitable with the image recording time. If the system evolution is too fast, it will be impossible to record several images and observe this evolution. At the contrary, if the reaction kinetic is low, the time necessary for image recording will be too long and incompatible with experimentation. The high and low limits can be estimated ranging between 2 minutes and 48 hours.
- The system must remain stable under the environmental conditions and/or irradiation by the electron beam during the time necessary for image recording. In the case of easily degradable samples, it is necessary to adjust the imaging conditions (high voltage, beam current, aperture, working distance, detector bias…) constantly, as the sample environmental conditions are modified during the experiment. Thus, the effect of the electron beam on the sample morphology modifications must be verified. Some authors report that it can act as an accelerator (Popma, 2002) or inhibitor (Courtois et al., 2011) of the observed reactions.
- The image resolution must fit well with the size of details to be observed. Improvements in the image resolution have been achieved in the last decade thanks to the field effect emission guns. However, the presence of gas in the VP-SEM/ESEM chamber contributes to the incident electron beam scattering and subsequent degradation of the image resolution. Thus, the acquisition conditions must be adapted to the sample to be studied depending on the higher magnification to be reached.
- The gaseous environmental conditions in which the studied system evolutes (or can be stabilized) must be reproduced in the SEM/LV-SEM/ESEM chamber. The development of the ESEM offers real new opportunities in term of composition of the atmosphere surrounding the sample. The large field detector and the gaseous secondary electron detector (Stokes, 2008) have been developed specifically for imaging under "high pressure" conditions (up to 300Pa and 3000Pa respectively) whatever the gas composition (air, water, He, He+H_2 mixtures, O_2). Other detectors have been developed for very specific applications (high temperature under vacuum (Nakamura et al., 2002), EBSD at high temperature (Fielden, 2005)).

- The constraint in which the studied system evolutes (or can be stabilized) must also be reproduced in the microscope chamber. Some devices are commercialized by official sellers. Among them, we must report the Peltier stage for temperature control in the -10 to 60°C range, hot stages for temperature control up to 1500°C, stages for mechanical tests (Figure 1). Some authors have developed their own specific stages adapted to the problem to be treated (Fielden, 2005; Bogner et al., 2007). However, the development of miniaturized stages that can be positioned in the SEM chamber without creating perturbations on the incident electron beam can be really challenging. This will probably be a key in the development of *in situ* experimentation in the next years (Torres & Ramirez, 2011).

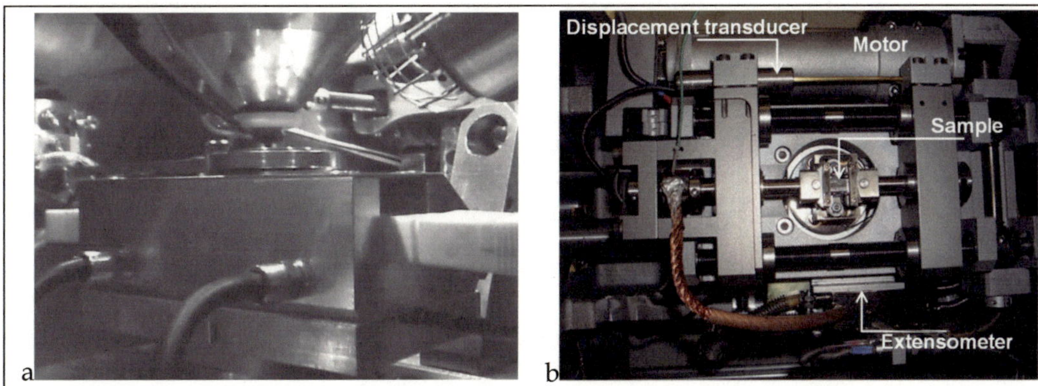

Fig. 1. a) hot stage (FEI) b) Hot tension/compression stage integrated into an SEM (Kammrath & Weiss Co.) (After Biallas & Maier, 2007 ; Gorkaya et al., 2010).

The basis of *in situ* experimentation in the SEM is the study of the morphological modifications of the sample under constraint. Thus, this requires recording of numerous high quality images for image post treatment and data extraction in order to characterize the reaction or matter properties. The sample size can vary from 1µm to 50mm, and the image resolution is in the 1-10nm range, depending on recording conditions. The images are SEM images, i.e. with a large depth of field and with grey level contrasts. In-SEM experimentation can be extended to a wide range of applications, corresponding to very different materials (plants (Stabentheiner et al., 2010), food (Thiel et al., 2002 ; James, 2009), paper (Manero et al., 1998), soft matter, polymers, metals, ceramics, solids, liquids…) or problems (plant behaviour, chemical reactivity, properties characterization, sintering, grain growth, corrosion…). In the literature, the main part of the data reported has been acquired using an environmental scanning electron microscope.

3. *In situ* mechanical tests

Boehlert (2011) have recently underlined the interest of performing *in situ* mechanical tests in the SEM and summarized it as follows. "*In situ* scanning electron microscopy is now being routinely performed around the world to characterize the surface deformation behavior of a wide variety of materials. The types of loading conditions include simple tension, compression, bending, and creep as well as dynamic conditions including cyclic fatigue with dwell times. These experiments can be performed at ambient and elevated temperatures and in different environments and pressures. Most modern SEMs allow for the adaptation of heating and mechanical testing assemblies to the SEM stage, which allows for tilting and rotation to optimal imaging conditions as well as energy dispersive spectroscopy X-ray capture. Perhaps some of the most useful techniques involve acquisition of electron backscatter diffraction (EBSD) Kikuchi patterns for the identification of crystallographic orientations. Such information allows for the identification of phase transformations and plastic deformation as they relate to the local and global textures and other microstructural features. Understanding the microscale deformation mechanisms is useful for modeling and simulations used to link the microscale to the mesoscale behavior. In turn, simulations require verification through *in situ* microscale observations. Together simulations and *in situ* experimental verification studies are setting the stage for the future of material science, which undoubtedly involves accurate prediction of local and global mechanical properties and deformation behavior given only the processed microstructural condition".

As a direct consequence of the great interest of the collected information, many different works from several scientific domains have been published for long. Thiel & Donald (1998) and Stabentheiner et al. (2010) describe the deformation of plants (carrots and leaves respectively) during room temperature tensile tests performed in the ESEM chamber. Similar tests are also reported with food (Stokes & Donald, 2000) and they are regularly performed on polymers (Poelt et al., 2010; Lin et al., 2011), composites (Schoßig et al., 2011) and metals (Boehlert et al., 2006; Gorkaya et al., 2007). Mechanical tests on metals, alloys and ceramics can also be performed at high temperature (Biallas & Maier, 2007; Chen & Boehlert, 2010). High temperature EDSB, developed by Seward et al. (2002), offers the possibility to observe phase transformations in materials as a function of temperature, as well as the direct visualization of the associated microstructural modifications (Seward et al., 2004).

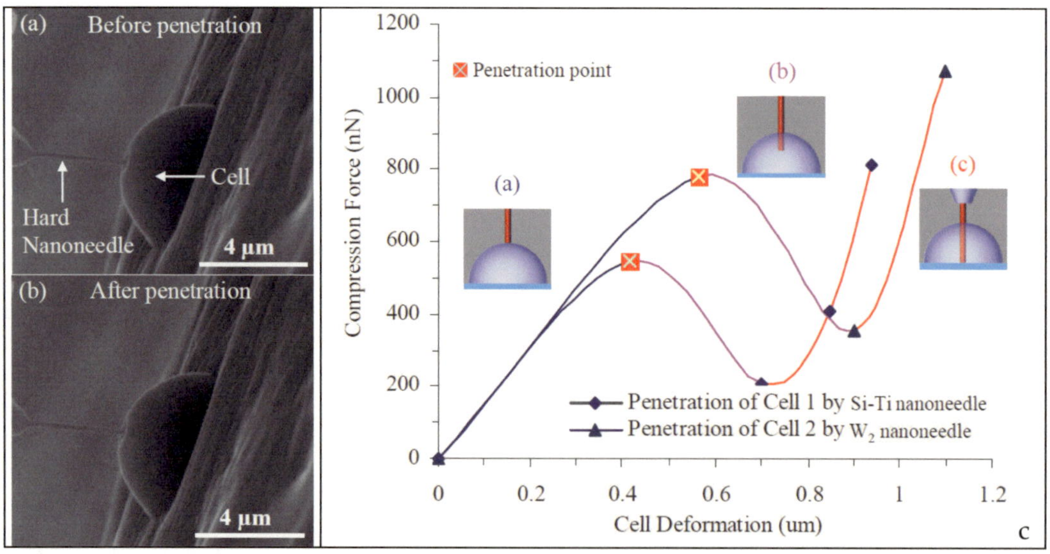

Fig. 2. (a) & (b) Single cell surgery without cell bursting using Si-Ti nanoneedle , (c) Force-cell deformation curve using Ti-Si and W2 nanoneedles at three different stages, i.e. (a) before penetration, (b) after penetration and (c) touching the substrate. (Ahmad et al., 2010).

Several recently developed techniques allow characterizing materials at the nanometer scale through both technological miniaturization and advancements in imaging and small-scale mechanical testing. Ahmad et al. (2010) have developed a coupled ESEM-atomic force microscope to characterize single cells mechanical properties (Figure 2). This ESEM-nanomanipulation system allowed determining effects of internal influences (cell size and growth phases) and external influence (environmental conditions) on the cell strength. Gianola et al. (2011) reports the development of a quantitative *in situ* nanomechanical testing approach adapted to a dualbeam focused ion beam and scanning electron microscope. *In situ* tensile tests on 75 nm diameter Cu nanowhiskers as well as compression tests on nanoporous Au micropillars fabricated using FIB annular milling are reported, the scientific question being the mechanical behaviour of nanosize materials. Both examples probably represent what will be the future of *in situ* mechanical tests using scanning electron microscopes.

4. *In situ* experimentation under wet conditions

4.1 Conditions for experimentation

Combination of the use of the ESEM and a Peltier stage with the development of specific detectors allows the possibility to control both specimen temperature and water pressure around the sample (Leary & Brydson, 2010). Water can be condensed or evaporated on the demand from the sample (Figure 3). This allows performing *in situ* experiments in a temperature-pressure domain that is reported on Figure 3a (dot zone). An easy to perform experiment, illustrated by a 6 images series, corresponding to the NaCl dissolution (during the increasing of the water pressure in the ESEM chamber and consecutive water condensation, at constant temperature) in water followed by the crystallization of NaCl (decrease of the water pressure) is reported on Figure 3b. This example corresponds to an "isothermal experiment". Another ways to work are to perform isobar experiments or to heat or cool a sample using a constant relative humidity (iso-RH experiments). These techniques allow the characterization of structural transitions of hydrated samples as a function of temperature (Bonnefond, 2011).

4.2 Biology and soft matter applications

This technique is particularly well adapted for the observation or experimentation on biological samples (Muscariello et al., 2005). Images of small and highly hydrated samples such as liposomes have been obtained by several authors (Perrie et al., 2007 ; Ruozi et al;, 2011) without any particular sample preparation. Perrie et al. (2007) have also been able to dynamically follow the hydration of lipid films and changes in liposome suspensions as water condenses onto, or evaporates from, the sample in real-time. The data obtained provides an insight into the resistance of liposomes to coalescence during dehydration, thereby providing an alternative assay for liposome formulation and stability (Perrie et al., 2010). However, Kirk et al. (2009) report that ESEM imaging of biological samples must remain combined with the classical techniques for sample preparation. Several works are specifically dedicated to *in situ* experimentation. Stabentheiner et al. (2010) state that "one unrivaled possibility of ESEM is the *in situ* investigation of dynamic processes that are impossible to access with CSEM where samples have to be fixed and processed". These authors have studied the anther opening that is a highly dynamic process involving several tissue layers and controlled tissue desiccation. This phenomenon can be observed because the sample is very stable under the ESEM conditions (Figure 4). Another recent study is relative to the closure of stomatal pores by Mc Gregor & Donald (2010). Even if the possibility for experimentation on biological samples is clearly demonstrated, the authors outline the fact that the electron beam damages are important even at low accelerating voltage (Zheng et al., 2009). Another surprising example that can be reported is the direct observation of living acarids available online: in the movie, colonies of acarids are directly observed in the ESEM chamber under several conditions (FEI movie).

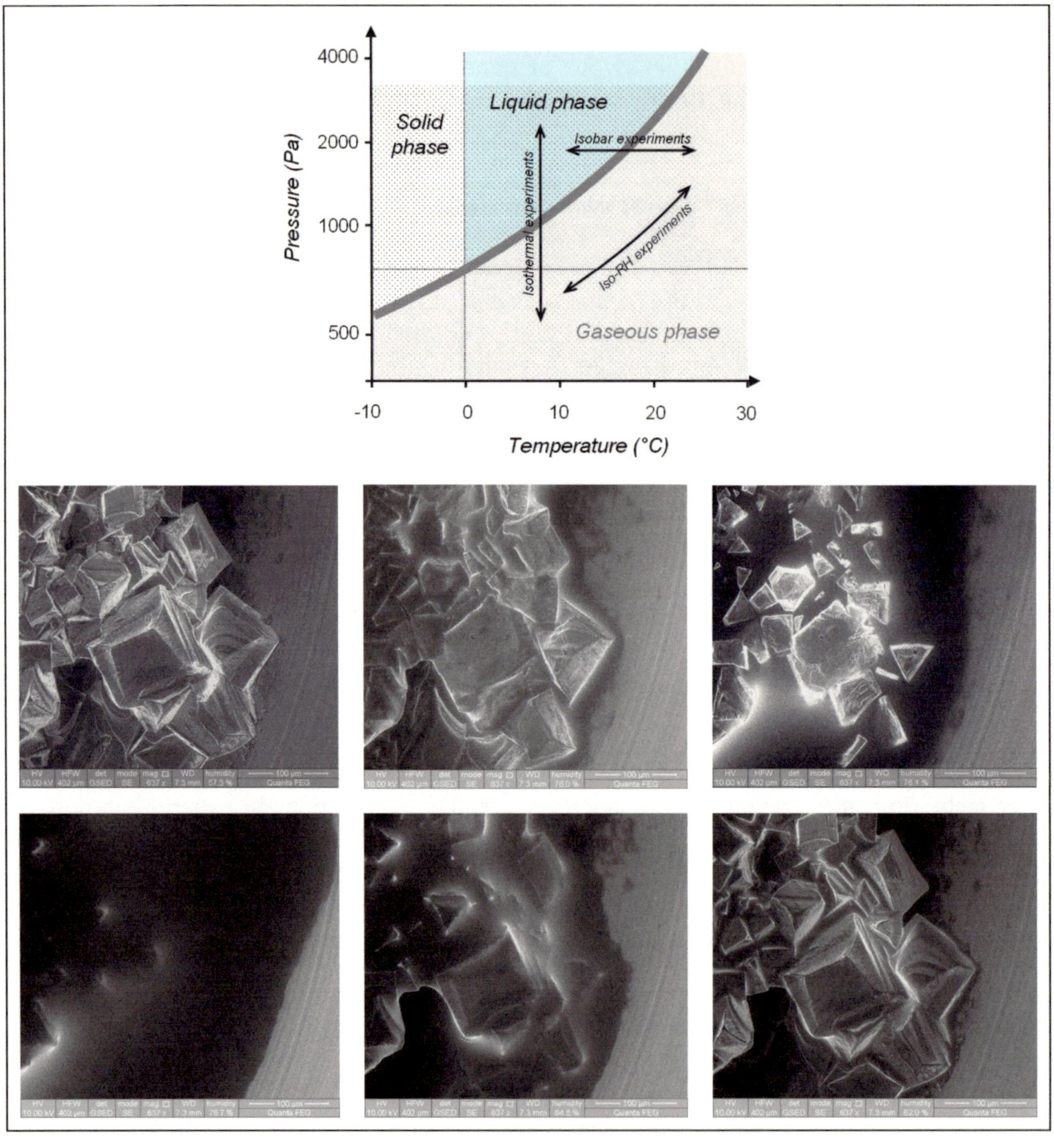

Fig. 3. (a) Simplified phase diagram for water indicating the ESEM domain (dot zone) and schemes to understand how isothermal or isobar experiments are performed.
(b) Solubilisation and crystallization of NaCl directly observed in the ESEM chamber.

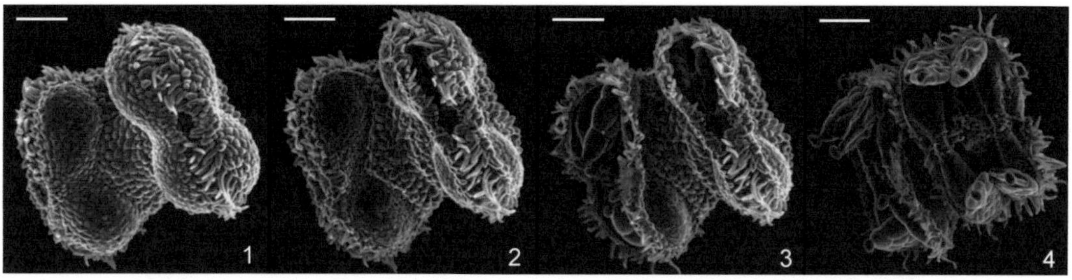

Fig. 4. *In situ* anther opening of C. angustifolia observed in LV-ESEM. 1) At the beginning, the valves of the anther are closed; 2) opening starts at the end of the stomium; 3) polyads are already seen; 4) opening proceeds till the valves are completely bent back and all eight polyads are presented (scale bar = 100µm). Time span from 1) to 3) was 25 min; 4) imaged 1 h after the start of the opening process (after Stabentheiner et al., 2010)

4.3 Applications on cements

Several works have been performed in order to study the reactivity of cement materials versus humidity. Hydration or dehydration (Sorgi & De Gennaro, 2007; Fonseca & Jennings, 2010; Camacho-Bragado et al., 2011) of phases have been followed and used to extract kinetic parameters (Montes-Hernandez, 2002 ; Montes & Swelling, 2005 ; Maison et al., 2009), as reported on Figure 5. In this work, the author uses ESEM image series to determine a three-step mechanism for bentonite aggregates evolution with relative humidity corresponding to an arrangement of particles followed by a particle swelling and a full destructuration. In SEM experiments are also used to characterize chemical reactivity (Camacho-Bragado et al., 2011). It has been recently used to characterize reaction of fly ash activated by sodium silicate by Duchene et al. (2010). These authors have determined very accurately the different steps of the reaction determining that the sodium silicate activator dissolves rapidly and begins to bond fly ash particles. Open porosity was observed and it was rapidly filled with gel as soon as the liquid phase is able to reach the ash particle. The importance of the liquid phase is underlined as a fluid transport medium permitting the activator to reach and react with the fly ash particles. The reaction products had a gel like morphology and no crystallized phase was observed.

4.4 Hydration and dehydration experiments

As previously reported for liposomes, new opportunities for the study of polyelectrolyte microcapsules versus their resistance to relative humidity and temperature modifications are opened and under consideration. The image series reported on Figure 6 clearly illustrate the possibility to image the native soft capsule at high relative humidity without any deformation. When decreasing the water pressure near the capsule, the object is deformed and do not shrink as observed when it is heated in water at temperature higher than 25°C (Basset et al., 2010). Thus, the walls of the object do not rearrange but collapse when submitted to a relative humidity decrease.

Similar tests have been performed on self-organized metal-organic framework compounds (Bonnefond, 2011). According to the image series reported on Figure 7, when the water pressure decreases, the size of sample remains constant up to a given water pressure (i.e. relative humidity) and for a transition pressure, the sample size decreases regularly. This can be associated to a local reorganisation in the sample that corresponds to a water loss associated to the sample collapsing The enthalpy of water ordering in the sample can be derived from the recorded image series as reported by Sievers et al.

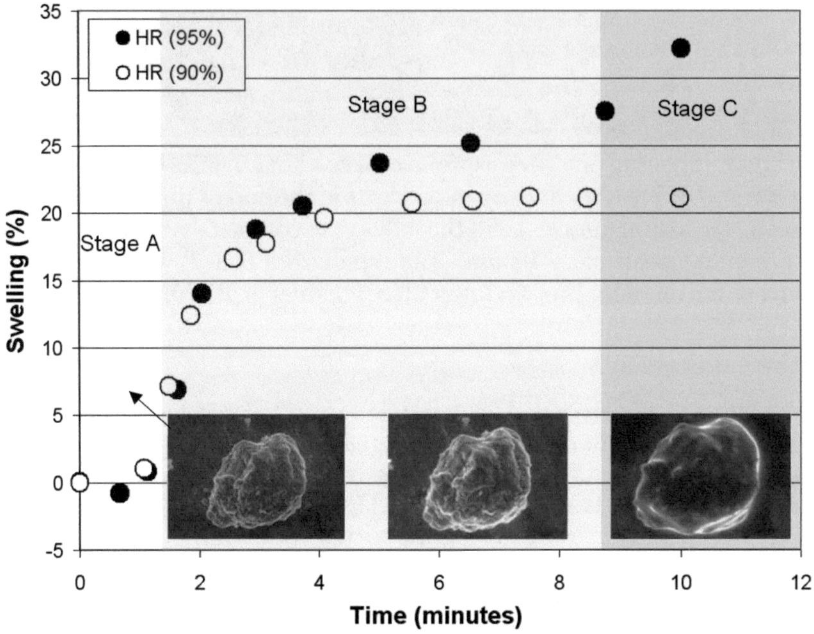

Fig. 5. Swelling kinetics of raw bentonite aggregates scale using ESEM-digital image analyses coupling (after Montes & Swelling, 2005).

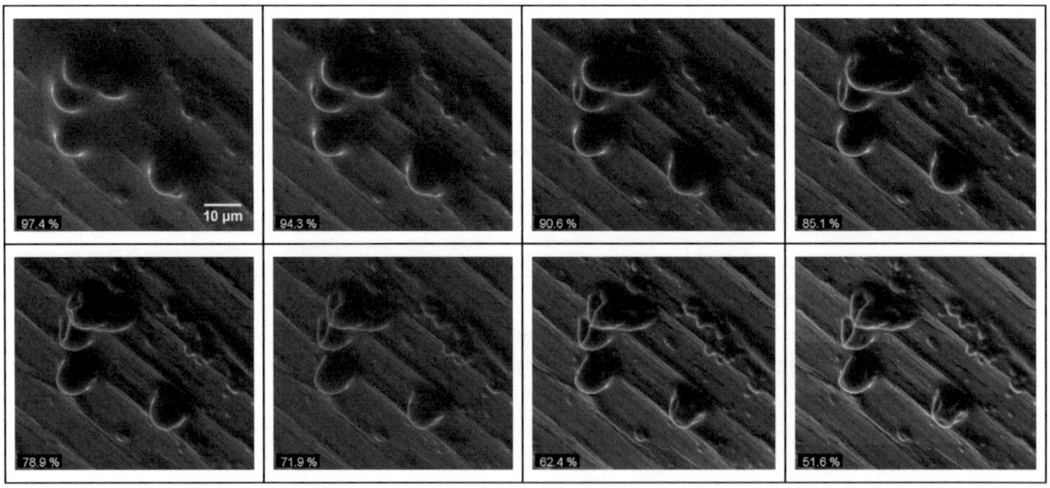

Fig. 6. ESEM micrographs of polyelectrolyte microcapsules suspended in double distilled water. Microcapsules were subjected to controlled dehydration in the ESEM sample chamber at T=5°C. At an operating pressure of 800Pa, vesicles appeared as spherical structures. (a) Gradual decrease of the operating pressure to 350 Pa showed regular deformation of the microcaspsules (b to h)

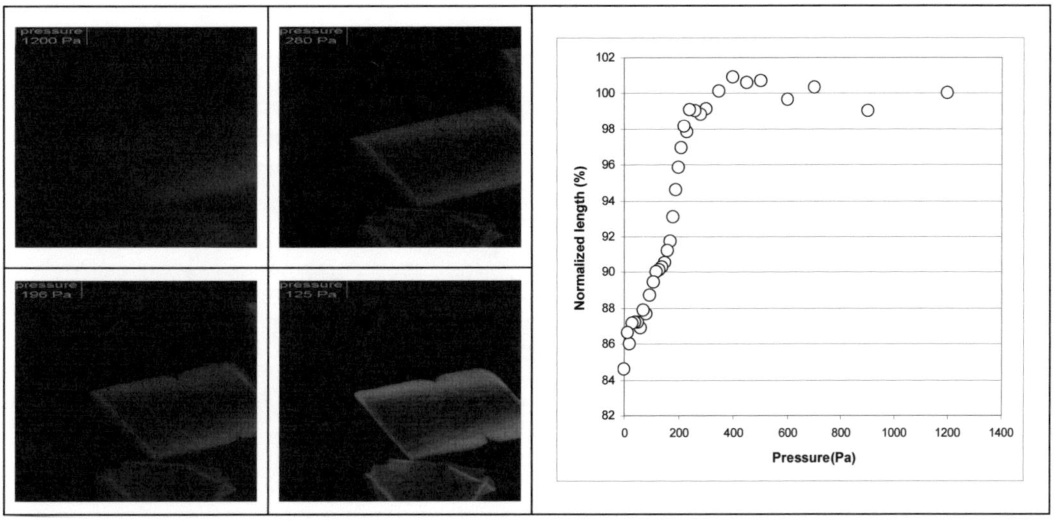

Fig. 7. Dehydration experiments performed on self-assembled organo-metallic compounds at T=22°C and corresponding size modification *versus* water vapour pressure (Bonnefond, 2011).

The effect of dehydration on lamellar bones was also studied by *in situ* ESEM experiments (Utku et al., 2008). The obtained results indicate that dehydration affects the dimensions of lamellar bone in an anisotropic manner in longitudinal sections, whereas in transverse sections the extent of contraction is almost the same in both the radial and tangential directions.

An original work on the heterogeneous ice nucleation on synthetic silver iodide, natural kaolinite and montmorillonite particles has been performed using the "increasing water pressure at constant temperature" (Zimmermann et al., 2007) in the temperature range of 250–270 K. Ice formation was related to the chemical composition of the particles. The obtained data are in very good agreement with previous ones obtained by diffusion chamber measurements (Figure 8).

4.5 Characterization of surface wetting properties

Characterization of the wetting properties of surfaces through the formation of microdroplets or nanodroplets is another important investigation field that can be explored using the ESEM. A recent review by Mendez-Vilas et al. (2009) has highlighted the main fundamental and applied results. Several strategies for the contact angle between water and the surface determination are reported (Stelmashenko et al., 2001; Stokes, 2001; Lau et al., 2003; Wei, 2004; Yu et al., 2006; Jung & Bhushan, 2008; Rykaczewski & Scott, 2011). The investigation of the hydrophobicity and/or hydrophilicity of a catalyst layer have been performed using ESEM for the first time by Yu et al. (2006). These authors have determined the micro-contact angle distribution as a function of the catalyst microstructure. Microdroplets growing and merging process was observed directly in the ESEM chamber by Lau et al. (2003).

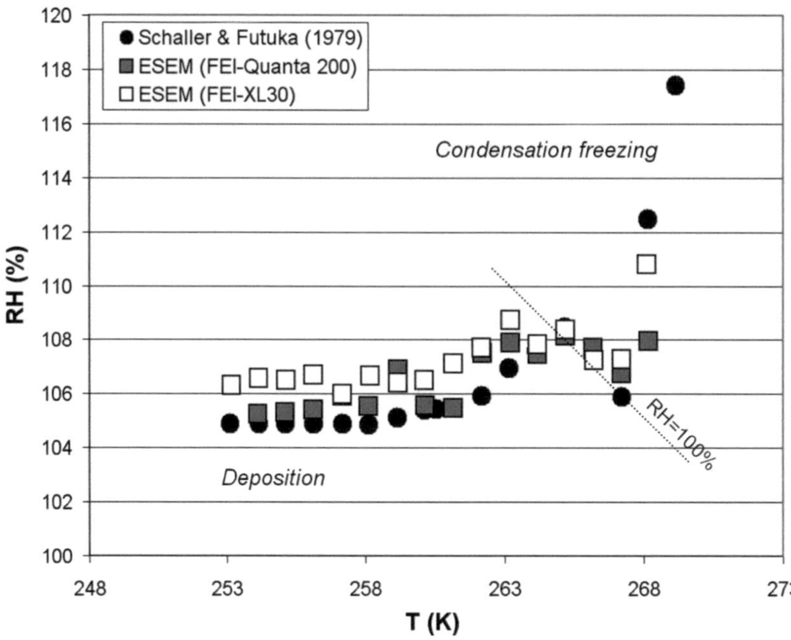

Fig. 8. Supersaturation *versus* temperature diagram for silver iodide (After Zimmermann et al., 2007).

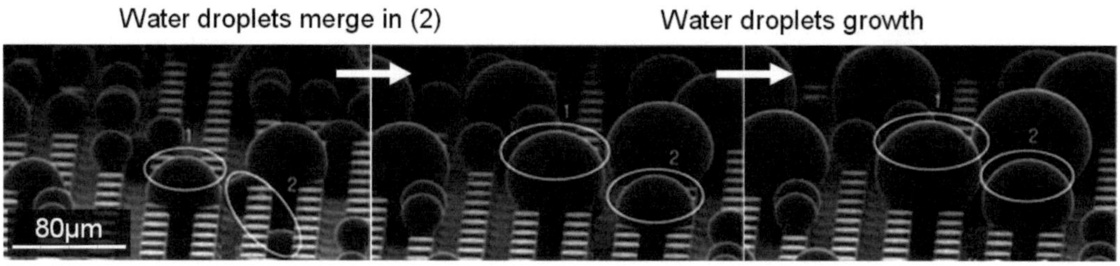

Fig. 9. Microdroplets growing and merging process under ESEM during increasing condensation by decreasing temperature. (After Jung & Bhushan, 2008)

4.6 Using the Wet-STEM mode

The development of the Wet-STEM by Bogner et al. (2005, 2007) allows observing samples in the transmission mode in the ESEM chamber, and more particularly, it offers the possibility to image directly nanoparticles dispersed in a few micrometer thin water film (Bogner et al., 2008), emulsions or vesicles (Maraloiu et al., 2010), without removing the liquid surrounding the objects of interest. One must keep in mind that images with soft matter, and more generally sample sensitive to the electron beam are very hard to obtain. Nevertheless, this technique also opens new research fields using *in situ* experimentation that only begin to be explored for wettability or deliquescence studies. By combining Wet-STEM imaging with Monte-Carlo simulation (Figure 10), Barkay (2010) have studied the initial stages of water nanodroplet condensation over a nonhomogeneous holey thin film. This study has shown a preferred water droplet condensation over the residual water film areas in the holes and has provided corresponding droplet shape and contact angle. On a similar way, Wise et al. (2008) have studied water uptake by NaCl particles prior to deliquescence by varying the relative humidity in the Wet-STEM environment (Figure 11).

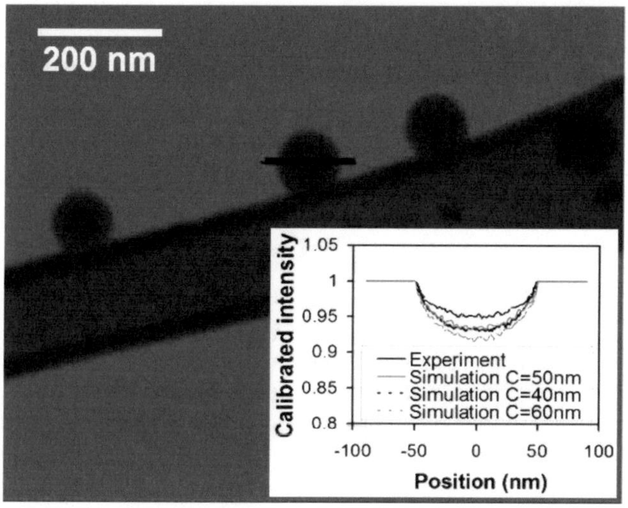

Fig. 10. Bright field image of 100 nm polystyrene latex spheres. Insert is the calibrated intensity corresponding to the dark line in the image (After Barkay (2010))

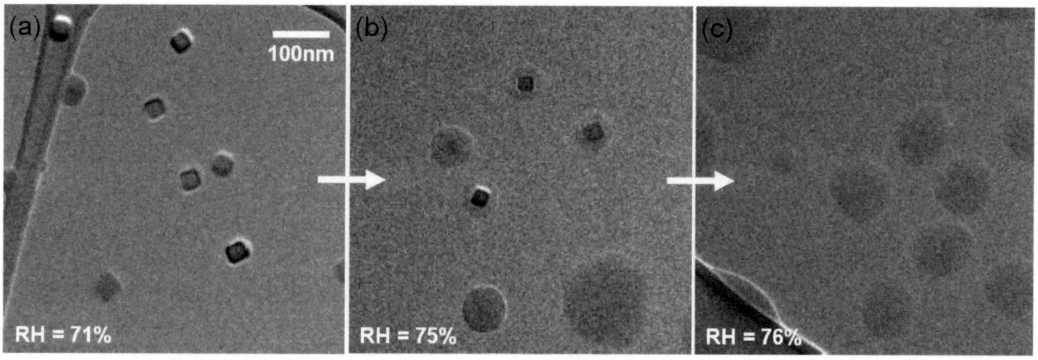

Fig. 11. ~40 nm NaCl particles as the RH was increased past the deliquescence point. Water uptake [(a) → (b)] prior to full deliquescence (c) is clearly observed. (After Wise et al., 2008)

4.7 Development of specific materials for experimentation

Several specific devices have been developed to characterize specific properties or reactions. Two of them will be shortly described below.

Chen et al. (2011) have developed an experimental platform that can be used to investigate chemical reaction pathways, to monitor phase changes in electrodes or to investigate degradation effects in batteries. They have performed *in situ* experiment runs inside a scanning electron microscope (SEM) and tracked the morphology of an electrode including active and passive materials in real time. This work has been used to observe SnO_2 during lithium uptake and release inside a working battery electrode.

Direct imaging of micro ink jets inside the ESEM chamber has been achieved using a specific device developed by Deponte et al. (2009), using a two-fluid stream consisting of a water inner core and a co-flowing outer gas sheath. ESEM images of water jets down to 700 nm diameter have been recorded. Details of the jet structure (the point of jet breakup, size and shape of the jet cone) can be measured. The authors conclude that ESEM imaging of liquid jets offers a valuable research tool for the study of aerosol production, combustion processes, ink-jet generation, and many other attributes of micro- and nanojet systems.

5. High temperature in the SEM

5.1 Application domains of HT-(E)SEM

Specific stages (and associated detectors) have been developed to heat samples up to 1500°C directly in the microscope chamber (Knowles & Evans, 1997; Gregori et al., 2001). The environmental scanning electron microscope (ESEM) equipped with this heating stage is an excellent tool for the *in situ* and continuous observation of system modifications involved by temperature. It allows recording image series of the morphological changes of a sample during a heat treatment with both high magnification and high depth of focus. The experiments can be carried out to observe the influence of all these parameters on the studied phenomenon under various conditions (heating rates, atmosphere compositions, variable pressure, final temperature and heating time). Images have been recorded up to 1400°C, with a decrease of the image resolution when the sample temperature increases (Podor et al., 2012). It is possible to work under vacuum (classical SEM) or under controlled atmosphere (H_2O, O_2, He+H_2, N_2, air...). Different types of studies have been reported, relative to corrosion of metals (Jonsson et al., 2011), oxidation of metals (Schmid et al., 2001a, 2001b ; Oquab & Monceau, 2001 ; Schmid et al., 2002 ; Abolhassani et al., 2003 ; Reichmann et al., 2008 ; Jonsson et al., 2009 ; Mège-Revil et al., 2009 ; Quémarda et al., 2009 ; Delehouzé et al., 2011), reactivity at high temperature (Maroni et al., 1999 ; Boucetta et al., 2010), phase changes (Fischer et al., 2004 ; Hung et al., 2007 ; Beattie & McGrady, 2009), hydrogen desorption (Beattie et al., 2009, 2011), redox reactions (Klemensø et al., 2006), microstructural modifications (Bestmann et al., 2005 ; Fielden, 2005 ; Yang, 2010), magnetic properties (Reichmann et al., 2011), sintering (Sample et al., 1996 ; Srinivasan, 2002 ; Marzagui & Cutard, 2004 ; Smith et al., 2006 ; Subramaniam, 2006 ; Courtois et al., 2011 ; Joly-Pottuz et al., 2011 ; Podor et al., 2012), thermal decomposition (Gualtieri et al., 2008 ; Claparède et al., 2011 ; Goodrich & Lattimer, 2011 ; Hingant et al., 2011), crystallisation (Gomez et al., 2009) in melts (Imaizumi et al., 2003 ; Hillers et al., 2007) and study of self-repairing – self-healing – properties of materials (Wilson & Case, 1997 ; Coillot et al., 2010a, 2010b, 2011) …

Even if numerous researchers are invested in HT-ESEM, only few of them have been successful in pursuing dynamic experiments at temperatures higher than 1100°C. Two recent studies report experiments performed at T=1350°C (Subramaniam, 2005) and 1450°C (Gregori et al., 2002). However, the resolution of the images remains poor (more than 1μm) mainly due to water cooling induced vibrations. Furthermore, the precision on the measure of the sample temperature remains poor (temperature differences up to 150°C with the expected temperature are sometimes measured). A recent device has been proposed by Podor et al. (2011) to overcome this difficulty.

A complete review specifically dedicated to *in situ* high temperature experimentation in the ESEM will be available soon. Several examples of *in situ* studies performed at high temperature in the ESEM chamber will be reported below, on the basis of original data acquired in our laboratory.

5.2 Investigation of the crystallization behaviour in silicate melts

The crystal growth and morphology during isothermal heating of glass melts can be directly observed using the hot stage associated with the ESEM. The image series reported on Figure 12 have been recorded during 10 minutes while heating the borosilicate melt sample isothermally at T=740°C. The development of large crystals in the melt rapidly yields to the complete crystallization of the melt. The crystal morphology presents cells filled with a second phase and the crystal formation yields to the deformation of the sample surface. Hillers et al. (2007) have used such data to quantify the variation of crystal length with time. They have established that the growth is only linear during the first minutes; afterward the growth rate decreases progressively with time.

This technique can also be used to determine the temperature of formation of the first crystals at the melt surface and to observe their formation. In the case of glass-ceramics, the density of nuclei as well as their size and shape development can be directly observed and used for crystallization kinetic determination (Vigouroux et al., 2011, in prep).

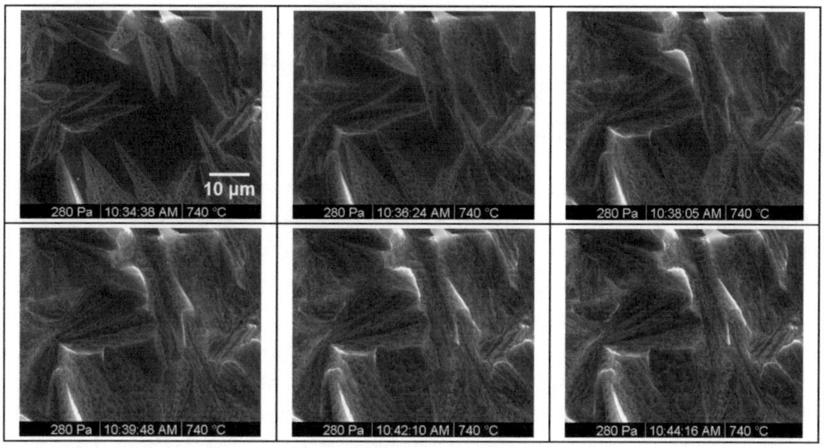

Fig. 12. Growth of crystals in a borosilicate melt during 10 minutes isothermal heat treatment at 740°C observed using the hot stage associated with the ESEM.

5.3 Decomposition of compounds

In situ thermal decomposition of composites, oxalates, oxides have been reported by several authors. Images of the heat treatment of a mixed uranium-cerium oxalate grain from 25°C to 1235°C are gathered on Figure 13. Morphological changes with temperature are directly linked with the oxalate decomposition as stated by Hingant et al. (2011) in the temperature range 25-500°C. The sample shrinkage observed when T>500°C is probably related with the first stage of the sintering process - i.e. beginning of bond formation between the nanograins and with the oxide grain growth (that can not be directly observed at this stage by HT-ESEM, but that is confirmed by X-Ray diffraction). Such a process has also been recently reported by Claparede et al. (2011) and Joly-Pottuz et al. (2011).

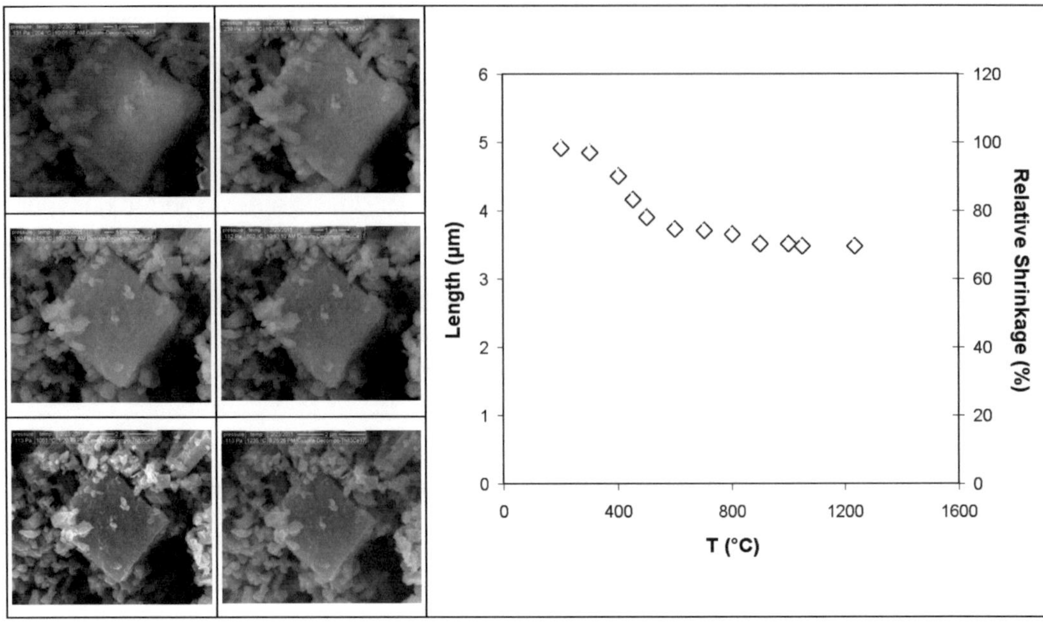

Fig. 13. Decomposition of a uranium-cerium mixed oxalate observed during *in situ* heating in the ESEM chamber and relative size and shrinkage modifications.

5.4 Study of sintering and grain growth

Several studies are relative to the sintering and grain growth processes in metals and ceramics. Depending on the system, the experiments have been performed in the temperature range 300-1450°C. The main interest of these studies is the possibility of direct observation of the individual grain behaviour during heat treatment. The example that is reported on Figure 14a corresponds to the heat treatment of the grain decomposed *in situ* (Figure 13). The image resolution is high enough to observe the nanograins growth inside the square plate agglomerate. Consequently, relative shrinkage and average grain diameter are extracted by image processing (Figure 14b). Assuming that the final density of the agglomerate is 99%, the sintering map is directly derived from these experimental data (Figure 14c). Thus, *in situ* sintering experiments can allow the establishment of the trajectories of theoretical sintering. Such data have never been already reported in previous studies, mainly due to the poor resolution of the recorded images.

The effect of the electron beam on sintering is controversy. Indeed, Popma (2002) noted that a local sintering stop was achieved by focusing the electron beam at a certain position during the *in situ* sintering experiments in the ESEM (performed on ZrO_2 nanolayers). On the contrary, Courtois et al (2011) performed experiments on the sintering of a lead phosphovanadate and concluded that the electric current induced by the electron beam was found to reduce the effective temperature of sintering by 50 to 150°C as well as to accelerate the kinetics of shrinkage of a cluster composed of sub-micrometric grains of material. Such effects were not evidenced in our study: the local sintering on sample surface zones that were not observed (i.e. exposed to the electron beam) was identical to the local sintering determined on the observed zone.

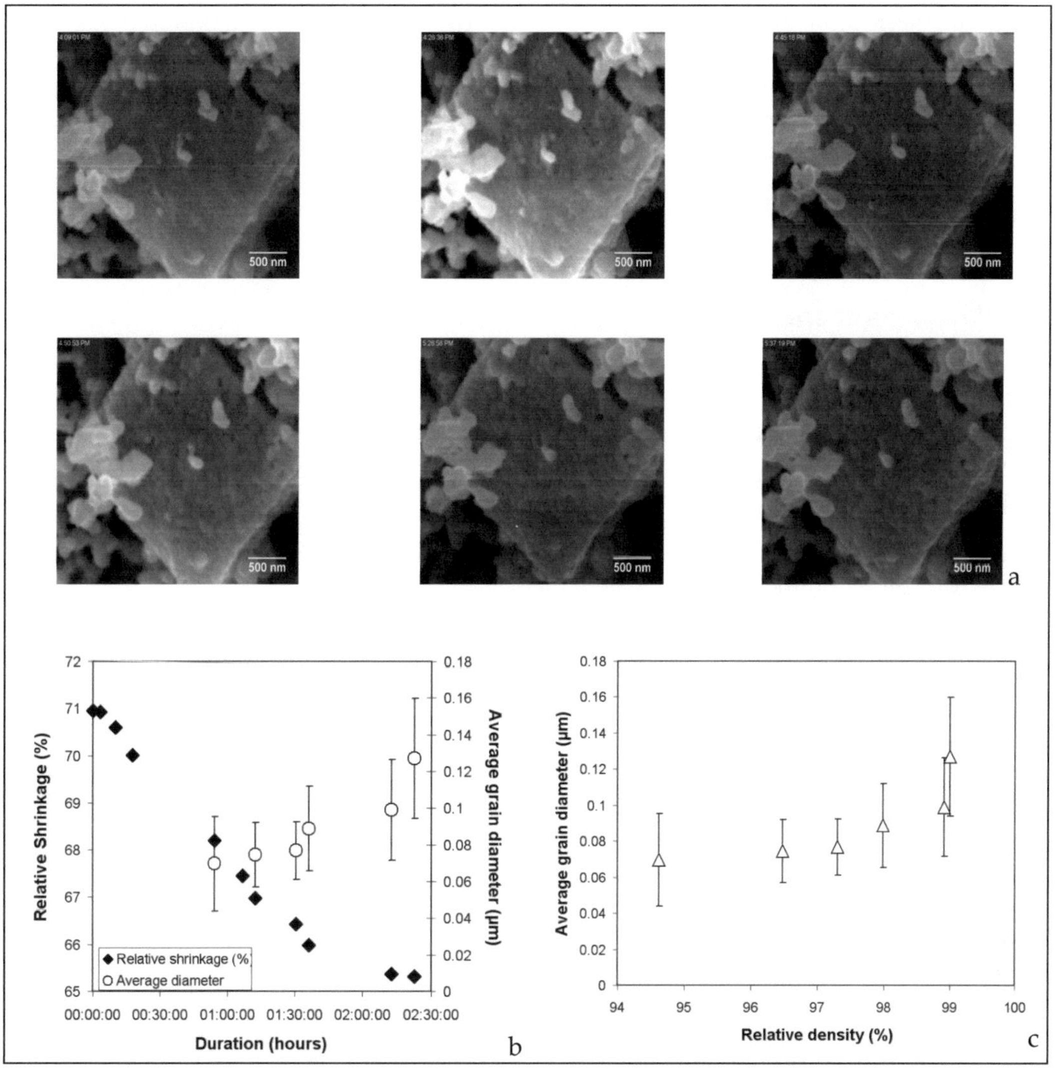

Fig. 14. (a) Sintering and grain growth of a uranium-cerium mixed oxide observed *in situ* in the ESEM chamber at T=1235°C, after 55′, 70′, 90′, 95′, 130′, 140′ (a). Corresponding Relative (b) Shrinkage and Average grain diameter versus duration and (c) derived sintering map - Grain growth versus densification rate –

6. Conclusions and perspectives

In situ scanning electron microscopy experimentation, that is generally associated with the use of the ESEM, allows the study of very different problems, the main limit being the availability of specific devices. Torres & Ramirez (2011) have written the best conclusion indicating that "the new generation of SEMs shows innovative hardware and software solutions that result in improved performance. This progress has turned the SEM into an extraordinary tool to develop more complex and realistic *in situ* experiments, achieving even at the subnanometer scale". In the near future, new SEM imaging modes, nanomanipulation and nanofabrication technologies (Miller & Russell, 2007 ; Romano-Rodriguez & Hernandez-Ramirez, 2007 ; Wich et al., 2011) will make possible to replicate more closely the conditions as the ones associated to the problems to be treated. *In situ* ESEM will probably be used to overcome technical and fundamental challenges in many scientific domains. The recent developments of a high temperature stage in the FIB (Fielden, 2008), a new tomography mode in the ESEM (Jornsanoh et al., 2011) and of the atmospheric scanning electron microscope (Nishiyama et al, 2010 ; Suga et al, 2011) can be cited as examples for this future.

7. Acknowledgment

The authors want to thank all the co-workers of the studies cited in this chapter, and more particularly F. Bonnefond, H. Boucetta, C. Dejugnat, T. Demars, A. Monteiro and L. Claparède for providing the samples and challenging projects.

8. References for videos

Reactivity of a salt with silicate melt at high temperature	http://www.dailymotion.com/icsmweb#videoId=xjknrt
Sintering of CeO_2 at T=1200°C	http://www.youtube.com/watch?v=4ijIUdQe3M4
Self-healing of a metal-glass composite at high temperature	http://www.dailymotion.com/icsmweb#videoId=xjknpp
Deformation of vesicles during dehydration	http://www.dailymotion.com/icsmweb#videoId=xjk75u
NaCl solubility and precipitation in water	http://www.dailymotion.com/icsmweb#videoId=xk22i9

9. References

Abolhassani, S., Dadras, M., Leboeuf, M. & Gavillet, D. (2003). In situ study of the oxidation of Zircaloy-4 by ESEM. *Journal of Nuclear Materials* 321, 70-77.

Ahmad, M.R., Nakajima, M., Kojima, S, Homma, M. & Fukada, T. (2010). *Single cell analysis inside ESEM – (ESEM)-nanomanipulator system* , InTech, ISBN 978-953-7619-93-0, "Cutting Edge Nanotechnology" 413-438.

Barkay, Z. (2010). Wettability study using transmitted electrons in environmental scanning electron microscope. *Applied Physic Letters* 96, 183109.

Basset, C., Harder, C., Vidaud, C. & Déjugnat, C. (2010). Design of Double Stimuli-Responsive Polyelectrolyte Microcontainers for Protein Soft Encapsulation. *Biomacromolecules* 11, 806–814.

Beattie, S.D. & McGrady, G.S. (2009). Hydrogen desorption studies of $NaAlH_4$ and $LiAlH_4$ by in situ heating in an ESEM. *International Journal of Hydrogen Energy* 34, 9151-9156.

Beattie, S.D., Langmi, H.W. & McGrady, G.S. (2009). In situ thermal desorption of H_2 from LiNH2-2LiH monitored by environmental SEM. *International Journal of Hydrogen Energy* 34, 376-379.

Beattie, S.D., Setthanan, U. & McGrady, G.S. (2011). Thermal desorption of hydrogen from magnesium hydride (MgH_2): An in situ microscopy study by environmental SEM and TEM. *International Journal of Hydrogen Energy* in press.

Bestmann, M., Piazolo, S., Spiers, C.J. & Prior, D.J. (2005). Microstructural evolution during initial stages of static recovery and recrystallization: new insights from in-situ heating experiments combined with electron backscatter diffraction analysis. *Journal of Structural Geology* 27, 447–457.

Biallas, G. & Maier, H.J. (2007). In-situ fatigue in an environmental scanning electron microscope - Potential and current limitations. *International Journal of Fatigue* 29, 1413–1425.

Boehlert, C.J., Cowen, C.J., Tamirisakandala, S., McEldowney, D.J. & Miracle, D.B. (2006).In situ scanning electron microscopy observations of tensile deformation in a boron-modified Ti-6Al-4V alloy. *Scripta Materialia* 55, 465–468.

Boehlert, C. J. (2011). In situ scanning electron microscopy for understanding the deformation behaviour of structural materials. Seminarios Internacionales de Fronteras de la Ciencia de Materiales. April 11th, 2011 (http://www.youtube.com/watch?v=wH3EYxT_ysM)

Bogner, A., Guimarães, A., Guimarães, R.C.O., Santos, A.M., Thollet, G., Jouneau, P.H. & Gauthier, C. (2008). Grafting characterization of natural rubber latex particles: wet-STEM imaging contributions. *Colloid and Polymer Science* 286, 1049–1059.

Bogner, A., Jouneau, P.H., Thollet, G., Basset, D. & Gauthier C. (2007). A history of scanning electron microscopy developments: Towards "wet-STEM" imaging. *Micron* 38, 390–401.

Bogner, A., Thollet, G., Basset, D., Jouneau, P.H. & Gauthier, C., (2005). Wet STEM: A new development in environmental SEM for imaging nano-objects included in a liquid phase *Ultramicroscopy* 104, 290-301.

Bonnefond, F. (2011). *Etude in situ de la déshydratation de composés organométalliques*. Master 1 thesis (30p.)

Boucetta, H., Schuller, S., Ravaux, J & Podor, R. (2010). *Etude des mécanismes de formation des phases cristallines RuO_2 dans les verres borosilicate de sodium*. Proceeding of Matériaux 2010 (18-22 oct Nantes, France)

Camacho-Bragado, G.A., Dixon, F. & Colonna, A. (2011). Characterization of the response to moisture of talc and perlite in the environmental scanning electron microscope. *Micron* 42, 257-262.

Chen, D., Indris, S., Schulz, M., Gamer, B. & Mönig, R. (2011). In situ scanning electron microscopy on lithium-ion battery electrodes using an ionic liquid. *Journal of Power Sources* 196, 6382–6387.

Chen, W., Boehlert, C.J. (2010). The 455°C tensile and fatigue behavior of boron-modified Ti–6Al–2Sn–4Zr–2Mo–0.1Si(wt.%). *International Journal of Fatigue* 32, 799-807.

Claparède, L., Clavier, N., Dacheux, N., Moisy, P., Podor, R. & Ravaux, J. (2011). Influence of crystallization state and microstructure on the chemical durability of cerium-neodynium mixed dioxides. *Inorganic Chemistry*, 50[18], 9059–9072.

Coillot, D., Podor, R., Méar, F.O. & Montagne, L. (2010a). Characterisation of self-healing glassy composites by high-temperature environmental scanning electron microscopy (HT-ESEM). *Journal of Electron Microscopy* 59, 359-366.

Coillot, D., Méar, F.O., Podor, R. & Montagne, L. (2010b). Autonomic Self-Repairing Glassy Materials. *Advanced Functional Materials* 20(24), 4371-4374.

Coillot, D., Méar, F.O., Podor, R. & Montagne, L. (2011). Influence of the Active Particles on the Self-Healing Efficiency in Glassy Matrix. *Advanced Engineering Materials* 13, 426-435.

Courtois, E., Thollet, G., Campayo, L., Le Gallet, S., Bidault, O. & Bernard, F. (2011). In situ study of the sintering of a lead phosphovanadate in an Environmental Scanning Electron Microscope. *Solid State Ionics* 186, 53–58.

Delehouzé, A., Rebillat, F., Weisbecker, P., Leyssale, J.M., Epherre, J.F., Labrugère C. & Vignoles G.L. (2011). Temperature induced transition from hexagonal to circular pits in graphite oxidation by O_2. *Applied Physics Letters* 99, 044102.

DePonte, D.P., Doak, R.B., Hunter, M., Liu, Z., Weierstall, U. & Spence, J.C.H. (2009). SEM imaging of liquid jets. *Micron* 40, 507–509.

Donald, A.M. (2003). The use of environmental scanning electron microscopy for imaging wet and insulating materials. *Nature Materials* 2, 511-516.

Duchene, J., Duong, L., Bostrom, T. & Frost, R. (2010). Microstructure study of early in situ reaction of fly ash geopolymer observed by ESEM. *Waste Biomass Valorisation* 1, 367–377.

FEI movie (1998) http://www.dailymotion.com/video/xirinx_acariens-cannibales-les-envahisseurs-invisibles_animals

Fielden, I.M. (2005). *Investigation of microstructural evolution by real time SEM of high temperature specimens.* PhD thesis Sheffield Hallam University (170p).

Fielden, I.M. (2008). In-Situ Focused Ion Beam (FIB) microscopy at high temperature. *Electron Microscopy and Analysis Group.*

Fischer, S., Lemster, K., Kaegi, R., Kuebler, J. & Grobety, B. (2004). In situ ESEM observation of melting silver and inconel on an Al_2O_3 powder bed. *Journal of Electron Microscopy* 53, 393-396.

Fonseca, P.C. & Jennings, H.M. (2010). The effect of drying on early-age morphology of C–S–H as observed in environmental SEM. *Cement and Concrete Research* 40, 1673–1680.

Gianola, D.S., Sedlmayr, A., Mönig, R., Volkert, C.A., Major, R.C., Cyrankowski, E., Asif, S.A.S., Warren, O.L., & Kraft, O. (2011). In situ nanomechanical testing in focused ion beam and scanning electron microscopes. *Review of Scientific Instruments* 82, 063901.

Gómez, L.S., López-Arce, P., Álvarez de Buergo, M. & Fort, R. (2009). Calcium hydroxide nanoparticles crystallization on carbonates stone: dynamic experiments with heating/cooling and Peltier stage ESEM. *Acta Microscopica* 18, 105-106.

Goodrich, T.W. & Lattimer, B.Y. (2011). Fire Decomposition Effects on Sandwich Composite Materials. *Composites A*, doi:10.1016/j.compositesa.2011.03.007

Gorkaya, T., Burlet, T., Molodov, D.A. & Gottstein, G. (2010). Experimental method for true in situ measurements of shear-coupled grain boundary migration. *Scripta Materialia* 63, 633–636.

Gregori, G., Kleebe, H.J., Siegelin, F. & Ziegle, G.(2002). In situ SEM imaging at temperatures as high as 1450°C. *Journal of Electron Microscopy* 51, 347-52.

Gualtieri, A.F., Lassinantti Gualtieri, M. & Tonelli, M. (2008). In situ ESEM study of the thermal decomposition of chrysotile asbestos in view of safe recycling of the transformation product. *Journal of Hazardous Materials* 156, 260-266.

Hillers, M., Matzen, G., Veron, E., Dutreilh-Colas, M. & Douy, A. (2007). Application of In Situ High-Temperature Techniques to Investigate the Effect of B_2O_3 on the

Crystallization Behavior of Aluminosilicate E-Glass. *Journal of the American Ceramic Society* 90, 720-726.

Hingant, N., Clavier, N., Dacheux, N., Hubert, S., Barré, N., Podor, R. & Aranda, L. (2011). Preparation of morphology controlled $Th_{1-x}U_xO_2$ sintered pellets from low-temperature precursors. *Powder Technology* 208, 454–460.

Hung, J.H.H., Chiu, Y.L., Zhu, T. & Gao, W. (2007). In situ ESEM study of partial melting and precipitation process of AZ91D. *Asia-Pacific Journal of Chemical Engineering*, Special Issue: Special issue for the Chemeca 2006 John A Brodie Medal Nominated Papers. Volume 2, Issue 5, pages 493–498, September/October 2007.

Imaizumi, K., Matsuda, N. & Otsuka, M. (2003). Coagulation/phase separation process in the silica/inorganic salt systems (1)—observation of state transformation— *Journal of Materials Science* 38, 2979 - 2986.

James, B. (2009). Advances in "wet" electron microscopy techniques and their application to the study of food structure. *Trends in Food Science & Technology* 20, 114-124.

Joly-Pottuz, L., Bogner, A., Lasalle, A., Malchere, A., Thollet, G. & Deville, S. (2011). Improvements for imaging ceramics sintering in situ in ESEM. *Journal of Microscopy*, 244, 93-100.

Jonsson, T., Folkeson, N., Svensson, J.E., Johansson, L.G., & Halvarsson M. (2011). An ESEM in situ investigation of initial stages of the KCl induced high temperature corrosion of a Fe–2.25Cr–1Mo steel at 400 °C. *Corrosion Science* 53, 2233-2246.

Jonsson, T., Pujilaksono, B., Hallström, S., Ågren, J., Svensson, J.E., Johansson, L.G. & Halvarsson, M. (2009). An ESEM in situ investigation of the influence of H_2O on iron oxidation at 500°C. *Corrosion Science* 51, 1914-1924.

Jornsanoh, P., Thollet, G., Ferreira, J., Masenelli-Varlot, K., Gauthier, C. & Bogner, A. (2011). Electron tomography combining ESEM and STEM: A new 3D imaging technique *Ultramicroscopy*, doi:10.1016/j.ultramic.2011.01.041

Jung, Y.C. & Bhushan, B. (2008). Wetting behaviour during evaporation and condensation of water microdroplets on superhydrophobic patterned surfaces *Journal of Microscopy* 229, 127-140.

Kirk, S.E., Skepper, J.N. & Donald, A.M. (2009). Application of environmental scanning electron microscopy to determine biological surface structure. *Journal of Microscopy* 233, 205–224.

Klemensø, T., Appel, C.C. & Mogensen, M. (2006). In Situ Observations of Microstructural Changes in SOFC Anodes during Redox Cycling. *Electrochemical and Solid-State Letters* 9, A403-A407

Knoll, M. (1935). Aufladepototentiel une Sekündaremission elektronbestrahlter Körper. Zeitschrift fur technische Physik 16, 467-475.

Knowles, R. & Evans, B. (1997). *High temperature specimen stage and detector for an ESEM.* Patent WO 97/07526

Lau, K.K.S., Bico, J., Teo, K.B.K., Chhowalla, M., Amaratunga, G.A.J., Milne, W.I., McKinley, G.H. & Gleason, K.K. (2003). Superhydrophobic Carbon Nanotube Forests. *Nano Letters* 3, 1701-1705.

Leary, R. & Brydson, R. (2010). Characterisation of ESEM conditions for specimen hydration control. *Journal of Physics: Conference Series* 241, 012024.

Lin, T., Jia, D. & Wang, M. (2010). In situ crack growth observation and fracture behavior of short carbon fiber reinforced geopolymer matrix composites. *Materials Science and Engineering* A 527, 2404–2407.

Maison, T., Laouafa, F., Fleureau, J.M. & Delalain, P. (2009). *Analyse aux échelles micro et macroscopique des mécanismes de dessiccation et de gonflement des sols argileux.* Proceeding of the 19ème Congrès Français de Mécanique Marseille, 24-28 août 2009.

Manero, J.M., Masson, D.V., Marsal, M. & Planell, J.A. (1998). Application of the Technique of Environmental Scanning Electron Microscopy to the Paper Industry. *Scanning* 21, 36–39.

Maraloiu, V.A., Hamoudeh, M., Fessi, H. & Blanchin, M.G. (2010). Study of magnetic nanovectors by Wet-STEM, a new ESEM mode in transmission. *Journal of Colloid and Interface Science* 352, 386–392.

Maroni, V.A., Teplitsky, M. & Rupich M.W. (1999). An environmental scanning electron microscope study of the AgrBi-2223 composite conductor from 25 to 840°C. *Physica* C 313, 169–174.

Marzagui, H. & Cutard, T. (2004). Characterisation of microstructural evolutions in refractory castables by in situ high temperature ESEM. *Journal of Materials Processing Technology* 155-156, 1474-1481.

McGregor, J.E. & Donald, A.M. (2010). ESEM imaging of dynamic biological processes: the closure of stomatal pores. *Journal of Microscopy* 239, 135–141.

Mège-Revil, A., Steyer, P., Thollet, G., Chiriac, R., Sigala, C., Sanchéz-Lopéz, J.C. & Esnouf, C. (2009). Thermogravimetric and in situ SEM characterisation of the oxidation phenomena of protective nanocomposite nitride films deposited on steel. *Surface & Coatings Technology 204*, 893–901

Mendez-Vilas, A., Belen Jodar-Reyes, A., & Gonzalez-Martin, M.L. (2009). Ultrasmall Liquid Droplets on Solid Surfaces: Production, Imaging, and Relevance for Current Wetting Research. *Small* 5(12), 1366–1390.

Miller, M.K. & Russell, K.F. (2007). Atom probe specimen preparation with a dual beam SEM/FIB miller. *Ultramicroscopy* 107, 761-766.

Montes-H., G. (2005). Shrinkage measurements of bentonite using coupled environmental scanning electron microscopy and digital image analysis. *Journal of Colloid and Interface Science* 284, 271–277.

Montes-Hernandez, G. (2002). *Etude expérimentale de la sorption d'eau et du gonflement des argiles par microscopie électronique à balayage environnementale (ESEM) et l'analyse digitale d'images*. Thèse de 3éme cycle 162pp.

Muscariello, L., Rosso, F., Marino, G., Giordano, A., Barbarisi, M., Cafiero, G. & Barbarisi, A. (2005). A Critical Overview of ESEM Applications in the Biological Field. *Journal of Cellular Physiology* 205, 328–334.

Nakamura, M., Isshiki, T., Tamai, M. & Nishio, K. (2002). *Development of a new heating stage equipped thermal electron filter for scanning electron microscopy*. Proceeding of the 15th International Congress on Electron Microscopy Durban, South Africa.

Nishiyama, H., Suga, M., Ogura, T., Maruyama, Y., Koizumi, M., Mio, K., Kitamura, S. & Sato, C. (2010). Atmospheric scanning electron microscope observes cells and tissues in open medium through silicon nitride film. *Journal of Structural Biology* 169, 438-449.

Oquab, D. & Monceau, D. (2001). In-situ SEM study of cavity growth during high temperature oxidation of β-(Ni, Pd)Al. *Scripta Materialia* 44, 2741-2746.

Perrie, Y., Ali, H., Kirby, D.J., Mohammed, A.U.R., McNeil, S.E. & Vangala A. (2010). *Environmental Scanning Electron Microscope Imaging of Vesicle Systems in "Liposomes: Methods and Protocols, Volume 2: Biological Membrane Models"*, Methods in Molecular Biology 606, 319-331.

Perrie, Y., Mohammed, A.U.R., Vangala, A. & McNeil, S.E. (2007). Environmental Scanning Electron Microscopy Offers Real-Time Morphological Analysis of Liposomes and Niosomes. *Journal of Liposome Research* 17, 27-37.

Podor, R., Clavier, N., Ravaux, J., Claparéde, L., Dacheux, N. & Bernache-Assollant, D. (2012). Dynamic aspects of cerium dioxide sintering: HT-ESEM study of grain growth and pore elimination. *Journal of the European Ceramic Society*, 32, 353-362.

Podor, R., Pailhon, D., Ravaux, J. & Brau, H.P. (2011). *Porte-échantillon à thermocouple intégré.* Demande de brevet français déposée le 21 juillet 2011 sous le n° 11 56612.

Poelt, P., Zankel, A., Gahleitner, M., Ingolic, E. & Grein, C. (2010). Tensile tests in the environmental scanning electron microscope (ESEM) - Part I: Polypropylene homopolymers. *Polymer* 51, 3203-3212.

Popma, R.L.W. (2002). *Sintering characteristics of nano-ceramics*, PhD thesis, University of Groningen.

Proff, C., Abolhassani, S., Dadras, M.M. & Lemaignan, C. (2010). In situ oxidation of zirconium binary alloys by environmental SEM and analysis by AFM, FIB, and TEM. *Journal of Nuclear Materials* 404, 97–108.

Quémarda, L., Desgranges, L., Bouineau, V., Pijolat, M., Baldinozzi, G., Millot, N., Nièpce, J.C. & Poulesquen, A. (2009). On the origin of the sigmoid shape in the UO_2 oxidation weight gain curves. *Journal of the European Ceramic Society* 29, 2791–2798.

Reichmann, A., Poelt, P., Brandl, C., Chernev, B. & Wilhelm, P. (2008). High-Temperature Corrosion of Steel in an ESEM With Subsequent Scale Characterisation by Raman Microscopy. *Oxidation of Metals* 78, 257-266.

Reichmann, A., Zankel, A., Reingruber, H., Pölt, P. & Reichmann, K. (2011). Direct observation of ferroelectric domain formation by environmental scanning electron microscopy. *Journal of the European Ceramic Society* , 31[15], 2939-2942.

Romano-Rodriguez, A. & Hernandez-Ramirez, F. (2007). Dual-beam focused ion beam (FIB): A prototyping tool for micro and nanofabrication. *Microelectronic Engineering* 84,789–792.

Ruozi, B., Belletti, D., Tombesi, A., Tosi, G., Bondioli, L., Forni, F. & Vandelli, M.A. (2011). AFM, ESEM, TEM, and CLSM in liposomal characterization: a comparative study. *International Journal of Nanomedicine* 6, 557–563.

Rykaczewski, K. & Scott, J.H.J. (2011). Methodology for Imaging Nano-to-Microscale Water Condensation Dynamics on Complex Nanostructures. *ACSNano* 5[7], 5962-5968.

Sample, D.R., Brown, P.W. & Dougherty, J.P. (1996). Microstructural evolution of copper thick films observed by environmental scanning electron microscopy. *Journal of the American Ceramic Society* 79, 1303-1306.

Schaller, R.C., Fukuta, N. (1979). Ice nucleation by aerosol particles: Experimental studies using a wedge-shaped ice thermal diffusion chamber. *Journal of Atmospheric Sciences* 36, 1788–1802.

Schmid, B., Aas, N., Grong, Ø. & ØDegard, R. (2001). High-temperature oxidation of nickel and chromium studied with an in-situ environmental scanning electron microscope. *Scanning* 23, 255-266.

Schmid, B., Aas, N., Grong, Ø. & ØDegard, R. (2001). In situ environmental scanning electron microscope observations of catalytic processes encountered in metal dusting corrosion on iron and nickel. *Applied Catalysis* A 215, 257-270.

Schmid, B., Aas, N., Grong, Ø. & ØDegard, R. (2002). High-Temperature Oxidation of Iron and the Decay of Wüstite Studied with in situ ESEM. *Oxidation of Metals* 57, 115-130.

Schoßig, M., Zankel, A., Bieröge, C., Pölt, P. & Grellmann, W. (2011). ESEM investigations for assessment of damage kinetics of short glass fibre reinforced thermoplastics – Results of in situ tensile tests coupled with acoustic emission analysis. *Composites Science and Technology* 71, 257-265

Seward, G.G.E., Prior, D.J., Wheeler, J., Celotto, S., Halliday, D.J.M., Paden, R.S. & Tye, M.R. (2002). High-Temperature Electron Backscatter Diffraction and Scanning Electron Microscopy Imaging Techniques: In-situ Investigations of Dynamic Processes. *Scanning* 24, 232–240.

Seward, G.G.E., Celotto, S., Prior, D.J., Wheeler, J. & Pond R.C. (2004). In situ SEM-EBSD observations of the hcp to bcc phase transformation in commercially pure titanium. *Acta Materialia* 52, 821-832.

Sievers, T.K., Bonnefond, F., Demars, T., Genre, C., Meyer, D., Podor, R. Vapour pressure dependent size of coordination polymer network meso-particles. *Advanced Materials* (submitted)

Smith, A.J., Atkinson, H.V., Hainsworth, S.V. & Cocks, A.C.F. (2006). Use of a micromanipulator at high temperature in an environmental scanning electron microscope to apply force during the sintering of copper particles. *Scripta Materialia* 55, 707-710.

Sorgi, C. & De Gennaro, V. (2007). Analyse microstructurale au MEB environnemental d'une craie soumise à chargement hydrique et mécanique. *Comptes-rendus Geosciences* 339, 468–481.

Srinivasan, N.S. (2002). Dynamic study of changes in structure and morphology during the heating and sintering of iron powder. *Powder Technology* 124, 40-44.

Stabentheiner, E., Zankel, A. & Pölt, P. (2010). Environmental scanning electron microscopy (ESEM)—a versatile tool in studying plants. *Protoplasma* 246:89–99.

Stelmashenko, N.A., Craven, J.P., Donald, A.M., Terentjev, E.M. & Thiel, B.L. (2001). Topographic contrast of partially wetting water droplets in environmental scanning electron microscopy. *Journal of Microscopy* 204, 172-183.

Stokes, D.J. & Donald A.M. (2000). In situ mechanical testing of dry and hydrated breadcrumb in the ESEM. *Journal of Materials Science* 35, 599-607.

Stokes, D.J. (2001). Characterisation of Soft Condensed Matter and Delicate Materials Using Environmental Scanning Electron Microscopy (ESEM). *Advanced Engineering Materials* 3, 126-130.

Stokes, D.J. (2008). *Principles and practice of variable pressure/environmental scanning electron microscopy (VP/ESEM)*, John Wiley & Sons Ltd, The Atrium, Southern Gate, Chichester, West Sussex, UK.

Subramaniam S. (2006). *In Situ High Temperature Environmental Scanning Electron Microscopic Investigations of Sintering Behavior in Barium Titanate*. PhD thesis, University of Cincinnati, Cincinnati USA.

Suga, M., Nishiyama, H., Konyuba, Y., Iwamatsu, S., Watanabe, Y., Yoshiura, C., Ueda, T. & Sato, C. (2011). The Atmospheric Scanning Electron Microscope with open sample space observes dynamic phenomena in liquid or gas. *Ultramicroscopy*, doi:10.1016/j.ultramic.2011.08.001

Thiel, B.L. & Donald, A.M. (1998). In situ Mechanical Testing of Fully Hydrated Carrots (Daucus carota) in the Environmental SEM. *Annals of Botany* 82: 727-733.

Thiel, B.L., Stokes, D.J. & Donald, A.M. (2002). Application of Environmental Scanning Electron Microscopy to the Study of Food Systems. *Microscopy and Microanalysis* 8, 960-961.

Torres, E.A. & Ramirez, A.J. (2011). In situ scanning electron microscopy. *Science and technology of welding and Joining* 16(1), 68-78.

Utku, F.S., Klein, E., Saybasili, H., Yucesoy, C.A. & Weiner, S. (2008). Probing the role of water in lamellar bone by dehydration in the environmental scanning electron microscope. *Journal of Structural Biology* 162, 361-367.

Vigouroux, H., Fargin, E., Le Garrec, B., Dussauze, M., Rodriguez, V., Adamietz, F., Ravaux, J., Podor, R., Vouagner, D., De Ligny, D. & Champagnon, B. (2011). *Phase Separation and Crystallization Mechanism in $LiNbO_3$-SiO_2 Glasses*. International conference on the chemistry of glasses, 4-8 sept 2011, Oxford (UK).

Vigouroux, H., Fargin, E., Le Garrec, B., Dussauze, M., Rodriguez, V., Adamietz, F., Ravaux, J., Podor, R., Vouagner, D., De Ligny, D. & Champagnon, B. In-Situ Study of $LiNbO_3$ crystallization in lithium niobium Silicate glass ceramic. (in prep).

von Ardenne, M. (1938a). Das Elektronen-Rastermikroskop. Praktische Ausführung. *Zeitschrift fur technische Physik* 19, 407-416.

von Ardenne, M. (1938b). Das Elektronen-Rastermikroskop. Theoretische Grundlagen. *Zeitschrift fur Physik* 109, 553-572.

Wei, Q.F. (2004). Surface characterization of plasma-treated polypropylene fibers. *Materials Characterization* 52, 231-235.

Wich, T., Stolle, C., Luttermann, T. & Fatikow, S. (2011). Assembly automation on the nanoscale. *CIRP Journal of Manufacturing Science and Technology*, doi:10.1016/j.cirpj.2011.03.003

Wilson, B.A. & Case, D.E. (1997). In situ microscopy of crack healing in borosilicate glass. *Journal of Materials Science* 32, 3163-3175.

Wise, M.E., Martin, S.T., Russell, L.M. & Buseck, P.R. (2008). Water uptake by NaCl particles prior to deliquescence and the phase rule. *Aerosol Science and Technology* 42(4), 281-294.

Yang, J. (2010). In-situ High Resolution SEM Imaging with Heating Stage. Scanning Electron Microscopes (SEM) from Carl Zeiss.

Yu, H.M., Ziegler, C., Oszcipok, M., Zobel, M. & Hebling, C. (2006). Hydrophilicity and hydrophobicity study of catalyst layers in proton exchange membrane fuel cells. *Electrochimica Acta* 51, 1199-1207.

Zheng, T, Waldron, K.W. & Donald, A.M. (2009). Investigation of viability of plant tissue in the environmental scanning electron microscopy. *Planta* 230, 1105–1113.

Zimmermann, F., Ebert, M., Worringen, A., Schutz, L. & Weinbruch, S. (2007). Environmental scanning electron microscopy (ESEM) as a new technique to determine the ice nucleation capability of individual atmospheric aerosol particles. *Atmospheric Environment* 41, 8219–8227.

4

Some Applications of Electron Back Scattering Diffraction (EBSD) in Materials Research

Zhongwei Chen[1], Yanqing Yang[1,*] and Huisheng Jiao[2]
[1]State Key Laboratory of Solidification Processing,
Shaanxi Materials Analysis & Research Center,
Northwestern Polytechnical University, Xi'an,
[2]Oxford Instruments Shanghai Office, Shanghai,
P.R. China

1. Introduction

Electron Back Scattering Diffraction (EBSD) is a technique based on the analysis of the Kikuchi pattern by the excitation of the electron beam on the surface of the sample in a scanning electron microscope (SEM). The crystal structure, orientation and correlative information can be acquired by the technique. EBSD has a unique advantage in the determination of the crystal orientation and microstructure compared with the traditional analysis methods. It can observe the grain boundary types, misorientations, and the distribution of them, and the statistical measurement and quantitative analysis also can be carried out. Therefore, the quantitative relationship between grain boundary structure, orientation, texture and material properties can be established. Consequently, it has been a very important experimental technique in materials science and engineering.

This chapter presents a few examples of applying EBSD to characterize the microstructure of different materials including steels and molybdenum sheets after rolling and heat treatment, and casting aluminum alloys in order to reveal the formation mechanism of microstructure during solidification.

2. Applications of EBSD in steel and molybdenum

In this section, we summarize the applications of EBSD in microstructure characterization, including second phase identification, texture analysis of steel and molybdenum after different heat treatments and rolling, in understanding the microstructure change during the cold and hot work processing, and the properties of the materials. EBSD also has been used to measure the Kurdjumov-Sachs orientation relationships between austenite and ferrite in stainless steel. The orientation relationships between the particles and the matrix show a spread around the Kurdjumov-Sachs relationship; the close packed planes in the FCC and BCC phases are usually parallel to or nearly parallel to each other. A model has

* Corresponding Author

been proposed for the interfacial structure in a duplex stainless steel based on the topological theory to explain the deviation angles. In order to verify the prediction of the distribution of the orientation relationship between BCC/FCC structures, EBSD provides a convenient approach for orientation relationship determination and fast data collection.

2.1 Microstructure characterization of steel after different heat treatments and rolling

Steel's mechanical properties and corrosion resistance are seriously affected by inclusions and precipitates at grain boundaries and inside grains. For example, in ferritic stainless steel, the precipitation of chromium carbides and nitrides at grain boundaries causes local depletion of chromium in the surrounding material, resulting in a much-reduced corrosion resistance [Kim *et al.*, 2010; Park *et al.*, 2006]. This will lead to a dramatic drop in the strength and stiffness of the materials. In order to prevent the formation of chromium carbide and nitride, stronger nitride and carbide formation elements are added to form more stable precipitates.

It is shown that, in this research, a number of precipitates have been formed in a ferritic stainless steel intended for a high temperature application during the steel making process. The precipitates vary in size from 50μm down to a few 10s of nanometers. The larger precipitates can be easily identified using a combination of EBSD and EDS chemical analyses, but the smaller ones make any chemical analysis problematical. However the superior resolution of the EBSD technique, coupled with a knowledge of the probable composition of the precipitates, makes their identification using EBSD alone relatively simple.

The sample was mechanical polished and the final polish with colloidal silica. Zeiss Supra 55 VP SEM was used for EBSD and EDS analysis with beam current around 1-2nA and 20kV. A combination of imaging and "point and click" phase identification (with an integrated EDS-EBSD system) has been used to identify and then map the larger particles; however, these are commonly intragranular particles and, as such, are less damaging to the properties of the steel. The finest particles lie on the grain boundaries - these have the potential to cause intergranular corrosion, and EBSD analysis alone was necessary for their identification. Different mapping step size was used for mapping coarse inclusions in grains and fine precipitates at grain boundaries, 0.5μm and 20nm respectively.

2.1.1 Phase identification of coarse precipitates

Imaging using the Nordlys forescatter system allows both orientation contrast images and atomic number contrast images to be collected, shown in Fig.1, simply by switching from the bottom to the top forescatter diodes. These clearly show 2 types of precipitates within the grains - a large precipitate (type 1) and a number of smaller precipitates (type 2).

The integrated EBSD+EDS system allows 1-click phase identification. From a single point, an EDS spectrum is collected along with the diffraction pattern (EBSP); the spectrum peaks are identified and this chemistry is used to search a phase database (or several databases) to find all matching phases. The EBSP is then indexed and the best matching phase is determined, shown in Fig.2. Here was used to identify the 2 types of precipitate: type 1 precipitates are aluminum nitrides (AlN - hexagonal), and the type 2 precipitates are chromium carbides ($Cr_{23}C_6$ - cubic).

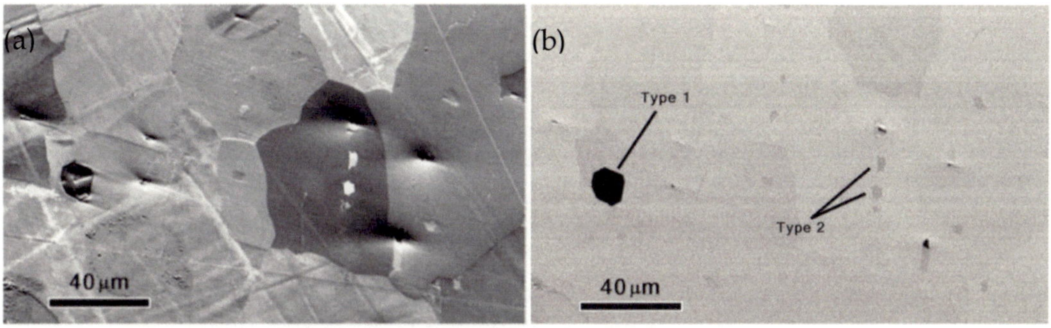

Fig. 1. Forescatter images. (a) orientation contrast image collected using the lower forescatter diodes, showing grains and surface topography; (b) atomic number contrast image collected using the top diodes, showing 2 types of precipitate in the ferrite matrix - types 1 and 2.

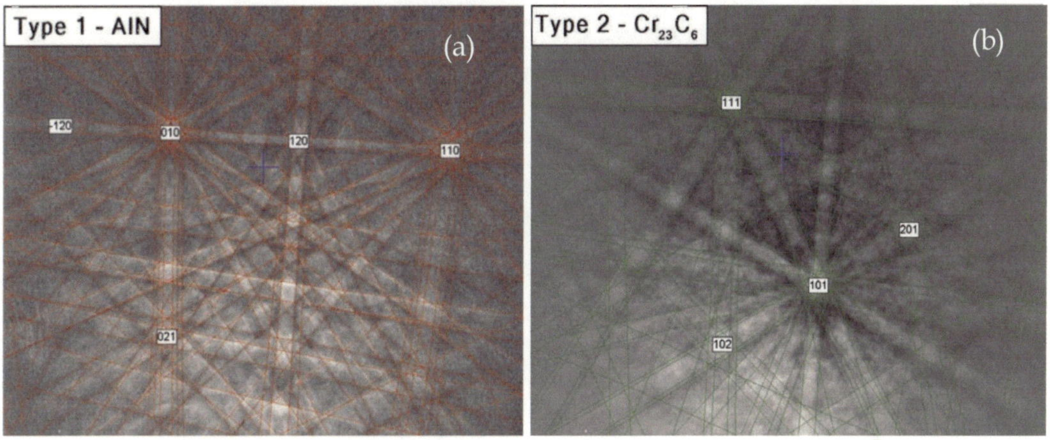

Fig. 2. Indexed EBSPs from the 2 types of precipitate. (a) type 1, indexed as hexagonal AlN; (b) type 2, indexed as cubic $Cr_{23}C_6$.

With the identity of the precipitates now known, it is possible to map the area and to show the distribution of the phases. The CHANNEL5 EBSD system can discriminate between the 2 precipitates and the ferrite matrix, on the basis of crystallography alone. The resulting phase map is shown in Fig.3.

2.1.2 Identification of small grain boundary precipitates

A closer look at the microstructure shows that many of the grain boundaries have small, elongated precipitates, less than 200 nm across. Unlike the coarse AlN and $Cr_{23}C_6$ precipitates, their location at the boundaries could cause intergranular corrosion, and as such it is important to identify them (see Fig. 4).

The size of these particles makes chemical analysis by EDS problematic, as the signal will predominantly originate from the steel matrix. Therefore EBSD is the ideal technique to identify such precipitates. It is expected that these precipitates are either carbides or nitrides, and so matching phases that fit the chemistry (Fe, Cr)(N, C) were used to index the EBSPs. In all cases the precipitates were identified as having a hexagonal M_7C_3 structure –

$(Fe,Cr)_7C_3$. An automated EBSP map along one of these boundaries was collected, and this showed that some of the precipitates have a distinct crystallographic relationship with one of the neighbouring grains: $(0001)_{carbide} || (111)_{ferrite}$ – a basal orientation relationship.

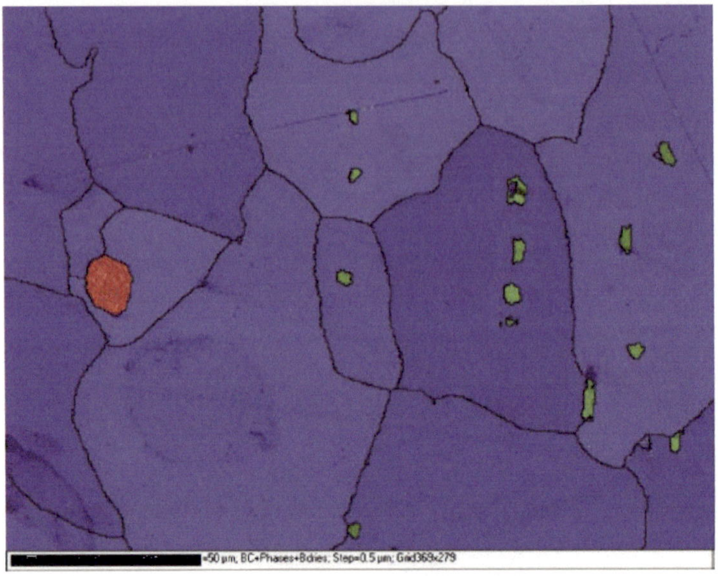

Fig. 3. Phase map showing the same area imaged in Fig. 1. Blue = ferrite, Red = AlN and Green = $Cr_{23}C_6$. Black lines represent grain boundaries.

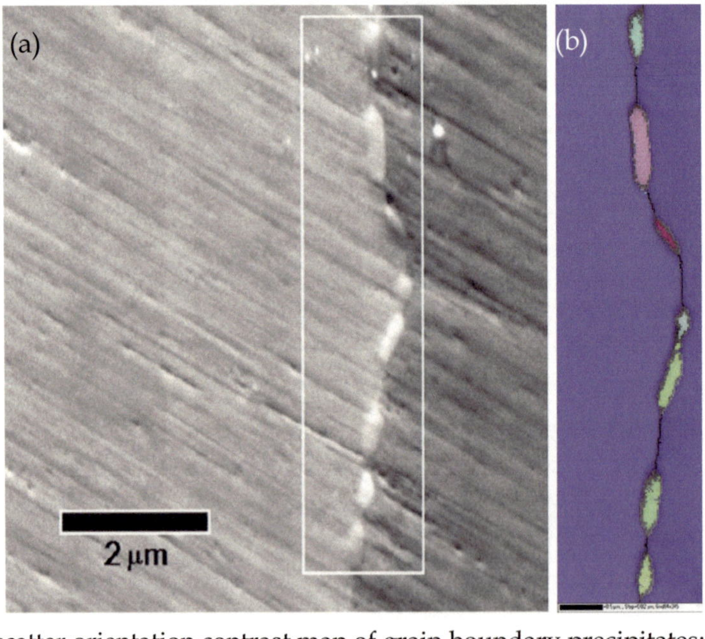

Fig. 4. (a) Forescatter orientation contrast map of grain boundary precipitates: white box marks EBSD analysis area; (b) Phase and orientation map of the boundary zone. Ferrite is shown in blue, with the $(Fe,Cr)_7C_3$ precipitates colored according to their orientation.

2.2 Microtexture analysis in molybdenum sheets

EBSD is an ideal technique for microtexture analysis. With the development of the speed of detectors, macro texture analysis is also possible. Comparison with XRD, in characterizing texture EBSD provides not only the types and of percentages of textures, but also the microstructure information. In this section, microtextures of a cross-rolled molybdenum were analyzed using EBSD technique. Molybdenum and its alloys are used in a variety of markets, including the electronics, materials processing and aerospace industries. There are a number of different properties that make molybdenum so attractive, notably its strength at high temperature, high stiffness, excellent thermal conductivity and low coefficient of thermal expansion [Cockeram *et al.*, 2005].

As with any metal, the physical characteristics of molybdenum can be tailored to suit particular applications. This is done using specific machining or metalworking procedures. One example of this is found in the aerospace industry, where molybdenum's strength and stiffness at high temperatures make it the ideal material for space satellite components. Sheets of molybdenum are cross-rolled in order to further enhance its properties [Oertel *et al.*, 2008]. Cross-rolling involves rolling the original sheet both parallel and perpendicular to its length, producing a specific texture (defined as the {001}<110> texture).

This research looks at a specific case in which, during the production of a molybdenum dish, undesirable surface ripples were observed. Obviously these ripples would have a damaging effect on the dish's performance, and therefore it was decided that the microstructural and crystallographic textural characteristics of the molybdenum sheet should be investigated in order to deduce the cause of the ripple formation. Samples were mechanically polished; final electro-polishing with 10% sulphuric acid/methanol electrolyte at -25°C, 55V applied voltage.

The microstructure of the sample is shown in Fig.5, a low magnification backscattered electron micrograph. It is clear that the sample has partially recrystallized, producing large, strain free grains. The recrystallized and unrecrystallized (i.e. deformed) fractions are arranged in alternate bands parallel to the final rolling direction (RD). The scale bar represents 50 μm.

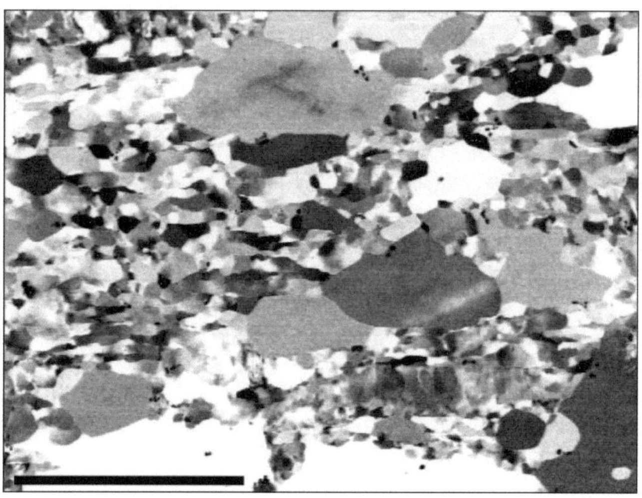

Fig. 5. Forescatter orientation contrast image

The results of the EBSD analysis are shown in this orientation map, Fig.6. The color scheme reflects the orientation (see the inverse pole figure, inset), with green color showing points with the crystallographic <110> direction aligned with the rolling direction. Grain boundaries (>10°) are marked in black, with subgrain boundaries (2-10°) in grey. Note that in this and all other maps, the final rolling direction is horizontal and the scale bar represents 300 μm.

There are many ways that the CHANNEL 5 data processing software can be used to characterize the extent of deformation and recrystallization in a sample such as this.

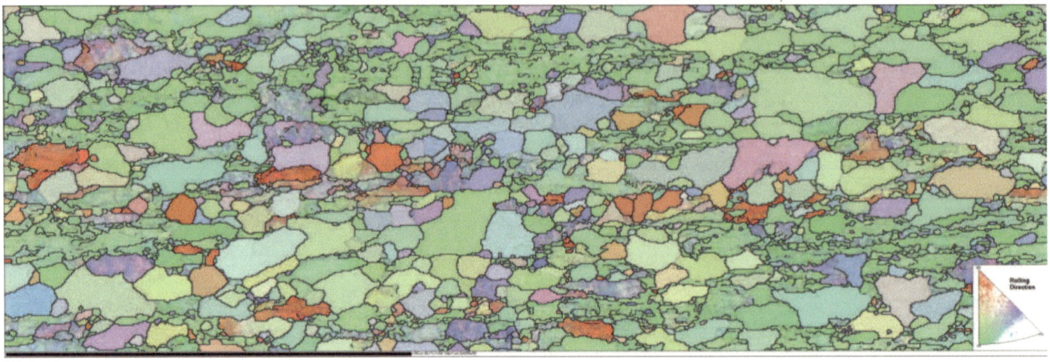

Fig. 6. Orientation map showing the orientation of each grain

In average misorientation map of Fig. 7(a) the misorientation between all the points in each grain has been calculated, and assigned a color (as defined in the legend, inset). The blue color indicates where grains have very little internal misorientation - these typically have been recrystallized. The green, yellow and red colors represent progressively increasing levels of internal misorientation, indicating no recrystallization. The software can also determine the recrystallized fraction automatically, as shown in Fig.7(b). Here the recrystallized grains are marked in blue and make up 54% of the total area. Unrecrystallized grains are colored red. It is clear from the orientation map (Fig.6) that this sample has a strong texture. This can be represented in pole figures or in orientation distribution functions (ODFs) in Fig.8.

In Fig. 8(a) the contoured pole figures show the complex nature of this texture. As observed in the orientation map, there is a strong alignment of the <110> axes with the rolling direction. However, there are also significant {100} and {111} textures. In Fig. 8(b) plotting the texture in an ODF clearly illustrates the texture characteristics. There is a strong fibre texture that splits into 2 branches, as well as other less important texture components. The colors are the same as in Fig. 7(b), showing that there is little difference between the orientation of recrystallized (blue) and non-recrystallized grains (red).

With a complex texture such as this, there are many individual "texture components" that can be used to describe parts of the overall texture. There are 2 main texture components that account for most of the texture in this sample, but if this sample has been fully cross-rolled, then a strong {001}<110> texture would be expected. In Fig.9, the inset shows the color scheme (up to a maximum 20° deviation) and the background grains are colored according to the pattern quality.

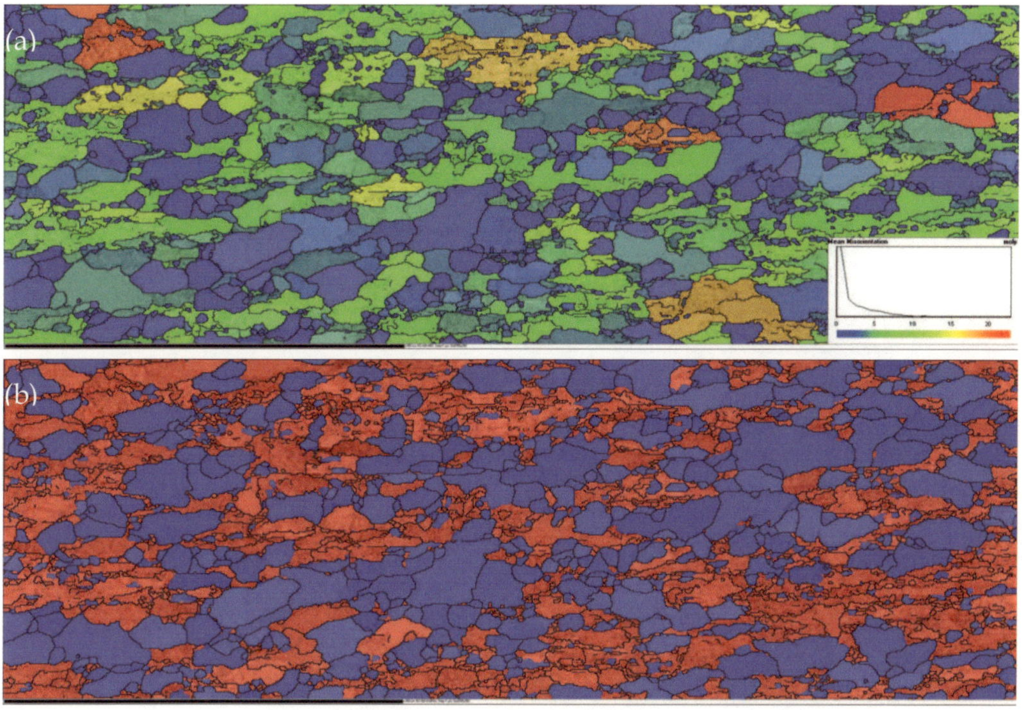

Fig. 7. (a) The average misorientation map and (b) recrystallised fraction map.

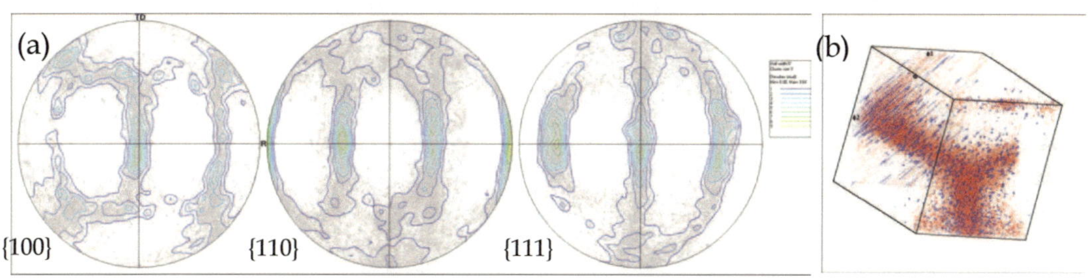

Fig. 8. (a) The contoured pole figures and (b) ODF showing texture characterisitics.

Cross-rolling should produce a dominant {001}<110> texture in BCC-metals (such as molybdenum). However this texture accounts for less than 25% of the area in this sample, with {110} and {111}-fibre textures more dominant. This indicates that the sheet has not been fully cross-rolled, and this would explain the formation of undesired ripples on the sheet surface.

2.3 Measuring the Kurdjumov-Sachs orientation relationships between austenite and ferrite in stainless steel

The interface between FCC and BCC crystals can be found in many important metallic alloys. The orientation relationships between these two phases show a spread around the Kurdjumov-Sachs relationship (K-S OR), Nishiyama-Wassermann relationship or other relationship; the close packed planes in the FCC and BCC phases are usually parallel to or nearly parallel to each other.

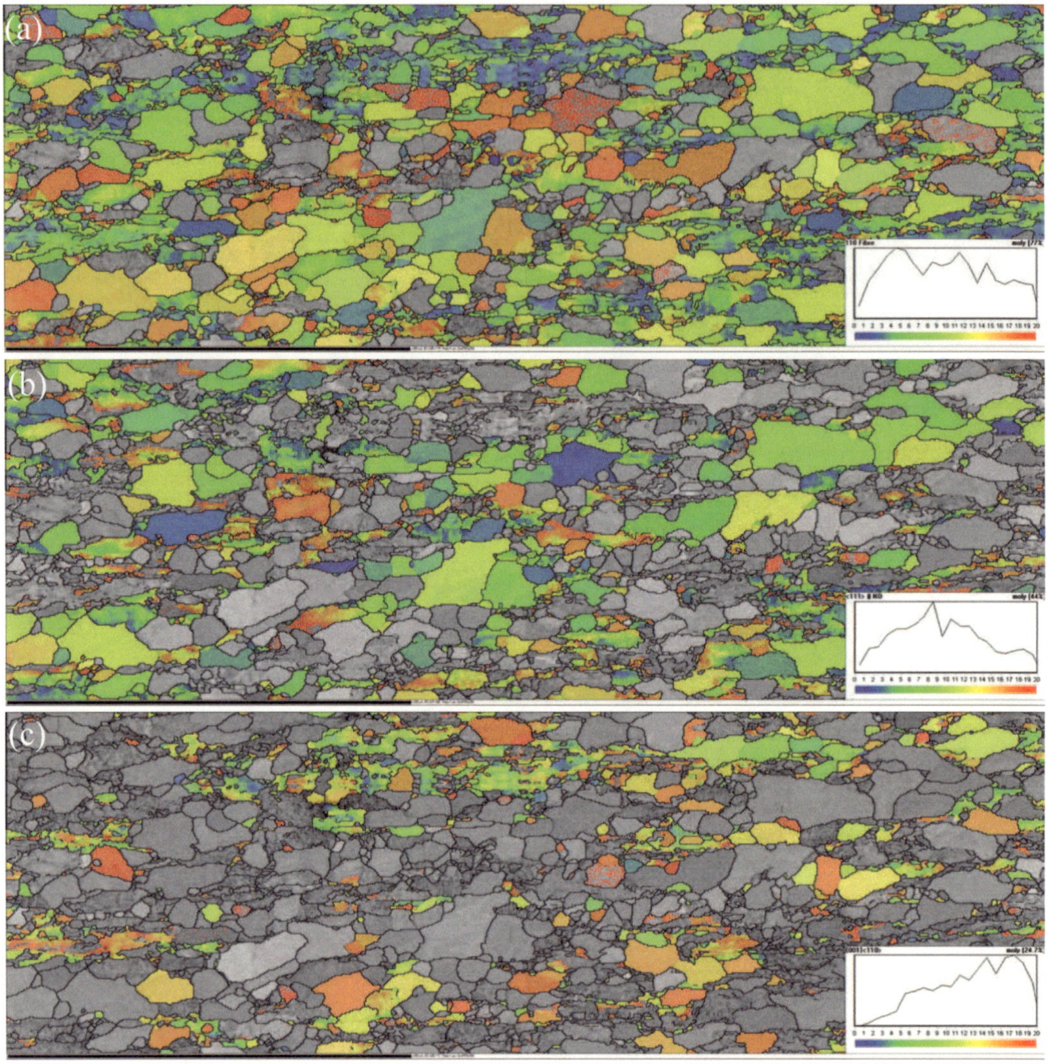

Fig. 9. (a) {110}-fibre texture (<110> parallel to RD) - 77% of the area; (b) {111}-fibre texture (<111> parallel to normal direction) - 44% of the area; (c) {001}<110> texture (a subset of the {110}-fibre texture) - 24.7% of the area.

In order to understand the morphology and structure of such interphase boundaries, different theoretical models have been proposed, for example, the O-lattice theory [Bollmann,1970], Invariant line model [Dahmen, 1982], CSL/DSC model [Balluffi, 1982], and the structural ledge model [Hall *et al.*, 1972]. However, these models can only explain or predict interfacial structure in part; in particular, interfacial defects have not always been accurately characterized in previous studies. Recently, by considering the symmetry of the bicrystals, the topological theory was developed for characterizing the parameters of interfacial defects [Pond, 1989]. A model has been proposed for the interfacial structure in a duplex stainless steel based on the topological theory [Jiao *et al.*, 2003]. The FCC/BCC interfaces have been characterized as arrays of interfacial defects superimposed on reference bicrystal structures. This model predicted deviation angle ranged from 0.3° to 4.97° from the ideal K-S OR. In order to verify the prediction of the distribution of the orientation relationship between BCC/FCC structures, EBSD has been used for orientation relationship determination and fast data collection.

In this research, a sample of Zeron-100 duplex stainless steel was heat treated at 1400°C for 30 minutes and followed by water quenching, and 10 seconds at 1000°C for precipitating. Sample was electropolished with a solution of 10 wt% oxalic acid in H_2O with a voltage of 10~15V at room temperature for 1~3 minutes.

Fig. 10 shows the forescatter orientation contrast image of the duplex stainless steel. The EBSD data was collected from the center area of the image. In Fig.11, it can be seen there are two phases, FCC γ phase is in red and BCC α matrix is in blue. For the γ phase we can see that there are three types: small particles, large particles in α grain and at grain boundaries. From the pole figures in Fig.12 the orientation relationship (OR) between FCC and BCC phases was determined as K-S OR. In Fig.11(b) the K-S OR interface boundaries are plotted in white. If the deviation from K-S OR over 7° the interface boundaries are in black. From the map we can see that the most of particles shows a K-S OR to matrix. However, the particles at the grain boundaries only show K-S OR with one side, which is because of the particle is nucleated from one grain. The large particles inside the α grain show non K-S OR with the matrix; that because these particles are retained during homogenization.

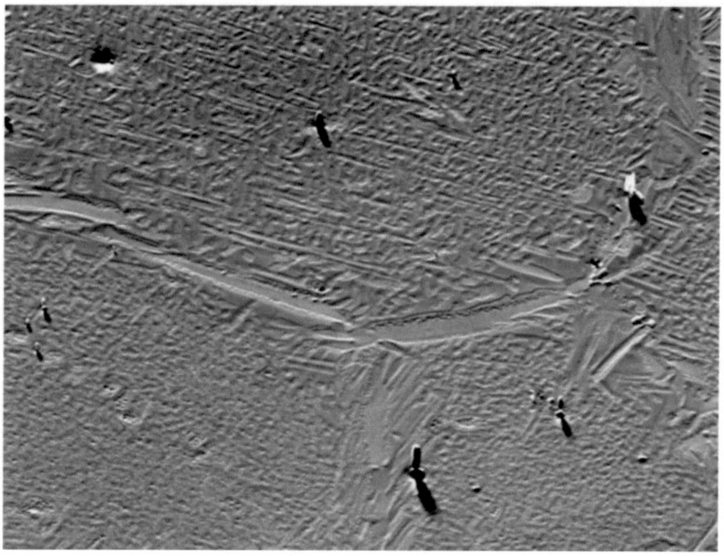

Fig. 10. Forescatter orientation contrast image of the duplex stainless steel. The EBSD data was collected from the center area of the image.

Although these two phases show a good consistent with K-S OR, there is always a small deviation from the ideal OR; that means the {111} plane in γ phase is not exactly parallel to {110} plane in BCC matrix. From the above distribution, it is found that most the particles show a 1.5° away the K-S OR and with a range of 0.2° to 5.4°, as shown in Fig. 13, which is in a good agreement with the prediction from the topological interfacial model.

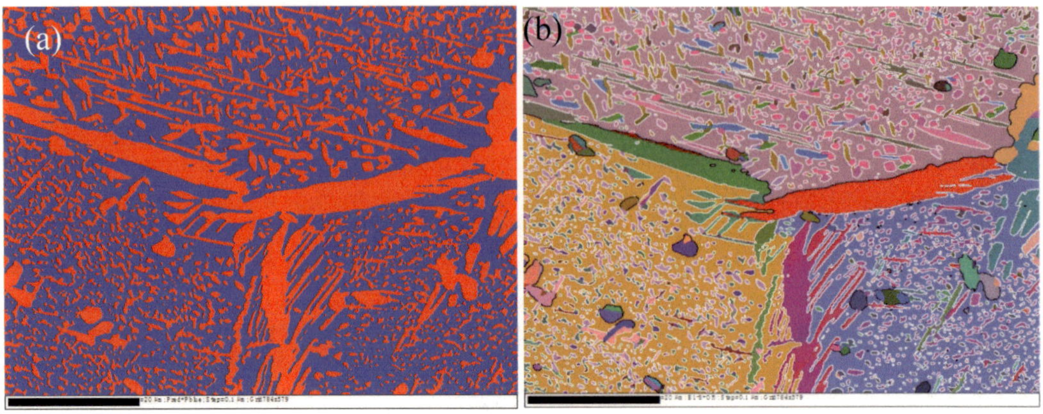

Fig. 11. (a) Phase map and (b) orientation map of the sample.

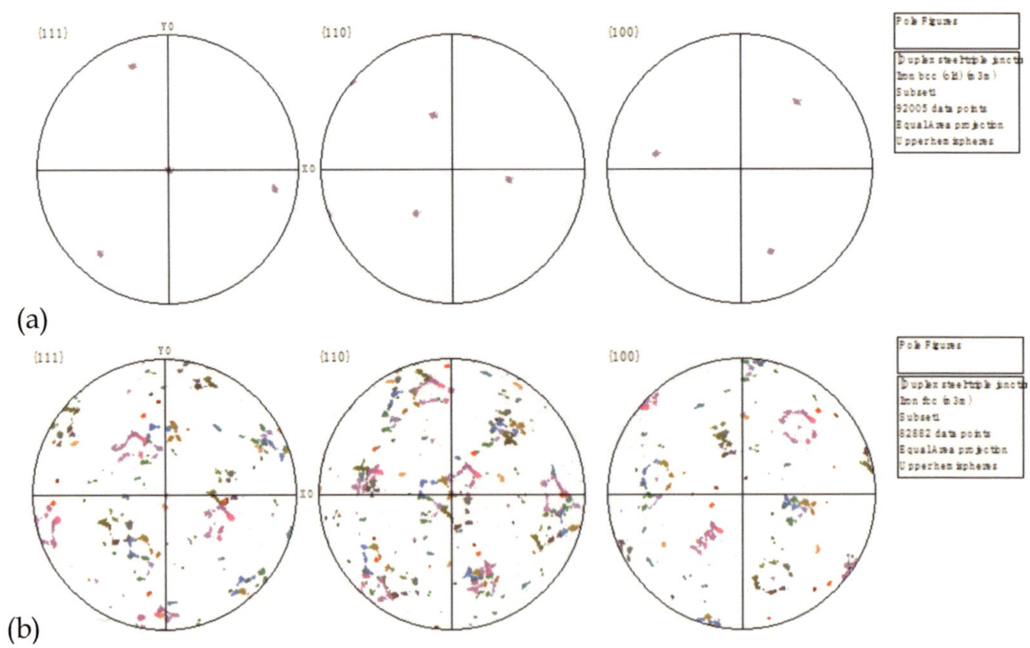

Fig. 12. Pole figures from one α grain (a) and γ particles in this grain (b). From these pole figures it is found that these two phases fall into K-S OR.

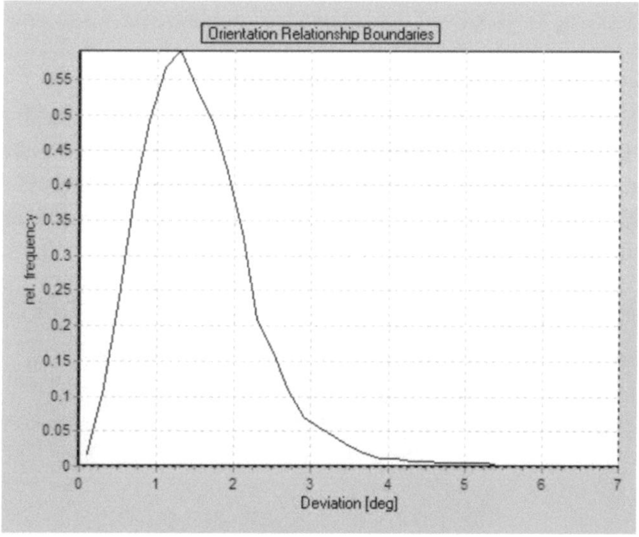

Fig. 13. The distribution of the deviation angle from K-S OR.

3. EBSD analysis in the field of solidification

3.1 Preparation of EBSD samples by ion etching

Both electrolytic polishing and ion etching can be used to prepare EBSD samples, however, the electrolyte for Al-Si alloy is relatively difficult to be prepared. On the other hand, for A357 alloy, electrolytic polishing of α-Al dendrites is faster than that of the eutectic Si due to the difference in electrochemical property, and then the prominency of eutectic Si phase is visible, which influences the surface roughness awfully. As a result, the reflection of the backscatter electron can not be received by the screen, inducing a low calibration rate, consequently the sample could not be analyzed [Nogita & Dahle, 2001]. Ion etching, as a new technique for EBSD sample preparation, is suitable for eliminating the surface stress layer of most materials, and the etching speed could be selected according to the etching voltage, ion beam current, geometrical shape and materials of the samples. Therefore, ion etching is chosen for EBSD sample preparation of particular materials which contain some hard brittle phases in microstructure.

Fig.14 shows the Kikuchi pattern of A357 alloy sample prepared by mechanical buffing and ion etching respectively. The difference between these two pictures is obvious. Rheology on the surface of the sample took place due to mechanical buffing, therefore no Kikuchi pattern could be observed, as shown in Fig. 14(a). Since the electron beam effects only 1-2μm deep on the surface of the sample, so ion etching must be performed to remove the surface stress layer in order that Kikuchi pattern could be observed. Fig.14(b) shows clear Kikuchi pattern.

3.2 Misorientation of secondary eutectic phase with primary phase in modification alloys

A357 aluminum alloy is casting alloy with the coarse α-Al dendrites and plate-like eutectic silicon in cast microstructure, of which the volume fraction of the eutectic silicon phase is more than 50% [Heiberg & Arnberg, 2001] and influences its mechanical properties. The coarse primary α-Al dendrite could be refined and equiaxed by adding proper quantities of Ti and B into the alloy [Easton & St John, 2001; Shabestari & Malekan, 2010], while Sr is a good modificator, which improves the mechanical properties of Al-Si alloy by changing the morphology of the eutectic Si [Chen & Zhang, 2010; Martínez *et al.*, 2005]. In the section, EBSD investigation of A357 alloy has been used to analyze the misorientation of eutectic phase with primary phase in A357 aluminum alloy with and without Sr modification and to study the nucleation and growth mechanism of the eutectic solidification by Sr modification.

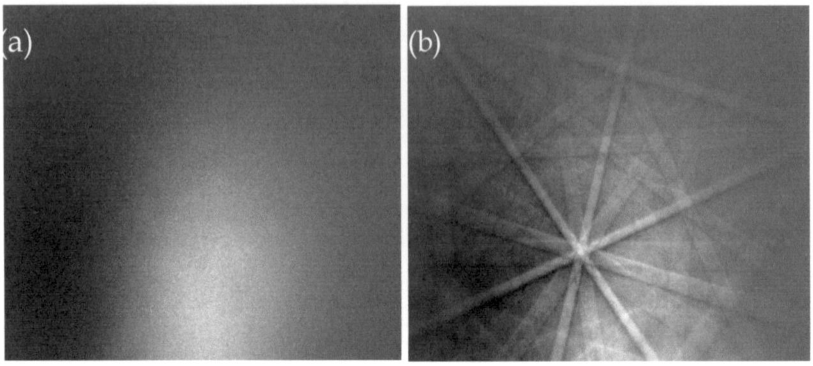

Fig. 14. Kikuchi pattern of unmodified A357 alloy samples prepared by (a) mechanical polishing and (b) ion etching.

For EBSD samples of the cast A357 alloys, the ion etching was carried out for the EBSD sample preparation by Gatan 682 ion etching equipment. EBSD tests were performed on ZEISS SUPRA55 field emission gun scanning electron microscope (FEG-SEM) with HKL channel 5 backscattered electron diffraction camera. The parameter of FEG-SEM is set as following: acceleration voltage of 20kV, working distance of 21.0 mm.

Fig.15 shows the microstructure of the unmodified and modified A357 alloy. The coarse α-Al dendrites and the plate-like Al-Si eutectic both can be observed clearly, as shown in Fig.15 (a). The coarse plate-like eutectic Si phases were transformed to fine fibrous eutectic by Sr modification, as shown in Fig.15 (b). It is concluded that there are approximately 480 grains in a visual field of the unmodified A357 sample, of which the average grain size is 21.0μm, while about 1362 grains can be found in the same scanning region of the sample which was modified and the average grain size is 7.4μm, according to the EBSD statistical results. The average grain size of the sample decreases by nearly two times, indicating that the microstructure can be refined by Sr modification.

Fig.16 shows the crystal misorientation and its distribution with and without Sr modification. Low-angle grain boundaries (LAGBs) are dominate in the unmodified sample and few high-angle grain boundaries (HAGBs) can be observed, while both of them in the modified sample can be found that they distribute evenly and occupy certain proportion respectively, as shown in Fig.16(d). Fig.17 shows the polar figures of the samples with and without Sr modification. Crystal orientation of the sample without Sr modification is relatively concentrated, which basically tends to be two kinds of crystal orientation, represented as red and purple respectively, as shown in Fig.17 (a), while crystal orientation in the modified sample is changed and tends to be scattered, as shown in Fig.17 (b).

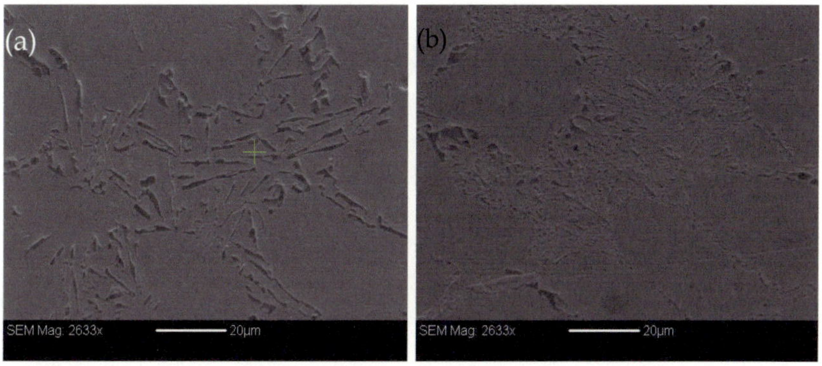

Fig. 15. Microstructure of samples (a) unmodified and (b) modified.

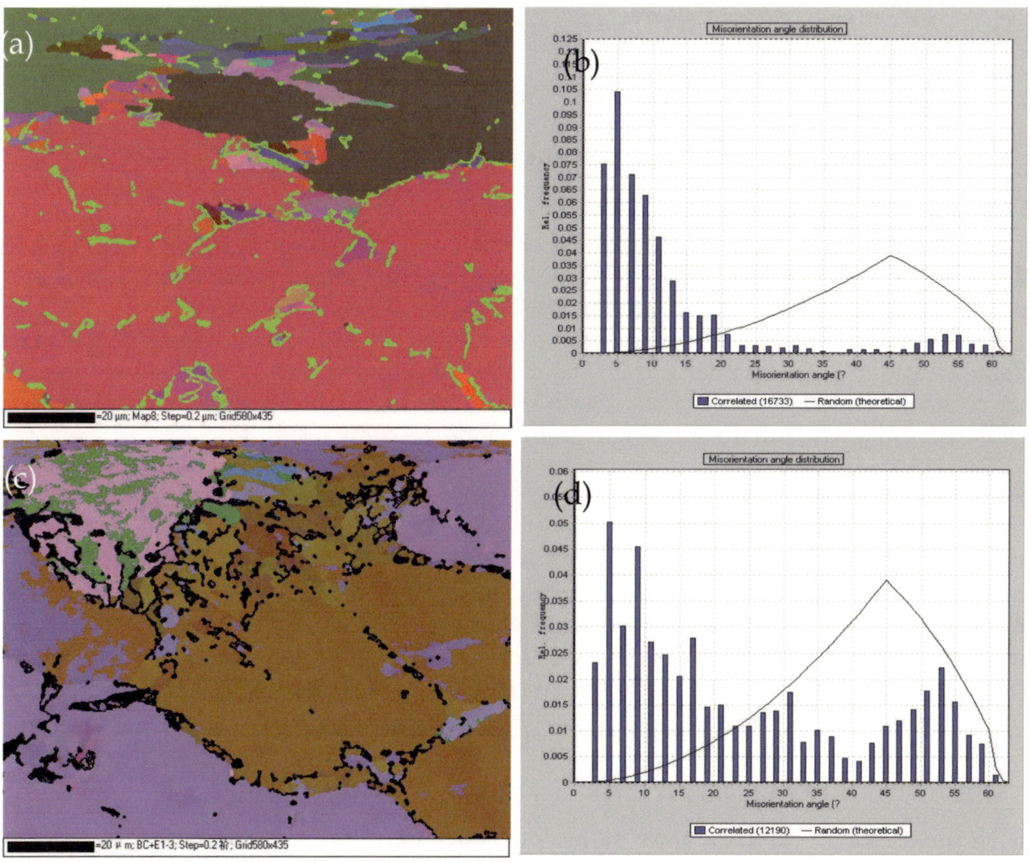

Fig. 16. Misorientation angle distribution for the Al grains. (a) and (b) unmodified; (c) and (d) modified.

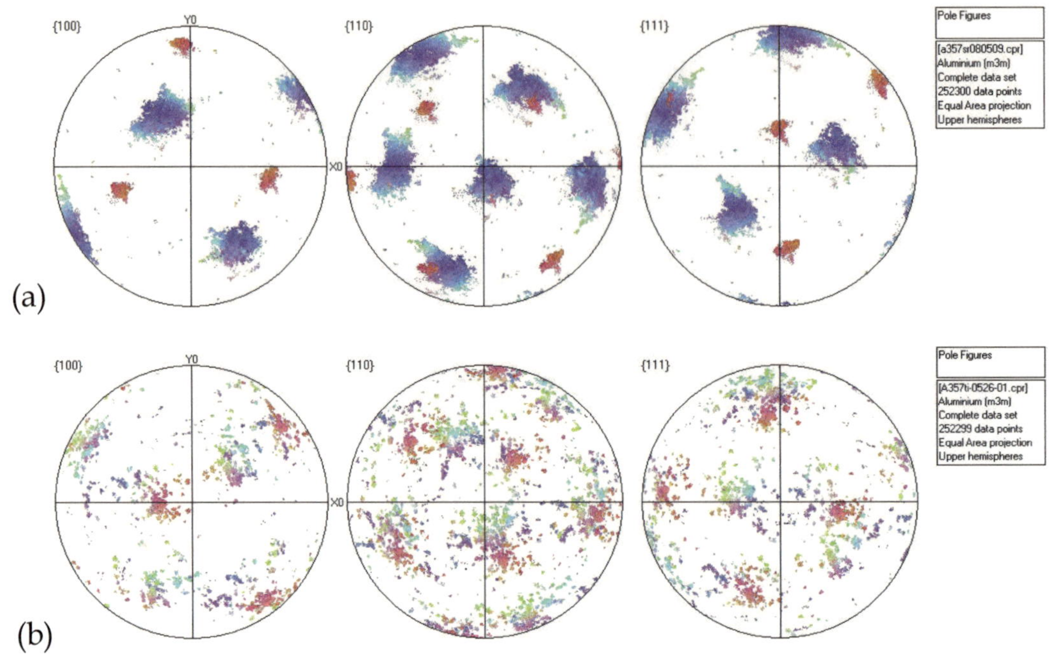

Fig. 17. Pole figure maps of samples. (a) unmodified and (b) modified.

A357 alloy is hypoeutectic Al-Si alloy, and heterogeneous nucleation takes place during the solidification process of the primary α-Al dendrites and then the nuclei grows up. For A357 alloy, recent work has indicated that there are three different possible eutectic nucleating and growth modes depending on the solidification conditions [Dahle *et al.*, 1997]: (1) nucleates on the casting wall and grows up in the opposite direction of thermal gradients, (2) nucleates on the primary α-Al dendrites and grows up, (3) nucleates on the heterogeneous nuclei located in the region between primary α-Al dendrites and grows up. As shown in Fig.16, with the low-angle grain boundary being responsible for the nucleation of the eutectic Si on the primary α-Al dendrites, the high-angle grain boundary is responsible for the nucleation of the eutectic Si on the heterogeneous nuclei located in the region between primary α-Al dendrites. It is obvious that the eutectic Si in the sample without Sr modification nucleates and grows up in mechanism (2) and those in the sample modified nucleates and grows up in mechanism (3). The results indicate that nucleation mechanism of the eutectic Si phase changes due to the modification, leading to the change of the growth patterns. Therefore, misorientation of secondary eutectic phase with primary phase in casting alloys can be detected by EBSD and used to study the nucleation and growth mechanism of modification.

3.3 Agglomeration of primary crystals during solidification

In suction casting, the extra force is provided by the differential pressure between the melting chamber and the suction chamber. Thus, suction casting is successful in preventing casting defects by means of suction force, and a high cooling rate in suction casting is generated due to the use of the Cu-mold. Although the primary solidification under the intensive force has been comprehensively understood [Chen *et al.*, 2009], not much attention has been paid to the solidification of the remaining liquid in the die-casting mould. However, the secondary solidification of the remaining liquid plays an important role in determining the final microstructure and corresponding mechanical properties [Stangleland *et al.*, 2004; Vernède *et al.*, 2006]. The secondary solidification was also found in rheo-diecasting (RDC) process of semi-solid metal (SSM) technology [Fan *et al.*, 2005; Hitchcock *et al.*, 2007]. Therefore, it is necessary to understand the secondary solidification behavior of the remaining liquid in relation to the final microstructure.

Agglomeration of primary crystals in the mushy zone during solidification is an important phenomenon influencing many aspects of casting processes and often affects the rheology of partially solid alloys and the microstructure of the as-cast component. Microstructural characterization of AlFeSi specimens produced by suction casting and die casting has been performed using EBSD. EBSD is used to study the agglomeration of equiaxed crystals in suction casting by analyzing the grain misorientation.

Typical as-cast microstructures produced by the two casting techniques are shown in Figs. 18(a) and 18(b). The primary grains have quasi-equiaxed morphology in two cases. From Fig. 18(b), it can be seen that in located region of the relatively large primary α-Al dendrites, there are fine α-Al particles, contributed by the solidification of the remaining liquid in the located region of α1 dendrites, which is referred to as "secondary solidification".

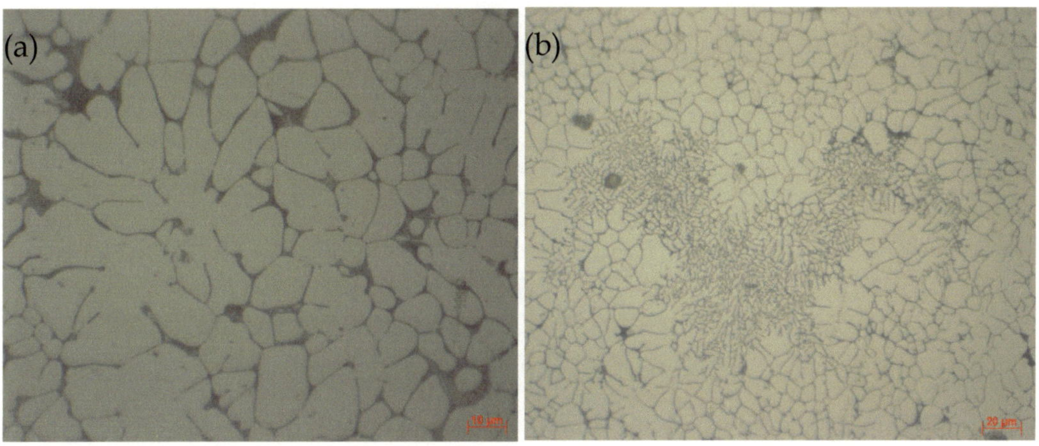

Fig. 18. Optical metallographs of AA8011 alloy. (a) die casting sample and (b) central region in 2 mm diameter sample in suction casting.

Similar to previous research [Otarawanna *et al.*, 2010], low-angle grain boundaries (LAGBs) and high-angle GBs (HAGBs) are defined here as boundaries with a misorientation between 5° and 15°, and more than 15°, respectively. Three types of GBs – LAGBs, HAGBs and coincidence-site-lattice GBs (CSL-GBs) – were determined after result extrapolation in each EBSD map. Fig. 19 shows extrapolated EBSD maps by LAGBs and HAGBs in each of the samples investigated. It shows that the fraction of low-energy GBs (LAGBs) is higher in the suction casting sample than in the die casting sample.

By producing samples with different casting methods where the solidifying alloy experiences different levels of external mechanical stresses, the effects of mechanical stresses applied during solidification can be assessed. The length percent of low-energy GBs shown in Fig. 19 is associated with crystal agglomeration during solidification. The high pressure applied on the solidifying alloy during suction casting is likely to promote crystal collisions and result in a strong degree of crystal agglomeration.

When two growing primary crystals impinge with one another, GB formation depends on the interfacial energy of the potential new GB. Coalescence or bridging is the transformation of two impinging solidification fronts into a solid bridge [Rappaz *et al.*, 2003] and in this case a new agglomerate can form. Therefore, coalescence can be considered as the disappearance of two solid–liquid interfaces, each with interfacial energy $\gamma_{s/l}$, and the formation of a GB with interfacial energy γ_{gb} [Mathier *et al.*, 2004]. When two solidifying crystals impinge on one another, bridging occurs readily if $\gamma_{gb} < 2\gamma_{s/l}$ [Rappaz *et al.*, 2003]. On the other hand, some energy is required to form a new boundary if $\gamma_{gb} > 2\gamma_{s/l}$. γ_{gb} is a function of the misorientation between the two impinging crystals. $\gamma_{gb} < 2\gamma_{s/l}$ occurs if the misorientation is less than 15° (LAGBs) [Mathier *et al.*, 2004]. In this case, there is an attractive force to bring the two crystals together and coalescence occurs as soon as the two interfaces are close enough. Only GBs with $\gamma_{gb} < 2\gamma_{s/l}$ are thought to form after a collision of two crystals [Sannes *et al.*, 1996]. If not, a liquid film is stable and the colliding crystals bounce back. In suction casting samples, the fraction of low-energy GBs among in-cavity solidified grains is significantly higher than in diecasting specimens. This is attributed to the increased number of crystal collisions during suction casting, which promotes agglomeration of favorably oriented crystals.

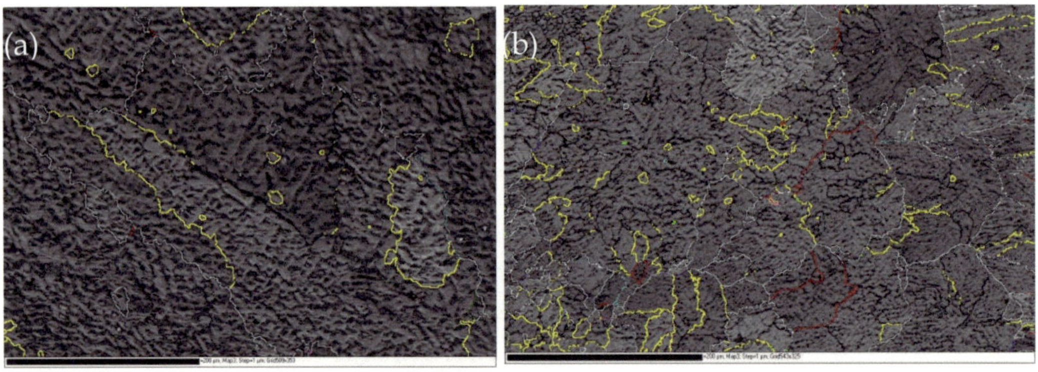

Fig. 19. EBSD micrographs by GB lines in color. White—HAGBs, Yellow—LAGBs, other color-- CSL-GBs. (a) die casting sample, (b) central region in 2 mm diameter sample in suction casting

4. Conclusions

Some examples of applications of EBSD to materials processing are presented in the chapter. It is shown that combined with SEM imaging and EDS composition analyzing, EBSD is very powerful for materials research. Phase identification, microtexture characterization and crystallographic orientation relationship determination are carried out in different steels and molybdenum sheet after heat treating and rolling in order to understand the microstructure evolution during the hot processing. It is also shown that EBSD analysis on casting aluminum alloys provides invaluable insight into mechanisms of nucleation and growth in modification during solidification, and agglomeration of equiaxed crystals in secondary solidification.

5. Acknowledgement

The authors gratefully acknowledge the financial support from the Research Fund of the State Key Laboratory of Solidification Processing (No. 42-QP-009) and the Fundamental Research Fund of Northwestern Polytechnical University (No. JC200929).

6. References

Balluffi, R. W.; Brokman, A. & King, A. H. (1982). CSL/DSC Lattice model for general crystal boundaries and their line defects. *Acta Metallurgica*, Vol.30, No.8, (August 1982), pp. 1453-1470, ISSN 0001-6160

Bollmann, W. (1970). *Crystal Defects and Crystalline Interfaces*, Springer, ISBN 978-0387050577, New York, UAS

Chen, Z. W. & Zhang, R. J. (2010). Effect of strontium on primary dendrite and eutectic temperature of A357 aluminum alloy. *China Foundry*, Vol.7, No.2, (May 2010), pp. 149-152, ISSN 1672-6421

Chen, Z. W.; He, Z. & Jie, W. Q. (2009). Growth restriction effects during solidification of aluminium alloys. *Transactions of Nonferrous Metals Society of China*, Vol.19, No.2, (April. 2009), pp. 410-413, ISSN 1003-6326

Cockeram, B. V.; Smith, R. W. & Snead, L. L. (2005). Tensile properties and fracture mode of a wrought ODS molybdenum sheet following fast neutron irradiation at temperatures ranging from 300°C to 1000°C. *Journal of Nuclear Materials*, Vol.346, No.2-3, (November 2005), pp. 165-184, ISSN 0022-3115

Dahmen, U. (1982). Orientation relationships in precipitation systems. *Acta Metallurgica*, Vo.30, No.1, (January 1982), pp.63-73, ISSN 0001-6160

Dahle, A. K.; Hjelen, J. & Arnberg, L. (1997). Formation of eutectic in hypoeutectic Al-Si alloys, *Proceedings of 4th Decennial International Conference on Solidification Processing*, pp. 527-530, ISBN 0-9522507-2-1, Sheffield, UK, July 7-10, 1997

Easton, M. A. & St John, D. H. (2001). A model of grain refinement incorporating alloy constitution and potency of heterogeneous nucleant particles. *Acta Materialia*, Vol.49, No.10, (June 2001), pp. 1867-1878, ISSN 1359-6454

Fan, Z.; Liu, G. & Hitchcock, M. (2005). Solidification behaviour under intensive forced convection, *Materials Science and Engineering: A*, Vol.413-414, (December 2005), pp. 229-235, ISSN 0921-5093

Hall, M. G.; Aaronson, H. I. & Kinsma, K. R. (1972).The structure of nearly coherent fcc:bcc boundaries in a CuCr alloy, *Surface Science*, Vol.31, (June 1972), pp. 257-274, ISSN 0039-6028

Heiberg, G. & Arnberg L. (2001). Investigation of the microstructure of the Al–Si eutectic in binary aluminum–7 wt% silicon alloys by electron backscatter diffraction (EBSD). *Journal of Light Metals*, Vol.1, No.1, (February 2001), pp. 43-49, ISSN 1471-5317

Hitchcock, M.; Wang, Y. & Fan, Z. (2007). Secondary solidification behavior of the Al–Si–Mg alloy prepared by the rheo-diecasting process. *Acta Materialia*, Vol.55, No.5, (March 2007), pp. 1589-1598, ISSN 1359-6454

Jiao, H. S.; Aindow, M. & Pond, R. C. (2003). Precipitate orientation relationships and interfacial structures in duplex stainless steel Zeron-100. *Philosophical Magazine*, Vol.83, No.16, (June 2003), pp. 1867-1887, ISSN 1478-6435

Kim, J. K.; Kim, Y. H.; Lee, J. S. & Kim K. Y. (2010). Effect of Chromium Content on Intergranular Corrosion and Precipitation of Ti-stabilized Ferritic Stainless Steels. *Corrosion Science*, Vol.52, No.5, (May 2010), pp. 1847-1852, ISSN 0010-938X

Martínez, D. E. J.; Cisneros, G. M. A. & Valtierra, S. (2005). Effect of strontium and cooling rate upon eutectic temperatures of A319 aluminum alloy. *Scripta Materialia*, Vol.52, No.6, (March 2005), pp. 439-443, ISSN 1359-6462

Mathier, V.; Jacot, A. & Rappaz, M. (2004). Coalescence of equiaxed grains during solidification. *Modeling and Simulation in Materials Science and Engineering*, Vol.12, No.3, (May 2004), pp. 479-490, ISSN 0965-0393

Nogita, K. & Dahle, A. K. (2001). Eutectic solidification in hypoeutectic Al–Si alloys: electron backscatter diffraction analysis. *Materials Characterization*, Vol.46, No.4, (April 2001), pp. 305-310, ISSN 1044-5803

Oertel, C. G.; Huensche, I.; Skrotzki, W.; Knabl, W.; Lorich, A. & Resch, J. (2008). Plastic anisotropy of straight and cross rolled molybdenum sheets. *Materials Science and Engineering: A*, Vol.483-484, (June 2008), pp. 79-83, ISSN 0921-5093

Otarawanna, S.; Gourlay, C. M.; Laukli, H. I. & Dahle, A. K. (2010). Agglomeration and bending of equiaxed crystals during solidification of hypoeutectic Al and Mg alloys. *Acta Materialia*, Vol.58, No.1, (January 2010), pp. 261-271, ISSN 1359-6454

Park, C. J.; Ahn, M. K. & Kwon, H. S. (2006). Influences of Mo substitution by W on the precipitation kinetics of secondary phases and the associated localized corrosion and embrittlement in 29% Cr ferritic stainless steels. *Materials Science and Engineering: A*, Vol.418, No.1-2, (February 2006), pp. 211-217, ISSN 0921-5093

Pond, R. C. (1989). Line defects in interfaces, In: *Dislocations in Solids, Vol.8*, Nabarro, F.R.N. & Hirsh, J. P., (Eds.), pp. 1-62, Elsevier, ISBN 0-444-70515-5, Amsterdam, the Netherland

Rappaz, M.; Jacot, A. & Boettinger, W. J. (2003). Last-stage solidification of alloys: Theoretical model of dendrite-arm and grain coalescence. *Metallurgical and Materials Transactions A*, Vol.34, No.3, (March 2003), pp. 467-479, ISSN 11661-003-0083-3

Sannes, S.; Arnberg, L. & Flemings, M. C. (1996). Orientational relationships in semi-solid Al-6.5wt%Al, In: *Light Metals 1996*, Hale, W., (Ed.), pp. 795-798, TMS, ISSN 1096-9586, Warrendale, PA, USA

Shabestari, S. G. & Malekan, M. (2010). Assessment of the effect of grain refinement on the solidification characteristics of 319 aluminum alloy using thermal analysis. *Journal of Alloys and Compounds*, Vol.492, No.1-2, (March 2010), pp. 134-142, ISSN 0925-8388

Stangleland, A.; Mo, A.; Nielsen, Ø.; M'Hamdi, M. & Eskin, D. (2004). Development of thermal strain in the coherent mushy zone during solidification of aluminum alloys. *Metallurgical and Materials Transactions A*, Vol.35, No.9, (September 2004), pp. 2903-2915, ISSN 11661-004-0238-X

Vernède, S.; Jarry, P. & Rappaz, M. (2006). A granular model of equiaxed mushy zones: Formation of a coherent solid and localization of feeding. *Acta Materialia*, Vol.54, No.15, (September 2006), pp. 4023-4034, ISSN 1359-6454

5

Dopant Driven Electron Beam Lithography

Timothy E. Kidd
Physics Department, University of Northern Iowa, Cedar Falls, IA, USA

1. Introduction

The scanning electron microscope (SEM) can be used for far more than just obtaining images. It has a long tradition of being used to directly manipulate a sample to create various surface structures. The scanning coils within the microscope can be utilized for directing the electron beam in a controlled manner rather than simply raster across the surface as is used in imaging. By focusing the electron beam on a given area of the sample, it can be used to induce various localized changes to the surface of a material with a high degree of precision. There are several established techniques by which an electron beam can be used to create patterned structures upon a surface, the most common of which is electron beam lithography. Electron beam lithography is a multi-step process in which a sacrificial polymer layer is first deposited onto the sample that can achieve feature sizes down to ten nanometer length scales (Broers et al. 1996; Liu et al. 2002). The electron beam can also be used to locally induce or break bonds to pattern nanostructures (Mendes et al. 2004) or simply burn material away from selected areas of the sample (Egerton et al. 2004). In essentially every case, the electron beam interacts with the surface to locally alter or break chemical bonds to form patterned surface structures with very high precision.

In the present study, we have developed a new method by which the electron beam can be used to create patterned surfaces (Kidd et al. 2011). The discovery was quite by accident, occurring during some standard studies of layered dichalcogenide crystals. These highly two dimensional materials have intriguing electronic and chemical bonding characteristics which have made them of great interest for the study of novel electronic phase transitions (Wilson et al. 1975; Sipos et al. 2008) and potential use in a variety of alternative energy applications (Whittingham 1976; Kline et al. 1981; Chen et al. 2003). The systems we had chosen for study were doped crystals, as doping these materials can be used to induce dramatic changes in their electronic and chemical behaviors (Levy 1979; Friend and Yoffe 1987). The particular sample of interest at the time was Cu_xTiSe_2, owing to a recent discovery of a superconducting phase in the system in competition with the charge density wave ground state of pure $TiSe_2$ (Morosan et al. 2006).

The SEM was being used to examine the microstructure of single crystal samples while energy dispersive x-ray spectrometry (EDX), via a spectrometer attached to our microscope, was used for measuring the chemical composition of the sample. These measurements were meant to do nothing more than determine the homogeneity of our samples and how much copper was successfully incorporated as a dopant. Interesting anomalies were quickly discovered, however. The samples had an inhomogeneous appearance in the SEM even after surface layers were removed, although EDX measurements showed the chemical composition was essentially homogenous. The EDX measurements were then taken over longer time periods, to try to detect any subtle inhomogeneity in the sample stoichiometry. These longer measurements induced further changes in the appearance of the sample, and there were signs that the copper concentration might actually be varying with the duration of the EDX measurement itself. After some time, we hypothesized that perhaps the samples were contaminated in some way, so we attempted to drive off any contamination by heating the measured crystals in a vacuum oven. This resulted in the creation of a multitude of sub-micron and nanoscale copper iodide crystallites upon the sample surface, except for where the EDX measurements were performed. In areas of the sample that had been exposed for long durations to the electron beam, the surface was unchanged.

Through further experimentation, the process was refined so that the size and density of the crystallites could be controlled by varying exposure dosages and the temperature at which samples were heated after exposure. The iodine was found to come from within the sample itself. Iodine is used as a catalyst for growing large single crystals of $TiSe_2$ and other dichalcogenides, and some remains trapped in various nooks or cracks long after the samples are grown. This iodine is then released during heating and draws copper ions from within the bulk of the $TiSe_2$ to its surface to react and form the CuI crystallites (Jaegermann et al. 1996). The electron beam radiation serves to reduce the amount of copper arriving at the surface in the exposed portions of the surface, allowing one to devise a patterned array of sub-micron structures. Unlike traditional electron beam lithography, with its required sacrificial polymer coating, no intermediate processing is required. The sample is simply exposed to the electron beam in a controlled fashion and then reacted with some quantity of a halogen gas like iodine in a simple two-step process. This lithographic technique is essentially one derived solely from controlling the mobility of the dopant ions stored within the sample to reach the surface for reaction.

2. Experimental methods

The Cu_xTiSe_2 samples used in this study were synthesized using a technique proven for growing large single crystals of pure $TiSe_2$ (Balchin 1976; Kidd et al. 2002). The basic process is to heat the starting elemental powders (Alfa Aesar, >99.5% purity) with iodine as a catalyst in a sealed and evacuated silica ampoule. The original intent for these samples was to study the superconducting ground state which emerges upon doping with copper (Morosan, Zandbergen et al. 2006). In an attempt to dope the samples with a higher concentration of Cu than was found in the literature, the copper was included with the initial growth of Ti, Se, and I powders in a single step growth process rather than the standard multi-step method of first growing $TiSe_2$ powders and then later incorporating the copper dopants.

To put it mildly, high doping concentrations were not achieved using this modified single-step method. Instead, the copper reacted strongly with iodine in the growth ampoule to form a CuI film which coated the surfaces of very lightly doped $TiSe_2$ (Figure 1). The samples had a stoichiometry of $Cu_{0.04}TiSe_2$, with copper concentrations varying by ±0.01 as determined by EDX. This was much less than the goal of doping concentrations greater than 10%, and in fact our samples showed no signs of superconductivity down to 3K.

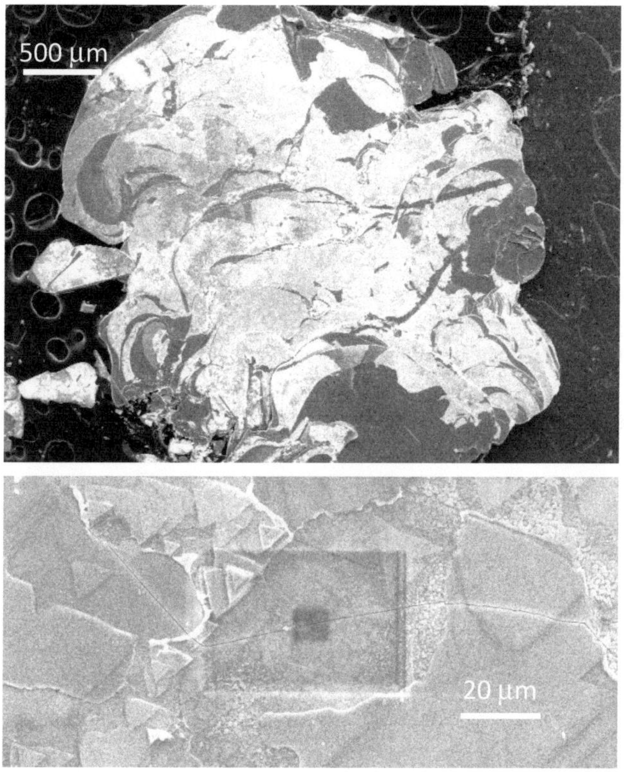

Fig. 1. SEM image of the CuI film coating as a grown Cu_xTiSe_2 single crystal. Most of the surface is covered with a CuI film, which appears much brighter in the SEM image than surrounding $TiSe_2$. The lower panel reveals triangular structures typical of the CuI coating. The darkened rectangular regions are areas in which higher magnification imaging was attempted. The focused beam burned away the CuI to reveal the underlying $TiSe_2$ surface.

The CuI surface layer was not a completely uniform coating, with portions seemingly chipped away during sample handling. It appeared much brighter than uncovered areas, and typically was composed of triangular features aligned over a local area of the sample. This CuI overlayer was very sensitive to electron beam exposure. High magnification imaging would often burn away a portion or all of the CuI layer within less than a minute of imaging. Rectangular areas in which the CuI layer was thus burned away can be seen in the lower panel of Figure 1.

The SEM measurements were performed using a TESCAN Vega II microscope with an attached Bruker Quantax EDX spectrometer. SEM images and EDX data shown here were taken using 20kV beam voltages. The samples were manipulated in air for short periods and stored long term in a dry box. The inert $TiSe_2$ crystals showed almost no signs of chemical decomposition months after they were grown. To expose clean portions of the samples and remove the CuI films, samples were exfoliated with Scotch tape to remove the uppermost surface layers. This process had to be repeated several times in some cases, as CuI inclusions could be found within some crystals. These inclusions, which were typically very thin CuI films, grew as the crystals were formed creating weak points within the crystals where exfoliation would occur but not provide a clean dichalcogenide surface.

3. Metastable surface inhomogeneity

SEM and EDX measurements were performed primarily on exfoliated Cu_xTiSe_2 single crystals. Interestingly, even in samples which showed no evidence of CuI via EDX after this process, the crystals appeared inhomogeneous in SEM images as in Figure 2. The apparent brightness of the surface varied in different locations. This was quite unexpected as pure $TiSe_2$ samples never showed any such inhomogeneity, and x-ray diffraction measurements indicated the samples were of a single phase. EDX measurements were then taken over different sections of the sample to try to correlate the apparent differences in brightness to differences in the local chemical composition. According to EDX measurements, the samples were essentially homogenous. Furthermore, high magnification SEM measurements did not reveal any substantial differences between the brighter and darker areas of the sample.

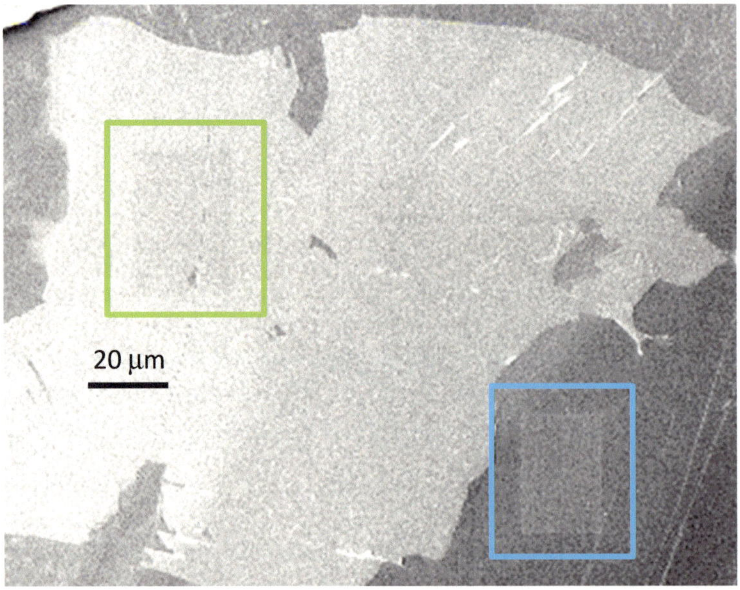

Fig. 2. SEM image taken from the inhomogeneous appearing surface of a freshly exfoliated single crystal of $Cu_{0.04}TiSe_2$. Various areas of the surface appeared more or less bright in the SEM. The colored boxes outline two areas exposed to relatively high electron beam radiation during long duration EDX measurements. The green square outlines an area from the brighter appearing portion of the surface that became darker after exposure. The blue square outlines a section from the darker area of the surface that became brighter after electron beam exposure.

To obtain more precise stoichiometry values, EDX measurements were made over longer exposure times than usual. Measurements were taken for half an hour or more, in which a relatively small portion of the surface was continuously exposed to electron beam radiation. While these measurements did not provide any conclusive results regarding the local stoichiometry, they did have a significant impact on the appearance of the sample surface (Figure 2). Essentially, long term electron beam exposure caused areas which initially appeared brighter to become darker and areas which were initially darker to become brighter.

The process could even be reversed to some extent in that an area could be switched back to its original appearance by a second exposure as seen in Figure 3. In this image, the sample areas appeared relatively dark until the portion in the upper half of the picture was exposed to long term electron beam radiation and became brighter. After this, a smaller portion of the sample was re-exposed to the beam. At this time, the brighter section reverted to its original appearance and the bottom half of the surface, which was not initially exposed to the beam for a long duration, was not changed at all. The edge of this second exposure is visible in the image, however. In general, it was often possible to effectively burn away a region made brighter by the beam, but we were not able to make a darkened area return to a more bright state by a second exposure.

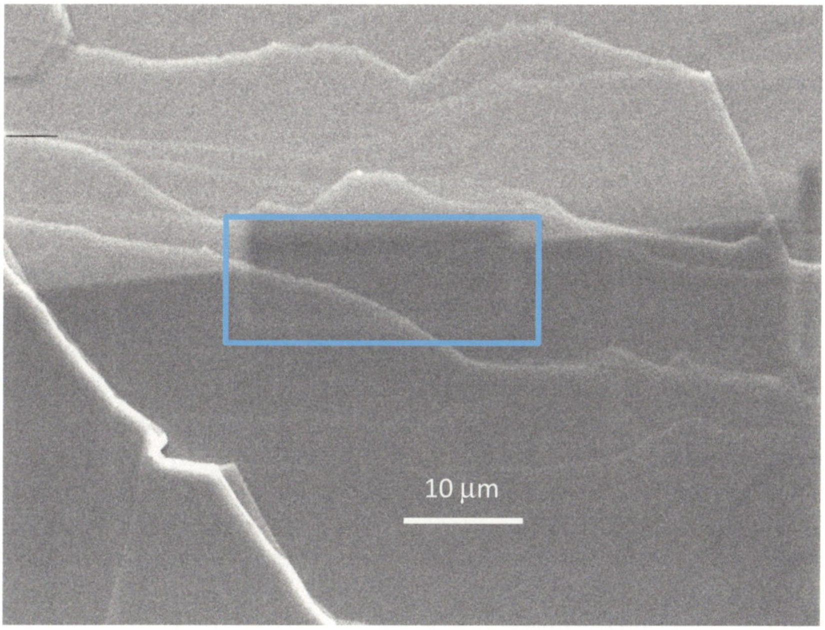

Fig. 3. SEM images of an area exposed to long term electron beam radiation multiple times. This area of the sample was originally one of the less bright regions of the surface. After long term EDX measurements, the top half of this image became brighter. A second long term EDX measurement was taken at the border between the original surface and the brightened area, as highlighted by the colored rectangle. Within this region, the area that became brighter in the initial exposure reverted to its original darker appearance after the second exposure.

These results indicate that the apparent brightness of the sample is most likely related to a phenomenon confined to the immediate surface of the sample. EDX measurements include x-rays generated by inelastically scattered electrons which can easily penetrate several hundred nanometers into the sample. Therefore, stoichiometry variations confined to a few surface layers are essentially undetectable by EDX. Furthermore, it is obvious that electron beam radiation can alter the apparent brightness of the sample as viewed in the SEM. When the beam is focused on a small area of the sample in high magnification measurements, this area would receive a relatively high dose of radiation even for short duration measurements. Therefore it is not surprising that significant differences could not be found when comparing different portions of the sample as the surface is likely changing during the measurement process itself.

These measurements in themselves show that the variation in brightness is a surface phenomenon sensitive to electron beam exposure, but do not reveal the source of the surface inhomogeneity. The apparent brightness in an SEM image can be influenced by many factors. The relative height, density, and conductivity of a given sample region can all play a role in determining how bright a particular area appears in an SEM image.

In this case, it appears that the brightness could be due to local changes in the sample density or surface conductivity. Copper ions have significant mobility within the $TiSe_2$ solid (Gunst et al. 2000), so that any local heating or other beam interactions could potentially cause the copper ions to migrate and/or aggregate to alter the local electron density in response to beam exposure. On the other hand, insulating copper iodide films appear not only on the surface of as-grown crystals, but could also be found after exfoliations interspersed within the crystal bulk. Even though EDX measurements did not detect significant iodine levels, the brighter areas could arise from an extremely thin CuI layer. This appears to be the more likely explanation. Copper ions assume a positive oxidation state as dopants within the layers of $TiSe_2$. If the beam caused them to migrate and cluster together in some manner, this would not be a stable configuration as the positive ions would repulse each other through Coulomb interactions to return to their equilibrium positions. The changes induced by beam exposure were stable for days if not weeks indicating equilibrium must be attained soon after the beam exposure is ended.

If the apparent changes in brightness arise to a thin CuI layer, there must be a source of iodine somewhere throughout the sample. In fact, a small concentration of iodine does remain interspersed throughout the sample from the growth process, where iodine is used to catalyze the growth of large single crystals. Also, iodine was found in significant amounts within cracks or at the edges of the crystals. These areas also tended to contain larger concentrations of copper, making them a mix of copper iodide and pure iodide in areas of the sample less affected by exfoliation. Due to charging effects in these insulating materials, the iodine rich particles appear much brighter than the surrounding dichalcogenide surface.

The presence of microscale iodide crystals was unexpected given the element's volatility. The samples used in this study were grown several weeks prior to the SEM measurements, making it likely that any surface iodine should have already sublimated. Therefore, the iodine seen here must have been stable while trapped inside various sample defects. This iodine would then be exposed after the exfoliation process to become more reactive.

Assuming the bright areas appear so due to a thin coating of CuI, they would become darker as this insulating material is burned away during the electron beam exposure. At the same time, the radiation from the electron beam would enhance the mobility of copper ions to flow towards the surface through local heating and/or the accumulation of negative charge. Another effect of the electron beam exposure would be to volatilize any exposed iodine. Any copper ions migrating to the surface would react with this iodine to become trapped as a very thin CuI layer, creating a region appearing more bright after exposure. This hypothesis is consistent with the fact that the surface was mostly stabilized within a day or so after exfoliation. While it was still possible to burn away brighter regions, it was no longer possible to reverse this process or induce darker regions to become bright on the aged samples. This is exactly what one would expect as the iodine available for reaction immediately after exfoliation would simply sublimate away so that there would be nothing left for copper ions to react with to form apparently bright regions as seen by SEM.

4. Formation of CuI crystals

The initial EDX measurements indicated that there could be variation in the copper concentration related to electron beam exposure, although these variations were not very significant given the precision of the measurement. If this were true, it could mean that the electron beam exposure could be used to create localized areas of superconductivity within the Cu_xTiSe_2 surface and induce interesting interfacial properties between charge density wave and superconducting regions of the sample. (Morosan, Zandbergen et al. 2006; Barath et al. 2008). To better attain information about the copper distribution, it was deemed important to first remove the iodine. Copper reacts strongly with any iodine present in the system, making iodine a contaminant that can strongly affect the distribution of the dopant copper ions within the $TiSe_2$ layers. Given the volatility of iodine, the decision was made to gently heat one of our exfoliated samples in a vacuum oven. For convenience and consistency, a sample in which EDX measurements had been performed was chosen. A low temperature, only 60°C, was used as this should be sufficient to vaporize any exposed iodine and it was not so high as to damage the carbon tape used to mount the sample.

The results were quite surprising, as seen in Figure 4. Not all of the iodine exposed by exfoliation sublimed from the sample during heating. In fact, it appeared as if the majority reacted with dopant copper ions to form a surface covered in CuI. That copper and iodine reacted was not completely unexpected. Halogen gas exposure has been shown to draw out high mobility dopants like copper or silver (Jaegermann, Pettenkofer et al. 1996). However, the relatively small amount of iodine seen in our measurements of the exfoliated surface did not lead us to expect the near totality with which the surface was covered after heating.

There are additional oddities concerning the copper iodine reaction of this surface. The region shown in Figure 4 is the same area as shown in Figure 3 after heat treatment. The portions of the sample that had turned relatively bright after e-beam exposure were almost completely free of CuI after heating. Other exposed areas can be seen as rectangular features in the image. The smaller region formed by the secondary exposure also had less CuI formation. However, some rectangular regions appear to have a more uniform CuI coverage than unexposed areas of the sample. These regions received a relatively small dose of electron beam radiation during the time in which images were taken, such as that shown in Figure 3.

The truly novel discovery was that rectangular portions of the surface remained CuI free. These rectangular areas could be identified as the regions on the sample which received very high electron beam exposure levels during the EDX studies. This discovery led to a more systematic series of measurements designed to illuminate whether controlled electron beam exposure could be used to create a patterned array of CuI features in a process similar to, but simpler than, standard electron beam lithography.

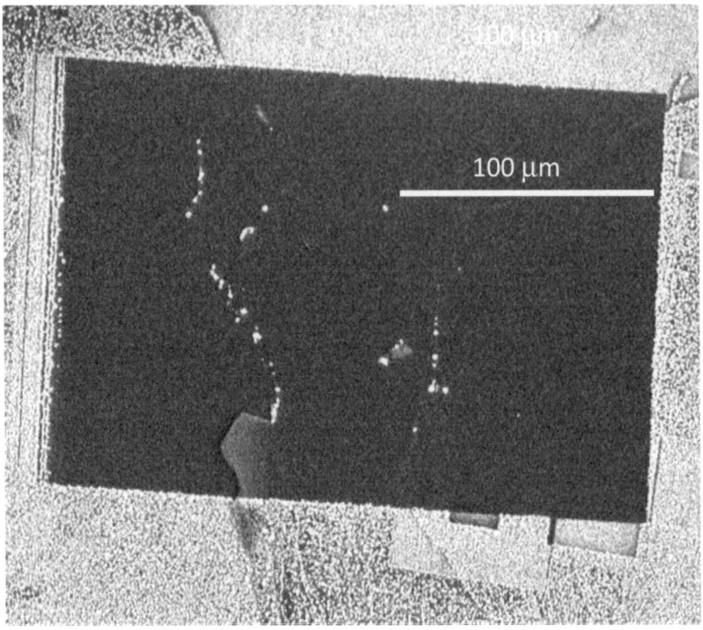

Fig. 4. SEM images taken from the area shown in Figure 3 after the sample was heated in a vacuum oven at 60°C for 24 hours. Iodine released from the sample during heating reacted with copper from within the sample to form CuI crystallites over the majority of the sample surface. The formation of CuI was reduced or eliminated in areas of the sample in which long term EDX measurements were taken.

5. Dopant driven lithography

To determine the effects of electron beam exposure, comparisons were made between a series of different exposure levels made upon a single Cu_xTiSe_2 crystal. After the exposures were performed, the samples were transferred to the vacuum oven and heated to temperatures between 60-100°C. The iodine exposure levels were varied by including various amounts of iodine in the vicinity of the sample when it was loaded into the vacuum oven.

Within a given sample, the results were very consistent. Figure 5 shows SEM images taken from a Cu_xTiSe_2 crystal that had been exposed to four different electron radiation doses before being heated in a vacuum oven. The surface was exposed to three ring patterns with dosages of 2.5, 5, and 10 mC/cm^2. Near the center of each ring a small circular area was also exposed to a dosage of roughly $50mC/cm^2$. This surface was exfoliated multiple times to attain a mostly homogenous surface to minimize any effects from a potential pre-existing CuI film upon the surface. These efforts were successful in that no significant changes were induced by the electron beam exposure itself. However, the radiation exposure patterns were still easily identifiable after the heat treatment. Even at low magnification levels, it can be seen that the CuI coverage is significantly reduced by the electron beam exposure.

The influence of the electron beam exposure is more obvious at higher magnification as seen in the area exposed with a 5 mC/cm^2 dose in Figure 6. After heating, the surface is coated with small CuI crystallites, which are reduced in size and density in areas exposed to higher levels of electron radiation. Significant changes were difficult to detect in doses less than $1 mC/cm^2$.

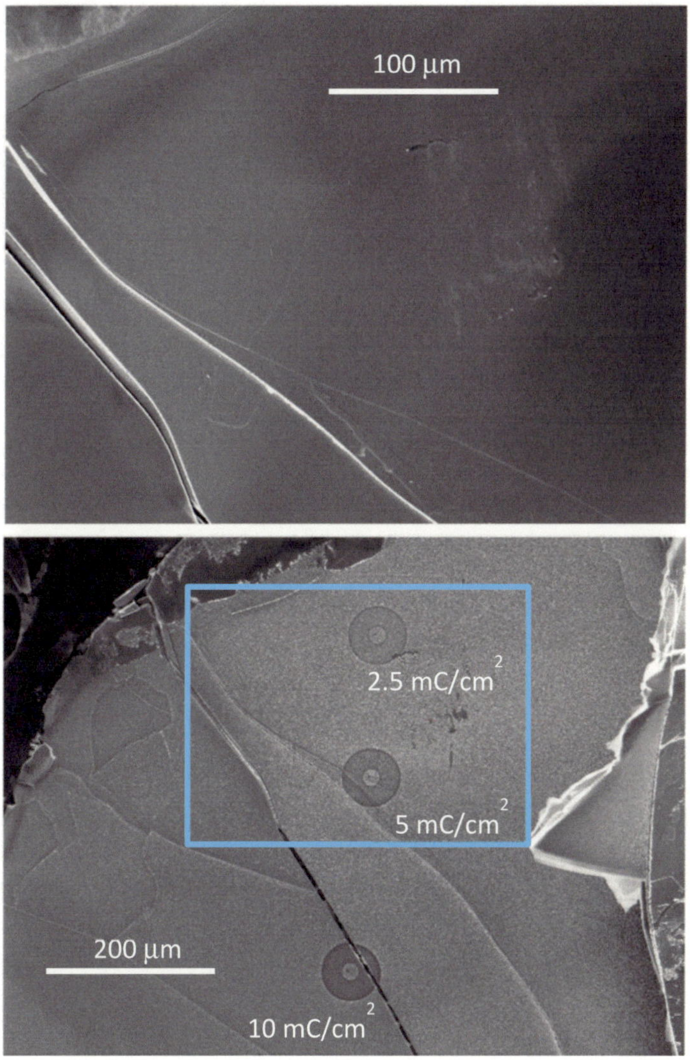

Fig. 5. SEM images before and after an irradiated sample was heated to 60°C in a vacuum oven for 24 hours. The top panel was taken immediately after selected areas of the surface were exposed to electron beam radiation. The surface showed no signs of damage from the electron beam exposure. The image in the bottom panel was taken after the heat treatment induced CuI formation. The blue rectangle indicates the area initially imaged in the upper panel, and the radiation exposure levels for each pattern are indicated on the image. Near the center of each ring, small circular areas were exposed to approximately 50 mC/cm^2 of electron beam radiation.

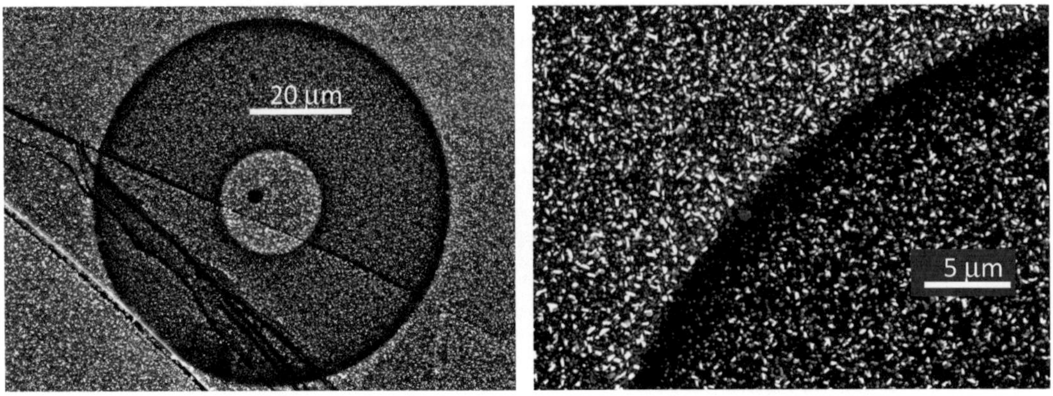

Fig. 6. SEM images taken after irradiation and vacuum heat treatment at 60°C. The left panel is an overview of an area exposed to 10mC/cm^2 of radiation in a ring pattern. The right panel is a magnified view of the exterior ring edge. CuI crystallites are smaller and fewer in number in the irradiated area. The boundary region of the exposed area received a larger dose of electron radiation, resulting in even less CuI crystallites at the edge of the pattern.

The edges of the exposed ring were relatively dark in the SEM images, with much less CuI crystallites. This is attributed to systematic errors in the coil control program which cause the beam to linger longer at edges in the prescribed pattern. The small circle offset from the center of the ring was exposed to a much higher dosage of electron beam radiation. Through additional measurements concerning multiple samples and treatment temperatures, it was found that a dosage of 50mC/cm^2 was consistently sufficient to completely suppress CuI formation.

It was difficult to determine how the chemical composition of the surface was altered by various exposure doses or iodine reaction temperatures. The signal for iodine was small and copper consistently measured between three and four percent of the overall composition. Iodine could not be detected at all in areas exposed to high enough doses to completely suppress CuI formation, but even in these areas the copper concentration was not significantly different from the rest of the sample. These results indicate the CuI formation was confined to the surface and that the overall copper doping concentration was not substantially altered within the exposed areas.

As shown in Figure 7, the CuI crystallites are typically sub-micron structures. The size and density of these crystallites are inversely proportional to the amount of electron radiation exposure. These results show that not only can the exposure process be used to direct where the CuI crystallites are formed, but also to control their size and average spacing.

Depending on the radiation dosage and iodine reaction temperature, crystallite sizes ranged from ten nanometers to a few microns. While not particularly uniform, the size of most crystallites within a given exposure region did not deviate by more than a factor of two. The crystallite size was dependent on both the exposure level and temperature at which the iodine exposure occurred within the vacuum oven. Higher processing temperatures led to the formation of larger crystallites (Figure 8). While the smaller crystallites appeared more rounded and randomly shaped, larger structures often showed clear symmetry respective of an underlying crystal structure. The variation in appearance is a size-effect related phenomenon. As the size of the crystallite decreases, atoms residing at the surface represent a larger and larger percentage of the material. Surface atoms have a relatively high energy state, and when the crystallite becomes small enough, minimizing the total number of surface atoms by attaining a more rounded structure becomes more energetically favorable than maintaining the proper bonding angles representing the crystal symmetry found in larger features.

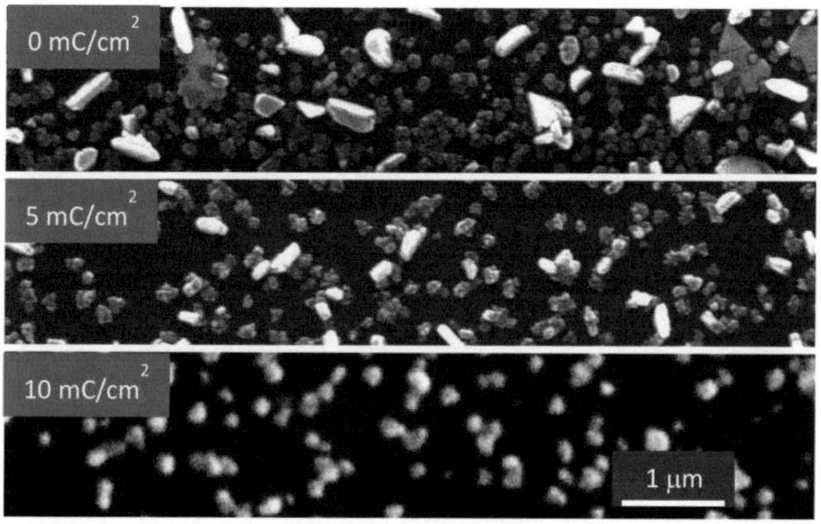

Fig. 7. SEM images comparing the effects of radiation exposure on the form of CuI crystallites formed after heating at 60°C. Each panel is shown at the same scale and was taken from different areas of the same sample with the indicated radiation exposure dose.

The lateral resolution of the process is theoretically expected to be on the order of the electron beam spot used during the exposure process. However, in practice the beam spot can be made to be much smaller than the average CuI crystallite size. Thus the lateral resolution of this technique is defined more by the size of the crystallites to be obtained rather than intrinsic exposure parameters.

There were significant variations found in the size and density of crystallites formed on different $TiSe_2$ crystals. This is attributed to variations in the amount of iodine available for reaction. It was found that including excess iodine seemed to have a much smaller effect than variations seen from sample to sample, indicating that the CuI reaction is dominated by iodine emerging from the $TiSe_2$ sample itself. While iodine is a volatile element in ambient conditions, it can remain present for years trapped between layers of inside defects of a $TiSe_2$ crystal. When the sample is exfoliated, trapped iodine becomes available for reaction, especially when the sample is heated, even at temperatures below 100°C. Iodine loaded in proximity to the crystal likely becomes volatile and spreads quickly throughout the vacuum oven, minimizing its effectiveness in forming CuI. Iodine trapped in the nooks and crannies of the sample will take more time to fully volatilize, and be in closer proximity to the surface, giving it more time to react with the copper dopant ions before it dissipates away from the sample. This hypothesis also supports the relationship between crystallite size and reaction temperature. At higher temperatures, the copper ions have a higher mobility which allows a larger number to reach the surface in a given time. As there is likely a limit to how long the chemical reaction takes place, heating to a higher temperature would allow more copper ions to take part in the reaction to form larger structures.

Fig. 8. SEM images comparing the effects of reaction temperature on the size of CuI crystallites. The top panel was heated in the vacuum oven at 100°C. The bottom panel was heated to 60°C. The bare area on the left of the bottom panel was initially dosed with about 50 mC/cm^2 of electron beam radiation, completely suppressing the formation of CuI.

It was also found that CuI formation was inhibited for samples which had been exfoliated for many days prior to beam exposure. Even though the samples looked almost identical to fresh surfaces in the SEM, including the presence of trapped iodine, CuI crystallites would rarely grow upon these aged samples. These results were reminiscent of an AFM study in which the copper ions could be drawn to the surface of a dichalcogenide crystal, but only for freshly exfoliated surfaces (Gunst, Klein et al. 2000).

6. Conclusions and future directions

The experimental results show that it is possible to use electron beam exposure to control the formation of CuI crystallites on the surface of a $TiSe_2$ crystal. This is a two-step process in which a fresh surface is first exposed to electron beam radiation in a controlled fashion, and then heated in the presence of iodine. Due to inconsistencies between samples, exact empirical parameters concerning the size and density of CuI formation have not been determined as yet, although qualitative relationships have been established. The size and density of CuI crystallites formed in this process are inversely proportional to the amount of electron beam radiation. The formation of crystallites can be completely eliminated in areas receiving relatively high exposure levels. It is evident this process can be used to create an array of CuI crystallites with diameters sufficiently small to be considered nanoparticles at certain processing conditions. The lateral resolution is essentially limited by the size of the crystallites, making this technique suitable for forming a pattern array of nanocrystallites with nanoscale precision.

In itself, this if of interest as CuI nanoparticles show some promise for applications in dye-sensitized solar cells (Perera and Tennakone 2003) or gas sensing applications for CO and NO (Wolpert et al. 2009). The process might also be useful for controlling the local copper dopant concentration, important for exploring competition between superconducting and charge density wave ground states in the material. However, the technique would be of far greater importance if it could be extended to other dopants and/or dichalcogenide crystals. For this to occur, one must have a better understanding of the process itself. While a complete quantitative understanding remains elusive given inconsistencies between samples, a qualitative explanation is forthcoming from our observations.

The second step of the process is relatively straightforward. Copper ions have a high mobility within dichalcogenides and can be reversibly incorporated into their structure using electrical or chemical gradients.(Gunst, Klein et al. 2000) Iodine present at the sample surface will react with surface copper ions, leading to a concentration gradient near the surface. Copper ions from deeper within the sample will be driven to the surface layers by this depletion, and thus in turn react themselves. This process will continue until the iodine is depleted, the concentration gradient is reduced below some threshold value depending on the overall percentage of copper ions in the sample, or the surface is completely coated with CuI. The increased size of the CuI crystallites at higher reaction temperatures could be due to the increased mobility of Cu ions at higher temperatures or the fact that more iodine is released from the sample. Either process could lead to the formation of more CuI at the surface before the iodine dissipates away from the sample.

The first step of the process is perhaps not as intuitive. It was not initially clear why exposure of the sample could be used to influence the reaction of copper and iodine at the sample surface. The exposure must in some way be inhibiting the copper ions from traveling to the surface during the iodine reaction. This is most likely due to the electron beam radiation inducing a localized surface chemical reaction. When $TiSe_2$ is exposed to wet or humid environments it will oxidize at the surface. Titanium diselenide is composed of molecular $TiSe_2$ layers as shown in Figure 8. Hydrogen from the water will react with Se to form gaseous H_2Se leaving the exposed titanium atoms to react with the remaining oxygen to form titanium oxide. Owing to the inert nature of the material and its layered structure, it can take days or even weeks for a visible oxide layer to be seen. However, it would not be surprising for the uppermost molecular layer to oxidize within even a few hours of exposure to ambient conditions. A surface layer of titanium oxide would be an effective barrier to Cu ion migration to the surface, which would explain why CuI would not form on the surface of samples that had been exfoliated for several days before electron beam exposure was initiated.

Within the SEM, the sample resides in a vacuum. However, the samples are exfoliated in air before being inserted into the microscope. This means that there will be a residual water layer of some thickness on the sample even after the measurement chamber is evacuated. In truth, such a water layer coats the entire interior surface of this and any other SEM that is not heated to at least 100°C before measurement. This minor amount of water would normally react very slowly with the $TiSe_2$ surface. However, the energy provided by the electron beam radiation could easily break up the water molecules into various radical ions that would be much more reactive. This would in turn create a titanium oxide surface layer to inhibit copper ion migration. The uniformity and thickness of such an oxide layer would depend on the radiation exposure level, making it possible to fine tune how much copper ions migrate to the surface during the time in which the sample reacts with iodine in the second step of the process. It is true that other chemical reactions could occur on the surface, such as with remnant gas molecules within the SEM, however the low pressure ($<10^{-2}$ Pa) makes this unlikely.

These results make it apparent which systems could be best utilized with this technique. A wide range of ions and molecules can be reversibly stored within various dichalcogenides (Levy 1979). This makes a wide range of surface features possible, given the proper method by which materials could be induced to leave the sample. The most straightforward extension of this technique would be to utilize silver rather than copper ions. Silver and copper both have similar mobility within a given dichalcogenide and silver also is highly reactive with halogen gasses. Thus it should not be difficult to create a patterned array of silver iodide crystallites on the surface of $TiSe_2$. AgI has been used in photographic and other optical applications for decades, and an array of AgI nanocrystallites could easily be reduced to form silver nanoparticles which are of high interest for their antibacterial properties.

Current research is being directed towards refining this technique to enable the formation of various copper and silver halide nanoparticle arrays. Titanium diselenide is used as the principle host, but Ta based dichalcogenides are also of interest for their similar chemical characteristics. It is hoped that this technique can be coupled with traditional electron beam and/or scanning probe lithographic techniques to enable to the fabrication of prototype device structures.

7. Acknowledgements

The author would like to thank Dusty Klein and Tyler Rash for assisting with the SEM/EDX measurements and Dr. Laura Strauss for assisting with crystal growth. This work was supported by Battelle and the Iowa Office of Energy Independence grant #09-IPF-11. The author was also supported by a UNI summer fellowship during portions of this research.

8. References

Balchin, A. A. (1976). Growth and the Crystal Characteristics of Dichalcogenides Having Layer Structures. Crystallography and Crystal Chemistry of Materials with Layered Structures. F. Levy. Dordrecht, D. Reidel. 2: 1-50.

Barath, H., M. Kim, J. F. Karpus, S. L. Cooper, P. Abbamonte, E. Fradkin, E. Morosan and R. J. Cava (2008). "Quantum and Classical Mode Softening Near the Charge-Density-Wave–Superconductor Transition of Cu_{x}TiSe_{2}." Physical Review Letters 100(10): 106402.

Broers, A. N., A. C. F. Hoole and J. M. Ryan (1996). "Electron beam lithography--Resolution limits." Microelectronic Engineering 32(1-4): 131-142.

Chen, J., S.-L. Li, Z.-L. Tao, Y.-T. Shen and C.-X. Cui (2003). "Titanium Disulfide Nanotubes as Hydrogen-Storage Materials." Journal of the American Chemical Society 125(18): 5284-5285.

Egerton, R. F., P. Li and M. Malac (2004). "Radiation damage in the TEM and SEM." Micron 35(6): 399-409.

Friend, R. H. and A. D. Yoffe (1987). "Electronic properties of intercalation complexes of the transition metal dichalcogenides." Advances in Physics 36(1): 1.

Gunst, S., A. Klein, W. Jaegermann, Y. Tomm, H. Crawack and H. Jungblut (2000). "Intercalation and deintercalation of transition metal dichalcogenides: Nanostructuring of intercalated phases by scanning probe microscopy." Ionics 6(3): 180-186.

Jaegermann, W., C. Pettenkofer, O. Henrion, Y. Tomm, C. Papageorgopoulos, M. Kamaratos and D. Papageorgopoulos (1996). "Surface science investigations of Cu intercalation in 1T $TaSe_2$ and $TiSe_2$ and its deintercalation by adsorbed Br_2." Ionics 2(3): 201-207.

Kidd, T. E., D. Klein, T. A. Rash and L. H. Strauss (2011). "Dopant based electron beam lithography in CuxTiSe2." Applied Surface Science 257(8): 3812-3816.

Kidd, T. E., T. Miller, M. Y. Chou and T. C. Chiang (2002). "Electron-Hole Coupling and the Charge Density Wave Transition in TiSe2." Physical Review Letters 88(22): 226402.

Kline, G., K. Kam, D. Canfield and B. A. Parkinson (1981). "Efficient and stable photoelectrochemical cells constructed with WSe2 and MoSe2 photoanodes." Solar Energy Materials 4(3): 301-308.

Levy, F., Ed. (1979). Intercalated Layer Materials. Physics and Chemistry of Materials with Layered Structures. Dordrecht, D. Reidel.

Liu, K., P. Avouris, J. Bucchignano, R. Martel, S. Sun and J. Michl (2002). "Simple fabrication scheme for sub-10 nm electrode gaps using electron-beam lithography." Applied Physics Letters 80(5): 865-867.

Mendes, P. M., S. Jacke, K. Critchley, J. Plaza, Y. Chen, K. Nikitin, R. E. Palmer, J. A. Preece, S. D. Evans and D. Fitzmaurice (2004). "Gold Nanoparticle Patterning of Silicon Wafers Using Chemical e-Beam Lithography." Langmuir 20(9): 3766-3768.

Morosan, E., H. W. Zandbergen, B. S. Dennis, J. W. G. Bos, Y. Onose, T. Klimczuk, A. P. Ramirez, N. P. Ong and R. J. Cava (2006). "Superconductivity in CuxTiSe2." Nature Physics 2(8): 544-550.

Perera, V. P. S. and K. Tennakone (2003). "Recombination processes in dye-sensitized solid-state solar cells with CuI as the hole collector." Solar Energy Materials and Solar Cells 79(2): 249-255.

Sipos, B., A. F. Kusmartseva, A. Akrap, H. Berger, L. Forro and E. Tutis (2008). "From Mott state to superconductivity in 1T-TaS2." Nature Materials 7(12): 960-965.

Whittingham, M. S. (1976). "Electrical Energy Storage and Intercalation Chemistry." Science 192(4244): 1126-1127.

Wilson, J. A., F. J. D. Salvo and S. Mahajan (1975). "Charge-density waves and superlattices in the metallic layered transition metal dichalcogenides." Advances in Physics 24(2): 117-201.

Wolpert, B., O. S. Wolfbeis and V. M. Mirsky (2009). "Gas sensing properties of electrically conductive Cu(I) compounds at elevated temperatures." Sensors and Actuators B: Chemical 142(2): 446-450.

6

Palmtop EPMA

Jun Kawai, Yasukazu Nakaye and Susumu Imashuku
Department of Materials Science and Engineering,
Kyoto University, Sakyo-ku, Kyoto,
Japan

1. Introduction

We have been developping palmtop electron probe X-ray microanalyzers (EPMA) for these several years [1-4] and succeeded to make such an instrument recently [3], and in the present chapter, we describe how to make an instrument in detail. The EPMA is an instrument to perform microanalysis (micrometer area elemental analysis) of any kinds of samples, such as metals, alloys, minerals, environmental and biological samples, by using an electron beam. Usually 10-30 keV kinetic energy electron beam is focused less than 1 μm in order to irradiate a sample, and consequently to excite characteristic X-rays such as Kα (2p → 1s) or Lα (3d → 2p) lines. From the energy and intensity of the characteristic X-ray lines, the kind (qualitative analysis) and concentration (quantitative analysis) of the elements in the specimen can be analyzed. Commercially available EPMA instruments are usually large (need a room of at least 3 m × 5 m to install) and expensive instruments (a few 10^5 USD). The palmtop EPMA we describe in the present chapter has features as follows and different from the conventional EPMA.

1. Small size. The size of the main part (sample holder, X-ray emission part, and the electron gun) is palmtop size. Typically less than 3 cm diameter and 5 cm length (**Fig.1**), but can be smaller than this size. The limitation of the size is due to the high voltage discharge distance.
2. Electric battery driven. The electron gun is driven by two 1.5 V electric D-batteries (**Fig. 2**), i.e. 3 V is enough for high energy (>10 keV) electron beam in order to excite characteristic X-rays.
3. The X-ray detector is Amptek Si-PIN detector (**Fig. 3**). Thus the size of the detector is also small. We use analog type X-ray detector amplifier. The size of the detector pre-amplifier is typically around 7 cm × 4 cm × 3 cm. The temperature control of the detector unit and bias power supply from the Amptek Co. is needed (**Fig.3**). Thus we need power supply for the Si-PIN controller to cool down the detector and bias voltage. However usually low voltage (5-12 V) DC is enough.
4. The pulse height analyzer (PHA) commercially available is not used, but we use a musician's amplifier and Windows computer as an alternative to the commercially available PHA [5-8]. This part is usually called DSP (digital signal processor). Usually a DSP is an expensive device which costs between 5000 and 10000 USD. However the musician's amplifier (**Fig. 4**) is typically less than 500 USD.

5. Vacuum in the sample chamber is of the order of 10^{-2} Torr (or ~1 Pa). This vacuum level is equivalent to the vacuum reached by a small rotary pump. Higher the vacuum than 10^{-2} Torr, in other words, too good or too high vacuum is not effective for high intensity X-ray emission. Too high vacuum down to 10^{-6} Torr needs expensive turbo molecular pump or oil diffusion pump, however, the residual gas is an electron source for the electron gun, thus such a high vacuum reduces the intensity of X-rays emitted. Too low vacuum up to 10^{-1} Torr will make electric discharge and 10^{-2} Torr is suitable for the present instrument. This vacuum level is suitable for oil sealed rotary pump **(Fig. 5)**.
6. The sample exchange is easy because the vacuum seal is usually rubber O-ring **(Fig. 6)** and can be easily opened after air leaked.
7. The electron beam size is not micrometer, but as wide as the vacuum vessel diameter. It irradiates whole part of the sample and thus sample holder as well as vacuum vessel materials are excited. However the sample size can be as small as possible depending on the signal intensity. A typical size of the specimen is 50 μm diameter × 5 mm length metal wire. Thus single sand particle can be measured.
8. The palmtop EPMA can be made in laboratory, even by students without special experience.

a

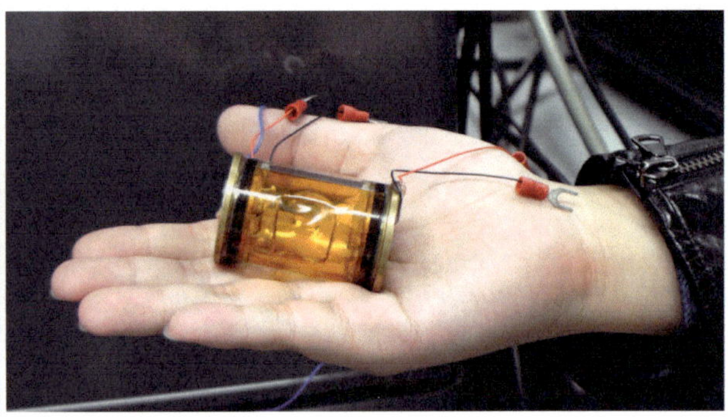

b

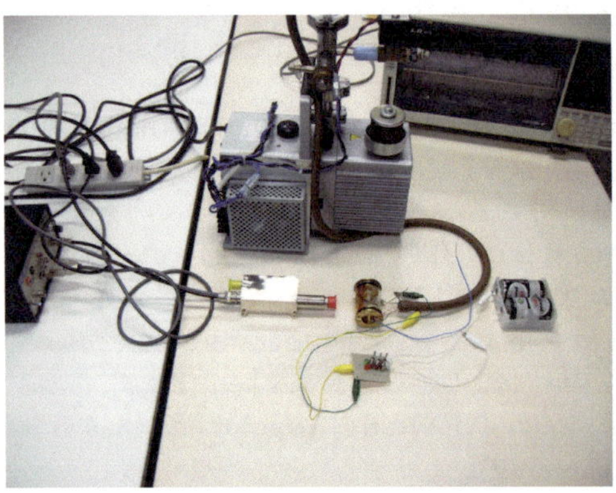

c

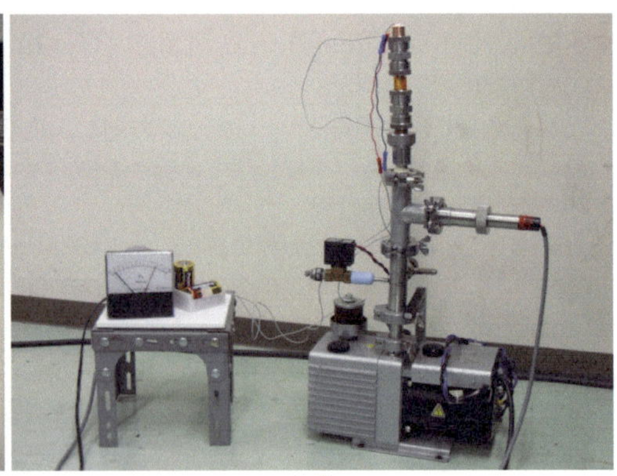

d

Fig. 1. Various kinds of palmtop EPMA made in our laboratory. (a) NW25 nipple is used for vacuum chamber, a hole is sealed by Kapton tape. (b) Glass type vessel is on the palm. (c) The whole system of the palmtop EPMA. (d) Another type of palmtop EPMA, with Pirani vacuum gauge.

Fig. 2. Electric D-batteries.

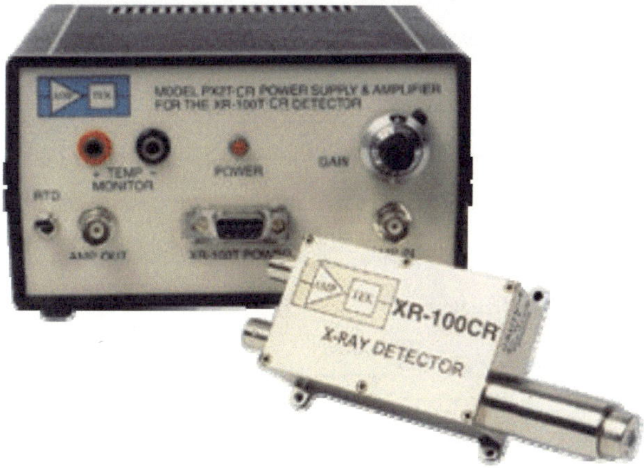

Fig. 3. Amptek Si-PIN X-ray detector (taken from Amptek Web page)

Fig. 4. Musician's amplifier for DSP.

2. The parts prepared before to build a palm-top EPMA

The following mechanical or electronic parts should be prepared to build up the instrument.

1. Rotary pump. Oil sealed rotary pump to reach to the vacuum level of the order of 10^{-2} Torr or around 1 Pa is needed. The smaller size rotary pump is preferable. Single phase 100 V power supply is enough. A typical rotary pump used in the present work is shown in **Fig. 5**.

Fig. 5. A small rotary vacuum pump.

2. Si-PIN X-ray detector. Amptek X-ray detector (**Fig. 3**) is needed. It costs usually less than 4000 USD including the preamplifier. The controller of Peltier device inside the Si-PIN detector to cool the Si-PIN device is needed. Usually the power supply associated with the Si-PIN detector is preferable. If Si-PIN X-ray detector is not available, Geiger-Müller counter is alternatively used for check the X-ray emission. The Geiger-Müller counter is available less than 500 USD (**Fig. 7**).
3. Notebook size computer. Windows computer to control the Amptek Si-PIN detector, to display the X-ray spectra, and to control and analyze the digital X-ray signal is needed as is shown in **Fig. 4**. Several USB devices are connectable at the same time.
4. $LiTaO_3$ single crystal (one piece). The size is around 3 mm (x) × 3mm (y) × 5 mm (z) (**Fig. 8**). The z direction should be known. From Shin-Etsu Chemical Co. Ltd., Japan, this single crystal is available by less than 200 USD. Several other companies treat $LiTaO_3$ single crystal of the size of 10 mm length and 3 mm diameter with similar price. Other alternative is $LiNbO_3$. These materials are called pyroelectric crystals. If these single crystals are not available, at the first stage, you can use PZT stone used in the cigarette lighter to ignite.
5. Peltier device. The size of Peltier device in **Fig. 9** is 8 mm × 8 mm. To drive this Peltier device, 3 V electric D-batteries are needed. **(Fig. 2)**
6. Glass vessel or steel nipple. Single crystal, peltier device, and sample should be inside of the vessel or nipple to evacuate by the vacuum pump (**Fig. 1**).
7. If possible, Pirani vacuum gauge is helpful to build up the instrument.

Fig. 6. O-ring, with brass specimen holder. Graphite is better than brass.

Fig. 7. Geiger-Müller counter

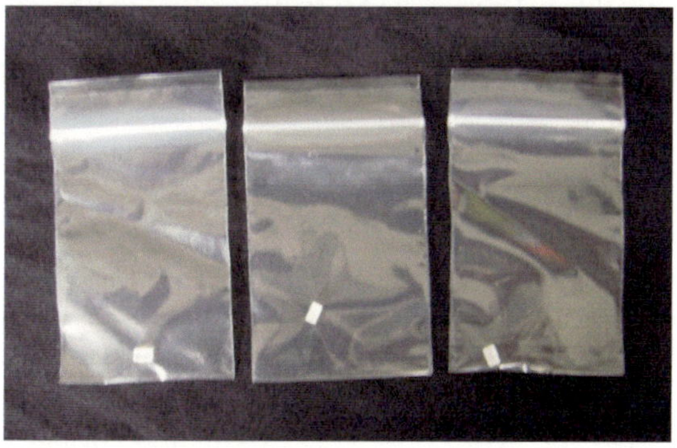

Fig. 8. $LiTaO_3$ single crystals in plastic bags.

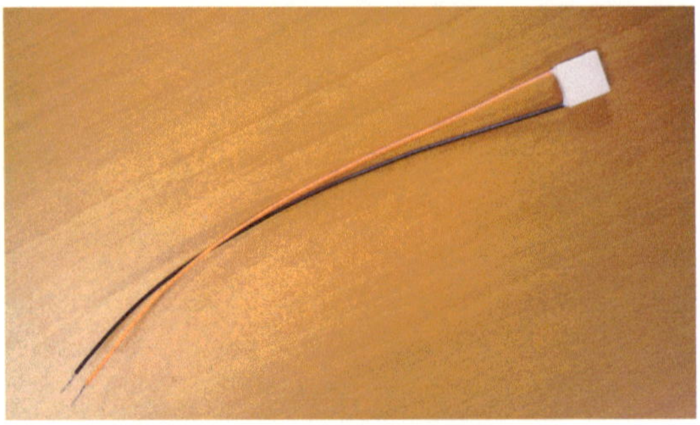

Fig. 9. Peltier device of size 8 mm × 8 mm.

3. Principle

The pyroelectric crystal is a material usually called as ferroelectric material. A mineral named tourmaline has similar characteristics. This is because the center of gravity of the plus charge and minus charge ions are not the same place in the crystal, but has a distance along z direction. Thus the extension or compression of these materials will produce electric high voltage. If a single crystal of the thickness of 1 mm in z direction of $LiTaO_3$ changes the temperature (usually the heating will expand, and cooling will compression) from room temperature (25 °C) to say 50 or 100 °C, then 10 kV high voltage will be produced. Thus when the crystal with thickness of 5 mm will produce 50 kV and 10 mm will produce 100 kV. When this single crystal is put into low vacuum such as 10^{-2} Torr, the electrons in the residual gas will be accelerated by this high voltage to hit the surface of the pyroelectric crystal or counter electrode of the pyroelectric crystal [9-19], as shown in **Fig. 12** below. If a specimen is attached on the counter electrode, the electrons accelerated by the high voltage between the pyroelectric (-HV) and counter electrode (0 V, grounded to earth) will hit the specimen. Consequently the X-rays are excited by the ionization of electrons by the bombardment of the accelerated electrons. Then the ionized electrons will contribute next instance to be accelerated to hit the specimen again and again, until the surface charge is neutralized. The vacuum should be not too good. Usually better side of 10^{-2} Torr is the suitable vacuum for this experiment. The heating and cooling of the pyroelectric crystal should be performed by a Peltier device. The temperature control of 50 °C from the room temperature (25 °C) is possible. The polarity of the surface changes when heated and cooled. Thus the electrons are moved inversely when heated and cooled. Usually it is heated for a few minutes, then next few minutes the pyroelectric crystal should be cooled. When the electron hits the pyroelectric surface, the X-rays are not from the specimen but Ta X-rays are observable from $LiTaO_3$.

4. How to build the main part of the palmtop EPMA

The Peltier device and the pyroelectric crystal are on the copper rod as shown in **Fig. 10**. They are glued by silver paste. The reason using the copper rod is its good heat conductivity as well as the electric conductivity. The one side of pyroelectric crystal glued to the Peltier device should be grounded to the earth. The heat created by the Peltier device should be diffused through the copper rod. Thus both silver paste and copper rod are suitable for this purpose.

Fig. 10. Pyroelectric crystal, Peltier device, and copper rod, glued by silver paste.

The anode is graphite rod as shown in **Fig. 11**. The specimen will be put on the graphite rod by double sided carbon adhesive tape. This is also because of the electric conductivity. The graphite rod and copper rod should be electrically connected to make the same ground potential. The X-ray intensity becomes weak if the electric connection between the copper and graphite rods is removed. Here the graphite is used as the specimen holder. The reason we use graphite and carbon adhesive tape is because the specimen and holder are hit by the electron beam from the pyroelectric crystal, and the carbon Kα X-rays from graphite as well as carbon adhesive tape are negligible because the C Kα energy is about 300 eV and thus such a low energy soft X-rays are strongly absorbed by the air and window of the vessel (Kapton), resulting the X-rays from graphite rod and carbon adhesive tape are not detectable.

Fig. 11. Graphite anode as specimen holder.

The lead wire from the Peltier device is inside of the vacuum vessel, and should go through the vacuum boundary to the outside of the vessel without the leak of the vacuum by using epoxy glue. If possible addition of thermocouple is helpful to measure the temperature of the Pettier device.

Both of the electrodes are sealed inside the glass or stainless pipe by rubber O-ring. Viton rubber O-ring is preferable. If NW-25 type quick coupling flange is directly used, this is quite easy to seal the vacuum. The stainless steel nipple of NW-25 or glass pipe should have ca 5 mm diameter through-hole to go through the X-rays, and this through-hole should be sealed by thin Kapton film. Adhesive type Kapton tape is commercially available. Polyester, PET (polyethylene terephthalate), or Mylar films glued by epoxy resin adhesives are alternative to the Kapton tape. The structure of the palmtop part is illustrated in **Fig. 12**.

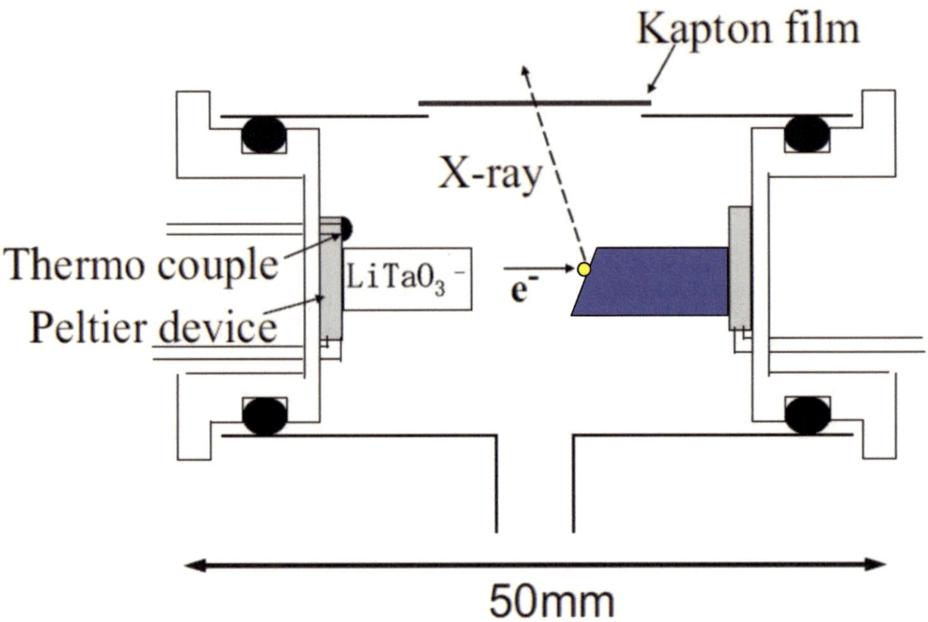

Fig. 12. Illustration of the palmtop EPMA.

5. How to operate

Putting stainless steel small plate, the size of which is typically 5 mm × 5 mm, on the graphite rod, by double sided carbon adhesive tape, then the vacuum vessel should be evacuated by the rotary pump. The connection between the vessel and the pump should be thick rubber tube. At the first experiment, the stainless steel as the specimen should be as large as possible to get enough X-ray intensity. The stainless steel is usually composed of 18 % Cr, 8 % Ni, and the rest Fe. Thus we can observe Cr, Fe, and Ni $K\alpha$ and $K\beta$ lines. After one or two minutes evacuation by the rotary pump, the vacuum reached to better than 2×10^{-2} Torr. Then the Peltier lead wires should be connected to 3 V D-batteries. Waiting for 10-20 seconds, the Geiger-Müller counter reacts the X-rays, and the intensity is not very strong but we can hear the X-ray counts sounds from the Geiger-Müller counter as almost continuous sounds from the counter. It is more than a few tens of cps (counts per seconds). The polarity of D-battery is no problem. If the polarity is for the pyroelectric crystal being +HV, the end of crystal is hit by the electron and X-rays of Ta are emitted.

If you confirm the emission of X-rays from the vessel, the next step is to measure the spectra by the Si-PIN detector. It is better to become accustomed to use the Si-PIN detector to measure the X-ray spectra by using other method, such as weak radio isotopes. Then the Si-PIN detector should be close to the Kapton window of the palmtop EPMA to measure the spectra. Usually the time interval to change the polarity of the D-battery from heat to cool the crystal is 1 or 2 minutes. A typical temperature change and X-ray intensity decay are shown in **Fig. 13**.

The spectra measured by the Si-PIN detector is shown in **Fig. 14**, where the peak intensity is several thousands of counts for a few minutes one cycle. The X-ray intensity is not very strong, but it is better to protect the exposure to the X-rays. Usually lead containing acrylic plate is good for protecting from the X-rays, but steel plate of the thickness of 1 or 2 mm is enough. The Geiger-Müller counter is not saturated during this experiment, but if the X-ray intensity is too strong when Geiger-Müller counter is directly irradiated by an 1 watt X-ray tube, then the Geiger-Müller counter will be saturated, and we cannot see the difference between no X-rays. Thus the present experiment should be performed with an expert of X-ray experiments.

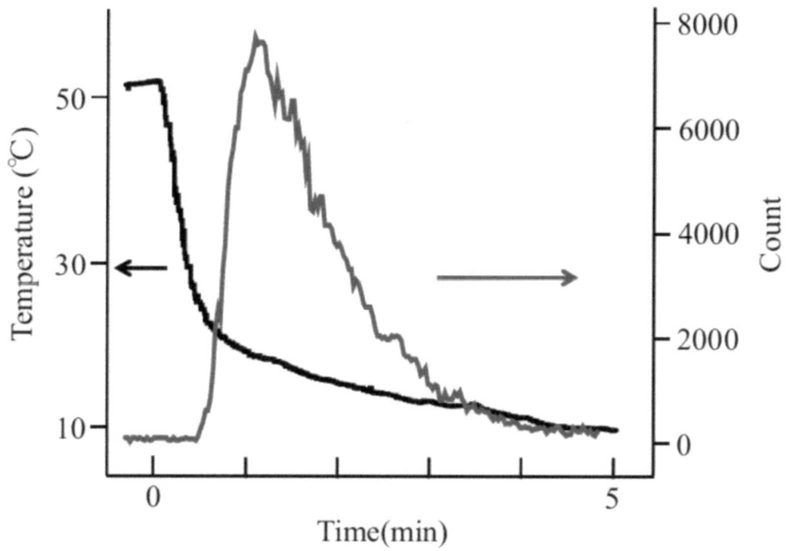

Fig. 13. Time dependent temperature and X-ray intensity after the crystal became to be cooled.

6. Digital signal processor

The X-ray signal from the Si-PIN X-ray detector is voltage signal. If the energy of X-rays is high, the voltage increase step-like according to the energy of the X-rays. Thus when the X-ray energy is high, the step height is larger. If the X-ray intensity is strong, then the step frequency in unit time increases. The step height increases again and again, and finally the voltage is larger than the voltage of the power source, then the step signal is reset and starts again. Thus if the time dependent voltage increase is recorded as voice signal in the memory of a computer as the voice recorder, we can plot the X-ray spectra by differentiating the step-like voltage signal and plotting the frequency of the pulses against the height of the peak. This is the X-tray spectra. One of the X-ray spectra obtained in such a way is shown in **Fig. 14**. The X-ray intensity of our palmtop EPMA is not very strong (less than 10000 counts per minute), thus the response of analog/digital (A/D) converter for music purpose is enough (**Fig. 15**). The important points to use the music A/D convertor, or musician's amplifier are as follows.

1. The input of music amplifier is usually biased by a DC voltage to drive a microphone, and thus direct DC connection between X-ray pre-amplifier and music A/D converter will cause to destroy the pre-amplifier, and thus AC coupling should be used.
2. Impedance should be matched.
3. Any kind of notebook computer has microphone input, and can be used for the similar purpose, but the notebook computer inside is full of digital noises. Consequently too high level of digital noises makes it impossible to measure the X-ray spectra. This is the reason we use separate musician's amplifier.

The above function is identical to the digital oscilloscope and if you have a digital oscilloscope, you can connect your oscilloscope to the computer by a USB and import the X-ray signal into your computer. After recording or during the recording, you can differentiate the X-ray signal numerically, and plot the X-ray spectra on the computer display. When you record the X-ray spectra, the peaks are not assigned at all, and the peaks should be assigned using linear relation between the voltage and X-ray energy, and the energy of the spectra of elements contained in the specimen should be assigned by an X-ray database. The resulted X-ray spectrum is shown in **Fig. 14**. An example of the software is provided from X-ray Precision Co. Ltd. Kyoto by the price of around 200 USD by CD-ROM, but you can make such a program by yourself.

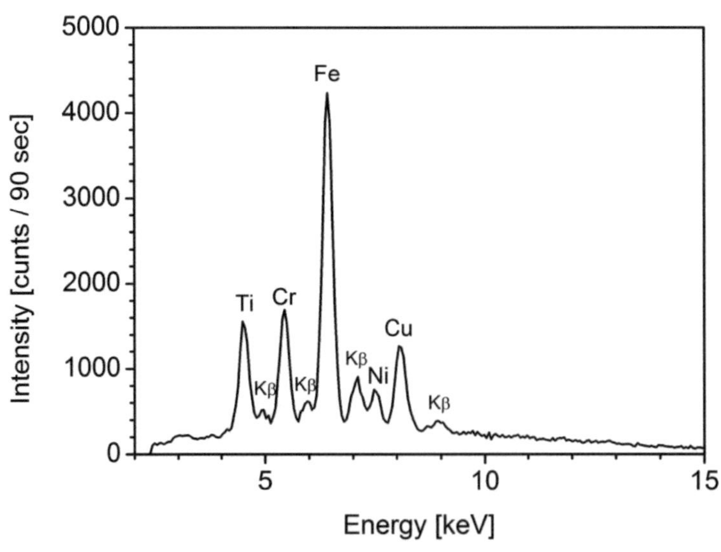

Fig. 14. X-ray spectra of steel and Ti (about 1 minute measurement).

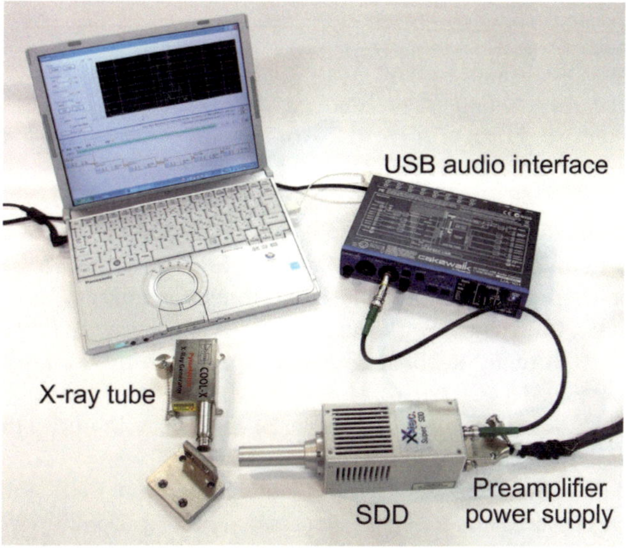

Fig. 15. Musician's amplifier connected to X-ray detector and computer.

7. Maintenance and safety

The surface of the pyroelectric crystal becomes dark because of the sputtered particles by the high voltage discharge will be deposited on the surface. These black fine particles are electric conductive, and thus the high voltage is not accumulated in the pyroelectric crystal. In such a case, the pyroelectric surface should be cleaned by cotton stick with ethanol to remove the dark sputtered particles. Then the X-rays will come back again.

The earth electric line is important to avoid the electric shock. Also the electric connection between the two electrodes is important to emit strong X-rays, as mentioned in the text. The X-ray is dangerous to be exposed even it is very weak. Thus monitoring of the X-ray intensity and shielding the X-rays are important for experiments.

After the experiment, the rotary pump should be leaked to the atmospheric pressure. If one forgets to leak the air into the vacuum system, the oil in the rotary pump will rise up to the vacuum vessel of the palmtop EPMA to fill the vessel, and consequently the experiment is not possible any more because of the contamination by the oil. To avoid this, auto-leak valve is preferable to attach just above the rotary pump. Since the auto-leak valve, manual air leak valve, and Pirani gauge are attached to the vacuum system, the instrument shown in **Fig. 1d** is not a simple one but complicated by vacuum parts. All these vacuum parts are connected to the vacuum system by the NW-25 flanges.

8. Acknowledgements

E. Hiro and S. Terada, who contributed at the early stage of the palmtop EPMA development, are acknowledged. The authors would like to thank Asahi Glass Foundation for financial support. The computer program for music amplifier has been developed by the JST support. The first version of the palmtop EPMA was a by-product of JST Sentan project.

9. References

[1] E. Hiro, T. Yamamoto, and J. Kawai: Applying pyroelectric crystal to small high energy X-ray source, *Adv. X-Ray Chem. Anal., Japan*, 41, 195-200 (2010).

[2] J. Kawai, Y. Nakaye, E. Hiro, and H. Ida: Application of pyroelectric crystal---a safety X-ray source, *Radioisotopes*, 60 (6), 249-263 (2011).

[3] S. Imashuku, A. Imanishi, and J. Kawai: Development of miniaturized electron probe X-ray microanalyzer, *Anal. Chem.* 83 (15), 6011-6017 (2011).

[4] S. Imashuku, A. Imanishi, and J. Kawai: Palmtop EPMA by electric battery, submitted to the Proceedings of the 21st ICXOM (Intern. Congress on X-ray Optics and Microanalysis, Campinas, September, 2011), *AIP Conference Proceedings* (submitted).

[5] Y. Nakaye, and J. Kawai: Recording X-ray spectra with an audio digitizer, *X-Ray Spectrom.*, 39 (5), 318–320 (2010).

[6] L. Ze, Y. Nakaye, Y. Morikawa and J. Kawai: SEM-EDX with audio digitizer, *Adv. X-Ray Chem. Anal. Japan*, 42, 111-114 (2011).

[7] Y. Nakaye and J. Kawai: Observation of pulsed electron field emission driven by a pyroelectric crystal, *Adv. X-Ray Chem. Anal. Japan*, 42, 249-253 (2011).

[8] Y. Nakaye and J. Kawai: X-Ray measurement using an audio A/D converter, *Adv. X-Ray Chem. Anal. Japan*, 42, 255-259 (2011).

[9] J. D. Brownridge: Pyroelectric X-ray generator, *Nature*, 358, 287-288 (1992).

[10] J. D. Brownridge, and S. Raboy: Investigations of pyroelectric generation of x rays, *J. Appl. Phys.*, 86, 640-647 (1999).

[11] S. M. Shafroth, W. Kruger, and J. D. Brownridge: Time dependence of X-ray yield for two crystal X-ray generators, *Nucl. Instrum. Meth. Phys. Res.*, A422, 1-4 (1999).

[12] J. D. Brownridge, and S. M. Shafroth: Electron and positive ion beams and X-rays produced by heated and cooled pyroelectric crystals such as $LiNbO_3$ and $LiTaO_3$ in dilute gases: Phenomenology and applications, in "Trends in Laser and Electro-Optics Research", Ed. W. T. Arkin, Nova Science Pub. (2006) pp. 59-95.

[13] B. Naranjo, J. K. Gimzewski, and S. Putterman: Observation of nuclear fusion driven by a pyroelectric crystal, *Nature*, 434, 1115-1117 (2005).

[14] J. Geuther, Y. Danon, F. Saglime, and B. Sones: Electron acceleration for X-ray producing using paired pyroelectric crystals, Abstracts of the 6th Intern. Meeting on Nuclear Applications of Accelerator Technology (AccApp'03), San Diego, June 1-5 (2003) pp.124-128. (http://www.rpi.edu/~danony/Publications.htm).

[15] J. A. Geuther and Y. Danon: Electron and positive ion acceleration with pyroelectric crystals, *J. Appl. Phys.*, 97, 074109 (2005).

[16] J. A. Geuther and Y. Danon: High-energy X ray production with pyroelectric crystals, *J. Appl. Phys.*, 97, 104916 (2005).

[17] J. A. Geuther, Y. Danon, and F. Saglime: Nuclear reaction induced by a pyroelectric accelerator, *Phys. Rev. Lett.*, 96, 054803 (2006).

[18] J. A. Geuther and Y. Danon: Application of pyroelectric particle accelerators, *Nucl. Instrum. Meth. Phys. Res.*, B261, 110-113 (2007).

[19] J. A. Guther and Y. Danon: Enhanced neutron production from pyroelectric fusion, *Appl. Phys. Lett.*, 90, 174103 (2007).

7

Adhesive Properties

Anna Rudawska
Lublin University of Technology,
Poland

1. Introduction

Adhesive joints function in multiple branches of technical engineering in which the phenomenon of adhesion appears: creating adhesive joints, sealing, applying protective or decorative coating (paint or varnish), printing, decorating and many others. Among adhesive bonding techniques these are adhesive joints which are used most often in various machine structure joints.

Surface phenomena, such as adhesion, cohesion and wettability, play an exceptionally important role in creating adhesive joints, as they influence the possibility of creating such a joint and its quality. Adhesive properties are fundamental in processes in which the phenomenon of adhesion appears. These properties are referred to as the whole of physical-chemical properties heavily influencing adhesion. Adhesive properties are a crucial indicator determining, for instance, whether the surface layer is properly prepared for permanent or temporary adhesive joints to be formed. The surface layer is the external layer of the material, limited by the real surface of the object, including this surface and the outer part of the material together with its real surface. It demonstrates different physical and chemical properties or qualities when compared with the core of the material.

When analysing the issue of constituting adhesive properties, exceptional importance is ascribed to the first two groups of technological operations aimed at preparing and obtaining specific properties of the surface and the surface layer of the material, as well as a special improvement (modification) of the aforementioned. They allow, for instance, obtaining proper energy and geometric properties of the surface layer of joined materials, which positively influence adhesion.

These operations are considered crucial in terms of constituting these properties in reference to forming and the quality of hybrid adhesive joints, as they are composed of materials of different physical, mechanical and chemical properties.

Surface preparation, conducted according to the requirements, is one of the methods of constituting adhesive properties of a surface. Depending on the characteristics and required properties of adhesive joints it is possible to increase or decrease adhesion, i.e. to improve or lower adhesive properties.

The selection of a surface preparation method (including appropriate technological operations allowing to achieve desired structure and energy properties) depends on many factors, among which the most important ones is the type of materials creating the adhesive joint.

2. Surface layer

In geometry, the surface is a two-dimensional geometric figure that limits the space filled with matter, i.e. surface in a theoretical sense. In mechanics, the surface is defined as the edge of a material body, which may be analysed in a molecular scale, micro- and macrosize, at the same time distinguishing different surfaces: material, nominal, real, observed, under machining, machined (Burakowski & Wierzchoń, 1995; Sikora,1997). The real surface may be defined as the surface separating the object from the surrounding environment. However, from the point of view of adhesion, the most important concept of surface is presented in the physical-chemical sense, as it involves the phases. In physical chemistry, surface is a boundary of two touching phases, i.e. interfacial surface or an interface, where an abrupt change of properties occurs together with the phase transition. Interfacial surfaces are surfaces between bodies of different states of aggregation (Hebda & Wachal, 1980). Surface in physical-chemical sense is analysed in three dimensions, despite the difficulties in determining the thickness/depth of the interface due to its small dimensions.

The physical space is not a homogenous area between two phases. Atomically clean surface is extremely active physically and chemically, therefore, each contact with another body results in the adsorption of the substances. Newly adsorbed substances may initiate formation a new phase. Another aspect is that under a physically clean surface there may be various deformations and defects resulting from surface formation. Consequently, different properties may be observed in the physically clean surface compared to the core of the object. As a result, different layers constituting the surface layer may occur: below the surface, surface and above the surface.

2.1 The surface layer structure

At present a number of definitions of the surface layer exist (Sikora, 1996,1997; Rożniatowski, Kurzydłowski, & Wierzchnoń,1994). One of the alternatives states that the surface layer is the external layer of the material, limited by the real surface of the object, including this surface and the outer part of the material under its real surface, which demonstrates different physical and, occasionally, chemical properties when compared with the core of the material. The articles (Kuczmaszewski, 2006; Sikora, 1996,1997) contain the description and the model of the surface layer of the material resulting from the adhesive failure. The surface layer has zonal structure. The proportions and the thickness of different zones vary, in addition the zones may interpenetrate, changing into one another or occupying the same space.

The structure and properties of the surface layer depend on the type and course of multiple phenomena and processes, including physical-chemical phenomena, such as adhesion.

2.2 The non-saturated surface force field

The surface of any body consists of atoms, particles or ions, which are in different conditions than the ones inside the body. In the volume phase the particles are subject to equal forces of interaction. In the interface, however, the particles come into contact with their own phase as well as another one, which leads to the occurrence of asymmetric forces of interaction (Fig. 1, Burakowski & Wierzchnoń, 1995). The particles on the surface are more forcefully drawn into the volume phase, and as a result the surface has higher energy than the inside of the body. Such a surface is active and is able to adsorb other atoms or particles in its vicinity (Burakowski & Wierzchnoń, 1995; Dutkiewicz 1998).

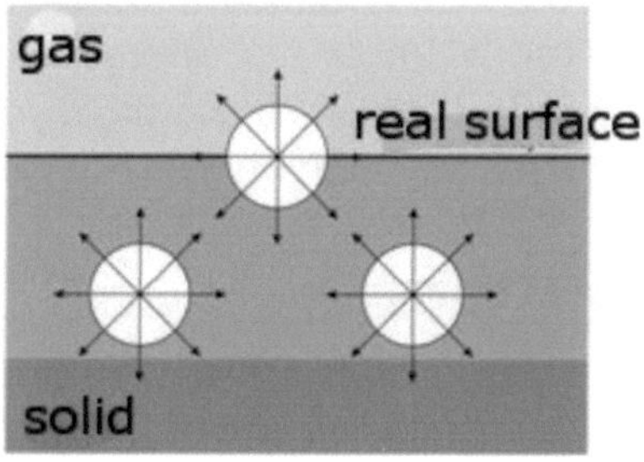

Fig. 1. A model of forces interacting with particles inside the solid and on its surface

What is equally important is the degree to which the surface particles are surrounded by other particles, i.e. whether the surface is flat or porous (Fig. 2, Dutkiewicz, 1998). The degree of non-saturation of forces is higher for a porous surface than for a flat one, therefore, the former is more active physically and chemically.

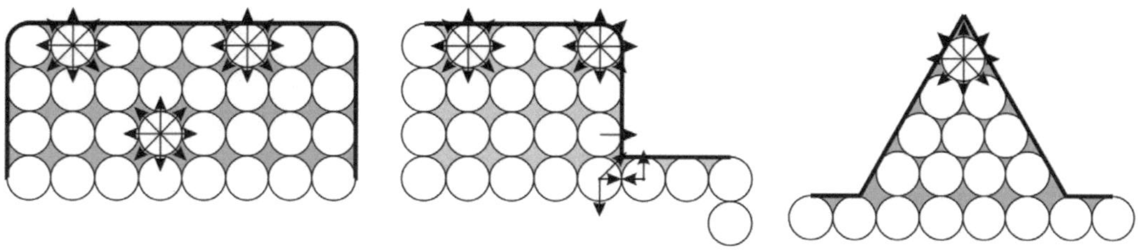

Fig. 2. The influence of porosity on the force field of various surfaces

It is the surface free energy, characteristic of solids, which is the measure of the interactions between the particles on the surface and inside the body.

3. Adhesion

3.1 Adhesion and adhesive properties definitions

The literature on the analysed subject is to some extent inconsistent in terms of contradicting terminology defining adhesion. Etymologically, 'adhesion' is derived from Latin *adhaesio* and stands for clinging or linkage. The adhesion is defined as a surface phenomenon, consisting in binding bodies in close contact as a result of force field interactions (Harding & Berg, 1997; Kuczmaszewski, 2006; Mittal, 1978, 1980; Żenkiewicz, 2000). The force field, induced by the charges of atoms constituting the surface layer (particles, ions), decreases exponentially with the distance to the surface (van der Waals interaction forces are negligible for the gap over 1-2 nm). Therefore for the adhesion to take place, the close contact of surfaces is required.

Knowledge of the adhesive propriety plays important role in processes in which appears the occurrence of the adhesion. To such processes we can number the bonding, the painting, the decoration, the printing, the lacquer finish, etc. The adhesive properties characterise the surface of the materials taking into account their applicability in the adhesive processes. Good adhesive properties have a positive influence on the strength of the adhesive joint obtained, low properties significantly lower this strength or even prevent the bonding. Knowing the properties allows as well to constitute them properly by means of required surface preparation treatment of the analysed materials (Rudawska, 2010).

3.2 Geometric structure and SEM technique

Ggeometric structure and adhesive properties are extremely important in the technology of creating adhesive joints. The geometrical structure of the material surface has an influence on the adhesive joints strength obtained, and that is the reason why it should be carefully analysed before bonding. Surface roughness is important in view of the part the mechanical adhesion plays in general adhesion; consequently, it is beneficial to know the structure of the material surface layer that will be used in the adhesion process.

A scanning electron microscope (SEM) is a type of electron microscope that images a sample by scanning it with a high-energy beam of electrons in a raster scan pattern. The electrons interact with the atoms that make up the sample producing signals that contain information about the sample's surface topography, composition, and other properties such as electrical conductivity. Due to the very narrow electron beam, SEM micrographs have a large depth of field yielding a characteristic three-dimensional appearance useful for understanding the surface structure of a sample (http://en.wikipedia.org/wiki/Scanning_electron_ microscope). Scanning electron microscopy (SEM) is generally considered micro-analytical techniques which are able to *image* or *analyze* materials we can not generally observe with the resolution offered by visible techniques. By *image* we mean photograph an object much smaller than we can see, even with the aid of an optical microscope (http://epmalab.uoregon.edu/epmatext.htm). SEM technique is very useful to analysis geometric structure of material for which is described adhesive properties (for example wettability or surface free energy).

Below there are some of example of materials for which it was determined the geometric structure (Rudawska, 2009 b, 2010).

The tests were conducted on aramide-epoxy composite samples. The composite consisted of two layers (2 x 0.3 mm) of aramide material marked KV-EP 285 199-46-003. The materials were arranged at 90 degree angle and subjected to the polymerisation process.

The geometric structure of the analysed composite was defined by means of SEM images. The results are shown in Fig. 3.

SEM images of the surface of analysed composite, show distinct differences in the surface structure, that are the result of specific character of the surface of the measures composites.

The pleat and the direction of the materials arranged at 90 degrees angle can be easily noticed.

The next tests were conducted on CP1 and CP3 titanium sheets samples. The samples of titanium sheets are made from:

1. CP1 (Grade 1- ASTM B265) and thickness 0.4 mm,
2. CP3 (Grade 3 - ASTM B265) and thickness 0.8 mm.

The results of SEM images of titanium sheets geometric structure are shown in Fig. 4 and Fig. 5.

a) b)

Fig. 3. Example of a surface topography SEM of the aramid/epoxy composite, magnification x250, a) spatial view, b) surface view (Rudawska, 2010)

a) b)

Fig. 4. Example of a surface topography SEM of CP1 titanium sheets surface, magnification x500, a) spatial view, b) surface view (Rudawska, 2009 b)

a) b)

Fig. 5. Example of a surface topography SEM of CP3 titanium sheets surface, magnification x500, a) spatial view, b) surface view (Rudawska, 2009 b)

SEM images of the titanium sheets surface show differences in the surface structure of analysed titanium sheets.

The following samples are concern the SEM images of aluminium sheets surface. The samples used were aluminium clad (plated) sheets type 2024-T3 (sheet thickness: 0.64 mm) The results of SEM images of aluminium sheets geometric structure are shown in Fig. 6 (own research).

a) b)

Fig. 6. Example of a surface topography SEM of aluminium 2024-T3 sheets surface, magnification x750, a) spatial view, b) surface view

The analysis of geometric structure of the analysed sheets considered in relation to adhesion technology is extremely important since these factors influence the obtained the adhesive joints strength.

4. Wetting phenomenon and contact angle

4.1 Wettability

The wetting phenomenon is a significant issue in various technological processes (Birdi & Vu, 1993; Norton, 1992; Parsons, Buckton & Chacham, 1993; Sommers & Jacobi, 2008; Qin &Chang, 1996) . Wetting is a surface phenomenon consisting in substituting the surface of the solid and the liquid with a boundary surface, characterised by certain tension (σ), which results from the difference in the surface tension between the solid and liquid in the gaseous medium (Fig. 7), (Hay, Dragila & Liburdy 2008; Żenkiewicz, 2000).

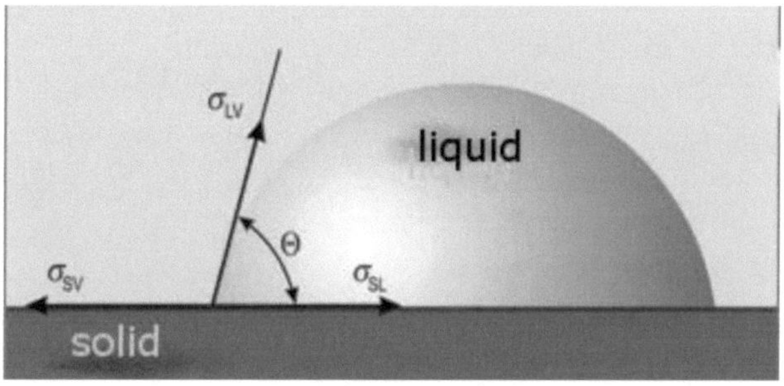

Fig. 7. Wetting a solid by a liquid

Wetting is a procedure that determines the diffusion of a liquid (adhesive) over a solid surface (substrate), creating an intimate contact between them. The air displacement caused by this physical attraction minimises the interfacial flaws. Good wettability of a surface is a prerequisite for a good adhesive bonding. Wettability is a crucial issue in the case of forming adhesive joints, because it directly affects the phenomenon of adhesion, increasing or decreasing adhesion forces.

4.2 Contact angle

The contact angle Θ provides the measure of wettability. This is the angle formed between the wetted solid surface and the tangent to the wetting liquid surface (to the meniscus of the wetting liquid), at the contact point of the liquid and the solid surface (Comyn, 1992; Hebda & Wachal 1980; Lee, 1993; McCarthy, 1998; Żenkiewicz, 2000).

When wetting the surface of a solid, the contact angle value will be lower than 90^0 (Fig. 8). The case when the contact angle $\Theta = 0^0$, indicates that the liquid spreads over the surface evenly and, furthermore, represents complete wetting of a solid surface by a liquid. If the contact angle $\Theta = 180^0$, then the result is absolute non-wetting (McCarthy, 1998).

The literature offers various tips on surface wetting, which account for the differences in size and interdependencies (as for the contact angle). In order for the liquid to wet the surface of the solid favourably, its surface tension should be lower than the surface tension of the liquid.

The contact angle can provide the measure of wettability of solids by liquids, it can determine critical surface tension, moreover, it can be used for determining surface free energy, as well as for the analysis of surface layer changes occurring when the surface is modified (Żenkiewicz, 2000).

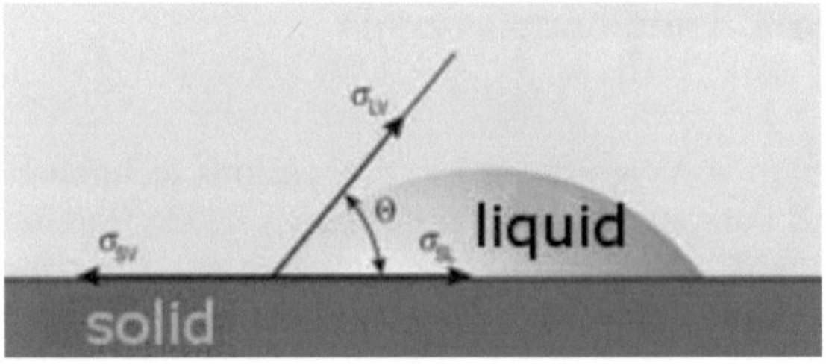

a)

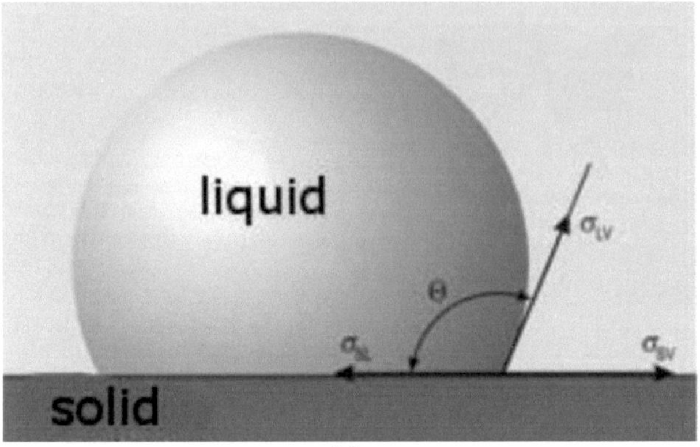

b)

Fig. 8. Wetting of a solid surface by the liquid in the case of: a) favourable wettability $\Theta < 90^0$ and insufficient wettability $\Theta > 90^0$

There are a number of factors significantly influencing the value of contact angle and the correctness of the angle measuring process, which include: the longitudinal modulus of elasticity (surface rigidity), surface porosity, chemical and physical homogeneity of the surface (and the surface layer), surface contaminants, the type of a measuring liquid, drop volume or humidity.

The aforementioned factors contribute to disturbing the measuring of the contact angle, hinder the interpretation of results and are the cause of various metastable states of the drop itself. What is more, these phenomena result in the contact angle hysteresis (Chibowski & González-Caballero, 1993; Diaz, M. Fuentes, Cerro & Savage, 2010; Żenkiewicz, 2000).

The hysteresis is assumed to consist of two basic components: thermodynamic and dynamic. The sources of the former can be found in porosity and heterogeneity of the surface and the surface layer of the analysed material. This component of the hysteresis is independent of the surface age of the drop, provided the volume of the drop remains unchanged throughout. The other component, the dynamic hysteresis, results from the wetting liquid – test material chemical interaction, as well as from penetration of the gaps in the material by the measuring liquid. The dynamic hysteresis depends on the surface age of the drop (Żenkiewicz, 2000).

There are a number of methods for measuring the contact angle, and the most common include such techniques as: the bubble measure method, geometric method (where the contact angle is measured from the dimensions of the drop), the capillary rise method (such as Wilhelmy plate method) or the direct measurement method (Ahadian, Mohseni & Morawian, 2009; Shang, Flury, Harsh & Zollars, 2008; Mangipudi, Tirrell &Pocius, 1994; Volpe& Siboni, 1998; Żenkiewicz, 2000).

At present, this is the direct measurement of the contact angle which is a commonly applied method, and the measurement is conducted by means of specialised instruments called goniometers or contact angle analysers (Żenkiewicz, 2000).

4.3 Factors influencing the contact angle

There is a number of factors substantially affecting the contact angle and the correctness of its measurements, which include: the longitudinal modulus of elasticity (surface rigidity), surface porosity, surface (and surface layer) physical and chemical homogeneity, surface contamination, the type of measuring liquid, drop volume, humidity, etc (Ajaev, Gambaryan-Roisman & Stephan 2010; Brown, 1994; Chibowski & González-Caballero, 1993; Extrand, 1998; Thompson, Brinckerhoff & Robbins, 1993; Żenkiewicz, 2000).

One of the factors influencing the contact angle is the *drop volume*. The impact of this factor is by no means certain, since there are no prevailing conclusions, due to the fact that the contact angle measurement methods and calculating models applied in tests were different. In his work (Żenkiewicz, 2000) M. Żenkiewicz included a lot of information both on the measuring drop volumes as well as contact angle measuring methods. M. Zielecka (Zielecka, 2004) observed the influence of the size of the drop on the contact angle measurement, and arrived at a drop volume range of 2-6 mm, within which the size of the drop bears no influence on the measurement of the contact angle. X. Tang, J. Dong, X. Li (Tang, Dong & Li, 2008). conducted contact angle measurements for distilled water drops in the volume range of 3-6 μl. In their tests, K. B. Borisenko and others (Borisenko, Evangelou, Zhao & Abel, 2008). used the diiodomethane drop volume of 5 μl. Although, in the tests conducted by M. Żenkiewicz (Żenkiewicz, 2005), Q. Bénard, M. Fois and M. Grisel (Bernard, Fois, & Grisel, 2005). the measuring liquids applied were different (distilled water, glycerol, formamide, diiodomethane, α-bromonaphthalene), the volume of the drop was identical - 3 μl. In the case of many works (Hołysz, 2000; Serro, Colaço & Saramago, 2008; Żenkiewicz, 2000) the measuring drop volume ranges from 2-5 μl (2 μl, 4 μl), e.g. J. Shang and others (Shang, Flury, Harsh & Zollars, 2008). apply a 2 μl drop for static contact angle measurements and larger 5 μl in the case of dynamic contact angle measurements. According to the data collected from the literature (Żenkiewicz, 2000), the size of the drop should range between 28mm^3 and 0.5 mm^3.

The surface age of the drop, i.e. the time between the application of a drop and the measurement, is one another contact angle affecting factor. M. Żenkiewicz (Żenkiewicz, 2000) notes that the time between the application and the measurement should be as short as possible, and moreover, identical for all the drops of the test series. Following this procedure should ensure a small influence of the drop-surface interaction and reduction of the drop volume as a result of evaporation.

X. Tang, J. Dong and X. Li (Tang, Dong & Li, 2008) deal with the phenomena of wetting and contact angle and additionally present test results of the influence of the surface age of the drop on the contact angle volume for different (wet and dry) surface states.

Another factor taken into consideration is *temperature*. M. Żenkiewicz (Żenkiewicz, 2000) mentions in his paper that within the range of 80⁰C, any changes in temperature only to a small degree trigger changes in the surface free energy, and natural temperature fluctuations, possible during laboratory tests, have a negligible impact on the samples contact angle measurements results. N. Zouvelou, X. Mantzouris, P. Nikolopoulos (Zouvelou, Mantzouris & Nikolopoulos, 2007) compared their tests observations with the literature data and drew a linear dependence of the surface free energy and the contact angle of certain materials on the temperature (nevertheless for high temperatures of approx. 800^0 C – 1173 K, 1500^0 C- 1773 K).

The longitudinal modulus of elasticity (surface rigidity) is yet another factor which should be considered when measuring the contact angle. M. Żenkiewicz, J. Gołębiowski and S. Lutomirski (Żenkiewicz, Gołębiewski & Lutomirski, 1999). stress that the surface of the test material where measuring drops are placed should be appropriately rigid. Therefore, the longitudinal modulus of elasticity of the material should be higher than 10 kPa, as it would prevent any drop deformations, resulting from the weight of the measuring drop.

One of the components of the thermodynamic hysteresis, *surface porosity*, is the next factor in question. R.D. Hazlett (Hazlett, 1992) describes and presents opinions of other researchers on the influence of surface porosity on the hysteresis of the contact angle, to conclude that the influence of porosity is beyond a shadow of a doubt. It can be, however, assumed that if Ra < 0,5 µm, then the impact of porosity on the contact angle is insignificant.

A.P. Serro, R. Colaço and B. Saramago (Serro, Colaço & Saramago, 2008) present test results for two samples made of UHMWPE (ultra-high-molecular-weight polyethylene) of different porosity, characterised by the Ra parameter of 3.9 and 1.0 nm, and the distance between the micropores of 23 and 6 nm respectively. They note that the wettability for given cases is irrespective of surface porosity, and that the contact angles measured for water and hexadecane are identical. However, J. Xian (Xian, 2008) points out that the wettability and the contact angle for a porous surface, e.g. analysed steel and polymers, is different for a smooth surface, adding that the change of the contact angle on a porous surface depends on the contact angle of a smooth surface of the analysed materials.

The physical and chemical homogeneity of the surface (and the surface layer) – i.e. physio-chemical homogeneity, which is the second source of the thermodynamic hysteresis, is another aspect taken into consideration when measuring the contact angle. Moreover, a considerable influence on the contact angle value may be observed on the part of the following: additive migration, diverse supermolecular structure, along with surface inhomogeneity - the result of different functional groups of different size and character formed on that surface.

What cannot be disregarded when measuring the contact angle is the analysis of *the type of the measuring liquid*. The measuring liquid penetration of the gaps in the surface layer of the material as well as of the intermolecular spaces is one of the causes of the dynamic hysteresis. The molar volume of the liquid plays an important role in the process as well – the rate of water penetration processes becomes slower and limited when the volume rises. Owing to its low molar volume water easily penetrates the structure of certain materials, therefore the importance of proper measuring liquid selection.

Other factors significantly disturbing the measurement of the contact angle are the *surface contaminants* and *air humidity* at the time of a test. Furthermore, *the sample should be firmly fixed* in order to prevent any measuring drop deformations as a result of vibrations.

Publications include plenty of information on the aspects of drop dispersion, along with the model of phenomena occurring when the contact angle measurement is taken for different liquid contact models, not to mention the characteristics of static and dynamic contact angle measurements. Some articles highlight the practical importance of wetting and wettability of different liquids in various processes, such as impregnation.

The factors mentioned in the preceding paragraphs hinder the measurement of the contact angle and the analysis of tests results, in addition they lead to different metastable states of the drop itself. These phenomena result in the contact angle hysteresis.

4.4 The contact angle hysteresis

Among many issues connected with the contact angle (the type of angle, measurements and values used in calculations) special importance is attributed to *the contact angle hysteresis*, which is the result of phenomena associated with metastable states of the measuring drop placed on the analysed surface of a solid (Bayer, Megaridis, Hang, Gamota & Biswas, 2007; Vedantam & Panchagnula, 2008; Zielecka, 2004; Żenkiewicz, 2000).

The first significant research on the contact angle hysteresis began in the middle of the 1970s and was conducted for example by R.J. Good (Good,1979). E. Chibowski and F. González-Caballero (Chibowski & González-Caballero, 1993). presented theoretical information on the contact angle hysteresis, factors causing it and the description of the observed contact angle hysteresis connected with chemical interactions. C.W. Extrand (Extrand, 1998). characterised some of the contact angle hysteresis theoretical models and presented the study of the contact angle hysteresis thermodynamic model based on the research on polymers.

The Young equation constitutes the basis for theories related to the phenomenon of wettability. This equation comprises a measurable geometric parameter - the contact angle with three thermodynamic indices, which allow explaining the properties of interactions in the interface. The Young equation (also called Young-Laplace equation) was formed in 1805 and since then its principles and description have been used in multiple publications (Diaz, Fuentes, Cerro & Savage 2010; Faibish, Yoshida & Cohen, 2002; Żenkiewicz, 2006,2000).

The Young equation describes an ideal system, which meets specific requirements of the contact angle measurement, geometric properties and qualities of the analysed surfaces (e.g. porosity, rigidity, physical and chemical homogeneity or the lack of surface contaminants). These requirements have been described in subsection 4.3.

If the surface meets the Young equation principles, the drop placed on it remains in equilibrium, which is accompanied by the lowest energy state. In such a situation, the contact angle is referred to as an equilibrium contact angle and its value does not depend on the changes of the drop volume. If the surface fails to meet the principles of Young equation, the measuring drop placed on it is in a metastable state, and then the contact angle of this drop may be higher or lower than the equilibrium angle. Initially, the gradual increase of the drop volume causes the increase of the contact angle until it reaches the limit, called *the advancing contact angle* Θ_A (Chibowski & González-Caballero, 1993; Żenkiewicz, 2000).

After this volume has been exceeded, an abrupt change of the drop position occurs - an abrupt change of the drop contour (decrease in height, increase in the contact area) and decrease in the volume of the contact angle. If the volume of the drop is gradually decreased, the value of the contact angle will initially decrease until it reaches the value called the *receding angle* Θ_R. After this value has been exceeded, the contour of the drop abruptly recedes (the height increases, the contact area decreases) and the value of the contact angle increases. A new metastable state of the drop location is, characterised by the contact angle is higher than the receding angle. Therefore, the contact angle hysteresis is defined as the difference between the advancing angle Θ_A and the receding angle Θ_R on the tree-phase contact line (Chibowski & González-Caballero, 1993; Faibish, Yoshida & Cohen, 2002; Vedantam & Panchagnula, 2008; Volpe & Siboni 1998; Żenkiewicz, 2000).

The contact angles Θ_A and Θ_R, along with their corresponding drop volumes: maximum (for Θ_A) and minimum (for Θ_R) with a constant diameter (D) of the circle created by the drop lying on the tested material, are shown in the Fig. 9 (Żenkiewicz, 2000).

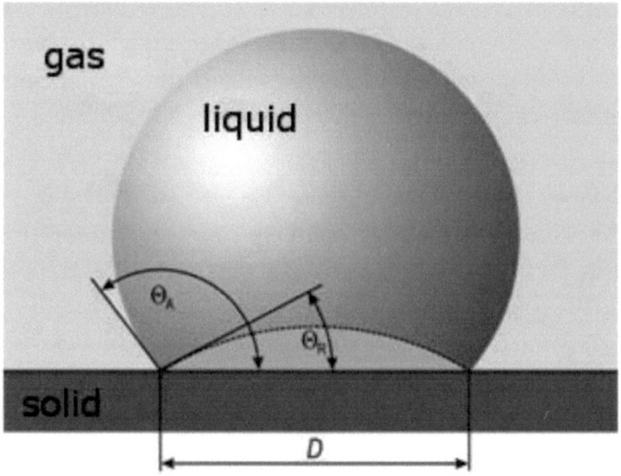

Fig. 9. The contact angles with a constant diameter D (D=const): Θ_A – the advancing angle, Θ_R – the receding angle, 1 – the maximum volume drop, 2 – the minimum volume drop

The hysteresis is assumed to consist of two basic components. First is the so called *thermodynamic hysteresis*, which results from porosity and heterogeneity of the surface and the surface layer of the tested material. This component is independent of the surface age of the drop (provided that the drop volume remains unchanged while measured). The other component is *the dynamic hysteresis*. It results from, among other things, chemical interactions of the measuring liquid with the tested material, and the measuring liquid penetration of the gaps in the material. The dynamic hysteresis depends on the surface age of the drop (Żenkiewicz, 2000).

The research on the hysteresis is extremely important from the practical point of view of, for instance, the surface free energy calculations. It is mostly connected with the question of which contact angle should be adopted in the simplified equation (3) in order to obtain the correct result. The contact angle used in calculations is the angle θ_A called the advancing angle.

5. Surface free energy

Surface free energy (SFE) is one of the thermodynamic quantities describing the state of atom equilibrium in the surface layer of materials (Hołysz, 2000; Żenkiewicz, 2000,2005). This quantity is characteristic for each substance. It reflects the specific state of unbalance in intermolecular interaction which is present at the phase boundary of two mediums.

Surface free energy is of equal number to the work necessary for creating a new surface unit while separating two phases in equilibrium, in a reversible isothermal process. It is measured in [mJ/m^2] (Żenkiewicz, 2000).

5.1 Young equation

The basis for methods of calculating surface free energy from the measurements of the contact angle is the Young equation (Fig. 7) (Chibowski & González-Caballero, 1993; Lee, 1993; Thompson, Brinckerhoff & Robbins, 1993; Żenkiewicz, 2000).

It was derived from the condition of equilibrium of forces which represent surface tensions at the contact point of three phases - solid, liquid and gas.

$$\sigma_{SV} = \sigma_{SL} + \sigma_{LV}\cos\Theta_V \tag{1}$$

where Θ_V is the equilibrium contact angle, and σ_{LV}, σ_{SV}, and σ_{SL} are the surface free energies of liquid-vapour, solid-vapour and solid-liquid interfaces, respectively.

The Young equation may also be derived from the energy balance for the triple point (Chibowski & González-Caballero, 1993; Michalski, Hardy & Saramago, 1998; Zouvelou, Mantzouris & Nikolopoulos, 2007). In this case, the equation is of the following form (Żenkiewicz, 2000):

$$\gamma_{SV} = \gamma_{SL} + \gamma_{LV}\cos\Theta_V \tag{2}$$

where: γ denotes surface free energy and the other symbols have the same meaning as in the equation (1).

It is impossible to determine surface free energy directly from the equation (2) because of the two unknowns: γ_{SV} and γ_{SL}. For calculation purposes, the following form of the equation (2) is commonly used to determine the surface free energy of solids (Chibowski & González-Caballero, 1993; Żenkiewicz, Gołębiewski & Lutomirski, 1999):

$$\gamma_S = \gamma_{SL} + \gamma_L\cos\Theta \tag{3}$$

where: γ_S -surface free energy of solids in a vacuum,
γ_{SL} - surface tension on the solid - liquid phase boundary,
γ_L -surface free energy of the measured liquid,
Θ_V - contact angle measured on the examined true surface.

The main drawback of the equation (1) is that it refers to an ideal system because it has been based on theoretical considerations, to a large extent not confirmed empirically. Still, this is the contact angle measurement which is the most often used method to determine energy properties of solids.

5.2 Surface free energy determination methods

The various SFE determination methods are based on specific relations, and involve the measurement of contact angles of various liquids. A number of factors have a substantial influence on the correctness of the contact angle measurement (subchapter 3.3). Some issues related to contact angle measurements and wettability have been highlighted shown in the literature.

Determination of surface free energy of solid objects involves indirect methods - direct methods can only be used in the case of liquids. Among the various indirect methods are the approaches due to Fowkes, Owens-Wendt, van Oss-Chaudhury-Good, Zisman, Wu, and Neumann (Ahadian, Mohseni & Morawian, 2009; González-Martín, Labajos-Broncano, Jańczuk & Bruque, 1999; Greiveldinger & Shanahan,1999; Hołysz, 2000; Jańczuk, Białopiotrowicz & Zdziennicka, 1999; Lee, 1993; Lugscheider & Bobzin, 2001; Żenkiewicz, 2000, 2006).

5.2.1 The Owens-Wendt (Kaelble-Owens-Wendt) method (OW)

The Owens-Wendt method (sometimes referred to as Kaelble-Owens-Wendt method) is a frequently applied method for determining the surface free energy of, e.g. polymers (Jańczuk & Białopiotrowicz, 1987; Rudawska & Kuczmaszewski, 2006; Rudawska, 2008). This method consists in determining dispersive and polar components of SFE based on Berthelot principle (Żenkiewicz, 2000) , which assumes that interaction between molecules of two bodies in their surface layers equals the geometric mean of the cohesion work between the molecules of each body.

This method assumes that the surface free energy (γ_S) is a sum of two components: polar (γ_S^p) and dispersive (γ_S^d), and that there is a relation between the three quantities:

$$\gamma_S = \gamma_S^d + \gamma_S^p \qquad (4)$$

The dispersive element is the sum of components derived from such intermolecular interactions as: polar, hydrogen, induction and acid-base, with the exception of dispersive interactions. Dispersive interactions constitute the dispersive component of the surface free energy.

The work of adhesion between the solid and the liquid can be described by means of the Dupré equation:

$$W_a = \gamma_{SV} + \gamma_{LV} - \gamma_{SL} \qquad (5)$$

By combining the equations 2 with 3, the Young–Dupré equation is obtained:

$$W_a = \gamma_{LV}(1+\cos\Theta) \qquad (6)$$

However, Owens and Wendt propose the following form of the work of adhesion between interacting solid and liquid.

$$W_a = 2(\gamma_S^d\, \gamma_{LV}^d)^{0,5} + 2(\gamma_S^p\, \gamma_{LV}^p)^{0,5} \qquad (7)$$

If we compare and combine equations (6) and (7), the following equation is obtained:

$$\gamma_{LV}(1+\cos\Theta) = 2(\gamma_S{}^d \gamma_{LV}{}^d)^{0,5} + 2(\gamma_S{}^p \gamma_{LV}{}^p)^{0,5} \qquad (8)$$

This equation allows determining the surface free energy of a solid and its SFE components.

In order to determine the polar and the dispersive components of the surface free energy, the measurements of the contact angle of the analysed samples need to be conducted with two measuring liquids. The surface free energy of the measuring liquids used in test is known, including its polar and dispersive components. One of the liquids is non-polar and the other is bipolar. Most frequently, the tests include distilled water as the polar liquid and diiodomethane as the non-polar one.

The SFE γ_S is calculated using the adjusted dependence describing the dispersive component of the surface free energy (Jańczuk& Białopiotrowicz, 1987; Rudawska & Kuczmaszewski, 2005; Rudawska, 2008; Rudawska & Jacniacka 2009).

$$\left(\gamma_s{}^d\right)^{1/2} = \frac{\gamma_d\left(\cos\Theta_d + 1\right) - \sqrt{\frac{\gamma_d{}^p}{\gamma_w{}^p}}\gamma_w\left(\cos\Theta_w + 1\right)}{2\left(\sqrt{\gamma_d{}^d} - \sqrt{\gamma_d{}^p\frac{\gamma_w{}^d}{\gamma_w{}^p}}\right)} \qquad (9)$$

and the polar component of the surface free energy

$$\left(\gamma_S{}^p\right)^{0,5} = \frac{\gamma_w\left(\cos\Theta_w + 1\right) - 2\sqrt{\gamma_S{}^d\gamma_w{}^d}}{2\sqrt{\gamma_w{}^p}} \qquad (10)$$

where: $\gamma_S{}^d$ – the dispersive component of the test material surface free energy, $\gamma_S{}^p$ – the polar component of the test material surface free energy, γ_d – the surface free energy of diiodomethane, $\gamma_d{}^d$ – the dispersive component of the surface free energy of diiodomethane, $\gamma_d{}^p$ – the polar component of the surface free energy of diiodomethane, γ_w – the surface free energy of water, $\gamma_w{}^d$ – the dispersive component of the surface free energy of water, $\gamma_w{}^p$ – the polar component of the surface free energy of diiodomethane, Θ_d – the contact angle of diiodomethane, Θ_w – the contact angle of water.

There is one of example of materials for which it was determined the surface free energy after various surface treatment (Rudawska, 2008, 2009).

The surface free energy of the material presented below was calculated with the Owens-Wendt method. This is a structural material applied in e.g. aircraft industry. The tests were to determine the influence of a surface preparation method on the SFE of the sample material.

The tests were conducted on glass-epoxy composite samples consisting of two layers (2x0.30 mm) of glass fibre 3200-7781. The fabric layers were arranged at a right angle and cured conforming to the technology standards.

The composite samples were tested for four surface preparation variants:

1. variant I – no surface preparation;
2. variant II – degreasing with Loctite 7036 (a detailed description of this method can be found in e.g. (Rudawska & Kuczmaszewski, 2005));

3. variant III - mechanical surface preparation with P320 abrasive tool;
4. variant IV - mechanical surface preparation with P320 abrasive tool, followed by degreasing with Loctite 7036.

The surface free energy values as well as the components of the SFE for four tested glass-epoxy composite surface preparation variants are presented in Fig. 10-13 (Rudawska, 2008, 2009).

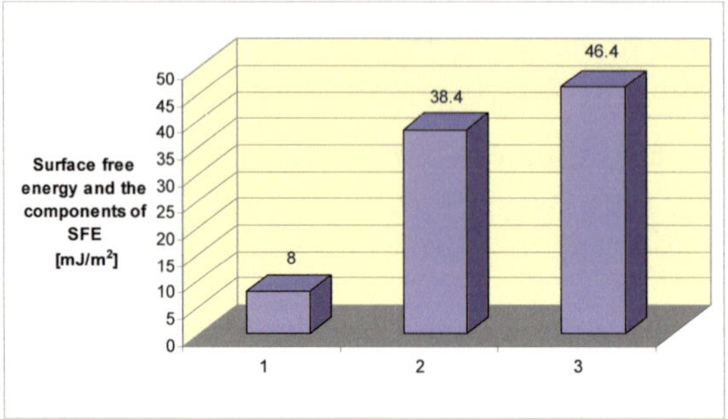

Fig. 10. Surface free energy and the components of SFE - the surface of glass/epoxy composite without surface treatment (variant I): 1 - polar component of SFE, 2 - dispersive component of SFE, 3 - surface free energy (SFE)

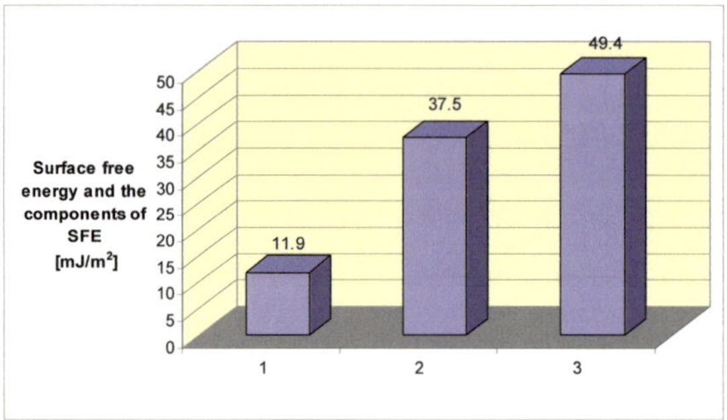

Fig. 11. Surface free energy and the components of SFE - the surface of glass/epoxy composite after degreasing (variant II): 1 - polar component of SFE, 2 - dispersive component of SFE, 3 - surface free energy (SFE)

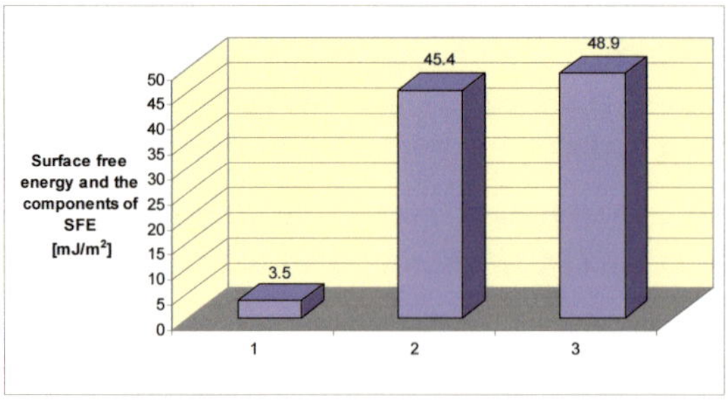

Fig. 12. Surface free energy and the components of SFE - the surface of glass/epoxy composite after the P320 grinding tool processing (variant III): 1 - polar component of SFE, 2 - dispersive component of SFE, 3 - surface free energy (SFE)

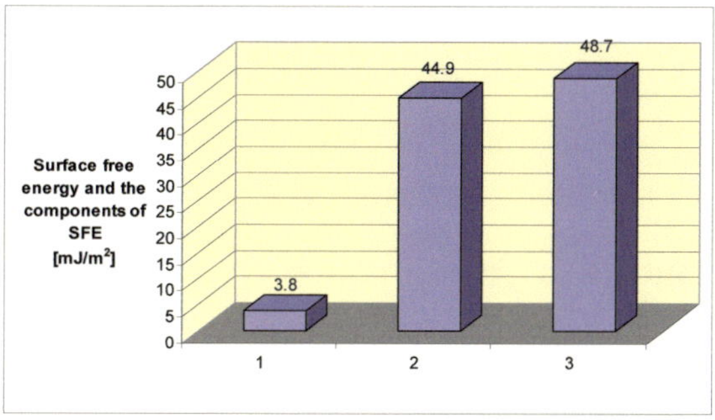

Fig. 13. Surface free energy and the components of SFE - the surface of glass/epoxy composite after the P320 grinding tool processing and degreasing (variant IV):
1 – polar component of SFE, 2 – dispersive component of SFE, 3 – surface free energy (SFE)

The results demonstrate that the highest values of the surface free energy were obtained in the case of degreasing, while the lowest were observed for variant I, with no surface preparation. Consequently, it appears that mechanical surface preparation and mechanical surface preparation followed by degreasing both increase the surface free energy. Additionally, no statistically relevant difference in the γ_S values of the two variants in question was observed.

Taking into consideration the polar component of the SFE, its highest value was noted in the case of surface preparation variant II, in which the surface free energy value was the highest as well. Additionally, the polar component constituted 24% of the total SFE. In the case of variant I, with the lowest γ_S value in the tests, the polar component constitutes 17% of the SFE. For the other two variants, III and IV, the polar component represented scant 7% and 8% respectively.

The analysis of the SFE values leads to the observation that degreasing the surface of the glass-epoxy composite has beneficial effect on the surface free energy value. It results in the increase of the SFE as compared to the surfaces with no prior surface preparation.

To conclude, it must be mentioned that, firstly, forming an adhesive joint should be preceded by certain surface preparation methods, and secondly, that this is degreasing which produces the best results in terms of adhesive properties of the analysed glass-epoxy composite.

5.2.2 The van Oss-Chaudhury-Good method (OCG)

In the case of the van Oss-Chaudhury-Good method the surface free energy is a sum of two components (Adão, Saramago & Fernandes, 1999; Żenkiewicz, 2000). While the first component γ_i^{LW} is connected with long-range interactions (dispersive, polar and inductive, referred to as Lifshitz-van der Waals electrodynamic interactions), the second component γ_i^{AB} describes the acid-base interactions (Hołysz 2000; Jansen, 1991):

$$\gamma_i = \gamma_i^{LW} + \gamma_i^{AB} \tag{11}$$

Good R.J. and van Oss C.J. (Good & van Oss, 1992) separate the acid component (electron-acceptor: γ_L^+, γ_S^+) and the base component (electron-donor: γ_L^-, γ_S^-) of the surface free energy.

Moreover, the γ_i^{AB} component can be described by means of equation for bipolar compounds (showing properties of both Lewis acids and bases), (Elftonson, Ström, Holmberg & Olsson, 1996):

$$\gamma_i^{AB} = 2(\gamma_i^{+}\gamma_i^{-})^{0,5} \tag{12}$$

where: γ_i^+ - Lewis acid surface free energy component, γ_i^- - Lewis base surface free energy component, index *i* - subsequent measuring solids or liquids.

Determining the SFE of test materials will consist in measuring their surfaces contact angle with three different measuring liquids and calculating the γ_S of the system of three equations:

$$(\gamma_S^{LW}\gamma_{Li}^{LW})^{0,5} + (\gamma_S^{+}\gamma_{Li}^{-})^{0,5} + (\gamma_S^{-}\gamma_S^{+})^{0,5} = \gamma_{Li}(1+\cos\Theta_i)/2 \tag{13}$$

where: i=1,2,3.

Measuring the contact angle requires the application of two polar and one non-polar liquids; nevertheless, solving the equation (3) requires additional information - particular values for the applied measuring liquids. Polar liquids applied in tests are water, glycerol, formamide or ethylene glycol, and non-polar liquids (not showing properties of either Lewis acids or bases) diiodomethane or α-bromonaphthalene.

A detailed description of this method is provided in the publications (Shen, Sheng, & Parker, 1999, Żenkiewicz, Gołębiewski & Lutomirski, 1999; Żenkiewicz 2000).

Determining the SFE with the van Oss-Chaudhury-Good method is uncomplicated, nevertheless, the test results should be carefully analysed. This method is burdened with a few problems, including e.g. the fact that the test results depend heavily on the applied measuring liquids configuration. This issue has been described by e.g. C. Della Volpe and S. Siboni (Volpe & Siboni, 1998). who in addition present the Drago theory, concerning, among other issues, the properties of Lewis acids and bases.

5.2.3 The comparison OW and OCG methods

Due to the fact that the methods of calculating the surface free energy presented in the previous chapters are most frequently applied, a comparison of selected structural materials SFE values calculated with the Owens-Wendt and the van Oss-Chaudhury-Good methods should be conducted (Kuczmaszewski & Rudawska, 2002).

The structural material under analysis was electrolytic zinc coated and hot dip zinc coated sheets, which find application in such industries as automotive, construction or machine-building. The zinc coated sheets were 0.7 mm thick, the hot dip zinc coating equalled 18 μm and electrolytic zinc coating equalled 7.5 μm (following the PN-89/H-92125 and PN-EN 10152 standards).

The sample material surface was degreased with degreasing agents: Loctite 7061 and acetone. Degreasing was conducted in ambient temperature between 18 and 20 ^{0}C with relative humidity oscillating between 38% and 40%.

The method applied for measuring the contact angle was the direct measurement of the angle between the measuring drop and the tested surface.

For calculating the surface free energy with the Owens-Wendt method relationships (9) and (10) were applied. The values of both the surface free energy and its components for the applied measuring liquids are presented in Table 1 (Jańczuk& Białopiotrowicz, 1987).

No.	Measuring liquid	Surface free energy and its components [mJ/m²]		
		γ_L	γ_L^p	γ_L^d
1	Distilled water	72.8	21.8	51.0
2	Diiodomethane	50.8	2.3	48.5

Table 1. The values of the surface free energy and its components for the applied measuring liquids

The SFE components values used in the van Oss-Chaudhury-Good method are presented in Table 2 (Żenkiewicz, Gołębiewski & Lutomirski, 1999).

No.	Measuring liquid	Surface free energy and its components [mJ/m²]				
		γ_L	γ_L^{LW}	γ_L^{AB}	γ_L^+	γ_L^-
1	Distilled water	72.8	21.8	51.0	34.2	19.0
2	Glycerol	64.0	34.0	30.0	5.3	42.5
3	Diiodomethane	50.8	50.8	0	0	0

Table 2. The values of the surface free energy and its components for the applied measuring liquids

The values of the surface free energy and its components were calculated with the van Oss-Chaudhury-Good method using the data presented in Table 2 as well as relationships (3) and (4).

The surface free energy values of the electrolytic zinc coated and hot dip zinc coated sheets calculated with the Owens-Wendt method are presented in Table 3 (Kuczmaszewski & Rudawska, 2002).

No.	The type of zinc coated sheets	The type of the degreasing agent	Surface free energy and its components [mJ/m²]		
			γ_S	γ_S^p	γ_S^d
1	Electrolytic zinc coated sheets	Loctite 7061	42.0	17.8	24.2
		Acetone	35.4	10.7	24.7
2	Hot dip zinc coated sheets	Loctite 7061	44.7	9.7	35.0
		Acetone	43.8	11.6	32.2

Table 3. The zinc coated sheets surface free energy calculated with the Owens-Wendt method

The results demonstrate that the dispersive component of the surface free energy for hot dip zinc coated sheets is higher (even three times) than its polar component. In the case of electrolytic zinc coated sheets degreased with Loctite7061, this difference is less significant.

The values of the surface free energy of the electrolytic zinc coated and hot dip zinc coated sheets calculated with the van Oss-Chaudhury-Good method are presented in Table 4 (Kuczmaszewski & Rudawska, 2002). The results were obtained from the tested sheet surface layer contact angle measurement taken with distilled water, glycerol and diiodomethane as measuring liquids.

No.	The type of zinc coated sheets	The type of the degreasing agent	Surface free energy and its components [mJ/m^2]				
			γ_S	γ_S^{LW}	γ_S^{AB}	γ_S^+	γ_S^-
1	Electrolytic zinc coated sheets	Loctite 7061	43.7	32.5	11.2	13.8	2.3
		Acetone	38.6	30.5	8.1	4.3	3.9
2	Hot dip zinc coated sheets	Loctite 7061	45.0	41.5	3.5	0.3	11.6
		Acetone	41.4	39.2	2.2	0.1	15.8

Table 4. The zinc coated sheets surface free energy calculated with the van Oss-Chaudhury-Good method

It can be observed that the component of the surface free energy connected with long range interactions γ_s^{LW} (polar, dispersive and inductive) is higher than the component describing acid-base interactions γ_s^{AB}. The γ_S^{AB} component is scant in hot dip zinc coated sheets. Drawn from the analysis of the acid-base interactions component γ_s^{AB}, certain regularity may be observed. Lewis acid (γ_s^+) surface free energy component is higher than Lewis base (γ_s^-) surface free energy component for electrolytic zinc coated sheets, whereas for hot dip zinc coated sheets the γ_s^+ value was negligible when compared with the γ_s^- component. Owing to the insignificant γ_s^{AB} value it may be presumed that these surfaces will show properties of monopolar or non-polar substances.

A comparison of the surface free energy calculated with both the Owens-Wendt method and the van Oss-Chaudhury-Good method for hot dip zinc coated and electrolytic zinc coated sheets degreased with Loctite 7061 is presented in Fig.14 (Kuczmaszewski & Rudawska, 2002).

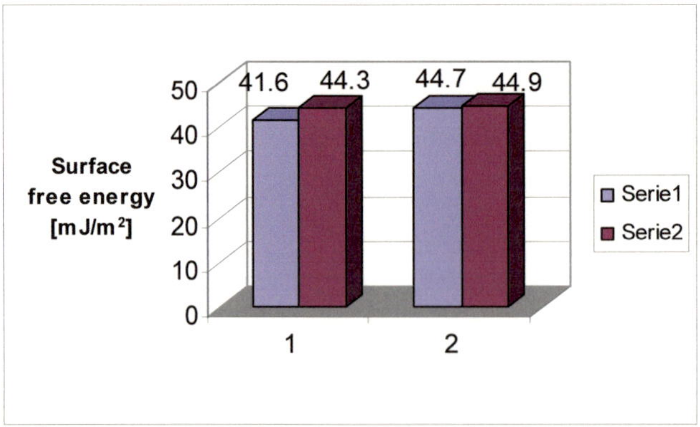

Fig. 14. The surface free energy values calculated with the Owens-Wendt method (series 1) and the van Oss-Chaudhury-Good method (series 2) for: 1- electrolytic zinc coated sheets, 2- hot dip zinc coated sheets after degreasing with Loctite 7061

A comparison of the surface free energy calculated with the Owens-Wendt method and the van Oss-Good method for hot dip zinc coated and electrolytic zinc coated sheets degreased with acetone is presented in Fig. 15 (Kuczmaszewski & Rudawska, 2002).

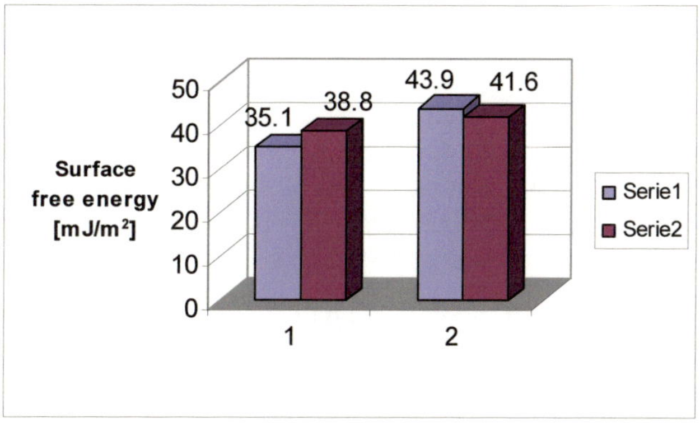

Fig. 15. The surface free energy values calculated with the Owens-Wendt method (series 1) and the van Oss-Chaudhury-Good method (series 2) for: 1- electrolytic zinc coated sheets, 2- hot dip zinc coated sheets after degreasing with acetone

The research results were subsequently analysed statistically using statistical models used for statistical verification (Krysicki et al., 1999). The statistical model – Student's t-test consisted in comparing means when the test variances were equal (Fisher - Snedecor distribution), with a predetermined level of significance $\alpha = 0.05$. The statistical analysis provided basis for formulating the following conclusions.

The analysis proved that there are no statistically significant differences in the values of the SFE calculated with either the Owens-Wendt or the van Oss-Good method when the sheets are degreased with Loctite 7061. This holds true for both electrolytic zinc coated sheets and hot dip zinc coated sheets.

When degreasing with acetone operation was applied on the sheet surface, statistically significant differences in the SFE calculated for the hot dip zinc coated sheets were observed. The γ_s value calculated with the van Oss-Chaudhury-Good method was higher. However, this difference is not too significant (lower than 10%). Still, there were no statistically significant differences in the SFE calculated with the van Oss-Good method for electrolytic zinc coated sheets.

When analysing the SFE values calculated with the Owens-Wendt method, it may be assumed that the surface layer of the electrolytic zinc coating would most likely demonstrate higher affinity with the polar substance than the hot dip zinc coating would.

6. Conclusion

Adhesion and concurrent phenomena, e.g. wettability, are present in numerous fields of engineering and life in general. Determining the factors influencing the quality of adhesion and finding technology that can increase or decrease it is of utmost importance when it comes to constituting adhesive joints. What cannot be disregarded is the structure of the surface layer of analysed materials or methods of determining adhesive properties, which assess materials suitability for adhesive processes. The existence of many methods for measuring the surface free energy stems from the fact that certain methods are suitable in particular circumstances. Existing methods describe the thermodynamic state of the surface layer differently yet all, through subsequent analyses of the surface free energy and its components, expand our knowledge of the phenomenon of adhesion.

SEM technique is very useful to analysis geometric structure of material for which is described adhesive properties. SEM micrographs have a large depth of field yielding a characteristic three-dimensional appearance useful for understanding the surface structure of a sample. The information of geometric structure is extremely important for the progress of adhesive processes like gluing, sealing, painting, coating.

In the subchapter devoted to a comparative analysis of the surface free energy measuring methods, the selection of the OW and the OCG methods was dictated by the fact that, on the one hand, these are the most frequently applied methods for measuring the surface free energy, on the other hand, due to relatively uncomplicated measurement of the contact angle with standard measuring liquids. The statistical analysis of the results evidences that, in most of the analysed cases, there are no statistically relevant differences between the values of surface free energy measured with either the Owens-Wendt or van Oss-Chaudhury-Good method.

Based on the statistical analysis it may be concluded that the choice of the surface free energy measurement method in the case of the analysed zinc coated sheets is basically of no relevance. Nevertheless, in ordinary working conditions it is the Owens-Wendt method which should be selected as a more efficient and less complicated tool for measuring the surface free energy of materials. The van Oss-Chaudhury-Good method, however, could be applied when a more detailed evaluation of the thermodynamic state of a surface (or a surface layer) is required.

Recent developments in the field of materials engineering contribute to creating structural materials or coatings, which are increasingly modern and specific – designated for particular applications. This creates the demand for continuous research into determining and describing their adhesive properties when adhesively bonding or joining such materials.

7. Acknowledgment

The some scientific study was funded from education finance for 2006-2009 as research project no. 3T10C02730, Poland

The some scientific study was funded from education finance for 2010-2013 as research project no. N N507 592538 The Ministry of Science and Higher Education, Poland

8. References

Adão, M.H.V.C.; Saramago, B.J. & Fernandes, A.C. (1999). Estimation of the Surface Properties of Styrene-Acrylonitrile Random Copolymers From Contact Angle Measurements. *Journal of Colloid Interface Science,* Vol. 217, No.1, pp. 319-328, ISSN 0021-9797

Ahadian, S.; Mohseni, M. & Morawian, S. (2009). Ranking proposed models for attaining surface free energy of powders using contact angle measurements. *International Journal of Adhesion and Adhesives* Vol.29, No. 4, pp. 458-469, ISSN 0143-7496

Ajaev, V.S.; Gambaryan-Roisman, T. & Stephan. P. (2010). Static and dynamic contact angles of evaporating liquids on heated surfaces. *Journal of Colloid Interface Science,* Vol.342, No.2, pp.550-558, ISSN 0021-9797

Bayer, I.S.; Megaridis, C.M.; Hang, J.; Gamota, D. & Biswas, A. (2007). Analysis and surface energy estimation of various model polymeric surfaces using contact angle hysteresis. *Journal of Adhesion Science and Technology*, Vol.21, No.15, pp. 1439-1467 (29), ISSN 0169-4243

Birdi, K.S. & Vu, D.T. (1993). Wettability and Evaporating Rates of Fluids from Solid Surfaces. *Journal of Adhesion Science and Techno*logy, Vol.7, No.6, pp. 485-493 (9), ISSN 0169-4243

Bernard, Q.; Fois, M. & Grisel M. (2005). Influence of Fibre Reinforcement and Peel Ply Surface Treatment Towards Adhesion of Composite Surface. *International Journal of Adhesion and Adhesives*, Vol. 25, No.5, pp.404-409, ISSN 0143-7496

Borisenko, K.B.; Evangelou, E.A.; Zhao, Q. & Abel, E.W. (2008). Contact Angle of Diiodomethane on Silicon-Doped Diamond-Like Carbon Coatings in Electrolyte Solutions. *Journal of Colloid Interface Science*, Vol.326, No.2, pp. 329-332. ISSN 0021-9797

Brown, S.D. (1994). Adherence Failure and Measurement: Some Troubling Question. *Journal of Adhesion Science and Technology*, Vol.8, No.6., pp. 687-711(25), ISSN 0169-4243

Burakowski, T. & Wierzchnoń, T. (1995). *Inżynieria powierzchni metali.* WNT, ISBN 83-204-1812-7, Warsaw, Poland

Comyn, J. (1992). Contact Angels and Adhesive Bonding. *International Journal of Adhesion and Adhesives*, Vol.12, No.3, pp.145–149, ISSN 0143-7496

Chibowski, E. & González-Caballero, F. (1993). Interpretation of Contact Angle Hysteresis. *Journal of Adhesion Science and Techno*logy, Vol.7, No. 11, pp. 1195-1209(15), ISSN 0169-4243

Diaz, M.E.; Fuentes, J.; Cerro, R.L. & Savage, M.D (2010). Hysteresis During Contact Angles Measurement. *Journal of Colloid Interface Science,* 343(2), pp.574-583, ISSN 0021-9797

Dutkiewicz, E.T. (1998). *Fizykochemia powierzchni.* WNT, ISBN 83-204-22-66-3, Warsaw, Poland

Elftonson, J.E.; Ström G.; Holmberg K. & Olsson J. (1996). Ashesion of Streptococcus Sanguis to porous and non-porous substrates with well-defined surface energies. *Journal of Adhesion Science and Technology,* Vol. 10, No.8, pp. 761-770, ISSN 0169-4243

Extrand, C.W. (1998). A Thermodynamic Model for Contact Angle Hysteresis. *Journal of Colloid Interface Science,* Vol.207, No.1, pp.11-19, ISSN 0021-9797

Faibish, R.S.; Yoshida, W. & Cohen, Y. (2002). Contact Angle Study on Polymer-Grafted Silicon Wafers. *Journal of Colloid Interface Science,* 256, pp.341-350, ISSN 0021-9797

Good, R.J. (1979). in: Surface and Colloid Science, R.J. Good, R.R. Stromberg (Eds.), vol.11. Plenyum Press, New York.

Good, R. J. & van Oss, C.J. (1992). in: Modern Approaches to Wettability, M.E. Schrader and G. Loeb (Eds), pp. 1-27, Plenum Press, New York.

González-Martín, M.L.; Labajos-Broncano, L.; Jańczuk, B. & Bruque, J.M. (1999). *Wettability and Surface Free Energy of Zirconia Ceramics and Their Constituents. Journal of Materials Science,* Vol.34, pp.5923-5926, ISSN 0022-2461

Greiveldinger, M. & Shanahan E.R. (1999). A Critique of the Mathematical Coherence of Acid/Base Interfacial Free Energy Theory. *Journal of Colloid Interface Science,* Vol.215, pp. 170-178, ISSN

Harding, H. & Berg, J.C. (1997). The role of adhesion in the mechanical properties of filled polymer composites. *Journal of Adhesion Science and Techno*logy, Vol.12, No.4, pp.471-493, ISSN 0169-4243

Hazlett, R.D. (1992). On Surface Roughness Effects in Wetting Phenomena. *Journal of Adhesion Science and Technology,* Vol.6, No.6, pp.625-633, ISSN 0169-4243

Hay, K.M.; Dragila, M.I. & Liburdy, J. (2008). Theoretical Model for the Wetting of a Rough Surface. *Journal of Colloid Interface Science,* Vol.325, No.2, pp.472-477, ISSN 0021-9797

Hebda, M. & Wachal, A. (1980). *Trybologia.* WNT, ISBN 83-204-0043-0, Warsaw, Poland

Hołysz, L. (2000). Investigation of the effect of substrata on the surface free energy components of silica gel determined by layer wicking method. *Journal of Material Science,* 35, pp.6081-6091, ISSN 0022-2461

Jansen, W.B. (1991). Overview Lecture. The Lewis Acid-Base Concepts: Recent Results and Prospects for the Future. *Journal of Adhesion Science and Technology,* Vol.5, No.1, pp.1-21, ISSN 0169-4243

Jańczuk, B.& Białopiotrowicz, T. (1987). Swobodna Energia Powierzchniowa Niektórych Polimerów. *Polimery,* pp.269-271, ISSN 0032-2725

Jańczuk, B.; Białopiotrowicz, T. & Zdziennicka A. (1999) Some Remarks on the Components of the Liquid Surface Free Energy. *Journal of Colloid Interface Science,* Vol.211, pp.96-103, ISSN 0021-9797

Krysicki, W.; Bartos. J.; Dyczka W.; Królikowska K. & Wasilewski, M. (1999). *Rachunek prawdopodobieństwa i statystyka matematyczna w zadaniach.* WNT, ISBN 83-204-2442-9, Warsaw, Poland

Kuczmaszewski, J. & Rudawska A. (2002). Porównanie wybranych metod określania swobodnej energii powierzchniowej na przykładzie blach ocynkowanych. *Farby i Lakiery,* No.6, pp.22–26, ISSN 1230-3321

Kuczmaszewski, J. (2006). Fundamentals of metal-metal adhesive joint design. Lublin University of Technology, Polish Academy of Sciences, Lublin Branch, ISBN 83-89293-11-0, Lublin, Poland

Lee, L.H.. (1993). Roles of molecular interactions in adhesion, adsorption, contact angle and wettability. *Journal of Adhesion Science and Technology,* Vol.7, No.6, pp.538–634, ISSN

Lugscheider, E. & Bobzin, K. (2001). The Influence on Surface Free Energy of PVD-Coatings. *Surface and Coatings Technology,* Vol.142-144, pp.755-760, ISSN 0257-8972

Mangipudi, V.; Tirrell, M. & Pocius, A.V. (1994). Direct Measurement of Molecular Level Adhesion Between Poly(ethylene terephthalate) and Polyethylene Films: Determination of Surface and Interfacial Energies. *Journal of Adhesion Science and Technology,* Vol.8, No.11, pp.1251–1270, ISSN 0169-4243

McCarthy, S.A. (1998). Dynamic Contact Angle Analysis and Its Application to Paste PVC Product. *Polimery* Vol.43, pp.314 - 319, ISSN 0032-2725

Michalski, M.-C.; Hardy, J. & Saramago, J.V. (1998). On the surface free energy of PVC/EVA polymer blends: comparison of different calculation methods. *Journal of Colloid Interface Science,* Vol.208, pp.319-328, ISSN 0021-9797

Mittal, K.L. (1978). (Ed.) in Adhesion Measurement of Thin Films, Thick Films, and Bulk Coatings, ASTM STP640, pp.5-17, American Society for Testing and Materials., Philadelphia, PA.

Mittal, K.L. (1980). Interfacial Chemistry and Adhesion: Recent Developments and Prospects. *Pure and Applied Chemistry,* Vol.52, pp.1295-1305, ISSN 0033-4545

Norton, M.G. (1992). The influence of contact angle, wettability, and reactivity on the development of indirect-bonded metallizations for aluminum nitride. *Journal of Adhesion Science and Technol*ogy ,Vol.6, No.6, pp.635-651, ISSN 0169-4243

Parsons, G.E.; Buckton, G. & Chacham, S.M. (1993). Comparison of measured wetting behaviour of material with identical surface energies, presented as particles and plater. *Journal of Adhesion Science and Technol*ogy, Vol.7, No.2, pp.95-104, ISSN 0169-4243

Qin, X. & Chang, W.V. (1996). The Role of Interfacial Free Energy in Wettability, Solubility, and Solvent Crazing of Some Polymeric Solids. *Journal of Adhesion Science and Technology,* Vol.10, No.10, pp. 963–987, ISSN 0169-4243

Rożniatowski, K.; Kurzydłowski, K.J. & Wierzchnoń T. (1994). Geometryczny opis cech mikrostrukturalnych warstwy powierzchniowej. *Inżynieria Materiałowa,* No.5, pp.141-149, ISSN 0208-6247

Rudawska, A. & Jacniacka E. (2009). Analysis of Determining Surface Free Energy Uncertainty with the Owens-Wendt method. *International Journal of Adhesion and Adhesives,* Vol.29, pp. 451-457, ISSN 0143-7496

Rudawska, A. & Kuczmaszewski, J. (2005). *Klejenie blach ocynkowanych.* Wyd. Uczelniane PL, ISBN 83-89246-43-0, Lublin, Poland

Rudawska, A. & Kuczmaszewski, J. (2006). Surface Free Energy of Zinc Coating After Finishing Treatment. *Material Science-Poland,* Vol.24, No.4, pp. 975 - 981, ISSN 2083-134X

Rudawska, A. (2008). Swobodna Energia Powierzchniowa i Struktura Geometryczna Powierzchni Wybranych Kompozytów Epoksydowych. *Polimery,* 53, No.6, pp.452-456, ISSN 0032-2725

Rudawska, A. (2009). Właściwości adhezyjne kompozytów szklano-epoksydowych po różnych sposobach przygotowania powierzchni, In: *Polimery i kompozyty konstrukcyjne,* G. Wróbel, (Ed.), pp. 97-105, ISBN 978-83-60917-40-4, Cieszyn, Poland

Rudawska, A. (2009). Wytrzymałość połączeń klejowych blach tytanowych po różnych sposobach przygotowania powierzchni. *Inżyniera Materiałowa,* No. 5, pp. 341 - 345, ISSN 0208-6247

Rudawska, A. (2010). *Geometric structure and surface layer adhesive properties of aramid-epoxy composite.* Conferencing mat. Polymer Processing Society PPS-26. Regional Meeting Istanbul, Turkey, October 20-24. 2010, pp.159.

Sekulic, A. & Curnier, A. (2010). Experimetation on Adhesion of Epoxy. *International Journal of Adhesion and Adhesives,* Vol.30 pp. 89-104, ISSN 0143-7496

Serro, A.P.; Colaço, R. & Saramago, B. (2008). Adhesion Forces in Liquid Media: Effect of Surface Topography and Wettability. *Journal of Colloid Interface Science,* Vol.325, pp.573-579, ISSN 0021-9797

Shang, J.; Flury, M.; Harsh, J.B. & Zollars, R.L. (2008). Comparison of Different Methods to Measure Contact Angles of Soil Colloids. *Journal of Colloid Interface Science*, Vol.328, pp.299-307, ISSN 0021-9797

Shen, W.; Sheng, Y.J. & Parker, I.H. (1999). Comparison of the Surface Energetics Data of Eucalypt Fibers and Some Polymers Obtained by Contact Angle and Inverse Gas Chromatography Methods. *Journal of Adhesion Science and Technology*, Vol.13, No.8, pp.887-901, ISSN 0169-4243

Sikora, R. (1996). Warstwa wierzchnia tworzyw wielkocząsteczkowych. *Polimery*, pp.96-113, ISSN 0032-2725

Sikora, R. (1997). Konstytuowanie warstwy wierzchniej tworzyw wielkocząsteczkowych. *Inżynieria Materiałowa*, No. 4, pp.160-164, ISSN

Sommers, A.D. & Jacobi, A.M. (2008). Wetting Phenomena on Micro-Grooved Aluminium Surfaces and Modeling of the Critical Droplet Size. *Journal of Colloid Interface Science*, Vol.328, pp.402-411, ISSN 0021-9797

Tang, X., Dong, J.& Li, X.(2008). A Comparison of Spreading Behaviors Silwet L-77 on Dry and Wet Lotus Leaves. *Journal of Colloid Interface Science*, Vol.325, pp.223- 227, ISSN 0021-9797

Thompson, P.A., Brinckerhoff, W.B. & Robbins, M.O.J. (1993). Microscopic Studies of Static and Dynamic Contact Angles. *Journal of Adhesion Science and Technology*, Vol.7, No.6, pp. 535-554, ISSN 0169-4243

Vedantam, S. & Panchagnula, M.V. (2008). Constitutive modeling of contact angle hysteresis. *Journal of Colloid Interface Science*, Vol.321, pp. 393-400, ISSN 0021-9797

Volpe, C.D. & Siboni, S. (1998). Analysis of dynamic contact angle on discoidal samples measured by the Wilhelmy method. *Journal of Adhesion Science and Technology*. Vol.12, pp.197–224, ISSN 0169-4243

Xian, J. (2008). Wettability of rough polymer, metal and oxide surfaces as well as of composite surface. *Journal of Adhesion Science and Technology*, Vol.22, No.15, pp.1893-1905, ISSN 0169-4243

Zielecka, M. (2004). Methods of Contact Angle Measurement as a Tool for Characterization of Wettability of Polymers. *Polimery* Vol.49, pp. 327-332, ISSN 0032-2725

Zouvelou, N.; Mantzouris, X. & Nikolopoulos, P. (2007). Interfacial energies in oxide/liquid metal systems with limited solubility. *International Journal of Adhesion and Adhesives* 27, pp.380-386, ISSN 0143-7496

Żenkiewicz, M.; Gołębiewski, J. & Lutomirski, S. (1999). Doświadczalna Weryfikacja Niektórych Elementów Metody van Ossa-Gooda. *Polimery*, Vol.44, No.3 ,pp. 212–217, ISSN 0032-2725

Żenkiewicz, M. (2000). *Adhezja i modyfikowanie warstwy wierzchniej tworzyw wielkocząsteczkowych*, WNT, ISBN 83-204-2547-6, Warsaw, Poland

Żenkiewicz, M. (2005). Wettability and Surface Free Energy of a Radiation-Modified Polyethylene Film. *Polimery*, Vol.50, No.5, pp.365-370, ISSN 0032-2725

Żenkiewicz, M. (2006). New Method of Analysis of the Surface Free Energy of Polymeric Materials Calculated with Owens-Wendt and Neumann Method. *Polimery*, Vol.51,

http://en.wikipedia.org/wiki/Scanning_electron_microscope

http://epmalab.uoregon.edu/epmatext.htm

Part 2

Biology, Medicine

8

Contribution of Scanning Electron Microscope to the Study of Morphology, Biology, Reproduction, and Phylogeny of the Family Syllidae (Polychaeta)

Guillermo San Martín and María Teresa Aguado
Departamento de Biología (Zoología), Facultad de Ciencias, Calle Darwin 2, Universidad Autónoma de Madrid, Canto Blanco, Madrid, Spain

1. Introduction

Syllidae is a highly diverse family of the polychaetes (Annelida, Phyllodocida), with 72 described genera and almost 700 species (San Martín, 2003; Aguado & San Martín, 2009; Aguado et al., in press), and continuously new taxa are being described. They are small marine worms, usually of few mm long, although some species can reach up to 90 mm. Contrariwise to their small size, they are very complex, with a body exhibiting numerous structures, external and internal, some of them difficult to examine properly under light microscope, even using higher magnifications and Nomarsky system of polarized light. Description of most species before around the year 2000 was based only on examinations and drawings made with camera lucida under light microscopes. The use of SEM to study syllids is relatively recent, but produced the discover and descriptions of a number of new, unknown structures, or only incompletely known before, whose physiology and significance open a new field of research. The oldest SEM picture of a syllid is from 1980 in which Heacox showed the head of a *Chaetosyllis* stolon and a larval compound chaeta in a study of the life cycle of *Syllis pulchra* Berkeley & Berkeley, 1948 (Heacox, 1980); somewhat later, Pocklington & Hutchenson (1983) reported the viviparity of the interstitial species *Parexogone hebes* (Webster & Benedict, 1884) showing excellent and surprising SEM photographs of juveniles emerging through segmental apertures (probably nephridial pores) and also some characteristic crenulations of the ventral surface of female's body after releasing juveniles; in the same year, Pawlick published the first SEM photos of general aspect of body, chaetae, and details of ciliation on the basis of dorsal cirri on the species *Branchiosyllis oculata* Ehlers, 1887. This is the first paper, in our knowledge, in which some previously overlooked structures were showed and described thanks to SEM; finally, Sardá & San Martín (1992) redescribed one species of syllid from East coast of USA, with some SEM pictures. However, few descriptions of new taxa with SEM and only few papers with SEM pictures of syllids were published during next years. Most of examinations came from our own relatively recent papers; one is the book "Fauna Ibérica. Syllidae", published in 2003, in which the 161 recorded species of the Iberian Peninsula are described and figured, with 122 SEM plates, showing morphology, and also details of the reproduction; the SEM study revealed some new, tiny structures, unknown before (San Martín, 2003). However, since that book was published in Spanish, several of these discoveries could be overlooked by some non-Spanish speakers. Same kind of study and descriptions, with both light microscope and SEM photos was followed since then for descriptions of syllids from other parts of the world (Aguado et al., 2006 and others), especially in the still unfinished series on Australian Syllidae (San Martín, 2005; San Martín & Hutchings, 2006; San Martín et al., 2008a, b; 2010, and others) and revisions of genera (Aguado & San Martín, 2008). In these papers, and others published by other authors profusely cited below, more unknown structures and new details on chaetae, ciliation pattern, pharyngeal armature, etc., were described, as well as important observations on the reproductive biology of many species. A review of these discoveries and their implications on some aspects of the knowledge of the family Syllidae, with especial relevance in taxonomy and systematics, will be analyzed in detail in this chapter.

Detailed morphological and reproductive observations under SEM had important relevance to the knowledge of the family Syllidae for two reasons.

- The first one is related with the phylogeny of the family; discoverment and descriptions of these structures and details of the reproductive modes provided more morphological features to analyze and, consequently, more robust hypothesis about the relationships among the different genera of the family.
- The second one is the great help to differentiate sibling complexes of species; species apparently identical morphologically can be differentiated by minute details, only perceptible under SEM.

Two books about polychaetes also includes excellent SEM pictures of Syllidae (as well as many others of different families of polychaetes): Rouse & Pleijel (2001), which shows 6 photos with details of morphology and reproduction, and Beesley et al. (2000) in which a couple of SEM photos about syllid reproduction are shown.

2. Material and methods

Syllids are generally of small size and hence, the process of preparation is usually complex and sometimes difficult, so, it is desirable to prepare several specimens for examination. At least, one to be examined dorsally and another one ventrally; specimen with exerted pharynx is also strongly recommended to examine and take photos of details of the anterior end of the pharynx and their armature, which are important characters for identification to genus level.

Minute specimens are very easily lost during the process of preparation, so it is important to be extremely careful, especially having a short number of specimens.

Fixation of the specimens is a very important process for taking good and sharp pictures. Specimens of syllids are usually dirty, especially those of the genera with dorsal papillae, which produce a sticky secretion which agglutinate debris. However, details of papillation is an important taxonomic trait for identification of species and therefore, it is necessary to clean up them, using a brush of a single, slender hair.

Techniques could be different depending upon the authors and the type of SEM used. The specimens for taxonomic studies are usually fixed in formalin, examined in pure water and finally stored in 70% ethanol. To be prepared for SEM, the selected specimens experience a series of progressive baths in more concentrated alcohol (80%, 90%, and pure ethanol), then dried on critical point and covered of a coat of gold. Rouse & Pleijel (2001) recommend Osmium tetroxide (OsO_4) as preferred fixative, although its high toxicity (see Rouse & Pleijel, 2001, p. 7).

3. Results

3.1 Morphology

3.1.1 Ciliation

Many syllids are provided with numerous cilia, whose arrangement has taxonomic importance, but they are difficult to see under light microscope; sometimes, it is possible to observe the presence of tufts of cilia in some appropriate areas, but a detailed description of the arrangement of these cilia is almost impossible or extremely difficult. However, well prepared specimens show, under SEM, a system of transversal rows of cilia, which can be single (fig. 1A) or double (fig. 1B), sometimes some anterior segments with single and from one segment backwards being double (fig. 1B), sometimes only tufts of cilia on some areas; in some species only the peristomium is dorsally provided with a single band of cilia and remaining segments lack them. These details were not included in the descriptions until recently, and they are certainly useful for segregation of species and even similar genera. Descriptions arrangement of ciliary bands, based on SEM examinations are in San Martín (2003) for the species *Paraehlersia ferrugina* (Langerhans, 1881), *Odontosyllis fulgurans* (Audouin & Milne Edwards, 1834), *Myrianida benazzi* (Cognetti, 1953), *Myrianida convoluta* (Cognetti, 1953), *Myrianida edwarsi* (Saint-Joseph, 1887), and *Myrianida dentalia* (Imajima, 1966); San Martín & Hutchings (2006) for the species *Eusyllis kupfferi* Langerhans, 1879, *Odontosyllis polycera* Schmarda, 1863, *O. australiensis* Hartmann-Schröder, 1979; *Paraehlersia weissmannioides* (Augener, 1913), *P. ehlersiaeformis* (Augener, 1913), *Perkinsyllis koolalya* (San Martín & Hutchings, 2006), and *P. serrata* (Hartmann-Schröder, 1984); and Nogueira & Fukuda (2008) for *Trypanosyllis zebra* (Grube, 1860) and *T. aurantiacus* Nogueira & Fukuda, 2008. Also some ciliated areas on bases of dorsal cirri, lateral of segments, palps, ventral surface or in ventral cirri were described and figured by Pawlik (1983) for *B. oculata;* Licher & Kuper (1998) for *Syllis tyrrhena* (Licher & Kuper, 1998); López et al. (2001) for *Pionosyllis magnifica* Moore, 1906; San Martín (2003) for *Xenosyllis scabra* (Ehlers, 1864), and *Trypanosyllis aeolis* Langerhans, 1879; San Martín & Hutchings (2006) for *O. australiensis;* Aguado & San Martín (2008) for *Brachysyllis infuscata* (Ehlers, 1901); Ramos et al. (2010) for *Streptodonta exsulis* Ramos, San Martín & Sikorski, 2010; and Salcedo-Oropeza et al. (2011) for *Trypanosyllis microdenticulata* Salcedo-Oropeza, San Martín & Solís-Weiss, 2011.

3.1.2 Nuchal organs

Nuchal organs are only present in polychaetes and are thought to be a synapomorphy for the group (Rose & Fauchald, 1997; Rouse & Pleijel, 2001), although they show different chemoreceptor structures and are different in shape among the families. The most typical shape appears as two semicircular, densely ciliated pits between prostomium and peristomium; these are the kind of nuchal organs most common in the Syllidae, in fact most genera have the typical nuchal organs (fig. 2A), sometimes extending laterally to prostomium, forming two semicircular ciliated areas (fig. 2B) as in the genera *Eusyllis* Malmgren, 1867, *Odontosyllis* Claparède, 1863, and *Trypanosyllis* Claparède, 1864; however, some genera show other kind of nuchal organs, called nuchal lappets, forming two dorsal, longitudinal evaginations over a number of segments, sometimes only one, more or less spherical or elongate, straight (fig. 2C) or sinuous (fig. 2D), ciliated or lacking of cilia. Numerous descriptions of species include SEM photos of the nuchal organs, as those in Lanera et al. (1994); Lewbart & Riser (1996); Licher & Kuper (1998); Martín et al. (2002); Nogueira & San Martín (2002); San Martín (2003; 2005); San Martín & Hutchings (2006), Aguado & San Martín (2007, 2008); Lattig et al. (2007; 2010 a, b); Nogueira & Fukuda (2008); Lattig & Martin (2009); Ramos et al. (2010); San Martín et al. (2008 a, b; 2010); Salcedo-Oropeza et al. (2011); Lattig & Martín (in press a, b). San Martín & López (2003) described a new genus and species form Australia with nuchal organs laterally located, protected by two lips.

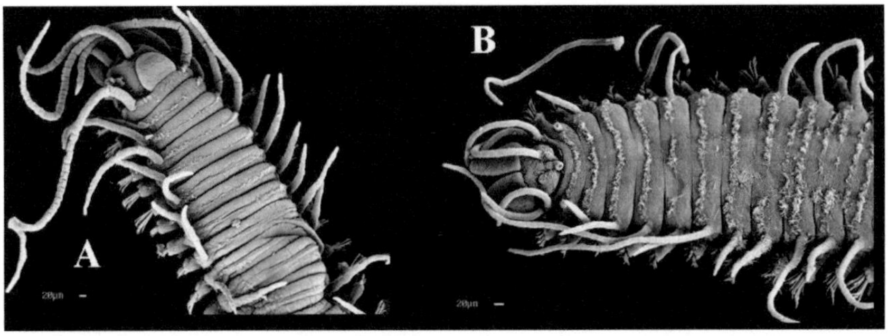

Fig. 1. A, Anterior end, dorsal view of *Odontosyllis australiensis*, showing single row of cilia on each segment. B, Anterior end, dorsal view of *Perkinsyllis serrata* showing single row of cilia on anterior segments and double row form chaetiger 9. San Martín & Hutchings (2006).

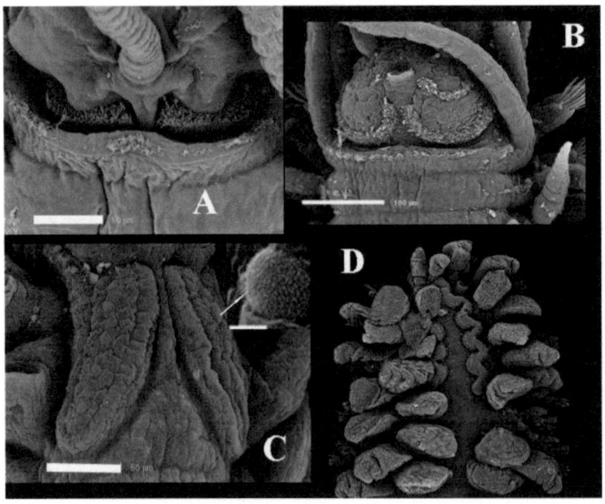

Fig. 2. Nuchal organs of: A *Syllis corallicola*, B, *Eusyllis kupfferi*, showing the ciliation extending to lateral sides of prostomium; C, *Proceraea aurantiaca*, and detail (arrow, right) of the rugose bulks; D, *Clavisyllis alternata* Knox, 1957. A, C, San Martín (2003); B, San Martín & Hutchings (2006); D, Aguado & San Martín (2008).

In the species *Proceraea aurantiaca* Claparède, 1868, San Martín (2003) found and described an enigmatic structure on the nuchal lappets, as semicircular, rugose bulks (fig. 2C, right, up, arrow) provided with pores and tiny hairs, whose function is totally unknown.

3.1.3 Papillation

Some genera of the family Syllidae, as *Sphaerosyllis* Claparède, 1863, *Prosphaerosyllis* San Martín, 1984, *Erinaceusyllis* San Martín, 2005, *Rhopalosyllis* Augener, 1913; and *Paraopisthosyllis* Hartmann-Schröder, 1991, as well as some species of other genera, as *Opisthosyllis* Langerhans, 1879, *Branchiosyllis* Ehlers, 1887, and *Trypanosyllis* Claparède, 1864, have the dorsal, and also sometimes the ventral, surface covered by conspicuous papillae (figure 3), sometimes also extending through parapodia, antennae, cirri and palps. These papillae can be scarce and scattered or densely distributed, in different sizes or all similar, hazardous distributed or arranged in rows. Information on size, shape and distribution of papillae has a very important taxonomic meaning, and it is very useful to differentiate species; although presence of papillae is easily perceptible under light microscope, the arrangement and sizes of papillae are much better evaluable under SEM. In *Sphaerosyllis, Prosphaerosyllis* and *Erinaceusyllis,* the papillae produce an adhesive secretion which sticks detritus for mask the individuals; so, it is sometimes difficult to see them because they are covered by detritus and the specimens requires a previous cleaning by a minute brush and washing of water. Numerous descriptions of species of these three genera are in San Martín (2003; 2005), Nogueira et al. (2004); Musco et al. (2005); Álvarez & San Martín (2009); and Olivier et al. (2011). Details of papillae of the species *Syllis papillosus* (Tovar-Hernández, Granados-Barba & Solís-Weiss, 2002) in Tovar-Hernández et al. (2002); *Trypanosyllis troll* Ramos, San Martín & Sikorski, 2010 are in Ramos et al. (2010), *Paraopisthosyllis alternocirra* San Martín & Hutchings, 2006 in San Martín & Hutchings (2006); *Opisthosyllis viridis* Langerhans, 1879 in San Martín et al. (2008 a); and *Branchiosyllis verruculosa* (Augener, 1913) in San Martín et al. (2008b).

3.1.4 Pharyngeal armature

The pharyngeal armature in syllids is an important diagnostic character at different taxonomic levels; for this reason, a careful examination of the pharyngeal armature is always necessary for identification. SEM photos help considerably in these observations, and also contributed greatly to the discover of some overlooked details, as ciliation and presence of pores on the pharyngeal papillae, coats of cilia on the pharyngeal opening, secondary crown of papillae, total absence of papillae, etc. The pharynx of syllids can be unarmed or provided with a single middorsal tooth (fig. 4A), usually surrounded by a crown of papillae, and sometimes with a complete or incomplete crown of teeth (trepan) (fig. 4B) sometimes there is a trepan but not a pharyngeal middorsal tooth; teeth of trepan are usually of the same size, but sometimes they are of different sizes (figs. 4C, D). Detailed observations under SEM of the pharyngeal armature, papillae and ciliation are in numerous papers: Capa et al. (2001); Tovar-Hernández et al. (2002); Nogueira & San Martín (2002); Martin et al. (2002; 2003; 2009); San Martín (2003; 2005); San Martín & Hutchings (2006); Aguado & San Martín (2007,2008); Nogueira & Fukuda (2008); San Martín et al. (2008a, b; 2010); Lattig & Martín (2009; in press a, b); Lattig et al. (2007; 2010 a, b); Ramos et al. (2010); Olivier et al. (2011); Salcedo-Oropeza et al. (2011).

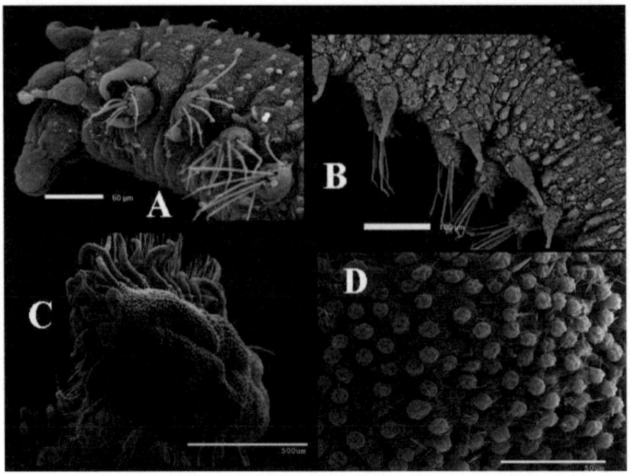

Fig. 3. Papillae: A, anterior end, lateral view,*Sphaerosyllis hirsuta*. B, midbody of the same. C, anterior end, dorsal view, *Opisthosyllis papillosa*. D, details of the papillae of the same. A, B, San Martín (2005). C, D, San Martín et al. (2008 a).

3.1.5 Chaetae

Chaetae have a great value for taxonomic identifications; small details of chaetae can be crucial to assign one specimen to one species. Therefore, a careful examination of chaetae is decisive. They are usually the most photographied structures of syllids; together with the pharyngeal armature. The chaetae are the unique hard structures in syllids, so they are relatively easy to be examined and photographied by SEM; however, they are not always clean enough. Additionaly, length and shape can vary from anterior to posterior part of body and from dorsal to ventral. In our knowledge, the oldest publications showing chaetae of any syllid are those of Heacox (1980) and Pawlik (1983) cited above. Later, Westheide (1990) showed the chaetae and also a special genital chaeta of a new species (*Sphaerosyllis hermaphrodita*) from Thailand. There are many variations on the chaetal types in syllids, and since then, numerous papers showing details of chaetae, many of them profusely cited in this chapter, and also others, like Licher et al (1995) or Martínez et al (2002), not mentioned in any other sections of this chapter. Relative length and orientation of spines on margin of blades of compound chaetae are difficult to see properly under light microscope but they are absolutely precise under SEM. In the figure 5, there is a selection of photos of different kinds of chaetae in syllids (see also figure 11). One interesting is the Australian species *Odontosyllis freycinetensis* Augener, 1913; compound chaetae of that species are apparently unidentate under light microscope, even examined with Normarsky system and high magnifications, but in fact they are bidentate, with a minute, spine-like distal tooth and a big, curved proximal tooth. The use of SEM for study of chaetae in syllids also clarify some other details; for instance, hyaline hoods on the margin of blades appear in few some species, but SEM examinations showed that in fact they are several rows of minute spines, instead of one row, with more or less well defined spines (fig. 6).

Images of the same chaeta under light microscope or SEM can be remarkably different, being the latter much more precise and useful for descriptions and phylogenetic inferences. Also study under SEM of special chaetae for reproduction and brooding of eggs produce interesting discoveries (see below).

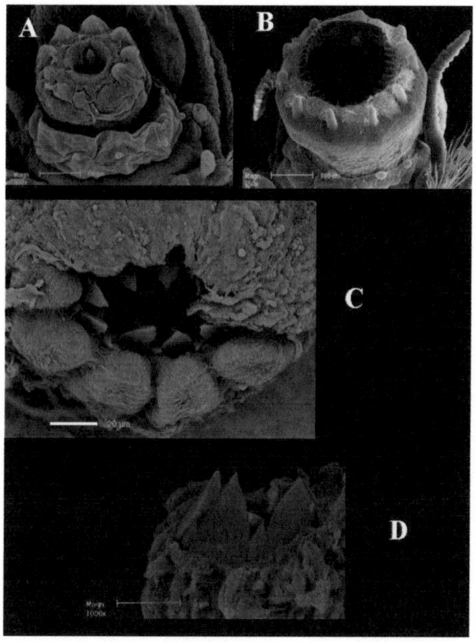

Fig. 4. Everted pharynx of: A, *Syllis gerundensis*, showing the pharyngeal tooth, and two crowns of papillae. B, *Eusyllis assimilis*, showing the middorsal tooth, incomplete trepan, and two crowns of papillae. C, D, frontal and lateral view of the trepan of *Proceraea picta*, showing the different sizes of teeth. San Martín (2003).

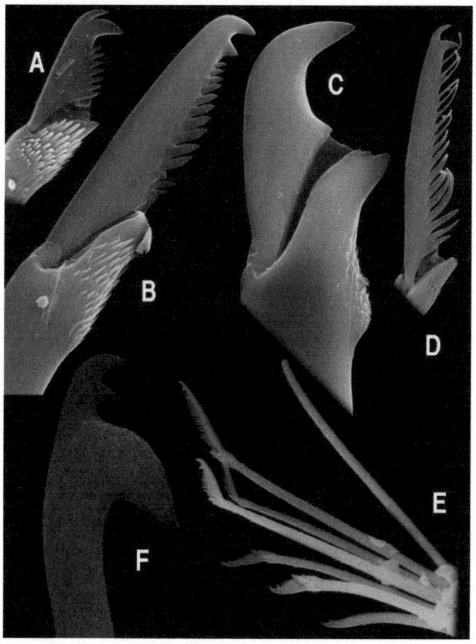

Fig. 5. Some examples of Syllid chaetae. A, *Paraehlersia elersiaeformis*; B, C, *Branchiosyllis maculata*; D, *Perkinsyllis heterochaetosa*; E, chaetal fascicle of *Syllis rosea*; F, *Parahaplosyllis brevicirra*. A, D, San Martín & Hutchings (2006); B, C, San Martín et al.(2008a); E, San Martín (2003); F, San Martín, et al. (2010).

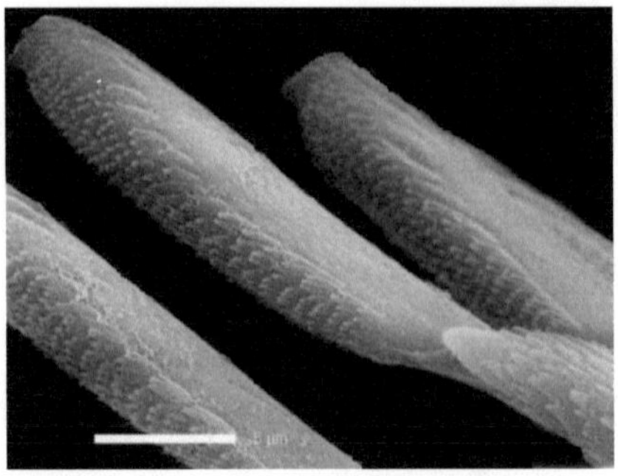

Fig. 6. Compound chaetae of *Streptodonta pterochaeta*, showing the rows of minute spines. San Martín (2003).

3.1.6 Glands

Syllids are provided with numerous kinds of glands, whose function is mostly unknown; some of them were known from long time ago but others were discovered and described recently using SEM. Some of the previously known glands were studied under SEM, and some details of their structures were showed after SEM observations. Most of the different kinds of glands in Syllidae are present in the subfamilies Exogoninae Langerhans, 1879 and Syllinae Grube, 1850, especially in interstitial, minute species, but also in some large syllids. Members of the genus *Syllis* Lamarck, 1818 (including *Typosyllis* Langerhans, 1879) have convoluted, refringent glands within the articles of cirri. They are easily visible under light microscope, even under compound microscope. These glands are usually opened by means of minute pores, only perceptible under SEM (fig. 7A). The first picture of these pores are in Licher & Kuper (1998), for *Syllis thyrrena*, and later San Martín (2003) for *Trypanosyllis zebra*, *Trypanosyllis aeolis*, and *Syllis amica* Quatrefages, 1865, but they are probably present in many other species. San Martín et al. (2008b) showed these pores in the species *Parasphaerosyllis indica*. San Martín (2003) also discovered and published some photos of pores in some anterior ventral cirri of some species, as *Eurysyllis tuberculata*, *Plakosyllis brevipes*, *Xenosyllis scabra*, and *Trypanosyllis zebra*; especially interesting are those of *P. brevipes*, since they are half-moon like, with an arranged alveolar organization (fig. 7B). Since that species is strictly interstitial in sands, these glands could be adhesive; however, they are also present, although less developed, in other non-interstitial species, as *T. zebra*. Pores on dorsum of several species have been also described, as in *P. brevipes* (San Martín, 2003). San Martín et al. (2008b) described a species, *Branchiosyllis carmenroldanae*, from Australia, with the granular dorsum covered by numerous pores. Pores on the dorsal tubercles have been reported in *E. tuberculata* by San Martín (2003) and in the tips of dorsal crests (see below) in *Xenosyllis moloch* San Martín, Hutchings & Aguado, 2008 (San Martín et al., 2008b) (fig. 7C). Also dorsum densely provided of granules, opened by pores was described in the Anoplosyllinae *Syllides fulvus* (Marion & Bobretzky, 1875) by San Martín (2003). Dorsal pores have been reported in *P. ferrugina* by the same author, *P. ehlesiaeformis*, *P. weissmannioides*, by San Martín & Hutchings (2006). Especially interesting are the pores described and figured by the same authors for the species *Brevicirrosyllis mariae* (San Martín & Hutchings, 2006), located on bases of dorsal cirri, similar to the pores of parapodial glands of some Exogoninae (see below) (fig. 7D), and also at tips of dorsal cirri. Pores in ventral cirri have been also reported for other few species, as *Amblyosyllis madeirensis* Langerhans, 1879, and *Eusyllis lamelligera* Marion & Bobretzky, 1875.

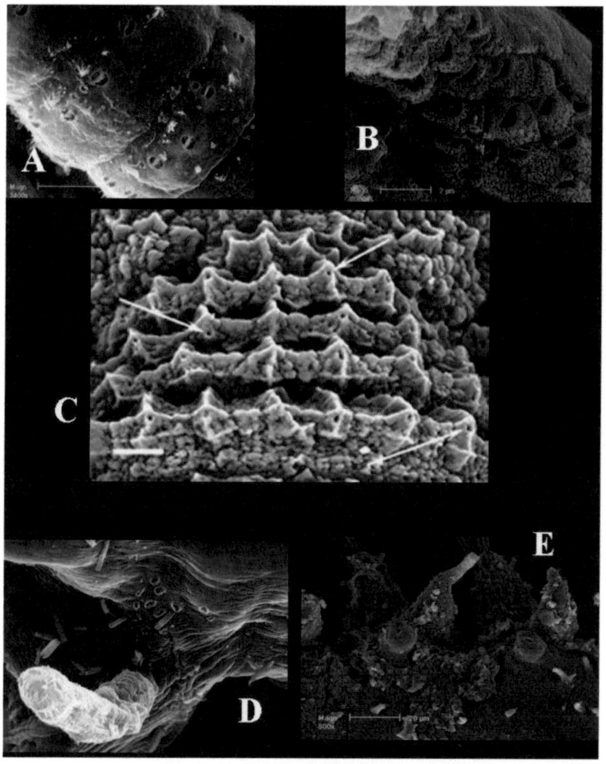

Fig. 7. Opening of different glands. A, articles of dorsal cirri, *Trypanosyllis zebra*; B, ventral cirri, *Plakosyllis brevipes*; C, dorsal crests, *Xenosyllis moloch*; D, base of dorsal cirri, *Brevicirrosyllis mariae*; E, parapodial glands, *Sphaerosyllis capensis*. A, B, San Martín (2003); C, San Martín et al. (2008a); D, San Martín & Hutchings (2006); E, San Martín (2005).

In some genera of the subfamily Exogoninae (*Brania* Quatrefages, 1865, *Sphaerosyllis* Claparède, 1863, *Parapionosyllis* Fauvel, 1923) there is a typical kind of glands, the parapodial glands, which might show rod, granular, or hyaline material. In *Sphaerosyllis*, the parapodial glands are connected to a special papilla, opened by one pore (fig. 7E) described for *S. capensis* Day, 1953 by San Martín (2005). San Martín (2003) took some detailed SEM pictures of pores of parapodial glands in some species of *Parapionosyllis*, as *P. labronica* Cognetti, 1965, *P. brevicirra* Day, 1954, *P. elegans* (Pierantoni, 1903), *P. minuta* (Pierantoni, 1903), and *P. cabezali* (Parapar, Moreira & San Martín, 2000). These pores are very well defined, with a thickened, circular area around, and a digitiform, eversible and contractile structure inside, even one SEM image shows one of these digitiform structures with glandular material emerging from the body (fig. 8, below).

3.1.7 Other structures

Additionally, SEM, has provided the possibility to find some enigmatic structures, such as transversal lines of spines or papillae in some species of *Trypanosyllis* (San Martín, 2003; Nogueira & Fukuda, 2008; San Martín et al, 2008a; Ramos et al., 2010; Salcedo-Oropeza et al., 2011) (fig. 9A), a detailed arrangement of crests in *Xenosyllis* (San Martín, 2003; San Martín et al., 2008b) (fig. 7C), dorsal glands in *Syllis pulvinata* (Langerhans, 1881), ciliated areas close to prostomium in *Sacconereis* stolons (San Martín, 2003) (fig. 9D), terminal papillae in palps of some species of *Syllides* and *Streptosyllis*, (San Martín, 2003; San Martín & Hutchings, 2006), subcirral papillae in some species of *Paraehlersia* (figs. 9B, C), etc. Furthermore, SEM has been used to take detailed photos of internal structures, as the proventricle, a distinct muscular and glandular structure of the gut, after dissection, formed by columns ending in hexagons (fig. 9E), which were photographed in SEM by San Martín (2003), San Martín et al., 2008a, 2010, Martín et al., 2009).

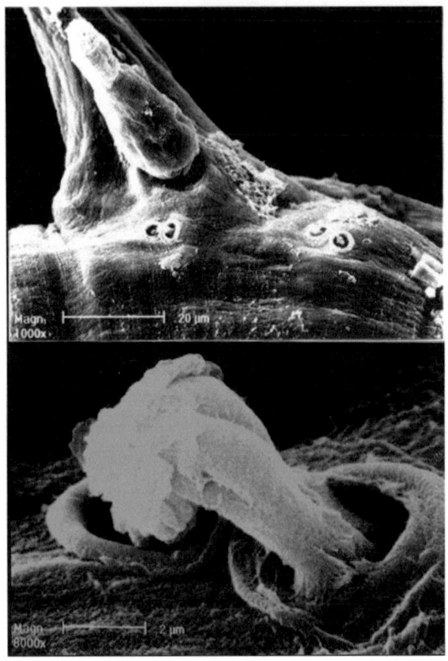

Fig. 8. Parapodial glands of *Parapionosyllis brevicirra* (above), detail of a papilla with glandular material (below). San Martín (2003).

3.2 Biology and reproduction

Some aspects of the biology has been also remarked using SEM, such is the case of specimens of the genus *Haplosyllis* crawling through the galleries inside sponges (Magnino et al., 1999 a, b), or specimens produced by asexual reproduction (Lattig et al., in press b), or in *Procerastea*. However, most important contributions of SEM to the knowledge of syllids are related with reproductive aspects, well documented by means of numerous pictures in several papers cited below. We have already mentioned the SEM photos of young *E. hebes* emerging from the mother's body (Pocklington & Hutchenson, 1983), stolons (Heacox, 1980), and genital especial chaetae (Westheide, 1990); after these papers, others also dealt with same or similar topics, as Qian & Chia (1989) for the larval development of one species of *Myrianida*.

Many interesting contributions were made on the external gestation of Exogoninae; Küper & Westheide (1998) demonstrated, including SEM photos, that in the genera *Prosphaerosyllis* and *Salvatoria*, the eggs are attached by means of tiny, simple notochaetae which penetrate into the eggs, also examined posteriorly in other species of the same genera by ourselves, and in species of the genera *Erinaceusyllis* San Martín, 2005 (San Martín, 2003; 2005) (figs. 10 A, B). Especially interesting were the SEM photos taken to the new Australian genus and species *Nooralia bulgannabooyanga* San Martín, 2002, because it is up to date, the unique syllid which develops compound notochaetae for dorsal brooding of eggs (San Martín, 2002). Ventral brooding of eggs and development of juveniles attached to females in some species were recorded and photographied by San Martín (2003), Mastrodonato et al. (2003), Nogueira et al. (2004), and Böggemann & Purschke (2005) (figs. 10C, D). SEM pictures were also very useful to the study of reproductive stolons in the subfamily Syllinae, as in San Martín (2003), San Martín & Nishi (2003), San Martín et al. (2008a,b, 2010), and Nogueira & Fukuda (2008).

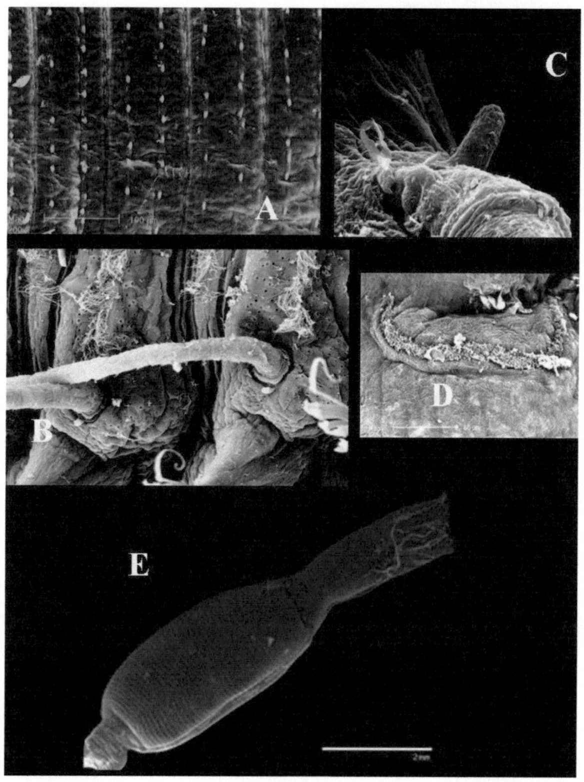

Fig. 9. A, transversal rows of spines, *Trypanosyllis aeolis*. B, subcirral papilla (under bases of dorsal cirri), *Paraehlersia ehlersiaeformis*. C, detail of subcirral papilla, *Paraehlersia weissmannioides*. D, ciliated area (nuchal organs?) of a *Sacconereis* stolon. E, pharynx and proventricle, *Megasyllis inflata*. A, D, San Martín (2003); B, C, San Martín & Hutchings (2006); E, San Martín et al. (2008a).

Especially interesting were some photos taken to stolons of the Australian species *Megasyllis corruscans* (Haswell, 1885), showing pores on ventral bases of parapodia, never reported before (fig. 10E). Also of great interest was the information taken on stolons in the complex of species of the genus *Haplosyllis* (see below).

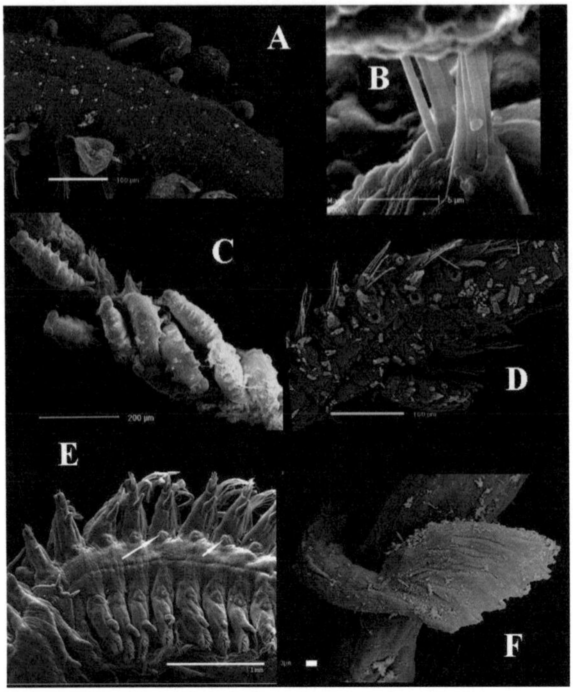

Fig. 10. A, dorsal brood of eggs by means of capillary notochaetae, *Prosphaerosyllis sexpapillata*. B, detail of capillary notochaetae inserting into an egg, *Salvatoria vieitezi*. C, juveniles attached to a female of *Exogone verugera*. D, juvenile attached to a female of *Exogone africana*, showing nephridial pores. E, ventral view of an stolon of *Megasyllis corruscans*, showing pores on parapodial bases. F, distal end of a dorsal cirri, *Murrindisyllis kooromundroola*. A, D, San Martín (2005); B, C, San Martín (2003); E, San Martín et al. (2008 a). F, San Martín et al. (2007).

3.3 Taxonomy

All these observations, discoveries, and more detailed descriptions of characters, thanks to SEM had important consequences, not only descriptions of new structures, opening new topics for future research, but also in the taxonomy and systematics of this complicate family of polychaetes. Detailed descriptions, using SEM, and comparison of different populations of supposed species of large distribution can help to differentiate cryptic species. The genus *Haplosyllis* has been recently revised, comparing specimens from all around the world, and using SEM techniques for all characters, especially the chaetae, apparently identical or very slightly different; however, the comparison of SEM photos show important differences (fig. 11). Before these studies, most of the reports all around the world of this genus were attributed to the species *Haplosyllis spongicola*, but that species is actually only present in the European seas, and with doubts in Australia. This study produced a number of papers (Martín et al., 2002, 2003; Lattig & Martín, 2009, in press a, b; Lattig et al., 2007, 2010 a, b) and the descriptions of numerous new species. Similarly, the genus *Haplosyllides* was also revised with similar results (Martín et al., 2009). Revisions of other genera of Syllidae are currently in process.

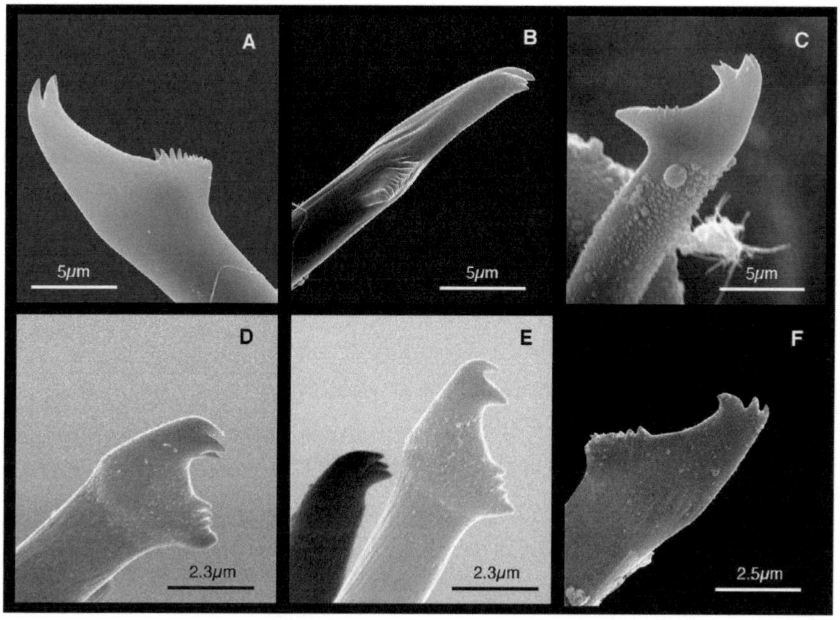

Fig. 11. Details of chaetae in different species of the genus *Haplosyllis*, showing the minute details, very difficult to see under light microscope. Martin et al. (2003).

3.4 Phylogeny

All these observations, above explained, under SEM on numerous species of syllids made possible the description of a number of new characters, previously unknown, as well as a better understanding of other features already known. This had important consequences in taxonomy, as already explained, and also on the phylogeny of the family. Sharp and detailed SEM photos are extremely useful to illustrate the characters used for phylogenetic analysis, as done by Nygren (1999), Aguado & San Martín (2009), and Aguado et al. (in press). Nygren (1999) used one plate with six SEM photos to illustrate 8 characters, and Aguado & San Martín (2009) 5 plates with a total of 33 SEM photos, in which ciliation of segments, body shape, ornamentation (papillae, tubercles, crests, among others) nuchal organs and details of nuchal lappets, details of peristomium and prostomium, size and shape of antennae and cirri, palps, parapodia, chaetae, several kinds of glands, pharyngeal armature, kinds of reproduction, etc., were profusely illustrated. The increasement on the number of characters and detailed descriptions supported more robust hypotheses about evolutionary relationships of syllids. The information of new morphological traits or new details about well known structures has been extremely useful to provide evidence for synapomorphies of some clades. For instance, the glands in anterior ventral cirri highly contributed to support a clade with the genera *Eurysyllis*, *Plakosyllis* and *Trypanosyllis*; a character that is only perceptible under SEM.

4. Perspectives and conclusions

The use of SEM in the study of syllids has been proved of high usefulness, showing characters not previously described and for giving appropriate descriptions of other already known. However, since it is also possible the discoverment of new taxa in wide unprospected areas of the world, and a detailed study under SEM still lacks for numerous syllid genera, new morphological structures may still be undescribed. For instance, San Martín et al. (2007) described a new genus and species from Australia (*Murrindisyllis kooromundroola*) with numerous autopomorphic characters. The most shocking characteristic is the tip of dorsal cirri, ending in a structure amazingly similar to a hand, with five "fingers" jointed by a membrane, giving a frog leg appearance (fig. 10F). These and other unusual characters of this genus were carefully detailed with SEM photos. At this point, the use of SEM is absolutely necessary for providing good and detailed descriptions of new taxa and numerous already known genera for which a detailed examination is still lacking. However, the use of SEM for the study of syllids is not the unique tool for improving the knowledge of the family; other techniques are absolutely necessary for phylogenetic studies in this family, such as molecular information, as shown by Aguado et al. (in press) who reorganizes the classification of the family using combined molecular and morphological data. Similarly, a worldwide catalogue for all the species of the family Syllidae is currently in process, in which SEM pictures acquire crucial importance.

5. Acknowledgements

We wish express our gratitude to the Editorial Board of InTech, Open Access Publisher, and especially to Ms. Daria Nahtigal, for their kind offer to participate in this book and for all the given facilities. We wish to acknowledge to Fauna Ibérica, JMBA, Recods of the Australian Museum, Zootaxa, and Springer Verlag, for allow us to reproduce the SEM photos, previously published. Contribution to the project no. CGL2009-12292 BOS funded by the Ministerio de Ciencia e Innovación of the Spanish Government.

6. References

Aguado, M. T. & San Martín, G. 2007. Syllidae (Polychaeta) from Lebanon with two new reports for the Mediterranean Sea. *Cahiers of Biologie Marine*, 48: 207-224.

Aguado, M. T. & San Martín, G. 2008. Re-description of some genera of Syllidae (Phyllodocida: Polychaeta). *Journal of Marine Biological Association United Kingdom* 88(1): 35-56.

Aguado, M.T. & San Martín, G. 2009. Phylogeny of Syllidae (Annelida, Phyllodocida) based on morphological data. *Zoologica Scripta*, 38 (4): 379-402.

Aguado, M. T., San Martín, G. & Nishi, E. 2006. Two new species of Syllidae (Polychaeta) from Japan. *Scientia Marina*, 70: 9-16.

Aguado, M. T., San Martín, G. & Siddall, M. E. in press. Systematics and evolution of syllids (Syllidae, Annelida). *Cladistics.*

Álvarez, P. & San Martín, G. 2009. A new species of *Sphaerosyllis* (Annelida: Polychaeta: Syllidae) from Cuba, with a list of syllids from the Guanahacabibes Biosphere Reserve (Cuba). *Journal of Marine, Biological Association United Kingdom*, 1-9.

Beesley, P. L., Ross, G. J. B. & Glasby, C. J. 2000. *Polychaetes & Allies: the Southern Synthesis. Fauna of Australia. Vol. 4A. Polychaeta, Myzostomida, Pogonophora, Echiura, Sipuncula.* CSIRO Publishing: Melbourne XII, 465 pp.

Böggemann, M. & Purschke, G. 2005. Abyssal benthic Syllidae (Annelida: Polychaeta) from the Angola Basin. *Organism, Diversity & Evolution* 5: 221-226.

Capa, M., San Martín, G. & López, E. 2001. Syllinae (Syllidae: Polychaeta) del Parque Nacional de Coiba (Panamá). *Revista de Biología Tropical*, 49 (1): 103-115.

Heacox, A. E. 1980. Reproduction and Larval Development of *Typosyllis pulchra* (Berkeley & Berkeley) (Polychaeta: Syllidae). *Pacific Science*, 34 (3): 245-259.

Küper, M. & Westheide, W., 1997. Ultrastructure of the male reproductive organs in the interstitial annelid *Sphaerosyllis hermaphrodita* (Polychaeta, Syllidae). *Zoomorphology (Berlin)*, 117: 13-22.

Lanera, P., Sordino, P. & San Martín, G., 1994. *Exogone (Parexogone) gambiae*, a new species of Exogoninae (Polychaeta, Syllidae) from the Mediterranean Sea. *Bolletino di Zoologia*, 61: 235-240.

Lattig, P. & Martín, D. 2009. A taxonomic revision of the genus *Haplosyllis* Langerhans, 1887 (Polychaeta: Syllidae: Syllinae). *Zootaxa*, 2220 : 1-40.

Lattig, P. & Martín, D. in press a. Two new endosymbiotic species of *Haplosyllis* (Polychaeta : Syllidae) from the Indian Ocean and Red Sea, with new data on *H. djiboutienesis* from the Persian Gulf. *Italian Journal of Zoology*.

Lattig, P. & Martín, D. in press b. Sponge associated *Haplosyllis* (Polychaeta : Syllidae : Syllinae) from the Caribbean Sea, with the description of four new species. *Scientia Marina*.

Lattig, P., San Martin, G. & Martín, D. 2007. Taxonomic and morphometric analyses of the *Haplosyllis spongicola* complex (Polychaeta: Syllidae: Syllinae) from the Spanish seas, with re-description of the type species and description of two new species. *Scientia Marina* 71(3): 551-570.

Lattig, P., Martín, D. & San Martín, G. 2010a. Syllinae (Syllidae: Polychaeta) from Australia. Part 4. The genus *Haplosyllis* Langerhans, 18789. *Zootaxa* 2552: 1-36.

Lattig, P., Martín, D. & Aguado, M. T., 2010b. Four new species of *Haplosyllis* (Polychaeta : Syllidae : Syllinae) from Indonesia. *Journal of the Marine Biological Association of the United Kingdom*, 90 (4): 789-798.

Lewbart, G. A. & Riser, N. W. 1996. Nuchal organs of the polychaete *Parapionosyllis manca* (Syllidae). *Invertebrate Biology* 115 (4): 286-298.

Licher, F. & Küper, M., 1998. *Typosyllis tyrrhena* (Polychaeta, Syllidae, Syllinae), a new species from the island of Elba, Tyrrhenian Sea. *The Italian Journal of Zoology (Modena)*, 65: 227-233.

Licher, F., Ding, Z., Fiege, D. & Sun, R. 1995. Redescription of *Typosyllis magnipectinis* (Storch, 1967) from the South China Sea. *Senckenbergiana maritime*, 25 (4/6): 107-113.

López, E., Britayev, T., Martín, D. & San Martín, G. 2001. New symbiotic associations involving Syllidae (Annelida: Polychaeta), with taxonomic and biological remarks on *Pionosyllis magnifica* and *Syllis* cf. *armillaris*. *Journal of the Marine Biological Association of the United Kingdom*, 81: 399-409.

Magnino, G., Pronzato, R., Sarà, A. & Gaino, E. 1999a. Fauna associated with the horny sponge *Anomoianthella lamella* Pulitzer-Finali & Pronzato, 1999 (Ianthellidae, Demospongiae) from Papua New-Guinea. *Italian Journal of Zoology*, 66: 175-181.

Magnino, G., Sarà, A., Lancioni, T. & Gaino, E. 1999b. Endobionts of the coral reef sponge *Theonella swinhoei* (Porifera, Demospongiae). *Invertebrate Biology*, 118 (3): 213-220.

Martín, D., Núñez, J., Riera, R. & Gil, J. 2002. On the association between *Haplosyllis* (Polychaeta: Syllidae) and gorgonians (Cnidaria: Octocorallia), with the description of a new species. *Biological Journal of the Linnean Society*, 77: 455-477.

Martín, D., Britayev, T., San Martín, G. & Gil, J., 2003. Inter-population variability and character description in the sponge associated *H. spongicola* species-complex (Polychaeta: Syllidae). *Hydrobiologia*, 496:145-162.

Martin, D., Britayev, T.A. & Aguado, M.T. 2009. Review of the Symbiotic genus *Haplosyllides* (Polychaeta, Syllidae), with description of a New Species. *Zoological Science*, 26: 646-655.

Martínez, J., I. Adarraga & San Martín, G. 2002. *Exogone (Exogone) mompasensis* (Exogoninae: Syllidae: Polychaeta), a new species from the Iberian Peninsula. *Proceedings of the Biological Society of Washington*, 115 (3): 676-680.

Mastrodonato, M., Sciscioli, M., Lepore, E., Gherardi, M., Giangrande, A., Mercati, D., Dallai, R. & Lupetti, P. 2003. External gestation of *Exogone naidina* Örsted, 1845 (Polychaeta, Syllidae): ventral attachement of eggs and embryos. *Tissue & Cell*, 35: 297-305.

Musco, L., Çinar, M. E. & Giangrande, A. 2005. A new species of *Sphaerosyllis* (Polychaeta, Syllidae, Exogoninae) from the coasts of Italy and Cyprus (eastern Mediterranean Sea). *Italian Journal of Zoology*, 72: 161-166.

Nogueira, J. M. & Fukuda, M. V. 2008. A new species of *Trypanosyllis* (Polychaeta: Syllidae) from Brazil, with a redescription of Brazilian material of *Trypanosyllis zebra*. *Journal of the Marine Biological Association of the United Kingdom*, 88 (5): 913-924.

Nogueira, J. M. & San Martín, G., 2002. Species of *Syllis* Lamarck, 1818 (Polychaeta: Syllidae) living in corals in the state of Sao Paulo, southeastern Brazil. *Beaufortia*, 52 (7): 57-93.

Nogueira, J. M., G. San Martín & Fukuda, M. V. 2004. On some exogonines (Polychaeta, Syllidae, Exogoninae) from the northern coast of the State of São Paulo, southeastern Brazil Results of BIOTA/FAPESP/Bentos Marinho Project. *Meiofauna Marina*, 13: 45-77.

Nygren, A. 1999. Phylogeny and reproduction in Syllidae (Polychaeta). *Zoological Journal of the Linnean Society*, 126: 365-386.

Olivier, F., Grant, C., San Martín, G., Archambault, P. & McKindsey, C. V. 2011. Syllidae (Annelida: Polychaeta: Phyllodocida) from the Chausey Archipelago (English Channel, France), with a description of two new species of the Exogoninae *Prosphaerosyllis*. *Marine Biodiversity*, 12526: 92.101.

Pawlick, J. R., 1983. A sponge-eating worm from Bermuda: *Branchiosyllis oculata* (Polychaeta, Syllidae). *Marine Ecology*, 4: 65-79.

Pocklington, P. & Hutchenson, M. S. 1983. New record of viviparity for the dominant benthic invertebrate *Exogone hebes* (Polychaeta: Syllidae) from the Grand Banks of Newfoundland. *Marine Ecology Progress Series*, 11: 239-244.

Qian, P.Y. & Chia, F. S. 1989. Larval development of *Autolytus alexandri* Malmgren, 1867 (Polychaeta, Syllidae). *Invertebrate Reproduction and Development*, 15: 49-56.

Ramos, J., San Martín, G. & Sikorski, A. 2010. Syllidae (Polychaeta) from the Arctic and sub-Arctic regions. *Journal of the Marine Biological Association of the United Kingdom*: 1-10.

Rouse, G. & Fauchald, K. 1997. Cladistics and polychaetes. *Zoologica Scripta*, 23: 271-312.

Rouse, G. & Pleijel, F. 2001. *Polychaetes.* Oxford University Press, 354 pp.

Salcedo-Oropeza, D. L., San Martín, G. & Solís-Weiss, V. 2011. Two new species of Syllidae (Annelida: Polychaeta) from the Southern Mexican Pacific. *Zootaxa*, 2800: 41–52.

San Martín, G. 2002. A new genus and species of Syllidae (Polychaeta) from Australia brooding eggs dorsally by means of compound notochaetae. *Proceedings of the Biological Society of Washington* 115(2): 333-340.

San Martín, G. 2003. *Annelida Polychaeta II. Syllidae.* IN: Fauna Ibérica, vol. 21. Ramos, M. A. *et al.* (EDS). Museo Nacional de Ciencias Naturales. CSIC. Madrid, 554 pp.

San Martín, G. 2005. Exogoninae (Polychaeta: Syllidae) from Australia, with the description of a new genus and twenty-two new species. *Records of the Australian Museum*, 57 (1): 39-152.

San Martín, G. & Hutchings, P. 2006. Eusyllinae (Polychaeta, Syllidae) from Australia, with the description of a new genus and fifteen new species. *Records of the Australian Museum* ,58: 257-370.

San Martín, G. & López, E. 2003. A new genus of Syllidae (Polychaeta) from Western Australia. *Hydrobiologia* 496: 191-197.

San Martín, G. & Nishi, E. 2003. A New species of *Alcyonosyllis* Glasby and Watson, 2001 (Polychaeta: Syllidae: Syllinae) from Shimoda, Japan. Commensal with the Gorgonian *Melithaea flabellifera. Zoological Science*, 20: 371-375.

San Martín, G., Aguado, M. T. & Murray, A. 2007. A new genus and species of Syllidae (Polychaeta) from Australia with unusual morphological characters and uncertain systematic position. *Proceedings of the Biological Society of Washington*, 120 (1): 39-48.

San Martín, G., Hutchings, P. & Aguado, M. T. 2008a. Syllinae (Polychaeta, Syllidae) from Australia. Part. 2. Genera *Inermosyllis, Megasyllis* n. gen., *Opisthosyllis,* and *Trypanosyllis. Zootaxa*, 1840: 1-53.

San Martín, G., Hutchings, P. & Aguado, M. T. 2008b. Syllinae (Polychaeta, Syllidae) from Australia. Part. 1. Genera *Branchiosyllis, Eurysyllis, Karroonsyllis,Parasphaerosyllis, Plakosyllis, Rhopalosyllis, Tetrapalpia* n. gen., and *Xenosyllis. Records of Australian Museum,* 60(2): 119–160.

San Martín, G., Hutchings, P. & Aguado, M. T. 2010. Syllinae (Polychaeta: Syllidae) from Australia. Part 3. Genera *Alcyonosyllis*, Genus A, *Parahaplosyllis*, and *Trypanosyllis* (*Trypanobia*). *Zootaxa*, 2493: 35-48.

Sardá, R. & San Martín, G. 1992. Systematic description of *Syllides verrilli* as *Streptosyllis verrilli* (Moore, 1907). Life cycle, population of a salt marsh in Southern England (MA, USA). *Bulletin of Marine Science*, 51 (3): 407-419.

Tovar-Hernández, M. A., Granados-Barba, A. & Solís-Weiss, V. 2002. *Typosyllis papillosus*, a new species (Annelida: Polychaeta: Syllidae) from the southwest Gulf of Mexico. *Proceedings of the Biological Society of Washington*, 115 (4): 760-768.

Westheide, W. 1990. A hermaphroditic *Sphaerosyllis* (Polychaeta: Syllidae) with epitokous genital chaetae from intertidal sands of the Island of Phuket (Thailand). *Canadian Journal of Zoology*, 68: 2360-2363.

9

Diversity of Lips and Associated Structures in Fishes by SEM

Pinky Tripathi and Ajay Kumar Mittal
Banaras Hindu University,
India

1. Introduction

Lips are specialized structures that cover the jawbones, and border the anterior orifice of alimentary canal, the mouth. In general, lips and structures associated with them in different fish species may be considered as mainly concerned with the selection, capture, deglutition and pre digestive preparation of food. The effectiveness of these structures is dependent on modifications in relation to food and feeding habits of the fishes and environmental niches inhabited by them.

Morphological data are also key to understanding fish nutrition in ecology and aquaculture, and during development as well as mechanisms for physiological adaptations to a changing environment. A number of the multifunctional roles of the fish lips and associated structures that are discussed incorporate distinctive morphological features that will be highlighted in this chapter. The lips and associated structures represent a significant vertebrate innovation and are highly diversified.

Therefore, present work was undertaken to investigate diversity of the epithelia of lips and associated structures in different fresh-water fish species with the aim to elucidate the surface architecture using Scanning Electron Microscope. The functional aspects of the lips and associated structures in family Gobiidae, Cobitidae, Belontiidae and few species of Cyprinidae show considerable variation and exhibit unique morphological modifications associated with their lips and other structures around the mouth regarding information on the level of surface architecture as seen under SEM in relation to various food and feeding habits and ecological niches.

When we started to survey, collect and organize the current knowledge on lips and associated structures for the invited chapter on SEM, we soon realized that such a study would lead to a greater understanding only if the lips were discussed as incorporate distinctive morphological features.

The successful maintenance of fish populations in challenging environments requires responsive adjustments in their behaviour, morphology and physiology and these have been reflected by modifications at the level of their organ systems, organs and tissues. The lips are no exception to this. The importance of food in daily life of a fish is obvious, and is reflected in the form of the mouth, lips, jaws and so on. These structures present more diverse modifications than any other organ of the body.

Lips and the structures intimately associated with them in different fish species are in direct contact with a complex ever-changing aquatic environment and ecological conditions in which fish inhabits. They are highly sensitive, serve a variety of functions and are characteristically modified in different groups of fishes. These modifications may be associated in some way either with the diet or the method of feeding.

In general, the upper and the lower jaws bear relatively simple and thin lips. These may be thick, fleshy and fimbriated or even unculiferous. In some cases one or both of the lips may regress or fail entirely to develop. The rostral cap is probably present in most fish species, although it may be so reduced as to be overlooked easily. In many forms it lies well above the upper lip and plays no direct role in feeding, while in others it is greatly enlarged, partially or completely overlies the upper lip and plays a major role in gathering food from the substrate. In the older literature the rostral cap and horny jaw sheaths frequently are confused with the lips.

The gross and fine structure of the lips, the rostral caps and the horny jaw sheaths is extremely varied. This involves, among other things, formation of unculiferous fimbriae, tubercles, unculi, papillae, or ridges and grooves of variable height and distribution on the lips and the rostral cap, and sharp cutting edge, cone shaped structure or unculi on the horny jaw sheaths.

Literatures pertaining to the morpho-anatomical structures of the lip in freshwater teleosts are fragmentary and many authors while studying the alimentary canal, briefly described the morphology and structural organisation of the lips of different fish species (Vanajakshi, 1938; Kapoor, 1958; Khanna, 1961, 1962; Pasha, 1964 a, b, c; Saxena & Bakhshi, 1964; Lal *et al.*, 1964; Chitray, 1965; Sehgal, 1966; Moitra & Bhowmik, 1967; Lal, 1968; Sehgal & Salaria, 1970; Moitra & Sinha, 1971; Sinha, 1975; Sinha & Moitra, 1975, 1976, 1978; Kapoor *et al.*,1975). Suzuki (1956) described the histological organisation of lips of bottom feeding scythe fish *Pseudogobio esocinus*. At the surface of lips, he reported the presence of characteristic processes, with numerous taste buds, playing an important role as food finders. Miller & Evans (1965) studied relationship between the external morphology of brain and lips with emphasis on the distribution of taste buds. Branson & Hake (1972) described morphological organisation of the lips of *Piaractus nigripinnis* and reported that the lips in this fish are adapted for accessory respiratory function. Kiyohara *et al.* (1980) gave an account of the distribution of taste buds on the lips of a minnow *Pseudorasbora parva*. Ono (1980) reported epidermal projections associated with taste buds on the lips in some loricariid catfishes.

Agrawal & Mittal (1991, 1992 a, b, c) reviewed the literature and described the structural organisation of the epithelia of lips and associated structures of three Indian major carps - a surface plankton and detritus feeder, *Catla catla*; a herbivorous column feeder, *Labeo rohita* and an omnivorous bottom feeder, *Cirrhina mrigala;* and a sluggish, bottom dwelling, carnivorous catfish, *Rita rita*. Mittal & Agrawal (1994) reported the structural organisation of the epithelia of lips and associated structures of an active predatory fish *Channa striata*.

Scanning Electron microscope (SEM) reveals the details of surface architecture of tissues to an extent not possible by other procedures. In spite of this fact, the review of literature reveals that not much attention has hitherto been paid to study the surface architecture of fish lips and structures associated with them using SEM. In view of this, Roberts (1982) who examined a variety of fish species using scanning electron microscope, reported that differences in morphology of the lips and associated structures include:

1. Degree of development and specialisation of the lips,
2. Degree of development and specialisation of the rostral cap,
3. Presence or absence of horny jaw sheaths on the jaws in addition to, or in place of the normal lips, and
4. Form and distribution of unculi on the rostral cap, lips and horny jaw sheaths.

Recently, Ojha & Singh (1992), using SEM described functional morphology of the anchorage system and food scrapers of *G. lamta*. Again, Pinky et al. (2002) made a detailed report on lips and associated structures of the same fish *G. lamta*. Yashpal et al. (2009) briefly reported the presence of unculi on the upper jaw epithelium of *Cirrhinus mrigala*. More recently, Tripathi & Mittal (2010) made a detailed report on lips and associated structures of the fish *Puntius sophore*.

Our current knowledge on aspects of modifications in lips and associated structures in fish arises from too little data to arrive at general trends without running the risk of confusing variability, both real and experimental with adaptive phenomenon. What will be required to remedy this situation is a more extensive examination of a larger variety of species, particularly species adapted to different living conditions. This is an area of research for which the extreme biodiversity of fishes will be a powerful tool. Recently, the lips and associated structures of the Labeonini cyprinids have traditionally been identified as important characters in their classification (Yang & Mayden 2010).

The present work has, therefore, been undertaken with the aim to make a comparative study of the organisational pattern of lips and structures associated with them, at scanning electron microscopic levels, in certain fresh-water fish species having different feeding habits and inhabiting varied ecological niches, to address following specific questions:

Do the lips and the structures associated with them in fish species with different feeding habits and inhabiting varied ecological niches show modifications in their organisational pattern?

Does the surface architecture of epidermis of lips and associated structures of the fish species show adaptive modifications in relation to their ecophysiological status and varied feeding habits?

This chapter treats the morphology of lips and associated structures of the family Gobiidae, Cobitidae, Belontiidae and few species of Cyprinidae. The fresh-water fish species inhabiting different ecological niches and having different feeding habits selected for this study are *Glossogobius giuris, Noemacheilus botia, Colisa fasciata, Garra lamta, Puntius sophore* and *Cyprinus carpio.*

Glossogobius giuris (Hamilton, 1822), belongs to the Family Gobiidae, suborder Gobioidei and Order Perciformes. It is predominantly a fresh water fish found throughout the plains of India and sometimes also found in brackish waters. The members of this genus, in general, are bottom fishes and food chiefly consists of small bottom living animals (Günther, 1989). *G. giuris* is a carnivorous surface feeder and feeds on small crustaceans, insects, molluscs, smaller fishes and tadpoles (Hora & Mukerji, 1953; Khanna, 1993).

Noemacheilus botia (Hamilton, 1822) belongs to the family Cobitidae, sub-order Cyprinoidei, and order Cypriniformes. It is a bottom dweller fish, which usually live under stones, and in currents swim from stone to stone (Nikolsky, 1963). It is an omnivorous very undemanding species, which accepts any kind of food and feeds on algal films as well.

Colisa fasciata (Bloch & Schneider, 1801) belongs to the family Belontiidae, suborder Anabantoidei and order Perciformes (Graham, 1997). It is an air breathing fish and the natural habitats of the members of this genus include very weedy rivers, streams and ponds, irrigation ditches, flooded rice fields and also very dirty accumulations of water such as drains (Günther, 1989). It is an omnivorous fish and feeds on nearly equal amount of plant and animal material (Khanna, 1993). It has been regarded a good larvicidal form and is a very efficacious fish (Hora & Mukerji, 1953).

Garra lamta (Hamilton, 1822) belongs to the family Cyprinidae, sub-order Cyprinoidei and order Cypriniformes. It is a hill stream fish and is predominantly adapted to life in swift-flowing waters. Behind the ventral mouth there is a sucking disc which enables the fish to hold fast in strong currents, mountain streams and rapids. Riverbeds comprising mainly rocks, boulders, stones and gravel form a useful hiding and anchoring substratum for the fish. These fishes have "stone clinging" and "stone licking" habit. Food mainly consists of algal felts, mats and periphyton that they scrape off stones.

Puntius sophore (Hamilton, 1822) belongs to the family Cyprinidae, sub-order Cyprinoidei and order Cypriniformes. It inhabits in shallow ponds and streams and is often found in large numbers in the polluted waters where large drains enter the main river (Hora & Mukerji, 1953). It is an omnivorous column feeder (Khanna, 1993) and likes to grub about the bottom (Günther, 1989). Food of the fish consists of much larger amount of plant material than that of animal.

Cyprinus carpio var. communis (Linnaeus, 1758) belongs to the family Cyprinidae, sub-order Cyprinoidei and order Cypriniformes. It is a cold water fish, but being very hardy, easily adapts to warm water. It is an omnivorous feeder. It browses on the shallow bottom and margins, takes in vegetable debris, insects, worms, crustaceans and also planktonic algae (Alikunhi, 1957). The species has been implicated in degradation of the aquatic environment mainly through its habit of rooting in the bottom that causes turbidity and deoxygenated conditions (Welcomme, 1988). In searching for worms and insect larvae, it burrows hole into the embankments (Hora & Pillay, 1962).

2. Materials and methods

Live specimens of *Glossogobius giuris* (approximately 60 ± 10 mm in length), *Noemacheilus botia* (approximately 35 ± 5 mm in length), *Colisa fasciata* (approximately 55 ± 5 mm in length), *Puntius sophore* (approximately 50 ± 10 mm in length), *Cyprinus carpio var communis* (approximately 40 ± 5 mm in length) were collected from river Ganges and ponds at Varanasi, Uttar Pradesh, India. *Garra lamta* (approximately 60 ± 10 mm in length) were collected from hill streams, Jonha Falls at Ranchi, Bihar, India. The fishes were maintained in the laboratory conditions at controlled room temperature (25 ± 2^{0}C). The fishes were cold anaesthetised following (Mittal & Whitear, 1978). Pieces of lips and associated structures were excised and rinsed in physiological saline, dipped briefly in 0.1% solution of S-carboxymethyl- L-cysteine to remove mucus (Whitear and Moate, 1994) and fixed in 3%

glutaraldehyde in 0.1M sodium cacodylate buffer, at pH 7.4 for 4h at 4°C. Following fixation, the tissues were washed in 0.1M sodium cacodylate buffer (pH 7.4) and dehydrated at 4°C with graded ethyl alcohol in ascending concentrations. The tissues were then treated with ethyl alcohol and acetone in the ratios 3:1, 1:1 and 1:3, anhydrous acetone and critical point dried using a critical point dryer (BIO-RAD, England) with liquid carbon dioxide as the transitional fluid. Tissues were glued to stubs, using conductive silver preparation (Eltecks Corporation, India), coated with gold using a sputter coater (AGAR, B7340, England) and examined with a Scanning Electron Microscope (Leo, 435 VP, England).

3. Results

In all the fish species investigated the upper jaws, in general, show a variable degree of protrusion. The upper lip (UL), borne on upper jaw, is associated with the rostral cap (RC) (rostralkappe = Minzenmay, 1933), through a fold of skin (FSUR), that in turn continues with the dorsal head skin. The FSUR, in general, is thin and membranous, and shows remarkable capacity of extendibility. The FSUR, when mouth is closed, lies in a deep groove between the UL and the RC.

The lower jaws in the fish species, in contrast, are only slightly protrusive. The lower lip (LL) is borne on lower jaw. Generally, it continues with the ventral head skin directly at the narrow middle region and through a fold of skin (FSLS) at the lateral sides of the lower jaw. In *G. lamta*, however, a specialised structure - the adhesive pad (AP) is located between the LL and the ventral head skin. The AP is separated from the LL by a narrow groove lined by a thin fold of skin (FSLA) and from the ventral head skin by a deep cleft lined by a thin and extensive fold of skin (FSAV).

In *N. botia, C. fasciata, G. lamta, P. sophore* and *C. carpio* the UL and the LL, on the side facing the mouth opening, are associated with the horny upper jaw sheath (HUJS) and the horny lower jaw sheath (HLJS) respectively. The horny jaw sheaths are absent in *G. giuris*. Several papilliform teeth are also observed in the jaws of *G. giuris* and *C. fasciata.*

In the fish species investigated, in general, the epithelia of the UL, the RC, the FSUR, the LL, the FSLS, the AP, the FSLA and the FSAV are mucogenic. The epithelia of the HUJS and the HLJS, in contrast, are keratinized. In the epithelia of the RC and the AP in *G. lamta* and those of the HUJS and the HLJS in *C. fasciata*, however, both mucogenic and keratinized regions are observed.

In the present account unculi or keratinization are recorded in a total of 5 fishes, and SEM observations are recorded for all of the above mentioned 6 fishes. These observations may be resulted as follows:

4. *Glossogobius giuris*

In *G. giuris* the mouth is terminal (Fig 1 a). The lower jaw remains slightly projected beyond the upper jaw. The upper jaw, as compared to the lower jaw, is highly protrusive (Fig. 1 b, c). The UL and the LL covering the upper jaw and the lower jaw respectively are thick and prominent.

The epithelia of the UL, the LL, the RC, the FSUR and the FSLS are mucogenic and are covered by a mosaic of irregularly polygonal epithelial cells of varied dimensions (Fig. 2 a, b, c). The surface architecture of the epithelial cells is characterised by the presence of a series of micro-ridges. The boundaries between adjacent epithelial cells are demarcated by smooth well-defined uninterrupted double row of closely approximated micro-ridges.

The oral cavity of *G. giuris* is prominent in the strong dentition of their jaws, armored by several papilliform teeth (Fig.1c).

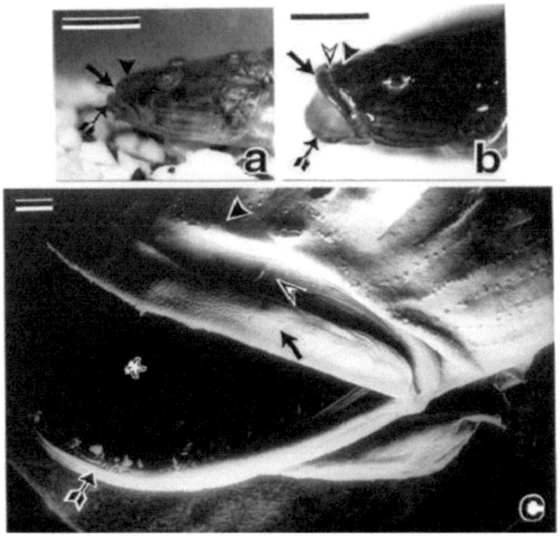

Fig. 1.

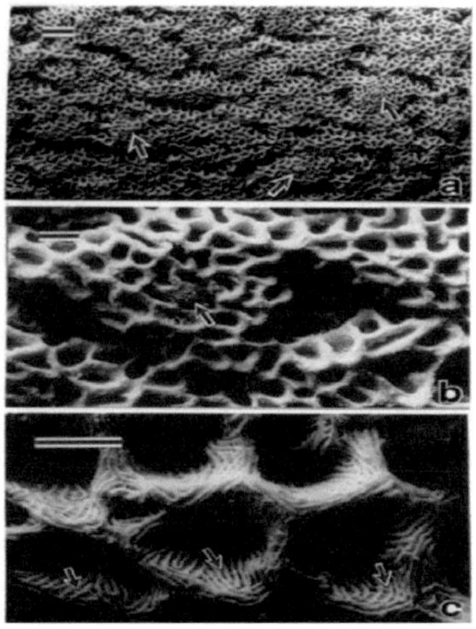

Fig. 2.

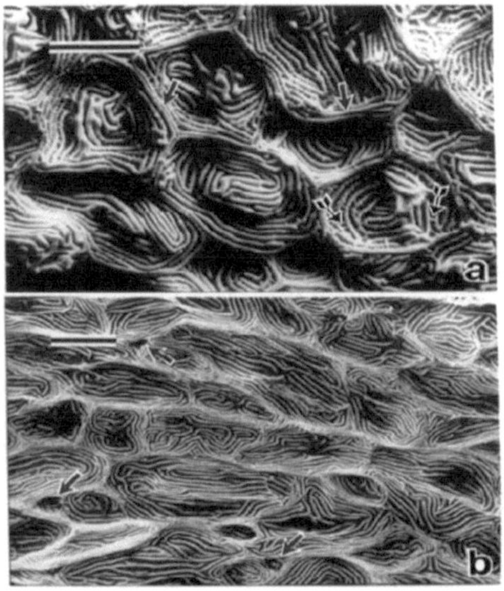

Fig. 3.

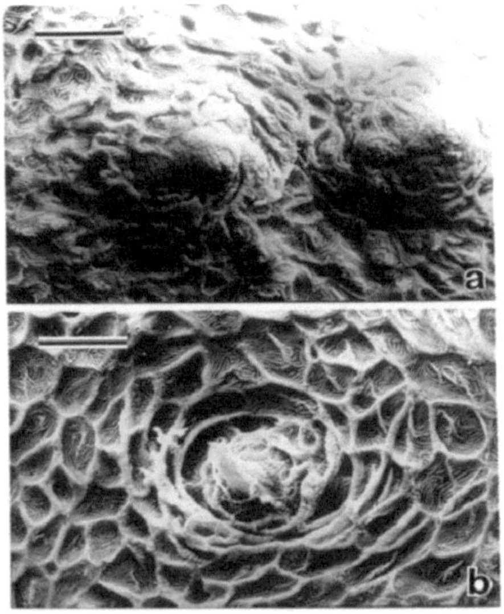

Fig. 4.

4.1 Epithelial cells

The surface of the epithelia of the UL and the LL characteristically appears like that of honeycomb. The epithelia appear folded and differentiated in to wide ridges separated by shallow and narrow furrows. The epithelial cells show a variable degree of invagination and thus their surfaces characteristically appear as concave depressions of varied depth (Fig. 2 a, b, c).

The micro-ridges on the surface of the epithelial cells in the epithelia of the UL and the LL appear smooth, extensive, uninterrupted, and are separated by wide furrows. In general, these appear systematically arranged parallel to each other often traversing towards the deeper regions of concave depressions in each cell (Fig. 2 b, c). The micro-bridges interconnecting the adjacent micro-ridges are prominent and are often located close to each other (Fig. 2 c).

The surfaces of the epithelial cells in the epithelium of the RC, in contrast, show only a slight concavity and appear as shallow depressions. The micro-ridges appear smooth, extensive, uninterrupted, at times branched and are separated by wide furrows. In general these are arranged systematically in a concentric manner, traversing almost parallel to the boundary of the cell forming intricate patterns. The micro-bridges interconnecting the adjacent micro-ridges are prominent similar to those in the epithelia of the UL and the LL (Fig. 3 a).

The epithelial cells in the epithelia of the FSUR and the FSLS, unlike those in the epithelia of the UL, the LL and the RC, appear flattened and do not show depressions at their surfaces. The micro-ridges on the surface of the epithelial cells in these regions though are extensive and often traverse parallel to each other are not interconnected by micro-bridges (Fig. 3 b).

4.2 Mucous cells

Interspersed between the epithelial cells in the epithelia of the FSUR and the FSLS mucous cell apertures of varied dimensions are observed similar to those in *P. sophore, C. carpio, N. botia* and *C. fasciata* (Fig. 3 b). In the epithelia of the UL, the LL and the RC, however, the mucous cell openings could not be clearly distinguished.

4.3 Taste buds

There are two types of TBs in *G. giuris,* type I-minute taste buds could be located on careful search in the epithelia of the UL and the LL. The epithelial cells around each taste bud are concentrically arranged (Fig. 2 a, b).

On the other hand type II-big taste buds on the surface of the epithelium of the RC are observed. At intervals, rounded mounds of epithelial cells bearing this type of comparatively big taste buds (Fig. 4 a). In the central region, the microvilli are arranged in the form of characteristic rosettes (Fig. 4 a). The microvilli of each rosette probably represent sensory hairs originating from sensory cells of the taste buds (Fig. 4 a).

4.4 Superficial neuromasts

Superficial neuromasts sunk slightly in the epithelium are observed. The central region of each such superficial neuromast is characterised by the presence of a characteristic structure consisting of tall-elongated closely approximated projections. This could represent the cupula of the superficial neuromast. The epithelial cells surrounding the superficial neuromasts are concentrically arranged to form a characteristic ring like pattern. (Fig. 4 b).

5. *Noemacheilus botia*

In *Noemacheilus botia* the mouth is small, inferior or sub-terminal, transverse and semicircular (Fig. 5 a). The upper jaw is highly protrusive (Fig. 5 b, c) and the lower jaw is only slightly protrusive. The HUJS and HLJS in this fish are prominent and are characteristically modified. Further, the UL and the LL in this fish are thick and plicate (Fig. 5 a, b, c).

5.1 Mucogenic epithelia

The epithelia of the UL, the LL, the RC, the FSUR and the FSLS are mucogenic. The epithelia of the UL and the LL are thrown in to distinctive protuberances of variable dimensions delineated by narrow furrows (Fig. 6 a, b). Further, the epithelia of the FSUR and the FSLS are characteristically pleated (Fig. 6 a, b, c).

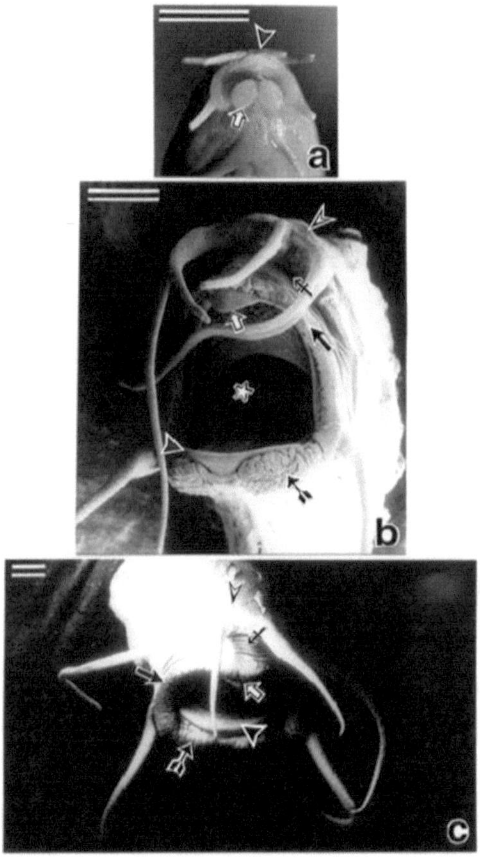

Fig. 5.

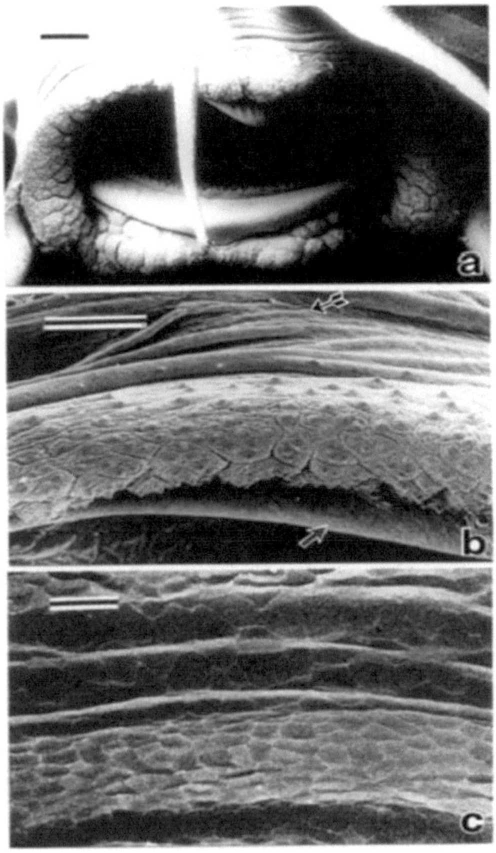

Fig. 6.

5.1.1 Epithelial cells

The epithelia of the UL, the LL, the RC, the FSUR and the FSLS in *N. botia* like those of *G. lamta, P. sophore* and *C. carpio* are covered by a mosaic of irregularly polygonal epithelial cells of varied dimensions (Fig. 6 c). The surface architecture of the epithelial cells is characterised by the presence of a series of micro-ridges separated by prominent irregular spaces. The micro-ridges in general appear sinuous having roughed surface, short with abrupt ends and irregularly interwoven to form web like patterns (Fig. 7 a, b). The boundaries between adjacent epithelial cells are demarcated by a well-defined double row of micro-ridges, which are often connected by transverse strands to give a braided appearance (Fig. 7 a, b).

5.1.2 Mucous cells

Mucous cell openings, seen as wide, rounded apertures or crypts, often containing blobs of mucus, are interspersed between the epithelial cells. Generally, such apertures occur where the boundaries of 3 or more epithelial cells meet (Fig. 7 a, b).

5.1.3 Taste buds

The most conspicuous surface feature of the epithelia of the UL and the LL in *N. botia* is the presence of distinct protuberances each studded with a large number of small, protrusions or elevations (Fig. 6 a, b; Fig. 8 a, b, c). These elevation appear conical and extend beyond the general epithelial surface (Fig. 8 a). The slopes of each elevation or papilla are covered by a mosaic pavement of concentrically arranged epithelial cells. The epithelial cells covering the surface of these papillae and at their vicinity, in general, are arranged concentrically (Fig. 8 b, c). The apex of each elevation is characterised by the presence of numerous closely packed microvilli (Fig. 8 b, c). These microvilli represent the taste hairs originating from taste cells of the buds and are projected through the rounded taste pore at the summit of these elevations.

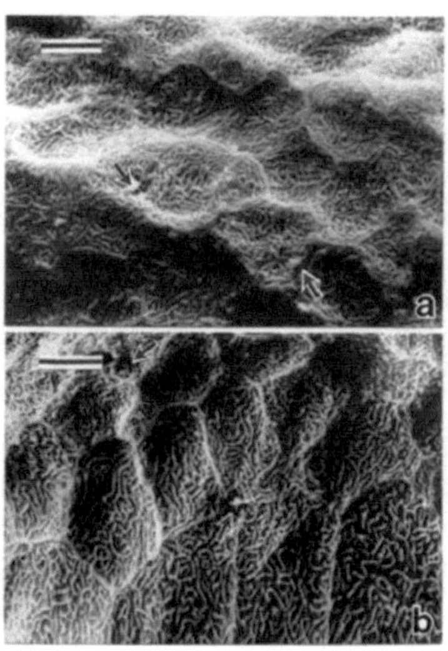

Fig. 7.

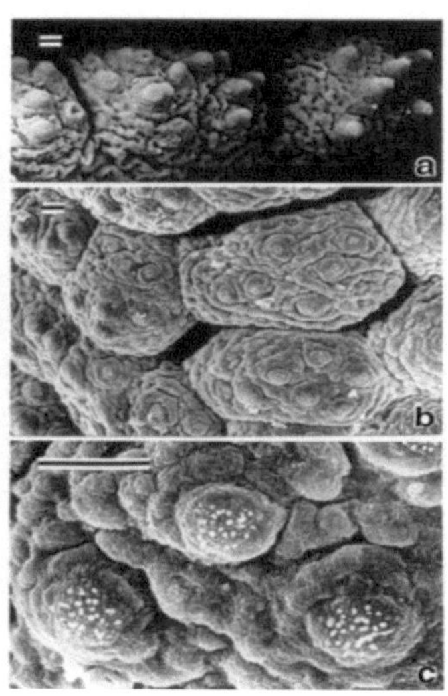

Fig. 8.

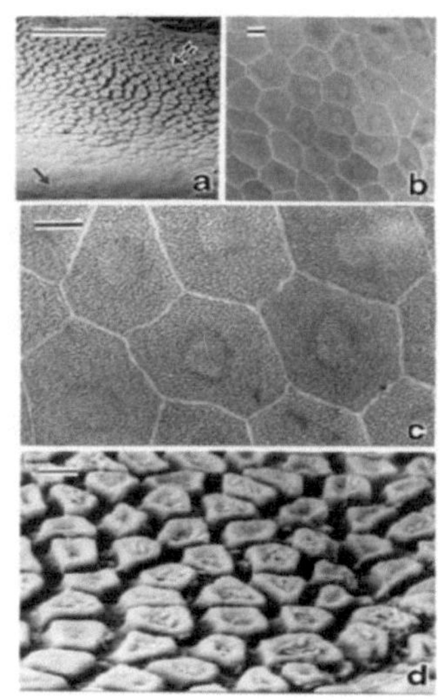

Fig. 9.

5.2 Keratinized epithelia

The epithelia of the HUJS and the HLJS are keratinized and resemble with each other in their surface architecture. The surface epithelial cells of the HUJS and the HLJS, in general, towards the proximal regions gradually get transformed in to truncated specialised structures - the unculi. Thus the surface architecture of the horny jaw sheaths at their distal regions i.e. the regions boarding the mouth is strikingly distinct from those at the proximal regions i.e. towards the buccal cavity (Fig. 9 a).

The surface of the epithelia at the distal regions of the HUJS and the HLJS are covered by a mosaic pavement of irregularly polygonal epithelial cells of varied dimensions (Fig. 9 b). The free surface of these cells is characterised by the presence of compactly arranged micro-ridges separated by narrow irregular spaces. The micro-ridges, in general, are short, sinuous, beaded, branched with abrupt ends and are irregularly interwoven to form maze like pattern (Fig. 9 c). The boundaries between adjacent epithelial cells are very prominent and appear slightly raised from the general surface of the epithelia (Fig. 9 c). These are demarcated by well-defined double rows of micro-ridges, which appear either lying very close to each other or fused. The central regions of these epithelial cells, in general, show rounded bulge at the surface. Each bulge is further demarcated by a narrow depression around them in the form of a ring (Fig. 9 b, c). These bulges could represent the nuclei of these epithelial cells, which appear greatly flattened in cross sections.

The surface of the epithelia at the proximal regions of the HUJS and the HLJS are studded with characteristic truncated, polygonal unculi (Fig. 9 d). The unculi, in general, appear uniform in dimensions and remain projected at the free surface. Each unculus represents modified surface relief of fine projections of a superficial layer epithelial cell. The unculi appear irregularly arranged and the central part of each unculus is wide and often irregularly distributed projections. The boundaries of the adjacent epithelial cells bearing the unculi are distinctly demarcated.

6. *Colisa fasciata*

In *C. fasciata* the mouth is oblique or slightly upturned (Fig. 10 a). The upper jaw is highly protrusive and extends forward to a great extent (Fig. 10 b, c). The lower jaw, however, is only slightly protrusive. Further, the UL and the LL, in general, are thin. The UL is, however, relatively less conspicuous than the LL. The lips are relatively thick and appear lobular at the lateral sides of the mouth. The lobes are separated by shallow grooves. The HUJS and the HLJS are very conspicuous and their distal regions are modified into prominent sharp cutting edges. At intervals papilliform teeth like structures are observed protruding on the surface of the jaw sheaths (Fig. 10 d).

The epithelia of the UL, the LL, the RC, the FSUR and the FSLS are mucogenic. In contrast, the epithelia of the HUJS and HLJS are keratinized and the dead keratinized epithelial cells at the surface are commonly visualised to be lifted up from the underlying tissues. They are probably in the process of being exfoliated. Further, the jaw sheaths are characterised by the presence of papilliform teeth like structures, which protrude at intervals from their surfaces facing the mouth opening (Fig. 11 a, b, c). The surface of the UL epithelium shows slight infoldings, which are visible even in stretched conditions. The folds on the surface of the LL, in contrast, are more distinct and are distinguished in to prominent ridges separated by shallow gutter-like depressions. Generally, these ridges run parallel to each other along the surface bordering the mouth. (Fig. 11 a, c).

6.1 Mucogenic epithelia

6.1.1 Epithelial cells

The epithelia of the UL, the LL, the RC, the FSUR and the FSLS in *C. fasciata* like those of *G. lamta, P. sophore, C. carpio* and *N. botia* are covered by a mosaic of irregularly polygonal epithelial cells of varied dimensions (Fig. 12 a). The surface architecture of the epithelial cells is characterised by the presence of a series of micro-ridges. The boundaries between adjacent epithelial cells are demarcated by smooth well-defined uninterrupted double row of closely approximated micro-ridges (Fig. 12 a, b, c).

The micro-ridges on the surface of the epithelial cells in the epithelia of the UL and the LL are generally short, straight or sinuous and smooth often arranged in the form of small groups (Fig. 12 c; Fig. 13 b). Several such groups of micro-ridges may be observed on the surface of each cell. Adjacent groups of micro-ridges are delineated from each other by extensive micro-ridges, which are often branched and encircle each group. The micro-ridges within a group are generally arranged parallel to each other either linearly or concentrically. The adjacent micro-ridges are interconnected with each other by fine transverse connections, the micro-bridges (Fig. 12 b).

The micro-ridges on the surface of the epithelial cells in the epithelium of the RC, in contrast, appear smooth, extensive, uninterrupted, at times branched and are separated by wide furrows. In general these are arranged systematically in a concentric manner, traversing almost parallel to the boundary of the cell forming intricate patterns. In the narrow central region of these cells, the micro-ridges are often either indistinct or fragmented (Fig. 12 a, b). The micro-ridges on the surface of the epithelial cells in the epithelia of the FSUR and the FSLS are relatively few, extensive and are located parallel to each other at long intervals (Fig. 13 a). Further micro-bridges could not be located.

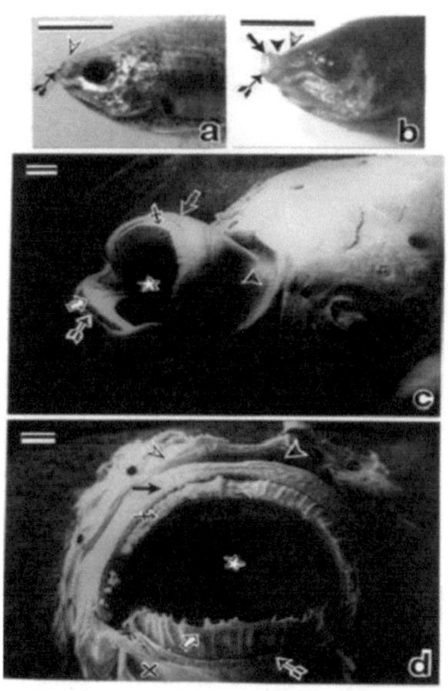

Fig 10.

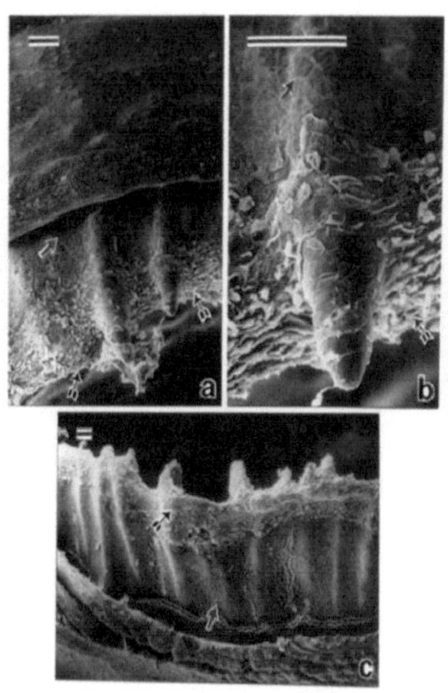

Fig. 11.

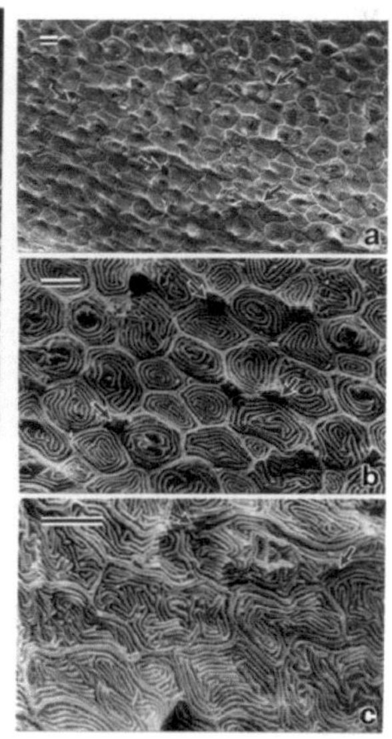

Fig. 12.

6.1.2 Mucous cells

Interspersed between the epithelial cells in the epithelia of the UL, the LL, the RC, the FSUR and the FSLS mucous cell apertures of varied dimensions are observed similar to those in *G. lamta, P. sophore, C. carpio* and *N. botia* (Fig. 12 a, b, c).

6.1.3 Taste buds

Taste buds are located on the ridges at the surface of the UL and the LL. Further, the epithelial surface in the regions where taste buds are located is thrown into papillae like projections protruding beyond the general surface of the epithelia. At the summit of each such papilla several microvilli representing the taste hairs of the taste buds are located. The arrangement of the epithelial cells at and around each papilla bearing a taste bud is concentric and the appearance of the taste buds at the summit of these papillae are similar to those in *G. lamta, P. sophore, C. carpio* and *N. botia* (Fig. 13 a, b).

6.2 Keratinized epithelia

The HUJS and HLJS are covered with a pavement of epithelia of close packed polygonal cells of irregular shape and size. The epithelial cells, however, show regional variations in their surface architecture.

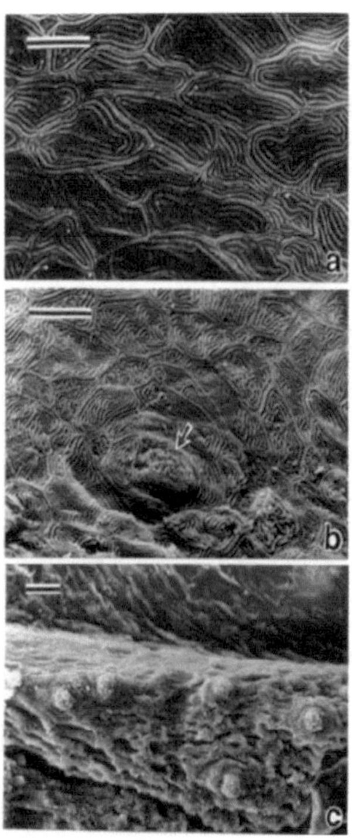

Fig. 13.

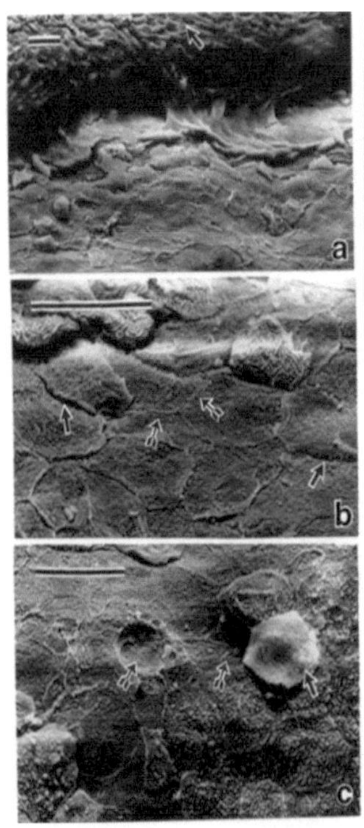

Fig. 14.

Fig.15.

At the proximal regions of the jaw sheaths, the surface of the epithelial cells appears scrawly. The micro-ridges are small, low, irregular and ill defined. Further, the pattern formed by these at the cell surfaces seems indistinctive. The boundary between the adjacent epithelial cells is delineated either by shallow separating clefts or by a double row of micro-ridges separated by distinct spaces (Fig. 14 a, b).

At the distal regions of the jaw sheaths the micro-ridges at the surface of the epithelial cells are frequently punctated (Fig. 14 c) and are separated by wide spaces. In addition short, sinuous, branched micro-ridges interwoven to form characteristic patterns are also observed (Fig. 15 a). The boundaries between adjacent epithelial cells are demarcated by double row of closely approximated micro-ridges.

At and near the apical margins of the horny jaw sheaths the epithelial cells generally, exhibited a surface relief of fine closely approximated micro-ridges such being often prominent in the central parts of the cells. The micro-ridges at the narrow peripheral portions of the cells were relatively short being more widely spaced and irregularly located. The boundaries of the epithelial cells were demarcated by prominent continuous marginal elevations of adjacent cells, sometimes with an inconspicuous gap between them.

The epithelial cells at the distal regions including the apical margins are frequently observed to be lifted up from the underlying tissues. They were probably in the process of being sloughed (Fig. 14 c; Fig. 15 a, b).

7. *Garra lamta*

In *Garra lamta* mouth is sub-terminal and is situated on the ventral side of the head. In this fish the upper jaw and the lower jaw are only slightly protrusive. The UL and the LL are rudimentary and are represented by slight thickening of the epithelia, at narrow regions, covering the upper jaw and the lower jaw respectively. In the regions bordering the lateral margins of the mouth, however, the lips are distinguished as small, stumpy, papillae like structures (Pinky et al., 2002, 2008).

7.1 Mucogenic epithelia

The RC is very prominent and greatly enlarged. Its epithelium may be distinguished in to a keratinized belt towards the mouth opening and a major mucogenic region towards dorsal head skin. At the apical margins of the RC mucogenic islands are observed between the keratinized regions. These non-keratinized and keratinized regions show characteristic alternate arrangements.

The epithelium of the AP like that of the RC is distinguished in to mucogenic and keratinized regions. The epithelium of the major central region of the AP is mucogenic. The narrow peripheral regions of the AP are, however, keratinized.

7.1.1 Epithelial cells

The surface of the mucogenic epithelium of the RC is covered by a mosaic pavement of irregularly polygonal epithelial cells of varied dimensions. The free surface of each epithelial cell is characterised by the presence of a series of compactly arranged micro-ridges separated by narrow irregular spaces. The micro- ridges, in general, appear sinuous, having smoothed surface, short with abrupt ends and irregularly interwoven to form maze like patterns. The boundaries between adjacent epithelial cells are demarcated by well-defined double row of micro- ridges, which are often interwoven to give a braided appearance.

7.1.2 Mucous cells

Interspersed between the epithelial cells rounded or irregular shaped crypts could be observed. These crypts often contain blobs of mucus and represent mucous cell openings. Generally, these apertures are located at the points where the boundaries of three or more epithelial cells meet.

7.1.3 Taste buds

A large number of epithelial protrusion or elevations that extends beyond the epithelial surface are located at irregular intervals. Each epithelial elevation is characteristically associated with a taste bud. The epithelial cells covering the surface of the elevations and at their vicinity, in general, are arranged concentrically. At the apical surface of each epithelial elevation, closely packed microvilli are observed. These microvilli appear to represent the taste hairs originating from the sensory cells of the taste buds.

The mucogenic islands in between the keratinized regions at the apical margin of the RC are characterised by the presence of several stumpy epithelial protrusions lying close to each other. Each epithelial protrusion is associated with a taste bud.

The surface relief of the epithelial cells in the mucogenic region of the AP, in general, is similar to that of the epithelial cells in the mucogenic epithelium of the RC. Further, the mucogenic epithelium of the AP resembles with that of the RC in the distribution of a large number of taste buds and in the presence of mucous cell openings.

7.2 Keratinized epithelia

7.2.1 Rostral cap

The surface of the epithelium of the RC in the keratinized regions, in contrast, to that in mucogenic regions appears shaggy. In general, it is matted with rounded projections or excrescencies in an organised array that are separated by shallow grooves.

The surface of each excrescence is represented by a cluster of several (15-25 or even more) prominent somewhat curved spine like unculi each having a broad base and a narrow apical end. Each unculus is projected at the free surface and represents modified surface relief of fine projections of a superficial layer epithelial cell.

These projections show a gradual increase in their height from the peripheral margin to the centre of the cell and in general appear compactly arranged or fused. The apical end of an unculus is either blunt or conical and the surface is rough with vertically oriented micro-villous projections. Between the unculi, the boundaries of the adjacent epithelial cells demarcated by distinct rows of micro-ridges may be observed.

The epithelial cells in shallow grooves between the excrescencies also show modified surface relief of fine projections. These are, however, less prominent and are not differentiated in to unculi like structures.

7.2.2 Adhesive pad

The keratinized epithelium at the posterior and lateral margins of the AP is characterized by the presence of rounded projections or excrescencies similar to those in the keratinized epithelium of the RC. In the keratinized epithelium at the anterior margin of the AP, in contrast, these excrescencies appear relatively prominent and tall. In general, these appear inverted cone shaped or basket like, each with a narrow proximal base, which gradually becomes relatively wide at the distal region. Like in the keratinized epithelium of the RC, the distal surfaces of these projections are characterised with the presence of a cluster of unculi, which represent modified surface relief of fine projections of superficial layer epithelial cells. The boundaries of the adjacent epithelial cells are often clearly demarcated by well-defined uninterrupted rows of micro-ridges.

7.2.3 Horny upper jaw sheath & Horny lower jaw sheath

The surface sculpture of the epithelia of the HUJS and the HLJS is similar to each other and are characteristically studded with tall, truncated, polygonal unculi. These unculi, in general, appear uniform in dimensions and shape and remain projected at the free surface. The unculi are arranged diagonally in parallel rows in an organised manner to form a characteristic pattern on the surface of the horny jaw sheaths. Each unculus, like that of the RC and the AP, represents modified surface relief of fine projections of a superficial layer epithelial cell. In contrast, these projections appear more developed smooth and prominent at the margins of the cells and show a gradual decline in their height towards the central part of the cell. This results in the formation of a characteristic sharp edge at the margin and a deep depression at the central region of each unculus. Each unculus thus appear very much like a tooth.

The rudimentary UL and the LL, and the delicate FSUR, the FSLA, the FSACAP and the FSAV remain concealed and thus the surface architecture could not be visualized because these regions are deeper in position.

8. *Puntius sophore*

In *Puntius sophore* the mouth is terminal (Fig. 16 a, b, c). The upper jaw is highly protrusive. In contrast, the lower jaw is only slightly protrusive. Further, the UL and the LL are prominent. The UL is however, thin and the LL, in contrast, is very thick (special permission for figures 16-19 have been taken from Tissue and Cell).

8.1 Mucogenic epithelia

8.1.1 Epithelial cells

Surface architecture of the mucogenic epithelia of the UL, the LL, the RC, the FSUR and the FSLS, in general, resembles with each other. The surface of the epithelia is covered by a mosaic pavement of irregularly polygonal epithelial cells of varied dimensions. The free surface of each epithelial cell is characteristically thrown in to a series of micro-ridges having smooth surface. The micro-ridges, in contrast to those of *G. lamta*, are separated by wide furrows and are extensive, at times branched and traverse almost parallel to the boundary of the cell forming intricate patterns (Fig. 17 a, b). Further, the adjacent micro-ridges are interconnected with each other by fine transverse connections, the micro-bridges.

The boundaries between adjacent epithelial cells are demarcated by smooth well-defined double row of closely approximated micro-ridges (Fig. 17 a, b). In addition, epithelial cells with compactly arranged sinuous, short micro-ridges with abrupt ends to form maze like pattern or with micro-ridges giving a punctated appearance to the surface are frequent in the epithelia on the apical side of the UL and the LL (Fig. 17 c).

8.1.2 Mucous cells

Crypts representing the mucous cell openings, often containing blobs of mucus, are frequently observed at the borders of 3 or 4 epithelial cells (Fig. 17 a). The crypts are relatively conspicuous, large, rounded and frequent in the epithelia of the UL, the LL, and the RC, than those of the FSUR and the FSLS.

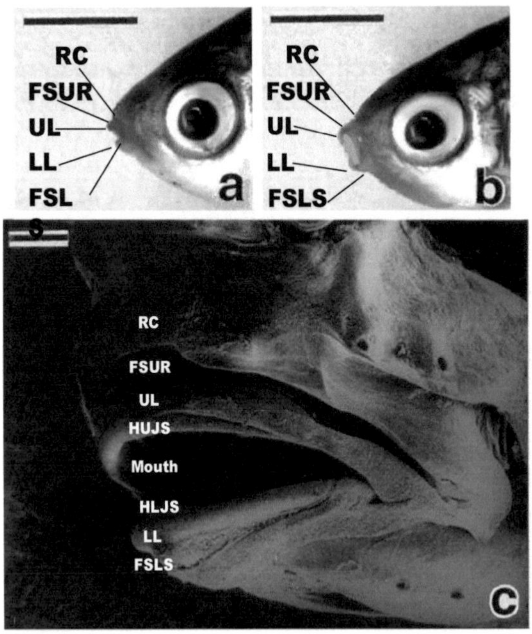

Fig. 16.

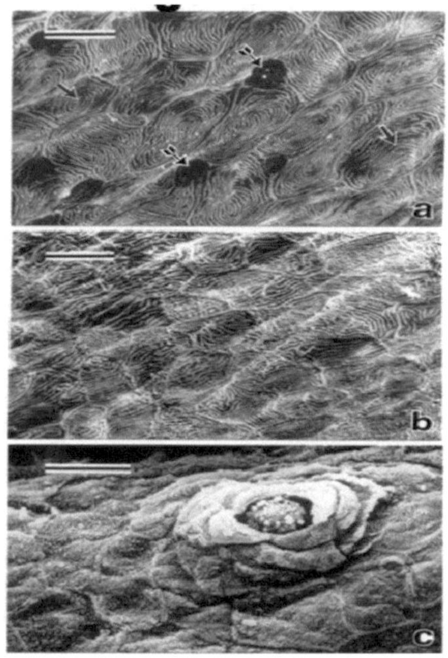

Fig. 17.

8.1.3 Taste buds

In the epithelia of the UL and the LL a large number of taste buds are observed (Fig. 18 a, b). Each taste bud is situated on a small epithelial papilla projecting at the surface. The epithelial cells covering the surface of these papillae and at their vicinity, in general, are arranged concentrically (Fig. 17 c; Fig. 18 a, b). At the summit of each papilla closely packed microvilli are observed (Fig. 17 c). These microvilli appear to represent the taste hairs originating from the sensory cells of the taste buds.

8.2 Keratinized epithelia

The epithelial surface of the HUJS and the HLJS at their distal regions are studded with characteristic truncated, polygonal unculi (Fig. 18 b). The unculi, in general, appear uniform in dimensions and remain projected at the free surface (Fig. 18 c; Fig. 19 a). Each unculus represents modified surface relief of fine projections of a superficial layer epithelial cell and resemble in their shape and organisation to those in the epithelia at the proximal regions of the HUJS and the HLJS of *N. botia*. The unculi in this fish appear relatively less orderly arranged. Further, the central part of each unculus is much wide and often shows the presence of characteristic irregularly distributed projections (Fig. 19 a, b).

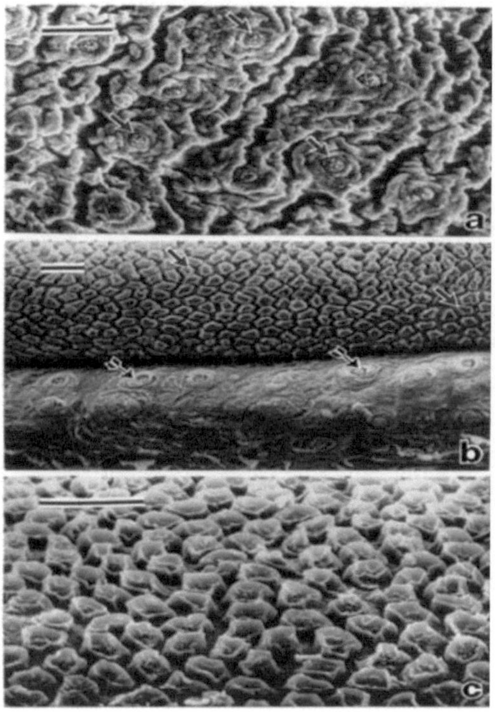

Fig. 18.

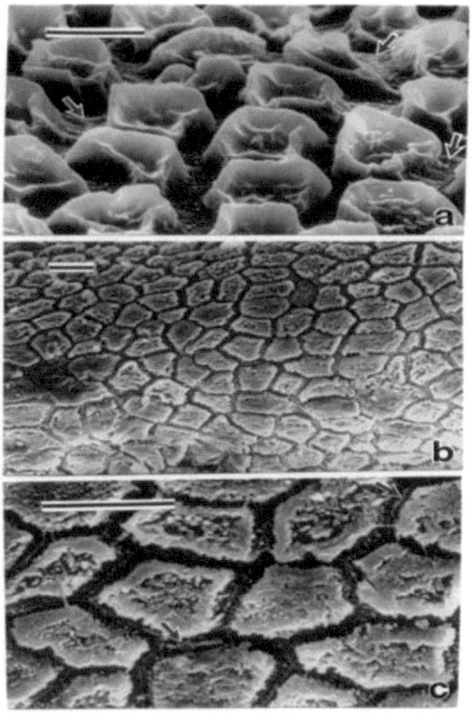

Fig. 19.

The unculi show a gradual decrease in their height towards the proximal regions of the HUJS and the HLJS. At these regions the surface relief of each unculus in contrast to those at the distal regions appears scraggy (Fig. 19 b, c). The micro-villous projections at the peripheral region on the surface of each cell appear fused to form the outer boundary of each unculus with scrawly surface (Fig. 19 c). The major central part of each unculus is occupied with micro-villous projections, which often appear fused awkwardly to give a scrambled or frothy appearance to the surface (Fig. 19 c).

Between the unculi both at the distal and proximal regions of the HUJS and the HLJS the boundaries of the adjacent epithelial cells demarcated by distinct rows of micro-ridges are observed (Fig. 19 a, c).

9. *Cyprinus carpio*

In *C. carpio* the mouth is terminal (Fig. 20 a, b). The upper jaw is highly protrusive (Fig. 20 c, d) and the lower jaw is slightly protrusive. The UL and the LL covering the upper jaw and the lower jaw respectively, in contrast, are thick and prominent and are associated with relatively inconspicuous HUJS and the HLJS.

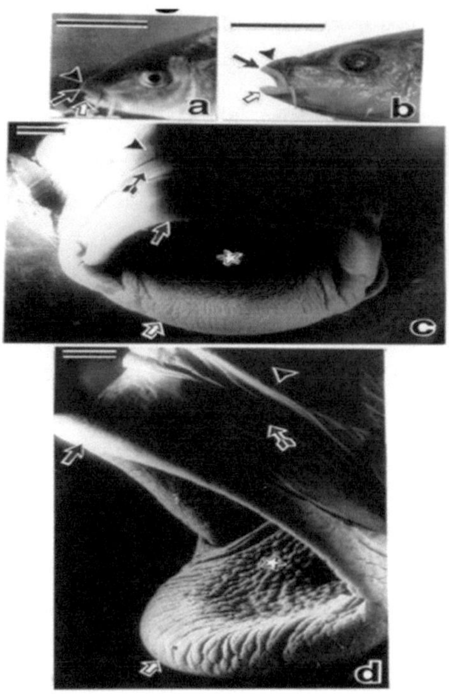

Fig. 20.

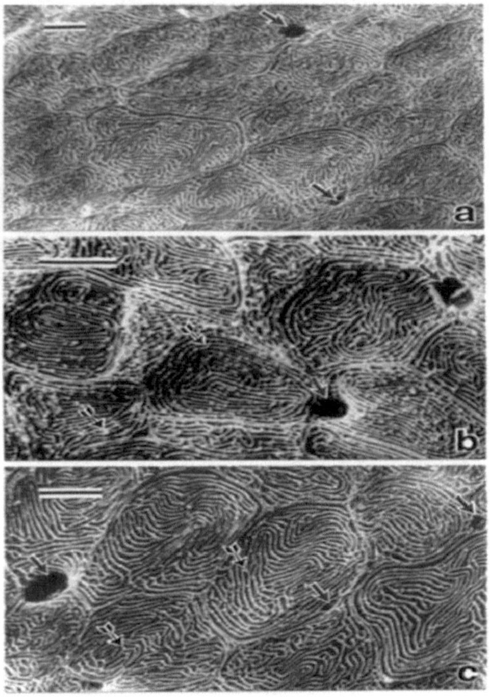

Fig. 21.

9.1 Mucogenic epithelia

9.1.1 Epithelial cells

The surface of the mucogenic epithelia of the UL, the LL, the RC, the FSUR and the FSLS, like those of *P. sophore,* is covered by a mosaic pavement of epithelial cells. The micro-ridges on the surface of the epithelial cells of these epithelia are separated by wide furrows and are extensive. In contrast, the micro-ridges in this fish often appear beaded, branched and interlocked to form intricate patterns (Fig. 21 a, b, c). Further, the micro-bridges interconnecting the adjacent micro-ridges are more prominent and frequent. The boundaries between adjacent epithelial cells are demarcated by well-defined double row of closely approximated micro-ridges often interconnected by transverse connections (Fig. 21 b, c).

9.1.2 Mucous cells

Interspersed between the epithelial cells in the epithelia of the UL, the LL, the RC, the FSUR and the FSLS are observed wide, rounded crypts or pores of varied dimensions, like in *P. sophore.* Generally these crypts or pores occur where the boundaries of three or four epithelial cells meet. These could represent the openings of the mucous cells. This is further confirmed by the presence of blobs of mucus in most of these crypts (Fig. 21 a, b, c).

9.1.3 Taste buds

In the RC epithelium, small taste buds are located individually at long intervals (Fig. 22 a). In the epithelia of the UL and the LL, however, a large number of taste buds are observed, each located on a small epithelial papilla projecting at the surface (Fig. 22 a, b). The taste buds at the major distal potion of the lips are characteristically located in-groups arranged in parallel rows (Fig. 22 b, c). The arrangement of the epithelial cells at and around each papilla bearing a taste bud and the appearance of the taste buds at the summit of these papillae are similar to those in *P. sophore* and *G. lamta*.

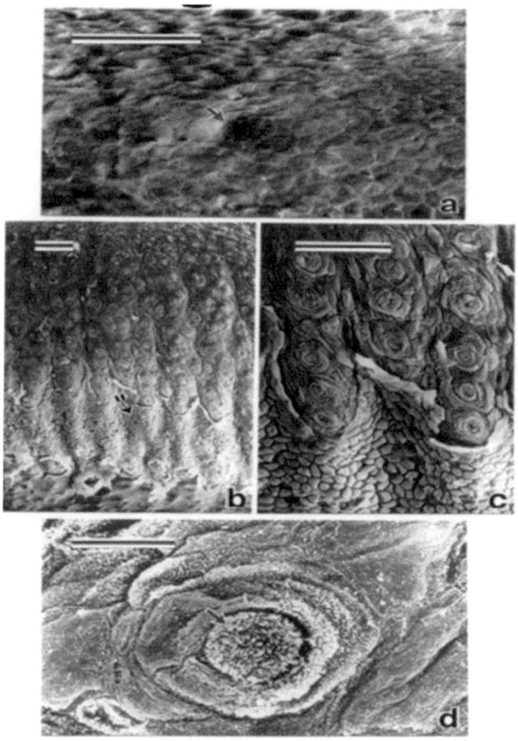

Fig. 22.

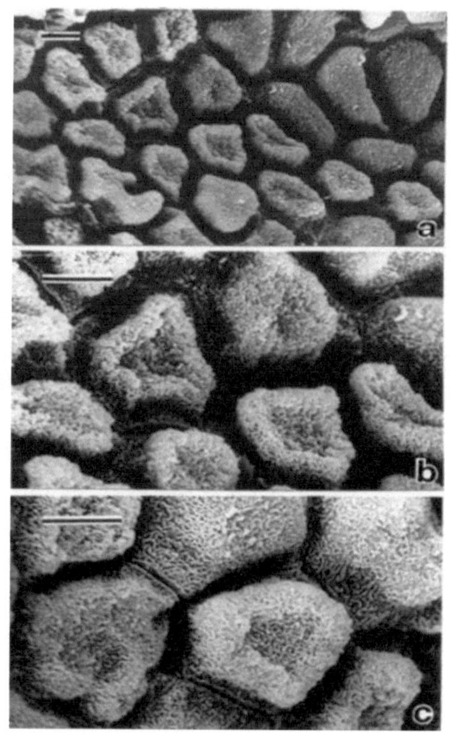

Fig. 23.

9.2 Keratinized epithelia

The epithelial cells at the surface of the HUJS and the HLJS are studded with characteristic polygonal unculi (Fig. 23 a, b, c). The surface relief of each unculus appears truncated. Each unculus is raised significantly from the general surface as a narrow bend at the periphery and has a shallow depression at its major central region (Fig. 23 a, b, c). Further, the surface relief appears scraggy similar to those at the proximal regions of the HUJS and the HLJS of *P. sophore*. In contrast, the micro-ridges at the surface of each unculus are prominent, separated by wide spaces and do not fuse with each other. These micro-ridges are interwoven forming web-like patterns (Fig. 23 b, c). The boundaries of the adjacent epithelial cells, modified as unculi, are demarcated by distinct double rows of micro-ridges (Fig. 23 c). The space between these micro-ridges is relatively prominent and wide.

10. Explanation of figures

Figure 1 (a - b) Photographs of a part of the dorso-lateral side of the head of *G. giuris,* with closed mouth (a) and with open mouth (b) showing the RC (arrowhead), the UL (arrow), the FSUR (white arrowhead) and the LL (winged arrow) (Scale bar = 5 mm) **(c)** Scanning electron photomicrograph of a part of the dorso-lateral side of the head of *G. giuris,* showing the RC (arrowhead), the FSUR (white arrowhead), the UL (arrow), the LL (winged arrow) and the mouth (white star). Several papilliform teeth are also observed inside the mouth. (Scale bar = 400 μm)

Figure 2 (a – c) Scanning electron photomicrographs showing surface architecture of the epithelia of the UL of *G. giuris.* (Scale bar = (a) 30 μm, (b, c) 10 μm). **(a)** Showing the epithelial cells having a variable degree of invagination and small type I taste buds (arrows). **(b)** Same as (a) in higher magnification. **(c)** Same as (b) in higher magnification, showing the surface of the epithelial cells characterised with smooth, extensive, uninterrupted micro-ridges traversing towards the deeper regions of concave depressions and are separated by wide furrows. Note fine micro-bridges interconnecting the adjacent micro ridges (arrows).

Figure 3 (a - b) Scanning electron photomicrographs showing surface architecture of the epithelia of the RC (a), the FSUR (b) of *G. giuris.* (Scale bar = 10 μm). **(a)** Showing the surface of the epithelial cells is characterised with smooth, extensive, uninterrupted sometimes branched micro-ridges separated by wide furrows. Note boundary is demarcated by double row of micro-ridges (arrows) and fine micro-bridges interconnecting the adjacent micro-ridges (winged arrows). Note each epithelial cell shows slight concavity, compare with plate 51 (b, c). **(b)** Showing micro-ridges forming characteristic pattern on the epithelial cell surface and small mucous cell apertures (arrows).

Figure 4 (a - b) Scanning electron photomicrographs showing surface architecture of the epithelia of the RC of *G. giuris.* (Scale bar = 20 μm) **(a)** Showing big type II taste buds located on the mounds of epithelial cells. Note the epithelial cells at each mound are characteristically arranged concentrically**. (b)** Showing superficial neuromast sunk slightly in the epithelium with characteristic cupula like structure.

Figure 5 (a) Photograph of the head region of *N. botia* showing the UL (arrow), the LL (white arrow) and the RC (arrowhead). (Scale bar = 5 mm). **(b - c)** Scanning electron photomicrograph of a part of the head of *N. botia* showing the RC (white arrowhead), the FSUR (barred arrow), the UL (arrow), the HUJS (white arrow), the HLJS (arrowhead), the LL (winged arrow) and mouth (white star) (Scale bar = 400μm).

Figure 6 (a) Scanning electron photomicrograph of a part of the head of *N. botia* showing the RC, the FSUR, the UL, the HUJS, the HLJS and the LL. On the UL and the LL protuberances of various dimensions separated by furrows are discernible. (Scale bar = 200 μm). **(b)** Scanning electron photomicrograph of the epithelium of the UL showing protuberances of variable dimensions delineated by narrow furrows. A part of the HUJS (arrow) and the FSUR (winged arrow) is also discernible. (Scale bar = 200 μm). **(c)** Scanning electron photomicrograph of the epithelium of the FSUR showing characteristic pleats and a mosaic pavement of surface epithelial cells. (Scale bar = 20 μm).

Figure 7 (a, b) Scanning electron photomicrographs of the epithelium of the RC (a) and the FSUR (b) of N. *botia*. (Scale bar = 5 µm). Showing intricate patterns of micro-ridges at the surface of the epithelial cells. Note the presence of mucous cell apertures (arrows).

Figure 8 (a - c) Scanning electron photomicrographs of LL of *N. botia*. (Scale bar = 10 µm). **(a)** Showing distinct protuberances separated by furrows. Each protuberance is studded with a large number of epithelial elevations. **(b)** Same as (a) in higher magnification. Showing taste buds at the apex of each epithelial elevation. **(c)** Same as (b) in still higher magnification. Showing the apex of each epithelial elevation characterised by the presence of numerous closely packed microvilli, which represent the taste hairs originating from taste cells of the taste buds. Note the epithelial cells at the epithelial elevations are concentrically arranged.

Figure 9 (a – d). Scanning electron photomicrographs of HLJS of *N. botia*. (Scale bar = (a) 50 µm, (b, c, d) 5µm). **(a)** Showing the superficial layer epithelial cells at the distal region (arrow) that gradually get transformed into truncated unculi towards the prox imal region on the buccal cavity side of the HLJS (winged arrow). **(b)** Showing a mosaic pavement of irregularly polygonal epithelial cells at the distal region of the HLJS. Note distinct boundaries between the epithelial cells and characteristic rounded bulge in the central region of the epithelial cells. **(c)** Same as (b) in higher magnification, showing double row of micro-ridges demarcating the boundaries of the adjacent cells. Note the micro-ridges on the surface of the epithelial cells are short, sinuous and compactly arranged. **(d)** Proximal region of the HLJS, towards buccal cavity showing the epithelial cells modified into truncated unculi.

Figure 10 (a – b) Photographs of the head region of *C. fasciata* with closed mouth (a) and with open mouth (b) showing the RC (white arrowhead), the FSUR (arrowhead), the UL (arrow) and the LL (winged arrow). (Scale bar = 5 mm). **(c - d)** Scanning electron photomicrograph of a part of head region of *C. fasciata* in dorso-lateral view (c) and in front view (d), showing the RC (white arrowhead), the FSUR (arrowhead), the UL (arrow), the HUJS (barred arrow), the HLJS (white arrow), the LL (winged arrow), the FSLS (cross) and mouth (white star). (Scale bar = 400 µm).

Figure 11 (a – c) Scanning electron photomicrograph showing the surface architecture of the HUJS (a, b) and the HLJS (c) of *C. fasciata*. (Scale bar = 50 µm) **(a)** Note the difference in the surface architecture of the epithelia at the proximal region (arrow) and the distal region (winged arrows) of the HUJS. The epithelial cells in the distal region exfoliate at the surface. Note the presence of papilliform teeth like structures protruding at the surface. **(b)** Same as (a) in higher magnification. **(c)** Note the difference in the surface architecture of the epithelium at the proximal region (arrow) and the distal region (winged arrow) of the HLJS. The epithelial cells in the distal region exfoliate at the surface. Note the presence of papilliform teeth like structures protruding at the surface.

Figure 12 (a – c) Scanning electron photomicrograph showing the surface architecture of the RC (a, b), the UL (c) of *C. fasciata*. (Scale bar = 5 µm). **(a)** Showing mosaic pavement of irregularly polygonal epithelial cells interspersed with mucous cell apertures (arrows). Epithelial cells with a series of micro-ridges and the boundaries demarcated by smooth well-defined un-interrupted double row of micro-ridges. **(b)** Same as (a) in higher magnification. **(c)** The micro-ridges are short, straight or sinuous and smooth often arranged in the form of small groups.

Figure 13 (a - c) Scanning electron photomicrograph showing surface architecture of the FSUR (a), the LL (b, c) of *C. fasciata*. (Scale bar = 10 μm). **(a)** Showing mosaic pavement of irregularly polygonal epithelial cells. **(b)** The micro-ridges are short, straight or sinuous and smooth often arranged in the form of small groups. Note a taste bud at the summit of a papilla like projection (arrow). The epithelial cells are arranged concentrically around the taste bud. **(c)** Showing taste buds at the summit of papilla like projections on the surface of the LL thrown in to papillae like projections protruding beyond the general surface of the epithelia. At the summit of each such papilla several microvilli representing the taste hairs of the taste buds are located.

Figure 14 (a - c) Scanning electron photomicrographs showing surface architecture of the epithelia of the HLJS of *C. fasciata.* (Scale bar = 20 μm) **(a)** Showing a mosaic pavement of irregularly polygonal epithelial cells. A part of the lower lip is also visible (arrow). **(b)** Showing ill-defined micro-ridges on the surface of the epithelial cells at the proximal region of the HLJS. Note the boundaries between the adjacent epithelial cells are delineated either by shallow clefts (arrows) or by double row of micro-ridges (winged arrows). **(c)** Showing the epithelium at the transitional zone between the proximal region and the distal region of the HLJS. Note an epithelial cell separated from the underlying epithelial cell before its exfoliation (arrow). Impressions left on the surface (winged arrows) of the epithelium is due to the exfoliated epithelial cells. The surface of the epithelial cells at the distal region is characterised with punctated micro-ridges.

Figure 15 (a – b) Scanning electron photomicrographs showing surface architecture of the epithelia of the distal region of the HUJS of *C. fasciata.* (Scale bar = 5 μm) **(a)** Micro-ridges at the surface are either punctated or are short and sinuous. Note an exfoliated epithelial cell (arrow). **(b)** Showing prominent and raised micro-ridges in the central region of the epithelial cells (arrows).

Figure 16 (a – b) Photographs of the head region of *P. sophore* with closed mouth (a) and with open mouth (b), showing the upper lip (UL), the lower lip (LL), the rostral cap (RC) ,the fold of skin between UL and RC (FSUR) and the fold of skin between LL and ventral head skin (FSLS). (Scale bar = 5 mm). **(c)** Scanning electron photomicrograph of a part of the lateral side of mouth of *P. sophore* (Scale bar = 400 μm).

Figure 17 Scanning electron photomicrographs showing surface architecture of the epithelia of the UL **(a),** the FSUR **(b)** and the LL **(c)** of *P. sophore.* (Scale bar = 10 μm). **(a)** Showing micro-ridges arranged characteristically (arrows) at the surface of the epithelial cells. Note wide mucous cell apertures (winged arrows). **(b)** Showing characteristic pattern of micro-ridges at the surface epithelial cells. **(c)** Showing a taste bud at the apex of a mound of epithelial cells. Note the epithelial cells surrounding the taste bud are arranged concentrically.

Figure 18 Scanning electron photomicrographs showing surface architecture of the epithelia of the UL **(a),** distal region of HLJS and the LL **(b)** and distal region of the HLJS **(c)** of *P. sophore.* (Scale bar = 10 μm). **(a)** Showing a large number of taste buds (arrows). **(b)** Showing truncated unculi at the HLJS (arrows) and taste buds at the LL (winged arrows). **(c)** Showing truncated unculi with central concavity and raised margins.

Figure 19 Scanning electron photomicrographs showing surface architecture of the epithelia at the distal region of the HLJS **(a)** and at the proximal region of the HLJS **(b, c)** of *P. sophore.* (Scale bar = 5 µm). **(a)** Same as 21 c, in higher magnification, showing the superficial layer epithelial cells modified in to truncated unculi. Note the boundaries of the adjacent epithelial cells are demarcated by distinct micro-ridges (arrows). **(b)** Showing scraggy surface relief of the unculi. **(c)** Same as (b) in higher magnification, showing the scrawly surface of the unculi. The microvillar projections occupying the central part of each unculus give a frothy appearance to the surface. Note, between the unculi the boundaries of the adjacent epithelial cells are demarcated by distinct micro-ridges (arrows).

Figure 20 (a – b) Photographs of the head region of C. carpio showing the UL (arrow), the LL (white arrow), the RC (arrowhead). (Scale bar = 5 mm). **(c – d)** Scanning electron photomicrograph of a part of the head with open mouth showing the UL (arrow), the LL (white arrow), the RC (arrowhead) and the FSUR (winged arrow) and mouth (white star) of *C. carpio.* (Scale bar = 400 µm). **(c)** front view **(d)** lateral view.

Figure 21 (a - c). Scanning electron photomicrographs showing surface architecture of the epithelia of the UL (a, c) and the RC (b) of *C. carpio.* (Scale bar = 5 µm). **(a)** Showing mosaic pavement of irregularly polygonal epithelial cells with micro-ridges forming intricate patterns. Note mucous cell apertures (arrows) at the boundary of the epithelial cells. **(b)** Showing micro-ridges forming intricate patterns at the surface of the epithelial cells. Note fine transverse connections in between the micro-ridges (winged arrows) and mucous cell apertures (arrows). **(c)** Showing micro-ridges forming intricate patterns at the surface of the epithelial cells. Note fine transverse connections in between the micro-ridges (winged arrows) and mucous cell apertures (arrows).

Figure 22 (a - d) Scanning electron photomicrographs showing surface architecture of the epithelia of the RC (a) and the LL (b, c, d) of *C. carpio.* (Scale bar = (a, b, c) 50 µm and (d) 10 µm). **(a)** Showing a mosaic pavement of the epithelial cells and a taste bud at the apex of a mound of epithelial cells (arrow). **(b)** Showing major mucogenic region characterised with the presence of a large number of taste buds arranged in parallel rows. Note a narrow keratinized region, the surface of which appears rough and studded with pebble like structures (winged arrows). **(c)** Same as (b) in higher magnification. **(d)** Showing a taste bud (arrow) at the apex of a mound of epithelial cells.

Figure 23 (a – c) Scanning electron photomicrographs showing surface architecture of the epithelium of HLJS of *C. carpio.* (Scale bar = 5 µm). **(a)** Showing truncated polygonal unculi with concavity in the central region of each unculus. **(b)** Same as (a) in higher magnification. Note prominent micro-ridges at the surface of the unculi, separated by wide spaces. **(c)** Same as (b) in higher magnification. Note the boundaries of the adjacent unculi are demarcated by double row of micro-ridges.

11. Discussion

Research on lips and associated structures began about 200 years ago, as described by Anson, 1929 in his manuscript "The comparative anatomy of the lips and labial villi of vertebrates". He made an attempt to define lips and on Danforth's interpretation of homology, homologous lips are found at certain stages of development in some representatives of all classes of vertebrates. The primary lips characteristic of selachians, after the maxillary and premaxillary bones have developed within the territory of the upper lip (toadfish, cod), may disappear (trout, Spelerpes), accompanied by a forward migration of the lower jaw. The secondary lips of higher forms are first indicated in certain teleosts and amphibians. Lips vary in structure to accord with their physiological functions, whether sensory, prehensile, or adhesive (Anson, 1929). By precise comparative morphology and gene expression analyses, a possibility was inferred that ammocoete lips may not be identical to gnathostome jaws (Kuratani, 2003).

The surface architecture of the superficial layer of epithelial cells in the lips and associated structures is characterised by specialised structures, the micro-ridges forming different patterns in different fish species investigated in this study. These structures in other fishes, have been described as cytoplasmic folds (Merrilees, 1974), microvilli (Harris & Hunt, 1975), microfolds (Hunter & Nayudu, 1978) or ridges (Iger *et al.*, 1988). Insofar as these structures appeared as micro-ridges under SEM and microvilli under TEM. The term "microridges" is used in this study following Whitear & Mittal (1986), Whitear (1990), Suzuki (1992) and Whitear & Moate (1998) and seems appropriate.

The micro-ridges on the surfaces of the mucogenic epithelia form characteristic maze like patterns in different fish species. Fishelson, (1984) correlated the variations in micro-ridge pattern with the locomotory activity of the fish. He suggested that in faster swimming fishes, micro-ridges are most developed and serve to trap mucus on the epithelial surface. The present study, however, is not in support of this since micro-ridges are well developed and conspicuous on the free surface of the lips and associated structures in all the six fish species investigated showing significant difference in their habits and habitats.

The retention of secretion has been the most popular hypothesis of micro-ridge function (Hughes & Wright, 1970; Hughes, 1979; Tillman *et al.*, 1977; Meyer-Rochow, 1981; Fishelson, 1984). Modifications in the pattern of micro-ridges can also be caused by various intrinsic, e.g. hormonal (Schwerdtfeger, 1979 a, b), or extrinsic factors e.g. temperature (Ferri, 1982), salinity (Ferri, 1983), mercury salts (Pereira, 1988), organic pollutants, (Iger *et al.*, 1988), handling and ectoparasites (Whitear, 1990). Some speculations about the function of micro-ridges have centred on mechanical considerations (Lanzing & Higginbotham, 1974; Hawkes, 1974; Sibbling & Uribe, 1985). The provision of reserve apical membrane to allow for distortion was postulated by Zeiske *et al.*, (1976) but Sperry & Wassersug (1976) found no change of pattern after stretching fish oesophageal epithelium and suggested that spread of mucus from goblet (mucous) cells might be guided by the direction of ridges.

Presence of conspicuous micro-ridges on the surfaces of the mucogenic epithelia in the fish species investigated could be considered to reflect high secretory activity of the surface epithelial cells in these regions. Secretion of glycoproteins (GPs), shown histochemically (Pinky et al., 2008, Tripathi & Mittal, 2010), in the surface epithelial cells in the mucogenic regions is in support of this. Further, Whitear (1990) proposed that the form of micro-ridges correlate with the type and rate of secretion at the cell apex. The development of micro-ridges would then be a consequence of arrival of new membranes as vesicles carrying the secretion fuse with the apical plasmalemma and high ridges would indicate a rapid sequence of arrival of secretory vesicles at the surface.

In the epithelia of lips and associated structures of all fish species studied, the adjacent micro-ridges are often interconnected with each other by fine cross connections i.e. micro-bridges. It should be pointed out that such specific structures connecting micro-ridges have also been reported previously in fish epidermis (Whitear, 1990) and have been variously named as interconnections (Reutter *et al.*, 1974), micro-bridges (Karlsson, 1983), cross-bridges (Avella & Ehrenfeld, 1997). Previous workers have not commented on the functional significance of these structural peculiarities. It is possible that these structures may provide mechanical strength to the micro-ridges. However, it is open to other interpretations.

Secretions elaborated by the epithelial cells and the mucous cells in the mucogenic epithelia could be regarded as an adaptation to lubricate and protect the epithelia from abrasion (Pinky et al., 2002). The role of mucus was likewise postulated previously to inhibit the invasion and proliferation of pathogenic micro-organisms and to prevent their colonisation in fish epidermis (Nigrelli, 1937; Nigrelli et al., 1955; Hildemann, 1962; Liguori et al., 1963; Lewis, 1970).

In the oral cavity the lining mucous membrane becomes keratinized to varying degrees in different animals and also in different areas of the mouth (Adams, 1976). Some of the most dramatic advances made over the past 2-3 decades in epidermal research have come about through the utilization of newly developed biochemical investigative techniques, examples of which include the use of gene cloning to study the organization of the keratin gene family, and the use of immuno-fluorescence with monoclonal antibodies to discern when various keratin proteins appear during differentiation. In SEM studies, the Keratinized surfaces of the fishes studied are shaggy and are matted with an organised array of horny projections separated by shallow grooves. The boundaries of the adjacent epithelial cells (the surfaces of which are modified into unculi) are demarcated by well-defined and distinct rows of microridges.

Horny projections from single cells of lips and associated structures have been reported in a wide variety of fishes by various workers (Leydig, 1895; Rauther, 1911; 1928; Hora, 1922; Minzenmay, 1933; Saxena, 1959; Thys, 1961; Kaiser, 1962; Lal *et al.*, 1966; Saxena & Chandy, 1966; Roberts, 1982). Girgis (1952), in a herbivorous bottom feeder *Labeo horie*, observed horny protuberances on lips and two sharp horny cutting edges lying in the upper and lower borders of mouth immediately inside the lips. Mester (1971) made a morpho-histological analysis of the buccopharyngeal cavity in *Noemacheilus barbatulus*, and reported horny plates on the inner surface of the lips that are used by the fish for trituration of nutrients. Verighina (1971) while describing the structure of the digestive tract, reported the presence of horny cutting edge on the lower lip of periphyton eater fish *Chondrostoma nasus variabile*. Agrawal & Mittal (1992 a) reported the presence of keratinized unculi on the ventral side of the upper lip and keratinized cone like structure with sharp cutting edge on the horny lower jaw sheath associated with the lower lip of an omnivorous bottom feeder, *Cirrhina mrigala*. Agrawal & Mittal (1992 b) observed keratinized unculi on the ventral side of the upper lip and on the dorsal side of the lower lip facing the mouth opening. In addition, they observed keratinized cone like structure on the horny upper jaw sheath and on the horny lower jaw sheath associated with the lips of a herbivorous column feeder, *Labeo rohita*.

The present study shows that in *N. botia*, *C. fasciata*, *G. lamta*, *P. sophore* and *C. carpio* the UL and the LL, on the side facing the mouth opening, are associated with the horny upper jaw sheath (HUJS) and the horny lower jaw sheath (HLJS) respectively. The horny jaw sheaths are absent in *G. giuris* having few villiform teeth.

Keratinization occurs in the structures associated with lips of fish *Garra lamta* by histochemical investigations (Pinky et al., 2004). The unculi also observed in the upper jaw of an herbivorous fish, *Cirrhinus mrigala* by SEM studies. More recently, in *Puntius sophore*, the HUJS and the HLJS are keratinized. The surface epithelial cells in these regions are modified to form single cell modification—the unculi, give positive results for keratin with histochemical reactions. The unculi appear functionally significant on the lips and associated structure in this fish species lacking jaw teeth (Tripathi & Mittal, 2010).

In the epithelia of the RC and the AP in *G. lamta* and those of the HUJS and the HLJS in *C. fasciata*, however, both mucogenic and keratinized regions are observed. In *C. fasciata* the apical regions of the horny jaw sheaths are modified into cone like structures each consisting of several superimposed keratinized epithelial cells. The spine like or conical unculi in the keratinized regions of the RC and AP of *G. lamta* may be considered to facilitate clinging or adherence of the fish to the substratum by engaging irregularities on the surface of rocks or stones. Truncated unculi on the surface of HUJS and HLJS in 4 fish species investigated except *G. giuris* and apical cone like structure of the horny jaw sheaths in *C. fasciata* may have a function to assist the fish in scooping mud from the bottom in search of food or to act as sharp cutting edge assisting the fish to browse upon or scrape the food attached with the substratum for feeding.

The keratinized epithelia are mainly composed of the epithelial cells only. The mucous cells and the taste buds are not observed. The absence of the gland cells in the keratinized epithelia suggests an inverse relationship between the degree of keratinization and slime secretion.

In most vertebrates the sense of taste is used as a close range receptor for food item discrimination. Fish are unique among vertebrates in having taste buds widely distributed over various regions. The present study shows that, the taste buds in the epithelia of the lips and structures associated with them are conspicuous in *C. fasciata* and are small and inconspicuous in *G. giuris*. In both these fish species, however, the taste buds are few and could be located at long intervals. This indicates that the cutaneous gustatory function in these fish species is probably of less importance. Low density of taste buds, in *G. giuris* having active predatory habit could be associated with the presence of superficial neuromasts and canal neuromasts in the localisation of its prey. This is supported by reports that the taste buds are absent on the lips of the ox-eye herring *Megalops cyprinoides* (Pasha, 1964 c), are poorly developed on lips of *Channa striata* (Agarwal & Mittal, 1994) and are relatively few on the outer surface of the body than those on the lips and palatal organs and gills in *Pseudorasbora parva* (Kiyohara *et al.*, 1980) all these fishes are active predators.

In the lips and associated structures of *G. lamta, P. sophore, C. carpio* and *N. botia* taste buds, as compared to those in *C. fasciata* and *G. giuris,* are very prominent and are distributed in large numbers. This indicates the development of acute gustatory function, an adaptation to the peculiar mode of life of these fishes. A similar correlation between the distribution of taste buds and gustatory feeding has been shown in two races of the Mexican characin, *Astyanax mexicanus;* one lives in caves and is sightless, and the other lives in river and is visually normal. Schemmel (1967) has shown that the cave-dwelling race has a larger number and a more extensive distribution of external buds than the river-dwelling race, corresponding to acute gustatory feeding behaviour.

Presence of large number of taste buds in *C. carpio* and *N. botia* could be due to its habit to live at the muddy bottom of water bodies. *C. carpio* browses on the shallow bottom and margins, takes in vegetable debris, insects, worms, crustaceans and also planktonic algae. *N. botia* is a bottom dwelling fish, which accepts any kind of food and feeds on algal films as well. At muddy bottoms the visibility, in general, is poor owing to (i) depth and (ii) increase in turbidity caused by the disturbance of the bottom mud due to the movements of the fish, which is in habit of suddenly burrowing into the mud or sand on the bottom to protect itself from predators. The presence of a large number of taste buds increases the probability of detecting and locating accurately prey concealed by darkness or turbidity and may also permit the accurate location of small food particles, which would be missed otherwise.

The taste buds in the lips and associated structures of all the six fish species investigated remain encircled by characteristic concentric whorls of epidermal cells. Harvey & Batty (1998), suggested that it was sometimes possible to locate and count taste buds by the presence of the characteristic ring of epidermal cells surrounding the sensory apex, even when the apex itself was damaged or missing.

Earlier SEM studies reported that fish taste buds fall into three categories based on their external surface morphology (Reutter *et al.*, 1974; Ezeasor, 1982). In addition to the three types of taste buds previously described from various teleost fish, a fourth type comprising very small buds, was found in some cardinal fish (Fishelson, 2004). The taste buds in the mouth cavity of *Rita rita* are of three types which are elevated from the epithelium at different levels, which may be useful for ensuring full utilization of the gustatory ability of the fish, detection and analysing of taste substances, as well as for assessing the quality and palatability of food, during its retention in the mouth cavity (Yashpal et al., 2006). In *C. mrigala*, there is only one type of taste buds observed in mouth cavity (Yashpal et al., 2009).In the present study only one type of taste buds are observed in the 5 fish species whereas in *G. giuris* there are two types of taste buds are observed in different regions of lips and associated epithelia.

In *G. giuris*, in addition to the taste buds, specialised sensory structures the superficial neuromasts are also observed in the epithelium of the RC and in the epithelia of the FSLS close to the ventral head skin. Neuromasts can be found on the entire body surface including the tail with the cupulae extending into the water (Schellart & Wubbels, 1998; Eastman & Lannoo, 2003; Tarby and Webb, 2003; Sane et. al., 2009).

The fishes studied are characterised by the peculiar trophic niche they occupy: many scrape epilithic or epiphytic algae and other food items from submerged substrates. This specialized feeding type is possible thanks to the remarkably formed, ventrally placed suckermouth of *G. lamta* that allows itself to attach to a surface while scraping and eating the food attached to it. In spite of this highly specialized feeding apparatus, diversity in both thickness of the different regions of lips and in shape of lips exists, and these fishes actually feed on a broad range of food. As such, Cyprinidae are the most specialized and successful fish family within the order Cypriniformes. More basal families within the teleosts like Gobiidae, Cobitidae and Belontiidae mostly display a more general (non-specialized) feeding mode, with a typical feeding apparatus suitable for finding and processing insects and other food items that abound in the water column or in the bottom. Some knowledge exists on the trophic, evolutionary trend in the group, but detailed studies dealing with the morphology of lips and associated structures or the feeding apparatus are few and often fragmentary.

12. References

Adams, D. (1976) Keratinization of the oral epithelium. *Annals of the Royal College of Surgeons of England* 58, 351-358.

Agrawal, N., Mittal, A. K. (1991) Epithelium of lips and associated structures of the Indian major carp, *Catla catla. Japan. J. Ichthyol.* 37, 363-373.

Agrawal, N., Mittal, A. K. (1992 a) Structural modifications and histochemistry of the epithelia of lips and associated structures of a carp - *Labeo rohita. Eur. Arch. Biol.* 103, 169-180.

Agrawal, N., Mittal, A. K. (1992 b) Structural organisation and histochemistry of the epithelia of lips and associated structures of a carp - *Cirrhina mrigala. Can. J. Zool.* 70, 71-78.

Agrawal, N., Mittal, A. K. (1992 c) Structure and histochemistry of the epithelia of lips and associated structures of a catfish *Rita rita. Japan. J. Ichthyol.* 39, 93-102.

Alikunhi, K. H. (1957). *Fish culture in India, Farm Bulletin No.* 20. New Delhi: Indian Council of Agricultural Research.

Anson, B. J. (1929) The comparative anatomy of the lips and labial villi of vertebrates. *J Morph. Physiol.* 47, 2, 335-413.

Avella, M. & Ehrenfeld, J. (1997). Fish gill respiratory cells in culture: A new model for Cl^- -secreting epithelia J. *Membrane Biol.* 156, 8-97.

Bloch, M. E. & Schneider, J. G. (1801). *Syst. Ichthyologie.* i-ix, 1-584.

Branson, B. A. & Hake, P. (1972). Observation on an accessory breathing mechanism in *Piaractus nigripinnis* (Cope) (Pisces: Teleostomi: Characidae). *Zool. Anz.* Leipzig. 189, 292-297.

Chitray, B. B. (1965). The anatomy and histology of the alimentary canal of *Puntius sarana* (Ham.) with a note on feeding habits. *Ichthyologica* 4, 53-62.

Eastman, J. T. and Lannoo, M. J. (2003) Diversification of Brain and Sense Organ Morphology in Antarctic Dragonfishes (Perciformes: Notothenioidei: Bathydraconidae). J Morphol 258:130–150.

Ezeasor, D. N. (1982). Distribution and ultrastructure of taste buds in the oropharyngeal cavity of the rainbow trout, *Salmo gairdneri* Richardson. *J. Fish Biol.* 20, 53-68.

Ferri, S. (1982). Temperature induced transformation of teleost (*Pimelodus maculatus*) epidermal cells. *Gegenbaurs Morph. Jahrb.* Leipzig. 128, 712-731.

Ferri, S. (1983). Modification of microridge pattern in teleost (*Pimelodus maculatus*) epidermal cells induced by NaCl. *Gegenbaurs Morph. Jahrb.* Leipzig. 129, 325-329.

Fishelson, L. (1984). A comparative study of ridge-mazes on surface epithelial cell/membranes of fish scales (Pisces, Teleostei). *Zoomorphologie.* 104, 231-238.

Fishelson, L. Delarea, Y. and Zverdling, A. (2004) Taste bud form and distribution on lips and in the oropharyngeal cavity of cardinal fish species (Apogonidae, Teleostei), with remarks on their dentition. J. Morph. 259:316–327.

Girgis, S. (1952). On the anatomy and histology of the alimentary tract of an herbivorous bottom-feeding Cyprinoid fish, *Labeo horie* (Cuvier). *J. Morph.* 90, 317-362.

Graham, J. B. (1997). *Air-breathing Fishes*. Academic Press. California, USA.

Günther, S. (1989). *Fresh water fishes of the world Vol. I & II* (Translated and revised by Tucker, D. W.), New Delhi: Falcon books, Cosmo Publications.

Hamilton, F. B. (1822). *An account of fishes found in the river Ganges and its branches*. Edinburg and London, pp. VIII+405 39 pls.

Harris, J. E. & Hunt, S. (1975). The fine structure of the epidermis of two species of salmonid fish, the Atlantic salmon (*Salmo salar* L.) and the brown trout (*Salmo trutta* L.). I. General organisation and filament containing cells. *Cell Tissue Res.* 157, -553-565.

Harvey, R. & Batty, R. S. (1998). Cutaneous taste buds in cod. *J. Fish Biol.* 53, 138-149.

Hawkes, J. W. (1974). The structure of fish skin. I. General organisation. *Cell Tissue Res.* 149, 147-158.

Hildemann, W. H. (1962). Immunogenetic studies of poikilothermic animals. *Am. Nat.* 96, 195-204.

Hora S. L. & Mukerji, D. D. (1953). Table for the identification of Indian fresh water fishes,with description of certain families and observation on the relative utility of the probable larvivorous fishes of India. (Revised by T.J. Job) *Health bulletin* No.12. *Malaria bureau*, No. 4. Simla, Govt of India Press.

Hora, S. L. & Pillay, T. V. R. (1962). *Handbook of fish culture in the Indo-Pacific region*. FAO Fish. Biol. Tech. Pap. No. 14. Fisheries Division, Biology Branch, Food and Agriculture Organisation of the United Nations, Rome.

Hora, S. L. (1922). Structural modifications in the fish of mountain torrents. *Rec. Indian Mus.* 24, 31-61.

Hughes, G. M. (1979). Scanning electron microscopy of the respiratory surfaces of trout gills. *J. Zool. Lond.* 188, 443-453.

Hughes, G. M. & Wright, D. E. (1970). A comparative study of the ultrastructure of the water/blood pathway in the secondary lamellae of teleost and elasmobranch fishes - benthic forms. *Z. Zellforsch. mikrosk. Anat.* 104, 478-493.

Hunter, C. R. & Nayudu, P. L. (1978). Surface folds in superficial epidermal cells of three species of teleost fish. *J. Fish Biol.* 12, 163-166.

Iger, Y., Abraham, M., Dotan, A., Fattal, B. & Rahamim, E. (1988). Cellular responses in the skin of carp maintained in organically fertilised water. *J. Fish Biol.* 33, 711-720.

Kaiser, P. (1962). Hornzahnchen als Lippenbewaffnug bei Jungfischen von Cypriniden. *Zool Anz.*, Leipzig 169, 158-161.

Kapoor, B. G. (1958). The anatomy and histology of the alimentary tract of a plankton-feeder, *Gadusia chapra* (Ham.). *Ann. Mus. Stor. nat.* Geneva. 70, 8-32.

Kapoor, B. G., Smit, H & Verighina, I. A. (1975). The alimentary canal and digestion in teleosts. *Adv. Mar. Biol.* 13, 109-239.

Karlsson, L. (1983). Gill morphology in the zebra fish, Brachydanio *rerio* (Hamilton-Buchanan). *J. Fish Biol.* 23, 511-524.

Khanna, S. S. (1961). Alimentary canal in some teleostean fishes. *J. Zool. Soc. India.* 13, 206-219.

Khanna, S. S. (1962). A study of bucco-pharyngeal region in some fishes. *Ind. J. Zool.* 3, 21-48.

Khanna, S. S. (1993). *An introduction to fishes*, Allahabad: Central Book Depot.

Kiyohara, S., Yamashita, S. & Kitoh, J. (1980). Distribution of taste buds on the lips and inside the mouth in the minnow, *Pseudorasbora parva*. *Physiol. Behav.* 24, 1143-1147.

Kuratani, S. (2003) Evolution of the vertebrate jaw: homology and developmental constraints. *Paleontological research* 7, 1, 89-102.

Lal, M. B. (1968). Studies on the anatomy and histology of the alimentary canal of a carp, *Tor putitora* (Ham). *Proc. Nat. Acad. Sci.* India. 38B, 127-136.

Lal, M. B., Bhatnagar, A. N. & Kailc, R. K. (1964). Studies on the morphology and histology of the digestive tract and associated structures of *Chagunius chagujnio* (Ham). *Proc. Nat. Acad. Sci. India.* 34B, 160-172.

Lal, M. B., Bhatnagar, A. N. & Uniyal, J. P. (1966). Adhesive modifications of a hill stream fish *Glyptothorax pectinopterus* (McClelland). *Proc. natl. Acad. Sci.* India (B) 36, 109-116.

Lanzing, W. J. R. & Higginbotham, D. R. (1974). Scanning microscopy of surface structures of *Tilapia mossambica* (Peters) scales. *J. Fish Biol.* 6, 307-310.

Lewis, R. W. (1970). Fish cutaneous mucus: a new source of skin surface lipids. *Lipids* 5, 947-949.

Leydig, F. (1895). Integument und Hautsinnesorgane der Knochenfishc. *Zool. Jb. Anat.* 8, 1-152.

Liguori, V. R., Ruggieri, G. D., Baslow, S. J. M. H., Stempien, M. F. & Nigrelli, R. F. (1963). Antibiotic and toxic activity of the mucus of the pacific golden striped bass *Grammistes sexlineatus*. *Am. Zool.* 3, 546.

Linnaeus, C. (1758). *Systema Naturae.* 10th edn. Vol. 1. Regnum Animale. Stockholm: Salvius. (Facsimile reprint (1956). London: British Museum (Natural History).)

Merrilees, M. J. (1974). Epidermal fine structure of the teleost *Esox americanus* (Esocidae, Salmoniformes). *J. Ultrastruct. Res.* 47, 272-283.

Mester, L. (1971). Studiul cavitatii buco-faringiene, La *Noemacheilus barbatulus* L. (Pisces, Cobitidae). *St. si. cerc. Biol. seria Zoologie Bucuresti*. 23, 439-444.

Meyer-Rochow, V. B. (1981). Fish tongues - surface fine structures and ecological considerations. *Zoo. J. Linn. Soc.* 71, 413-426.

Miller, R. J. & Evans, H. E. (1965). External morphology of the brain and lips in Catostmid fishes. *Copeia.* 4, 467-487.

Minzenmay, A. (1933). Die Mundregion der Cypriniden. *Zool. Jb. Anat.* 57, 191-286.

Mittal, A. K. & Agrawal, N. (1994). Modifications in the epithelia of lips and associated structures of the predatory murrel (*Channa striata*). *J. Appl. Ichthyol.* 10, 114-122.

Mittal, A. K. & Whitear, M. (1978). A note on cold anaesthesia of poikilotherms. *J. Fish Biol.* 13, 519-520.

Moitra, S. K. & Bhowmik, M. L. (1967). Functional histology of the alimentary canal of the young *Catla catla* (Ham.), an omnivorous surface-feeding fish of Indian fresh-waters. *Vestnik Cs. spol. Zool.* 31, 41-50.

Moitra, S. K. & Sinha, G. M. (1971). Studies on the morphohistology of the alimentary canal of a carp, *Chagunius chagunio* (Ham.) with reference to the nature of taste buds and mucous cells. *Inland Fish.Soc India.* 3, 44-56.

Nigrelli, R. F. (1937). Further studies on the susceptibility and acquired immunity of marine fishes to *Epibdella melleni*, a monogenetic trematode. *Zoologica, N.Y.* 22, 185-192.

Nigrelli, R. F., Jakowska, S. & Padnos, M. (1955). Pathogenicity of epibionts in fishes. *J. Protozool.* 2 (suppl.) 7.

Nikolsky, G.V. (1963). *The Ecology of Fishes.* (Translated by L. Birkett). Academic Press, London and New York.

Ojha, J. & Singh, S. K. (1992). Functional morphology of the anchorage system and food scrapers of a hill stream fish, *Garra lamta* (Ham.) (Cyprinidae, Cypriniformes). *J. Fish Biol.* 41, 159-161.

Ono, D. R. (1980). Fine structure and distribution of epidermal projections associated with taste buds on the oral papillae in some Loricariid catfishes (Siluroidei: Loricariidae). *J. Morph.* 164, 139-159.

Pasha, S. M. K. (1964 a). The anatomy and histology of the alimentary canal of an omnivorous fish, *Mystus gulio*. *Proc. Ind. Acad. sci.* 59B, 211-221.

Pasha, S. M. K. (1964 b). Anatomy and histology of the alimentary canal of a herbivorous fish, Tilapia *mosambica*. *Proc. Ind. Acad. Sci.* 59B, 340-349.

Pasha, S. M. K. (1964 c). The anatomy and histology of the alimentary canal of a carnivorous fish, *Megalops cyprinoides*. *Proc. Ind. Acad. Sci.* 60B, 107-115.

Pereira, J. J. (1988). Morphological effects of mercury exposure on windowpane flounder gills as observed by scanning electron microscopy. *J. Fish Biol.* 33, 571-580.

Pinky, Mittal, S., Ojha, J. Mittal, A. K., 2002. Scanning electron microscopic study of the structures associated with lips of an Indian hill stream fish *Garra lamta* (Cyrinidae, Cyriniformes) *European Journal of Morphology* 40, 161-169.

Pinky, Mittal S, Yashpal M, Ojha J, Mittal AK. 2004. Occurrence of keratinization in the structures associated with lips of a hill stream fish *Garra lamta* (Hamilton) (Cyprinidae. Cypriniformes). *J Fish Biol.* 65, 1165–1172.

Pinky, Mittal S, Mittal AK. 2008. Glycoproteins in the epithelium of lips and associated structures of a hill stream fish *Garra lamta* (Cyprinidae. Cypriniformes): A histochemical investigation. *Anat Histol Embryol* 37,101–113.

Rauther, M. (1911). Beiträge zur Kenntnis der Panzerwelse. *Zool. Jb. Anat.* 31, 497-528.

Rauther, M. (1928). Der Saugmund von *Discognathus. Zool. Jb. Anat.* 45, 45-76.

Reutter, K., Breipohl, W. & Bijvank, G. J. (1974). Taste bud types in fishes. II Scanning electron microscopical investigations on *Xiphophorus helleri* Heckel (Poeciliidae, Cyprinodontiformes, Teleostei). *Cell Tissue Res.* 153, 151-165.

Roberts, T. R. (1982). Unculi (Horny projections arising from single cells), an adaptive feature of the epidermis of Ostariophysan fishes. *Zoologica Scripta.* 11, 55-76.

Sane, S. P. and McHenry, M. J. (2009) The biomechanics of sensory organs. *Integrative and Comparative Biology* 1–16.

Saxena, D. B. & Bakshi, P. L. (1964). Functional anatomy of the alimentary canal of a torrential stream fish *Botia birdi* (Choudhari). *Kashmir sci.* 1, 76-86.

Saxena, S. C. & Chandy, M. (1966). Adhesive apparatus in certain Indian hill stream fishes. *J. Zool.* 148, 315-340.

Saxena, S. C. (1959). Adhesive apparatus of a hill stream cyprinid fish *Garra mullya* (Sykes). *Proc. Natn. Inst. Sci. India.* 25, 205-214.

Schellart, N. A. M. & Wubbels, R. J. (1998). The auditory and mechanosensory lateral line system. In *The physiology of fishes.* (Evans, D. H. ed.) pp. 283-312. New York: CRC Press.

Schemmel, C. (1967). Vergleichende Untersuchungen an den Hautsinnesorganen ober-und unterirdisch lebeder Astyanax-Formen. *Z. Morph. Okol. Tiere* 61, 253-316.

Schwerdtfeger, W. K. (1979 a). Morphometrical studies of the ultrastructure of the epidermis of the guppy, *Poecilia reticulata* Peters, following adaptation to sea-water and treatment with prolactin. *Gen.Comp. Endocrinol.* 38, 476-483.

Schwerdtfeger, W. K. (1979 b). Qualitative and quantitative data on the fine structure of the guppy (*Poecilia reticulata* Peters) epidermis following treatment with thyroxine and testosterone. *Gen. comp. Endocr.* 38, 484-490.

Sehgal, P. (1966). Anatomy and histology of the alimentary canal of *Labeo calbasu* (Ham). *Res. Bull. Punjab Univ. Sci.* 17, 257-266.

Sehgal, P. & Salaria J. (1970). Functional anatomy of histology of the digestive organs a *Cirrhina mrigala* (Cuvie and Val.) *Proc. nat. Acad. Sci. India.* 40B, 212-222.

Sibbing, F. A. & Uribe, R. (1985). Regional specialisations in the oropharyngeal wall and food processing in the carp (*Cyprinus carpio* L.). *Neth. J. Zool.* 35, 377-422.

Sinha, G. M. (1975). On the origin development and probable function of taste buds in the lip and bucco-pharyngeal epithelia of an Indian freshwater major carp, *Cirrhinus mrigala* (Hamilton) in relation to food and feeding habits. *Z. mikrosk -anat Forsch,* Leipzig. 82, 294-304.

Sinha, G. M. & Moitra, S. K. (1975). Functional morpho-histology of the alimentary canal of an Indian fresh water major carp *Labeo rohita* (Hamilton) during its different life history stages. *Anat. Anz.* 138, 222-239.

Sinha, G. M. & Moitra, S. K. (1976). Studies on the morpho-histology of the alimentary canal of fresh water fishes of India. Part I. The alimentary canal of young *Cirrhinus reba* (Ham.) with a comparison with that of the adult in relation to food. *Vest. Cs. Spol. Zool.* 40, 221-231.

Sinha, G. M. & Moitra, S. K. (1978). Studies on the comparative histology of the taste buds in the alimentary tract of a herbivorous fish, *Labeo calbasu* (Ham.) and a carnivorous fish, *Clarius batrachus* (Linn.) in relation to food and feeding habits. *Zool. Beitr.* 24, 43-57.

Sperry, D. G. & Wassersug, R. J. (1976). A proposed function for microridges on epithelial cells. *Anat. Rec.* 185, 253-258.

Suzuki, N. (1992). Fine structure of the epidermis of the mudskipper, *Periophthalmus modestus* (Gobiidae). *Japan. J. Ichthyol.* 38, 379-396.

Suzuki, Y. (1956). A histological study of the granular processes on the lips of scythe fish *Pseudogobio esocinus* (T. et S.) *Jap. J. Ichthyol.* 5, 12-14.

Tarby ML, Webb JF. 2003. Development of the supraorbital and mandibular lateral line canals in the cichlid, *Archocentrus nigrofasciatus*. J Morphol 255, 44–57.

Thys (van den Audenaerde), D. F. E. (1961). L' anatomie de phractolaemus ansorgei Blgr et la position Systematique des phractolaemidae. *Anuls. Mus. r. Afr. Cent. Ser. 8vo (Zool).* 103, 99-167.

Tillmann, B., Pietzsch-Rohrschneider, I. & Huenges H. L. (1977). The human vocal cord surface. *Cell Tissue Res.* 185, 279-283.

Tripathi, P. and Mittal A. K. (2010) Essence of Keratin in Lips and Associated Structures of a Freshwater Fish *Puntius sophore* in Relation to its Feeding Ecology: Histochemistry and Scanning Electron Microscope Investigation. *Tissue and Cell.* 42, 223-233.

Vanajakshi, T. P. (1938) Histology of the digestive tract of *Sacchobranchus fossilis* and *Macrones vittatus*. *Proc. Indian Acad. Sci.* 7 (B), 61-79.

Verighina, I. A. (1971). The structure of the digestive tract of the Volga under mouth *Chondrostoma nasus* variable Jak. *Voprosy Ikhtiologi.* 11, 311-318.

Welcomme, R.L. (1988). *International introductions of inland aquatic species.* FAO fisheries technical paper 294, pp 1-318. Food and Agriculture Organisation of the United Nations, Rome.

Whitear, M. (1990). Causative aspects of microridges on the surface of fish epithelia. *J. Submicrosc. Cytol. Pathol.* 22, 211-220.

Whitear, M. (1986). Epidermis. *In Biology of the Integument.* Vol. 2, *Vertebrates* (Bereiter-Hahn, J. Matoltsy, A. G. & Richards, K. S., eds.), pp. 8-38. Berlin: Springer-Verlag.

Whitear, M. & Moate, R. (1998). Cellular diversity in the epidermis of *Raja clavata* (Chondrichthyes). *J. Zool., Lond.* 246, 275-285.

Whitear, M. & Moate, R. M. (1994). Microanatomy of taste buds in the dogfish, *Scyliorbinus canicula. J. Submicrosc. Cytol. Pathol.* 26, 357-367.

Yang, L. and Mayden, R. L. (2010) Phylogenetic relationships, subdivision, and biogeography of the cyprinid tribe Labeonini (sensu Rainboth, 1991) (Teleostei: Cypriniformes), with comments on the implications of lips and associated structures in the labeonin classification. *Molecular Phylogenetics and Evolution* 54, 254–265.

Yashpal, M., Kumari, U., Mittal, S., Mittal, A.K. (2006) Surface architecture of the mouth cavity of a carnivorous fish *Rita rita* (Hamilton, 1822) (Siluriformes, Bagridae). *Belg. J. Zool.* 136 (2), 155-161.

Yashpal, M., Kumari, U., Mittal, S., Mittal, A.K. (2009) Morphological specialization of the buccal cavity in relation to the food and feeding habit of a carp *Cirrhinus mrigala*: A scanning electron microscopic investigation. *J. Morphol.* 270, 714 - 728.

Zeiske, E., Melinkat, R., Breucker, H. & Kux J. (1976). Ultrastructural studies on the epithelia of the olfactory organ of Cyprinodonts (Teleostei, Cyprinodontoide). *Cell Tissue Res.* 172, 245-267.

10

Effects of Er:YAG Laser Irradiation on Dental Hard Tissues and All-Ceramic Materials: SEM Evaluation

Bülent Gökçe
Ege University, School of Dentistry, Department of Prosthodontics, Turkey

1. Introduction

A reliable bond to dental hard tissues and materials has always been one of the most significant contributions for restorative dentistry (Leinfelder, 2001). A durable and stable bond between resins and dental hard tissues and restorative materials which has to integrate all parts of the system into one coherent structure is fundamental for the long-term retention and clinical success of the restorations. However, micromechanical attachment is one of the key mechanisms for a reliable adhesion to dental hard tissues and restorative materials (Matinlinna & Vallitu, 2007; Van Noort, 2002b; Fabienelli, et al., 2010). Advances in adhesive dentistry have resulted in the recent introduction of modern surface conditioning methods in order to achieve high bond strengths through increased surface roughness of both dental hard tissues and the restorative materials (Matinlinna & Vallitu, 2007; Van Noort, 2002b).

The use lasers in dentistry has evolved since their development in 1962. Researches have been carried out on effects of lasers on dental hard tissues and materials and applications of different wavelengths as they become available (Roberts-Harry, 1992; Convissar & Goldstein, 2001; White, et al., 1993; Frentzen, et al., 1992; Arima & Matsumoto, 1993; Wilder-Smith, et al., 1997; Cernavin, 1995; Keller & Hibst, 1989; Burkes, et al., 1992; Wigdor, et al., 1993; Visuri, et al., 1996b). According to current literature there is no optimum wavelength for all dental applications. Each wavelength has distinct treatment advantages and offers various treatment options. Understanding the differences between laser wavelengths will help to choose the adequate wavelength for each application in the dental office (Kutsch, 1993).

Laser light has properties such as being coherent, monochromatic and collimated. Laser light travels in specific wavelengths in a predictable pattern (coherent) and parallel (collimated) and it has one color (monochromatic). Lasers and target tissues interact in four ways. When a laser light hits the target it can be reflected, absorbed, scattered throughout the target or transmitted into the target (Kutsch, 1993). During laser application light energy is converted into heat and energy absorption on the target surface causes the vaporization. This process is called ablation or photoablation by vaporization (Cardoso, et al., 2008; Esteves-Oliveira, et al., 2007; Tachibana, et al., 2008; Lee, et al., 2007). Among currently available lasers, the erbium:yttrium-aluminum-garnet (Er:YAG) and Erbium,Chromium:Yttrium-Scandium-Gallium-Garnet (Er,Cr:YSGG) lasers have been proposed for different dental applications, including carious dentin removal, cavity preparation, surface conditioning, and as a surface treatment method for indirect restorations (Trajtenberg, et al., 2004; Atsu, et al., 2006; Bottino, et al., 2005; Gökçe, et al., 2007; Harashima, et al., 2005).

2. Morphological analysis of Er:YAG laser treated enamel and dentin

Etching of enamel with phosphoric acid was first recommended by Buonocore in 1955. (Buonocore, 1955). Resin bonding to tooth ensured by acid etching of enamel and/or dentin with total etch or self-etching techniques and followed by the use of a dentin adhesive (Fusayama, et al., 1979). Phosphoric acid removes the matrix phase of enamel and increases the surface area as well as creating high-energy hydrophilic surface with honey-comb-like structure (Sharpe, 1967; Reynold, 1975). Acid etching results in dissolution of the hydroxyapatite and enhances the penetration of adhesive monomers (Van Meerbeek, et al., 2003) forming resin tags in situ after polymerization (Barkmeier & Cooley, 1992; Leinfelder, 2001).

Conversely bonding to dentin is more complex due to its hydrated biological structure. To obtain intimate association of adhesive and dentin is hard when dentin is conditioned with total etch technique (Marshall, et al., 1997; Pashley, 1992). Etching dentin results in smear-free surface, open dentinal tubules with widened orifices due to removal of peritubular dentin, increased permeability by the loss mineralized dentin within the collagen matrix and exposed collagen web (Marshall, et al., 1997; Pashley, 1992; Pashley & Carvalho, 1997; Schein, et al., 2003).

Micromechanical retention is still the key factor for bonding to dentin. Monomers containing hydrophilic radicals infiltrate through the collagen fibrils and polymerized to develop the micromechanical retention. Many efforts have been spent to promote this dentin-resin interdiffusion zone, hybrid layer, since its description in 1982 (Nakayabashi, et al., 1982).

Air abrasion has also been introduced for enamel pretreatment by Olsen et al., in 1940. It was used for cavity preparation (Olsen, et al., 1997a). In this method, alumina particles were applied under air pressure to roughen the enamel surface (Zachrisson & Buyukyılmaz, 1993).

Etching dental hard tissues with laser has recently been proposed and may enable strong bonds with the restorative materials. Pulsed Nd:YAG lasers are sometimes used to etch enamel in preparation for bonding of restorative materials but some studies suggest that Nd:YAG etching alone results weaker bonds compared with acid etching (Roberts-Harry, 1992). It was suggested that to use the Nd:YAG laser efficiently for surface roughening a topical absorber must be applied to enamel surfaces and low pulse energies (100 mj or less) should be used (Roberts-Harry, 1992). SEM evaluation of the surface of Nd:YAG laser treated dentin was partially obliterated due to resolidification of molten dentin with grooves, fissures and concavities but without smear layer (Ariyaratnam, et al., 1999). They also stated that lased dentin surfaces produced a rougher surface compared to untreated dentin. This difference was suggested to maintain the micromechanical interlocking with the dentin adhesive. It was concluded that although laser irradiation with Nd:YAG laser produced a favorable surface for bonding, the bond strength to dentin did not differ from the conventionally treated dentin.

Both enamel and carious dentine were suggested to be removed with Nd:YAG and excimer lasers without signs of thermal damage (White, et al., 1993; Frentzen, et al., 1992; Arima & Matsumoto, 1993; Wilder-Smith, et al., 1997). When compared with Nd:Yag laser, Ho:YAG laser was shown to remove dental hard tissues more effectively with less cracks (Cernavin, 1995).

Some investigations suggest that CO_2 laser etching results in bonds of comparable strength on enamel and higher bond on dentin surfaces, compared to acid etching (Cooper, et al., 1988; Liberman, et al., 1984). Therefore CO_2 lasers can be recommended for enamel etching prior to composite restorations and fissure sealants without need of an absorber (Walsh, 1994). However excessive heat generated by some lasers may cause pulpal damage (Akova, et al., 2005). Adequate laser parameters can supply limited pulpal temperature increases within safety limits (Obata, et al., 1999). Controversially CO_2 laser at high fluencies and in continuous wave mode may cause cracking, flaking, crater formation, charring, melting and recrystallization of dental hard tissues (Stern, et al., 1972; Boehm, et al., 1997; McCormack, et al., 1995; Malmström, et al., 2001).

Other pulsed lasers whose wavelengths are strongly absorbed by dental hard tissues and hydroxyapatite, e.g. erbium lasers (Er:YAG and Er,Cr:YSGG), can successfully be used for dental hard tissue procedures including conditioning or etching without any side effects. Again no absorber is required (Liberman, et al., 1984; Keller & Hibst, 1989; Burkes, et al., 1992; Wigdor, et al., 1993; Visuri, et al., 1996b).

The water and the hydrated components of dental hard tissues absorb the high energy of erbium lasers and evaporate with micro explosions resulting in particle removal (ablation). (Cardoso, et al., 2008; Esteves–Oliveira, et al., 2007; Tachibana, et al., 2008; Lee, et al., 2007). This thermomechanical effect of erbium lasers on dental hard tissues can vary according to the tissue composition and mainly the water concentration. The mechanism of ablation of dental hard tissues with erbium lasers is still unclear but it was proposed that it takes place by the expansion of subsurface water resulting in microexplosions. This microexplosion induce strong mechanical separation of the calcified tissue (Kayano, et al., 1989). This constitutes the major principle of erbium laser ablation and produce non-uniform tissue removal with ejection of both organic and inorganic tissue microparticles, creating the micro-crater like appearance typical of lased surfaces (Corona, et al., 2007)

Erbium lasers have a shallow thermal penetration depth and can ablate sound and carious enamel and dentine (Keller & Hibst, 1989; Burkes, et al., 1992; Wigdor, et al., 1993; Visuri, et al., 1996b). Besides rough and irregular surface with sharp edged craters without color changes indicative of thermal damage (burning or carbonization) of surrounding tissues and/or the pulp have been reported. Concave and convex surfaces caused by microablation have been observed (Harashima, et al., 2005; Oelgiesser, et al., 2003). Er:YAG laser with appropriate parameters proposed to can selectively remove enamel hydroxyapatite crystals resulting in irregular surface that would enhance the micromechanical retention (Hibst & Keller, 1989; Hossain, et al., 1999).

Sasaki, et al., (2008) made a structural analysis of acid and Er:YAG laser etched enamel. They stated that acid etching exhibited a more homogenous etching pattern whereas Er:YAG alone showed areas of ablation. Er:YAG laser irradiation followed by acid etching resulted in more homogenous surface pattern than the only lased surfaces.

Harashima, et al., (2005) reported that cavities prepared by Er:YAG laser showed characteristic rough surface similar to an acid etched surface with open dentinal tubules and stripped surfaces. They also stated very clean surfaces, almost free of debris when the laser tip was aligned perpendicular to the surface. Scratched appearance with interspersed open dentinal tubules at areas covered by melted surfaces was found with angulated laser application (Harashima, et al., 2005). Unlike acid etching it was shown that the collagen fibrils were not found forming a porous network responsible for the increased porosity of dentin surface and subsurface. The morphological analysis of resin-dentin interface of acid etched dentin revealed triangular hybridization with resin tags in different lengths at the transition between peri- and intertubular dentin. But little or no hybridization zones with fewer and thinner tags at the intertubular dentin areas could be observed due to scarcity and discontinuity of the interdiffusion area at the resin-dentin interface (Schein, et al., 2003).

Literature review also states crater formations, mineral meltdowns and enamel melting, cracks, fissuring in enamel and smooth edged voids (Frentzen & Koort, 1992; Olsen et al., 1997b). Parameter factors and wavelength specificity relate to the degree of change that can be induced to enamel. Varying pulse width, pulse mode and spot size can produce significant changes in enamel and dentin surface morphology (Frentzen & Koort, 1992).

Erbium lasers also denatures the organic content and reduces the solubility of hydroxyapatite (Keller & Hibst, 1989; Hibst & Keller, 1989; Bader & Krejci, 2006). The interaction of erbium lasers with dental hard tissues results in negatively effected bond between the composite resins and dentin and collagen fibrils (Moretto, et al., 2010; Ceballo, et al., 2002; Ramos, et al., 2010, Oliveira, et al., 2010). Carvalho, et al., (2011) suggested that removal of laser irradiated dentin with phosphoric acid gel and sodium hypochlorite had increased the bond strength to dentin.

In a recent study phosphoric acid etching of enamel was compared with Er:YAG laser and Er:YAG laser+acid etching, and it was concluded that Er:YAG laser+acid group exhibited the highest bond strength, followed by acid and laser groups. The lower bond strength with only laser group was attributed to the non-homogenous laser application leaving untouched areas on the surface. Laser application followed by acid etching effectively conditioned the non-lased spots remained within the irradiated area (Sasaki, et al., 2008).

On the other hand some authors reported that the microretentive pattern resulting from laser irradiation could be favorable to bonding procedures (Hossain, et al., 2001; Li, et al., 1992; Visuri, et al., 1996a). Some studies suggest that laser irradiated dentinal tissue resulted in lower bond strength than does non-irradiated dentin. Visuri, et al. (1996a) reported a significantly higher shear bond strength of composite to dentin prepared with an Er:YAG laser. In contrast, Sakakibara, et al. (1998), Ceballo, et al. (2002) and Dunn, et al. (2005) reported a decrease in bond strength to laser-irradiated dentin, and Armengol, et al. (1999) and Kataumi, et al. (1998) found no difference between laser- irradiated and non-irradiated specimens.

Treating dentin erbium lasers (Er:YAG and Er,Cr:YSGG) creates a rough, smear layer-free surface with open dentinal tubules. SEM observations of Carvalho, et al., (2011) revealed irregular and rugged dentinal surfaces, following Er,Cr:YSGG laser. Harashima, et al., (2005) observed smaller width and stripped surfaces on the cavities prepared by Er:YAG laser. They may also cause fissures and cracks that can be considered as drawbacks of using erbium lasers for surface pretreatment (Aoki, et al., 1998; Hossain, et al., 1999; De Munck, et al., 2002; De Oliveira, et al., 2007; Moretto, et al., 2010). Increase in acid resistance of dental hard tissues after laser irradiation was also been reported by some authors (Fried, et al., 1996; Hossain, et al., 2000; Apel, et al., 2002; Liu, et al., 2006).

SEM evaluation of Er:YAG laser treated enamel and dentin revealed different surface morphologies in accordance with literature reviewed depending on the laser parameters.

2.1 Morphological analysis of Er:YAG laser treated enamel

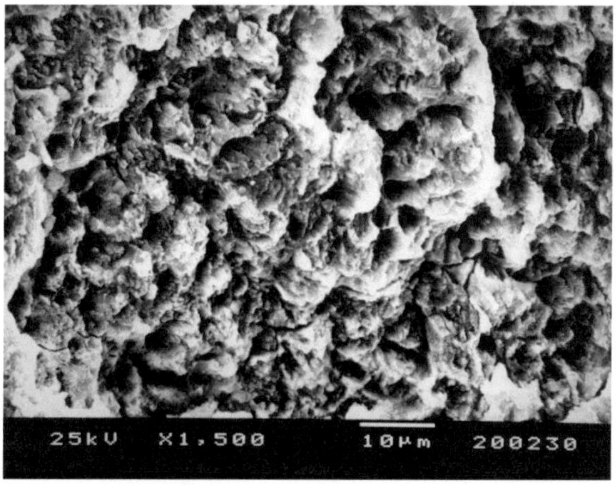

Fig. 1. Enamel. 100 mj. 10 Hz. With water cooling. Honey-comb appearance can be seen but not throughout the surface which is due to non-homogenous application of the laser.

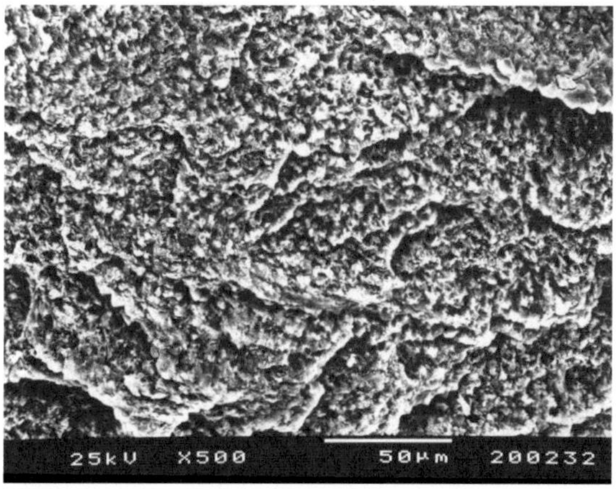

Fig. 2. Enamel. 100 mj. 10 Hz. With water cooling. Honey-comb appearance can be seen on the surface similar to acid etching.

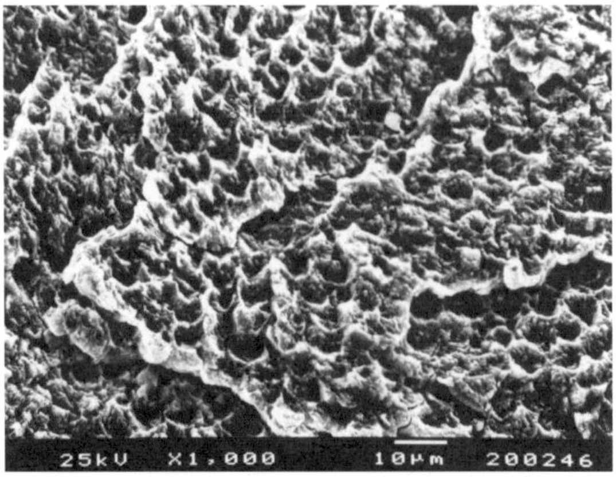

Fig. 3. Enamel. 100 mj. 10 Hz. With water cooling. Higher magnification of the surface in Fig. 2. No signs of thermal damage. Honey-comb appearance.

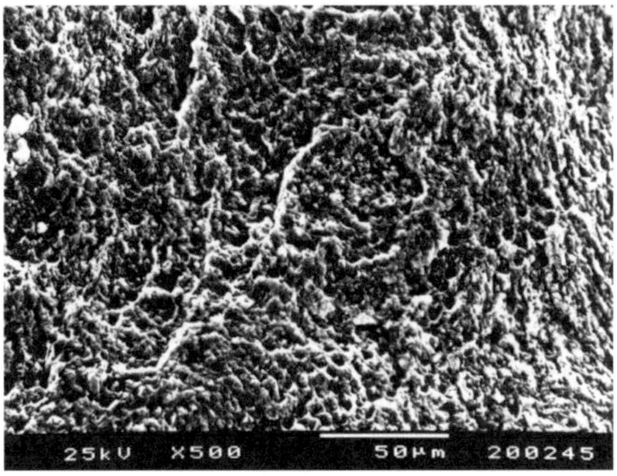

Fig. 4. Enamel. 250 mj. 10 Hz. With water cooling. Serrated surface with honey-comb appearance.

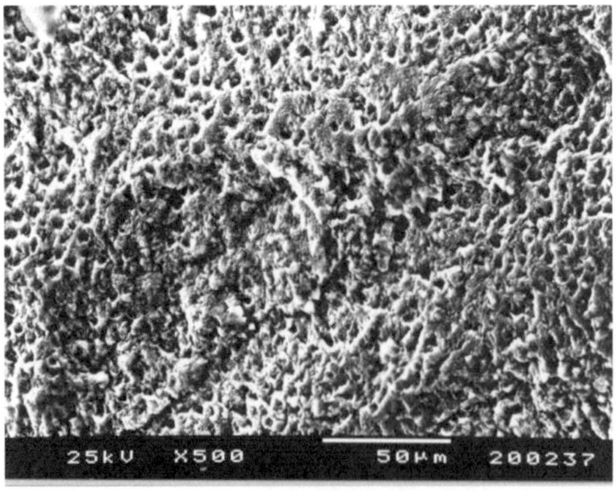

Fig. 5. Enamel. 500 mj. 10 Hz. With water cooling. Interprismatic matrix has been removed. Similar to acid etching but some melting points probably due to repeated shots at the same point can be observed.

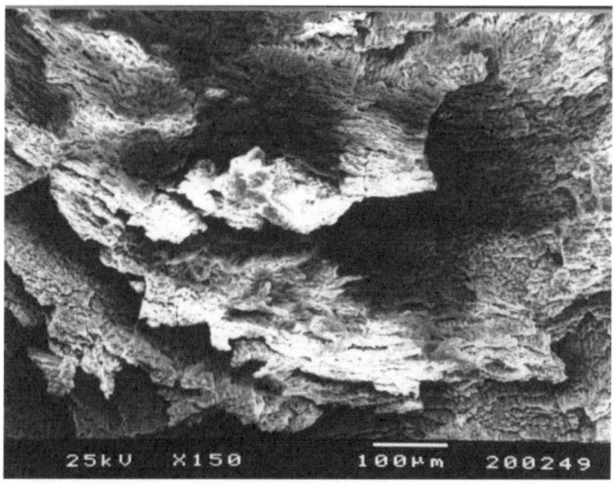

Fig. 6. Enamel. 600 mj. 10 Hz. Without water cooling. Layered enamel surface possibly due to dehydration of enamel during laser application.

Fig. 7. Enamel. 750 mj. 10 Hz. Without water cooling. Higher magnification of the previous Fig. Layered enamel surface.

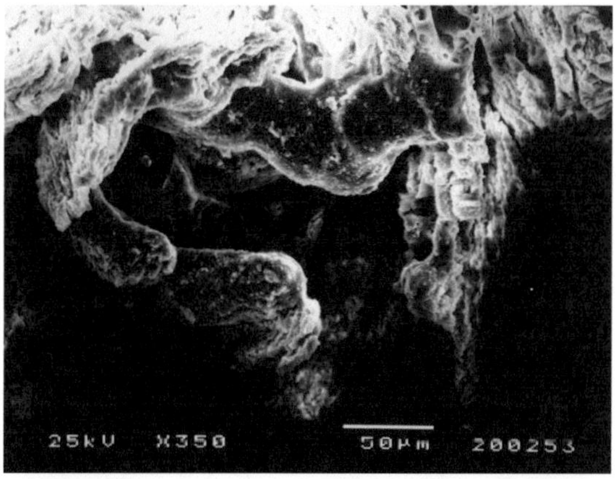

Fig. 8. Enamel. 800 mj. 5 Hz. Without water cooling. Melted and resolidified enamel. This texture is highly acid resistant.

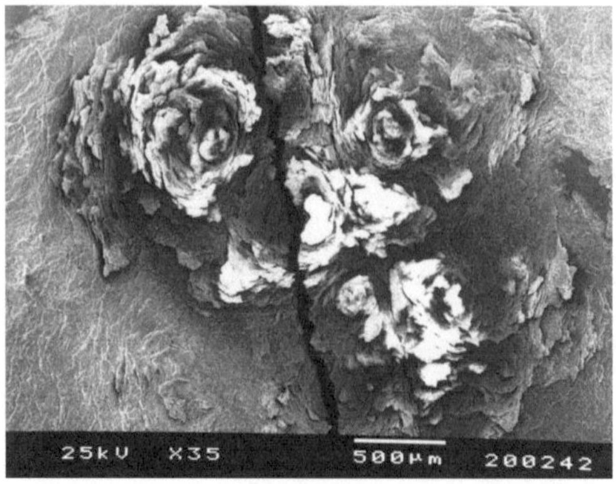

Fig. 9. Enamel. 1000 mj. 10 Hz. Without water cooling. Rose-bud like appearance. Clear evidence of over destruction of enamel with high energy intensity. Enamel lost its integrity in layers around the lased point. (The crack at midline is a result of dehydration during preparation of the specimen for SEM evaluation).

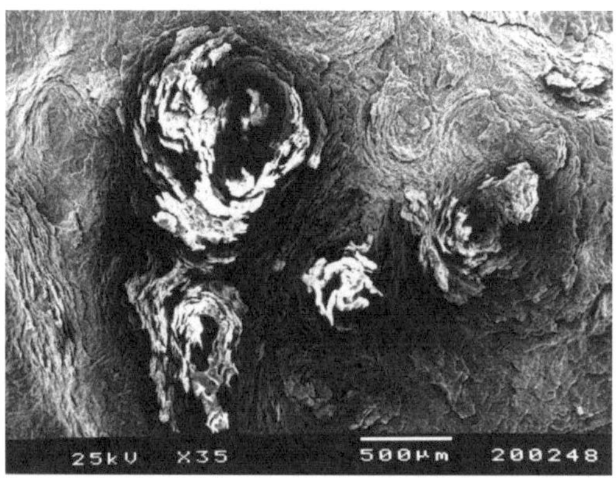

Fig. 10. Enamel. 1000 mj. 10 Hz. Without water cooling. Similar appearance with Fig. 13. Overdestructed and layered surface as a result of excessively heated enamel.

2.2 Morphological analysis of Er:YAG laser treated dentin

Fig. 11. Dentin. 250 mj. 10 Hz. Without water cooling. Swollen dentin orifices.

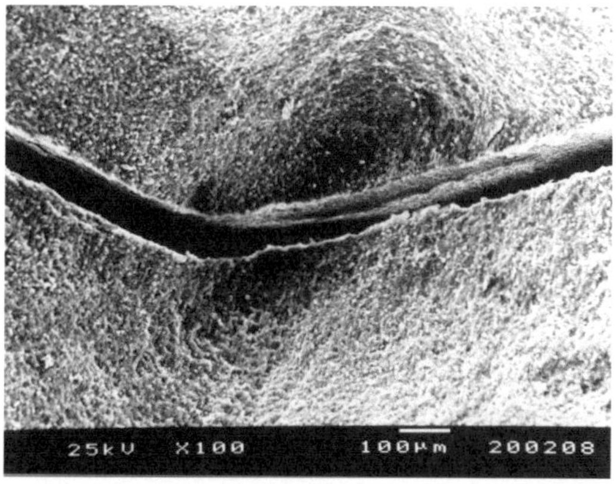

Fig. 12. Dentin. 400 mj. 10 Hz. Without water cooling. Cavitation with charring. (The crack at midline is a result of dehydration during preparation of the specimen for SEM evaluation).

Fig. 13. Dentin. 250 mj. 10 Hz. With water cooling. Partially open dentinal tubules with crater formations.

Fig. 14. Dentin. 500 mj. 10 Hz. Without water cooling. Pop-corn like appearance. One exploded (right) an done over swollen dentin orificies. Evidence of thermal destruction of dentin.

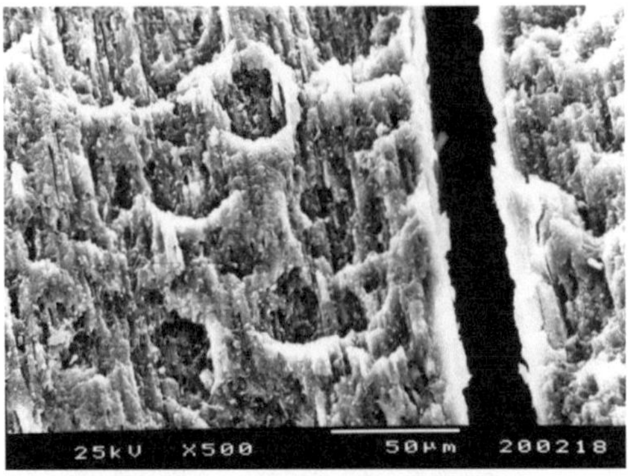

Fig. 15. Dentin. 500 mj. 5 Hz. With water cooling. intertubular Apperent evidence of intertubular dentin being affected dramatically by laser. (The crack on the right is a result of dehydration during preparation of the specimen for SEM evaluation)

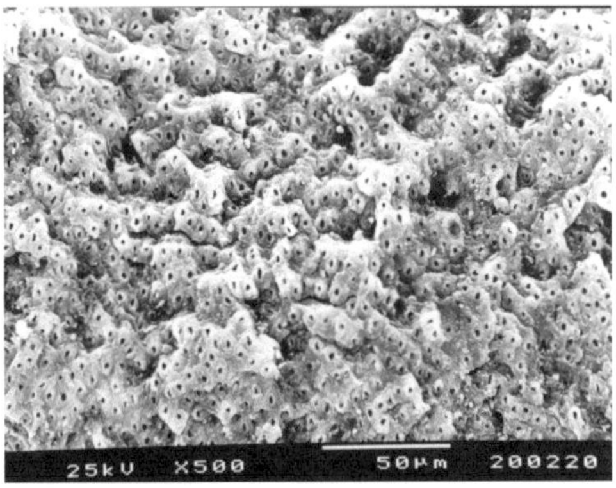

Fig. 16. Dentin. 250 mj. 5 Hz. With water cooling. Nearly all dentinal tubules are open. Adequate surface for bonding procedures. Stratified surface due to non-homogenous application of laser. Calcospherite areas which are usually seen following Na(OH) were observed.

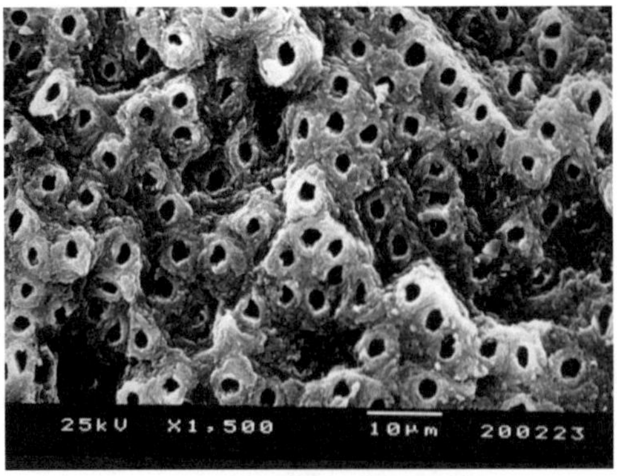

Fig. 17. Dentin. 250 mj. 5 Hz. With water cooling. Higher magnification of the surface in Fig. 16. No signs of thermal damage. No melted and swollen dentin. All dentinal tubules are open. Adequate surface for bonding procedures.

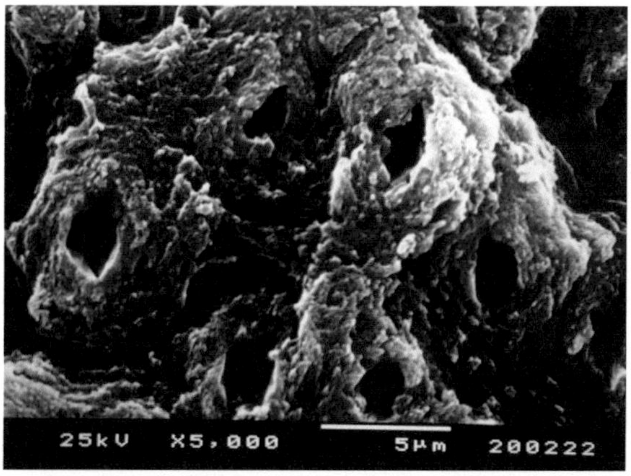

Fig. 18. Dentin. 250 mj. 5 Hz. With water cooling. Higher magnification of the surface in Fig. 17. No signs of thermal damage.

3. Morphological analysis of Er:YAG laser treated all-ceramic materials

A new class of dental framework materials have been introduced to the market for crown and fixed partial denture fabrication such as high-aluminium trioxide (alumina) ceramics, leucite reinforced feldspathic ceramics, castable glass-ceramics, machining and CAD/CAM ceramic systems and yttrium tetrogonal zirconia polycrystal (Y-ZTP; zirconia) (Atsu, et al., 2006; Amaral, et al., 2006; Bottino, et al., 2005; Kim, et al., 2005; Kern & Wegner, 1998). Alumina and zirconia demonstrate high clinical success due to their high cristalline content and are potential substitutes for traditional materials (Cavalcanti, et al., 2009a; Jacobsen, et al., 1997; Haselton, et al., 2000; Toksavul & Toman, 2007; Fradeani & Redemani, 2002).

The tetragonal to monoclinic phase transformation capability of zirconia results in high mechanical properties (Guazzato, et al.,2004). External stresses such as sandblasting, grinding, impact, and thermal aging can trigger this phase transformation mechanism (Karakoca & Yılmaz, 2009).

The clinical succes and survival rates of these restorations depend on several factors such as cementation procedure. To maintain a micromechanical bond, a key factor between restoration and the resin, luting surfaces of the restorations should be conditioned (Awliya, et al., 1998; Özcan, et al., 2001). To achieve reliable adhesion to these new materials, surface pre-treatments usually followed by silanization are required (Atsu, et al., 2006; Amaral, et al., 2006; Bottino, et al., 2005; Kim, et al., 2005; Kern & Wegner, 1998).

To obtain high mechanical bond strength to newer restorations, the inner surfaces are roughened by numerous techniques to increase the luting surface area. Among several methods that have been investigated for surface modification dental restorative materials, grinding, abrasion with diamond rotary instruments, airborne particle abrasion with aluminum oxide particles (sandblasting), chemical etching with different concentrations of hydrofluoric acid (HF), silica coating (Cojet, Rocatec), Silicoater MD, PyrosilPen silanization, selective infiltration-etching technique and combinations of any of these methods are the most common conditioning techniques prior to luting procedures (Amaral, et al., 2006; De Oyague, et al., 2009; Özcan, et al., 2001; Özcan, 2002; Kern & Thompson, 1994; Aboushelib, et al., 2007). Although surface treatments are used to micromechanical retention, they might affect the mechanical properties of zirconia (Sato, et al., 2008).

For chemical etching, different concentrations of HF acid, acidulated phosphate fluoride and ammonium bifluoride are used to condition the restorations (Blatz, et al. 2003; Clauss, 2000; Janda, et al. 2003). Etching dissolves the low fusing glass matrix exposing the cristalline structure and creates a micromechanically retentive surface but also promotes hydroxyl group formation on the etchable ceramic materials (Matinlinna & Vallitu, 2007; Özcan, 2003; Van Noort, 2002a; Özcan, et al., 2001). But some new materials such as zirconia and alumina are non-etchable because of they do not have glassy phase at the cristalline border and it is difficult to form microretentive surfaces to obtain strong and durable bonds with chemical etching techniques (Blatz, et al., 2003; Clauss, 2000; Janda, et al., 2003; Awliya, et al., 1998). Therefore different surface conditioning methods such as sandblasting and silica coating have been suggested for surface pretreatments of alumina and zirconia frameworks to modify the surface properties (Della Bona, et al., 2004; Phark, et al., 2009; Ersu, et al., 2009; Jacobsen, et al., 1997).

Different sizes of alumina particles between 25 and 250 µm are used (Blatz, et al., 2003; Kern & Wegner, 1998; Hummel & Kern, 2004; Curtis, et al., 2006). Sand blasting the surface with aluminum oxide particles cleans the ceramic surface and creates adequate bonding with micromechanical mechanisms to alumina- and zirconia based frameworks (Matinlinna & Vallitu, 2007; Phark, et al., 2009; Blatz, et al., 2003; Kern & Wegner, 1998; Hummel & Kern, 2004; Blatz, et al., 2004). The abrasive process removes loose contaminated layers, increases surface area and improves the wettability (Amaral, et al., 2006; Kümbüloğlu, et al., 2006).

Large abrasive particles result in rougher surface since the abrasion of the surface increases in proportion to the square of the diameter of the particle. Particle size variations and and the high pressure during sandblasting may cause flaws and phase transformation that expedites micro-crack formation and lead to altered mechanical properties of zirconia (Zhang, et al., 2004; Zhang, et al., 2006). Mechanical grinding and sandblasting may create subcritical microcracks and phase transformation within zirconia surface which might negatively affect the mechanical properties (Karakoca & Yılmaz, 2009; Ayad, et al., 2008).

Sandblasting is not recommended to roughen In-Ceram Zirconia frameworks as the aluminum oxide particles used to condition the surface have a hardness similar to that of the aluminum oxide crystals present in the target material (Borges, et al., 2003). Alternatively use of synthetic diamond particles 1-3 µm in size have been advocated to roughen the aluminous ceramics (Sen, et al., 2000).

Another method to inrease the surface energy of ceramic materials is tribochemical silica coating that is based on forming a SiO_2 layer followed by silane application with accelerated silica coated alumina particles on to the ceramic surface, including non-etchable alumina and zirconia (Matinlinna & Vallitu, 2007; Kramer, et al., 1996; Sindel, et al., 1996; Özcan, 2002). Silica coating method also provides micromechanical retention like sandblasting and silica deposition on the luting surface (Kern & Thompson, 1995; Matinlinna & Vallitu, 2007; Özcan, et al., 2001). In a recent study AFM results revealed irregular and heterogeneous surfaces following silica coating and sandblasting of zirconia with the formation of high peaks and shallows while SEM observations showed microretentive grooves in conjuction with Atomic Force Microscope (AFM) results (Subaşı & İnan, 2011).

In addition to currently used conditioning methods, laser-induced modifications of dental materials have also been studied. Lasers have been proposed to modify the surface of materials in relatively safe and easy means (Ersu, et al., 2009; Gökçe, et al., 2007; Akova, et al., 2005; Spohr, et al., 2008; Cavalcanti, et al., 2009b; Jacobsen, et al., 1997). Implant surfaces treated with lasers exhibit high degree of purity with adequate surface roughness (Gaggl, et al., 2000; Cho & Jung, 2003).

Among the several applications of lasers, surface conditioning for bonding have also been reported. Various laser types such as Nd:YAG, Er:YAG, Er,Cr:YSGG and CO_2 have been studied for surface alterations of dental materials (Convissar & Goldstein, 2001). But only limited studies are available on the laser treatment of all ceramic materials (Ersu, et al., 2009; Gökçe, et al., 2007; Akova, et al., 2005; Cavalcanti, et al., 2009b; Jacobsen, et al., 1997; Cavalcanti, et al. (2009a).

Ceramics do not effectively absorb some certain wavelengths such as 1064 nm (Nd:YAG). To increase the energy absorption of this laser the surface of ceramic material can be covered with graphite powder prior to laser irradiation. During laser application the graphite is removed from the surface with microexplosions (Spohr, et al., 2008).

Some lasers are also used for other applications such as forming a glazed surface layer on ceramics, the removal of resin composite filling materials, laser welding of ceramics and metal alloys, including titanium, and increasing the corrosion resistance of metal alloys (Ersu, et al., 2009; Schmage, 2003). Focussed CO_2 laser causes in conchoidal tears (result of surface warming) on ceramic surface that provides mechanical retention between resin composite and ceramics. But sudden temperature changes could create internal tensions that might affect the bond strength (Ersu, et al., 2009). The authors concluded that CO_2 laser surface modification demonstrated higher bond strength than control, sandblasted and chemical etching.

Results of studies that compared the bond strength of resins to CO_2 laser and chemically etched zirconia vary. Obata, et al., (1999) stated that laser etching produced lower bond strength compared to acid ecthing whereas Ural, et al., (2010) proposed higher bond strength. They attribute the high bond strength values to power levels of the laser used in their study. Increased power settings caused micro-cracks and high bond strength (Ural, et al., 2010).

Watanabe, et al., (2009) suggested that Nd:YAG laser irradiation improved the mechanical properties of cast titanium. Nd:YAG laser as an etchant was also used to enhance the bond strength of low-fusing ceramic to titanium (Kim & Cho, 2009).

Nd:YAG laser was also used to roughen In-Ceram Zirconia and feldspathic ceramic (Li, et al., 2000; Spohr, et al., 2008). Li, et al., (2000) reported that SEM images of Nd:YAG laser applied specimens was fovarable to mechanical retention between the feldspathic ceramic and the resin cement and both laser and HF acid etched groups exhibited same shear bond strength. Nd:YAG laser treatment of In-Ceram Zirconia caused surface changes characterized with material removal due to the micro-explosions resulting in formation of voids and fusing and melting of the most superficial ceramic layer followed by solidification to a smooth blister-like surface (Spohr, et al., 2008). Nd:YAG laser irradiation of zirconia causes color change to black with many cracks and reduced oxygen content (Noda, et al., 2010).

Recently roughening capacity of the Er:YAG laser for the inner surfaces of the lithium disilicate material has been introduced (Gökçe, et al., 2007). Ceramic specimens laser etched with low energy levels exhibited similar bond strength that of chemicaly etched specimens. But as the energy level increased bond strength values decreased dramtically. They concluded that their results could be explained by insufficient micro depths of the irregularities formed by high Er:YAG laser power settings, which resulted in limited penetration of silane and low bond strength. Higher power settings resulted in low bond strengths which might be due to over destruction (disassociation) of the crystal and/or matrix phases or heat damaged layer which was poorly attached to the infra layers or increased luting agent thickness due to craters caused by laser pulses (Gökçe, et al., 2007).

Erbium lasers are absorbed mainly by water and their absorption by water-free materials are compromised. To increase the effect of erbium lasers, covering the zirconia surfaces with graphite or hydroxapatite powder was recommended (Cavalcanti, et al., 2009b). Akın, et al., (2011) irradiated zirconia surface with Er:YAG laser and found increased surface roughness and surface irregularities compared to the untreated specimens. The authors used low power settings with water cooling and did not observe microcracks. They concluded that altering the zirconia surface with Er:YAG laser increased the shear bond strength of ceramic to dentin and found to be effective for decreasing microlekage in the adhesive-ceramic interface. Their results were in accordance with the study of Cavalcanti, et al. (2009a). Erdem & Erdem, (2011) studied the effect of Er:YAG laser irradiation with water cooling on zirconia and unlike forementioned resarchers they suggested that laser treatment decreased the bond strength of resin composite to zirconia framework. They observed microcracks throughout the surface in contrast with Akın, et al., (2011). They attributed the low bond strength values to excessively affected surfaces and crack formation which was possibly a result of laser irradiation. Stepped local temperature changes and pressurized water followed by thermocycling could be responsible for low temperature degredation of zirconia resulting in low bond strengths (Erdem & Erdem, 2011). They might have also induced phase transformation (Cavalcanti, et al., 2009a). The microcrack formation and sizes enlarged as the laser intensity increased (Cavalcanti, et al., 2009b). Stübinger, et al., (2008) demonstrated that Er:YAG and CO_2 lasers adversely affected the zirconia implant surfaces. They found crack formations up to 100 µm depth and large grains in blackened areas under SEM evaluation. Excessive power settings shown to be deterious to zirconia and their use for zirconia surface conditioning was questionable (Cavalcanti, et al. 2009b; Navarro, et al., 2010).

Subaşı & İnan, (2011), evaluated the Er:YAG laser treated zirconia with AFM and SEM. AFM and SEM results of lased surfaces revealed similar texture to that of the control group with the exception that sharp peaks formations of the lased surfaces. Cavalcanti, et al. (2009b) also demonstrated that increased laser energy levels increased surface roughness of zirconia. Melting, excessive loss of mass, and the presence of smooth areas surrounded by cracks were observed. Lower energy intensities (200 mj) had milder effect with smaller cracks along with melting, solidification and color changes without loss of structure compared to higher intensities (400 and 600 mj) (Cavalcanti, et al., 2009b). 200 mj irradiation also provided alterations similar to sandblasting. Effect of Nd:YAG laser (100 mj) and Er:YAG laser (200 mj) exhibited similar topographies although the Nd:YAG laser had a totally different target interaction compared with the Er:YAG laser (Da Silveira, et al., 2005; Cavalcanti, et al., 2009b).

There is no consensus about energy levels of Er:YAG laser that could be used to modify the zirconia surface. 400 mj at 10 Hz (Subaşı & İnan, 2011), 150 mj at 10 Hz (Akın, et al., 2011), 200 mj at 10 Hz (Erdem & Erdem, 2011), 200 mj at 10 Hz (Cavalcanti, et al., 2009a); 200 mj, 400 mj, 600 mj at 10 Hz (Cavalcanti, et al., 2009b), 300 mj at 10 Hz (Şen & Ceylan, 2010) were chosen to roughen zirconia surfaces. Besides different methods have been chosen to evaluate the bond strength and surface topography. Therefore it is difficult to compare the results of the studies reviewed.

3.1 Morphological analysis of Er:YAG laser treated and hydroflouric acid etched Li-Disilicate material

SEM evaluations of 9.5% Hydrofluoric acid and Er:YAG laser Li-disilicate material revealed different surface morphologies, depending on the surface conditioning methods.

3.1.1 SEM evaluation before shear bond strength testing

Fig. 19. The untreated surface showing intact glassy phase without any apparent crystals.

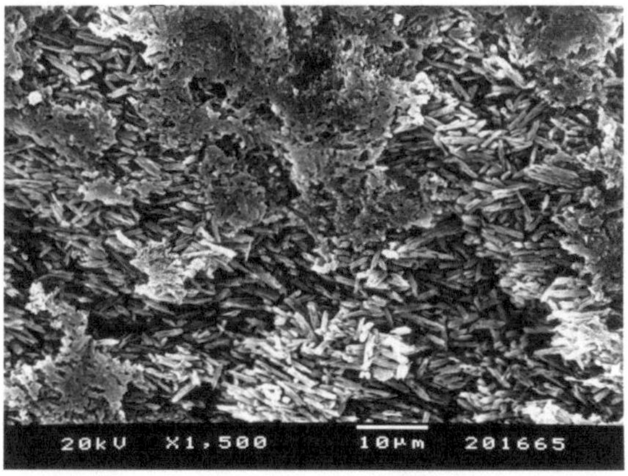

Fig. 20. 9.5% HF, 30 seconds. The surface has both apparent Li-disilicate crystals and glassy matrix. Glass matrix phase could not be completely removed if not applied homogenously and might lead to ill penetration of silane and the adhesive.

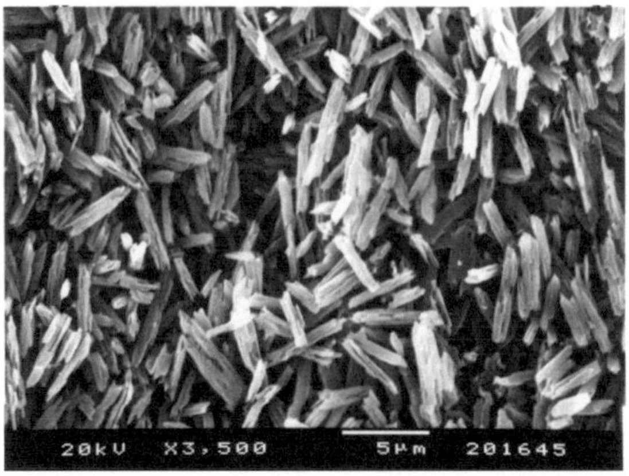

Fig. 21. 9.5% HF, 30 seconds. Visible Li-disilicate crystals. Completely removed glassy matrix. Appropriate etching pattern and surface for adhesive cementation (Gökçe et al., 2007).

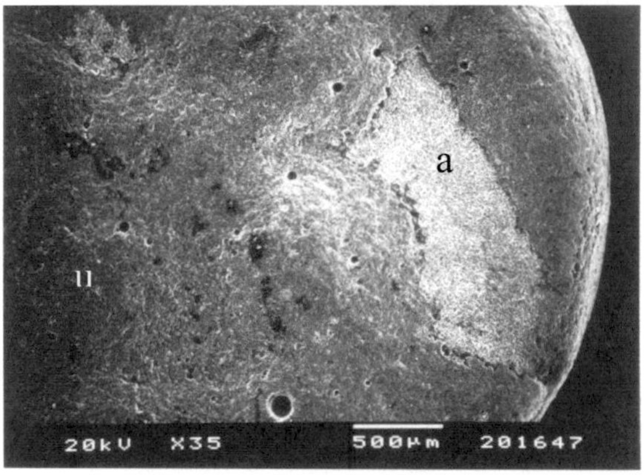

Fig. 22. Er:YAG laser, 300 mj, 10 Hz. Affected (a) and unaffected (u) areas of lased surface.

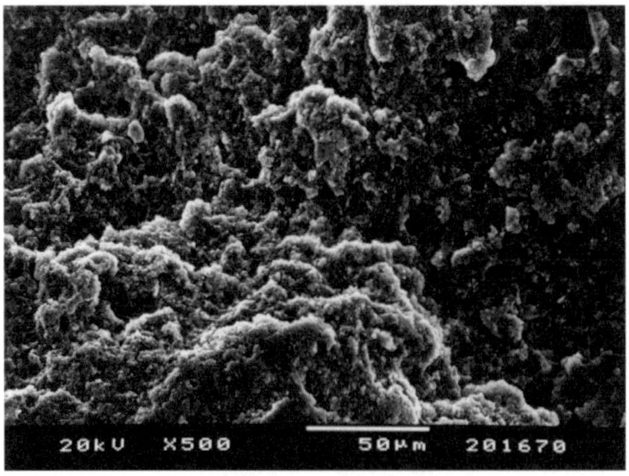

Fig. 23. Er:YAG laser, 300 mj, 10 Hz. Irregular Li-disilicate crystals in smaller sizes (Gökçe et al., 2007).

Fig. 24. Er:YAG laser, 600 mj, 10 Hz. Increased surface irregularities with severely affected and disassociated Li-disilicate crystals (Gökçe et al., 2007).

Fig. 25. Er:YAG laser, 900 mj, 10 Hz. Severely affected and disassociated Li-disilicate crystals (Gökçe et al., 2007).

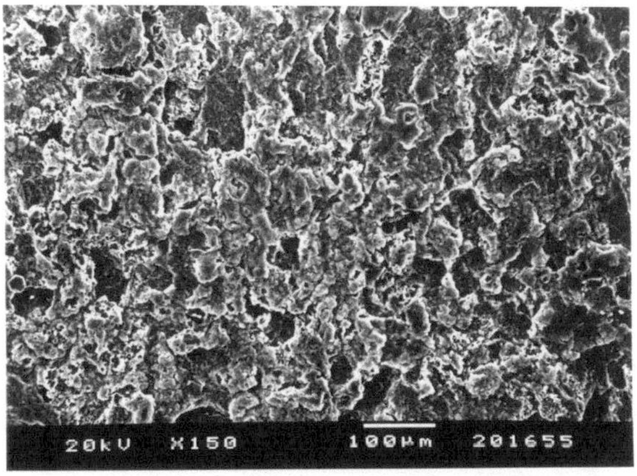

Fig. 26. Er:YAG laser, 1000 mj, 10 Hz. Melted and resolidified surface. This layer is poorly attached to the underlying intact phase.

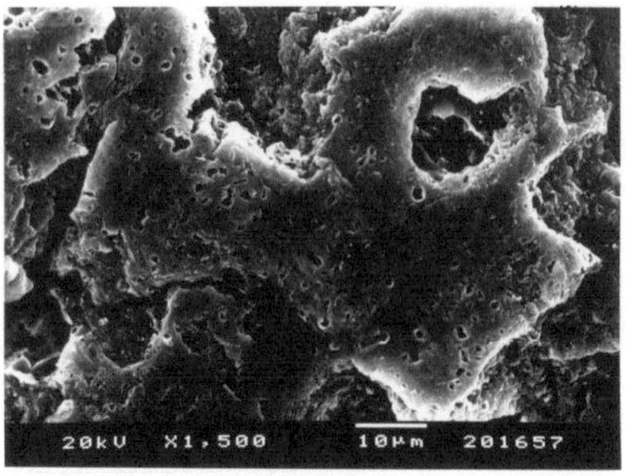

Fig. 27. Er:YAG laser, 1000 mj, 10 Hz. Higher magnification of the surface in Fig. 26.

3.1.2 SEM evaluation after shear bond strength testing

SEM evaluation following shear bond strength of the untreated, HF acid etched and Er:YAG laser conditioned Li-Disilicate material exhibited different failure modes, indicatives of adhesion of the bonding agent and the luting cement (Variolink II).

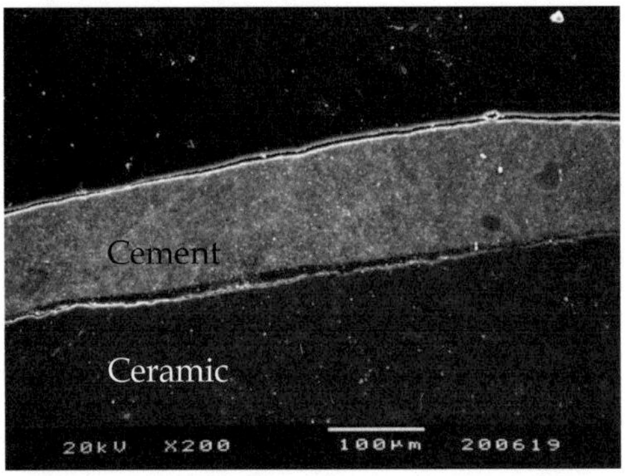

Fig. 28. Untreated ceramic surface. Adhesive failure inbetween the ceramic and the cement. No rough surfaces were noted on the ceramic (Gökçe et al., 2007).

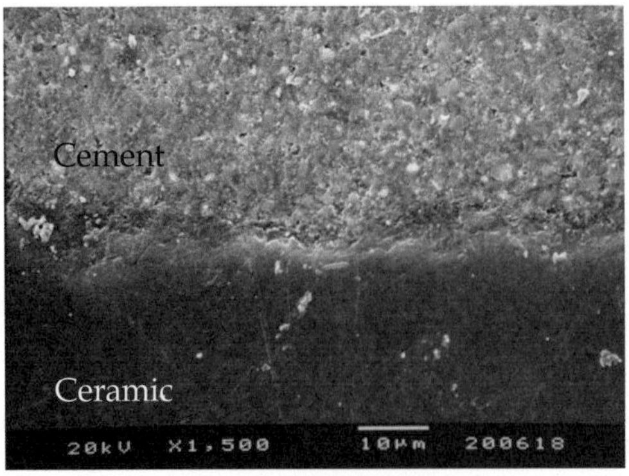

Fig. 29. 9.5% HF, 30 seconds. Good adhesion at the cement-ceramic interface with increased surface roughness. Mainly cohesive failures within the cement (Gökçe et al., 2007).

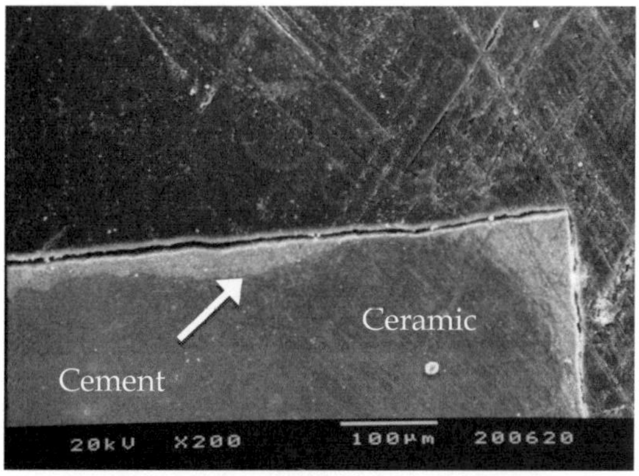

Fig. 30. Er:YAG laser, 300 mj, 10 Hz. No visible cement on the margins, while a cement remnant at the center of the specimen with adhesive+cohesive failures were observed (Gökçe et al., 2007).

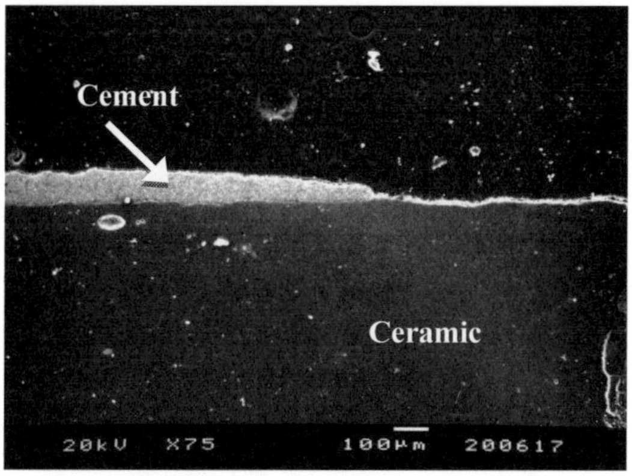

Fig. 31. Er:YAG laser, 600 mj, 10 Hz. Partially delaminated cement surfaces can be observed with adhesive failures (Gökçe et al., 2007).

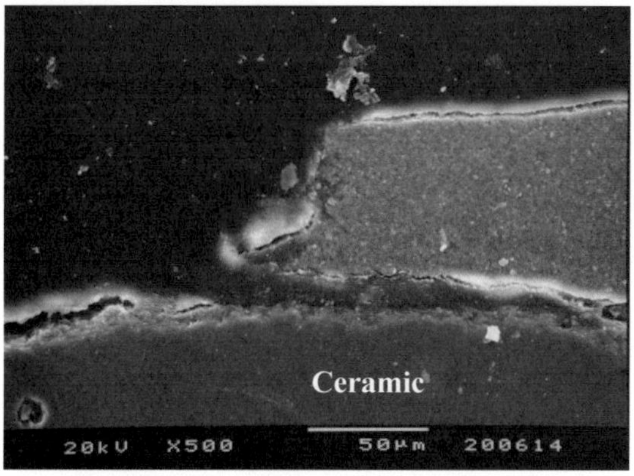

Fig. 32. Er:YAG laser, 900 mj, 10 Hz. Adhesive failure between cement and ceramic. Decreased irregularities and severe effects of laser on the ceramic surface (Gökçe et al., 2007).

3.2 Morphological analysis of sandblasted and Er: YAG laser-roughened alumina material

Fig. 33. Untreated In-Ceram Alumina (Şen, 2010).

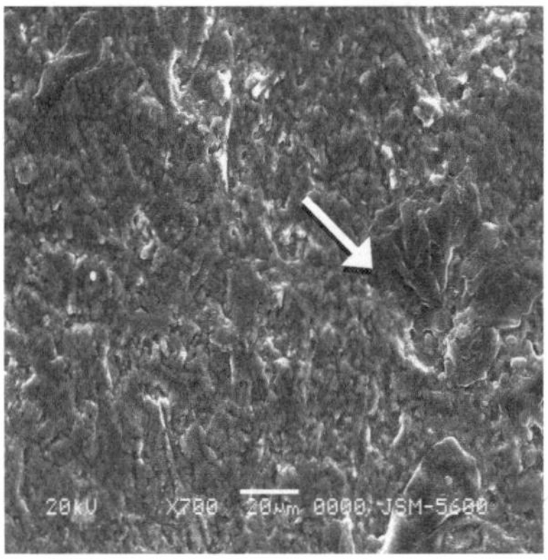

Fig. 34. In-Ceram Alumina. Airborne particle abrasion (110µm Al_2O_3). Affected and rougher surface compared to untreated surface with shallow pits (Şen, 2010).

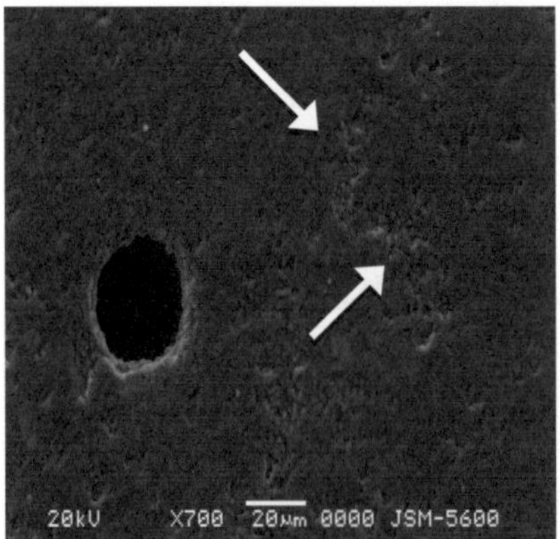

Fig. 35. In-Ceram Alumina. Er:YAG laser, 150 mj at 10 Hz with water cooling. Locally affected points on the surface due ton on homogenous application of the laser (Şen, 2010).

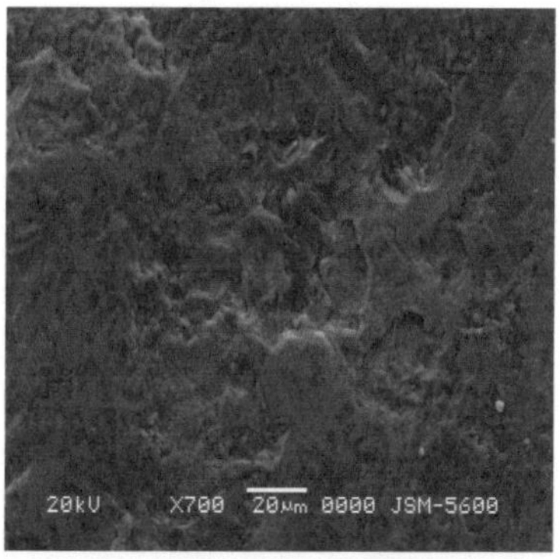

Fig. 36. In-Ceram Alumina. Er:YAG laser, 250 mj at 10 Hz with water cooling. Generalized effect of laser rougher surface compared to untreated and 150 mj laser applied surfaces (Şen, 2010).

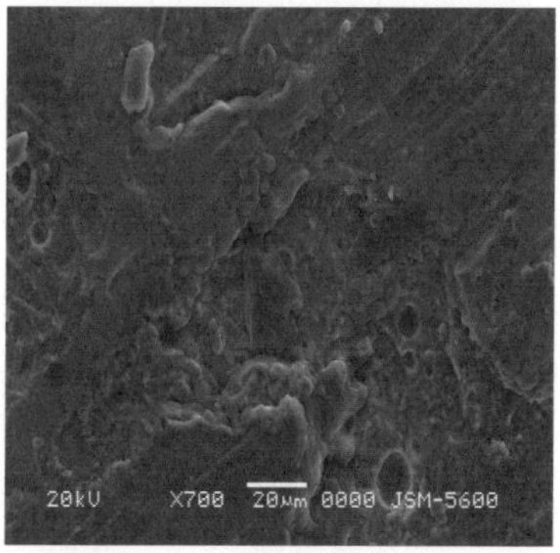

Fig. 37. In-Ceram Alumina. Er:YAG laser, 400 mj at 10 Hz with water cooling. Serrated and smoothened surface by resolidification of melted areas. This resolidified layer might be poorly attached to the underlying material (Şen, 2010).

3.3 Morphological analysis of sandblasted, silica coated and Er: YAG laser-roughened zirconia

3.3.1 SEM evaluation before shear bond strength testing

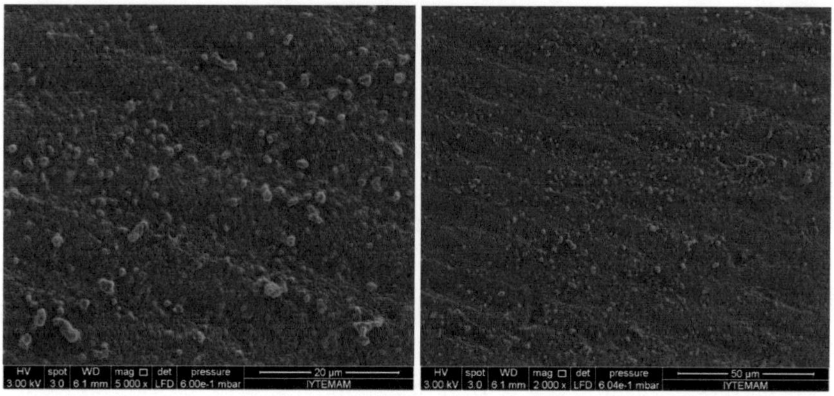

Fig. 38. Untreated zirconia. 2000x (left) and 5000x (right) magnifications (Erdem & Erdem, 2011).

Fig. 39. Sandblasted (particle size 110 µm) zirconia. 2000x (left) and 5000x (right) magnifications. Increased roughness compared to untreated zirconia (Erdem & Erdem, 2011).

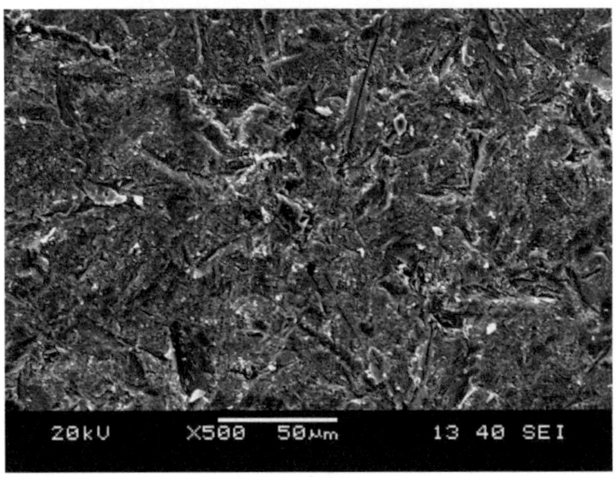

Fig. 40. Sandblasted (particle size 180 µm) zirconia. 500x magnification. Similar texture with the 110 µm air abraded surface.

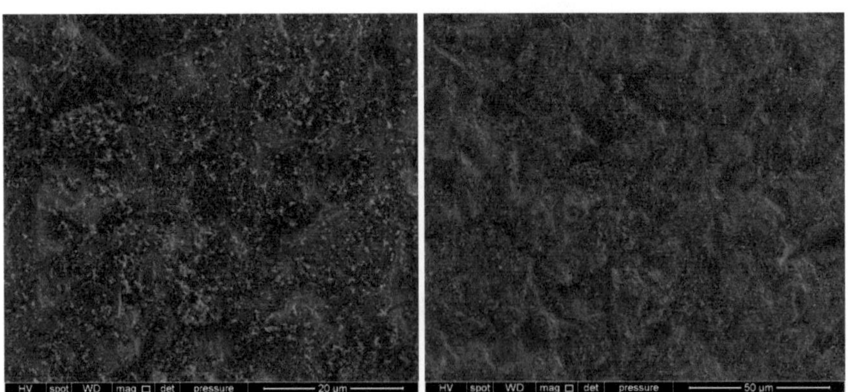

Fig. 41. Silica coated (Rocatec Pre110 µm and Rocatec Soft 30 µm) zirconia. 2000x (left) and 5000x (right) magnifications. Increased roughness similar to sandblasting and silica deposition on the surface can be observed (Erdem & Erdem, 2011).

Fig. 42. Silica coated (Rocatec Pre110 µm and Rocatec Soft 30 µm) zirconia. 500x magnification. Increased roughness similar to particle abrasion with aluminum oxide (Şen, 2010).

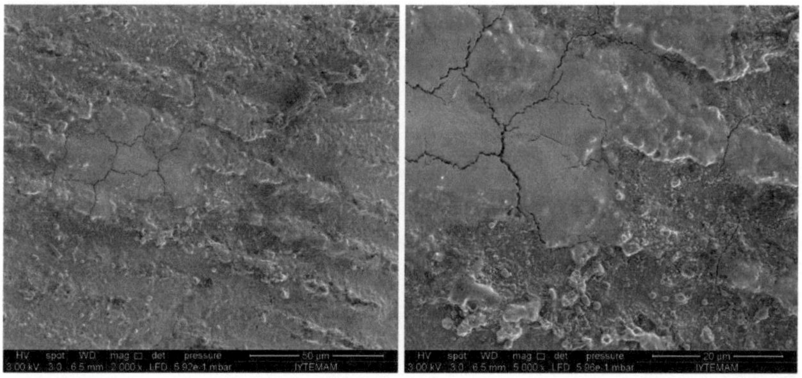

Fig. 43. Graphite coated and lased (200 mj, 10 Hz) zirconia. 2000x (left) and 5000x (right) magnifications. Rough and severely affected appearance with irregular surface (Şen, 2010) with micro cracks (Erdem & Erdem, 2011).

3.3.2 SEM evaluation after shear bond strength testing

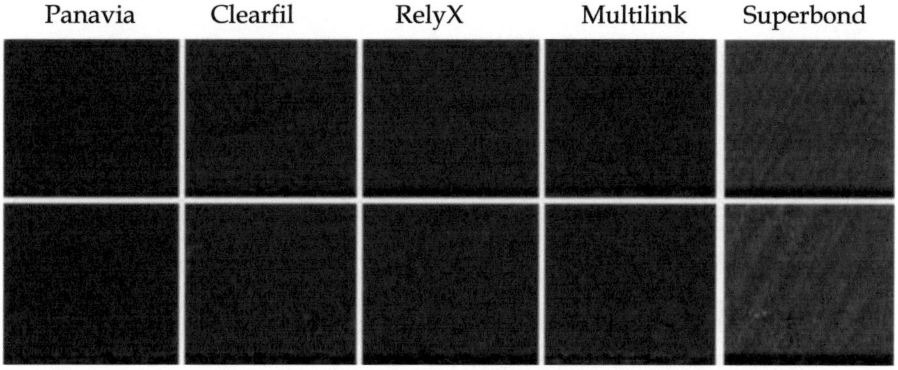

Fig. 44. Untreated zirconia. 2000x (first line) and 5000x (second line) magnifications. No cement retention was observed on untreated zirconia (Erdem & Erdem, 2011).

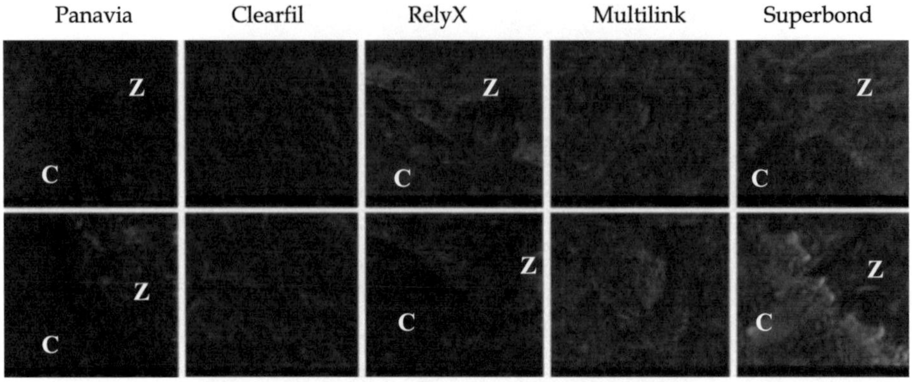

Fig. 45. Sandblasted (particle size 110 μm) zirconia. 2000x (first line) and 5000x (second line) magnifications. Two of the cements tested exhibited both adhesive and cohesive failures. (Z: Zirconia, C: Cement) (Erdem & Erdem, 2011).

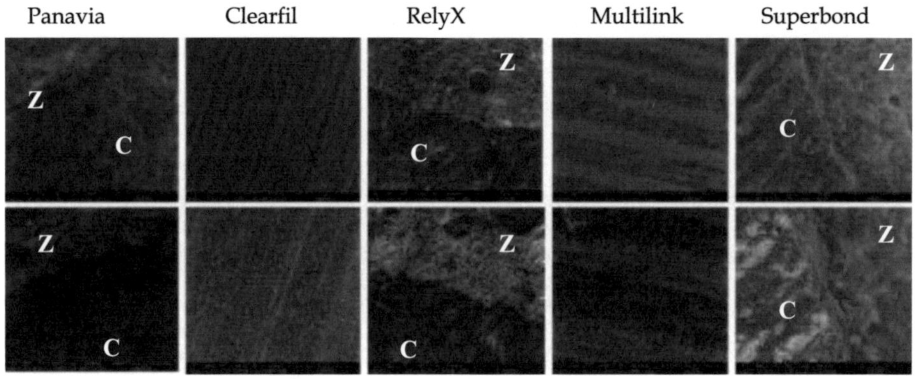

Fig. 46. Silica coated (Rocatec Pre110 μm and Rocatec Soft 30 μm) zirconia. 2000x (first line) and 5000x (second line) magnifications. Similar results with sandblasting was oserved after shear bond strength testing (Z: Zirconia, C: Cement) (Erdem & Erdem, 2011).

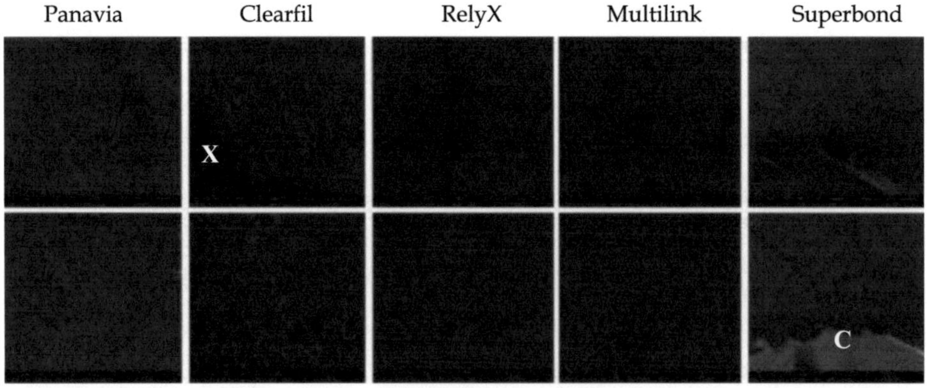

Fig. 47. Graphite coated and lased (200 mj, 10 Hz) zirconia. 2000 (first line) and 5000 (second line) magnifications. Adhesive failures observed in all cement groups. No cement retention on any of the groups. (X: severely affected area, C: Cement) (Erdem & Erdem, 2011).

4. Conclusion

There are many techniques to condition dental hard tissues and luting surfaces of indirect restorations prior to bonding. Operators find it difficult to decide which technique offers better results, and are also uncertain about the factors that might influence their techniques of choice. However micromechanical retention of luting materials to acid etched conditioned dental hard tissues is currently seems to be the most successful and reliable approach for dental bonding. But surface characteristics of Er:YAG lased enamel and dentin are responsible for considering this surface adequate for resin bonding.

It is assumed that the ablation rate of lasers on the dental materials is strongly influenced by the differences in composition and microstructure of the material and the presence of water. In spite of its great potential for ablation, Er:YAG laser effectiveness and safety is also directly related to adequate setting parameters. Power settings, frequency and durations of laser irradiation play an important role to obtain optimum bond strength and roughness values.

Future studies are needed to evaluate the superficial and sub-superficial layers of irradiated dental hard tissues and materials in order to develop new agents that can interact properly with lased substrate. In my opinion, in the near future, 9.6 µm CO_2 laser with an adequate delivery system that has the absorption peak in hydroxyapatite will replace many dental hard tissue lasers, which are currently being used. In the presented chapter, the morphological assessment of Er:YAG lased dental hard tissues and materials have been discussed under the light of the current literature.

5. References

Aboushelib, MN.; Kleverlaan, CJ. & Feilzer, AJ. (2007). Selective infiltration-etching technique for a strong and durable bond of resin cements to zirconia-based materials. *Journal of Prosthetic Dentistry*, Vol.98, No.5, pp. 379-388.

Akin, H.; Tugut, F.; Emine, A.G.; Guney, U. & Mutaf, B. (2011). Effect of Er:YAG laser application on the shear bond strength and microleakage between resin cements and Y-TZP ceramics. *Lasers in Medical Science,* Vol 21. [Epub ahead of print]

Akova, T.; Yoldas, O.; Toroglu, MS. & Uysal, H. (2005). Porcelain surface treatment by laser for bracket-porcelain bonding. *American Journal of Orthodontics & Dentofacial Orthopedics,* Vol.128, pp. 630–637.

Amaral, R.; Özcan, M.; Bottino, MA. & Valandro, LF. (2006). Microtensile bond strength of a resin cement to glass infiltrated zirconia-reinforced ceramic: the effect of surface conditioning. *Dental Materials,* Vol.22, pp. 283–290.

Aoki, A.; Ishikawa, I.; Yamada, T.; Otsuki, M.; Watanabe, H.; Tagami, J.; Ando, Y. & Yamamoto, H. (1998). Comparison between Er:YAG laser and conventional technique for root caries treatment in vitro. *Journal of Dental Research,* Vol.77, pp. 1404–1414.

Apel, C.; Meister, J.; Schmitt, N.; Gra¨ber, H.G. & Gutknecht, N. (2002). Calcium solubility of dental enamel fol- lowing sub-ablative Er:YAG and Er:YSGG laser irradiation in vitro. *Lasers in Surgical Medicine,* Vol.30, pp. 337–341.

Arima, M. & Matsumoto, K. (1993). Effects of ArF excimer laser irradiation on human enamel and dentin. *Lasers in Surgical Medicine,* Vol.13, pp. 97–105.

Ariyaratnam, MT.; Wilson MA. & Blinkhorn, AS. (1999). An analysis of surface roughness, surface morphology and composite/dentin bond strength of human dentin following the application of the Nd:YAG laser. *Dental Materials,* Vol.15, No.4, pp. 223-8.

Armengol, V.; Jean, A.; Rohanizadeh, R. & Hamel, H. (1999). Scanning electron microscopic analysis of diseased and healthy dental hard tissues after Er:YAG laser irradiation: in vitro study. *Journal of Endodontics,* Vol.25, No.8, pp. 543–546

Atsu, SS.; Kilicarslan, MA.; Kucukesmen, HC. & Aka, PS. (2006). Effect of zirconium-oxide ceramic surface treatments on the bond strength to adhesive resin. *Journal of Proshtetic Dentistry,* Vol.95, pp. 430–436.

Awliya, W.; Oden, A.; Yaman, P.; Dennison, JB. & Razzoog, ME. (1998). Shear bond strength of a resin cement to densely sintered high-purity alumina with various surface conditions. *Acta Odontologica Scandinavica,* Vol.56, pp. 9–13.

Ayad, MF.; Fahmy, NZ. & Rosenstiel, SF. (2008). Effect of surface treatment on roughness and bond strength of a heat-pressed ceramic. *Journal of Prosthetic Dentistry,* Vol.99, pp. 123–130

Bader, C. & Krejci, I. (2006). Marginal quality in enamel and dentin after preparation and finishing with an Er:YAG laser. *American Journal of Dentistry,* Vol.19, pp. 337–342.

Barkmeier, WW. & Cooley, RL. (1992). Laboratory evaluation of adhesive systems. *Operative Dentistry,* Supplementary 5, pp. 50-61

Blatz, MB.; Sadan, A. & Kern, M. (2003). Resin-ceramic bonding: a review of the literature. *Journal of Prosthetic Dentistry,* Vol.89, pp. 268–274.

Blatz, MB.; Sadan, A.; Martin, J. & Lang, B. (2004). In vitro evaluation of shear bond strengths of resin to densely-sintered high-purity zirconium-oxide ceramic after long-term storage and thermalcycling. *Journal of Prosthetic Dentistry,* Vol.91, pp. 356–362.

Boehm, R.; Rica, J.; Webster, J. & Janke, S. (1997). Thermal stress effects and surface cracking associated with laser use on human teeth. *Journal of Biomechanical Engineering,* Vol.99, pp. 189-194.

Borges, G.A.; Spohr, A.M.; Goes, M.F.; Correr Sobrinho, L. & Chan, J.D. (2003). Effect of etching and airborne parti- cle abrasion on the microstructure of different dental ceramics. *Prosthetic Dentistry,* Vol.89, pp. 479–487.

Bottino, MA.; Valandro, LF.; Scotti, R. & Buso, L. (2005). Effect of surface treatments on the resin bond to zirconium-based ceramic. *International Journal of Prosthodontics,* Vol.18, pp. 60–65.

Buonocore, MG. (1955). A simple method of increasing the adhesion of acrylic filling materials to enamel surfaces. *Journal of Dental Research,* Vol.34, No.6, pp. 849-853.

Burkes, EJ. Jr.; Hoke, J.; Gomes, E. & Wolbarsht, M. (1992). Wet versus dry enamel ablation by Er:YAG laser. *Journal of Prosthetic Dentistry,* Vol.67, No.6, pp. 847–851.

Cardoso, M.V.; Coutinho, E., Ermis, RB.; Poitevin, A.; Van Landuyt, K.; De Munck, J.; Carvalho, RC.; Lambrechts, P. & Van Meerbeek, B. (2008). Influence of Er,Cr:YSGG laser treatment on the microtensile bond strength of adhesives to dentin. *Journal of Adhesive Dentistry,* Vol.10, No.1, pp. 25–33.

Carvalho, AO.; Reis AF de Oliveira, MT.; de Freitas, PM.; Aranha, AC.; Eduardo, CD. & Giannini, M. (2011). Bond Strength of Adhesive Systems to Er,Cr:YSGG Laser-Irradiated Dentin. *Photomedicine and Laser Surgery,* Vol.16, (Epub ahead of print)

Cavalcanti, AN.; Foxton, RM.; Watson, TF.; Oliveira, MT.; Giannini, M. & Marchi, GM. (2009a). Bond strength of resin cements to a zirconia ceramic with different surface treatments. *Operative Dentistry,* Vol.34, No.3, pp. 280–287

Cavalcanti, AN.; Pilecki, P.; Foxton, RM.; Watson, TF.; Oliveira, MT.; Gianinni, M. & Marchi, GM. (2009b). Evaluation of the surface roughness and morphologic features of Y-TZP ceramics after different surface treatments. *Photomedicine and Laser Surgery,* Vol.27, No.3, pp. 473-479.

Ceballo, L.; Toledano, M.; Osorio, R.; Tay, FR. & Marshall, GW. (2002). Bonding to Er-YAG-laser-treated dentin. *Journal of Dental Research,* Vol.81, No.2, pp. 119–122

Cernavin, I. (1995). A comparison of the effects of Nd:YAG and Ho:YAG laser irradiation on dentine and enamel. *Australian Dental Journal,* Vol.40, pp. 79–84.

Cho, SA. & Jung, SK. (2003). A removal torque of the laser-treated titanium implants in rabbit tibia. *Biomaterials,* Vol.24, pp. 4859–4863

Clauss, C. (2000). All-ceramic restoration based on milled zirconia. *Zahn Mund Kieferheilkd Zentralbl,* Vol.18, pp. 436–442.

Convissar, RA. & Goldstein, EE. (2001). A combined carbon dioxide/ erbium laser for soft and hard tissue procedures. *Dentistry Today,* Vol.20, pp. 66–71.

Cooper, LF.; Myers, ML.; Nelson, DG. & Mowery, AS. (1988). Shear strength of composite bonded to laser pre-treated dentin. *J Prosthetic Dentistry,* Vol.60, pp. 45-49.

Corona, S.A.; de Souza, A.E.; Chinelatti, M.A.; Borsatto, M.C.; Pecora, J.D. & Palma-Dibb, R.G. (2007). Effect of energy and pulse repetition rate of Er: YAG laser on dentin ablation ability and morphological analysis of the laser-irradiated substrate. *Photomedicine and Laser Surgery,* Vol.25, pp. 26–33.

Curtis, A.R.; Wright, A.J. & Fleming, G.J. (2006). The influence of surface modification techniques on the performance of a Y-TZP dental ceramic. *Journal of Dentistry,* Vol.34, pp. 195–206.

Da Silveira, B.L.; Paglia, A.; Burnett, L.H.; Shinkai, R.S.; Eduardo, C.P. & Spohr, A.M. (2005). Micro-tensile bond strength between a resin cement and an aluminous ceramic treated with Nd:YAG laser, Rocatec System, or aluminum oxide sandblasting. *Photomedicine and Laser Surgery,* Vol.23, pp. 543–548.

De Munck, J.; Van Meerbeek, B.; Yudhira, R.; Lambrechts, P. & Vanherle, G. (2002). Microtensile bond strength of two Erbium:YAG-lased vs. bur-cut enamel and dentin. *European Journal of Oral* Sciences, Vol.110, pp. 322–329.

De Oliveira, M.T.; de Freitas, P.M.; de Paula Eduardo, C.; Ambrosano, G.M. & Giannini, M. (2007). Influence of di- amond sono-abrasion, air-abrasion and Er:YAG laser irradiation on bonding of different adhesive systems to dentin. *European Journal of Oral* Sciences, Vol.1, pp. 158–166.

De Oyague, RC.; Monticelli, F.; Toledano, M.; Osorio, E.; Ferrari, M. & Osorio, R. (2009). Influence of surface treatments and resin cement selection on bonding to densely-sintered zirconium-oxide ceramic. *Dental Materials,* Vol.25, pp. 172–179

Della Bona, A.; Shen, C. & Anusavice, KJ. (2004). Work of adhesion of resin on treated lithia disilicate-based ceramic. *Dental Materials,* Vol.20, No.4, pp. 338-344.

Dunn, WJ.; Davis, JT. & Bush, AC. (2005). Shear bond strength and SEM evaluation of composite bonded to Er:YAG laser-prepared dentin and enamel. *Dental Materials,* Vol.21, No.7, pp. 616-624

Erdem A, Erdem A, (2011) Evaluation of bond strength of luting resins to a Y-TZP framework material processed with different surface treatments. PhD Thesis. Ege University School of Dentistry Department of Prosthodontics, Izmir, Turkey

Ersu, B.; Yuzugullu, B.; Ruya, Y.A. & Canay, S. (2009). Surface roughness and bond strengths of glass-infiltrated alumina- ceramics prepared using various surface treatments. *Journal of Dentistry,* Vol.37, pp. 848–856.

Esteves–Oliveira, M.; Zezell, DM.; Apel, C.; Turbino, ML.; Aranha, AC.;Eduardo Cde, P. & Gutknecht, N. (2007). Bond strength of self-etching primer to bur cut, Er,Cr:YSGG, and Er:YAG lased dental surfaces. *Photomedicine and Laser Surgery,* Vol.25, No.5, pp. 373–380.

Fabianelli, A.; Pollington, S.; Papacchini, F.; Goracci, C.; Cantoro, A.; Ferrari, M. & Van Noort, R. (2010). The effect of different surface treatments on bond strength between leucite reinforced feldspathic ceramic and composite resin. *Journal of Dentistry,* Vol.38, No.1, pp. 39–43.

Fradeani, M. & Redemagni, M. (2002). An 11-year clinical evaluation of leucite-reinforced glass-ceramic crowns: a retrospective study. *Quintessence International,* Vol.33, pp. 503–510.

Frentzen, M. & Koort, H.J. (1992). The effect of Er:YAG laser irradiation on enamel and dentin. *Journal of Dental Research,* Vol.71, pp. 571-577.

Frentzen, M.; Koort, H.J. & Thiensiri, I. (1992). Excimer laser in dentistry: future possibilities with advanced technology. *Quintessence International,* Vol.23, pp. 117–133.

Fried, D.; Featherstone, J.D.B.; Visuri, S.R.; Seka, W.D. & Walsh, J.T. (1996). The caries inhibition potential of Er:YAG and Er:YSGG laser radiation. *SPIE Proceedings,* 2672, pp. 73–78.

Fusayama, T.; Nakamura, M.; Kurosaki, N. & Iwaku, M. (1979). Non-pressure adhesion of a new adhesive restorative resin. *Journal of Dental Research,* Vol.58, pp. 1364–1370.

Gaggl, A.; Schultes, G.; Müller, WD. & Kärcher, H. (2000). Scanning electron microscopical analysis of laser-treated titanium implant surfaces–a comparative study. *Biomaterials,* Vol.21, pp. 1067–1073

Gökçe, B.; Özpınar, B.; Dündar, M.; Çömlekoglu, E.; Sen, BH. & Güngör, MA. (2007). "Bond Strengths of All Ceramics: Acid vs Laser Etching". *Operative Dentistry,* Vol.32, pp. 168-173.

Guazzato, M.; Albakry, M.; Ringer, SP. & Swain, MV. (2004). Strength, fracture toughness and microstructure of a selection of all-ceramic materials. Part II. Zirconia-based dental ceramics. *Dental Materials,* Vol.20, pp. 449–456.

Harashima, T.; Kinoshita, J. & Kimura, Y.; Brugnera, A.; Zanin, F.; Pecora JD. & Matsumoto, K. (2005). Morphological comparative study on ablation of dental hard tissues at cavity preparation by Er:YAG and Er,Cr:YSGG lasers. *Photomedicine and Laser Surgery,* Vol.23, No.1, pp. 52–55.

Haselton, DR.; Diaz-Arnold, AM. & Hillis, SL. (2000). Clinical assess- ment of high-strength all-ceramic crowns. *Journal of Prosthetic Dentistry,* Vol.83, pp. 396–401.

Hibst, R. & Keller, U. (1989). Experimental studies of the application of the Er:YAG laser on dental hard substances: I. Measurement of the ablation rate. *Laser in Surgery and Medicine,* Vol.9, No.4 , pp. 338- 344.

Hossain, M.; Nakamura, Y.; Yamada, Y.; Kimura, Y.; Nakamura, G. & Matsumoto, K. (1999). Ablation depths and morphological changes in human enamel and dentin after Er:YAG laser irradiation with or without water mist. *Journal of Clinical Laser in Medical Surgery,* Vol.17, pp. 105- 109.

Hossain, M.; Nakamura, Y.; Kimura, Y.; Yamada, Y.; Ito, M. & Matsumoto, K. (2000). Caries-preventive effect of Er:- YAG laser irradiation with or without water mist. *Journal of Clinical Laser in Medical Surgery,* Vol.18, pp. 61–65.

Hossain, M.; Nakamura, Y.; Yamada, Y.; Suzuki, N.; Mur- akami, Y. & Matsumoto, K. (2001). Analysis of surface roughness of enamel and dentin after Er,Cr:YSGG laser ir-radiation. *Journal of Clinical Laser in Medical Surgery,* Vol.19, pp. 297–303.

Hummel, M. & Kern, M. (2004). Durability of the resin bond strength to the alumina ceramic Procera. *Dental Materials,* Vol.20, No.5, pp. 498–508

Jacobsen, NL.; Mitchell, DL.; Johnson, DL. & Holt, RA. (1997). Lased and sandblasted denture base surface preparations affecting resilient liner bonding. *Journal of Prosthetic Dentistry,* Vol.78, No.2, pp. 153–158

Janda, R.; Roulet, JF.; Wulf, M. & Tiller, H-J. (2003). A new adhesive technology for all-ceramics. *Dental Materials,* Vol.19, No.6, pp. 567–573

Karakoca, S. & Yılmaz, H. (2009). Influence of surface treatments on surface roughness, phase transformation, and biaxial flexural strength of Y-TZP ceramics. *Journal of Biomedical Materials Research Part B: Applied Biomaterials,* Vol.91, No.2, pp. 930–937

Kayano, T.; Ochiai, S.; Kiyono, K.; Yamamato, H.; Nakajima, S. & Mochizuki, T. (1989). Effects of Er:YAG laser irradiation on human extracted teeth. *The Journal of the Stomatological Society, Japan,* Vol.56, No.2, pp. 381–392

Kataumi, M.; Nakajima, M.; Yamada, T. & Tagami, J. (1998). Tensile Bond strength and SEM evaluation of Er:YAG laser irradiated dentin using dentin adhesive. *Dental Materials Journal,* Vol.17, pp. 125–138

Keller, U. & Hibst, R. (1989). Experimental studies of the application of the Er:YAG laser on dental hard substances: II. Light microscopic and SEM investigations. *Lasers in Surgery and Medicine,* Vol.9, No.4, pp. 345–351

Kern, M. & Thompson, V.P. (1994). Sandblasting and silica coating of a glass-infiltrated alumina ceramic: volume loss, morphology, and changes in the surface composition. *Journal of Prosthetic Dentistry,* Vol.71, No.5, pp. 453–461

Kern, M. & Thompson, V.P. (1995). Bonding to glass infiltrated alumina ceramic: adhesive methods and their durability. *Journal of Prosthetic Dentistry,* Vol.73, No.3, pp. 240–249

Kern, M. & Wegner, SM. (1998). Bonding to zirconia ceramic: adhesion methods and their durability. *Dental Materials,* Vol.14, No.1, pp. 64–71

Kim, BK.; Bae, HE.; Shim, JS. & Lee, KW. (2005). The influence of ceramic surface treatments on the tensile bond strength of composite resin to all-ceramic coping materials. *Journal of Prosthetic Dentistry,* Vol.94, No.4, pp. 357–362

Kim, JT. & Cho, SA. (2009). The effects of laser etching on shear bond strength at the titanium ceramic interface. *Journal of Prosthetic Dentistry*, Vol.101, No.2, pp.101–106

Kramer, N.; Popp, S.; Sindel, J. & Frankenberger, R. (1996). Einfluss der Vorbehandlung von ompositinlays auf die Verbundfestigkeit. *Deutsch Zahnarztl Z*, Vol.51, pp. 598–601

Kumbuloglu, O.; Lassila, L.V.; User, A. & Vallittu, P.K. (2006). Bonding of resin composite luting cements to zirconium oxide by two air-particle abrasion methods. *Operative Dentistry*, Vol.31, No.2, pp. 248–255

Kutsch, VK. (1993). Lasers in dentistry: comparing wavelengths. *Journal of the American Dental Association,* Vol.124, No.2, pp.49-54

Lee, B.S.; Lin, P.Y.; Chen, M.H.; Hsieh, TT.; Lin, CP.; Lai, JY. & Lan, WH. (2007). Tensile bond strength of Er,Cr:YSGG laser-irradiated human dentin and analysis of dentin–resin interface. *Dental Materials,* Vol.23, No.5, pp.570–578.

Leinfelder, KF. (2001). Dentin adhesives for the twenty-first century. *The Dental Clinics of North America*, Vol.45, No.1, pp. 1-6

Li, R.; Ren, Y. & Han, J. (2000). Effects of pulsed Nd:YAG laser irradiation on shear bond strength of composite resin bonded to porcelain. *Hua Xi Kou Qiang Yi Xue Za Zhi,* Vol.18, No.6, pp. 377–379.

Li, Z.Z.; Code, J.E. & Van De Merwe, W.P. (1992). Er:YAG laser ablation of enamel and dentin of human teeth: Determination of ablation rates at various fluencies and pulse repetition rates. *Lasers in Surgery and Medicine,* Vol.12, No.6, pp. 625–630

Liberman, R.; Segal, TH.; Nordenberg, D. & Serebro, LI. (1984). Adhesion of composite materials to enamel: Comparison between the use of acids lasing as pretreatment. *Lasers in Surgery and Medicine*, Vol.4, No.4, pp. 232-237

Liu, J.F.; Liu, Y. & Stephen, H.C. (2006). Optimal Er:YAG laser energy for preventing enamel demineralization. *Journal of Dentistry,* Vol.34, No.1, pp. 62–66

Malmström, HS.; McCormack, SM.; Fried, D. & Featherstone, JD. (2001). Effect of CO_2 laser on pulpal temperature and surface morphology: an in vitro study. *Journal of Dentistry,* Vol.29, No.8, pp. 521-529

Marshall Jr., GW.; Marshall, SJ.; Kinney, JH. & Balooch, M. (1997). The dentin substrate: structure and properties related to bonding. *Journal of Dental Research*, Vol.25. No.6, pp. 441-458

Matinlinna, JP. & Vallitu, PK. (2007). Bonding of resin composites to etchable ceramic surfaces - an insight review of the chemical aspects on surface conditioning. *Journal of Oral Rehabilitation,* Vol.34, No.8, pp. 622-630

McCormack, SM.; Fried, D.; Featherstone, JD.; Glena, RE. & Seka, W. (1995). Scanning electron microscope observations of CO2 laser effects on dental enamel. *Journal of Dental Research*, Vol.74, No.10, pp. 1702-1708

Moretto, SG.; Azambuja, N.Jr.; Arana–Chavez, V.E; Reis, AF.; Giannini M.; Eduardo C de P. & De Freitas, PM. (2010). Effects of ultramorphological changes on adhesion to lased dentin-Scanning electron microscopy and transmission electron microscopy analysis. *Microscopy Research Technique*, Vol.74, No.8, pp. 720–726

Nakabayashi, N.; Kojima, K. & Mashura E. (1982). The promotion of adhesion by resin infiltration of monomers into tooth structure. *Journal of Biomedical Materials Research*, Vol.16, pp. 265-273

Navarro, RS.; Gouw-Soares, S.; Cassoni, A.; Haypek, P.; Zezell, DM. & Eduardo, CP. (2010). The influence of erbium:yttrium-aluminum-garnet laser ablation with variable pulse width on morphology and microleakage of composite restorations. *Lasers in Medical Science*, Vol.25, No.6, pp. 881–889

Noda, M.; Okuda, Y.; Tsuruki, J.; Minesaki, Y.; Takenouchi, Y. & Ban, S. (2010). Surface damages of zirconia by Nd:YAG dental laser irradiation. *Dental Materials Journal*, Vol.29, No.5, pp. 536-541

Obata, A.; Tsumura, T.; Niwa, K.; Ashizawa, Y.; Deguchi, T. & Ito, M. (1999). Super pulse CO2 laser for bracket bonding and debonding. *European Journal of Orthodontics*, Vol.21, No.2, pp. 193–198

Oelgiesser, D.; Blasbalg, J. & Ben-Amar, A. (2003). Cavity preparation by Er-YAG laser on pulpal temperature rise. *American Journal of Dentistry*, Vol.16, No.2, pp. 96-98

Oliveira, MT.; Arrais, CA.; Aranha, AC.; Paula Eduardo, C.; Miyake, K.; Rueggeberg, FA. & Giannini, M. (2010). Micromorphology of resin–dentin interfaces using one-bottle etch & rinse and self-etching adhesive systems on laser- treated dentin surfaces: a confocal laser scanning microscope analysis. *Lasers in Surgery and Medicine*, Vol.42, No.7, pp. 662–670

Olsen, ME.; Bishara, SE.; Damon, P. & Jakopsen, JR. (1997a) Comparison of shear bond strength and surface structure between conventional acid etching and air abrasion of human enamel. *American Journal of Orthodontics and Dentofacial Orthopedics*, Vol.112, No.5, pp. 502-506

Olsen, ME.; Bishara, SE.; Damon, P. & Jakopsen, JR. (1997b). Evaluation of Scotchbond multipurpose and maleic acid as alternative methods of bonding orthodontic brackets. *American Journal of Orthodontics and Dentofacial Orthopedics*, Vol.111, No.5, pp. 498-501

Özcan, M.; Alkumru, HN. & Gemalmaz, D. (2001). The effect of surface treatment on the shear bond strength of luting cement to a glass-infiltrated alumina ceramic. *International Journal of Prosthodontics*, Vol.14, No.4, pp. 335–339

Özcan, M. (2002). The use of chairside silica coating for different dental applications: a clinical report. *The Journal of Prosthetic Dentistry*, Vol.87, No.5, pp. 469–472

Özcan M. (2003). Adhesion of resin composites to biomaterials in dentistry: an evaluation of surface conditioning methods. PhD Thesis, University of Groningen, Groningen, The Netherlands

Parker, S. (2004). The use of lasers in fixed prosthodontics. *Dental Clinics of North America*, Vol.48, No.4, pp. 971-998

Pashley, DH. (1992). The effects of acid etching on the pulpodentin complex. *Operative Dentistry*, Vol.17, pp. 229-242

Pashley, DH. & Carvalho, RM. (1997). Dentin permeability and dentin adhesion. *Journal of Dentistry*, Vol.25, pp. 355-372

Phark, JH.; Duarte S, Jr.; Blatz, M. & Sadan, A. (2009). An in vitro evaluation of the long-term resin bond to a new densely sintered high-purity zirconium-oxide ceramic surface. *The Journal of Prosthetic Dentistry*, Vol.101, No.1, pp. 29–38

Ramos, A.C.; Esteves-Oliveira, M., Arana-Chavez, V.E. & de Paula Eduardo, C. (2010). Adhesives bonded to erbium: yttrium–aluminum–garnet laser-irradiated dentin: transmission electron microscopy, scanning electron microscopy and tensile bond strength analyses. *Lasers in Medicine and Science*, Vol.25, No.2, pp. 181–189

Reynold, IR. (1975). A review of direct bonding. *British Journal of Orthodontics*, Vol.2, pp. 171-180

Roberts-Harry, DP. (1992). Laser etching of teeth for orthodontic bracket placement: A preliminary clinical study. *Lasers in Surgery and Medicine*, Vol.12, No.5, pp. 467-470

Sakakibara, Y.; Ishimaru, K. & Takamizu, M. (1998). A study on bond strength to dentin irradiated be Erbium:YAG laser. *The Japanese Journal of Conservative Dentistry*, Vol.41, pp. 207-219

Sasaki, LH.; Lobo, PD.; Moriyama, Y.; Watanabe, IS.; Villaverde, AB.; Tanaka, CS.; Moriyama, EH. & Brugnera A, Jr. (2008). Tensile bond strength and SEM analysis of enamel etched with Er:YAG laser and phosphoric acid: a comparative study in vitro. *Brazilian Dental Journal*, Vol.19, No.1, pp. 57-61

Sato, H.; Yamada, K.; Pezzotti, G.; Nawa, M. & Ban, S. (2008). Mechanical properties of dental zirconia ceramics changed with sandblasting and heat treatment. *Dental Materials Journal*, Vol.27, No.3, pp. 408–414

Schein, MT.; Bocangel, JS.; Nogueira, GE. & Schein, PA. (2003). SEM evaluation of the interaction pattern between dentin and resin after cavity preparation using ER:YAG laser. *Journal of Dentistry*, Vol.31, No.2, pp. 127-135

Schmage, P.; Nergiz, I.; Herrmann, W. & Özcan, M. (2003). Influence of various surface-conditioning methods on the bond strength of metal brackets to ceramic surfaces. *American Journal of Orthodontic Dentofacial Orthopedic*, Vol.123, No.5, pp. 540-546

Sen, D.; Poyrazoglu, E.; Tuncelli, B. & Goller, G. (2000). Shear bond strength of resin luting cement to glass-infiltrated porous aluminum oxide cores. *The Journal of Prosthetic Dentistry*, Vol.83, No.2, pp. 210–215

Şen S & Ceylan G. (2010) The Effects of Different Surface Treatments on the Bond Strength of Zirconium-Oxide Ceramic and Adhesive Resin. PhD Thesis. Ondokuz Mayıs University, School of Dentistry Department of Prosthodontics, Samsun, Turkey.

Sharpe, AN. (1967). Influence of the crystal orientation in human enamel on its reactivity to acid as shown by high resolution microradiography. *Archieves of Oral Biology*, Vol.12, No.5, pp. 583-591

Sindel, J.; Gehrlicher, S. & Petschel, A. (1996). Untersuchungen zur Haftung von Kompositan VMK-Kerakim. *Deutsch Zahnarztl Z.*, Vol.51, pp. 712–716

Spohr, AM.; Borges, GA.; Júnior, LH.; Mota, EG. & Oshima, HM. (2008). Surface modification of In-Ceram Zirconia ceramic by Nd:YAG laser, Rocatec system, or aluminum oxide sandblasting and its bond strength to a resin cement. *Photomedicine and Laser Surgery*, Vol.26, No.3, pp. 203-208

Stern, RH.; Vahl, J. & Sognnaes, RF. (1972). Lased enamel: ultrastructural observations of pulsed carbon dioxide laser effects. *Journal of Dental Research*, Vol.51, No.2, pp. 455-460

Stu̍binger, S.; Homann, F.; Etter, C.; Miskiewicz, M.; Wieland, M. & Sader, R. (2008). Effect of Er:YAG, CO2 and diode laser irradiation on surface properties of zirconia endosseous dental implants. *Lasers in Surgery and Medicine*, Vol.40, No.3, pp. 223-228

Subaşı, MG. & Inan, O. (2011). Evaluation of the topographical surface changes and roughness of zirconia after different surface treatments. *Lasers in Medical Science*, July 24. [Epub ahead of print]

Tachibana, A.; Marques, MM.; Soler, JM. & Matos, AB. (2008). Erbium, chromium:yttrium scandium gallium garnet laser for caries removal: influence on bonding of a self-etching adhesive system. *Lasers in Medical Science*, Vol.23, pp. 435–441

Toksavul, S. & Toman, M. (2007). A short-term clinical evaluation of IPS Empress 2 crowns. *The International Journal of Prosthodontics*, Vol.20, No.2, pp. 168–172

Trajtenberg, CP.; Pereira, PN. & Powers JM. (2004). Resin bond strength and micromorphology of human teeth prepared with an Erbium:YAG laser. *American Journal of Dentistry*, Vol.17, No.5, pp. 331-336

Ural, Ç.; Külünk, T.; Külünk, Ş. & Kurt, M. (2010). The effect of laser treatment on bonding between zirconia ceramic surface and resin cement. *Acta Odontologica Scandinavica*, Vol.68, No.6, pp. 354-359

Van Meerbeek, B.; De Munck, J. & Yoshida, Y. (2003). Buonocore memorial lecture. Adhesion to enamel and dentin: current status and future challenges. *Operative Dentistry*, Vol.28, pp. 215–235

Van Noort, R. (2002a). Dental Ceramics, In: *An introduction to dental materials*, Van Noort R, (Ed.), 231–246, Elsevier Science, Hong Kong

Van Noort, R. (2002b). Principles of adhesion. In: *An introduction to dental materials*, Van Noort R, (Ed.), 68-78, Elsevier Science, Hong Kong

Visuri, SR.; Gilbert, JL.; Wright, DD.; Wigdor, HA. & Walsh, JT. Jr. (1996a). Shear strength of composite bonded to Er:YAG laser-prepared dentin. *Journal of Dental Research*, Vol.75, pp. 599–605

Visuri, SR.; Walsh, JT.; & Wigdor, HA. (1996b). Erbium laser ablation of dental hard tissue: Effect of water cooling. *Lasers in Surgery and Medicine*, Vol.18, No.3, pp. 294–300

Walsh, LJ. (1994). Clinical evaluation of dental hard tissue applications of carbon dioxide lasers. *Journal of Clinical Laser Medicine & Surgery*, Vol.12, No.1, pp. 11-15

Watanabe, I.; McBride, M.; Newton, P. & Kurtz, KS. (2009). Laser surface treatment to improve mechanical properties of cast titanium. *Dental Materials*, Vol.25, No.5, pp. 629–633

White, JM.; Goodis, HE.; Setcos, JC.; Eakle S.; Hulscher BE. & Rose CL. (1993). Effects of pulsed Nd:YAG laser energy on human teeth: a three-year follow- up study. *Journal of American Dental Association*, Vol.124, No.7, pp. 45–51

Wigdor, H.; Abt, E.; Ashrafi, S. & Walsh JT. Jr. (1993). The effect of lasers on dental hard tissues. *Journal of American Dental Association,* Vol.124, No.2, pp. 65–70

Wilder-Smith, P.; Lin, S.; Nguyen, A.; Liaw LH.; Arrastia AM.; Lee JP. & Berns MW. (1997). Morphological effects of ArF excimer laser irradiation on enamel and dentin. *Lasers in Surgery and Medicine*, Vol.20, No.2, pp. 142–148

Zachrisson, BU. & Buyukyılmaz, T. (1993). Recent advances in bonding to gold, amalgam and porcelain. *Journal of Clinical Orthodontics*, Vol.27, pp. 661-675

Zhang, Y.; Lawn, BR.; Rekow, ED. & Thompson, VP. (2004). Effect of sandblasting on the long-term performance of dental ceramics. *Journal of Biomedical Materials Research. Part B Applied Biomaterials*, Vol.71, No.2, pp. 381–386

Zhang, Y.; Lawn, BR.; Malament, KA.; Van Thompson, P. & Rekow, ED. (2006). Damage accumulation and fatigue life of particle-abraded ceramics. *The International Journal of Prosthodontics*, Vol.19, No.5, pp. 442–448

11

The Application of Scanning Electron Microscope (SEM) to Study the Microstructure Changes in the Field of Agricultural Products Drying

Hong-Wei Xiao and Zhen-Jiang Gao
College of Engineering, China Agricultural University, Qinghua Donglu, Beijing, China

1. Introduction

The objective of this part: Highlight the significance of microstructure investigation in the field of agricultural product drying or dehydration.

It is a common sense that structure of material determines its function. The change of macroscopic properties of materials is caused by the changes of its microstructure. For example, a porous structure as a honeycomb would facilitate rapid water diffusion or promote a rapid water uptake during drying or cooking. On the contrary, a compact structure or fewer pores at the surface of the product can cause a slower moisture migration during drying or water penetration into the interior during rehydration or cooking. Therefore, microstructure investigation can help quantifying product changes during processing and may also improve the understanding of mechanisms and changes in quality factors, especially the changes in food texture (**Aguilera & Stanley, 1999; Xiao et al., 2009**). For example, the pore sizes and the number of pores can significantly influence the texture of food. Smaller number of pores and small sizes led to the dense structure. While, larger number of pores and large pore size can cause a decrease of the hardness of the product.

2. The classification of the specific research

The objective of this part: Through concrete examples to introduce the main content and conclusions of the microstructure investigation in the field of agricultural materials drying.

2.1 Observing the microstructure of the material surface before and after the processing to find out the effect of drying on the product microstructure

The scanning electron micrographs on the surface of fresh and dried samples can be used to analyze the microstructure changes during drying or dehydration process. Fresh apple tissue has a well-organized structure consisting of cells and intercellular spaces, however, the breakdown of cell walls, a decreased intercellular contact and collapse of cell structure were found in the dried apple tissues (Deng &Zhao, 2008 a and b).

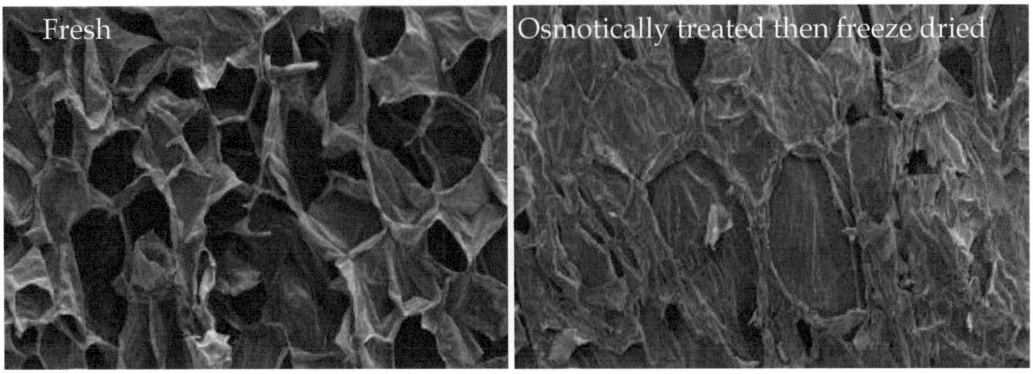

Fig. 1. Scanning electron micrograph of fresh and dried apple (Fuji)
(Deng & Zhao, 2008 a and b)

Take the pressure pulsed osmotic dehydration (PPOD) of salt eggs using NaCl solution for another example. Osmotic dehydration of salt eggs is an ancient method of egg preservation, which can be traced back to several hundreds of years. With special taste and flavour, salt eggs is one of the most popular egg food in Asian countries especially in China. Whereas, presently in China the osmotic dehydration of egg is done manually, the process of which is tedious, time consuming and labor intensive. The traditional osmotic dehydration process involves submerging eggs in NaCl solution in a static situation which last about 30 days at room temperature. During this process the NaCl transfer from salt solution through eggshell into the egg white and yolk. In order to increase the automation and decrease the processing time, the PPOD technology has been applied in salting eggs, which has been proved more efficient reducing the osmotic time from 30 days to 2 or 3 days (Chen and Gao, 2006; Wang and Gao, 2010). However, the mechanisms of PPOD of salt eggs hasn't be explored.

The authors try to find the mechanism using the microstructures of the eggshell before and after PPOD, as shown in Figure 2. From Figure 2, it can be found that the pore sizes and the number of pores in the surface of eggshell was increased after the processed of PPOD, which could significantly facilitate the NaCl immigration from the salt solution to the egg inside. As regards the cross section of eggshell, it can be observed that after PPOD the eggshell become less dense and a few pore channels were formed, which enabled the NaCl to penetrate easily from solution to egg inside and increased the moisture transport from egg to the solution, and thus accelerated the osmotic dehydration process. This phenomenon may be due to the tunneling effect of PPOD, which can create canals in the microstructure of eggshell during PPOD by stretching and enlarging the pore sizes. The membrane of eggshell is the most resistance for mass transfer during osmotic dehydration of salt egg (Chen et al., 1999). From Figure 2, it can be found that the eggshell membrane comprised of multi-layers of "fibers" as a bird's nest built of sticks. It can be also found that after PPOD the microstructure of eggshell membrane become looser. Certainly, such a loose structure can promote mass transport and improve the overall process rate compared to the dense structure.

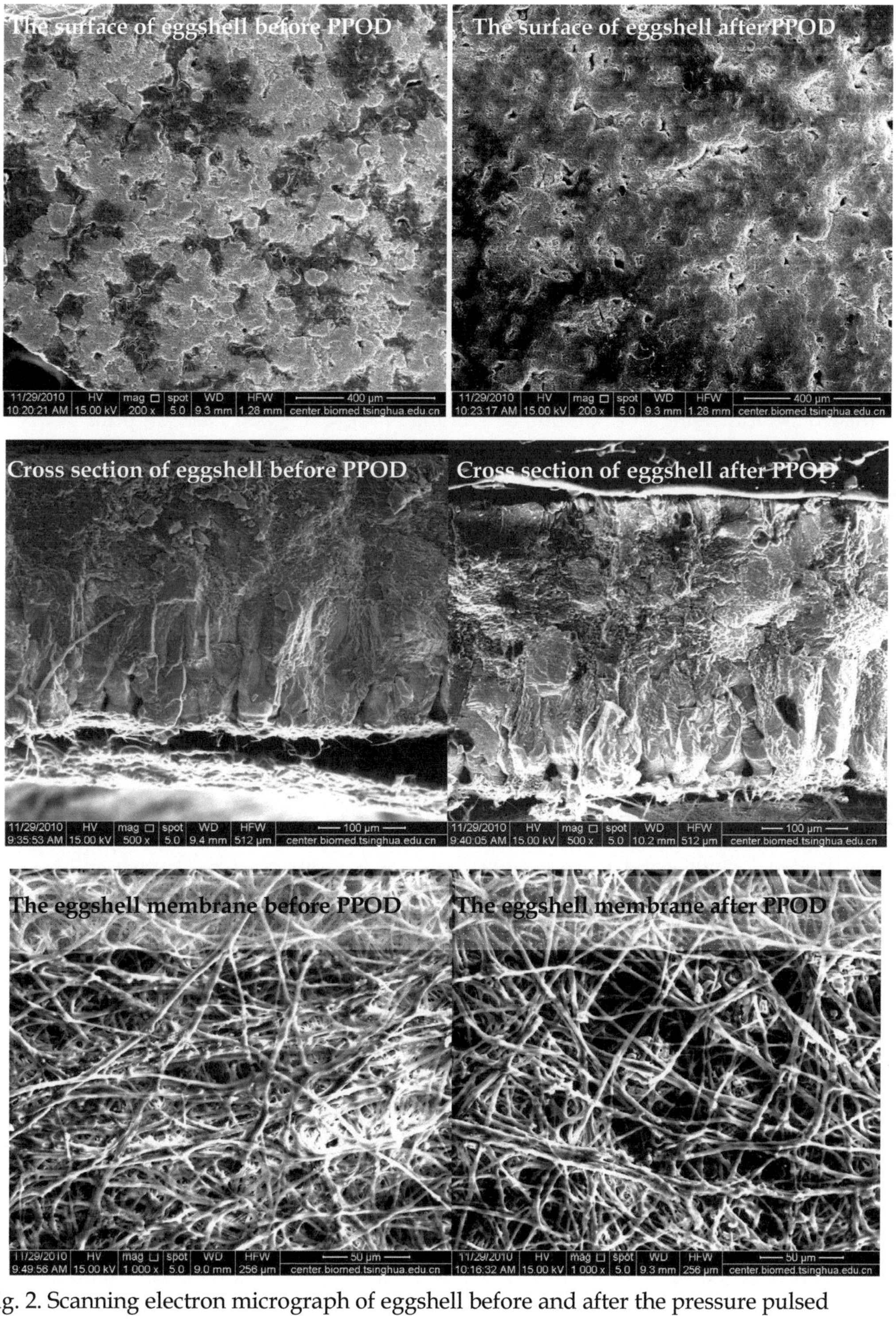

Fig. 2. Scanning electron micrograph of eggshell before and after the pressure pulsed osmotic dehydration (PPOD).

2.2 To investigate the effect of different pretreatment methods and drying conditions on the microstructure of the samples

The information on microstructure changes is essential for enabling better process control and improvement in the appearance by optimizing the pretreatment and drying parameters. The microstructure observation of the product's surface can be carried out to explore the effects of different pretreatment methods and drying temperatures on the microstructure changes of the samples. Recently, many researchers have investigated the microstructure changes of various fruits, vegetables and other food materials during pretreatment and drying process.

Vega-Gálvez et al. (2008) investigated the influence of pretreatment and air drying temperature on the quality and microstructure properties of rehydrated dried red bell pepper. Microscopic evaluation of the rehydrated pepper samples, as shown in **Figure3**, illustrated that the damage to cellular structure was minimized when the samples was pretreated by immersing in a solution containing NaCl, $CaCl_2$ and $Na_2S_2O_5$ prior to drying in comparison with the no pretreated samples. They also found that the drying temperature had an significant effect on the microstructure of the dried sample and the damage to cellular structure could be alleviated by decreasing the drying temperature.

Xiao et al. (2009) studied the effect of different pretreatments on drying kinetics and quality of sweet potato bars in terms of textural properties, microstructure, and colour undergoing air impingement drying. Microstructure observation of the surface of dried sweet potato bars, as shown in **Figure4**, was carried out to evaluate the effects of different pretreatments on the microstructure changes of the dried samples. It was found that when the sweet potato bars were subjected to hot water blanching and superheated steam blanching pretreatments, the dried samples had a homogeneous compact structure. In addition, no pores and starch granules were found on the surface of the samples. As a result, the structure would slower water transfer or penetration during drying or rehydration process. However, the samples subjected to citric acid pretreated for 30 min had large with non uniform pores and lots of starch granules on its surface. Absolutely, such porous structure would facilitate rapid water migration during drying. In terms of no pretreated ones, the dried sweet potato tissues showed more numerous starch granules and fewer pores than the citric acid pretreated samples on its surface. Therefore it is interesting to note that different pretreatments cause various changes of microstructure of the samples and lead to product properties varied differently.

Bondaruk, Markowski, Blaszczak (2007) investigated the effect of drying conditions on the quality of vacuum-microwave dried potato cubes in terms of colour, starch content, sugar content, mechanical properties and microstructure. Concerning the microstructure, it was observed that compared with the forced convection air drying the application of microwave energy led to different physical changes in the sample microstructure. It was also found that in the case of hot-air drying the intensity of structural changes depended on the drying temperature. In addition, a higher temperature causes greater damage to the microstructure of potato cubes.

Pimpaporn et al. (2007) reported that potato chips dried at 80 and 90°C had more uniform pore size and pore distribution compared with the chips dried at 70°C and more extensive surface shrinkage was found on the samples dried at 70°C. However, Fang et al. (2011)

reported that lower drying temperature led to relatively uniform size and shape with smooth particle surface, whereas higher drying temperature resulted in size variations and wrinkled particle surfaces when they carried out the milk spray drying under different drying temperatures. The SEM micrographs of milk powder under different spray drying temperatures were shown in **Figure5**.

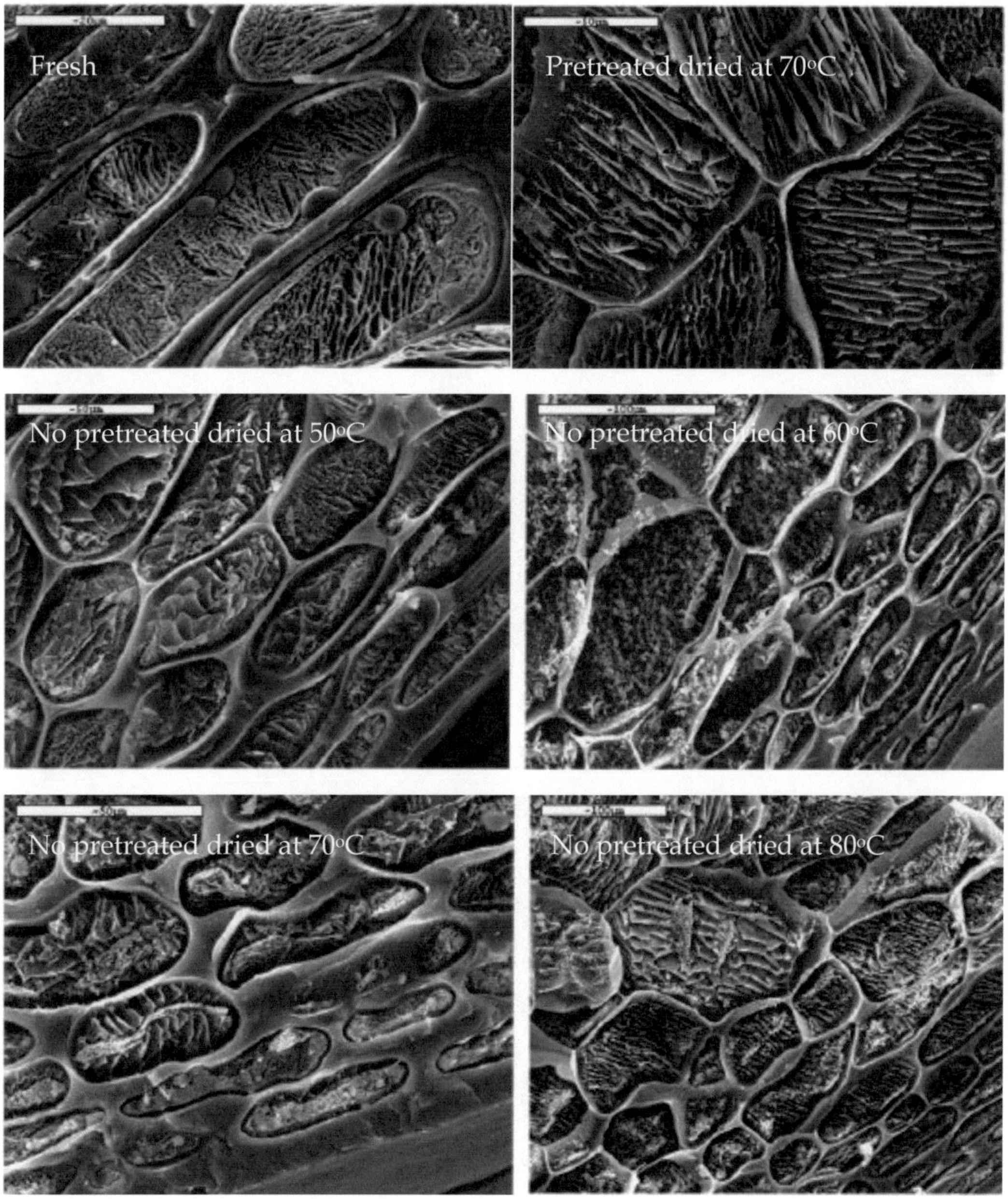

Fig. 3. SEM micrographs of fresh and rehydrated red pepper samples with and without pretreatment dried at different temperatures (Vega-Gálvez et al., 2008).

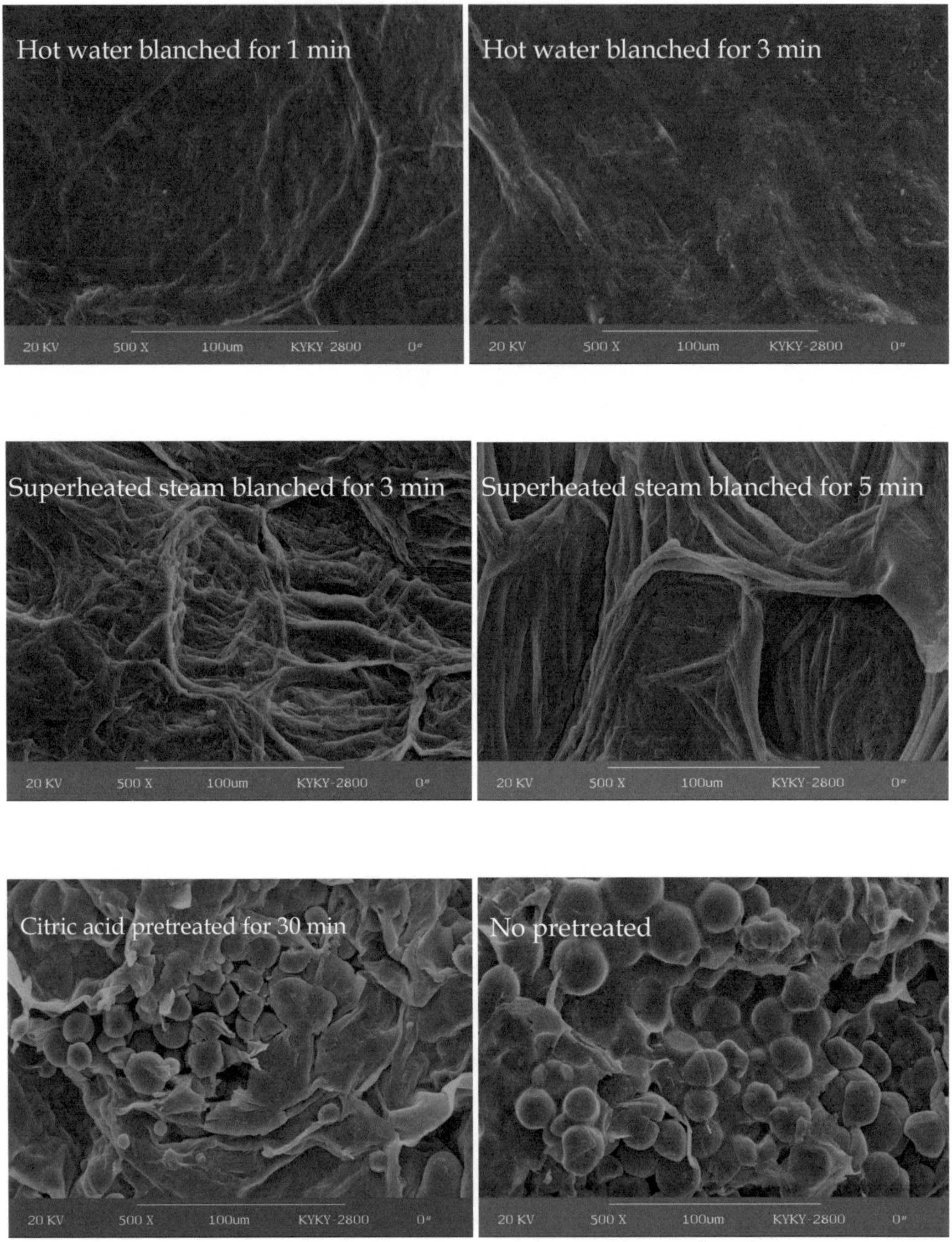

Fig. 4. SEM micrographs of the surface of dried sweet potato bars underwent different pretreatments (Xiao et al., 2009).

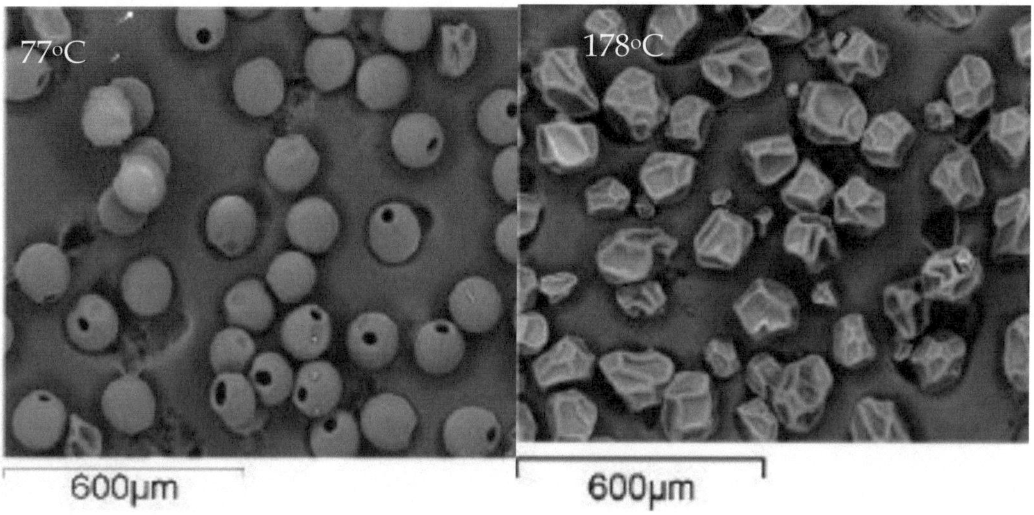

Fig. 5. SEM micrographs of milk powder under different spray drying temperatures (Fang, e t al., 2011).

2.3 Using SEM micrographs of materials to analyze the moisture transfer mechanisms during drying process and interpret the rehydration characteristics or the texture properties of the dried products

In general, most of the drying occurs in the falling rate period and moisture migration controls the whole process. Due to the limited information on the mechanism of moisture movement during drying and the complexity of the process that may involve molecular diffusion, capillary flow, Knudsen flow, hydrodynamic flow, surface diffusion and all other factors which affect drying characteristics, the moisture transfer mechanism hasn't been described completely (Madamba et al., 1996). Knowledge of the microstructure in which moisture and heat transfer take place may assist in finding the mechanisms and their relative contributions to the transport phenomena. Yang et al. (2010) made a try when they carried out an experiment to investigate the influence of glutinous components in plant tissue on the drying characteristics of plant materials taking Chinese angelica and Astragalus slices as the samples.

From the SEM micrographs of the sample (as shown in **Figure6**), they found that the trachea with relative large pore diameter was surrounded by massive parenchyma cells. It implied that during the drying process there was two parallel ways for moisture transfer from parenchyma cells to surrounding drying media: the direct moisture diffusion through some pores or open structures on the surface layer and the moisture emigrate from the parenchyma cells inside matrix to the surrounding drying media. Further more, they also reported that the process of moisture transfer from the parenchyma cells inside matrix to the surrounding drying media included three steps: firstly from the parenchyma cell to the adjacent cell via plasmodesma; secondly from the parenchyma cell to the adhered trachea via aperture; thirdly from tranchea to the surrounding drying media.

Rehydration ratio and texture of the samples, which is the macro performance of the material microstructure, is strongly dependent on the product microstructure. Therefore, the SEM micrographs can be used to analyze, interpret or even predict the rehydration characteristics or the texture properties of the dried products.

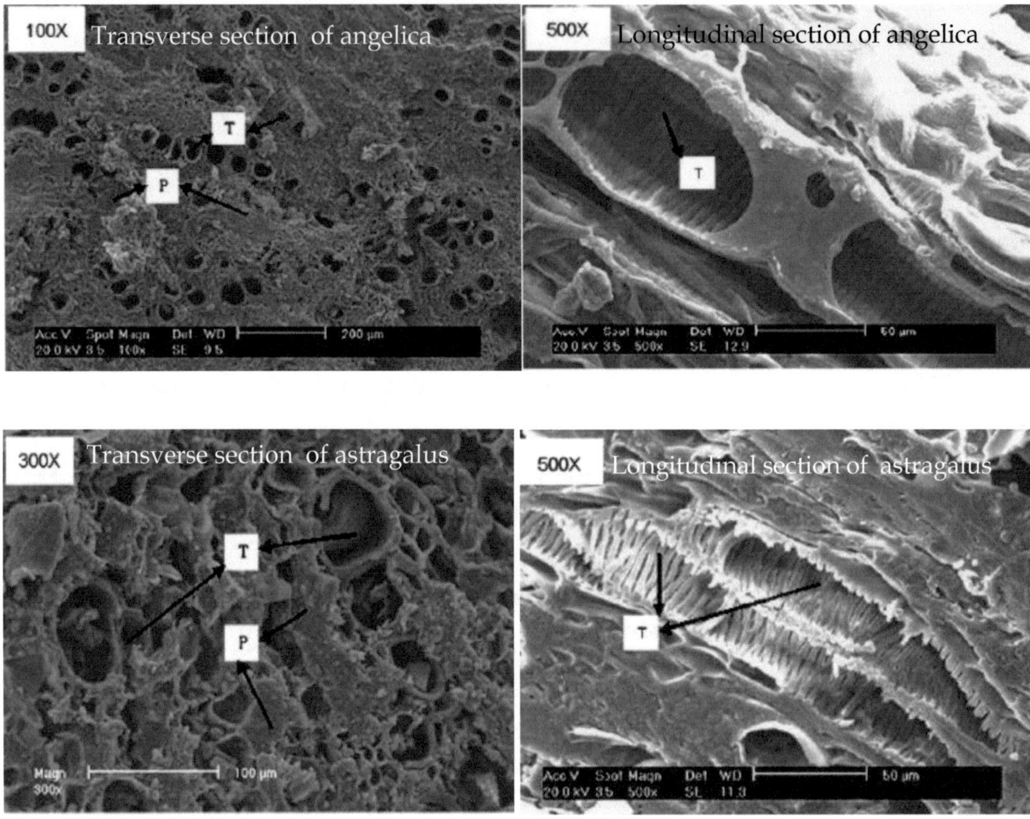

Fig. 6. The SEM micrographs of transverse section and longitudinal section of Chinese angelica slices and Astragalus sices (T part: trachea; P part: parenchyma cells). (Yang et al., 2010)

Thuwapanichayanan et al. (2011) studied the influence of drying temperatures on the moisture diffusivity and quality attributes of the dried banana slices in terms of volatile compound, shrinkage, colour, texture and microstructure. On the subject of microstructure, as shown in Figure7, they found that the drying temperature strongly affected the microstructure of dried banana and on the surface the pore sizes and the number of pores increased with increasing drying temperature, which significantly influenced the product texture in terms of hardness. The hardness of the samples dried at 90°C was lower than that dried at 70°C but was not significantly different from that dried at 80°C. This might be due to the effect of puffing that occurred more at higher temperatures and probably increased the porosity and resulted in a decrease of hardness and less shrinkage of the samples.

Brown et al. (2008) evaluated the microstructure characteristics of carrot pieces that had been dried using different techniques. They found that samples dried in ethano-modified supercritical carbon dioxide possessed many pores which could facilitate the movement of water into the internal structure and decrease the rehydrated and cooked time compared with the air-dried samples. Recently, similar results has been reported by Yang et al. (2010), who revealed that the larger porosity and total volume of the sample, the higher rehydration ratio of the sample.

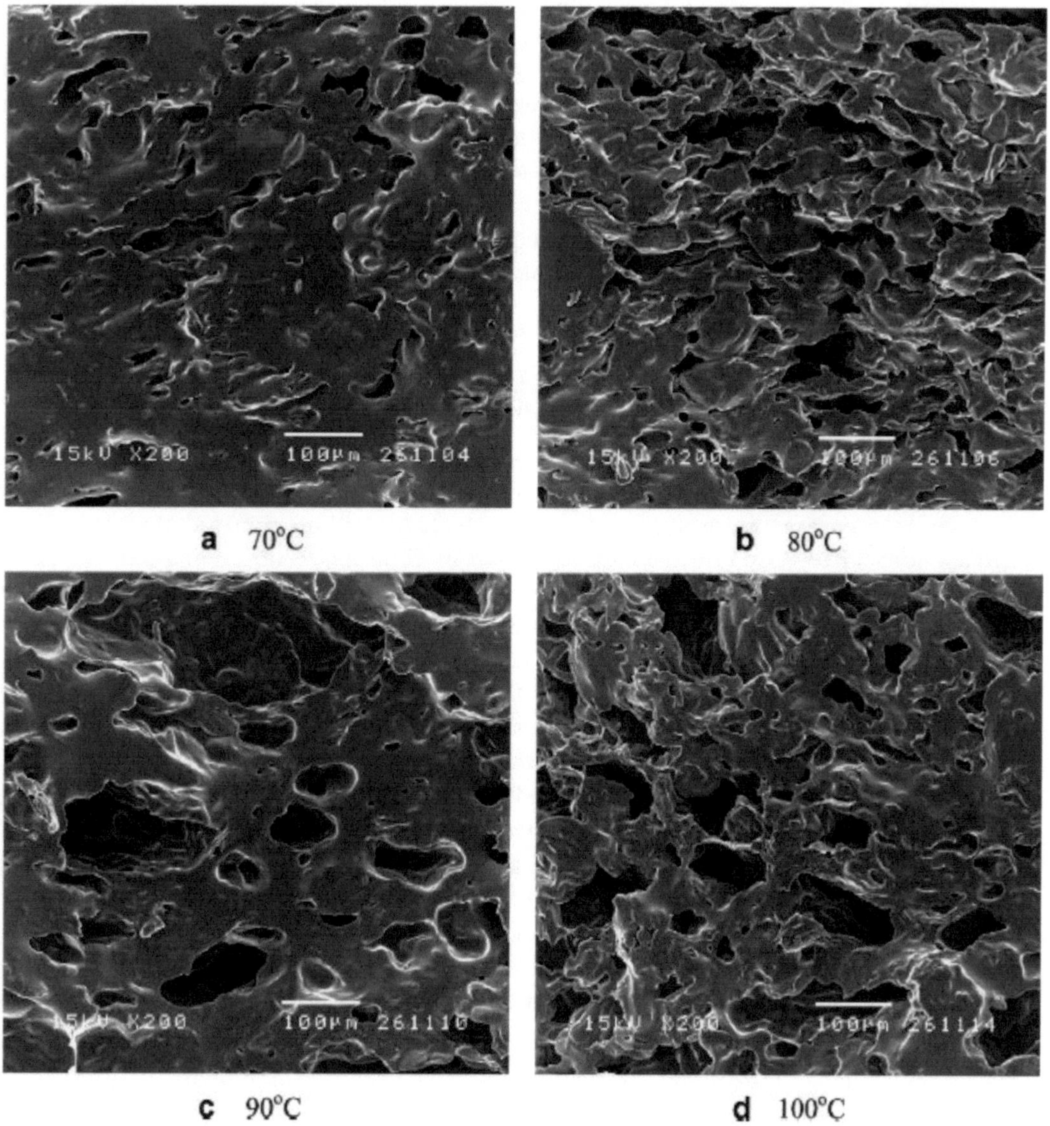

Fig. 7. SEM micrographs of banana slices dried at different temperatures (Thuwapanichayanan et al., 2011).

3. The causes leading to microstructure changes

The objective of this part: Analysis of the reasons behind the microstructure change phenomenon through specific case

3.1 Thermal and moisture gradients can cause cell wall disruption, deformation and folding during drying process

Deng & Zhao (2008b) explored the effect of different osmoconcentration pretreatment on glass transition temperature, texture, microstructure and calcium penetration of dried apples. In terms of microstructure, which was observed using SEM, it was found that it impossible to distinguish between cells and other spaces because of the disruption of cell walls, a decreased intercellular contact and the collapse of cell structure. Furthermore, they pointed that this phenomenon might be due to the fact that during drying process the transient thermal and moisture gradients causing cell wall breakdown, deformation and folding. Additionally, they also reported that the structural deformation, folding and collapses of cell structure might also associated with surface tension, environment pressure, moisture transport mechanism, and generation of internal pressure in samples during drying.

3.2 The microstructure changes of samples is closely associated with the stress developed in the tissue, which may be set up by shrinkage

Drying is a simultaneous heat and mass transfer process, shrinkage takes place when the materials are heated and moisture is lost. Since the composition of materials is very difference, take corn as an example which is constituted of soft floury endosperm, hard vitreous endosperm, germ and pericarp as shown in **Figure 8**. In addition, each substance has different shrinkage characteristics. These fact leads to the occurrence of non-uniform shrinkage during drying process. The non-uniform shrinkage within the product results in two types of internal stresses: the thermal stress which is due to temperature gradients within the material and the hydro stress which is due to moisture gradients within the product. When the combination of thermal and hydro stress exceeds the binding force between cells of the material stress crack or burst phenomena occur, which can change the macrostructure and microstructure of the product during drying process. As Wang and Brennan (1995) pointed that during drying internal cracks are formed and shrinkage stresses pull the tissue apart. Lewicki & Pawlak (2003) also demonstrated that physical changes are mostly due to stress and are pronounced by macro and micro alterations of size, shape and internal structure.

3.3 Stress in cell walls, phase changes in membrane lipids and chemical changes can also cause structural modifications of the product

Vega-Gálvez et al. (2011) explored different pretreatments such as high hydrostatic pressure, blanching, enzymatic and microwaves on the microstructure of Aloe vera gel during convective drying at 70°C, as shown in **Figure9**. It was found that the intact cellular structure of aloe was transformed into a more separated and ruptured cellular structure with non-distinct middle lamella. They ascribed this effect to the degradation of pectinacious material and the damage of most cell wall due to excessive strain in membranes and stress in cell walls during processing.

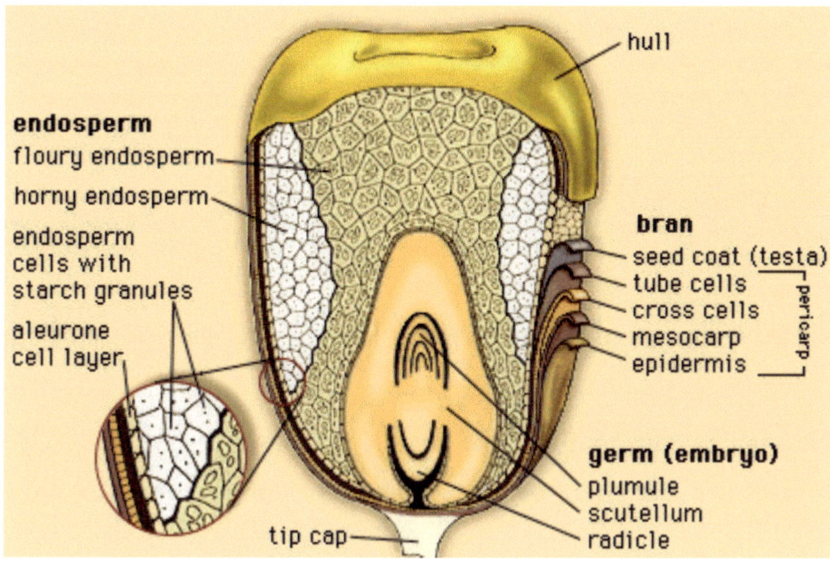

Fig. 8. The structure and compositon of corn kernel.

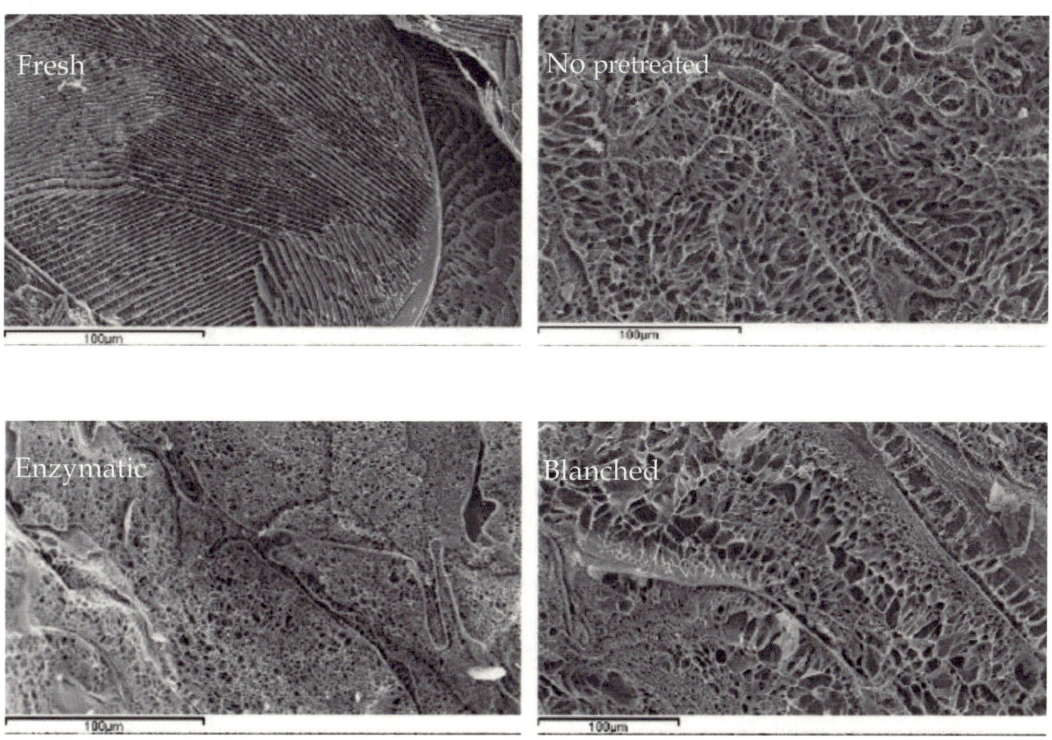

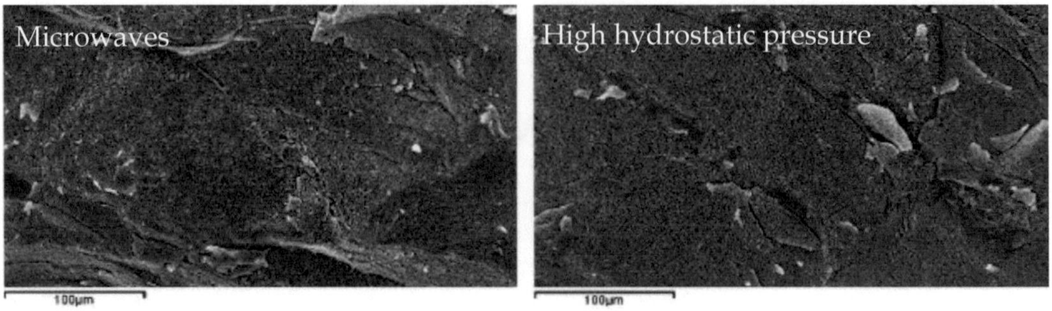

Fig. 9. The SEM micrographs of fresh and dried Aloe vera gel under different pretreatments (Vega-Gálvez et al., 2011).

4. Further research suggestion

The objective of this part: Point out the inadequacy of current research and the future research directions.

The microstructure of fresh and dried samples can provide a powerful tool and strong evidence for analyzing the properties changes of the samples during drying process. In this sense, research about the relationship between microstructure and the properties of the dried samples should be enhanced. In detail, more work needs to be performed on how the microstructure changes of dried food affects the food properties such as texture, rehydration ratio as well as its functionality, or even the availability of bioactive components.

Since information on microstructure change kinetics during products processing would be useful in predicting the quality changes during drying such as the texture and the surface shape of the product, thereby enabling better process control and improvement in the appearance by optimizing the process parameters. In addition, understanding the microstructure changes of the product is very important to clarify the change mechanism of quality during it processing. Therefore, the change kinetics of the microstructure during processing should be carried out in the further research.

5. References

Aguilera, J.M., & Stanley, D.W.(1999). Microstructural principles of food processing and engineering (second ed.). Gaithersburg: Aspen Publishers.

Bondaruk, J., Markowski, M., Blaszczak, W. (2007). Effect of drying conditions on the quality of vacuum-microwave dried potato cubes. *Journal of Food Engineering*, 81, 306-312.

Brown, Z.K., Fryer, P.J., Norton, I.T., Bakalis, S., Bridson, R.H. (2008). Drying of foods using supercritical carbon dioxide-Investigations with carrot. *Innovative Food Science and Emerging Technologies*, 9, 280-289.

Chen, S.T., Gao, Z.J. (2006). Study on the process of salted eggs under pulsed pressure. *Transaction of the CSAE*, 22, 163-166. (in Chinese with English abstract)

Chen, X.D., Freeman, Y., Guo, F., Chen, P. (1999). Diffusion of sodium chloride through chicken eggshell in relation to an ancient method of egg preservation. *Trans IchemE*, 77, Part C, 40-46.

Deng, Y., Zhao, Y.Y. (2008a). Effects of pulsed-vacuum and ultrasound on the osmodehydration kinetics and microstructure of apples (*Fuji*). *Journal of Food Engineering*, 85, 84-93.

Deng, Y., Zhao, Y.Y. (2008b). Effect of pulsed vacuum and ultrasound osmopretreatments on glass transition temperature, texture, microstructure and calcium penetrationof dried apples (Fuji). LWT-Food Science and Technology, 41, 1575-1585.

Fang, Y., Rogers, S., Selomulya, C., Chen, X.D. (2011). Functionality of milk protein concentrate: effect of spray drying temperature. *Biochemical Engineering Journal*, doi:10.1016/j.bej.2011.05.007

Lewicki, P.P., & Pawlak, G. (2003). Effect of drying on microstructure of plant tissue. *Drying Technology*, 21, 657-683.

Madamba, P. S., Griscoll, R. H., & Buckle, K. A. (1996). The thin layer drying characteristics of garlic slices. *Journal of Food Engineering*, 29, 75 - 97.

Pimpaporn, P., Devahastin, S. and Chiewchan, N. (2007). Effects of combined pretreatments on drying kinetics and quality of potato chips undergoing low-pressure superheated steam drying. *Journal of Food Engineering*, 81, 318-329.

Thuwapanichayanan, R., Prachayawarakorn, S., Kunwisawa, J., Soponronnarit, S. (2011). Determination of effective moisture diffusivity and assessment of quality attributes of banana slices during drying. *LWT-Food Science and Technology*, 44, 1502-1510.

Vega-Gálvez, A., Lemus-Mondaca, R., Bilbao-Sáinz, C., Fito, P., Andrés, A. (2008). Effect of air drying temperature on the quality of rehydrated dried red bell pepper (var. Lamuyo). *Journal of Food Engineering*, 85, 42-50.

Vega-Gálvez, A., Uribe, E., Perez, M., Tabilo-Munizaga, G., Vergara, J., Garcia-Segovia, P., Lara, E., Scala, K.D. (2011). Effect of high hydrostatic pressure pretreatment on drying kinetics, antioxidant activity, firmness and microstructure of Aloe vera (Aloe barbadensis Miller) gel. LWT-Food Science and Technology, 44, 384-391

Wang, N., Brennan, J.G. (1995). Changes in structure density and porosity of potato during dehydration. *Journal of Food Engineering*, 24, 61 - 76.

Wang, X.T., Gao, Z.J. (2010). Technologic optimization of pickled salted eggs under pulsed pressure. *Food Science*, 31, 97-101.(in Chinese with English abstract)

Xiao, H.W., Lin, H., Yao, X.D., Du, Z.L., Lou, Z., Gao, ZJ. (2009). Effects of different pretreatments on drying kinetics and quality of sweet potato bars undergoing air impingement drying. *International Journal of Food Engineering*, 5, Article 5.

Yang, J.H., Di, Q.Q., Zhao, J. (2010). Effect of glutinous components on matrix microstructure during the drying process of plant porous materials. *Chemical Engineering and Processing: Process Intensification*, 49, 286-293.

/

12

Scanning Electron Microscopy Imaging of Bacteria Based on Nucleic Acid Sequences

Takehiko Kenzaka and Katsuji Tani
Osaka Ohtani University, Faculty of Pharmacy, Japan

1. Introduction

Scanning electron microscopy (SEM) has been widely used in environmental microbiology to characterize the surface structure of biomaterials and to measure cell attachment and changes in morphology of bacteria. Moreover, SEM is useful for defining the number and distribution of microorganisms that adhere to surfaces. Traditionally, inability to provide phylogenetic or genetic information about microorganisms has been one limitation of SEM in environmental microbiology.

Modern molecular studies based on DNA and RNA sequence analysis have led to an understanding of the microbial diversity and composition of bacterial communities in various environments. Introduction of the concept of in situ hybridization (ISH) with RNA- or DNA-targeted fluorescent probes has led to important research regarding the identification and quantification of individual cells and has demonstrated great potential in the analysis of the composition of bacterial communities in the environment. Combining morphological study with SEM and ISH techniques (SEM-ISH) has provided new insights into the understanding of the spatial distribution of target cells on various materials.

2. Application of ISH techniques to SEM

2.1 Concepts of ISH

ISH is a method of detecting and localizing specific nucleic acid sequences in morphologically preserved tissues or cell preparations by hybridizing the complementary strand of a nucleotide probe to the sequence of interest. To detect target cells, the permeability of the cell and the visibility of the nucleotide sequence to the probe must be increased without destroying the structural integrity of the cell. The type of probe to be used and how to label it to yield better resolution with the highest accuracy should be taken into consideration.

In environmental microbiology, fluorescent in situ hybridization (FISH) with rRNA-targeted oligonucleotide probes has provided information about the absolute abundance, morphology, and cell size of bacteria with defined phylogenetic affiliations and had been applied to the investigation of community composition in lakes, river, oceans, activated sludge, drinking water, etc. (Amann et al., 1995). FISH analysis of bacterial communities with mRNA- or DNA-targeted probes presents a unique challenge because of its low sensitivity and resolution (Moraru et al., 2010; Pernthaler & Amann, 2004). The concepts of ISH can be incorporated into scanning electron microscopy imaging of microorganisms (Kenzaka et al., 2005a, 2009).

2.2 Metal labeling of target cells for SEM-ISH

To identify cells of interest by SEM, cells carrying specific DNA or RNA sequences must be labeled with a metal instead of fluorescent molecules. Target cells that were hybridized with oligonucleotide or polynucleotide probes can be labeled with gold or platinum (Fig. 1). In case of gold labeling of the target sequence, hybridization was performed with biotin-labeled oligonucleotide probes, and then target microbes were identified by nanogold-labeled streptavidin (Hacker, 1998; Fig. 1a). In case of the platinum probe complex, hybridization was performed with platinum-labeled oligonucleotide probes (Fig. 1b). Platinum-labeled probes can be prepared by allowing the platinum complex to bind to the probe at guanine-N7 atoms (Brabec & Leng 1993; Dalbiès et al., 1994). To amplify the signal in the both cases, gold enhancement was performed to enlarge the gold particles. In this reaction, gold ions in solution are catalytically deposited onto nanogold particles as metallic gold (Au^0). These particles grow in size as development time elapses. Consequently, hybridized cells contain gold up to tens of nm in size inside the hybridized cells. These hybridized cells release a strong backscatter electron signal (BSE) because of the accumulation of gold atoms inside cells (Kenzaka et al., 2005a, 2009).

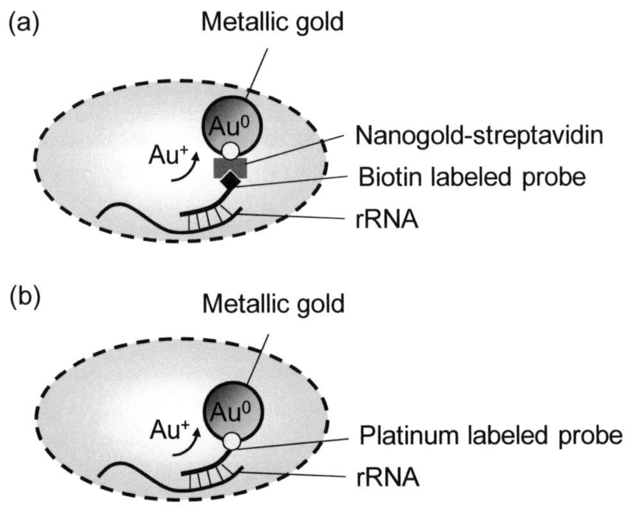

Fig. 1. Systematic representation of metal labeling for SEM-ISH. (a) Biotin labeled probes are hybridized with rRNA, followed by streptavidin-gold labeling, gold enhancement.
(b) Platinum labeled probes are hybridized with rRNA, followed by gold enhancement.

High-vacuum SEM images of a mixture of bacterial strains after ISH with an rRNA-targeted probe and labeling with gold are shown in Fig. 2. Secondary electron (SE) image demonstrate the surface topography of all cells (Fig. 2a). In BSE image, *Escherichia coli* cells hybridized with the ES445 probe which was targeted for *Escherichia-Shigella* (Kenzaka et al., 2001) were identified as bright spots because of the enhanced signal as a result of gold labeling (Fig. 2b). In hybridized cells, the amount of gold in the cells was greater than that on the cell surface; thus, this labeling resulted in a higher BSE signal. In the same microscopic field, both images could be viewed side by side.

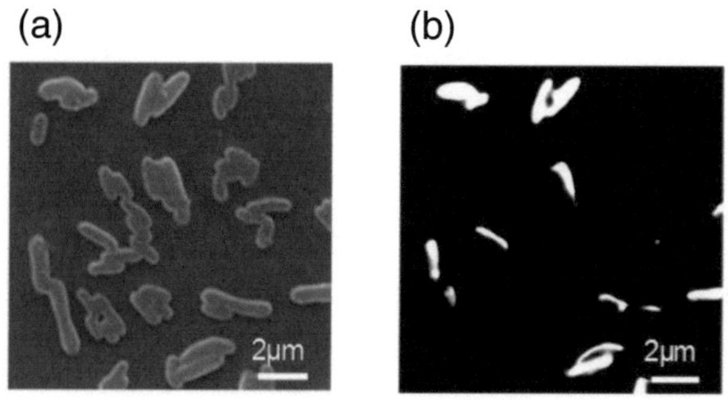

Fig. 2. High-vacuum SEM images of bacterial cells after ISH: Mixture of *Escherichia coli* and *Aeromonas sobria* cells hybridized with ES445 probe (targeted for *Escherichia-Shigella*). The same microscopic fields are shown with SE (a) and BSE images (b).

3. Experimental protocols for SEM-ISH

Experimental protocols for SEM-ISH are similar to FISH except metal labeling, and both techniques have the same challenges and limitations. For success in ISH, several issues need to be considered before proceeding with experiments: permeabilization and pretreatment, hybridization condition (composition of the hybridization solution, temperature, sodium concentration, and presence of organic solvents), washes, controls, etc. Here we discuss the general protocols for ISH in comparison to FISH.

3.1 Preparation of materials for ISH

Before processing for ISH, the specimens must be fixed and permeabilized to allow the penetration of the probes and reagents into the cells and protect the target RNA or DNA from degradation by nucleases. Fixation conditions may vary according to the target bacteria and type of sample. Optimal fixation results in good material penetration as well as the maintenance of cell integrity and morphological detail. For FISH, 3%–4% (v/v) formaldehyde or paraformaldehyde is generally suitable for gram-negative bacteria. In case of gram-positive bacteria, cells were fixed with 50% ethanol (Amann et al., 1995).

Before or after fixation, cells are usually prepared on glass slides or trapped on a polycarbonate filter. For aggregated samples or biofilms, species are fixed in formalin and then embedded in paraffin before being sectioned (Sekiguchi et al., 1999). For SEM-ISH, similar protocols with paraformaldehyde can be employed.

For better attachment of the specimens to the slide or polycarbonate filter, the surfaces were treated with a coating agent such as gelatin (Amann et al., 1990b), poly-L-lysine (Lee et al, 1999), or agarose (Pernthaler et al., 2002a). If peroxidase-labeled molecules were used for signal enhancement, an additional enzymatic treatment with lysozyme was required (Pernthaler et al., 2002b; Schönhuber et al., 1997). Some cases require further enzymatic treatment to open the peptidoglycan layer. Lysozyme and pancreatic lipase for enterococci (Waar et al., 2005), lysozyme and lysostaphin for staphylococci (Kempf et al., 2000), and lysozyme and achromopeptidase for actinobacteria (Sekar et al., 2003) have been previously employed. When peroxidase-labeled probes and antibodies were used, diethyl pyrocarbonate treatment or other additional treatments were required to inactivate intracellular peroxidase (Pernthaler et al., 2002a). Microbial communities in the natural environment are complex, and the permeability of their cell walls is not uniform. The application of mixed enzymes or other chemical treatments may be required (Pernthaler et al., 2002b), but it remains difficult to sufficiently permeabilize the cell walls of all complex bacterial communities.

3.2 Probes types

Probes are sequences of nucleotide bases complimentary to the specific DNA or RNA sequence of interest. These probes can be as small as 15–30 nucleotides or up to 1000 nucleotides. The strength of the binding between the probe and the target molecules is crucial in hybridization. This strength is affected by the various hybridization conditions described below.

Several types of probes can be used in performing ISH: oligonucleotide DNA probes, polynucleotide DNA probes, polyribonucleotide probes, peptide nucleic acid (PNA) probes, locked nucleic acid (LNA) probes, etc. Stability, availability, speed, expense, ease of use, specificity, cell wall penetration ability, and reproducibility should be considered for selecting a probe type.

3.2.1 Oligonucleotide DNA probes

An oligonucleotide DNA probe is a short sequence of nucleotides that are synthesized to match a target. They are synthetically produced, commercially available, and economical. The target nucleotide sequence must be known. The probes used to detect bacteria are small, generally approximately 15–30 bp. A small-sized probe allows for easy penetration into the bacterial cells of interest. For effective hybridization, the thermodynamics of nucleic acid hybridization based on Gibbs free energy change should be considered. The affinity of the probe to the target site is defined as the overall Gibbs free energy change for intramolecular DNA and RNA interactions that take place during ISH (Yilmaz & Noguera, 2004).

3.2.2 Polynucleotide DNA probes

Polynucleotide DNA probes have similar advantages to oligonucleotide DNA probes, but they are much larger, typically 50–1000 base long. These are synthetically produced by reverse transcription of RNA, or by PCR, or by fragmentation of the PCR product (Niki & Hiraga, 1998). Fragmented chromosomal DNA can be used as a probe (Lanoil & Giovannoni, 1997). However, synthesizing this type of probe requires time and expensive reagents. Polynucleotide DNA probes can be labeled at multiple sites with fluorescent dyes, digoxigenin, or biotin and thus are used to amplify probe-derived signals.

3.2.3 Polyribonucleotide probes

RNA probes have the advantage that RNA–RNA hybrids are considerably thermostable and resistant to digestion by RNases. In vitro transcription of plasmid DNA with RNA polymerase can be used to produce RNA probes (Delong et al., 1999; Pernthaler et al., 2004; Zwirglmaier et al., 2004). These probes, however, can be very difficult to work with as they are highly sensitive to ubiquitous RNases.

3.2.4 Peptide nucleic acid (PNA) probes

PNAs are the synthetic analogs of DNA. DNA and RNA have deoxyribose and ribose sugar backbones, respectively, whereas the backbone of a PNA backbone comprises repeating N-(2-aminoethyl)-glycine units that are linked by peptides (Nielsen et al., 1991). The backbone of PNAs contains no charged phosphate groups. Less electrostatic repulsion occurs when the PNA probe hybridizes to DNA or RNA sequences. The PNA–DNA or PNA–RNA complex is more stable than the natural nucleic acid complexes. PNA is not easily identified by either nucleases or proteases, making them resistant to enzyme degradation. Because of their higher binding strength, it is not necessary to design long PNA oligomers. Such oligomers are chemically synthesized and commercially available. PNA-FISH also has broad applications in clinical microbiology (Stender, 2003).

3.2.5 Locked nucleic acid (LNA) probes

LNAs are a class of analogs that contain an extra bridge connecting the 2' oxygen and 4' carbon (Obika et al., 1997). LNA nucleotides can be mixed with DNA or RNA residues in the oligonucleotide whenever desired. The LNA–DNA heteroduplex is thermostable. DNA probes with LNA have the advantage of higher affinity and specificity than normal DNA probes, and greater design flexibility and lower costs than PNA probes (Silahtaroglu et al., 2003). Such oligomers are chemically synthesized and commercially available. LNA nucleotides are used to increase the sensitivity and specificity of expression in DNA microarrays, FISH probes, real-time PCR probes, and other molecular biology techniques.

3.3 Hybridization

Hybridization must be performed under stringent conditions to allow the binding of the probe to the target sequence. The representative components in a hybridization buffer are shown in Table 1. For hybridization of oligonucleotide DNA probes to rRNA, sodium chloride (NaCl), tris(hydroxymethyl)aminomethane-HCl (Tris-HCl), sodium dodecyl sulfate (SDS), and formamide were the major components. For hybridization of polyribonucleotide probes to rRNA, the components were slightly modified. For polynucleotide DNA probes, NaCl, sodium citrate and formamide were major components. A buffer solution containing NaCl, Tris-HCl, and SDS or SSC buffer (containing NaCl and sodium citrate) was used as the wash solution.

The preheated hybridization buffer was applied to the sample containing probes complementary to the target sequence. In case of rRNA-targeted ISH, stringency can be adjusted by varying either the formamide concentration or the hybridization temperature. Formamide decreases the melting temperature by weakening the hydrogen bonds, thus enabling lower temperatures to be used with high stringency. The salt immobilizes hybrid molecules and is used instead of formamide to reduce toxic waste. The general hybridization conditions used for ISH with rRNA-targeted probes are shown in Table 2. The concentration of NaCl ranged from 0 to 900 mM and that of formamide ranged from 0% to 50%, and the hybridization temperature ranged from 37° to 55°C. Hybridization time ranged from 30 min to 16 h. After hybridization, the slides or filters are rinsed with the appropriate buffer to remove the unbound probe.

Components in hybridization buffer	Probe type	Target molecule	References
NaCl, Tris-HCl, SDS	Oligonucleotide DNA	rRNA	Amann et al., 1990a, 1995
NaCl, Tris-HCl, SDS, formamide	Oligonucleotide DNA	rRNA	Amann et al.,1995; Manz et al., 1996
NaCl, Tris HCl, dextran sulfate, SDS, formamide, blocking reagent	Oligonucleotide DNA	rRNA	Pernthaler et al., 2002b
NaCl, Tris-HCl, EDTA, poly(A), formamide, dextran sulfate	Polyribonucleotide	rRNA	Delong et al., 1999
NaCl, Tris-HCl, dextran sulfate, SDS, formamide, *E.coli* tRNA, salmon sperm DNA, blocking reagent	Polyribonucleotide	rRNA	Pernthaler et al., 2002a
NaCl, sodium citrate, formamide, dextran sulfate, blocking reagent, Denhardt's solution, yeast RNA, salmon sperm DNA	Polyribonucleotide	mRNA	Pernthaler & Amann, 2004.
NaCl, Tris-HCl, SDS, formamide	Polyribonucleotide	DNA	Zwirglmaier et al., 2004
NaCl, sodium citrate, formamide, salmon sperm DNA	Polynucleotide DNA (fragmented PCR product)	DNA	Niki & Hiraga, 1998
NaCl, sodium citrate, formamide, dextran sulfate	Polynucleotide DNA (fragmented genomic DNA)	DNA	Lanoil & Giovannoni, 1997

Table 1. Representative components of hybridization buffer.

	Formamide (%)	NaCl (mM)	Temperature	Time (h)
Condition	0-50	0-900	37-55	0.5 -16

Table 2. General hybridization conditions for rRNA-targeted ISH.

Bouvier & del Giorgio (2003) investigated factors that influenced the detection of bacterial cells in rRNA-targeted FISH. They collected experimental conditions for FISH and environmental factors based on published reports and found that both NaCl and formamide in the hybridization buffer and wash solution significantly influence the performance of rRNA-targeted FISH. These two chemicals are used to adjust the stringency conditions of hybridization and wash steps.

Appropriate controls play an important role in optimizing hybridization and wash conditions. In general, target bacteria include perfect match sequence as positive control. As a negative control, non-target bacteria include a known mismatched sequence.

4. Problems and pitfalls in ISH

4.1 False positive results

The accuracy and reliability of ISH is highly dependent on the specificity of the probe. The sequence design and evaluation of the new probe are essential. Appropriate positive and negative control strains should be included in every ISH experiment. Probe sequences should be carefully examined using the latest version of sequencing data. The stringency conditions of hybridization and wash steps affect both false positive and negative results The mild conditions result in nonspecific probe binding to mismatched sequences or cell structures. Newly designed probes should be evaluated in laboratories by molecular microbiological methods such as ISH, dot blot hybridization, melting curve analysis etc.

4.2 False negative results

Low signal intensity may be a result of insufficient penetration of the probe into the target cells. It depends on the structure of the cell wall or membrane of the bacterial cells. Permeabilization conditions need to be optimized so that all reagents can penetrate the cell. The permeability of the bacterial cell wall structures is not uniform, and different permeabilization procedures have been employed for different cells. In general, gram-negative bacteria tend to be permeable under well-known permeability conditions. For gram-positive bacteria, special fixation and pretreatment is required as described above.

In case rRNA is the target molecule, loop and hairpin formation of target RNA hampers hybridization, leading to differential accessibility of these probes. Self-annealing and self-hairpin formation of the probe itself can also lead to low signal intensity (Fuchs et al., 1998). In addition, the rRNA content of bacterial cells may vary considerably within species as well as strains depending on the physiological state in the given environment. Low rRNA content may result in low signal intensity or false negative results. Various strategies have been used to overcome this difficulty, including FISH combined with direct viable counting, use of multiply-labeled polynucleotide DNA or RNA probes, enzymatic signal amplification, and in situ DNA amplification (see 5. Enhancement of signal intensity).

To test false negative results because of methodological problems, universal probes such as EUB338 (Amann et al., 1990a) have been commonly used as positive control probes. If the control with the universal probe yields good results in ISH, then fixation, probe permeabilization, and rRNA content of the target cells are not the problem. A commonly used negative control probe is NON338, which is complimentary to EUB338 (Wallner et al.,

1993). Using this probe, the non-specific binding of the probe to cell structures other than target nucleic acids can be evaluated.

5. Enhancement of signal intensity

The signal from hybridized cells is highly dependent on the content of target molecules. Despite its potential, the application of ISH in targeting rRNA of resident bacteria in oligotrophic environments is hampered by the low copy number of target molecules. To increase signal intensity, two major environmental microbiological approaches were employed, signal amplification and target nucleic acids amplification. The representative methods are shown in Table 3.

Category	Methods	References
Signal amplification	Tyramid signal amplification	Schönhuber et al., 1997
	Two-pass Tyramid signal amplification	Kubota et al., 2006
	HNPP/Fast Red	Yamaguchi et al., 1996
	Multiply-labeled polyribonucleotide	Delong et al., 1999
Target DNA amplification	In situ PCR	Hodson et al., 1995
	LAMP	Maruyama et al., 2003
	CPRINS	Kenzaka et al., 2005b
	In situ RCA	Maruyama et al., 2005
Target RNA amplification	Direct viable counting	Kenzaka et al., 2001

Table 3. Signal amplification and target nucleic acids amplification methods.

5.1 Signal amplification

Enzymatic signal amplification using a tyramide signal amplification system (TSA) or HNPP/Fast Red was combined with oligonucleotide/polynucleotide probes to increase the sensitivity of ISH (Kenzaka et al., 1998; Kubota et al., 2006; Schönhuber et al., 1997). In the alkaline phosphatase-HNPP/Fast Red system, a digoxigenin-labeled oligonucleotide probe was detected by an alkaline phosphatase-conjugated anti-digoxigenin antibody. Fluorescent molecules accumulate in the target cells because of the activity of alkaline phosphatase (Yamaguchi et al., 1996). In horseradish peroxidase (HRP)-TSA system, oligonucleotide probes were labeled with HRP that generates fluorescent molecules in cells when fluorescent tyramide was used as a substrate (Schönhuber et al., 1997). These enzymatic amplification systems resulted in an 8–20-fold amplification of signal intensity. In case of a two-pass TSA, tyramide tagged with dinitrophenyl was generated around the probe/target site by the activity of HRP, and then HRP-labeled anti-dinitrophenyl antibody further accumulated in cells. Tyramide-Cy3 was deposited by the activity of accumulated HRP (Kawakami et al., 2010).

The signal in SEM–ISH can be amplified using a similar system to detect low copy number target DNA sequences in individual cells (Kenzaka et al., 2009). In the study, digoxigenin-labeled polynucleotide probes were hybridized with plasmid DNA. Peroxidase-labeled anti-digoxigenin antibody was bound to digoxigenin. By using tyramide signal amplification, biotin molecules were accumulated inside target cells. Target cells were identified by streptavidin bound to a gold immunoprobe. Gold particle enhancement was performed to amplify probe signals from hybridized cells. Low vacuum SEM images of a mixture of *E. coli* JM109 harboring plasmid pT7GFP (tagged with *gfp* and *bla* genes) cells and *E. coli* Okayama O27 cells after ISH with *gfp* and *bla* probes are shown in Fig. 3a, 3b, respectively. *E. coli* JM109 harboring pT7GFP were approximately 5-μm long and rod shaped. The long rod-shaped cell with the target gene showed a strong signal because of the high density of gold in one portion of the cell (shown as arrows in Fig. 3a and 3b).

(a) (b)

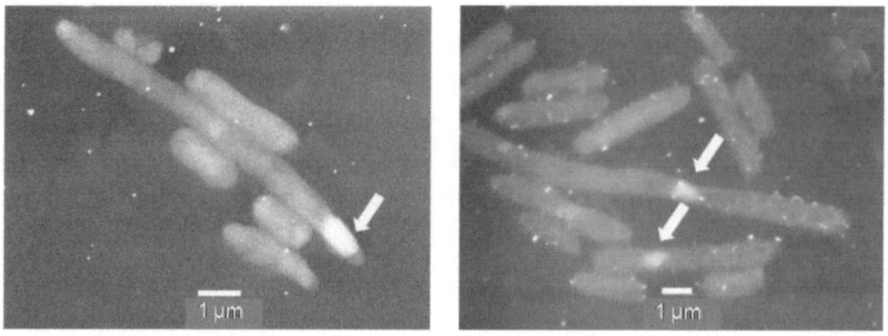

Fig. 3. Low-vacuum SEM images of a mixture containing *E. coli* JM109 harboring pT7GFP cells and *E. coli* Okayama O27 cells after ISH with *gfp* probe (a) and *bla* probe (b). Arrow indicates target *E. coli* JM109 cells harboring pT7GFP.

5.2 Target nucleic acids amplification

In situ DNA amplification techniques based on fluorescent labeling have been successfully applied to identify individual genes in a single bacterial cell. The basic approach is in situ PCR in which target DNA sequences are amplified inside cells (Hodson et al., 1995; Tani et al., 1998). The application of in situ PCR with functional probes provides a powerful tool for detection of genes or gene products in individual cells. This method, however, cannot be applied to diverse species in the natural environment mainly because of permeability, the leakage of amplified products, and less effective concentration of target cells. Longer amplified products after in situ DNA amplification would give better results because these products are less likely to leak out from the cell.

In situ loop-mediated isothermal amplification (LAMP) generates long tandem repeats of the target sequence, preventing amplicons from leaking outside the cell (Maruyama et al., 2003). The mild permeabilization conditions and low isothermal temperature used in the in situ LAMP method causes lesser cell damage than in situ PCR.

Cycling primed in situ amplification (CPRINS) uses a single primer and results in linear amplification of the target DNA. The amplicons are long, single-stranded DNA and are thus retained within the permeabilized microbial cells. ISH with a multiply labeled probe set enabls significant reduction in a nonspecific background while maintaining high signals of target bacteria (Kenzaka et al., 2005b).

In situ rolling circle amplification (RCA) require one short target sequence (less than 40 mer) and generate large, single-stranded, and tandem repeats of target DNA as amplicons. The circularizable probes are approximately 90 mers, comprising short complementary sequences of the target DNA at the 3' and 5' ends, respectively, with an arbitrary sequence in the middle. These probes are labeled with a phosphate group at the 5' end and circularized by ligation when hybridized to the target sequences. The RCA primers amplify the complementary sequence of the circularized probe by hybridization to a specific region of the probe. The amplicons are a single-stranded tandem repeats of the circularized probe sequence. It can be detected with labeled oligonucleotide probes. (Maruyama et al., 2005). CPRINS and in situ RCA can be performed on polycarbonate filters, which allow the effective concentration of target cells from aquatic samples and enhance the quantitative analysis.

Target rRNA molecules can be increased by the direct viable counting (DVC) method (Kenzaka et al., 2001). This method is based on the incubation of samples with antimicrobial agents and nutrients. The antibiotic cocktail acts as a specific inhibitor of DNA synthesis and prevents cell division without affecting other metabolic activities. The resulting cells can continue to metabolize nutrients and elongate and/or become fattened after incubation. By employing these techniques, SEM-ISH will lead to further improvements in sensitivity.

6. Applicability of SEM-ISH to complex microbial communities

Fluorescence microscopy and confocal laser scanning microscopy are important tools in effectively examining complex microbial communities attached to various materials in the natural environment. Potential problems with these fluorescent techniques include autofluorescence, which results from natural substances within plant tissue, organic debris, soil particles, etc. This may hamper the observation of target microbes in complex microbial communities.

SEM allows the visualization of cells attached to materials (e.g., sediment particles) without the interference of autofluorescence and without requiring the ultrathin sectioning of materials. In our experiments, *E. coli* JM109 cells expressing green fluorescent protein (GFP) were introduced into natural river water samples and subjected to FISH with an ES445 probe. Figs. 4a and 4b show the fluorescence micrographs of bacterial cells hybridized with a Cy3-labeled probe. Under blue excitation, the inoculated cells expressing GFP were identified (Fig. 4a).

Although these cells were expected to show bright Cy3 fluorescence under green excitation, the *E. coli* cells attached to organic debris in river water samples were masked by the nonspecific binding of the probe to organic debris in river water samples (Fig. 4b). Consequently, this hampered the identification and accurate enumeration of target cells by FISH.

To enhance the reliability of enumerating target cells, SEM-ISH was employed for the same river water samples containing inoculated *E. coli* JM109 cells. Fig. 4c shows the SE image of organic debris with *E. coli* cells on the surface. The advantage of SEM-ISH is to enable clear observation of the cell surface structure, detritus, etc., under higher magnification. Fig. 4d shows the portion of Fig. 4c that was magnified. Inoculated *E. coli* cells were approximately 5-µm long and rod shaped. The magnification of the SE image allowed clear differentiation between the target cells (shown as arrow in Fig. 4d) and the other cells. By comparing the

BSE image from the same microscopic field, the probe signal was detected from the hybridized cells (Fig. 4e). Even when SEM-ISH was applied to the river water samples, the problem of nonspecific binding of probes to organic debris as shown in Fig. 4b was not completely resolved. However, high magnification allowed target cells to be detected and distinguished from others, even with a high background. As a result, SEM-ISH proved better than fluorescence-based methods in discriminating between target cells and others in the water samples (Fig. 4d and 4e).

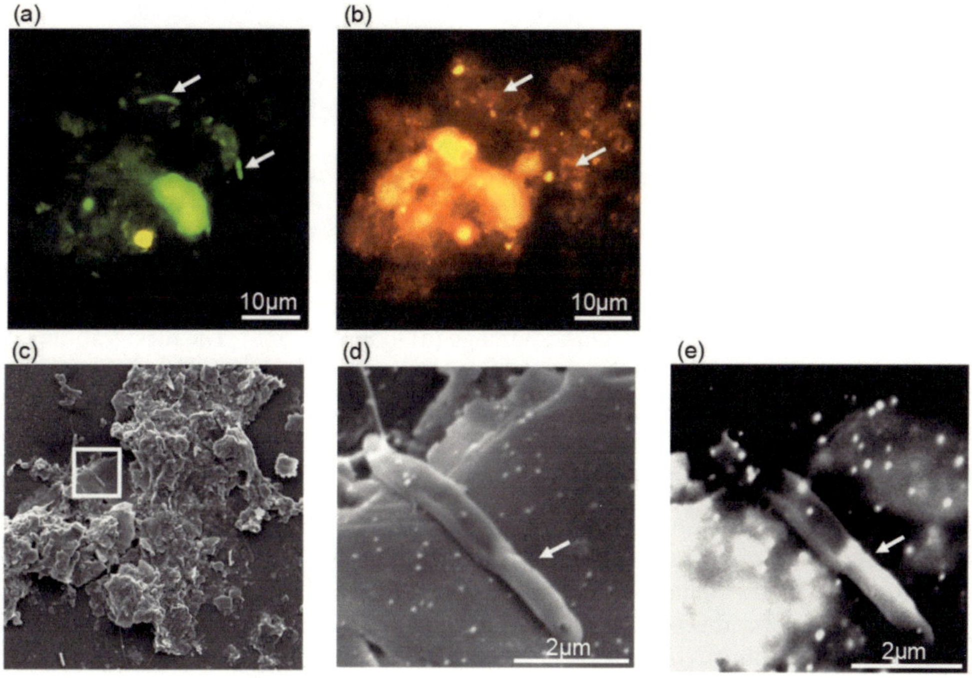

Fig. 4. Organic debris masks the probe signal in river water. *E. coli* cells expressing GFP were inoculated into river water samples, and subjected to FISH with Cy3 labeled probe ES445 (a, b) or SEM-ISH with biotin labeled probe ES445 (c, d, e). *E. coli* cells expressing GFP became attached to organic debris in the river water sample and delineated under blue excitation (a), but probe signals were masked under green excitation (b). High vacuum SEM imaging of *E. coli* cells attached to organic debris in the river water sample (c) was magnified (d). The topographic information was obtained with SE images (c and d), and cells hybridized with ES445 probe were detected with the BSE image (e). Arrow indicates introduced *E. coli* cell.

Certain bacterial cells happen to be buried in the surface of materials such as sediments or soil particles, and cell boundaries were unclear under SEM. In this case, it was difficult to distinguish individual cells within the particle structure using only SE signals. BSE images provided the probe signals from the hybridized cells in the same microscopic fields and clarified the existence of buried cells. We applied SEM-ISH with rRNA-targeted probes to examined bacteria communities on surface of sediment samples (Kenzaka et al., 2005a). SEM-ISH revealed the significant abundance of the *Cytophaga–Flavobacterium* cluster (detectable by probe CF319) on the surface of sediment particles and confirmed a wide distribution over the particle surface. When observed at high magnification, certain bacterial cells were found to be buried in the particles (Fig. 5a and 5b). SEM-ISH with rRNA-targeted probes identified the buried cells based on rRNA sequence.

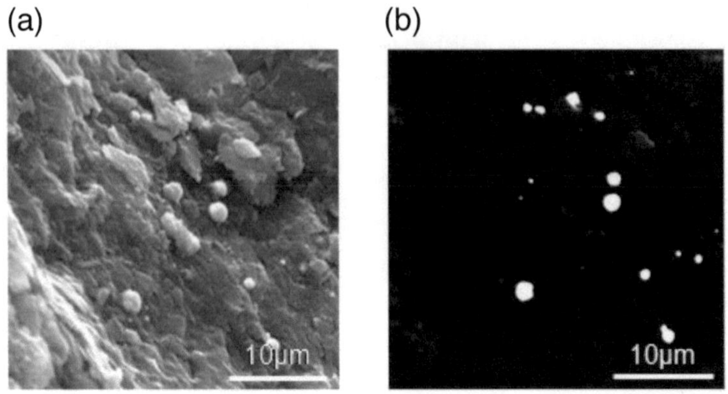

Fig. 5. High-vacuum SEM images of bacteria attached on surface of river sediment particles detected by ISH with the probe CF319 (targeted for *Cytophaga-Flavobacterium* phylum). The same microscopic fields are shown with SE image (a), BSE image (b). The topographic information was obtained with SE images (a), and cells hybridized with the probe CF319 were detected with the BSE image (b).

7. Access to yet-to-be-cultured bacteria with ISH

In the environment, more than 90% of the bacterial communities cannot be cultured by standard techniques, and the yet-to-be-cultured fraction includes diverse microorganisms that are only distantly related to the cultured ones. Culture-independent methods are essential to understand the genetic diversity, population structure, and ecological roles of the bacterial communities. The use of PCR-based clone libraries or metagenomics of an assemblage of microorganisms has great potential for the exploring novel species and sequences and for the understanding the composition and function of microbial communities and their dynamics in the environment (Handelsman, 2004).

Once novel RNA or DNA sequences are obtained by such approaches, ISH with newly designed probes targeting the novel sequences would be a valuable tool for the investigation of their distribution and abundance in the given environment. To validate probe specificity, ISH of clones with target sequences as inserts into plasmids (clone-ISH) would play important roles (Schramm et al., 2002).

8. Conclusion

SEM has been widely used in environmental microbiology to study cell and biofilm morphology. Combining morphological study with SEM and ISH with RNA- or DNA-targeted probes has demonstrated great potential in visualization of cells attached to materials without interference by autofluorescence and requiring no ultrathin sectioning of materials. SEM-ISH addresses some of the limitations of FISH alone and enhances the reliability of monitoring target cells in environmental samples in which the application of fluorescence-based methods is limited. The concepts of ISH in electron microscopic studies can lead to a new understanding of the spatial distribution of target cells as well as of the extent of cell heterogeneity on plant, metal, alloy, bioreactor, or the three-dimensional structures of the attachment matrix.

9. References

Amann, R.I., Binder, B.J., Olson, R.J., Chisholm, S.W., Devereux, R. & Stahl, D.A. 1990a. Combination of 16S rRNA-targeted oligonucleotide probes with flow cytometry for analyzing mixed microbial populations. *Applied and Environmental Microbiology,* Vol.56, No.6, pp. 1919-1925.

Amann, R.I., Krumholz, L. & Stahl, D.A. (1990b). Fluorescent-oligonucleotide probing of whole cells for determinative, phylogenetic, and environmental studies in microbiology. *Journal of Bacteriology,* Vol.172, No.2, pp. 762–770.

Amann, R.I., Ludwig, W. & Schleifer, K.H. 1995. Phylogenetic identification and in situ detection of individual microbial cells without cultivation. *Microbiological Reviews,* Vol.59, No.1, pp. 143-169.

Bouvier, T. & del Giorgio, P.A. (2003). Factors infuencing the detection of bacterial cells using fuorescence in situ hybridization (FISH): A quantitative review of published reports. *FEMS Microbiology Ecology,* Vol.44, No.1, pp. 3-15.

Brabec, V. & Leng, M. (1993). DNA interstrand cross-links of trans-diamminedichloroplatinum(II) are preferentially formed between guanine and complementary cytosine residues. *Proceedings of the National Academy of Sciences of the United States of America.* Vol.90, No.11, pp. 5345-5349.

Dalbiès, R., Payet, D. & Leng, M. (1994). DNA double helix promotes a linkageisomerization reaction in trans-diamminedichloroplatinum(II)-modified DNA. *Proceedings of the National Academy of Sciences of the United States of America.* Vol. 91, No.17. pp. 8147–8151.

DeLong, E.F., Taylor, L.T., Marsh, T.L. & Preston, C.M. (1999). Visualization and enumeration of marine planktonic archaea and bacteria by using polyribonucleotide probes and fluorescent in situ hybridization. *Applied and Environmental Microbiology,* Vol.65, No.12, pp. 5554-5563.

Fuchs, B.M., Wallner, G., Beisker, W., Schwippl, I., Ludwig, W. & Amann, R. (1998). Flow cytometric analysis of the in situ accessibility of *Escherichia coli* 16S rRNA for fluorescently labeled oligonucleotide probes. *Applied and Environmental Microbiology,* Vol.64, No.12, pp. 4973–4982.

Hacker, G.W. (1998). High performance Nanogold-silver in situ hybridisation. *European journal of histochemistry,* Vol.42, No.2, pp. 111-120.

Handelsman J. (2004). Metagenomics: application of genomics to uncultured microorganisms. *Microbiology and Molecular Biology Reviews,* Vol.68, No.4, pp. 669-685.

Hodson, R.E., Dustman, W.A., Garg, R.P. & Moran, M.A. (1995). In situ PCR for visualization of microscale distribution of specific genes and gene products in prokaryotic communities. *Applied and Environmental Microbiology,* Vol.61, No.11, pp. 4074-4082.

Kawakami, S. Kubota K., Imachi, H., Yamaguchi, T. Harada, H. & Ohashi, A. (2010). Detection of single copy genes by two-pass tyramide signal amplification fluorescence in situ hybridization (Two-Pass TSA-FISH) with single oligonucleotide probes. *Microbes and Environments*, Vol.25, No.1, pp. 15–21.

Kempf, V.A.J., Trebesius, K. & Autenrieth, I.B. (2000). Fluorescent in situ hybridization allows rapid identification of microorganisms in blood cultures. *Journal of Clinical Microbiology*, Vol.38, No.2, pp. 830-838.

Kenzaka, T., Yamaguchi, N., Tani, K. & Nasu, M. (1998). rRNA-targeted fluorescent in situ hybridization analysis of bacterial community structure in river water. *Microbiology*, Vol.144, No.8, pp. 2085-2093.

Kenzaka, T., Yamaguchi, N., Prapagdee, B., Mikami, E. & Nasu, M. (2001). Bacterial community composition and activity in urban rivers in Thailand and Malaysia. *Journal of Health Science*, Vol.47, No.4, pp. 353-361.

Kenzaka, T., Ishidoshiro, A., Yamaguchi, N., Tani, K. & Nasu, M. (2005a). rRNA sequence-based scanning electron microscopic detection of bacteria. *Applied and Environmental Microbiology*,Vol.71, No.9, pp. 5523-5531.

Kenzaka, T., Tamaki, S., Yamaguchi, N., Tani, K. & Nasu, M. (2005b). Recognition of individual genes in diverse microorganisms by cycling primed in situ amplification. *Applied and Environmental Microbiology*, Vol.71, No.11, pp. 7236-7244.

Kenzaka, T., Ishidoshiro, A., Tani, K. & Nasu, M. (2009). Scanning electron microscope imaging of bacteria based on DNA sequence. *Letters in Applied Microbiology*, Vol.49, No.6, pp. 796-799.

Kubota, K., Ohashi, A. Imachi, H. & Harada, H. (2006). Visualization of mcr mRNA in a methanogen by fluorescence in situ hybridization with an oligonucleotide probe and two-pass tyramide signal amplification (two-pass TSA-FISH). *Journal of Microbiological Methods*, Vol.66, No.3, pp. 521–528.

Lanoil, B.D. & Giovannoni, S.J. (1997). Identification of bacterial cells by chromosomal painting. *Applied and Environmental Microbiology*, Vol.63, No.3, pp. 1118-1123.

Lee, N., Nielsen, P.H. Andreasen, K.H., Juretschko, S., Nielsen, J.L., Schleifer, K.H. & Wagner, M. (1999). Combination of fluorescent in situ hybridization and microautoradiography-a new tool for structure-function analyses in microbial ecology. *Applied and Environmental Microbiology*, Vol.65, No.3, pp. 1289-1297.

Manz, W., Amann, R., Ludwig, W., Vancanneyt, M. & Schleifer, K.H. (1996). Application of a suite of 16S rRNA-specific oligonucleotide probes designed to investigate bacteria of the phylum Cytophaga-Flavobacter-Bacteroides in the natural environment. *Microbiology*, Vol.142, No.5, pp. 1097-1106.

Maruyama, F., Kenzaka, T., Yamaguchi, N., Tani, K. & Nasu, M. (2003). Detection of bacteria carrying the *stx*2 gene by in situ loop-mediated isothermal amplification. *Applied and Environmental Microbiology*, Vol.69, No.8, pp. 5023-5028.

Maruyama, F., Kenzaka, T., Yamaguchi, N., Tani, K. & Nasu, M. (2005). Visualization and enumeration of bacteria carrying a specific gene sequence by in situ rolling circle amplification. *Applied and Environmental Microbiology*, Vol.71, No.12, pp. 7933-7940.

Moraru, C., Lam, P., Fuchs, B.M., Kuypers, M.M.M. & Amann, R. (2010). GeneFISH – an *in situ* technique for linking gene presence and cell identity in environmental microorganisms. *Environmental Microbiology*, Vol.12, No.11, pp. 3057–3073.

Nielsen, P.E., Egholm, M, Berg, R.H. & Buchardt, O. (1991). Sequence-selective recognition of DNA by strand displacement with a thymine-substituted polyamide *Science*, Vol.254, No.5037, pp. 1497-1500.

Niki, H. & Hiraga, S. (1998). Polar localization of the replication origin and terminus in *Escherichia coli* nucleoids during chromosome partitioning. *Genes and Development*, Vol.12, No.7, pp. 1036-1045.

Obika, S. Nanbu, D., Hari, Y., Morio, K., In, Y., Ishida, T. & Imanishi, T. (1997). Synthesis of 2'-O,4'-C-methyleneuridine and -cytidine. Novel bicyclic nucleosides having a fixed C3'-endo sugar puckering. *Tetrahedron Letters*. Vol.38, No.50, pp. 8735-8738.

Pernthaler, A., Pernthaler, J. & Amann, R. (2002a). Fluorescence in situ hybridization and catalyzed reporter deposition for the identification of marine bacteria. *Applied and Environmental Microbiology*, Vol.68, No.6, pp. 3094-3101.

Pernthaler, A., Pernthaler, J., Schattenhofer, M. & Amann. R. (2002b). Identification of DNA-synthesizing bacterial cells in coastal North Sea plankton. *Applied and Environmental Microbiology*, Vol.68, No.11, pp. 5728-5736.

Pernthaler, A. & Amann, R. (2004). Simultaneous fluorescence in situ hybridization of mRNA and rRNA in environmental bacteria. *Applied and Environmental Microbiology*, Vol.70, No.9, pp.5426-5433.

Schönhuber, W., Fuchs, B., Juretschko, S. & Amann, R. (1997). Improved sensitivity of whole-cell hybridization by the combination of horseradish peroxidase-labeled oligonucleotides and tyramide signal amplification. *Applied and Environmental Microbiology*, Vol.63, No.8, pp. 3268-3273.

Schramm, A., Fuchs, B.M., Nielsen, J.L., Tonolla. M. & Stahl, D.A. (2002). Fluorescence in situ hybridization of 16S rRNA gene clones (Clone-FISH) for probe validation and screening of clone libraries. *Environmental Microbiology*, Vol.4. No.11, pp. 713-720.

Sekar, R., Pernthaler, A., Pernthaler, J., Warnecke, F., Posch, T. & Amann, R. (2003). An improved protocol for the quantification of freshwater actinobacteria by fluorescence in situ hybridization. *Applied and Environmental Microbiology*, Vol.69, No.5, pp. 2928-2935.

Sekiguchi, Y., Kamagata, Y., Nakamura, K., Ohashi, A. & Harada, H. (1999). Fluorescence in situ hybridization using 16S rRNA-targeted oligonucleotides reveals localization of methanogens and selected uncultured bacteria in mesophilic and thermophilic sludge granules. *Applied and Environmental Microbiology*, Vol.65, No.3, pp. 1280-1288.

Silahtaroglu, A.N., Tommerup, N. & Vissing, H. (2003). FISHing with locked nucleic acids (LNA): Evaluation of different LNA/DNA mixmers. *Molecular and Cellular Probes*, Vol.17, No.4, pp. 165–169.

Stender., H. (2003). PNA FISH: an intelligent stain for rapid diagnosis of infectious diseases. *Expert Review of Molecular Diagnostics*. Vol.3, No.5, pp. 649-655.

Tani, K., Kurokawa, K. & Nasu, M. (1998). Development of a direct in situ PCR method for detection of specific bacteria in natural environments. *Applied and Environmental Microbiology*, Vol.64, No.4, pp. 1536-1540.

Waar, K., Degener, J.E., van Luyn, M.J. & Harmsen, H.J.M. (2005) Fluorescent in situ hybridization with specific DNA probes offers adequate detection of *Enterococcus faecalis* and *Enterococcus faecium* in clinical samples. *Journal of Medical Microbiology*, Vol.54, No.10, pp. 937-944.

Wallner, G., Amann, R. & Beisker, W. (1993). Optimizing fluorescent in situ hybridization with rRNA-targeted oligonucleotide probes for flow cytometric identification of microorganisms. *Cytometry*, Vol.14, No.2,. pp. 136–143.

Yamaguchi, N., Inaoka, S., Tani, K., Kenzaka, T. & Nasu, M. (1996). Detection of specific bacterial cells with 2-hydroxy-3-naphthoic acid-2′-phenylanilide phosphate and Fast Red TR in situ hybridization. *Applied and Environmental Microbiology,* Vol.62, No.1, pp. 275-278.

Yilmaz, L. S. & Noguera, D.R. (2004). Mechanistic approach to the problem of hybridization efficiency in fluorescent in situ hybridization. *Applied and Environmental Microbiology,* Vol.70, No.12, pp. 7126-7139.

Zwirglmaier, K., Ludwig, W. & Schleifer, K.H. (2004). Improved method for polynucleotide probe-based cell sorting, using DNA-coated microplates. *Applied and Environmental Microbiology,* Vol.70, No.1, pp. 494-497.

13

Ionizing Radiation Effect on Morphology of PLLA: PCL Blends and on Their Composite with Coconut Fiber

Yasko Kodama[1,*] and Claudia Giovedi[2]
[1]*Instituto de Pesquisas Energéticas e Nucleares – IPEN–CNEN/SP,*
[2]*Centro Tecnológico da Marinha em São Paulo – CTMSP,*
Brazil

1. Introduction

The problem of non-biodegradable plastic waste remains a challenge due to its negative environmental impact. In this sense, poly(L-lactic acid) (PLLA) and poly(ε-caprolactone) (PCL) have been receiving much attention lately due to their biodegradability in human body as well as in the soil, biocompatibility, environmentally friendly characteristics and non-toxicity (Tsuji & Ikada, 1996; Kammer & Kummerlowe, 1994; Dell'Erba et al., 2001; Yoshii et al., 2000; Zhang et al., 2005). The controlled degradation of polymers is sometimes desired for biomedical applications and environmental purposes (Michler, 2008).

PLLA is a poly(α-hydroxy acid) and PCL is a poly(ω-hydroxy acid) (Tsuji & Ikada, 1996). PLLA is a hard, transparent and crystalline polymer (Mochizuki & Hirami, 1997). On the other hand, PCL can be used as a polymeric plasticizer because of its ability to lower elastic modulus and to soften other polymers (Kammer & Kummerlowe, 1994). Both polymers, PLLA and PCL, can be used in biomedical applications, which require a proper sterilization process. Nowadays, the most suitable sterilization method is high energy irradiation. Ionizing radiation exposure induces to crosslinking or scission of polymer main chain (Broz et al., 2003), besides other chemical alterations. Nature of those alterations is affected by chemical structure of polymer, and also by gaseous compounds present, as oxygen. Irradiation in the presence of air or oxygen leads to oxidized products formation that normally are undesirable, being less thermally stable and decreasing crosslinking degree by reaction of polymeric radicals (Charlesby et al., 1991).

The market for biodegradable polymers had shown strong growth from 2001 up to 2005. A number of major plant expansions for commercial scale production had been planned. The major classes of biopolymers, polylactic acid and aliphatic-aromatic co-polyesters has been used in a wide variety of niche applications, particularly for manufacture of rigid and flexible packaging, bags and sacks and foodservice products. In 2005, starch-based materials were the largest class of biodegradable polymer and polylactic acid (PLA) was the second

---

* Corresponding Author

largest material class followed by synthetic aliphatic-aromatic co-polyesters. Product development and improvement has a crucial role to play in the further development of the biodegradable polymers market. Biodegradable polymers can be found in a wide range of end use markets. Continued progress in terms of product development and cost reduction will be required before they can effectively compete with conventional plastics for mainstream applications. The main markets for PLA are thermoformed trays and containers for food packaging and food service applications. In 2005, packaging was the largest sector with 39% of total biodegradable polymer market volumes. Loose-fill packaging was the second largest sector, followed by bags and sacks. Fibers or textiles is an important sector for PLA, and accounted for 9% of total market volumes. Others include a wide range of very small application areas, the most important of which are agriculture and fishing, medical devices, consumer products and hygiene products. While the cost of some biodegradable plastics is high compared to conventional polymers, from a marketing perspective, it is important not only to consider the material cost, but also all associated costs, including the costs of handling and disposal, which are of course lower for biodegradable plastics. Users of biodegradable plastics can differentiate themselves from the competition by demonstrating how innovative and proactive they are for the benefit of the environment (Platt, 2006).

Polylactic acid is a biodegradable polymer derived from lactic acid. It is a highly versatile material and is made from 100% renewable resources like corn, sugar beet, wheat and other starch rich products. The homopolymer of L-lactide is a semicrystaline polymer. Due to high costs, the focus was initially on the manufacture of medical grade sutures, implants and controlled drug release applications. PLA has many potential uses, including many applications in the textile and medical industries as well the packaging industry.

From a commercial point of view the most important synthetic biodegradable aliphatic polyester was traditionally polycaprolactone (PCL). The ring opening polymerization of ε-caprolactone yields a semicrystalline polymer, which is regarded as tissue compatible and was originally used in the medical field as a biodegradable suture in Europe. Polycaprolactone aliphatic polyesters have long been available commercially for use as adhesives, compatibilizers, modifiers and films, as well as medical applications. Caprolactone limits moisture sensitivity, boosts melt strength, and helps to plasticize the starch (Platt, 2006).

In order to improve some desirable properties two or more polymers can be mixed to form polymeric blends (Utracki, 1989). The original reasons for preparing polymer blends are to reduce costs by combining high-quality polymers with cheaper materials (although this approach is usually accompanied by a drastic worsening of the properties of the polymer) and to create a polymer that has a desired combination of the different properties of its components (Michler, 2008). However, according to Michler (2008), usually different polymers are incompatible. Improved properties can be only realized if the blend exhibits optimum morphology. According to Sawyer (2008), in polymer science, the term morphology generally refers to form and organization on a size scale above the atomic arrangement, but smaller than the size and shape of the whole sample. Thus, improving compatibility between the different polymers and optimizing the morphology are the main issues to address when producing polymer blends.

In general, the morphology results from the complex thermomechanical history experienced by the different constituents during processing. So, parameters like the composition, viscosity ratio, shear rate/shear stress, elasticity ratio and interfacial tension among the component polymers, as well as processing conditions such as time, temperature and type of mixing, rotation speed of rotor of mixing determine the final size, shape and distribution of dispersed phase during the melting process (Dell'Erba et al., 2001; Michler, 2008; Nakayama & Tanaka, 1997) and are very important to define the characteristics of the obtained material. One common feature of semicrystalline polymers is a hierarchical morphology. The scales of structural details within them range from nanometers to millimeters. Under certain conditions, macromolecules are able to form periodic structures that involve adjacent chains or chain segments. By repeated folding of a flexible polymer chain results in densely packed and highly ordered domains. Polymers normally are partially crystalline, highly ordered domains will coexist with regions of an amorphous phase and a crystalline fraction with identical chemical compositions but divergent physical properties (Michler, 2008). Moreover, the structural modifications induced by ionizing radiation may alter the morphology of the samples.

PLLA:PCL blends have attracted great interest as temporary absorbable implants in human body, but they suffer from poor mechanical properties due to macro phase separation of the two immiscible components, and to poor adhesion between phases (Dell'Erba et al., 2001). Chemical structure influences the biodegradation of solid polymers. Enzymatic and non enzymatic degradations occur easier in the amorphous region (Mochizuki & Hirami, 1997; Tsuji & Ishizaka, 2001). The crystallinity and the resulting morphology are usually controlled by using either different proportions of stereoisomers (enantiomers, e.g. L-lactide and D-lactide) of the same monomer or a defined ratio of comonomers of related polyesters (Michler, 2008).

The morphology of the blends affects also the biodegradation of the polymers. So the control of the morphology of an immiscible polymer processed by melting is important for the tailoring of the final properties of the product (Dell'Erba et al., 2001; Michler, 2008). Kikkawa et al. (2006) cited that one of the approaches used to generate biodegradable materials with a wide range of physical properties is blending, and miscibility of blends is one of the most important factors affecting the final polymer properties. In particular, surface structure and morphology of biodegradable polymer blends have a great impact on the enzymatic degradation behavior by enzymes. According to Michler (2008), a high degree of crystallinity yields low degradation rates since it has been shown that hydrolytic degradation (cleavage of ester bonds) preferentially occurs in the amorphous regions.

Plastic solid waste has become a serious problem recently concerning environmental impact. In this scenario, preparation of polymers and composites based on coconut husk fiber would lead to a reduction on the cost of the final product. Additionally, it will reduce the amount of agribusiness waste disposal in the environment. In Brazil, coconut production is around 1.5 billion fruits by year in a cultivated area of 2.7 million hectares, but the coconut husk fiber has not been used much for industrial applications. According to Sawyer (2008), composites can contain short or continuous fibers, inorganic or organic. Even though according to Michler (2008), the word "composite" should only be applied to polymers with inorganic components, in this chapter it will be applied to the mixture of polymeric blend and natural coconut fiber.

It is worthy of note that the use of natural fibers as reinforcement in polymeric composites is an important research field that has been growing in the last decades, (Kapulskis et al, 2006;

Martins et al., 2006; Tomczak et al., 2007). Thus, by incorporating fibers of low cost to the polymeric blend, it is possible to obtain an improvement of the mechanical properties without loss of the original characteristics of polymeric components.

Regarding the irradiation effects, vegetable fiber, like as coconut fiber, is composed by cellulose and lignin, which suffer chemical alteration by irradiation such as scission or cross-linking. In the case of natural polymers, such as cellulose, main chain scission occurs predominantly due to irradiation and as a result molecular weight decrease (Chmielewski, 2005).

The structures and morphologies of polymers have been under investigation for more than 60 years. Scanning electron microscopy (SEM) was introduced in the 1960´s, and has been used to investigate fracture of surfaces, phase separation in polymer blends and crystallization of spherulites. Several improvements have been made in electron generation and electron optics that have enhanced the resolution power. Also, computerized SEM led to the introduction of the digital scanning generator for digital image recording and processing. Consequently, SEM is at present the most popular of the microscopic techniques. This is because of the user-friendliness of the apparatus, the ease specimen preparation, and the general simplicity of image interpretation. The limitation is that only surface features are easily accessible. Furthermore, chemical analysis of different elements is usually possible with SEM (energy dispersive or wavelength dispersive analysis of X-rays, EDXA, WDXA) (Michler, 2008). According to Sawyer (2008), SEM is used to evaluate features as the size, distribution, and adhesion of the fibers or particles, their adhesion to the matrix play a major role in revealing the strength and toughness enhancement. Regarding to the studied material, coarse structures such as larger particles in a matrix can easily be studied at low temperature (brittle) fracture surfaces in the SEM, since the fracture path follows the shape of the particle (Michler, 2008) considering that from a practical point of view, a good indicator of the degree of compatibility is the heterogeneity of the polymer system (e.g how the sizes of the dispersed domains depend on the processing conditions).

The objective of this chapter is to present the effect of ionizing radiation on the morphology of PLLA:PCL blend and composites containing coconut fiber. SEM and field emission (FE) SEM micrographs were taken of non irradiated and irradiated samples with gamma rays and electron beam.

2. Experimental

Coconut fiber

Coconut fibers were from three different origins: Embrapa- Empresa Brasileira de Pesquisa Agropecuária, Paraipaba region, Ceará; Projeto Coco Verde - Quissamã, Rio de Janeiro; Poematec - Ananindeua, Pará.

Size reduction of the coconut fibers was carried out using helix mill Marconi - model MA 680, from Laboratório de Matéria-prima Particulados e Sólidos Não Metálicos - LMPSol, Departamento de Engenharia de Materiais of Escola Politécnica/USP.

The fiber size distribution was measured using sieves of the Tyler series 16, 20, 35 and 48, fiber sizes of 1.0mm, 0.84mm, 0.417mm, and 0.297mm, respectively. The 0.417mm fiber size was used for the assays. The triturated material was separated using a sieve shaker Produtest, for 1 min.

In order to remove lignin from coconut fiber surface, fibers were soaked with Na_2SO_3 2% aqueous solution for 2h using ultrasound. Fibers from Embrapa were washed several times with tap water and finally, tree times with deionized water, as described by Calado et al. (2000).

Fiber acetylation was performed as described by d'Almeida et al. (2005). As received fibers from Embrapa were soaked in a solution of acetic anhydrate and acetic acid (1.5:1.0, w:w). It was used as a catalyst, 20 drops of sulfuric acid in 500mL solution. Set were submitted to ultrasound for 3h, then for more 24h rest at the same solution. Fibers were washed using tap water and for more 24h rested in deionized water. Fibers were separated from water and washed with acetone, after that, were evaporated at room temperature.

Fiber residue of non irradiated and irradiated samples were obtained using a furnace at 600°C, air atmosphere, by 2 hours. Residues were analyzed in a Scanning Electron Microscope, SEM, Jeol, JXA-6400, at Centro Tecnológico da Marinha em São Paulo (CTMSP). It was used Energy Dispersion Spectroscopy, EDS, to analyze chemical elements present in the residues.

Preparation of blend sheets

PLLA pellets were dried in a vacuum oven at 90°C and PCL pellets were dried at 40°C overnight to avoid hydrolysis of polymers during the melt-processing. Sheets of PCL and PLLA homo-polymers and blends with PCL:PLLA weight ratio of 25:75, 50:50 and 75:25 were prepared using a twin screw extruder (Labo Plastomill Model 150C, Toyoseki, Japan) equipped with a T-die (60mm width and 1.05mm thickness). T-die temperature was set at 205°C for PLLA homo-polymer and its blends, and at 90°C for PCL. Extruded sheets were quenched using a water bath set at room temperature. The take up speed was selected at 0.35 m min^{-1}. As the take up speed was set at slightly higher than the extrusion out-put speed, finally obtained thickness of films was around 1 mm.

Preparation of composite pellets and sheets

PCL (pellets, $\bar{M}_W$=2.14 10^5 g mol^{-1}; $\bar{M}_W/\bar{M}_n$= 1.423), PLLA (pellets, $\bar{M}_W$=2.64 10^5 g mol^{-1} $\bar{M}_W/\bar{M}_n$= 1.518 - Gel Permeation Chromatographic values) and dry coconut fiber (from Embrapa, Ceará, Brazil) were used to prepare blends and composites. A Labo Plastomil model 50C 150 of Toyoseiki twin screw extruder was used for pellets preparation. Pellets of PLLA:PCL 80:20 (w:w) blend and composites containing 5 and 10% of untreated and chemically treated coconut fiber were prepared at AIST.

Sheets (150mm x 150mm x 0.5mm) of PCL, PLLA, PLLA:PCL 80:20 (w:w) blend and composites containing 5 and 10% untreated and chemically treated coconut fiber were prepared using Ikeda hot press equipment of JAEA– Japan Atomic Energy Agency. Mixed pellets of samples were preheated at 195°C for 3 min and then pressed by under heating at the same temperature for another 3 min under pressure of 150 kgf cm^{-2}. The sample was then cooled in the cold press using water as a coolant for 3 min.

Gamma irradiation

Samples were irradiated at IPEN–CNEN/SP (Brazil) using a Co-60 irradiator Gammacell model 220, series 142 from Atomic Energy of Canada Limited. Doses of 25, 50, 75, 100 and

500 kGy were applied at a dose rate of 4.3 kGy h^{-1}. Samples were cut 10 × 100 cm^2, and irradiated at room temperature in air.

Electron beam irradiation

Hot pressed sheets were irradiated using a electron beam accelerator (E=2 MeV, 2 mA) applying radiation doses from 10 kGy to 500 kGy with a dose rate of 0.6 kGy s^{-1}, at JAEA, Takasaki, Japan.

Scanning Electron Microscopy (SEM)

Morphology of the fractured surfaces of the non-irradiated homopolymers and blends was examined by a scanning electron microscope (SEM) DS-720 TOPCON Co. The photomicrographs of the cryogenic fractured surface of the blends sheet were taken after 4-5 min gold coating.

SEM micrographs of the irradiated homopolymers and blends sheets; and coconut fibers surfaces from cryogenic fractured samples were obtained using a scanning electron microscope model JXA-6400 (JEOL) at Centro Tecnológico da Marinha em São Paulo.

Field Emission Scanning Electron Microscopy (FE-SEM)

Photomicrographs of the, cryogenic fractured, non-irradiated and irradiated samples were taken using a field emission scanning electron microscope, JEOL, JSM - 7401F, acceleration voltage 1.0 kV at Central Analítica IQ-USP.

3. Results and discussion

Figure 1 shows scanning electron micrographs of coconut fiber surface, as received samples from Embrapa.

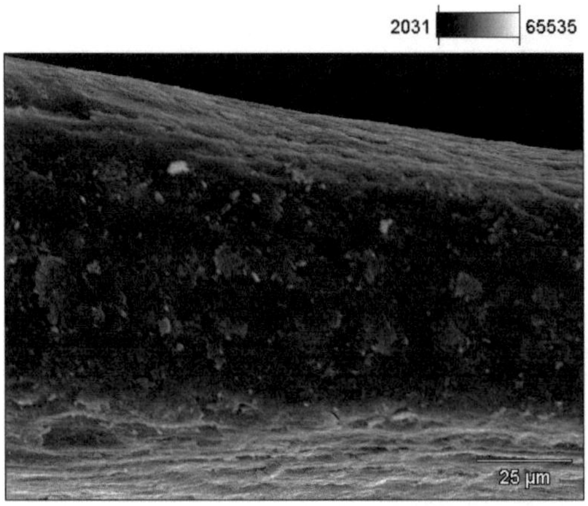

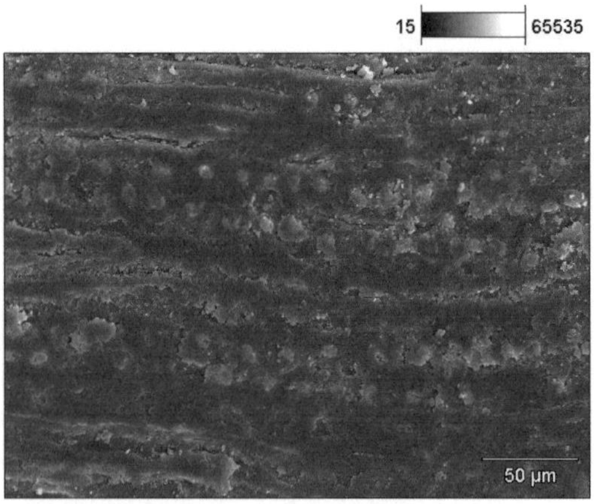

Acceleration voltage: 20.0 kV Magnification: 700× Acceleration voltage: 20.0 kV Magnification: 350×

Fig. 1. Scanning electron micrographs of coconut fiber surface, as received from Embrapa.

There are some studies involving chemical treatment of vegetal fibers to improve its compatibility to polymers (Kapulskis et al, 2006; Abdul-Khalil & Rozman, 2000; Lee et al., 2004). Cell wall of a plant in the dry state consist mainly of carbohydrates combined to lignin, and few quantities of protein, starch and inorganic compounds, chemical composition varies from plant to plant and through different parts of the same plant (Rowell et al., 2000). The lignocellulosic fibers characteristics vary considerably with the place where they are produced, and possess different chemical compositions that affect their physical and chemical properties (Tomczak et al., 2007).

According to Calado et al. (2000), coconut coir fiber treated with Na_2SO_3 suffers lignin removal from surface and its roughness increases. This increase of surface roughness improves their adhesion due to promotion of mechanical interaction between fibers and polymeric matrix. In this sense, scanning electron micrographs of coconut fibers surface, chemically treated with Na_2SO_3 are shown in Figure 2.

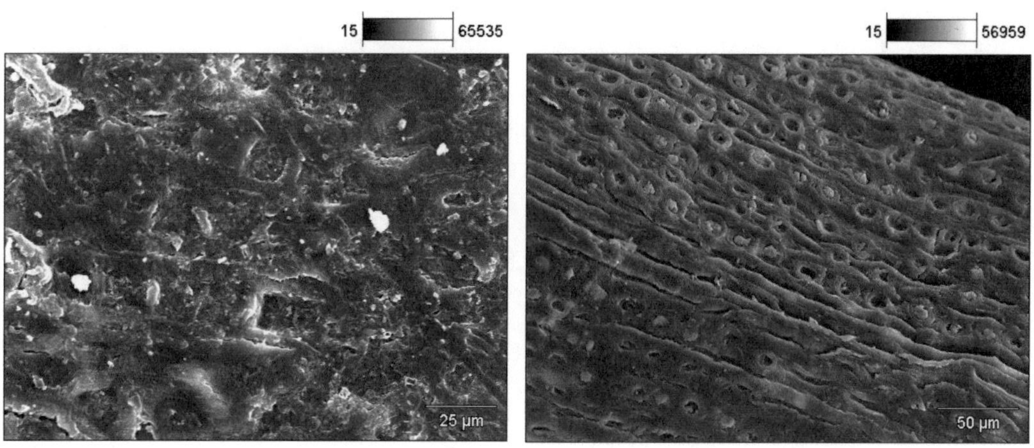

Acceleration voltage: 20.0 kV Magnification 600× Acceleration voltage: 20.0 kV Magnification: 350×

Fig. 2. Scanning electron micrographs of coconut fibers surface from Embrapa , chemically treated with Na_2SO_3.

Chemical treatment with anhydride acetic reduces the number of hydroxyl radicals, as shown in Figure 3. This treatment would reduce cellulose molecules polarity and then would improve the compatibility with thermo rigid matrix used in composites.

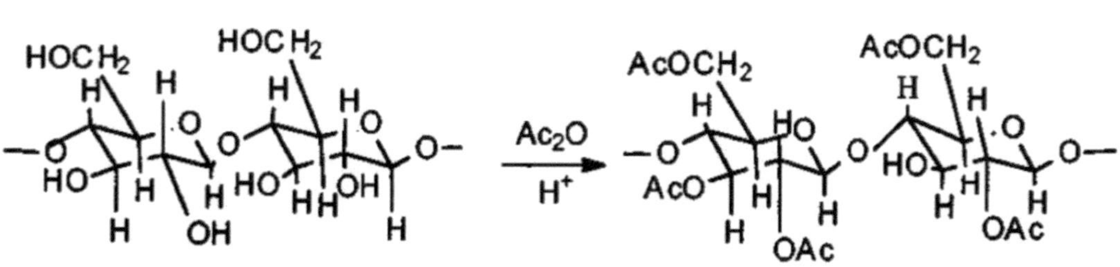

Fig. 3. Cellulose acetylation reaction (Calado et al., 2000).

Calado et al. (2000) observed clear difference on the morphologies of non treated and chemical treated surfaces of coir fibers. Same behavior was observed on the fibers studied in this chapter.

It was possible to observe roughness increase on the surface treated with Na_2SO_3 and anhydride acetic, as shown in Figure 4.

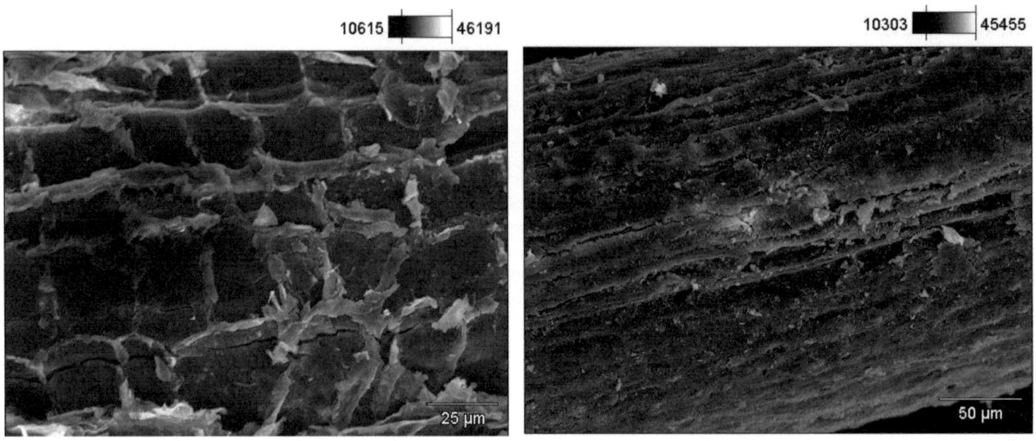

Acceleration voltage: 20.0 kV Magnification: 600× Acceleration voltage: 20.0 kV Magnification: 350×

Fig. 4. Scanning electron micrographs of coconut fibers surface from Embrapa, chemically treated with Na_2SO_3 and anhydride acetic, respectively.

X rays spectra by EDS of Embrapa coconut fiber, non irradiated with 0.297 mm up to 0.417 mm size, are shown in Figure 5. It was possible to identify that chemical composition of the residue varied for the same fiber. It can be explained by the fact that chemical composition varies from plant to plant and also through different parts of the same plant (Rowell et al., 2000; Severiano et al., 2010).

In the residue of thermal decomposition in air of fibers from three different origins, most common elements found were K, Na, Si, Ca and, in some cases, Fe. In the literature, they found P, Mg and , in small concentrations, Cu, Zn, Mn (Rosa et al., 2001), in addition to Cl, S, Br e Rb (Mothé & Miranda, 2009) that were not observed in samples in this study. This variation can be attributed to soil where coconut tree was grown. Peak that had attributed to Zr, in fact is due to P, as they appear at the same channel of energy and, it is more probable to find P in higher quantity in soil than Zr. Peak that was attributed to As observed in some spectra was due to coating process used to allow image formation.

Scanning electron micrographs of surfaces of cryogenic fractured non irradiated as extruded blend samples were taken. As extruded PCL micrograph shows a homogeneous morphology, as shown in Figure 6. The as extruded PCL:PLLA 50:50 micrograph shows spheres with different sizes and shapes, as shown in Figure 7. According to Michler (2008), when a polymer is cooled down from the melt, the (primary) crystallization starts from initial points that are randomly distributed in the volume. Such starting points are either homogeneous or heterogeneous nuclei (i.e. nucleating agents, impurities, or filler particles). This radial growth results in a characteristic arrangement of lamellae. The superstructures (spherulites and sheaf-like boundless of lamellae) come in a variety of forms depending on the polymer and its crystalline structure. These superstructures generally form a texture consisting of one or more spherulite types with a characteristic spherulite size distribution.

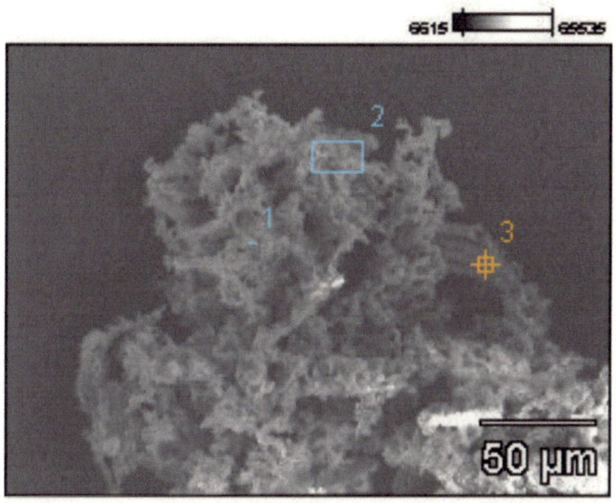

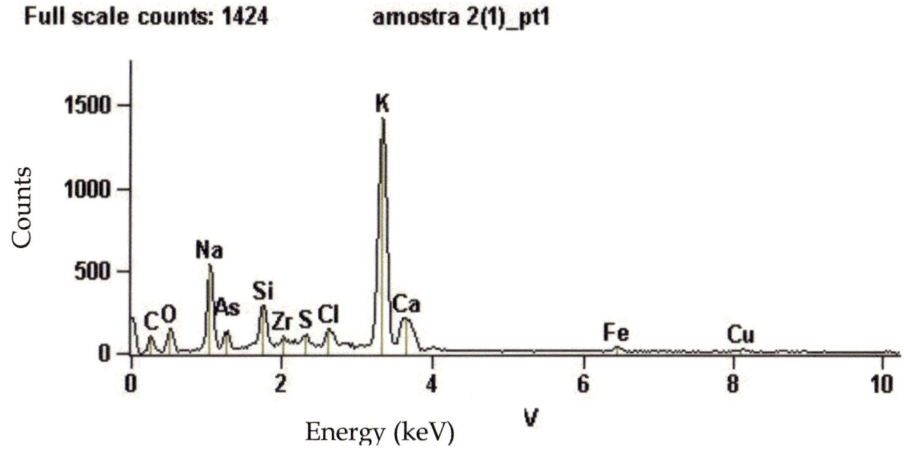

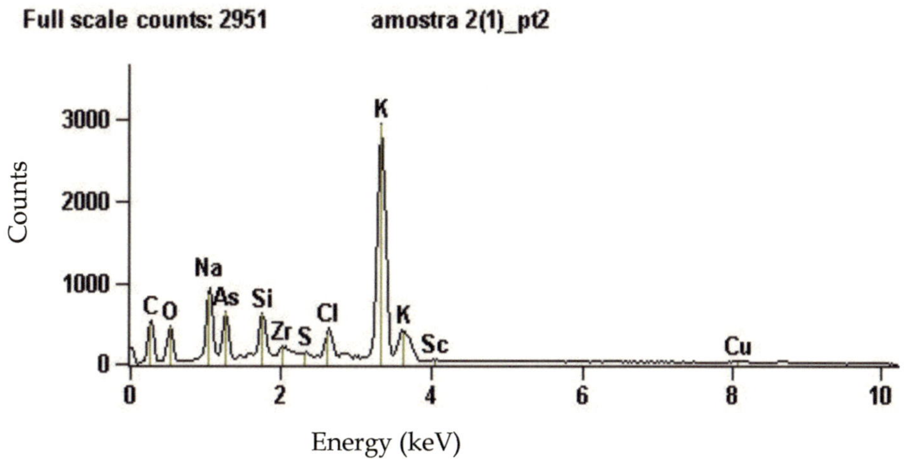

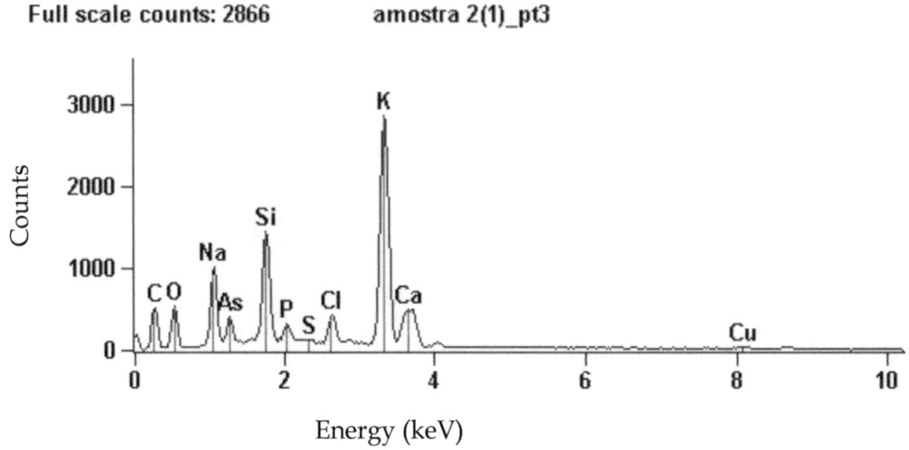

Fig. 5. Scanning electron micrograph of non irradiated Embrapa coconut fiber (0.297 and 0.417 mm) residue and X ray spectra by EDS (points 1, 2 and 3).

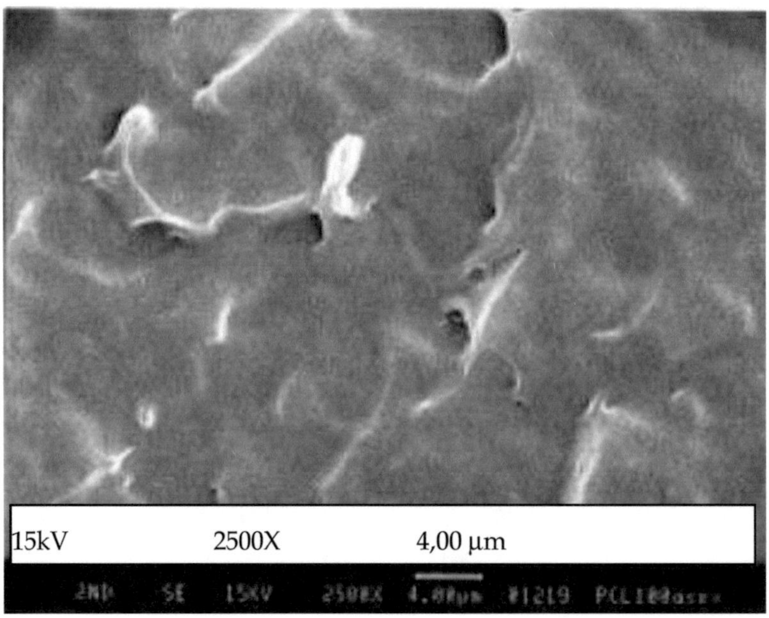

Fig. 6. Scanning electron micrograph of as extruded PCL, cryogenic fractured sample.

It is possible to observe the continuous PLLA-rich phase and the PCL-rich dispersed phase with a maximum domain size of 1.5 µm, as visualized before by Tsuji et al. (2001) in PCL:PLLA solution-cast blends. The PCL-rich phase is homogeneously dispersed in the PLLA matrix. One can observe some cracks or voids in the PLLA film probably caused by the temperature of processing, as shown in Figure 8. It has mentioned before that the degradation of aliphatic polyesters can occur because of melting at high temperatures (Yoshii et al., 2000; Kodama et al., 1997).

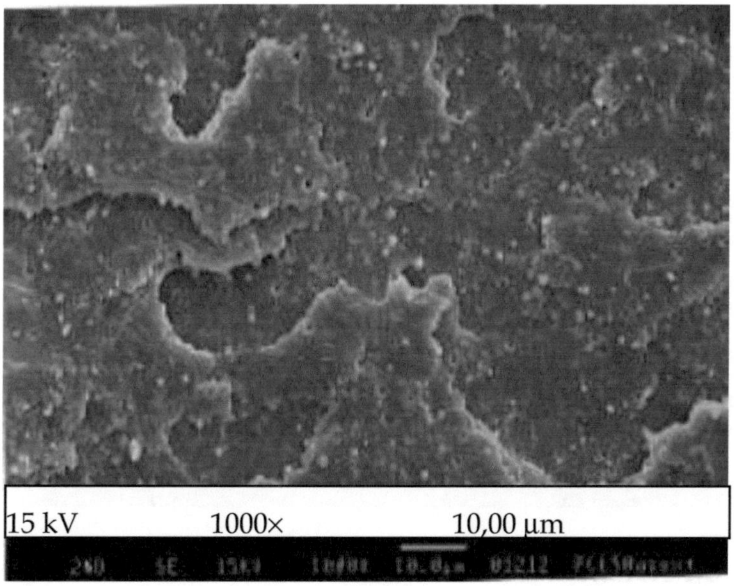

Fig. 7. Scanning electron micrograph of as extruded PCL:PLLA 50:50 (w:w), cryogenic fractured sample.

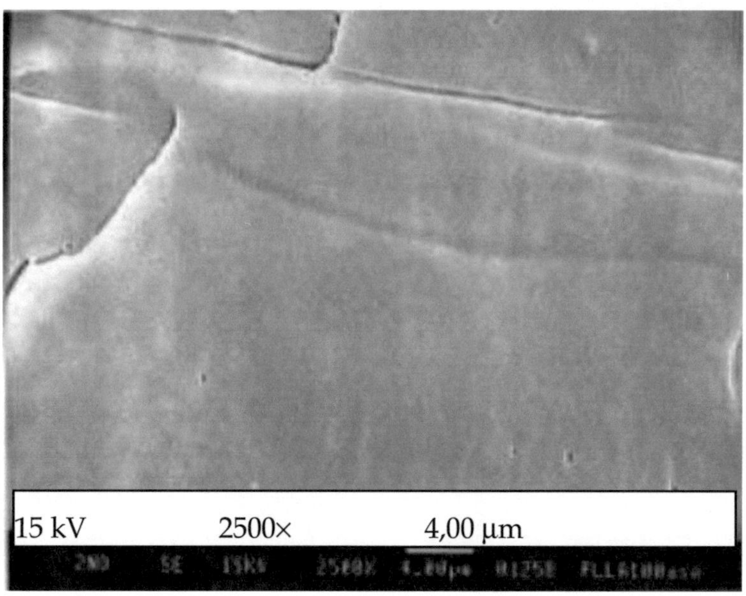

Fig. 8. Scanning electron micrograph of as extruded PLLA, cryogenic fractured sample.

Micrographs of annealed PLLA (Figure 9) and PCL:PLLA 50:50 (w:w) (Figure 10) shows changes in the morphology due to the crystallization of PLLA. Utracki (1989) explained that depending on the crystallization conditions various types of morphology can be obtained, which proceeds through melt, nucleation, lamellar growth and, spherulitic growth.

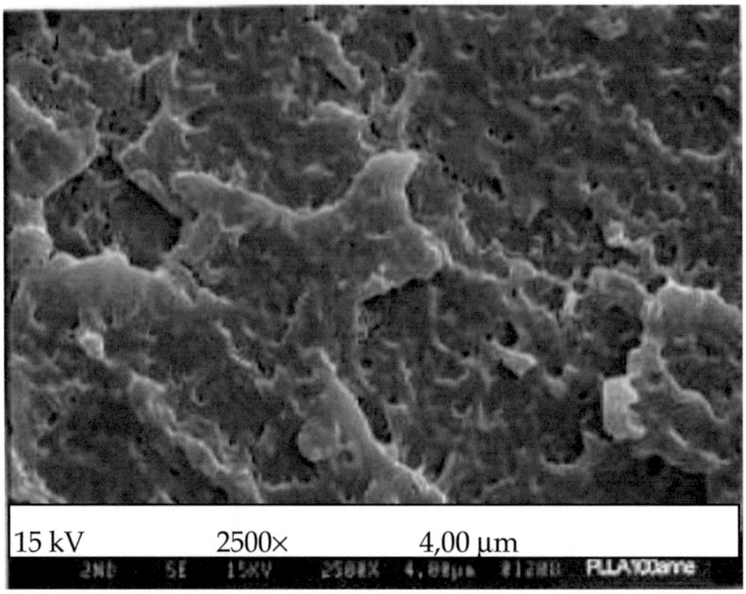

Fig. 9. Scanning electron micrograph of *annealed* PLLA, cryogenic fractured sample.

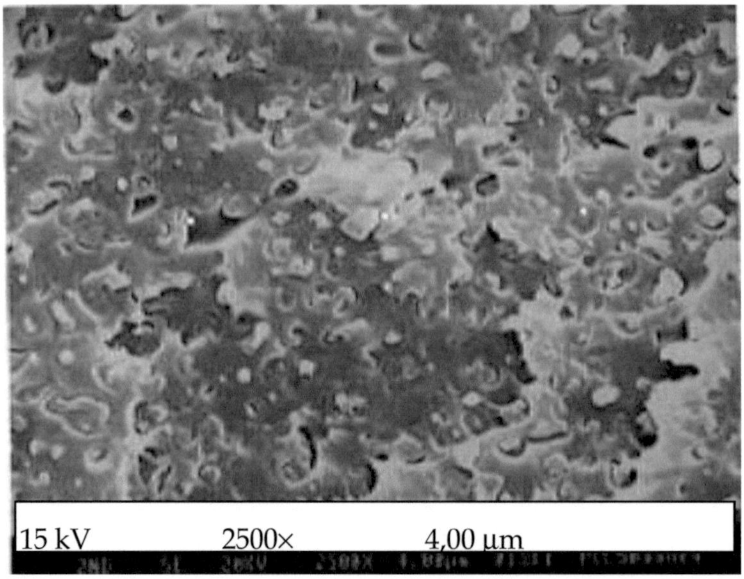

Fig. 10. Scanning electron micrograph of *annealed* PCL:PLLA 50:50 (w:w), cryogenic fractured sample.

Although the temperature of annealing was lower than the PLLA melting temperature (T_m), the thermal treatment allowed the crystallization of PLLA. Moreover, even though as extruded and annealed samples temperatures of processing were the same, one can observe the reduction of the cracks on the PLLA annealed sample. It is also possible to verify some morphological changes. The spheres are apart from the matrix, and in addition the matrix was changed due to the crystallization of PLLA. The blends are not miscible and the after extrusion cooling from the melt to room temperature causes the phase separation due to the difference between the melting temperatures of both blend components, PCL and PLLA.

In preliminary studies by differential scanning calorimetry (DSC) no change in the PLLA melting temperature was observed by increasing the PCL content in the blends. It was observed that, as PCL amount increased, PCL T_m peak increased in the region of the glass transition temperature of PLLA (Kodama et al., 1997). Although the immiscibility occurs, it is possible to observe by SEM some interfacial interaction, as the spheres seem to be covered by a thin layer of the polymeric matrix of the blends.

It should be noted that the blends were well mixed during extrusion as shown by the distribution and the size of the spherulites in the matrix. Dell'Erba et al. (2001) have found that it was reasonable to assume that low interfacial tensions were obtained in PLLA:PCL blends because of their similar chemical nature of the blends components, which allows interpolymer polar interactions across phase boundaries, thus favoring a well-dispersed morphology.

Preliminary studies have shown that although both are semi-crystalline polymers, only PCL crystallizes during extrusion. PLLA is amorphous and crystallizes after annealing, which was observed by x-ray diffraction of the non-irradiated samples (Broz et al., 2003). Also the orientation of crystallites in the blends was observed by x-ray diffraction, PLLA crystallizes in the α form with 10_3 helical conformation (Zhang et al., 2005). The thermal treatment increases the quantity of spheres, it was possible to notice some ellipsoids. This suggests that the new spherulites were formed due to the crystallization of amorphous PLLA, as observed previously (Broz et al., 2003). In this case, the segregation is also clear. It is possible to observe the separated spheres from the matrix and the cavities. It was discussed in literature (Dell'Erba et al., 2001) that the PLLA spherulites growth mechanism does not change when different amounts of PCL are present in the blend. Additionally, the presence of PCL enhances the PLLA crystallization rate, suggested to likely occur through the increase in the nucleation rate, it was observed that the presence of PCL domains in the PLLA matrix causes a small lowering in the half time of crystallization (Dell'Erba et al., 2001).

Furthermore, the results indicated that even though PLLA and PCL are immiscible, revealed by the presence of two glass transition temperatures for the blends very close to those found for pure PLLA and PCL, they are not highly incompatible (Dell'Erba et al., 2001). The binary mixture of (two) polymers is considered a compatible blend, when a homogeneous solid system is formed, without phase separations. It means a complete mutual solubility of the two polymers in molten state as well. This compatibility is reflected in, among other physical and mechanical properties, the fact that the system will have one single glass transition temperature (T_g) (IAEA-TECDOC-1420, 2004).

According to Michler (2008), γ or electron irradiation initiates pronounced crosslink in the amorphous parts of semi-crystalline polymers, whereas the structure inside the lamellae (the crystallinity) is not destroyed, so long as critical doses are not used. In this study, micrographs of PCL and PLLA homopolymers and PCL:PLLA 50:50 (w:w) blend irradiated with 100 kGy and 500 kGy, respectively, are shown in Figure 11 up to Figure 16.

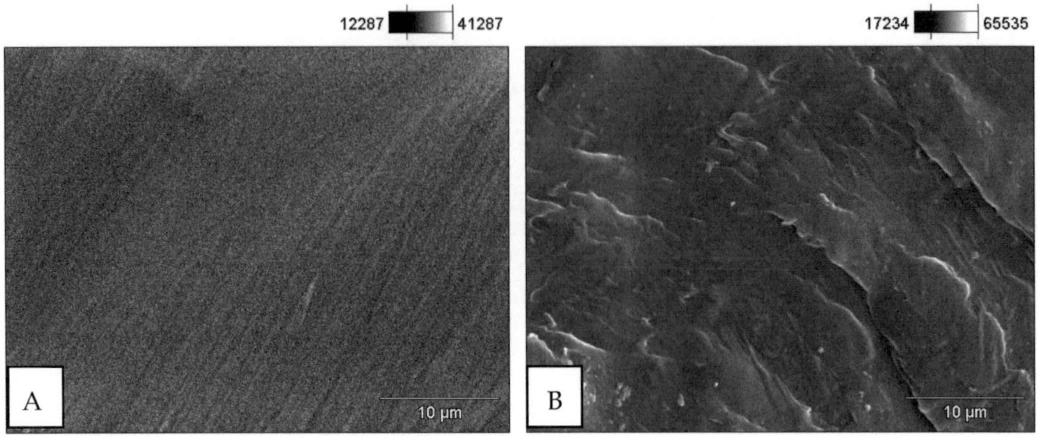

Acceleration voltage: 20.0 kV Magnification: 2500

Fig. 11. Scanning electron micrographs of as extruded PCL, cryogenic fractured sample: A) non irradiated and B) irradiated with 100 kGy.

In Figure 12 it is possible to observe that lamellar structures increase on the fractured surface of PCL irradiated with 500 kGy, indicating that possibly significant crosslinking occurred.

Comparing the scanning electron micrographs, it is observed very few changes on the surface of ruptured samples. In earlier studies by size exclusion chromatography of electron beam irradiated PCL in air, it was observed a small increase followed by a decrease of crosslinking degree up to 5 kGy and after that an increase of gel-content of 15 wt %, indicating the enhance of crosslinking degree up to 200 kGy radiation dose (Södergard & Stolt, 2002). Even though PCL crosslinking predominates at radiation doses higher than 5 kGy and random chain-scission at lower doses (Södergard & Stolt, 2002; Ohrlander et al.,2000). In this chapter, only few changes could be seen by SEM for the irradiated PCL up to 100 kGy. However, the ruptured sample surface of irradiated PCL with 500 kGy became full of scales suggesting that the increase of crosslinking density induced by the ionizing radiation caused this alteration.

Some differences observed on micrographs A and B on Figure 13 probably are related to different regions analyzed. Apparently, an increase of granulation occurred on the polymeric matrix for 100 kGy irradiated sample.

Comparing Figures 13 and 14, some cracks appear, and polymeric surface seems to become smoother, probably due to significant scission of polymeric chains.

The surface of PLLA sample became rough. This fact is correlated to chain scissions promoted by gamma radiation. In the literature it was observed that PLA mainly undergoes chain-scissions at doses below 250 kGy. At higher doses of radiation, crosslinking reactions increase as a function of the increasing radiation dose. The reactions occur in the amorphous phase of the polymer (Södergard & Stolt, 2002). Samples submitted to doses in the range of 30 up to 100 kGy showed a marked depression in mechanical properties attributed to oxidative chain-scissions in amorphous region (Nugroho et al., 2001). Apparently no changes are visible by SEM on the irradiated PLLA with 500 kGy radiation dose, in contrast to other properties observed previously in the literature (Södergard & Stolt, 2002; Ohrlander et al., 2000; Nugroho et al., 2001).

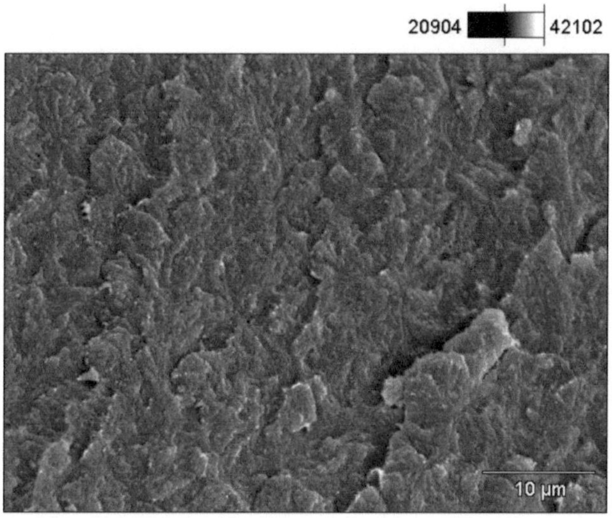

Acceleration voltage: 10.0 kV Magnification: 2000×

Fig. 12. Scanning electron micrograph of as extruded PCL, cryogenic fractured sample irradiated with 500 kGy.

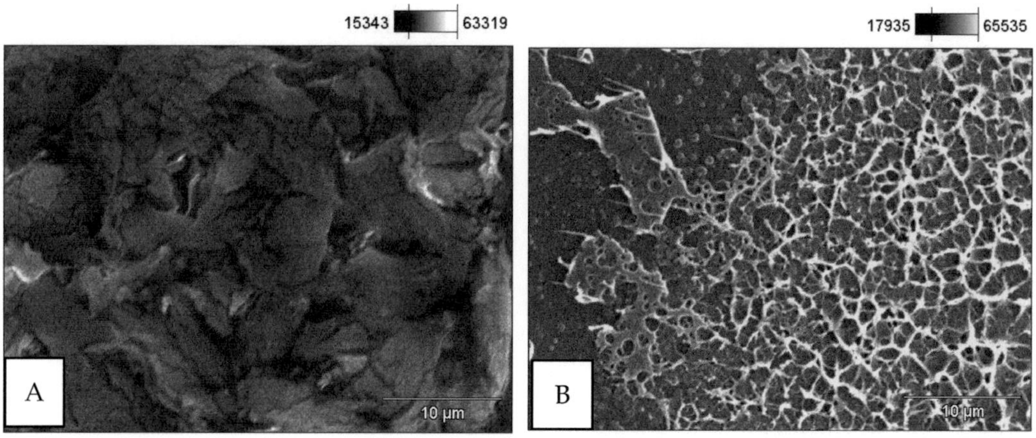

Acceleration voltage: 20.0 kV Magnification: 2500×

Fig. 13. Scanning electron micrographs of as extruded PLLA, cryogenic fractured sample: A) non irradiated and B) irradiated with 100 kGy.

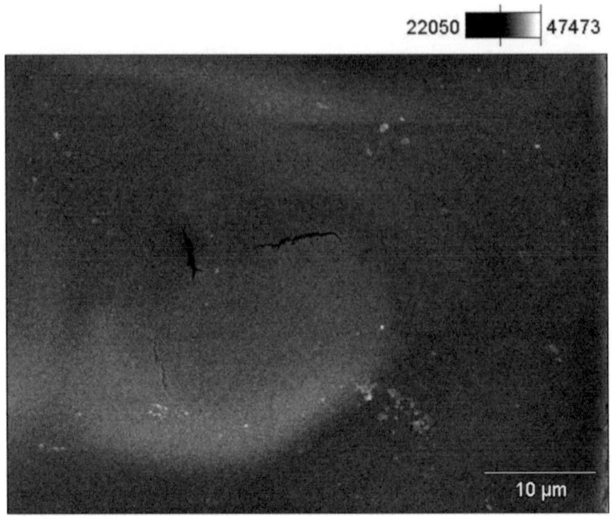

Acceleration voltage: 10.0 kV Magnification: 2000×

Fig. 14. Scanning electron micrograph of as extruded PLLA, cryogenic fractured sample irradiated with 500 kGy.

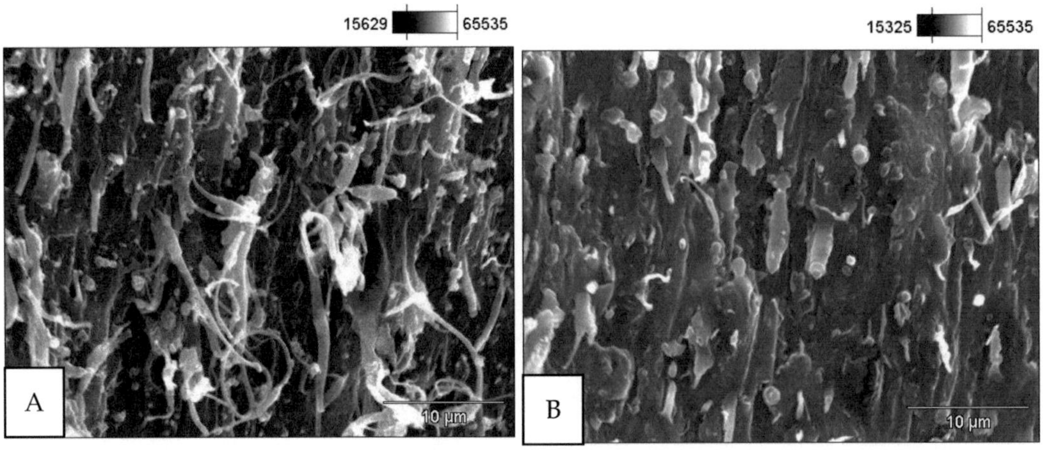

Acceleration voltage: 20.0 kV Magnification: 2500

Fig. 15. Scanning electron micrographs of as extruded PCL:PLLA 50:50 (w:w), cryogenic fractured sample: A) non irradiated and B) irradiated with 100 kGy.

Broz et al. (2003) observed that microstructure of PLLA 0.4 in PCL was characterized by relatively wide quantity of spherical inclusions of PLLA on PCL matrix. Particles had sizes varying from 5 µm to 100 µm, apparently isolated on the matrix. Similar micrograph was observed in Figure 15A. Irradiated blends micrographs, shown in Figures 15B and 16, suffered scales formation similar to that observed for irradiated PCL. The interface between spherical inclusions and polymeric matrix seemed to be clean, suggesting that exists a weak adhesion between two phases, in consonance with absence of thermal transitions dislodgement observed by DSC. This lack of adhesion is unexpected since both polymers had been stated as miscible in the molten state.

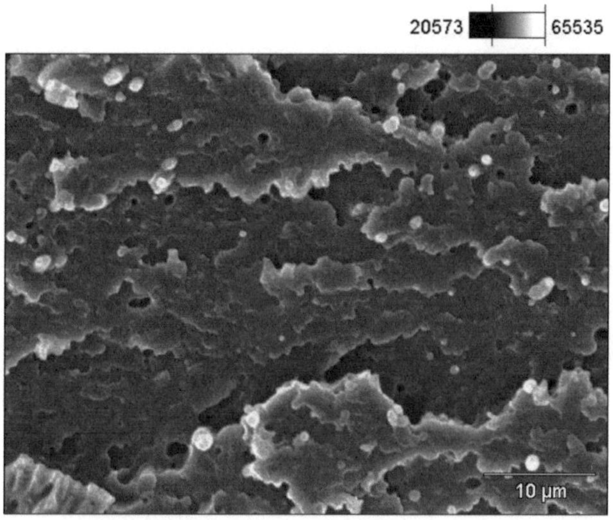

Acceleration voltage: 10.0 kV Magnification: 2000××

Fig. 16. Scanning electron micrograph of as extruded PCL:PLLA 50:50 (w:w), cryogenic fractured sample irradiated with 500 kGy.

It seems that the ionizing radiation induced some shape alteration in the PCL dispersed phase in blends that were irradiated with 100 kGy. Likewise, samples submitted to irradiation processing up to dose of 500 kGy present the matrix with decreased PCL spherulites disperse phase, suggesting that some interaction between both polymeric phases had been promoted by the ionizing radiation. Previous studies demonstrated that gamma radiation does not affect significantly thermal properties of the blends when doses were kept bellow 75 kGy. A small decrease of PLLA T_m occurred probably due to PLLA main chain-scission. Thermal treatment induces PLLA T_m variation on irradiated blends with high concentration of PLLA. The crystallinity of PCL homopolymer and PLLA homopolymer as well as in the studied blends was not significantly affected by irradiation up to 100 kGy (Kodama et al., 2005). On the other hand, after irradiation with higher doses, PCL samples were more thermally stable than PLLA and blend (Nugroho et al., 2001). Thermal properties of PLLA were not affected by gamma radiation up to 100 kGy (Kodama et al., 2006a). Other results obtained previously by Kodama et al.(2006b) have shown that both, gamma and EB radiation, at doses up to 500 kGy, do not cause sample degradation to any significant extent to be detectable by FTIR (Fourier Transform Infrared Spectroscopy). As well, the miscibility of the polymeric blends was not affected by the irradiation process (Kodama et al., 2006b).

It can be observed in Figure 17A and B that fractured surface of non irradiated blend presented several needles like structures orthogonal to the surface, apparently related to PCL, due to the proportion o this component in the blend. Radiation dose seemed to induce these structures to diminish, probably related to PCL crosslinking that occurred with 100 kGy radiation dose.

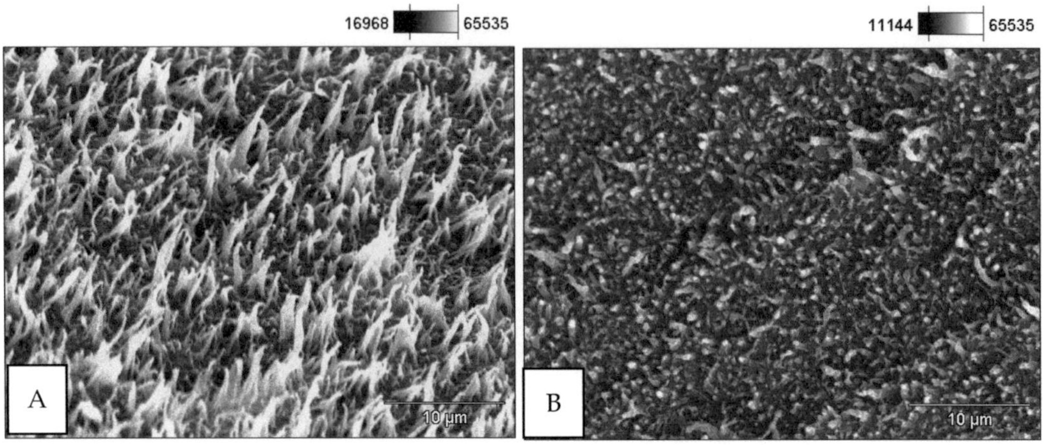

Acceleration voltage: 20.0 kV Magnification: 2500×

Fig. 17. Scanning electron micrographs of as extruded PCL:PLLA 75:25 (w:w), cryogenic fractured sample: A) non irradiated and B) irradiated with 100 kGy.

Following micrographs were obtained using field emission scanning electron microscopy (FE-SEM) that dispense the use of Au^0 coating and energy for image obtaining is lower. Micrograph of PCL:PLLA 20:80 (w:w), non irradiated is shown in Figure 18. It was observed a rough surface.

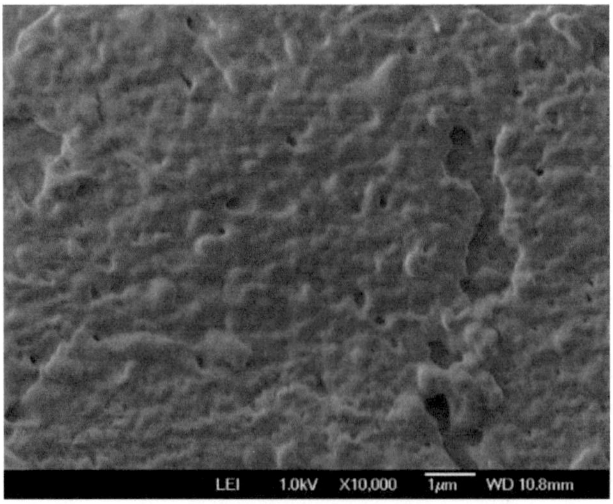

Fig. 18. Field emission scanning electron micrograph of as extruded PCL:PLLA 20:80 (w:w), cryogenic fractured sample, non irradiated.

In Figure 19 is observed surface micrographs of samples of PCL:PLLA 20:80 (w:w) prepared by hot press process , non irradiated and irradiated with 20 kGy radiation dose. This dose look as if do not affect significantly blend surface. Considering that conventional radiation dose for sterilization is 25 kGy, it is not expected that significant alteration occurs on the blend morphology sterilized by ionizing radiation.

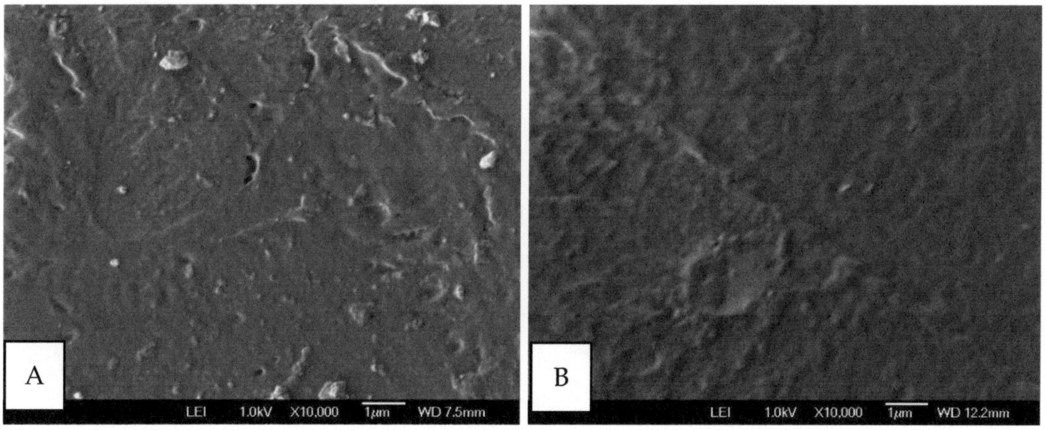

Fig. 19. Field emission scanning electron micrographs of hot pressed PCL:PLLA 20:80 (w:w) blends: A) non irradiated and B) irradiated with 20 kGy.

Arbelaiz et al. (2006) studied composites of linen fiber and PCL, and observed by SEM that fibers were clean and almost without adhesion points with PCL polymeric matrix, that indicated low wettability of fibers and lack of adhesion between phases. In Figure 20 it was no possible to observe in this sample regions containing fibers as observed by the authors mentioned before, probably it is related to its low concentration on the composite (5 or 10%) studied in this chapter. It was just observed what looks like a fragment of fiber.

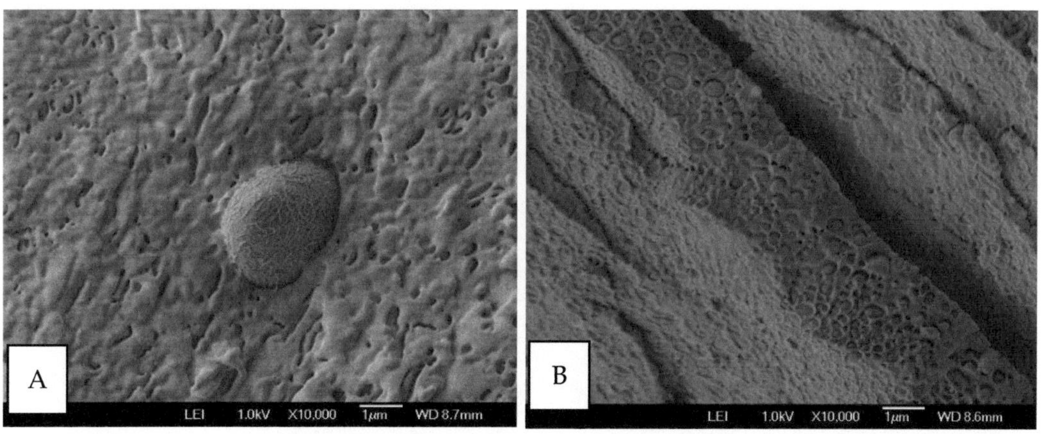

Fig. 20. Field emission scanning electron micrographs of composite with 10% non chemically treated coconut fiber: A) non irradiated; B) irradiated with 100 kGy, cryogenic fractured.

It was no possible to observe in Figure 21 significant alteration between structure existent and blend matrix. Apparently, acetylating process was not effective referring to the adhesion. However, it seems that ellipsoidal structures apart from polymeric matrix increased. Visually, structures suffered elongation and size reduction. Irradiated sample have smoother surface, probably related to scission process prevail of PLLA with doses above 100 kGy.

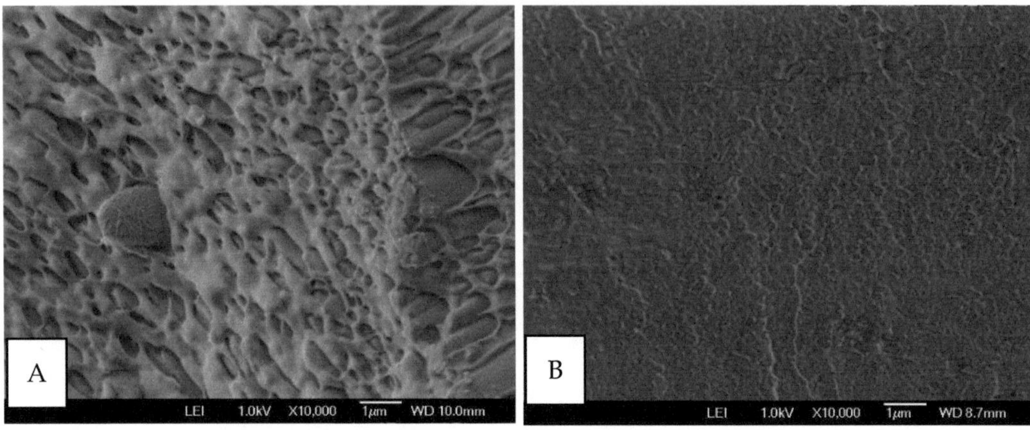

Fig. 21. Field emission scanning electron micrographs of composite with 10% acetylated coconut fiber: A) non irradiated; B) irradiated with 100 kGy, cryogenic fractured.

Micrographs of hot pressed sheets surface of composites containing 10% non chemically treated coconut fiber, non irradiated, and EB irradiated with 50 kGy and 100 kGy radiation doses, respectively, are shown in Figure 21. It was not possible to observe fiber presence on the surface analyzed. Neither any significant alteration on the irradiated surface, in the dose range studied. It suggests that cryogenic fractured surfaces allows the observation of irradiation effect on the polymeric bulk that otherwise could not be observed on sample surface. Probably it occurs due to the fact that species formed by energy deposition of radiation through polymeric matrix reacts mainly on the bulk than on the surface.

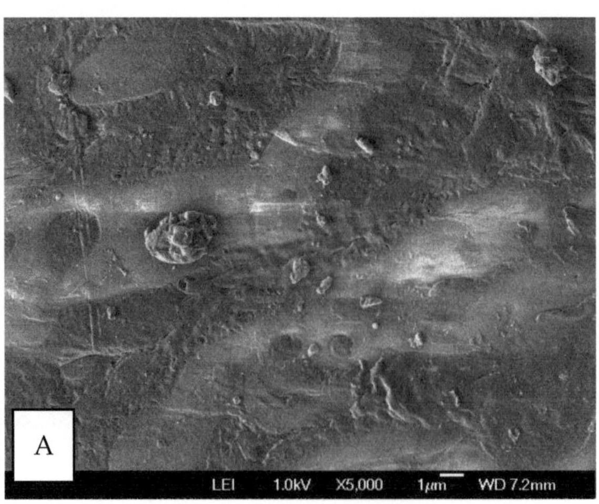

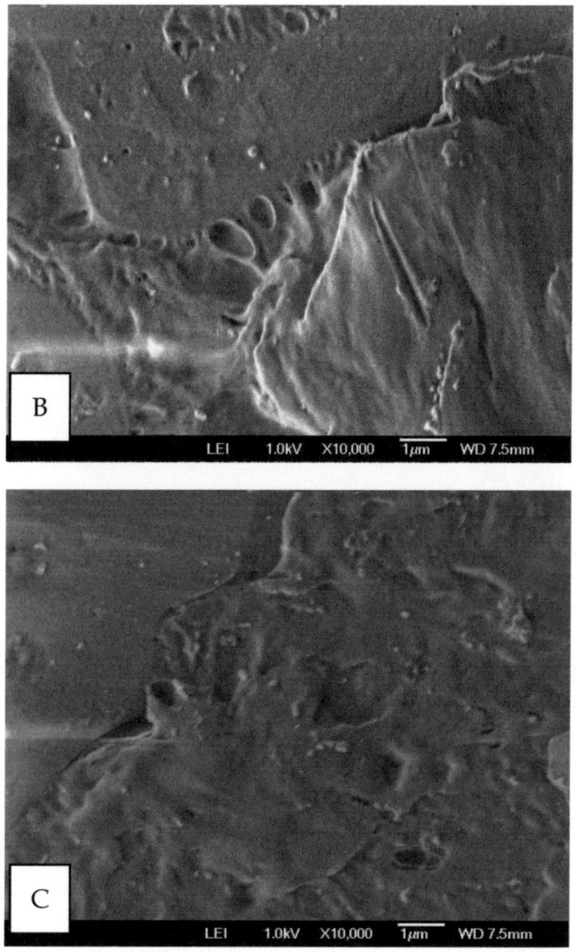

Fig. 22. Field emission scanning electron micrograph surface of hot pressed sheet of composite containing 10% non chemically treated coconut fiber: A) non irradiated; B) EB irradiated with 50 kGy, and C) EB irradiated with 100 kGy.

4. Conclusion

Due to coalescence effect it was possible to observe spherical inclusions of PLLA in PCL:PLLA blend. Increasing radiation dose induced elongation of inclusions, as well, lamellar structures increase in the PCL matrix. Radiation doses higher than 100 kGy altered morphologies of samples surfaces, that became smoother, attributed to the presence of smaller fractions of PLLA, as a result of long chain scission, and high crosslinking reaction in PCL phase. Radiation processing and chemical acetylating did not promote measurable interaction between fibers with polymeric matrix.

The SEM micrographs of the fractured homopolymers and blends have shown their immiscibility. The crystallization of PLLA could be observed on the annealed samples. Samples irradiated with 100 kGy presented little variation on the morphology, even supposing that the structural modifications induced by ionizing radiation may alter the morphology of the samples. It seems that some shape alteration in the PCL dispersed phase in blends occurred. Likewise, samples submitted to irradiation processing up to dose of 500 kGy presented the matrix with decreased PCL spherulites disperse phase, suggesting that some interaction between both polymeric phases had been promoted by the ionizing radiation. However, in PCL homopolymer and PCL:PLLA 50:50 irradiated with 500 kGy samples it was possible to observe significant alteration. The ruptured sample surface of irradiated PCL with 500 kGy became full of scales probably due to an increase of crosslinking density induced by the ionizing radiation. The surface of PLLA sample became rough with 100 kGy radiation dose correlated to chain scissions promoted by gamma radiation. On the other hand, apparently no changes are visible by SEM on the irradiated PLLA with 500 kGy radiation dose, in contrast to the observed previously in the literature. It was also studied blends and composites based on PCL, PLLA, and coconut fiber. Acetylation of fibers was not effective in order to induce any interaction between fibers and polymeric matrix, as expected. Ionizing radiation neither promoted detectable interaction between polymeric matrix and fibers.

5. Acknowledgements

We are grateful to the financial support from JICA and IAEA. Additionally, to Dr. Akihiro Oishi and Dr. Kazuo Nakayama, from National Institute of Advanced Industrial Science and Technology - AIST, Japan, for samples preparation and valuable discussion; to Dr. Naotsugu Nagasawa and Dr. Masao Tamada , from Japan Atomic Energy Agency - JAEA, Japan, for samples preparation and irradiation. We also would like to thank Dr. Morsyleide Freitas Rosa from Embrapa for providing coconut fiber; to Prof. Dr. Hélio Wiebeck, and Mr. Wilson Maia from Laboratório de Matérias-Primas Particuladas e Sólidos Não Metálicos - LMPSol, Departamento de Engenharia de Materiais, Escola Politécnica da USP (EPUSP) for coconut fiber size reduction and segregation; also to Eng. Elisabeth S.R. Somessari, Eng. Carlos G. da Silveira, and Mr. Paulo de Souza Santos, from IPEN, for blends and composites irradiation. In addition, to Centro Tecnológico da Marinha em São Paulo - CTMSP, for SEM and SEM EDS micrographs. We would like also to thank Dr. Luci Diva Brocardo Machado for helpful discussion.

6. References

Abdul Khalil, H.P.S., Rozman, H.D. (2000). Acetylated plant-fiber-reinforced polyester composites: a study of mechanical, hygrothermal, and aging characteristics, *Polymer-Plastics Technology and Engineering* Vol. 39 (No. 4) : 757-781.

Advances in Radiation Chemistry of Polymers, IAEA-TECDOC-1420, IAEA, Vienna, 2004.

Broz, M.E., VanderHart, D.L., Washburn, N.R. (2003). Structure and mechanical properties of poly(D,L-lactic acid)/poly(epsilon -caprolactone) blends, *Biomaterials* Vol. 24: 4181-4190.

Calado, V., Barreto, D.W., D'Almeida, J.R.M., (2000). The effect of a chemical treatment on the structure and morphology of coir fibers, *Journal of Materials Science Letters* Vol. 19: 2151-2153.

Charlesby, A., Clegg, D.W. & Collyer, A.A. (Ed.). (1991). *Irradiation Effects on Polymers*, Elsevier Applied Science, London and New York.

Chmielewski, A.G. New Trends in radiation processing of polymers, In: International Nuclear Atlantic Conference; Encontro Nacional de Aplicações Nucleares, 7th, ago. 28 - set. 2, 2005, Santos, SP. *Anais...* São Paulo: ABEN, 2005.

D'Almeida, A.L.F.S.; Calado, V.; Barreto, D.W. (2005). Acetilação da fibra de bucha (Luffa cylindrica), *Polímeros: Ciência e Tecnologia* Vol.15(No. 1): 59-62.

Dell'Erba, R., Groeninckx, G., Maglio, G., Malinconico, M., Migliozzi, A., (2001). Imiscible polymer blends of semicrystalline biocompatible components: thermal properties and phase morphology analysis of PLLA/PCL blends, *Polymer* Vol. 42: 7831-7840.

Kammer, H.W., Kummerlowe, C. (1994). Poly (ε-caprolactone) Comprising Blends - Phase Behavior and Thermal Properties, *in* Finlayson, K. (ed.) *Advances in Polymer Blends and Alloys Technology*, Technomicv, USA, 5, pp. 132-160.

Kantoğlu, Ö., Güven, O., (2002). Radiation induced crystallinity damage in poly(L-lactic acid), *Nuclear Instruments and Methods in Physics Research B* Vol. 197: 259-264.

Kapulskis, T.A., de Jesus, R.C., Innocentini-Mei L.H. Modificação química de fibras de coco visando melhorar suas interações interfaciais com matrizes poliméricas biodegradáveis. "XIII Congresso Interno de Iniciação Científica da UNICAMP - PIBC 2005,
www.prp.unicamp.br/pibic/congressos/xiiicongresso/cdrom/html/area3.html.
Accessed in 18/09/06

Kikkawa, Y., Suzuki, T., Tsuge, T., Kanesato, M., Doi, Y., Abe, H., (2006). Phase structure and enzymatic degradation of poly(L-lactide)/atactic poly(3-hydroxybutyrate) blends: an atomic force microscopy study, *Biomacromolecules* Vol. 7: 1921-1928.

Kodama, Y., Machado, L.D.B., Nakayama, K. (2005). Thermal Properties of Gamma Irradiated Blends Based on Aliphatic Polyesters. In: International Nuclear Atlantic Conference - INAC 2005, August 28 to September 2, 2005, Santos, SP, Brazil, Associação Brasileira de Energia Nuclear - ABEN ISBN 85-99141-01-5. 1CD-ROM.

Kodama, Y., Machado, L.D.B., Nakayama, K. Effect of Gamma Rays on Thermal Properties of Biodegradable Aliphatic Polyesters Blends, In: V Congresso Brasileiro de Análise Térmica e Calorimetria, April 02 to 05, 2006a, Poços de Caldas, MG, Brazil, Brazilian Association of thermal analysis and calorimetry - ABRATEC.

Kodama, Y., Machado, L.D.B., Giovedi, C., Nakayama, K. FTIR Investigation of Irradiated Biodegradable Blends. In: 17° Congresso Brasileiro de Engenharia e Ciência dos Materiais - CBECIMAT, 2006b, Foz do Iguaçu, PR, Brazil.

Lee, S.H., Ohkita, T., Kitagawa, K. (2004). Eco-composite from poly (L-lactic acid) and bamboo fiber, *Holzforschung* Vol. 58: 529-536.

Martins, M.A., Forato, L.A., Mattoso, L.H.C., Colnago, L.A. (2006). A solid state ^{13}C high resolution NMR study of raw and chemically treated sisal fibers, *Carbohydrate Polymers* Vol. 64: 127-133.

Michler, G.H. (2008). *Electron Microscopy of Polymers*, Springer-Verlag.

Mochizuki, M., Hirami, M., (1997). Structural effects on the biodegradation of aliphatic polyesters, *Polymers for AdvancedTechnology* Vol. 8: 203-209.

Mothé, C.G.; de Miranda, I.C. (2009) Characterization of sugarcane and coconut fibers by thermal analysis and FTIR, *Journal of Thermal Analysis and Calorimetry* Vol. 97: 661-665.

Nakayama K. & Tanaka, K. (1997). Effect of heat treatment on dynamic viscoelastic properties of immiscible polycarbonate-linear low density polyethylene blends, *Advanced Composite Materials* Vol. 6 (No. 4): 327-339.
Nugroho, P., Mitomo, H., Yoshii, F., Kume, T. (2001). Degradation of poly(l-lactic acid) by γ -irradiation, *Polymer Degradation and Stability* Vol. 72: 337-343.
Ohrlander, M., Erickson, R., Palmgren, R., Wisén, A., Albertsson, A.-C. (2000). The effect of electron beam irradiation on PCL and PDXO-X monitored by luminescence and electron spin resonance measurements, *Polymer* Vol. 41: 1277-1286.
Platt, D.K. (2006). *Biodegradable Polymers: Market Report*, Smithers Rapra Limited, United Kindom.
Rosa, M.F.; Santos, F.J.S.; Montenegro, A.A.T.; Abreu, F.A.P; Correia, D.; Araújo, F.B.S.; Norões, E.R.V. (2001). *Caracterização do pó da casca de coco verde usado como substrato agrícola*, Embrapa, Comunicado Técnico, Vol. 54: 1-6.
Sawyer, L.C., Grubb, D.T. & Meyers, G.F. (2008). *Polymer Microscopy 3rd ed*, Springer.
Severiano, L.C.; Lahr, F.A.R.; Bardi, M.A.G. Machado; L.D.B. (2010). Estudo do efeito da radiação gama sobre as propriedades térmicas de madeira usadas em patrimônios artísticos e culturais brasileiros. In: VII CONGRESSO DE ANÁLISE TÉRMICA E CALORIMETRIA, 2010, São Paulo, SP. *Anais...* São Paulo: 25 a 28 de abril. 1 CD-ROM.
Rowell, R.M., Han., J.S., Rowell, J.S., Characterization and Factors Effecting Fiber Properties, In: Frollini, E.; Leão, A.L.; Mattoso, L.H.C. (Ed.). *Natural Polymers and Agrofibers Composites: preparation, properties and applications*, São Carlos: USP-IQSC/Embrapa Instrumentação Agropecuária, Botucatu: UNESP, São Paulo, 2000.
Södergard, A. & Stolt, M. (2002). Properties of lactic acid based polymers and their correlation with composition, *Progress in Polymer Science* Vol. 27: 1123-1163.
Tomczak, F., Sydenstricker, T.H.D., Satyanaryana, K.G. (2007). Studies on lignocellulosic fibers of Brazil. Part II: Morphology and properties of Brazilian coconut fibers, *Composites: part A* Vol. 38: 1710-1721.
Tsuji, H. & Ikada, Y. (1996) Blends of aliphatic polyesters. I. Physical properties and morphologies of solution-cast blends from poly (DL-lactide) and poly(ε-caprolactone), *Journal of Applied Polymer Science* Vol. 60: 2367-2375.
Tsuji, H., Ishizaka, T., (2001). Blends of aliphatic polyesters. VI. Lipase-catalyzed hydrolysis and visualized phase strucuture of biodegradable blends from poly(e-caprolactone) and poly(L-lactide). International Journal of Biological Macromolecules Vol. 29: 83-89.
Utracki, L.A. (1989). *Polymer Alloys and Blends: Thermodinamics and Rheology*, Hanser, New York.
Yoshii, F., Darvis, D., Mitomo, H., Makuuchi, K. (2000). Crosslinking of poly (ε-caprolactone) by radiation technique and its biodegradability, *Radiation Physics and Chemistry* Vol. 57: 417-420.
Zhang, J., Duan, Y., Sato, H., Tsuji, H., Noda, I., Yan, S., Ozaki, Y. (2005). Crystal modifications and thermal behavior of poly (L-lactic acid) revealed by infrared spectroscopy, *Macromolecules* Vol. 38: 8012-8021.

14

Study of Helminth Parasites of Amphibians by Scanning Electron Microscopy

Cynthya Elizabeth González[1], Monika Inés Hamann[1] and Cristina Salgado[2]
[1]*Centro de Ecología Aplicada del Litoral,*
Consejo Nacional de Investigaciones Científicas y Técnicas,
[2]*Servicio de Microscopia Electrónica de Barrido,*
Universidad Nacional del Nordeste (UNNE), Corrientes,
Argentina

1. Introduction

Amphibians, like all other animals, are subject to a variety of parasites and diseases, including viral, bacterial and fungal infections as well as some forms of cancer and tuberculosis (Hoff et al., 1984). Various viruses and bacteria such as *Pseudomonas* or *Salmonella,* and fungi such as *Candida,* are recorded as common infectious agents in amphibians, but currently the focus of studies are the fungi of genus *Batrachochytrium,* agents of the disease known as chytridiomycosis, which is considered as one of the factors responsible for the decline of amphibian populations in many parts of the world (Berger and Speare, 1998). In addition, protozoans of the genera *Opalina* and *Entamoeba* in the digestive tract, and trypanosomes in the circulatory system, as well as coccidian protozoa have been recorded in amphibians (Duellman and Trueb, 1986; Duszynski et al., 2007).

However, helminths are the most common invertebrate parasites of amphibians. One well known example among trematodes is the monogenean genus *Polystoma,* which infects the urinary bladder of adult amphibians around the world. Parasitic digenean trematodes include both larval stages (metacercariae) and adults. Cestodes are not common parasites in the gastrointestinal tract of amphibians, but when present may persist for a long time. Adult acanthocephalans adhere to the mucosa of the stomach or intestine. Finally, nematodes are particularly abundant in the digestive tract, lungs and blood vessels of these vertebrates (Pough et al., 2001). Amphibians are also hosts to other groups of less common parasitic invertebrates, such as annelids, pentastomids and arthropods (copepods, ticks, insects) (Tinsley, 1995).

Of these, analyses using scanning electron microscopy techniques have been mainly applied to digenean trematodes (flatworms), nematodes (roundworms) and acanthocephalans (thorny or spiny headed), particularly to their adult stages (Fig. 1-3).

The study of parasitic nematodes and trematodes by scanning electron microscopy began in the 1970s and involved mainly those organisms that produced diseases in humans and livestock, as well as parasites of crops (Halton, 2004). In particular, studies in amphibian hosts were first made by Nollen and Nadakavukaren (1974) and Nadakavukaren and

Nollen (1975) who provided details of the tegument of the trematodes *Megalodiscus temperatus* (Paramphistomatidae) and *Gorgoderina attenuata* (Gorgoderidae) from *Rana pipiens*. Regarding nematodes, Navarro et al. (1988) provided details of the cuticle of the species *Cosmocerca ornata, Oxysomatium brevicaudatum* (Cosmocercidae) and *Seuratascaris numidica* (Ascarididae) collected in ranid hosts from different geographical areas of the Iberian Peninsula.

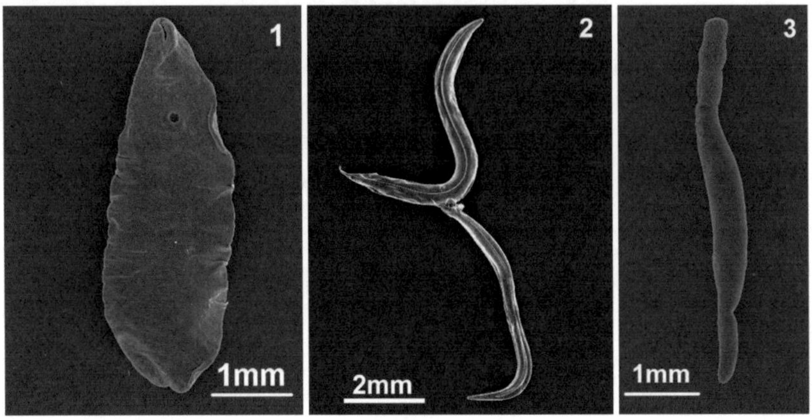

Fig. 1-3. Helminth parasites found in amphibian hosts. 1. Flatworms (Trematoda), general view. 2. Roundworms (Nematoda) male and female mating. 3. Thorny or spine headed (Acanthocephala), general view.

This chapter presents scanning electron micrographs taken during diverse studies undertaken to determine the helminth fauna of Argentinean anurans, especially those living in Northeastern Argentina. The survey includes the classes Trematoda (specifically subclass Digenea) and Nematoda (specifically subclass Secernentea) and the phyllum Acanthocephala. A total of five families of amphibians (Bufonidae, Cycloramphidae, Hylidae, Leptodactylidae, Leiuperidae) were analyzed, both at larval (tadpole) and adult stages, to study their helminth parasites. At the end of the chapter we present a summary of the present-day advances in this topic, including new contributions presented in this work; finally, we discuss possible future lines of research in this field of Parasitology.

The classification of helminths follows Anderson et al. (2009) and Gibbons (2010) for class Nematoda; Gibson et al. (2002) and Jones et al. (2005) for class Trematoda and Amin (1985) for Acanthocephala.

2. Preparation of helminth parasites of amphibians for observation by scanning electron microscopy

2.1 Collection of hosts and obtaining of helminth parasites

Adult frogs were hand captured, mainly at night, using the sampling technique defined as visual encounter survey (Crump and Scott, 1994). The individuals were transported live to the laboratory and killed in a chloroform solution ($CHCl_3$). The abdominal cavity of each frog was opened and the oesophagus, stomach, gut, lungs, liver, urinary bladder, kidneys, body cavity, musculature, integument and brain examined for parasites under a dissecting microscope (Fig. 4). Tadpoles were captured with a 45-cm-diameter dip net and kept alive in the laboratory until their dissection. They were killed using a chloroform solution, and subsequently all organs, musculature and body cavity were examined for parasites.

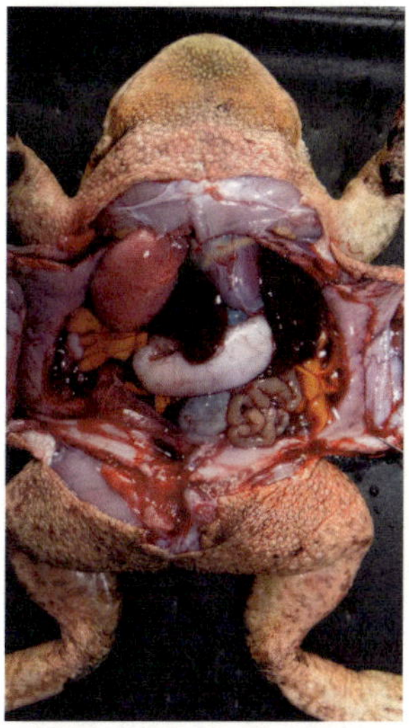

Fig. 4. Ventral view of adult amphibian showing all organ systems.

The analized amphibian species were: *Rhinella bergi, R. fernandezae, R. granulosa, R. schneideri* (Bufonidae), *Odontophrynus americanus* (Cycloramphidae), *Dendropsophus nanus, D. sanborni, Hypsiboas raniceps, Pseudis limellum, Phyllomedusa hypochondrialis, Scinax acuminatus, S. nasicus, Trachycephalus venulosus, Pseudis paradoxa* (Hylidae), *Physalaemus albonotatus, P. santafecinus, Pseudopaludicola boliviana, P. falcipes* (Leiuperidae), *Leptodactylus bufonius, L. chaquensis, L. elenae, L. latinasus, L. latrans, L. podicipinus* (Leptodactylidae), *Lepidobatrachus laevis* (Ceratophryidae). Anurans were identified using different guides and keys (Faivovich et al., 2005; Frost et al., 2004).

2.2 Procedure applied to helminth. Complications

In 1972, Allison et al. proposed a simplified four-step procedure that resulted in excellent preservation, support in the high vacuum and dissipation of surface charging of nematodes. The procedure involved: fixation, dehydration, treatment with an antistatic agent and gold-palladium coating. These authors obtained best results using fixation with 4% paraformaldehyde (phosphate-buffered) and AFA (acetic acid-formalin-alcohol); specimens were dehydrated in an ascending series of ethanol solutions to 70%, then transferred to 5% glycerine-95% ethanol from which the alcohol was allowed to evaporate, and cleared in 96.6% glycerol-0.05% potassium chloride-3.35% distilled water, 24 to 48 hours prior to examination. Subsequently, specimens were mounted on metal specimen stubs with Duco cement, outgassed in a vacuum evaporator for 1 hour or more and rotary-coated with gold-palladium.

In this study we basically followed the aforementioned procedure, with some modifications. Helminths were observed *in vivo*, counted, fixed and stored. Adult nematodes and trematodes were fixed in hot 4% formaldehyde; larval nematodes and trematodes were removed from cysts with the aid of preparation needlees and fixed in hot 4% formaldehyde; acanthocephalan larvae were placed in distilled water for 24 hours at 4° C for the proboscis to evert, and then fixed in hot 4% formaldehyde.

Another technique used for the study of nematodes and acanthocephalans by SEM consists of transferring specimens for 2 hours into 1% osmium tetroxide and dehydrate them in an ethanol series for 2 hours in each bath (Mafra and Lanfredi, 1998). In the case of trematodes, fixation can be made with paraformaldehyde, glutaraldehyde or a formaldehyde-glutaraldehyde mix such as Karnovsky's fixative in phosphate or cacodylate buffer. Postfixation is usually done for 2 to 3 hours at 4°C with cacodylate- or phosphate- buffered 1% osmium tetroxide (Karnovsky, 1965).

For processing of helminths, it should be taken into account that a hypertonic solution will cause shrinkage and almost complete disappearance of inflation in specimens, whereas a hypotonic solution may produce artificial inflations. In some cases, in spite of careful processing of samples for SEM study, good results are not achieved. Very frequently, the samples contain bacteria or debris, or tears of the cuticle of nematodes or tegument of trematodes (Fig. 5-9).

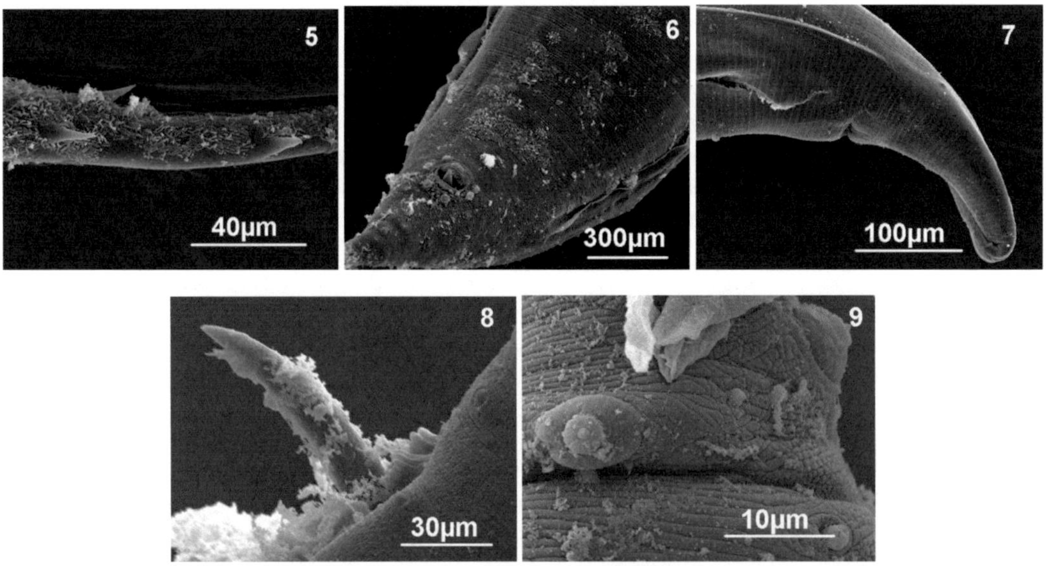

Fig. 5-9. Some complications in samples for SEM studies. 5. 6. Bacteria in posterior end of *Skrjabinodon* sp. (5) and in ventral surface of *Cosmocerca parva* (6). 7. Broken cuticle in posterior end of female of *Aplectana* sp. 8. 9. Debris in spicules and adanal papillae of *Cosmocerca podicipinus* (8) and in posterior end of *Falcaustra* sp. (9).

2.3 Characteristic of the scanning electron microscope used for this study

The Microscopy department at Universidad Nacional del Nordeste possesses a Jeol 5800LV scanning electron microscopy. Specimens are critical-point dried using a Denton Vacuum DCP-1 critical point drying apparatus, and sputter-coating is made using a Denton Vacuum Desk II sputter-coating unit.

For examination by scanning electron microscopy (SEM), specimens were dehydrated through an alcohol 70° and acetone series (70%, 85% and 100%; 15 minutes in each solution) and then subjected to critical point drying using CO_2; in the case of larval digenean trematodes, critical point drying time is shortened because these individuals are more fragile than adults. Samples were mounted on a metal sheet of copper or aluminum using double-sided tape. Then the specimens were sputter-coated with gold or gold-palladium for three minutes. Helminth parasites were observed in high vacuum in all cases.

Adequate fixation contributes to specimen preservation and stability within the microscope; dehydration allows outgassing and drying of specimens in the vaccum evaporator prior to coating without subsequent "bubbling" or shrinkage artifacts; glyceration provides an antistatic surface coating over all surface irregularities, even those which might not be adequately coated with 200 a of gold-palladium. The thin metal coating does not obscure fine structural detail but, in combination with the glycerol-KCl coating, ensures that the specimen can be exposed for extended periods of time to the electron beam without surface changes (Allison et al., 1972).

3. What can SEM studies tell us about helminth parasites of amphibians?

3.1 Nematoda

The body surface of nematodes is covered by a truly inert cuticle of extracellular material in the form of cross-linked collagens and insoluble proteins that are synthesised and secreted by the underlying epidermis (= hypodermis). In addition to proteins and collagen, the cuticle possesses glycoproteins, fibrin and keratin. It consists of three main layers and the epicuticle. The external layer is divided into internal and external cortex; the middle layer varies from having a granular uniform structure to presenting skeletal rods, fibers or channels; the internal basal layer can be laminated or grooved. The thin epicuticle may have a coating of quinone (Lee, 2002).

The above discussion refers specifically to the cuticle of the external body surface. In other body parts, such as the buccal cavity, excretory pore, vagina, cloaca and rectum, as well as the eggs and spicules in males, the cuticle is of a different nature.

The particular composition of the cuticle along the whole body of nematodes, and especially the ornamentations that they present in the anterior and posterior extremity, are the focus of primary study of these helminths by scanning electron microscopy.

The nomenclature for the structures detailed here is based on Chitwood and Chitwood (1975), Gibbons (1986) and Willmott (2009). We describe the modifications of the cuticle along the body surface, at the anterior and posterior body ends, and finally the cuticle of the eggs, vulva and spicules of these helminths.

3.1.1 Cuticle of body

The features analyzed along the body surface are the morphology and disposition of somatic papillae as well as the striation of the cuticle -longitudinal, transverse or oblique-, and the presence of annulations, punctuations, longitudinal ridges, alae -lateral, cervical or caudal-, inflation and spination.

There are several types of gross cuticular markings, namely, *transverse, longitudinal* and *oblique* markings. Of these, *transverse* and *longitudinal* markings are very common in nematode parasites of amphibians.

Tranverse markings: the superficial markings are grooves or ridges. *Striations* are defined as fine transverse grooves occurring at regular intervals; the distance between two striae is the intestrial region (*Aplectana hylambatis, Cosmocerca* spp.) (Figs. 10, 11). Deep striae are very commonly and are known as *annulations;* the distances between them are termed *annules* (*Aplectana delirae, Falcaustra* sp.) (Figs. 12, 13). In nematode parasites of the family Pharyngodonidae, annulations are much broader and more prominent (Figs. 14, 15, 34).

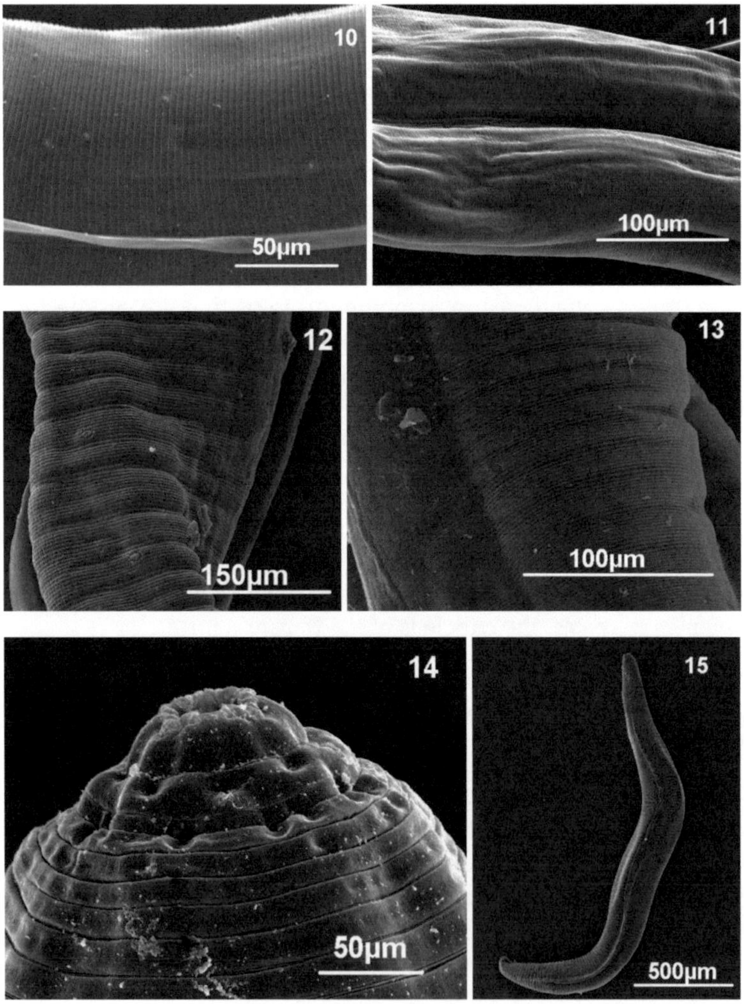

Fig. 10-15. Transverse marking in cuticle of nematode parasites of amphibians. 10. 11. Striations (*Aplectana hylambatis, Cosmocerca podicipinus,* respectively). 12. 13. Striations and annules (*A. delirae, Falcaustra* sp., respectively). 14. 15. Annulations (*Skrjabinodon* sp.).

Striations can be distributed uniformly along the whole body of the parasite (*Cosmocerca* spp.), or they may become less evident (*A. delirae*) or wider (*A. hylambatis*) at the body ends, or more marked, as for example, in the posterior body (*Skrjabinodon*) (Figs. 10-15).

Longitudinal markings: these may take the form of ridges or alae, or they may be merely the result of gaps in transverse markings.

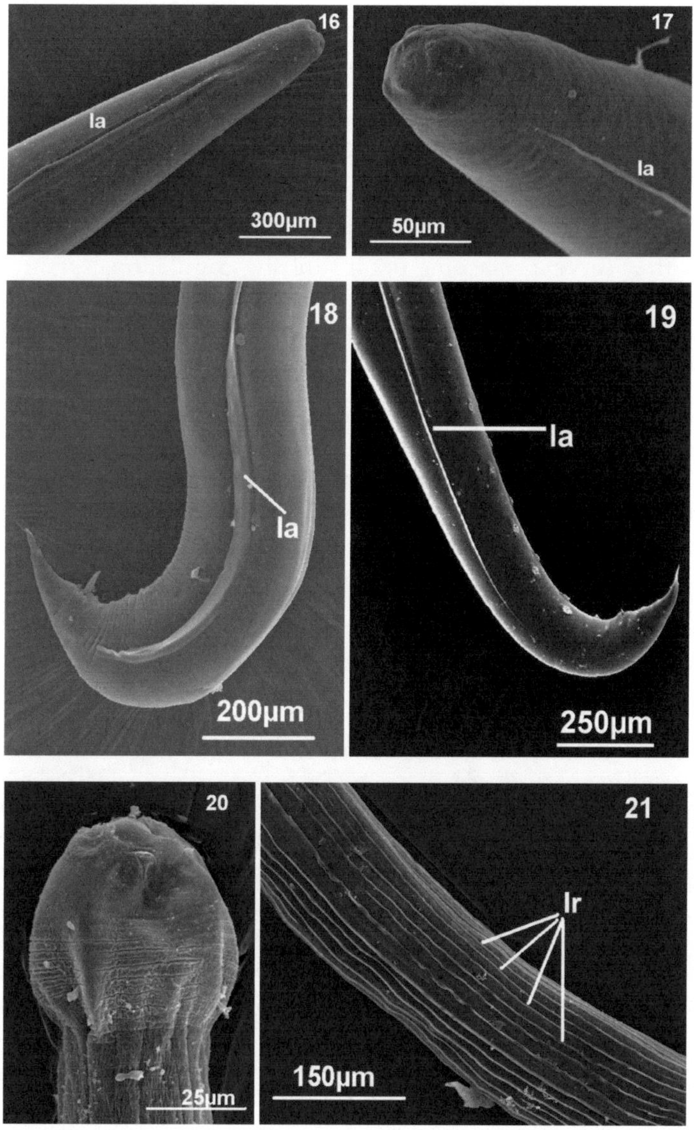

Fig. 16-21. Longitudinal markings in nematode parasites of amphibians. 16. 17. Lateral alae in anterior end of body in *Paraoxyascaris* sp. (16) and in *Porrocaecum* sp. (larvae) (17). 18. 19. Lateral alae in posterior end of body in *Cosmocerca* spp. 20. Cephalic cuticular vesicle in *Oswaldocruzia* sp. 21. Longitudinal ridges in *Oswaldocruzia* sp. la: lateral alae; lr: longitudinal ridges.

Alae: these are usually lateral or sublateral cuticular thickenings or projections. There are three types of alae: longitudinal, cervical and caudal.

Longitudinal alae: these are usually lateral or sublateral and occur in both sexes; extending along the length of the body. They occur in the families Cosmocercidae (*Aplectana, Cosmocerca, Cosmocercella, Paraoxyascaris*) (Figs. 16, 18, 19), Ascarididae (*Ortleppascaris, Porrocaecum*) (Fig. 17); in *Gyrinicola* sp. (Family Pharyngodonidae) lateral alae are present in males only.

Cervical alae: these structures are confined to the anterior part of the body. This modification of the cuticle does not occur in nematode species that parasitize amphibians.

In some parasitic nematodes, the cervical alae is modified as a *cephalic cuticular vesicle* as in the genus *Oswaldocruzia;* in some cases this cuticular vesicle is divided into a larger anterior part and a smaller posterior part (Fig. 20).

Caudal alae: these alae are confined to the caudal region of the body and limited to the males only; apparently they serve as clasping organs during copulation. Among nematodes that parasitize amphibians, they occur in the family Pharyngodonidae (*Parapharyngodon*). On the other hand, modified caudal alae occur in the males of some families such as the Molineidae, where they are called *bursa copulatrix* or *copulatory bursa* (see below).

Longitudinal ridges: these are raised areas that extend along the length of the body and are present on the submedian as well as on the lateral surfaces. *Oswaldocruzia* (Family: Molineidae) has longitudinal ridges throughout the body length that can disappear or appear along its body (Fig. 21). The system of longitudinal cuticular ridges or *synlophe* in this genus is very important as a taxonomic character.

Inflation: when the cuticle is swollen in a blister-like manner, this condition may be termed inflation (eg. *Rhabdias*; Figs. 22-24). Besides being inflated, the cuticle can be also striated. Inflation of the cuticle may be generalized over the whole body surface instead of restricted to certain areas.

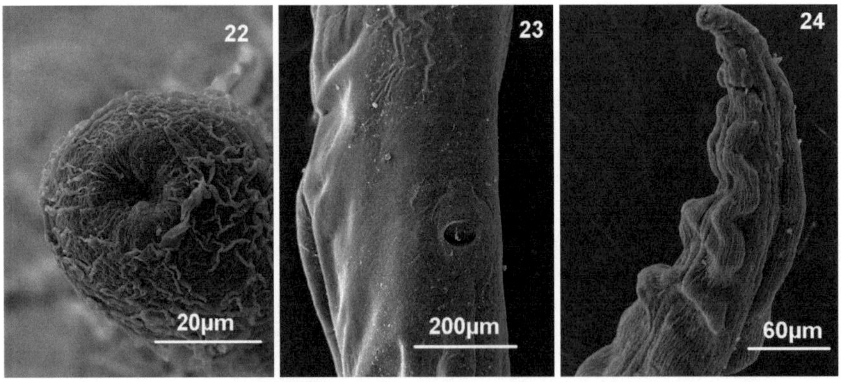

Fig. 22-24. Inflation in the cuticle of nematode parasites of genus *Rhabdias*. 22. Cephalic end. 23. Vulvar region. 24. Anterior end of body.

Papillae: these structures are nerve endings, some of which have a tactile function and others are chemoreceptors; they appear as cuticular elevations of different shapes and sizes (Figs. 25-27). According to their position, they are divided into labial, cephalic, cervical and genital papillae. In this section, we deal with the distribution and structure of somatic papillae, i.e. those that are distributed throughout the body of the nematode. These papillae are generally arranged in two subdorsal and two subventral rows along the body of the nematode.

Finally, in some genera of nematode parasites of amphibians, the excretory pore has conspicuous cuticularized walls, surrounded by a rough cuticular area (Pharyngodonidae); in others, the excretory pore is located in a depression of the cuticula (*Falcaustra*) (Figs. 28, 29).

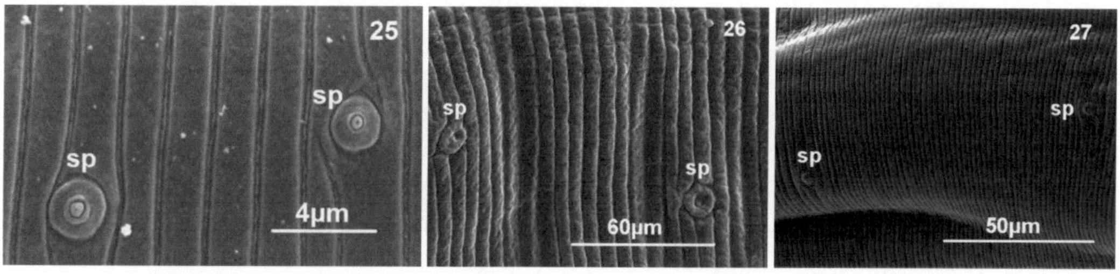

Fig. 25-27. Somatic papillae in nematode parasites of amphibians. 25. *Aplectana hylambatis*. 26. *Aplectana delirae*. 27. *Cosmocerca podicipinus*. sp: somatic papillae.

Fig. 28-29. Excretory pore in nematode parasites of amphibian. 28. *Falcaustra* sp. 29. *Skrjabinodon* sp.

Punctuations: this type of marking is frequent and appears as minute round areas of the cuticle which are arranged in definite patterns for each species. In nematode parasites of amphibians, this type of marking appears in males of the family Cosmocercidae, specifically in the genus *Cosmocerca* as part of the rosette papillae (see below; Figs. 38-44) and in the genus Cosmocercella (Figs. 46, 48).

3.1.2 Cuticle of anterior end of body

The cuticular modifications that occur in the anterior end of the body of nematode parasites of amphibians are: head papillae - cephalic papillae, externo-labial papillae, interno-labial papillae-, interlabia, deirids and amphids. These two latter structures are not modifications of the cuticle itself, but their opening in the cuticle can present diverse shape and structure.

Head papillae: these are tactile sensory organs usually located on the lips or labial region, including two circles of six labial papillae and one circle of four cephalic papillae. This arrangement has been proposed for ancestral nematodes considering a radial type of symmetry (De Coninck, 1965). The head papillae are divided into: *cephalic papillae*: outer circle of four head papillae (or latero-ventral and latero-dorsal papillae); *externo-labial papillae*: median circle of six head papillae, and *interno-labial papillae*: inner circle of six head papillae.

The basic structure proposed by De Coninck (1965) shows modifications in the nematodes that parasitize amphibians. For example, cosmocercid nematodes present three lips. The genus *Aplectana* has a circle of four internal labial papillae, 1 on each subventral lip and 2 on the dorsal lip, and a circle of six external labial papillae; the amphids are large, one on each subventral lip. The anterior end of the oesophagus presents three tooth-like projections covered with a thick cuticle, also called cuticular flap (Fig. 30). The genus *Cosmocerca* presents the same arrangement in the internal and external labial papillae and, as in the previous genus, the anterior end of the oesophagus bears three tooth-like projections (cuticular flap); the amphids are very prominent in some cases (Fig. 31). In the nematodes of family Atractidae, specifically in the genus *Schrankiana*, each lip has a cuticular flange overhanging the mouth opening (Fig. 32).

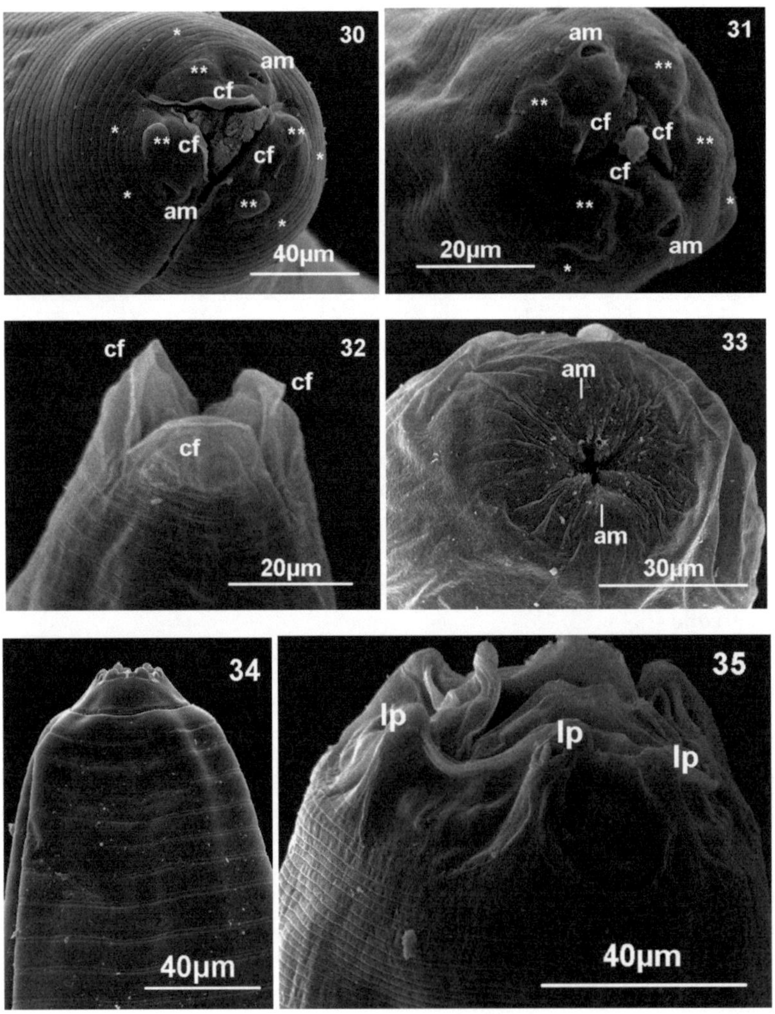

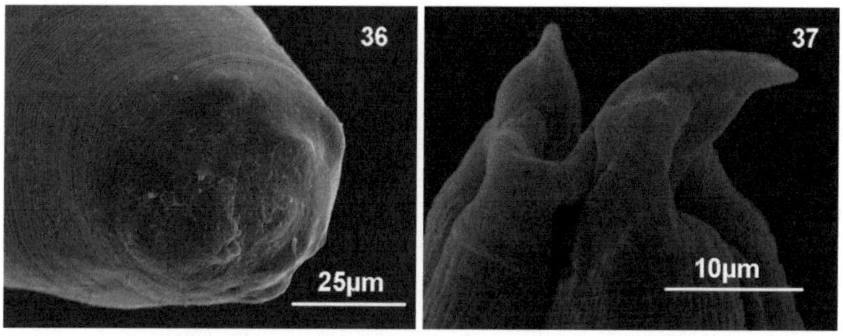

Fig. 30-37. Modifications of the cuticle in the anterior end of body. 30. *Aplectana* sp. 31. *Cosmocerca* sp. 32. *Schrankiana* sp. 33. *Rhabdias* sp. 34. *Pharyngodon* sp. 35. *Falcaustra* sp. 36. *Porrocaecum* sp. (larvae). 37. *Brevimulticaecum* sp. (larvae). am: amphids; cf: cuticular flap; lp: labial papillae; one asterisk: external labial papillae; two asterisk: internal labial papillae.

In the lung nematodes of genus *Rhabdias,* which have an inflated cuticle, the cephalic structures are in most cases very difficult to observe (Fig. 33). In oxyurid nematodes, eg. *Parapharyngodon,* there are three lips, each one bilobed with one small papillae (Fig. 34). In genus *Falcaustra* the mouth is surrounded by 3 large lips, each with 2 forked papillae, and one amphid on each ventrolateral lip (Fig. 35)

On the other hand, in nematode parasites found in larval stage, the cephalic structures of adult forms are generally not present. For example, *Porrocaecum,* a genus found in the liver of the host, has a very little developed lip anlagen (Fig. 36); *Brevimulticaecum,* found encapsulated in various organs of the hosts, presents two toothlike prominences, 1 dorsal and 1 ventral (Fig. 37).

Interlabia: these are cuticular outgrowths (neoformations) originating at the base of the lips or pseudolabia and extending between them, occurring in some ascarids and spirurids. These modifications of the cuticle are present in the larval stage of genus *Physaloptera;* this genus is characterized by the presence of four teeth on the internolateral face of each pseudolabium, an internal group of three teeth, two sublateral and one lateral, and a single externolateral tooth.

Deirids: a pair of sensory organs found laterally in the cervical region and usually protruding above the surface of the cuticle. That may be simple or claw-like and forked, and are situated laterally in the vicinity of the nerve ring in most species (Rhabdochonidae).

Amphids: a pair of glandular sensory organs situated laterally in the cephalic region and opening through the cuticle; according to some authors, they are chemoreceptors. They have various shapes and sizes but usually occur as two lateral pores (Figs. 30, 31).

3.1.3 Cuticle of posterior end of body

In the case of nematode parasites of amphibians, the special modifications of the cuticle in the posterior end of males, which are generally associated with the copula, are particularly important. These comprise: plectanes, rosettes papillae, vesiculated rosette, papillae, bursa, phasmids, spines suckers and caudal lateral alae. Phasmids are not modifications of the cuticle such as the deirids and the amphids, but their opening onto the cuticle can present different shapes and structures.

Plectanes: these are cross striated cuticular plates functioning as supports for the genital papillae in some males (Fig. 38). This structure is very common in the genus *Cosmocerca*, which is widely distributed in amphibians. In some species, these supports are very marked, as in *C. podicipinus*, in this case, the plectanes of each row are even fused to each other (Fig. 39). In other species, this support is not so developed (eg. *C. parva*); on the other hand, the same species in different hosts can present different degree of development of these structures (Figs. 40-43); indeed, in some immature specimens, the plectanes are imperceptible (Fig. 44).

These structures are arranged in two longitudinal rows on the ventral surface of the males; the number of pairs of plectanes varies among species (*C. podicipinus*, *C. cruzi*, *C. travassossi*: 5 pairs; *C. chilensis* and *C. rara*: 6 pairs; *C. longispicula*, *C. uruguayensis*, *C. vrcibradici*: 7 pairs). Furthermore, some species show wide intraspecific variation regarding the number of pairs of plectanes, even in the same host species (*C. parva*: 5-7 pairs; *C. brasiliensis*: 8-11 pairs; *C. paraguayensis*: 4-5 pairs) (Figs. 18, 19) (González and Hamann, 2010b).

Table 1 shows characteristics of the posterior end of males for species of *Cosmocerca* that parasitize amphibians, observed under SEM.

Cosmocerca spp.	Plectanes + rosette papillae	Adanal papillae*	Caudal papillae	Puncatations of rosette papillae	References
C. ornata	5 pairs	Not stablished	Not stablished	Not stablished	Navarro et al. (1988)
C. ornata	5 pairs	Not stablished	Not stablished	6-7 pad-like protuberances around the posterior border only	Grabda-Kazubska and Tenora (1991)
C. commutata	7 pairs	Not stablished	Several pairs	Interior and exterior rosette: 15	Grabda-Kazubska and Tenora (1991)
C. parva	5-7 pairs + 1 unpaired	1 + 2-4	3 pairs	Interior and exterior rosette: 12-16	Mordeglia and Digiani (1998)
C. podicipinus	5 pairs	3 pairs	Not stablished	Interior rosette: 11-12; exterior rosette: 12-15	González and Hamann (2010b)
C. parva	5-7 pairs	1 + 3 pairs	Not stablished	Interior rosette: 10-11; exterior rosette: 12-14	González and Hamann (2010b)
C. parva	4-5 pairs	1 + 3 pairs	Not stablished	Interior rosette: 12-15; exterior rosette: 12-15	González and Hamann (2008)

Table 1. Characteristics of the caudal region of males of *Cosmocerca* spp. that parasitize amphibian hosts, studied under SEM. *Arrangement of adanal papillae: unpaired papillae anterior to anus + pairs of adanal papillae.

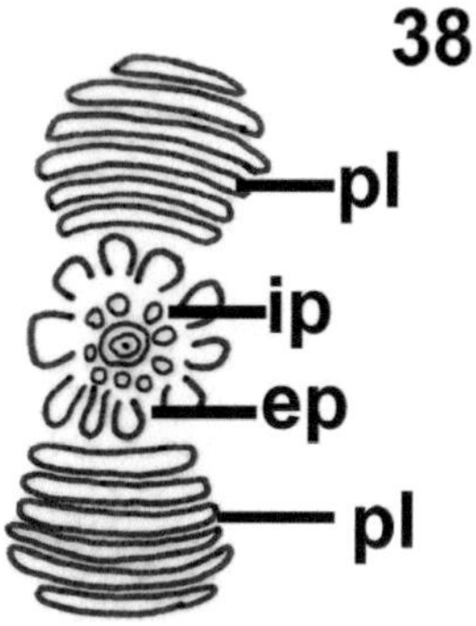

Fig. 38. Plectanes and rosette papillae in nematode parasites of genus *Cosmocerca*. pl: plectanes; ip: internal circle of punctations; ep: external circle of punctations.

Rosette papillae: these structures consist of papillae surrounded by punctuations. This modification is present in the genus *Cosmocerca*. In this genus, the plectanes are located outside the rosette papillae and are directed toward the anterior and posterior end (Fig. 38).

These rosette papillae are formed by two circles of punctuations, one internal and one external. The number of punctuations in each circle varies among species and within the same species for individuals collected from different hosts (González and Hamann, 2008; 2010b) (Figs. 38-44). In the genus *Cosmocercoides*, the caudal rosette papillae are not raised above the cuticular surface.

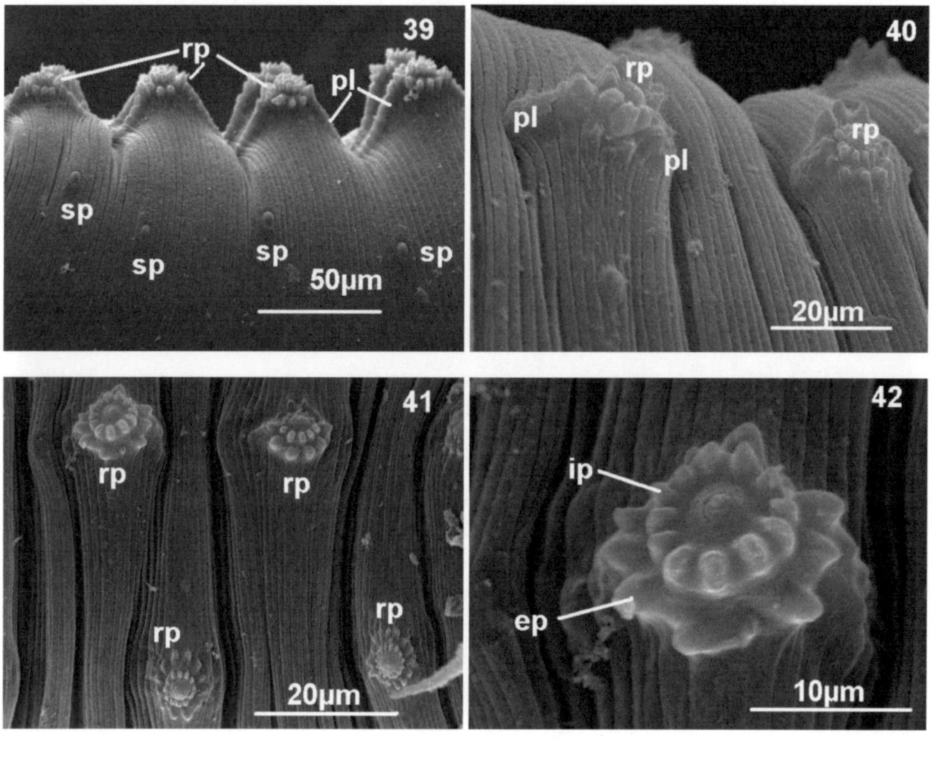

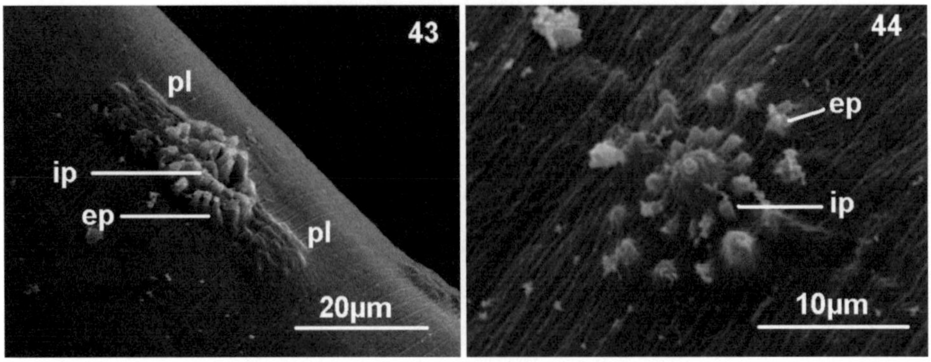

Fig. 39-44. Plectanes and rosette papillae in nematode parasites of genus *Cosmocerca*. 39. *Cosmocerca podicipinus* collected in *Pseudopaludicola falcipes* with fused plectanes. 40. *Cosmocerca parva* collected in *Rhinella granulosa* with very marked plectanes. 41. 42. *Cosmocerca parva* collected in *Rhinella schneideri*. 43. *Cosmocerca parva* collected in *Leptodactylus bufonius*. 44. *Cosmocerca parva*, immature specimen, collected in *Rhinella fernandezae* with imperceptible plectanes. sp: somatic papillae; pl: plectanes; rs: rosette papillae; ip: internal circle of punctations; ep: external circle of punctations.

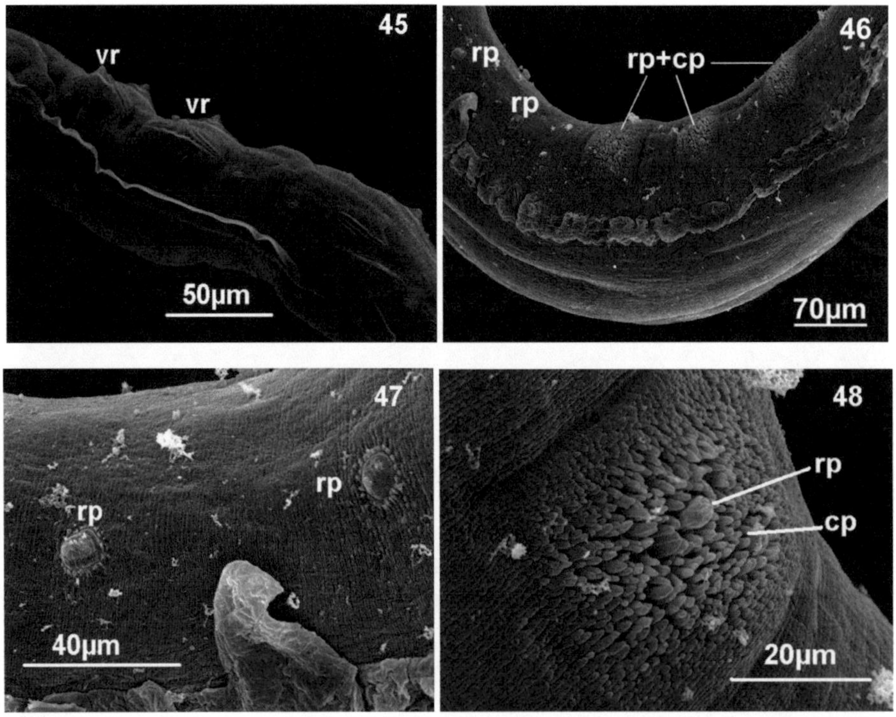

Fig. 45-48. Modifications of the cuticle in the posterior end of males of the genus *Cosmocercella*. 45. Vesiculated rosette in *C. minor*. 46. Combination of different structures in *C. phyllomedusae*. 47. Detail of small rosette papillae. 48. Detail of rosette papillae surrounded by wide areas of cuticular punctations. vr: vesiculated rosette; rp: rosette papillae; cp: cuticular punctation; rp+cp: rosette papillae surrounded by areas of cuticular punctations.

Vesiculated rosette: caudal rosette papillae raised on the surface of a clear vesicle. These modifications are present in the genus *Cosmocercella*. In *C. minor*, for example, there are 4 pairs of vesiculated papillae (Fig. 45) and in *C. phyllomedusae* there is a combination of different structures; this species has small rosette papillae in the preanal subventral surface (Figs. 46, 47), rosette papillae surrounded by wide areas of cuticular punctuations (Fig. 46, 48), and large unpaired vesiculated papillae extending almost to the level of the oesophagus.

Papillae: some genera of nematode parasites, such as *Aplectana, Raillietnema, Falcaustra* or *Paraoxyascaris* do not possess conspicuous structures in the posterior end such as plectanes, rosette papillae or vesiculated papillae; these genera have simple papillae with variable number and arrangement (Figs. 49-56). These structures can be divided into caudal papillae (located in the tail, posteriorly to the anus) and cloacal papillae (surrounding the cloaca; these can be precloacal, postcloacal and adcloacal). These papillae may be sessile (*Aplectana, Schrankiana, Falcaustra*) (Figs. 49-52, 54-56) or pedunculate (*Parapharyngodon*) (Fig. 53). On the other hand, other genera such as *Cosmocercella* and *Cosmocerca* have this type of papillae in addition to plectanes, rosettes and vesiculated rosette. These papillae are commonly surrounded by one or two small rosettes of punctuations (Fig. 51). In some cases they protrude above the surface of the cuticle (Fig. 52).

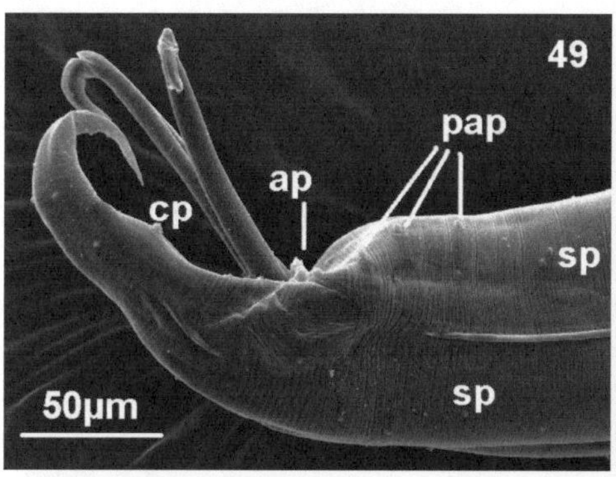

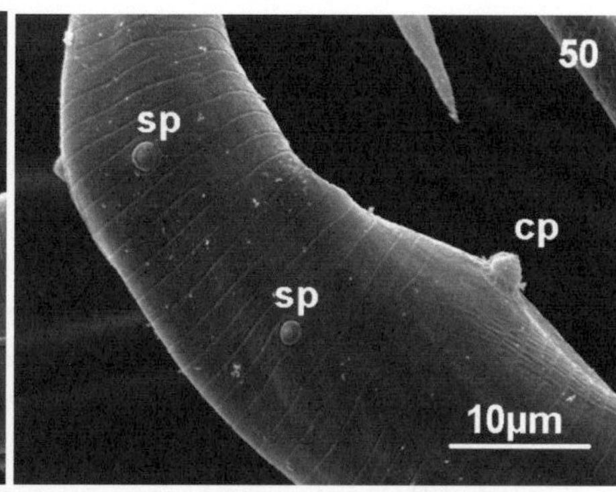

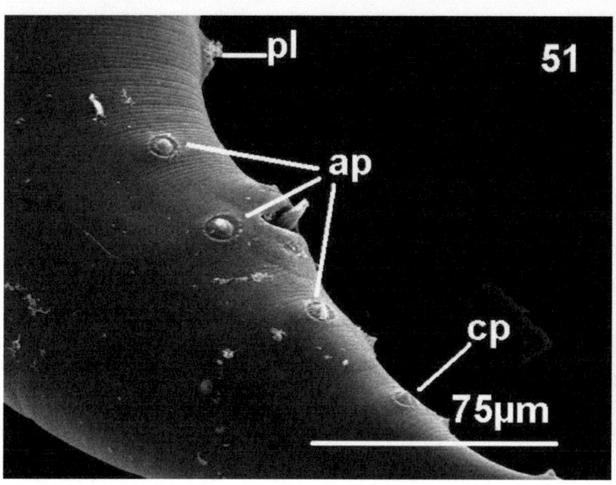

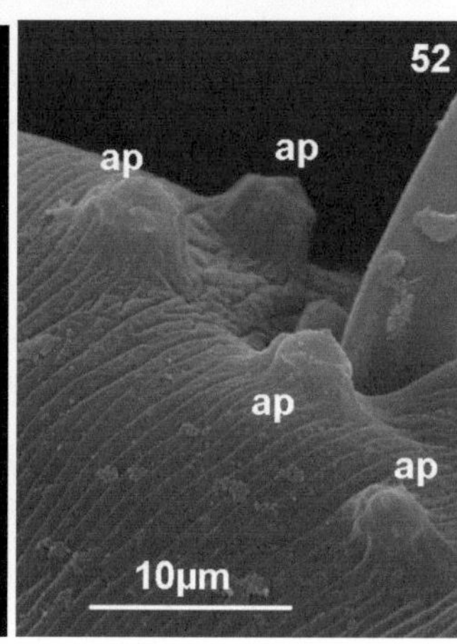

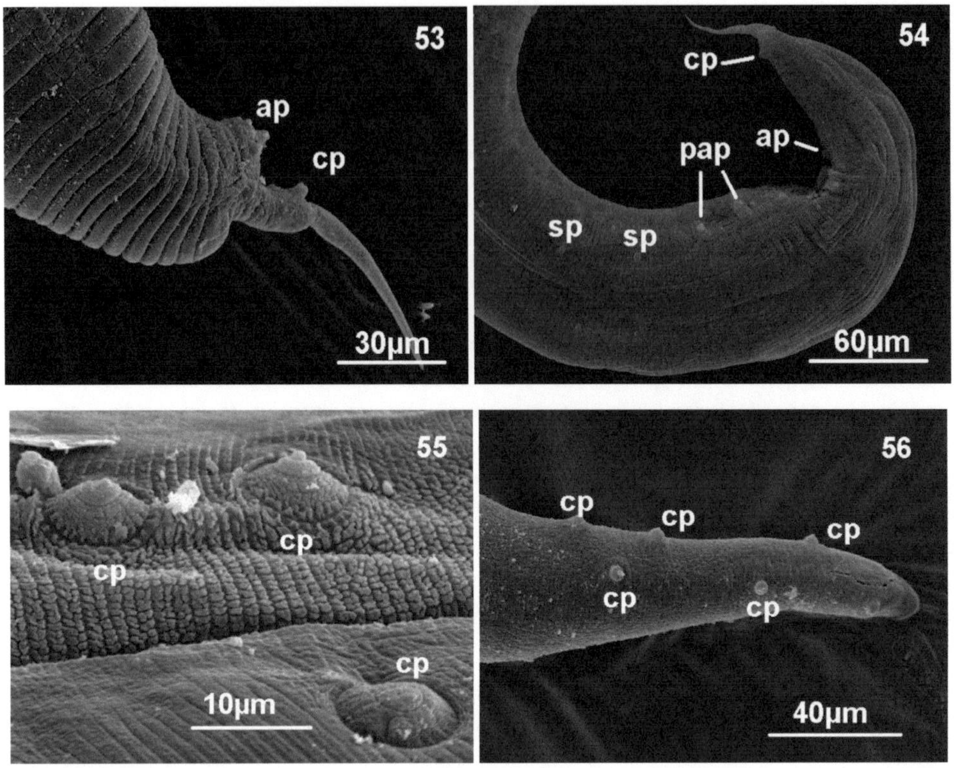

Fig. 49-56. Papillae in the posterior end of body of males nematode parasites of amphibians. 49. 50. Arrangement of caudal papillae in *Aplectana hylambatis*. 51. Arrangement of precloacal, adcloacal and postcloacal papillae in genus *Cosmocerca*. 52. Detail of precloacal, adcloacal and postcloacal papillae in *Cosmocerca*. 53. Pedunculate papillae in tail of *Parapharyngodon*. 54-56. Sessile papillae in *Schrankiana* (54), *Falcaustra* (55) and *Cosmocerca* (56). sp: somatic papillae; pl: plectane; pap: preanal papillae; ap: adanal papillae; cp: caudal papillae.

Bursa: this structure is present in nematodes of the genus *Oswaldocruzia*. This structure may be circular or oval, often divided into two symmetrical or asymmetrical lateral lobes, separated by a dorsal lobe and supported by rays or papillae. The rays of the bursa are visualized well under light microscope because they are not a part of the cuticle but embedded in the lobes.

Caudal lateral alae: these are sublateral or lateral longitudinal wings of the cuticle that occur on the male tail. Among nematode parasites of amphibians, they occur in the genus *Physaloptera*; however, these caudal alae develop in the adult stage, while it is typically the larval stage of *Physaloptera* that occurs as a parasite of amphibians.

Spines: some male and female nematode parasites, such as those of the genus *Skrjabinodon*, have cuticular spines on the tail filament (Fig. 57).

Suckers: this is a sucker-like pre-cloacal structure. A series of stages in sucker development occurs in some kathlaniids, indicating that there is first a concentration of copulatory muscles in this area which later becomes a sucker through modification of the cuticle. This modification of the cuticle is found in some species of genus *Falcaustra*.

Phasmids: these are paired glandular sensory organs situated laterally in the caudal region and opening to the surface by a slit or pore (Fig. 58).

As in the previous case, the structures and modifications of the cuticle that are observed in the posterior body end of adult specimens do not occur in larvae (Fig. 59).

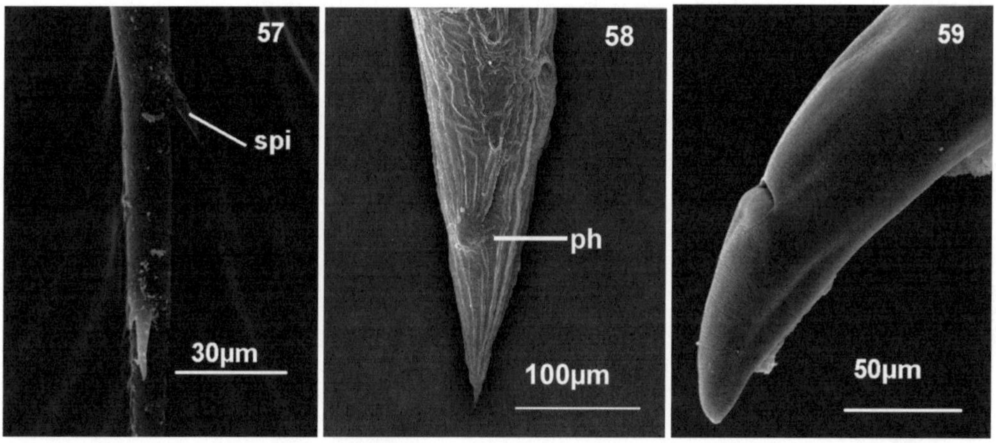

Fig. 57-59. Modifications of the cuticle in posterior end of body of nematode parasites. 57. Spines in the tail of *Skrjabinodon* sp. 58. Phasmid in *Rhabdias* sp. 59. Posterior end of *Physaloptera* sp. spi: spines; ph: phasmid.

3.1.4 Eggs

The eggs are variable in size, shape and structure; they usually have a many-layered shell with either smooth or rough, sometimes sculptured, external surface, and their poles may bear a characteristic operculum or plug. Among the nematode parasites of amphibians, eggs may present punctuations (Pharyngodonidae) (Figs. 60, 61), an operculum (*Gyrinicola*) or a thin membrane that has no special features, as in the Cosmocercidae.

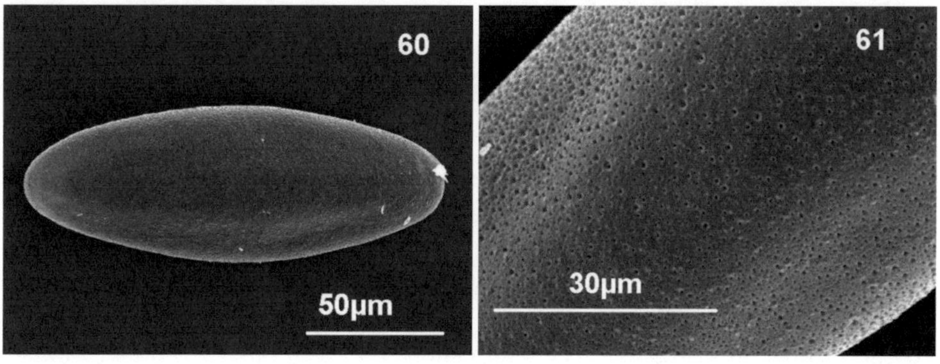

Fig. 60-61. Egg cuticle of nematode parasites of amphibians. 60. General view. 61. Detail of punctations.

3.1.5 Vulva

The area of the body inmediately anterior and posterior to the vulvar opening is called vulvar region. In most females of nematode parasites of amphibians, this may be simply an opening transversal to the longitudinal body axis without any special striation, as in the genus *Rhabdias* (Fig. 62), or with a striation that differs slightly from that of the rest of the body as in the genus *Cosmocercella* (Fig. 63); or it may present more complex structures as in *Aplectana* (Fig. 64); in this latter case, the cuticle forms an extension in the anterior side of the vulvar opening that can be observed as a vulvar flap (Gibbons, 1986).

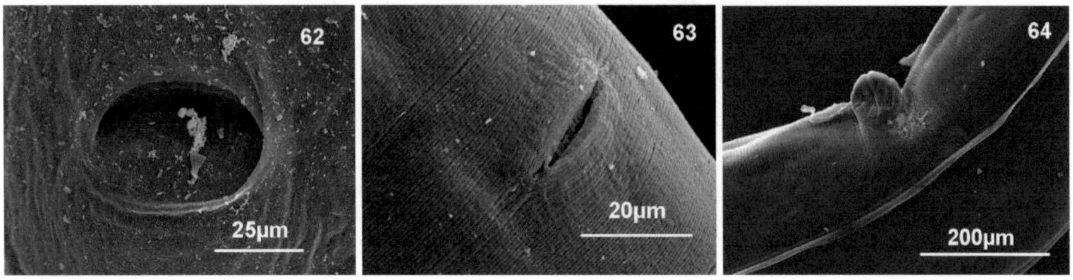

Fig. 62-64. Vulvar cuticle in nematode parasites of amphibians. 62. *Rhabdias* sp. 63. *Cosmocercella* sp. 64. *Aplectana* sp.

3.1.6 Spicules

Nematodes usually have two spicules; each one is essentially a tube covered by a sclerotized cuticle and containing a central protoplasmic core. In terms of the taxonomy of nematode parasites of amphibians, the importance of the spicules lies in their morphology and size, and not in the presence of ornamentation on the cuticle of these structures. In this case the spicules can be studied with SEM only when they are outside the individual, i.e., when protruding from the cloaca (Figs. 65-67).

Fig. 65-67. Spicules in nematode parasites of amphibians. 65. *Aplectana hylambatis.* 66.*Cosmocerca* sp. 67. *Falcaustra* sp.

3.2 Trematoda

The tegument of trematodes is syncytial and consists of a tegumental outer membrane (trilaminate), a matrix (with discoid bodies, membranous bodies and usually mitochondria) and a basal tegumental membrane. The tegument is variously interrupted by cytoplasmatic projections of gland cells and by openings of excretory pores. The tegumental surface often contains ornamentations such as spines between the outer and basal membranes; these are often present in different areas of the body; there are also numerous sensory papillae, pits and ridges of various configurations (Fried, 1997; Schmidt and Roberts, 2000). The surface topography of the cirrus of digenetic trematodes also shows spine-shape protrusions and papillae (Bušta and Našincová, 1987).

3.2.1 Tegument of suckers

The rim of the oral and ventral suckers in some species of digenean parasites of amphibians (both larval and adult) shows sensory papillae with variable morphology and distribution. Thus, papillae may appear as button-like structures and can be distributed as single and double papillae on the oral and ventral surfaces of the sucker (Hamann and González, 2009; Mata-López, 2006; Nadakavukaren and Nollen, 1975). In other digeneans, the oral sucker has a spongy surface with numerous pores (Whitehouse, 2002). Figures 68-70 show some characteristics of the tegument of the oral and ventral suckers of digenean trematodes found in Argentinean amphibians.

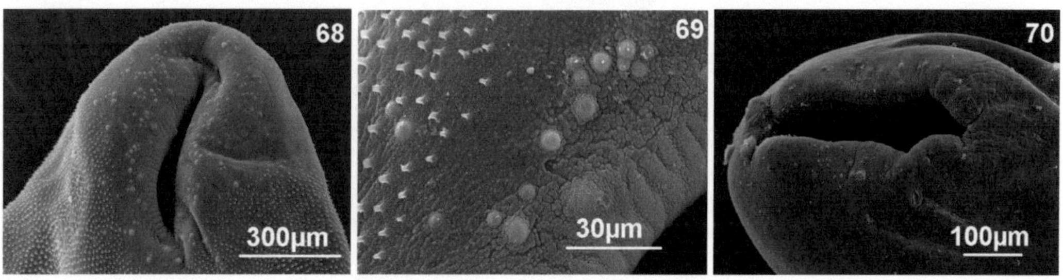

Fig. 68-70. Tegument of suckers of digenean parasites of amphibians. 68. Papillae on oral sucker of Macroderoididae. 69. Detail of papillae on oral sucker. 70. Detail of papillae on ventral sucker of Paramphistomatidae.

3.2.2 Tegument of the ventral surface

The tegument of digenean trematodes (larval and adults) that occur in amphibian hosts shows spines with varied morphology (e.g. scale-like spines) and variable distribution; they generally extend from the anterior end to variable levels of the posterior region (Hamann and González, 2009; Razo-Mendivil et al., 2006). In other digeneans, the surface of the tegument possesses regular ridges and interspersed protuberances (Nadakavukaren and Nollen, 1975). Figures 71-72 show the shape and distribution of spines found in the tegument of some digenean parasites of Argentinean amphibian.

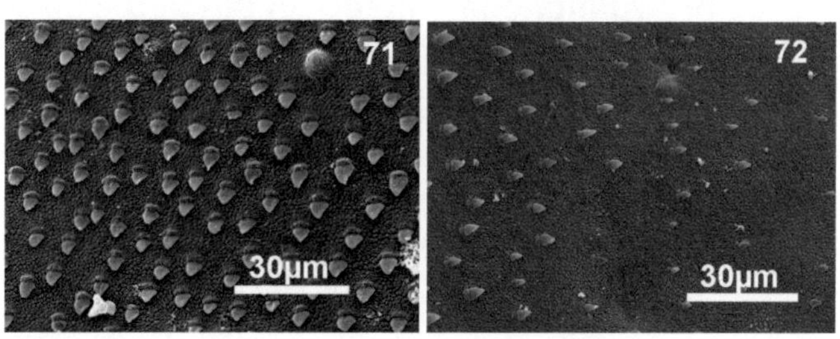

Fig. 71-72. Tegument of the ventral surface of Macroderoididae. 71. Spines of the anterior third of the body. 72. Spines of the posterior third of body.

3.3 Acanthocephalan

The body surface of acanthocephalans has 5 layers. The outermost layer is the epicuticle, followed by the cuticle which is composed mainly of lipoproteins; the third layer has a homogeneous nature; the fourth layer possesses fibrous bands besides mitochondria, bladders and lacunar channels, and the fifth layer contains scarce fibres but larger and more abundant lacunar channels than in the previous layer (Olsen, 1974).

Regarding this group of helminth parasites, most SEM studies are focused on the hooks that they possess in the proboscis, as well as their body spines. Likewise, they present sensory structures with diverse ornamentations in the posterior part of the bursa.

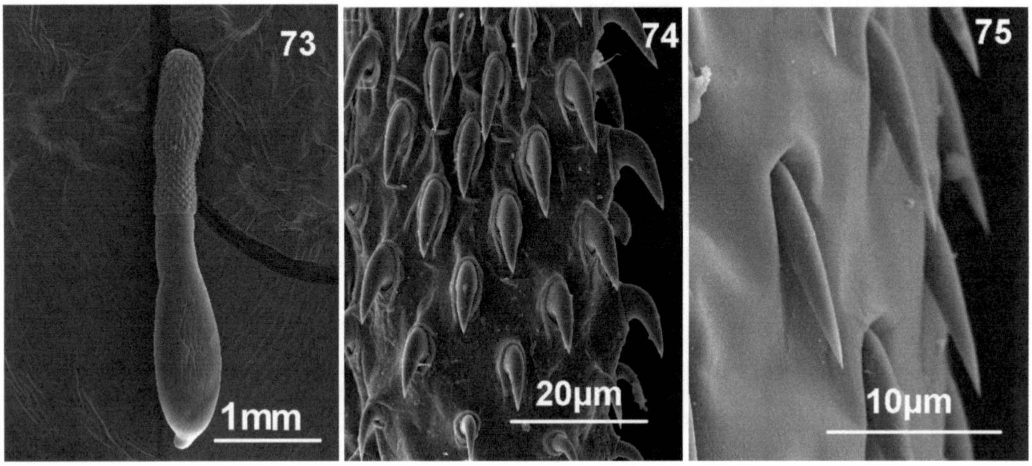

Fig. 73-75. Acanthocephalan parasites of amphibians. *Centrorhynchus* sp. 73. General view. 74. Detail of hooks of the proboscis. 75. Detail of spines of the proboscis.

In our study we found larval stages belonging to the genus *Centrorhynchus*. This genus is characterized for possessing an unarmed trunk, the proboscis divided into two regions (an anterior portion with hooks and a posterior portion with spines), and for having subterminal genital pores. Because the genital complex was not fully developed, specific determination was not possible. The proboscis presented 28 to 30 longitudinal rows of 20 to 23 hooks, 8 to 11 rooted hooks and 10 to 13 rootless spines (Figs. 73-75).

4. Helminth parasites of Argentinean amphibian studied with SEM: Synthesis and new contributions. Research perspectives

Up to the present, studies performed with scanning electron microscopy techniques on helminth parasites of Argentinean amphibians have included three families of nematodes: Rhabdiasidae, *Rhabdias füelleborni* (González and Hamann, 2008), Cosmocercidae, *Cosmocerca parva*, *C. podicipinus*, *Aplectana hylambatis*, *A. adaechevarriae* (González and Hamann, 2010b; Mordeglia and Digiani, 1998; Ramallo et al., 2008) and Physalopteridae, *Physaloptera* sp. (González and Hamann, 2010b), and one family of trematodes: Diplostomidae, *Lophosicyadiplostomum* aff. *nephrocystis* and *Bursotrema tetracotyloides* (Hamann and González, 2009).

New contributions presented in this work include, for the Class Nematoda, the families Molineoidae (*Oswaldocruzia* spp.), Pharyngodonidae (*Parapharyngodon, Pharyngodon, Skrjabinodon*), Cosmocercidae (*Aplectana* spp., *Cosmocerca* spp., *Cosmocercella* spp., *Paraoxyascaris* sp.), Kathlaniidae (*Falcaustra* sp.), Atractidae (*Schkrankiana* sp.) and Ascarididae (*Porrocaecum* sp., *Brevimulticaecum* sp.), and for the Phyllum Acanthocephala, family Centrorhynchidae (*Centrorhynchus* sp.).

Future research in this topic should focus on extending the geographical areas studied while at the same time, expanding the examination to other possible amphibian hosts.

Reports about helminth parasites of Argentinean amphibians studied under SEM refer mainly to specimens collected in host from the Northeast region, specifically Corrientes province, and the Northwest region, with only one record for Salta province so; thus, there is still a vast portion of the Argentinean territory that has not yet been studied (Fig. 76).

Fig. 76. Reports of helminth parasites of amphibians studied using SEM in Argentina. ¶: González and Hamann (2008); *: González and Hamann (2010b); +: Hamann and González (2009); £: Mordeglia and Digiani (1998); §: Ramallo et al. (2008).

Lavilla et al. (2000) reported a total of 271 amphibian species for Argentina (167 anurans and 4 gymnophions); of these, only 32 (11.8%) have been cited as hosts for helminth parasites (González and Hamann, 2004, 2005, 2006a, 2006b, 2007a, 2007b, 2008, 2009, 2010a, 2010b, 2011; Hamann and Pérez, 1999; Hamann and González, 2009; Hamann et al., 2006a, 2006b, 2009a, 2009b, 2010; Lajmanovich and Martinez de Ferrato, 1995; Lunaschi and Drago, 2007; Ramallo et al., 2007a, 2007b, 2008). Of all the anuran families, the most studied for helminth parasites are Hylidae, Bufonidae, Leiuperidae and Leptodactylidae. Nine species of hylids have been studied for helminth parasites, but only one of these studies included SEM: *Scinax nasicus* (González and Hamann, 2008; Hamann and González, 2009); similarly, seven species of bufonids have been studied for helminth parasites, but SEM was employed in only two cases: *Rhinella schneideri* and *R. granulosa* (González and Hamann, 2008; Mordeglia and Digiani, 1998; Ramallo et al., 2008), finally, six species of leiuperids and leptodactylids have been studied for helminth parasites, but only one of them was analyzed with SEM, the leiuperid *Physalaemus santafecinus* (González and Hamann, 2010b).

The SEM study of the tegument of helminth species (e.g. morphology of spines) collected in different localities could detect possible intraspecific variation related to geographical location; this phenomenon has been highly documented in both trematode (Grabda-Kazubska and Combes, 1981; Kennedy 1980a) and nematode (Chitwood, 1957) parasites of amphibians. Similarly, variations related to occurrence in a wide range of phylogenetically unrelated hosts, i.e. the cases of generalist helminths could detect possible intraspecific variation related with the host age or diet, previous exposure to the parasite, presence of another parasite and number of specimens present (Chitwood, 1957; Haley, 1962; Kennedy, 1980b; Watertor, 1967).

5. Importance of the use of scanning electron microscope for the study of helminth parasites of amphibians

In helminth parasites, all morphological aspects must be studied under light microscope, because these structures are very important in the context of their systematic classification. Some examples of these traits include, in the case of parasitic nematodes, the type of esophagus (oxyuroid, rhabditoid, strongyloid), presence of ventriculus and its shape, and the caecum and its shape; in the females, the arrangement of ovaries (prodelphic, amphidelphic or opisthodelphic), number of uteri (monodelphic or didelphic), structure of the ovoyector and the vagina and, in the males, the structure and measurements of the gubernaculum. In the case of trematodes, the distribution of vitelline follicles, the position of the testes and ovary, size of the eggs, position of oral and ventral suckers, reproductive structures, among others, are characteristics of taxonomic importance. Finally, regarding the internal anatomy of acanthocephalans, some particularly relevant structures are the proboscis receptacle, lemnisci, retractor muscle, testis, seminal vesicle, cement gland, Saefftigen's pouch, etc. Thus, the scanning electron microscope represents an additional tool for the study of this group of organisms. The importance of SEM lies in its ability to provide three-dimensional images with high magnification that allow understanding the spatial relationships among surface structures. It could be used to separate species that appear morphologically identical when examined under light microscope, validate species and demonstrate differences between populations or races (Gibbons, 1986; Hirschmann, 1983).

6. Acknowledgments

We thank Secretaría General de Ciencia y Técnica of Universidad Nacional del Nordeste, Corrientes, Argentina, for supporting partially this work.

We are grateful Dr. Graciela T. Navone, Dr. Julia I. Díaz, Dr. María del Rosario Robles at the Centro de Estudios Parasitológicos y de Vectores, La Plata, Argentina, Licentiate Rodrigo Cajade at Centro de Ecología Aplicada del Litoral, Corrientes, Argentina, Dr. Lorena Sereno at Centro de Energia Nuclear na Agricultura, Universidade de São Paulo, Brazil, Dr. Viviane Gularte Tavares dos Santos at Instituto de Biociências, Universidade Federal do Rio Grande do Sul, Brazil, for helping with literature search.

We are grateful Dr. Marta I. Duré and Dr. Eduardo F. Schaefer at Centro de Ecología Aplicada del Litoral for the photographs of the host and for helping with the edition of the map.

We thank to Graphic Designer Cecilia Rios Encina for help in photograph edition.

7. References

Anderson, R.; Chabaud, A. & Willmont, S. (2009). *Keys to the Nematode Parasites of Vertebrates. Archival Volumen.* CAB International, ISBN 978-1-84593-572-6, Wallingford, United Kingdom.

Allison, V.; Ubelaker, J.; Webster Jr., R. & Riddle, J. (1972). Preparations of Helminths for Scanning Electron Microscopy. *Journal of Parasitology*, Vol. 58, No. 2, (April 1972), pp. 414-416, ISSN 0022-3395.

Amin, O. (1985). Classification, In: *Biology of the Acanthocephala*, D. Crompton and B. Nickol (Ed.), 27-72, Cambridge University Press, ISBN 0-521-24674-1, Cambridge, United Kingdom.

Berger, L. & Speare, R. (1998). *Chytridiomycosis*: a New Disease of Wild and Captive Amphibians. Australian & New Zealand Council for the Care of Animal in Research and Teaching. *Newsletter*, 11 (4), (December 1998), pp. 1-3.

Bušta, J. & Našincová, V. (1987). Ultrastructure of the Surface of External Genitals of six Species of Digenetic Trematodes Studied by Scanning Electron Microscopy. *Folia Parasitologica*, Vol. 34, No. 2, pp. 137-143, ISSN 0015-5683.

Chitwood, M. (1957). Intraespecific Variation in Parasitic Nematodes. *Systematic Zoology*, Vol. 6, No. 1, (March 1957), pp. 19-23, ISSN: 0039-7989

Chitwood, B. & Chitwood, M. (1975). *Introduction to Nematology* (Second Revised Edition). University Park Press, ISBN 0-8391-0697-1. Maryland, United State of America.

Crump, M. & Scott Jr, N. (1994). Visual Encounters Surveys. In: *Measuring and Monitoring Biological Diversity - standard methods for amphibians,* W. Heyer, M. Donnelly, R. McDiarmid, L. Hayek and M. Foster (Ed.), 84-91, Smithsonian Institution Press, ISBN-10 1560982845, Washington, United State of America.

De Coninck, L. (1965). Classe de Nématodes. In: *Traité de Zoologie. Anatomie, Systématique, Biologie*, P. Grassé (Ed.), pp. 1-217, Vol 4(2) Némathelminthes, pp. 1-217. Masson et Cie, ISBN : 9782225585104, Paris, France.

Duellman, W. & Trueb, L. (1986). *Biology of Amphibians*. The Johns Hopkins University Press, ISBN 0-8018-4780-X, London, United Kingdom.

Duszynski, D.; Bolek, M. & Upton, S. (2007). Coccidia (Apicomplexa: Eimeriidae) of Amphibians of the World. *Zootaxa*, Vol. 1667, (December 2007), pp. 1-77, ISBN 978-1-86977-183-6.

Faivovich, J.; Haddad, C.; García, P.; Frost, D.; Campbell, J. & Wheeler, W. (2005). Systematic Review of the Frog Family Hylidae, with Special Reference to Hylinae: Phylogenetic Analysis and Taxonomic Revision. *Bulletin of the American Museum of Natural History*, Vol. 294, (June 2005), pp. 1-240, ISSN 0003-0090.

Fried, B. (1997). An Overview of the Biology of Trematodes. In: *Advances in Trematode Biology*. Fried, B. and Graczyk T. (Ed.). CRC Press, ISBN 0-8493-2645-1, New York, United State of America.

Frost, D.; Grant, T.; Faivovich, J.; Bain, R.; Haas, A.; Haddad, C.; De Sá, R.; Channing, A.; Wilkinson, M.; Donnellan, S.; Raxworthy, C.; Campbell, J.; Blotto, B.; Moler, P.; Drewes, R.; Nussbaum, R.; Lynch, J.; Green, D. & Wheeler, W. (2004). The Amphibian Tree of Life. *Bulletin of the American Museum of Natural History*, Vol. 297, pp. 1-370, ISSN 0003-0090.

Gibbons, L. (1986). *SEM Guide to the Morphology of Nematode Parasites of Vertebrates*. CAB International, ISBN 085198-569-6, Oxu, United Kingdom.

Gibbons, L. (2010). *Keys to the Nematode Parasites of Vertebrates. Supplementary Volume*. CAB International/The Natural History Museum, ISBN 978-1-84593-571-9, Wallingford, United Kingdom.

Gibson, D.; Jones, A. & Bray, R. (2002). *Keys to the Trematoda. Vol 1*. CAB International/The Natural History Museum, ISBN 0-85199-547-0, London, United Kingdom.

González, C. & Hamann, M. (2004). Primer Registro de *Cosmocerca podicipinus* Baker y Vaucher, 1984 (Nematoda, Cosmocercidae) en *Pseudopaludicola falcipes* (Hensel, 1867) (Amphibia, Leptodactylidae) en Argentina. *FACENA*, Vol. 20, (December 2004), pp. 65-72, ISSN 0325-4216.

González, C. & Hamann, M. 2005. *Gyrinicola chabaudi* Araujo & Artigas, 1982 (Nematoda: Pharyngodonidae) in Tadpoles of *Scinax nasicus* (Cope, 1862) (Anura: Hylidae) from Corrientes, Argentina. *FACENA*, Vol. 21, (December 2005), pp. 145-148, ISSN 0325-4216.

González, C. & Hamann, M. (2006a). Nematodes Parásitos de *Chaunus granulosus major* (Müller & Hellmich, 1936) (Anura: Bufonidae) en Corrientes, Argentina. *Cuadernos de Herpetología*, Vol. 20, No. 1, (September 2006), pp. 43-49, ISSN 0326-551X.

González, C. & Hamann, M. (2006b). Helmintos Parásitos de *Leptodactylus bufonius* Boulenger, 1894 (Anura: Leptodactylidae) de Corrientes, Argentina. *Revista Española de Herpetología*, Vol. 20, pp. 39-46, ISSN 0213-6686.

González, C. & Hamann, M. (2007a). Nematode Parasites of two Species of *Chaunus* (Anura: Bufonidae) from Corrientes, Argentina. *Zootaxa*, Vol. 1393, (January 2007), pp. 27-34, ISSN 1175-5326

González, C. & Hamann, M. (2007b). The First Record of Amphibians as Paratenic Host of *Serpinema* larvae (Nematoda: Camallanidae). *Brazilian Journal of Biology*, Vol. 67, No. 3, (August 2007), pp. 579-580, ISSN 1519-6984.

González, C. & Hamann, M. (2008). Nematode Parasites of two Anurans Species, *Rhinella schneideri* (Bufonidae) and *Scinax acuminatus* (Hylidae), from Corrientes, Argentina. *Revista de Biología Tropical*, Vol. 56, No. 4, (December 2008), pp. 2147-2161, ISSN 0034-7744.

González, C. & Hamann, M. (2009). First Report of Nematodes in the Common Lesser Escuerzo *Odontophrynus americanus* (Duméril and Bibron, 1841) (Amphibia: Cycloramphidae) from Corrientes, Argentina. *Comparative Parasitology*, Vol. 76, No. 1, (January 2009), pp. 122-126, ISSN 1525-2647.

González, C. & Hamann, M. (2010a). Larval Nematodes Found in Amphibians from Northeastern Argentina. *Brazilian Journal of Biology*, Vol. 70, No. 4, (November 2010), pp. 1089-1092, ISSN 1519-6984.

González, C. & Hamann, M. (2010b). First Report of Nematode Parasites of *Physalaemus santafecinus* (Anura: Leiuperidae) from Corrientes, Argentina. *Revista Mexicana de Biodiversidad*, Vol. 81, No. 3, (December 2010), pp. 677-687, ISSN 1870-3453.

González, C. & Hamann M. (2011). Cosmocercid Nematodes of Three Species of Frogs (Anura: Hylidae) from Corrientes, Argentina. *Comparative Parasitology*, Vol. 78, No. 1, (January 2011), pp. 212-216, ISSN 1525-2647.

Grabda-Kazubska, B. & Combes, C. (1981). Morphological Variabily of *Haplometra cylindracea* (Zeder, 1800) (Trematoda, Plagiorchiidae) in Populational and Geografic Aspects. *Acta Parasitologica Polonica*, Vol. 28, No. 5, pp. 39-65, ISSN 1230- 2821.

Grabda-Kazubska, B. & Tenora, F. (1991). SEM Study on *Cosmocerca ornata* (Dujardin, 1845) and *C. commutata* (Diesing, 1851) (Nematoda, Cosmocercidae. *Acta Parasitologica Polonica*, Vol. 36, No. 1, pp. 45-50, ISSN 0065-1478.

Haley, A. (1962). Role of Host Relationship in the Systematics of Helminth Parasites. *Journal of Parasitology*, Vol. 48, No. 5, (October 1962), pp. 671-678, ISSN 0022-3395.

Halton, D. (2004). Microscopy and the Helminth Parasite. *Micron*, Vol. 35, No. 5, (July 2004), pp. 361-390, ISSN 0968-4328.

Hamann, M. & González, C. (2009). Larval Digenetic Trematodes in Tadpoles of six Amphibian Species from Northeastern Argentina. *Journal of Parasitology*, Vol. 95, No. 3, (June 2009), pp. 623-628, ISSN 0022-3395.

Hamann, M. & Pérez, V. (1999). Presencia de *Haematoloechus longiplexus* Stafford, 1902 (Trematoda, Haematoloechidae) en Anfibios de Argentina. *FACENA*, Vol. 15, No. (December 1999), pp. 157-162, ISSN 0325-4216.

Hamann, M., González, C. & Kehr, A. (2006a). Helminth Community Structure of the Oven Frog *Leptodactylus latinasus* (Anura, Leptodactylidae) from Corrientes, Argentina. *Acta Parasitologica*, Vol. 51, No. 4, (December 2006), pp. 294-299, ISSN 1230-2821.

Hamann, M.; Kehr, A. & González, C. (2006b). Species Affinity and Infracommunity Ordination of Helminths of *Leptodactylus chaquensis* (Anura: Leptodactylidae) in two Contrasting Environments from Northeastern Argentina. *Journal of Parasitology*, Vol. 92, No. 6, (April 2006), pp. 1171-1179, ISSN 0022-3395.

Hamann, M.; Kehr, A.; González, C.; Duré, M. & Schaefer, E. (2009a). Parasite and Reproductive Features of *Scinax nasicus* (Anura: Hylidae) from a South American Subtropical Area. *Interciencia*, Vol. 34, No. 3, (March 2009), pp. 214-218, ISSN 0378-1844.

Hamann, M.; Kehr, A. & González, C. (2009b). Niche Specificity of two *Glypthelmins* (Trematoda) Congeners Infecting *Leptodactylus chaquensis* (Anura: Leptodactylidae) from Argentina. *Journal of Parasitology*, Vol. 95, No. 4, (August 2009), pp. 817-822, ISSN 0022-3395.

Hamann, M.; Kehr, A. & González, C. (2010). Helminth Community Structure of *Scinax nasicus* (Anura: Hylidae) from a South American Subtropical Area. *Diseases of Aquatic Organisms*, Vol. 93, No. 1, (December 2010), pp. 71-82, ISSN 0177-5103.

Hirschmann, H. (1983). Scanning Electron Microscopy as a Tool in Nematode Taxonomy, In: *Concepts in Nematode Systematics*, A. Stone, H. Platt and L. Khalil (Ed.), 95-111, Academic Press, New York, United State of America and London, United Kingdom.

Hoff, G.; Frye, F. & Jacobson, E. (1984). *Diseases of Amphibians and Reptiles*. Plenum Press, ISBN 978-0306417115. New York, United State of America.

Jones, A.; Bray, R. & Gibson, D. (2005). *Keys to the Trematoda. Vol 2.* CAB International/The Natural History Museum, ISBN 0-85199-547-0, London, United Kingdom.

Karnovsky, M. (1965). A Formaldehyde-glutaraldehyde Fixative of High Osmolality for use in Electron Microscopy. *The Journal of Cell Biology*, Vol. 27, pp. 137A-138A (Abstr.), ISSN 0021-9525.

Kennedy, J. (1980a). Geographical Variation in Some Representatives of *Haematoloechus* Looss, 1899 (Trematoda: Haematoloechidae) from Canada and the United State. *Canadian Journal of Zoology*, Vol. 58, No. 6, (June 1980), pp. 1151-1167, ISSN 0008-4301.

Kennedy, J. (1980b). Host-induced Variations en *Haematoloechus buttensis* (Trematoda: Haematoloechidae). *Canadian Journal of Zoology*, Vol. 58, No. 3, (March 1980), pp. 427-442, ISSN 0008-4301.

Lajmanovich, R. & Martinez de Ferrato, A. (1995). *Acanthocephalus lutzi* (Hamman 1891) Parasite de *Bufo arenarum* en el Río Paraná, Argentina. *Revista de la Asociación de Ciencias Naturales del Litoral*, Vol. 26, No. 1, pp. 19– 23, ISSN 0325-2809.

Lavilla, E.; Richard, E. & Scrocchi, G. (2000). Categorización de los Anfibios y Reptiles de la República Argentina. Edición Especial Asociación Herpetológica Argentina. Argentina: 1- 97, ISBN 987-98331-0-4.

Lee, D. (2002). Cuticle, moulting and exsheathment. In: *The Biology of Nematodes*, D.L. Lee, (Ed.), 171-209, Taylor and Francis, ISBN 0-415-27211-4, London, United Kingdom.

Lunaschi, L. & Drago, F. (2007). Checklist of Digenean Parasites of Amphibians and Reptiles from Argentina. *Zootaxa*, Vol. 1476, (May 2007), pp. 51–68, ISSN 1175-5326.

Mafra, A. & Lanfredi, R. (1998). Reevaluation of *Physaloptera bispiculata* (Nematoda: Spiruroidea) by Light and Scanning Electron Microscopy. *Journal of Parasitology*, Vol. 84, No. 3, (June 1998), pp. 582–588, ISSN 0022-3395.

Mata-López, R. (2006). A New Gorgoderid Species of the Urinary Bladder of *Rana zweifeli* from Michoacán, Mexico. *Revista Mexicana de Biodiversidad*, Vol. 77, No. 2, (December 2006), pp. 191-198, ISSN 1870-3453.

Mordeglia, C. & Digiani, M. (1998). *Cosmocerca parva* Travassos, 1925 (Nematoda: Cosmocercidae) in Toads from Argentina. *Memorias do Instituto Oswaldo Cruz*, Vol. 93, No. 6, (November/December 1998), pp. 737-738, ISSN 0074-0276.

Nadakavukaren, M. & Nollen, M. (1975). A Scanning Electron Microscopic Investigation of the Outer Surfaces of *Gorgoderina attenuata*. *International Journal for Parasitology*, Vol. 5, No. 6, (December 1975), pp. 591-595, ISSN 0020-7519.

Navarro, P.; Izquierdo, S.; Pérez-Soler, P.; Hornero, M. & Lluch, J. (1988). Contribución al Conocimiento de la Helmintofauna de los Herpetos Ibéricos. VIII. Nematoda Ascaridida Skrjabin et Schultz, 1940 de Rana *spp*. *Revista Ibérica de Parasitología*, Vol. 48, No. 2, (April, May, June 1988), pp. 167-173, ISSN 0034-9623.

Nollen, M. & Nadakavukaren, M. (1974). *Megalodiscus temperatus*: Scanning Electron Microscopy of the Tegumental Surfaces. *Experimental Parasitology*, Vol. 36, No. 1, (August 1974), pp. 123-130, ISSN 0014-4894.

Olsen, O. (1974). *Animal parasites, their life cycles and ecology*, Dover Publications, Inc., ISBN 0839106432, New York, United State of America.

Pough, F.; Andrews, R.; Cadle, J.; Crump, M.; Savitzky, A. & Wells, K. (2001). *Herpetology* (Second Edition), Prentice Hall, ISBN 0-13-030795-5, New Jersey, United State of America.

Ramallo, G; Bursey, C. & Goldberg, S. (2007a). Two New Species of Cosmocercids (Ascaridida) in the Toad *Chaunus arenarum* (Anura: Bufonidae) from Argentina. *Journal of Parasitology*, Vol. 93, No. 4, (August 2007), pp. 910-916, ISSN 0022-3395.

Ramallo, G; Bursey, C. & Goldberg, S. (2007b). Primer Registro de *Oswaldocruzia proencai* (Nematoda: Molineoidae), Parásito de *Rhinella schneideri* (Anura: Bufonidae) en Salta, Argentina. *Acta Zoológica Lilloana*, Vol. 51, No. 1, pp. 91–92, ISSN 0065-1729.

Ramallo, G; Bursey, C. & Goldberg, S. (2008). New Species of *Aplectana* (Ascaridida: Cosmocercidae) in the Toads, *Rhinella granulosa* and *Rhinella schneideri* (Anura: Bufonidae) from Northern Argentina. *Journal of Parasitology*, Vol. 94, No. 6, (December 2008), pp. 1357-1360, ISSN 0022-3395.

Razo-Mendivil, U.; León-Règagnon, V. & Pérez-Ponce de León, G. (2006). Monophyly and Systematic Position of *Glypthelmins* (Digenea), Based on Partial lsrDNA Sequences and Morphological Evidence. *Organisms, Diversity & Evolution*, Vol. 6, No. 4, (November 2006), pp. 308–320, ISSN 1439-6092.

Schmidt, G. & Roberts, L. (2000). *Foundations of Parasitology* (Sixth Edition), Mcraw-Hill, ISBN: 0071168966, New York, United State of America.

Tinsley, R. (1995). Parasitic Diseasse in Amphibians: Control by the Regulation of Worm Burdens. *Parasitology*, Vol. 111, No. Suplement 1, (January 1995): pp. 153-178, ISSN 0031-1820.

Watertor, J. (1967). Intraespecific Variation of Adult *Telorchis bonnerensis* (Trematoda: Telorchiidae) in Amphibian and Reptilian Hosts. *Journal of Parasitology*, Vol. 53, No. 5, (October 1967), pp. 962-968, ISSN 0022-3395.

Whitehouse, C. (2002). A Study of the Frog Lung Fluke *Haematoloechus* (Trematoda: Haematoloechidae) Collected from Areas of Kentucky and Indiana. *Proceedings of the Indiana Academy of Science*, Vol. 1, No. 1, pp. 67–76, ISSN 0073-6767.

Willmott, S. (2009). Glossary and Keys to Subclasse, In: *Keys to the Nematode Parasites of Vertebrates. Archival Volume*, R. Anderson, A. Chabaud and S. Willmott (Ed.), pp. 1-17. CAB International, ISBN: 978-1-84593-572-6, Wallingford, United Kingdom.

15

Pathogenic Attributes of Non-*Candida albicans Candida* Species Revealed by SEM

Márcia Cristina Furlaneto[1], Célia Guadalupe Tardeli de Jesus Andrade[2], Luciana Furlaneto-Maia[3], Emanuele Júlio Galvão de França[1] and Alane Tatiana Pereira Moralez[1]

[1]Department of Microbiology,
[2]Electronic Microscopy and Microanalysis Laboratory,
State University of Londrina (UEL),
[3]Technological Federal University of Paraná (UTFPR), Londrina-PR,
Brazil

1. Introduction

The advent of microscopy provided an expressive progress on the knowledge of the biological world. Particularly important was the development of the electron microscopy at 1930s, making possible to find out a universe of unimaginable dimensions. Its great highlight is the much shorter wavelength of the electron that increases the resolution power of the equipment (Lee, 1993). Currently the electron microscopy is considered a specialized field of science (Bozzola and Russel, 1999). Although electron microscopy is useful to answer important questions about the ultrastructure of biological materials, it is also represent an additional tool that may be used as an ally in several research fields.

The scanning electron microscope (SEM) is useful to analyze microstructural features of solid bodies' surfaces, such as yeast cells. Besides, it leads to the formation of a three-dimensional image as a direct result of the great depth of field (Lee, 1993). Even samples observed by naked eye may be analyzed at low magnifications with great depth of focus, making possible to obtain images with a pronounced resolution using detectors of secondary electrons. Thus, the electron microscopy may contribute to reveal this nanometer´s world including fungal cell structure and the interaction between fungal cells and their microenviroment.

Most of the ultrastructural studies of *Candida* are based on the polymorphic species *Candida albicans*. In a pioneering study, the employment of SEM allowed the analyses of the surface features of different morphologies (budding yeast cells, germ tubes, hyphae and chlamydospores) of *C. albicans* (Barnes et al., 1971). Since then, many studies have applied SEM to elucidate several biological features of *Candida*. Recently, an updated useful review of the ultrastructural biological features of superficial candidiasis was presented (Jayatilake, 2011).

The evolution of intensive care medicine prolongs life expectancy leading populations to high susceptibility for candidal infection. Although ubiquitous in nature, *Candida* species can cause various infections that can vary from a relatively mild skin mycoses to life-threatening systemic disease. Over the past two decades, an increase in the number of cases caused by non-*Candida albicans Candida* (NCAC) species has been reported. In this context, *Candida parapsilosis* and *Candida tropicalis* are among the commonest species of *Candida* responsible for nosocomial blood infection worldwide (Krcmery & Barnes, 2002, Almirante et al., 2006, Colombo et al., 2006, Nucci & Colombo, 2007). Besides, *C. parapsilosis* has gained increasing recognition as the most common etiological agent causing *Candida* nail infections (reviewed in Trofa et al., 2008).

Yeasts belonging to genus *Candida* produce daughter cells by budding that readily separate at sites of septation. However other morphologies also occur, including pseudohyphal cells that also grow by budding and display distinct constrictions at septa, although they are more elongated and do not readily separate, and true hyphal cells formed as long thin tubes with parallel cell walls that lack septal constrictions.

With reference to morphological characteristics, *Candida* budding cells display oval, round, or cylindrical shapes. *C. parapsilosis* does not form true hyphae and exists in either a yeast phase or a pseudohyphal form. Differently, *C. tropicalis* can exist in multiple morphogenetic forms, including yeast phase, pseudohyphae and true hyphae, being considered a polymorphic fungi.

For pathogenic yeasts, it is widely accepted that yeast form cells are essential for efficient dissemination through the body, whereas the filamentous forms are required for tissue invasion. For *C. albicans*, strains lacking hyphal formation exhibited lower ability to invade tissue compared with wild-type strains (Jayatilake et al., 2006).

Most pathogenic *Candida* species have developed a wide range of putative virulence factors to assist in their ability to colonize host tissues, cause disease, and overcome host defenses. Despite intensive research to identify pathogenic factors in yeast, particularly in *C. albicans*, relatively little is known about the virulence attributes associated with NCAC species. Although non-*albicans* species seem to share common virulence determinants with *C. albicans* it is believed that they have a particular repertoire of specific virulence traits (Haynes, 2001). Currently, we still have much to learn about the virulence of NCAC species, particularly *C. parapsilosis* and *C. tropicalis*.

The purpose of this chapter is to sum up some of the recent ultrastructural findings of *C. parapsilosis* and *C. tropicalis* virulence-associated characteristics that are thought to contribute in their process of pathogenesis.

2. Ultrastructural features of pathogenic attributes

Multiple characteristics have been proposed to be putative virulence factors related to the pathogenesis of *C. parapsilosis* and *C. tropicalis*, including adherence to host surfaces (cells and tissues) and medical devices, formation of filamentous forms, biofilm formation, production of hydrolytic enzymes and phenotypic switching (reviewed in Trofa et al., 2008, reviewed in van Asbeck et al., 2009, reviewed in Silva et al., 2011).

Recently, França et al. (2010) described a correlation between *in vitro* haemolytic and proteinase activities in clinical isolates of *C. parapsilosis* and *C. tropicalis* and site of fungal isolation. According to these authors, anatomical sites of isolation seem to be correlated with these activities, particularly for *C. parapsilosis* isolates.

Currently, we are employing the SEM to evaluate many events related to pathogenicity of clinical strains of *C. parapasilosis* and *C. tropicalis,* including adherence to biotic substrates, agar invasion capability, switching morphogenesis and morphological alterations of *Candida* cells by compounds from natural resources.

Precise imaging of yeasts depends on the adequate preservation. To SEM analysis, yeast samples are well fixed by immersion on glutaraldehyde in phosphate buffer at proper concentrations (Hayat, 2000). The use of buffered osmium tetroxide at 1%, for 1 hour, is recommended as post fixation to preserve cellular content and surfaces. In our studies, samples of planktonic cells or biofilms are critical point dried after ethanolic dehydration. In order to preserve the colonies organization some steps were optimized, such as omission of osmium tetroxide, ethanolic dehydration and critical point dried. As a routine in our laboratory colonies are freeze-dried to avoid distortions and to maintain their architecture. Besides, sputtering was performed using a thick layer (50 nm) of gold.

2.1 Adherence patterns *in vitro*

Adherence is essential for members of the genus *Candida* to develop their pathogenic potential since it triggers the process that leads to colonization and allows their persistence in the host. Furthermore, different intra-species adherence ability has been reported for *Candida* species (reviewed in Silva et al., 2011). For instance, *C. tropicalis* exhibited higher ability to colonize reconstituted human epithelium (RHE) than did *C. parapsilosis* and *Candida glabrata* (Jayatilake et al., 2006). Adherence of *C. albicans* to epithelial cells is greater than that of *C. parapsilosis* (Lima-Neto et al., 2011).

For *C. parapsilosis* its emergence as a major opportunistic and nosocomial pathogen may relate to an ability to colonize the skin and adhere to inert polymeric surfaces and forms biofilms on these surfaces, such as catheters, prosthetic valves, artificial dentures and others (Douglas, 2003). According to Panagoda et al. (2001) the initial adherence of *C. parapsilosis* to surfaces is associated with cell surface hydrophobicity.

As cited previously, *C. parapsilosis* is a common etiological agent causing *Candida* nail infections. In Brazil, *C. parapsilosis* is the first or second most common cause of onychomycosis lesions (Figueiredo et al., 2007, Martins et al. 2007). Candidal onychomycosis, infection affecting nails, is increasingly found especially in immunocompromised patients. Far more than being a simple esthetics problem, infected nail serves as a reservoir of infections of the skin and mucous membrane. Multiple virulence mechanisms of *Candida* are involved in the pathogenesis of nail infections (reviewed in Jayatilake et al., 2009).

Therefore, ultrastructural investigations of the interface of *C. parapsilosis* and the keratinised substrates from human source reveal important features, which may help to clarify the pathogenesis of superficial candidiasis.

We have recently initiated experiments to verify the *in vitro* adherence pattern of *C. parapsilosis sensu stricto* (formerly *C. parapsilosis* group I) isolates obtained from candidal onychomycosis with keratinous substrates from human source. In a recent work, SEM was employed to verify the capability of *C. parapsilosis* cells to adhere and grow as biofilm on human natural substrates (nail and hair). In addition, the adherence pattern of isolates exhibiting distinct colonies phenotypes (smooth and crepe) was compared (Oliveira et al. 2010). This analysis allowed us to observe, for the first time, extracellular material and biofilm formation by *C. parapsilsois* on keratinised substrates.

In the present study, we compared ultrastructural features related to adhesion of *C. parapsilosis* cells of isolates obtained from distinct clinical sources (nail infection and tracheal secretion), on soft keratin (cutaneous stratum corneum - the outermost layer of skin) and hard keratin (nail and hair) substrates, following growth on these substrates as sole nitrogen source. In general, the surfaces of the budding cells (blastoconidia) and pseudohyphae were generally smooth except for occasional bud scars. Based on the SEM images, a different pattern of adhesion was observed for the isolates tested (Figure 1). For the onychomycosis (finger nail lesion) isolate the cell population attached to keratin substrates consisted mainly of cells in the budding-yeast phase of growth (blastoconidia) (Figure 1 A1, B1 and C1). Besides, on stratum corneum keratin short hyphal form was observed (Fig. 1A1). Differently, the isolate obtained from tracheal secretion (colonization site) the cellular population consisted mostly of pseudohyphae, particularly on stratum corneum (Figure 1 A2), a pattern that could indicate that this situation favors cellular morphologies with capacity for tissue invasion. Furthermore, SEM analysis also revealed that the tracheal secretion isolate presented different morphological pattern according to the substrate that they were in contact. For instance, cells adhered to hair keratin, consisted mainly of blastoconidia (Figure 1 C2). Overall, there was a loose association between yeast cells and keratinous substrates. However, on stratum corneum extracellular material was seen evolving cells from the onychomycosis isolate by forming a biofilm-like structure (Fig. 1A1). This feature was not observed on the other two sources of human keratin (nail and hair). These results extend our knowledge about the course of adhesion of *C. parapsilosis* on keratinized substrates which may help to clarify the pathogenesis of superficial candidiasis.

2.2 Invasion capability

Morphogenesis between yeast and hyphal growth, which facilitates fungal tissue invasion and enables the fungus to evade the defense system of the host is generally accepted as virulence traits of *C. albicans*. For instance, Jayatilake et al. (2008, 2009) have demonstrated that multiple host-fungal interactions such as cavitations, thigmotropism, and morphogenesis take place during candidal tissue invasion.

For non-*albicans Candida* species it is suggested that filamentous forms (hyphae and/or pseudohyphae) also assist in the invasive penetration of physical barriers (reviewed in Jayatilake, 2011). However, relatively little is known about the invasive potential regarding NCAC species. According to the literature, the invasiveness of non-*albicans Candida* species is variable among species. For instance, in RHE model *C. tropicalis* exhibited higher ability to invade this tissue than did *C. parapsilosis* and *C. glabrata* (Jayatilake et al., 2006).

Although it is well established that *C. tropicalis* is a polymorphic fungus few studies have analyzed the importance of its morphology on virulence. Recently, Silva et al. (2010) demonstrated that only filamentous forms of *C. tropicalis* were able to invade an oral epithelium.

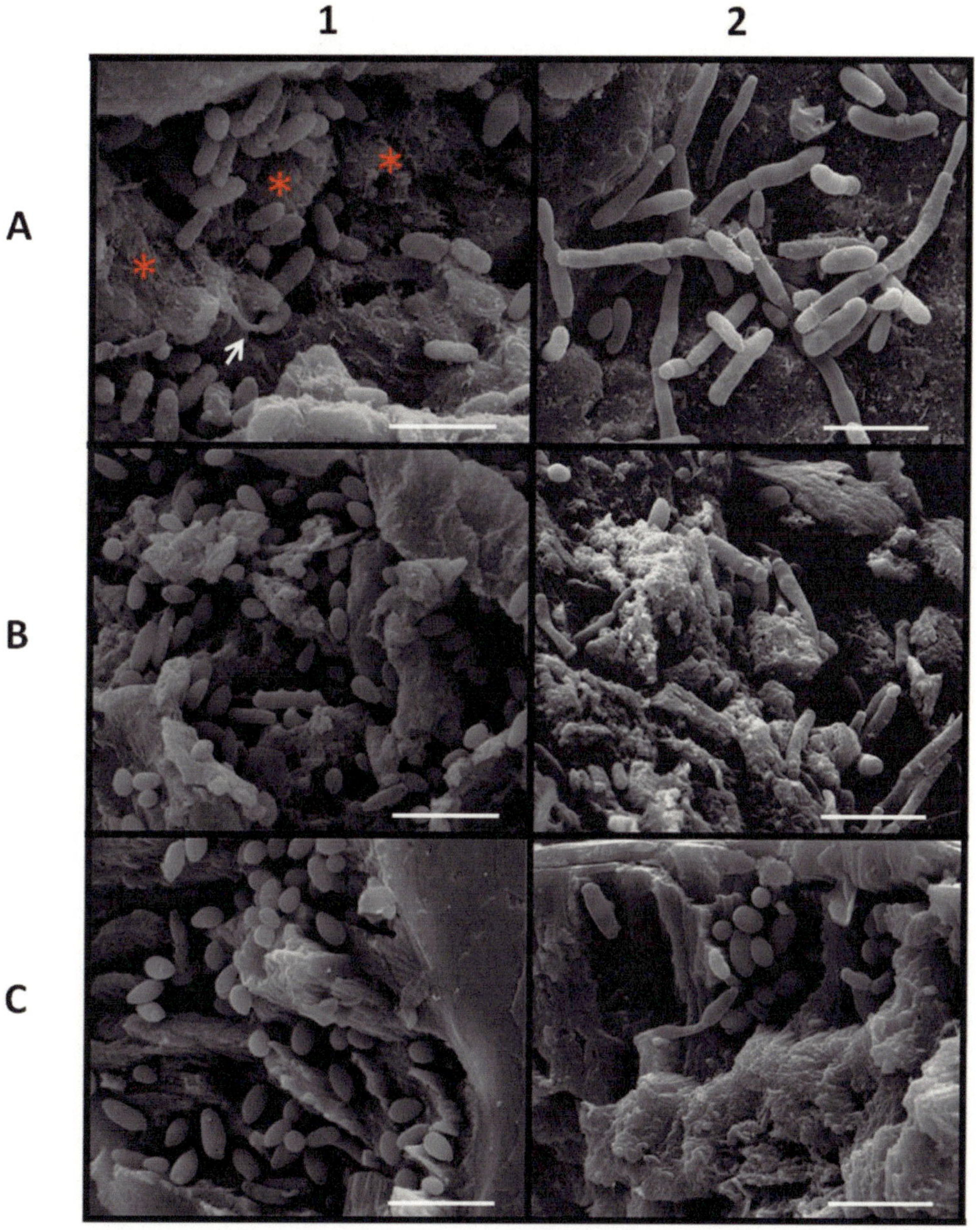

Fig. 1. These figures illustrate the *in vitro* adherence pattern of *Candida parapsilosis* isolates recovered from onychomycoses (1) and tracheal secretion (2) to human keratinous substrates. (A) Stratum corneum. On this substrate the cellular shape and the extend of extracellular material (*) were isolate-dependent. Note that cells from the onychomycosis isolate exhibits an oval shape while that cells from the tracheal secretion isolate displays pseudohyphae forms. (B) Nail. The pattern of cell morphology on nail keratin was also isolate-dependent. (C) Hair. The cellular features were independent to the site of yeast isolation. Short hyphal form (arrow). Scale bars = 100μm.

According to Brown et al. (1999), *C. albicans* cells respond to growth in contact with agar (semi solid matrix) medium by producing filaments that invade the agar. Production of invasive hyphae during growth in synthetic medium may occur by the same mechanism that is involved in production of invasive lesions during candidiasis. Recently, it has been showed that invasive filamentation of *C. albicans* into agar medium is promoted by a cell wall-linked protein (Zucchi et al., 2011). Thus, agar invasion tests enable sorting strains by their degree of invasiveness.

In this study, we employed the agar invasion assay to determine the invasive potential of a switch variant strain, exhibiting a crepe morphotype (Figure 2A), obtained from a clinical *C. tropicalis* isolate recovered from tracheal secretion. For this, cells were grown on the surface of YPD medium for 4 days a 37°C. After colony were washed off the agar surface, with a stream of water and gentle rubbing, cells that had invaded the agar remained as macroscopically visible microcolonies on the surface of the washed plate (Fig. 2B). The parental isolate also invade, although not to the same extent (data not shown). This data suggests that colonies of *C. tropicalis* exhibiting distinct morphologies differ in their capabilities to invade agar. Similar data were observed by *C. parapsilosis* (Laffey and Bluter, 2005) using the agar invasion assay.

Further we analyzed the invaded agar at ultrastructural level, by the employment of fracture technique. Fracture is valuable to reveal internal surfaces and it is performed by dipping the fixed samples in liquid nitrogen and breaking with a sharp scalpel. Growth in YPD shows filamentous forms invading the agar at different planes and angles, as well as yeast cell forms (Figure 3). Thus, SEM may be useful for the detailed analysis of extend and pattern of yeast cells in the course of the invasion process. This is the first report of the employment of SEM to examine the pattern of agar invasion by *Candida* cells.

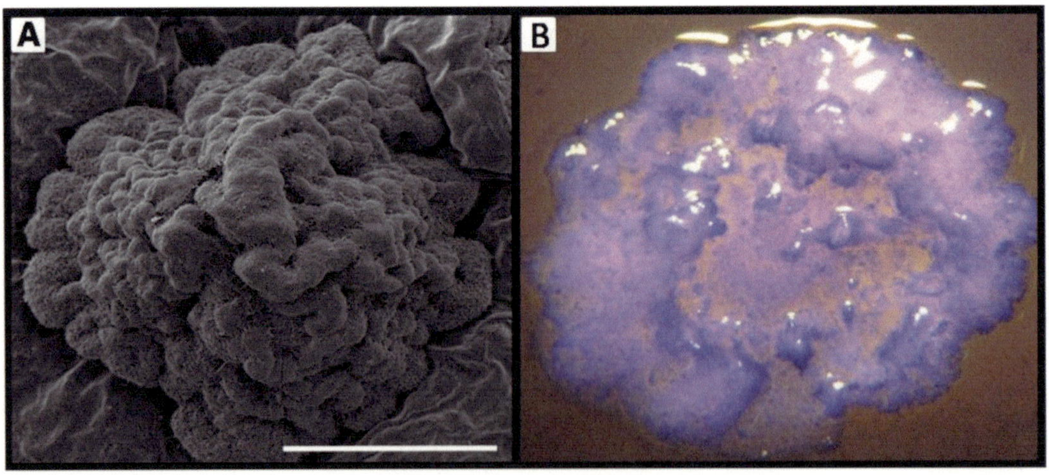

Fig. 2. (A) Scanning electron micrograph shows *C. tropicalis* morphotype crepe following 96 h growth on YPD agar. (B) Photomicrograph of footprint of attached cells after washing out the colony. Scale bar = 1mm (A). 2.5X (B).

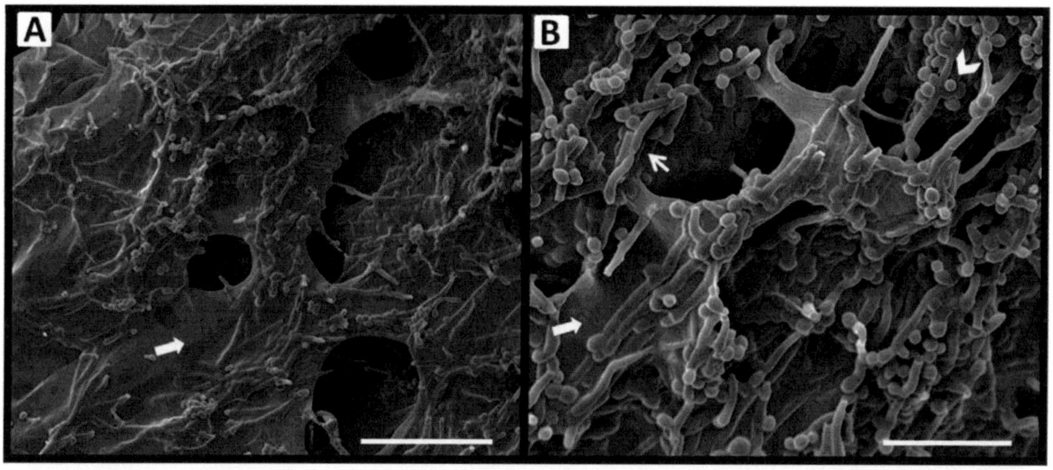

Fig. 3. Scanning electron micrographs of agar invasion by *C. tropicalis* crepe after cells were washed off the agar surface. Cells includes on agar substrate (large arrow) appear like filamentous form (thin arrow) and blastoconidia (head arrow) Scale bars = 100μm (A) and 50μm (B).

2.3 Switching morphotypes: Ultrastructure and morphological types

Phenotypic switching represents an epigenetic state that occurs in a small fraction of the population, is random and reversible. This biological phenomenon is related to the occurrence of spontaneous emergence of colonies with different morphologies that enables the microorganism to undergo rapid microevolution and to adapt to different environments, including various anatomical sites in the human body (reviewed in Soll, 1992). Thus, the switching phenotype event has also being considered a candidal virulence factor (Segal, 2004).

Furthermore, switching has been demonstrated to regulate virulence-associated characteristics in *C. albicans*, such as adhesion, expression of cell surface hydrophobicity and biofilm formation (Kennedy et al., 1988, Antony et al., 2009, Lohse and Johnson, 2009). Concerning NCAC species, *C. parapsilosis* distinct switch phenotypes exhibited differential ability to form biofilm on polystyrene surfaces (Laffey and Butler, 2005). For *C. tropicalis*, França et al. (2010) also found a correlation in phenotypic switching and biofilm formation.

For fungi this event is defined as the reversible change manifested as altered colony morphology at a rate higher than the somatic mutation rate (reviewed in Soll, 1992).

In yeast, phenotypic switching was originally described in *C. albicans* strains (Soll, 1992), but is also known to exist in other *Candida* species, such as *C. tropicalis* (Soll et al., 1988, França et al., 2010).

Ultrastructural investigations revealed a relationship between *C. albicans* switched variant colonies and microstructure (Radford et al., 1994, 1997). In a pioneer study we report the presence of extracellular material, resembling a biofim-like colony, throughout the development of *C. tropicalis* switch colonies, suggesting that its presence is correlated with the complex architecture of colonies (França et al. 2010).

SEM was successfully employed for the analyses of whole *Candida* colonies architecture (França et al. 2010, Furlaneto et al., 2012). Additional studies on switching event in clinical isolates of *C. tropicalis* are in progress. For instance, different architectures exhibited by *C. tropicalis* colonies morphotypes are shown in Figure 4. The smooth phenotype colony (Figure 4A) showed a hemispherical shape character, while the rough phenotype exhibited more complex architecture and was characterized by the presence of deep central and peripheral depressions areas (Figure 4B). The irregular wrinkled colony was characterized by a highly wrinkled centre and an irregular periphery (Figure 4C). Crepe colony was characterized by the presence of aerial hyphae on the colony surface (Figure 4D). The preparation of colonies by a freeze-drying technique allowed their architecture preservation with maintenance of the phenotypes observed at lower magnitude (data not shown).

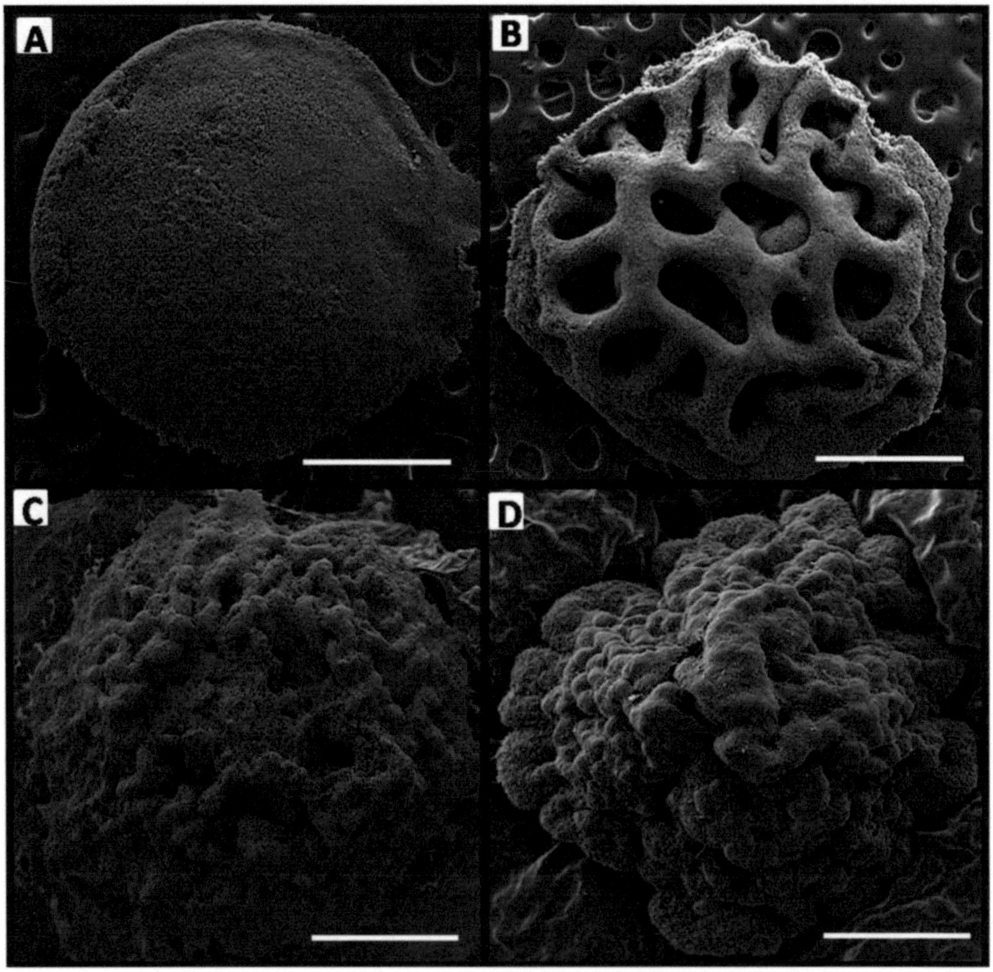

Fig. 4. Different architectures exhibited by *C. tropicalis* colonies morphotypes following 96 h incubation on YPD agar. (A) Smooth colony. (B) Rough colony, shows cells in a tridimensional disposition. (C) Cells are establishing an irregular wrinkled colony and (D) crepe colony shows the homogeneous substance coating its surface, besides filamentous forms. Scale bars = 1mm.

The ultrastructural analysis allowed the observation of the arrangement of individual cells within the colonies. After 4 days of colony development, the whole smooth and irregular wrinkled colonies consisted entirely of yeast cells (not shown). The crepe colony phenotype also comprised mainly yeast cells as observed at depressions areas (Figure 5A). Most interesting was the presence of extracellular material forming a biofilm-like colony where many of the cells were almost hidden by this material. It was observed as fibrils, with enlarged structures, connecting neighbouring cells (Figure 5B).

A *C. tropicalis* variant exhibiting a myceliated phenotype is shown in Figure 6. The whole colony surface is formed by aerial mycelia with a prominent centre (Figure 6A). The aerial hyphae showed a compact nature that is composed by hyphae and blastoconidia (Figure 6B).

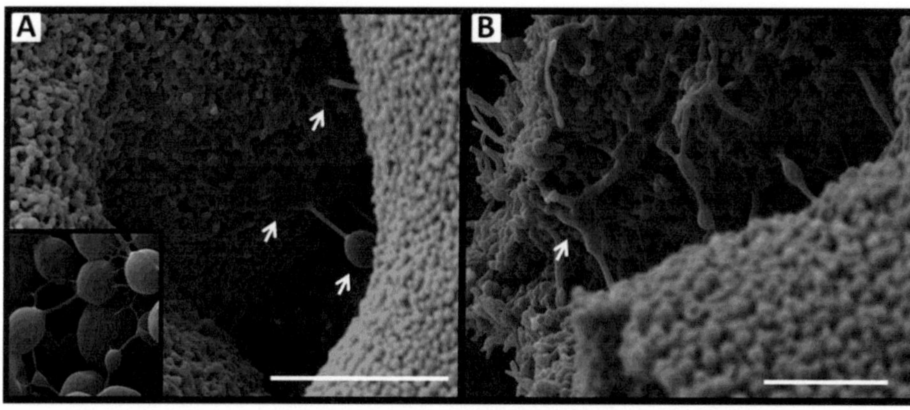

Fig. 5. *C. tropicalis* crepe morphotype following 96 h incubation on YPD agar.
(A) Extracellular material (arrows) is seeing forming a biofilm-like colony. (B) Fractured colonies reveal details of extracellular material recovering and connecting cells. Scale bar = 100μm (A), 50μm (B). Inset shows fibrilar extracellular material connecting cells (5000x)

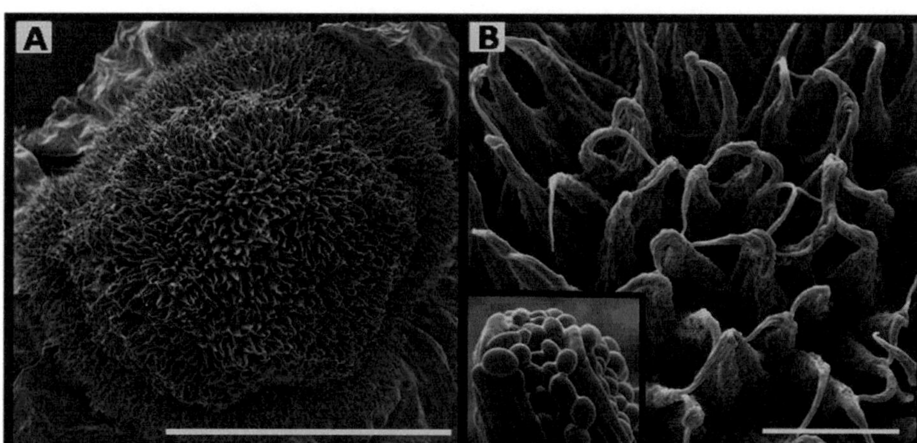

Fig. 6. Electron micrographs of the *C. tropicalis* myceliated morphotype following 96 h incubation on YPD agar. (A) Whole myceliated colony, (B) In higher magnification aerial hyphae showing compact nature. Scale bars = 1mm (A), 100μm (B). Inset shows detail of a distal end of an aerial hyphae that is composed by hyphae and blastoconidia (5000x)

2.4 Effect of antifungal compounds on yeast morphology

The therapy of deep fungal infections, particularly those caused by opportunistic pathogens, including *Candida* species, remains a difficult medical problem. Besides, compared with antibiotics, the development of antifungal agents has been relatively limited. Widespread use of antifungal agents could be an explanation for the emergence of the more resistant non-*albicans* species of *Candida* (Pfaller & Diekema, 2004). Fluconazole is a systemic antifungal drug effective against most of the *Candida* species; however, the emergence of fluconazole resistance has been reported in NCAC species, particularly *C. tropicalis* and *C. parapsilosis* (Yang et al., 2004, Pereira et al., 2010, Oxman et al., 2010, Bruder-Nascimento et al., 2010). The emergence of yeast species with decreased susceptibility to contemporary antifungal regimens demonstrates the need for new antifungal agents.

Many studies have addressed the search for natural compounds with antifungal activity. As an example, Duarte et al. (2005) screened 35 medicinal plants commonly used in Brazil for anti-*Candida albicans* activity. In this context, the flavonoid baicalein, originally isolated from the roots of *Scutellaria baicalensis* Georgi (a Chinese herb) has been tested against *C. albicans* (Cao et al., 2008, Huang et al., 2008). According to these authors, antifungal activity was observed on free cells as well as on biofilm.

Ultrastructural investigations of the effect of natural compounds on morphology are limited. However, SEM analysis allowed the observation of irregular budding patterns and pseudohyphae formation in *C. albicans* type strain treated with compounds isolated from pomegranate peels (Endo et al., 2010). In contrast, the exposure of cells from the same type strain to berberine (alkaloid found in medicinal herbs) did not affect cell morphology (Iwazaki et al., 2010).

We employed the scanning electron microscopy to evaluate the effect of baicalein alone and in combination with fluconazole on the morphology of *C. parapsilosis* and *C. tropicalis*. For *C. parapsilosis*, SEM analyses showed control cells (untreated cells) with a normal budding profile where no extracellular material was seen (not shown). After exposure to baicalein alone, the general aspect of the cells was not modified, however, a profusely flocculent extracellular material was seeing connecting yeast cells (Fig. 7A). Similar pattern was observed for cells exposed to baicalein in combination with fluconazole, although, the amount of extracellular was visible higher (Fig. 7B). SEM images also showed markedly reduced number of organisms due to baicalein.

On the other hand, the data obtained in this study showed that *C. tropicalis* underwent morphological alterations visible by SEM when treated with subinhibitory concentration of baicalein alone and in association with fluconazole.

Untreated cells (control) consisted of blastoconidia and pseudohyphae (not shown). For cells exposed to baicalein alone we observed the presence of elongated cells as well as a great capacity for producing pseudohyphae (Fig. 8A). Cells exposed to baicalein in combination with fluconazole showed an oval shape with profusely flocculent extracellular material connecting yeast cells (Fig. 8B). These data, suggest different inter-species response to baicalein alone as well as to in association with fluconazole.

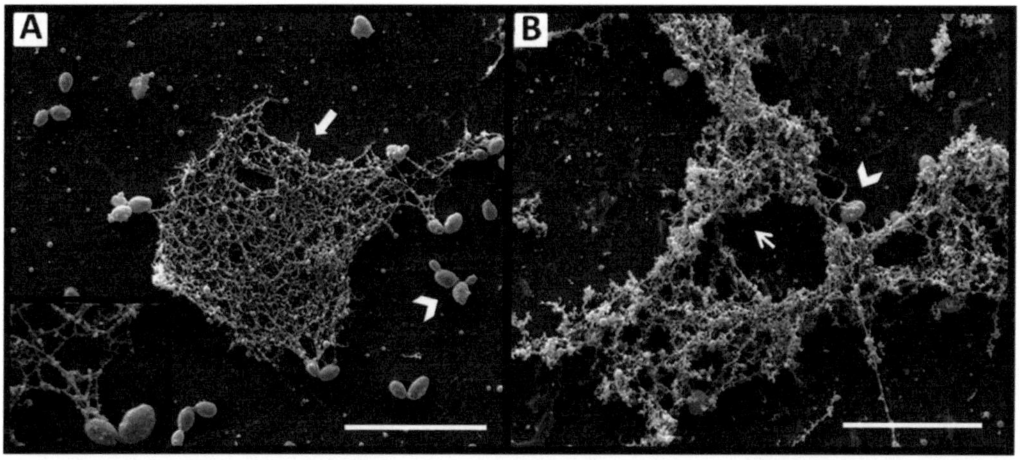

Fig. 7. *C. parapsilosis* treated with baicalein alone (MIC50) and in combination with fluconazole. (A) baicalein, (B) baicalein plus fluconazole. *C. parapsilosis* cells display a typical oval shape (heads arrow). Some of them are involved by an extracellular material constituted by irregular fibrils disposed as a network (large arrow). Note that fibrils appear like beads on a string (inset). In B, a flocculent extracellular material (thin arrow) is seeing connecting cells. Scale bar = 20μm. Inset =12,000 X

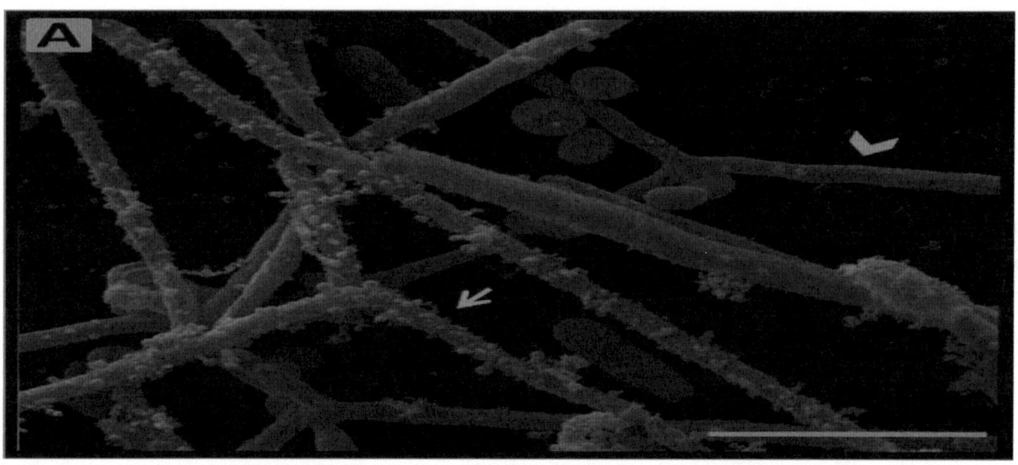

Fig. 8. *C. tropicalis* treated with baicalein alone (A) and in combination with fluconazole (B). Hyphae form is predominant in treatment with baicalein (head arrow), while in combination with antifungal blastoconidia prevails. Flocculent extracellular material (arrows) is seen surrounding hyphae and a prominent one is connecting blastoconidia. Scale bar = 20μm.

3. Conclusion

Yeast pathogenicity arises through complex interactions between the organism's virulence characteristics and the host's response. *Candida* species can exhibit several virulence factors such as adherence, biofilm formation and phenotypic switching that increase their persistence within the host as well as allow adapting to different anatomical sites in the human body. Therefore, the increase in the incidence and antifungal resistance of NCAC species, specifically, *C. parapsilosis* and *C. tropicalis*, and the unacceptably high morbidity and mortality associated with these species, make it essential to further enhance our knowledge on the virulence and resistance mechanisms associated with these species. An understanding of the virulence determinants of these species would provide insight into their pathogenic mechanisms. Our studies have shown that ultrastructural investigations at SEM of some of these virulence traits may help to elucidate general mechanisms of fungal virulence. Another approach that is in progress in our laboratory is the processing of yeast samples to analyse at transmission electron microscopy that may reveal inner ultrastructure at higher resolution.

4. Acknowledgments

The author gratefully acknowledges to all colleagues and students joined to the medical mycology group at State University of Londrina and Technological Federal University of Paraná in Londrina, Paraná-Brazil. Special thanks to the group members: Ana Flávia Leal Specian and Rosapa Serpa that contributed with results obtained during their master thesis work. The authors also thank Osvaldo Capello for technical support. Team´s members using original results made all the figures´ composition. This work was partially supported by Conselho Nacional de Desenvolvimento Científico e Tecnológico –CNPq-Brazil.

5. References

Anthony, G.; Saralaya, V.; Gopalkrishna Bhat, K; Shalini Shenoy, M. & Shivananda, P.G. (2009). Effect of phenotypic switching on expression of virulence factors by *Candida albicans* causing candidiasis in diabetic patients. *Revista Iberoamericana de Micologia,* Vol.26, No.3 (September 2009), pp. 202-205, ISSN 1130-1406.

Almirante, B.; Rodríguez, D. & Cuenca-Estrella, M. (2006). Epidemiology, risk factors, and prognosis of *Candida parapsilosis* bloodstream infections: case-control population-based surveillance study of patients in Barcelona, Spain, from 2002 to 2003. *Journal of Clinical Microbiology*, Vol.44, No.5 (May2006), pp. 1681-1685, ISSN 0095-1137.

Barnes, W.G.; Flesher, A.; Berger, A.E. & Arnold, D.J. (1971). Scanning electron microscopic studies of *Candida albicans*. *Journal of Bacteriology*, Vol.106, No.1 (April 1971), pp. 276–280, ISSN 0021-9193.

Bozzola, J.J. & Russel, L.D. 1999. *Electron microscopy. Principles and Techniques for Biologists.* (2nd ed.), Jones and Bartlett Publishers, ISBN 0-7637-0192-0, Sudbury, Massachusetts, USA.

Branchini, M.L.; Pfaller, M.A.; Rhine-Chalberg, J.; Frempong, T. & Isenberg, H.D. (1994). Genotypic variation and slime production among blood and catheter isolates of *Candida parapsilosis*. *Journal of Clinical Microbiology*, Vol.32, No.2 (February 1994), pp. 452–456, ISSN 0095-1137.

Brown, D.H.Jr.; Giusani, A.D.; Chen, X. & Kumamoto, C.A. (1999). Filamentous growth of *Candida albicans* in response to physical environmental cues and its regulation by the unique *CZF1* gene. *Molecular Microbiology*, Vol.34, No.4 (November 1999), pp. 651–662, ISSN 1365-2958.

Bruder-Nascimento, A.; Camargo, C.H.; Sugizaki, M.F.; Sadatsune, T.; Montelli, A.C.; Mondelli, A.L.; Bagagli, E. (2010). Species distribution and susceptibility profile of

Candida species in a Brazilian public tertiary hospital. *BCM Research Notes*, Vol.3, No.1 (January 2010), pp. 1-5, ISSN 1756-0500.

Cao, Y.Y.; Dai, B.D.; Wang, Y.; Huang, S.; Xu, Y.G.; Cão, Y.B.; Gao, P.H.; Zhu, Z.Y. & Jiang, Y.Y. (2008). In vitro activity of baicalein against *Candida albicans* biofilms. *International Journal of Antimicrobial Agents*, Vol.32, No.1 (July 2008), pp. 73–77, ISSN 0924-8579.

Colombo, A.L.; Nucci, M.; Park, B.J.; Nouer, A.S.; Arthington-Skaggs, B.; da Matta, D.A.; Warnock, D. & Morgan, J. (2006). Epidemiology of candidemia in Brazil: a nationwide sentinel surveillance of candidemia in eleven medical Centers. *Journal of Clinical Microbiology*, Vol.44: No.8 (August 2006), pp. 2816-23, ISSN 0095-1137.

De Bernardis, F.; Mondello, F.; San Millàn, R.; Ponón J. & Cassone, A. (1999). Biotyping and virulence properties of skin isolates of *Candida parapsilosis*. *Journal of Clinical Microbiology*, Vol.37, No.11 (November 1999), pp. 3481-3486, ISSN 0095-1137.

Douglas, J. (2003). *Candida* biofilms and their role in infection. *Trends in Microbiolgy*, Vol.11, No.1 (January 2003), pp.: 30–36, ISSN 0966-842X.

Duarte, M.C.T.; Figueira, G.M.; Sartoratto, A.; Rehder, V.L.G. & Delarmelina C. (2005). Anti-*Candida* activity of Brazilian medicinal plants. *Journal of Ethnopharmacology*, Vol.97, No.28 (February), pp. 305-311, ISSN 0378-8741.

Endo, E.H.; Cortez, D.A.G.; Ueda-Nakamura, T.; Nakamura, C.V. & Filho, B.P.D. (2010). Potent antifungal activity of extracts and pure compound isolated from pomegranate peels and synergism with fluconazole against *Candida albicans*. *Research in Microbiology*, Vol.161, No.7 (September 2010), pp. 534-540, ISSN 0923-2508.

Figueiredo, V.T.; Santos, D.A.; Resende, M.A. & Hamdan, J.S. (2007). Identification and in vitro antifungal susceptibility testing of 200 clinical isolates of *Candida* spp. responsible for fingernail infections. *Mycopathologia*, Vol.164, No.1 (June 2007), pp.27-33, ISSN 0301-486X.

França, E.J.G.; Andrade, C.G.T.J.; Furlaneto-Maia, L.; Serpa, R.; Oliveira, M.T.; Quesada, R.M.B. & Furlaneto, M.C. (2011). Ultrastructural architecture of colonies of different morphologies and biofilm produced by phenotypic switching of *Candida tropicalis*. *Micron*, Vol.42, No.7 (October 2011), pp. 726-732, ISSN 0968-4328.

França, E.J.G.; Furlaneto-Maia, L.; Quesada, R.M.B.; Favero, D.; Oliveira, M.T. & Furlaneto, M.C. (2010). Haemolytic and proteinase activities in clinical isolates of *Candida parapsilosis* and *Candida tropicalis* with reference to the isolation anatomic site. *Mycoses*, Vol.54, No.4 (July 2011), pp. e44-e51, ISSN 1439-0507.

Furlaneto, M.C.; Andrade, C.G.T.J.; Aragão, P.H.A.; França, E.J.G.; Moralez, A.T.P. & Ferreira, L.C.S. (2012). Scanning Electron Microscopy as tool for the analyses of colonies architecture of different morphologies produced by phenotypic switching of a human pathogenic yeast *Candida tropicalis*. *Journal of Physics. Conference Series*, ISSN 1742-6588, in press.

Hayat, M.A. 2000. *Principles and Techniques of Electron Microscopy. Biological Applications*. (4th ed.), Cambridge University Press, ISBN 0-521-63287-0, USA.

Haynes, K. (2001). Virulence in *Candida* species. *Trends in Microbiology*, Vol.9, No.12 (December 2001), pp. 591-596, ISSN 0966-842X.

Huang, S.; Cao, Y.Y.; Dai, B.D.; Sun, X.R.; Zhu, Z.Y.; Cao, Y.B.; Wang, Y.; Gao, P.H. & Jiang, Y.Y. (2008). In vitro synergism of fluconazole and baicalein against clinical isolates of *Candida albicans* resistant to fluconazole. *Biological Pharmaceutical Bulletim*, Vol.31, No.12 (December 2008), pp. 2234-2236, ISSN 1347-5215.

Iwazaki, R.S.; Endo, E.H.; Ueda-Nakamura, T.; Nakamura, C.V.; Garcia, L.B. & Filho, B.P.D. (2010). In vitro antifungal activity of the berberine and its synergism with fluconazole. *Antonie van Leeuwenhoek*, Vol.97, No.2 (November 2009), pp. 201-205, ISSN 0003-6072.

Kennedy, M.J.; Rogers, A.L.; Hanselman, L.R.; Soll, D.R. & Yancey, R.J. (1988). Variation in adhesion and cell surface hydrophobicity in *Candida albicans* white and opaque phenotypes. *Mycopathologia*, Vol.102, No.3 (June 1988), pp. 149-156, ISSN 0301-486X.

Jayatilake, J.A.M.S. (2011). A review of the ultrastructural features of superficial candidiasis. *Mycopathologia*, Vol.171, No.4 (October 2010), pp. 235-250, ISSN 0301-486X.

Jayatilake, J.A.; Samaranayake, Y.H.; Cheung, L.K. & Samaranayake, L.P. (2006). Quantitative evaluation of tissue invasion by wild type, hyphal and SAP mutants of *Candida albicans*, and non-albicans *Candida* species in reconstituted human oral epithelium. *Journal of Oral Pathology and Medicine*, Vol.35, No. 8 (August 2006), pp. 481–491, ISSN 1600-0714.

Jayatilake, J.A.M.; Samaranayake, Y.H. & Samaranayake, L.P. (2008). A comparative study of candidal invasion in rabbit tongue mucosal explants and reconstituted human oral epithelium. *Mycopathologia*, Vol.165, No.6 (March 2008), pp. 373–380, ISSN 0301-486X.

Jayatilake, J.A.M.; Tilakaratne, W.M. & Panagoda, G. J. (2009). Candidal onychomycosis: A Mini-Review. *Mycopathologia*, Vol.168, No.4 (May 2009), pp. 165–173, ISSN 0301-486X.

Krcmery, V. & Barnes, A. (2002). Non-*albicans Candida* spp. causing fungaemia pathogenicity and antifungal resistance. *Journal of Hospital Infection*, Vol.50, No.4 (April 2002), pp. 243-260, ISSN 0195-6701.

Laffey, S.F. & Butler, G. (2005). Phenotype switching affects biofilm formation by *Candida parapsilosis*. *Microbiology*, Vol.151, No.4 (January 2005), pp. 1073-1082, ISSN 0002-7739.

Lee, R.E. 1993. *Scanning Electron Microscopy and X-Ray Microanalysis*. PTR Prentice-Hall, ISBN 0-13-813759-5, New Jersey, USA.

Lima-Neto, R.G.; Beltrão, E.I.; Oliveira, P.C. & Neves, R.P. (2011). Adherence of *Candida albicans* and *Candida parapsilosis* to epithelial cells correlates with fungal cell surface carbohydrates. *Mycoses*, Vol.54, No.1 (January 2011), pp. 23-39, ISSN 0933-7407.

Lohse, M.B. & Johnson, A.D. (2009). White–opaque switching in *Candida albicans*. *Current Opininion in Microbiology*, Vol.12, No. 6 (December 2009), pp. 650-654, ISSN 1369-5274.

Martins, E.A.; Guerrer, L.V.; Cunha, K.C.; Soares, M.M.C.; & Almeida, M.T.G. (2007). Onychomycosis: clinical, epidemiological and mycological study in the municipality of São José do Rio Preto. *Revista da Sociedade Brasileira de Medicina Tropical*, Vol.40, No.5 (October 2007), pp. 596-598, ISSN 0037-8682.

Manzano-Gayosso, P.; Hernández-Hernández, F.; Méndez-Tovar, L.J.; Palacios-Morales, Y.; Córdova-Martínez, E.; Bazán-Mora, E. & Martinez-López, R. (2008). Onychomycosis incidence in type 2 diabetes mellitus patients. *Mycopathologia*, Vol.166, No.1 (March 2008), pp. 41-45, ISSN 0301-486X.

Nucci, C. & Colombo, A.L. (2007). Candidemia due to *Candida tropicalis*: clinical, epidemiologic, and microbiologic characteristics of 188 episodes occurring in tertiary care hospitals. *Diagnostic Microbiology and Infectious Diseases*, Vol.58, No.1 (May 2007), pp. 77-82, ISSN 0732-8893.

Oliveira, M.T.; Specian, A.F.; Andrade, C.G.T.J.; França, E.J.G.; Furlaneto-Maia, L. & Furlaneto, M.C. (2010). Interaction of *Candida parapsilosis* with human hair and nails surfaces revealed by scanning electron microscopy analysis. *Micron*, Vol.41, No.6 (August 2010), pp. 604-606, ISSN 0868-4328.

Oxman, D.A.; Chow, J.K.; Frendl, G.; Hadley, S.; Hershkovitz, S.; Ireland, P.; McDermott, L.A.; Tsai, K.; Marty, F.M.; Kontouiannis, D. & Golan, Y. (2010). Candidaemia associated with decreased in vitro fluconazole susceptibility: is *Candida* speciation predictive of the susceptibility pattern? *Journal of Antimicrobial Chemotherapy*, Vol.65, No.7 (April 2010), pp. 1460-1465, ISSN 0305-7453.

Panagoda, G.J.; Ellepola, A.N. & Samaranayake, L.P. (2001). Adhesion of *Candida parapsilosis* to epithelial and acrylic surfaces correlates with cell surface hydrophobicity. *Mycoses*, Vol.44, No.1-2 (March 2001), pp. 29–35, ISSN 1439-0507.

Pereira, G.H.; Mulles, P.R.; Szeszs, M.W.; Levin, A.S. & Melhem, M.S.C. (2010). Five-year evaluation of bloodstream yeast infections in a tertiary hospital: the predominance of non-*C. albicans Candida* species. *Medical Mycology*, Vol.48, No.6 (September 2010), pp. 839-842, ISSN 1369-3786.

Pfaller, M.A. & Diekema, D.J. (2004). Twelve years of fluconazole in clinical practice: global trends in species distribution and fluconazole susceptibility of bloodstream isolates of *Candida*. *Clinical Microbiology and Infection*, Vol.10, No.1 (January 2004), pp. 11–23.

Radford, D.R.; Challacombe, S.J. & Walter, J.D. (1994). A scanning electron microscopy investigation of the structure of colonies of different morphologies produced by phenotypic switching of *Candida albicans*. *Journal of Medical Microbiology*, Vol.40, No.6 (June, 1994), pp. 416-423, ISSN 1473-5644.

Radford, D.R.; Challacombe, S.J. & Walter, J.D. (1997). Scanning electron microscopy of the development of structured aerial mycelia and satellite colonies of phenotypically switched Candida albicans. *Journal of Medical Microbiology*, Vol.46, No. 4 (April, 1997), pp. 326-332, ISSN 1473-5644.

Segal, E. (2004). *Candida*, still number one – what do we know and where are we going from there? *Mycoses*, Vol.48, No.1 (April, 2005), pp. 3–11, ISSN 1439-0507.

Silva, S.; Hooper, S.J.; Henriques, M.; Oliveira, R.; Azeredo, J. & Williams, D.W. (2010). The role of secreted aspartyl proteinases in *Candida tropicalis* invasion and damage of oral mucosa. *Clinical Microbiology and Infection*, Vol.17, No. 2 (April 2010), pp. 264–272, ISSN 1469-0691.

Silva, S.; Negri, M.; Henriques, M.; Oliveira, R.; Williams, D.W. & Azeredo, J. (2011). Adherence and biofilm formation of non-*Candida albicans Candida* species. *Trends in Microbiology*, Vol.19, No.5 (May 2011), pp. 241-247, ISSN 0966-842X.

Slutsky, B.; Buffo, J. & Soll, D.R. (1985). High frequency switching of colony morphology in *Candida albicans*. *Science*, Vol.230, No. 4723 (November, 1985), pp. 666-669, ISSN 1535-9778.

Soll, D.R. (1992). High-frequency switching in *Candida albicans*. *Clinical Microbiology Reviews*, Vol.5, No.2 (April, 1992), pp. 183-203, ISSN 0893-8512.

Soll, D.R.; Staebell, M.; Langtimm, C.; Pfaller, M.; Hicks, J. & Gopala Rao, T.V. (1988). Multiple *Candida* strains in the course of a single systemic infection. *Journal of Clinical Microbiology*, Vol.26, No. 8 (August, 1988), pp. 1448-1459, ISSN 0095-1137.

Trofa, D.; Gácser, A. & Nosanchuk, J.D. (2008). *Candida parapsilosis*: an emerging fungal pathogen. *Clinical Microbiology Review*, Vol.21, No.4 (October 2008), pp. 606-625, ISSN 0893-8512.

van Asbeck, E.C.; Clemins, K.V. & Stevens, D.A. (2009) *Candida parapsilosis*: a review of its epidemiology, pathogenesis, clinical aspects, typing and antimicrobial susceptibility. *Critical Review in Microbiology*, Vol.35, No.4 (November, 2009), pp. 283–309, ISSN 1040-841X.

Yang, Y.L.; Ho, Y.A.; Cheng, H.H.; Ho, M. & Lo, H.J. (2004). Susceptibilities of *Candida* species to amphotericin B and fluconazole: the emergence of fluconazole resistance in *Candida tropicalis*. *Infection Control Hospital Epidemiology*, Vol.25, No.1 (January, 2004), pp. 60-64, ISSN 0899-823X.

Zucchi, P.C.; Davis, T.R. & Kumamoto, C.A. (2010). A *Candida albicans* cell wall-linked protein promotes invasive filamentation into semi-solid medium. *Molecular Microbiology*, Vol.76, No.3 (May 2010), pp. 733-748, ISSN 1365-2958.

Part 3

Material Science

16

Multimodal Microscopy for Ore Characterization

Otávio da Fonseca Martins Gomes and Sidnei Paciornik
Centre for Mineral Technology – CETEM,
Dept. of Materials Engineering, Catholic University of Rio de Janeiro,
Brazil

1. Introduction

The recent developments in electronics and computing have brought a radical change to the microstructural characterization of materials. The integration of digital image acquisition and digital image analysis with microscope automation methods is giving rise to a rich set of techniques in the new field of Digital Microscopy (Paciornik & Maurício, 2004).

Modern microscopes of all kinds (optical, electron, scanning probe) are controlled by software and have digital image acquisition. This setup allows many integrated tasks to be run under the control of automation routines like, for instance, specimen scanning and automatic focusing. Additionally, some microscopes can be fully automated. Thus, it is possible to integrate specimen scanning, image acquisition and storage, processing, analysis and report generation in a single routine.

Besides the automation of routine tasks in the microscopes, Digital Microscopy really opens new possibilities for microstructural characterization. In this context, multimodal microscopy emerges as a promising trend. Multimodal microscopy aims at combining complementary types of information from a given sample in order to build a multidimensional data set. It generates multi-component images combining layers obtained from different microscopy modalities, or from the same microscope in diverse conditions. For instance, multimodal microscopy may consider different signals in scanning electron microscopy (SEM) and different contrast modes in optical microscopy. Sometimes, multimodal microscopy is also referred as co-site, correlative or collaborative microscopy.

The key step of a multimodal microscopy methodology is the registration between images from a given field and/or set of fields. Image registration is the process of overlaying two or more images of the same scene taken at different conditions or by different sensors. Actually, registration is a crucial procedure in all image analysis tasks in which the final information is obtained from the combination of various data sources. Typically, registration is required to combine or compare images in remote sensing and medical imaging applications.

Once the multimodal set of images is acquired and registered, image segmentation can be employed to discriminate phases, regions or objects of interest. Due to the nature of this problem, multidimensional pattern recognition techniques arise as potential methods for image segmentation. Then, after segmentation, one is able to measure size, shape, intensity, and position parameters, leading to the possibility of automatic characterization of microstructural features.

The present chapter presents a multimodal methodology that combines images obtained by reflected light microscopy (RLM) and SEM. The so-called RLM-SEM co-site microscopy (Gomes & Paciornik, 2008a, 2008b) was developed to solve some ore microscopy problems that cannot be solved by either RLM or SEM.

2. Ore microscopy

Ore microscopy is an essential tool for ore characterization. It was generally employed in its various modalities (stereoscopic, transmitted and reflected light, SEM, etc.) for mineral identification and quantification, and in the determination of mineral texture and liberation analysis. In certain conditions, ore microscopy is the single approach to access this kind of information. In the mining industry, it is extensively used to provide parameters to the Geometallurgy procedures for exploration, production planning, and processing plant design and optimization purposes.

Transmitted and reflected light microscopy, respectively for transparent and opaque minerals, are probably the most traditional techniques of mineralogical identification. During the last two centuries, diverse analytical methods based on various properties of minerals were developed and refined. Referring to reflected light microscopy, it is worth to mention properties such as reflectivity, colour, reflection pleochroism, internal reflections, hardness, preferential polishing, chemical reactivity, crystalline habit, and crystalline texture, among others. There are some classical text-books that cover both theoretical and practical aspects of ore microscopy such as Galopin & Henry (1972), Gribble & Hall (1992), Criddle & Stanley (1993), and Craig & Vaughan (1994).

However, these traditional methods generally require an expert mineralogist and only few of them can be applied in automated systems. Thus, optical microscopy was being left aside in favour of SEM in ore characterization methodologies. In fact, in the last decades, research and development of microscopy in Applied Mineralogy field were focused on SEM.

The SEM is a very versatile analytical instrument. It builds images through synchronization of the electron beam scanning and one of the many signals that come from the interaction between the electron beam and the specimen. Thus, the pixels present intensities proportional to the signal measured by one of the SEM detectors such as, for instance, back-scattered electrons (BSE) or secondary electrons (SE) detectors. If the SEM has a coupled energy dispersive X-ray spectrometer (EDS), it becomes even more versatile, and can also perform elemental chemical analysis with a resolution down to approximately 1 μm on the surface. This is the great advantage of SEM - a large variety of electron-specimen interactions can be used to form images and to provide information with different physical meanings (Reimer, 1998; Goldstein et al., 2003).

The most used signal for ore characterization is BSE that can furnish topography information and atomic number contrast. Nevertheless, if the specimen is plane, each pixel is proportional to the average atomic number of its corresponding region on the specimen surface (Jones, 1987). Therefore, BSE images of polished samples are indirectly compositional images, in which mineral phases can be correlated to characteristic intensities or grey levels that are proportional to their average atomic numbers. Figure 1 shows a BSE image of iron ore in which four phases can be recognized: the embedding resin (the black background), quartz (dark grey), goethite (the grey particle at centre), and hematite (white). Table 1 presents chemical formula, colour on RLM, and average atomic number of epoxy resin and some minerals.

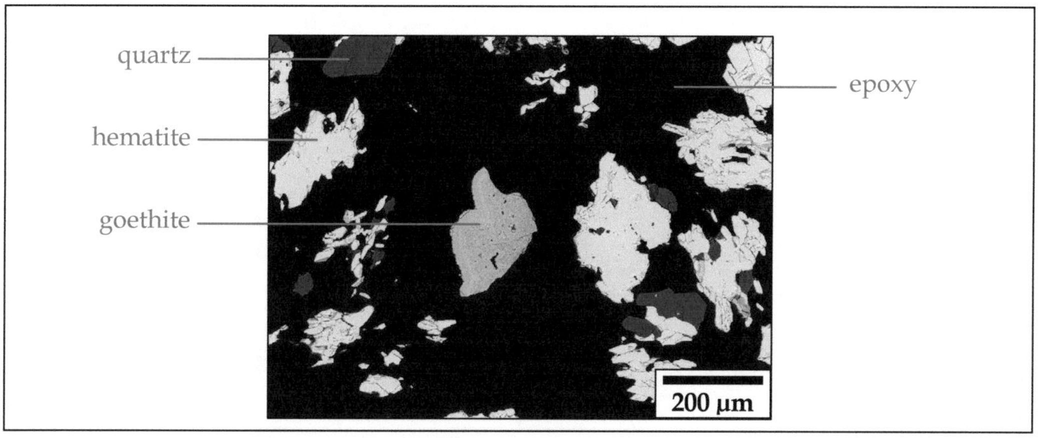

Fig. 1. BSE image of iron ore: the embedding resin (the black background), quartz (dark grey), goethite (the grey particle at centre), and hematite (white).

Based on BSE and EDS techniques, some automated systems for ore characterization were developed and commercially launched (Petruk, 1988; Sutherland & Gottlieb, 1991; Gu, 2003). These systems are SEM's especially dedicated to quantitative mineral analysis. They can identify minerals using BSE and EDS signals, and perform quantification routines through integrated image analysis software. Their capabilities may include particle-by-particle analysis, mineral phase classification and quantification, and mineral liberation analysis. Therefore, they became dominant for ore characterization, both in academy and industry, due to their enormous analytical capacity and relative simplicity of use.

Nevertheless, in recent years, there was a growing use of optical microscopy applied to ore characterization. Basically, three facts contributed to this trend: better optics, better digital image acquisition devices (Pirard et al., 1999), and the advent of Digital Microscopy. The progress in microscope optics, mainly due to infinity correct tubes and new advanced objective lenses, provided images with reduced spherical aberration and free of colour distortions (Davidson & Abramowitz, 1999), which are more suitable to image analysis and consequently to quantitative microscopy.

The colour has always been one of the most important properties used for mineral identification under a microscope (Piller, 1966). Moreover, there are some relevant minerals that are not distinguishable in the SEM, but can be discriminated through their colours in the reflected light microscope, such as, for instance, hematite and magnetite, which are the major iron ore minerals. Hematite and magnetite have similar average atomic numbers, respectively 20.59 and 21.02, and consequently show similar grey levels in BSE images, preventing their discrimination. The segmentation of hematite and magnetite in such kind of images requires a strong image contrast. However, this contrast condition avoids the segmentation of other present phases. In practice, not even SEM-based systems for automated mineralogy can discriminate hematite and magnetite, because the discrimination of these minerals is not possible through EDS due to their similar chemical composition (Gomes & Paciornik, 2008b).

On the other hand, transparent minerals and the embedding resin generally cannot be distinguished by their specular reflectances. For instance, quartz and epoxy resin have practically the same reflectance through the visible light spectrum (Neumann & Stanley, 2008). Actually, this is a classical problem in ore microscopy that renders unfeasible this kind of analysis through reflected light microscopy.

Phase	Chemical formula	Colour on RLM	Average atomic number
Epoxy resin	$C_{21}H_{25}ClO_5$	Dark grey	7.90
Quartz	SiO_2	Dark grey	10.80
Goethite	FeO.OH	Grey / Brown	19.23
Hematite	Fe_2O_3	Light grey	20.59
Pyrite	FeS_2	Pale yellow	20.66
Magnetite	Fe_3O_4	Pinkish grey	21.02
Pentlandite	$(Fe,Ni)_9S_8$	Pale yellow	23.36
Chalcopyrite	$CuFeS_2$	Brass yellow	23.54
Covelline	CuS	Blue	24.64
Bornite	Cu_5FeS_4	Purple	25.34
Sphalerite	ZnS	Grey	25.39
Chalcocite	Cu_2S	Light grey	26.38
Native copper	Cu	Bright yellow	29.00

Table 1. Chemical formula, colour on RLM, and average atomic number of epoxy resin and some minerals.

Figure 2 shows a pair of images of an iron ore sample acquired by reflected light microscopy and SEM. Comparing them, one can observe that the segmentation between quartz and epoxy resin in the BSE image is easy, but it is not viable in the optical image. On the other hand, hematite and magnetite present distinct colours, respectively light grey and pinkish grey, in the optical image, but have practically the same grey level in the BSE image.

Another example of minerals of difficult discrimination can be observed in Figure 3. It shows images of the same field of a copper ore sample acquired by reflected light microscopy and SEM. In the optical image, chalcopyrite can be easily identified by its characteristic brass yellow colour, but pyrite and pentlandite present a very similar colour (pale yellow). On the other hand, in the BSE image, chalcopyrite and pentlandite are practically indistinguishable, due to their similar average atomic numbers (23.54 and 23.36, respectively). Nevertheless, pyrite is slightly darker than pentlandite, because pyrite has a lower average atomic number (20.66).

The RLM-SEM co-site microscopy was developed to overcome these challenges. This methodology can improve the SEM analytical capacity adding specular reflectance (colour) information from RLM. The methodology was employed with some mineral samples, aiming at the discrimination of phases that are not distinguishable by either RLM or SEM, but can be discriminated through the combined use of both techniques.

Fig. 2. Images of an iron ore sample acquired on (a) reflected light microscope and (b) SEM.

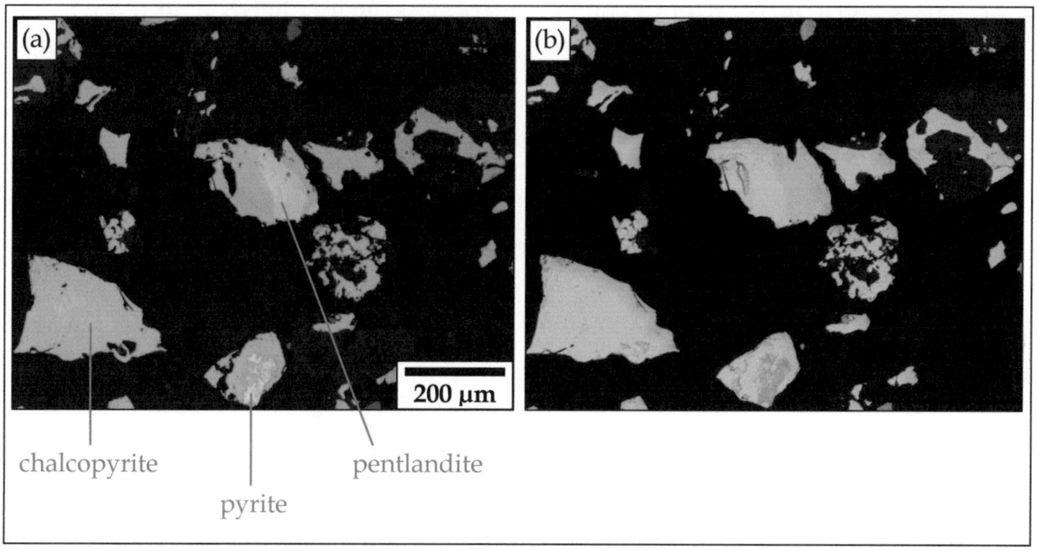

Fig. 3. Images of a copper ore sample acquired on (a) reflected light microscope and (b) SEM.

3. Image registration

Image registration comprises the operation to determine the correspondence point to point between two or more images of the same area (or volume) obtained by different sensors or in different conditions, and the subsequent process of overlaying them.

Image registration is a fundamental procedure in all image analysis tasks in which the final information is gained from the combination of various data sources. Only after the registration, a multi-component image that represents a multimodal database can be properly composed and analyzed.

Typically, image registration is employed for composition and comparison of multi-spectral images in Remote Sensing (Schowengerdt, 1983). It also has several applications in Medicine, such as diagnosis, preparation of surgeries, treatment evaluation, etc. It is used, for instance, for fusion of anatomical and functional information, which are usually obtained through different medical imaging techniques (van den Elsen et al., 1993; Maintz & Viergever, 1998).

There is in the literature a wide variety of image registration methods based on different principles and employed for diverse applications (Zitova & Flusser, 2003; Goshtasby, 2005). Anyway, registration consists in the determination and implementation of a geometric operation (spatial transformation) between images in order to correlate the spatial coordinates of both images. Therefore, the fundamental aspect of any registration method is the spatial transformation used to correctly overlay images. Although many types of variations may be present in images, a suitable transformation must remove only spatial distortions between them (Brown, 1992). Other differences, due to the diversity of information that each image represents, must be maintained, since these are the interesting characteristics that one aims to expose.

In fact, image registration is more complex than a simple image alignment. It is not limited to translation and rotation of images. It may be composed of a combination of six distinct basic transformations: translation, rotation, scale, shear, projection, and other non-linear and local distortions. Figure 4 presents the six basic transformations, showing their effects in a sample base image.

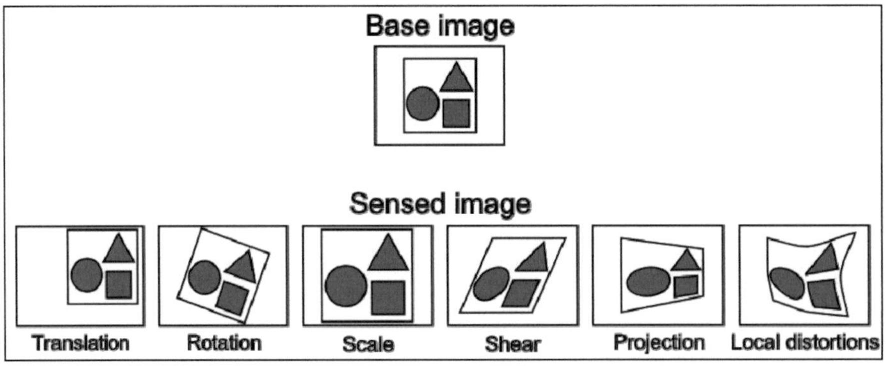

Fig. 4. Basic spatial transformations.

Spatial transformations can convert the coordinates of the sensed image to the coordinates of the base or reference image. Thus, they correlate these digital images pixel by pixel, allowing the assemblage of a multi-component image.

The transformations that involve only translation and rotation are generally called rigid body or Euclidean transformations, since the Euclidean distances within images are preserved (Szeliski, 2004). In contrast, the other ones are classified as non-rigid or elastic. Although this nomenclature is the most commonly found in the literature, including the present text, it is not a consensus. Some authors consider scale as a rigid body transformation too, and there are still others that also include shear and projection in the class of rigid transformations (Crum, et al., 2004).

In this context, multimodal microscopy procedures that are performed intrinsically in a unique microscope constitute probably the simplest cases, generally involving only rigid body transformations. Sometimes it is even possible to acquire images that are directly registered.

Multimodal microscopy methodologies on reflected light microscope can be carried out with or without specimen removal from the stage. In the first case, specimen removal, for instance, for chemical etchings like in a classical metallographic approach, generally imply some displacement between images. Therefore, translation and occasionally rotation corrections are required. Soto et al. (2004) and Paciornik & Gomes (2009) present case studies of multimodal methods that involve specimen removal for chemical etchings.

On the other hand, optical methodologies without specimen removal can be sometimes performed without translation corrections (De-Deus et al., 2007). Nevertheless, there are some exceptions. For instance, Pirard (2004) proposed a multispectral imaging technique applied to ore characterization in which shifts of the order of several pixels occur between images obtained from different wavelengths; and Iglesias et al. (2011) developed a multimodal microscopy methodology based on the combination of cross-polarized and bright field images in which there were small misalignments between them.

The SEM forms an image through scanning its electron beam in a raster across the specimen and then it synchronizes the scanning with a signal from one of its detectors. Thus, in a given field, it can acquire several different images, which are ready to compose a multi-component image without the need for a registration procedure. However, in practice, older equipments usually exhibit some translation between images from different detectors. In this case, a translation correction is not enough to properly register the images, because SEM's generally present complex and non-linear distortions (Goldstein et al., 2003) that must be considered.

In the RLM-SEM co-site microscopy, the registration procedure involves rigid and non-rigid transformations. The specimen handling between the microscopes and the different stages imply that translation and rotation adjustments are necessary. Besides, non-rigid transformations are required due to the complex distortions that occur in images from SEM. Even a fine calibration of the equipment is not capable of preventing them.

A registration procedure consists of a sequence of mathematical operations that determine the suitable spatial transformation and then defines and applies the geometric operation that properly performs the registration. The base of a registration procedure is the kind of information used by its algorithm. Therefore, as stated in the already classical review paper of Zitova & Flusser (2003), there are two main classes of algorithms according to their nature: area-based and feature-based.

Area-based algorithms, also called template matching, estimate the correspondence between images (or parts of them) in order to determine which transformations provide the best correspondence. The correlation between two signals (cross-correlation) is a standard approach to template matching algorithms that can be particularly efficient if it is computed in the frequency domain using the fast Fourier transform (Lewis, 1995). Area-based algorithms are in general simpler than feature-based ones. They are applied mostly in cases involving only rigid and scale transformations. Besides, they are more sensitive to noise in images. For instance, the multimodal methodologies presented by Soto et al. (2004), Paciornik & Gomes (2009), and Iglesias et al. (2011) employed cross-correlation in the frequency domain for the registration of their optical images.

Feature-based algorithms consist of four steps: feature detection, feature matching, mapping function design, and image transformation and resampling (Zitova & Flusser, 2003). Two sets of features, which are salient and distinctive objects such as corners, line intersections, edges, etc., are manually or automatically detected in both base and sensed images. These features are represented by the so-called control points (points themselves, centers of gravity, line endings, etc.). The aim is to find the pairwise correspondence between control points and then to map a suitable transformation from them. Therefore, the sensed image is transformed through the determined mapping function and an appropriate interpolation technique is employed in order to treat non-integer coordinates.

The detection of control points and the determination of their correspondence in base and sensed images are crucial and difficult tasks. The method named Scale Invariant Feature Transform (SIFT), proposed by Lowe (2004), has been shown computationally efficient and robust upon diverse distortions and multisensor cases.

In contrast to the area-based methods, the feature-based ones do not work directly with image intensity values. Control points constitute higher level information. This fact makes feature-based methods suitable for applications in which diverse sensors with different data structures and physical meanings are involved. Besides, it allows registering images with any nature of distortions, including non-linear and local ones (Zitova & Flusser, 2003).

Furthermore, in multimodal microscopy methodologies, an alternative approach can facilitate the determination of control points. By introducing indentation marks in the sample through a microdurimeter, the control points can be properly defined as their centers of gravity. In fact, this method can be useful for the registration of one or few fields in which specific microstructural features are of interest. However, it becomes impractical when the number of fields is large, as usually occurs in ore characterization procedures.

The RLM-SEM co-site microscopy methodology employs a feature-based method for registration that is described in the section 5.3.

4. Image processing and analysis

A typical image processing and analysis sequence comprises the steps of image acquisition, digitization, pre-processing, segmentation, post-processing, feature extraction and classification (Gonzalez & Woods, 2007).

Pre-processing, or image enhancement, is the first step after image digitization and is used to correct basic image defects, normally created during the image acquisition step. Typical operations, at this step, are background correction, edge enhancement and noise reduction. Pre-processing is useful for qualitative reasons, as it improves the visibility of relevant features in the image, but it is even more important to prepare the image for the following step of segmentation.

Segmentation is the technical term used for the discrimination of objects in an image. Segmentation is probably the most complex step in the sequence because it tries to represent computationally a cognitive process that is inherent to the human vision. When we look at an image we use many different inputs to distinguish the objects: brightness, boundaries, specific shapes or textures. Our brains process this information in parallel at high speed, using previous experience. Computers, on the other hand, do not have the same associative power. The recognition of objects in an image is made through the classification of each pixel of the image as pertaining or not to an object.

There are many segmentation methods based on different principles such as thresholding, edge detection, texture analysis, mathematical morphology, etc. Each one is generally more suitable for a specific application. Categorically, there is not an ideal generic method that is always the best one. Some classical references in this area are: Haralick et al. (1973), Otsu (1979), Haralick (1979), Beucher & Lantuéjoul (1979), Marr & Hildreth (1980), Pun (1981), Canny (1986), and Adams & Bischof (1994).

The most common segmentation method is thresholding. It is based on the assumption that pixels pertaining to a given class of objects (e.g. a specific mineral phase) have similar colour or greyscale intensity, and this colour is different from the background and from other classes of objects. In other words, there must be sufficient contrast between different phases in the material. If that is the case, then the segmentation is based on selecting colour/intensity thresholds that represent the various phases.

Noise, uneven illumination, edge effects contribute to degrade the discrimination between phases, and that is why a pre-processing step may be so relevant. Evidently, phase contrast maybe too low, depending on the microscope used, as described before, and that is where combining information from different types of signals becomes critical.

In many situations the results of segmentation contain artefacts, such as spurious objects, touching or partially overlapping objects, etc. A very common artefact in mineralogical images is segmenting a phase together with the edges of a different phase that share the same intensity range. Some of these defects can be minimized with an appropriate pre-processing step, such as delineation (edge enhancement), but many must be corrected after segmentation, in the so-called post-processing step.

Post-processing makes intense use of morphological operators such as erosion, dilation, opening, closing, and more sophisticated functions like the watershed separation method for touching objects (Serra, 1982, 1988). Ideally, the final result is an image in which just the relevant objects are present and separated in groups that correspond to each phase present in the sample. However, as described below, this is rarely the case, and further analysis of the segmented objects must be done to complete the discrimination.

Given a segmented post-processed image containing a set of objects, several measurements are available. Field features such as number of objects and area fraction are some of the simplest ones. Object specific features are more sophisticated and include measurements of size, shape, position, intensity and texture of each object in the image (Friel, 2000). These features are critical for the classification step.

4.1 Multi-component image analysis

Common colour images, generated by either scientific or general-purpose digital cameras, are generally 24-bit RGB images. Actually, they consist of multi-component images composed by three images that represent the primary colours (red, green and blue) with 8-bit quantization each (Orchard & Bouman, 1991). The RGB system is the most common colour representation system employed by cameras and displays. It was developed to be similar with the *tri-stimulus* response of human vision. Figure 5 shows a sample RGB image above its three components, respectively, R, G and B. In this figure, one can observe some samples of how the primary colours are mixed to compose other colours and grey levels.

Fig. 5. A RGB image and its components.

The components of a multi-component image can represent information of any source and physical meaning. It is not necessary that the data within components are correlated. They just must have spatial correspondence pixel-to-pixel, i.e., they must be registered.

Multi-component images with up to three components have the advantage that they can be viewed as RGB colour images in standard image viewer software. In this case, the so-called pseudo-colours denote the properties which are represented in the components. Pseudo-colour images constitute a useful approach for data visualization.

Each component image can be singly processed and analyzed as a common intensity or grey level image. However, this processing should be carefully performed so that the spatial relations within images are not undesirably modified. Geometric operations must be especially avoided.

A multi-component image can also be conceived as a matrix in which each element is a vector, not a scalar value. Each vector represents a pixel, and each element of vector is the value of this pixel in one of the component images. In a RGB image, each pixel consists of a three-element vector that represents the intensity of the three primary colours. Therefore, in a multi-component image, the probability density function becomes multivariate and consequently its histogram of intensities becomes multidimensional. Figure 6 shows a RGB image of a copper ore sample in the reflected light microscope and its bi-dimensional RG histogram.

Image segmentation by thresholding can be generalized to an n-dimension problem. In this case, a threshold becomes a decision boundary, whose form depends on the number of components. One component leads to decision boundaries that are scalar values (thresholds); two components imply that decision boundaries are straight lines; three components involve decision boundaries that are planes; and so on. Nevertheless, there are more sophisticated segmentation methods that offer more complex decision boundaries, such as curves, surfaces, etc. Besides, they can also be discontinuous. In this context, multidimensional pattern recognition techniques arise as potential methods for segmentation of multi-component images.

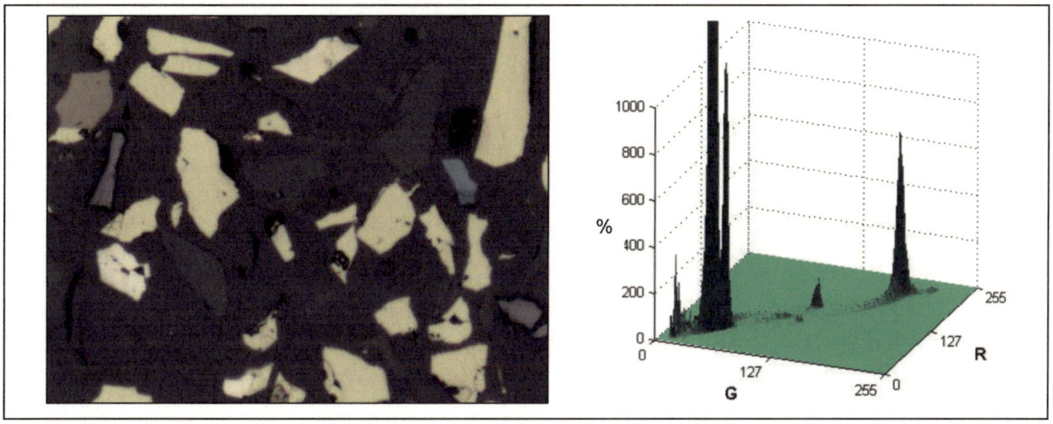

Fig. 6. A RGB image and its bi-dimensional RG histogram.

Pattern recognition is the scientific discipline whose goal is the classification of objects (patterns) into a number of classes from the observation of their characteristics (Theodoridis & Koutroumbas, 2003). It aims to build a simpler representation of a data set through its more relevant features in order to perform its partition into classes (Duda et al., 2001).

Pattern recognition techniques can be used to classify objects (pixels, regions, etc.) within images. Any part of image that has at least one measurable property can be considered as an object and consequently it can be classified. In multi-component images, a pixel consists of a vector in which the elements represent its values in the components.

The classification of pixels is actually an image segmentation procedure. Each class of pixels can properly represent a phase or mineral in an ore microscopy case. If the phases of interest are known, this problem becomes supervised (Duda et al., 2001). In a supervised classification procedure, this known information is exploited so that the classification system learns how the different classes can be recognized.

A supervised classification procedure involves three stages: training, validation and classification. The training stage comprises sampling of known pixels of each class in order to compose the so-called training set that is used as knowledge base. Therefore, the classifier is trained, i.e., it is designed based on the known information. Following, in the validation stage, another known set of pixels, the validation set, is classified aiming to estimate the performance of the classifier, considering its generalization capacity (Toussaint, 1974). If the validation stage indicates that the training was successful, the classification is then possible and consequently the segmentation procedure is implemented.

It is worth to mention that although the RGB system is vastly predominant in image acquisition devices, it is not generally the most appropriate colour representation system for classification purposes because its three components (R, G and B) are very correlated, due to their strong dependence from the light intensity (Littmann & Ritter, 1997). Besides, it doesn't represent colours in a uniform scale, preventing measurements of similarity between colours through their distance in RGB space (Cheng et al., 2001). There are many colour systems described in the literature (Sharma & Trussell, 1997). Actually, one can define any colour system from linear or non-linear transformations of RGB (Vandenbroucke et al., 2003). Systems like rgb, HSI, L*a*b* and L*u*v* present less correlated features and consequently tend to be more suitable for classification (Gomes & Paciornik, 2008a).

5. Combining reflected light microscopy and SEM

This section describes the RLM-SEM co-site microscopy methodology by reviewing two case studies, in which it was applied for the characterization of a copper ore and an iron ore. These case studies were originally presented by Gomes & Paciornik (2008a and 2008c, respectively).

The RLM-SEM co-site microscopy comprises four sequential steps: image acquisition in RLM; image acquisition in SEM; registration; and image processing and analysis. The three first steps consist of generic routines of the methodology, but the image processing and analysis procedure depends on the case study. The segmentation of minerals in both case studies was performed through supervised classification of pixels, exploiting their multidimensional nature. However, used features and classification methods differ.

Ore microscopy procedures commonly involve acquisition and analysis of tens to hundreds of images per cross-section in order to provide a representative sampling. Therefore, in the development of the RLM-SEM co-site microscopy, automatic routines for field scanning and image acquisition were implemented for both used microscopes.

5.1 Image acquisition in reflected light microscopy

A motorized and computer controlled microscope with a digital camera (1300 x 1030 pixels) was used. A routine was implemented for microscope and camera control, and for image acquisition. It integrates and automates many common procedures such as specimen x-y scanning, automatic focusing, background correction and imaging.

The following image acquisition procedures and conditions were employed:

a. Before image acquisition, a SiC reflectivity standard was used to generate background images for each objective lens, which were subsequently employed for automatic background correction (Pirard et al., 1999) of every acquired image.
b. Illumination was kept constant by direct digital control of the lamp voltage.
c. Camera sensitivity, exposure and white balance were optimized initially for a representative field of the sample and kept constant there on.
d. Objective Lenses: 5X (NA 0.13); 10X (NA 0.20); 20X (NA 0.40), leading to resolutions of 2.11, 1.05, and 0.53 μm/pixel, respectively.
e. Single fields regularly spaced on the sample were imaged through specimen scanning with a motorised stage and automatic focusing.
f. Each field position was recorded in a database for subsequent image acquisition on in SEM.
g. All images were acquired at 24 bit RGB colour quantisation.

5.2 Image acquisition in SEM

A digital SEM was used to acquire a BSE image (1024 x 768 pixels) of each field imaged on RLM. In this procedure, the sample must be placed in the SEM stage at a similar arrangement as positioned in the RLM stage. It is unnecessary and impractical to place the sample in the exact same way, but a similar arrangement can make image registration easier and faster.

The magnification of the SEM was set to keep similar resolutions to optical images. Other SEM operational parameters were manually tuned for a representative field of the sample and then kept constant. After that, the field positions database was loaded with the acquisition routine developed in the SEM control software. It converts RLM stage coordinates to SEM stage coordinates and subsequently performs automatic specimen scanning and image acquisition. Thus, respectively for copper ore and iron ore samples, 121 (11 x 11) and 81 (9 x 9) fields per cross-section were imaged with the RLM and the SEM. Figure 7 presents a field of the copper ore sample as imaged on RLM and SEM.

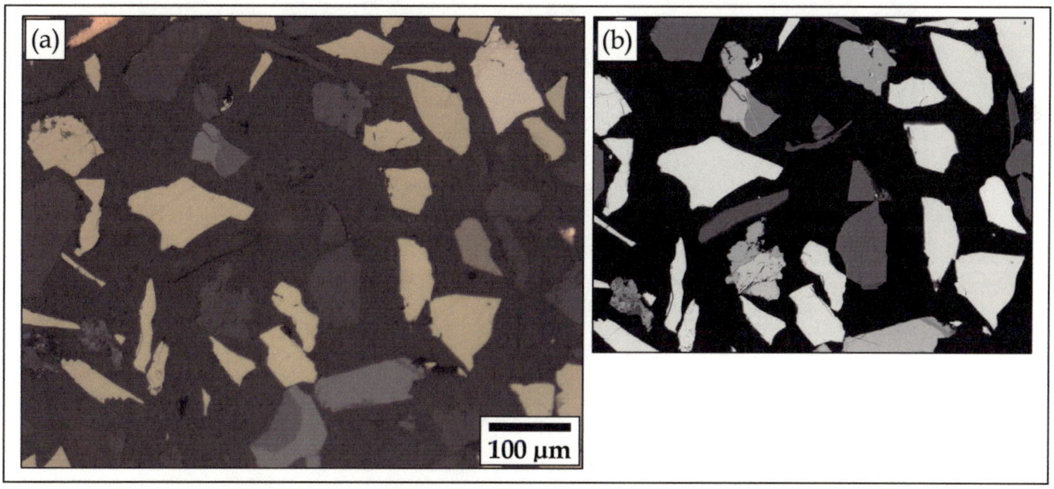

Fig. 7. Images of a field of the copper ore sample obtained by (a) RLM and (b) SEM.

5.3 Image registration

An automatic method for the registration of RLM and SEM images was developed. The distortions were considered according to their sources, and the registration procedure was accomplished through sequential steps. The first step comprises distortions from the SEM, such as astigmatism and local distortions. The second step adjusts rotation and the third one corrects translation. At the end, the registered images are cropped to represent exactly the same field.

The first step is carried out through a feature-based registration algorithm. It maps the SEM characteristic distortions based on several control points that are automatically detected. These distortions do not depend on samples. They are a function of SEM operational parameters, such as magnification and working distance. Therefore, this step was employed only one time for each SEM set-up, i.e., once for each magnification whose pixel size corresponds to the pixel size obtained through one of the objective lenses of the RLM.

A standard specimen with regular distributed and easily extractable control points must be imaged on RLM and SEM. In the present case, two copper grids (200 mesh for 5X and 10X lenses, and 400 mesh for 20X lens) were used. These images were analyzed by an automatic routine in order to determine the centroid (center of gravity) of each grid hole, the control points. Then, a suitable spatial transformation was computed from the pairs of control points using the local weighted mean method proposed by Goshtasby (1988). Therefore, this spatial transformation was applied to just remove distortions in every SEM image of the ore sample.

Figure 8 shows the pairs of control points superimposed to the RLM image of the grid (10X). The white circles represent the control points extracted from the RLM image, and the white dots are the points obtained from the corresponding SEM image. As can be observed, circles and points are not aligned. Besides, the misalignment varies, evidencing the complexity of these distortions.

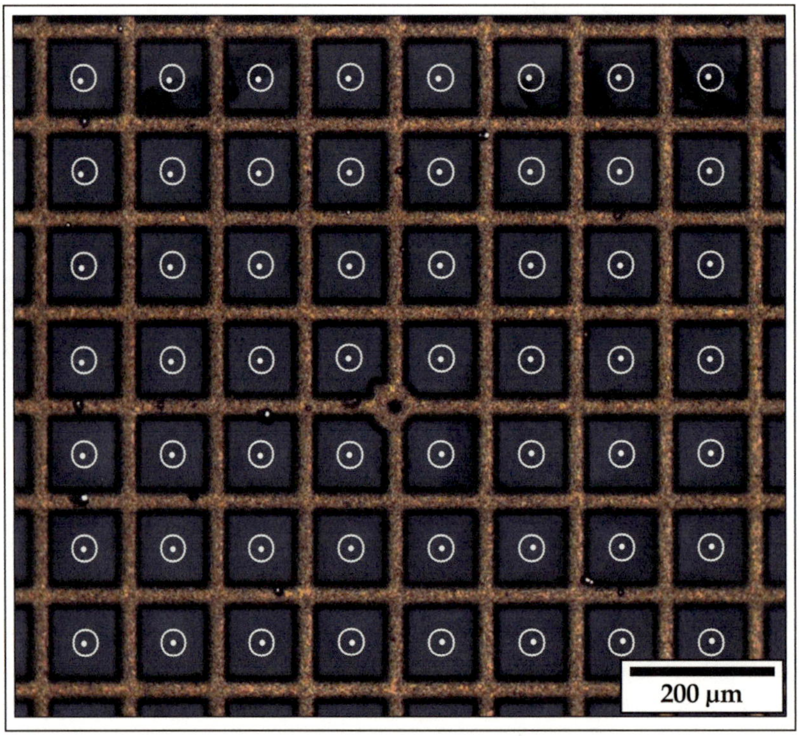

Fig. 8. Pairs of control points superimposed on the RLM image of the grid (10X). The white circles represent the control points extracted from the RLM image, and the white dots are the control points obtained from the corresponding SEM image.

The second step of the registration method aims at finding the rotation angle between images and subsequently to adjust rotation. This angle is due to sample manipulation and its different arrangement in the sample holders of the microscopes. Thus, this rotation is constant in all fields of a sample in a given experiment.

An iterative algorithm is used to determine the angle that maximizes the normalized cross correlation between a pair of images. The algorithm uses coarse-to-fine approach. It evolves in order to adjust the angle down to 0.01° of precision. This procedure is applied for one pair of images and the obtained angle is used to correct rotation in all SEM images.

After the second step, the SEM images are free of distortions and they are put in the same coordinate system of the RLM images. Therefore, only translation problems remain. Thus, in the third step, the RLM and the SEM images are finally registered through the maximization of normalized cross correlation. At the end, they are cropped to represent exactly the same field. Figure 9 shows the images present in Figure 7, after registration.

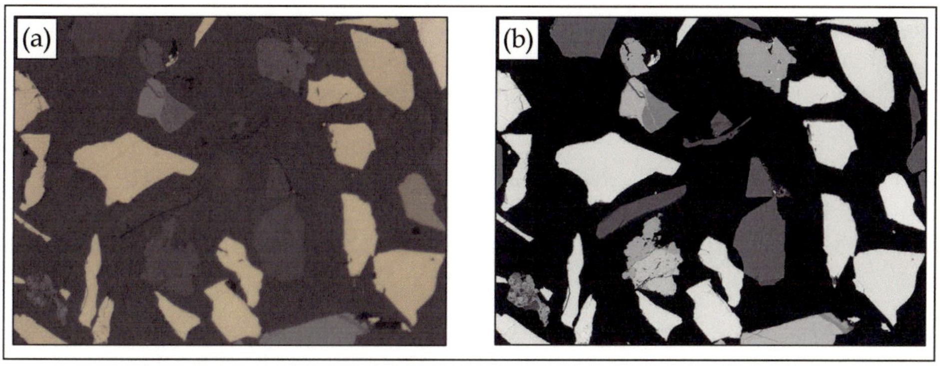

Fig. 9. (a) RLM and (b) SEM images from Figure 7, after registration.

5.4 Image processing and analysis

The registered SEM and RLM images went through a pre-processing step of delineation to reduce the well-known halo effect, making them more suitable for the subsequent segmentation procedures. Figure 10 shows a detail of an image obtained by RLM, before and after delineation operation. Comparing them, one can observe that delineation improves phase transitions and consequently allows better segmentation results.

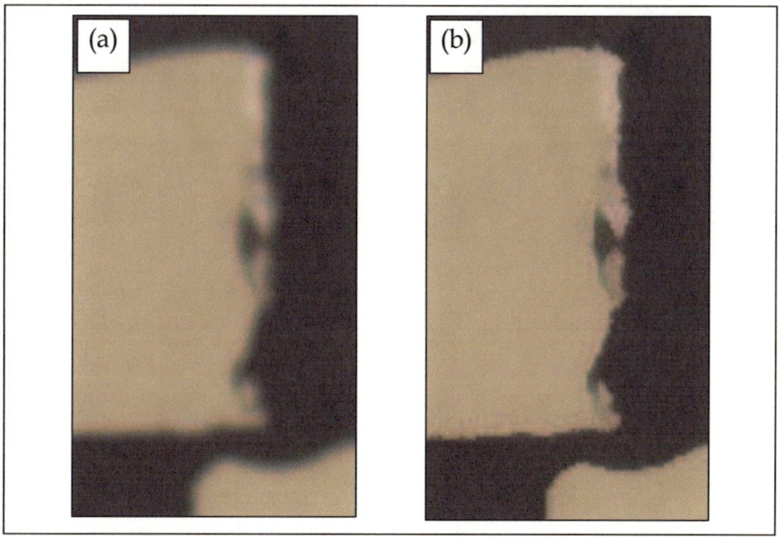

Fig. 10. Delineation of a RLM image: (a) detail of the original image; (b) after delineation.

5.4.1 Copper ore

The copper ore sample had a complex mineralogy mainly composed of thirteen minerals, which were then taken as individual classes (quartz, three different silicates, apatite, magnetite, pentlandite, chalcopyrite, covelline, bornite, sphalerite, chalcocite, and native copper). Besides mineral phases, epoxy resin was taken as a class too. Thus, the training stage involved sampling of pixels from the fourteen classes. In practice, 6000 pixels of each phase were interactively selected from several images.

The delineated RLM images were converted from RGB to the rgb, HSI, L*a*b* and L*u*v* colour systems with the goal of revealing colour information hidden by the correlated RGB system. These conversion operations generate ten new components (r, g, b, H, S, I, a*, b*, u*, v*), increasing the system dimensionality from four to fourteen.

A Bayes classifier (Duda et al., 2001) was employed and the fourteen components (BSE intensity and the thirteen colour components) were used as features. The validation was carried out through holdout estimate (Toussaint, 1974), reaching a success rate larger than 99.5%.

The classification result was a grey level image per field in which each phase was represented by a distinct grey level. Thus, pixels classified as phase one have intensity one, pixels recognised as phase two have intensity two, and so on. Therefore, a suitable look-up table can be applied in order to attribute a different pseudo-colour to each phase and consequently to make their visualisation easier. Figure 11 presents the classification result for the images shown in Figure 9 and its look-up table.

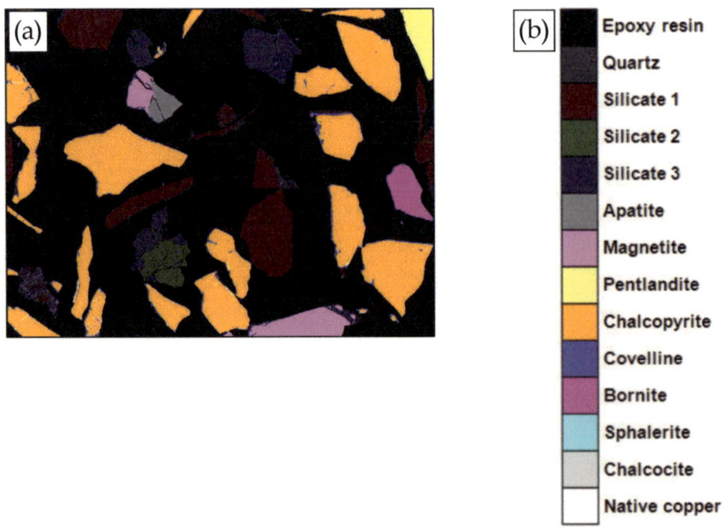

Fig. 11. The classification result for the images shown in the Figure 9: (a) segmented image; and (b) look-up table.

5.4.2 Iron ore

The mineralogical assemblage of the iron ore sample was simple. It was mainly composed by hematite, magnetite, goethite and quartz. Therefore, in this case study, five classes were considered (epoxy resin, quartz, goethite, haematite, and magnetite).

The segmentation process was split in two supervised classification procedures. The first one recognised epoxy resin, quartz, goethite, and a hematite-magnetite composed phase through the classification of pixels in SEM images, using the BSE intensity as feature in a Bayes classifier. Then, the second classification procedure was able to discriminate hematite and magnetite. It was carried out through the classification of pixels in RLM-SEM composed images. In this case, their four components (R, G, B, and BSE intensity) were used as features and a Bayes classifier was employed. The training stage for both classification procedures comprised interactive sampling of 1000 pixels of each one of the five classes from a RLM-SEM composed image. Figure 12 presents the segmentation result for the images shown in Figure 2 and its applied look-up table.

Fig. 12. The classification result for the images shown in the Figure 2: (a) segmented image; and (b) look-up table.

5.5 Discussion

The information from RLM and SEM presents different structures and physical meanings. RLM data consists of a vector of three 8-bit values, which denote specular reflectance in the RGB colour system. On the other hand, SEM data is represented as 8-bit values of BSE intensity that provides average atomic number contrast. Thus, it is very difficult to compute suitable measurements of similarity between patterns (pixels) that can be used to recognise the classes. In this kind of problem, Valev and Asaithambi (2001) pointed out that different classifiers can be used to complement one another. This is the approach employed for the segmentation in the iron ore case study.

The increase of dimensionality carried out in the copper ore case study can make the classification task easier, since hidden information is discovered and consequently patterns are better described. However, the training data must grow exponentially with the dimensionality in order to prevent the so-called curse of dimensionality (Marques de Sá,

2001). In practice, this is not an issue for pixels classification as a typical image of 1024 x 768 pixels, for instance, has about 0.8 million of pixels, and it is easy to obtain several thousands of pixels of each class.

The resulting segmented images in both case studies reveal small amounts of misclassified pixels in borders between phases. It occurs mainly due to little cracks and relief, in spite of the good sample preparation and the delineation pre-processing. This misclassification is quite small and it can be negligible in mineralogical identification and quantification procedures. However, it becomes more significant for microstructural characterisation purposes, such as mineral liberation analysis. Therefore, post-processing routines should be developed.

6. Conclusion

Multimodal microscopy extends the capabilities of traditional microscopy techniques, improving the discrimination of mineral phases in ores. By combining Optical Microscopy and Scanning Electron Microscopy it takes advantage of the complementary contrasts provided by these techniques.

This method relies on microscope automation, digital image acquisition, processing and analysis. Over the last years many of these techniques have become readily available in both commercial and free software environments.

The use of supervised classification methods relies on operator experience, during the training stage, but once the classifier is optimized and validated, the effective classification of unknown samples is fully automatic and fast.

The developed method is applicable to other materials for which individual microscopy techniques do not provide enough discrimination between the relevant phases.

7. Acknowledgment

One of the authors (S. Paciornik) acknowledges the support of CNPq, the Brazilian Research Council.

8. References

Adams, R. & Bischof, L. (1994). Seeded region growing. *IEEE Transactions on Pattern Analysis and Machine Intelligence*, Vol. 16, No. 6, (June 1994), pp. 641–647, ISSN 0162-8828

Beucher, S. & Lantuéjoul, C. (1979). Use of watersheds in contour detection, *Proceedings of International Workshop on Image Processing, Real-time Edge and Motion detection/estimation*, pp. 2.1-2.12, Rennes, France, September 17-21, 1979

Brown, L.G. (1992). A Survey of Image Registration Techniques. *ACM Computing Surveys*, Vol. 24, No. 4, (December 1992), pp. 325-376, ISSN 0360-0300

Canny, J. (1986). A computational approach to edge detection. *IEEE Transactions on Pattern Analysis and Machine Intelligence,* Vol. 8, No. 6, (November 1986), pp. 679–698, ISSN 0162-8828

Cheng, H.D.; Jiang, X.H.; Sun, Y. & Wang, J. (2001). Color image segmentation: advances and prospects. *Pattern Recognition*, Vol. 34, No. 12, (December 2001), pp. 2259-2281, ISSN 0031-3203

Criddle, A.J. & Stanley, C.J. (1993). *Quantitative Data File for Ore Minerals* (3rd Edition), Chapman & Hall, ISBN 041246750X, London, UK

Craig, J.R. & Vaughan, D.J. (1994). *Ore microscopy and ore petrography* (2nd Edition), John Wiley & Sons, ISBN 0471551759, New York, USA

Crum, W.R.; Hartkens, T. & Hill, D.L.G. (2004). Non-rigid image registration: theory and practice. *The Britsh Journal of Radiology*, Vol. 77, (December 2004), pp. S140-S153, ISSN 0007-1285

Davidson, M.W. & Abramowitz, M. (1999). *Optical microscopy,* The Florida State University, Retrieved from: <http://micro.magnet.fsu.edu/primer/opticalmicroscopy.html>

De-Deus, G.; Reis, C.M.; Fidel, R.A.S.; Fidel, S.R.; Paciornik, S. (2007). Co-site digital optical microscopy and image analysis: an approach to evaluate the process of dentine demineralization. *International Endodontic Journal*, Vol. 40, No. 6, (June 2007), pp. 1365-2591, ISSN 1365-2591

Duda, R.O.; Hart, P.E. & Stork, D.G. (2001). *Pattern classification* (2nd Edition), Wiley-Interscience, ISBN 0-471-05669-3, New York, USA

Friel, J.J. (2000). Measurements, In *Practical guide to image analysis*, pp. 101-128, ASM International, ISBN 0871706881, Materials Park, USA

Galopin, R. & Henry, N.F.M. (1972). *Microscopic study of opaque minerals*, W. Heffer and Sons Ltd., ISBN 0852700474, Cambridge, UK

Goldstein, J.I.; Newbury, D.E.; Echlin, P.; Joy, D.C.; Lyman, C.E.; Fiori, C.; Lifshin , E.; Sawyer, L. & Michael, J.R. (2003). *Scanning Scanning Electron Microscopy and X-ray Microanalysis* (3rd Edition), Springer, ISBN 0306472929, New York, USA

Gomes, O.D.M. & Paciornik, S. (2008a). Co-site microscopy: combining reflected light and scanning electron microscopy to perform ore mineralogy, *ICAM 2008 - Ninth International Congress for Applied Mineralogy Conference Proceedings*, pp. 695-698, ISBN 9781920806866, Brisbane, Australia, September 8-10, 2008

Gomes, O.D.M. & Paciornik, S. (2008b). Iron ore quantitative characterization through reflected light-scanning electron co-site microscopy, *ICAM 2008 - Ninth International Congress for Applied Mineralogy Conference Proceedings*, pp. 699-702, ISBN 9781920806866, Brisbane, Australia, September 8-10, 2008

Gomes, O.D.M. & Paciornik, S. (2008c). RLM-SEM co-site microscopy applied to iron ore characterization, *3rd International Meeting on Ironmaking and 2nd International Symposium on Iron Ore Conference Proceedings*, pp. 218-224, ISBN 9788577370320, São Luís, Brazil, September 22-26, 2008

Gonzalez, R.C. & Woods, R.E. (2007). *Digital Image Processing* (3rd Edition), Prentice-Hall, ISBN 013168728X, Upper Saddle River, USA

Goshtasby, A. (1988). Image Registration by Local Approximation Methods. *Image and Vision Computing*, Vol. 6, No. 4, (November 1988), pp. 255-261, ISSN 0262-8856

Goshtasby, A. (2005). *2-D and 3-D image registration for medical, remote sensing, and industrial applications*, John Wiley & Sons, ISBN 0471649546, Hoboken, USA

Gribble, C. & Hall, A.J. (1992). *Optical Mineralogy: Principles and Practice,* UCL Press, ISBN 185728013X, London, UK

Gu, Y. (2003). Automoated Scanning Electron Microscope Based Mineral Liberation Analysis - An Introduction to JKMRC/FEI Mineral Liberation Analyser. *Journal of Minerals and Materials Characterisation and Engineering,* Vol. 2, No. 1, pp. 33-41, ISSN 1539-2511

Haralick, R.M.; Shanmugam, K.; Dinstein, I. (1973). Textural features for image classification. *IEEE Transactions on Systems, Man, and Cybernetics,* Vol. 3, No. 6, (November 1973) pp. 610-621, ISSN 0018-9472

Haralick, R.M. (1979). Statistical and Structural Approaches to Texture. *Proceedings of the IEEE,* Vol. 67, No. 5, (May 1979), pp. 786-808, ISSN 0018-9219

Iglesias, J.C.A.; Gomes, O.D.M. & Paciornik, S. (2011). Automatic recognition of hematite grains under polarized reflected light microscopy through image analysis. *Minerals Engineering,* Vol. 24, No. 12, (October 2011), pp. 1223-1378, ISSN 0892-6875

Jones, M.P. (1987). *Applied mineralogy: a quantitative approach,* Graham and Trotman Ltd., ISBN 0860105113, London, UK

Lewis, J.P. (1995). Fast Template Matching, *Vision Interface 95,* pp. 120-123, Quebec City, Canada, May 15-19, 1995

Littmann, E. & Ritter, H. (1997). Adaptive color segmentation - a comparison of neural and statistical methods. *IEEE Transactions on Neural Networks,* Vol. 8, No. 1, (January 1997), pp. 175-185, ISSN 1045-9227

Lowe, D.G. (2004). Distinctive Image Features from Scale-Invariant Keypoints. International Journal of Computer Vision, Vol. 60, No. 2, (November 2004), pp. 91–110, ISSN 1573-1405

Maintz, J.B.A. & Viergever, M.A. (1998). A survey of medical image registration. *Medical Image Analysis,* Vol. 2, No. 1, (March 1998), pp. 1-36, ISSN 1361-8415

Marques de Sá, J.P. (2001). *Pattern Recognition: Concepts, Methods and Applications,* Springer, ISBN 3540422978, Berlin, Germany

Marr, D. & Hildreth, E. (1980). Theory of Edge Detection. *Proceedings of the Royal Society of London. Series B, Biological Sciences,* Vol. 207, No. 1167, (February 1980), pp. 187-217, ISSN 00804649

Neumann, R. & Stanley, C.J. (2008). Specular reflectance data for quartz and some epoxy resins: implications for digital image analysis based on reflected light optical microscopy, *ICAM 2008 - Ninth International Congress for Applied Mineralogy Conference Proceedings,* pp. 703-706, ISBN 9781920806866, Brisbane, Australia, September 8-10, 2008

Orchard, M.T.; Bouman, C.A. (1991). Color Quantization of Images. *IEEE Transactions on Signal Processing,* Vol. 39, No. 12, (December 1991), pp. 2677-2690, ISSN 1053-587X

Otsu, N. (1979). A Threshold Selection Method from Gray-Level Histograms. *IEEE Transactions on Systems, Man, and Cybernetics,* Vol. 9, No. 1, (January 1979), pp. 62-66, ISSN 0018-9472

Paciornik, S. & Gomes, O.D.M. (2009). Co-site Microscopy: Case Studies. *Praktische Metallographie,* Vol. 46, No.9, (September 2009), pp. 483-498, ISSN 0032-678X

Paciornik, S. & Maurício, M.H.P. (2004). Digital Imaging, In *ASM Handbook, Volume 9: Metallography and Microstructures*, G.F. Vander-Voort (Ed.), pp. 368-402, ASM International, ISBN 0871707063, Materials Park, USA

Petruk, W. (1988). The capabilities of the microprobe Kontron image analysis system: application to mineral benefication. *Scanning Microscopy*, Vol. 2, No. 3, pp. 1247-1256, ISSN 0891-7035

Piller, H. (1996). Colour measurements in ore-microscopy. *Mineralium Deposita*, Vol. 1, No. 3, (December 1966), pp. 175-192, ISSN 0026-4598

Pirard, E.; Lebrun, V. & Nivart, J.-F. (1999). Optimal Acquisition of Video Images in Reflected Light Microscopy. *Microscopy and Analysis*, Vol. 60, (July 1999), pp. 9–11, ISSN 0958-1952

Pirard, E. (2004). Multispectral imaging of ore minerals in optical microscopy. *Mineralogical Magazine*, Vol. 68, No. 2, (April 2004), pp. 323–333, ISSN 0026-461X

Pun, T. (1981). Entropic thresholding, a new approach. *Computer Graphics and Image Processing*, Vol. 16, No. 3, (July 1981), pp. 210-239, ISSN 0146-664X

Reimer, L. (1998). *Scanning Electron Microscopy: Physics of Image Formation and Microanalysis* (2nd Edition), Springer-Verlag, ISBN 3540639764, Berlin, Germany.

Schowengerdt, R.A. (1983). *Techniques for image processing and classification in remote sensing*, Academic Press, ISBN 0126289808, Orlando, USA

Serra, J. (1982). Image Analysis and Mathematical Morphology, Academic Press, ISBN 0126372403, London, UK

Serra, J. (1988). Image Analysis and Mathematical Morphology, Volume 2: Theoretical Advances, Academic Press, ISBN 0126372411, London, UK

Sharma, G. & Trussell, H.J. (1997). Digital color imaging. *IEEE Transactions on Image Processing*, Vol. 6, No. 7, (July 1997), pp. 901-932, ISSN 1057-7149

Soto, O.A.J.; Gomes, O.D.M.; Pino, G.A.H. & Paciornik, S. (2004). Native Copper Analysis through Digital Microscopy, In *Applied Mineralogy: Developments in Science and Technology*, M. Pecchio, F.R.D. Andrade, L.Z. D'Agostino, H. Kahn, L.M. Sant'Agostino, M.M.M.L. Tassinari (Eds.), Vol. 2, pp. 1043-1046, ISBN 859865602X

Sutherland, D. & Gottlieb, P. (2001). Application of automated quantitative mineralogy in mineral processing. *Minerals Engineering*, Vol. 4, No. 7-11, pp. 753-762, ISSN 0892-6875

Szeliski, R. (2004). *Image alignment and stitching: a tutorial*, Technical Report MSR-TR-2004-92, Microsoft Research, December 2004.

Theodoridis, S. & Koutroumbas, K. (2003). *Pattern Recognition* (2nd edition), Academic Press, ISBN 0-12-685875-6, London, UK

Toussaint, G.T. (1974). Bibliography on Estimation of Misclassification. *IEEE Transactions on Information Theory*, Vol. 20, No. 4, (July 1974), pp. 472-479, ISSN 0018-9448

Valev, V. & Asaithambi, A. (2001). Multidimensional pattern recognition problems and combining classifiers. *Pattern Recognition Letters*, Vol. 22, No. 12, (October 2001), pp. 1291-1297, ISSN 0167-8655

Vandenbroucke, N.; Macaire, L. & Postaire, J.G. (2003). Color image segmentation by pixel classification in an adapted hybrid color space. Application to soccer image analysis. *Computer Vision and Image Understanding,* Vol. 90, No. 2, (May 2003), pp. 190-216, ISSN 1077-3142

van den Elsen, P.A.; Pol, E.J.D. & Viergever, M.A. (1993). Medical Image Matching - A Review with Classification. *IEEE Engineering in Medicine and Biology,* Vol. 12, No. 1, (March 1993), pp. 26-39, ISSN 0739-5175

Zitova, B. & Flusser, J. (2003). Image registration methods: a survey. *Image and Vision Computing,* Vol. 21, No. 11, (October 2003), pp. 977-1000, ISSN 0262-8856

17

SEM Analysis of Precipitation Process in Alloys

Maribel L. Saucedo-Muñoz,
Victor M. Lopez-Hirata and Hector J. Dorantes-Rosales
Instituto Politecnico Nacional (ESIQIE),
Mexico

1. Introduction

The microstructural characterization of the precipitation process in alloys is a very important aspect in order to understand the formation mechanism and growth kinetics of precipitated phases during its heating because of either the heat treating process or the operation-in-service conditions. Additionally, the microstructure control is a key point to know the degree of hardening after heat treating of the alloys and to assess their mechanical properties after a prolonged exposure at high temperature during the operation of an industrial component. There are different characterization techniques for microstructure; however, the use of the scanning electron microscopy, SEM, has been very popular for the microstructural observation and it has become a power tool for characterization of the phase transformations. Besides, the application of energy-dispersed-spectra, EDS-SEM system to the microstructural characterization has permitted to know not only the morphology of phases, sizes, distribution and then growth kinetics, but also their chemical composition and thus element distribution of the formed phases. Thus this chapter shows the application of SEM-EDS system to the characterization of microstructural of precipitation process in different alloy systems such as Fe-Ni-Al alloy, austenitic stainless steels and Mg-Zn-Al alloy.

2. Precipitation in alloys

Phase separation in alloys usually consists of the formation of a supersaturated solid solution by heating the alloy at temperatures higher than the equilibrium solvus line and subsequently quenched rapidly. This supersaturated solid solution can usually be separated in two or more phases as a result of the isothermal aging at temperatures lower than that of equilibrium. Phase separation can mainly take place by two mechanisms, nucleation and growth, and spinodal decomposition (Porter, 2009). The former mechanism consists of the formation of a stable nucleus with a nucleation barrier to overcome and it is characterized by an incubation period. In contrast, the latter one is initiated by the spontaneous formation and subsequent growth of coherent composition fluctuations. The formation of fine second-phase dispersion in a matrix promotes its hardening, known as precipitation hardening. If the aging of alloys continues, it is expected that larger precipitates will grow at the expense of smaller ones which dissolve again given rise to a change in the precipitate size distribution (Kostorz, 2005).

2.1 Coarsening process in Fe-Ni-Al alloys

The precipitation of the β´phase is important for strengthening at high temperatures in different engineering ferritic alloys such as, PH stainless steels, nitralloy, Fe-Cr-Ni-Al based alloys, etc. These alloys are used in industrial components which require good mechanical strength and oxidation resistance at high temperatures. The β´phase is an ordered phase of the B2 type crystalline structure (Sauthoff, 2004). The coarsening resistance of precipitates is a key factor to keep the high strength at high temperatures in this type of alloys. An alternative to have a good coarsening resistance, it is to have a low value of lattice misfit which maintains a coherent interface between the precipitate and matrix (Kostorz, 2005). Thus, this section shows the effect of structural and morphological characteristics of the β´ precipitates on the coarsening behavior during the isothermal aging of an Fe-10Ni-15Al alloy.

2.1.1 Experimental details

An Fe-10Ni-15Al alloy (wt. %) was melted using pure metallic elements in an electrical furnace under an argon atmosphere. The ingot of 30 x 10 x 10 mm was encapsulated in a quartz tube with argon gas and then homogenized at 1100 °C for one week. Specimens were solution treated at 1100 °C for 1 h and subsequently aged at temperatures of 750, 850 and 920 °C for times from 0.25 to 750 h. These samples were also observed with a SEM analysis with EDS detector at 20 kV. Vickers hardness was tested for the aged specimens using a load of 100 g.

2.1.2 Microstructural evolution of coarsening

SEM micrographs of precipitates are shown for the sample aged at 750 and 920 °C for different times in Figs. 1 (a-c) and (d-f), respectively. The shape of the β´ precipitates was round without any preferential alignment for the aging at 750 °C up to 75 h and 920 °C up to 0.5 h, Figs 1 (a). A further aging changed the shape of the β´ precipitates to cuboids with a preferential alignment on the <100> directions of the ferritic α phase, Figs. 1 (c-e). A prolonged aging at 920 °C promoted the change of shape to rectangular plates also aligned in the <100> directions, Fig. 1 (f). The volume percentage of precipitation was determined to be about 30, 25 and 20 % for the samples aged at 750, 850 and 920 °C, respectively.

2.1.3 Growth kinetics of coarsening

The variation of the β´ precipitates size expressed as r^3-r_o^3 with aging time for the sample aged at 750, 850 and 920 °C is shown in Fig. 2. It can be noticed that the experimental data fit to a straight line for each temperature. Thus the growth kinetics of coarsening followed the behavior predicted by the Lifshitz-Slyozov-Wagner (LSW) theory for coarsening controlled by volume diffusion. This fact shows a good agreement with the modified theory for the diffusion-controlled coarsening in ternary alloys (Kostorz, 2005) which predicts that growth kinetics is similar to that of LSW theory. The size distribution of precipitates is shown in Figs. 3 (a-c) for the sample aged at 750, 850 and 920 °C for 200 h, respectively. It can be seen that the size distribution is broader and lower than that predicted by the LSW theory because of the high volume fraction of precipitates, which has been reported in the coarsening process of several alloy systems. It has been observed that the growth or shrinkage rate of an individual particle depends not only on its normalized radii but also on its local environment. That is, a particle surrounded by several larger particles will grow slower, or shrink faster, than a particle of the same size whose neighbors are smaller. Thus, as the volume fraction increased, the particle size distribution widened increasing the coarsening rate at the same time. It was also observed that the higher aging temperature, the faster coarsening kinetics of the β´ precipitates because of the increase in volume diffusion (Ratke & Vorhees, 2002).

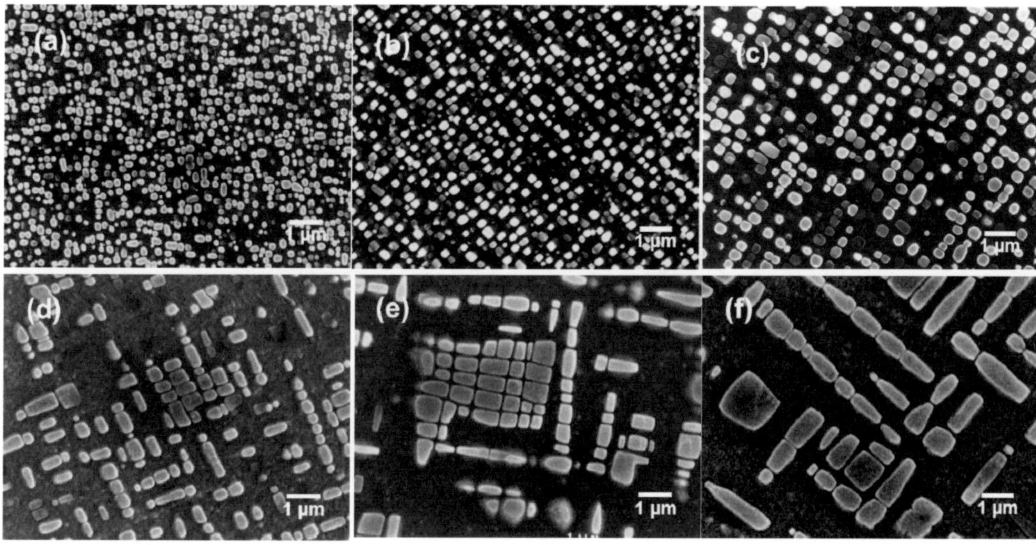

Fig. 1. SEM micrographs for the Fe-10Ni-15Al alloy aged at 750 °C for (a) 75, (b) 250, and (c) 500 h, and at 920 °C for (d) 25, (e) 100 and (f) 200 h.

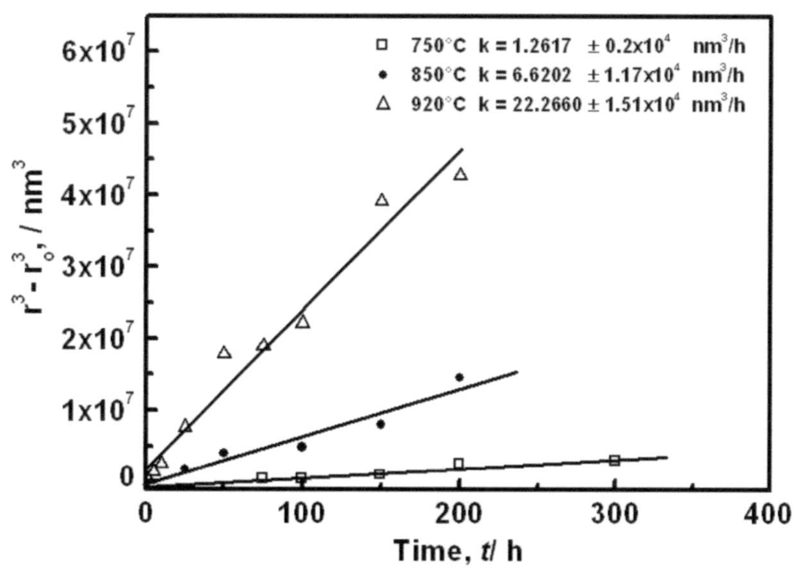

Fig. 2. Plot of r^3-r_o^3 vs. aging time for the Fe-10Ni-15Al alloy aged at 750, 850 and 920 °C.

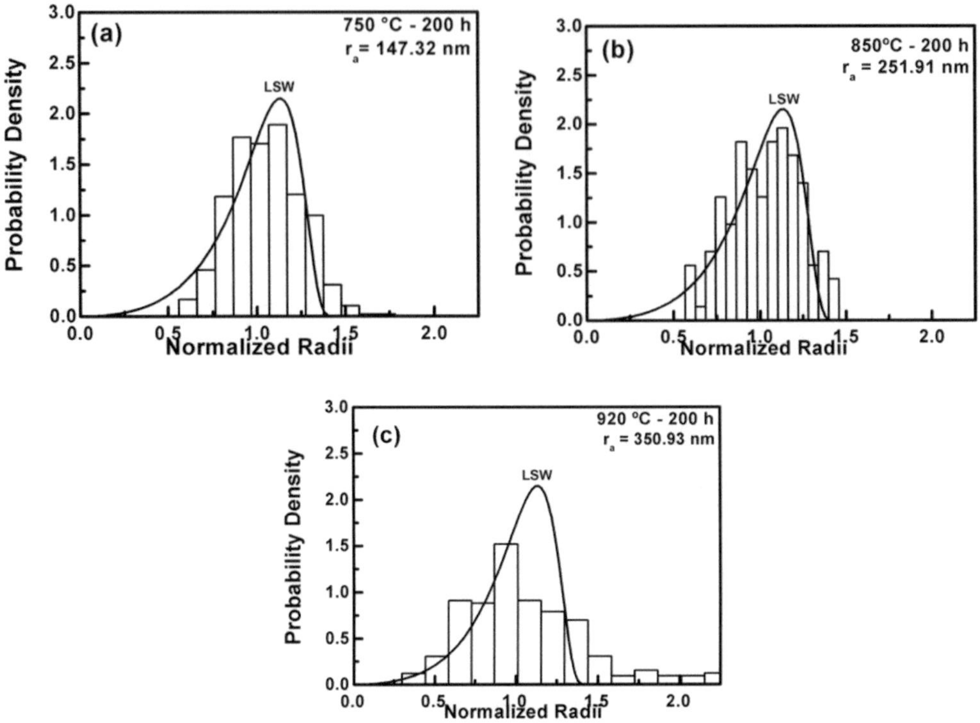

Fig. 3. Size distribution of precipitates for the Fe-10Ni-15Al alloy aged at (a) 750, (b) 850 and (c) 920 °C for 200 h.

2.1.4 Hardening behavior

Figure 4 shows the aging curves for the sample aged at 750 and 920 °C. A higher hardness can be noticed in the sample aged at 920 °C. This can be attributed to the morphology and alignment of β´ precipitates. That is, they are cuboids aligned in the <100> crystallographic directions of the ferritic matrix. A similar hardening behavior was observed in Fe-Ni-Al alloys aged at lower temperatures, 500 °C (Soriano-Vargas et al. 2010, Cayetano-Castro et al. 2008). In contrast, the precipitates are rounded particles without any preferential crystallographic alignment for aging at 750 °C up to 75 h. Besides, the size of β´ precipitates is much smaller than that of the sample aged at 920 °C. It can also be observed that the hardness peak is first reached in the aging at 750 °C than that at 920 °C. Additionally, the overaging started first for the aging at 750 °C. Besides, the hardness is almost the same value for prolonged aging at both temperatures. All the above facts suggest that even the coarsening process at 920 °C is the fastest one, the cuboid morphology and alignment of β´ precipitates causes a higher hardness peak and a slower overaging process than those corresponding at 750 °C.

In summary, the aging process of the Fe-10Ni-15Al alloy promoted the precipitation of the β´(Fe(NiAl)) precipitates with the B2 type crystalline structure. The morphology of β´ precipitates was rounded at the early stages of aging and then it changed to cuboids aligned in the <100> directions of the ferritic matrix. A prolonged aging caused the formation of rectangular plates also aligned in this direction. The coarsening process followed the growth kinetics predicted by the LSW theory. Nevertheless, the hardness peak was higher and the overaging process occurred later in the sample aged at 920 °C than those of the sample aged at 750 °C. This behavior can be attributed to the fast formation of cuboid morphology and alignment in the <100> direction due to the higher lattice misfit between the ferritic matrix and β′ precipitate at this aging temperature.

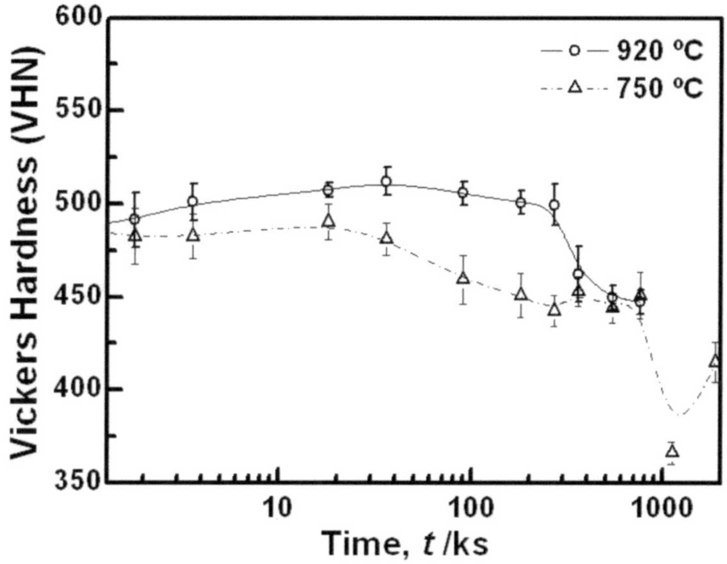

Fig. 4. Aging curves for the Fe-10Ni-15Al alloy aged at 750 and 920 °C.

2.2 Precipitation in austenitic stainless steels

The austenitic stainless steels are construction materials for key corrosion-resistant equipment in most of the major industries, particularly in the chemical, petroleum, and nuclear power industries (Marshal, 1984). These steels are iron alloys containing a minimum of approximately 12 % chromium. This content of chromium allows the formation of the passive film, which is self-healing in a wide variety of environments. Nitrogen as an alloying element in iron-based alloys is known since the beginning of the last century having been profoundly studied during the last three decades (Nakajima et al., 1996). Nevertheless, nitrogen steels are now not widely used. The reason for the comparatively narrow industrial application lies in the old customer skepticism in relation to nitrogen as an element causing brittleness in ferritic steels, some technical problems involved with nitrogen into steel, and the insufficient knowledge of the physical nature of nitrogen in iron and its alloys. In the case of austenitic stainless steels, the main driving force in the development of nitrogen-containing steels is due to the higher yield and tensile strengths achieved, compared with conventionally-processed austenitic stainless steels without sacrificing toughness. Nitrogen stainless steels have yield and tensile strengths as much as 200-350 % of the AISI 300 series steels. It is also important to notice that, in contrast to carbon, nitrogen-containing austenitic stainless steels retain high fracture toughness at low temperatures. Therefore, the higher mechanical properties of nitrogen-containing austenitic stainless steels have made very attractive its application in the power-generation industry, shipbuilding, railways, cryogenic process, chemical equipment, pressure vessels and nuclear industries (Nakajima et al. 1989). These stainless steels are also susceptible to the precipitation of different phases because of the aging for long exposition at high temperatures or during continuous cooling after a welding process. Therefore, it is important to evaluate the degree of microstructural degradation due to the precipitation phenomenon which may affect the cryogenic toughness in this type of steels. In this section, three types of austenitic stainless steels, JJ1, JN1 and JK2 developed for applications to the superconducting magnets of fusion experimental reactor by JAERI, were selected to study the microstructure evolution during isothermal aging.

2.2.1 Experimental details

Materials used in this work were forged-steel plates of 200 mm thick and their chemical compositions are shown in Table 1. The solution treatment of JN1, and JJ1 and JK2 was carried out at 1075 and 1050 °C, respectively, for 1 hour under an argon atmosphere, and then water-quenched. The aging temperatures and times were 600, 700, 800 and 900 °C and from 10 to 1000 minutes, respectively. The aged samples were prepared metallographically and etched with Vilella´s reagent. The precipitates in the aged samples were extracted electrolitically by dissolution of the austenitic matrix in a solution of 10 vol. %HCl-CH_3OH at 4 volts. The X-ray diffraction pattern of extracted precipitates was measured in a diffractometer using Kα Cu radiation. The SEM/EDX microanalysis of precipitates was also conducted using the extraction replica technique.

Material	C	Si	Mn	Ni	Cr	Al	N	Mo
JN1	0.040	0.97	3.88	15.07	24.32	0.023	0.32	---
JJ1	0.025	0.48	10.13	11.79	12.01	---	0.236	4.94
JK2	0.05	0.39	21.27	9.15	12.97	---	0.247	0.97

Table 1. Chemical composition (wt.%) of materials.

2.2.2 Microstructural evolution

An intergranular precipitation can be observed for all cases. The highest and lowest volume fraction of intergranular precipitates corresponded to the aged JN1 and JK2 steels, respectively, Figs. 5 (a-b) and (e-f). The presence of an intergranular cellular precipitation of Cr_2N was observed to occur in the JN1 steel sample aged at 900 °C. No intergranular precipitation was practically detected for the JK2 steel aged at 700 °C. The intragranular precipitates can be classified into two types: cellular or discontinuous precipitation and plate-like precipitates, which have a preferred alignment with the austenitic matrix. The morphology of cellular precipitates is similar to that of pearlite in carbon-steels, Fig. 6. The formation of this lamellar microstructure initiated at grain boundaries and grew into the austenite γ matrix, according to the following reaction:

$$\gamma \rightarrow \gamma + Cr_2N \quad (1)$$

The volume fraction of the discontinuous precipitation increased with time and the maximum value was determined by the point-count grid method, to be about 0.04. This value seems to be reasonable, since a volume fraction of 0.1 was reported in an austenitic stainless steel containing 0.42 % N, after a long heat treatment (Kikuchi et al., 1991). Some small intragranular precipitates were present in the JN1 and JJ1 steels aged at 700 and 800 °C for 5 h, Figs. 5 (a-d). The volume fraction of intragranular precipitates for the aged JJ1 steel was slightly higher than that of the aged JN1 steel. This tendency became higher by increasing the aging temperature. Almost no intragranular precipitation was observed in the aged JK2 steel. The precipitation of particles was also observed to occur on twin boundaries for the aged JN1 steel. The X-ray diffraction patterns of residues extracted from the JN1, JJ1 and JK2 steels aged at 700 and 800 °C for 5 h are shown in Fig. 7. The extracted precipitates of the JN1 steel, aged at 700 and 800 °C for 5 h, were identified as $Cr_{23}C_6$ and Cr_2N. The Cr_2N and $Cr_{23}C_6$ phases were also detected in the aged JJ1 steel. Besides, the presence of the (Fe_2Mo) η phase was also noted in the samples aged at 800 and 900 °C. The precipitated particles of JK2 steel were mainly composed of $Cr_{23}C_6$. According to the chemical composition, shown in Table 1, the JN1 steel has the highest and lowest contents of interstitial solutes (C and N), and Mn, respectively. This suggests that the highest volume fraction of precipitation for carbides and nitrides must have occurred in this steel. In contrast, the JK2 steel has an interstitial solute content lower than that of the JN1 steel, but it has the highest content of manganese, which maintains nitrogen in solid solution, avoiding its precipitation. That is, it is only expected the precipitation of carbides for this steel. This fact showed a good agreement with the above results.

2.2.3 Precipitation kinetics

All the above results are summarized in the Time-Temperature-Precipitation (TTP) diagrams of JN1, JJ1 and JK2 steels, as shown in Figs. 8 (a-c), respectively. In general, it can be noticed that the kinetics of precipitation for JN1 steel is faster than that of JJ1 steel, because of its higher interstitial solute content. The TTP diagrams show that the intergranular precipitation of $Cr_{23}C_6$ and Cr_2N preceded to the intragranular precipitation of Cr_2N, and Cr_2N and η phase in JN1 and JJ1 steels, respectively.

2.2.4 Fracture toughness

In contrast, Figs. 9 (a), (b) and (c) show the plots of CVN fracture energy at -196 °C versus aging time for the JN1, JJ1 and JK2 steels aged at 700, 800 and 900 °C, respectively. All the steels showed a monotonotic decrease in the CVN fracture energy with aging time at the three temperatures. It is also evident that the drop of fracture toughness of JN1 steel is always faster than that of JJ1 steel. The fastest drop of fracture toughness occurred in the JN1 steel samples aged at 900 °C. This fact may be attributed to the higher content of C and N in JN1 steel, which can lead to faster kinetics in intergranular precipitation during the aging process, as discussed in a later section. The CVN fracture energy of solution treated JK2 steel was lower than that corresponding to the other two steels. The lowest decrease in the CVN fracture energy was for the aged JK2 steel. Furthermore, the JK2 steel, aged at 900 °C, showed almost no change in the fracture energy with time. All the JN1, JJ1 and JK2 steels fractured in a ductile manner in the solution treated condition. Intergranular facets were found in all the aged samples, although the area fraction of intergranular facets to ductile surface was strongly dependent on aging conditions. The fraction of intergranular brittle fracture increased with time and temperature, and it seemed consistent with the CVN fracture energy value. Nevertheless, the fracture surface of the JK2 steel, aged at 900 °C, showed almost a complete ductile- fracture mode. These results are in agreement with the fracture mode observed in the tested SP test specimens.

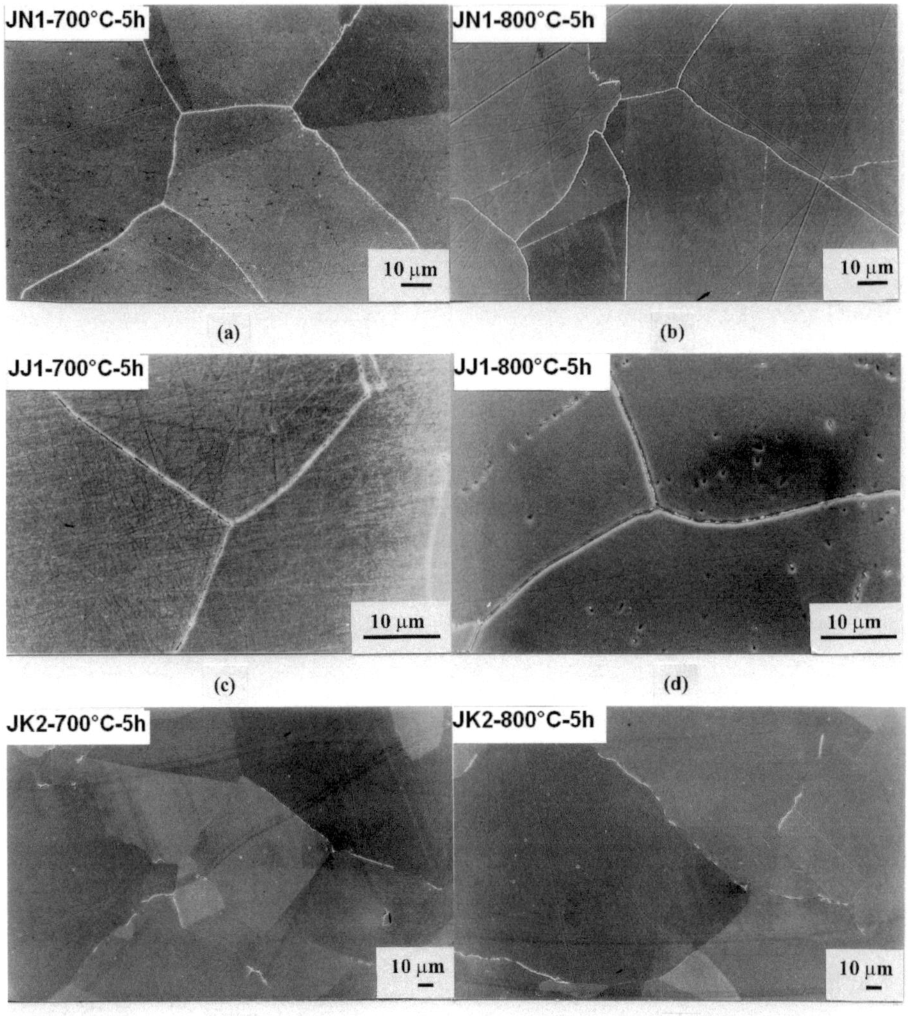

Fig. 5. SEM micrographs of JN1, JJ1 and JK2 steels aged at 700 and 800 °C for 5 h.

In summary, the highest and lowest degradation in toughness for JN1 and JK2 steel, respectively, is associated with the volume fraction of intergranular precipitation formed during the thermal aging. An abundant presence of intergranular precipitates was reported to causes the reduction of cohesive strength of grain boundaries (Saucedo et al., 2001). This is also confirmed by the increase in intergranular brittle fracture as the thermal aging progresses.

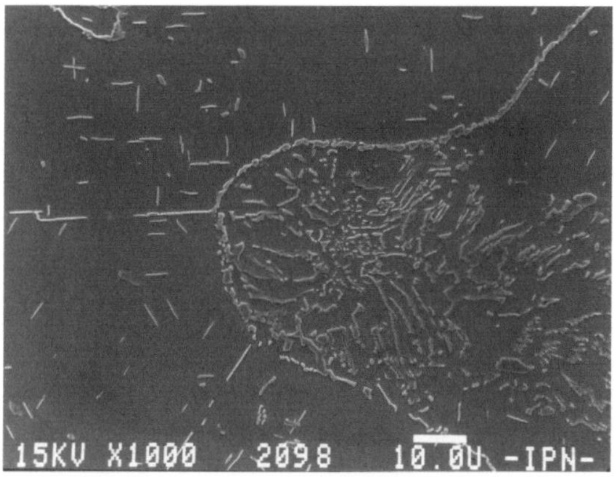

Fig. 6. SEM micrograph of the cellular precipitation in the JN1 steel aged at 700 °C for 1000 h.

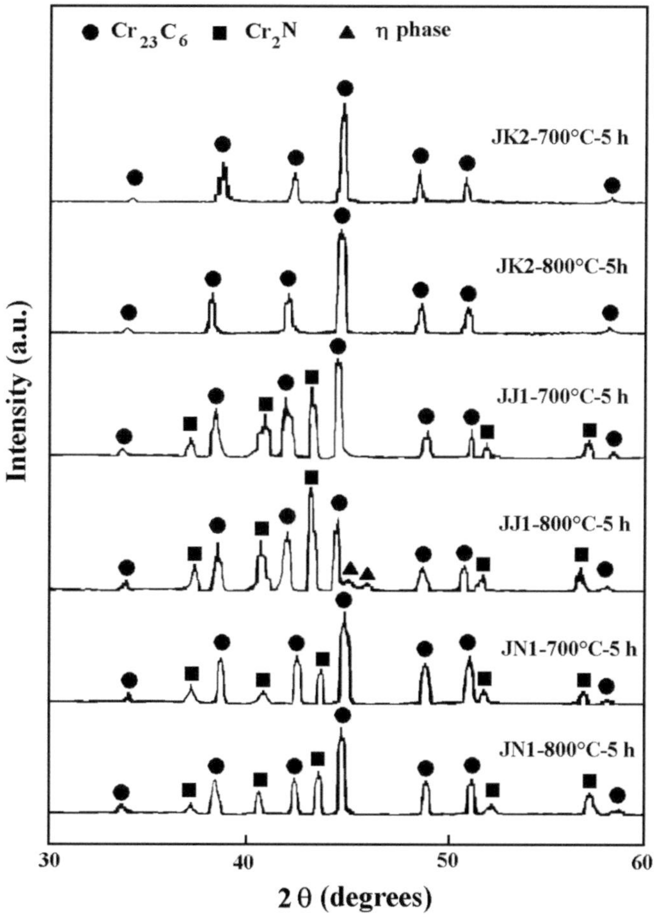

Fig. 7. X-ray diffraction patterns of extracted residues for JN1, JJ1 and JK2 steels agedat 700 and 800 °C for 5 h.

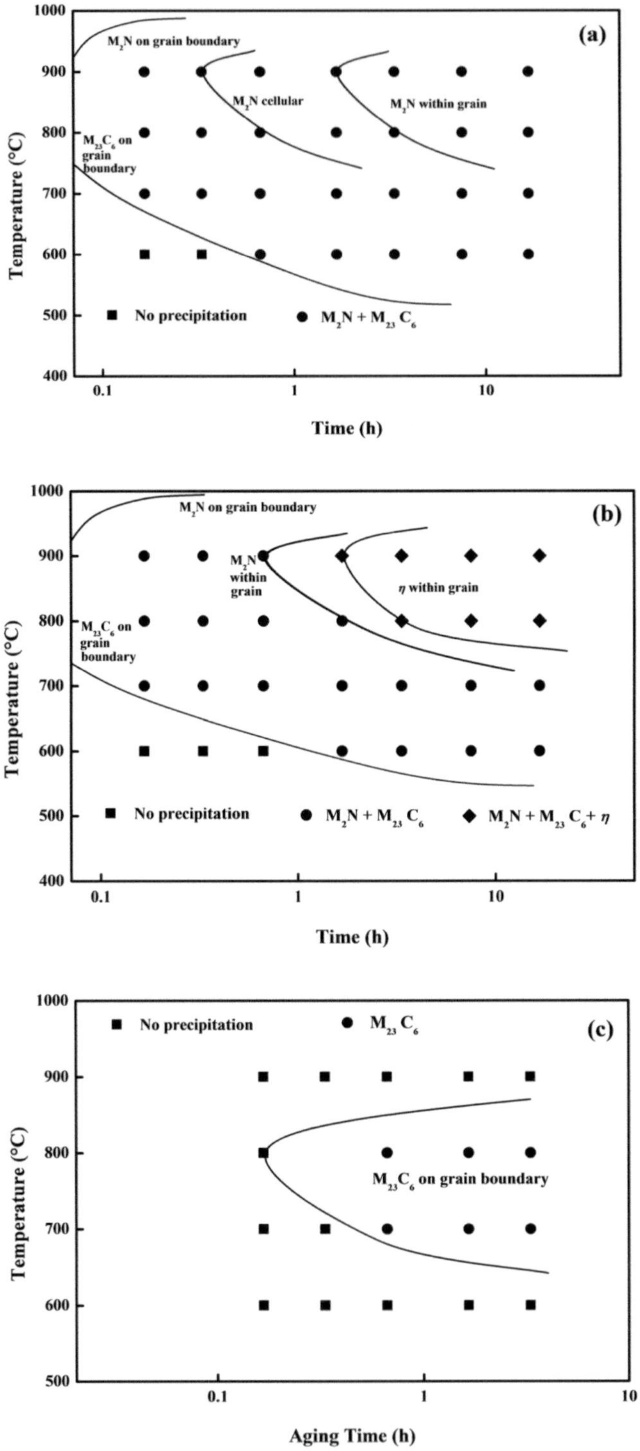

Fig. 8. TTP diagrams of the (a) JN1, (b) JJ1 and (c) JK2 steels.

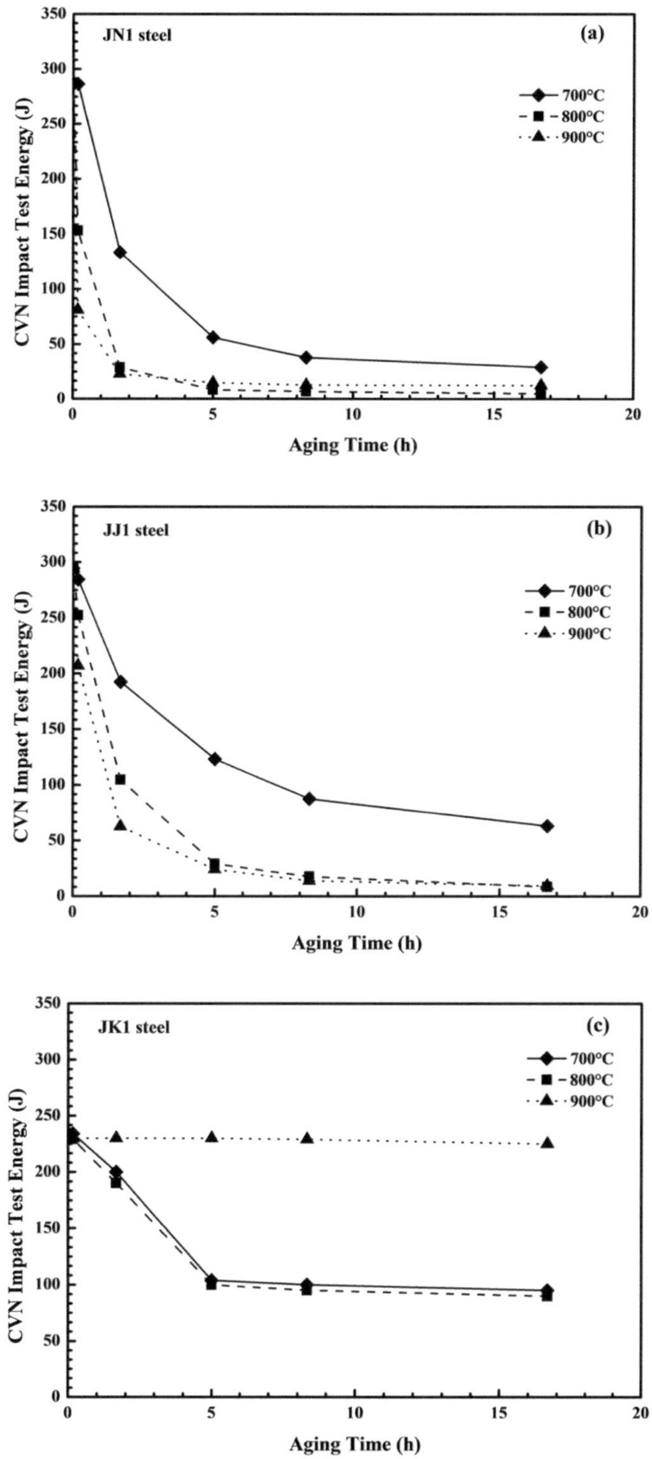

Fig. 9. Plot of CVN impact energy at – 196 °C vs. aging time for tested steels.

2.3 Cellular precipitation in a Mg-8.5Al-0.5Zn-0.2Mn alloy

Mg-Al-Zn alloys have become one of the most important light alloys with a wide range of applications in the automotive industry. This is attributed to the best combination of castability, mechanical strength and ductility (Kainer, 2003). The AZ series of magnesium alloys are mainly based on the Mg-Al binary alloys system. According to the equilibrium Mg-Al alloy phase diagram, the equilibrium phases are the hcp Mg-rich α phase and $Mg_{17}Al_{12}$-γ phase with a complex bcc structure. During the aging process of the Mg-Al based alloys, two types of precipitation are present. That is, discontinuous precipitation takes place on grain boundaries. One of these, intergranular precipitations occurs forming a lamellar structure and it is also known as cellular precipitation. Additionally, continuous precipitation takes place in an intragranular manner and it exhibits more complicated morphologies and orientation relationships than the cellular precipitation. It has been shown in several works (Lai et al., 2008) that these alloys have a poor response to precipitation hardening, compared with precipitation-hardenable Al alloys. Furthermore, the aging hardness is strongly influenced by the morphology, the size and the distribution density of $Mg_{17}Al_{12}$ precipitates. Besides, it has been reported that both discontinuous and continuous precipitations have an effect on the hardness of these alloys (Contreras-Piedras, et al., 2010). Thus, this section shows the mechanism and growth kinetics of cellular precipitation in a Mg-8.5Al-0.5Zn-0.2Mn (wt.%) alloy aged isothermally at 100, 200 and 300 °C for different time periods.

2.3.1 Experimental details

A Mg-8.5Al-0.5Zn-0.2Mn (wt.%) alloy was melted using pure metallic elements under a protective argon atmosphere. Table 1 shows the chemical analysis corresponding to this alloy. Specimens of 10 mm x 10 mm x 10 mm were cut from the ingot and encapsulated in a Pyrex tube under an argon atmosphere. These were homogenized at 430 °C for 3 days and subsequently water-quenched. Homogenized and solution-treated specimens were aged at 100, 200 and 300 °C for different times. The heat-treated specimens were analyzed by X-ray diffraction with copper Kα radiation. These specimens were prepared metallographically and etched with an etchant composed of 19 ml distilled water, 60 ml ethylene glycol, 20 ml glacial acetic acid and 1 ml nitric acid. Etched specimens were observed at 25 kV with a scanning electron microscope equipped with EDS analysis. Vickers hardness was measured in all the heat-treated samples with a load of 100 g. The volume fraction of the discontinuous precipitation was determined from SEM images using a commercial image analyzer.

2.3.2 Microstructural characterization

The X-ray diffraction patterns of the specimens in the conditions of solution-treated and aged at 300 °C for 150 h are shown in Fig. 10. A single-phase is confirmed in the solution-treated specimen, while the appearance of XRD peaks corresponding to the $Mg_{17}Al_{12}$-γ phase are evident in the XRD pattern of the specimen aged at 300 °C for 150 h. No other phases were detected. The presence of these phases for each case is in agreement with the equilibrium Mg-Al phase diagram. Figures 11 (a-i) show the SEM micrographs for the specimens aged at 100, 200 and 300 °C for different time periods. There is a clear competition between the discontinuous and continuous precipitation from the early to the late stages of aging (see, for instance Fig. 11 (e)). Some intragranular precipitates are also observed in these micrographs, Fig. 11 (h). In general, there is a precipitate coarsening as the aging process progresses, Figs. 11 (e-f) and (h-i). The morphology of cellular precipitation at 100 and 200 °C mainly consisted of an S-shape and double-seam morphologies. In contrast, the shape corresponding to 300 °C was mainly a single-seam. It has been reported (Aaronson et al., 2010) that the first morphology occurs at a low temperature ($T<T_m/2$) and it is associated with the free-boundary mechanism and the second one takes place at lower temperature and it is related to the precipitate-assisted mechanism.

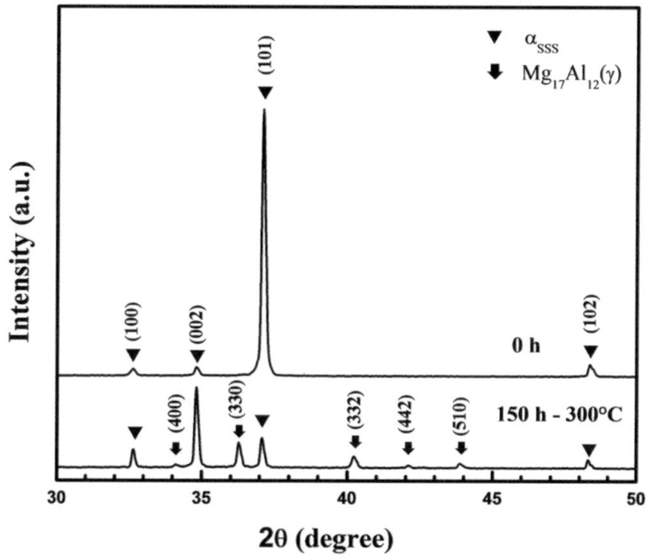

Fig. 10. XRD patterns of the specimens solution-treated and aged at 300 °C for 150 h.

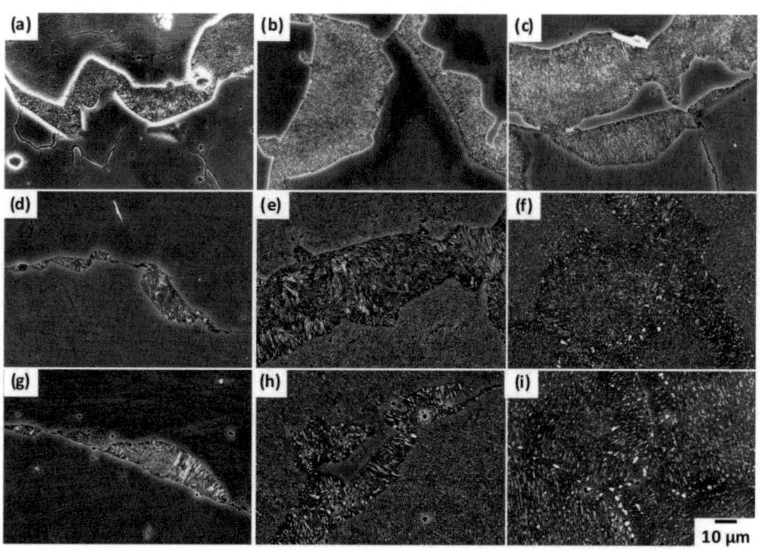

Fig. 11. SEM micrographs of the alloy aged at 100°C for (a) 550, (b) 1500 and (c) 3000 h, At 200 °C for (d) 1, (e) 10 and (f) 250 h, and at 300°C for (g) 0.9, (h) 1 and (i) 25 h.

2.3.3 Growth kinetics of cellular precipitation

The plot of volume fraction of cellular precipitation vs. aging time is shown in Fig. 12. The highest volume fraction occurred for the lowest aging temperature, 100 °C. This fact suggests that continuous precipitation extends more rapidly within grains limiting the growth of cellular precipitation. The analysis of the plot of the volume fraction X_f vs. aging time *t*, Fig. 12, was carried out using the Johnson-Mehl-Avrami-Kolmogorov equation (Cahn, 1975):

$$X_f = 1 - \exp\left(-kt^n\right) \tag{2}$$

The time exponent *n* was determined to be about 1.1, 0.85 and 0.87 for 100, 200 and 300 °C, respectively. These exponent values are close to 1 and it is associated with the dimensionality of the saturation site. That is, it corresponds to a boundary (Cahn, 1975). The lamellar structure always nucleates at grain boundaries and grows perpendicularly to them. The cellular growth stops only if the volume fraction of continuous precipitation is significant to impede its growth. In addition, the activation energy for the cellular precipitation was determined to be about 64.6 kJ mol^{-1}. It was obtained by the slope of the straight line in the Arrhenius plot of the time for a volume fraction of 0.6 vs. the reciprocal value of the absolute temperature as shown in Fig. 5. This energy value seems to be reasonable because it is much lower than the self-diffusion of Mg, 135 kJ mol^{-1} (Mehrer, 1990). That is, it seems to correspond to a grain boundary diffusion process. Additionally, an energy value of 84 kJ mol^{-1} was reported for the cellular precipitation in the binary Al-Zn alloy system (Contreras et al., 2010), which is also a low energy value as that found in present work. Figure 6 shows the variation of interlamellar spacing, S, of discontinuous precipitation as a function of temperature. An increase in lamellar spacing is observed with the increase in temperature. A similar behavior was reported for the discontinuous precipitation in Al-Zn alloys (Contreas et al. 2010). According to the Turnbull theory for cell growth kinetics, the interlamellar spacing S is defined as follows (Aaronson et al. 2010):

$$S=-4\gamma V/\Delta G \tag{3}$$

Where γ is the interfacial energy, V the molar volume, and ΔG the free energy associated with the cellular reaction. ΔG has an inverse relation with undercooling, temperature. Thus, the lower temperature corresponds to the shorter interlamellar spacing. Moreover, the interlamellar spacing remains constant with the increase in aging time for all aging temperatures. These facts seem to be in agreement with the Turnbull theory, which predicts constant lamellar spacing and lamellae growth rate according to the following equation (Aaronson et al., 2010):

$$G \approx \sim 4\delta D_\delta / S^2 \qquad 4)$$

Where G is the lamellae growth rate, D_b is the solute diffusivity along the cell boundary and δ is the cell boundary thickness.

2.3.4 Hardenin behavior

The aging curves for 100, 200 and 300 °C are shown in Fig. 7. The lowest and fastest hardness peak was observed in the aging at 300 °C. This behavior can be attributed to the rapid coarsening of the $Mg_{17}Al_{12}$-γ precipitates either in the discontinuous or continuous precipitations. In contrast, the highest and slowest hardness peaks occurred in the alloy aged at 100 °C. This fact seems to be related to the fine continuous precipitation due to the slow diffusion process at this temperature.

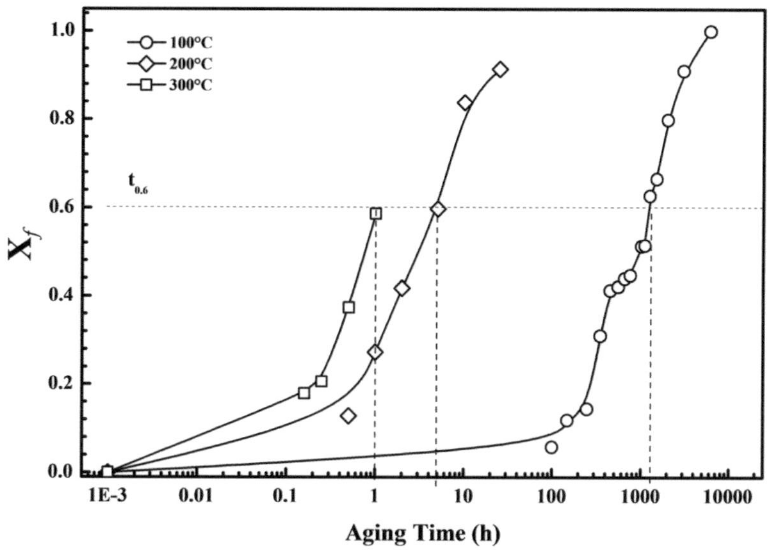

Fig. 12. Volume fraction of cellular precipitation vs. aging time of the alloy aged at 100, 200 and 300 °C.

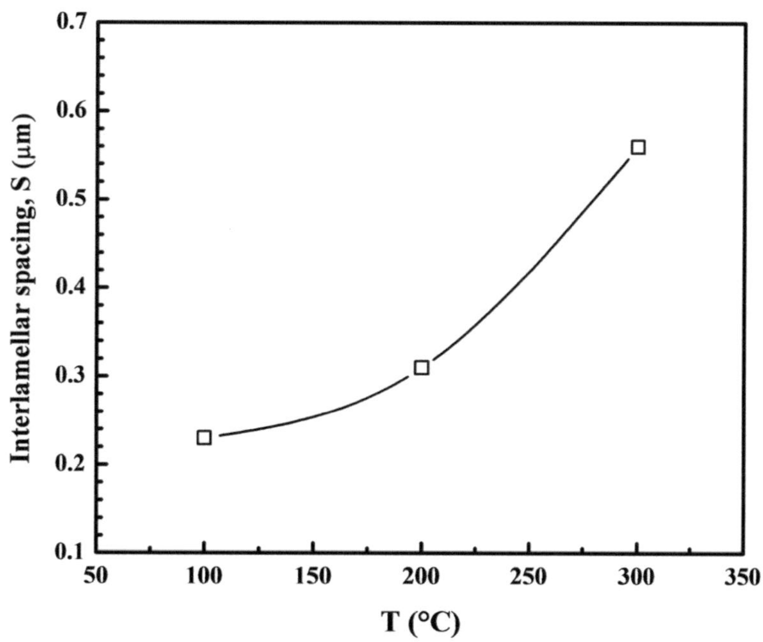

Fig. 13. Interlamellar spacing as a function of aging temperature.

In summary, the microstructural evolution and growth kinetics were studied in an isothermally-aged Mg-8.5Al-0.5Zn-0.2Mn (wt%) alloy and the growth kinetics of cellular precipitation was evaluated using the Johnson-Mehl-Avrami-Kolmogorov equation analysis (Cahn, 1975), which gives a time exponent close to 1. This value confirms that cellular precipitation takes place on the saturation sites corresponding to grain boundaries. Additionally, the activation energy for the cellular precipitation was determined to be about 64.6 kJ mol^{-1}. This also indicates a grain boundary diffusion process. The variation of cellular spacing with temperature follows the behavior expected by Turnbull theory. The highest hardness peak corresponded to the lowest aging temperature and it is associated with a fine continuous precipitation, while the lowest hardness peak was detected at the highest aging temperature and it is attributed to the rapid coarsening process of both precipitations.

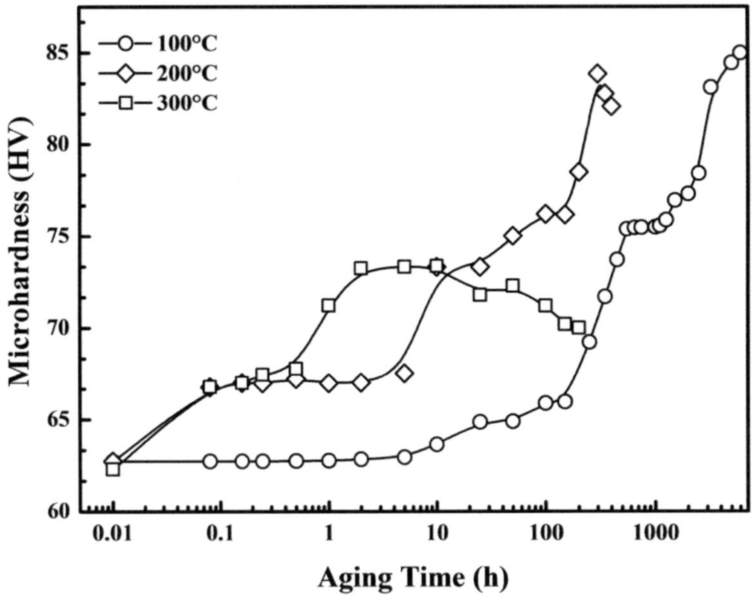

Fig. 14. Aging curves for 100, 200 and 300 °C.

3. Conclusion

This chapter showed three applications of SEM characterization for the analysis of different phase transformations in ferrous and nonferrous alloys, as well as its effect on their mechanical properties. The analysis of these phase transformations enables us to characterize the growth kinetics of these transformations which can be useful either to design heat tretments in order to obtain better mechanical properties or to analyze the microstructural evolution in order to asses the mechanical properties of a component-in-service. Besides, it was shown that the SEM characterization parameters can be used along with the phase transformation theories permitting a better understanding of the transformation behavior in materials after heating.

4. Acknowledgment

The authors wish to acknowledge the financial support from Instituto Politecnico Nacional (ESIQIE), SIP-IPN and CONACYT 100584.

5. References

Aaronson, H.I, Enomoto M. & Lee, J.K. (2010). *Mechanism of Diffusional Phase Transformations in Metals and Alloys*, CRC Press, ISBN 978-1-4200-6299-1, NW, USA

Cayetano-Castro, N.; Dorantes-Rosales H., Lopez-Hirata, V.M., Cruz-Rivera, J. & Gonzalez-Velazquez, J.L. (2008). Cinética de Engrosamiento de Precipitados Coherentes en la Aleación Fe-10%Ni-15%Al. *Revista de Metalurgia de Madrid*, Vol. 44, No. X, (Month, 2008) pp. 162-169, ISSN 1582-2214

Christian J.W. (1975), *The Theory of Transformations in Metals and Alloys*, Pergamon Press, ISBN 0-08-018031-0, Oxford, UK

Contreras-Piedra, E., Esquivel-Gonzalez, R., Lopez-Hirata, V.M., Saucedo-Muñoz, M.L., Paniagua-Mercado, A.M. & Dorantes-Rosales, H.J. (2010). Growth Kinetics of Cellular Precipitation in a Mg-8.5Al-0.5Zn-0.2Mn (wt.%) Alloy, *Materials Science Engineering A*, Vol. 527, pp. 7775-7778, 2010. ISSN 0921-5093

Kainer, K.U. (2003), Magnesium- Alloys and Technologies, Wiley-VCH, ISBN 3-527-30570-X, Germany

Kikuchi M., Kajihara M. & Choi S. (1991). Cellular Precipitation Involving both Substitutional and Interstitial Solutes: Cellular of Cr_2N in Cr-Ni Austenitic Steels. *Materials Science Engineering A*, Vol. 146, pp. 131-150, ISSN 0921-5093

Kostorz, G. (2001). *Phase Transformations in Materials*, Wiley-VCH, ISBN 3-527-30256-5, Weinheim, Germany

Lai W.J.; Lai, Y.Y. Lu, Y.F. Hsu, S. Trong, W.H. Wang. (2009). Aging behaviour and precipitate morphologies in Mg–7.7Al–0.5Zn–0.3Mn (wt.%) alloy, *Journal of Alloys Compounds*, Vol. 476, pp.118-124, ISSN 0925-8388

Marshal, P. (1984). *Austenitic Stainless Steels Microstructure and Properties*, Elsevier Applied Science Publisher, ISBN 0267-0836, NY, USA

Mehrer, H. (1990), *Numerical Data and Functional Relationship in Science and Technology*, Landolt-Borstein New Series III/26, ISBN 0-387-50886-4, Springer-Verlag, Berlin, Germany

Nakajima H., Nunoya Y., Nozawa M., Ivano O., Takano K., Ando S. & Ohkita S. (1996). Development of High Strength Austenitic Stainless Steel for Conduit of Nb_3Al Conductor, *Advances in Cryogenic Engineering*, Vol. 42 A, pp. 323-330, ISSN 0065-2482

Porter D.A.; Easterling, K.E. & Sherif, M.Y (2009). *Phase Transformations in Metals and Alloys*, CRC Press, ISBN 978-1-4200-6210-6, NW, USA.

Ratke, L. & Vorhees, P.W. (2002). *Growth and Coarsening: Ripening in Materials*, Springer, Berlin, Germany, ISBN 3-540—42563-2

Sauthoff, G. (1995). *Intermetallics*, Wiley-VCH, ISBN 3-527-29320-5, Weinheim, Germany

Soriano-Vargas, O.; Saucedo-Muñoz, M.L., Lopez-Hirata, V.M. & Paniagua Mercado, A. (2010). Coarsening of β´ Precipitates in an Isothermally-Aged Fe_{75}-Ni_{10}-Al_{15} Alloy, *Mater. Trans. JIM*, Vol. 51, No. x, (Month, 2010), pp.442-446, ISSN 1345-9678

Saucedo-Muñoz, M.L.;, Watanabe Y., Shoji T. & Takahashi H. (2001), Effect of Microstructure Evolution on Fracture Toughness in Isothermally Aged Austenitic Stainless Steels for Cryogenic Applications. *Journal of Cryogenic Materials*, Vol. 40, pp. 693-700. ISSN 011-2275

18

Cutting Mechanism of Sulfurized Free-Machining Steel

Junsuke Fujiwara
Osaka University,
Japan

1. Introduction

In order to improve efficiency of cutting process in production industry, development of new steel which has good machinablity is desired. The work material which has good surface roughness, easy breakable chip and small tool wear as the good machinablity is expected. And the free-machining steel was developed owning to adding elements which could make the machinablity better. Of all others, leaded free-machining steel and sulfurized free-machining steel are famous. The leaded free-machining steel and sulfurized free-machining steel are well used in the production industry. However the use of the leaded free-machining steel is limited from an environmental problem. In order to develop new environmental friendly free-machining steel, it is necessary to find out the behavior of the inclusion in the work material for the improvement of the machining performance.

There are a lot of studies about the behavior of the inclusion in the free-machining steel (Narutaki et al., 1987), (Yaguchi, 1991), (Usui et al., 1980). There are some papers about the role of the lead and the manganese sulfide which are the representative inclusions. The Pb inclusion acts as lubricant and reduces cutting resistance (Akazawa, 1997). As the MnS is harder than steel, the MnS acts as an internal stress concentration source when the work material reforms into a chip at the cutting edge. And the MnS produced the micro-cracks at shear deformation zone. This is the cause that the shear area became small and reduces the cutting stress (Yamamoto, 1971). Although these results are almost reasonable, we must think over the role of the inclusion again in order to produce new free-machining steel. The experiment was carried out to find out the mechanism of the sulfurized inclusion on the machinablity, using some kinds of steels which have different size of the inclusion. The observation of the deformation behavior near the cutting edge was carried out to investigate the effect of the inclusion in detail.

2. Experimental method

In this experiment, two kinds of the sulfurized free-machining steels (Steel A and Steel B) which have different size of the inclusion were used. Figure 1 show optical microphotographs of microstructure and size distribution of MnS in the Steel A and Steel B, respectively. The area fraction of equivalent circle diameter of the inclusions was also shown in these figures. The steel A contains larger inclusions than the steel B. These sulfurized free-machining steels contain 0.42% S, and chemical compositions of these materials are almost the same as shown in Table 1. The inclusions tend to be slender parallel to rolling direction.

In an orthogonal cutting at low speed, the cutting forces were measured. The cutting width of the work material was 2 mm. The surface of work materials was polished to observe deformation of the inclusions. An orthogonal cutting was performed using table feeding system of a horizontal milling machine as shown in Fig. 2. Table 2 shows cutting conditions in the orthogonal cutting. The cutting speed was 16mm/min and the depth of cut was 0.1mm. The tool material was high speed steel and its rake angle was 10 degree.

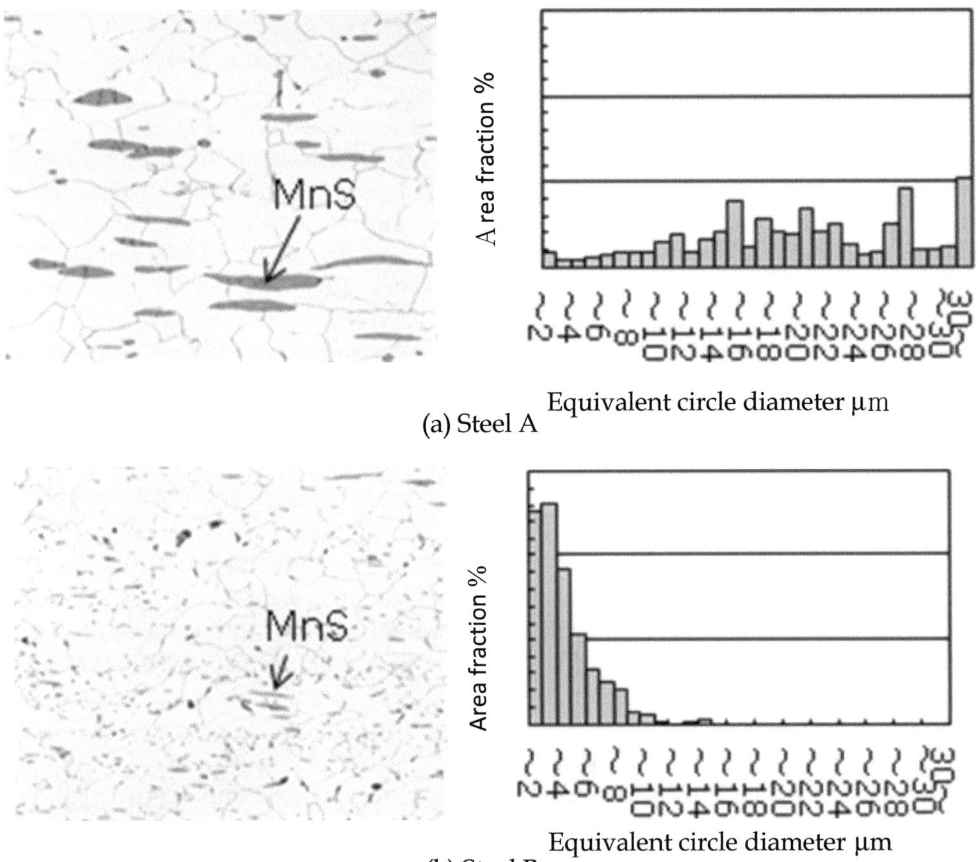

Fig. 1. Optical micrographs of microstructure and size distribution of MnS

Mass %	C	Si	Mn	S	Al	O_2
Steel A	0.03	0.01	1.44	0.42	0.001	0.0175
Steel B	0.03	0.01	1.7	0.43	0.001	0.0044

Table 1. Chemical compositions of work materials

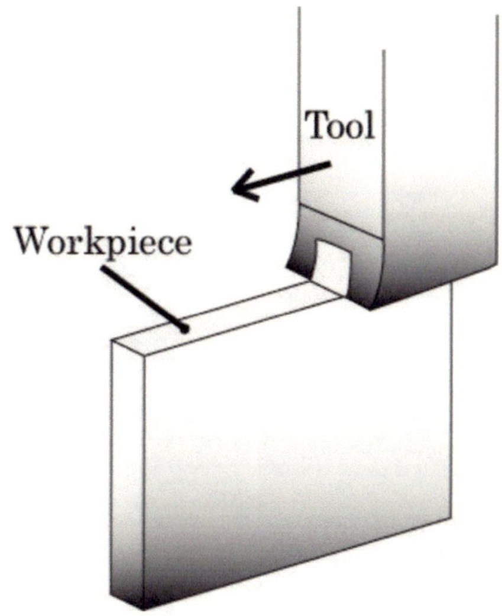

Fig. 2. Method of orthogonal cutting

Cutting speed	0.016 m/min
Depth of cut	0.1 mm
Tool	SHK4
Rake angle	10°
Clearance angle	17°

Table 2. Cutting conditions in orthogonal cutting

3. Experimental results and discussions

3.1 Orthogonal cutting

The Cutting forces were measured in the orthogonal cutting. These results are shown in Fig. 3. On the whole, the cutting force in the Steel A was bigger than that in the Steel B. The cutting force in the Steel A was more stable than that in the Steel B. This fact led smooth surface roughness.

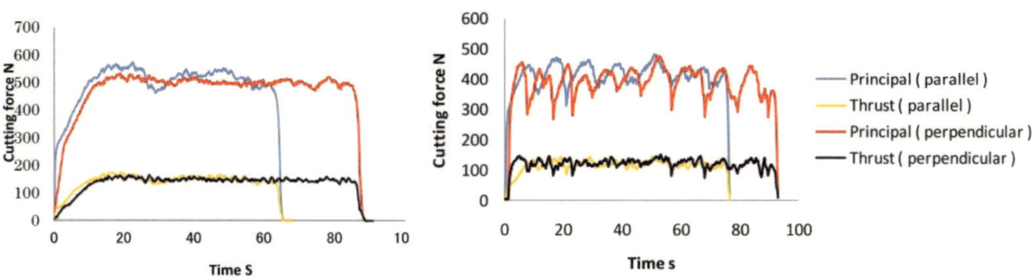

(a) Steel A (b) Steel B

Fig. 3. Cutting forces in orthogonal cutting

In order to investigate the flow state of around the shear zone, a quick stop test was carried out during the orthogonal cutting. Figure 4 shows the enlarged photographs around the shear zone. In case of the Steel A, the large crack parallel to the shear plane was found. In case of Steel B, the chip is thin, and there are small cracks near the rake face.

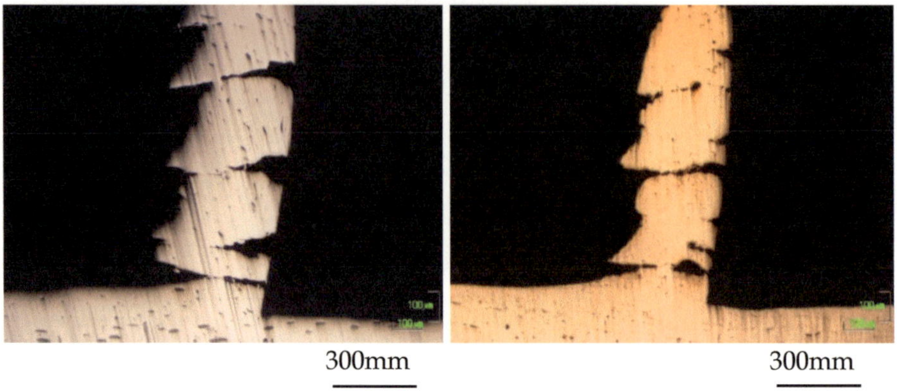

(a) Steel A (b) Steel B

Fig. 4. Microphotographs of partially formed chip

3.2 Micro-cutting in SEM

A small orthogonal cutting equipment as shown in Fig. 5 was mounted into the Scanning Electrical Microscope (SEM) (Iwata 1977). The deformation behavior around the shear zone was observed in detail with the SEM. An example of the micro-cutting in SEM is also shown in this figure. The cutting speed was 0.27 mm/s and the depth of cut was 20 - 50 µm. Table 3 shows the cutting conditions. The work material was cut from the test piece as shown in Fig.6.

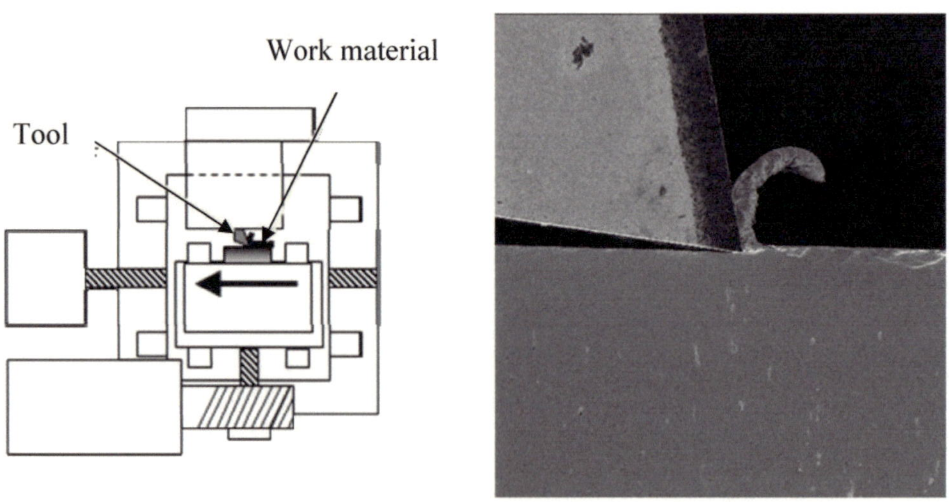

(a) Cutting equipment (b) Micro-cutting

Fig. 5. Cutting equipment in SEM and micro-cutting

Cutting speed	0.27 mm/s
Depth of cut	20~50 μm
Tool	SKH4
Rake angle	6°
Clearance angle	3°

Table 3. Cutting conditions in SEM

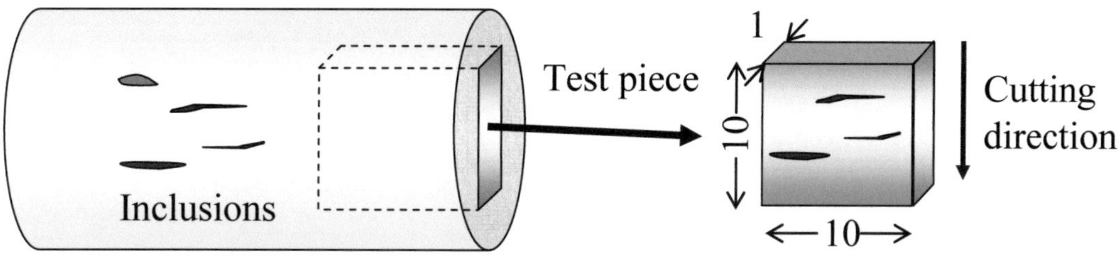

Fig. 6. Work material for micro-cutting in SEM

Figure 7 shows the sequential photographs during cutting of the Steel A. As the Steel A has the spindle shaped inclusions, the inclusion MnS which was extended perpendicular to the cutting direction can be found. This inclusion turned in counterclockwise and broke to several pieces around the shear zone. These pieces create voids around them, and flowed to the chip in the direction parallel to the shear plane. As shown in this figure, the void was formed at the upper of the inclusion and the micro-crack was formed along the primary shear plane.

Figure 8 shows the sequential photographs during cutting of the Steel B. The inclusion of higher aspect ratio than that in the Steel A can be found. As the Steel B had long slender inclusions, this inclusion broke into smaller pieces than that in the Steel A. These pieces create very small voids between them. The inclusions in the Steel B are well dispersed, so these very small voids are created at various places in the work material. It causes the reduction of the cutting force.

3.3 Image analysis for stress distribution

It is very important to know strain and stress distribution in shearing zone. The sequential images could be taken during micro-cutting in SEM. Using with image processing, the strain increase and stress distribution around MnS can be calculated. That is to say, as comparison with two sequential SEM images after micro movement of the tool, the displacement within observed zone could be measured by tracing a same point. Moreover the strain increase and stress distribution could be calculated from the displacement.

In order to measure the displacement from the sequential SEM images, PIV (Particle Image Velocimetry) method was used (Raffel etal., 2007). The moving distance was calculated from gray level pattern between SEM image A at t in time and SEM image B at t+△t in time as shown in Fig.9.

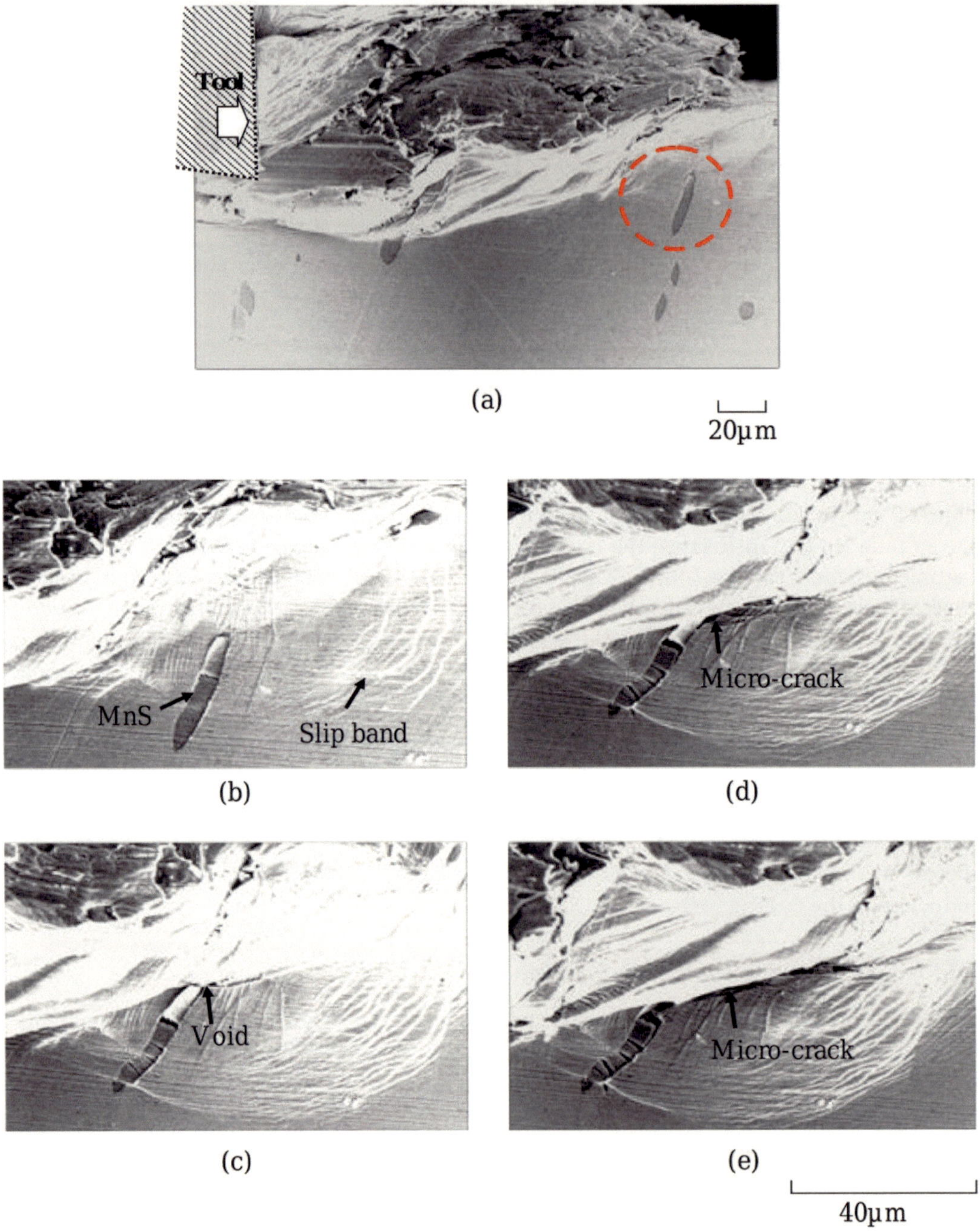

Fig. 7. Deformation behavior of large spindle shaped MnS inclusion (Steel A)

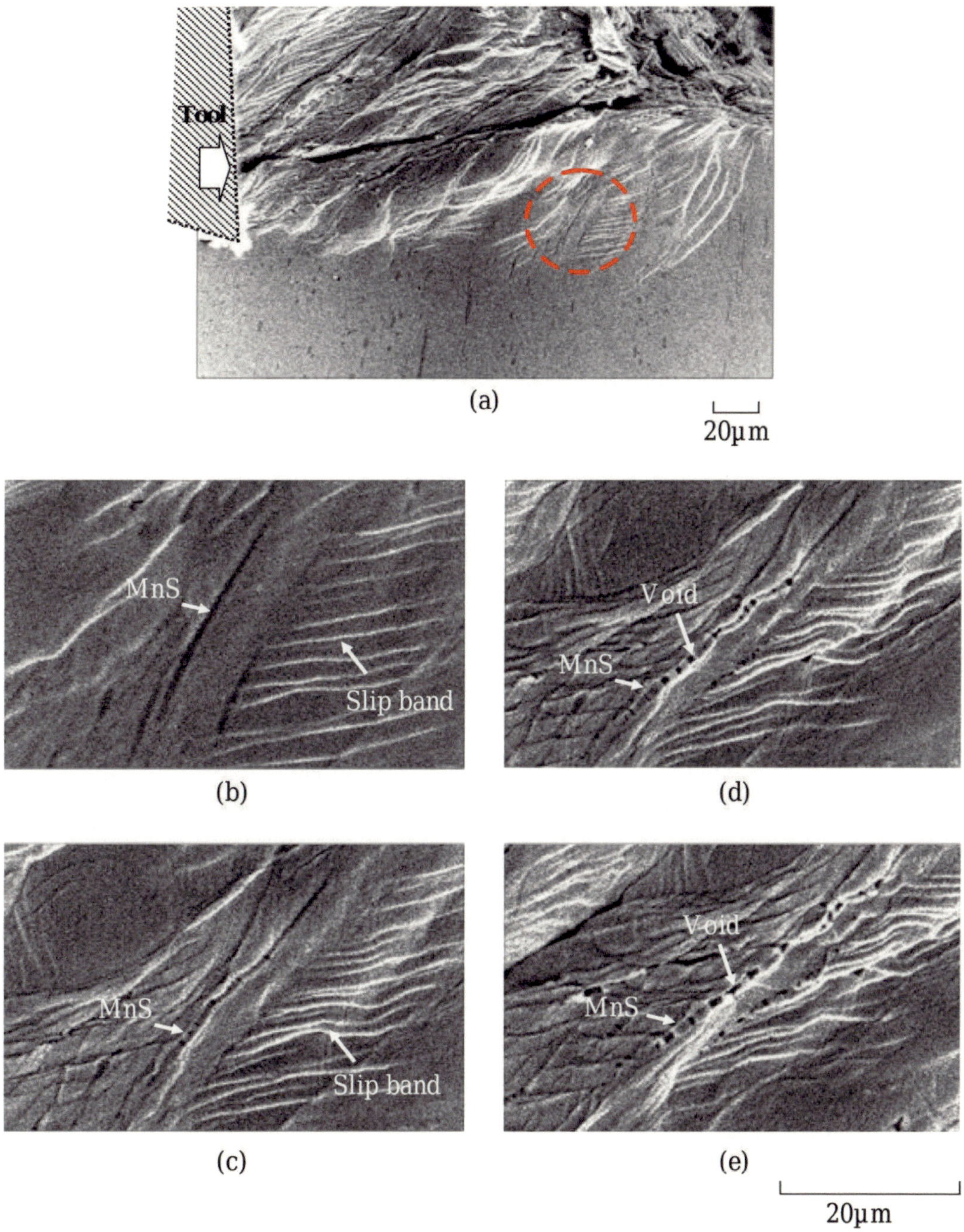

Fig. 8. Deformation behavior of long slender MnS inclusion (Steel B)

The strain and stress distribution was calculated from the displacement increase measured with the PIV method using with FEM as shown in Fig.10 (Usui et al., 1990). In the calculation, the mechanical properties as shown in Table 4 were used.

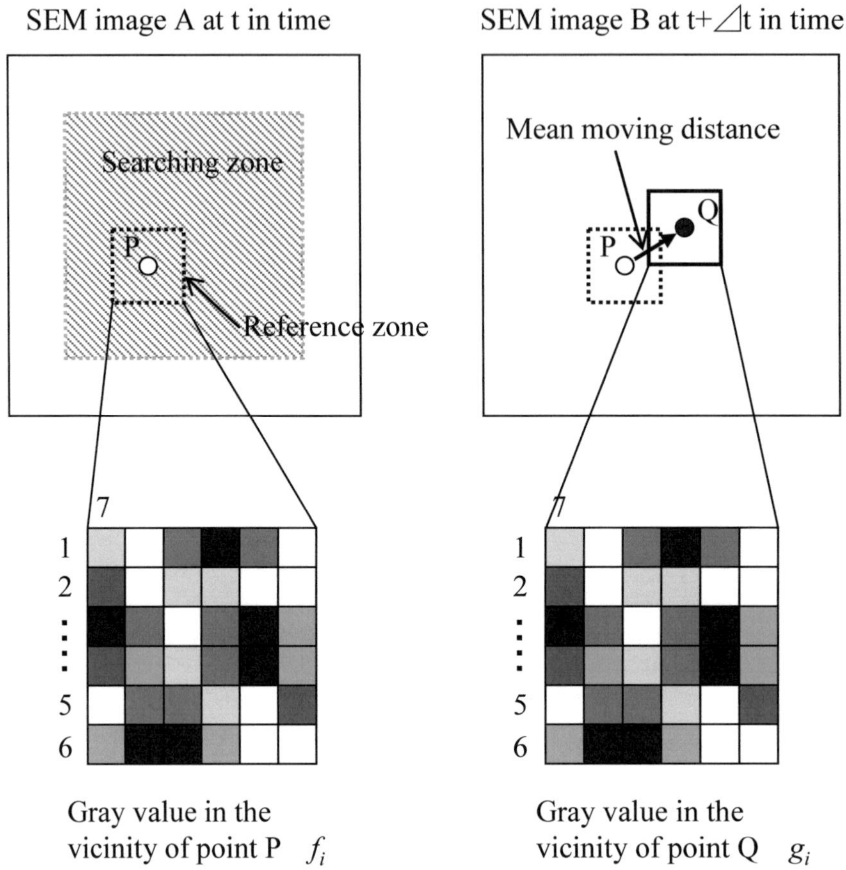

Fig. 9. Outline of correlation method in particle image velocimetry

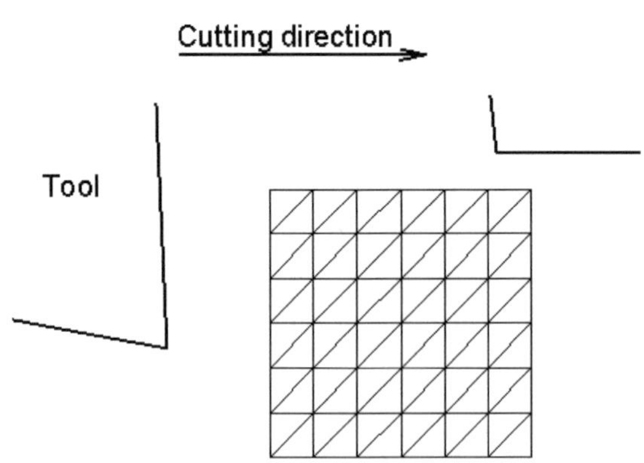

Fig. 10. Model of triangulated zone for FEM calculation

	Steel A	Steel B
Yield strength MPa	522	560
Tensile strength MPa	525	563
Elongation %	13.2	12.0
Reduction of area %	52.1	48.0

Table 4. Mechanical properties of work materials

The strain increase distribution was calculated from the two sequential SEM Images as shown in Fig. 7. Figure 11 shows the strain increase distribution in micro-cutting of the Steel A. The moving distance of two images was 1.8 μm. As shown in Fig 11 (d), the shear strain increase was large in shear zone but another large strain increase was found around MnS.

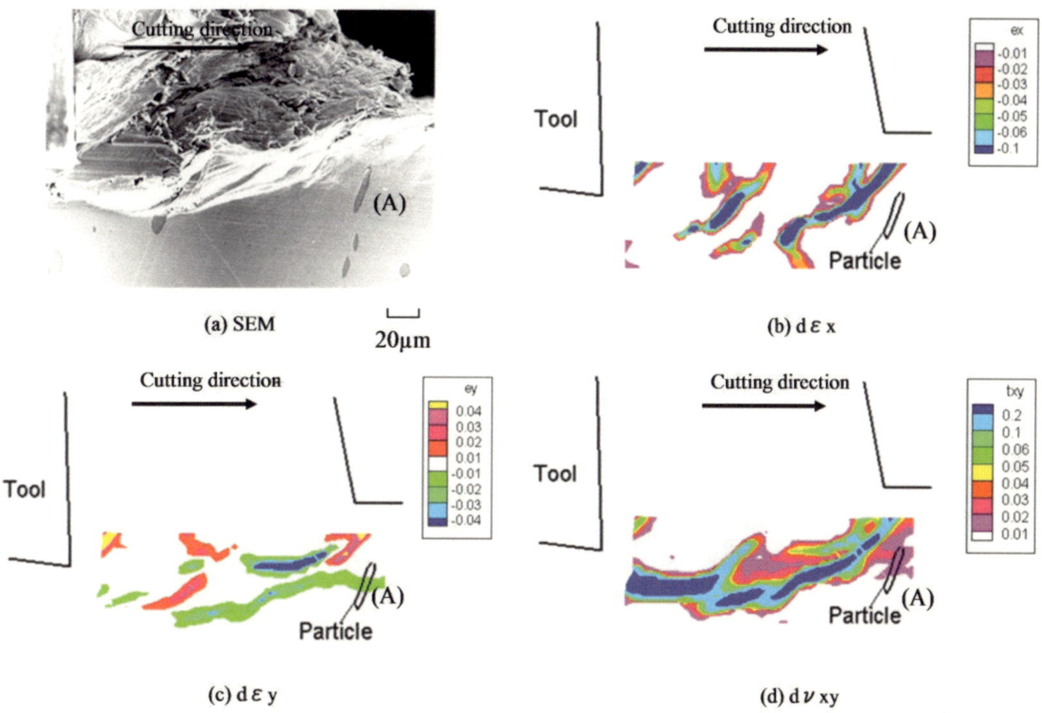

Fig. 11. SEM image and strain increase distribution in micro-cutting of Steel A

The stress distribution was calculated from the strain increase distribution. Figure 12 shows the stress distribution in micro-cutting of the Steel A. It shows that the stress was big at the upper of the MnS because of the stress concentration.

The strain increase distribution was calculated from the two sequential SEM Images as shown in Fig. 8. Figure 13 shows the strain increase distribution in micro-cutting of the Steel B. The moving distance of two images was 4.3 μm. As shown in Fig 13 (d), the strain increase along shear zone was large in and no large strain increase was found around MnS.

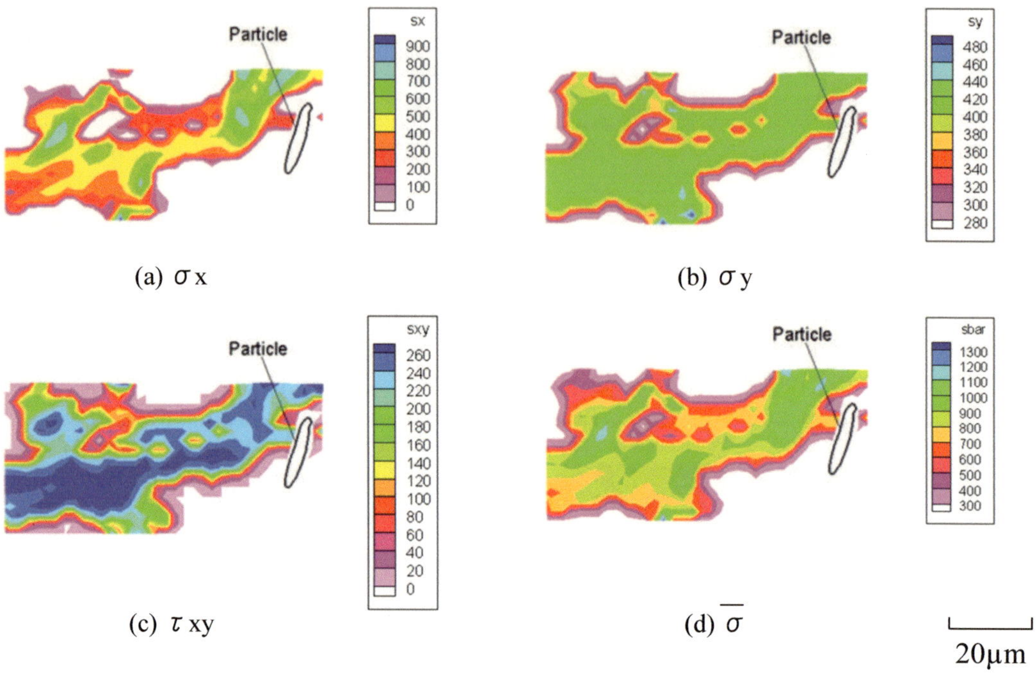

Fig. 12. Stress distribution in micro-cutting of Steel A

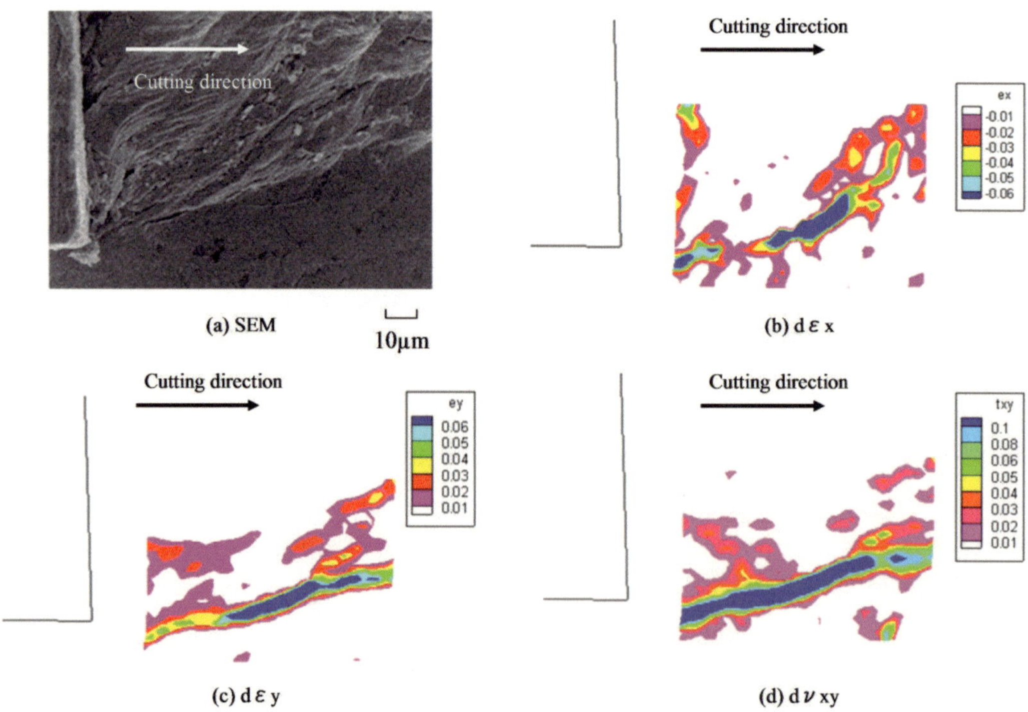

Fig. 13. SEM image and strain increase distribution in micro-cutting of Steel A

The stress distribution was calculated from the strain increase distribution. Figure 14 shows the stress distribution in micro-cutting of the Steel B. As the Steel B had long slender inclusions, the inclusion MnS broke into small pieces and the stress distributed along the shear plane. There is little stress concentration around the MnS.

Consequently, in the Steel A which has large spindle type MnS, the micro-crack is easily formed. As this micro-crack affects the breakage of a build-up edge (BUE) and chip, the BUE could not become big and the good finish surface roughness was gained in the Steel A. As shown in Fig. 15, large micro-crack affected the separation between the chip and the BUE.

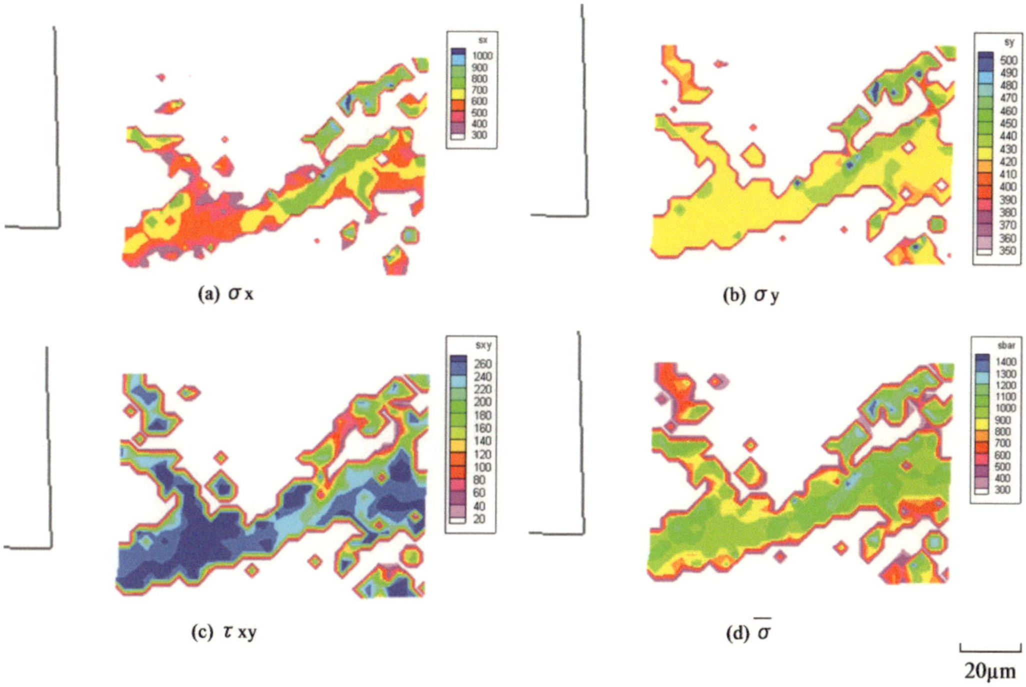

Fig. 14. Stress distribution in micro-cutting of Steel A

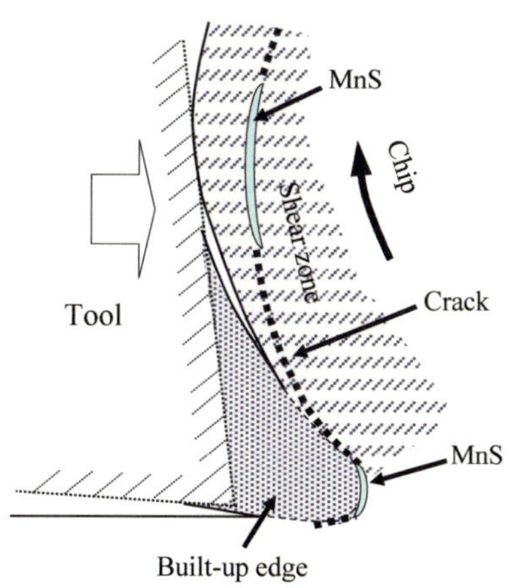

Fig. 15. Effect of spindle-type MnS to suppress BUE growth in the vicinity of BUE

3.4 Quick stop test during turning.

A quick stop experiment during turning was carried out with the quick stop system which was attached to a conventional lathe. Figure 16 shows the equipment of the quick stop test. The tool was rotated down at the moment when pulling a pin which fixed the tool and the cutting state was stopped quickly. The deformation behavior around the shear zone was observed from the workpiece with a chip. The cutting speed was 62 m/min and the depth of cut was 0.2 mm. Table 5 shows the cutting conditions.

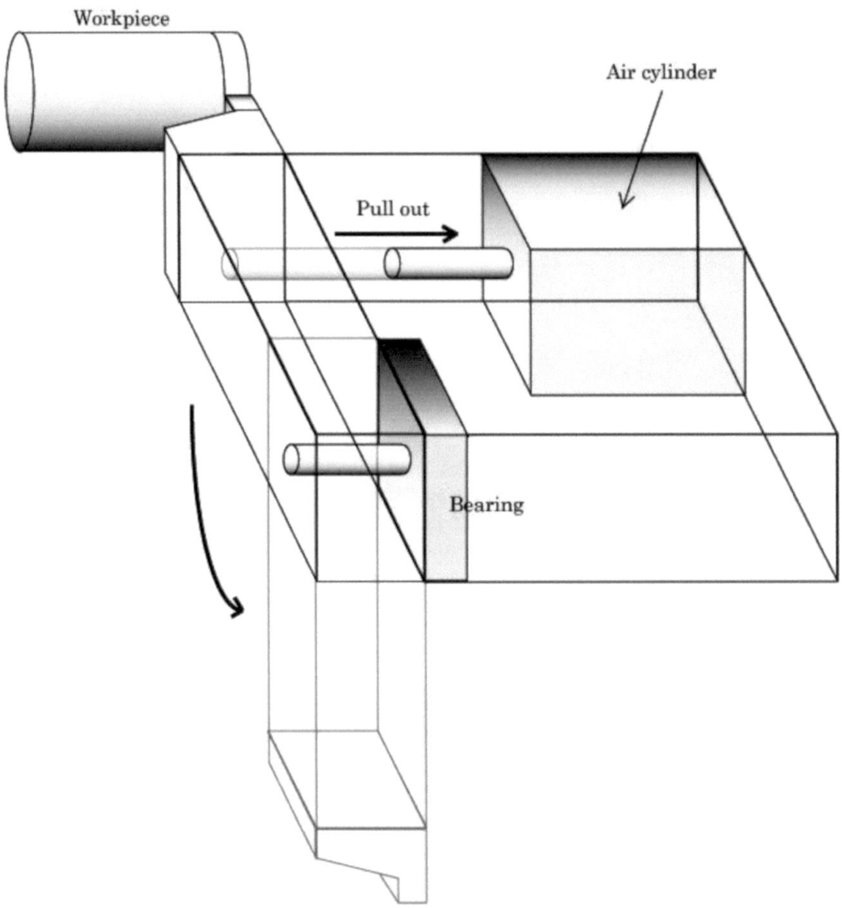

Fig. 16. Equipment of quick stop test

Cutting speed	62 m/min
Depth of cut	0.2 mm
Tool	Cemented carbide P10
Rake angle	20°
Clearance angle	6°

Table 5. Cutting condition in quick stop test

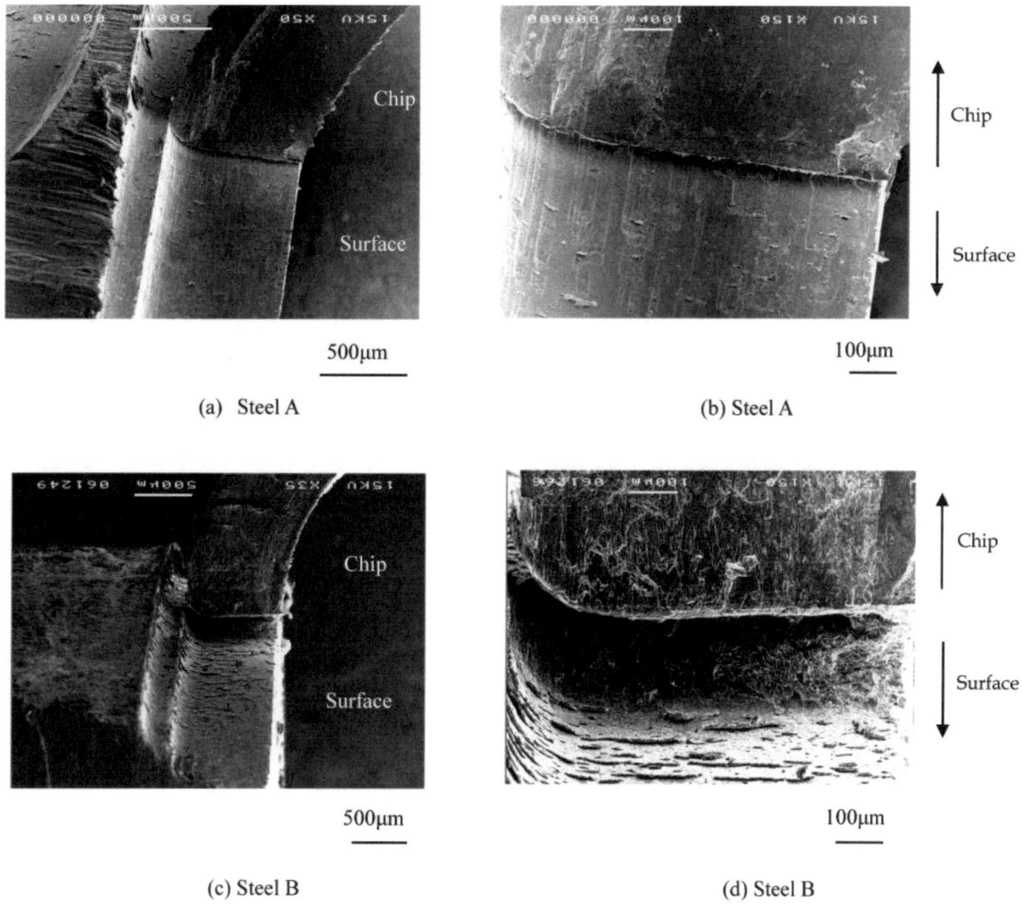

Fig. 17. Chip and finish surface in quick stop test

In quick stop test, the chips in the vicinity of the tool face and machined surface was observed as shown in Fig. 17. In case of the Steel A which contains large spindle shaped inclusions, the BUE could not be found. And machined surface had good surface roughness. In case of the Steel B which contains small slender inclusions, the BUE could be found on the rake face of the chip. There were many tears on the machined surface. The BUE partially separated and they leave on the machined surface. As a result, the surface roughness became bad. In this experiment, it was clear that the larger inclusions could reduce the formation of the BUE.

4. Conclusions

The main results obtained are as follows.

1. In machining of sulfurized free-machining, some inclusions creates voids around them, some break to several pieces depending on their conditions around the shear zone.
2. The larger inclusions can reduce the formation of the BUE.

5. References

Akazawa, T., Free cutting steels contributing to industry, *Journal of Special Steel,* Vol.46, No.5 (May 1997), pp.6-10.

Araki, T., Yamamoto, S., Machinability of Steel and Metallugical Factors, *Iron and Steel,* Vol.57, No.13 (Nov. 1971), pp.1912-1932.

Iwata, K., Ueda, K., Shibasaka, T, Study on Micro-machining Mechanics Based on Direct-SEM Observation, *Journal of Japan Society of Precision Engineering,* Vol. 43, No.3 (1977) pp.311-317.

Katayama, S., Toda, M., Hashimura, M., Growing Model of Build-up Edge in Relation to Inhomogeneities of Steel Microstructure, *Journal of Japan Society of Precision Engineering,* Vol. 62, No. 9 (1996) pp.1345-1349.

Maekawa, K., Kubo, A., Kitagawa, T., MachinabilityAnalysis of Free-machining Steel, *Journal of Japan Society of Precision Engineering,* Vol. 57, No. 12 (1991) pp.2193-2198.

Narutaki, N., Yamane, Y., Usuki, H., Yan, B., Kuwana, T., Machinability of Resulfurized Steels under High Cutting Speed, *Journal of Japan Society for Precision Engineering,* Vol.53, No.3 (March 1987), pp.455-466.

Raffel, M., Willert, C.E., Kompenhaus J., (2000) Particle Image Velocimetry, Springer, ISBN 3-540-63683-8 New York.

Usui, E., (1990) *Modern Cutting Theory* (1st), Kyouritsu Shuppan, ISBN 4-320-08054-8, Tokyo.

Usui, E., Obikawa, T., Shirakashi, T., Embrittle Action of Free-machining Additives, *Japan Society for Precision Engineering,* Vol.46, No.7 (July 1980), pp.849-855.

Yaguchi, H., Effect of MnS Inclusion Size on Machinability of Low-Carbon Leaded Resulfurized Free-Maching Steel, *Journal of Applied Metalworking* Vol. 4, No. 3 (1986) pp.214-220.

Yaguchi, H., Effect of soft assitives (Pb, Bi) on formation of Build-up edge, *Journal of Material Science Technology,* No. 4, (1988) pp.926-932.

Yaguchi, H., The Role of Liquid Metal Embrittlement on the Chip Disposability of Steel, I*ron and Steel* Vol.77, No.5 (May 1991), pp.683-690.

19

Catalyst Characterization with FESEM/EDX by the Example of Silver-Catalyzed Epoxidation of 1,3-Butadiene

Thomas N. Otto, Wilhelm Habicht,
Eckhard Dinjus and Michael Zimmermann
Karlsruhe Institute of Technology, IKFT,
Germany

1. Introduction

Ag catalysts are of outstanding importance in the field of heterogeneous catalysis. Optimum distribution and morphology of the Ag particles must be ensured by controlled, tailored catalyst synthesis. Hence, there is a growing demand for the characterization of Ag-dispersed fine particle systems requiring high-resolution surface observation of particles down to a few tens of nanometers and elemental analysis by *field emission scanning electron microscopy and energy-dispersive X-ray spectrometry* (FESEM/EDX). It is beneficial to characterize the particle morphology by comparison of different imaging methods like secondary electron (SE)-, backscattered electron (BSE)- and transmitted electron (TE) detection. In scanning electron microscopy surface topography becomes visible due to the dependency of the SE yield on the angle of electron incidence. Together with the large depth of field informative images of irregularly shaped particle structures are obtained. The increased BSE yield of high atomic numbers (Z) such as Ag catalysts and promoters (e.g. Cs) compared to a low-density matrix and the high penetration depth of 20-30 keV electrons also allows imaging and analysis of inclusions that would be obscured at low beam energies. Both SE and BSE detectors, in particular at low beam voltages, can additionally reveal interesting surface features of fine Ag particles. A well-known example for a Ag catalyzed reaction is the α-Al_2O_3 supported Ag-catalyzed epoxidation of 1,3-butadiene to 3,4-epoxybutene. The electrophilic addition of oxygen across the carbon-carbon double bond of 1,3-butadiene, resulting in a three-member ring structure that can undergo further chemical transformations to oxygenated products, such as ketones, alcohols, and ethers. Supported silver catalysts have been shown to epoxidize olefins with nonallylic hydrogen when an alkali promotor is doped on the surface. Thus, the direct kinetically controlled oxidation to the corresponding epoxide is preferred. The guiding hypothesis for this partially oxidation is that surface oxametallacycles are key intermediates for epoxidation on promoted Ag catalysts. Therefore, the preparative application of Ag and promoters (Cs, Ba) on the catalyst support material is of great importance. Another important aspect is sintering of Ag particles which may reduce the catalytically active surface and decreases the overall reaction performance. For this research, catalysts are produced by sequential impregnation of two mineralogically differing support materials (SC13, SLA2) with an aqueous active component solution ($AgNO_3$, $CsNO_3$). The two selected Ag catalyst systems are examined using mainly FESEM/EDX. The determination of metal amount were carried out with EDX area analysis and, if necessary, supplemented by EDX-spot analysis. In some cases, characterization of Ag-distributions by EDX mappings were made. Furthermore, for the Monte Carlo (MC)- simulations x-ray line scans of two different SEM preparation techniques (bulk specimen, thin-film supported specimen) were performed to underpin the relationships impressively. Temperature-programmed O_2 desorption (O_2-TPD) as well as N_2 sorption (BET) measurements are important analysis methods in catalysis chemistry, too. Both methods are used to support the FESEM/EDX investigation to provide complementary contributions with regard to the Ag distribution and the properties of the carrier surfaces.

2. Characterization and measurement methods

2.1 Ag catalyst systems

Huge varieties of materials are used in the preparation of heterogeneous catalysts, especially industrial catalysts. Catalysts can be divided into three groups of constituents, namely active catalytic agents, promoters, and supports [1]. Catalysts are manufactured by various methods, such as wet impregnation, leaching, drying and calcination. The major components of the catalyst system are the catalyst support (bulk material, e.g. Al_2O_3, TiO_2, SiO_2), which might influence the catalytic activity of the active components (metal-support interactions, MSI) [2] and the active metal e.g. Ag, Pd, Pt is the active agent. Increasing the surface area of the active agent is one function of the support. Maintaining a high dispersion of the active components is the other function. α-Al_2O_3 with its small specific surface area has proved to be a wear-resistant carrier material for Ag and to be highly suited for the selective oxidation of 1,3-butadiene [3].

2.2 Catalyst preparation

The Ag catalysts for this research are produced by sequential incipient wetness impregnation of two mineralogically different carrier materials with an aqueous solution of $AgNO_3$ and $CsNO_3$ as active components. The following catalyst supports are applied:

- SC13 (Almatis) mineralogical: 80 % alpha-Al_2O_3, 20 % γ-Al_2O_3, Chemical composition: 99 % Al_2O_3, 0.05 % CaO, 0.03 % Fe_2O_3, 0.02% SiO_2, 0.4 % Na_2O.
- SLA92 (Almatis) mineralogical: Main phase Ca hexa aluminate (CA6) hibonite, secondary phase alpha-Al_2O_3, chemical composition: 91% Al_2O_3, 8.5 % CaO, 0.04 % Fe_2O_3, 0.07 % SiO_2, 0.4 % Na_2O.

Dry SC13 and SLA92 are pre-sieved and subjected to wet sieving (Retsch laboratory sieves manufactured according to DIN 3310, mesh width 0.045 mm - 0.063 mm). The sieve fractions are dried at 120°C in a circulating air oven for 5 h. The desired active component solutions $AgNO_3$(aq) and $CsNO_3$(aq) are applied sequentially to the carrier materials and subjected to an ultrasound bath for 0.5 min (Sonorex RK 100H) [4]. The precursors are dried in a circulating air oven at 40°C for 5 h and then oxidized with 100 % O_2 (4.8) at 250 °C for 10 min (4000 ml_{STP} h^{-1}). A reduction with 100 % H_2 (6.0) follows at 200 °C for 10 min (1000 ml_{STP} h^{-1}). The concentrations of the active metal (oxidation number ± 0) components after conditioning are given by:

D1, SC13, 45/63 µm, 5 % Ag
D2, SC13, 45/63 µm, 5 % Ag, 1500 ppm Cs
D3, SC13, 45/63 µm, 10 % Ag
D4, SC13, 45/63 µm, 10 % Ag, 1500 ppm Cs
D5, SC13, 45/63 µm, 20 % Ag
D6 SC13, 45/63 µm, 20 % Ag, 1500 ppm Cs
MZ06, SLA 92, 45/63 µm, 5 % Ag, 1500 ppm Cs
MZ09, SLA 92, 45/63 µm, 10 % Ag, 1500 ppm Cs

2.3 Electron interactions with the specimen

Scanning electron microscopy in combination with energy-dispersive X-ray spectrometry (SEM/EDX) is a well-established and versatile method for the characterization of heterogeneous catalysts, especially Ag catalysts. They are predominantly composed of small metal particles dispersed onto a supporting material, generally a chemically inert oxide. Information about particle size distribution, deposition on the substrate surface, and elemental compositions can be obtained easily. In FESEM, a high brightness Schottky-type field emission (FE) cathode with its small beam diameter (spot size) enables imaging of features with high resolution and high contrast down to the nanoscale even on bulk substrates. Coincidently, the element composition of the specimen is available by excitation of inner shell electrons to collect characteristic X-rays with sufficient intensity for analytical information. Due to the high depth of field ($D \approx d^2 / \lambda$, $\lambda_{20\,keV} = 0.0086$ nm), where D is the depth of field, d is the apparent resolution, and λ is the electron wavelength, impressive images of differently shaped catalyst particles can be generated [5]. In the SEM, the primary electron beam creates different types of electron interactions, while penetrating the specimen. For this research, however, we utilize the following 4 important electron interactions with the specimen, namely the generation of

- secondary electrons (SE)
- backscattered electrons (BSE)
- transmitted electrons (TE)
- characteristic X-rays

Secondary and backscattered electrons are essentially for the topographical imaging of the specimen surface and therefore are described in more detail in chapter 2.4. Transmitted electrons (TE) interact less with the specimen. The specimen appears more or less "transparent" for electrons, depending on the *thickness and density* of the specimen. In energy-dispersive X-ray Spectroscopy (EDX), the X-rays are produced by inelastic scattering of primary beam electrons with bound inner shell electrons during their penetration into the matter. Subsequent deexcitation by transition of outer shell electrons to the inner shell vacancy results in emission of an element specific X-ray quantum [6]. These characteristic X-rays are essential for the determination of the elemental composition of a specimen. The electron penetration depth depends mainly on the primary beam energy and the target composition. The X-ray production range also depends on the critical excitation energy of the specific X-ray line (e.g. Kα or Lα) and is always smaller than the electron range. The electron range is the travelling distance from the primary beam electron incident at the target surface to the point where the electrons lost their energy by multiple interaction processes within the material. Figure 1 illustrates the different kinds of electron interactions with specimen and the according electron detectors.

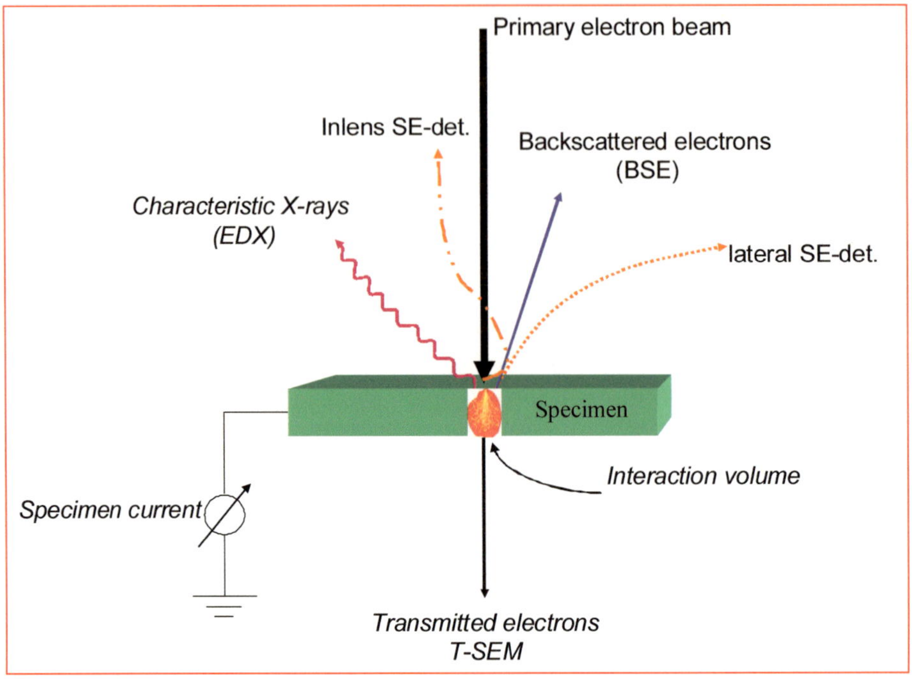

Fig. 1. Different types of electron interactions with specimen and related detection modes.

2.4 Imaging modes

Secondary electrons (SE), backscattered electrons (BSE) and transmitted electrons (TE) are responsible for three different imaging modes in SEM. The fourth type of electron interaction is the generation of X-rays and, hence, not a typical imaging mode.

The 3 imaging modes are explained as follows:

1. Backscattered electrons provide contrast based on atomic number (Z-contrast) and density. The BSE-detector will be used as a quick response to visualize materials heterogeneously composed and distributed. It will be applied often in combination with the EDX- unit to capture images for subsequent microanalysis. Depending on the primary beam energy, backscattered electrons are created inside the specimen by elastic scattering. They possess approximately the energy of the primary electrons. As a rule of thumb, the exit depth of the BSE is half the primary electron range. As a consequence, the imaging resolution achieved by backscattered electrons is worse than those achievable by secondary electrons.
2. Low energetic (< 50 eV) secondary electrons are used for true surface imaging, because they are created in the vicinity of the primary beam impact on the target surface. They are responsible for high-resolution imaging (in-lens SE detection).
3. Transmission-type images (also called T-SEM mode) are obtained by a special mounting device for TEM grids with a diode-type detector beneath. The detector is mounted like a specimen stub instead and also adjusted to the electron beam. The specimen must be sufficiently thin (< 200 nm) to permit penetration of beam electrons at 20-30 kV, which is the typical operation voltage in T-SEM mode. Imaging of mass-

thickness contrast is enabled. The application of the TE- detector can provide additional information on the sub-surface structure of many particles which cannot be resolved clearly in the corresponding SE and BSE images. The thin-film-supported specimen permits imaging with enhanced contrast due to an improved signal-to-noise ratio and X-ray analysis with reduced scattering background [7], [8].

2.5 Specimen preparation for SEM / T-SEM

The specimens are fixed with conductive tabs onto Al-stubs as bulk specimen for conventional SEM imaging. For T-SEM imaging, the samples are suspended in ethanol, and a droplet is poured onto a carbon-filmed TEM-grid (400 meshes Cu, ca. 8 nm C) for transmitted electron measurements.

2.6 SEM equipment

The electron microscopic investigations are carried out with a DSM 982 Gemini, Zeiss corp., Germany, equipped with a 4-quadrant solid-state BSE detector, a high brightness in-lens SE detector and a lateral SE detector. The DSM 982 GEMINI is applied with a thermal (Schottky-type) field-emission electron source (SFE). Element-specific quantification is performed by the dedicated EDX unit, equipped with a 30 mm^2 Si(Li) detector INCA Pentafet™, FWHM 129 eV @ MnKα (Oxford corp., England).

2.7 O_2-TPD

O_2-TPD analysis is performed using a BELCAT-B (BEL INC. Japan) system. It is coupled with a GAM 400 quadrupole mass spectrometer (In Process Instruments, Germany) as a mass-selective detector. The samples are subjected to a preliminary in-situ treatment in an O_2 / He test gas mixture. For this purpose, the sample is heated to 250°C in steps of 5 K/min and then kept at 250°C for 1 h. After this, it is cooled down to room temperature (RT). At RT, the non-adsorbed O_2 is removed by rinsing with He. In the subsequent O_2-TPD experiment the pre-treated samples are heated linearly in He. A thermal conductivity detector (WLD) is applied. The quadrupole mass spectrometer determines the composition of the desorbed products.

2.8 N_2-BET

The BET surface area (S. **B**runauer, P. H. **E**mmett, and E. **T**eller) is determined with N_2 (77 K) (BELSORP-mini II, BEL INC. Japan). All samples are subjected to a preliminary treatment in a vacuum at 200°C for 5 h. To determine the sorption isotherms, the amount of molecules adsorbed on the samples are measured as a function of the relative pressure p/p_0. The specific surfaces are calculated from the adsorption parts of the isotherms at 77 K in the relative pressure range from 0.01 to 0.35 p/p_0 with N_2.

3. Discussion

3.1 Electron microscopy affecting parameters and MC simulation

FESEM and the dedicated EDX unit are important aids in the determination of the relationship among catalyst particle (cluster) sizes, dispersion onto support, support morphology and the influence of promoters. For a better understanding it makes sense to calculate the practical electron range by the Monte Carlo simulation program MOCASIM™ [9]. Additionally, the depth of the X-ray generating region (X-ray range) is estimated for the materials under consideration. It is based on an analytical expression useful for most elements and is calculated by the equation 1 of Andersen-Hasler:

$$R = \frac{0.064}{\rho} \cdot \left(E_0^{\ 1.68} - E_c^{\ 1.68} \right) \tag{1}$$

Whereby R (μm) is the X-ray range, E_0 (keV) is the primary electron (beam) energy, E_c (keV) is the critical excitation energy for the characteristic (analytical) X-ray line and ρ is the density of the elements. The dimension of the primary X-ray generation volume is important for the information depth obtained. It depends on the beam energy and the X-ray line chosen for the measurement [10]. The dependency of the practical electron range (penetration depth) and the X-ray range on the material density and beam voltage (beam energy) of the catalyst materials Ag, Al, and Al_2O_3 is depicted in Tab. 1. The higher the density of the element or specimen, the less is the penetration depth of the electrons (practical electron range) and the X-ray range (information depth). As a consequence, the material should be ideally homogeneous over the electron range to minimize errors caused by specimen heterogeneity. The critical excitation energy is determined by the specific absorption edge for the electron shell of an element from which the analytical X-ray line will be emitted. Therefore, E_c is always slightly higher than the corresponding X-ray line.

Specimen	ρ / g cm^{-3}	X-ray line / keV	E_c / keV	HV / kV	pract. Electron range / nm	X-ray range / nm
Al	2.7	1.487 (Kα)	1.56	5	361	304
				10	1146	1085
				15	2252	2192
				20	3639	3585
Al_2O_3	3.5	-	-	5	278	235
				10	884	837
				15	1738	1691
				20	2807	2766
Ag	10.5	2.984 (Lα1,2)	3.35	5	93	45
				10	295	245
				15	579	530
				20	936	888

Table 1. Parameters affecting the electron range and X-ray range. E_c is the critical excitation energy for the corresponding analytical X-ray line. The practical (pract.) electron range is calculated by MOCASIM™.

The different penetration depths for heterogeneously composed materials will be overcome by using the T-SEM technique as it will be discussed later. As depicted in Figure 2 and Figure 3 the MC simulations elucidate the different excitation volumes for 5 keV and 20 keV beam energies. They interact with an Alox (Al_2O_3) support covered with an Ag cluster that is assumed to be 100 nm in thickness. It demonstrates the differences in the electron ranges as a function of beam energies and material density whereas the relative atomic number Ag (107.8) and the relative molecular weight Al_2O_3 (101.9) are close together. Remarkable is the beam spreading effect caused by multiple scattering in the bulk specimen, which leads to a pear-like shape of the electron trajectories (interaction volume). This will be drastically reduced by using thin supporting foils and thinning of the specimen or using sufficiently small particles, depending on the beam voltage and density of the specimen (see Figure 10).

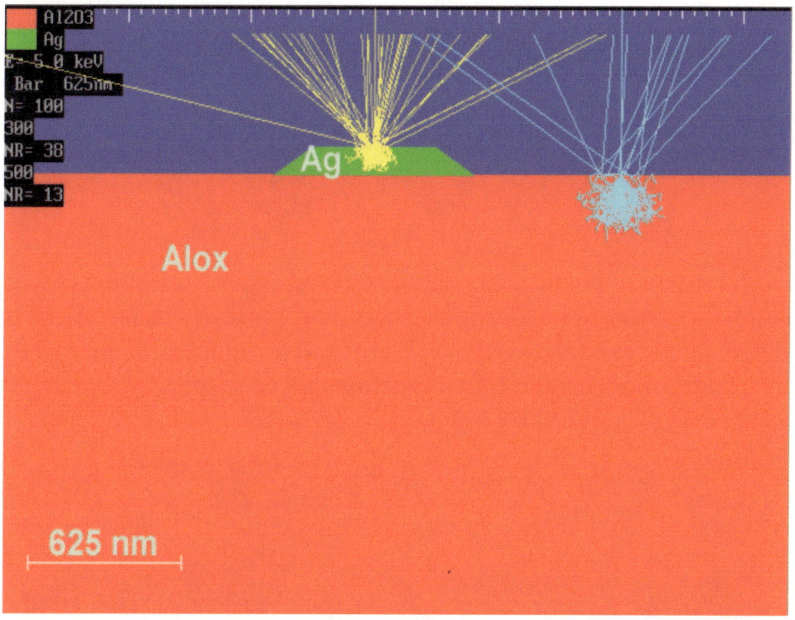

Fig. 2. MC simulation of electron trajectories (interaction volumes) created by primary beam energy of 5 keV. Note that N is equal to the number of used trajectories (100) for MC simulation. NR corresponds to the number of BS-electrons in case of Ag (38) and Alox (Al_2O_3, 13).

Obviously, specimens that are heterogeneously composed of materials strongly different in density and atomic number require a careful X-ray analysis. Nevertheless, due to the irregular shape of the analyzed particles, the systematic error of EDX measurements is increased compared to that of a homogeneous specimen with flat and smooth surfaces. Commonly, the commercially available "standard less analysis software" are adjusted to 'ideally' smooth samples. Usually, the EDX analysis results are expressed in weight-percent (wt.-%), in which the collected X-ray counts (intensity = f(element conc.)) will be converted into concentrations by means of an evaluation program taking into account fundamental X-ray parameters and the detector efficiency. The procedure is known as "standard less analysis" and is already established by the suppliers. To reduce errors resulting through different X-ray exit and scattering angles due to the different particle sizes, irregular dispersion, or agglomerated catalyst deposits, a huge population of particles should be chosen for elemental analysis of bulk specimen. To obtain e.g. the amount of metal covering the support, data collection at low magnification as demonstrated in Fig.4. should preferably be performed.

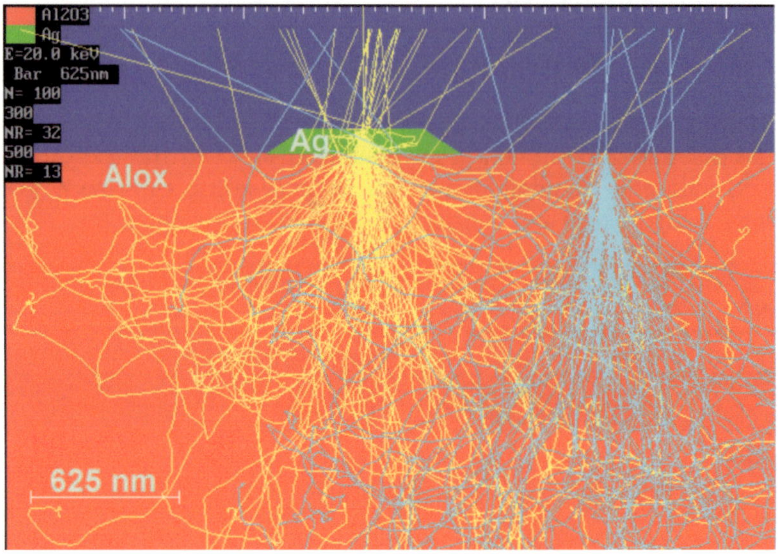

Fig. 3. MC simulation of electron trajectories created by a primary beam energy of 20 keV. Note the progress in the interaction volumes compared to Figure 2. N is equal to the number of used trajectories (100) for MC simulation. NR corresponds to the number of BS-electrons in case of Ag (32) and Alox (Al_2O_3, 13).

Fig. 4. Magnified (100 x) BSE image of a large population of Al_2O_3 particles differently covered with Ag, prepared on a conventional specimen stub.

Fig. 5. High-magnification (30,000) image of a single Al_2O_3 particle covered with different sized Ag clusters, prepared on a conventional specimen stub.

Therefore a great number of particles should be included in the analyzed region to average the influence of different shaped and sized particles. As an example, a low-magnified, large number of supporting alumina (Alox = Al_2O_3) particles partially covered with Ag is shown in Figure 4 and produced under conditions mentioned in chapter 2.2. The magnification of 100 x is recommended for an EDX analysis to determine the amount of Ag deposited. The alumina particle diameter varies between 30 µm and 80 µm. For comparison, a highly magnified single alumina particle (sample D4) with differently sized and shaped Ag deposits is presented in Figure 5. It represents a high-magnification (30,000 x) image of a single substrate particle, covered with differently sized Ag clusters. The size range of the deposited Ag is between 20 nm and 850 nm.

3.2 EDX line scans

Figure 6 shows an EDX line scan which is simulated by MC calculation using a bulk specimen geometry with an Ag deposit of 100 nm in thickness onto an alumina (Alox) support. It reveals the course of Ag-Lα, Al-Kα and O-Kα X-rays across the specimen. The X-ray signals of Al and O dropped drastically down over the whole Ag cluster, contrarily the Ag signal becomes dominant. The lack in signal strength (intensity) is caused by absorption of the weak Al and O X- rays in the dense Ag particle. The maximum of the Ag signal, observed at the ends of the Ag deposit (edges) in the simulation is conspicuous. This is not observed in practice, only in the simulation.

Figure 7 is an example of an EDX line scan across the Ag agglomerate (approximately 620 nm) on a selected surface image of sample D6, which is generated by backscattered electrons. The line scan is shown in detail in Figure 8. One can see the pronounced maximum of the Ag signal and the minimum of the Al signal, whereas the minimum of the O signal is less distinct as expected. It is a hint for the presence of subsurface oxygen or Ag bulk-dissolved oxygen, which will discussed in chapter 3.6 in detail by means of O_2-TPD and BET. The comparison between the experimental EDX line scan of sample D6 (Figure 8) and the simulated line scan (see Figure 6) is feasible.

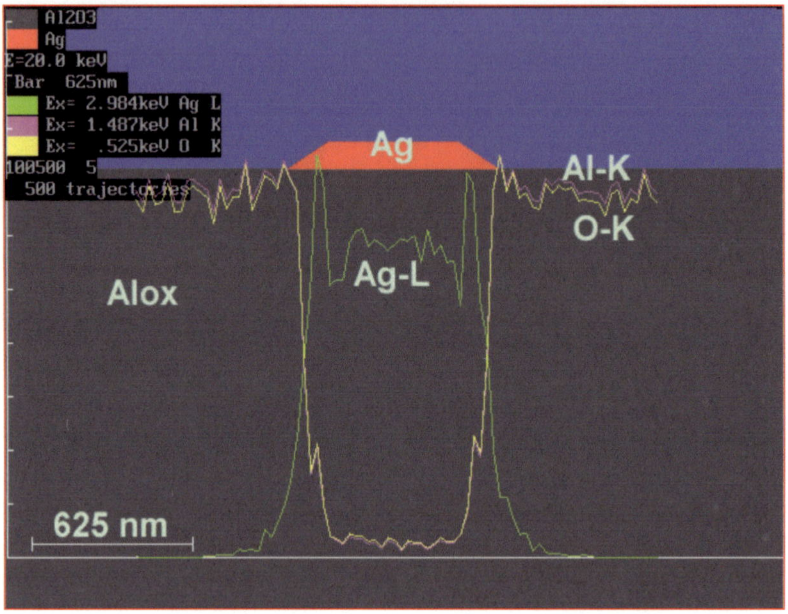

Fig. 6. MC simulation of an X-ray line scan (20 kV beam voltage) across a single particle deposited on an alumina (Alox) substrate.

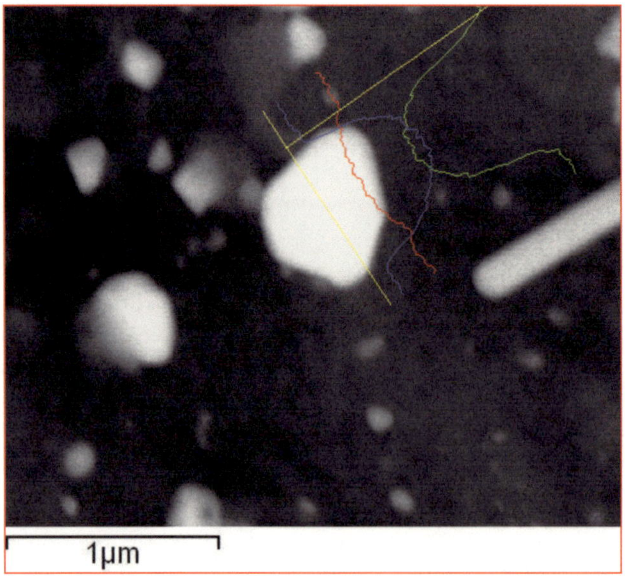

Fig. 7. BSE image of sample D6 and EDX line scan across a selected Ag deposit.

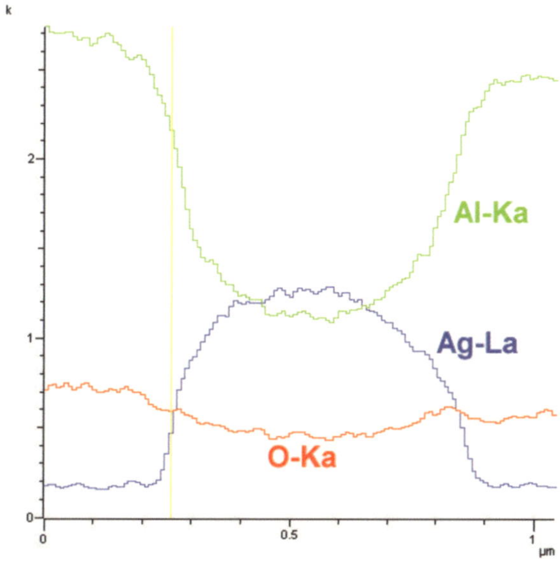

Fig. 8. Line scan in detail of Figure 7.

To demonstrate the imaging power of the T-SEM mode an example of a very tiny single alumina particle (D4) partially covered with Ag deposits in comparison to a extended particle (Figure 5) is shown in Figure 9. The dense Ag clusters appear as dark spots on the nearly "electron-transparent" alumina matrix. The graphical representation of an MC calculation based on an Al_2O_3 (Alox) substrate of 92 nm in thickness and covered with a 20 nm thick Ag deposit on a TEM-grid support is depicted in Figure 10. It illustrates the transmitted electron trajectories with a strongly reduced excitation volume in contrary to Figure 3.

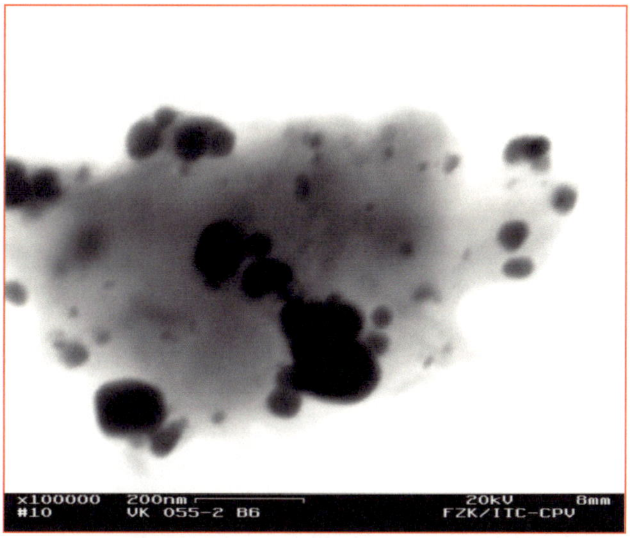

Fig. 9. Transmitted (TE) electron image (100,000 x) of an irregularly shaped single alumina particle (light grey) partly covered with Ag clusters (dark), prepared on a TEM-grid support.

The MC simulation of the electron beam trajectories in Figure 10 and the X-ray line scan in Figure 11 visualizes the interaction of primary electrons with matter of different densities and thickness on a thin-film support. The broadening of the beam (skirt) depends on the specimen thickness and density.

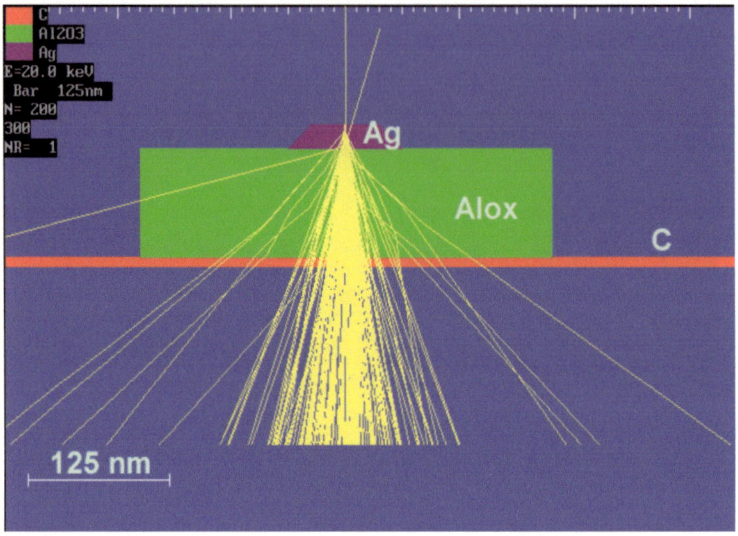

Fig. 10. MC simulation of the electron trajectories of an Ag deposit on an Alox (Al_2O_3) substrate, supported by an 8 nm carbon film (C) (TEM-grid). Note the broadening of the beam by passing the beam electrons through the matter. Backscattering events are drastically reduced (NR =1).

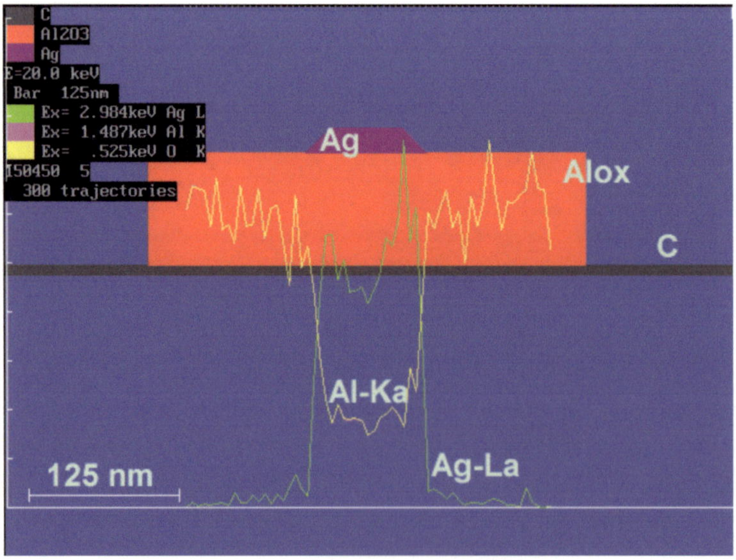

Fig. 11. X-ray line scan simulation of the thin film-supported (C) specimen. The course of the X-ray scan across the specimen is similar to that in Figure 6, but the intensity is decreased (indicated by the statistical fluctuations) due to the reduced beam interaction volume.

Thinner samples have better performance in resolution and contrast. In order to get highly magnified images even of tiny (sub nanometer) Ag particles, a TEM grid has to be prepared.

3.3 Ag distribution by EDX

In Figure 12 one can see the correlation of measured Ag concentration and the percentage of Ag area coverage visualized by image analysis (Figure 13). The image analysis is based on the intensity of the BSE signal.

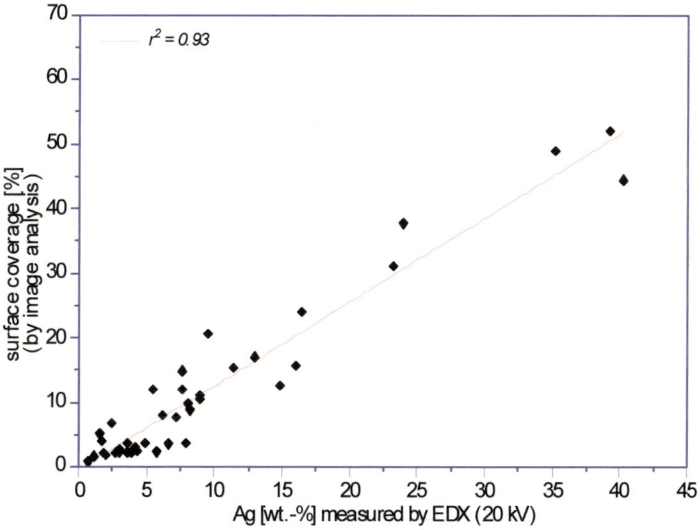

Fig. 12. Ag area coverage derived from the BSE signal and the correlation to the Ag concentration measured by EDX.

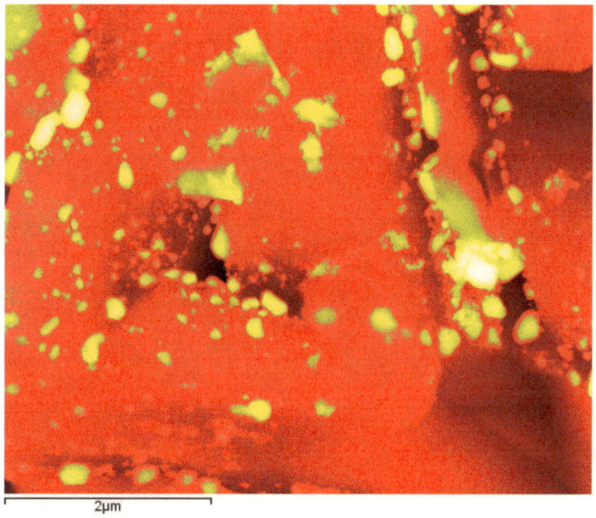

Fig. 13. Color rendering of a BSE image. The yellow areas represent Ag deposits. The red colored areas show the matrix (Al_2O_3).

The yellow spots represent the Ag deposits, approximately 13.5% from the total area. It corresponds to approximately 11 wt.-% Ag measured by EDX. This procedure will not replace the numerical calculation by the evaluation program, but visualizes in a first approximation the dependence of the Ag concentration on the BSE contrast. The Ag distribution on the carrier material is discussed with regard to the penetration depth of primary beam electrons and lower limits of detection for X-ray analysis. It is crucial to catalytic activity that Ag particles cover the carrier material almost homogeneously. Sintering of Ag particles to Ag agglomerates has to be prevented [11]. Using sample D6 (20 % Ag / 1500 ppm Cs) as an example, the FESEM images (Figure 14) and EDX analysis results (Tab. 2) shall be discussed. The image reveals uniform grain sizes of the carrier material and variable coverage (bright areas) by Ag particles or Ag clusters. In the quadrants Q1, Q2, Q3, and Q4 of the SEM image of sample D6 the coverage of the carrier grains by Ag particles (Ag-clusters) varies (bright areas). The apparently uniform catalyst grains are assumed to result from the wet sieving of the carrier material (SLA92 and SC13). The EDX spectrum in Figure 15 reveals all significant elements Ag, Al, O, and Cs. Obviously, some of the grains are densely covered with Ag, others not. Figure 16 shows the shape of huge Ag agglomerates of the D6 sample in a higher magnification (30,000 x). There is no significant difference in the measured Ag surface content between the 4 quadrants in Figure 14, which indicates an "apparently homogeneous" distribution of Ag.

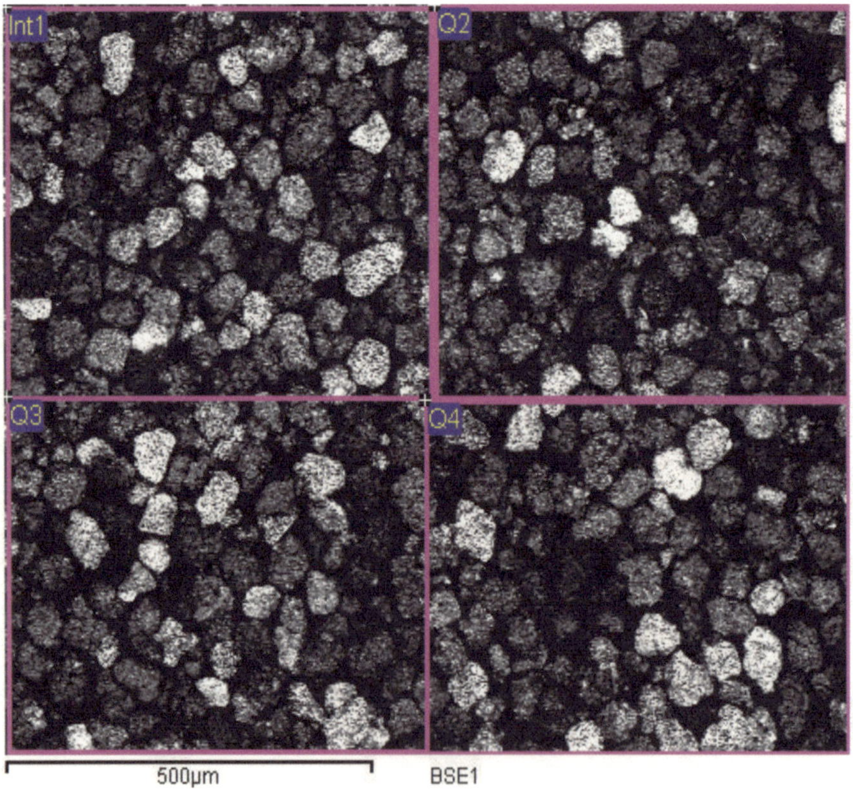

Fig. 14. SEM image of sample D6 (100 x, 20 kV) showing the 4 quadrants and the whole (integral) surface area analyzed by EDX (Tab.2).

EDX results sample D6	C	O	Al	Ag	Cs
Int1	3.0	51.8	29.8	15.1	0.12
Q1	2.6	51.3	30.2	15.9	0.00
Q2	5.4	51.2	30.3	13.1	0.00
Q3	1.9	52.2	30.6	15.2	0.15
Q4	3.1	51.4	29.1	16.4	0.00
average	3,3	51.6	30.0	15.1	0.01
standard deviation	1.3	0.40	0.60	1.30	0.07
max.	5.4	52.1	30.6	16.4	0.15
min.	1.9	51.2	29.1	13.1	0.00

Table 2. Results (wt %) of the EDX analysis of sample D6 shown in the FESEM image Figure 14 (100 x, 20 kV).

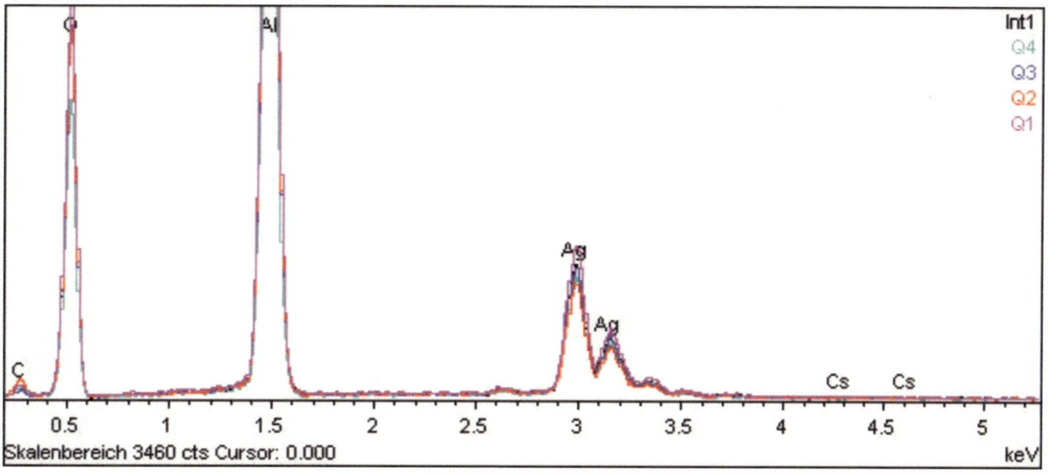

Fig. 15. EDX analysis of sample D6 (Figure 14) revealing Ag, Al, O, and Cs as significant elements.

According to Tab. 2, the average Ag concentration is 15.1% (average of Int1, Q1 to Q4) and the Ag concentrations in the quadrants do not vary significantly (Q1 = 15.9 % Ag, Q2 =13.1 % Ag, Q3 = 15.2 % Ag, and Q4 = 16.4 % Ag). But the Ag concentrations doesn't correlate well with the absolute amounts of Ag (20 wt %) used for catalyst preparation. In general, the heterogeneous catalysts are influenced considerably by the penetration depth of the electron beam because of layer thickness, Ag cluster size, surface coverage, preparation performance and, under certain conditions, Ag bulk-dissolved oxygen (see chapter 3.6). Hence, the Ag values determined by EDX could differ from the absolute Ag concentrations of the prepared samples. The accuracy is strongly influenced by morphological effects and heterogeneity of the Ag coverage of the analyzed samples. Availability of reliable standards in this respect would be beneficial. In either case, for an absolute Ag determination a quantitative digestion should be carried out (4 wt% HF) and analyzed by ICP-AES (Inductive coupled plasma - atom emission spectroscopy) under appropriate conditions.

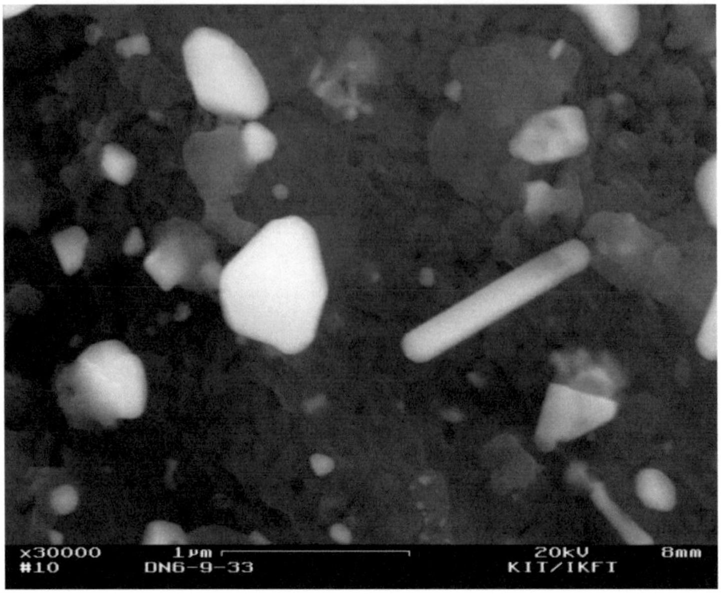

Fig. 16. Shape of huge Ag agglomerates of the D6 sample at 20 kV and 30,000 x.

3.4 EDX mapping

The dispersion of the Ag and Cs metal deposited on the catalyst support can also be characterized by the EDX mapping technique. Topographical and elemental imaging is possible at the same time. EDX mapping is a valuable tool to indicate the quality of the Ag dispersion. The electron beam is scanned pixel by pixel across a selected area of interest. In the following example, the element mappings for a selected specimen area (Figure 17, BSE image, bright spots indicating metal) revealed an inhomogeneous distribution of Ag (Figure 18) and Cs (Figure 19) of the catalyst support. Generally, the brighter the color appears, the higher is the concentration of the specific element. In both cases it correlates with the results in Tab. 3. Cs as a promoter facilitates Ag distribution on the α-Al_2O_3 surface based on the high coverage and lack of crystallites with large contact angles [12]. The presence of Cs in the catalyst improves the distribution of Ag over the support and the Ag/Al interfacial area. Area P1 (see Figure 17 and Tab. 3), for instance, shows a remarkably high silver content of 66%, which is about twice as high as the average. The deviation from the average amount of Cs is not noticeable, because Cs is only present in small quantities (ppm). Nevertheless, the distribution of Ag and Cs can be shown very impressively by the time-consuming mapping. Spectra of the analyzed regions are shown in Figure 20. In summary it can be said that the inhomogeneity of metal distribution on the catalyst carrier material can be shown very clearly also for small amounts of metal. This example demonstrates a less quality of the preparation. One can see that the texture of the substrate SLA2 is very rough in comparison to SC31, which means that a homogeneous distribution of Ag and Cs is not so easy to achieve and requires an improvement in preparation. In case of an excellent preparation, this is a optimal Ag distribution for both catalyst systems, the question is which of the two catalyst systems is more favourable for the Ag-catalyzed epoxidation of 1,3-butadiene (1,3-BD) to 3,4-epoxy-1-butene (EpB) in the reactor.

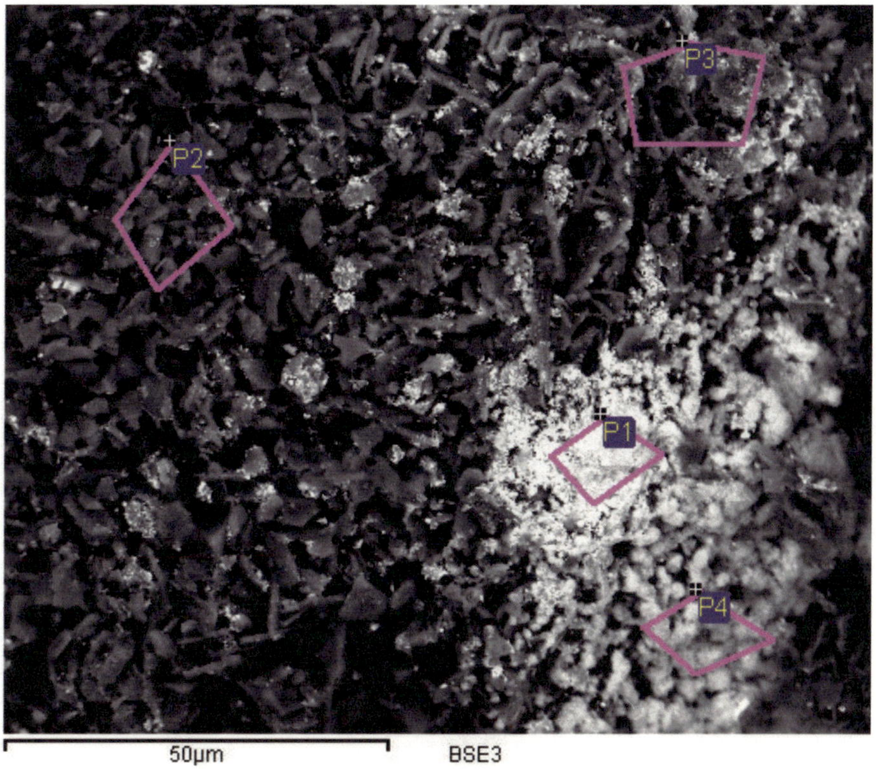

Fig. 17. BSE image of a selected area of an Ag catalyst particle (MZ06). The marked (red) areas P1, P2, P3 and P4 represent analyzed regions.

Fig. 18. Map of Ag-Lα showing the inhomogeneous distribution of Ag. The Ag concentration correlates with the intensity of the green area (measuring time about 5 h).

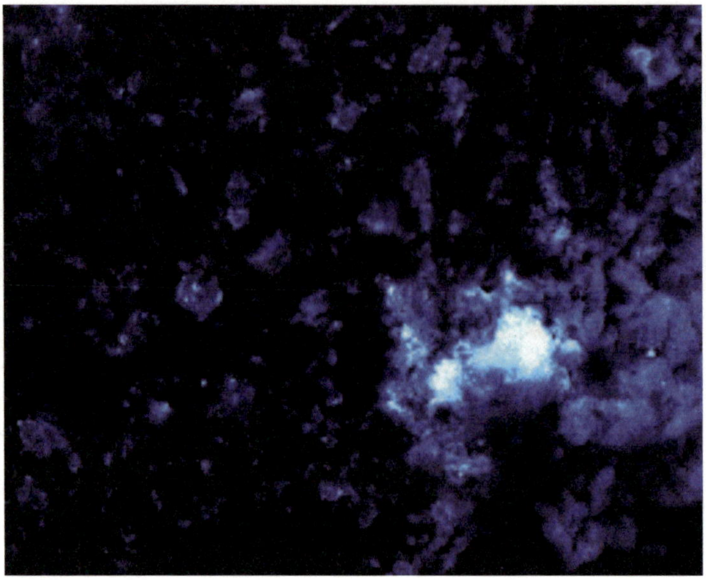

Fig. 19. Map of Cs-Lα showing the inhomogeneous distribution of Cs. The Cs concentration is correlated with the intensity of the blue area(measuring time about 5 h).

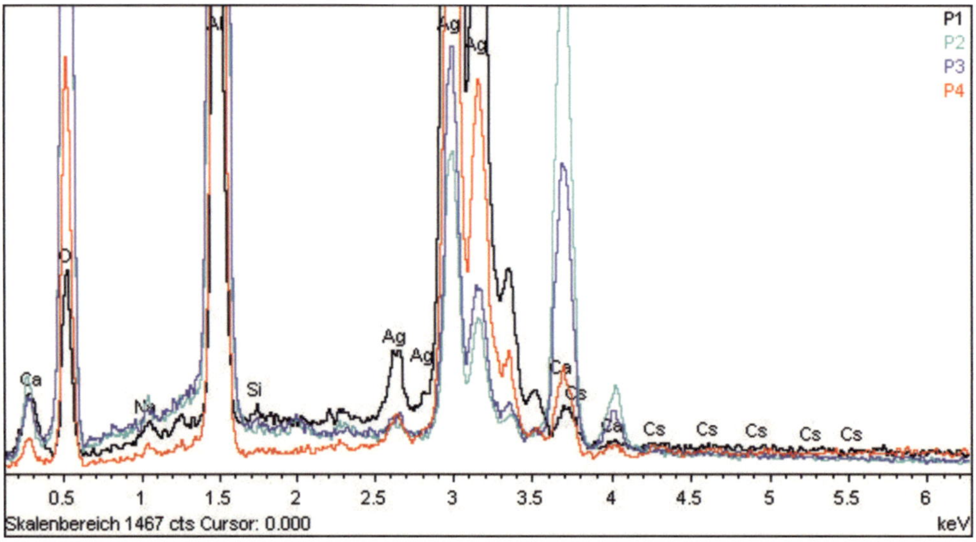

Fig. 20. Spectra of the analyzed regions (see Figure 17).

The rate of a catalyzed reaction should be proportional to the surface area of the active agent. Therefore it is desirable to have the active phase in form of the smallest possible particle. But the most undesired contribution to the reduction of the active surface (deactivation) is sintering (welding together of particles by applying heat below the melting point). The function of the support is to increase the active surface and to reduce the rate of sintering of the metal particles. On the other hand, the interaction between the lattice oxygen of the carriers and the metal particles also influences the behaviour of the active metal agent.

The support can modify the electronic character of the metal particle regarding to its adsorption and reactivity properties. Furthermore, the bond between the metal particle and the support can influence the shape of the metal particle (clusters). Both effects are so-called metal-support interactions (MSI). This effect decreases for supported Ag catalysts in the SiO_2 > Al_2O_3 > C sequence [2].

Spectrum	O	Na	Al	Si	Ca	Ag	Cs	Sum
P1	21.8	0.45	10.08	0.15	0.67	**66.33**	**0.45**	100
P2	54.1	0.32	31.16	0.00	6.62	**7.60**	**0.18**	100
P3	52.5	0.18	34.09	0.08	3.59	**9.53**	**0.00**	100
P4	38.6	0.28	19.59	0.04	2.16	**38.32**	**0.92**	100
average	41.8	0.31	23.73	0.07	3.26	**30.45**	**0.39**	100
max.	54.1	0.45	34.09	0.15	6.62	**66.33**	**0.92**	
min.	21.8	0.18	10.08	0.00	0.67	**7.60**	**0.00**	

Table 3. All results in wt.-%, derived from the BSE image in Figure 17.

Another tool to elucidate the distribution of elements is the so called CAMEO™ imaging. Figure 21 represents a CAMEO™ image from the examined specimen which is based on BSE detection. CAMEO™ is a tool to convert X-ray energies into visible wavelengths. In comparison to the X-ray mapping technique, the CAMEO™ procedure is much faster, but it may lead in some cases to a false color rendering (color overlap) caused by adjacent X-ray energies. As depicted in Figure 21 one can see that the brownish areas obscured the small Cs-spots. Therefore, Cs and Ag are not distinguishable because they are located at the same area of surface.

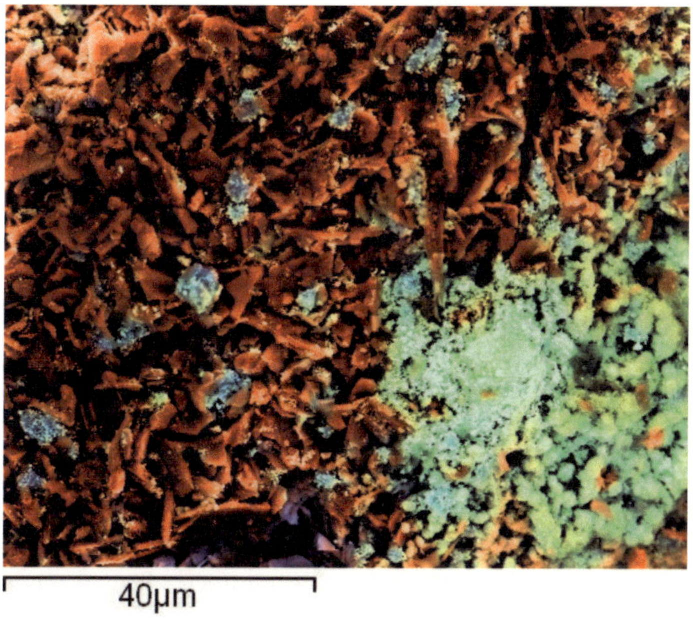

Fig. 21. CAMEO™ rendering of the area shown in Figure 17. The green colored areas represent Ag, whereas the brownish areas represent Al, respectively.

3.5 Remarks

All analytical results have been obtained at 20 kV acceleration voltages, which turned out to be the best choice for excitation conditions. Because the in-lens detector is switched off above 20kV, therefore high resolution SE imaging is disabled. The EDX results are determined by the manufacturer's spectrum evaluation software (Oxford corp.). Originally, all quantitative results are calculated with two significant fractional decimal digits provided with 1σ errors, which includes the errors resulting from spectra processing (background subtraction, filtered least squares fitting, peak overlap). Notably, this may not reflect uncertainties caused by surface topography and other systematic influences. According to experience, the actual uncertainties are considerably higher, especially for light elements. We suggest for C and O a relative uncertainty of 5 – 20%, for all other elements 1-5% [13]. Another important issue is the estimation of the *limit of detection* (LOD), which can be calculated from a synthetic spectrum by the equation 2.

$$LOD_{3\sigma} = 3 \cdot \sqrt{B} \cdot \frac{C}{P} \tag{2}$$

Where B is the number of background counts, $3 \cdot B^{1/2}$ represents 3σ error of background measurement, C is the concentration of the element, P corresponds to the number of counts in the X-ray line after background subtraction. In practice, calculation will be performed by a special program tool named '*spectrum synthesis*', which is provided with the INCA-Energy evaluation software [14].

LOD 3σ / w t.- %		I / nA	t_m / s	U / kV
Ag	Cs			
0.21	-	0.23	60	20
0.18	0.27	0.23	100	20
0.09	0.15	0.71	100	20
0.06	0.09	0.71	300	20

Table 4. Calculated 3σ LOD's regarding to equation 2 and [14]. The matrix composition, except for alumina, is assumed to be 1.8 wt.-% Ca, 0.3 wt.-% Na, 0.2 wt.-% P.

For the 3σ LOD estimation, the composition of the sample matrix and the acquisition parameters like measuring time t_m, beam voltage U, current I, detector parameters, X-ray take-off angle are required. Tab. 4 shows 3σ LOD's which are calculated for 3 different measuring times and aperture adjustments (beam current). For Ag, this is of minor importance, since the amounts of Ag are sufficiently high in comparison to the promoters.The pore size of the alumina support is in a similar order of magnitude as the X-ray generating range (see Fig. 1b), which leads to an uncertainty in the measurements in these regions. As a consequence, the EDX analysis is considered to be semi-quantitative on such Al_2O_3 supports. A loss of X-rays (Al-Kα and O-Kα) emitted from mesoporous media compared to that of dense monocrystalline alumina is described in literature [15]. It may be caused by charge effects due to the specific surface of particles and distribution of pores. In the course of our various measurements, no significant charging was observed. Compared to common *inductively coupled plasma atomic emission spectroscopy* (ICP-AES) the values of metal deposition determined by EDX are frequently higher. One explanation is a sometimes observable incomplete chemical digestion prior to ICP measurements. In the case of Al_2O_3 we recommend a digestion with HF (4%). As already mentioned, EDX with an electron penetration depth in the order of several nanometers up to microns is a technique for surface analysis, while the ICP technique is applicable to the quantitative determination of the bulk composition with the detection limits in the µg/l (ppb) range. Due to the incipient wetness impregnation technique applied here for surface preparation, the values of metal deposition analyzed by EDX are expected to be higher than those by ICP-AES [16]. The differences in the amounts of Ag are plausible. The reason therefore is that the D-samples and MZ-samples are completely chemically digested, which is a physical homogenization. In contrast, the sample preparation for FESEM/EDX is non-destructive, which means that the sample is not physically homogenized. Tab. 5 gives a selected overview about catalysts characterization methods applied on sample D1.

Promoter $CsNO_3$ / wt.-%	ICP Ag / wt.-%	EDX Ag / wt.-%	BET m^2g^{-1}
0.15	4.46	5.76	6.03

Table 5. A listing of commonly applied laboratory methods (ICP, EDX, BET) for characterization of a catalyst with nominal 5 wt.-% Ag, grain fraction 45-63 µm (sample D1).

3.6 O_2-TPD and BET measurement

To get a deeper insight it makes sense to combine SEM with O_2-TPD experiments and BET measurements, which additionally were carried out. The TPD experiment measures the temperature-dependent desorption rate of a molecule from the catalyst surface [17]. Typically, O_2 / Ag interactions are studied on the Ag monocrystal surfaces (110) and (111) [18]. This means that silver may storage O_2, with the amount adsorbed being dependent on temperature and O_2 partial pressure. O_2 dissociatively adsorbs on the Ag catalyst and may assume the following characteristic forms:

- Surface oxygen
- Subsurface oxygen
- Ag bulk-dissolved oxygen

Ag bulk-dissolved oxygen may act as a storage of converted surface oxygen and, hence, be supplied later on. Subsurface O_2 increases the coordination number of Ag surface atoms, which results in a smaller binding strength of surface oxygen and favourably influences the epoxidation reaction of 1,3-BD. The surface area has been corrected by the subsurface oxygen value. Comparison of the measured amount of O_2 desorbed with the amount theoretically required for an O_2 monolayer shows that the measured amount of desorbed O_2 is higher by a factor of 2. This may be explained by the presence of subsurface O_2 [19].

O_2-TPD studies reveal significant differences between MZ samples with an SLA92 carrier and D samples with an SC13 carrier for the same Ag contents. In general it can be said that the desorbed O_2 amounts of the MZ samples are below those of the D samples by a factor of 5. Moreover, the temperatures of the desorption maximums of the MZ samples are lower. This difference (binding strength of oxygen) results in a strongly variable reactivity of the adsorbed oxygen species. Figure 22 shows O_2 desorption and temperature of the samples D1, D3, and D5 with different Ag concentration on the support that is *not doped with Cs*. With increasing Ag concentration, the desorptions maxima are shifted towards lower temperatures (red line). This indicates that epoxidation already may start at lower temperatures. Figure 23 shows that the sample D6 doped with Cs reaches a higher temperature at the desorption maximum than the non-doped sample D5.

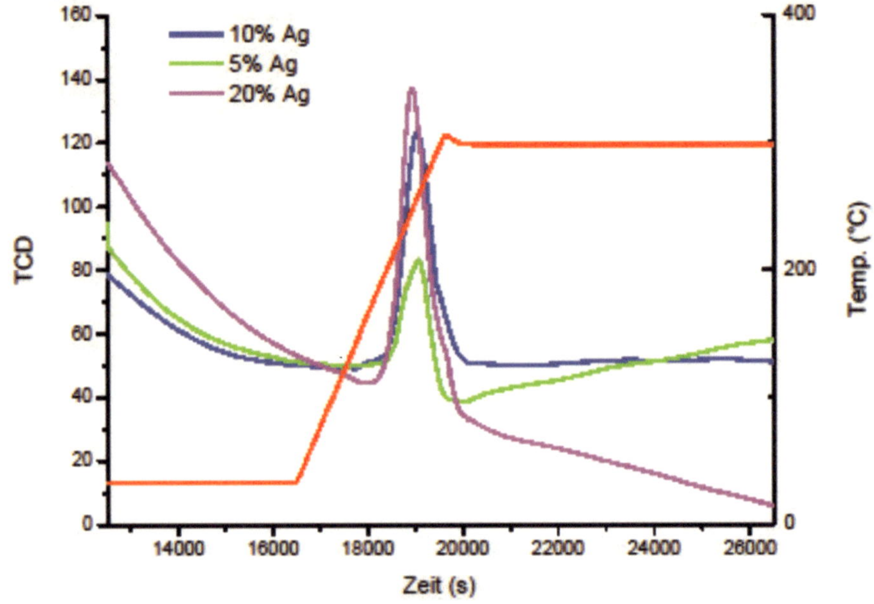

Fig. 22. O_2 detection (WLD) by O_2-TPD of the samples D1 (5 % Ag), D3 (10 % Ag), and D5 (20 % Ag) versus temperature (red line).

This indicates that the adsorbate is stabilized by the presence and the grade of distribution of Cs and Ag on the surface, which correlates with TPD results, FESEM, EDX mapping and the reaction performance. The determination of the specific surface areas of the catalysts D3, D4, D5, D6, MZ06 and MZ09 gives an important hint regarding to metal distribution of MSI effects. Tab. 6 lists the specific surface areas for the mentioned catalysts determined with N_2 as adsorptive. All isotherms are of the IUPAC type II ("s-shaped") and, hence, can be evaluated according to the BET theory [20], [21]. The Cs-doped samples (D4, D6) have a slightly higher surface than the non-doped samples (D3, D5) with the same Ag content, which correlates to a higher desorption rate for O_2-TPD. In case of the samples MZ06 and MZ09 one can see that the difference in BET surface is not remarkable, even in the presence of different Ag amounts. The total error for BET measurement is about 0.1 m^2/g.

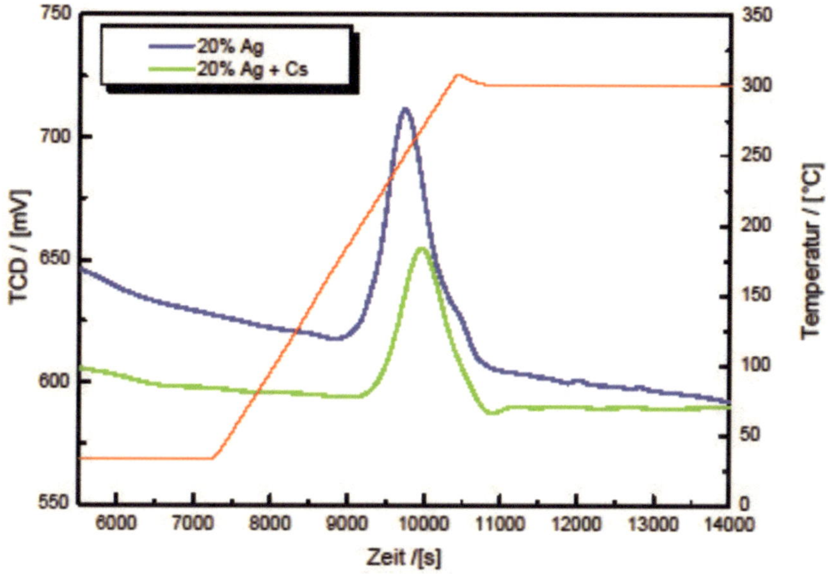

Fig. 23. O2 detection (WLD) by O2-TPD of samples D5 (20 % Ag) and D6 (20 % Ag + Cs) versus temperature (red line).

	D3 Ag 10%	D4 Ag 10% + Cs	D5 Ag 20%	D6 Ag 20% + Cs	MZ06 Ag 5% + Cs	MZ09 Ag 10% + Cs
$a_{s,BET}$ [$m^2 g^{-1}$]	6,0424	6,5521	7,6083	8,0461	0,80722	0,7009

Table 6. Specific surface areas from sample D3 - D6 and MZ06 and MZ09.

The BET reveals that there is a significant difference in the catalyst surface, which is expressed by the topological images of FESEM of the D samples (SC31) and MZ (SLA92) samples. Note: Very small amounts (0,1g) of sample are used for the investigations (FESEM, TPD). In the case of BET-Measurement the sample weight is in the range of 0.1 g up to 10 g, depending on the specific surface. The carrier material SC13 (high BET Surface) was not found to be suited for the epoxidation of 1,3-butadiene (Figure 24) on Ag particles. Already at 180°C a total oxidation of 1,3-butadiene to CO_2 and H_2O does occur, also when Cs-doped catalysts are used. In contrast to this, epoxidation of the Ag catalyst with SLA29 (less BET surface) carrier material results in an EpB selectivity of 74 % (200° C, SV = 2590 h^{-1}) at a 1,3-BD conversion rate of 15 % [22]. The constitution of the oxametallacycle intermediate is depicted in Figure 25. The guiding hypothesis is that surface oxametallacycles are key intermediate for epoxidation on Ag / Cs catalysts. The intermediate EpB(ads), finally leading to molecular EpB, is probably strongly adsorbed on the catalyst surface indicated by theoretically calculations which also support its identity as an oxametallacycle. The oxametallacycle intermediate is more thermodynamically stable than EpB by approx. 24 kcal/mol. Moreover, the transition state for EpB formation from the oxametallacycle intermediate is actually lower in energy than the reactants, butadiene and oxygen [23].

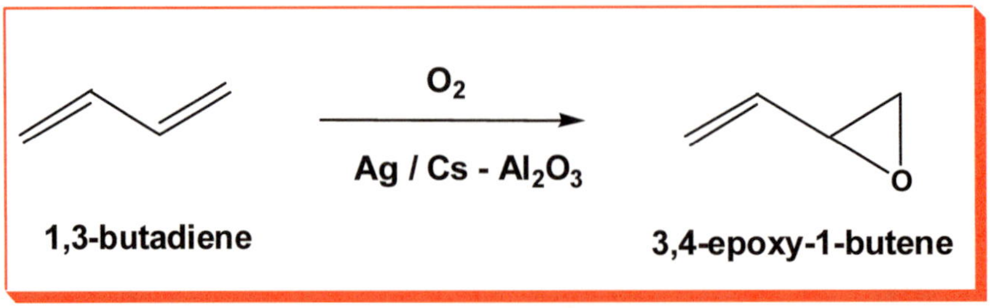

Fig. 24. Epoxidation of 1,3-butadiene over Ag / Cs catalyst on Al_2O_3 to 3,4-epoxy-1-butene.

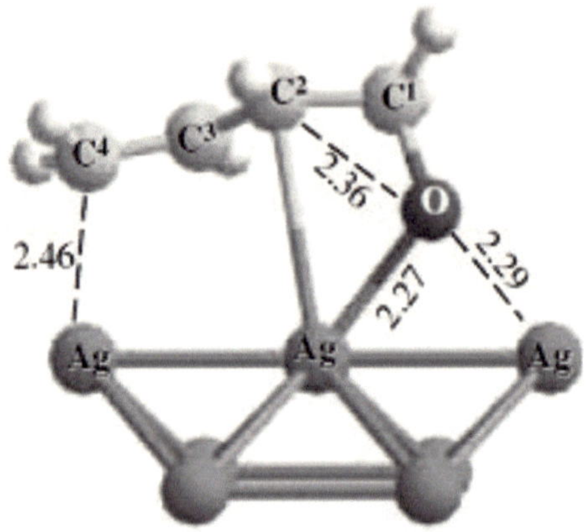

Fig. 25. The "intermediate EpB" finally leading to molecular EpB. The oxametallacycle intermediate is more thermodynamically stable than EpB. Moreover, the transition state for EpB formation from the oxametallacycle intermediate is actually lower in energy than the reactants, butadiene and oxygen

4. Conclusions and outlook

Production of Ag catalysts based on a corundum-containing (SC13) and a calcium hexa aluminate-containing (SLA92) carrier material is crucial to the selective epoxidation of 1,3-butadiene. Optimum distribution and morphology of the Ag particles must be ensured by controlled, tailored catalyst synthesis. An increase in the activity by enhanced Ag dispersion on a corundum-containing carrier material with a larger surface area leads to completely unselective catalysts. FESEM/EDX results provide major information with regard to the Ag distribution and the properties of the carrier surfaces. The carrier material SC13 was not found to be suited for the epoxidation of 1,3-butadiene on Ag particles. At 180°C already does a total oxidation of 1,3-butadiene to CO_2 and H_2O occur also when Cs-doped catalysts are used. In contrast to this, epoxidation of the Ag catalyst with SLA29 carrier material results in an EpB selectivity of 74 % (200 °C, SV = 2590 h^{-1}) at a 1,3-BD conversion rate of 15 %. All the analysis methods complement each other to form an overall impression, which is reflected in the product selectivity, catalyst activity and educts conversion during the reaction. The distribution and composition of metal particles on the surfaces can be seen and detected with EDX and FESEM. Also topological and morphological effects can be shown. BET measurements allow drawing a conclusion for successful metal loading. Furthermore, the efficiency of Ag-loading and promoters with TPD measurements can be determined. The analytical result reflects the final behaviour of the epoxidation regarding to the product selectivity and conversion rate of the educts.

5. Acknowledgement

Doreen Neumann-Walter, (preparation), KIT, Germany
Bernhard Powietzka, (preparation, reaction control), KIT, Germany
Sara Essig, (preparation), KIT, Germany
Dr. Angela Puls, (BET measurement), Rubotherm GmbH Bochum, Germany
Dr. Volker Hagen (O_2-TPD measurement), Rubokat GmbH Bochum, Germany

6. References

[1] R. J Wijngaarden, A. Kronberg, K. R. Westerterp, Industrial Catalysis, WILEY-VCH Verlag GmbH Weinheim, 1998
[2] B. Cornils, W. A. Hermann, R. Schlögl, Chi-Huey Wong, Catalysis from A to Z, WILEY-VCH Verlag GmbH, 2000
[3] J. R. Monnier, *Prepr. Pap. - Am. Chem. Soc.,Div. Fuel Chem.* 2007, 52 *(2)*, 163
[4] M. Pohl, S. Hogekamp, N. Q. Hoffmann, H. P. Schuchmann, Chem. Ing. Tech. 2004, 76 (4), 392
[5] J. M. Thomas, R. M. Lambert (Ed.), Characterisation of Catalysts, Chichester, Wiley 1980
[6] P. L. Gai, E. D. Boyes, Electron Microscopy in Heterogeneous Catalysis, Inst. of Physics Publishing, Bristol a. Philadelphia, 2003
[7] W. Habicht, N. Boukis, G. Franz, O. Walter, E. Dinjus, Microsc. Microanal. 2006, 12, 322-326
[8] C. E. Lyman et al.; Scanning Electron Microscopy, X-ray Microanalysis, Analytical Electron Microscopy: a Laboratory Workbook; Plenum Press N. Y. (1990)
[9] L. Reimer, Monte Carlo Simulation of Electron Diffusion, updated Version 3.1, Program and Handbook, Münster, Germany 1998
[10] J. I. Goldstein et al., Scanning Electron Microscopy and X-ray Microanalysis, 3rd ed., Kluwer Acdemic/Plenum Publishers, New York, 2003
[11] Handbook of Heterogeneous Catalysis, Vol.1, 2nd Ed. (Eds: G. Ertl, H. Knözinger, F. Schüth, J. Weitkamp) Wiley-VCH, Weinheim 2008, 561
[12] D. M. Minahan, G. B. Hoflund, W. S. Epling, D. W. Schoenfeld, J. of Catalysis 168, 1997, 393-399
[13] W. Habicht, N. Boukis, E. Hauer, E. Dinjus, X-ray Spectrometry 2011, 40, 69-73
[14] INCA Energy, tools for INCA users, provided by Oxford corp., or more profound: P. Duncumb, I. R. Barkshire, P. J. Statham, Microsc. Microanal. 2001, 7, 341-355
[15] L. Sorbier, E. Rosenberg, C. Merlet, X. Llovet, Mikrochim. Acta, 2000, 132, 189-199
[16] C. Xu, J. Zhu, Nanotechnology 2004, 15, 1671-1681
[17] M.I. Szynkoska, E. Lesniewska, T. Paryjczak, Pol. J. Chem. 2003, 77, 657
[18] F. M. Leibsle et al., Phys. Rev. Lett. 1994, 72, 569, 569-2572.
[19] G.W. Busser, O. Hinrichsen, M. Muhler, Cata. Lett. 2002, 79 (1 - 4), 49

[20] D. D. Do, Adsorption Analysis: Equilibria and Kinetics, Imperial College Press, London 1998
[21] P. Christopher and S. Linic, ChemCatChem 2010, (1), 2, 78
[22] T. N. Otto, P. Pfeifer, S. Pitter, B. Powietzka, *Chem. Ing. Tech.* 2009, 81 *(3)*, 349
[23] Mark A. Barteau, Topics in Catalysis Vol. 22, Nos. 1/2, January 2003

20

Fractal Analysis of Micro Self-Sharpening Phenomenon in Grinding with Cubic Boron Nitride (cBN) Wheels

Yoshio Ichida
Utsunomiya University,
Japan

1. Introduction

Self-sharpening phenomenon of the grain cutting edges during grinding is the main factor controlling the performance and the tool life of grinding wheels. Therefore, many studies on the relationship between the wear behavior and the self-sharpening of the grain cutting edges have been carried out (Yoshikawa, 1960; Tsuwa, 1961; Ichida et al., 1989, 1995; Malkin, 1989; Show, 1996). However, it is very difficult to evaluate this relation quantitatively because of the complexity in wear mechanism and the irregularity in shape and distribution of the grain cutting edges (Webster & Tricard, 2004). Especially, self-sharpening of the cutting edges in the grinding process with cBN wheels has not yet been clarified sufficiently (Ichida et al. 1997, 2006; Guo etal., 2007). To develop an innovative machining system using cBN grinding wheels, it is essential to clarify the self-sharpening mechanism due to the micro fracturing of the cutting edges that is the most important factor controlling the grinding ability of cBN wheel during the grinding process (Ichida et al. 2006; Kalpakjian, 1995; Comley et al., 2006).

The main purpose of this study is to evaluate quantitatively such a complicated self-sharpening phenomenon of the cutting edges in cBN grinding on the basis of fractal analysis. The changes in three-dimensional surface profile of cBN grain cutting edge in the grinding process are measured using a scanning electron microscope with four electron detectors and evaluated by means of fractal dimension.

2. Three-dimensional fractal analysis

There are several methods for calculating fractal dimension (Mandelbrot, 1983; Mandelbrot et al. 1984; Hagiwara et al., 1995; Itoh et al., 1990). In this report, we have used a 3D-fractal analysis that is expanded based on the idea in the fractal analysis using two-dimensional mesh counting method (Sakai et al., 1998). The analysis method is shown as follows. A 3D-profile under test is divided by cube grid with a mesh size of r. And then, the number of cubes intersected with 3D-profile $N(r)$ is counted. If there is a fractal nature in this 3D-profile, the relationship between $N(r)$, r and fractal dimension D_S is given by

$$N(r) = \alpha \cdot r^{-D_S} \qquad (1)$$

where α is constant number.

Area of square with mesh size r is expressed r^2. Therefore, the surface area of 3D-profile $S(r)$ based on $N(r)$ is given by

$$S(r) = r^2 \cdot N(r) = \alpha \cdot r^2 \cdot r^{-D_S} \tag{2}$$

If the logarithm of both sides is taken, eq. (2) is rewritten as follows;

$$\log S(r) = \log \alpha + (2 - D_S)\log r \tag{3}$$

Fractal dimension D_S is calculated by the following equation using the proportionality constant between log S(r) and log r in eq. (3).

$$D_S = 2 - \frac{d \log S(r)}{d \log r} \tag{4}$$

However, actual fractal analysis is conducted according to the following procedures by computer processing in this study. As shown in Fig. 1 (a), a square grid with mesh size r_1 is set on a 3D-profile of the top surface of grain cutting edge. It is divided to two triangular elements with mesh size r_1. Surface areas of each triangle $s_1(r_1)$ and $s_2(r_1)$ are evaluated using height coordinates in each grid point and $S(r_1)$ is decided by sum of these surface areas. Next, as shown in Fig. 1 (b), each triangle is divided with mesh size r_2 that is half a size of r_1. Surface areas of 8 triangles $s_1(r_2)$, ..., $s_8(r_2)$ are evaluated using height coordinates in 9 grid points and $S(r_2)$ is decided by sum of these surface areas. In addition, as shown in Fig. 1 (c), 8 triangles are divided with mesh size r_3 that is half a size of r_2. Surface areas of 32 triangles $s_1(r_3)$, ...,$S_{32}(r_3)$ are evaluated using height coordinates in 25 grid points and $S(r_3)$ is decided by sum of these surface areas. Afterward, mesh size r is scaled down and the surface area of 3D-profile is evaluated as follows;

$$S(r_n) = \sum_{i=1}^{2^{2n-1}} s_i(r_n) \tag{5}$$

On the basis of these equations, r is taken on the horizontal log axis, and $S(r)$ is taken on the vertical log axis. When data points are on a straight line in double log plot, fractal dimension D_S is given by a slope of the straight line. Figure 2 shows an example of relationship between $S(r)$ and r (Sample: surface profile of cBN cutting edge shown in Fig.1). Fractal nature is approved in a region of $0.4 < r < 4$ μm. From a slope of the straight line, it is decided that fractal dimension is 2.015.

3. Three-dimensional observation of wheel working surface

3.1 Experimental procedure

Grinding experiments were conducted with surface plunge grinding method on a horizontal spindle surface grinding machine. The schematic illustration of the experimental setup is shown in Fig. 3. A vitrified cBN wheel with a replaceable cBN segment shown in Fig. 3 was used to observe directly the profile of the wheel surface in the grinding process using a three-dimensional (3D) scanning electron microscope with four electron probes (3D-SEM/

EDM3000) (Ichida, 2008). The cBN segment detached for the observation can be precisely returned to former state again. It was confirmed experimentally that the cBN wheel with the replaceable cBN segment has almost the same grinding ability as the usual complete cBN wheel (Fujimoto, 2006).

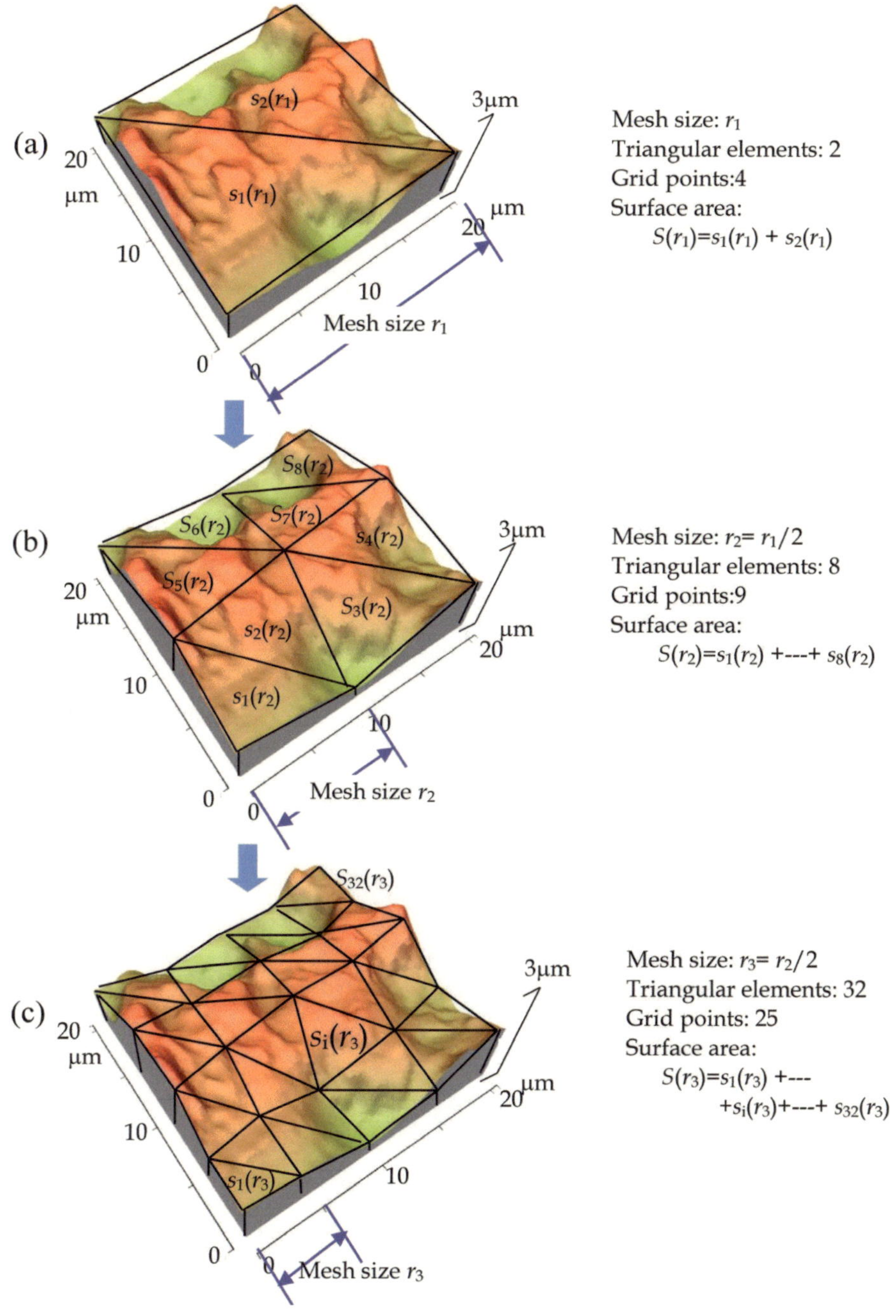

Fig. 1. Method of 3D-fractal analysis (Sample: surface profile of cBN abrasive grain).

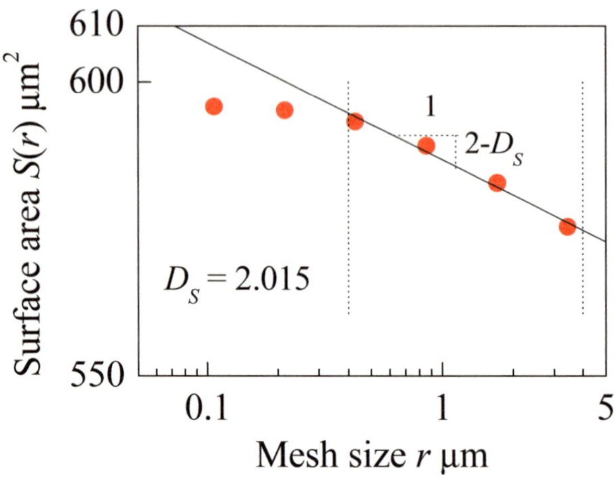

Fig. 2. Relationship between surface area $S(r)$ and mesh size r (Sample: surface profile shown in Fig.1).

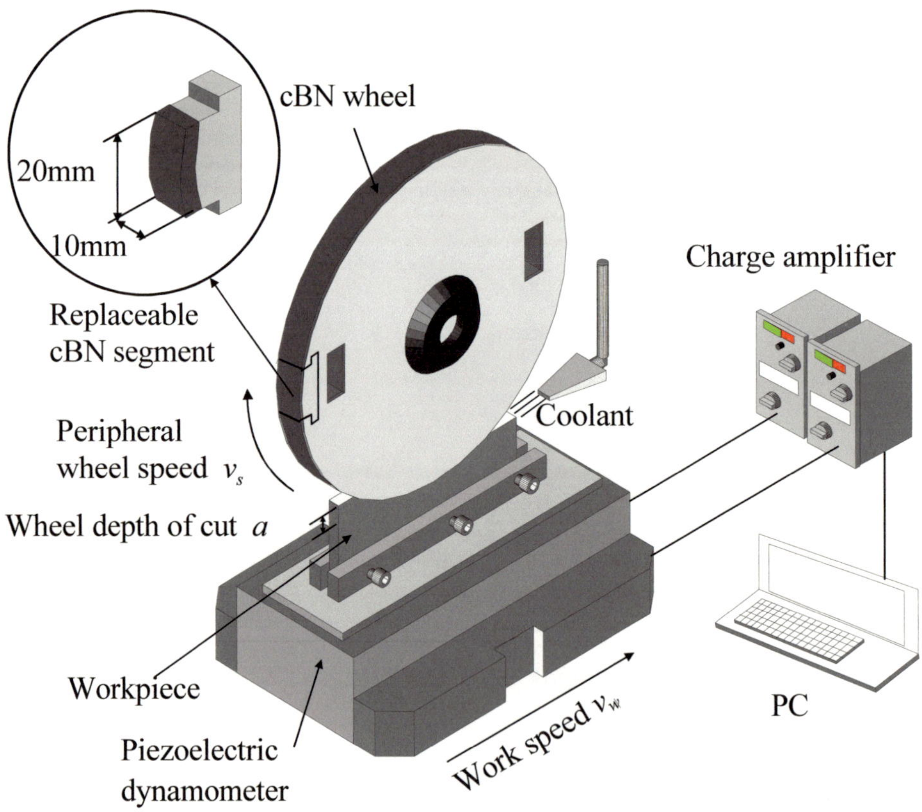

Fig. 3. Schematic illustration of the experimental setup.

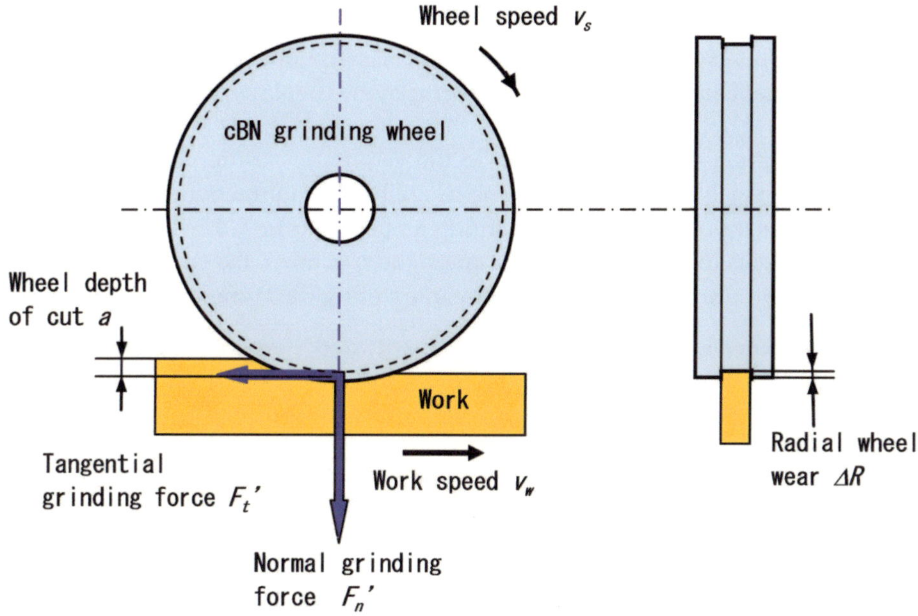

Fig. 4. Measurements of grinding characteristic palameters.

The expermental conditions are listed in Table 1. Representative single crystal cBN grain was used for cBN wheel. The dressing of cBN wheel was performed using a rotary diamond dresser (Dressing wheel: SD40Q75M) equipped with an AE sensor under the following dressing conditions: peripheral dressing speed 16.5 m/s, peripheral wheel speed ratio 0.5, dressing lead 0.1 mm/rev, dressing depth of cut 2μm×5 times. High speed steel SKH51/JIS (M2/ASTM) is used as the workpiece material.

Grinding method	Surface plunge grinding(Up-cut)
Grinding wheel	CBN80L100V Dimensions:200 ^{D}× 10^{T} [mm]
cBN grain	Single crystal cBN
Peripheral wheel speed v_s	33 [m/s]
Work speed v_w	0.1 [m/s]
Wheel depth of cut a	10 [μm]
Grinding fluid	Soluble type (JIS W-2-2) 2% dilution
Workpiece	High speed steel (JIS/SKH51) Hardness: 65HRC Dimensions:100^{l}×5^{t}×30^{h} [mm]

Table 1. Grinding conditions.

3.2 Measuring method of wheel surface profile with 3D-SEM

This 4-channel secondary electron (SE) detection system enables quantitative surface roughness measurements and enhances the topography by displaying the differential signal calculated from the 4 signals. The intensities of these detected signals are determined by the tilt angle of the specimen surface in relation to the geometric positioning of the detectors. The quantitative angular information can be obtained by the subtraction between the signal intensities of the detectors. By calculating 4 tilting angles (two in X-direction and two in Y-direction) on many spots in the X-Y matrix taken on the specimen, the surface topography of the specimen can be accurately re-constructed by integrating these angles over the matrix.

In this system, no eucentric tiling for stereo pairs is required, thereby simplifying operation and allowing much better precision and resolution than conventional SEMs using stereo photogrammetry. The vertical resolution in measuring a 3D profile using this 3D-SEM is 1 nm.

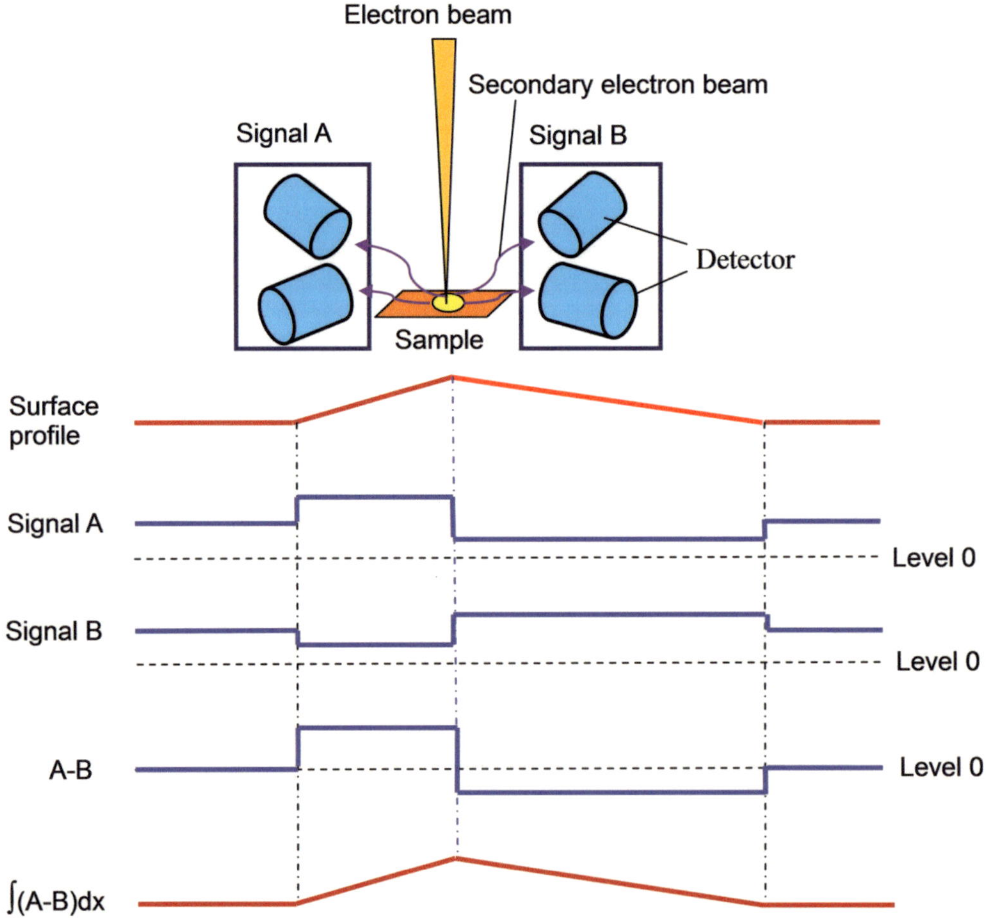

Fig. 5. Illustration of the 4-channel SE detector layout detailing the measurement principle of 3D-SEM.

4. Grinding wheel wear and wheel working surface

Grinding wheel wear is an important consideration because adversely affects the shape and accuracy of ground surface. Grinding wheel wear by three different mechanisms: attritious grain wear, grain fracture and, and bond fracture, as shown in Fig.6. In attritious wear, the cutting edges of a sharp grain dull by attrition, developing a wear flat. Wear is caused by the interaction of the grain with the workpiece material, involving both physical and chemical reactions. These reactions are complex and involve diffusion, chemical degradation or decomposition of the grain, fracture at a microscopic scale, plastic deformation, and melting. If the wear flat caused by attritious wear is excessive, the grain becomes dull and grinding becomes inefficient and produces undesirable high temperatures. Optimally, the grain should fracture or fragment at a moderate rate, so that new sharp cutting edges are produced continuously during grinding. This phenomenon is self-sharpening. However, self-sharpening by a large fracture is not suitable for precision grinding, because it gives large wheel wear and bad surface roughness during grinding. Therefore, self-sharpening due to micro fracture as shown in Fig.6 is suitable for effective precision grinding, because it offers small wheel wear and good surface roughness. We call this phenomenon 'micro self sharpening'(Ichida, 2008).

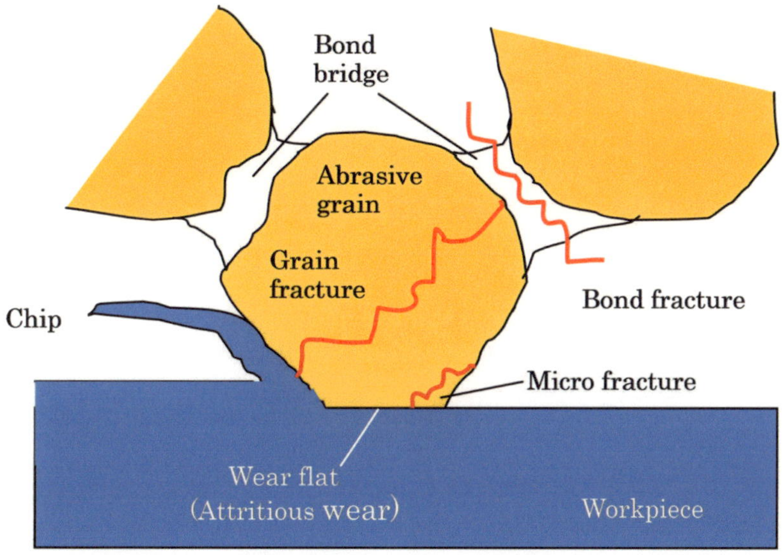

Fig. 6. Wear mechanisms of abrasive grain during grinding.

Fig. 7 shows the change in radial wheel wear ΔR with increasing the accumulated stock removal (cumulative volume of material removed per unit grinding width) V_w' when grinding under the conditions indicated in Table 1. At the same time, some typical sequential SEM images of the wheel working surface with an increase of stock removal are shown in this figure. The wear process of grinding wheel can be divided into two different regions: a) initial wear region over stock removal range from 0 to 1000 mm^3/mm, in which a rapid increase of wheel wear occurs with increasing stock removal, b) steady-state wear region over stock removal range larger than 1000 mm^3/mm, in which the wheel wear rate maintains a nearly constant value. In the initial wear region, a releasing of grain due to bond fracture and grain fracture are sometimes observed, as shown in grains C, D and so on. However, they are not observed so much in the steady-state wear region. The wheel wear in steady-state region dominantly occurs due to attritious wear and micro fracture, as shown in grains A, B and so on. As a typical example, a high magnification SEM image of grain A is shown in Fig.8 (a). Fig.8 (b) is its contour map. Wear flat developed due to attritiou wear and some brittle surfaces generated by micro fracture can be observed on the point of the grain.

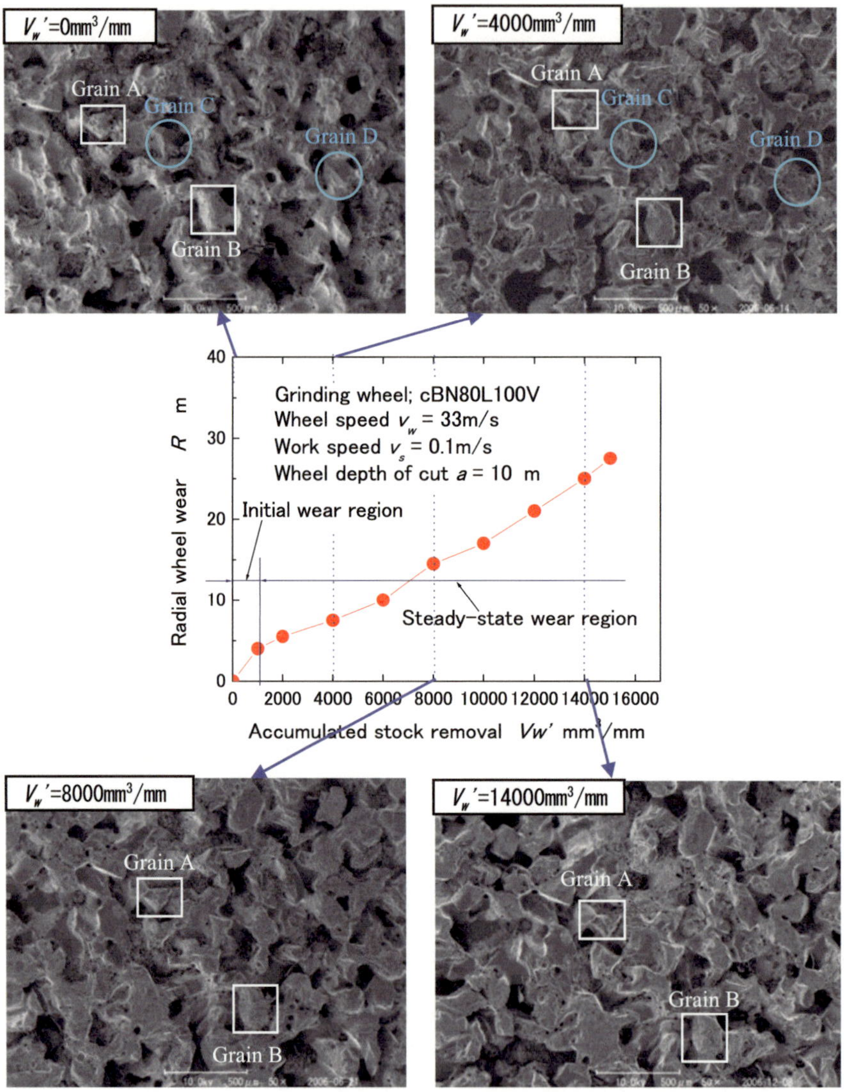

Fig. 7. Change of radial wheel wear with increasing stock removal and typical sequential SEM images of wheel working suraface.

There is little research that has quantitatively evaluated self-sharpening phenomenon of grinding wheel. We have tried to grasp the actual behavior of self-sharpening and evaluate it by the attritious wear flat area percentage of the grain cutting edge (Ichida, 2008, 2009).

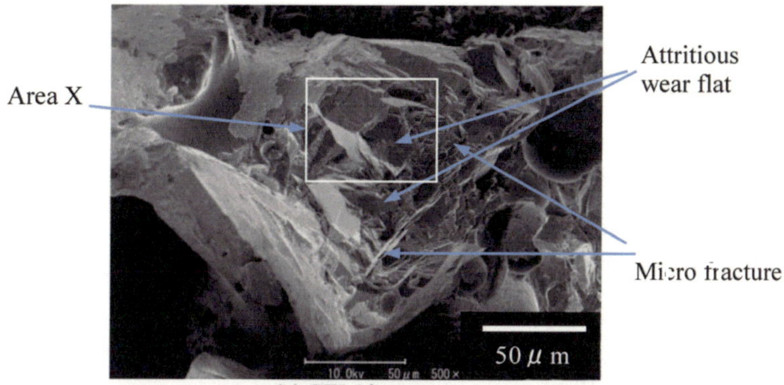

(a) SEM image

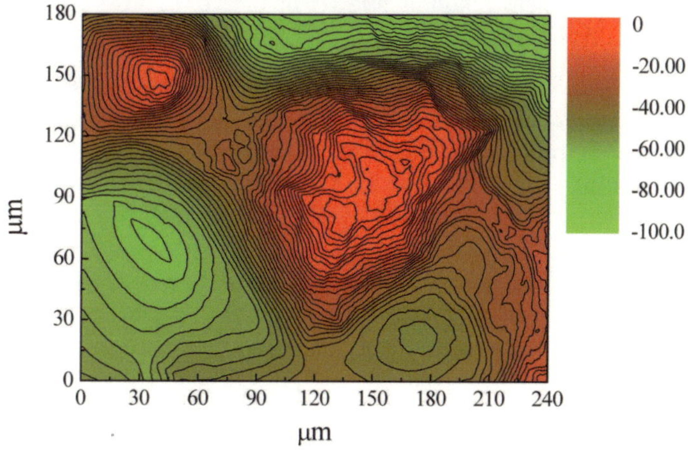

(b) Contour map of Fig. (a)

Fig. 8. High magnification SEM image and its contour map of grain A in Fig.7 (V_w'=4 000mm³/mm).

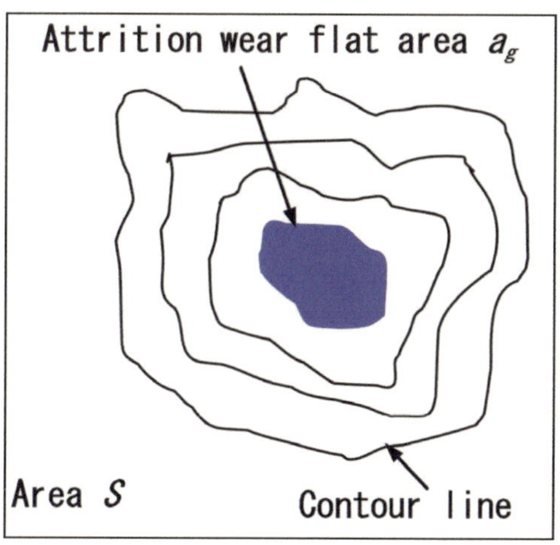

Fig. 9. Measuring method of attritious wear flat percentage.

Figure 9 shows the measuring method of the attritious wear flat percentage. Attritious wear flat area a_g in the area S to be observed is measured using SEM image and contour map made by 3D-profiles. Here, attritious wear flat percentage A_g is given by:

$$A_g = \frac{a_g}{S} \times 100[\%] \quad (6)$$

5. Self-sharpening phenomenon due to micro fracture of cutting edges

Grain cutting edges on the wheel surface change their shapes in various forms with the progress of wheel wear when the accumulated stock removal V_w' increases. Many sequential observations of the grain cutting edge with accumulated stock removal have been carried out using 3D-SEM. The typical result is shown in Fig. 10. Those are high magnification images of area X on grain A in Fig. 8(a).

The surface with a micro unevenness formed by the diamond dresser is observed on the tip of the grain cutting edge after dressing, as shown in Fig. 10 (a). And after grinding the stock removal V'_w = 500 mm^3/mm, an attritious wear flat is observed in the center part on the top surface of the grain cutting edge, as shown in Fig. 10 (b). Moreover, at V'_w = 2000 mm^3/mm, the wear flat becomes larger than that at V'_w = 500 mm^3/mm, as indicated in the comparison between Figs. 10 (b) and (c). The ductile attritious wear flat area takes the largest value at V'_w = 2000 mm^3/mm in the grinding process, as seen from all SEM images in Fig.4. However, between the stock removals from 2000 to 4000 mm^3/mm, some micro fractures take place at the lower left side part of cutting edge and consequently the wear flat area decreases a little, as indicated in the comparison between Figs. 10(c) and (d). Moreover,between the stock removals from 4000 to 10000 mm^3/mm, as many micro fractures take place repeatedly, the ductile attritious wear flat area is decreased and some new sharp edges are formed on the top surface of cutting edge, as shown in the comparison between Figs. 10 (e) and (f).

In addition, between the stock removals from 10000 to 12000 mm^3/mm, a small fracture takes place at the right side part of cutting edge and some new sharp edges are formed, and at the same time the wear flat is formed slightly in the center part on the cutting edge surface, as indicated in the comparison between Figs. 10 (f) and (g). Afterward, between the stock removals from 12000 to 14000 mm^3/mm, some new sharp edges due to the micro fracture are observed in the middle part on the cutting edge top surface, while the new attritious wear flat is formed again at the upper part of the cutting edge, as indicated in the comparison between Figs. 10 (g) and (h).

Thus, although the grain cutting edges become dull due to the ductile attritious wear, they can reproduce and maintain their sharpness due to the micro fractures occurred repeatedly on their top surfaces. Namely, an actual behavior of the self-sharpening phenomenon due to the micro fracture may be grasped on the basis on the sequential SEM observation method used in this study.

6. Evaluation of self-sharpening using fractal dimension

As mentioned above, the shape of the cutting edges on the wheel working surface is variously changed due to the fracture wear or the attritious wear when the accumulated stock removal increases. Such a complicated wear process is evaluated using 3D-fractal dimension. Fractal dimension is calculated in an area of 27.4 × 20.6 μm² enclosed with white frame in Fig. 10. The center of these areas is almost located in the top part of the cutting edge that acts as an effective edge. The range of mesh size r is $0.11 < r < 6.4$ μm.

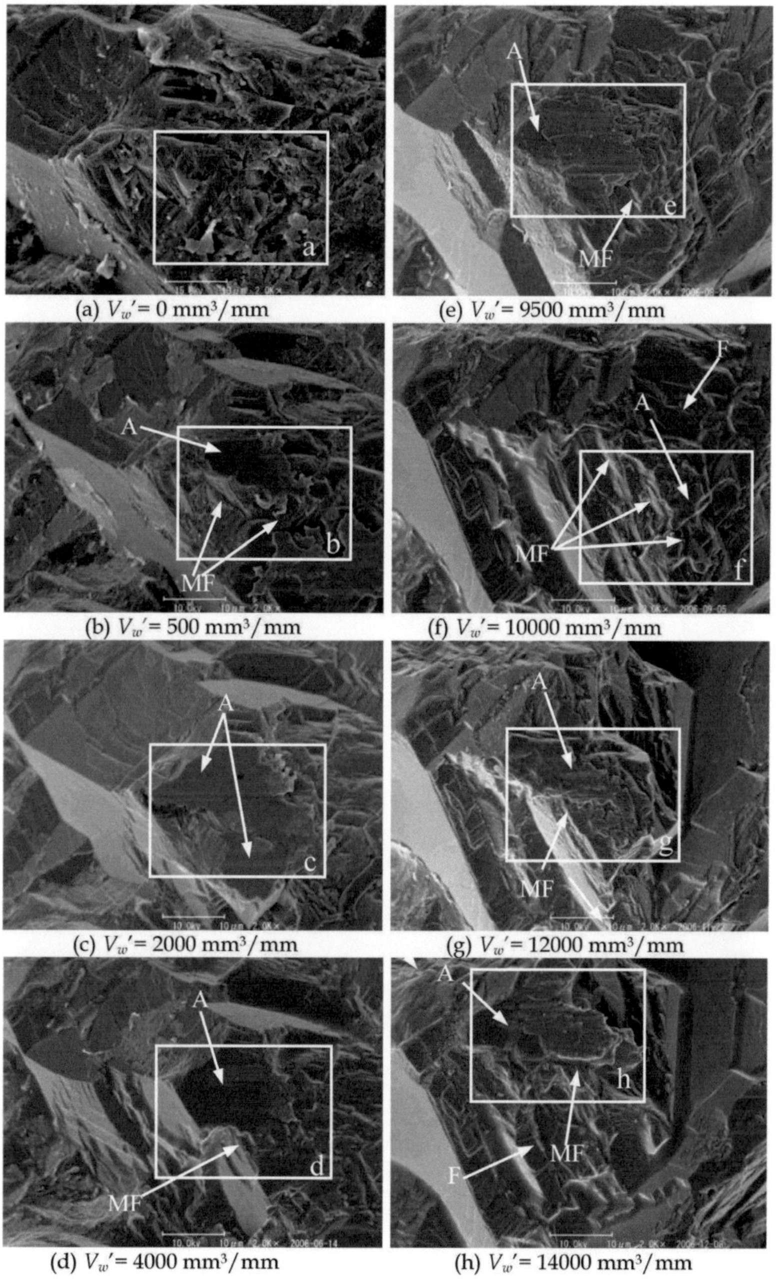

Fig. 10. Change in shape of grain cutting edge with accumulated stock removal (Area X on grain A in Fig.8(a)) (A: attritious wear, MF: micro fracture, F: fracture).

Figure 11 shows the 3D-profiles of the typical eight areas a, b, ----, h on the cutting edge used for fractal analysis (areas enclosed with white frame in Fig. 10). The relationships between mesh size r and surface area $S(r)$ obtained in these typical eight areas are shown in Fig. 12. This figure indicates that the fractal nature is approved in a region of $0.4 < r < 4$ µm. Using this relationship, fractal dimension is calculated. The results obtained are shown in Fig. 13. The fractal dimension changes complexly and randomly when the accumulated stock removal increases.

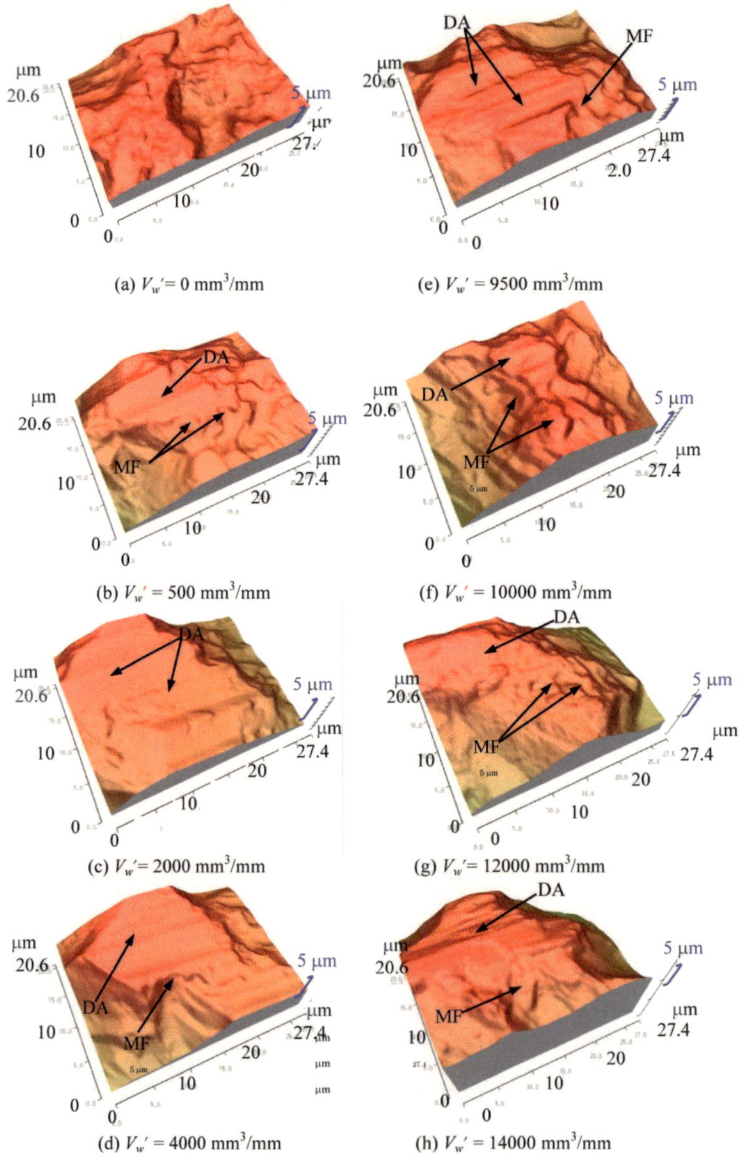

Fig. 11. Sequential 3D-profiles of typical area on grain cutting edge used for fractal analysis (DA: ductile attritious wear, MF: micro fracture).

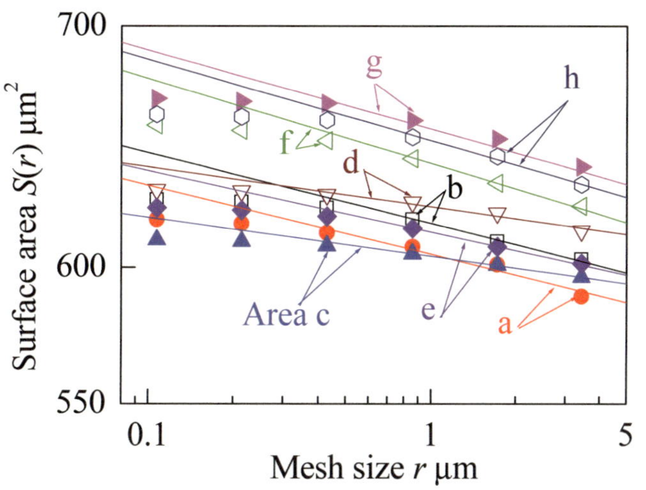

Fig. 12. Relationship between surface area $S(r)$ and mesh size r in areas shown in Fig.11.

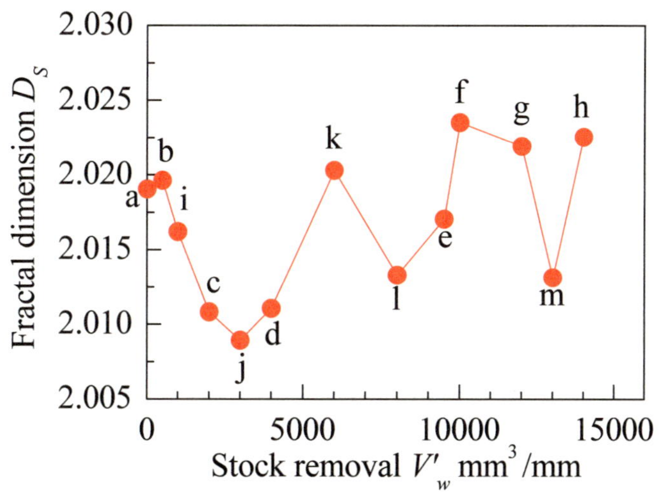

Fig. 13. Change in fractal dimension on top surface profile of grain cutting edge with accumulated stock removal.

To consider the reason for such complicated change of fractal dimension, the attritious wear flat area percentage of the cutting edge was measured. In this study, a percentage of ductile attritious wear area in same area of 27.4 × 20.6 μm² used for fractal analysis is measured and defined as attritious wear flat area percentage A_g. Figure 14 shows the change in the attritious wear flat area percentage A_g of the grain cutting edge with increasing accumulated stock removal.

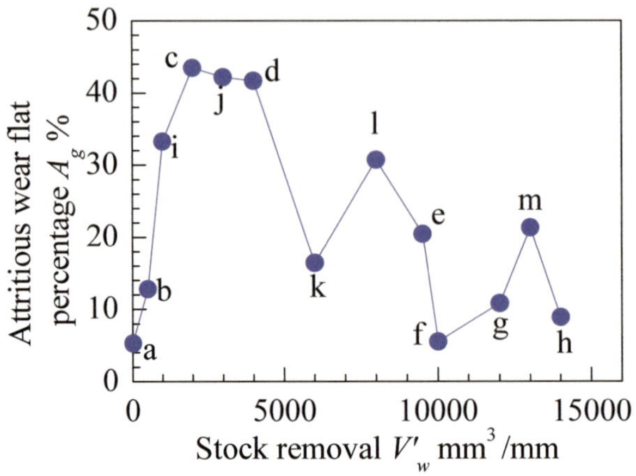

Fig. 14. Change in attritious wear flat percentage on top surface of grain cutting edge with accumulated stock removal.

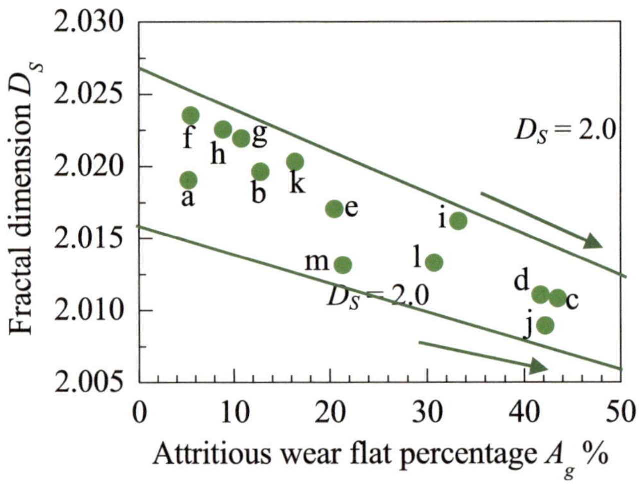

Fig. 15. Relationship between fractal dimension D_S and attritious wear flat percentage A_g

As shown in Figs. 13 and 14, the cutting edge after dressing comparatively takes a high fractal dimension (D_s =2.02), because it has complicated surface with a micro ruggedness formed by the diamond dresser. And then, between the stock removals from 500 to 3000mm^3/mm, fractal dimension decreases because the attritious wear flat increases with the accumulated stock removal. In addition, between the stock removals from 2000 to 4000 mm^3/mm, fractal dimension indicates the lowest value (D_s =2.01) because the attritious wear flat takes the highest value. Afterward, over a range of stock removals from 4000 to 6000 mm^3/mm, fractal dimension tends to increase because the attritious wear flat decreases and new sharp cutting edges are formed by self-sharpening due to micro fractures. However, between the stock removals from 6000 to 8000 mm^3/mm, fractal dimension decreases slightly because of a little increase in attritious wear flat area. Moreover, fractal dimension increases rapidly over a range of stock removals from 8000 to 10000 mm^3/mm because the attritious wear flat area decreases and micro fracture occurs repeatedly, i.e., self-sharpening due to micro fracture takes place actively. Afterward, although the fractal dimension decreases because of increasing in attritious wear flat at the stock removal 13000 mm^3/mm, it increases again because self-sharpening due to micro fracture takes place actively over a range of stock removals from 13000 to 14000 mm^3/mm.

As mentioned above, self-sharpening of the grain cutting edge can be characterized using fractal dimension. Especially, these results show that there is a close relationship between fractal dimension D_S and attritious wear flat percentage A_g. Figure 15 shows relationship between fractal dimension and attritious wear flat percentage. The alphabets in Fig.15 correspond to those in Figs.10, 11, 12, 13 and 14. As shown in this figure, fractal dimension decreases with increasing the attritious wear flat percentage and then becomes 2.0 at A_g = 100 % (perfect smooth surface) as a limit value. Thus, there is a negative correlation between fractal dimension and attritious wear flat percentage.

7. Effect of self-sharpening on grinding characteristics

Fig.16 and 17 show the changes of grinding forces and ground surface roughness with increasing accumulated stock removal, respectively. Under this grinding condition, grinding forces maintains a stable level in the steady-state wear region. Especially, tangential grinding force keeps a small variation between 4 and 6 N/mm in this wear region. On the other hand, although surface roughness increases with increasing stock removal, its increasing rate maintains comparatively low level. Thus, high grinding ability of cBN wheel is brought from such self-sharpening due to micro fracture of grain cutting edges, that is, micro self-sharpening phenomenon.

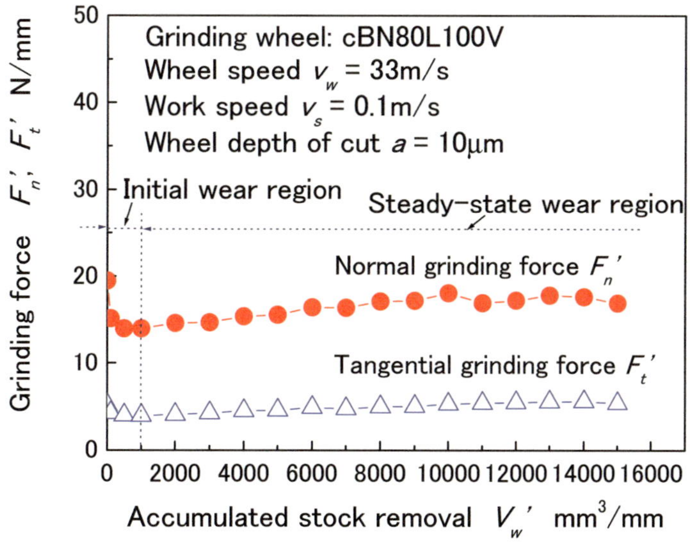

Fig. 16. Changes in grinding forces with increasing accumulated stock removal.

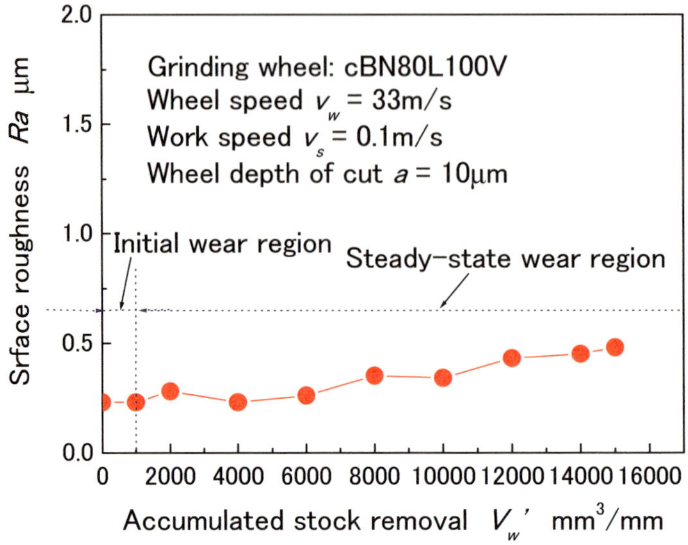

Fig. 17. Change of ground surface roughness with increasing accumulated stock removal.

8. Conclusions

The changes in three-dimensional surface profile of grain cutting edge in the grinding process with cBN wheels have been measured using a 3D-SEM and evaluated by means of fractal dimension. The main results obtained in this study are summarized as follows;

1. Actual behavior of self-sharpening phenomenon due to the micro fracture in the grinding process can be grasped using sequential observation method with 3D-SEM.
2. The fractal dimension for surface profile of the cutting edge formed by the micro fracture is higher than that of the cutting edge formed due to ductile attitious wear. An increase in ductile attritous wear flat area on the grain cutting edge results in a decrease in fractal dimension for its surface profile.
3. The complicated changes in shape of the cutting edge due to self-sharpening can be evaluated quantitatively using fractal dimension.

9. Acknowledgment

This research was supported in part by Grants-in-Aid for General Science Research (C) (No.19560106) from the Ministry of Education, Culture, Sports, Science and Technology of Japan.

10. References

Yoshikawa, H. (1960). Process of Wear in Grinding Wheel with Fracture of Bond and Grain, *Journal of the Japan Society for Precision Engineering*, Vol.26, No. 11, (1960), pp.691-700, ISSN0912-0289

Tsuwa, H. (1961). On the Behaviors of Abrasive Grains in Grinding Process (Part 4)-Microscopic Observations of Cutting Edges-, *Journal of the Japan Society for Precision Engineering*, Vol. 27, No. 11,(1961), pp. 719-725 , ISSN0912-0289

Kalpakjian, S. (1995), *Manufacturing Engineering and Technology*, Third Edition, Addison-Wesley Publishing Company Inc. , ISBN 0-201-53846-6, New York, pp.795-798.

Show, M. C. (1996), *Principles of Abrasive Processing*, Clarendon Press, Oxford, ISBN 0-19-859021-0, pp.55-62.

Malkin, S. (1989). *Grinding Technology: Theory and Applications of Machining with Abrasives*, Ellis Horwood Limited, Chichester, UK, PP.197-202. ISBN 0-85312-756-5

Ichida, Y.; Fredj, N. B. & Usui, N. (1995). The Micro Fracture Wear of Cutting Edges in CBN Grinding, *The Second International ABTEC Conference*, Vol. 11, (1995), pp. 501-504.

Ichida, Y.; Kishi, K.; Suyama, Y. & Okubo, J. (1989). Study of Creep Feed Grinding with CBN Wheels, -Characteristics of Wheel Wear-, Vol.55, No.8, pp.1468-1474, ISSN0912-0289

Ichida, Y. & Kishi, K. (1997). The Development of Nanocrystalline cBN for Enhanced Superalloy Grinding Performance, *Transactions of the ASME, Journal of Manufacturing, Science and Engineering*, Vol. 119, No. 1, (1997), pp.110-117. ISSN 1087-1357

Webster, J. & Tricard, M. (2004). Innovations in Abrasive Production for Precision Grinding, *Annals of the CIRP*, Vol.53, No.2, pp.597-617, ISSN 0007-8506

Ichida, Y.; Sato, R.: Morimoto, Y. & Inoue, Y. (2006). Profile Grinding of Superalloys with Ultrafine-Crystalline cBN Wheels, *JSME International Journal, Series C*, Vol.49, No.1, pp.94-99, ISSN 1344-7653.

Guo, C.; Shi, Z.; Attia, H. & Mclntosh, D. (2007). Power and Wheel Wear for Grinding Nickel Alloy with Plated CBN Wheels, *Annals of the CIRP*, Vol.56, No.1, pp.343-346, ISSN 0007-8506

Comley, P.; Walton, I.; Jin, T. & Stephenson, D. J. (2006). A High Material Removal Rate Grinding Process for the Production of Automotive Crankshafts, *Annals of the CIRP*, Vol.55, No.1, pp.347-350, ISSN 0007-8506

Mandelbrot, B. B. (1983). The Fractal Geometry of Nature, Freeman, W. H. and Company, New York, (1983), pp.109-111, ISBN4-532-06254-3

Mandelbrot, B. B.; Passoja, D. E. & Paullay, A. J. (1984). Fractal Characterization of Fracture Surfaces of Metals, *Nature*, Vol. 308, pp.1571-1572, ISSN 0028-0836.

Hagiwara, S.; Obikawa, T. & Yanai, H. (1995). Evaluation of Lapping Grains Based on Shape Characteristics, *Journal of the Japan Society for Precision Engineering*, Vol. 61, No.12, pp.1760-1764, ISSN0912-0289

Itoh, N.; Tsukada, T.& Sasajima, K. (1990). Three-Dimensional Characerization of Engineering Surface by Fractal Dimension, *Bulletin of the Japan Society for Prcision Engineering*, Vol.24, No.2, pp.148-149.

Sakai, T.; Sakai, T. & Ueno, A. (1998). Fractal Analysis of Metal Surface Mechanically Finished by Several Methods, *Transactions of the Japan Society of Mechanical Engineers, Series A*, Vol. 64, No. 620, (1998-4), pp.1104-1112, ISSN 1884-8338.

Fujimoto, M.; Ichida, Y.; Sato, R. & Morimoto, Y. (2006), Characterization of Wheel Surface Topography in cBN Grinding, *JSME International Journal, Series C*, Vol.49, No.1, pp.106-113, ISSN 1344-7653.

Ichida, Y.; Sato, R.; Fujimoto, M. & Tanaka, H. (2008), Fractal Analysis of Grain Cutting Edge Wear in Superabrasive Grinding, *JSME Journal of Advanced Mechanical Design, Systems, and Manufacturing*, Vol.2, No.4, pp.640-650 ISSN 1881-3054 .

Ichida, Y.; Sato, R.; Fujimoto, M. & Fredj, N. B. (2009). Fractal Analysis of Self-Sharpening Phenomenon in cBN Grinding, *Key Engineering*, Vols. 389-390, (2009), pp.42-47, ISBN-13978-0-87849-364-7

Ichida, Y.; Fujimoto, M.; Akbari, J. & Sato, R. (2008). Evaluation of Cutting Edge wear in cBN Grinding Based on Fractal Analysis, *6th International Scientific and technical Symposium on Manufacturing and Materials*, Monastir, Tunisia, pp.287-294.

21

Evolution of Phases in a Recycled Al-Si Cast Alloy During Solution Treatment

Eva Tillová, Mária Chalupová and Lenka Hurtalová
University of Žilina, Slovak Republic

1. Introduction

Aluminium has been acquiring increasing significance for the past few decades due to its excellent properties and diversified range of applications. Aluminium has been recognized as one of the best candidate materials for various applications by different sectors such as automotive, construction, aerospace, etc. The increasing demand for aluminium-based products and further globalization of the aluminium industry have contributed significantly to the higher consumption of aluminium scrap for re-production of aluminium alloys (Mahfoud et al., 2010).

Secondary aluminium alloys are made out of aluminium scrap and workable aluminium garbage by recycling. Production of aluminium alloys belong to heavy source fouling of life environs. Care of environment in industry of aluminium connects with the decreasing consumptions resource as energy, materials, waters and soil, with increase recycling and extension life of products. More than half aluminium on the present produce in European Union comes from recycled raw material. By primary aluminium production we need a lot of energy and constraints decision mining of bauxite so European Union has big interest of share recycling aluminium, and therefore increase interest about secondary aluminium alloys and cast stock from them (Sencakova & Vircikova, 2007).

The increase in recycled metal becoming available is a positive trend, as secondary aluminium produced from recycled metal requires only about 2.8 kWh/kg of metal produced while primary aluminium production requires about 45 kWh/kg produced. It is to the aluminium industry's advantage to maximize the amount of recycled metal, for both the energy-savings and the reduction of dependence upon overseas sources. The remelting of recycled metal saves almost 95 % of the energy needed to produce prime aluminium from ore, and, thus, triggers associated reductions in pollution and greenhouse emissions from mining, ore refining, and melting. Increasing the use of recycled metal is also quite important from an ecological standpoint, since producing aluminium by recycling creates only about 5 % as much CO_2 as by primary production (Das, 2006; Das & Gren, 2010).

Today, a large amount of new aluminium products are made by recycled (secondary) alloys. This represents a growing "energy bank" of aluminium available for recycling at the end of components' lives, and thus recycling has become a major issue. The future growth offers an opportunity for new recycling technologies and practices to maximize scrap quality; improve efficiency and reduce cost.

Aluminium-silicon (Al-Si) cast alloys are fast becoming the most universal and popular commercial materials, comprising 85 % to 90 % of the aluminium cast parts produced for the automotive industry, due to their high strength-to-weight ratio, excellent castability, high corrosion resistant and chemical stability, good mechanical properties, machinability and wear resistance. Mg or Cu addition makes Al-Si alloy heat treatable.

The alloys of the Al-Si-Cu system have become increasingly important in recent years, mainly in the automotive industry that uses recycled (secondary) aluminium in the form of various motor mounts, pistons, cylinder heads, heat exchangers, air conditioners, transmissions housings, wheels, fenders and so on due to their high strength at room and high temperature (Rios & Caram, 2003; Li et al., 2004; Michna et al., 2007). The increased use of these recycled alloys demands a better understanding of its response to mechanical properties.

The quality of recycled Al-Si casting alloys is considered to be a key factor in selecting an alloy casting for a particular engineering application. Based on the Al-Si system, the main alloying elements are copper (Cu) or magnesium (Mg) and certain amount of iron (Fe), manganese (Mn) and more, that are present either accidentally, or they are added deliberately to provide special material properties. These elements partly go into solid solution in the matrix and partly form intermetallic particles during solidification. The size, volume and morphology of intermetallic phases are functions of chemistry, solidification conditions and heat treatment (Li, 1996; Paray & Gruzleski, 1994; Tillova & Panuskova, 2007, 2008).

Copper substantially improves strength and hardness in the as-cast and heat-treated conditions. Alloys containing 4 % to 6 % Cu respond most strongly to thermal treatment. Copper generally reduces resistance to general corrosion and, in specific compositions and material conditions, stress corrosion susceptibility. Additions of copper also reduce hot tear resistance and decrease castability. Magnesium is the basis for strength and hardness development in heat-treated Al-Si alloys too and is commonly used in more complex Al-Si alloys containing copper, nickel, and other elements for the same purpose.

Iron considers the principal impurity and detrimental alloying element for Al-Si-Cu alloys. Iron improves hot tear resistance and decreases the tendency for die sticking or soldering in die casting. Increases in iron content are, however, accompanied by substantially decreased ductility. Iron reacts to form a myriad of insoluble phases in aluminium alloy melts, the most common of which are Al_3Fe, Al_6FeMn, and α-Al_5FeSi. These essentially insoluble phases are responsible for improvements in strength, especially at elevated temperature. As the fraction of insoluble phase increases with increased iron content, casting considerations such as flowability and feeding characteristics are adversely affected. Iron also lead to the formation of excessive shrinkage porosity defects in castings (Warmuzek, 2004a; Taylor, 2004; Shabestari, 2004; Caceres et al., 2003; Wang et al. 2001; Tillova & Chalupova, 2010).

It is clear that the morphology of Fe-rich intermetallic phases influences harmfully also fatigue properties (Taylor, 2004; Tillova & Chalupova, 2010). It is recognized that recycled Al-Si-Cu alloys are not likely to be suitable for fracture-critical components, where higher levels of Fe and Si have been shown to degrade fracture resistance. However the likelihood exists that they may perform quite satisfactorily in applications such as those listed where service life is determined by other factors (Taylor, 2004).

2. Experimental material and methodology

As an experimental material recycled (secondary) hypoeutectic AlSi9Cu3 alloy, in the form of 12.5 kg ingots, was used. The alloy was molten into the sand form (sand casting). The melting temperature was maintained at 760 °C ± 5 °C. Molten metal was before casting purified with salt AlCu4B6. The melt was not modified or grain refined. The chemical analysis of AlSi9Cu3 cast alloy was carried out using arc spark spectroscopy. The chemical composition is given in the table 1.

Si	Cu	Mn	Fe	Mg	Ni	Pb	Zn	Ti	Al
10.7	2.4	0.22	< 0.8	0.47	0.08	0.11	1.1	0.03	rest

Table 1. Chemical composition of the alloy (wt. %)

AlSi9Cu3 cast alloy has lower corrosion resistance and is suitable for high temperature applications (dynamic exposed casts, where are not so big requirements on mechanical properties) - it means to max. 250 °C. Experimental samples (standard tensile test specimens) were given a T4 heat treatment - solution treatment for 2, 4, 8, 16 or 32 hours at three temperatures (505 °C, 515 °C and 525 °C); water quenching at 40 °C and natural aging for 24 hours at room temperature. After heat treatment samples were subjected to mechanical test. For as cast state, each solution temperature and each aging time, a minimum of five specimens were tested.

Metallographic samples were prepared from selected tensile specimens (after testing) and the microstructures were examined by optical (Neophot 32) and scanning electron microscopy. Samples were prepared by standards metallographic procedures (mounting in bakelite, wet ground, DP polished with 3 μm diamond pastes, finally polished with commercial fine silica slurry (STRUERS OP-U) and etched by Dix-Keller. For setting of Fe-rich intermetallic phases was used etching by H_2SO_4. For setting of Cu-rich intermetallic phases was used etching by HNO_3.

Some samples were also deep-etched for 30 s in HCl solution in order to reveal the three-dimensional morphology of the eutectic silicon and intermetallic phases (Tillova & Chalupova, 2001, 2009). The specimen preparation procedure for deep-etching consists of dissolving the aluminium matrix in a reagent that will not attack the eutectic components or intermetallic phases. The residuals of the etching products should be removed by intensive rinsing in alcohol. The preliminary preparation of the specimen is not necessary, but removing the superficial deformed or contaminated layer can shorten the process. To determine the chemical composition of the intermetallic phases was employed scanning electron microscopy (SEM) TESCAN VEGA LMU with EDX analyser BRUKER QUANTAX.

Quantitative metallography (Skocovsky & Vasko, 2007; Vasko & Belan, 2007; Belan, 2008; Vasko, 2008; Martinkovic, 2010) was carried out on an Image Analyzer NIS - Elements 3.0 to quantify phase's changes during heat treatment. A minimum of 20 pictures at 500 x magnification of the polish per specimen were taken.

Hardness measurement was preformed by a Brinell hardness tester with a load of 62.5 kp, 2.5 mm diameter ball and a dwell time of 15 s. The Brinell hardness value at each state was obtained by an average of at least six measurements. The phases Vickers microhardness was measured using a MHT-1 microhardness tester under a 1g load for 10 s (HV 0.01). Twenty measurements were taken per sample and the median microhardness was determined.

3. Results and discussion

3.1 Microstructure of recycled AlSi9Cu3 cast alloy

Controlling the microstructure during solidification is, therefore, very important. The Al-Si eutectic and intermetallic phases form during the final stage of the solidification. How the eutectic nucleates and grows have been shown to have an effect on the formation of defects such as porosity and microporosity too. The defects, the morphology of eutectic and the morphology of intermetallic phases have an important effect on the ultimate mechanical properties of the casting.

As recycling of aluminium alloys becomes more common, sludge will be a problem of increasing importance due to the concentration of Fe, Mn, Cr and Si in the scrap cycle. During the industrial processing of the Al-Si alloys, these elements go into solid solution but they also form different intermetallic phases. The formation of these phases should correspond to successive reaction during solidification - table 2 (Krupiński et al., 2011; Maniara et al., 2007; Mrówka-Nowotnik & Sieniawski, 2011; Dobrzański et al., 2007, Tillova & Chalupova, 2009). Thus, control of these phases e. g. quantitative analysis (Vasko & Belan, 2007; Martinkovic, 2010) is of considerable technological importance. Typical structures of the recycled as-cast AlSi9Cu3 alloys are shown in Fig. 1. The microstructure consists of dendrites α-phase (1), eutectic (mixture of α-matrix and spherical Si-phases - 2) and variously type's intermetallic Fe- and Cu-rich phases (3 and 4).

Reactions	Temperature, °C
α - dendritic network	609
Liq. → α - phase + $Al_{15}Mn_3Si_2$ + Al_5FeSi	590
Liq. → α - phase + Si + Al_5FeSi	575
Liq. → α - phase + Al_2Cu + Al_5FeSi + Si	525
Liq. → α - phase + Al_2Cu + Si + $Al_5Mg_8Si_6Cu_2$	507

Table 2. Reactions occurring during the solidification of AlSi9Cu3 alloys

The α-matrix precipitates from the liquid as the primary phase in the form of dendrites and is nominally comprised of Al and Si. Experimental material was not modified and so eutectic Si particles are in a form of platelets (Fig. 2a), which on scratch pattern are in a form of needles – Fig. 2b (Skocovsky et al., 2009; Tillova & Chalupova, 2001; 2009).

Iron is one of the most critical alloying elements, because Fe is the most common and usually detrimental impurity in cast Al-Si alloys. Iron impurities can either come from the use of steel tools or scrap materials or be acquired during subsequent melting, remelting and casting, e.g. by contamination from the melting pot etc.

A number of Fe-rich intermetallic phases, including α (Al_8Fe_2Si or $Al_{15}(FeMn)_3Si_2$), β (Al_5FeSi), π ($Al_8Mg_3FeSi_6$), and δ (Al_4FeSi_2), have been identified in Al-Si cast alloys (Samuel et al., 1996; Taylor, 2004; Seifeddine, 2007; Seifeddine et al. 2008; Moustafa, 2009; Fang et al., 2007; Lu & Dahle, 2005).

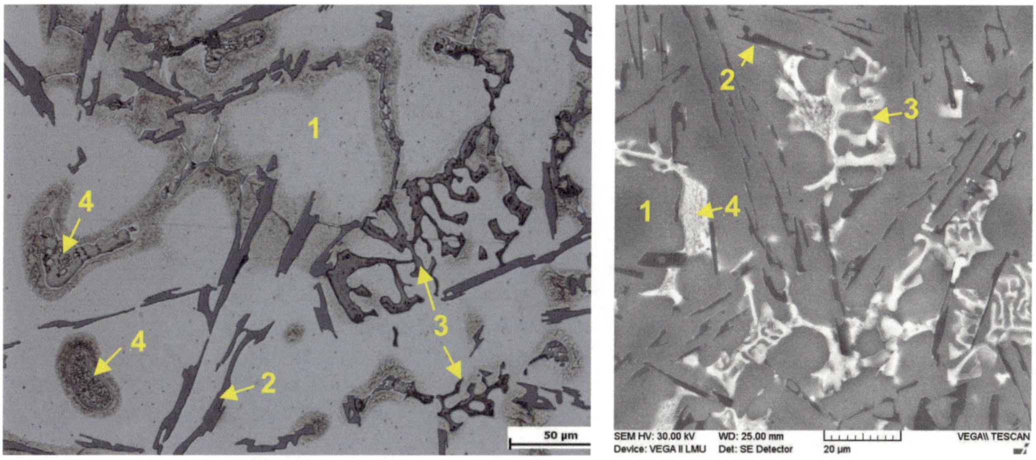

a) optical microscopy b) SEM

Fig. 1. Microstructure of recycled AlSi9Cu3 cast alloy (1 - α-phase, 2 - eutectic silicon, 3 - Fe-rich phases, 4 - Cu-rich phases), etch. Dix-Keller

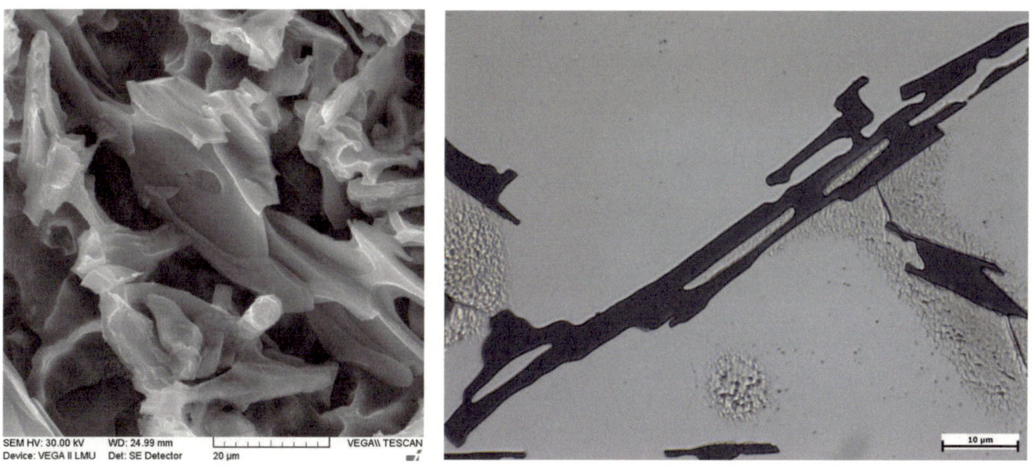

a) deep etch. HCl, SEM b) etch. Dix-Keller

Fig. 2. Morphology of eutectic silicon

In experimental AlSi9Cu3 alloy was observed the two main types of Fe-rich intermetallic phases, Al_5FeSi with monoclinic crystal structure (know as beta- or β-phase) and $Al_{15}(FeMn)_3Si_2$ (know as alpha- or α-phase) with cubic crystal structure. The first phase (Al_5FeSi) precipitates in the interdendritic and intergranular regions as platelets (appearing as needles in the metallographic microscope - Fig. 3). Long and brittle Al_5FeSi platelets (more than 500 μm) can adversely affect mechanical properties, especially ductility, and also lead to the formation of excessive shrinkage porosity defects in castings (Caceres et al., 2003). Platelets are effective pore nucleation sites. It was also shown that the Al_5FeSi needles can act as nucleation sites for Cu-rich Al_2Cu phases (Tillova et al., 2010).

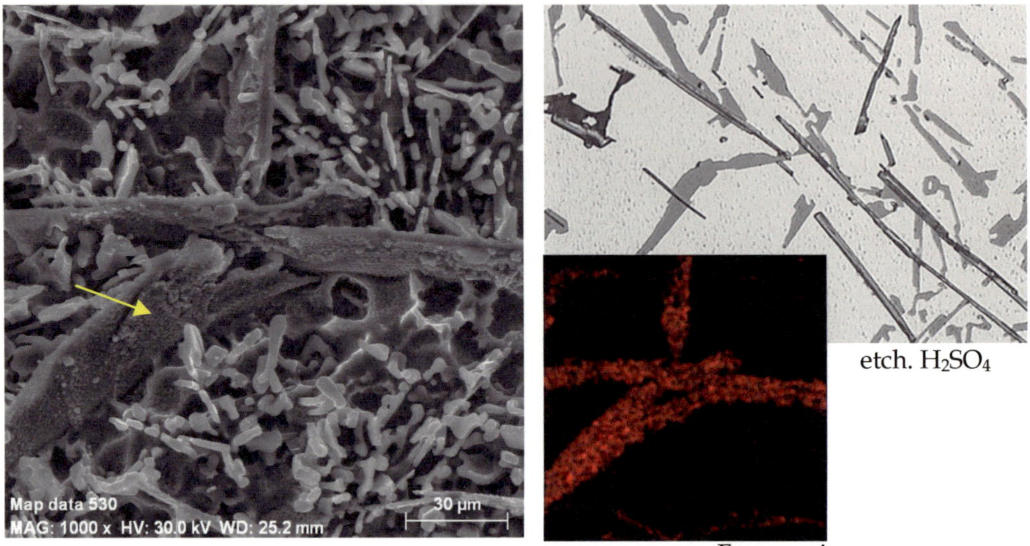

deep etch., SEM etch. H_2SO_4 Fe-mapping

Fig. 3. Morphology of Fe-phase Al_5FeSi

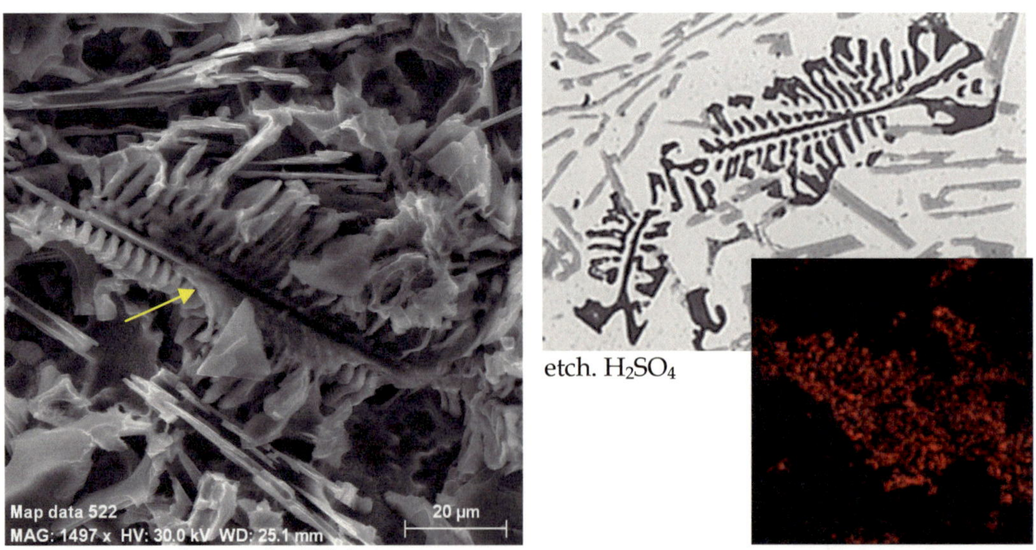

etch. H_2SO_4

deep etch., SEM Fe-mapping

Fig. 4. Morphology of Fe-phase $Al_{15}(FeMn)_3Si_2$

The deleterious effect of Al_5FeSi can be reduced by increasing the cooling rate, superheating the molten metal, or by the addition of a suitable "neutralizer" like Mn, Co, Cr, Ni, V, Mo and Be. The most common addition has been manganese. Excess Mn may reduce Al_5FeSi phase and promote formation Fe-rich phases $Al_{15}(FeMn)_3Si_2$ in form „skeleton like" or in form „Chinese script" (Seifeddine et al., 2008; Taylor, 2004) (Fig. 4). This compact morphology "Chinese script" (or skeleton - like) does not initiate cracks in the cast material to the same extent as Al_5FeSi does and phase $Al_{15}(FeMn)_3Si_2$ is considered less harmful to the mechanical properties than β phase (Ma et al., 2008; Kim et al., 2006). The amount of manganese needed to neutralize iron is not well established. A common "rule of thumb" appears to be ratio between iron and manganese concentration of 2:1.

Alloying with Mn and Cr, caution has to be taken in order to avoid the formation of hard complex intermetallic multi-component sludge, $Al_{15}(FeMnCr)_3Si_2$ - phase (Fig. 5). These intermetallic compounds are hard and can adversely affect the overall properties of the casting. The formation of sludge phases is a temperature dependent process in a combination with the concentrations of iron, manganese and chromium independent of the silicon content. If Mg is also present with Si, an alternative called pi-or π-phase can form, $Al_5Si_6Mg_8Fe_2$. $Al_5Si_6Mg_8Fe_2$ has a script-like morphology. The Fe-rich particles can be twice as large as the Si particles, and the cooling rate has a direct impact on the kinetics, quantities and size of Fe-rich intermetallic present in the microstructure.

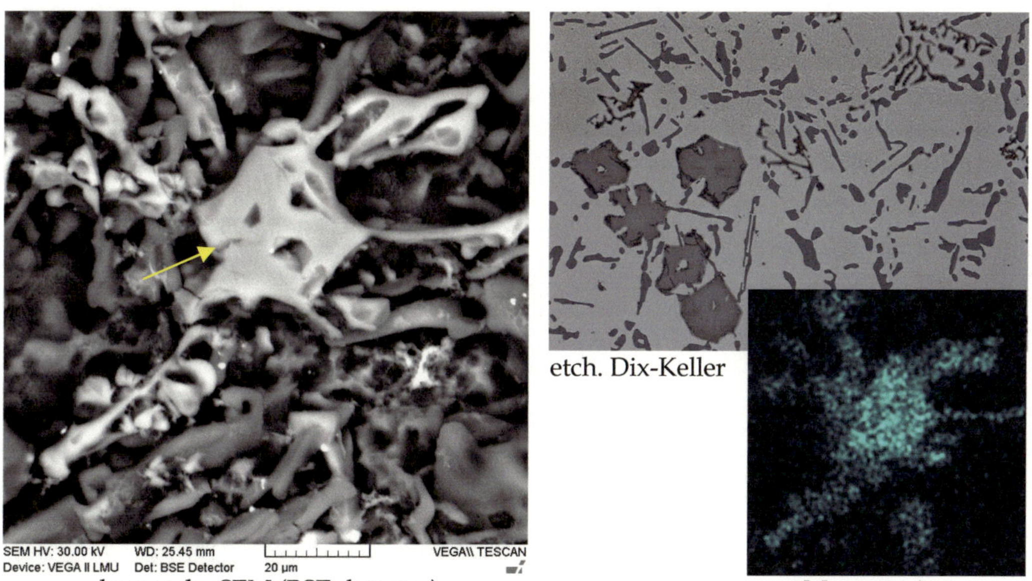

deep etch., SEM (BSE detector) Mn-mapping

Fig. 5. Morphology of sludge phase $Al_{15}(FeMnCr)_3Si_2$

Cu is in Al-Si-Cu cast alloys present primarily as phases: Al_2Cu, Al-Al_2Cu-Si or $Al_5Mg_8Cu_2Si_6$ (Rios et al., 2003; Tillova & Chalupova, 2009; Tillova et al.; 2010). The average size of the Cu-phase decreases upon Sr modification. The Al_2Cu phase is often observed to precipitate both in a small blocky shape with microhardness 185 HV 0.01. Al-Al_2Cu-Si phase is observed in very fine multi-phase eutectic-like deposits with microhardness 280 HV 0.01

(Tillova & Chalupova, 2009). In recycled AlSi9Cu3 alloy was analysed two Cu-phases: Al_2Cu and Al-Al_2Cu-Si (Fig. 6).

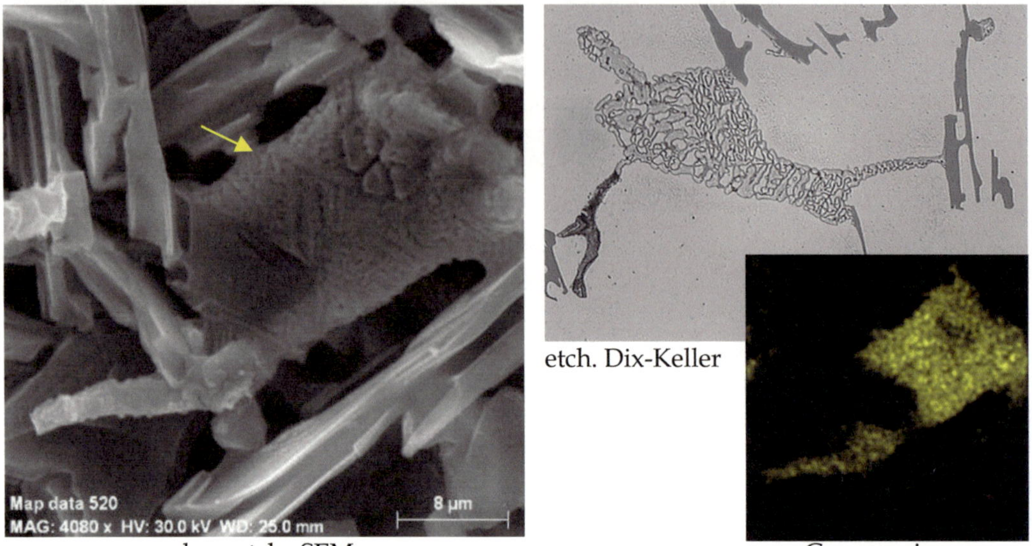

deep etch., SEM — Cu-mapping

Fig. 6. Morphology of Cu-phase - Al-Al_2Cu-Si

The microhardness of all observed intermetallic phases was measured in HtW Dresden and the microhardness values are indicated in table 3. It is evident that the eutectic silicon, the Fe-rich phase Al_5FeSi and the multicomponent intermetallic $Al_{15}(FeMn)_3Si_2$ are the hardest.

Intermetallic phases	HV 0.01	Chemical composition, wt. %					
		Al	Mg	Si	Fe	Cu	Mn
$Al_{15}(MnFe)_3Si_2$	483	61	-	10.3	13.4	2.6	13.6
Al_5FeSi	1 475	67.7	-	16.5	15.8	-	-
Al_2Cu	185	53.5	-	-	-	42.2	-
Al-Al_2Cu-Si	280	53	4.5	14.8	-	18.5	-
Si	1084	-	-	99.5	-	-	-

Table 3. Microhardness and chemical composition of intermetallic phases

Influence of intermetallic phases to mechanical and fatigue properties of recycled Al-Si cast alloys depends on size, volume and morphology this Fe- and Cu-rich phases.

3.2 Effect of solution treatment on the mechanical properties

Al-Si-Cu cast alloys are usually heat-treated in order to obtain an optimum combination of strength and ductility. Important attribute of a precipitation hardening alloy system is a temperature and time dependent equilibrium solid-solubility characterized by decreasing solubility with decreasing temperature and then followed by solid-state precipitation of second phase atoms on cooling in the solidus region (Abdulwahab, 2008; Michna et al., 2007; ASM Handbook, 1991). Hardening heat treatment involves (Fig. 7):

- Solution heat - treatment - it is necessary to produce a solid solution. Production of a solid solution consists of soaking the aluminium alloy at a temperature sufficiently high and for such a time so as to attain an almost homogeneous solid solution;
- Rapid quenching to retain the maximum concentration of hardening constituent (Al_2Cu) in solid solution. By quenching it is necessary to avoid slow cooling. Slow cooling can may the precipitation of phases that may be detrimental to the mechanical properties. For these reasons solid solutions formed during solution heat-treatment are quenched rapidly without interruption to produce a supersaturated solution at room temperature;
- Combination of artificial and over-ageing to obtain the desired mechanical properties in the casting. Generally artificial aging imparts higher strength and hardness values to aluminium alloys without sacrificing other mechanical properties.

The precipitation sequence for Al-Si-Cu alloy is based upon the formation of Al_2Cu based precipitates. The sequence is described as: $\alpha_{ss} \rightarrow$ GP zones $\rightarrow \theta' \rightarrow \theta$ (Al_2Cu). The sequence begins upon aging when the supersaturated solid solution (α_{ss}) gives way first to small coherent precipitates called GP zones. These particles are invisible in the optical microscope but macroscopically, this change is observed as an increase in the hardness and tensile strength of the alloy. As the process proceeds, the GP zones start to dissolve, and θ' begins to form, which results in a further increase in the hardness and tensile strength in the alloy. Continued aging causes the θ' phase to coarsen and the θ (Al_2Cu) precipitate to appear. The θ phase is completely incoherent with the matrix, has a relatively large size, and has a coarse distribution within the aluminium matrix. Macroscopically, this change is observed as an increase in the ductility and a decrease in the hardness and tensile strength of the alloy (Abdulwahab, 2008; Michna et al., 2007; Panuskova et al., 2008).

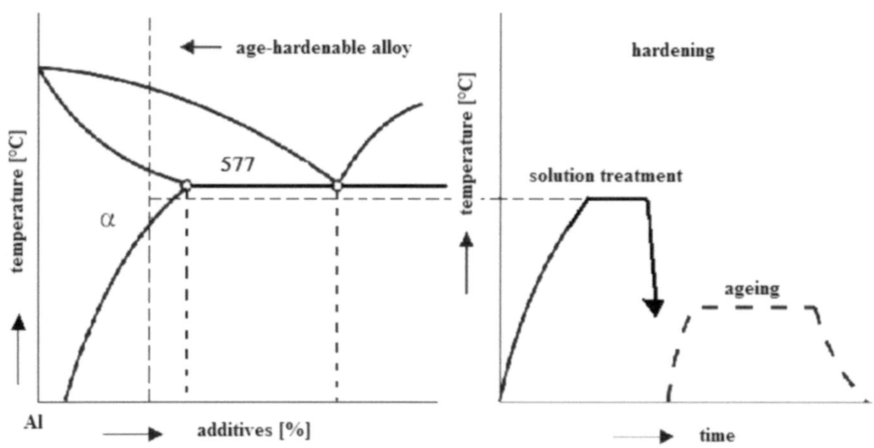

Fig. 7. The schematic diagram of hardening process for Al-Si-Cu cast alloy

Although the morphology, the amount and the distribution of the precipitates during aging process significantly influence the mechanical properties, an appropriate solution treatment is a prerequisite for obtaining desirable aging effect. From this point of view, the solution heat treatment is critical in determining the final microstructure and mechanical properties of the alloys. Thus, it is very important to investigate the effects of solution heat treatment on the alloys, before moving on to aging issues.

Solution treatment performs three roles (Li, 1996; Lasa & Rodriguez-Ibabe, 2004; Paray & Gruzleski, 1994; Moustafa et al., 2003; Sjölander & Seifeddine, 2010):

- homogenization of as-cast structure;
- dissolution of certain intermetallic phases such as Al_2Cu;
- changes the morphology of eutectic Si phase by fragmentation, spheroidization and coarsening, thereby improving mechanical properties, particularly ductility.

For experimental work heat treatment consisted of solution treatment for different temperatures: 505 °C, 515 °C and 525 °C; rapid water quenching (40 °C) and natural ageing (24 hours at room temperature) was used.

Influence of solution treatment on mechanical properties (strength tensile - R_m and Brinell hardness - HBS) is shown in Fig. 8 and Fig. 9.

After solution treatment, tensile strength, ductility and hardness are remarkably improved, compared to the corresponding as-cast condition. Fig. 8 shows the results of tensile strength measurements. The as cast samples have a strength value approximately 204 MPa. After 2 hours the solution treatment, independently of temperature of solution treatment, strength value immediately increases. By increasing the solution holding time from 2 to 4 hours, the tensile strength increased to 273 MPa for 515 °C. With further increase in solution temperature more than 515 °C and solution treatment time more than 4 hours, tensile strength decreases during the whole period as a result of gradual coarsening of eutectic Si, increase of inter particle spacing and dissolution of the Al_2Cu phase (at 525 °C).

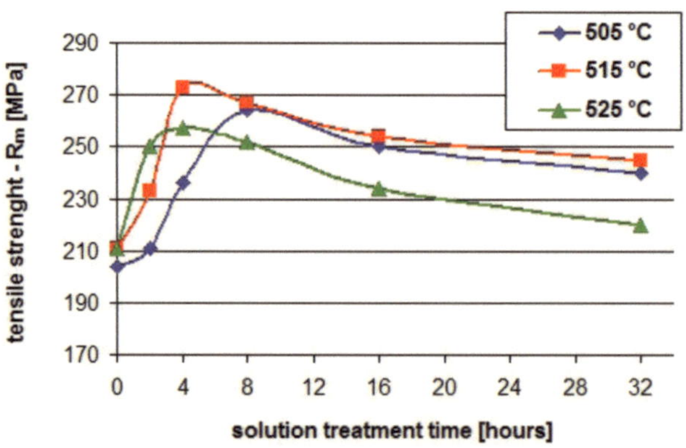

Fig. 8. Influence of solution treatment conditions on tensile strength

Fig. 9 shows the evolution of Brinell hardness value. Results of hardness are comparable with results of tensile strength. The as cast samples have a hardness value approximately 98 HB. After 2 hours the solution treatment, independently from temperature of solution treatment, hardness value immediately increases. The maximum was observed after 4 hours - approximately 124 HBS for 515 °C. However, after 8 hours solution treatment, the HB values are continuously decreasing as results of the coarsening of eutectic silicon, increase of inter particle spacing and dissolution of the Al_2Cu phase. After prolonged solution treatment time up to 16 h at 525 °C, it is clearly that the HB values strongly decrease probably due to melting of the $Al\text{-}Al_2Cu\text{-}Si$ phase.

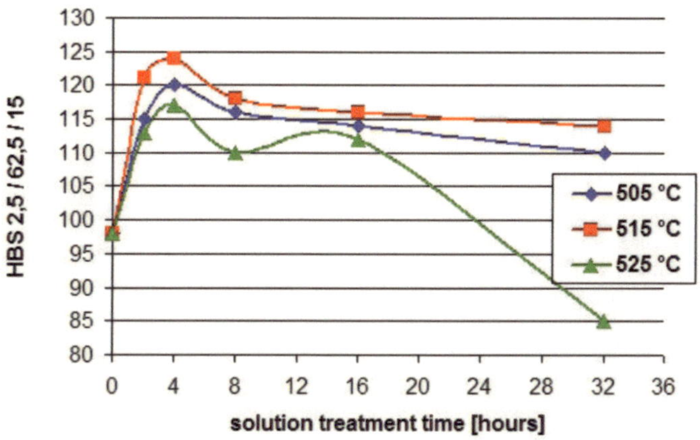

Fig. 9. Influence of solution treatment conditions on Brinell hardness

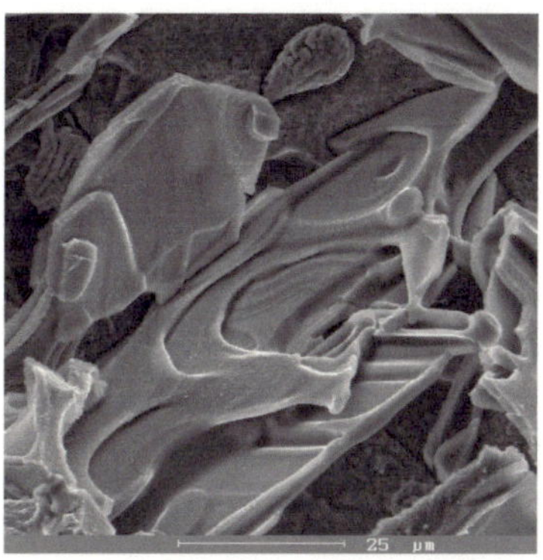

a) untreated state, deep-etch. HCl, SEM

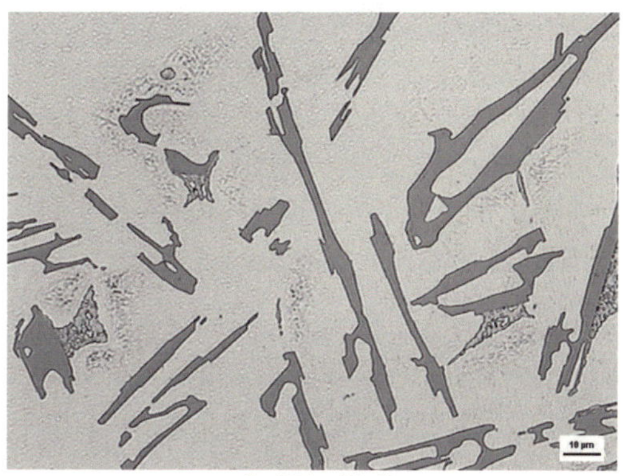

etch. Dix-Keller

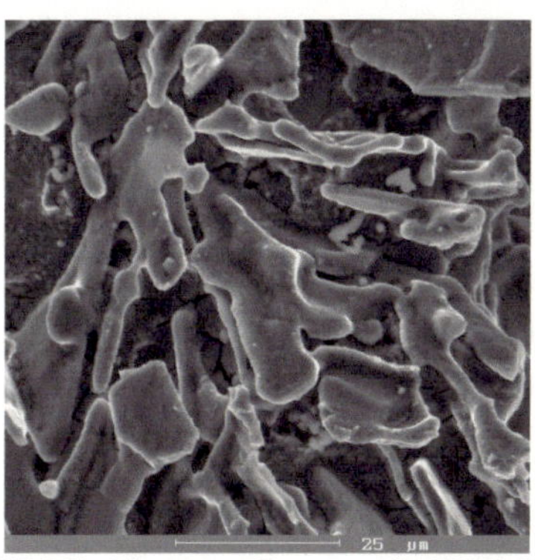

b) 505 °C, 4 hours, deep-etch. HCl, SEM

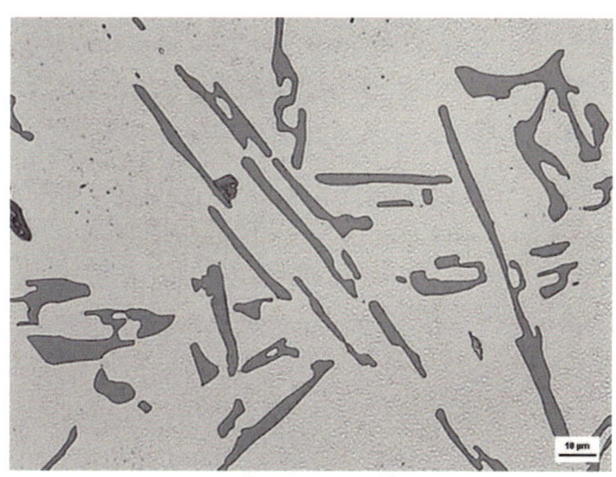

etch. Dix-Keller

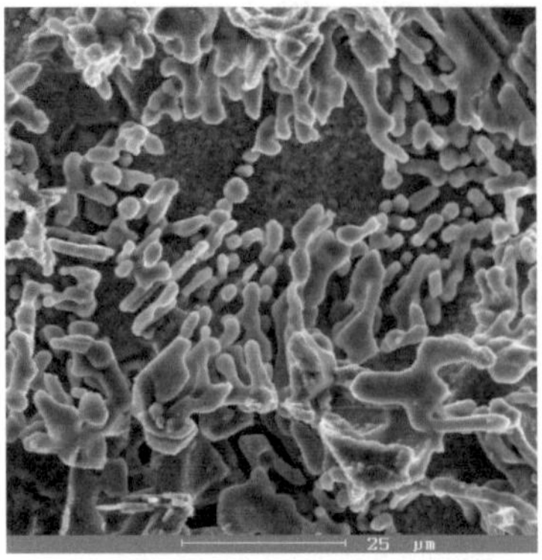

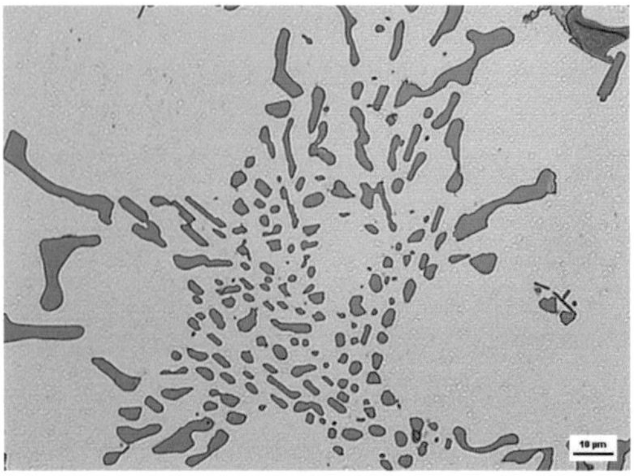

c) 515 °C, 4 hours, deep-etch. HCl, SEM etch. Dix-Keller

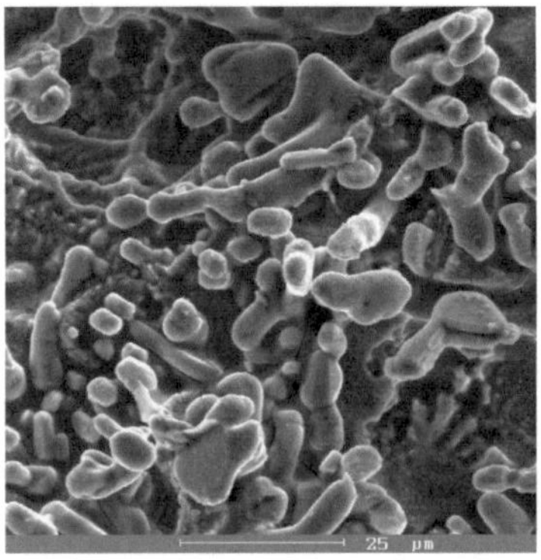

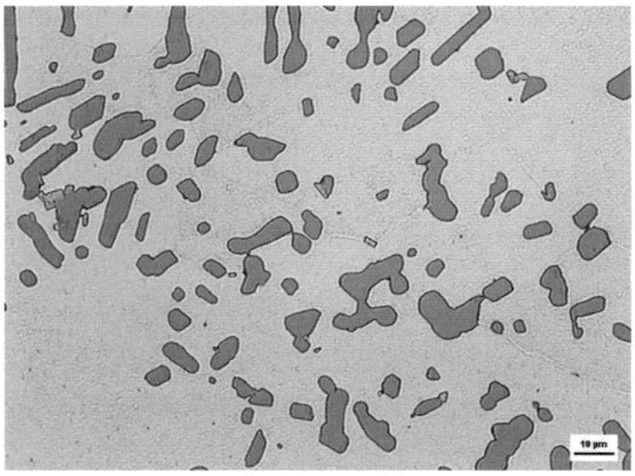

d) 525 °C, 4 hours, deep-etch. HCl, SEM etch. Dix-Keller

Fig. 10. Effect of solution treatment on morphology of eutectic Si

Obtained results (Fig. 8 and Fig. 9) suggests that to enhance the tensile strength or hardness of this recycled cast alloy by increasing of solution temperature more than 515 °C and by extending the solution time more than 4 hours does not seem suitable.

3.3 Effect of solution treatment on the morphology of eutectic silicon

The mechanical properties of cast component are determined largely by the shape and distribution of Si particles in the matrix. Optimum tensile, impact and fatigue properties are obtained with small, spherical and evenly distributed particles.

It is hypothesized (Paray & Gruzleski, 1994; Li, 1996; Tillova & Chalupova, 2009; Moustafa et al, 2010) that the spheroidisation process of the eutectic silicon throughout heat treatment takes place in two stages: fragmentation or dissolution of the eutectic Si branches and the spheroidisation of the separated branches. Experimental material was not modified or grain refined and so eutectic Si particles without heat treatment (untreated – as cast state) are in a form of platelets (Fig. 10a), which on scratch pattern are in a form of needles.

The solution temperature is the most important parameter that influences the kinetics of Si morphology transformation during the course of solution treatment. The effect of solution treatment on morphology of eutectic Si, for holding time 4 hour, is demonstrated in Figures 10b, 10c and 10d. After solution treatment at the temperature of 505 °C were noted that the platelets were fragmentized into smaller platelets with spherical edges (Fig. 10b) (on scratch pattern round needles). The temperature 505 °C is for Si-spheroidisation low.

The spheroidisation process dominated at 515 °C. Si platelets fragment into smaller segments and these smaller Si particles were spheroidised to rounded shape; see Fig. 10c. By solution treatment 525 °C the spheroidised particles gradually grew larger (coarsening) (Figures 10d).

Quantitative metallography (Skocovsky & Vasko, 2007; Vasko & Belan, 2007; Belan, 2008; Vasko, 2008; Martinkovic, 2010) was carried out on an Image Analyzer NIS-Elements to quantify eutectic Si (average area of eutectic Si particle) by magnification 500 x. Figure 11 shows the average area of eutectic Si particles obtained in solution heat treated samples. This graphic relation is in line with work Paray & Gruzleski, 1994.

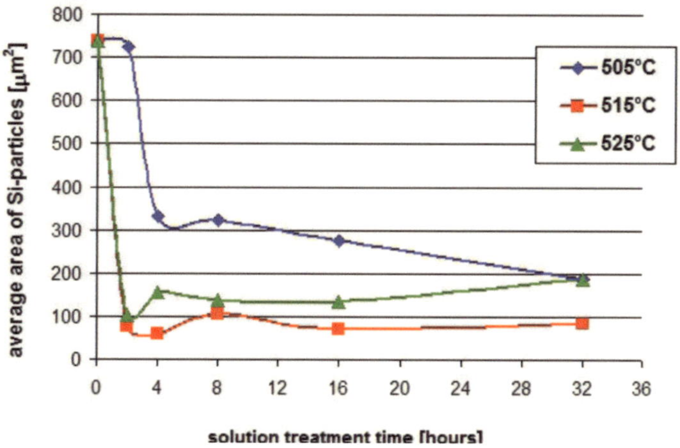

Fig. 11. Influence of solution treatment on average area of eutectic Si particles

Average area of eutectic Si particles decreases with increasing solution temperature and during the whole solution period. During the two hours, the area of Si-particles decreases which indicated that they undergo fragmentation and break into smaller segments.

Minimum value of average eutectic Si particles was observed by temperature 515 °C (approximately 89 μm^2). It's probably context with spheroidisation of eutectic silicon on this temperature. By solution treatment 525°C the spheroidised Si-particles in comparison with temperature 515 °C coarsen. The value of average eutectic Si particles at this temperature was observed from approximately 100 μm^2 (2 hour) till 187 μm^2 (32 hour). Prolonged solution treatment at 515°C and 525°C leads to a significant coarsening of the spheroidised Si particles.

3.4 Effect of solution treatment on the morphology of Fe-rich phases

The influence of iron on mechanical properties of aluminium alloys depends on the type, morphology and quantity of iron in the melt. Nevertheless, the shape of iron phases is more influential than the quantity of those iron compounds.

The evolution of the Fe-rich phases during solution treatment is described for holding time 4 hours in Fig. 12. Al_5FeSi phase is dissolved into very small needles (difficult to observe). The $Al_{15}(FeMn)_3Si_2$ phase was fragmented to smaller skeleton particles. In the untreated state $Al_{15}(FeMn)_3Si_2$ phase has a compact skeleton-like form (Fig. 12a). Solution treatment of this skeleton-like phase by 505°C tends only to fragmentation (Fig. 12b) and by 515°C or 525°C to fragmentation, segmentation and dissolution (Fig. 12c, Fig. 12d).

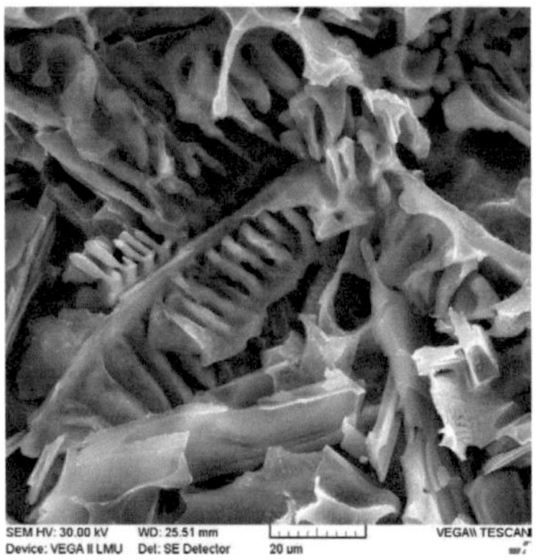

a) untreated state, deep-etch. HCl, SEM

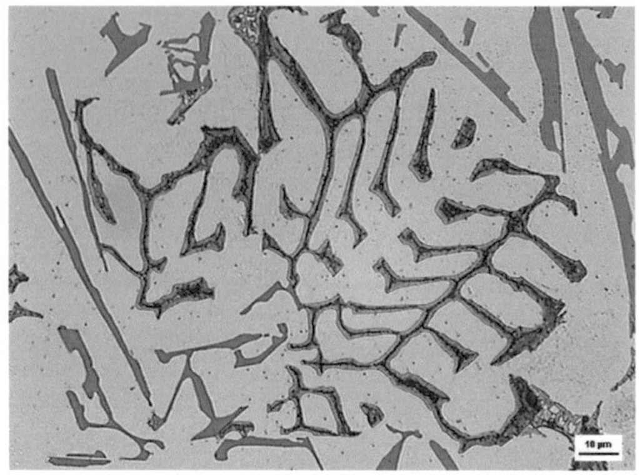
etch. H_2SO_4

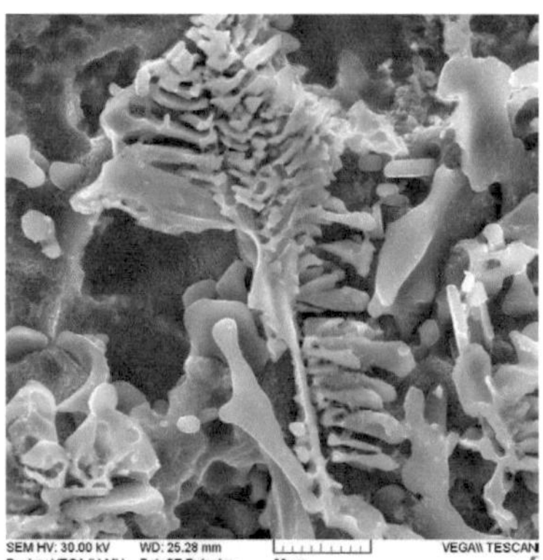

b) 505 °C, 4 hours, deep-etch. HCl, SEM

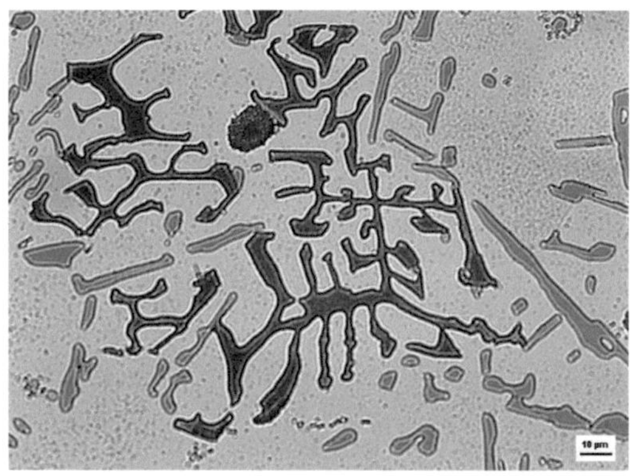
etch. H_2SO_4

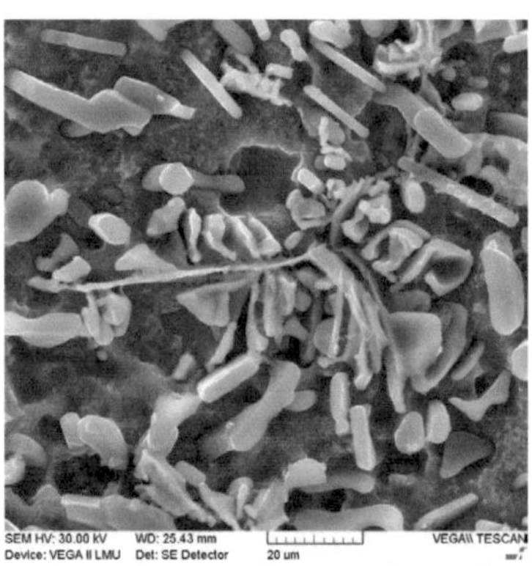

c) 515 °C, 4 hours, deep-etch. HCl, SEM

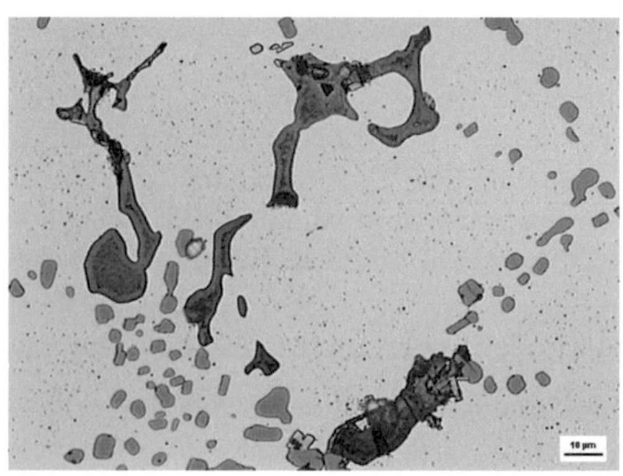
etch. H_2SO_4

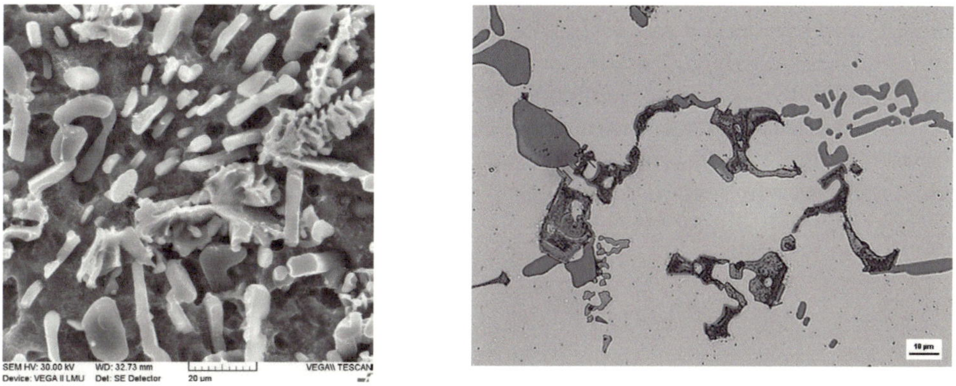

d) 525 °C, 4 hours, deep-etch. HCl, SEM etch. H_2SO_4

Fig. 12. Effect of solution treatment on morphology of Fe-rich phases

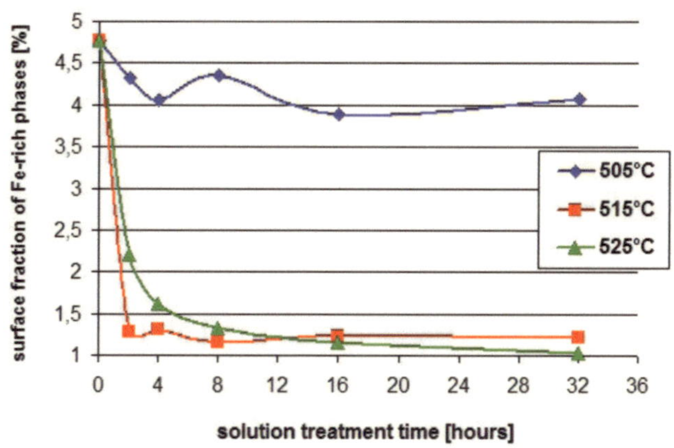

Fig. 13. Influence of solution treatment on surface fraction of Fe-rich phases

Quantitative metallography was carried out on an Image Analyzer NIS-Elements to quantify Fe-phases changes, during solution treatment. It was established that the temperature increase of solution treatment was attended not only by fragmentation of $Al_{15}(FeMn)_3Si_2$ phase, but also by decrease of surface fraction of all Fe-rich phases in AlSi9Cu3 alloy (Fig. 13). For the non-heat treated state the surface fraction of Fe-rich phase was c. 4.8 %, for temperature 515 °C c. 1.6 % and for 525 °C only c. 1.25 %. Solution treatment reduces its surface fraction rather than changes its morphology (Fig. 12 and Fig. 13).

3.5 Effect of solution treatment o the morphology of Cu-rich phases

The Cu-rich phase solidifies as fine ternary eutectic (Al-Al_2Cu-Si) - Fig. 6. Effect of solution treatment on morphology of Al-Al_2Cu-Si is demonstrated on Fig. 14. The changes of morphology of Al-Al_2Cu-Si observed after heat treatment are documented for holding time 4 hours. Al-Al_2Cu-Si phase without heat treatment (untreated state) occurs in form compact oval troops (Figures 14a and 15a).

After solution treatment by temperature 505 °C these phase disintegrated into fine smaller segments and the amount of Al-Al_2Cu-Si phase during heat treatment decreases. This phase is gradually dissolved into the surrounding Al-matrix with an increase in solution treatment time (Fig. 14b). By solution treatment by 515 °C is this phase observed in the form coarsened globular particles and these occurs along the black needles, probably Fe-rich Al_5FeSi phase (Figures 14c and 15b). By solution treatment 525 °C is this phase documented in the form molten particles with homogenous shape (Fig. 14d).

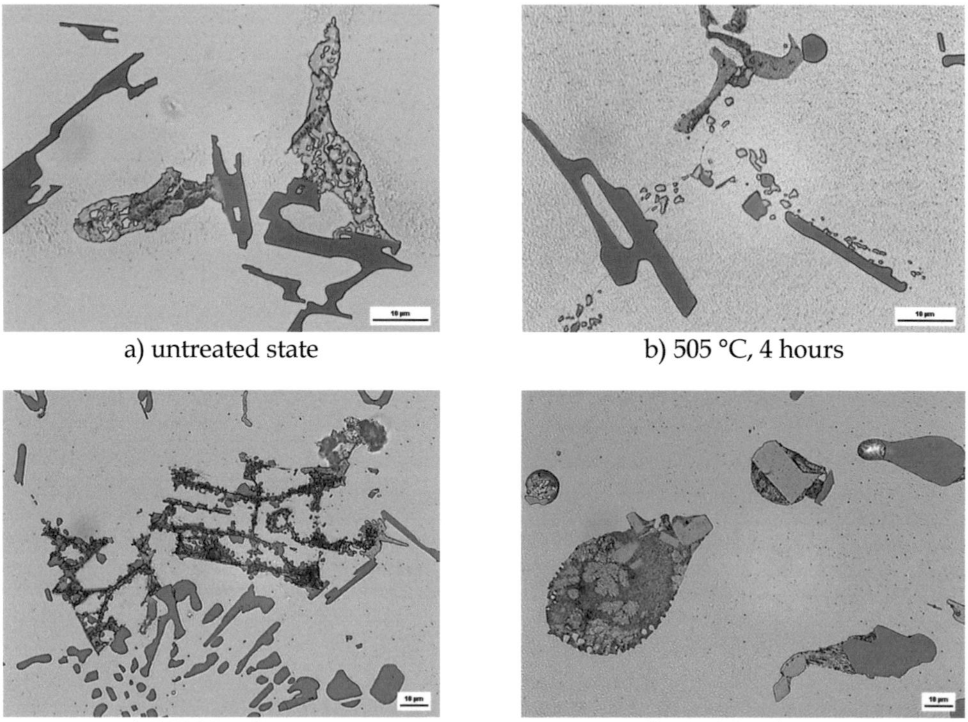

a) untreated state

b) 505 °C, 4 hours

c) 515 °C, 4 hours

d) 525 °C, 4 hours,

Fig. 14. Effect of solution treatment on morphology of Cu-rich phases, etch. HNO_3

Compact Al-Al_2Cu-Si phase disintegrates to fine separates Al_2Cu particles. The amount of these phases was not obvious visible on optical microscope. On SEM microscope we observed these phases in form very small particles for every temperatures of natural aging. By observation we had to use a big extension, because we did not see these elements.

Small precipitates (Al_2Cu) incipient by hardening were invisible in the optical microscope and electron microscope so it is necessary observation using TEM microscopy.

3.6 SEM observation of the fracture surface

Topography of fracture surfaces is commonly examined by SEM. The large depth of field is a very important advantage for fractographic investigations. Fracture surfaces of Al-Si-Cu cast alloys can be observed by means of SEM without almost any special preparation; nevertheless, if it is possible, the specimens should be examined immediately after failure because of the very fast superificial oxidation of Al-alloys. In some cases, the specimen should be cleaned mechanically by rinsing in ultrasonic cleaner, chemical reagents, or electrolytes (Michna et al., 2007; Tillova & Chalupova, 2009; Warmuzek, 2004b).

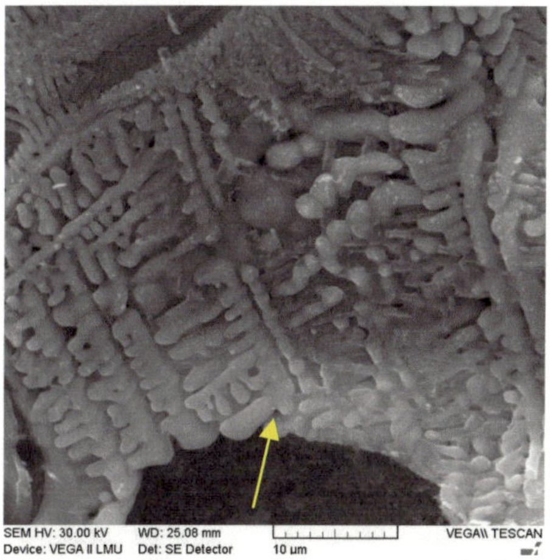

a) untreated state - compact morphology

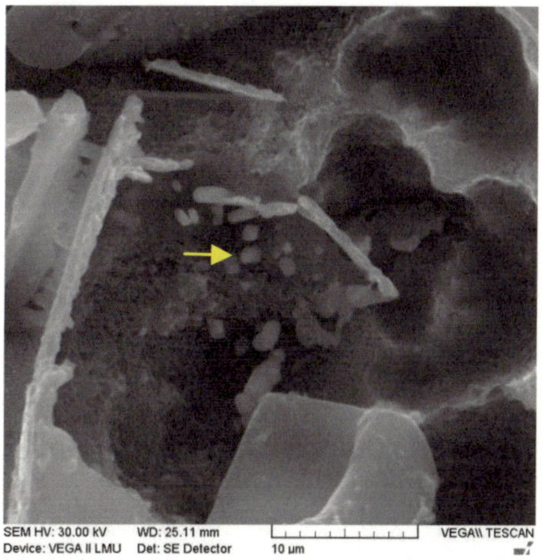

b) 515 °C, 4 hours - fine round particles

Fig. 15. Morphology of Cu-rich phases after deep etching, etch. HCl, SEM

Fractographs of the specimens in untreated state (as cast state) after impact test are documented in Fig. 16. As the experimental material was not modified and eutectic Si particles are in a form of platelets (Fig. 2), fracture surfaces are mainly composed of ductile fracture with cleavage fracture regions.

Fracture of the α-matrix is transcrystalline ductile with dimples morphology and with plastically transformed walls (Fig. 16a, b). The shape of walls depends on the orientation of Si particles on fracture surface. The brittle eutectic Si and Fe-rich phases (Figures 3-5) are fractured by the transcrystalline cleavage mechanism (Fig. 16c, d, e). Cu-phase (compact ternary eutectic Al-Al_2Cu-Si - Fig. 6) is fractured by transcrystalline ductile fracture with the very fine and flat dimples morphology (Fig. 16f). In some cases, to improve the contrast of the matrix/phase interface, detection of backscattered electrons (BSE) in a SEM is a very useful method (Fig. 16c). This method provides another alternative when phase attribution by morphology and/or colour, is not clearly.

Fractographs of the specimens after solution treatment are documented in Fig. 17. By temperature 505 °C of solution treatment were noted that the Si-platelets were fragmentized into smaller platelets with spherical edges (Fig. 10b). Spheroidisation process of eutectic silicon was not observed. The morphology from transcrystalline brittle fracture (cleavage) is mainly visible, but some degree of plastic deformation in the aluminium solid solution (α-matrix) also may be noticed in the form of shallow dimples and plastically transformed walls (Fig. 17a, b).

After solution treatment at the 515 °C eutectic silicon is completely spheroidised (Fig. 10c). Number of brittle Fe-phases decreases (Fig. 12c). Fracture is transcrystalline ductile with fine dimples morphology (Fig. 17c, d). The size of the dimples shows the size of eutectic silicon. Local we can observe little cleavage facets of Fe-rich phases.

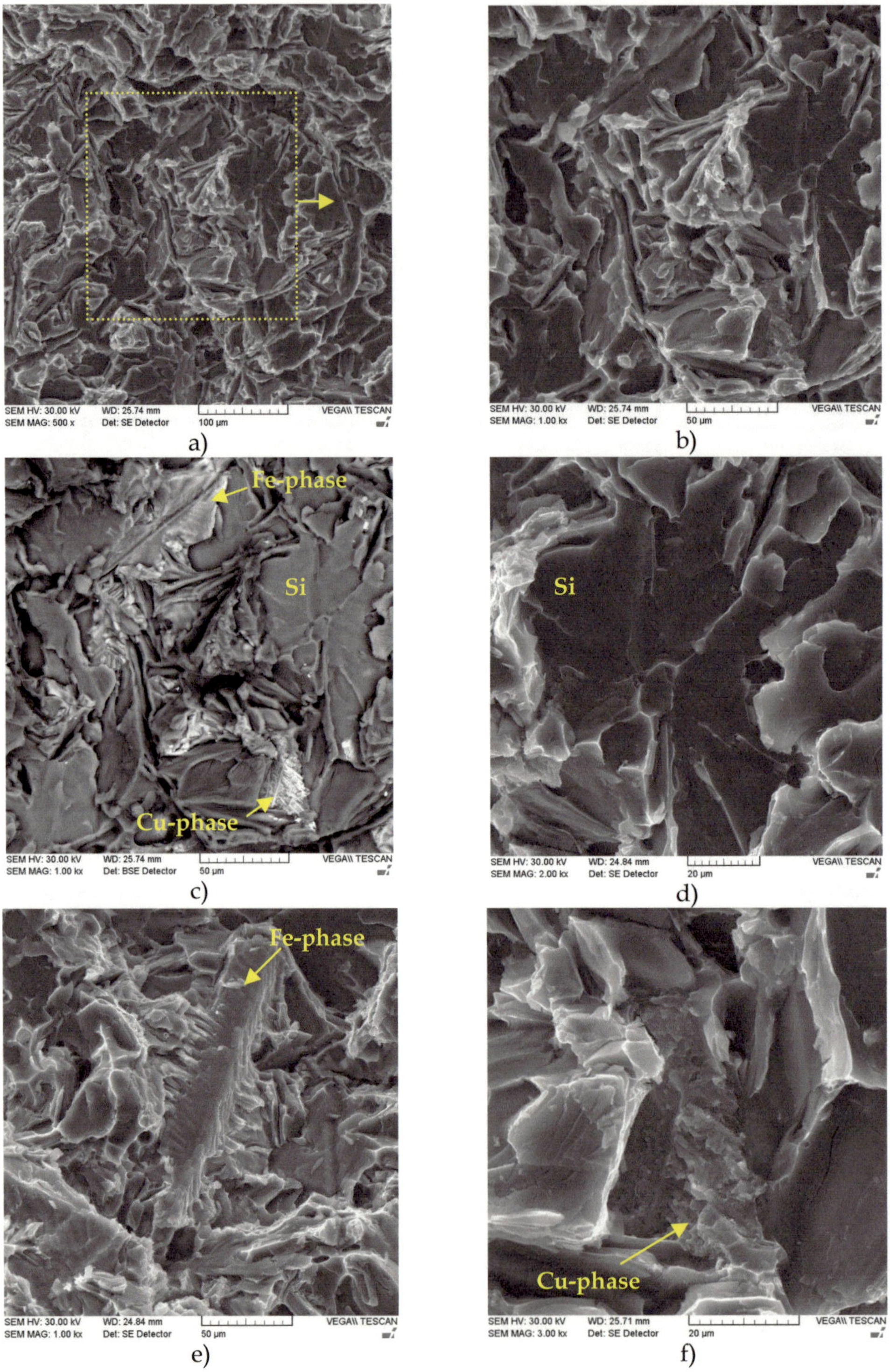

Fig. 16. Fractographs of the impact test specimen – as cast state, SEM

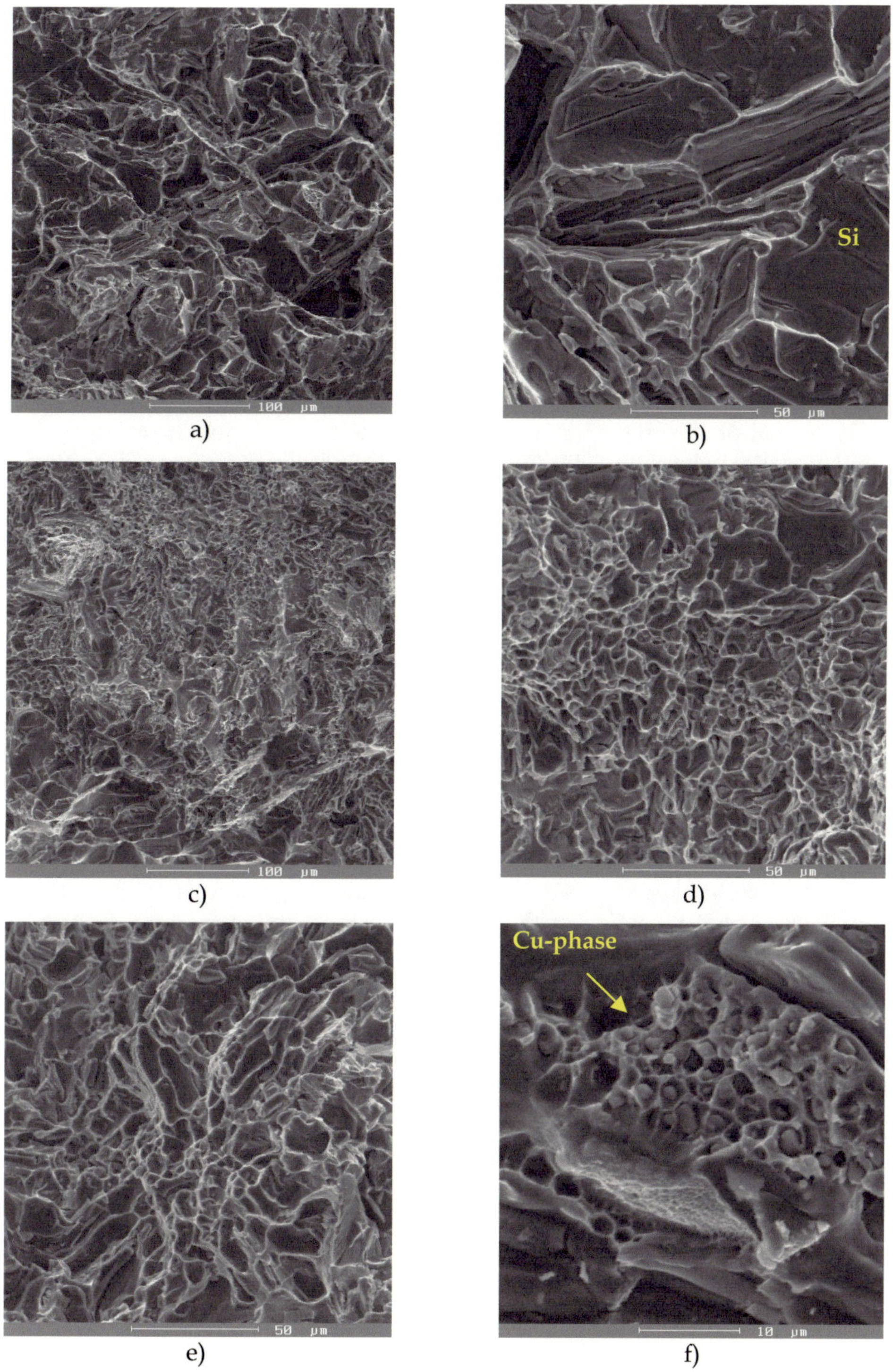

Fig. 17. Fractographs of the impact test specimen – after solution treatment, SEM

Fractograph of the specimen after solution treatment at the 525 °C is documented in Fig. 17e. Eutectic silicon is completely spheroidised too (Fig. 10d), but spheroidised Si-particles gradually grew larger. The fracture mechanism was identified as transcrystalline ductile with dimples morphology accompanied by plastically transformed walls (Fig. 17e). The size of the dimples shows the larger size of eutectic silicon as compared with fractograph Fig. 17d. Figure 17f is an example of a transcrystalline ductile fracture of Cu-rich phase after solution treatment at the 515 °C.

3.7 Influence of solution annealing on fatigue properties

To successfully utilize recycled Al-Si-Cu alloys in critical components, it is necessary to thoroughly understand its fatigue property too. Numerous studies have shown that fatigue property of conventional casting aluminium alloys are very sensitive to casting defects (porosity, microshrinkages and voids) and many studies have shown that, whenever a large pore is present at or near the specimen's surface, it will be the dominant cause of fatigue crack initiation (Bokuvka et al., 2002; Caceres et al., 2003; Moreira & Fuoco, 2006; Novy et al.; 2007). The occurrence of cast defects, together with the morphology of microstructural features, is strongly connected with method of casting too. By sand mould is the concentration of hydrogen in melt, as a result of damp cast surroundings, very high. The solubility of hydrogen during solidification of Al-Si cast alloys rapidly decreases and by slow cooling rates (sand casting) keeps in melt in form of pores and microshrinkages (Michna et al., 2007).

Fe is a common impurity in aluminium alloys that leads to the formation of complex Fe-rich intermetallic phases, and how these phases can adversely affect mechanical properties, especially ductility, and also lead to the formation of excessive shrinkage porosity defects in castings (Taylor, 2004; Tillova & Chalupova, 2009).

It is clear, that the morphology of Fe-rich intermetallic phases influences harmfully on fatigue properties too (Palcek et. al., 2003). Much harmful effect proves the cast defects as porosity and microshrinkages, because these defects have larger size as intermetallic phases. A comprehensive understanding of the influence of these microstructural features on the fatigue damage evolution is needed.

In the end heat treatment is considered as an important factor that affects the fatigue behaviour of casting Al-Si-Cu alloys too (Tillova & Chalupova, 2010).

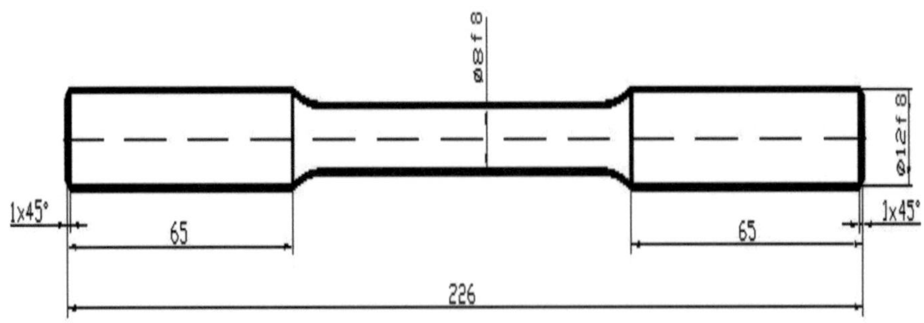

Fig. 18. Fatigue specimen geometry (all dimension in mm)

The fatigue AlSi9Cu3 tests (as-cast, solution heat treated at two temperatures 515 and 525 °C for times 4 hours, then quenched in warm water at 40 °C and natural aged at room temperature for 24 hours) were performed on rotating bending testing machine ROTOFLEX operating at 30 Hz., load ratio R = -1 and at room temperature 20 ± 5 °C on the air. Cylindrical fatigue specimens were produced by lathe-turning and thereafter were heat treated. Geometry of fatigue specimens is given in Fig. 18. The fatigue fracture surfaces of the fatigue - tested samples under different solution heat treatment condition were examined using a scanning electron microscope (SEM) TESCAN VEGA LMU with EDX analyser BRUKER QUANTAX after fatigue test.

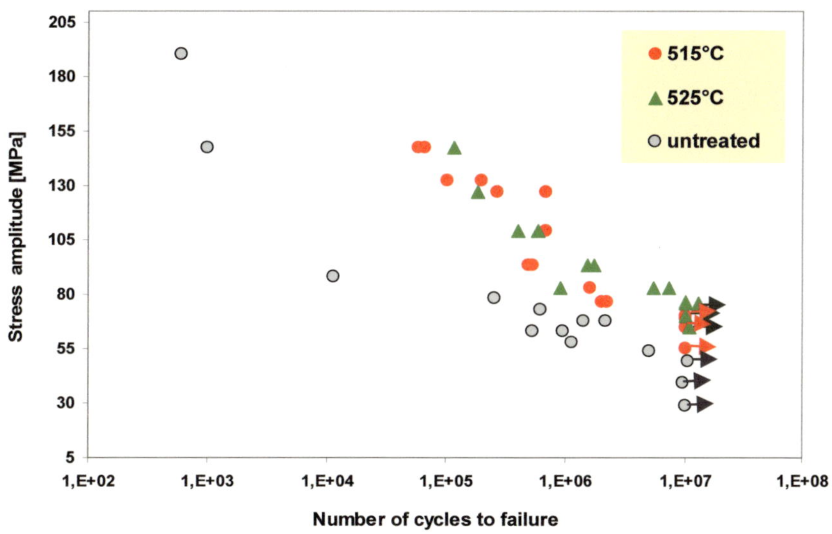

Fig. 19. Effect of solution treatment on fatigue behaviour of AlSi9Cu3 cast alloy

The untreated specimens were tested first to provide a baseline on fatigue life. In this study, the number of cycle, 10^7, is taken as the infinite fatigue life. Thus, the highest applied stress under which a specimen can withstand 10^7 cycles is defined as the fatigue strength of the alloy. The relationship between the maximum stress level (S), and the fatigue life in the form of the number of fatigue cycles (N), (S-N curves) is given in Fig. 19. Comparison on the fatigue properties of specimens with and without heat treatment was made. In heat untreated state has fatigue strength (σ) at 10^7 cycles the lowest value, only σ = 49 MPa. It is evident, that after solution treatment increased fatigue strength at 10^7 cycles. By the conditions of solution treatment 515 °C/4 hours the fatigue strength at 10^7 cycles increases up to value σ = 70 MPa. The solution treatment by 525 °C/4 hours caused the increasing of fatigue strength at 10^7 cycles to value σ = 76 MPa. The growths in fatigue strength at 10^7 cycles with respect to the temperature of solution treatment are 42 and 55 % respectively.

Fatigue fracture surfaces were examined in the SEM in order to find the features responsible for crack initiation. Typical fractographic surfaces are shown in Fig. 20, Fig. 21, Fig. 22 and Fig. 23. The global view of the fatigue fracture surface for untreated and heat treated specimens is very similar. The process of fatigue consists of three stages – crack initiation stage (I), progressive crack growth across the specimen (II) and final sudden static fracture of the remaining cross section of the specimen (III) (Bokuvka et al., 2002; Palcek et al., 2003; Novy et al., 2007; Tillova & Chalupova, 2009; Moreira & Fuoco, 2006).

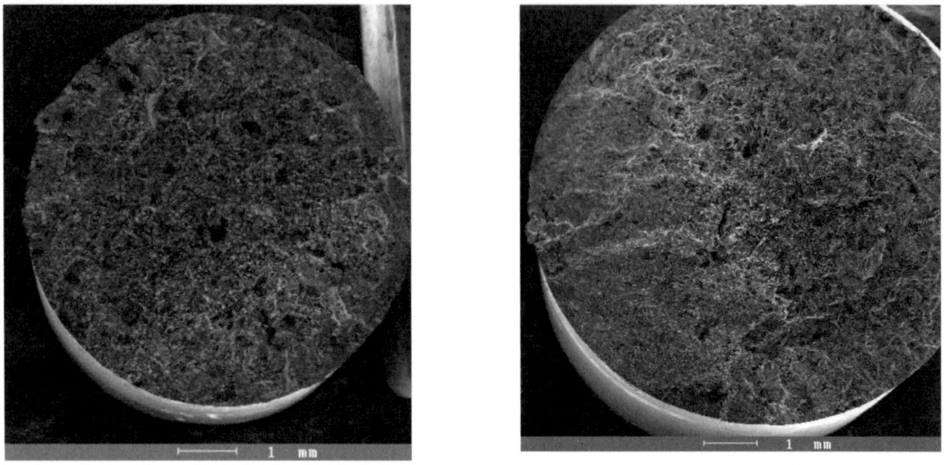

a) $\sigma = 88$ MPa, $N_f = 11\,560$ cycles   b) $\sigma = 54$ MPa, $N_f = 5.10^6$ cycles

Fig. 20. Complete fracture surfaces, SEM

Stage I and II is so-called fatigue region. The three stages are directly related to the macrographic aspects of the fatigue fractures (Fig. 20). Within the bounds of fatigue tests was established that, high stress amplitude caused small fatigue region (Fig. 20a) and large region of final static rupture. With the decreasing of stress amplitude increases the fatigue region of stable propagating of cracks (Fig. 20b) and the initiation places are more focused to one point simultaneously.

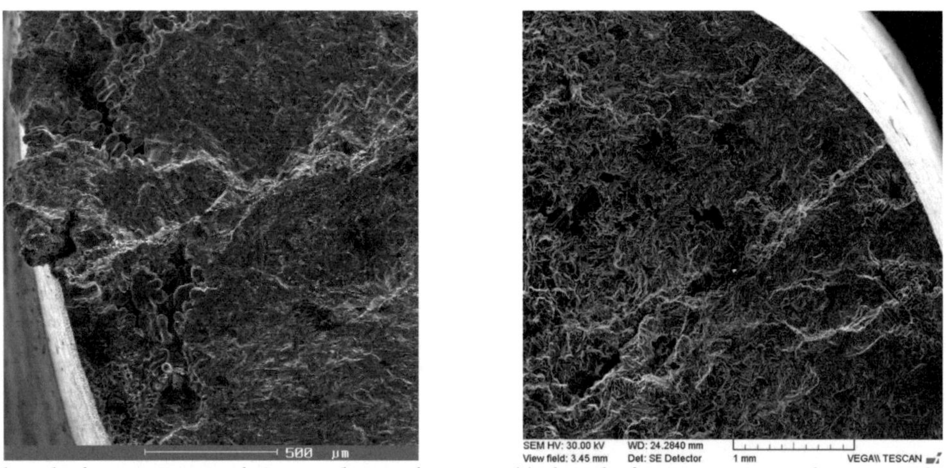

a) detail of one initiation' site on the surface   b) detail of more initiation' sites on the surface

Fig. 21. Fatigue crack nucleation - overview of a fracture surface

Important to the stress concentration and to fatigue crack nucleation is the presence of casting defects as microporosities, oxide inclusions and shrinkage porosities, since the size of these defects can by much larger than the size of the microstructure particles. It was confirmed, that if are in structure marked cast defects, then behaved preferentially as an initiation's places of fatigue damage.

The cast defects were detected on the surface of test fatigue specimens. Details of the initiations site are shown in Fig. 21. For low stress amplitudes were observed one initiation place (Fig. 21a). For high stress amplitudes existed more initiation places (Fig. 21b). The occurrence of these cast defects (Fig. 22) causes the small solubility of hydrogen during solidification of Al-Si alloys.

The main micrographic characteristics of the fatigue fracture near the initiating site are the tear ridges (Fig. 23a-c) in the direction of the crack propagation and the fatigue striation in a direction perpendicular to the crack propagation. The striations are barely seen (Figure 23d). Fig. 23b illustrates the same fatigue surface as Fig. 23a, near the initiating site, in BSE electron microscopy. The result of BSE observation presents the contrast improvement of brittle Fe-rich intermetallic phases $Al_{15}(FeMn)_3Si_2$.

Final rupture region for untreated and heat treated specimens is documented in Fig. 24. Fracture path is from micrographic aspect thus mostly transgranular and the appearance of the fracture surface is more flat. The fracture of the α-dendritic network is always ductile but particularly depends on morphology of eutectic Si and quantity of brittle intermetallic phases (e.g. $Al_{15}(FeMn)_3Si_2$ or Al_5FeSi).

The fracture surface of the as-cast samples revealed, in general, a ductile rupture mode with brittle nature of unmodified eutectic silicon platelets (Fig. 24a).

Fracture surface of heat treated samples consists almost exclusively of small dimples, with morphology and size that traced morphologhy of eutectic silicon (solution treatment resulted spheroidisation of eutectic silicon), such as those seen in Fig. 24b and Fig. 24c.

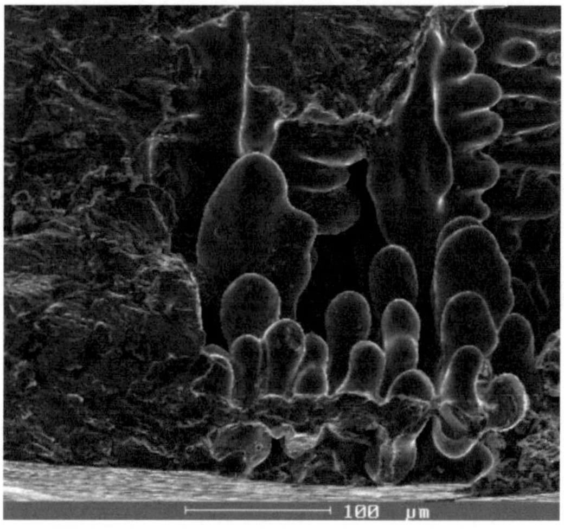

Fig. 22. Detail of cast defect

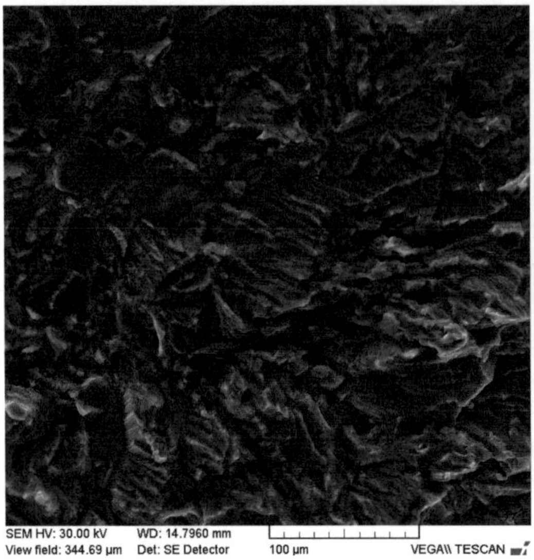

a) fatigue fracture surface near the initiating site - fine tear ridges

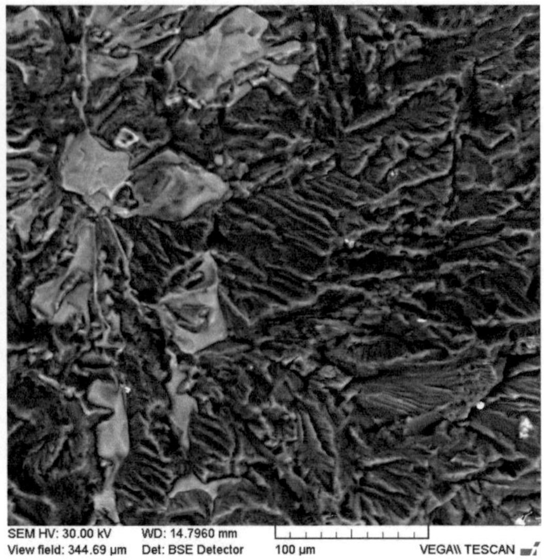

b) fatigue fracture surface near the initiating site - BSE

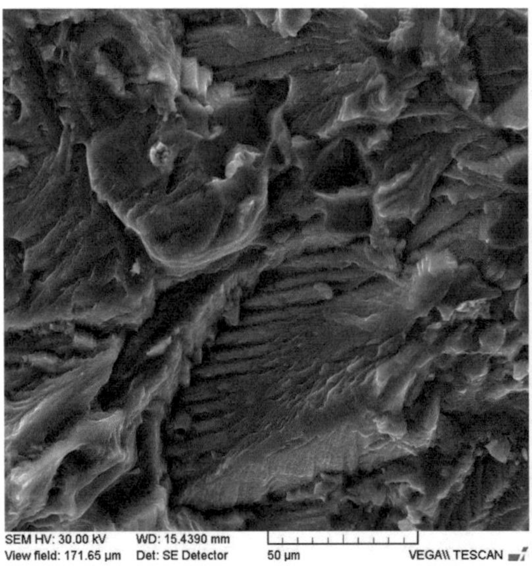

c) detail of fatigue fracture surface - tear ridges

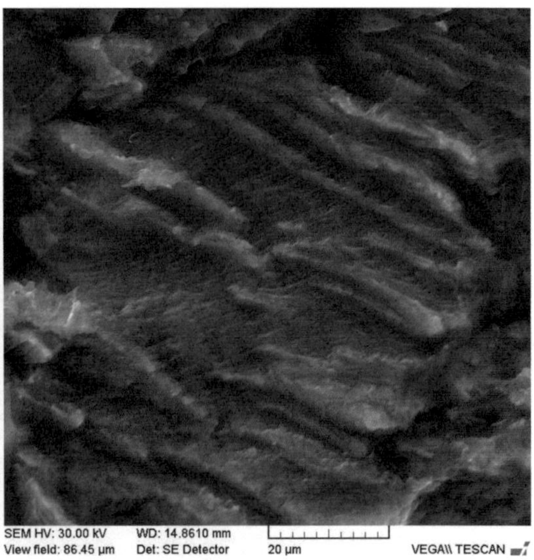

d) detail of the typical aspect of fatigue - extremely fine striations

Fig. 23. Typical fatigue fracture surface

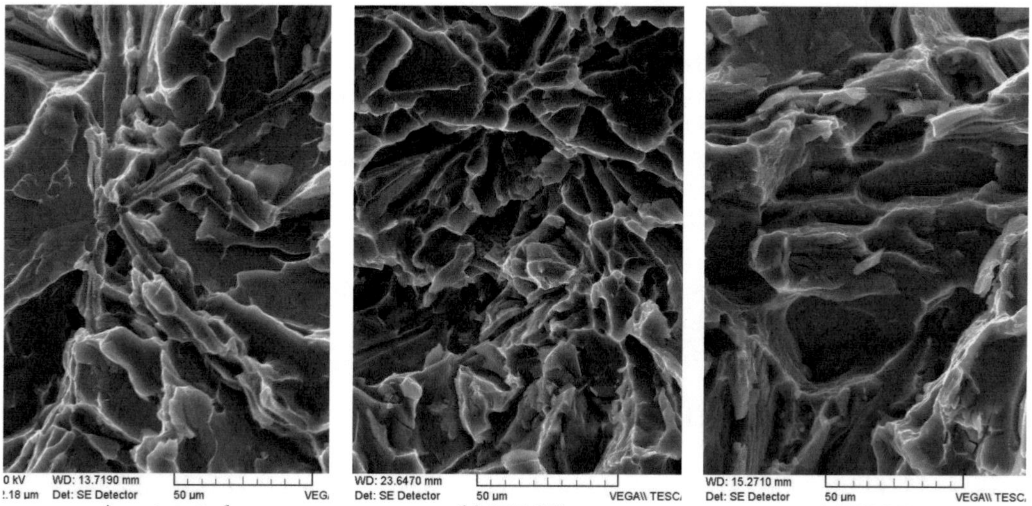

a) untreated b) 515 °C c) 525 °C

Fig. 24. Final rupture region - detail

4. Acknowledgment

The authors acknowledge the financial support of the projects VEGA No1/0249/09; VEGA No1/0841/11 and European Union - the Project "*Systematization of advanced technologies and knowledge transfer between industry and universities (ITMS 26110230004)*".

5. References

Abdulwahab, M. (2008). Studies of the mechanical properties of age-hardened Al-Si-Fe-Mn alloy. *Australian Journal of Basic and Applied Sciences*, Vol. 2, 4, pp. 839-843

ASM Handbook (1991). Volume 4, *Heat treating*. pp. 1861-1956, ASM International

Belan, J. (2008). Structural Analyses of Advanced Materials for Aerospace Industry. *Materials science/Medžiagotyra*, Vol.14, 4, pp. 315-318, ISSN 1392-1320

Bokuvka, O.; Nicoletto, G.; Kunz, L.; Palcek, P. & Chalupova, M. (2002). Low and high - frequency fatigue testing. CETRA, EDIS, Žilina, ISBN 80-8070-011-7

Caceres, C. H.; Svenson, I. L. & Taylor, J. A. (2003). Strenght-ductility Behaviour of Al-Si-Cu-Mg Casting Alloys in T6 temper. *Int. J. Cast Metals Res.*, No. 15, pp. 531-543, ISSN 1364-0461

Das, K. S. (2006). Designing Aluminum Alloys for a Recycle-Friendly World. *Materials Science Forum*, Vols. 519-521, pp. 1239-1244, ISSN 1662-9752

Das, K. S. & Gren J. A. S. (2010). Aluminum Industry and Climate Change-Assessment and Responses. *JOM*, 62, 2, pp. 27-31, ISSN 1047-4838

Dobrzański, L. A.; Maniara, R.; Krupiński, M. & Sokołowski, J. H. (2007). Microstructure and mechanical properties of AC AlSi9CuX alloys. *Journal of Achievements in Materials and Manufacturing Engineering*, Vol. 24, Issue 2, pp. 51-54, ISSN 1734-8412

Fang, X.; Shao, G.; Liu, Y. Q. & Fan, Z. (2007). Effect of intensive forced melt convection on the mechanical properties of Fe- containing Al-Si based alloy. *Materials science and engineering A*, Structural materials: properties, microstructure and processing. Vol. 445-446, pp. 65-72, ISSN 0921-5093

Krupiński, M.; Labisz, K.; Rdzawski, Z. & Pawlyta, M. (2011). Cooling rate and chemical composition influence on structure of Al-Si-Cu alloys. *Journal of Achievements in Materials and Manufacturing Engineering,* Vol. 45, Issue 1, pp. 13-22, ISSN 1734-8412

Kim, H. Y.; Park, T. Y.; Han, S. W. & Mo, L. H. (2006). Effects of Mn on the crystal structure of α-Al(Mn,Fe)Si particles in A356 alloys. *Journal of Crystal Growth*, Vol. 291, Issue 1, pp. 207-211, ISSN 0022-0248

Lasa, L. & Rodriguez-Ibabe, J. M. (2004). Evolution of the main intermetallic phases in Al-Si-Cu-Mg casting alloys during solution treatment. *Journal of Materials Science*, 39, pp. 1343-1355, ISSN 0022-2461

Li, R. (1996). Solution heat treatment of 354 and 355 cast alloys. *AFS Transaction*, No. 26, pp. 777-783

Li, R. X.; Li, R. D.; Zhao, Y. H.; He, L. Z.; Li, C. X.; Guan, H. R. & Hu, Z. Q. (2004) . Age-hardening behaviour of cast Al-Si base alloy. Materials Letters, 58, pp. 2096-2101, ISSN 0167-577X

Lu, L. & Dahle, A. K. (2005). Iron-Rich Intermetallic Phases and Their Role in Casting Defect Formation in Hypoeutectic Al-Si Alloys. *Metallurgical and Materials Transactions A*, Volume 36A, pp. 819-835, ISSN 1073-5623

Ma, Z.; Samuel, A. M.; Samuel F. H.; Doty, H. W. & Valtierra, S. (2008). A study of tensile properties in Al-Si-Cu and Al-Si-Mg alloys: Effect of β-iron intermetallics and porosity. *Materials Science and Engineering A*, Vol. 490, pp. 36-51, ISSN 0921-5093

Mahfoud, M.; Prasada Rao, A. K. & Emadi, D. (2010). The role of thermal analysis in detecting impurity levels during aluminum recycling. *J Therm Anal Calorim*, 100, pp. 847-851, ISSN 1388-6150

Maniara, R., Dobrzański, L. A., Krupiński, M. & Sokołowski, J. H. (2007). The effect of copper concentration on the microstructure of Al-Si-Cu alloys. *Archiwes of Foundry engineering,* Vol. 7, Issue 2, pp. 119-124, ISSN 1897-3310

Martinkovic, M. (2010). *Quantitative analysis of materials structures.* STU Bratislava, ISBN 978-80-2273-445-5 (in Slovak)

Michna, S.; Lukac, I.; Louda, P.; Ocenasek, V.; Schneider H.; Drapala, J,; Koreny, R. ; Miskufova, A. et. al. (2007). *Aluminium materials and technologies from A to Z.* Adin, s.r.o. Presov, ISBN 978-80-89244-18-8

Moreira, M. F. & Fuoco, R. (2006). Characteristics of fatigue fractures in Al-Si cast components. *AFS Transactions*, pp. 1-15

Mrówka-Nowotnik, G. & Sieniawski, J. (2011). Microstructure and mechanical properties of C355.0 cast aluminium alloy. *Journal of Achievements in Materials and Manufacturing Engineering*, Vol. 47, Issue 2, pp. 85-94, ISSN 1734-8412

Moustafa, M. A.; Samuel, F. H. & Doty, H. W. (2003). Effect of solution heat treatment and additives on the microstructure of Al-Si (A413.1) automotive alloys. *Journal of Materials Science*, 38, p. 4507-4522, ISSN 0022-2461

Moustafa, M. A. (2009). Effect of iron content on the formation of ß-Al_5FeSi and porosity in Al-Si eutectic alloys. *Journal of Materials Processing Technology*, 209, pp. 605-610, ISSN 0924-0136

Novy, F.; Cincala, M.; Kopas, P. & Bokuvka, O. (2007). Mechanisms of high-strength structural materials fatigue failure in ultra-wide life region. *Materials Science and Engineering A*, Vol. 462, No. 1-2, pp. 189-192, ISSN 0921-5093.

Palcek, P.; Chalupova, M.; Nicoletto, G. & Bokuvka, O. (2003). Prediction of machine element durability. CETRA, EDIS Žilina, ISBN 80-8070-103-2

Panuskova, M.; Tillova, E. & Chalupova, M. (2008). Relation between mechanical properties and microstructure of Al-cast alloy AlSi9Cu3. *Strength of Materials*, Vol. 40, No. 1, pp. 98-101, ISSN 1573-9325

Paray, F. & Gruzleski, J. E. (1994). Microstructure–mechanical property relationships in a 356 alloy. Part I: Microstructure. *Cast Metals*, Vol. 7, No.1, pp. 29-40

Rios, C. T. & Caram, R. (2003). Intermetallic compounds in the Al-Si-Cu system. *Acta Microscopica*, Vol.12, N°1, pp. 77-81, ISSN 0798-4545

Sencakova, L. & Vircikova, E. (2007). Life cycle assessment of primary aluminium production. *Acta Metallurgica Slovaca*, 13, 3, pp. 412-419, ISSN 1338-1156

Shabestari, S. G. (2004). The effect of iron and manganese on the formation of intermetallic compounds in aluminum-silicon alloys. *Materials Science and Engineering A*, 383, pp. 289-298, ISSN 0921-5093

Samuel, A. M.; Samuel, F. H. & Doty, H. W. (1996). Observation on the formation ß-Al_5FeSi phase in 319 type Al-Si alloys. *Journal of Materials Science*, 31, pp. 5529-5539, ISSN 0022-2461

Seifeddine, S. 2007. The influence of Fe on the microstructure and mechanical properties of cast Al-Si alloys. Literature review - Vilmer project. Jönköping University, Sweden

Seifeddine, S.; Johansson, S. & Svensson, I. (2008). The Influence of Cooling Rate and Manganese Content on the β-Al_5FeSi Phase Formation and Mechanical Properties of Al-Si–based Alloys. *Materials Science and Engineering A*, No. 490, pp. 385-390, ISSN 0921-5093

Sjölander, E. & Seifeddine, S. (2010). Optimisation of solution treatment of cast Al-Si-Cu alloys. *Materials and Design*, 31, pp. 44–49, ISSN 0261-3069

Skocovsky, P.; Tillova, E. & Belan, J. (2009). Influence of technological factors on eutectic silicon morphology in Al-Si alloys. *Archiwes of Foundry Engineering*, Vol. 9, Issue 2, pp. 169-172, ISSN 1897-3310

Skocovsky, P. & Vasko, A. (2007). *Quantitative evaluation of structure in cast iron*, EDIS Žilina, ISBN 978-80-8070-748-4 (in Slovak)

Taylor J. A. 2004. The effect of iron in Al-Si casting alloys. *35th Australian Foundry Institute National Conference*, pp. 148-157, Adelaide, South Australia

Tillova, E. & Chalupova, M. (2001). Využitie hlbokého leptania pri štúdiu morfológie eutektického kremíka. *Scientific papers of the University of Pardubice*, Series B - The Jan Perner Transport Faculty, 7, pp. 41-54, ISSN 1211-6610

Tillova, E. & Panuskova, M. (2007). Effect of Solution Treatment on Intermetallic Phase's Morphology in AlSi9Cu3 Cast Alloy. *Materials Engineering*, No. 14, pp. 73-76, ISSN 1335-0803

Tillova, E. & Panuskova, M. (2008). Effect of Solution Treatment on Intermetallic Phase's Morphology in AlSi9Cu3 Cast Alloy. *Mettalurgija/METABK*, No. 47, pp. 133-137, 1-4, ISSN 0543-5846

Tillova, E. & Chalupova, M. (2009). *Structural analysis of Al-Si cast alloys*. EDIS Žilina, ISBN 978-80-554-0088-4, Žilina, Slovakia (in Slovak)

Tillova, E.; Chalupova, M. & Hurtalova, L. (2010). Evolution of the Fe-rich phases in Recycled AlSi9Cu3 Cast Alloy During Solution Treatment. *Communications,* 4, pp. 95-101, ISSN 1335-4205

Tillova, E. & Chalupova, M. (2010). Fatigue failure of recycled AlSi9Cu3 cast alloy. *Acta Metallurgica Slovaca Conference*, Vol.1, No.2, pp. 108–114, ISSN 1335-1532

Vasko, A. & Belan, J. (2007). Comparison of methods of quantitative metallography, In: *Improvement of Quality Regarding processes and Materials*, S. Borkowski & E. Tillova, (Ed.), 53-58, PTM, ISBN 978-83-924215-3-5, Warszawa, Poland

Vaško, A. (2008). Influence of SiC additive on microstructure and mechanical properties of nodular cast iron. *Materials science/Medžiagotyra*, Vol.14, 4, pp. 311-314.,ISSN 1392-1320

Wang, Q. G.; Apelian, D. & Lados, D. A. (2001). Fatigue Behavior of A356/357 Aluminum Cast Alloys - part II. Efect of Microstructural Constituents. *J. of Light Metals*, 1, pp. 85-97, ISSN 1471-5317

Warmuzek, M. (2004a). Metallographic Techniques for Aluminum and Its Alloys. *ASM Handbook. Metallography and Microstructures*, Vol 9, pp. 711-751, ISBN 978-0-87170-706-2, ASM International

Warmuzek, M. (2004b). *Aluminum-Silicon-Casting Alloys: Atlas of Microfractographs.* ISBN 0-87170-794-2, ASM International, Materials Park

22

Strength and Microstructure of Cement Stabilized Clay

Suksun Horpibulsuk
Suranaree University of Technology, Thailand

1. Introduction

The soil/ground improvement by cement is an economical and worldwide method for pavement and earth structure works. Stabilization begins by mixing the in-situ soil in a relatively dry state with cement and water specified for compaction. The soil, in the presence of moisture and a cementing agent becomes a modified soil, i.e, particles group together because of physical-chemical interactions among soil, cement and water. Because this occurs at the particle level, it is not possible to get a homogeneous mass with the desired strength. Compaction is needed to make soil particles slip over each other and move into a densely packed state. In this state, the soil particles can be welded by chemical (cementation) bonds and become an engineering material (Horpibulsuk et al., 2006). To reduce the cost of ground improvement, the replacement of the cement by waste materials such as fly ash and biomass ash is one of the best alternative ways. In many countries, the generation of these waste materials is general far in excess of their utilization. A feasibility study of utilizing these ashes (waste materials) to partially replace Type I Portland cement is thus interesting.

The effects of some influential factors, i.e., water content, cement content, curing time, and compaction energy on the laboratory engineering characteristics of cement-stabilized soils have been extensively researched (Clough et al., 1981; Kamon & Bergado, 1992; Yin & Lai, 1998; Miura et al., 2001; Horpibulsuk & Miura, 2001; Horpibulsuk et al., 2003, 2004a, 2004b, 2005, 2006, 2011a). The field mixing effect such as installation rate, water/cement ratio and curing condition on the strength development of cemented soil was investigated by Nishida et al. (1996) and Horpibulsuk et al. (2004c, 2006 and 2011b). Based on the available compression and shear test results, many constitutive models were developed to describe the engineering behavior of cemented clay (Gens and Nova, 1993; Kasama et al., 2000; Horpibulsuk et al., 2010a; Suebsuk et al., 2010 and 2011). These investigations have mainly focused on the mechanical behavior that is mainly controlled by the microstructure. The structure is fabric that is the arrangement of the particles, clusters and pore spaces in the soil as well as cementation (Mitchell, 1993). It is thus vital to understand the changes in engineering properties that result from the changes in the influential factors.

This chapter attempts to illustrate the microstructural changes in cement-stabilized clay to explain the different strength development according to the influential factors, i.e., cement content, clay water content, fly ash content and curing time. The unconfined compressive strength was used as a practical indicator to investigate the strength development. The microstructural analyses were performed using a scanning electron microscope (SEM), mercury intrusion porosimetry (MIP), and thermal gravity (TG) tests. For SEM, the cement stabilized samples were broken from the center into small fragments. The SEM samples were frozen at -195°C by immersion in liquid nitrogen for 5 minutes and evacuated at a pressure of 0.5 Pa at -40°C for 5 days (Miura et al., 1999; and Yamadera, 1999). All samples were coated with gold before SEM (JEOL JSM-6400) analysis.

Measurement on pore size distribution of the samples was carried out using mercury intrusion porosimeter (MIP) with a pressure range from 0 to 288 MPa, capable of measuring pore size diameter down to 5.7 nm (0.0057 micron). The MIP samples were obtained by carefully breaking the stabilized samples with a chisel. The representative samples of 3-6 mm pieces weighing between 1.0-1.5 g were taken from the middle of the cemented samples. Hydration of the samples was stopped by freezing and drying, as prepared in the SEM examination. Mercury porosimetry is expressed by the Washburn equation (Washburn, 1921). A constant contact angle (θ) of 140° and a constant surface tension of mercury (γ) of 480 dynes/cm were used for pore size calculation as suggested by Eq.(1)

$$D = -(4\gamma\cos\theta)/P \tag{1}$$

where D is the pore diameter (micron) and P is the applied pressure (MPa).

Thermal gravity (TG) analysis is one of the widely accepted methods for determination of hydration products, which are crystalline $Ca(OH)_2$, CSH, CAH, and CASH, ettringite (Aft phases), and so on (Midgley, 1979). The CSH, CAH, and CASH are regarded as cementitious products. $Ca(OH)_2$ content was determined based on the weight loss between 450 and 580°C (El-Jazairi and Illston, 1977 and 1980; and Wang et al., 2004) and expressed as a percentage by weight of ignited sample. When heating the samples at temperature between 450 and 580°C, $Ca(OH)_2$ is decomposed into calcium oxide (CaO) and water as in Eq. (2).

$$Ca(OH)_2 \text{----------------------}> CaO + H_2O \tag{2}$$

Due to the heat, the water is lost, leading to the decrease in overall weight. The amount of $Ca(OH)_2$ can be approximated from this lost water by Equation (2), which is 4.11 times the amount of lost water (El-Jazairi and Illston, 1977 and 1980). The change of the cementitious products can be expressed by the change of $Ca(OH)_2$ since they are the hydration products.

2. Compaction and strength characteristics of cement stabilized clay

Compaction characteristics of cement stabilized clay are shown in Figure 1. The clay was collected from the Suranaree University of Technology campus in Nakhon Ratchasima, Thailand. It is composed of 2% sand, 45% silt and 53% clay. Its specific gravity is 2.74. The liquid and plastic limits are approximately 74% and 27%, respectively. Based on the Unified Soil Classification System (USCS), the clay is classified as high plasticity (CH). It is found that the maximum dry unit weight of the stabilized samples is higher than that of the unstabilized samples whereas their optimum water content is practically the same. This characteristic is the same as that of cement stabilized coarse-grained soils as reported by Horpibulsuk et al. (2006). The adsorption of Ca^{2+} ions onto the clay particle surface decreases the repulsion between successive diffused double layers and increases edge-to-face contacts between successive clay sheets. Thus, clay particles flocculate into larger clusters, which increases in the plastic limit with an insignificant change in the liquid limit (*vide* Table 1). As such, the plasticity index of the mixture decreases due to the significant increase in the plastic limit. Because the *OWC* of low swelling clays is mainly controlled by the liquid limit (Horpibulsuk et al., 2008 and 2009), the *OWC*s of the unstabilized and the stabilized samples are almost the same (*vide* Table 1). Figure 2 shows the compaction curve of the fly ash (FA) blended cement stabilized clay for different replacement ratios (ratios of cement to fly ash, C:F) compared with that of the unstabilized clay. Two fly ashes are presented in the figure: original, OFA (D_{50} = 0.03 mm) and classified, CFA (D_{50} = 0.009 mm) fly ashes. It is noted that the compaction curve of the stabilized clay is insignificantly dependent upon replacement ratio and fly ash particles. Maximum dry unit weight of the stabilized clay is higher than that of the unstabilized clay whereas their optimum water content is practically the same.

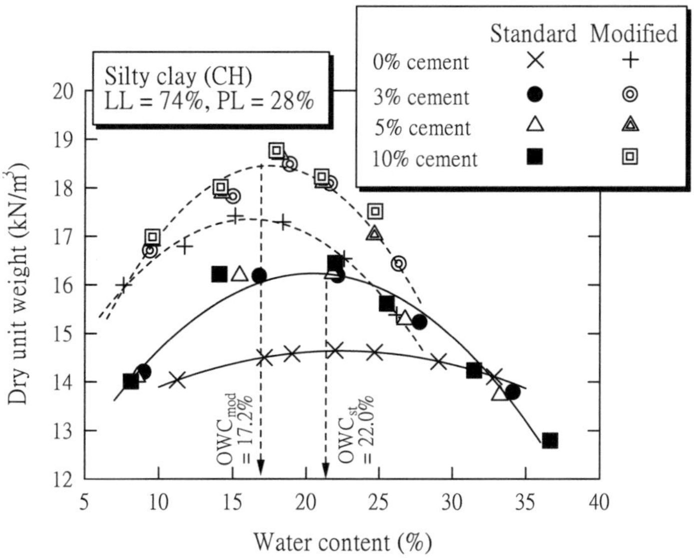

Fig. 1. Plots of dry unit weight versus water content of the uncemented and the cemented samples compacted under standard and modified Proctor energies (Horpibulsuk et al., 2010b)

Cement (%)	Atterberg's limits (%)			*OWC* (%)		γ_{dmax} (kN/m³)	
	LL	PL	PI	Std.	Mod.	Std.	Mod.
0	74.1	27.5	46.6	22.4	17.2	14.6	17.4
3	74.1	45.0	29.1	22.2	17.5	16.2	18.5
5	72.5	45.0	27.5	21.8	17.3	16.2	18.7
10	71.0	44.8	26.2	22.0	17.4	16.4	18.8

Table 1. Basic properties of the cemented samples

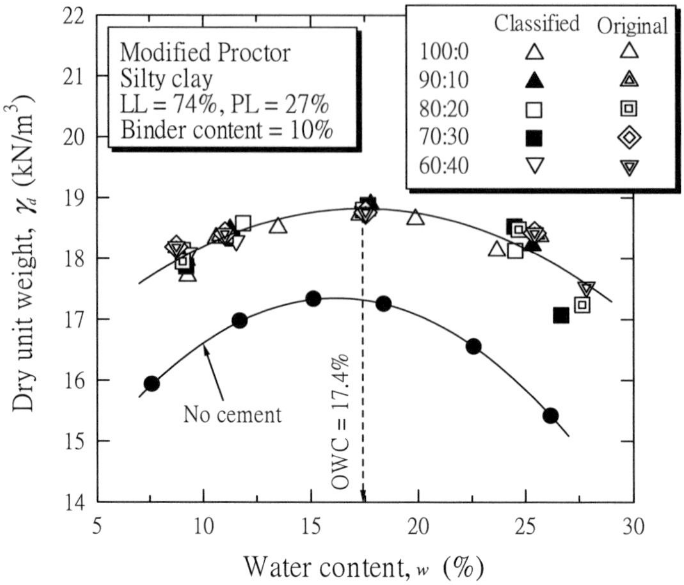

Fig. 2. Compaction curves of the OFA and CFA blended cement stabilized clay and the unstabilized clay (Horpibulsuk et al., 2009)

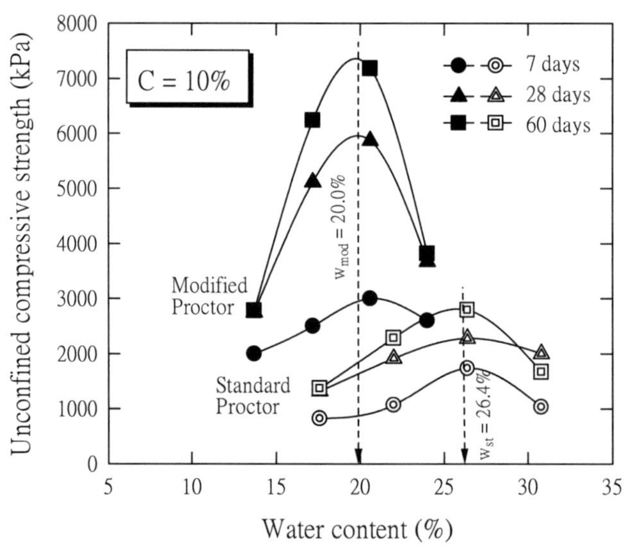

Fig. 3. Effect of compaction energy and curing time on strength development (Horpibulsuk et al., 2010b)

Typical strength-water content relationships for different curing times and compacton energies of the stabilized samples are shown in Figure 3. The strength of the stabilized samples increases with water content up to 1.2 times the optimum water content and decreases when the water content is on the wet side of optimum. At a particular curing time, the strength curve depends on the compaction energy. As the compaction energy increases, the maximum strength increases and the water content at maximum strength decreases. For the same compaction energy, the strength curves follow the same pattern for all curing times, which are almost symmetrical around 1.2*OWC* for the range of the water content tested. Figure 4 shows the strength versus water content relationship of the CFA blended cement stabilized clay at different replacement ratios after 60 days of curing compared with that of the unstabilized clay. The maximum strengths of the stabilized clay are at about 1.2*OWC* whereas the maximum strength of the unstabilized clay is at *OWC* (maximum dry unit weight). This is because engineering properties of unstabilized clay are mainly dependent upon the densification (packing).

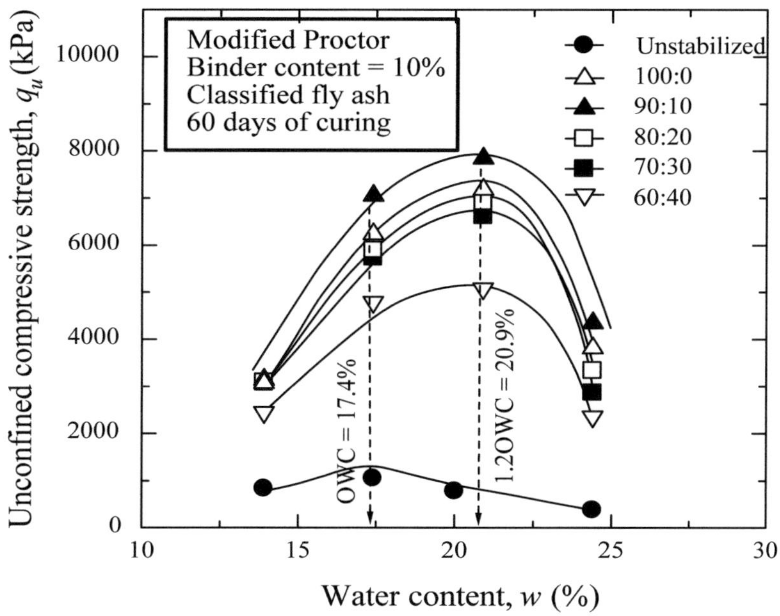

Fig. 4. Strength versus water content relationship of the CFA blended cement stabilized clay at different replacement ratios and 60 days of curing (Horpibulsuk et al., 2009)

Figure 5 shows the strength development with cement content (varied over a wide range) of the stabilized samples compacted under the modified Proctor energy at 1.2 *OWC* (20%) after 7 days of curing. The strength increase can be classified into three zones. As the cement content increases, the cement per grain contact point increases and, upon hardening, imparts a commensurate amount of bonding at the contact points. This zone is designated as the *active zone*. Beyond this zone, the strength development slows down while still gradually increasing. The incremental gradient becomes nearly zero and does not make any further significant improvement. This zone is referred to as the *inert zone* (C = 11-30%). The strength decrease appears when $C > 30\%$. This zone is identified as the *deterioration* zone.

Influence of replacement ratio on the strength development of the blended cement stabilized clay compacted at water content (w) of 1.2*OWC* (w = 20.9%) for the five curing times is presented in Figure 6. For all curing times, the samples with 20% replacement ratio exhibit almost the same strength as those with 0% replacement ratio. The 30 and 40% replacement samples exhibit lower strength than 0% replacement samples. The samples with 10% replacement ratio exhibit the highest strength since early curing time. The sudden strength development with time is not found for all replacement ratios. This finding is different from concrete technology where the role of fly ash as a pozzolanic material comes into play after a long curing time (generally after 60 days). In other words, the strength of concrete mixed with fly ash is higher than that without fly ash after about 60 days of curing.

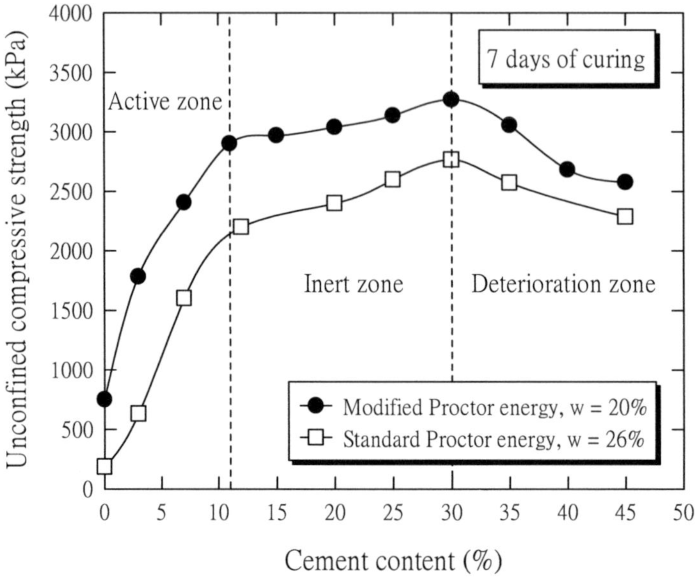

Fig. 5. Strength development as a function of cement content (Horpibulsuk et al., 2010b)

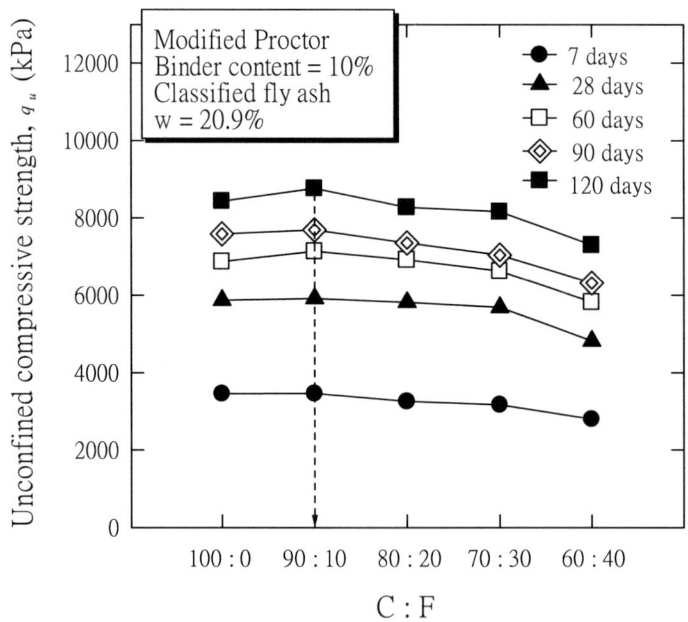

Fig. 6. Relationship between strength development and replacement ratio of the CFA blended cement stabilized clay at different curing times (Horpibulsuk et al., 2009)

3. Microstructure of cement stabilized clay

3.1 Unstabilized clay

For compacted fine-grained soils, the soil structure mainly controls the strength and resistance to deformation, which is governed by compaction energy and water content. Compaction breaks down the large clay clusters into smaller clusters and reduces the pore space. Figure 7 shows SEM photos of the unstabilized samples compacted under the modified Proctor energy at water contents in the range of 0.8*OWC* to 1.2*OWC*. On the wet side of optimum (*vide* Figure 7c), a dispersed structure is likely to develop because the quantity of pore water is enough to develop a complete double layer of the ions that are attracted to the clay particles. As such, the clay particles and clay clusters easily slide over each other when sheared, which causes low strength and stiffness. On the dry side of optimum (*vide* Figure 7a), there is not sufficient water to develop a complete double-layer; thus, the distance between two clay platelets is small enough for van der Waals type attraction to dominate. Such an attraction leads to flocculation with more surface to edge bonds; thus, more aggregates of platelets lead to compressible flocs, which make up the overall structure. At the *OWC*, the structure results from a combination of these two characteristics. Under this condition, the compacted sample exhibits the highest strength and stiffness.

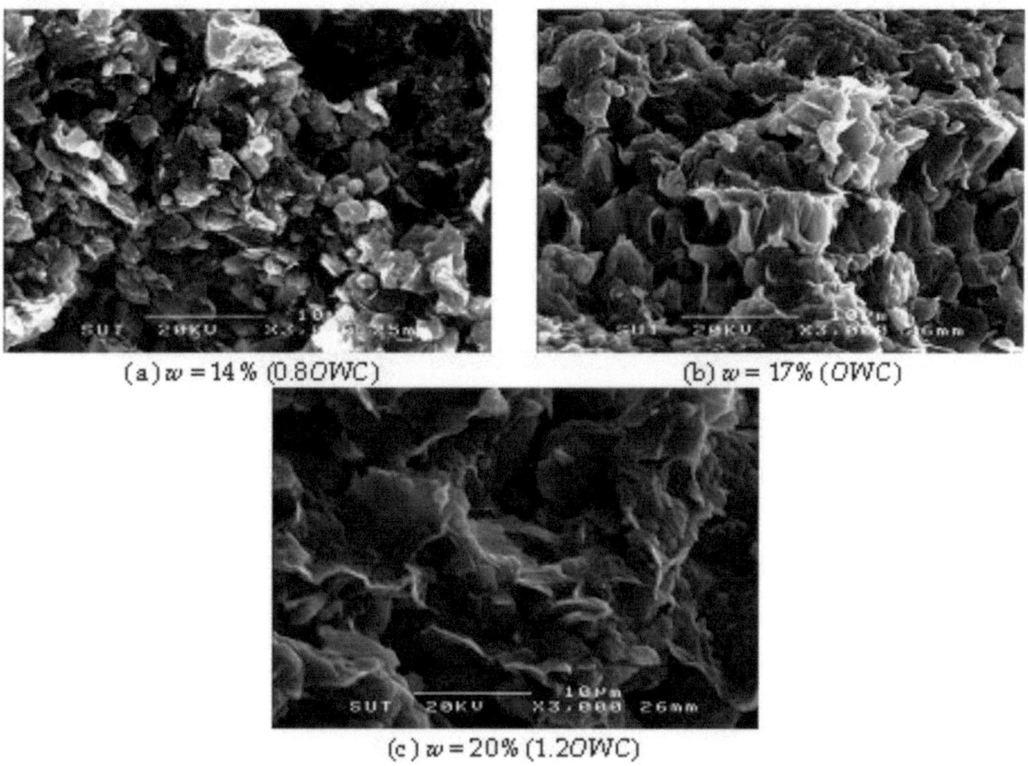

Fig. 7. SEM photos of the uncemented samples compacted at different molding water contents under modified Proctor energy (Horpibulsuk et al., 2010b)

3.2 Stabilized clay

3.2.1 Effect of curing time

Figure 8 shows SEM photos of the 10% cement samples compacted at w = 20% (1.2*OWC*) under the modified Proctor energy and cured for different curing times. After 4 hours of curing, the soil clusters and the pores are covered and filled by the cement gel (hydrated cement) (*vide* Figure 8a). Over time, the hydration products in the pores are clearly seen and the soil-cement clusters tend to be larger (*vide* Figures 8b through d) because of the growth of cementitious products over time (*vide* Table 2).

The effect of curing time on the pore size distribution of the stabilized samples is illustrated in Figure 9. It is found that, during the early stage of hydration (fewer than 7 days of curing), the volume of pores smaller than 0.1 micron significantly decreases while the volume of pores larger than 0.1 micron slightly increases. This result shows that during 7 days of curing, the cementitious products fill pores smaller than 0.1 micron and the coarse particles (unhydrated cement particles) cause large soil-cement clusters and large pore space. After 7 days of curing, the volume of pores larger than 0.1 micron tends to decrease while the volume of pores smaller than 0.1 micron tends to increase possibly because the cementitious products fill the large pores (larger than 0.1 micron). As a result, the volume of small pores (smaller than 0.1 micron) increases, and the total pore volume decreases.

a) 4 hours of curing

b) 7 days of curing

c) 28 days of curing

d) 60 days of curing

Fig. 8. SEM photos of the 10% cement samples compacted at 1.2*OWC* under modified Proctor energy at different curing times (Horpibulsuk et al., 2010b)

Curing time (days)	Weight loss (%)	$Ca(OH)_2$ (%)
7	1.52	6.25
28	1.65	6.78
60	1.85	7.63

Table 2. $Ca(OH)_2$ of the 10% cement samples compacted at 1.2*OWC* at different curing times under modified Proctor energy (Horpibulsuk et al., 2010b)

3.2.2 Effect of cement content

Figures 10 and 11 and Table 3 show the SEM photos, pore size distribution, and the amount of $Ca(OH)_2$ of the stabilized samples compacted at w = 20% under the modified Proctor energy for different cement contents after 7 days of curing. Figures 10a-c, 10d-g, and 10h-j show SEM photos of the cemented samples in the active, inert, and deterioration zones, respectively. The SEM photo of the 3% cement sample (Figure 10a) is similar to that of the unstabilized sample because the input of cement is insignificant compared to the soil mass. As the cement content increases in the active zone, hydration products are clearly seen in the pores (*vide* Figures 10b and c) and the cementitious products significantly increase (Table 3).

The cementitious products not only enhance the inter-cluster bonding strength but also fill the pore space, as shown in Figure 11: the volume of pores smaller than 0.1 micron is significantly reduced with cement, thus, the reduction in total pore volume. As a result, the strength significantly increases with cement. For the inert zone, the presence of hydration products (Figures 10d to g) and cementitious products (Table 3) is almost the same for 15-30% cement. This results in an insignificant change in the pore size distribution and, thus, the strength. For the deterioration zone (Figure 10h-j), few hydration products are detected. Both the volumes of the highest pore size interval (1.0-0.1 micron pores) and the total pore tend to increase with cement (Figure 11). This is because the increase in cement content significantly reduces the water content, which decreases the degree of hydration and, thus, cementitious products (Table 3).

3.2.3 Effect of fly ash

Figures 12 and 13 show SEM photos of the CFA blended cement stabilized clay compacted at w = 1.2OWC (w = 20.9%) and cured for 28 and 60 days at different replacement ratios. The fly ash particles are clearly shown among clay-cement clusters especially for 30% replacement ratio (C:F = 70:30) for both curing times (Figures 12a and 13a). It is noted that the hydration products growing from the cement grains connect fly ash particles and clay-cement clusters together. Some of the surfaces of fly ash particles are coated with layers of amounts of hydration products. However, they are still smooth with different curing times. This finding is different from concrete technology where the precipitation in the pozzolanic reaction is indicated by the etching on fly ash surface (Fraay et al., 1989; Berry et al., 1994; Xu and Sarker, 1994; and Chindapasirt et al., 2005). This is because the input of cement in concrete is high enough to produce a relatively high amount of $Ca(OH)_2$ to be consumed for pozzolanic reaction. Its water to binder ratio (W/B) is generally about 0.2-0.5, providing strength higher than 30 MPa (30,000 kPa) at 28 days of curing, whereas for ground improvement, the W/B is much lower. From this observation, it is thus possible to conclude that the pozzolanic reaction is minimal for strength development in the blended cement stabilized clay.

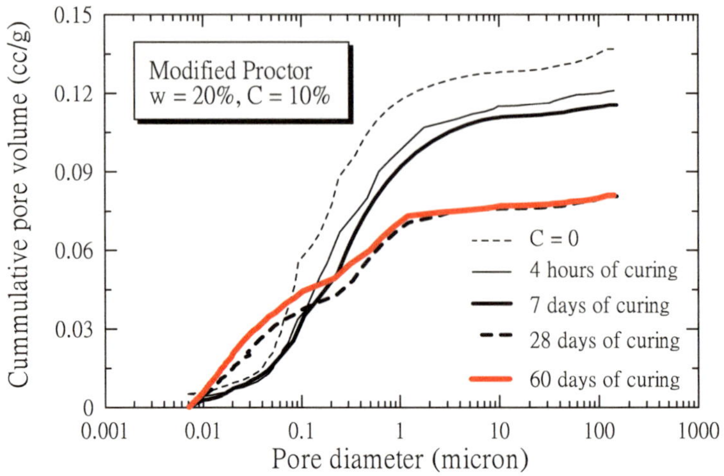

(a)

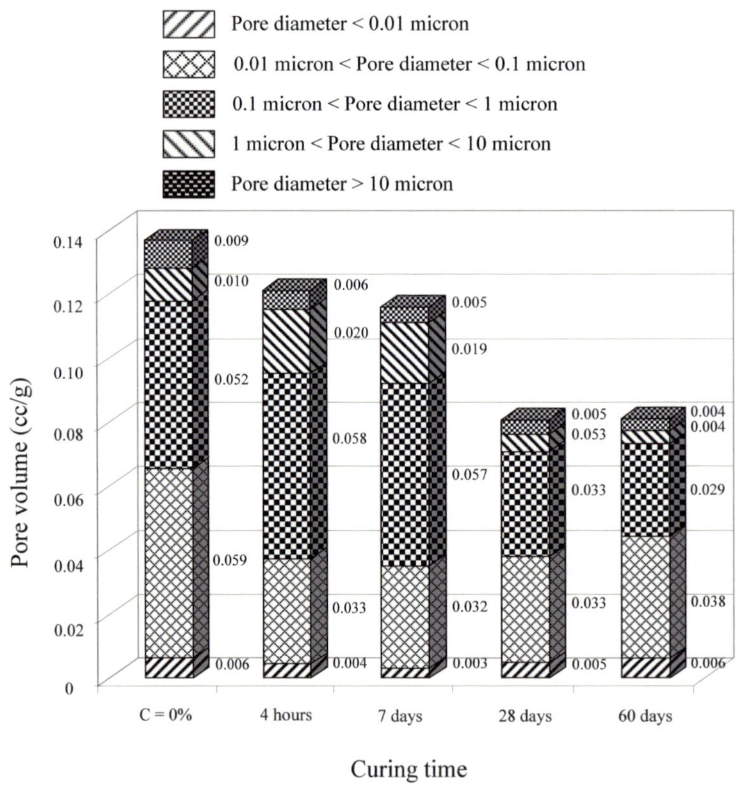

(b)

Fig. 9. Pore size distribution of the 10% cement samples compacted at different curing times under modified Proctor energy after 7 days of curing (Horpibulsuk et al., 2010b)

Fig. 10. SEM photos of the cemented samples compacted at different cement contents under modified Proctor energy after 7 days of curing (Horpibulsuk et al., 2010b)

(d)15% cement

(e)20% cement

(f)25% cement

(g)30% cement

Inert Zone

(h)35% cement

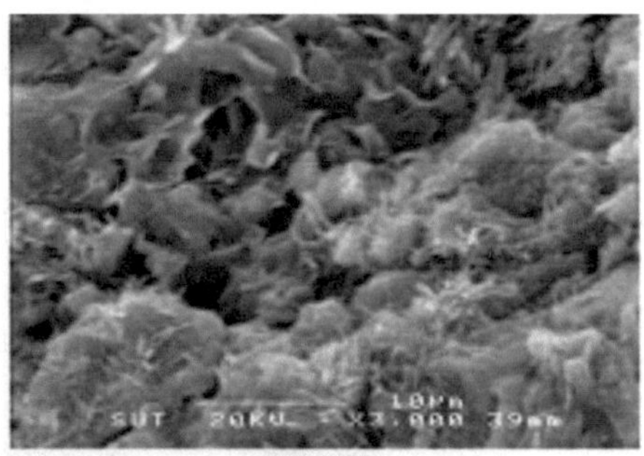

(i)40% cement

(j)45% cement

Decline Zone

Fig. 10. *(Continued)*

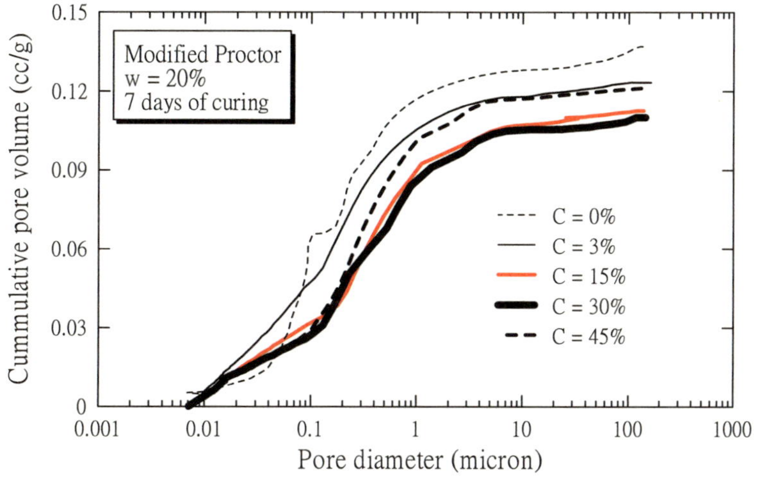

(a)

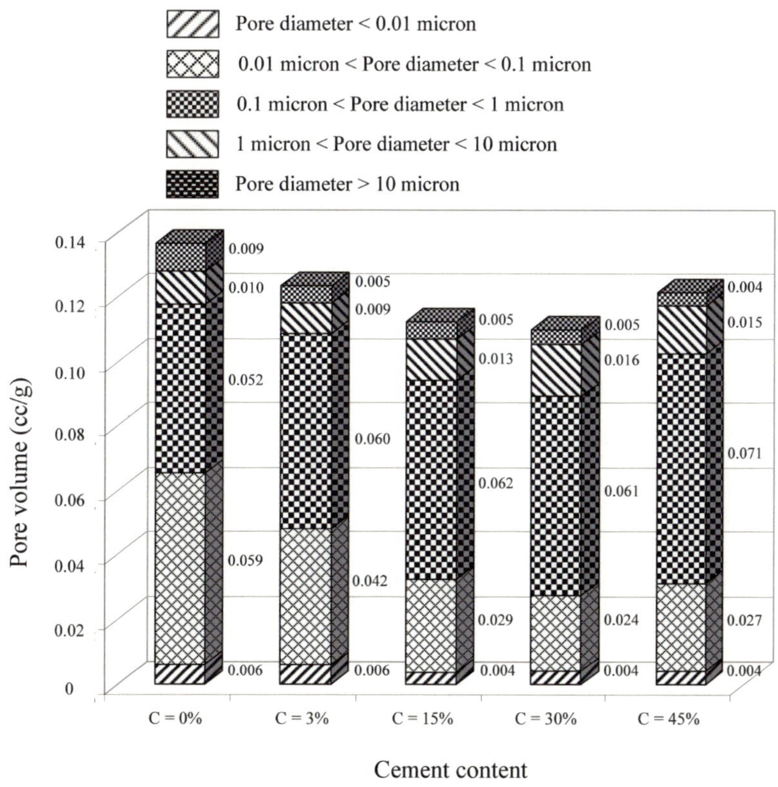

(b)

Fig. 11. Pore size distribution of the cemented samples compacted at different water contents and under modified Proctor energy after 7 days of curing time (Horpibulsuk et al., 2010b)

Improvement zones	Cement (%)	Weight loss (%)	$Ca(OH)_2$ (%)
	3	1.34	5.51
Active	7	1.50	6.17
	11	1.60	6.58
	15	1.62	6.66
Inert	20	1.65	6.78
	30	1.68	6.90
	35	1.54	6.33
Deterioration	40	1.48	6.08
	45	1.37	5.63

Table 3. $Ca(OH)_2$ of the cemented samples compacted at different cement contents under modified Proctor energy after 7 days of curing.

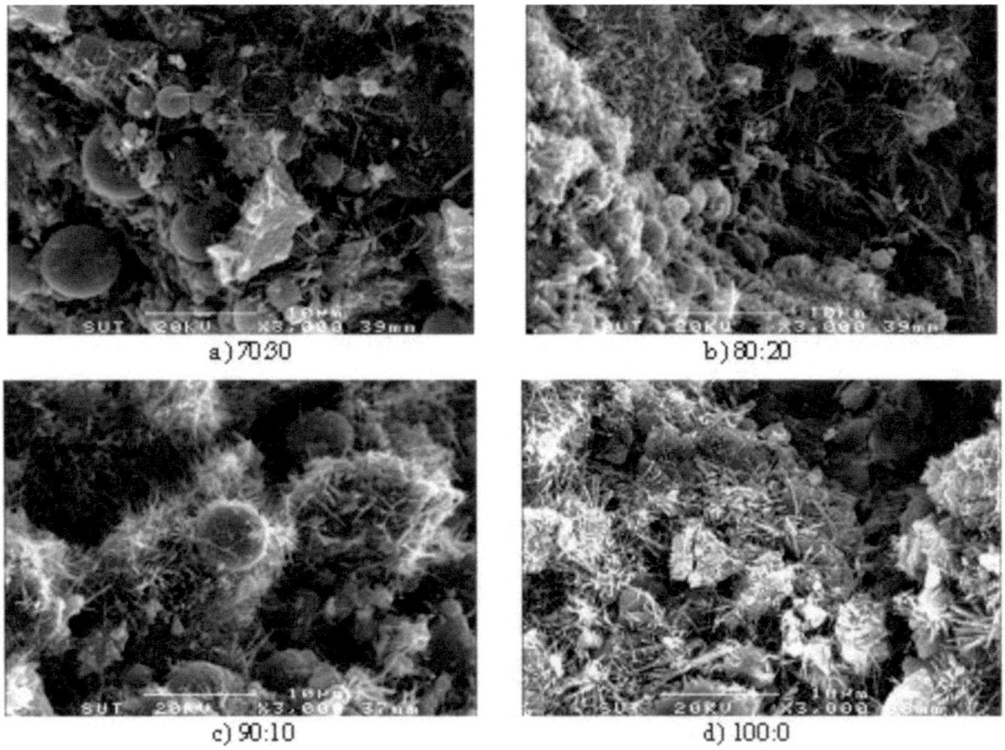

Fig. 12. SEM photos of the blended cement stabilized clay at different replacement ratios after 28 days of curing (Horpibulsuk et al., 2009)

Fig. 13. SEM photos of the blended cement stabilized clay at different replacement ratios after 60 days of curing (Horpibulsuk et al., 2010b)

Figure 14 shows the pore size distribution of the CFA blended cement stabilized clay at different curing times and replacement ratios. The pore size distribution for all replacement ratios is almost identical since the grain size distribution and D_{50} of PC and CFA are practically the same. This implies that the strength of blended cement stabilized clay is not directly dependent upon only pore size distribution. However, it might control permeability and durability. With time, the total pore and large pore (>0.1 micron) volumes decrease while the small pore (<0.1 micron) volume increases. This is due to the growth of cementitious products filling up pores.

Table 4 shows $Ca(OH)_2$ of the blended cement stabilized clay at $w = 1.2OWC$ for different curing times. For a particular water content and curing time, $Ca(OH)_2$ for the CFA blended cement stabilized clay decreases with replacement ratio only when the replacement ratio is in excess of a certain value. This finding is different from concrete technology in which $Ca(OH)_2$ decreases significantly with the increase in fineness and replacement ratio (Berry et al., 1989; Sybertz and Wiens, 1991; and Harris et al., 1987; and Chindapasirt et al., 2005 and 2006; and others) due to pozzolanic reaction. The highest $Ca(OH)_2$ is at 10% replacement ratio (C:F = 90:10) for all curing times. For replacement ratios higher than 10%, $Ca(OH)_2$ decreases with replacement ratio. $Ca(OH)_2$ at 20% replacement ratio is almost the same as that at 0% replacement ratio. This finding is associated with the strength test results that the 10% replacement ratio gives the highest strength and the strengths for 0% and 20% replacement ratios are practically the same for all curing times. It is thus concluded that cementitious products mainly control the strength development. In other words, the strengths of the blended cement stabilized clay having different mixing condition (binder content, replacement ratios, and curing time) could be identical as long as cementitious products are the same.

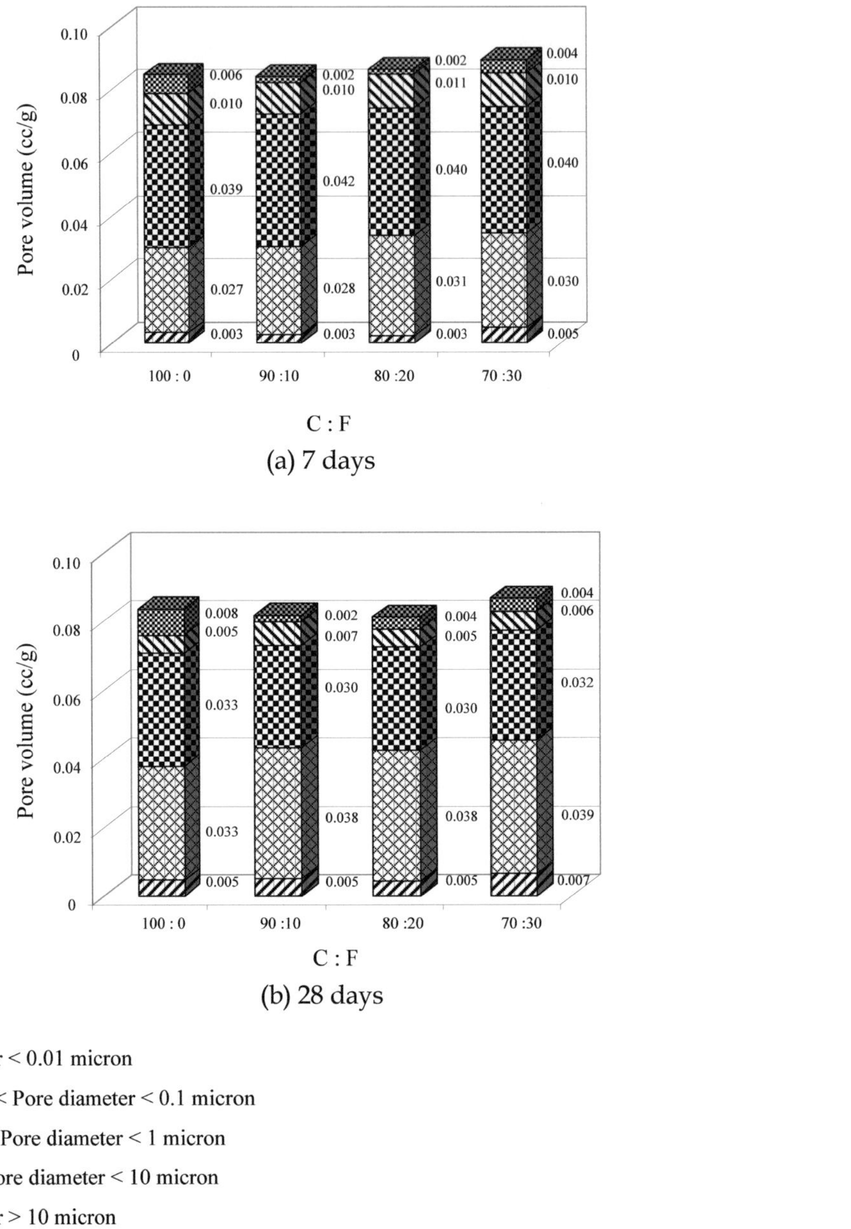

Fig. 14. Pore size distribution of the blended cement stabilized clay at different replacement ratios and curing times (Horpibulsuk et al., 2010b)

Curing time (days)	Replacement ratio C : F	Fly ash	$Ca(OH)_2$ (%)		
			Test (Combined effect)	Hydration	Induced (dispersion effect)
7	100:0	-	6.67	6.67	0.00
	90:10	CFA	6.97	6.00	0.97
	80:20	CFA	6.79	5.34	1.45
	70:30	CFA	6.39	4.67	1.72
28	100:0	-	6.79	6.79	0.00
	90:10	CFA	6.96	6.11	0.85
	80:20	CFA	6.81	5.43	1.38
	70:30	CFA	6.57	4.75	1.82
60	100:0	-	6.82	6.82	0.00
	90:10	CFA	7.16	6.14	1.02
	80:20	CFA	6.92	5.46	1.46
	70:30	CFA	6.68	4.77	1.91
90	100:0	-	7.07	7.07	0.00
	90:10	CFA	7.28	6.36	0.91
	80:20	CFA	6.94	5.66	1.28
	70:30	CFA	6.67	4.95	1.72
120	100:0	-	7.08	7.08	0.00
	90:10	CFA	7.29	6.37	0.92
	80:20	CFA	6.96	5.66	1.30
	70:30	CFA	6.70	4.96	1.74

Table 4. $Ca(OH)_2$ of the blended cement stabilized clay at different replacement ratios and curing times.

From SEM and MIP observation, it is notable that the small pore (<0.1 micron) volumes of the blended cement stabilized clay are higher than those of the cement stabilized clay. This implies that a number of large clay-cement clusters possessing large pore space reduce when fly ashes are utilized. In other words, the fly ashes disperse large clay-cement clusters into small clusters, resulting in the increase in small pore volume. The higher the replacement ratio, the better the dispersion. Consequently, the reactive surfaces increase, resulting in the increase in cementitious products as illustrated by dispersion induced $Ca(OH)_2$ (*vide* Table 4). It is the difference in $Ca(OH)_2$ of the blended cement stabilized clay due to the combined effect (hydration and dispersion) and due to hydration. $Ca(OH)_2$ due to combined effect is directly obtained from TG test on the blended cement stabilized sample. $Ca(OH)_2$ due to hydration is also obtained from TG test on the cement stabilized sample having the same cement content as the blended cement stabilized sample. For simplicity, $Ca(OH)_2$ due to hydration at any cement content can be estimated from known $Ca(OH)_2$ of cement stabilized clay at a specific cement content by assuming that the change in the cementitious products is directly proportional to the input of cement (Sinsiri et al., 2006). Thus, $Ca(OH)_2$ due to hydration (H) for any replacement ratio at a particular curing time is approximated in the form.

$$H = T \times (1 - F / 100) \tag{3}$$

where T is known $Ca(OH)_2$ of the cement stabilized clay (0% replacement ratio) obtained from TG test, and F is the replacement ratio expressed in percentage. Sinsiri et al. (2006) have shown that $Ca(OH)_2$ of the cement paste with fly ash is always lower than $Ca(OH)_2$ of the cement paste without fly ash, resulted from $Ca(OH)_2$ consumption for pozzolanic reaction. The same is not for the blended cement stabilized clay. It is found that $Ca(OH)_2$ due to combined effect is higher than that due to hydration for all replacement ratios and curing times. The dispersion induced $Ca(OH)_2$ increases with the replacement ratio for all curing times.

4. Conclusions

This chapter presents the role of curing time, cement content and fly ash content on the strength and microstructure development in the cement stabilized clay. The following conclusions can be advanced:

1. The strength development with cement content for a specific water content is classified into three zones: active, inert and deterioration. In the active zone, the volume of pores smaller than 0.1 micron significantly decreases with the addition of cement because of the increase in cementitious products. In the inert zone, both pore size distribution and cementitious products change insignificantly with increasing cement; thus, there is a slight change in strength. In the deterioration zone, the water is not adequate for hydration because of the excess of cement input. Consequently, as cement content increases, the cementitious products and strength decreases.
2. The flocculation of clay particles due to the cation exchange process is controlled by cement content, regardless of fly ash content. It results in the increase in dry unit weight with insignificant change in liquid limit. Hence, *OWCs* of stabilized and unstabilized silty clay (low swelling clay) are practically the same.
3. The surfaces of fly ash in the blended cement stabilized clay are still smooth for different curing times and fineness, suggesting that pozzolanic reaction is minimal. Fly ash is considered as a dispersing material in the blended cement stabilized clay. This is

different from the application of fly ash as a pozzolanic material in concrete structure in which $Ca(OH)_2$ from hydration is much enough to be consumed for pozzolanic reaction.

4. From the microstructural investigation, it is concluded that the role of fly ash as a non-interacting material is to disperse the cement-clay clusters with large pore space into smaller clusters with smaller pore space. The dispersing effect by fly ash increases the reactive surfaces, and hence the increase in degree of hydration as clearly illustrated by the increase in the induced $Ca(OH)_2$ with replacement ratio and fineness.
5. The increase in cementitious products with time is observed from the scanning electron microscope, mercury intrusion porosimetry and thermal gravity test. With time, the large pore (>0.1 micron) and total pore volumes decrease while the small pore (<0.1 micron) volumes increase. This shows the growth of the cementitious products filling up the large pores.

5. Acknowledgment

This work was a part of the author's researches conducted in the Suranaree University of Technology. The authors would like to acknowledge the financial support provided by the Higher Education Research Promotion and National Research University Project of Thailand, Office of Higher Education Commission, the Thailand Research Fund (TRF), and the Suranaree University of Technology. The author is indebted to Dr. Theerawat Sinsiri, School of Civil Engineering, Suranaree University of Technology for his technical advice in cement and concrete technology. The author is grateful to Mr. Yutthana Raksachon, ex-master's student for his assistance.

6. References

Berry, E.E., Hemmings, R.T., Langley, W.S. & Carette, G.G. (1989). Beneficiated Fly Ash: Hydration, Microstructure, and Strength Development in Portland Cement Systems, in: V.M. Malhotra (Ed.), Third CANMET/ACI, *Conference on Fly Ash, Silica Fume, Slag, and Natural Pozzolans in Concrete* (SP-114), Detroit, pp.241-273.

Chindaprasirt, P., Jaturapitakkul, C. & Sinsiri, T. (2005). Effect of Fly Ash Fineness on Compressive Strength and Pore Size of Blended Cement Plates. *Cement and Concrete Composites*, Vol.27, pp.425-258.

Clough, G.W., Sitar, N., Bachus, R.C. & Rad, N.S. (1981). Cemented Sands under Static Loading. *Journal of Geotechnical Engineering Division*, ASCE, Vol.107, No.GT6, pp.799-817.

El-Jazairi, B. & Illston, J.M. (1977). A Simultaneous Semi-Isothermal Method of Thermogravimetry and Derivative Thermogravimetry, and Its Application to Cement Plates. *Cement and Concrete Research*, Vol.7, pp.247-258.

El-Jazairi, B. & Illston, J.M. (1980). The Hydration of Cement Plate Using the Semi-Isothermal Method of Thermogravimetry. *Cement and Concrete Research*, Vol.10, pp.361-366.

Fraay, A.L.A, Bijen, J.M. & de Haan, Y.M. (1989). The Reaction of Fly Ash in Concrete: A Critical Examination. *Cement and Concrete Research*, Vol.19, pp.235-246.

Gens, A. & Nova, R. (1993). Conceptual Bases for Constitutive Model for Bonded Soil and Weak Rocks, *Geotechnical Engineering of Hard Soil-Soft Rocks*, Balkema.

Kamon, M. & Bergado, D.T. (1992). Ground Improvement Techniques, *Proceedings of 9th Asian Regional Conference on Soil Mechanics and Foundation Engineering*, pp.526-546.

Kasama, K., Ochiai, H. & Yasufuku, N. (2000). On the Stress-Strain Behaviour of Lightly Cemented Clay Based on an Extended Critical State Concept. *Soils and Foundations*, Vol.40, No.5, pp.37-47.

Harris, H.A., Thompson, J.L. & Murphy, T.E. (1987). Factor Affecting the Reactivity of Fly Ash from Western Coals. *Cement and Concrete Aggregate*, Vol.9, pp.34-37.

Horpibulsuk, S. & Miura, N. (2001). A new approach for studying behavior of cement stabilized clays, *Proceedings of 15th international conference on soil mechanics and geotechnical engineering* (ISSMGE), pp. 1759-1762, Istanbul, Turkey.

Horpibulsuk, S., Miura, N. & Nagaraj, T.S. (2003). Assessment of Strength Development in Cement-Admixed High Water Content Clays With Abrams' Law As a Basis. *Geotechnique*, Vol.53, No.4, pp.439-444.

Horpibulsuk, S., Bergado, D.T. & Lorenzo, G.A. (2004a) Compressibility of Cement Admixed Clays at High Water Content. *Geotechnique*, Vol.54, No.2, pp.151-154.

Horpibulsuk, S., Miura, N. & Bergado, D.T. (2004b). Undrained Shear Behavior of Cement Admixed Clay at High Water Content. *Journal of Geotechnical and Geoenvironmental Engineering*, ASCE, Vol.30, No.10, pp.1096-1105.

Horpibulsuk, S., Miura, N., Koga, H. & Nagaraj, T.S. (2004c). Analysis of Strength Development in Deep Mixing – A Field Study. *Ground Improvement Journal*, Vol.8, No.2, pp.59-68.

Horpibulsuk, S., Miura, N. & Nagaraj T.S. (2005). Clay-Water/Cement Ratio Identity of Cement Admixed Soft Clay. *Journal of Geotechnical and Geoenvironmental Engineering*, ASCE, Vol.131, No.2, pp.187-192.

Horpibulsuk, S., Katkan, W., Sirilerdwattana, W. & Rachan, R. (2006). Strength Development in Cement Stabilized Low Plasticity and Coarse Grained Soils : Laboratory and Field Study. *Soils and Foundations*, Vol.46, No.3, pp.351-366.

Horpibulsuk, S., Katkan, W. & Apichatvullop, A. (2008). An Approach for Assessment of Compaction Curves of Fine-Grained Soils at Various Energies Using a One Point Test. *Soils and Foundations*, Vol.48, No.1, pp.115-125.

Horpibulsuk, S., Katkan, W. & Naramitkornburee, A. (2009). Modified Ohio's Curves: A Rapid Estimation of Compaction Curves for Coarse- and Fine-Grained Soils. *Geotechnical Testing Journal*, ASTM, Vol.32, No.1, pp.64-75.

Horpibulsuk, S., Liu, M.D., Liyanapathirana, D.S. & Suebsuk, J. (2010a). Behavior of Cemented Clay Simulated via the Theoretical Framework of the Structured Cam Clay Model. *Computers and Geotechnics*, Vol.37, pp.1-9.

Horpibulsuk, S., Rachan, R., Chinkulkijniwat, A., Raksachon, Y., and Suddeepong, A. (2010b). Analysis of Strength Development in Cement-Stabilized Silty Clay based on Microstructural Considerations. *Construction and Building Materials*, Vol.24, pp.2011-2021.

Horpibulsuk, S., Rachan, R. & Suddeepong, A. (2011a). Assessment of Strength Development in Blended Cement Admixed Bangkok Clay. *Construction and Building Materials*, Vol.25, No.4, pp.1521-1531.

Horpibulsuk, S., Rachan, R., Suddeepong, A. & Chinkulkijniwat, A. (2011b). Strength Development in Cement Admixed Bangkok Clay: Laboratory and Field Investigations. *Soils and Foundations*, Vol.51, No.2, pp.239-251.

JEOL JSM-6400 (1989). Scanning Microscope Operations Manual, Model SM-6400 LOT No. SM150095, Copyright 1989, JEOL Ltd.

Midgley, H.G. (1979). The Determination of Calcium Hydroxide in Set Portland Cements. *Cement and Concrete Research*, Vol.9, pp.77-82.

Mitchell, J.K. (1993). *Fundamentals of Soil Behavior*. John Wiley&Sons, Inc., New York.

Miura, N., Yamadera, A. & Hino, T. (1999). Consideration on Compression Properties of Marine Clay Based on the Pore Size Distribution Measurement. *Journal of Geotechnical Engineering*, JSCE.

Miura, N., Horpibulsuk, S. & Nagaraj, T.S. (2001). Engineering Behavior of Cement Stabilized Clay at High Water Content. *Soils and Foundations*, Vol.41, No.5, pp.33-45.

Nishida, K., Koga, Y. & Miura, N. (1996). Energy Consideration of the Dry Jet Mixing method, *Proceedings of 2nd International Conference on Ground Improvement Geosystems*, IS-Tokyo '96, Vol.1, pp.643-748.

Sinsiri, T., Jaturapitakkul, C. & Chindaprasirt, P. (2006). Influence of Fly Ash Fineness on Calcium Hydroxide in Blended Cement Paste, *Proceedings of Technology and Innovation for Sustainable Development Conference (TISD2006)*, Khon Kaen University, Thailand.

Suebsuk, J., Horpibulsuk, S. & Liu, M.D. (2010). Modified Structured Cam Clay: A Constitutive Model for Destructured, Naturally Structured and Artificially Structured Clays. *Computers and Geotechnics*, Vol.37, pp.956-968.

Suebsuk, J., Horpibulsuk, S. & Liu, M.D. (2011). A Critical State Model for Overconsolidated Structured Clays. *Computers and Geotechnics*, Vol.38, pp.648-658.

Sybert, F. & Wiens, U. (1991). Effect of fly ash fineness on hydration characteristics and strength development, *Proceedings of International Conference on Blended Cement in Construction*, pp.152-165, University of Sheffield, UK.

Wang, K.S., Lin, K.L., Lee, T.Y. & Tzeng, B.Y. (2004). The Hydration Characteristics When C_2S Is Present in MSWI Fly Ash Slag. *Cement and Concrete Research*, Vol.26, pp.323-330.

Washburn, E.W. (1921). Note on Method of Determining the Distribution of Pore Size in Porous Material, *Proceedings of the National Academy of Science*, USA. Vol.7, pp.115-116.

Xu, A. & Sarker, S.L. (1994). Microstructure Development in High-Volume Fly Ash Cement System. *Journal of Material in Civil Engineering*, ASCE, Vol.6, pp.117-136.

Yamadera, A. (1999). *Microstructural Study of Geotechnical Characteristics of Marine Clays*. Ph.D. Dissertation, Saga University, Japan.

Yin, J.H. & Lai, C.K. (1998). Strength and Stiffness of Hong Kong Marine Deposit Mixed With Cement. *Geotechnical Engineering Journal*. Vol.29, No.1, pp.29-44.

Part 4

Nanostructured Materials for Electronic Industry

23

FE-SEM Characterization of Some Nanomaterial

A. Alyamani[1] and O. M. Lemine[2]
[1]*National Nanotechnology Research Centre, KACST, Riyadh,*
[2]*Physics Department, College of Sciences, Imam University Riyadh,*
Saudi Arabia

1. Introduction

In 1931 Max Knoll and Ernst Ruska at the university of Berlin built the first electron microscope that use accelerated electrons as a source instead of light source. However, the first scanning electron microscope (SEM) was built in 1938 due to the difficulties of scanning the electrons through the sample. Electron microscope is working exactly the same as the optical microscope expects it use a focused accelerated electron beam [1].

Since the invention of the electron microscope, it became one of the most useful instruments that has an impact in understanding scientific phenomena in different fields, such as physics, nanotechnology, medicine, chemistry biology..etc. Electron microscope has the ability to resolve objects ranging from part of nano-metre to micro-metre compared to light microscope that has a magnificationin the range of 1000 and resolution of 200 nm.

In the first part of the chapter, we will describe some of the basics of electronic microscope and its applications. The second part will be dedicated to the results obtained mainly by SEM.

2. Electron microscopy

2.1 Fundamental principles of electron microscopy

The principal of electron microscope is the same as a light microscope but instead of using visible light it use very energetic electrons as a source. However, the resolution of the optical microscope is limited by its wavelength compared to accelerated electrons which have very short wavelength.This is what makes it possible to see very small features.

In electron microscopes, electrons have very small wavelength λ. This wavelength can be changed according to the applied high voltage. Hence, according to Rayleigh's criterion the wavelength λ of an electron is related to the momentum $p=mv$ of the electron by: [2]

$$\lambda = \frac{h}{p} = \frac{h}{mv} \tag{1}$$

where h = 6x10^{-34} J s is the Planck constant, m and v are the mass and velocity of the electron respectively. Since the electron can reach nearly the velocity of light c then we can use the relativistic equations. In this case the electron mass is changing according to:

$$m = \frac{m_e}{\sqrt{1-(v/c)^2}} \quad (2)$$

where m_e is the rest mass of the electron. The energy eV transmitted to an electron is giving by:

$$eV = (m - m_2)c^2 \quad (3)$$

By using equations 1,2 and 3 the electron wavelength can be written as function of accelerated voltage: [3]

$$\lambda = \sqrt{\frac{1.5}{V\left(1+V*10^{-6}\right)}} nm \quad (4)$$

for example an accelrated voltage of 10 kV will yield a wavelength of 0.0122 nm. The extremely small wavelengths make it possible to see atomic structures using accelerated electrons.

2.2 Interaction of accelerated electrons with the specimen

The electron beam interacts with the specimen reveal useful information about the sample including: its surface features, size and shape of the features, composition and crystalline structure. The interaction of the electron beam with the specimen can be in different ways:

2.2.1 Secondary electrons

If the incident electronscome close enough to the atom then these electrons will give some of their energy to the specimen electrons mainly in the K-shell. As a result, these electrons will change their path and will ionize the electrons in the specimen atoms. These ionized electrons that escape the atoms are called secondary electrons. These electrons will move to the surface of the specimen and undergoing to elastics and inelastic collision until reaching the surface. However, due to their low energy $\sim 5eV$ only those electrons that are close to the surface (~ 10 *nm*) will escape the surface and then can be detected and can used for imaging the topography of the specimen.

2.2.2 Backscattered electrons

When the incident electrons hit an atom directly, then they will be reflected or back-scattered. Different atomic type of atoms will result in a different rate of backscattered electrons and hence the contrast of the image will vary as the atomic number of the specimen change, usually atoms with higher atomic number will appear brighter than those have lower atomic number.

2.2.3 Transmitted electrons

If the incident electrons pass through the specimen without any interaction with their atoms, then these electrons called transmitted electrons, these electrons are used to get an image of thin specimen. Another scattering mechanism called elastic scattered where electrons don't loss their energy these scattered electrons can be used to get information about orientation and arrangement of atoms.

2.2.4 Other interactions

When the atoms bombarded with incident electrons, electrons will released from these atoms and this will leave the atom in the excited state. In order for the atom to return to the ground state, it needs to release the excess energy Auger electrons, X-Rays,and cathodoluminescence are three ways of relaxation. The x-ray is used to identify the elements and their concentrations in the specimen by using a technique called Energy -dispersive X-ray analysis (EDX) technique. Chemical analysis can be done by using Auger electrons.

2.3 Types of electron microscopes

There are two types of electron microscopes. Scanning Electron Microscopes (SEM), and Transmission Electron Microscope (TEM), these types of microscopes detect electrons that emitted from the surface of the sample. The accelerated voltage is ranging from 10kV to 40kV for the SEM. The thickness of the specimen in this case is not important. In addition, the samples to be tested have to be electrically conductive; otherwise they would be overcharged with electrons. However, they can be coated with a conductive layer of metal or carbon.

In TEM the transmitted electrons are detected, and in this case the specimen thickness is important and typically should not exceed 150 nm. The accelerated voltage in this case $\geq$ 100kV.

Since the electrons are easily scattered in air all electron microscopes should operate under a high vacuum.

All types of electron microscopes are basically consist of three basic components:

Electron Gun which is used to provide and supply electrons with the required energy.There are different types of electron gun;the old type was a bent piece of Tungsten wire with 100 micro-metresin diameter. Higher performance electron emitters consist of either single crystals of lanthanum hexaboride (LaB6) or from field emission guns.

3. Experimental

3.1 Pulse Laser Deposition (PLD)

As a materials processing technique, laser ablation was utilized for the first time in the 1960's, after the first commercial ruby laser was invented [4]. Nevertheless, as a thin film growth method it did not attract much research interest until the late 1980′s [5], when it has been used for growing high temperature superconductor films. Since then, the development of the pulsed laser deposition (PLD) technique has been more rapid and the amount of research devoted to this topic has increased dramatically [6]. The growth and quality of the resulting film will generally depend on a number of fundamental parameters, including the choice of substrate, the substrate temperature and the absolute and relative kinetic energies and/or arrival rates of the various constituents within the plume.

The PLD process is shown in figure 1:

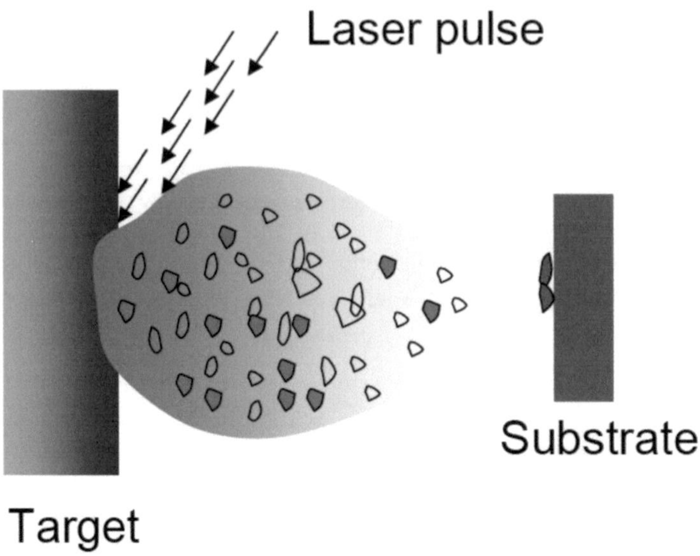

Fig. 1. Schematic presentation of the pulsed laser deposition process
a) Laser – target interaction, b) Plume expansion and c) Film deposition [6].

The growth and quality of the resulting thin film will generally depend on a number of fundamental parameters, including the choice of substrate, the substrate temperature, T_S, distance between target-substrate, pressure and laser energy.

In our case the laser energy was 300 mJ and the time was fixed at 60 minutes. For the others parameters (substrate, substrate temperature, pressure), different values were used.

3.2 Mechanical Alloying (MA)

The ball milling constitutes new promising methods to produce nanosized particles [7,8]. It has many advantages, e.g low cost, simple operation.The ball-milling is generally used as a mechanical co-grinding of powders, Initially different in nature, up to the preparation of a new powder, homogeneous in composition. The milling is done in cylindrical containers called vials and containing balls. The nature of the milling tools can be as diverse as steel, agate, tungsten carbide… The vials are generally filled under an inert atmosphere to avoid side reactions, since the particles are fractured during the milling process and, therefore, new highly reactive surfaces can react with the surrounding gases [8].

Several terms are used to call this technique: "Mechanical Alloying" when there is a chemical reaction between different powders, "Mechanical Grinding" or "Mechanical Milling" when the only goal is to modify the texture and/or the structure of a material (no chemical reaction is involved in the process).

Two kinds of milling systems were used to prepare our nanopowders (Vibrant and planetary milling) and different milling parameters were considered (milling times, balls to powders mass ratio, size of balls and rotation speed).

3.3 Filed Emission Scanning Electron Microscopy (FESEM)

The field emission scanning electron microscope (FE-SEM) is a type of electron microscope that images the sample surface by scanning it with a high-energy beam of electrons in a raster scan pattern. Electron emitters from field emission gun was used. These types of electron emitters can produce up to 1000x the emission of a tungsten filament. However, they required much higher vacuum conditions. After the electrons beam exit the electron gun, they then confined and focused into a thin focused, monochromatic beam using metal aperturesand magnetic lenses. Finally, Detectors of each type of electrons are placed in the microscopes that collect signals to produce an image of the specimen.

Particles morphology of our samples was investigated using Nova 200 NanoLab field emission scanning electron microscope (FE-SEM).

4. Results

4.1 Thin film prepared by Pulse Laser Deposition (PLD)

Fig. 2 shows FESEM micrographs of ZnO thin films grown on sapphire substrate by pulse laser deposition at growth temperature from 685 to 750 °C by using a ZnO powder target at high grade. The experimental parameters are summarized in table.1. It is seen that with the substrate temperature increasing the morphology of ZnO thin films have a little difference. The thickness of films decreases with the increase of substrate temperature.

The effect of the distance between target and substrate on the morphology was also studied. Fig. 3 shows the FESEM images of ZnO thin film with different distance between the target and thin film. It is clear that the distance affect the morphology of the film.

TEMPERATURE(°C)	THICKNESS (nm)	Distance between target and the film	Oxygen pressure	LASER ENERGY(mJ)	SUBSTRATE
750	510	37.5 mm	150 mTorr	350	**Sapphire**
700	1230	37.5 mm	150 mTorr	350	**Sapphire**
685	1115	37.5 mm	150 mTorr	350	**Sapphire**
400	-	10 mm	150 mTorr	350	**Sapphire**
400	-	23mm	150 mTorr	350	**Sapphire**

Table.1.Growth parameters of ZnO thin films

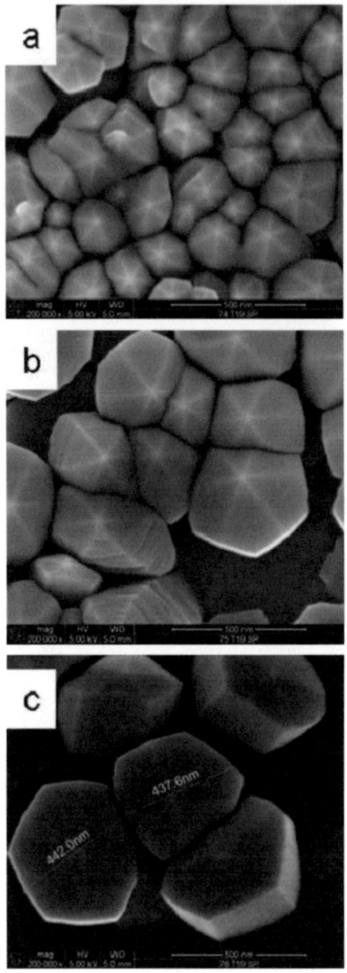

Fig. 2. FESEM images of thin film grown on sapphire at substrate temperature of: (a) 750 °C, (b) 700 °C and (c) 685°C.

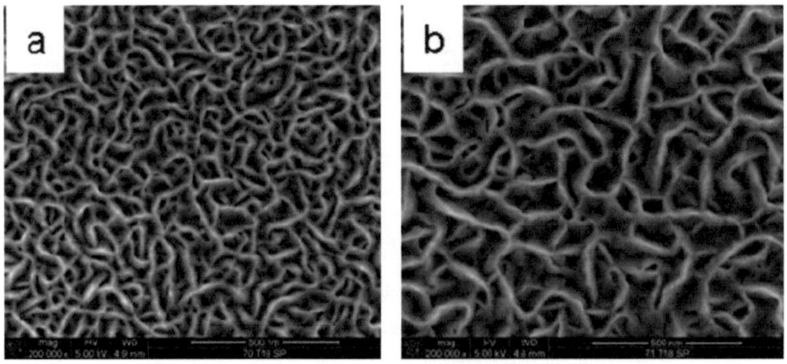

Fig. 3. **FESEM** images of thin film grown on sapphire at substrate temperature of 400°C: a) distance between target and thin film =10mm and b) distance between target and thin film =23 mm

4.2 Nanopwders obtained by mechanical alloying

4.2.1 Hematite (α-Fe_2O_3) nanocrystallines

The conditions for production of α-Fe_2O_3 nano-crystallines by dry milling was studied. [9,10] Commercial α-Fe_2O_3 powder was used as the starting material. The mechanical milling was carried out in a planetary ball mill Fritsch Pulverisstte 6. The powder was ground in vial with 200g of mixture 1:1 in weight of stainless steel balls (10 and 15 mm in diameter). Different milling times were considered (1, 6, 12, 24 and 48h) and the sample to balls weight ratio was fixed to 1:10. The milling intensity was 250 rpm. Fig.4 shows scanning electron micrographs before and after milling. It is clear that un-milled powder shows an in homogeneities regarding particle size distribution (Fig. 4a). After milling, a reduction of the particle size can be observed with relatively better homogeneity (Fig. 4b-d)). SEM images for increasing milling times reveal clearly that large particles are in fact agglomerates of much smaller particles.

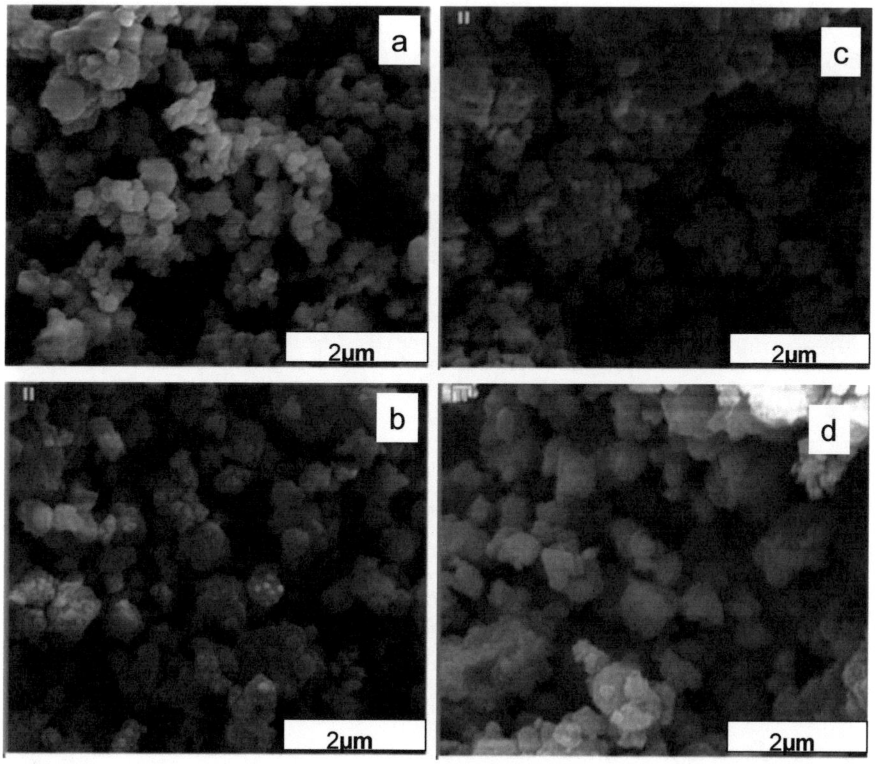

Fig. 4. FESEM images for the samples milled at different times : a) 0h, b) 12h, c) 24h and d) 48h.

4.2.2 Nanocrystalline zinc ferrite ($ZnFe_2O_4$)

Nanocrystalline zinc ferrite ($ZnFe_2O_4$) is synthesized by high-energy ball-milling from a powders mixture of zinc oxide (ZnO) and hematite (α-Fe_2O_3). [11] Commercially powders of hematite (α-Fe_2O_3) and zinc oxide (ZnO) are used with equal molar (1:1) and were introduced into a stainless steel vials with stainless steel balls (12 mm and 6 mm in diameter) in a high energy mill (SPEX 8000 mixer mill). Different milling times were considered (6, 12 and 24) and two values of the balls to powders mass ratio were used (10:1 and 20:1). SEM micrographs of the samples before and after milling are shown in Figure 5 It is clear that unmilled powder shows a different chap of powders due to zinc oxides and hematite powders (fig.5a, 5b). After milling, a reduction of the crystallite size can be observed fig.5c. High magnification images (fig.5d) reveal clearly the formation of a new nanocrystalline different from the started materials.

4.2.3 Zinc oxides Nanocrystalline (ZnO)

The effects of milling times on the mechanically milled ZnO powder are also studied [12]. Commercially ZnO powders with average particle size of about 1 μm and 99.9% of purity, were introduced into a stainless steel vials with stainless steel balls (12mm and 6mm in diameter) in a SPEX 8000 mixer mill, then milled for different milling periods of time. The balls to powder mass ratio was fixed to 10:1. SEM micrographs of the samples before after milling are shown in Figure.6. It is clear that un-milled powder shows un-homogeneities regarding particle size distribution, where the average size varies in the range 150 - 800 nm (Fig. 6a, 6b). After milling, a reduction of the particle size can be observed with relatively better homogeneity (Fig. 6c, 6e). High magnification images (Fig. 6d, 6f) reveal clearly that large particles are in fact agglomerates of much smaller particles. The average particle size after milling is less than 100nm.

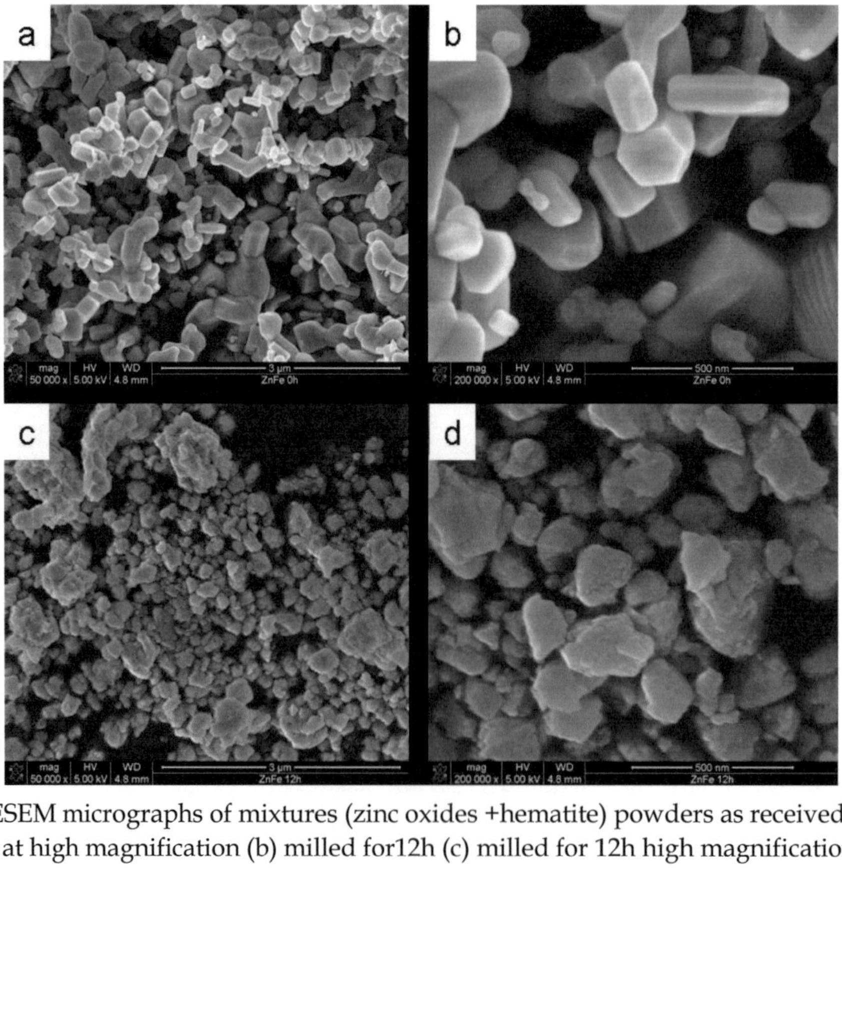

Fig. 5. FESEM micrographs of mixtures (zinc oxides +hematite) powders as received (a) as received at high magnification (b) milled for12h (c) milled for 12h high magnification (d)

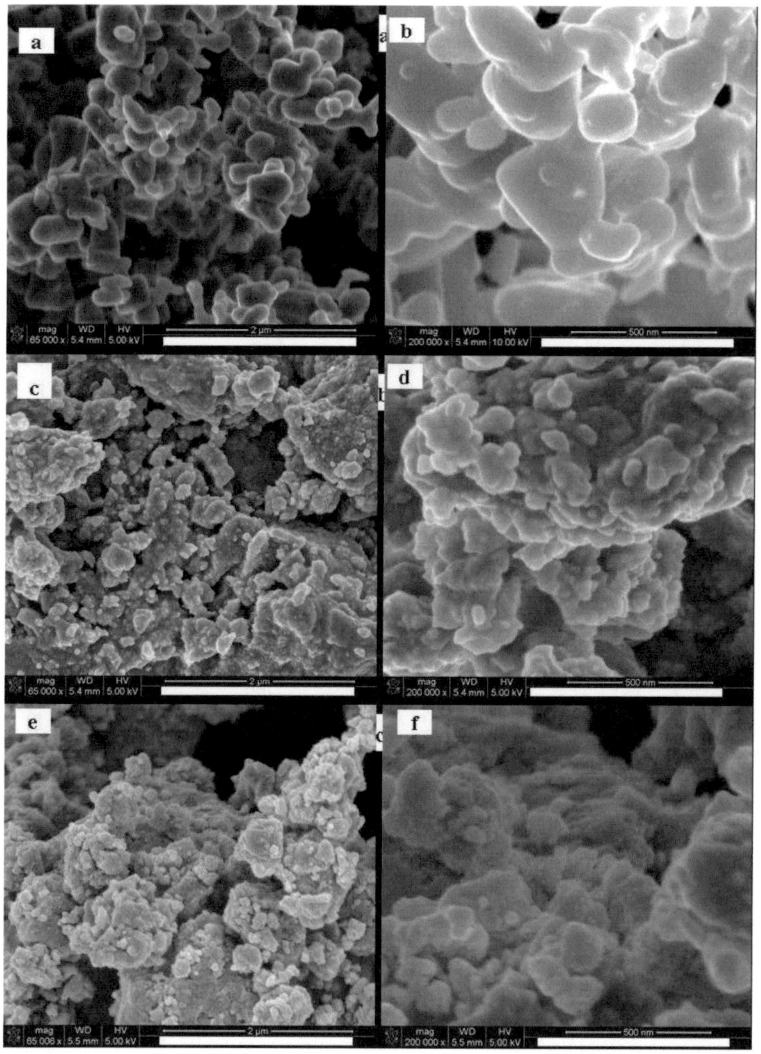

Fig. 6. SEM micrographs of ZnO powder for different milling time: a) as received; b) as received at high magnification; c) milled for 3h; d) milled for 3h high magnification; e) milled for 5h; f) milled for 5h at high magnification.

In summary, it is clear that scanning electron microscopy gives tremendous information about the microstructure of nanomaterials including thin film and nano-powders. In addition to that the signals coming from the sample can be used to get information about the composition of the materials and the structure.

5. References

[1] David B Williams and C Barry Carter, "Transmission Electron Microscopy", Springer 2009

[2] Nouredine Zietteli, " Quantum mechanic, concept and applications", Wiley 2001.

[3] Arthur Beiser, " Concept of modern physics", McGraw-Hill, Inc, 1995 H.M. Smith and A.F. Turner, Appl. Opt. 4 (1965) 147.

[4] D. Dijkkamp, T. Venkatesan, X.D. Wu, S.A. Shaheen, N. Jisrawi, Y.H. Min-Lee, W.L. McLean and M. Croft, Appl. Phys. Lett. 51 (1987) 619-621.

[5] D.B. Chrisey and G.K. Hubler (Eds.), "Pulsed Laser Deposition of Thin Films", Wiley, New York, 1994.

[6] Raphaël Janot and Daniel Guérard, Progress in Materials Science, Volume 50, Issue 1, January 2005, 1-92

[7] E. Petrovsky, M.D. Alcala, J.M. Criado, T. Grygar, A. Kapicka and J. Subrt, J. *Magn. Magn. Mater*. 210 (2000), p. 257.

[8] O.M.Lemine., A.Alyamani, M. Sajieddine and M.Bououdina,, Journal of alloys and compounds, 502 (2010), pp. 279-282

[9] O. M. Lemine , R. Msalam, M. Sajieddine , S. Mufti, A. Alyemani , A. F. Salem, Kh. Ziq and M. Bououdina, International Journal of Nanoscience, Vol. 8, No. 3 (2009) 1–8.

[10] O.M. Lemine , M. Bououdina, M. Sajieddine, A. M. Al-Saie, M. Shafi, A. Khatab, M. Al-hilali1 and M. Henini, Physica B 406 (2011) 1989–1994

[11] O. M. Lemine, A.Alyemani and M.Bououdina, Int. J. Nanoparticles, Vol. 2, 2009

A Study of the Porosity of Activated Carbons Using the Scanning Electron Microscope

Osei-Wusu Achaw
Department of Chemical Engineering, Kumasi Polytechnic, Kumasi, Ghana

1. Introduction

The earliest mention of the significance of porosity in the performance of activated carbons is generally attributed to the French chemist Antoine-Alexandre-Brutus Bussy who in a 1822 publication suggested that porosity was important to the adsorptive properties of activated carbons. Since then a lot of research has gone into elucidating the nature of porosity of activated carbons, its development and measurement. In particular, a great deal of research has been spent on understanding factors that affect the development of porosity and how to model the porosity in terms of these factors. Similarly, much effort has gone into identifying accurate methods and procedures for characterizing activated carbons in general and particularly its pore structure. The continued interest in these research is because of the continued use and importance of activated carbons in industry and an unrelenting pursuit to improve on its performance. Characterization of porosity is often done indirectly by measurement of secondary data from which the requisite pore parameters are estimated. But direct methods also exist for characterizing the pore structure of activated carbons. Methods such as optical microscopy and scanning electron microscopy (SEM), in view of their ability to directly view the micro-structure of activated carbons have demonstrated enormous potential for use in the study and characterization of activated carbons [Manocha et al., 2010; Lazslo et al., 2009; Achaw & Afrane, 2008]. However, this latter approach has only been applied in a very limited capacity in the past. Rather, industry and researchers alike continue to rely on the indirect methods to determine and quantify porosity in activated carbons. The indirect methods calculate activated carbon characteristics from measurement of other parameters that are generally thought to relate to the properties of interest. Adsorption measurements and related mathematical models wherein information regarding the pore structure of an activated carbon is determined are the most commonly used amongst the indirect methods. Porosity measurements using this approach extracts such pore characteristics as pore volume, surface area, pore size distribution and average pore diameter based on mathematical models of the adsorption process, information on the adsorbate and an adsorption isotherm. Besides adsorption measurements, several other indirect methods also exist to estimate the pore characteristics of activated carbons. Among these are immersion calorimetry, small angle scattering of X-rays (SAXS), small angle scattering of neutrons (SANS), and mercury porosimetry[Rigby & Edler, 2002; Stoeckli et al, 2002; Daley et. al., 1996].

The weakness of the indirect methods, is that they are based on models that do not always match with observed behavior of activated carbons. Others like mercury porosimetry are based on very simplified descriptions of the pore structure of activated carbons that are greatly deviated from pores observed directly using direct methods. Not surprisingly, pore characteristics estimated based on two different such models or methods rarely agree [Rodriguez-Reinoso and Linares-Solano, 1989]. The weaknesses notwithstanding, the indirect methods have thus far served a useful purpose of providing a framework for assessing and comparing activated carbons. In particular, they have provided a useful vehicle for predicting and evaluating the performance of these materials for industrial and other applications. The drawbacks of these methods, however, have meant that more consistent and reliable methods continue to be searched to measure the characteristics of activated carbons. The direct methods represent a viable option in that regard. Direct methods allow the direct viewing of the topography of the activated carbon surfaces which makes possible improved description of activated carbon properties such as pore shapes and pore orientation. When coupled with other methods or instrumentations, such as computerized image analysis, it is possible to estimate the pore characteristics of activated carbons more accurately. Again, these methods make possible a visual follow up of the stages of activated carbon manufacture which in turn makes possible the tracking of the changes that a precursor material goes through in forming an activated carbon. It thus offers enormous possibilities of shedding light on the pore development processes than hitherto known [Achaw & Afrane, 2008]. Already, in areas such as materials engineering, biology and medical sciences, the SEM has been extensively used to study and characterize the microstructure of substances [Chira et al., 2009; Vaishali et al., 2008; Chung, et al., 2008; Kamran, 1997]. The purpose of this chapter is to discuss the potential use of the SEM in understanding porosity development in activated carbon and pore structure characterization using micrographs of coconut shells at different stages during the manufacture of coconut shell-based activated carbons.

2. Tracking porosity development using the scanning electron microscope

To better control porosity in activated carbons, it is essential that its development during its preparation be well understood. It is now generally known that porosity in activated carbons is derived from three main sources, namely, the inherent cellular structure of the precursor material, the conditions extant during the preparation of activated carbons and the composition of the precursor material [Heschel & Klose, 1995; Raveendran et al., 1995; Evans & Marsh, 1979;]. How these factors combine to produce an activated carbon of a given specification has been and continues to be a subject of intense research. This continued search is borne out of the need to find newer applications for activated carbons and an unending desire to improve on the performance of activated carbons in such operations like filtration, gas and metal adsorption and separation, gas storage, and finally in water purification. In all these applications the performance of an activated carbon depends as much on the total pore volume as it does on the pore size distribution, the prevalence of a certain pore size regime, and the surface chemistry of the carbon. For instance, during operations involving molecular sieve activated carbons, that pore size characteristics is required that permits the separation of two or any number of molecules of differing molecular sizes. Achieving this kind of performance demands a special design of the pore structure of activated carbons. This in turn demands not only an understanding of how porosity in activated carbons is developed but an additional insight into how to control its development. The different pore sizes play unique roles during activated carbon application. Indeed, the classification of the pores in activated carbons into micropores, mesopores and macropores is based more on the varied behavior of admolecules in these pore regimes than on the actual sizes of the pores. Thus, more than the total pore volume or total surface area of the activated carbons, the fraction of the total pore volume or surface area due to the various sizes of pores is of utmost importance. Again, understanding this development is essential for the design of models to describe the performance of activated carbons and the prediction of activated carbon behavior. According to IUPAC nomenclature [Sing, et al., 1985], micropores are those pores with width less than 2 nm. The micropores play the key role of providing the bulk of the surface involved in adsorption, which is the basis of many applications of activated carbons. The mesopores are wider than the micropores and have pore widths in the range 2 nm to 50 nm. The mesopores also play a role in adsorption albeit on a reduced scale compared to the micropores. The role of the mesopores becomes more important during the adsorption of large molecules that cannot be accommodated in the micropores. Finally, there are the macropores which have much larger pore sizes and which play the important role of being the conduits through which access to the interior of the activated carbon and hence to the mesopores and micropores are achieved. They are generally considered as being part of the external surface of the activated carbon. The macropores have size greater than 50 nm.

Pores in activated carbons are areas of zero electron density in the carbon matrix. These constitute volume elements distributed throughout the particle and posses varied sizes and shapes. The individual volume elements are connected with each other through open channels which are themselves also volume elements. The volume elements are now known to originate from several sources. First, there are those whose source can be traced directly to the primary pore structure of the precursor material. Another group of pores are created as a result of the imperfections that arise from the arrangements of the lamellar constituent molecules (LCM) which are the building blocks of activated carbons. The LCM are layers of sheets which are made up of interconnected aromatic rings. They are formed when the precursor materials are subjected to heat treatment at the appropriate temperature and conditions [Bryne & Marsh, 1995; Evans & Marsh, 1995]. The imperfect arrangements of the LCM creates space in-between parallel layers of the molecules. Volume elements are also created when parts of the LCM are reacted away during contact with agents used in the activation process. The volume elements arising as a result of LCM arrangements and reactions constitute microporosity, and to a lesser extent mesoporosity in the activated carbon. The microporosity confers on activated carbons the unique ability to adsorb large quantities of a diverse range of molecules which makes activated carbons so useful in separation processes and other applications.

The LCM and the accompanying microporosity are formed when the original cellular structure of the precursor material undergoes molecular transformation and reconstitution. During heat treatment of the precursor material a number of physical and chemical processes occur that culminate in the final activated carbon pore structure. Among these, first, moisture and other volatile constituents of the precursor material escape leaving voids that may be later transformed or retain themselves in the final activated carbon product. Secondly, the macromolecules of the precursor material breakdown, lose mostly oxygen and hydrogen and reconstitute into aromatic rings which become the building blocks of the LCM. The new constituent molecules form to enclose the vacancies left by the escaping elements and molecules. These vacant lots also constitute porosity in activated carbons. The transformations are initiated during the pyrolysis of the precursor material and are continued and enhanced during the subsequent activation stage. The loss of volatile matter from the precursor materials occurs at all temperatures but aromatization of the material occurs at temperatures in excess of 700°C. Another important process occurs during the activation process to create new pores or enhance existing pores formed during the pyrolysis step. During activation, activation agents react with the carbon skeleton to create new pores or enlarge existing ones. It is also at the activation stage that other phenomena that facilitate pore creation manifest. For instance, inherent mineral matter such as alkali metals in the precursor material catalyze the pore formation process leading to such phenomena as pitting, channeling and pore enlargement [Bryne & Marsh, 1995]. Pores are also developed as a result of thermal stress on the cellular structure of the precursor material. This stress leads to the development of cracks, crevices, slits, fissures, and all manner of openings in the matrix of the ultimate carbon material. The events leading to the formation of pores occur mainly at the micro and sub micro levels and most of the products of the process such as LCM and associated carbon rings are hardly, directly, observable even with the most powerful of electron microscopes available today. As such these processes have most remained in the realm of theoretical discourse. However, there are other manifestations of these transformations that with the appropriate tools are observable. Using the SEM it has been possible to view images of some of the phenomena that engender porosity development in activated carbons. The ability of the SEM to distinguish objects as small as 1 nm makes it ideal for tracking the transformations happening in the precursor material during activated carbon formation.. This SEM has however not been fully exploited yet for the study of porosity development in activated carbons safe for the pioneering work of Achaw & Afrane, 2008. Figures 1-3 below, show SEM micrographs of sections of coconut shell at different stages during the preparation of coconut shell-based activated carbons. The images reveal details about these materials that shed useful light on aspects of porosity development in activated carbons.

Micrographs in Figures 1 - 3 reveal details of activated porosity development that, previously, has only being a matter of theoretical discourse. Samples for the SEM micrographs used in this study were prepared by cutting sections of well dried coconut shell (raw coconut shells, carbonized shells or coconut shell-based activated carbon) and mounting a on specimen stub with the help of a conductive silver adhesive. The specimen surfaces were thereafter sputter coated with a thin film of silver, placed in the sample holder and viewed with a Ziess DSM 962 electron microscope. A look at Figures 1 and 2 suggest that the original cellular structure of the coconut shells as seen from the transverse section are largely maintained albeit in a modified form following carbonization. In Figure 1 the largely isolated cylindrical units, see positions labeled A and B, has walls made up of layers of thin sheets. In Figure 2, these units seen in Figure 1 have joined together at the walls into a singular solid matrix interspersed with pores. The sheets of the walls are no longer visible in Figure 2. The joining of the walls and the fusing together of the sheets of the walls suggests a profound transformation at the molecular level in the shell during the pyrolysis.

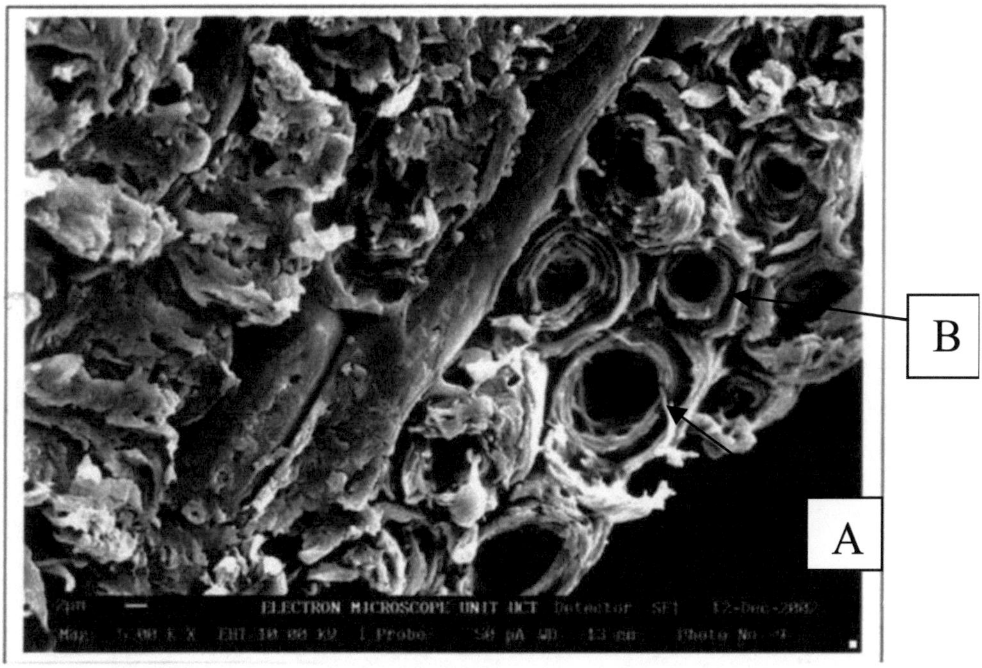

Fig. 1. Micrograph of surface of transverse section of raw coconut shell.
Source: *Achaw & Afrane, 2008*

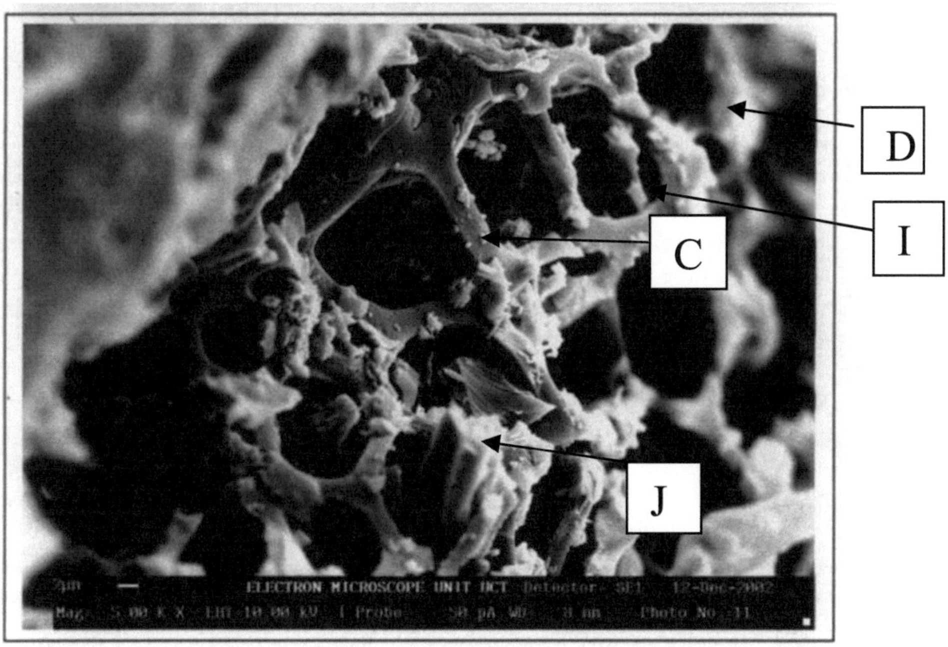

Fig. 2. Micrograph of surface of transverse section of carbonized coconut shell
Source: *Achaw & Afrane, 2008*

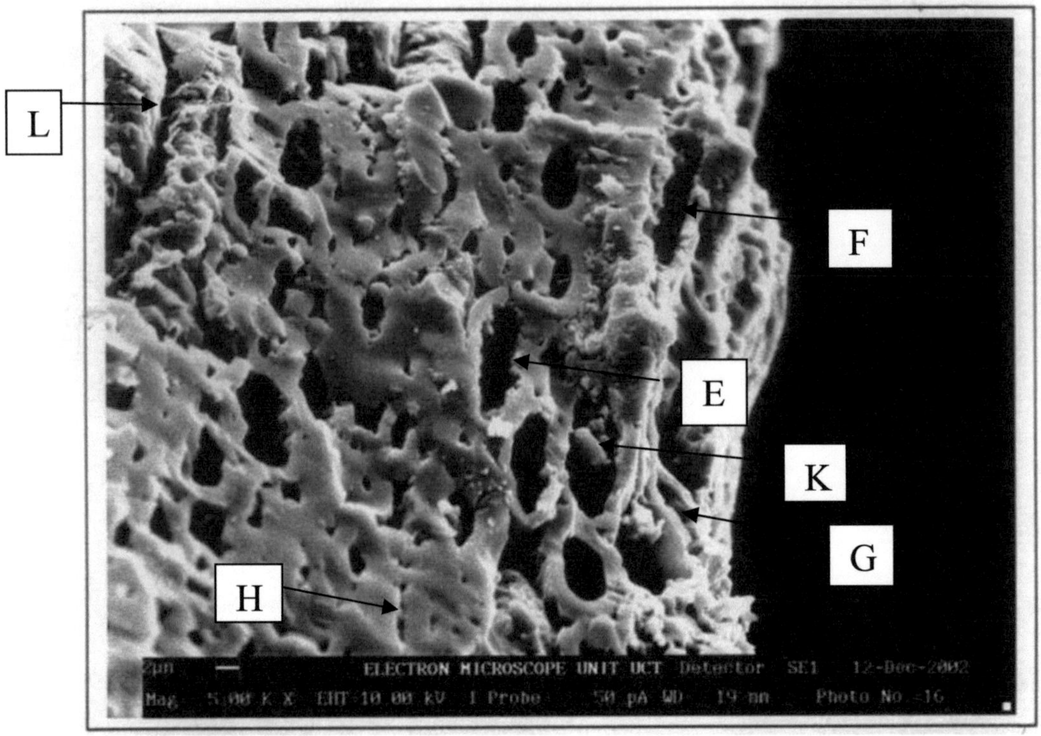

Fig. 3. Micrograph of transverse section of coconut shell-based activated carbon
Source: *Achaw & Afrane, 2008*

Notice that the cylindrical units in Figure 1 has deformed into all manner of shapes after the pyrolysis, see Figure 2, positions marked C and D. The transformation of the pores continues to the activation stage, see positions marked E and F in Figure 3, where what were originally cylindrical shape has now become partially flattened cylinder. The micrograph of the activated carbon further demonstrates transformation of the matrix of the shell as a result of continued heating during the activation. The narrowing of the pore widths suggest a kind of deformation where the matrix softens and walls of the pores give way and close in on each other. Notice further that as a result of this deformation, some of the walls have completely fallen in on each other resulting in a complete zipped up of the pores, see positions marked G and H. The transformations suggest that the carbon matrix passes through a plastic phase as a result of the thermal treatment. Another phenomena observable from these micrographs is the preponderance of foreign materials in the pores of the carbonized product, see positions marked I and J on Figure 2, and the almost lack of these materials in the activated carbon of Figure 3. This means that the activation process serves the additional purpose of cleaning foreign materials from the carbon besides the creation of pores. It is nonetheless noteworthy that even at the activated carbon stage some pores still remain blocked by foreign material, see position marked K in Figure 3. This last observation is an indication that the activation process was not complete. It is anticipated that these foreign materials would be completely cleared at the end of the activation process. Yet another important feature of these micrographs is the position marked L on Figure 3 which is a crack in the carbon matrix probably developed as a result of thermal stress on the carbon matrix due to temperature changes in the activation process. Such cracks contribute to the overall surface of the activated carbon and as such are important. Yet discourse on porosity development of activated carbons often ignore these cracks.

3. Characterization of activated carbons

Characterization of activated carbons is driven by the need to have qualitative and quantitative information which serve as the basis for comparison and selection of activated carbons for specific applications. Such data are also useful for modeling the behavior and performance of activated carbons. Furthermore, characterization provides feedback for use in the design and preparation of activated carbons. The characteristics often measured are density, abrasion resistance, surface area, average pore size, pore size distribution, pore shape, pore volume, and the surface chemistry of the carbon. Whilst there are well established standard methods for measuring the density and abrasion resistance, scientists and industry are still grappling with what accurate methods to use for measuring the others. Most current methods estimate these parameters indirectly from measurements of secondary data on the activated carbons. As a result there are still concerns with the accuracy of values determined for these parameters. The most popular method for characterizing activated carbons is through the measurement of adsorption data and application of mathematical models that relate the adsorption data to such characteristics as pore volume of the adsorbent and the properties of the adsorptive. Other indirect methods, namely, mercury porosimetry, immersion calorimetry, small angle scattering of X-rays (SAXS), neutrons (SANS), high-resolution transmission electron microscopy are also sometimes used to determine the characteristics of activated carbons. Then there are direct methods that hold enormous potential for characterizing activated carbons but which use are rarely mentioned in activated carbon literature. These latter methods are mainly the microscopic methods which enable the observation of micro- and sub-micro features of activated carbons and hence the direct measurement of these features. These microscopic methods are optical microscopy and the SEM.

3.1 Adsorption methods of characterizing activated carbons

These methods almost invariably combine the adsorption isotherm of a given adsorbate-adsorbent system and a theoretical or empirical model of the adsorption process to estimate the characteristics of activated carbons [Machnikowski, et al., 2010; Noor & Nawi, 2008; Lozano-Castello et al., 2004; Stoeckli et al., 2002; Stoeckli et al., 2001 Yuna et al., 2001; Rodriguez-Reinoso, 1989]. Most commercial sorption equipment estimate activated carbon data using in-built software based on one version or the other of this approach. Of historical importance is the Langmuir model [Gregg & Sing, 1982] which was first developed in 1916 to describe adsorption behavior on solid adsorbents in general. The model relates the adsorption of molecules in a gaseous medium onto a solid surface to the gas pressure above the solid at a fixed temperature and can be expressed mathematically as

$$\frac{P}{V}=\frac{P}{V_m}+\frac{1}{bV_m} \tag{1}$$

V is the equilibrium adsorbed amount (mmolg^{-1}) of the adsorbate per unit mass of the adsorbent at a pressure P. V_m is the amount of gas required for monolayer coverage of the adsorbent ($mmolg^{-1}$), and b is a constant whose value depends on the temperature. A linear plot of equation (1) allows V_m to be evaluated from the gradient and hence the adsorbents surface area from the relation

$$S = V_m L\sigma \tag{2}$$

S is the total surface area of the adsorbent (m^2g^{-1}), L is Avogadro's number and σ is the projected surface area of the adsorbate molecule. S is the sum total of pore surfaces and external (non-pore) surface of the adsorbent. Equation (1) is based on the assumption that i) there is a mono-layer adsorption, ii) there are no adsorbate-adsorbate interactions on the adsorbent surface, iii) the adsorbent has a homogeneous surface, iv) all adsorption sites on the adsorbent are equivalent and, v) the adsorbing gas adsorbs into an immobile state.

The Langmuir's model has been found to be of limited applications for activated carbons. In particular, activated carbon surfaces are rarely homogeneous. Generally, the assumptions have been found not to be consistent with observations therefore the Langmuir model is rarely used to characterize activated carbons. Consequently, other relatively more accurate models of adsorption are often used. One such model is the Braunnauer, Emmett and Teller (BET) model [Sing et. al., 1985]. The BET method has a much wider application and is most often used to interpret adsorption isotherms obtained using Nitrogen at 77K as the adsorbate. The model is an improvement on the Langmuir model in that it can account for multilayer adsorption. It relates the adsorption pressure and the volume of the adsorbed adsorbate according to the equation

$$\frac{p}{V(p^o - p)} = \frac{1}{V_{mc}} + \frac{c-1}{V_{mc}}\frac{p}{p^o} \tag{3}$$

where

$$c = \exp(\frac{(\Delta H_A - \Delta H_L)}{RT}) \tag{4}$$

In equation (3), V_{mc} is the monolayer capacity of the adsorbent, p^o is the saturation vapour pressure of the adsorbate gas, p is the pressure of the gas, and c is a constant which is exponentially related to the heat of first layer adsorption. ΔH_A is the heat of adsorption, ΔH_L is the heat of liquefaction of the adsorption fluid, T is the temperature, and R is the gas constant. A linear plot of equation (3) allows V_{mc} to be determined from the intercept and from which the surface area of the activated carbon can be estimated using equation (2). The BET method has found a number of applications in adsorption studies and is especially used in the determination of the surface areas of adsorbents including activated carbons. Even so, the BET equation is unable to account for adsorption in a number of instances. For activated carbons equation (3) is only linear at $p/p^o < 0.1$. This introduces error into the measurement of those pores for which adsorption is possible at pressures for which $p/p^o > 0.1$. Secondly, the calculation of the adsorbent surface area using equation (2) requires knowledge of the projected surface area of the adsorbate molecule, in this case Nitrogen. This in turn requires that the adsorbate molecules (Nitrogen) be in a close packed, monolayer coverage on the adsorbent. The use of the method implicitly assumes that the value estimated for V_{mc} is necessarily accurate and that σ is constant for the adsorbate under all conditions. Further, adsorption in micropores is characterized more by pore filling than by surface coverage. As such, the application of the BET method does not always yield the correct result for the surface area of activated carbons, especially if they are predominantly microporous.

Probably more accurate among the adsorption methods for the determination of pore characteristics of activated carbons is the Dubnin-Raduskevitch (DR) equation and its improved and more versatile version, the Dubnin-Astakov (DA) equation [Dubnin & Raduskekevich, 1947; Dubnin, 1989; Carrasco-Merin et al., 1996; Gil, 1998]. The DR equation is premised on the assumption that adsorption in micropores occurs by pore filling rather than by physical adsorption on the surface of the micropores. The equation relates the volume of pores, W, filled by an adsorbate at a given temperature T and relative pressure p/p^o and other parameters of the adsorption system as

$$W = W_o \exp[-(\frac{A}{\beta E_o})^2] \tag{5}$$

where W_o is the total volume of the micropores, Eo is the characteristic energy, and β is the affinity coefficient. Both Eo and β are system dependent. The deferential molar work of adsorption on the adsorbent, A, is further defined as

$$A = RT\frac{p}{p^o} \tag{6}$$

For slit-shaped pores, a relationship exists between E_o and the average micropore width $\overline{L}$ as

$$\overline{L}(nm) = 10.8(E_O - 11.4) \tag{7}$$

The DR equation has a narrow range of application as it corresponds to mostly Type I isotherms. The Dubnin-Astakov (DA) equation which is a modification of the DR equation and which is applicable to a wide range of microporous carbons is therefore preferred. The generalized form of the DA equation is

$$W = W_o \exp[-(\frac{A}{\beta E_o})^n] \tag{8}$$

When n=2, equation (8) becomes equal to the DR equation. Values of n between 1 and 4 are observed for most carbon adsorbents, n> 2 for molecular sieve carbons or carbons with highly homogeneous and small micropores, n < 2 for strongly activated carbons and heterogeneous micropore carbons. For monodisperse carbons, n=3 and for strongly heterogeneous carbons n= 2 [Carrasco-Merin et al., 1996].

3.2 Characterization of activated carbons with the scanning electron microscope

In the areas of porosity development and characterization of activated carbons, a number of issues still remain unresolved. The current state of knowledge has not been able to address all observed behavior and performance of activated carbons. For instance, to what extent does thermal stresses on the carbon matrix during pyrolysis affect porosity development.

How realistic is the often used slit-shaped pore model in describing activated carbons. Similarly, in characterizing activated carbons the often used methods such as adsorption measurement and mercury porosimetry all rely on secondary data and mathematical models to estimate pore characteristics. But these methods are fraught with a number of drawbacks. In these methods, only those pores can be characterized that the adsorbate molecules could have access to. Also the mathematical models of adsorption which are the basis of estimating pore characteristics are based on assumptions most of which do not match with observations. Adsorption measurements in particular have other drawbacks that affect the accuracy of parameters estimated. For instance, phenomena such as activated diffusion and gate effects introduce errors into adsorption based estimates of pore parameters. Equally, idealized pore models such as the cylindrical or slit-shape pore which are the basis of a number of methods for estimating average pore width of activated carbons are too simplistic in the face of the observed complex nature of porosity in activated carbons. Then again some of the pore parameters are not at all amenable to estimation by the indirect methods. Such parameters like pore shape, pore location and distribution, and pore orientation have all eluded estimation by the indirect methods. These parameters nonetheless have important consequences on modeling and prediction of performance characteristics of activated carbons and therefore are worth estimating or measuring.

Direct methods, particularly, microscopy offers an alternative approach to resolving most of the drawbacks of the indirect methods. Using microscopy, it is possible to observe the micro and submicro-features of activated carbons directly and therefore makes possible a proper qualitative and quantitative description of its characteristics [Ito & Aguiar, 2009; Daley et al., 1996; Tomlinson, et al., 1995; Hefter, J., 1987; Ball & McCartney, 1981]. There are two types of this method, namely, optical microscopy which has a resolution of about 1 μm and electron microscopy whose resolution is much greater and in the range of about 1.5 nm and which can achieve a magnification of about 2,000,000x. The SEM is one version of the electron microscopy which uses a beam of electrons to scan the surface of a specimen and makes possible the direct observation of its surface features at the micro and submicro levels. Due to the huge magnifications and impressive resolutions achievable with the SEM it has been used in many areas of science and industry, particularly in materials engineering, biological and medical sciences for the study and characterization of the micro-structure of substances. It use provides an avenue to resolve some of the yet unresolved issues in activated carbon porosity development and characterization. The SEM functions exactly as its optical counterparts except that it uses a focused beam of electrons instead of light to "image" a specimen and gain information about its structure and composition. The SEM can yield information about the topography (surface features of an object), morphology (shape and size of the particles making up the surface of an object), composition (the elements that the object is composed of and the relative amounts of these) and crystallographic information (how the atoms are arranged in the object). This ability makes the SEM a hugely useful instrument for the study of activated carbons. The topographic information attainable using the SEM allows that surface features such as pore characteristics, the description of which has been a major preoccupation of activated carbon chemists, to be studied and measured directly from SEM micrographs. Also SEM's ability to reveal compositional details of a specimen makes it a potent instrument for studying the surface chemistry of activated carbons. Other features of the SEM that make it a unique instrument for studying activated carbons include its ability to reveal details of a sample less than 1 nm in size. This means that, in principle, even micropores could be exposed for study by SEM micrographs. Again, due to the very narrow electron beam employed, SEM micrographs have a large depth of field that yields a pseudo three-dimensional appearance useful for understanding the surface structure of a sample. Figures 4-7 are examples of SEM micrographs that reveal details of activated carbons that only the direct methods can show.

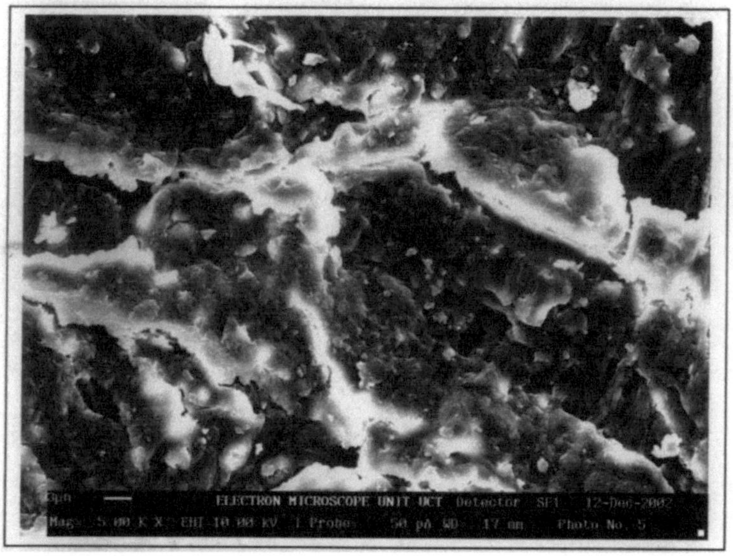

Fig. 4. Micrograph of outer surface of raw coconut shell
Source: *Achaw & Afrane, 2008*

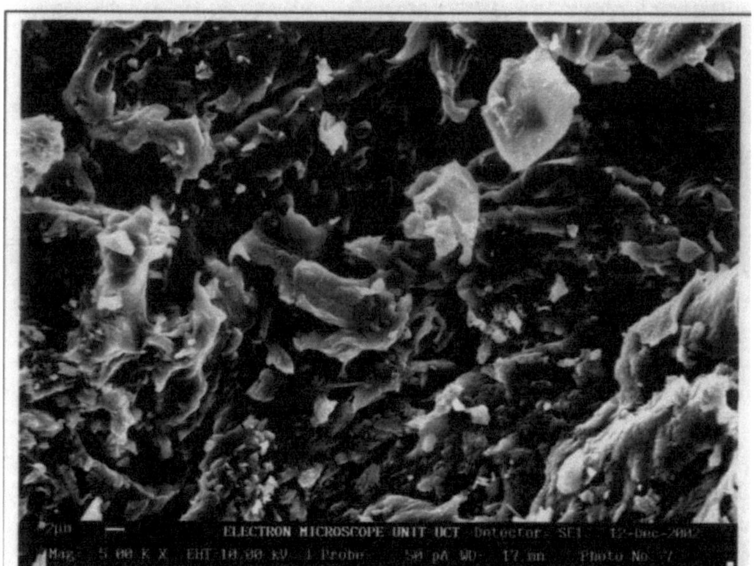

Fig. 5. Micrograph of outer surface of carbonized coconut shell.
Source: *Achaw & Afrane, 2008*

Fig. 6. Micrograph of outer surface of coconut shell-based activated carbon
Source: Achaw & Afrane, 2008

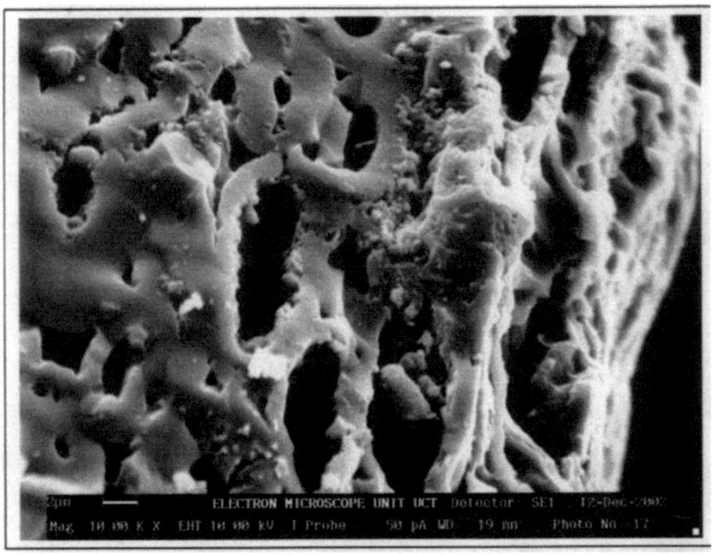

Fig. 7. A Micrograph of transverse section of coconut shell-based activated carbon
Source: This study

The micrographs in Figures 4-7 are all of coconut shells at different stages during the preparation of activated carbons. Figures 4-6 are micrographs of the surfaces of the outer sections of the shell. The surface features observed from these micrographs are significantly different from those seen from the corresponding micrographs of on Figures 1-3, indicating that the nature of pores seen of activated carbons depend on the sections used. Crucially, the nature of porosity observed in Figure 6 is totally different from any description of porosity previously described except in the work of Achaw and Afrane (2008). What is observed here is a network of cracks which are more the result of thermal stress on the carbon matrix. This is totally in contrast to what was observed of the micrograph in Figure 3 which is more likely the result of re-arrangement of molecules in the carbon matrix. The observed porosity in Figure 6 further show the difficulty of defining a generalized pore structure for activated carbons. It calls to question such often used concepts as average pore size, and pore size distribution. Further, it raises questions about models associated with pore structure such as the slit-shaped model which is the basis of a number of mathematical models of adsorption in activated carbons or the cylindrical models used by mercury porosimetry and other empirical methods for estimating pore characteristics. Figures 6 & 7 are micrographs of different sections of an activated carbon. The two images clearly demonstrate the extent of inhomogeneity of the surfaces of activated carbons. Equally noteworthy of the micrographs in Figures 4-6 is the fact that there is hardly any common trend linking the structures in these micrographs. Whilst, hardly, any pore is observable at all in Figure 4, the structure in Figure 5 is fuzzy and confusing, and hardly yielding to any definition at all. Finally, even though the structure in Figure 6 has some semblance of order, it also defies any exact definition. An important observation of these micrographs is that pore development in activated carbons is as a result of several phenomena. Particularly, it seems that thermal stress plays an important role in pore development in these materials than previously thought. The foregoing observations demonstrate the strength of SEM in studying activated carbons.

4. The scanning electron microscope

The main features of the SEM are an electron source which provides the electrons that interact with the material to be examined, an arrangement of metal apertures, magnetic lenses and scanning coils or deflectors plates that confines, focuses and turns the beam of electrons into a thin and focused monochromatic beam which is accelerated towards the sample and which irradiates the specimen in a raster fashion [Goldstein, J et al., 2003; Reimer, L, 1998]. The interaction of the electrons with the specimen initiates a number of reactions inside the sample which results in the generation of signals which are taken advantage of to gain information about the sample. The SEM imaging process involves four major steps. These include sample preparation, the specimen scanning process, image formation and image analysis. The kind of preparation required of the sample depends on whether it is electrically conducting or not. Electrically conducting samples, for instance metals, only require minimal sample preparation prior to mounting on a sample stub for scanning and imaging. Non-conductive specimens such as activated carbons, however, must first be made conducting before mounting for study. Otherwise, these tend to charge when scanned by the electron beam leading to scanning faults and other image artifacts. Non-conducting samples are therefore first sputter coated with an ultra-thin coating of an electrically-conducting material before imaging. Other reasons for coating the sample surface are to increase the signal and surface resolution, especially with samples of low atomic number. Some of the commonly used materials for coating samples include gold, graphite, platinum, chromium, tungsten, osmium, and indium. For biological materials it is possible to increase the conductivity without coating by impregnating them with osmium before imaging. It is also possible to image non-conducting specimen without coating by using the Environmental SEM(ESEM) or the field emission gun (FEG) SEM [Schatten & Pawley, 2008; Hardt, T. A. 1999;]. Samples for SEM study do not need to be as thin as it is the case in optical microscopy or transmission electron microscopy (TEM). The specimen size is dictated primarily by the size of the sample chamber and must be rigidly mounted on the specimen stub. Again, in view of the applied vacuum, it is important that the sample be completely dry. Hard, dry materials such as activated carbons can be examined with little further treatment besides sputter coating.

The signals of interest in the characterization of activated carbons using the SEM are the secondary electrons whose detection and imaging gives information on the surface topography and hence on the pore structure of the activated carbons, the backscattered electrons and the X-ray radiation which give complementary information of the chemical composition of the sample surface. Whilst secondary electrons can be detected and imaged in a conventional SEM instrument, harnessing signals from the backscattered electrons and X-ray radiation requires complementary instrumentation. The secondary electrons are produced when an incident electron in the primary electron beam excites an electron in an atom of the sample and loses most of its energy in the process. The excited electron moves to the surface of the sample where it can escape if it possesses sufficient energy. In view of their low energy, only secondary electrons that are very near the surface (<10 nm) can exit the sample and be detected. The secondary electrons are detected and accelerated onto a photomultiplier from which an amplified electrical signal output is displayed as a two-dimensional intensity distribution that can be viewed and photographed on an analogue video display or converted and displayed or stored as a digital image.

Backscattered electrons (BSE) are produced when the primary electron beam hits the sample and some of the electrons are reflected or scattered back out of the specimen. The production of backscattered electrons varies directly with the specimen's atomic number. High atomic number elements backscatter electrons more strongly than low atomic number elements, and thus appear brighter in an image. BSE are therefore used to detect contrast between areas on the sample surface with different chemical compositions. This provides opportunity for examination of the chemistry of the surface of activated carbons. Dedicated backscattered electron detectors, usually either scintillator or semiconductor types are positioned above the sample to detect the backscattered electrons.

As earlier intimated, inelastic scattering, places the atoms of the sample in an excited state. The tendency therefore is for an atom to return to its ground or unexcited state. To achieve this, the atom gives off the excess energy. This may result in the production of X-rays, cathodoluminescence and Auger electrons. The relaxation energy is the fingerprint of each element in the sample. Thus detection and analysis of the relaxation energies enables the identification of the specific elements in the surface of a sample. When an SEM is equipped with energy-dispersive X-ray spectroscopy (EDX) or wavelength dispersive X-ray spectroscopy (WDS), it is possible to get information on the elemental composition on the surface of the specimen. For an activated carbon, this method can be used to study the effect on porosity development of specific elements in the carbon and generally the chemistry of the activated carbon surface [Afrane & Achaw, 2008].

4.1 3D images in scanning electron microscope

In conventional (standard) SEM a pseudo-three dimensional (3D) view of the sample surface can be observed directly. The standard SEM image is, however, really a two-dimensional (2D) structure from which mostly qualitative data is possible regarding the microstructure of the specimen. In the study of activated carbons however, qualitative data though useful as for instance in understanding pore development phenomena, it is quantitative data on pore structure- pore sizes, pore shapes, pore surface area, and pore size distribution- that are most useful for characterization and modeling of performance and behavior. Getting quantitative data from 2D SEM images of activated carbons, however, poses a number of challenges. First, 2D SEM cannot determine three-dimensional porosity of the materials because the conventional SEM is unable to observe images of inner parts of a specimen. Then also there are the difficulties associated with getting precise descriptions or a representative pore structure in view of the otherwise complex, varied and numerous pores in the field of view of the microscope. To get quantitative data of a specimen from conventional SEM images, the practice is to convert the 2D images into 3D from which the requisite quantitative data can be measured or estimated often with the help of computerized image analysis software. A number of methods for getting 3D data from 2D images are available [Marinello et al., 2008; Spowart, 2006; Spowart et al. 2003; Alkemper and Voorhees, 2001; Lyroudia et al., 2000]. These include stereology, photogrammetry, photometric stereo, and the more useful serial sectioning method. An automated variation of the serial sectioning method called the focus ion beam-scanning electron microscopy (FIB-SEM) is increasingly being used in the areas of materials engineering and biological sciences to study the micro-features of substances. This last method appears to have enormous potential for use in the study and characterization of activated carbons.

In serial sectioning, 2D images of a sample are collected after a series of successive layers of equal width have been removed from the sample. Afterwards the stack of 2D data files (2D SEM images) is combined and processed in such a way that the microstructural features that are within the 3D data stack can be classified. This is most efficiently done using computerized image processing software. Serial sectioning is made up of two basic steps that are iteratively repeated until completion of the experiment. In the first step a nominally flat surface of the sample is prepared using any of a variety of methods such as cutting, polishing, ablating, etching, or sputtering. These processes remove a constant depth of material from the specimen between sections. The second step is to collect 2D characterization data after each section has been prepared, for instance by imaging with an SEM. Finally, computer software programs are used to construct a 3D array of the characterization data that can be subsequently rendered as an image or analyzed for morphological or topographical features of the sample. Using this method any micro and submicro features of the sample that can be distinguished by the SEM can also be characterized. Thus the capability of the methodology in characterizing topographical and morphological features of substances is limited only by the resolution of the SEM.

Automated serial sectioning techniques which facilitates the methodology exist for material removal and imaging. An example is the FIB-SEM which combines ion beam sectioning with SEM imaging to generate tomographic data that are well suited to characterize microstructural features of a sample in 3D via serial sectioning [Desbois et al., 2009; Orloff et al. 2003]. While less common, the FIB-SEM method has demonstrated the ability to complete 3D volumetric reconstruction at a resolution of 10 nm or better in all three dimensions. The method has been widely used for studies in materials engineering and life sciences and holds tremendous potential for characterizing the porosity of activated carbons.

5. Conclusion

In spite of the tremendous progress in development and use of activated carbons a number of questions still remain that the conventional indirect methods of studying activated carbon are still grappling to answer. The mode of porosity development is one such area. Another area is the characterization of the porosity of activated carbons, where the existing mathematical models and methods have still not succeeded in finding accurate ways to estimate pore parameters. Recent developments in scanning electron microscopy, especially in the conversion of 2D SEM to 3D and computerized image analysis has opened avenues for improved study of porosity development and characterization of activated carbon. This potential of the SEM has not really been adequately explored for the study of activated carbons as not much work exits in the literature in that regard. However, judging from the enormous strides researchers in the areas of materials engineering and the biological sciences have made in using this methodology to identify micro and submicro-features of substances, it is anticipated that its adaptation for use in the study of activated carbons would facilitate the study of porosity development and pore characterization. The SEM micrographs shown in this work clearly demonstrate this point. The major limitation of the SEM is the level of resolution achievable with it currently. At 1.5 nm, this poses difficulty in characterizing most micropores in activated carbons. It is nonetheless hoped that continued advances in SEM instrumentation will overcome this difficulty and facilitate the use of the SEM in the study of activated carbons.

6. References

Achaw, O-W. & Afrane, G. (2008). The evolution of pore structure of coconut shells during the preparation of cocnut shell-based activated carbons. *Microporous and Mesoporous Materials*, 112, pp. 284 - 290, ISSN 1387-1811

Afrane, G. & Achaw, O-W. (2008). Effect of the concentration of inherent mineral elements on the adsorption capacity of coconut shell-based activated carbons, *Bioresource Technology*, Vol. 99, No. 14, pp. 6678 - 6682, ISSN 0960-8524

Alkemper, J., Voorhees P. W. (2001). Quantitative serial sectioning analysis. *Journal of Microscopy*, Vol. 201, No.3, pp. 388–394, *ISSN: 1365-2818*

Ball, M. D. & McCartney, D. G. (1981). The measurement of atomic number and composition in a SEM using backscattered detectors. *J. Microsc.* 124, 57-68, ISSN: 1365-2818

Bryne, J. F. & Marsh, H. (1995). Origins and structure of porosity, in John W. Patrick (ed), *Porosity in carbons: Characterization and applications*, Edward Arnold, London, ISBN 0 340544732

Carrasco-Marin, F., Alvarez Merino, M. A., & Moreno-Castilla, C. (1996). Microporous activated carbons from a bituminous coal. *Fuel*, Vol. 75, No. 8, pp. 966-970, ISSN: 0016-2361

Chung, W., Sharifi, V. N., Swithenbank J, Osammor, O., Nolan, A. (2008).. Characterisation of Airborne Particulate Matter in a City Environment. *Modern Applied Science*, Vol.2, No. 4, p.17, ISSN 1913 - 1844

Dalye, M. A., Tandon, D., Economy, J. abd Hippi, E. J. (1996). Elucidating the porous structure of activated carbon fibers using direct and indirect methods. *Carbon*, Vol. 34, No. 10, pp 1191-1200, ISSN: 0008-6223

Desbois, G, Urai, J. L., & Kukla, P. A. (2009). Morphology of the pore space in claystones - evidence from BIB/FIB ion beam sectioning and cryo-SEM observations. *eEarth Discuss.*, 4, 1–19, 2009, www.electronic-earth-discuss.net/4/1/2009/

Dubnin, M. M. (1989). Fundamentals of the theory of adsorption in micropores of carbon adsorbents: Characteristics of their adsorption properties and microporous structures. *Pure & Applied Chemistry*, Vol. 10, pp. 1841 - 1843, ISSN 0033-4545

Evans, M. & Marsh, H. (1979). Origins of microporosity, mesoporosity, and macroporosity in carbons and graphites, in *Characterization of Porous Solids*, Unger, K. K., Rouguerol, J., Sing, K. S. W., Kral, H., (eds). London, SCI, ISBN 0 - 444 - 42953 -

Farkas, L., Major, N., Mihalko, A., Abraham, J. & Kozar, Z. (2009). Analysis of active carbon catalysis. *Material Sciences and Engineering, Miskolc,* Volume 34/2 , pp 41-51

Gil, A. (1998). Analysis of the micropore structure of various microporous materials from nitrogen adsorption at 77K. *Adsorption,* Vol. 4, Numbers 3-4, pp 197-206

Goldstein, J., Newbury D. E., Joy, D. C, Lyman, C. E., Echlin, P., Lifshin, E., Sawyer, L., and Michael, J. R. (2003). *Scanning Electron Microscopy and X-Ray Microanalysis.* Kluwer Academic /Plenum Publishers, New York, ISBN 978-0-306-47292-3

Gregg, S. J. and Sing, K. S. W. (1991). *Adsorption, Surface Area and Porosity.* Academic Press, 2nd Edition, London, 1991, ISBN 0123009561

Hardt, T. A. (1999). Environmental SEM and related applications. In Impact of electron and scanning probe microscopy on materials research. (eds). Rickerby, D. G. Giovanni Valdrè, G., & Valdrè, U, Kluwer Academic Publishers, Netherlands, p. 407, ISBN 0-7923-5939-9

Hefter, J. (1987). Morphological characterizations of materials using low voltage scanning electron microscopy. *Scanning Microsc.* 1(l), 13-21, ISSN: 1932-8745

Heschel, W. & Klose, E. (1995). On suitability of agricultural by-products for the manufacture of granular activated carbon. *Fuel,* Vol. 74, No. 12 , ISSN: 0016-2361

Ito, L. X. & Aguiar, M. L. (2009). A study of the porosity of gas filtration cakes. *Brazilian Journalof Chemical Engineering.* Vol. 26, No. 02, pp. 307 - 315. ISSN 0104-6632

Joshi, V. C., Khan, I. A., &. Sharaf, M. H. M. (2008). Use of Scanning Electron Microscopy in the Authentication of Botanicals. *Pharmacopeial Forum. Vol. 34, No.4, ISBN 0363-4655*

Kamran, M. N. (1997).Fracture analysis of concrete using scanning electron microscopy. *Scanning* Vol. 19, pp. 426 - 430, ISSN: 1932-8745

Lozano-Castello, D., Cazorla-Amoros,D., Linares-Solano, A. (2004).Usefulness of CO2 adsorption at 273 K for the characterization of porous carbons .*Carbon 42*, pp. 1231–1236, ISSN: 0008-6223

Lyroudia, K, Pantelidou, O. Mikrogeorgis, G., Nikopoulos, N., & Pitas,L. (2000). Three-dimensional reconstruction: A new method for the evaluation of apical microleakage. *Journal of Endodontics, Vol. 26, No. 1, pp. 36- 38,* ISSN: 0099-2399

Machnikowski, J., Kierzek, K., Lis, K., Machnikowska, H., and Czepirski, L. (2010).Tailoring porosity development in monolithic adsorbents made of KOH-activated pitch coke and furfuryl alcohol binder for methane storage. *Energy Fuels* , 24, 3410–3414,

Marinello, F., Bariani, P., Savio, E., Horsewell, A., & De Chiffre, L. (2008). Critical factors in SEM 3D stereo microscopy. *Measurement Science and. Technology.* Vol. 19, No. 6, pp. 1- 12, ISSN 0957-0233

Md Noor., A. A. B. & Nawi, A. B. M. (2008).. Textural characteristics of activated carbons prepared from oil palm shells activated with ZnCl2 and pyrolysis under nitrogen and carbon dioxide. *Journal of Physical Science,* Vol. 19(2), 93–104, 2008, ISSN: 1675-3402

Mfanacho, S. M. , Hemang, P. & Manocha, L. M. (2010). Enhancement of microporosity through physical activation. PRAJÑĀ - *Journal of Pure and Applied Sciences,* Vol. 18, pp. 106 - 109, ISSN 0975 - 2595

Orloff, J., Utlaut, M., Swanson, L., (2003). *High Resolution Focused Ion Beams: FIB and Its Applications*. Kluwer Academic/Plenum, New York, ISBN 0-306-47350-X

Raveendran, K., Ganesh, A., Khilar, K. C. (1995). Influence of mineral matter on boimasspyrolysis characteristics. *Fuel* Vol. 74, No. 12, ISSN: 0016-2361

Reimer, L. (1998). *Scanning Electron Microscopy: Physics of Image Formation and Microanalysis,* Springer-Verlag, New York, ISBN 3-540-63976-4

Rigby, S. P. & Edler, K. J. (2002). The influence of mercury contact angle, surface tension, and retraction mechanism on the interpolation of mercury porosimetry data. Journal of Colloid and Interface Science 250, pp. 175-190

Rodriguez-Reinoso, F. & Linares-Solano, A. (1989). In 'Chemistry and Physics of Carbon', . P. A. Thrower (ed), Dekker, Vol. 21 , New York, p.1, ISBN 0824781139

Rodriguez-Reinoso, F., An overview of methods of the characterization of activated carbons. *Pure & Applied Chemistry,* Vol. 61, No. 11, pp 1859 - 1867, 1989, ISSN 0033-4545

Schatten, H & Pawley, J. (2008). *Biological low voltage field emission scanning electron microscopy. Springer Science + Business Media, New York,* e-ISBN 978-0-387-72972-5

Sing., K. S. W., Everett, D. H., Haul, R. A. W., Moscou, L., Pierotti, R. A., Rouquerol, J. &

Siemieniewska. (1985). Reporting on physiosorption data for gas/solid systems with special reference to determination of surface and porosity. *Pure & Applied Chemistry,* Vol 57, p 603, ISSN 0033-4545

Spowart, J. E. (2006). Automated serial sectioning for 3-D analysis of microstructures. *Scripta Materialia,* Vol. 55, No.1, pp. 5–10, ISSN: 1359-6462

Spowart, J. E., Mullens H. M., Puchala, B. T. (2003) Collecting and analyzing microstructures in three dimensions: A fully automated approach. *JOM Journal of the Minerals, Metals and Materials Society,* Vol. 55, No. 10, pp. 35–37, ISSN 1047-4838

Stoeckli, F., Guillot, A., Slasli, A. M.,a, and Hugi-Cleary, D. (2002). Microporosity in carbon blacks. *Carbon 40,* issue 2, pp. 211-215, ISSN: 0008-6223

Stoeckli, F., Guillot, A., Slasli, A. M. & Hugi-Cleary, D. (2002). The comparison of experimental and calculated pore size distributions of activated carbons. *Carbon* Vol. 40, No. 3, pp. 383-388, ISSN: 0008-6223

Tomlinson, J. B., Freeman, J. J., Sing, K. S. W., and Theocaris, C. R. (1995). Rates of activation and scanning electron microscopy of polyaryllamide-derived chars. Carbon, Vol. 33, No. 6, pp. 789-793, ISSN: 0008-6223

Yuna, C. H., Parka, Y. H., Park, C. R.,. (2001). Effects of pre-carbonization on porosity development of activated carbons from rice straw, *Carbon 39,* pp.559–567, ISSN: 0008-6223

25

Study of Structure and Failure Mechanisms in ACA Interconnections Using SEM

Laura Frisk
Tampere University of Technology, Department of Electronics, Finland

1. Introduction

The trends in the electronics industry have for several decades been for smaller size combined with greater functionality. One enabler for this trend has been development of new packaging solutions which has required the development of new materials and also new interconnections technologies. In the development of these technologies it has been essential to have effective tools to study the structure of the packages and their failure mechanisms. Due to the versatility of electronics packages concerning materials, structures, and functions a plethora of different methods have been used. These include for example electrical characterization technologies, x-ray, scanning electron microscopy (SEM), scanning acoustic microscopy (SAM), optical microscopy, differential scanning calorimetry (DCS), and thermomechanical analysis (TMA) (Chan et al., 2000; Jang et al., 2008; Yim & Paik, 2001).

This chapter concentrates on flip chip technology, which is one of the technologies developed to miniaturise an electronics package. In this technology a bare chip is attached directly onto a substrate without wiring needed to connect the chip. As the attachment is done with the active side of the chip towards the substrate, the chip is flipped before bonding. Hence the name flip chip. Flip chip technology has several advantages as it enables the production of small, very high density packages. In addition, it has good electrical performance due to the short interconnection path (Lau, 2000). This method can be used to attach a chip directly onto a substrate, but also as an attachment method in single-chip packages, such as a ball grid array (BGA) and a chip scale package (CSP), and in multi-chip modules.

One problem with this technology is that the quality of the joints is relatively difficult to assess. Semiconductor chips used in this technology may have hundreds of contacts which form an area array below the chip. In order to study the joints a cross-sectioning is often needed. Although cross-sections give lots of valuable information, they are restricted to very small area of a package and thereby often several cross-sections are needed. Additionally, the information of the interconnections may need to be increased using other techniques. For example scanning acoustic microscopy may be used to study the amount of delamination in a package. The cross-sections may be studied by optical microscopy. However, for detailed information scanning electron microscopy is the preferred method of analysis.

Currently mainstream flip chip technology is based on solder bumps. These can be produced using both traditional tin-lead and new lead-free solders. However, mounting environmental concern has increased interest in electrically conductive adhesives, as they are environmentally friendly. In addition to being lead free, they can be used with substrate materials which do not withstand soldering temperatures. Thus they can be used to solve the problem caused by the high reflow temperature needed by most lead-free solders (Li & Wong, 2006).

Compared to solders adhesive materials are more complex as they are polymeric materials containing conductive particles. They have several advantages which makes their use profitable. However, due to their complex structure quality of the interconnections made by these materials needs to be determined carefully to attain good reliability. There are two types of electrically conductive adhesives. In isotropic conductive adhesives (ICA) the concentration of the conductive particles is high and they conduct in all directions. On the other hand, in anisotropic conductive adhesives (ACA) the concentration of conductive particles is low and the adhesive conducts in z-direction only after the bonding process. This chapter will concentrate on ACA materials used in flip chip applications. This chapter will discuss specifically how SEM may be utilised to study the quality and failure mechanisms of ACA interconnections.

2. Polymeric interconnections for electronics

Electrically conductive adhesives used in electronics consist of polymer binder and conductive particles. Polymer resins used in these adhesives are inherently insulators. To obtain an electrically conductive adhesive they must therefore be filled with electrically conductive fillers, such as metal particles. In the following properties and materials of two main types of electrically conductive adhesives are discussed. Isotropic conductive adhesives (ICA) have high concentration of the particles and they conduct in every direction. These materials may be used to replace solders. If the concentration of conducting particles is low, an anisotropic conductive adhesive (ACA) is formed. ACAs conduct electrically only in a vertical direction and thereby may be used in very high density applications.

2.1 Isotropic conductive adhesives

Isotropically conductive adhesives are formed by adding enough conductive filler to a polymer matrix to transform it from an insulator into a conductor. This transformation has been explained by a percolation theory. When the concentration of conductive filler is increased, the resistivity of the adhesive drops dramatically above a critical concentration, and this is called the percolation threshold. It is believed that at this concentration the conductive particles contact each other forming a three dimensional network, which enables the conductivity. After the percolation threshold the resistivity decreases only slightly with increased concentration of the conductive filler. (Lau et al., 2003) The mechanical interconnection of an ICA joint is provided by the polymer matrix. If too high a concentration of the conductive filler is used, it may impair this interconnection. Thus the amount of conductive filler needs to be large enough to ensure good conductivity without sacrificing the mechanical properties of the adhesive (Lu, 2006). A typical volume fraction of the conductive filler is approximately 25 to 30 percent (Licari, 2005).

Several materials can be used in ICAs. The most widely used ICAs in the electronics industry are silver-filled epoxies, which also provide a high level of thermal conductivity. The popularity of epoxies is due to their excellent properties as a conductive adhesive. They have good adhesive strength, thermal stability and dielectric properties. Furthermore, they have good retention of these properties under thermomechanical stresses and under demanding conditions such as high humidity. However, other thermoset resins, such as silicones, cyanate esters, and cyanoacrylates, can also be used. Another option is thermoplastic resins. (Licari, 2005) Silver is the most commonly used filler material (Lau, 2003). The popularity of silver is due to its excellent conductivity and chemical stability (Morris 2005). Moreover, its oxide is highly conductive.

ICAs have been used in the electronics industry mainly as die-attach adhesives. However, lately they have also been proposed as an alternative to solders in surface mount and flip chip applications. For use in flip chip applications ICAs need to be carefully applied only on those areas which need to be conductive. Additionally, spreading of the adhesive should be prevented during the bonding process. A separate underfilling step is needed to improve the reliability of the joints. (Lau, 2003; Li, 2006) A typical cross section of an ICA flip chip joint is presented in Figure 1.

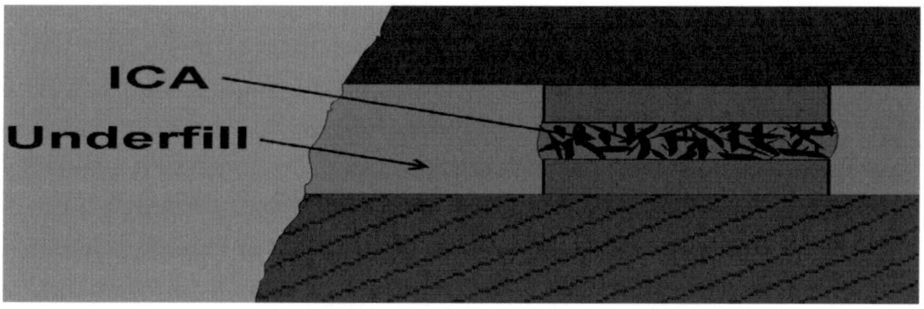

Fig. 1. Schematic illustration of an ICA flip chip joint with underfill.

2.2 Anisotropic conductive adhesives

In an anisotropic conductive adhesive (ACA) the concentration of conductive particles is below the percolation threshold (Lau, 1995; Licari, 2005) and the adhesive does not conduct before the interconnection is formed. Typically the number of particles is 0.5% - 5 % by volume (Licari, 2005) but depends largely on the size and shape of the conducting particles and on the application the ACA is used in (Watanabe, 2004). Normally the particles are randomly dispersed in the matrix, but adhesives having uniformly dispersed particles have also been developed (Ishibashi & Kimura, 1996; Jin et al., 1993; Sungwook & Chappell, 2010).

During the ACA attachment process the adhesive is placed between the mating contacts. The ACA interconnection is established by applying pressure and heat simultaneously to the interconnection. When the temperature is raised the adhesive matrix will transform into low viscosity fluid (Tan et al., 2004), which allows excess adhesive to flow from the joints and fill the spaces around the contacts forming a physical connection between the parts to be attached. The conductive particles are trapped between the contacts and deform forming an electrical connection. As a result, electrical conduction is restricted to the z-direction and the electrical insulation in x-y directions is maintained. During cooling residual stresses are formed as a result of contraction of the adhesive matrix. In addition, residual stresses form when the adhesive shrinks during curing. However, it has been shown that the residual stresses formed during cooling dominate (Kwon & Paik, 2004). This contraction builds up a sufficient force to create a stable, low-resistance connection. A typical cross section of an ACA flip chip joint is shown in Figure 2.

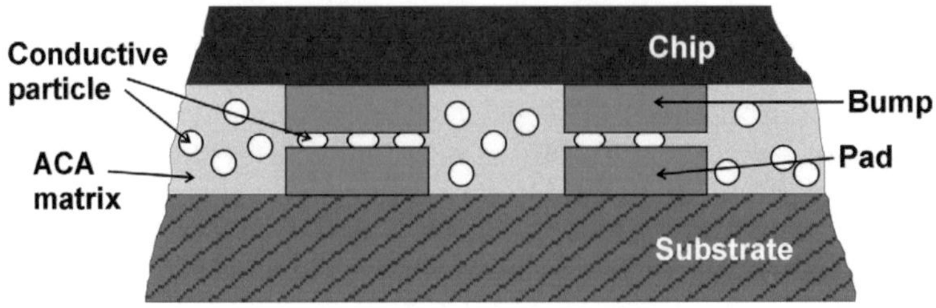

Fig. 2. Schematic illustration of an ACA joint between a chip and a substrate.

ACA joints have several advantages compared to underfilled solder interconnections. As the ACA process is solderless, there is no lead or alpha emission (Zhong, 2005). Moreover, the process is fluxless and no cleaning is required (Zhong, 2005). Furthermore, the process temperature is lower than that needed in soldering (Yim & Paik, 1998), which enables the use of heat sensitive or non-solderable substrate materials (Uddin et al., 2004). As the polymer matrix protects the contacts from mechanical damage and no underfilling is required (Lai & Liu, 1996), the ACA process costs less due to fewer processing steps (Yim & Paik, 2001). The ACA joining also enables very high interconnection density. On the other hand, the ACA joint has higher contact resistance and lower current capability than that made with solder (Jim & Paik, 1998; Zhong, 2005). Since the ACA has no self-alignment capability, a special bonding machine is needed for accurate alignment. During the bonding process heat and pressure also need to be applied simultaneously.

2.2.1 Materials used in ACAs

Both thermoplastic and thermoset materials and their mixtures have been used as an ACA matrix. Initially, the ACAs were made of thermoplastic materials, as they have better reworkability and pot life (Lau, 1995). However, their stability at high temperatures is not good and the thermoplastic material is not strong enough to hold the conducting particles in position, which increases the contact resistance of the joint (Asai et al., 1995; Kim et al., 2004). Thermoset adhesives were developed to overcome the problems with thermoplastic adhesives. Thermoset adhesives are stable at high temperatures and enable low joint resistance. Epoxies are commonly used as an ACA matrix due to their good properties. Epoxies have excellent adhesion to a variety of substrates, due to the highly polar hydroxyl and ether groups (Luo, 2002). In addition, they have high glass transition temperature (T_g) and favourable melt viscosity (Kim et al., 2004; Yim & Paik, 1999). Furthermore, epoxies give low contact resistance and by selecting suitable curing agent long self-life and fast-cure properties can be achieved. The epoxy resin forms a crosslinked structure during bonding with good mechanical properties. However, their reworkability is problematic, as they are not thermally reversible and do not dissolve in common organic solvents (Lau, 1995).

The electrical conduction in ACA is formed by the conductive particles. The size, concentration and material of the conductive particles depend on the application area and on the manufacturer. Typically the conductive particles are approximately 3-10 um in size. Nowadays, the most common conductive particles are nickel, which may be gold plated, and metal plated polymer particles. However, other materials, such as carbon fibres and solder balls, have also been used (Asai et al., 1995). The polymer particles are made of polystyrene cross-linked with divinyl benzene (Asai et al., 1995) and the metal plating on them may be of nickel, silver, or gold (Liu, 1996). The polymer particles are pliant and during the bonding process they deform, thereby forming the connection to the contacts. The deformation of the rigid nickel particles during the bonding process is less than that of the soft particles and the contact area formed is smaller. However, if the bonded contacts are made of softer metal, such as gold or copper, these contacts deform during bonding increasing the contact area with the rigid particles (Divigalpitiya & Hogerton, 2004; Yim & Paik, 1998; Frisk & Ristolainen, 2005; Frisk & Kokko, 2006).

In the flip chip process the interconnection is formed between pads on the substrate and bumps on the chip. However, as the cost of bumping may be unattractive in certain applications, bumpless chips are also used. The most commonly used bump materials are gold plated nickel, and gold. The gold bumps can be manufactured using an electroplating process. Copper bumps formed by similar electroplating techniques have also been considered as an alternative to gold because of their lower cost (Lau, 2000, Lau, 2003). However, copper oxidizes and corrodes easily, which may cause problems if it is used without plating. The nickel bumps can be made using an electroless plating process. This process has high potential for cost reduction, as it enables metal deposition directly on the aluminium pads on the chips. Thus the costly equipment needed in the electroplating process for sputtering, photoresist imaging, and electroplating is eliminated. The gold bumps may also be processed using a modified wire bonder to form stud bumps. The advantage of this process is that bumps can be formed on single chips in addition to whole wafers. (Lau, 1995)

2.2.2 The ACA bonding process

ACAs can be used either as films (ACF) or as pastes (ACA or ACP). The type of the ACA affects the bonding process and the equipment needed. ACFs are typically supplied in a reel and a dedicated in-line bonding machine is needed for cutting, aligning, and tacking to achieve high assembly speed. On the other hand, ACPs can be applied either by printing or by dispensing using a syringe. Even though the ACF process requires special equipment, ACFs are often used as they offer advantages compared to ACPs. The ACF process consumes less material than the ACP process. Moreover, the ACP process may destroy the randomness of the particle distribution leading to problems in process quality.

In the ACF bonding process the adhesive film is first cut to the correct size to cover the bonding area. After this the adhesive is aligned to the substrate and pretacked using light pressure and low temperature to attach the ACF to the substrate. After pretacking the carrier film on the ACF is removed. Next a chip is picked by a flip chip bonder. Typically a special flip chip bonder capable of simultaneously applying pressure and heat is used. The bumps on the chip and the pads on the substrate are aligned. The chip is pressed onto the substrate and heat is applied to the chip and the polymer matrix is cured. A schematic illustration of the bonding process used is presented in Figure 3. In case ACP material is used the steps a and b are replaced by deposition of the ACP. After this steps c and d are performed similarly to the ACF process.

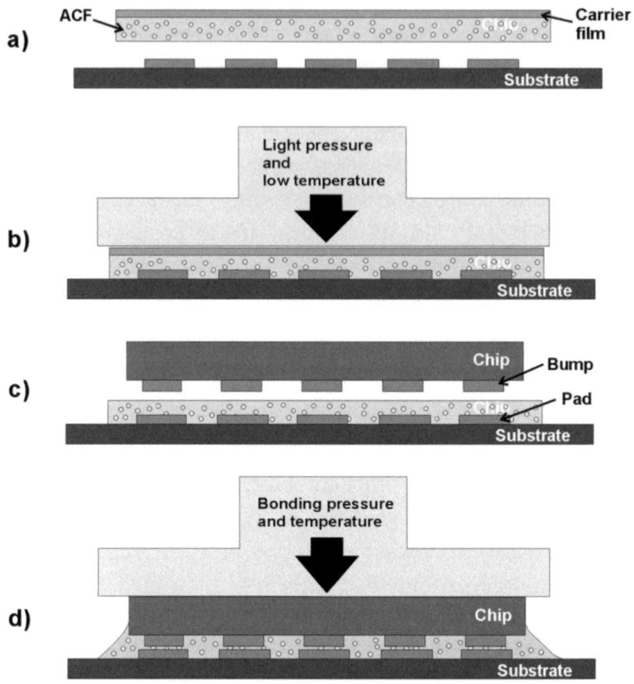

Fig. 3. Schematic illustration of the ACF flip chip bonding process: a) placement of ACF on the substrate, b) prebonding, c) alignment of the chip and the substrate and d) final bonding.

3. Evaluation of the quality of ACA interconnections

The quality of the ACA interconnections may be studied using several techniques. In general, a good quality ACA interconnection is characterized by low contact resistance and good mechanical properties. Therefore, electrical measurements may be done to assess the quality. However, often this is not possible due to the design of the semiconductor chip. Electrical measurements may show alignment and planarity problems as higher resistance values. However, this is not always the case, as it is possible that the electrical connection seems good even though there are problems in the joints causing reliability problems during use. The alignment of an interconnection may be studied through the substrate in the case of transparent substrates such as glass or thin polyimide film. However, often the most effective way to examine both alignment and the structure of the interconnection is to make a cross-section of the structure and study them using either optical or scanning electron microscope. SEM especially is often very a effective tool yielding a plethora of information of the joint, which is important for optimisation of the bonding process.

The mechanical properties of the ACA interconnections may be studied using adhesion testing, which will indicate how well the ACA material is attached to the substrate and the chip. In this technique both adhesion strength and failure mechanism during testing will give valuable information. In the following several different parameters affecting the quality of ACA interconnections are discussed. Special attention is paid to the information obtained from the interconnections using SEM analysis.

3.1 SEM analysis of ACA interconnections

As mentioned above, both optical and scanning electron microscopy may be used to study the cross-sections of ACA interconnections. Although, studying the cross-sections is typically very effective it has some drawbacks. The number of particles in ACA interconnections is typically quite low. Thus, when a cross-section of an interconnection is studied, the probability of seeing particles in interconnections is small even if there are sufficient particles in the interconnections to ensure proper joining. From a cross-section only one side of a chip is seen. Therefore, to determine planarity and alignment issues several cross-sections are needed. Furthermore, making a cross-section of a sample destroys it thereby gravely restricting further analysis. Optical microscopy may give valuable information especially when planarity and alignment are concerned. However, typically it does not give very good detailed information. Especially, if interfaces of different materials are studied an optical microscope does not give reliable information. On the other hand, SEM is often a very powerful tool to determine the quality and structure of an ACA interconnection. However, SEM analysis benefits from information of other analysis methods if they are available such as, for example, electrical characterisation and scanning acoustic microscopy.

For SEM analysis the quality of the cross-sections needs to be good and they need to be clean. Typically epoxy moulding is used followed by grinding and polishing. However, other materials are possible such as acrylics. Use of high temperature mould materials may cause problems depending on the structure studied and its materials. As the ACA structure has many different materials the analysis is often challenging and the parameters used for SEM need to be determined according to the samples studied. Additionally, the area of interest in the interconnection affects the parameters. Both thin gold and carbon layers may be used to make samples electrically conductive. However, gold gives better quality of analysis and is recommended if elemental analysis is not needed.

3.2 Bonding parameters

Bonding parameters are the key factors when the quality and the reliability of the ACA interconnections are considered. The ACA process makes the bonding parameters especially important as the complex mechanical, rheological, and chemical properties of the ACA materials need to be considered (Dou, 2006). The most important parameters in the ACA bonding process are time, temperature and pressure. However, other bonding parameters, such as application rates of pressure and temperature, also affect the quality and reliability of the joints (Ogunjimi et al., 1996; Whalley et al., 1997). Moreover, the parameters may interact with each other.

Finding optimum process parameters necessitates careful study. Quite often the quality of the bonding parameters cannot be determined on the basis of electrical or adhesion measurements only and cross-sections are needed for verification of the interconnection quality. For thermoset adhesives the bonding temperature together with the bonding time determine the degree of cure of the adhesive matrix. The higher the bonding temperature used the higher the degree of cure of the adhesive matrix is within the same amount of time (Chan & Luk, 2002a; Rizvi et al., 2005; Tan et al., 2004; Wu et al., 1997) as the higher temperature accelerates the crosslinking reaction (Uddin et al., 2004). Similarly, longer bonding time increases the degree of cure. Both bonding time and temperature needs to be high enough to achieve adequate curing of the adhesive matrix as the mechanical and chemical properties of the ACA have been found to depend heavily on the degree of cure (Wu et al., 1997). Too low degree of cure is often seen as high contact resistance values and also as inadequate adhesive strength.

The bonding pressure determines the deformation of the conductive particles. If the bonding pressure is too low the conductive particles cannot make good contact with the bonded surfaces and the contact resistance will be high (Chan & Luk, 2002b; Lau, 1995). In addition, the reliability of this kind of joint is poor (Frisk & Ristolainen, 2005; Lai & Liu, 1997). Examples of only slightly deformed particles are shown in Figure 4. In this case the variation is assumed to be caused by too fast curing of the adhesive during the bonding process. This adhesive has been designed to cure very quickly. Consequently it may have started to cure before the adhesive had flowed properly, leaving the particles insufficiently deformed.

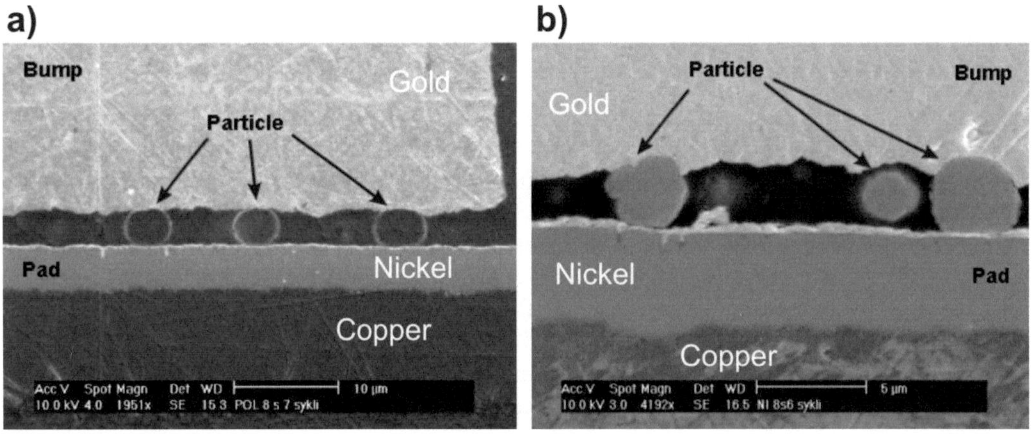

Fig. 4. An example of slightly deformed particles: a) gold coated polymer particles and b) nickel particles.

When the bonding pressure is increased, the contact resistance typically decreases sharply at first before evening out (Yim & Paik, 1998; Yin et al., 2003). This is caused by greater deformation of the conductive particles leading to a larger contact area between the particle and contacts (Kwon et al., 2006; Yin et al., 2003). Figure 5 shows examples of properly deformed particles. However, if the pressure is increased too much, the contact resistance may start to increase again (Chan & Luk, 2002b; Yim & Paik, 1998). If metal plated polymer particles are used, too high pressure may crush the particles (Wang et al., 1998) and lead to direct contact between the bump and the pad. The cracking of the metal plating may separate it from the polymer core and reduce the amount of conductive path between the pad and the bump (Yim & Paik, 1998). Another problem with too high bonding pressure is elastic stress formed in the chip or in the substrate (Frisk et al., 2010; Lai & Liu, 1996). In some cases, the high temperature used during the bonding process may soften the substrate material used and it will deform markedly during the bonding process.

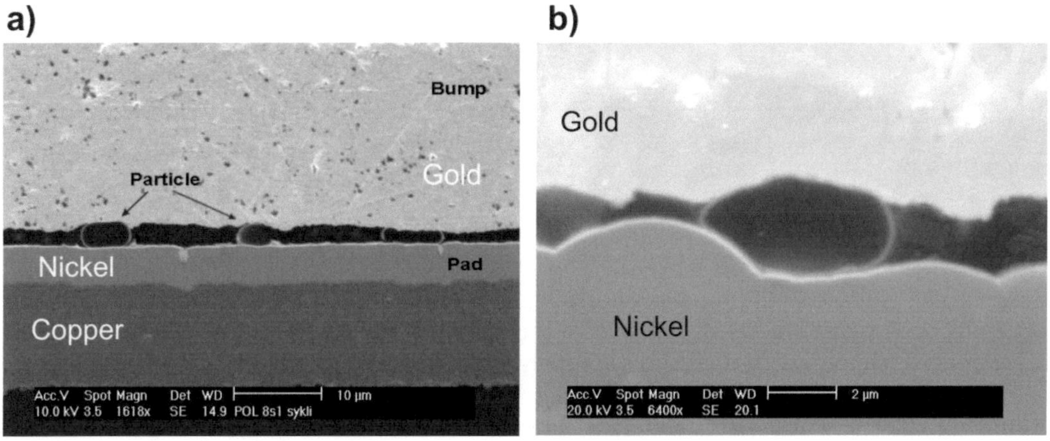

Fig. 5. a) and b) examples of marked deformation of gold plated polymer particles.

If rigid particles are used, the bump and the pad material also have a strong effect on the bonding pressure. With soft bump materials, such as gold or copper, the particles will sink into the bump forming a strong bond. If high bonding pressure is used, the particles will sink completely into the bumps and direct contact between the pad and the bump will be formed. An example of direct contact is presented in Figure 6. The hard nickel-plating on the pads prevented the penetration of the particles into the pads.

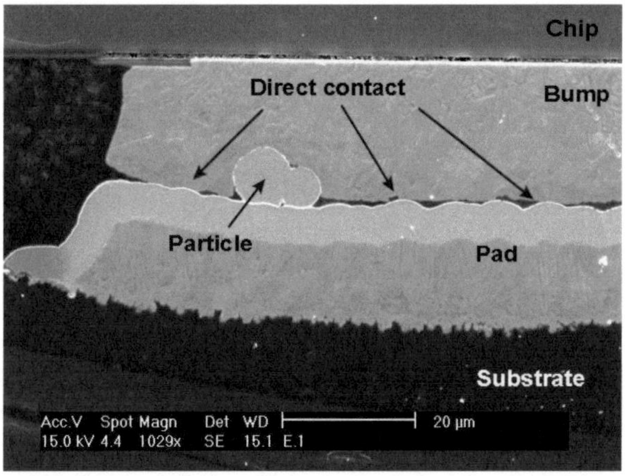

Fig. 6. Penetration of the nickel particles into the copper bump and direct contact between the bump and the pad.

3.3 Effect of the substrate material

Poor substrate and chip quality may cause coplanarity problems. This may increase the contact resistance, as not all joints have adequate deformation of the particles. For the flip chip process to be usable with the organic substrates, the planarity of the substrate is very important. Planarity issues often need to be determined using cross-sections. In the following examples of SEM analysis used for quality studies of ACA interconnections with different substrate materials are given.

3.3.1 Glass fibre reinforced substrates

Glass fibre reinforced materials are commonly used as substrates in electronics. The most widely used material is FR-4, which is a grade designated by the National Electrical Manufacturers Association (NEMA), which determines that the material is flame resistant, is primarily epoxy based, and has woven glass fibre reinforcement (Coombs, 2000). As the resin material between different FR-4s may vary, the properties of FR-4 materials are not identical. For example typical T_g of FR-4 is between 130 and 140 °C or between 170 and 180 °C. The popularity of the FR-4 is based on its good properties, availability, and low cost. When rigid glass fibre reinforced substrates are used in ACA applications, coplanarity problems may arise due to the woven structure of the substrate (Frisk & Cumini, 2009). During the bonding process the high temperature may soften the resin, which leads to its deformation under the pads. This deformation has been found to depend on the orientation of the glass fibres in the substrate and affect the electrical conductivity and reliability of the joints (Liu et al., 1999).

Deformation of substrate in ACA interconnections was studied using cross-sections and SEM (Frisk & Kokko, 2006; Frisk & Cumini, 2009; Frisk et al., 2010). SEM has proven to be a very effective method for such studies as the different materials and their interfaces may be clearly seen. Figure 7 shows micrographs of a FR-4 substrates after an ACA bonding process. Deformation of the substrates may be seen between contacts. Deformation is especially considerable in the areas where the glass fibres were far from the surface. This varying deformation of the substrate causes pressure variation in the joints (Pinardi, 1998). The varying pressure is important as it may result in different deformation of the particles leading to variation in the contact resistance and also impairing the reliability of the joints.

Quite often in flip chip applications with ACA materials very high wiring densities are needed. These are difficult to achieve with the traditional FR-4 substrates shown in Figure 7 One possibility to meet these demands is to use sequential build-up (SBU) processes. However, in this process conductor and dielectric layers are formed one after another on a rigid core board, which may be an FR-4 glass reinforced laminate (Tagagi et al., 2003). The electrical connection between the core board and the build-up layers is formed using microvia technologies. An example of a substrate made with the SBU process is presented in Figure 8. The typical dielectric materials used in SBU build-up layers are resin-coated copper foil (RCC or RCF), thermally cured resin, and photo-imageable resin (Tagagi et al., 2003). The most widely used dielectric material in the SBU process is RCC. The RCC is formed by adding a layer of resin to a thin copper foil, which is laminated to the core board. A typical resin material is epoxy. The RCC has several advantages and is suitable for processing in standard printed circuit board processes.

As the RCC layer does not have glass fibres, it is more pliable than an FR-4 substrate. Its effect on the interconnection structures was studied using cross-sections and SEM (Frisk & Kokko, 2006). During the bonding process the depression of the copper pads into the RCC was found to be much stronger than the depression into the FR-4 substrate. In Figure 9 an example of the depression is presented for test samples with the RCC test substrate. As can be seen, the RCC has deformed markedly more during the bonding process than the pure FR-4 substrates shown in Figure 7. Furthermore, some deformation of the FR-4 substrate beneath the RCC has occurred. Although the deformation of the RCC is greater, it is almost identical under every pad leading to more uniform distribution of pressure. This causes the deformation of the particles to be identical in the joints and increases reliability compared to the substrate without the RCC.

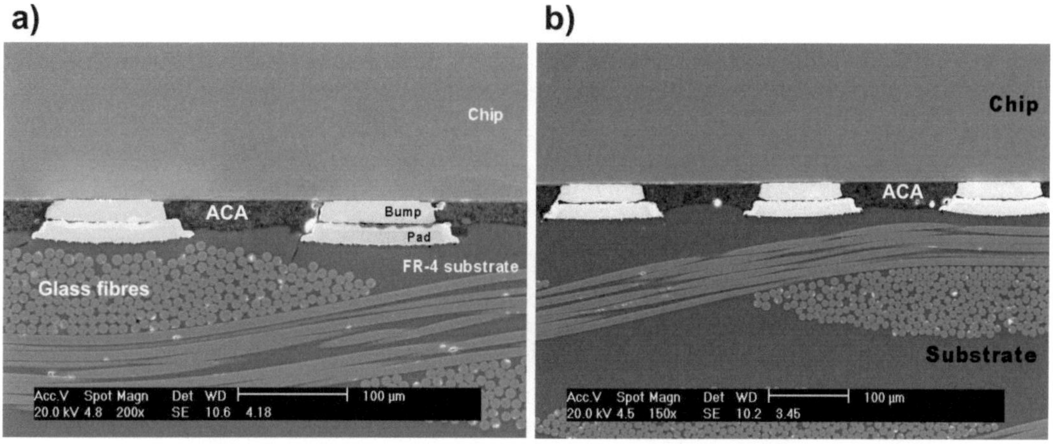

Fig. 7. a) and b) micrographs showing deformation of a FR-4 substrate

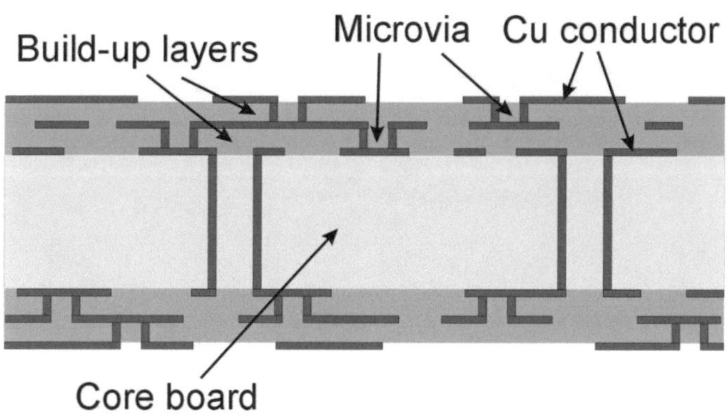

Fig. 8. Schematic cross-section of a printed circuit board made with SBU process.

Deformation of the particles between these test boards was also studied with SEM. With the test substrate, which did not have the RCC, the particles sank into the copper bump. However, the fairly thick nickel plating on the pads prevented the particles from sinking into it and caused deformation of the particles, as can be seen in Figure 10 a). In addition, due to relatively high magnification of the SEM micrograph the gold layers on both conducting particles and pad can be easily seen. The thinner nickel plating on the RCC test board enabled the particles to sink into the pad, as can be seen in Figure 10 b). Moreover, as the RCC gives in more under the pads during bonding than the FR-4 substrate, the deformation of the particles was less on the substrates with the RCC. Deformation of the particles is important for the reliability of the joints. Sufficiently deformed particles form a strong atomic interaction between the pad and the bump creating increased stability of the joint (Lai & Liu, 1996). Using SEM the interface between the particles and the pad or the bump may be studied and problems such as thin layers of polymer matrix may be detected. In both Figures 10 a) and 10 b) a good contact of particles is seen to both the pad and the bump .

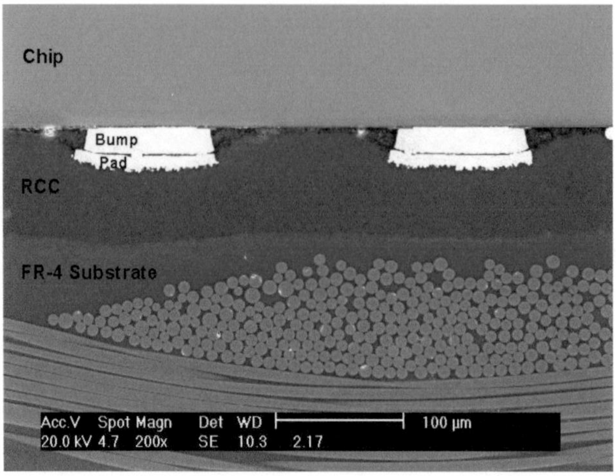

Fig. 9. Micrograph presenting the immersion of the pads in the RCC, when high bonding pressure is used.

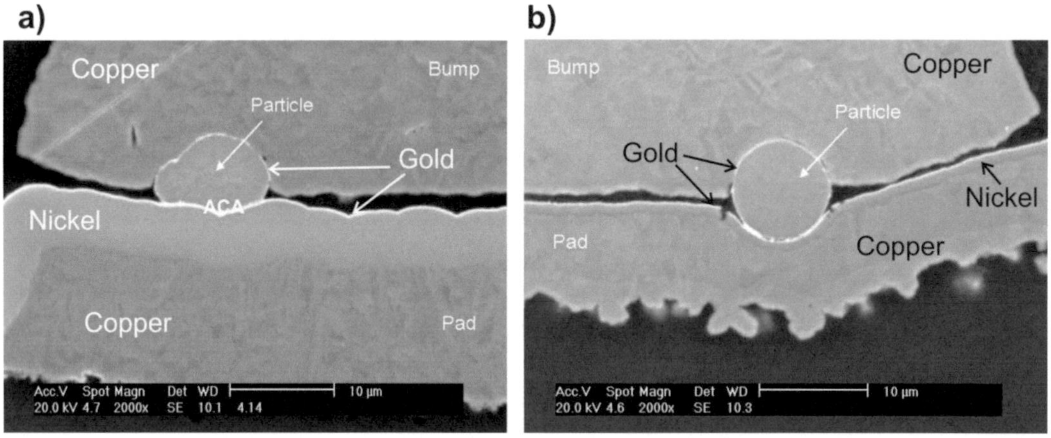

Fig. 10. a) Micrograph presenting the deformation of the rigid nickel particle, when a FR-4 substrate is used. b) Micrograph presenting the immersion of the rigid nickel particle in the pad and the bump, when the substrate with the RCC is used.

3.3.2 Flexible substrates

ACA materials are often used with flexible substrates, which are fabricated using pliable unreinforced polymeric materials. Flexible substrates have several advantages compared to fibre-reinforced substrates and lighter and thinner products can be produced using them. Flexible substrate may absorb stress, which may be important for the reliability of the interconnections, especially in flip chip applications. In addition, the thermal transfer through a thin substrate is more effective. Furthermore, very high density substrates are available with flexible substrates and this is often critical for ACA applications. On the other hand, the thinness of the substrate may cause problems in the stability of the construction. Moreover, the cost of the flexible substrates is higher than that of the rigid substrates. (Coombs, 2001)

The pliability of the flexible substrates may cause some problems during the bonding process. Figure 11 shows SEM micrographs of ACA interconnections with flexible liquid crystal polymer (LCP) substrates with two different pressures. A marked deformation of the liquid crystal polymer film with high pressure may be seen. Such deformation may cause problems. On the other hand, it may also even out some planarity problems as deformation can absorb the height variations (Connell, 1997; Savolainen, 2004) and thereby increase the quality of the interconnection. However, in some cases deformation may cause cracking of the wiring and thereby lead in to reliability problems. With the lower pressure the pressure exerted to the particles may not be high enough and therefore cause reliability problems.

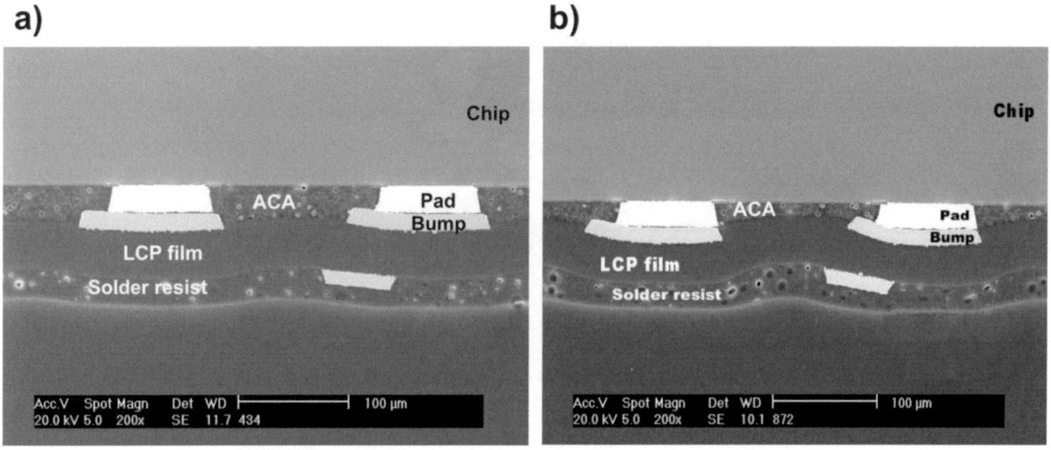

Fig. 11. An example of deformation of the LCP substrate when a) low bonding pressure and b) high bonding pressure was used.

With thin flexible substrates routing may also be critical to the distribution of the pressure, if the substrate has several conductive layers (Lai & Liu, 1996). If double sided flexible substrates are used they may have wiring on both sides of the bonding area. This may cause uneven distribution of the pressure and deformation of contact areas. Figure 12 shows an interconnection with a polyimide substrate having double sided wiring. As can be seen, the gold bump has markedly deformed during the bonding process because of the wiring on the other side of the substrate. A similar effect may be seen in the LCP substrates in Figure 11. However, as the solder resist on the LCP substrate evened out the effect of the wiring, there is clearly less deformation. Consequently, the design of flexible circuitry is very important for good quality interconnections. Furthermore, such quality problems are difficult to detect without cross-sectioning.

Another problem in a substrate may be overetching of the pads. As flexible substrates are often used in application where very high density substrates are needed such as attachment of driver chips is display application, overetching may cause marked problems. It reduces the contact area and may cause alignment and planarity problems. Figure 13 shows a SEM micrograph of a interconnection with an overetched polyimide (PI) substrate. In the original substrate layout the pad width was designed to be slightly greater than the bump. However, due to overetching the size of the pad is clearly less than that of the bump. Such overetching may be seen when substrates are quality checked before use. However, cross-section is a good way to evaluate the effect of overetching on an interconnect.

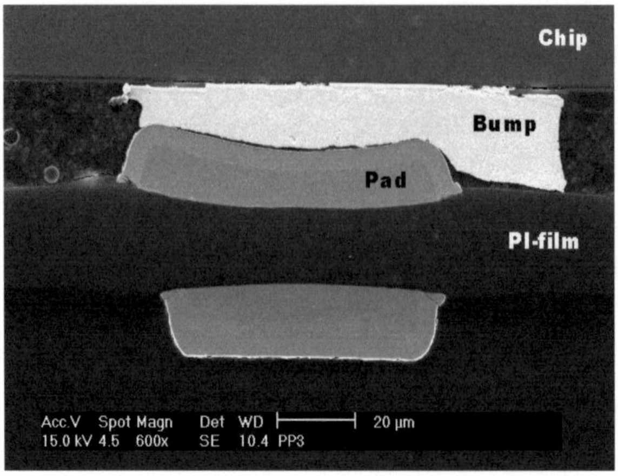

Fig. 12. Effect of the tracks on joint deformation on the PI substrate when the tracks are on both sides of the substrate.

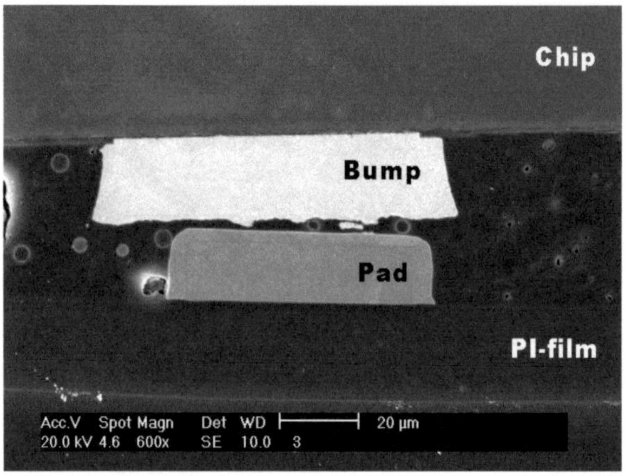

Fig. 13. An overetched pad on the PI substrate.

4. Failure mechanisms of the ACA flip chip joints

In general, a good ACA joint is characterized by low contact resistance. A key issue for the ACA to function properly is the retention of this contact resistance during the operational life of the product. Failure of a product can be defined as its inability to perform its intended function [For an ACA flip chip joint this typically means too great an increase in contact resistance. There is no single specific definition of failure of ACA joint resistance, as this depends largely on the application.

To improve the reliability of interconnections the reasons for their failure and the failure mechanisms must be understood. During the design phase reliability is typically assessed using accelerated environmental testing. The aim of such testing is to predict the future performance of a product in a shorter period of time than the service life of the product. Accelerated life tests can also be used to detect failure mechanisms occurring in products under different conditions of use. The acceleration is accomplished by using elevated stress levels or higher stress cycle frequency during testing compared to those under normal operational conditions of the product (Suhir, 2002). Depending on the condition failures may occur through several mechanisms. It is important that the testing conditions are determined so that the failure mechanisms during testing are similar to those occurring under normal conditions of use. The test conditions depend decisively on the application of the product. For example, the test conditions for products used in military and space applications are much more rigorous than those for consumer electronics.

Several different accelerated life tests have been used to study the reliability of ACA joints; see for example (Frisk & Ristolainen, 2005; Frisk & Cumini, 2006; Jang et al., 2008; Kim et al., 2004; Lai & Liu, 1996; Saarinen et al., 2011). These include high temperature aging tests, temperature cycling tests, high temperature and high humidity tests, and humidity and temperature cycling tests. The reliability and failure mechanisms of ACA interconnections depend on several factors, including the materials and bonding parameters. The materials used in the pads, bumps, and conductive particles need to be compatible with each other. The substrate material may also have a marked influence on the reliability of the joints. As the properties of the joints are much influenced by the bonding process, it is important that optimum bonding parameters are determined and used in the bonding process. In the following failure mechanisms and analysis of ACA interconnections are discussed.

4.1 Failures of ACA interconnections with rigid FR-4 substrates under thermal cycling

One of the major problems in ACA assembly using organic substrates is the great difference between the coefficient of thermal expansions (CTE) of chip and substrate. The bonding process is done at elevated temperatures. When the package is cooled down the contraction of the silicon chip is very small due to its low coefficient of thermal expansion and high Young's modulus (Kwon et al., 2005). On the other hand, the contraction of the substrate is much greater. At low temperatures this causes stresses to form between the chip and the substrate and the ACA flip chip package to warp downwards. At temperatures below its T_g the adhesive matrix holds the chip and the substrate together enabling this warpage (Kwon & Paik, 2006). However, when the T_g of the adhesive is exceeded, its mechanical strength decreases and it cannot provide mechanical support between the substrate and the chip. The warpage is evened out and both chip and substrate may expand with their inherent CTEs (Kwon & Paik, 2006). The warpage of the ACA package is presented in Figure 14. If the ambient temperature fluctuates, as in a temperature cycling test, the flip chip package warps repeatedly. The warping reduces the shear stress in the joints and a greater degree of warp has been reported to decrease the shear strain in the joints (Kwon et al., 2005).

The amount of warpage depends on the difference between the coefficient of thermal expansions of chip and substrate, but also on their stiffness. When a rigid fibre reinforced substrate is used, the shear stress caused by CTE mismatches is localised between the pad and the bump as the of deformation the substrate is less than that of the adhesive matrix (Lai et al., 1998). On the other hand, deformation of an unreinforced flexible substrate occurs much more easily due to its low modulus, and some of the shear stress may be absorbed by the substrate (Connell et al., 1997). This decreases the shear stress in the interconnection reducing the delamination and increasing the reliability.

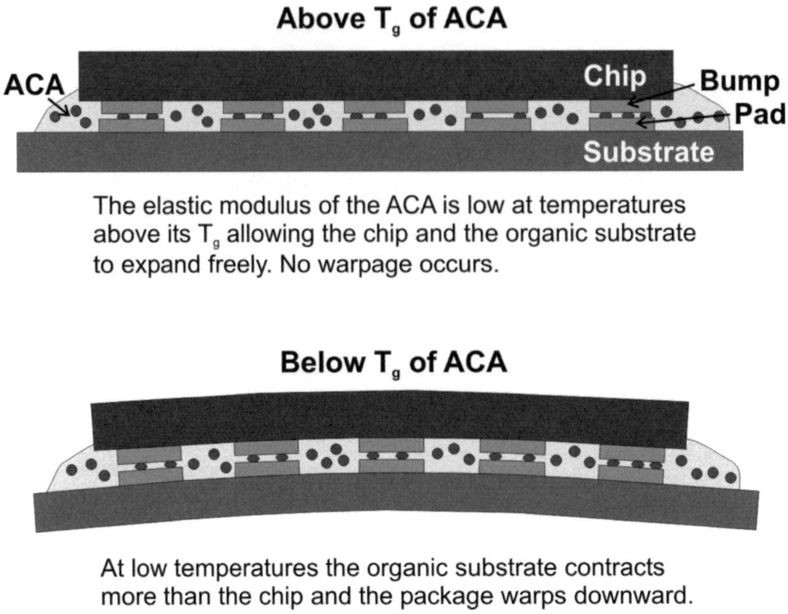

Fig. 14. Warpage of the ACA flip chip package during temperature cycling test.

ACA interconnections with rigid FR-4 substrate were exposed to thermal cycling between -40 °C and +125 °C (Frisk & Kokko, 2006). When the failed test samples were studied using SEM, delamination was found in several test samples after testing. This delamination is probably caused by shear stress between the pad and the bump formed due to differences in the coefficient of thermal expansion of the substrate and the chip. As shown in Figures 16 a), 16 b), and 16 c) with SEM the delamination is seen clearly and its place can be determined. In some cases delamination may be less pronounced and therefore harder to detect. In Figure 16 c) another example of delamination which is more difficult to detect is shown.

Another phenomenon seen in these samples was cracking of the substrate material. This is assumed to be caused by repeated warping of the substrate as the duration of testing was several thousand cycles. This type of cracking was typically found in areas where the glass fibres were far from the surface of the substrate. They started from the corners of the pads and often continued into the substrate until they reached the glass fibres. Examples of such cracking are presented in Figure 16. The formation of such cracks has probably facilitated delamination between pad and the bumb by providing sites for crack initiation, as the cracks often connected to the delamination between the pad and the bump. An example of this is presented in Figures 16 b) and 16 c).

It has been suggested that during thermal testing above the T_g of the ACA matrix sliding between the pad and the bump occurs (Kwon et al, 2006; Uddin et al., 2004). This may break the conductive particles and lead to failure. Furthermore, this phenomenon may cause fretting and thereby the formation of oxides on the conductive surfaces, which increases the resistance of the joints above an acceptable value. When ACA interconnections having nickel particles or gold coated polymer particles, were studied after thermal cycling clear indication of this phenomenon was seen. Figure 17 a) shows a SEM micrograph of polymer particles form a failed ACA interconnection. In the picture cracking of the gold layer in the particle can be clearly seen. Additionally, agglomeration of material under the particles can be seen indicating failure caused by this sliding. Figure 17 b) shows a similar situation for nickel particle and similar agglomeration of material.

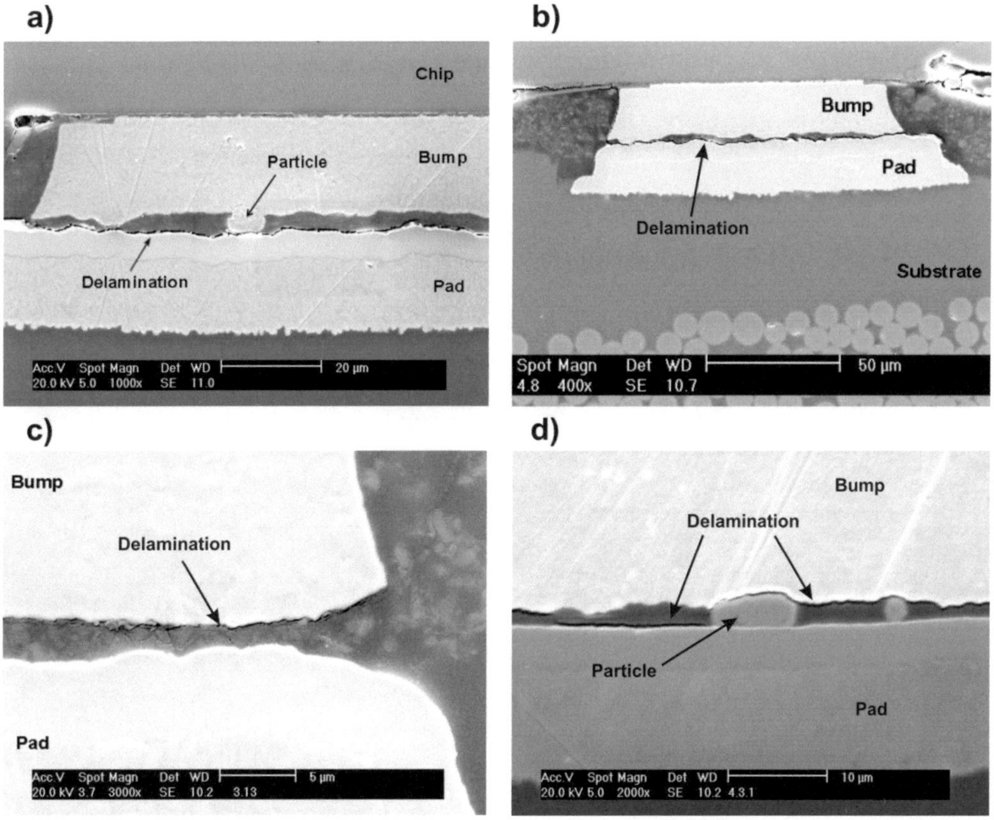

Fig. 15. Examples of delamination in test samples with FR-4 substrates after thermal cycling a) clear delamination between ACA and pad, b) delamination between bump and pad which continues to ACA-chip interface, c) less pronounced delamination, and d) delamination varying between the ACA-pad interface and the ACA-bump interface.

4.2 Failures of ACA interconnections with rigid FR-4 substrates under humidity testing

In addition to temperature changes, humidity has been found to have a major influence on the reliability of ACA interconnections. Under humid conditions the adhesive matrix may deform as it absorbs water. The adhesive matrix may also relax due to the increased temperature. The effect of water absorption depends on the structure of the adhesive and also on the duration of the exposure. The most common adhesive material is epoxy. It has been suggested that the water absorbed in the epoxy polymer has two states to which the water molecules can diffuse. The water may either fill the free volume in the polymer matrix or form hydrogen bonds with the epoxy polymer. If the water is hydrogen-bonded, it causes swelling of the adhesive matrix (Chiang & Fernandez-Garcia, 2002; Luo et al., 2002). This swelling due to humidity typically increases with temperature, but decreases sharply across the T_g of the polymer (Wong et al., 2000). Swelling due to moisture differs significantly between the materials used in flip chip packages causing the formation of hygroscopic stresses in the structure (Mercado et al., 2003; Wong et al., 2000).

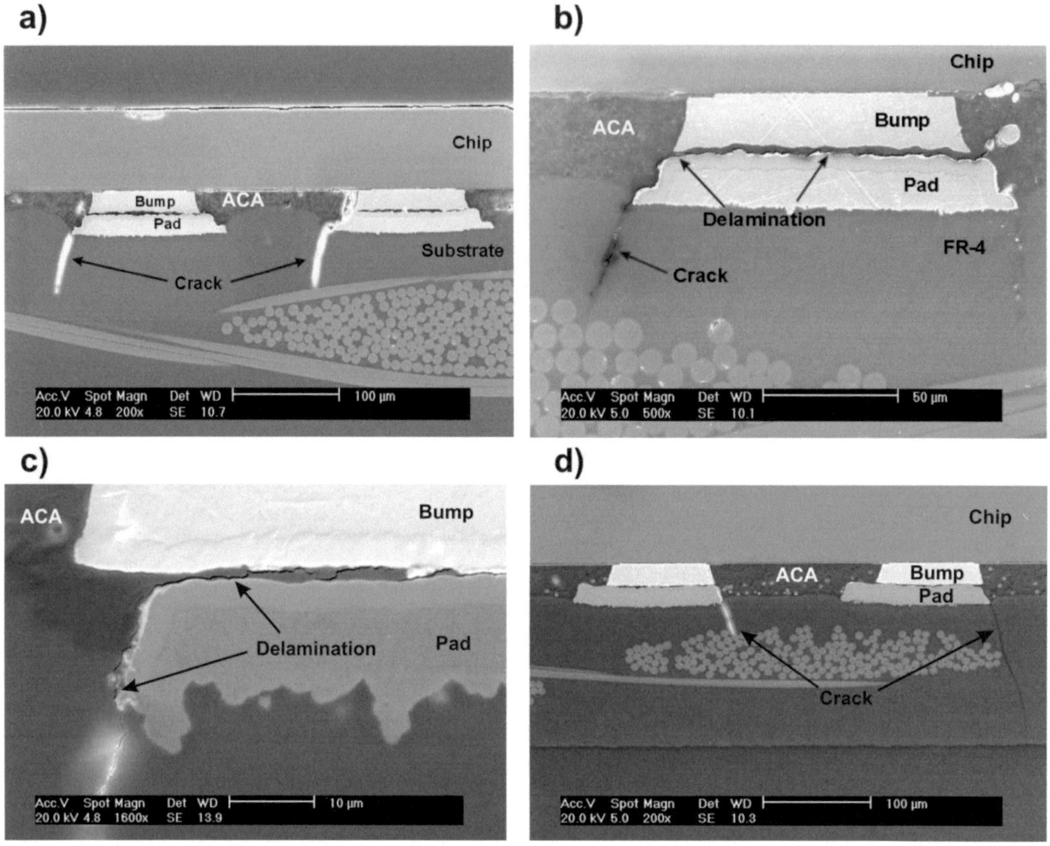

Fig. 16. Examples of cracks in FR-4 substrates: a) marked deformation and clear cracking of the epoxy matrix, b) cracking which continues between the pad and the bump, c) less pronounced cracking which continues between the pad and the bump, and d) cracking of thin FR-4 substrate

The swelling of the adhesive matrix may be marked (Mercado et al., 2003) and concurrent with thermal expansion may cause the conductive particles to lose contact. However, it has been reported that the absorbed water may also weaken the mechanical properties of the adhesive. Unlike the hydrogen-bonded water, the water filling the free volume in the epoxy polymer does not cause swelling, as it occupies a volume that already exists (Chiang & Fernandez-Garcia, 2002). However, it acts as a plasticizer affecting the mobility of the chains and increasing chain flexibility, which decreases the T_g of the polymer. As the water acts like a plasticizer it may also impair the mechanical strength of the adhesive.

The substrate material may also have a great influence on the reliability of the joints under humid conditions (Frisk & Cumini, 2006). If the substrate material absorbs water, it penetrates the interfaces more easily and may cause delamination. A large amount of moisture in the substrate also facilitates the moisture absorption of the adhesive matrix, and may accelerate the formation of moisture related failures. The effect of humidity at elevated temperatures on the ACA joints was studied using an 85°C/85RH test. Flexible polyimide was used as a substrate material. When studied after testing using SEM, every test sample with the PI substrate which showed an open interconnection after testing also showed delamination. An example of the delamination is presented in Figure 18. The moisture absorption of polyimide is marked and it is assumed to be the reason for the formation of delamination during testing.

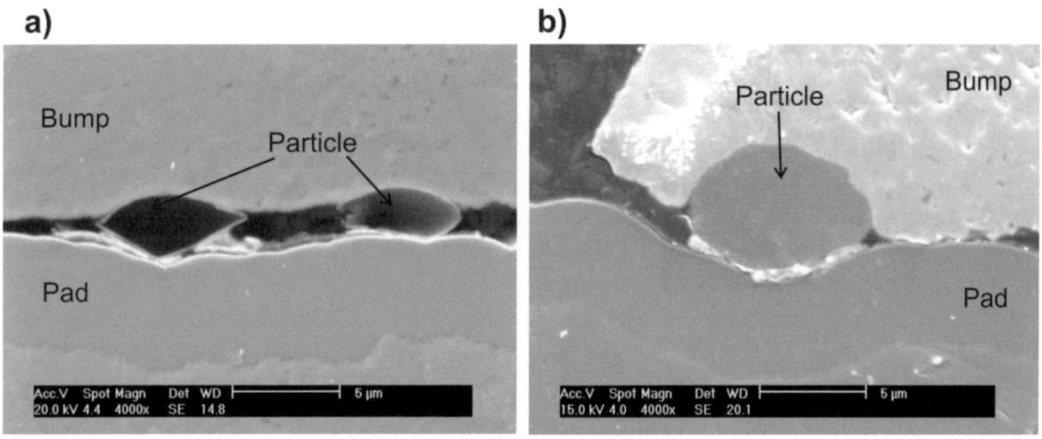

Fig. 17. Micrograph of a failed ACA interconnection after thermal cycling a) with polymer particles and b) with nickel particles.

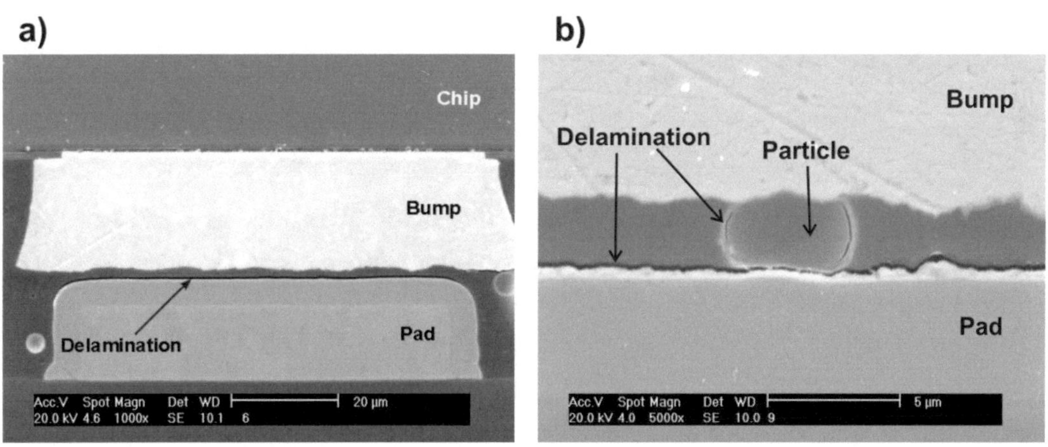

Fig. 18. a) An example of delamination after constant humidity testing on the PI substrate.b) close view of delamination after constant humidity testing on the PI substrate and on particle.

ACA technique was also studied with thinned silicon chips. When silicon chips are thinned below 100 um, they become pliant and can be used in solutions where they are bent. Consequently, they can be used in flexible electronics. Thinning also allows the chips to dissipate more heat, which is important when the densities of the packages increase.

However, thinned chips have certain drawbacks. They are more fragile than thicker chips, which needs to be taken into account when thinned chips are handled. Special tools may be needed since the fragile edges of the thinned chips are easily broken during handling. During the thinning and dicing processes a considerable amount of stress may be induced in the chips. It has been found that the thinning of the chips changes the shear stress distribution in a package (Frisk et al., 2011). In the analysis 50 μm thick chips were used instead of approximately 500 μm thick chips used in other studies. Very strong delamination was seen in these test samples which indicates failure mechanisms which is different from the one seen with the thicker chips. Examples of this delamination are shown in Figures 19 a) and 19 b). Furthermore cracking of the thinned chips was seen (Figure 19 c) and 19 d)). However, most of this occurred during the bonding process already before testing.

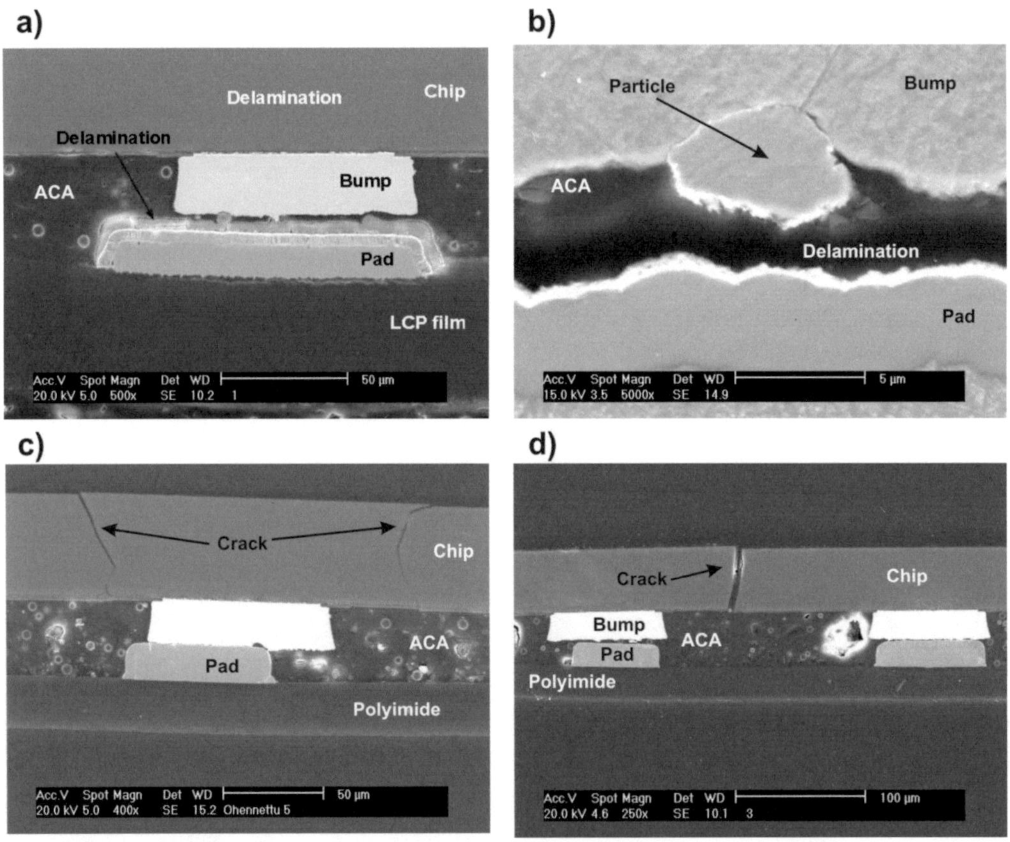

Fig. 19. Micrographs for ACA interconnections with thinned chips: a) marked delamination after humidity testing, b) close view of delamination under particle, c)and d) cracks in the thin chips.

5. Conclusion

Anisotropic conductive adhesives (ACA) are an interesting interconnection method for several applications. Due to the low cost of the ACA process and capability for high density they nowadays dominate many fields for instance attachment of chips in radiofrequency identification (RFID) tags and attachment of driver chips in display applications. ACA interconnections often show the typical trend in electronics for very small size with high functionality. In general this means high density of contacts in a chip and often also a large number of contacts per chip. Such applications are often very challenging both for studies of the interconnection process and the quality and reliability of the interconnections. Making cross-sections of the interconnections has proven to be efficient way to obtain detailed information about the interconnections structures, their quality and failure mechanisms, and this has been used effectively with other characterisation methods such as scanning acoustic microscope, DSC, and x-ray. Due to the small features currently common in electronics applications SEM is often the preferred method for examinations compared to optical microscopy.

In this chapter several SEM analyses of ACA interconnections were described. Many of these have been critical to both understanding the bonding process of these materials and also for the development of the ACA interconnection techniques and their reliability. Studying cross sections with SEM has been shown to be an effective way to analyse several of the failure mechanisms found in ACA interconnections. However, in general a good understanding of SEM analysis technique is needed for analysis due to the complexity of the ACA structure. ACA interconnection has many interfaces and the failure may occur on any of these or in the bulk materials of the structure. For effective analysis the critical parts need to be already understood when the analysis is made, and therefore, a good understanding of the technique and materials is needed. Lately the use of ACA technology has increased and it has been adopted on new areas for example high temperature electronics and sensor applications. In the future, this will increase the need for detailed knowledge of this interconnection technique. Therefore, it is critical that studies such as presented in this chapter are continued as they give vital information for both development and applicability of this technique.

Making cross-sections of studied samples is favoured in ACA flip chip applications due to the difficulties of studying the interconnections below a component. This also applies to many other interconnections and packaging technologies and similar methods have been used successfully in other applications. For example in the research of lead free solders cross-sections are systematically used for failure analysis and studies related to the microstructure of the interconnections. Additionally, in other techniques, such as flex on board attachments for example, in which flexible substrate is attached to a rigid substrate, and small packaging solutions such as Chip Scale Package (CSP) cross-sectioning and SEM analysis is often needed to determine the structure. In the future, the size of interconnections in electronics will decrease and their number will increase. As a consequence, the need for techniques capable for the analysis of such structures will increase markedly. SEM has proven to be an extremely useful tool for analysing electronics structures as it is relatively fast, typically easily available, and capable for analysing small features.

6. Acknowledgment

I would like to thank my colleagues Dr. Kati Kokko, M.Sc. Kirsi Saarinen, M.Sc. Janne Kiilunen, and M.Sc. Sanna Lahokallio, and, additionally, my former colleague Dr. Anne Cumini for their help in this research. Furthermore, I would like to thank the staff at the Institute of Material Science at Tampere University of Technology.

7. References

Asai, S., Saruta, U., Tobita, M., Takano, M., and Miyashita, Y. (1995) Development of an Anisotropic Conductive Adhesive Film (ACAF) from Epoxy Resins. *Journal of Applied Polymer Science*, Vol. 56, No. 7, 1995, pp. 769-77. ISSN 1097-4628.

Chan, Y., Hung, K., Tang, C., and Wu, C. (2000) Degradation Mechanisms of Anisotropic Conductive Adhesive Joints for Flip Chip on Flex applications. *Proceedings of 4th IEEE Conference on Adhesive Joining and Coating Technology in Electronics Manufacturing*, ISBN 0780364600, Espoo, Finland, June, 2000.

Chan, Y. C. and Luk, D. Y. (2002a) Effects of Bonding Parameters on the Reliability Performance of Anisotropic Conductive Adhesive Interconnects for Flip-chip-on-flex Packages Assembly I. Different Bonding Temperature. *Microelectronics Reliability*, Vol. 42, No. 8, 2002, pp.1185-94. ISSN 0026-2714

Chan, Y. C. and Luk, D. Y. (2002b) Effects of Bonding Parameters on the Reliability Performance of Anisotropic Conductive Adhesive Interconnects for Flip-chip-on-flex Packages Assembly II. Different Bonding Pressure. *Microelectronics Reliability*, Vol. 42, No. 8, 2002, pp.1195-1204. ISSN 0026-2714

Chiang, M. and Fernandez-Garcia, M. (2002) Relation of swelling and T_g depression to the apparent free volume of a particle-filled, epoxy-based adhesive. *Journal of Applied Polymer Science*, Vol. 87, No. 9, 2002, pp. 1436 - 44. ISSN 1097-4628

Connell, G., Zenner, R., and Gerber, J. (1997). Conductive Adhesive Flip Chip Bonding for Bumped and Unbumped Die. *Proceedings of the 47th IEEE Electronic Components and Technology Conference*, ISBN: 0-7803-3857-X, San Jose, CA, USA, May, 1997.

Coombs, C. (2001). *Coombs' Printed Circuits Handbook*. McGraw-Hill Professional, 5th edition, ISBN 0071350160.

Divigalpitiya, R. and Hogerton, P. (2004). Contact Resistance of Anisotropic Conductive Adhesives. *Journal of Microelectronics and Electronic Packaging*, Vol. 1, No. 3, 2004, pp. 194-9. ISSN 1551-4897.

Frisk, L. and Ristolainen, E. (2005) .Flip Chip Attachment on Flexible LCP-Substrate using an ACF. *Microelectronics Reliability*, Vol. 45, No. 3-4, 2005, pp. 583-8. ISSN 0026-2714

Frisk, L. and Kokko, K. (2006). The Effects of Chip and Substrate Thickness on the Reliability of ACA Bonded Flip Chip Joints. *Soldering and Surface Mount Technology*, Vol. 18, No. 4, 2006, pp. 28-37. ISSN 0954-0911

Frisk L. and A. Seppälä A. (2006). Reliability of Flip Chip Joints on LCP and PI Substrate. *Soldering and Surface Mount Technology*, Vol. 18, No. 4, 2006, pp. 12-20. ISSN 0954-0911

Frisk L. and Kokko K. (2007). Effect of RCC on the reliability of adhesive flip chip joints. *Journal of Electronic Packaging*, Vol. 129 No. 3, 2007, pp. 260-5. ISSN 1043-7398

Frisk, L. and Cumini, A. "Effect of Substrate Material and Thickness on Reliability of ACA Bonded Flip Chip Joints", Soldering and Surface Mount Technology, Vol. 21, No. 3, 2009 pp. 15-23.

Frisk L. and Cumini A. (2010). Reliability of ACF interconnections on FR-4 substrates. *IEEE Transactions on Components and Packaging Technologies*, Vol. 33, No. 1, 2010, pp. 138-147. ISSN 1521-3331

Frisk, L., Saarinen K., and Kokko K. (2011). Reliability of ACA joined thinned chips on rigid substrates under humid conditions. *Proceedings of European Microelectronics and Packaging Conference, EMPC*, Brighton, UK, September 2011.

Ishibashi, K. and Kimura, J. (1996). A New Anisotropic Conductive Film with Arrayed Conductive Particles. *IEEE Transactions on Components, Packaging, and Manufacturing Technology*, Vol. 19, No. 4, 1996, pp.752–7. ISSN 2156-3950

Jang K-W., C-K. Chung, W-S. Lee, K-W.Paik. (2008). Material properties of anisotropic conductive films (ACFs) and their flip chip assembly reliability in NAND flash memory applications. *Microelectronics Reliability*, Vol. 48, No. 7, 2008, pp. 1052-61. ISSN 0026-2714

Jin, S., Tiefel, T.H, Li-Ilan C., and Dahringer, D.W. (1993). Anisotropically Conductive Polymer Films with a Uniform Dispersion of Particles. *IEEE Transactions on Components, Hybrids, and Manufacturing Technology*, Vol. 16, No. 8, 1993, pp.972 - 977. ISSN 0148-6411

Kim, J., Kwon, S., and Ihm, D. (2004). Reliability and Thermodynamic Studied of an Anisotropic Conductive Adhesive Film (ACAF) Prepared from Epoxy/Rubber Resins. *Journal of Material Processing Technology*, Vol. 152, No. 3, 2004, pp.357-62. ISSN 0924-0136

Kwon, W. and Paik, K. (2004). Contraction stress build-up of anisotropic conductive films (ACFs) for flip-chip interconnection: Effect of thermal and mechanical properties of ACFs. *Journal of Applied Polymer Science*, Vol. 93, No. 6, 2004, pp. 2634-41. ISSN 1097-4628

Kwon, W., Yim, M., Paik, K., Ham, S., and Lee, S. (2005). Thermal Cycling Reliability and Delamination of Anisotropic Conductive Adhesives Flip Chip on Organic Substrates with Emphasis on the Thermal Deformation. *Journal of Electronic Packaging*, Vol. 127, No. 2., 2005, pp.86-90. ISSN 1043-7398

Kwon, W., Ham, S., and Paik, K. (2006). Deformation Mechanism and its Effect on Electrical Conductivity of ACF Flip Chip Package Under Thermal Cycling Condition: An experimental Study. *Microelectronics Reliability*, Vol. 46, No.2-4, 2006, pp.589-99. ISSN 0026-2714

Kwon, W. and Paik, K. (2006). Experimental Analysis of Mechanical and Electrical Characteristics of Metal-coated Conductive Spheres for Anisotropic Conductive Adhesives. *IEEE Transactions on Components and Packaging Technologies*, Volume 29, No. 3, 2006, pp. 528-34. ISSN 1521-3331

Lai, Z. & Liu J. (1996). Anisotropically conductive adhesive flip-chip bonding on rigid and flexible printed circuit substrate. *IEEE Transactions on Components, Packaging, and Manufacturing Technology, Part B: Advanced Packaging*, Vol. 19, No. 3, 1996, pp. 644-66. ISSN 2156-3950

Lai, Z., Lai, R., Persson, K., and Liu, J. (1998). Effect of Bump Height on the Reliability of ACA Flip Chip Joining with FR4 Rigid and Polyimide Flexible Substrate. *Journal of Electronics Manufacturing*, Vol. 8, No. 3-4, 1998, pp. 217-24. ISSN 0960-3131

Lau, J. (1995). *Flip Chip Technologies*, McGraw-Hill, ISBN 0070366098, New York, USA.

Lau, J. (2000). *Low Cost Flip Chip Technologies for DCA, WLCSP, and PBGA Assemblies*, McGraw-Hill, ISBN 0071351418, New York.

Lau, J., Wong, C.P., Lee, N., and Lee, R. (2003). *Electronics Manufacturing: with Lead-Free, Halogen-Free, and Conductive-Adhesive Materials*. McGraw-Hill Professional, ISBN 0071386246, New York.

Li, Y. and Wong, C.P. (2006). Recent Advances of Conductive Adhesives as a Lead-free Alternative in Electronic Packaging: Material, Processing, Reliability and Applications. *Materials Science and Engineering R 51*, 2006, ISSN 0927-796X.

Licari, J. J. and Swanson, D. W. (2005). *Adhesives Technology for Electronic Applications*, William Andrew Publishing, ISBN 0815515138, Norwich, New York.

Liu, J., Tolvgard, A., Malmodin, J., and Lai, Z. (1999). A Reliable and Environmentally Friendly Packaging Technology - Flip Chip Joining Using Anisotropically Conductive Adhesive", *IEEE Transactions on Components and Packaging Technology*, Vol. 22, No. 2, 1999, pp. 186-90, ISSN 1521-3331.

Liu, J. (2001). ACA Bonding Technology for Low Cost Electronics Packaging Applications – Current Status and Remaining Challenges. *Soldering & Surface Mount Technology*, Vol. 13, No. 3, 2001, pp. 39-57, ISSN 0954-0911.

Lu, D. (2006). Overview of Recent Advances on Isotropic Conductive Adhesives. *Proceeding of 6th Conference on High Density Microsystem Design and Packaging and Component Failure Analysis (HDP)*, ISBN 9781424404889, Shanghai, China, June 2006.

Luo, S., Leisen, J., and Wong, C. (2002). Study on Mobility of Water and Polymer Chain in Epoxy and Its Influence on Adhesion. *Journal of Applied Polymer Science*, Vol. 85, No. 1, 2002, pp.1-8. ISSN 1097-4628

Mercado, Lei., white, J., Sarihan, V., and Lee, T. (2003). Failure Mechanism Study of Anisotropic Conductive Film (ACF) Packages. *IEEE Transactions on Components and Packaging Technologies*, Vol. 26, No. 3, 2003, pp. 509-16. ISSN 1521-3331

Morris, J., Lee, J., and, Liu, J. (2005). Isotropic Conductive Adhesive Interconnect Technology in Electronics Packaging Applications. *Proceedings of the 5th International Conference on Polymers and Adhesives in Microelectronics and Photonics*, ISBN 0780395530, Wrocaw, Poland, January 2006.

Ogunjimi, A. O., Mannan, S. H., Whalley, D. C., & Williams, D. J. (1996). Assembly of Planar Array Components Using Anisotropic Conducting Adhesives-A Benchmark Study: Part I-Experiment. *IEEE Transactions on Components, Packaging, and Manufacturing Technology*, Vol. 19, No 4, 1996, pp. 257-63. ISSN 2156-3950

Pinardi, K., Liu, J., Haug, R., Treutler, C., and Willander, M. (1998). Deformation study of the PCB during the flip chip assembly processing anisotropically conductive adhesive (ACA) as a bonding agent. Proceedings of 3rd IEEE Conference on Adhesive Joining and Coating Technology in Electronics Manufacturing, Binghamton, ISBN 0780349342, NY, USA, September, 1998.

Rizvi, M. J., Chan, Y. C., Bailey, C., Lu, H., and Sharif, A. (2005). The Effect of Curing on the Performance of ACF Bonded Chip-on-flex Assemblies After Thermal Ageing. *Soldering and Surface Mount Technology*, 2005, Vol. 17, No. 2, pp.40-8. ISSN 0954-0911

Saarinen, K., Frisk, L. and Ukkonen, L. (2011). Effects of different combinations of environmental test on the reliability of UHF RFID tags. *Procedings of 17th European Microelectronics and Packaging Conference, EMPC*, Brighton, UK September, 2011.

Savolainen, P. (2004). Display driver packaging: ACF reaching the limits?. *Proceedings of the 9th International Symposium on Advanced Packaging Materials: Processes, Properties and Interfaces*, ISBN 0780384369, Atlanta, Georgia, USA, January, 2004.

Suhir, E. (2002). Accelerated Life Testing (ALT) in Microelectronics and Photonics: Its role, Attributes, Challenges, Pitfalls, and Interaction with Qualification Tests. *Journal of Electronic Packaging*, Vol. 124, No. 3, 2002, pp. 281-91. ISSN 1043-7398

Sungwook, M. and Chappell, W. J. (2010). Novel Three-Dimensional Packaging Approaches Using Magnetically Aligned Anisotropic Conductive Adhesive for Microwave Applications. *IEEE Transactions on Microwave Theory and Techniques*, vol. 58, No. 12, pp. 3815-23. ISSN 0018-9480

Takagi, K., Honma, H., and Sasabe, T. (2003). Development of Sequential Build-up Multilayer Printed Wiring Boards in Japan. *IEEE Electrical Insulation Magazine*, Vol. 19, No. 5, 2003, pp. 27 - 56. ISSN 0883-7554

Tan, S. C., Chan, Y. C., Chiu, Y. W., and Tan, C. W. (2004). Thermal Stability Performance of Anisotropic Conductive Film at Different Bonding Temperatures. *Microelectronics Reliability*, Vol. 44, No. 3, 2004, pp. 495-503. ISSN 0026-2714

Uddin, M. A., Alam, M. O., Chan Y. C., and Chan, H. P. (2004). Adhesion Strength and Contact Resistance of Flip Chip on Flex Packages - Effect of Curing Degree of Anisotropic Conductive Film. *Microelectronics Reliability*, Vol. 44, No. 3, 2004, pp.505-14. ISSN 0026-2714

Wang, X. Wang, Y., Chen, G., Liu, J., and Lai, Z. (1998). Quantitative Estimate of the Characteristics of Conductive Particles in ACA by using Nano Indenter. *IEEE Transactions on Components and Packaging, and Manufacturing Technology* , Vol. 21, no. 2, 1998, pp.248 - 51. ISSN 1070-9886

Watanabe, I. Fujinawa, T. Arifuku, M. Fujii, M. Yasushi, G. (2004). Recent advances of interconnection technologies using anisotropic conductive films in flat panel display applications. *Proceedings of the 9th IEEE International symposium Advanced Packaging Materials: Processes, Properties and Interfaces*, ISBN 0780384369, Atlanta, Georgia, USA, August 2004.

Whalley, D., Mannan, S., and Williams, D. (1997). Anisotropic Conducting Adhesives for Electronic Assembly. *Assembly Automation*, Vol. 17, No. 1, 1997, pp.66-74. ISSN 0144-5154

Wong, E. H., Chan, K. C., Rajoo, R., and Lin, T. B. (2000). The Mechanics and Impact of Hygroscopic Swelling of Polymeric Materials in Electronic Packaging. *Proceedings of the 50th IEEE Electronics Components and Technology Conference*, ISBN 0780359089, Las Vegas, NV, USA, May, 2000.

Wu, S. X., Zhang, C. Yeh, C, Wille, S., and Wyatt, K. (1997). Cure Kinetics and Mechanical Properties of Conductive Adhesive. *Proceedings of the 47th IEEE Electronics Components and Technology Conference*, ISBN 0-7803-3857-X, San Jose, CA, USA, May, 1997.

Yim, M. J. and Paik, K. W. (1998). Design and understanding of anisotropic conductive films (ACF's) for LCD packaging. *IEEE Transactions on Components and Packaging Technologies*, Vol. 21, No. 2, 1998, pp.226-34. ISSN 1521-3331

Yim, M. J. and Paik, K. W. (1999). The Contact Resistance and Reliability of Anisotropically Conductive Film (ACF). *IEEE Transactions on Advanced Packaging*, Vol. 22, No. 2, 1999, pp.166-73, ISSN 1521-3323.

Yim, M. J. and Paik, K. W. (2001). Effect of Nonconducting Filler Additions on ACA Properties and the Reliability of ACA Flip Chip on Organic Substrates. *IEEE*

Transactions on Components and Packaging Technologies, Vol. 24, No. 1, 2001, pp.24-32. ISSN 1521-3331

Yin, C.Y., Alam, M.O., Chan, Y.C., Bailey, C. and Lu, H. (2003). The Effect of Reflow Process on the Contact Resistance and Reliability of Anisotropic Conductive Film Interconnection for Flip Chip on Flex Applications. *Journal of Microelectronics Reliability*, Vol. 43, No. 4, 2003, pp. 625-33. ISSN 0026-2714

Zhong, Z. W. (2005). Various Adhesives for Flip Chips. *Journal of Electronic Packaging*, Vol. 127, No. 1, 2005, pp. 29-32. ISSN 1043-7398

26

Exploring the Superconductors with Scanning Electron Microscopy (SEM)

Shiva Kumar Singh[1,2,*], Devina Sharma[1], M. Husain[2], H. Kishan[1], Ranjan Kumar[3] and V.P.S. Awana[1]
[1]*Quantum Phenomena and Applications, National Physical Laboratory (CSIR), New Delhi,*
[2]*Department of Physics, Jamia Millia Islamia, New Delhi,*
[3]*Department of Physics, Panjab University, Chandigrah, India*

1. Introduction

The characterization of materials supports their development and in particular of superconductors, for their technological applications. Scanning electron microscopy (SEM) is one of these characterization techniques, whose data is used to estimate the properties, determine the shortcomings and hence improve the material. The phenomenon of superconductivity initially develops within the grain and eventually crosses over the grain boundaries, leading to the bulk. Hence SEM can be a useful tool to probe the microstructure of the superconductors and the properties related to it. Along with this the Energy-dispersive Spectroscopy (EDS) can tell about the chemical composition of compounds. Grain size and its connectivity can be seen through SEM and can be correlated with the corresponding properties. The superconducting materials developed for practical applications are some of the complex materials used today. These materials have large number of potential variables such as their processing conditions, composition, structure etc., whose dependence on the superconducting properties have to be analyzed critically. The characterization techniques are the tools that help to reveal and explore both the macro and microstructure of materials. It is known that the larger grains (reduction in grain boundaries) lead to increased pinning type behavior with enhanced J_c [1]. In contrast Rosko *et al.* [2] reported that J_c is determined by weak links and grain size has little role on it. Also, Smith *et al.* [3] interpreted reduction of J_c and activation of weak link type behavior with increasing grain size for $YBa_2Cu_3O_{7-\delta}$ (YBCO) polycrystalline samples in terms of micro-cracks in large grains. The superconducting parameters are broadly divided into two categories; first, the intrinsic parameters such as penetration depth (λ), which are intrinsic to the material and are not affected by, grain size. On the other hand, values such as shielding/Meissner fraction, the inter- and intra-grain critical current density and diamagnetic fraction depend upon particle size of bulk superconductors. Thus SEM can be very important to probe and in understanding the superconducting phenomena.

* Corresponding Author

In SEM electron beam is scanned across a sample's surface. When the electrons strike the sample, a verity of signals arises and produces elemental composition of the sample. *SEM* with *EDS* is a major tool for qualitative and quantitative analyses which is done by bombarding a finely focused electron beam (electron probe) on the specimen, and measuring the intensities of the characteristic X-ray emitted. The three signals in SEM are the secondary electrons, backscattered electrons and X-rays, provide the greatest amounts of information. Secondary electrons are emitted from the atoms occupying the top surface and produce interpretable image of the surface. The contrast in the image is determined by the sample morphology. Backscattered electrons are primary electrons which are "reflected" from atoms in the solid. The contrast in the image produced is determinate by the atomic numbers of the elements in the sample. Therefore the image shows the distribution of different chemical elements in the sample. Since these electrons are emitted from the depth of the sample, the resolution of the image is not as good as for secondary electrons.

This chapter deals with the ability of SEM in extracting the information from superconductors. Since, the microstructure and topology of the materials determine largely its properties in terms of its grains and their connectivity, the utility of SEM in studying the properties of various superconductors discovered till date will be reviewed. The limitations will also be discussed. How other characterization tools, can provide better information along with *SEM*, will be explored.

2. SEM: As a characterization tool for superconductors

Some of the important aspects with which *SEM* deals with, is the grain size, morphology and alignment, structural defects, chemical composition. While trying to optimize the transport properties, grain to grain alignment within the superconductor has to be considered. It is important to analyze the grain alignment and enhance it in order to relate it to the improvement of transport properties. Defects play an important role in determining the properties of superconductors, especially those of HTSc. While macro defects such as porosity, cracks, secondary phases etc. may adversely affect the transport critical super currents, on the other hand microscopic defects such as addition of nanoparticles, dislocation etc., can prove to be beneficial. As the size of the defects is smaller than the coherence length of the superconductor, they may act as pinning centers, thereby enhancing the critical current. Chemical composition of the material within the superconducting grain and at the grain boundaries has significant effect over its properties. The compositions and the change in the compositions can be measured or inferred from a variety of techniques such as energy dispersive spectroscopy. Backscattered electron imaging (BEI) can also be used to infer chemical compositional variations. In all, the complexity in the superconducting materials requires continuous research into their fundamental properties and evolution of new improved materials. For all these evaluations, various characterization tools have to be relied upon among which SEM and allied techniques such as EDS and BEI play an important role.

2.1 Determination of grain size from *SEM* micrographs

Grain size determination is perhaps one of the most commonly performed microstructural measurements from *SEM* micrographs. Standards organizations, including American Society for Testing and Materials (ASTM) [4] and some other national and international organizations, have developed standard test methods describing how to characterize microstructures quantitatively. The methods for grain size measurement are described in great detail in the ASTM Standard, E112, "Standard Test Methods for Determining Average Grain Size" [5]. The information below will provide cursory explanation to the methods for determining grain size in *ASTM* E112. The microstructural quantity known as the *ASTM* micro gain size number, G is defined as

$$n = 2^{G-1}$$

Where, n is the number of grains per square inch measured at a magnification of 100x. Grain size or the value of G is most commonly measured by (a) Planimetric method and by (b) Intercept method.

(a) Planimetric method

In the Planimetric method [Fig. 1 (a)] (developed by Zay Jeffries in 1916), a count is made of the number of grains completely within a circle of known area and half of the number of grains intersected by the circle to obtain N_A. Then, N_A is related to G. This method is slow when done manually because the grains must be marked when counted to obtain an accurate count. This method is described in the section 9 of *ASTM E112.*

The basic steps of the procedure are as follows:

i. A circle of known size is inscribed over the SEM image.
ii. The numbers of grains are counted that are completely within the area (n_{inside}).
iii. The number of grains is counted that are partially within the area ($n_{intercepted}$).
iv. The number of grains per sq. mm, N_A, is calculated from $N_A = f\,\{n_{inside} + ½(n_{intercepted})\}$
v. The multiplier f is calculated from (M^2/circle area), where M is the linear magnification of the image.
vi. From N_A, we can calculate the *ASTM* grain size number, G, using the following formula from E 112-96:$G = \{3.322\,(\log10\,N_A) - 2.954\}$

(b) Linear intercept method

The *ASTM* grain size can also be determined using the intercept method (developed by Emil Heyn in 1904) counting either the number of grains intercepted (N) or the number of grain boundaries intersected (P) with a test line. *ASTM* recommends using a grid with three concentric circles (as shown in the Fig. 1 (b)) with a 500mm total line length. The count of the number of grains intercepted by the circle is N. To calculate the number of interceptions per mm, N_L, we divide N by the true length of the circle. The true length (L_T) is obtained by dividing the circumference of the circle by the magnification, M. Hence, $N_L = N/L_T$ interceptions per mm. To calculate the grain size, we first determine the mean linear intercept length, *l*, which is the reciprocal of N_L (or of P_L, the number of grain boundary intersections per unit length). G is calculated from an equation from E 112-96:

$$G = \{-6.644\,(\log10\, l) - 3.288\},$$

where, *l* is in mm.

ASTM E112 provides table that relates grains/in^2 @ 100x and grains/mm^2 @ 1x to *ASTM* grain size number G. Since the two methods are sensing different geometric aspects of the three-dimensional grain structure, they will not give exactly the same value, but they will be close, generally within the experimental limitations of the measurements. In practice, these measurements are repeated on a number of fields in order to obtain a good estimate of the grain size.

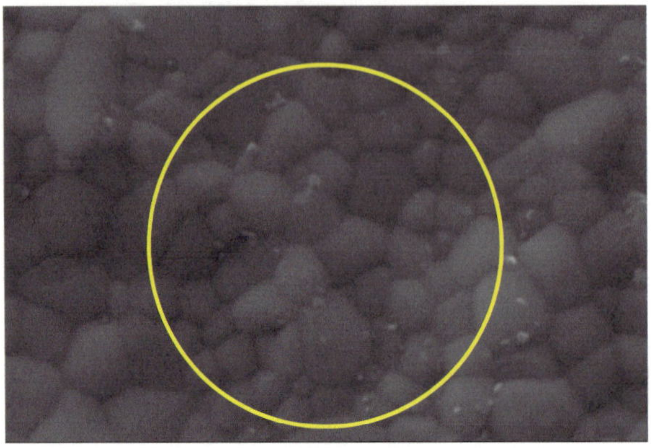

(a)

(b)

Fig. 1. **(a)** Planimetric method, **(b)** Linear Intercept method

2.2 Chemical composition using EDS

One of the most outstanding features of the *SEM-EDS* is that it allows elemental analysis and observation from an ultra micro area to a wide area on the specimen surface without destroying the specimen. Qualitative and quantitative analysis (by *EDS*) by electron probe takes advantage of the emission of characteristic X-radiation by electron interactions in the valence shell of atoms. Backscattered electron images in the *SEM* display compositional contrast that results from different atomic number and their distribution of elements. *EDS* allows one to identify what those particular elements are and their relative proportions, for example their atomic percentage. Initial EDS analysis usually involves the generation of an X-ray spectrum from the entire scan area of the SEM. In the X ray spectra generated from the entire scan, Y-axis shows the counts (number of X-rays received and processed by the detector) and the X-axis shows the energy level of those counts. The EDS software associates the energy level of the X-rays with the elements and shell levels that generated them.

3. Exploring superconductors using *SEM*

In this section we will discuss the *SEM* study of superconductors of different families. For high temperature superconductors (HTSc), *SEM* has been widely used to explore the superconducting behavior. Grain size matters a lot in deciding the superconducting parameters of cuprate HTSc. Decrease in shielding current is observed with decrease in particle size [6]. Magnesium diboride (MgB_2) represent an attractive alternative to low temperature superconductors. For most of the practical applications, high critical current density (J_c) in the presence of a magnetic field along with high upper critical field (H_{C2}) and high irreversibility field (H_{irr}) are required. Moderate impurities and nano (n) materials are being used to improve these parameters [7-14]. In particular, significant flux pinning enhancement in MgB_2 is observed with *n-SiC* addition [15]. In case of newly discovered pnictides superconductors, *SEM* is being used in an ingenious way. The spatially resolved electrical transport properties have been studied on the surface of optimally-doped superconducting $Ba(Fe_{1-x}Co_x)_2As_2$ single crystal by using a four-probe scanning tunneling microscopy [16]. Results will be discussed in following subtitles.

3.1 Grain size and gain connectivity in cuprate high temperature superconductors

One of the characteristics of high temperature superconductors (HTSc) is their small coherence length which is comparable with the unit cell. The coherence length is a key parameter for the performance of superconductors for applications, since this determines the size of the normally conducting core of the flux lines [17]. In order to control the motion of flux lines one needs a microstructure with defects as small as the coherence length. The extremely small coherence length of HTSc, which is for YBCO only 2.7 nm at 77 K within the ab-plane, is the reason that defects such as grain boundaries, which are very beneficial in low-temperature superconductors because they act as pinning defects, serve as weak links and limit the critical current, especially in the presence of an external magnetic field. HTSc bulk material can therefore be considered as a matrix of superconducting grains embedded in a non superconducting material.

The cuprate superconductors belong to the family of HTSc in which Cu-O chains and planes are responsible for the conduction of super currents. The size as well as the shape of the grains varies in different cuprate HTSc [18-20]. This variation in microstructure in turn leads to different superconducting behavior. Although the nature of the occurrence of superconductivity in cuprate HTSc is same, T_c varies from 38 K in LSCO to 110 K in BSCCO system. This wide range of T_c itself indicates that micro-structural parameters are of much importance. In this section various cuprate HTSc will be discussed in the chronological order of their discovery, in terms of the use of SEM for their characterization.

A. $La_{2-x}Sr_xC_uO_4$

Discovered by Bednorz and Muller in 1986 [21], $La_{2-x}Sr_xCuO_4$ was the trendsetter breakthrough in the history of superconductivity leading to the new era of High T_c superconductivity. It has T_c of 38 K which is beyond the BCS limit of 30 K. Although it has higher T_c than conventional superconductors, its critical parameters such as critical temperature, field and current density, which are applicable in practical applications, are weaker. Various attempts had been made earlier to enhance these parameters and to understand the physics behind it. The values such as shielding/Meissner fraction, the inter- and intra-grain critical current density and diamagnetization fraction depends upon particle size of bulk superconductors. One of the earliest reports on the effect of particle size on the physical properties of a superconductor was by Chiang et al. [22], who varied particle size from 1 to 10 μm and found significant changes in the superconducting and physical properties. Another fact that has well been established is that the critical current measurements done in HTSc have shown a much lower value for polycrystalline bulk samples [1] than single crystals of the same compound. This difference cannot be attributed alone to the intrinsic nature (anisotropy etc.) of the material. In fact, the quasi-insulating grain boundaries of HTSc play a detrimental role in limiting the critical current and other superconducting and magnetic properties [22-23]. In a recent report D. Sharma et al. [18] have investigated the influence of grain size (sintering temperature) on various superconducting parameters in $La_{1.85}Sr_{0.15}CuO_4$. From SEM micrographs [see Fig. 2 (a), (b) and (c)] it is clear that with the increase in the sintering temperature, there is a considerable increase in the grain size. Increment in the grain size brings in double boom to the superconductivity as it reduces the number of insulating grain boundaries (weak links) as well as increases the effective superconducting volume fraction. Thus both the inter and intra critical current density which is limited by the weak links is expected to enhance with increase in the grain size. Qualitative picture given by the SEM micrographs corroborates with the quantitative results obtained from various transport and magnetic measurements as given in the table 1.

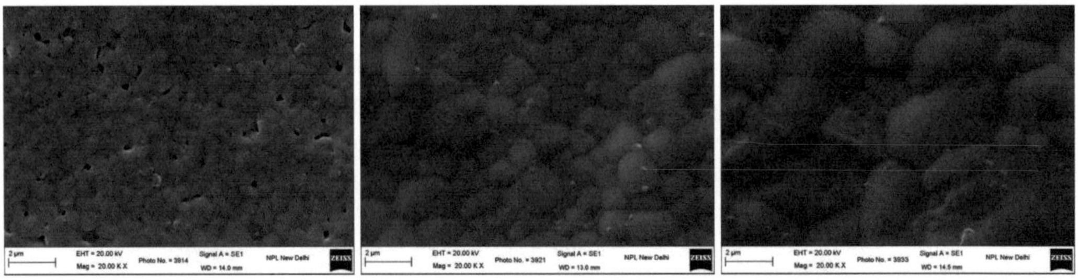

Fig. 2. SEM of $La_{1.85}Sr_{0.15}CuO_4$ samples sintered at (**a**) 900 °C, (**b**) 1000 °C, and (**c**) 1050 °C [Ref. 18].

Grain size (2*R*)	Critical current density J_c (A/cm^2)		Diamagnetization (at $T = 10$ K, $H = 10$ Oe)		Volume fraction not penetrated $(R-\lambda)^3/R^3$
	$J_c = (20 \times \Delta M)/a(1 - a/3b)$ Inter-granular J_c	$J_c = (15 \times \Delta M \times \rho)/R$ Intra-granular J_c	ZFC	FC	
1 μm	1.1×10^3	3.75×10^5	0.36%	0.28%	9.08%
2 μm	3.8×10^3	8.36×10^5	2.10%	1.59%	38.07%
4.5 μm	40.4×10^3	32.73×10^5	16.95%	4.08%	64.20%

Table 1. Critical current density, diamagnetization fraction and percentage volume fraction not penetrated by magnetic flux calculated from magnetic measurement data for different grain sizes of $La_{1.85}Sr_{0.15}CuO_4$ [Ref. 18].

In another particle size controlled study of non-superconducting $La_{1.96}Sr_{0.04}CuO_4$ was made with SEM and IR spectra by S. Zhou et al [24]. They observed that as the particle size reduces, the IR band at around 685 cm^{-1}, corresponding to in-plane Cu–O asymmetrical stretching mode, shifts to higher frequency and the magnetization exhibits a large enhancement at low temperature. A visible spin-glass transition was found under a relatively weak external field in the sample with the largest particle sizes. Whereas the sample with the smallest particle sizes exhibits no visible spin-glass transition. They suggested that surface effects play a dominant role in determining the magnetic properties as the particle size reduces.

B. $YBa_2Cu_3O_{7-\delta}$

Cuprate superconductors are very sensitive to oxygen content. Depending on oxygen content, $YBa_2Cu_3O_{7-\delta}$ crystallizes in two phases. Tetragonal *P4/mmm* with $\delta = 0.60$ results in non-superconducting and antiferromagnetic YBCO whereas orthorhombic *pmmm* with $\delta = 0.05$ phase leads to a superconducting YBCO with T_c 93 K [25]. Also, the intra-grain signal depends much on oxygen content of the composition. The structure of YBCO can be viewed as (Ba,Sr)O/CuO_2/RE/CuO_2/(Ba,Sr)O slabs interconnected through a sheet of Cu and O with variable composition of CuO_x.. Charge transport and high temperature superconductivity is believed to reside in the CuO_2 planes of all known HTSc cuprates, except that $CuO_{1+\delta}$ chains have been reported to participate in the *b*-axis transport of $YBa_2Cu_3O_{7-\delta}$ [26]. In $YBa_2Cu_3O_{7-\delta}$ ($CuBa_2YCu_2O_{7-\delta}$, Cu-1212) there are two different Cu sites, namely Cu1 and Cu2. Cu1 resides in CuO_x chains and Cu2 in superconducting CuO_2 planes. Even at macroscopic level, any contravene in integral CuO_2 stacks, affects superconductivity drastically [27-28]. The CuO_x chain acts as a charge reservoir and provides the mobile carriers to superconducting CuO_2 planes.

To understand the physics of the superconducting nature, investigation of doping various elements at Cu1 site was carried by some of us [29]. And it was found that the YBCO structure is versatile and changes with doped elements at Cu1 site. Single phase samples of 1212 type with different MOx layers showed the great flexibility of these rocksalt layers and variable structure formation. With different M, as the oxidation state and ionic state changes, carrier concentration and structure changes as well. While, Nb-, Fe-, Ru- and Al-1212 possess tetragonal *P4/mmm* space group structure, the Ga-1212 and Co-1212 are crystallized in orthorhombic *Ima2* space group.

The SEM images [Fig. 3] suggest that with the doping of variety of elements at the Cu1 site in Y-123 structure, the morphology also changes suggesting change in the structure of the new compounds formed. Change in structure was authenticated by the Rietveld analysis also. In another report Nalin *et al.* [30] studied effect of Zn doping at Cu1 site. With ac susceptibility inter and intra granular changes are studied. In the Zn-doped samples, the inter-grain peak got reduced dramatically. In the χ^{II} plots of Zn doped samples the inter-grain peak superposes with intra-grain peak and inter-grain peak depresses further. It was concluded that as (see *SEM* images [Fig. 4]) the average grain size is increasing with Zn doping. The increased grain size provides more area for the eddy currents loops to persist in the individual grains, thus systematic enhancing the intra-grain peak.

A combined study through *SEM* and *EDS* for the compounds $Y_{1-x}Ca_xBa_{1.9}Nd_{0.1}Cu_3Oy$ (YCBNCO) with $x \leq 0.40$ have been made [31]. Back-scattered electron SEM micrographs of samples with $x = 0.10$ and 0.30 have been taken. The *SEM* studies proved that the samples with x ≤ 0.20 are homogeneous and stone like grains with typical size of several microns. On the other hand the samples with x ≥ 0.20 are inhomogeneous. The *SEM* micrographs show that the stone-like grains and the sponge-like grains co-exist in the surface of the samples. The *EDS* results show that the constituted elemental ratios in both regions are different and hence the superconducting properties.

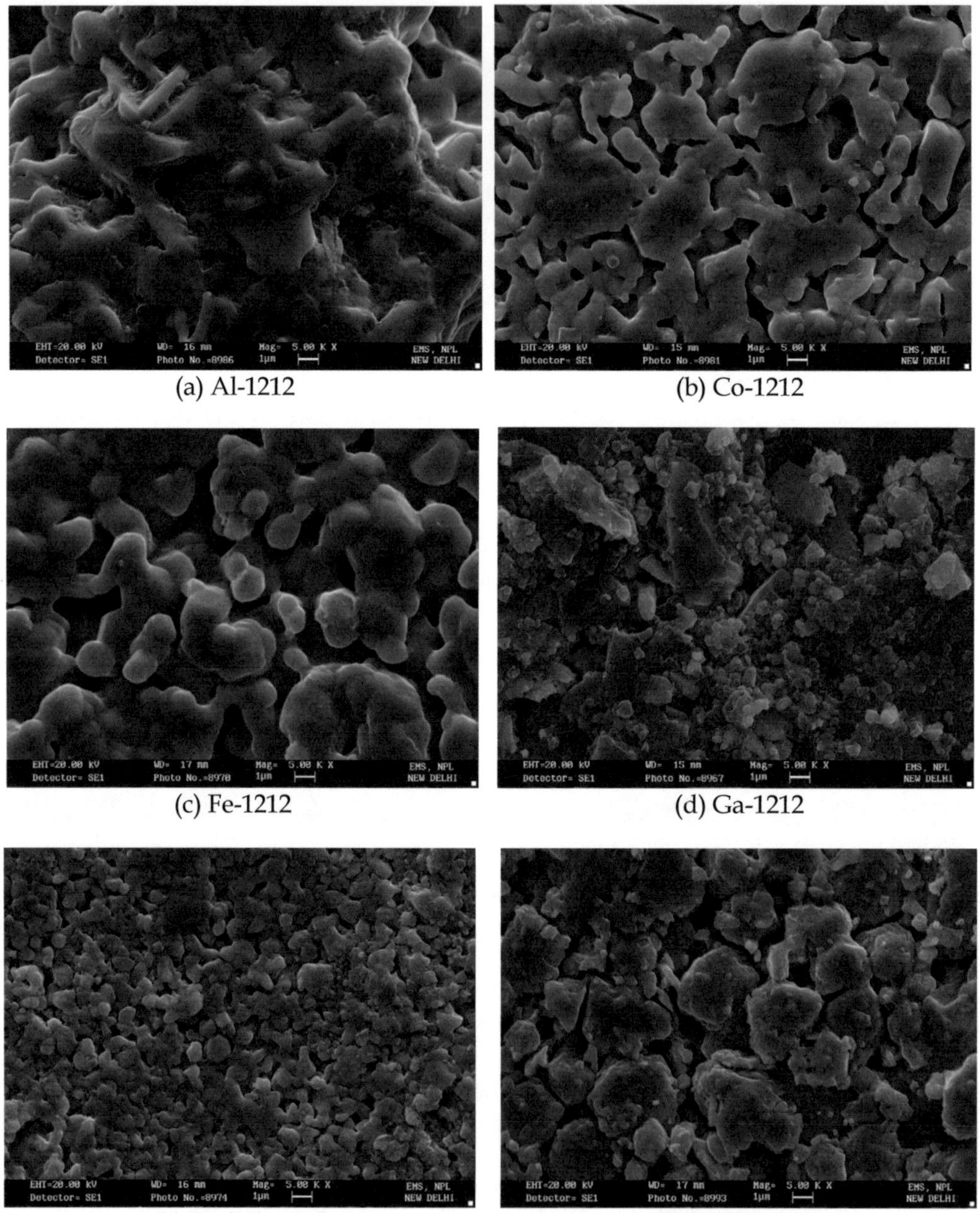

(a) Al-1212 (b) Co-1212

(c) Fe-1212 (d) Ga-1212

(e) Eu/Nb-1212 (f) Eu/Ru-1212

Fig. 3. SEM pictures of the M-1212: (a) $AlSr_2YCu_2O_{7+\delta}$, (b) $CoSr_2YCu_2O_7$, (c) $FeSr_2YCu_2O_{7+\delta}$, (d) $GaSr_2YCu_2O_7$, (e) $NbSr_2EuCu_2O_{7+\delta}$ and (f) $RuSr_2EuCu_2O_{7+\delta}$. [Ref. 29].

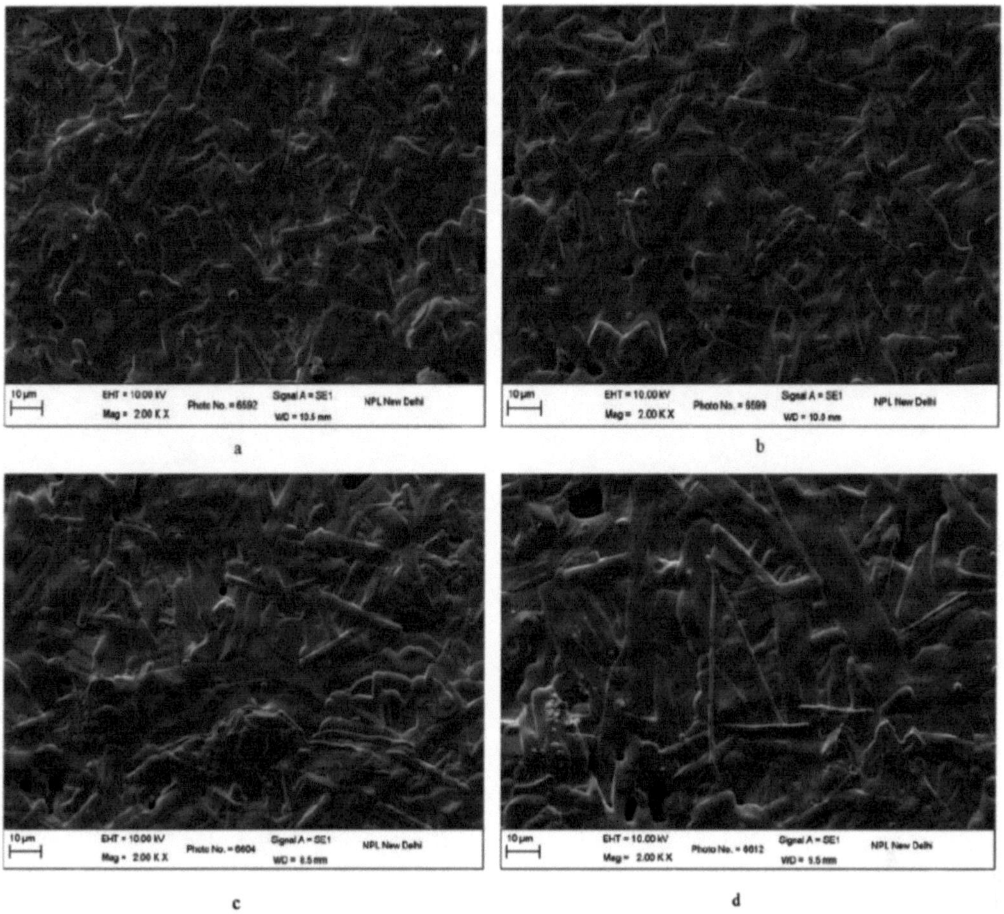

Fig. 4. SEM images of $YBa_2Cu_{3-x}Zn_xO_{7-\delta}$, (**a**) $x = 0.01$, (**b**) $x = 0.03$, (**c**) $x = 0.05$ and (**d**) $x = 0.10$ [Ref. 30].

C. $(Bi, Pb)_2Sr_2Ca_{n-1}Cu_nOy$

Bismuth based superconducting cuprates (in short named as BSCCO) is an another family of cuprate HTSc which are expressed by a general formula of $(Bi, Pb)_2Sr_2Ca_{n-1}Cu_nOy$. For ($n$ = 1, 2, 3), these are abbreviated as Bi2201, Bi2212 and Bi2223 phases, whose superconducting transition temperatures (T_c) are around 20, 85 and 110 K respectively [32]. Though the mechanism of superconductivity in HTSc superconductors has been extensively studied, it is still unclear. As a result of substitution experiments it is well known, that in HTSc's, there is a strong relationship between carrier concentration and transition temperature. In addition, intergrain carrier transportation is also a key factor in deciding the sharpness of the transition. It is a well known fact that intergrain region behaves as a non-conducting region. Thus grain connectivity becomes more important for the sharpness of transition and other critical parameters. The grain growth and its shape varies in $Bi_{2-x}Pb_xSr_2CaCu_2O_8$ (see Fig. 5) with substitution of Pb at Bi site ($0< x < 0.40$) [19]. From the flake type grain shape in pristine samples to needle type grain shape in $x = 0.40$ composition is observed from SEM micrographs. Moreover, improvement in the packing fraction and hence the inter-granular connectivity was seen in the samples for x = 0.0 up to x = 0.16, which degrades with further increase in x. Decrease in the grain alignment with increase in Pb content has also been seen in SEM micrographs.

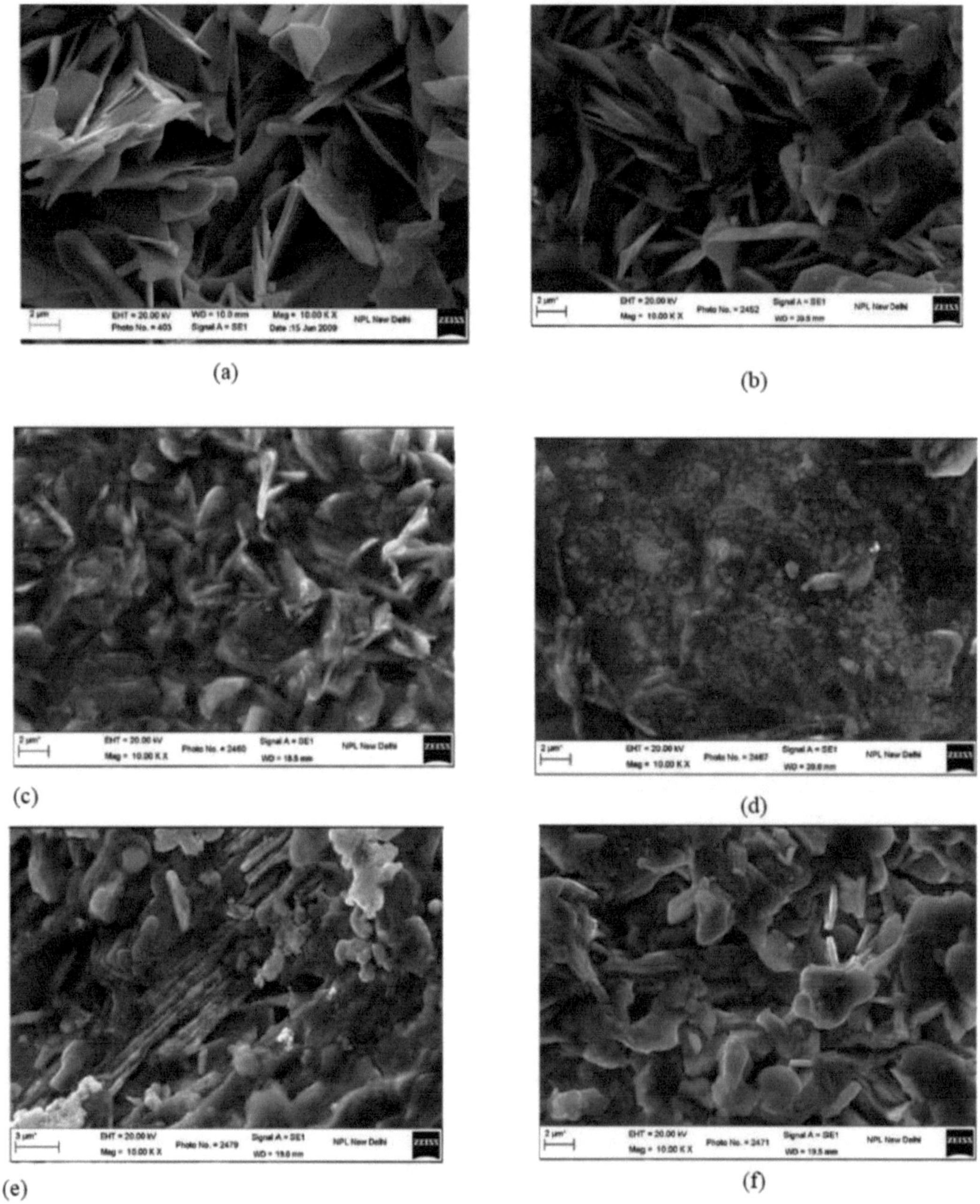

Fig. 5. SEM micrographs of $Bi_{2-x}Pb_x Sr_2CaCu_2O_{8+\delta}$ (**a**) x = 0.04, (**b**) x = 0.06, (**c**) x = 0.08 and (**d**) x = 0.16 (**e**) x = 0.20 and (**f**) x = 0.40[Ref. 19].

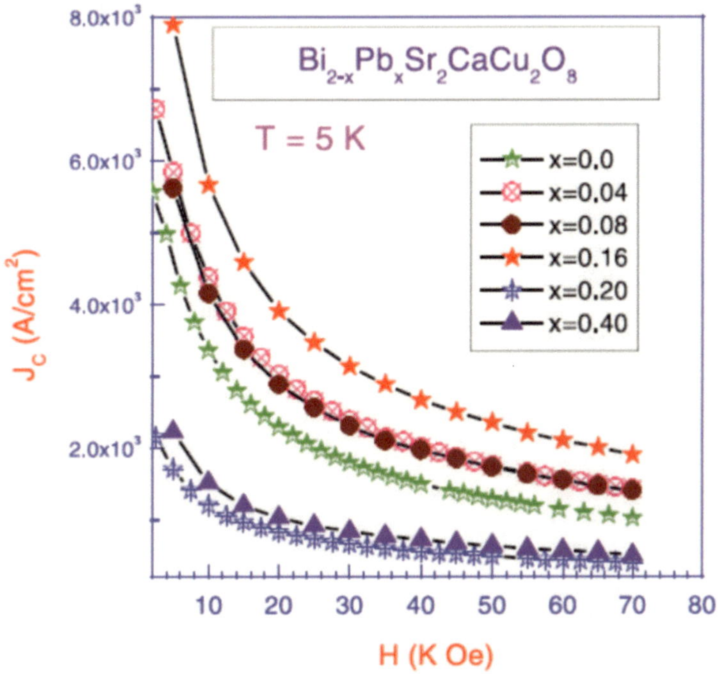

Fig. 6. *Jc* measurements for $Bi_{2-x}Pb_x Sr_2CaCu_2O_{8+\delta}$ ($x = 0$ to 0.40) [Ref. 19].

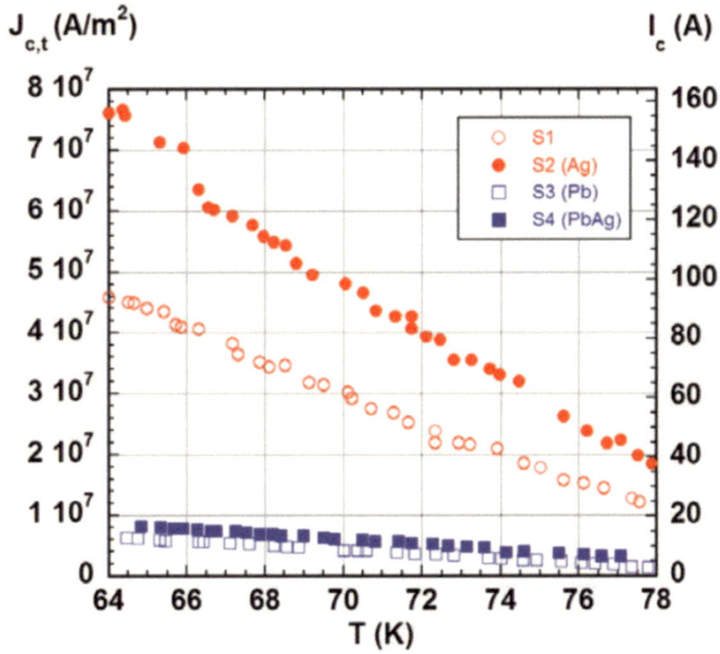

Fig. 7. Temperature dependence of the transport critical current densities, between 65 and 77 K and self-field, for the four samples.

Reprinted with permission from [Supercond. Sci. Technol. 22 (2009) 034012].

The critical current density data (see Fig. 6) for the same samples shows an increase in the current density with Pb substitution in pristine sample till x=0.16 after which it decreases with further increase in x. The decrease in conductivity for samples having $x > 0.16$ has been explained on the basis of the effects arising from decrease of the grain alignment, increase of porosity and secondary phases.

Also, A. Sotelo et al. [20] have studied the Lead (Pb) and Silver (Ag) doping of Bi-2212 samples. It was found that Pb doping results in the decrease of the transport critical current density (see Fig. 7), *Jc*,t (from 4.4 × 107 to 6 × 106 Am−2 at 65 K and self-field) as well as in the worsening of the mechanical properties, by about 35% compared to the undoped samples. In contrast, Ag doping results in the improvement of both the critical current density and mechanical strength.

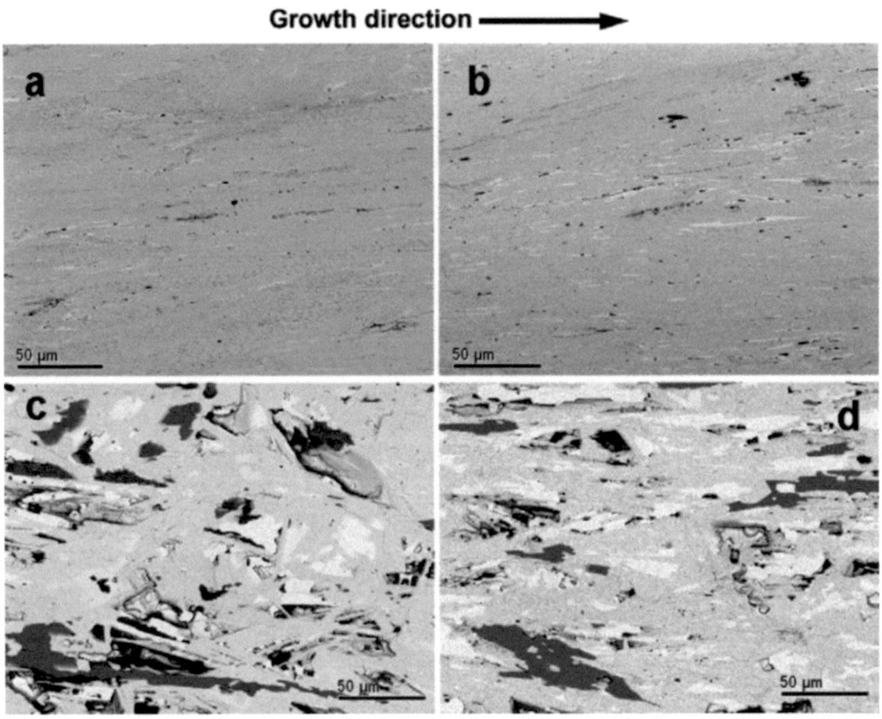

Fig. 8. Longitudinal SEM images obtained on annealed polished samples. (a) S1; (b) S2 (Ag); (c) S3 (Pb); and (d) S4 (PbAg). Phases can be identified as Bi-2201 (white contrast), Bi-free phases (dark grey; CuO, $Sr_{14}Cu_{24}O_{41}$, and $SrCuO_2$), plumbate-like phases and Bi-2212 (grey contrast), Ag (light grey contrast).
Reprinted with permission from [Supercond. Sci. Technol. 22 (2009) 034012].

These described effects are related to the microstructural observations (see Fig. 8) as, Pb doping dramatically reduces the texture, while Ag doping improves it. Moreover, for samples with Ag addition, an intergrowth of Bi-2223 inside the Bi-2212 grains is observed, which would explain the improved superconducting properties of these samples. The stability of these superconductors has been studied through the corrosion process in a moisture atmosphere [33]. By means of optical and SEM observations, several morphologies of the alteration products have been observed.

3.2 Flux pinning in MgB_2

Although cuprate superconductors exhibit very high transition temperatures, their in-field performance [34] is compromised by their large anisotropy, the result of which is to restrict high bulk current densities. On the other hand in-field performance (higher *Jc*), leads diboride of magnesium to much better candidate for application purposes besides its lower T_c than cuprates. With the magneto-optical (MO) and polarized light (PL), SEM was used to assess the issue of inhomogeneous and granular behaviour in MgB_2 [35]. It was speculated through SEM that the strongly shielding high-*Jc* regions are microstructurally subdivided on a scale of 100 nm. Also, in a darker central area a fine mixture of MgB_2 and a boron-rich phase was found through SEM [Fig. 9]. They concluded that the strongly shielding regions contain a large number of high-angle grain boundaries. Thus along with MO and PL, SEM suggests that MgB_2 is more similar to a low-*Tc* metallic superconductor than to a high-*Tc* copper oxide superconductor [35]. A sol-Gel synthesis of MgB_2 nanowires is reported Nath et.al [36]. SEM study reveals formation of a thick mesh of nanowires. The nanowires are found to be ca. 50–100 nm in diameter with very smooth surfaces having lengths up to at least 20 micrometer. It is observed that nanowires oriented vertically with respect to the electron beam. Thus SEM also revealed a hexagonal cross section for MgB_2 nanowires which is consistent with a degree of crystallinity. The Crystallinity of MgB_2 nanowires was also supported by their selected area electron diffraction (*SAED*) study on some individual nanowires. As MgB_2 has better candidature for practical applications various dopants has been added to improve its performance [6-14]. Arpita et al. [7] noted that with n-SiC addition though *Tc* decreases, but critical current density (*Jc*) and flux pinning improved significantly. Presence of Mg_2Si phase was also revealed through SEM and EDS [Fig. 10]. Dual reaction occurs with n-SiC addition first n-SiC reacts with Mg forming Mg_2Si and then free C is incorporated into MgB_2 at B site [37]. Thus both reactions help in the pinning of vortices which results in improved superconducting performance. Mg_2Si and excess carbon can be embedded within MgB_2 grains as nanoinclusions. They argued that due to the substitution of C at the B site the formation of a nanodomain structure takes place due to the variation of Mg-B spacing. These nanodomain defects, having the size of 2-3 nm, can also behave as effective pinning centers. So, highly dispersed nanoinclusions within the grains and the presence of nanodomain defects act as pinning centers and thus result in the improved *Jc(H)* behavior for the n-SiC doped samples.

3.3 Pnictides: Chemical composition and electrical transport

A. $REFeAsO_{1-x}F_x$ (1111)

Iron pnictides are the latest entrant in family of high temperature superconductors [38]. Superconductivity originates in parent pnictides *RE*FeAsO with doping of F at O site. The reactive nature of *REs* towards oxygen results a very critical synthesis condition for these compounds. Though the compounds are being synthesized in inert/oxygen controlled atmosphere, it is very hard to acquire the desired composition. Thus it is better to analyze the chemical composition of the synthesized compound before going insight and describing the physical properties. Thus *SEM* with *EDS* can be very useful in invoking the composition (especially effective F concentration) of the arsenides. The *SEM* analysis of the parent and non superconducting SmFeAsO compound after metallographic preparation reveals very small amounts of unreacted phases (iron arsenides), which are completely dissolved after

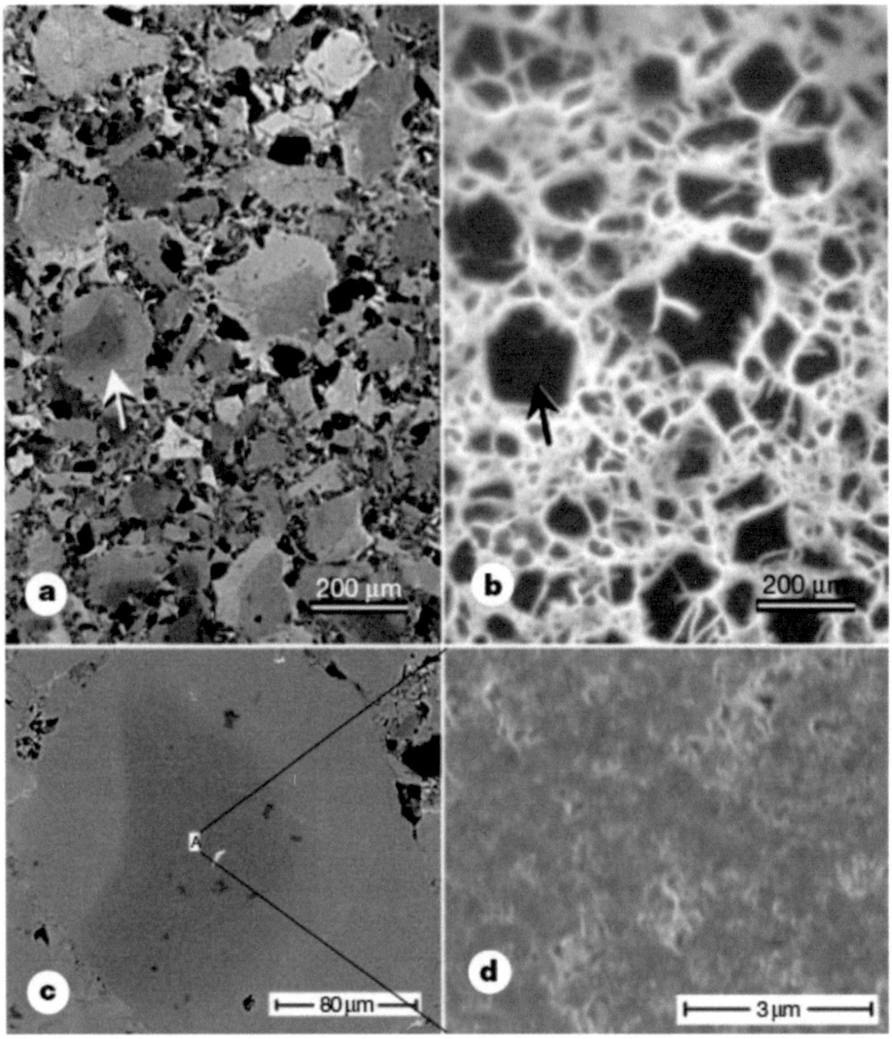

Fig. 9. Polarized light microscope and magneto-optical images of the same area of sample B are compared in **(a)** and **(b)**, respectively. Bright regions of **(b)** indicate areas where magnetic flux has penetrated the sample after a field of 120 mT was applied after cooling the sample in zero field to 11 K. Image **(c)** presents a magnified view using SEM backscattered electron imaging of the strongly superconducting region marked with an arrow in **(a)** and **(b)**. At higher resolution, image **(d)**, a secondary electron examination of the central region in **(c)**, reveals that the area marked by an arrow in **(a)** and **(b)** has, 100-nm, fine-scale structure [Reprinted by permission from Macmillan Publishers Ltd: NATURE 410 (2001) 186].

sintering. In general, sintering greatly increases the density of the samples, but favours the formation of Sm_2O_3 small particles. This feature reveals that at the sintering temperature the formation of Sm_2O_3 competes with the thermodynamic stability of the oxy-pnictide. S. Kaciulis et al. [39] studied $SmFeAsO_{0.85}F_{0.15}$ sample with *SEM*, *EDS* and *XPS*. *SEM* image after the fracture manifests the crystals appear clean at the surface, without any contamination of secondary phases. On the other hand another *SEM* image of the same sample after metallographic preparation reveals that crystals are aggregated within a matrix constituted of FeAs, which was also evidenced by their *EDS* analysis. However, their XPS study speculated that the formation of secondary phases, such as FeAs and SmOF. The discrepancy with EDS data, indicating only the presence of FeAs in the matrix, was explained by different analysis depth: up to a few micrometers for *EDS* and only a few nanometers for *XPS*. However, the aggregation of $REFeAsO_{0.85}$ crystals in FeAs matrix is also observed in Back Scattered *SEM* study [Fig. 11] of Ketnami et al. [40]. The absence of significant transport currents in polycrystalline samples has raised the concern that there is a significant depression of the superconducting order parameter at grain boundaries (GB) [41-43]. Remnant magnetization and *MO* studies of polycrystalline $NdFeAsO_{0.85}$ and $SmFeAsO_{0.85}$ uncovered that intergrain and intra-grain current densities had different temperature dependences and differed by three orders of magnitude, leaving open the possibility of an intrinsic GB blocking effect [40]. Moreover, the BSE-SEM images revealed that even the best $SmFeAsO_{0.85}$ bulk had non-superconducting Fe-As and RE_2O_3 occupying at least three quarters of the $REFeAsO_{0.85}$ GBs, making the active current path certainly much smaller than the geometrical cross-section of the sample [40]. Further to reveal the active local current paths, combined low temperature laser scanning microscopy (LTLSM) and SEM studies had been made [44]. With the *SEM* images they are able to show significant micro-structural differences between various regions of the sample. It is revealed that insulating Sm_2O_3 has a small surface to volume ratio and is mostly located within $SmFeAsO_{0.85}$ grains, so it has the smallest effect on current transport. On the other hand the dark gray Fe-As phase wets many GBs, thus interrupting grain to grain supercurrent paths, which are further degraded by extensive cracking, sometimes at GBs.

a

b

Fig. 10. (a), (b) SEM images of pure MgB_2 and 10 wt%n-SiC added samples.
Reprinted with permission from [Nanotechnology 19 (2008) 125708]

Fig. 11. BSE image of the (a) Sm and (b) Nd sample at high magnification. Although some grain boundaries are well connected, others are clearly obstructed by the Fe-As phase (dark contrast), Sm_2O_3 or Nd_2O_3 (white contrast) and cracks.
Reprinted with permission from [Supercond. Sci. Technol. 22 (2009) 015010].

Also, magnetic contamination is detected through *SEM* in the case of NdFeAsO single crystals grown out of NaAs flux under ambient pressure [45]. It is observed that some crystals show a lambda anomaly in the specific heat curve at ~12 K while the same is absent in others. They examined the cleaved (001) surfaces with *SEM* which were showing lambda anomaly. They turned to look at the edges of the crystals carefully. A 10-μm-thick layer was observed on the edges of the NdFeAsO crystals. With *EDS* it was found that some particles of TaAs, were surrounded by some very fine particles. After removal of that impurity the lambda anomaly disappeared. Thus with the help of *SEM* and *EDS* we can find out the actual cause which leads particular nature of a material.

B. Ba/Sr/K/Fe_2As_2 (122)

Soon after the discovery of 1111 family, other superconducting families based on FeAs layers (122, 111 and 11 structure) were reported, such as (Ba,K)Fe_2As_2 [46], LiFeAs [47], and FeSe [48]. Among all of these iron-based superconductors, the 122-type superconductors with a *Tc* of 38 K have a lower synthesis temperature and are oxygen free in comparison to the 1111-type. In addition, its *Tc* is much higher than those of the 111 and 11-type superconductors. SEM is used ingeniously to study the single crystal of Ba(Fe/Co)$_2$As$_2$ [16]. The topographic images of cleaved surface of optimally doped $BaFe_{1.8}Co_{0.2}As_2$ with *uniform* contrast have been observed in the secondary electron emission images acquired by SEM. They used secondary electron emission mode as it can register the contrast according to topography, chemical composition, and surface barrier (work function or ionization energy) of the sample [49] Small darker regions in the SEM image have been identified as marked by a rectangular box in Fig 12 (a). An SEM zoom-in image in the dark region reveals microscopic *domain* structures [Fig. 12]. They made resolved electrical transport measurements with use of SEM which have provided direct evidence of the coupling between superconductivity and local environment that is reflected by Co-concentration variation. In the uniform regions, the superconducting transition occurs at T_C = 22.1 K for 10% fixed percentage of the normal-state resistance. In the domain regions, although the onset superconducting transition temperature is found very close to that of the uniform regions, T_C varied over a broader range of 0.3-3.2 K. In addition, resistance of the domain regions above the transition onset temperature was noticed higher than that of the uniform region, indicating higher defect density in the domain regions.

Like HTSc's improvement in J_c is observed with increase of grain size in 122 systems also [50]. Effect of sintering temperature on the microstructure and superconducting properties of $Sr_{0.6}K_{0.4}Fe_2As_2$ bulk samples was made. It was found that the annealing temperature had little influence on the critical temperature *Tc*. However, the irreversibility field H_{irr} and J_c were significantly affected by the sintering temperature. The *SEM* images reveal although samples had similar microstructure, the grain size increases monotonically as the sintering temperature rises. The grain size was less influenced by temperatures over 850 C. It was concluded the J_c enhancement may result mainly from better grain connectivity due to the decrease of impurity phases.

4. Limitations

Although SEM is very useful in finding grain size, their connectivity and then revealing various microscopic properties with physics behind that but there are some limiting conditions for it. It is not a complete characterization in the sense that it needs extra characterization techniques such as TEM, PL, XPS and MO to support the results. The sensitivity of SEM is known to be relatively poor for lighter elements such as B, C and O. Thus through EDS results, the actual percentage ratio cannot be determined as very light elements boron and carbon are lesser sensitive in comparison to others elements [15]. In nano-TiO_2 doped samples [51] almost similar micrographs for all samples was found irrespective of whether they are doped with n-TiO_2 or not. In their study with HRTEM it was concluded that several black holes that appear in the image are presumably the n-TiO_2.

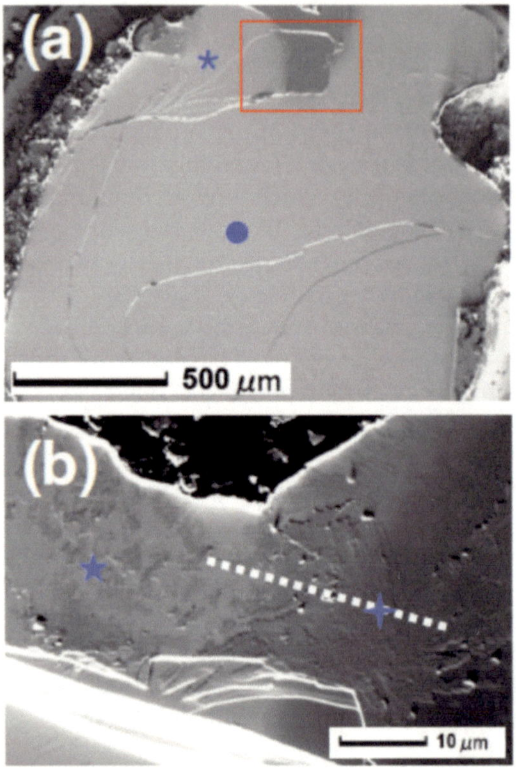

Fig. 12. Topographic images of cleaved surfaces of $BaFe_{1.8}Co_{0.2}As_2$ single crystal. (a) SEM image showing uniform contrast with some dark regions near the edge of crystal, as marked by a rectangular box. (b) Zoom-in SEM image showing domain structures. Marked regions by symbols and a dash line in (a) and (b) indicate where the transport measurements and composition probing are carried out.
Reprinted (Fig.) with permission from [T.-H. Kim, R. Jin, L. R. Walker, J. Y. Howe, M. H. Pan, J. F. Wendelken, J. R. Thompson, A. S. Sefat, M. A.McGuire, B. C. Sales, D. Mandrus, A. P. Li, Phys. Rev. B 80 (2009) 214518].

5. Summary

Summarily we can see that *SEM* has been widely used to explore the superconducting behavior. Grain size matters a lot in deciding the superconducting parameters of cuprate

HTSc. Some of the important aspects with which *SEM* deals with, is the grain size, morphology and alignment, structural defects, chemical composition. It has been used widely to explore from HTSc's, diborides to pnictides. With the time ways to use *SEM* and to extract information from superconductors got improved. Earlier for HTSc's it was used simply in finding grain size, grain connectivity and to figure out impurity regions. With these parameters superconducting behaviour was explained. Along with this in diborides it was also used to figure out the pinning centers and hence to enhance the applicable parameters. In Pnictides with spatially resolved electrical transport measurements it provided direct evidence of the coupling between superconductivity and local environment variation. We suppose in future SEM will be more plausible to understand so that superconductivity is better understood and improved.

6. Acknowledgements

Author S. K. Singh would like to acknowledge *CSIR*, India for providing fellowships. We are very much thankful NPG [Macmillan Publishers Ltd: NATURE 410 (2001) 186], IOP Publishing Ltd.{(SUST [Supercond. Sci. Technol. 22 (2009) 015010, Supercond. Sci. Technol. 22 (2009) 034012] and Nanotechnology **19** (2008) 125708)}, and APS [Phys. Rev. B 80 (2009) 214518] for providing permissions for the reprints of the images. We want to acknowledge the authors of the Ref. No. [15], [16], [20], [35], [40] for giving their consent to re-use the images.

7. References

[1] J. W Ekin, Adv. Ceram. Matter. 2 (1987) 586
[2] A Rosko, Y. M Chiang, J. S Moodera and D. A Rudman, American Ceram. Socie. (1988) 308
[3] D. S smith, S. Suasmoro, C.Gault, F. Caillaud and A. Smith, Revue Phys. Appl. 25 (1990) 61
[4] http://www.astm.org/
[5] http://www.astm.org/Standards/E112.htm
[6] T. Nagano, Y. Tomioka, Y. Nakayama, K. Kishio, K. Kitazawa, Phys. Rev. B 48 (1993) 9689
[7] S X Dou *et al.* Appl. Phys. Lett. 81 (2002) 3419
[8] W. K. Yeoh *et al.* Supercond. Sci. Technol. 19 (2006) 596
[9] H. Yamada, *et al.* Supercond. Sci. Technol. 19 (2006) 175
[10] C H Cheng, *et al.* Supercond. Sci. Technol. 16 (2003) 1182
[11] Senkowicz *et al.* Appl. Phys. Lett. 86 (2005) 202502
[12] R H T Wilke, *et al.* Phys. Rev. Lett. 92 (2004) 217003
[13] Xiang *et al.* Physica C 386 (2003) 611
[14] A. Matsumoto, *et al.* Supercond. Sci. Technol. 16 (2003) 926
[15] Arpita Vajpayee, V P S Awana, G L Bhalla and H Kishan Nanotechnology 19 (2008) 125708
[16] T.-H. Kim, R. Jin, L. R. Walker, J. Y. Howe, M. H. Pan, J. F. Wendelken, J. R. Thompson, A. S. Sefat, M. A. McGuire, B. C. Sales, D. Mandrus, A. P. Li, Phys. Rev. B 80 (2009) 214518
[17] Charles P. Poole, Jr., Handbook of superconductivity, Academic Press, California (2000)
[18] D. Sharma, Ranjan Kumar, H. Kishan and V.P.S. Awana, J Supercond Nov Magn 24 (2011) 205
[19] J. Kumar *et al.* J Supercond Nov Magn 23(2010) 493
[20] A Sotelo *et al.* Supercond. Sci. Technol. 22 (2009) 034012

[21] J.C. Bednorz and K.A. Müller, Zeitschrift für Physik B. 64(2) (1986) 189
[22] Chiang, Y.-M., Rudman, D.A., Leung, D.K., Ikeda, J.A.S., Roshko, A., Fabes, B.D., Physica C 152 (1988) 77
[23] J.W. Ekin *et al.* J. Appl. Phys. 62, (1987) 4821
[24] Shiming Zhou, Jiyin Zhao, Songnan Chu, Lei Shi Physica C 451 (2007) 38.
[25] R.J. Cava, Science 247 (1990) 656
[26] R. Gagnon, C. Lupien, and L. Taillefer, Phys. Rev. B 50 (1994) 3458
[27] M. Karppinen, V.P.S. Awana, Y. Morita, H. Yamauchi, Physica B, 312 (2003) 62
[28] P.R. Slater, C. Greaves: Physica C 180 (1991) 299
[29] Shiva Kumar, Anjana Dogra, M. Husain, H. Kishan and V.P.S. Awana, J. Alloys and compd. 352 (2010) 493
[30] N.P Liyanawaduge, Shiva Kumar Singh, Anuj kumar, V.P.S Awana and H.Kishan, J Sup. and Novel Magn doi: 10.1007/s10948-010-1063-7
[31] X.S. Wu, W.S. Tan, Y.M. Xu, E.M. Zhang,J. Du, A. Hu, S.S. Jiang, J. Gao Physica C 398 (2003) 131
[32] J. M. Tarascon *et al.* Phys. Rev. B 38 (1988) 8885
[33] O. Monnereau, Z.C. Kang, E. Russ, I. Suliga, G. Vacquier, T. Badéche, C. Boulesteix, A. Casalot Applied Superconductivity April (1995) 197
[34] Yeshurun, Y., Malozemoff, A. P. & Shaulov, A. Magnetic relaxation in high-temperature superconductors. Rev. Mod. Phys. 68 (1996) 911
[35] L. D. Larbalestier *et al.* NATURE 410 (2001) 186
[36] Manashi Nath and B. A. Parkinson Adv. Mater. 18 (2006) 1865
[37] S X Dou *et al.* Phys. Rev. Lett. 98 (2007) 097002
[38] Y. Kamihara, T. Watanabe, M. Hirano, H. Hosono, J. Am. Chem. Soc. 130 (2008) 3296
[39] S. Kaciulis *et al.* Surf. Interface Anal. 42 (2010) 692.
[40] F. Kametani, A. A. Polyanskii, A. Yamamoto, J. Jiang, E. E. Hellstrom, A. Gurevich, D. C. Larbalestier, Z. A. Ren, J. Yang, X. L. Dong, W. Lu, and Z. X. Zhao, Supercond. Sci. Technol. 22 (2009) 015010.
[41] A. Yamamoto, J. Jiang, C. Tarantini, N. Craig, A. A. Polyanskii, F. Kametani,F. Hunte, J. Jaroszynski, E. E. Hellstrom, D. C. Larbalestier, R.Jin, A. S. Sefat, M. A. McGuire, B. C. Sales, D. K. Christen, and D. Mandrus, Appl. Phys. Lett. 92 (2008) 252501
[42] B. Senatore, G. Wu, R. H. Liu, X. H. Chen, and R. Flukiger, Phys. Rev. B 78 (2008) 054514
[43] R. Prozorov, M. E. Tillman, E. D. Mun, and P. C. Canfield, New J. Phys.11 (2009) 035004
[44] F. Kametani, P. Li,1 D. Abraimov, A. A. Polyanskii, A. Yamamoto, J. Jiang, E. E. Hellstrom, A. Gurevich, D. C. Larbalestier, Z. A. Ren, J. Yang, X. L. Dong, W. Lu, and Z. X. Zhao Appl. Phys. Lett. 95 (2009) 142502
[45] J.-Q. Yan, Q. Xing, B. Jensen, H. Xu, K. W. Dennis, R. W. McCallum, and T. A. Lograsso Phys. Rev. B. 84 1250
[46] M. Rotter, *et al.* Phys. Rev. B 78 (2008) 020503
[47] X C Wang, *et al.* Solid State. Commun. 148 (2008) 538
[48] Hsu F C *et al.* Proc. Natl Acad. Sci. USA 105 (2008) 14262
[49] Ma Y W, Gao Z S, Wang L, Qi Y P, Wang D L and Zhang X P Chin. Phys. Lett. 26 (2009) 037401
[50] Zhiyu Zhang, Yanpeng Qi, Lei Wang, Zhaoshun Gao, DongliangWang, Xianping Zhang and Yanwei Ma Supercond. Sci. Technol. 23 (2010) 065009
[51] H. Kishan, V.P.S. Awana, T.M. de Oliveira, Sher Alam, M. Saito, O.F. de Lima Physica C 458 (2007) 1

27

Morphological and Photovoltaic Studies of TiO_2 NTs for High Efficiency Solar Cells

Mukul Dubey and Hongshan He*
Center for Advanced Photovoltaics, Department of Electrical Engineering & Computer Science South Dakota State University, Brookings, SD, USA

1. Introduction

Highly ordered nanostructures, especially TiO_2 NTs, have attracted considerable research interest in recent years due to their diverse applications in photocatalysis, photonic crystals, sensors, batteries and photovoltaic devices. The photophysical, photochemical, electrical and surface properties of these nanostructured materials depend highly on their morphology because of the quantum size effect. Hence it is critical to study the effect of morphology of the ordered nanostructures for device applications. In this chapter we will only focus on the TiO_2 NT morphology in context of their applications in dye-sensitized solar cells (DSCs).

DSC is an electrochemical device that converts sunlight to electricity. The major components of DSC are photoelectrode, counterelectrode and electrolyte sandwiched between them. The photoelectrode is a dye-coated wide band gap semiconductor, such as TiO_2, on a transparent conducting oxide (TCO) glass substrate. Dye molecules absorb sunlight and the electrons in the ground state are excited to the excited state. The electrons in the excited states inject into the conduction band of TiO_2. The injected electrons transports to the TCO electrode via diffusion through TiO_2 NPs. The electrons then flow through the external circuit to the counterelectrode, which is usually a platinized TCO glass. The redox species in the electrolyte, usually iodide, take the electron from counterelectrode, and are reduced to tri-iodide, which further gets oxidized by providing its electron to the ground state of dye molecule for its regeneration. There are several factors that affect the efficiency of DSC such as absorption band of dye molecule, electron injection efficiency from dye to TiO_2, redox potential of electrolyte and charge transport through TiO_2. The morphology of TiO_2 photoelectrode is one critical factor that plays a pivotal role in the conversion of sunlight to electricity in DSCs. Remarkable breakthrough in photoelectrode by changing the planar structure to randomly packed mesoporous structure of TiO_2 NPs improved the efficiency from less than 1 % to 8% by Grätzel *et al.* The mesoporous structures are promising due to their high surface area for the adsorption of photosensitzer leading to the improved light absorption and hence high efficiency. The photoelectrode was further optimized by introducing a compact layer with small TiO_2 NPs and a scattering layer with large TiO_2 NP underneath and at the top

* Corresponding Author

of normal TiO_2 NPs respectively. Both improved electrical and optical properties of photoelectrode and hence the device efficiency. With those structures and ruthenium bipyridine dyes, a respectable efficiency of 11.5% has been achieved rendering the DSCs as promising and cost-effective alternative to its otherwise expensive silicon technology.

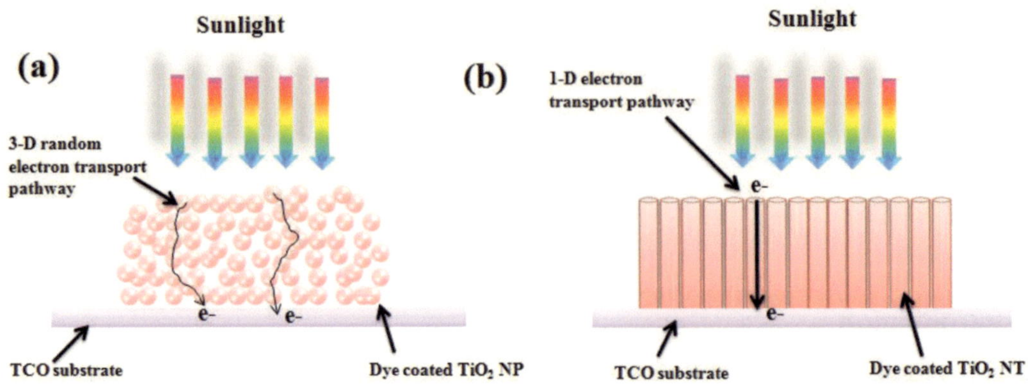

Fig. 1. **(a)** Schematic representation of electron transport in TiO_2 NPs based photoelctrode; **(b)** electron transport in TiO_2 NT based photoelectrode

The electron collection efficiency is a critical factor governing the overall photo conversion efficiency of solar cell. Various investigations suggest that the random morphology of polycrystalline TiO_2 NPs exhibits high defect density, which leads to the electron losses via recombination and the reduced electron collection efficiency. The presence of numerous defects, grain boundaries and surface states provides several trapping/detrapping and recombination sites in the electron transport pathway. The presence of defects reduces the electron mobility leading to increased recombination and hence reduced cell performance. In this regard anodic TiO_2 NTs proposed by Grimes *et al* is considered as an excellent electron acceptor for DSC. Architecturally, these NTs are well aligned in regular array perpendicular to the substrate leading to rapid unidirectional electron transport with reduced recombination. A schematic for difference in dimensionality of electron transport between random nanocrystalline particle network and one-dimensional NT is shown in Figure 1. The electron from dye molecules migrate directly from top of the NT to the bottom for electron collection without migration in a three dimensional network. A close to 100% electron collection efficiency at the bottom of the nantotube was observed. In addition, NTs also have strong light scattering behavior which increases the optical path length in the film and improve the light absorption efficiency for high solar cell efficiency.

Despite being promising both electrically and optically, the highest energy conversion efficiency obtained from NT based DSCs is only ~ 7%, which is much lower than the conventional NP based DSC. One of the disadvantages identified was the back illumination geometry of devices due to the presence of non-transparent Ti metal underneath the TiO_2 NT arrays. The TiO_2 NT arrays are usually grown directly from a thin layer of Ti metal, which is difficult to remove. This requires photo illumination from the counterelectrode (a platinum coated transparent conducting electrode) side as shown in Figure 2. The back illumination leads to significant loss in the photon flux by reflection from the platinum and absorption in the electrolyte. It was difficult to realize front illumination since the NTs were grown on titanium substrate and no technique was known to either grow or transfer the NT films on to the transparent conducting substrate.

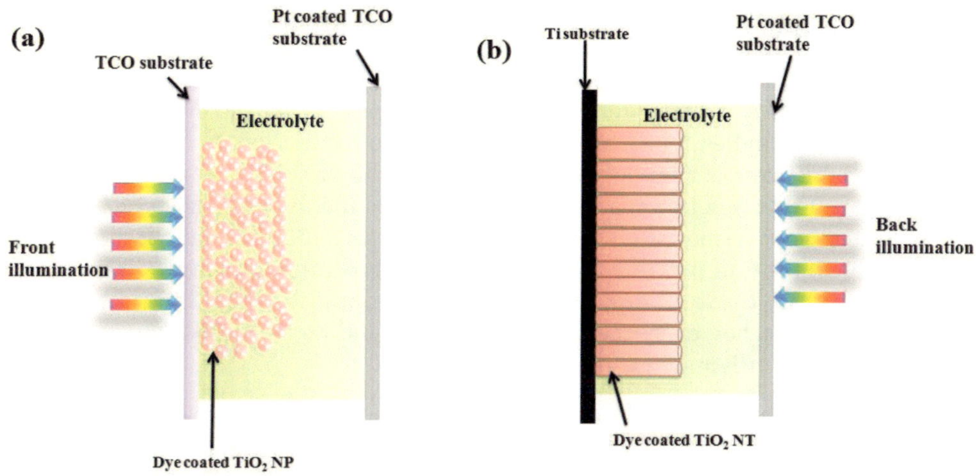

Fig. 2. **(a)** TiO_2 NP based DSC with front illumination geometry from photo-electrode side; **(b)** TiO_2 NT based DSC with back illumination geometry from counter-electrode side

Front illumination in TiO_2 NT-based DSCs can be realized through several recently reported methods. The first method is the growth of NTs on glass substrate with sputtered Ti metal on top. The sputtering must be performed at high temperatures to prevent peeling after anodization. Grimes *et al* recently reported a new method for sputtering Ti on FTO glass at low temperature that produced TiO_2 NTs with lengths up to 33 μm after anodization. A cell with a 17 μm NT array achieved a conversion efficiency of 6.9%. Two concerns emerge with this process: (1) the time- consuming nature of sputtering several tens of micrometer Ti may increase cost, and (2) the FTO layer on the glass could be damaged during anodization.

The second method is to remove the NT array from the Ti foil and attach it to FTO glass. In 2008, Jong Hyeok Park *et al* put anodized Ti foil in 0.1 M HCl aqueous solution for 1 hour, obtained an NT membrane, and attached it to FTO glass with the help of titanium isopropoxide. They achieved 7.6% efficiency with 8 μm NT arrays. Although the team claimed that NT membranes could be handled with tweezers, optical images in their publication suggested that these NT membranes were very fragile. In 2009, Qinwei Chen *et al* reported a re-anodization process that was followed by immersing the foil in 10% aqueous H_2O_2 solution for 24 hours and resulted in large sized NT membranes. The NTs were then attached to FTO glass with the help of a TiO_2 NP paste to achieve a conversion efficiency of 5.5%. Long-time immersion in solution diminishes the attractiveness of this mild process.

He *et al* also developed a method that can lift off the NT arrays in less than four minutes. The yellow membrane could easily be transferred to other substrates without any fracturing. He *et al* also developed a unique low temperature method to tightly plant the NT membranes on FTO glass. The NTs were embedded inside the NP layer. The DSCs with these films exhibited 6.1% efficiency using N719 as dye. It was found that the geometry of NT orientation on the glass substrate also plays a significant role in determining the efficiency of DSC. The test tube geometry of NTs with one end open and other end closed provides freedom in choosing the configuration of the freestanding NT fixation on substrate with either closed end or the open end on to the substrate. This finding suggests that both optically and electrically open end of the NT on to the substrate is superior to the other orientation and hence can help significantly in improving DSC efficiency.

Another challenge for the effective use of NT for DSC application is how to grow highly ordered TiO_2 NT arrays. Many researchers have reported that the NT tends to cluster together and form bundles which not only inhibits the infiltration of dye and electrolyte throughout the thickness of film but also increases recombination by incorporating disorder induced defects. It was reported that fine polishing of the titanium substrate prior to growth minimized the cluster formation. Several reports also indicated that the bundle and micro crack formation in the film was due to the capillary stress during the sample drying process. The supercritical CO_2 oxide drying technique was introduced, which indeed reduced the formation of clusters; however, the complete understanding of cluster formation is still elusive and requires further study.

To summarize the morphology of TiO_2 NT plays a critical role in dye-sensitized solar cell. Study of the effect of morphology of TiO_2 NT on DSC performance is therefore worthy of pursuit for achieving high conversion efficiency of the DSCs. In the following sections, we will discuss the growth mechanism of TiO_2 NTs and approaches for highly ordered TiO_2 NT array of NTs for DSC applications. We will also discuss how the effect of orientation of the NT on the TCO glass affects the photovoltaic properties of DSC.

2. Growth mechanism of TiO_2 NTs

This section reviews the growth mechanism of TiO_2 NTs by potentiostatic anodization technique in fluoride-containing electrolyte. The NT formation in acidic electrolyte containing F^- ion is generally agreed to occur via the field assisted formation and dissolution of oxidized titanium surface. It involves two critical steps that occur simultaneously: formation of TiO_2 on the titanium surface and the dissolution of oxide. The process can be described by following two reactions:

$$Ti + 2H_2O \rightarrow TiO_2 + 4e^- + 4H^+ ....(1)(Oxidation)$$

$$TiO_2 + 6F^- + 4H^+ \rightarrow [TiF_6]^{2-} + 2H_2O...........(2)(Dissolution)$$

In this two-electrode setup, titanium serves as anode and platinum as cathode. The electrolyte is composed of ethylene glycol, ammonium fluoride and water. A constant DC voltage is applied across the electrodes as shown in Figure 3. After some time, a layer of TiO_2 NTs will form on the surface of Ti metal. Figure 4 shows schematically how the TiO_2 NTs are formed. When pristine Ti is immersed into electrolyte solution, it is surrounded by various ionic species such as OH^- and F^-. (a) Once the DC voltage is applied these ionic species tends to oxidize the surface of titanium substrate (b) forming a thin barrier layer of TiO_2 as depicted in the equation 1 of reaction mechanism. Simultaneously the process of dissolution of TiO_2 layer in presence of F^- ion occurs leading to the formation of random pores during the initial stage of growth process (c). The F^- ions localize to the bottom of the pore i.e. at the oxide/metal interface which further undergoes oxidation and dissolution processes. Since the concentration of F^- ion is more at the bottom of the pore due to the external electric field; the effective dissolution of TiO_2 is more pronounced at the pore bottom leading to vertical cavity formation (d to f). The formation of round shape at the bottom of tube is still a topic of debate. It is proposed that this is results of volume expansion of TiO_2 compared to the space available from metal loss leading to high stress at the interface, high electric field distribution density at pore bottom and enhancement in acidity at the pore bottom due to the external electric field.

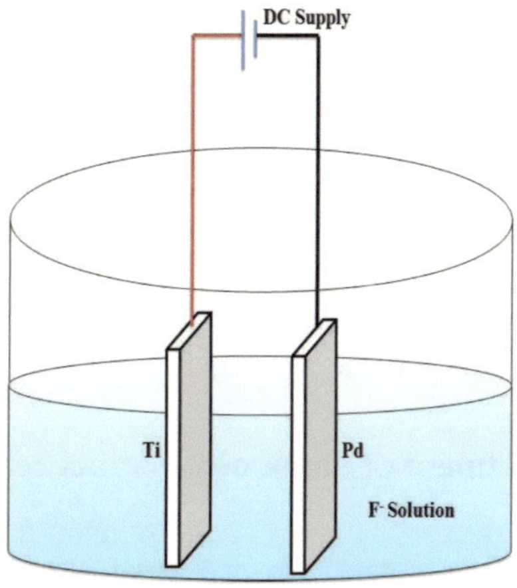

Fig. 3. Electrochemical anodization set up.

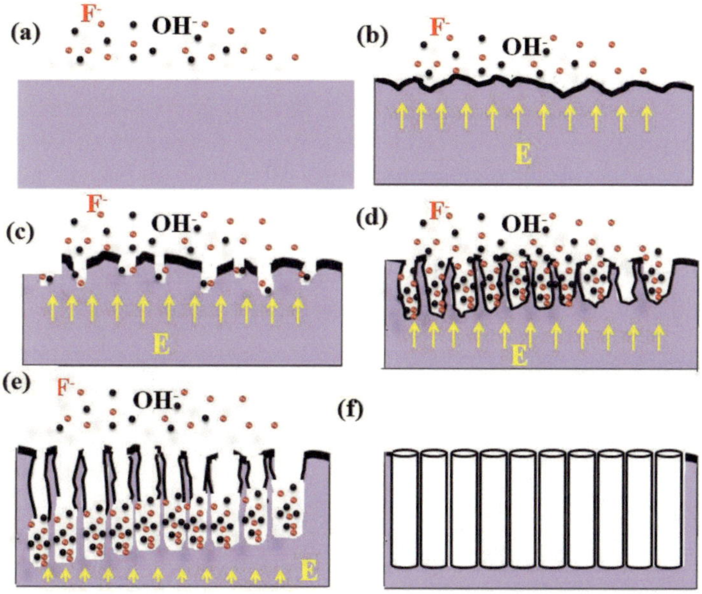

Fig. 4. **(a)** Titanium substrate in the ionic environment of electrolyte; **(b)** Formation of porous oxide layer on exposed surface of titanium right after field is switched on; **(c)** Initial random pore growth by dissolution; **(d)** elongation of pore geometry after few minutes of anodization; **(e)** development of regular array of pore geometry in the field direction; **(f)** fully developed NT array. Red and Black dots represent the fluoride & hydroxide ions respectively.

3. Effect of substrate morphology on growth of TiO_2 NTs

The formation of NTs largely depends on the type and concentration of ionic species present in the electrolyte as well as the extrinsic parameters such as anodization voltage, time and temperature. By controlling these factors, TiO_2 NTs having different length, diameter, and wall thickness can be obtained. However, it should be noted that field assisted directional dissolution of the oxide layer formed on titanium foil is a crucial step towards the formation of NTs which so far have been shown to depend on many variables such as electrolyte composition, concentration, anodization voltage and time, but least importance was given to the effect of substrate morphology on the growth of NTs which is discussed in the next section. We found that the morphology of titanium substrate also plays a key role in the morphological order of the NT thus formed. This section highlights the effect of morphological features of titanium substrate on NT growth which is further connected with the microscopic morphology drawing outline for the plausible reasons for the clustering of NTs and cost effective way to deal with it.

3.1 Effect of mechanical treatment of titanium substrate on TiO_2 NT growth

Commercial Ti foil with thickness ~ 250 μm is usually used for the growth of the TiO_2 NTs arrays. Before the anodization the Ti foil is cleaned by detergent, ethanol, toluene, and deionized water sequentially to remove any impurities on the surface. There are several commercial providers for Ti foil with high purity; however, the surface morphology of these as-purchased Ti foils is quite different. It was found the as-purchased Ti foil has many crack sites distributed throughout the surface of the substrate. Figure 5 shows the typical SEM image of the surface of one sample from Sigma-Aldrich. Many cracks were observed on the surface. The size of the cracks ranges from several hundred nanometer to several micrometer. The presence of such cracks leads to the formation of vertical gaps on the substrate leading to the absence of material up till certain depth. In addition there are several submicron range heterogeneous morphologies present in the vicinity of crack sites which render high degree of roughness to the substrate. The existence of cracks on the Ti surface leads to high degree of non-uniformity in the morphology of NTs thus formed resulting in the cluster and bundle formation of NT. Figure 6 (a) shows the SEM image of

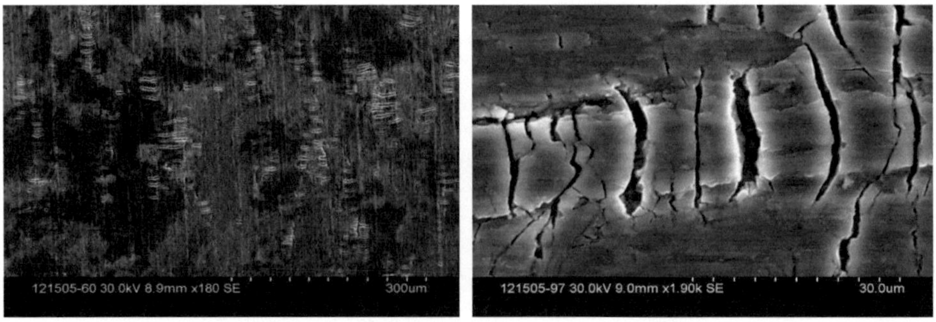

(a) (b)

Fig. 5. **(a)** Cracks or vertical gaps present on the surface of as purchased commercially pure titanium substrate; **(b)** magnified image of crack showing the absence of material up till certain depth.

surface morphology for NTs grown on as purchased commercially pure titanium foil for 15 minutes. The fingerprint of substrate crack structures and submicron heterogeneously distributed morphology near crack site were clearly observed on the NT film.

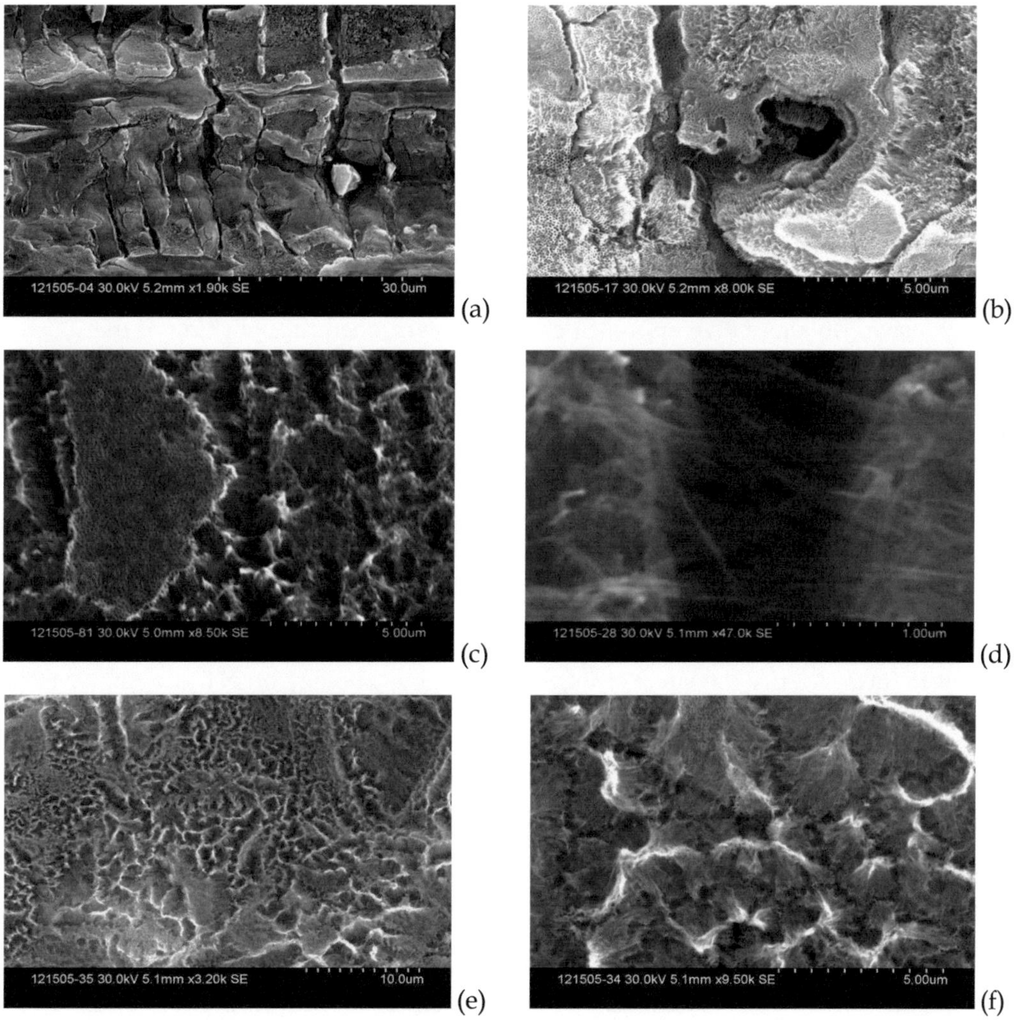

Fig. 6. **(a)** TiO_2 NTs after 30 min anodization; **(b)** TiO_2 NT with a whirlpool geometry at crack site; **(c)** TiO_2 NT cluster formed at crack lines; **(d)** Collapsed TiO_2 NTs at crack lines; **(e)** TiO_2 bundles throughout sample and **(f)** TiO_2 bundles under higher magnification

We further investigated the local morphology of NTs near the crack sites which is shown in higher magnification SEM image of Figure 6 (b). Whirlpool geometry of NT distribution at the crack site was observed, which shows the strong influence of substrate morphology on the initial growth of NTs. This effect was more pronounced in the NT under short anodization time. The clusters are formed near the crack lines of the substrate. Uniformed NTs are observed on the surface without any cracks. We also observed that the tubes over the edges of cracks tended to collapse on each other forming intercrossed tubes as shown in

Figure 6 (d). The collapsing of the NTs on each other can potentially lead to the cluster formation which can be seen from Figure 6 (e & f).

Based on the results of anodization on commercially purchased Titanium, it can be observed that smooth surface for anodization is very crucial to obtain highly ordered morphology of NTs. Han et al and Lee et al reported two step anodization processes to obtain ordered morphology of NTs. In their report first anodization was performed for shorter time followed by removal of the first NT layer. The surface of Ti after removal of first layer was very smooth leading to highly ordered morphology of NT formed in the second step. On the other hand Kang et al reported electropolishing technique in which Ti substrate was electropolished to render it a smooth surface followed by anodization to form ordered NT structure. Both electropolishing and two step anodization processes were found promising to obtained highly ordered NT array.

However, these processes involves complex two step processes which is time consuming and expensive. An alternative approach could be the mechanical polishing of the substrate to remove cracks. To this end we have tried to polish the Ti substrate using fine sand paper. However, our SEM results shows that even with very fine sand paper the micron size scratches are developed on the surface scratches are developed on the surface. It can be clearly seen that there It can be clearly seen that there were significant clumping and clustering of the NTs. Additionally at many other places NTs were found to be completely broken. Based on the results it can be inferred that even the fine mechanical polishing can form micron level roughness which cannot be used to grow highly ordered NTs.

3.2 Effect of chemical treatment of titanium substrate on TiO_2 NT morphology

In order to further verify the effect of local substrate morphology on NT growth, we etched the titanium substrate for 30 minutes in 0.75 M hydro fluoric acid (HF) introducing high degree of surface roughness to the substrate. Figure 8 (a) shows the morphology of rough surface of titanium after etching. TiO_2 NTs were then grown on the etched substrate for 15 minutes. It was observed that the initial pore formation for NT growth takes the local geometry of the substrate as shown in Figure 8 (b). The local pore formation might largely depend on the direction of local electric field was further confirmed by the NT formation in the etched substrate. Figure 8 (c) shows the SEM image of a large pit formed on the substrate due to etching. The pit shown in the image can be visualized to have three different planes i.e. x-y, y-z and x-z. It is interesting to note that the pore formation can be seen on all these three planes with their cross-sections perpendicular to the respective plane clearly indicating that the initial pore formation does depend on the direction of local electric field at the breakdown site this further depends on the local morphology of the substrate as shown in Figure 8 (d). The dependence of NT growth associated with the local electric field distribution corresponding to the substrate morphology can be a profound reason for the bundle and cluster formation in NTs which was further confirmed from SEM results. Figure 8 (e) shows the SEM image of NT at one of the crack sites of the NT film grown on etched substrate. It can be clearly observed that the NTs at crack site grew in different direction. Considering x-y plane to be the plane of substrate and z as direction normal to the substrate which is the preferred direction of NT growth, it can be clearly seen that the cross-sectional plane of NTs are facing in two different directions, one parallel to x-y plane highlighted with red circle and other in z- direction highlighted with yellow circle. The NTs facing x-y direction bends toward the z- direction. The initial bending followed by z growth of NTs was further confirmed in Figure 6 (f) where it can be observed from one of the pits that the initial pore formation on the walls of the pit is in all three directions. However as the NTs grew longer they start bending in one direction which latter completely follows one directional growth. Interestingly it can be seen that the initial bending ranging to several microns leads to the collapse of NTs on each other leading to the formation of clusters. Hence formation of highly ordered NTs can be severely influenced by the substrate morphology.

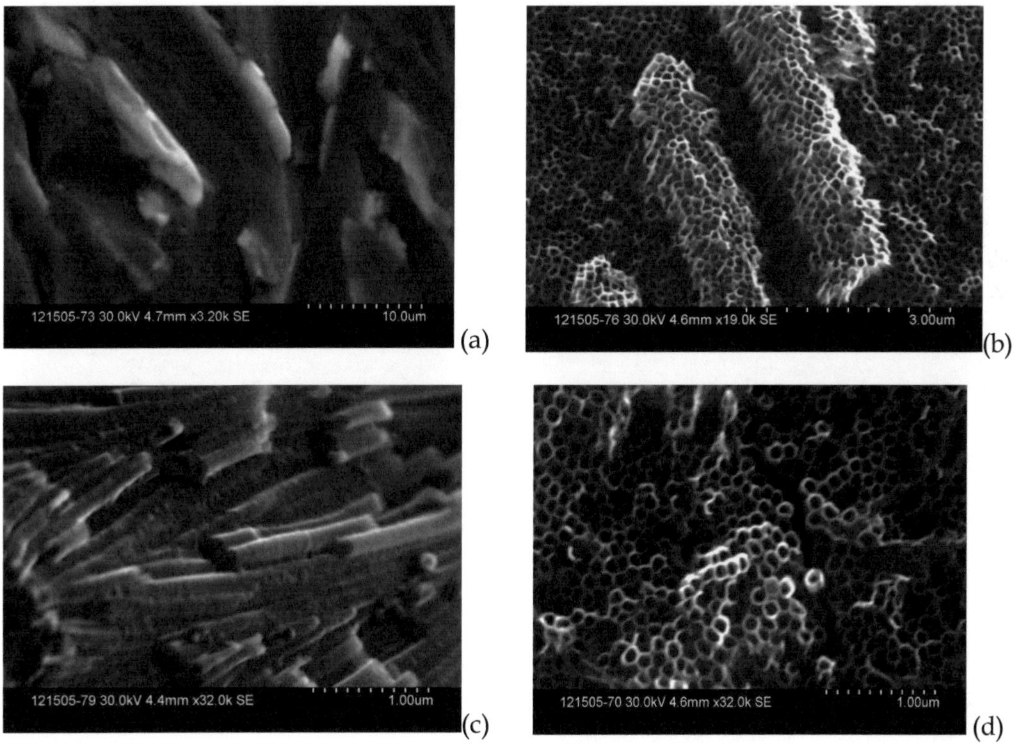

Fig. 7. **(a)** SEM image of polished Ti substrate; **(b)** bundle formation and non-uniformed TiO2 NT morphology; **(c)** side view showing different length of NTs and the bundle formation; **(d)** unevenly packed TiO2 NTs

Removing structural disorder from NTs was recently a key concern in the area of DSC. Some techniques including post growth ultrasonic treatment and supercritical CO_2 drying of NT samples showed promise in removing of the structural disorder. These techniques are very useful if the disorder in NT morphology is induced through impurities in the electrolyte, viscosity of the electrolyte or during drying of NTs after growth. Their applications to remove substrate induced disorder are limited. We employed a chemical etching process to solve this problem. The Ti substrates were immersed in 0.75 M HF ranging from 1 to 15 minutes. The cracks present on the substrate were removed completely in 10 minutes of etching time. Figure 9 (a) shows the SEM image of titanium foil etched for 5 minutes in 0.75 M HF where the crack features could still be observed. Figure 9 (b) shows

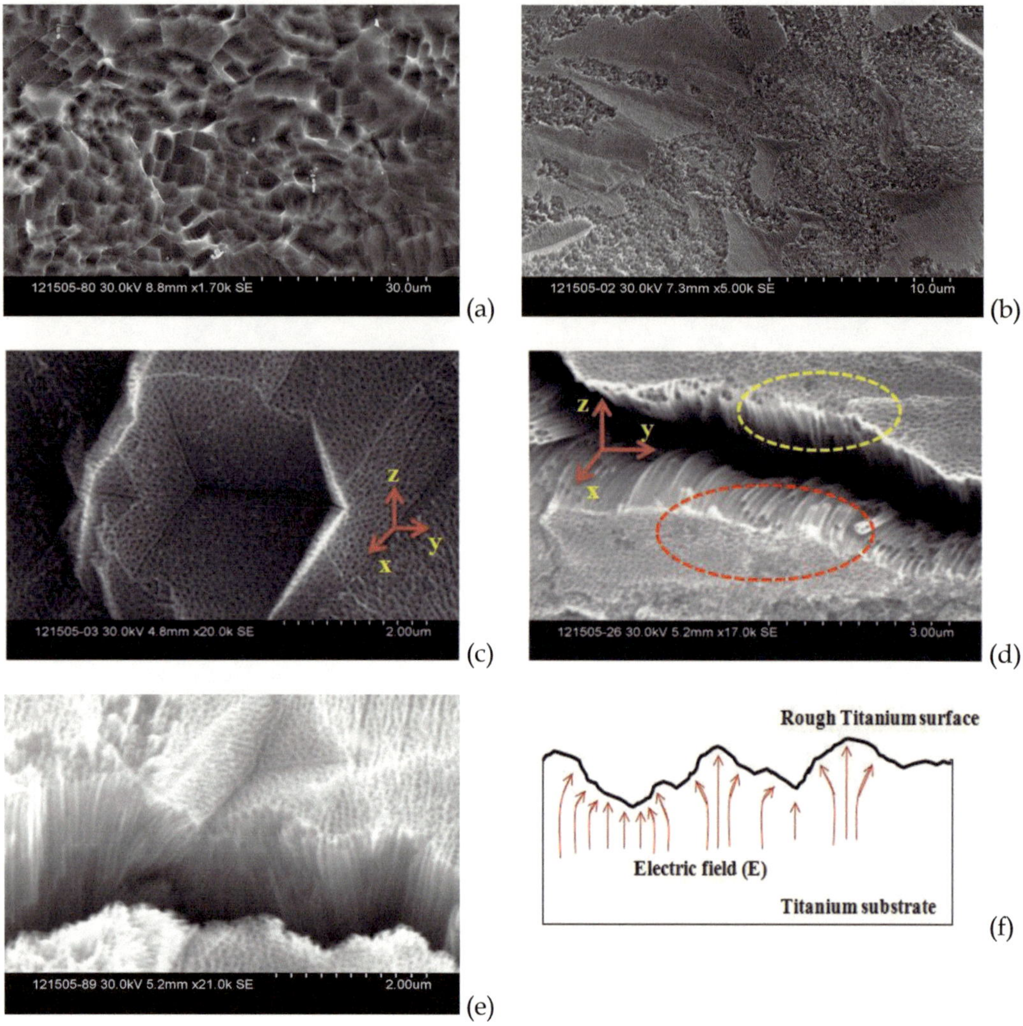

Fig. 8. SEM images of **(a)** etched Ti substrate; **(b)** TiO2 NTs grown on etched Ti; **(c)** large pit of TiO2 NTs on etched Ti; **(d)** TiO2 NT at a crack site on etched Ti; **(e)** TiO2 NT at the edge of one pit **(f)** non uniform local electric field distribution near the rough surface of titanium.

the cracks or vertical gaps completely disappeared after 10 minutes of etching but also introducing high degree of surface roughness induced on the substrate. Further etching the substrate for 15 minutes led to highly disordered coarse surface as can be seen in Figure 9 (c). A closer investigation of individual pits formed after 10 minutes etching of the substrate as shown in Figure 9 (d) revealed that these pits offer a very smooth concave shaped surface with average size of 5 - 10 μm. This observation suggested that highly oriented NTs can be grown over these smooth surfaces with short range of order on the surface of the substrate. Further concavity of the pit structure can lead to small bending in the NTs with cross-section plane facing towards the center of conic cross-section. The small bending of NTs can further help preventing the NTs to interact and collapse over the NTs formed in the neighboring pits, providing global order in the overall morphology of NTs. In order to verify our assumption we performed 30 minutes of anodization to grow shorter NTs on the titanium substrate etched for 10 minutes in 0.75 M HF. Figure 10 (a) shows the SEM image of NTs grown on etched substrate for 30 minutes anodization time. The image clearly shows that the NTs followed the local morphology of each pit taking the overall geometry of the substrate. In addition clustering or collapse of NTs was also not observed anywhere on the surface suggesting that overall order in the morphology can be achieved by this process. However, the method can find its applicability only when longer NTs can be successfully grown with long range order which is the essential need for solar cells. To investigate the morphology of longer NTs, we performed anodization of the etched substrate for 5 hrs which can lead to the formation of ~ 20 μm long NTs.

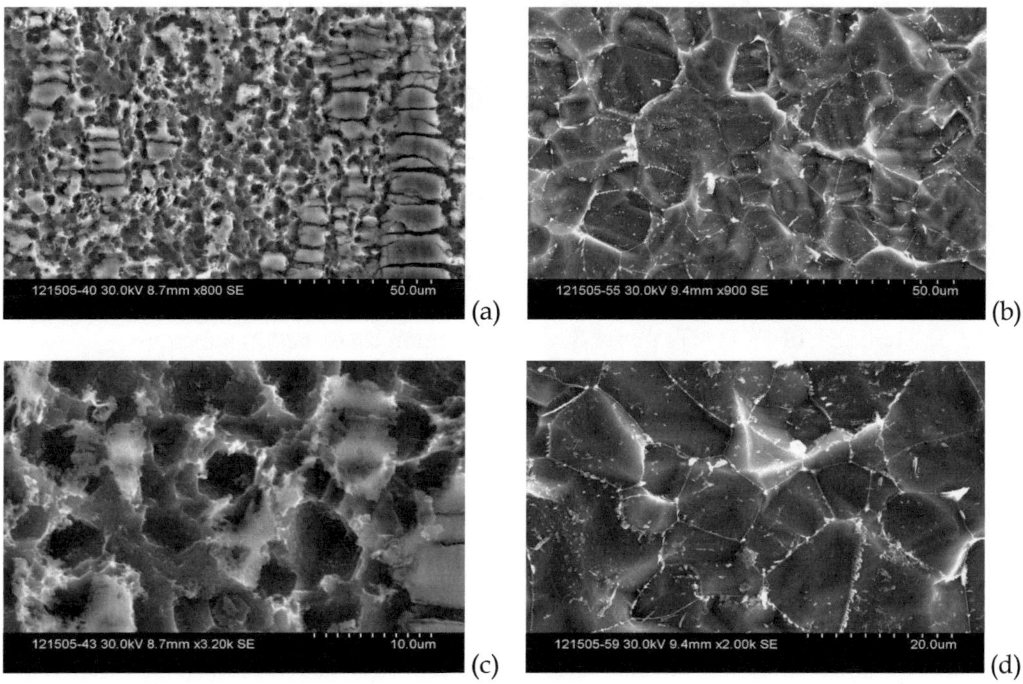

Fig. 9. SEM images of titanium substrate etched in 0.75 M HF under different etching time. (**a**) 5 minutes; (**b, d**) 10 minutes; (**c**) 15 minutes.

Interestingly the SEM image of Figure 10 (b) shows that the NTs even after 5hrs of anodization time followed highly ordered morphology without cluster formation anywhere on the substrate. It was also evidenced that the NTs retained the concave geometry of the substrate shown highlighted in yellow circle of Figure 10 (c). The overall morphology of the NTs were observed to be comprised of several small concave shaped honeycomb structure grouped together to form structured NT film which can be seen from SEM image of Figure 10 (d). Thus it can be seen that the morphology of the NTs significantly depends on both the morphology of the substrate and simple chemical pretreatment of the substrate can prove to be useful in growing oriented NTs which might further help in improving the efficiency of DSC.

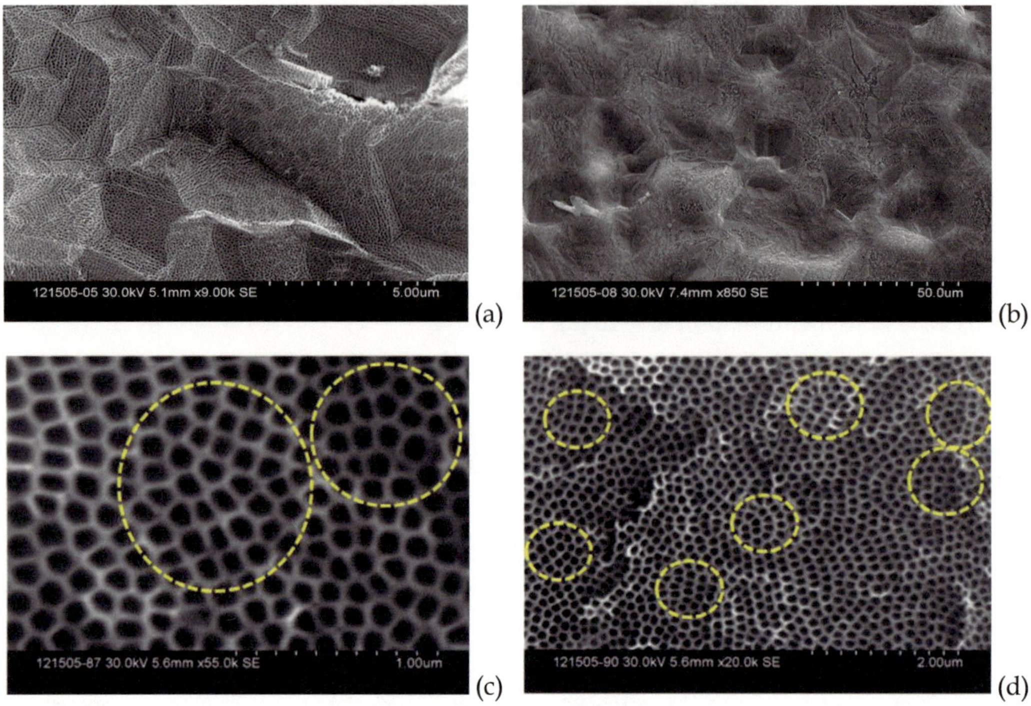

Fig. 10. SEM images of TiO2 NTs under 5 hrs anodization on etched Ti substrate.
(a) anodized surface at higher magnification; **(b)** anodized surface at lower magnification; **(c)** highly ordered NT with local concavity shown highlighted in yellow circle; **(d)** several concave geometries highlighted in yellow circles

4. Effect of TiO_2 NT morphology on PV performance of DSC

The TiO_2 NTs on the Ti substrate can be used directly for the fabrication of DSCs. The Ti metal will function as same as TCO layer in conventional Gratzel type DSCs. Due to the non-transparency of Ti metal to the sunlight, the cell has to be illuminated from counter electrode (back illumination). In 2007, Grimes *et al* reported 6.89 % conversion efficiency of this type of cell using ruthenium dye (N719) as light absorber, 20 μm long TiO_2 NT arrays for dye adsorption, and iodide/tri-iodide as electrolyte. Several other groups who fabricated DSCs with this configuration achieved efficiencies ~3% under similar conditions. In 2009, Grätzel *et al* reported a 3.59% conversion efficiency of DSCs using ruthenium dye (N719) as a light absorber, 14 μm long TiO_2 NT array for dye adsorption, and ionic liquid as electrolyte. He *et al* also achieved an efficiency of 3.45% with this configuration. Since TiO_2 NT arrays are often attached on the Ti foil and difficult to lift off, the NT arrays with Ti foil were used directly for cell fabrication. Sunlight must come from the rear of the cell. The absorption and reflection of sunlight by electrolyte and Pt counterelectrode respectively lead to reduction in photon flux reaching the dyes. Various techniques were reported from 2008 – 2010 for the growth, liftoff and fixation of NTs on transparent conducting substrate but they either lacked reproducibility or was time consuming.

In our work Freestanding NT films were obtained by preferential etching of the TiO_2/Ti interface followed by its fixation on TCO with colloidal TiO_2 paste as adhesive layer. The SEM image of freestanding NT film reveals that one end of the NT is open while other end is closed rendering it to be like a test tube structure. Figure 11 (a) shows the morphology of open end of NT while Figure 11 (b) shows the surface morphology of the closed end side of NT. The freestanding NTs can be used in two different orientations for fixation on TCO substrate; one with open end of NT facing the substrate while other with closed end of NT facing the substrate as shown in Figure 12. This section thus tends to highlight the effect of NT orientation on DSC performance. It was reported earlier that the closed end of NT facing the substrate might be helpful in improving the efficiency of DSC by serving as a barrier layer in between substrate and TiO_2 active layer improving the charge transport by minimizing the substrate/TiO_2 interface recombination analogous to the compact layer in NP based DSC.

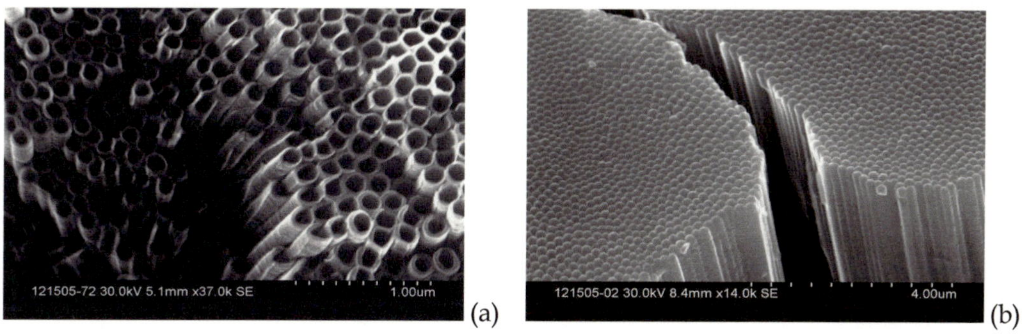

Fig. 11. **(a)** Top view of NT showing one end to be open; **(b)** bottom view of NT showing other end to be closed

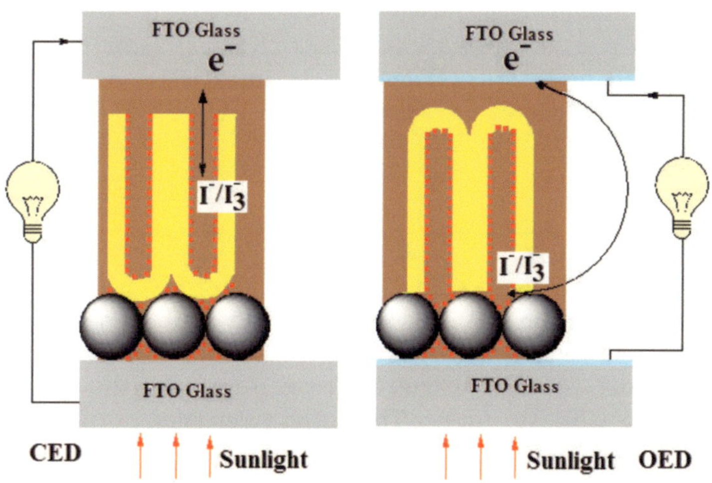

Fig. 12. Simplified DSC structure with CED and OED orientation of NT on TCO

In contrast it was also reported that ~ 2 - 3 μm thick layer of TiO_2 at the closed end of NT might serve as an insulating layer between TCO/TiO_2 layers which can be detrimental for effective charge transport from active layer to the electrode. In order to investigate the effect of closed end layer on PV performance, we fabricated DSC with two different orientations i.e. closed end facing the substrate and open end facing the substrate, hereafter referred to as CED and OED respectively. The DSCs fabricated with these two structures have apparently shown a big difference in their PV performance as can be seen from the J-V characteristics shown in Figure 13 (a).

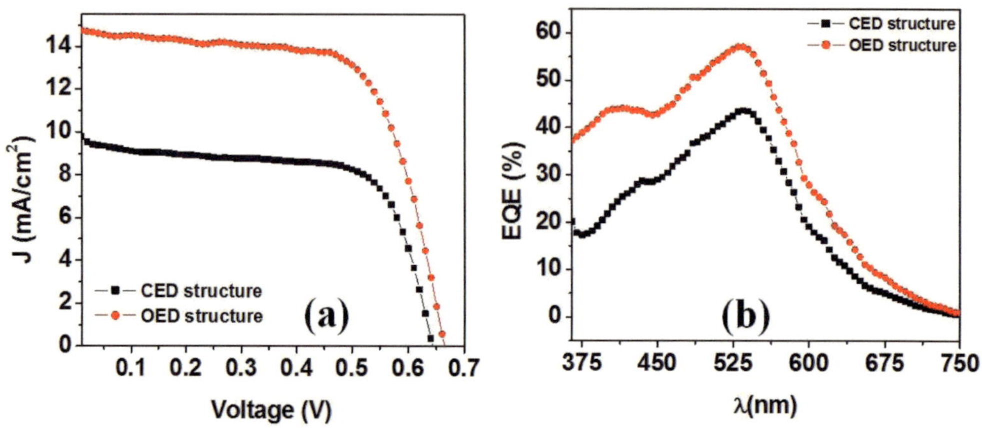

Fig. 13. **(a)** J-V curve under illumination for cells with OED and CED structures; **(b)** EQE curves for cells with OED and CED structures

It was found that the OED structure had higher efficiency of 6.58% as opposed to 4.17% efficiency of CED structure. It was found that cell with OED structure exhibited higher values of short circuit current density (J_{SC}), open circuit voltage (V_{OC}) and fill factor (FF) compared to CED structure. The J-V data for the photovoltaic performance of two cells is provided in Table 1.

Orientation of NT	NP layer thickness (μm)	NT length (μm)	J_{SC} (mA/cm²)	V_{OC} (mV)	FF (%)	η (%)
OED	3	22	14.75	666	67.05	6.58
CED	2.7	23	9.5	642	68.45	4.17

Table 1. J-V data for cells with OED and CED orientation of NTs.

In order to further support our J-V data we performed the external quantum efficiency (EQE) measurements on two cells as shown in Figure 13 (b). The EQE data was found to be very consistent with our J-V data where OED structure have shown greater quantum efficiency compared to CED structure. The current densities calculated from the EQE measurements were found to be ~ 15 and 10 mA/cm² for OED and CED structures respectively which were in close agreement with the J-V data. Overall the cell performance indicated the superiority of the OED over CED orientation.

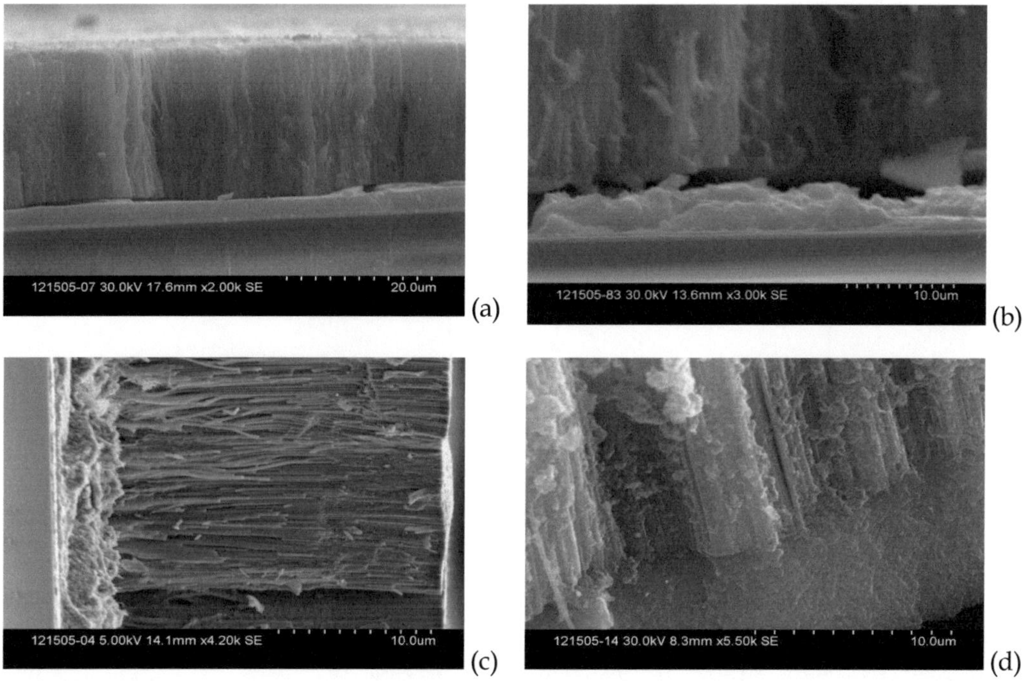

Fig. 14. Shows the cross-sectional SEM image of TiO2 NTs on FTO glass (**a**) CED orientation; (**b**) CED NT/TiO2 NP interface; (**c**) OED orientation; (**d**) OED NT/ TiO2 NP interface

In order to investigate the reason for difference in the PV performance of two structures we performed the cross-sectional SEM imaging of CED and OED structures shown in Figure 14. The interface between colloidal TiO_2 NP layer and the NT for CED structure (shown in Figure 14 (a & b)) can be seen to have gaps in between these two layers which suggest that the electron transfer between these two layers is not efficient leading to excessive slow down of the electrons at this interface increasing the recombination probability. We attribute the poor interface quality of this structure to the round shaped closed end of the NT which might have prevented the colloidal particles to partially penetrate into the tube leading to weak interface formation which upon high temperature sintering of the film might have introduced gaps at the interface. Interestingly this feature was not observed in the case of OED structure as can be seen from the cross-sectional image of Figure 14 (c & d). The NTs were found to have formed very good interface by embedding itself into the NP matrix leaving behind no gaps. It can be seen from the image that even after sintering at high temperature the interface retained its good morphology.

In order to investigate the reason for higher photocurrent in OED structure we performed dye loading measurements for two cells. The dye loading densities for cells with OED and CED structures were found to be ~ 7.16×10^{-6} mol g^{-1} and 3.58×10^{-6} mol g^{-1} respectively which indicates higher dye loading for OED compared to CED structure and hence higher photocurrent. In addition we also anticipate that the improved photocurrent can also be a result of higher confinement of light in the active layer of TiO_2 due to the nano-dome structure of closed end being on top leading to the increase in optical path length and hence improved absorption. A schematic for light confinement effect for CED & OED structures are shown in Figure 15 (a & b) respectively. Overall it can be seen that orientation of the NTs for cell fabrication also plays a critical role in determining the efficiency of DSC.

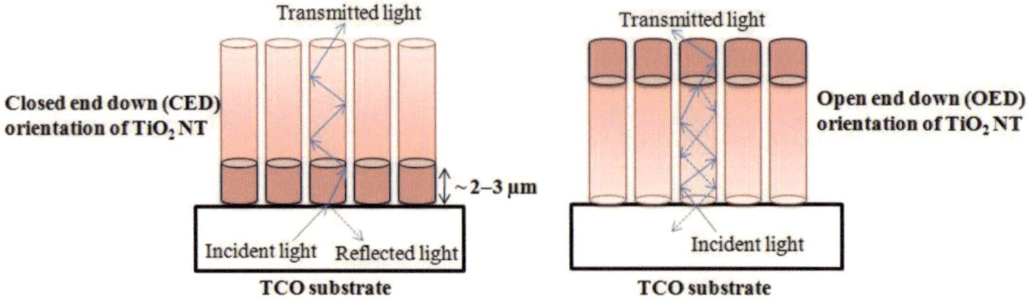

Fig. 15. (**a**) Schematic of light propagation through NT photoelectrode on FTO with (**a**) CED structure; and (**b**) OED structure

5. Conclusions

We found that morphology of NTs largely depends on the macro and microstructural topology of the substrate. Removal of substrate induced disorder in the morphology might be difficult by using simple ultrasonication or drying processes. A simple chemical pretreatment of substrate leads to substantial change in the morphology of grown NTs that can help in obtaining highly oriented and ordered TiO_2 NT arrays. The chemical pretreatment technique can find potential utility for being simple, cost effective and less time consuming. In addition we also found that the orientation of the NTs was critical in determining the efficiency of DSC. Hence a meticulous choice of NT orientation along with surface texturing of substrate can significantly help in engineering NT morphology for its successful implementation as a promising material for solar cells as well as other optoelectronic device applications.

6. References

Albu, S.P.; Ghicov, A., Macak, J.M., Hahn, R. & Schmuki, P. (2007). Self-organized , free-standing TiO_2 NT membrane for flow through photocatalytic applications. (April, 2007). Nano Letts., 7, 5, 1286-1289

Ali, G.; Chen, C., Yoo, S.H., Kum, J.M. & Cho, S.O. (2011). Fabrication of complete titania nanoporous structures via electrochemical anodization of Ti. *Nanoscale Research Letts.* 2011, 6:332

Anta, J.A.; Casanueva, F. & Oskam, G. (2006). A numerical model for charge transport and recombination in dye-sensitized solar cells. (March, 2006). *J. Phys. Chem. B.*, 110, 5372-5378

Bisquert, J. & Vikhrenko, V. S. (2004). Interpretation of the time constant measured by kinetic techniques in nanostructured semiconductor electrodes and dye-sensitized solar cells. (January, 2004). *J. Phys. Chem. B.*, 108, 2313-2322

Bisquert, J. & Zaban, A (2003). The trap limited diffusivity of electrons in nanoporous semiconductor networks permeated with a conductive phase (November, 2002). *Appl. Phys. A.*, 77, 507-514

Burke, A.; Ito, S, Snaith, H., Bach, U., Kwiatkowski, J. & Grätzel, M (2008). The function of a TiO compact layer in dye-sensitized solar cells incorporating "planar" organic dyes (March, 2008). *Nano Letts.* 8, 4, 976-981

Cameron, P.J. & Peter, L.M. (2003). Characterization of titanium dioxide blocking layers in dye-sensitized nanocrystalline solar cells (October, 2003). *J. Phys. Chem. B.*, 107, 14394-14400

Cameron, P.J. & Peter, L.M. (2005). How important is the back reaction of electrons via the substrate in dye-sensitized nanocrystalline solar cells? (October, 2004). *J. Phys. Chem. B.*, 109, 930-936

Cao, C.; Zhang, G., Song, X. & Sun, Z. (2011). Morphology and microstructure of as synthesized anodic TiO_2 NT arrays. *Nanoscale Res. Lett* (2011), 6, 64

Cao, F.; Oskam, G., Meyer, G.J. & Searson, P.C. (1996). Electron transport in porous nanocrystalline TiO_2 photoelectrochemical cells (October, 1996). *J. Phys. Chem*, 100, 17021-17027

Cass, M. J.; Qiu, F.L., Walker, A.B., Fisher, A.C. & Peter, L.M. (2003). Influence of grain morphology on electron transport in dye sensitized nanocrystalline solar cells (October, 2002). *J. Phys. Chem. B*, 107, 113-119

Chen, Q. & Xu, D. (2009). Large-scale, noncurling, and free-standing crystallized TiO_2 NT arrays for dye-sensitized solar cells. (March, 2009). *J. Phys. Chem. C.*, 113, 15, 6310-6314

Chen, Q.; Xu, D., Wu, Z. & Liu, Z. (2008). Free-standing TiO_2 NT arrays made by anodic oxidation and ultrasonic splitting. (July, 2008). *Nanotechnology*, 19, 365708

Crawford, G.A. & Chawla, N. (2009). Tailoring TiO_2 NT growth during anodic oxidation by crystallographic orientation of Ti. (February, 2009)., *Scripta Materialia*, 60, 874-877

Dor, S.; Grinis, L., Ruhle, S. & Zaban. A. (2009). Electrochemistry in mesoporous electrodes: Influence of nanoporosity on the chemical potential of the electrolyte in dye sensitized solar cells (January 2009). *J. Phys. Chem. C*, 113 (5), 2022-2027

Dubey, M. & He, H. (2009). Morphological studies of vertically aligned TiO_2 NTs for high efficiency solar cell. *Proc. 34th IEEE Photovolt. Conf.* (2009), 002130

Dubey, M.; Shrestha, M., Zhong, Y., Galipeau, D. & He, H. (2011). TiO_2 NT membranes on transparent conducting glass for high efficiency dye-sensitized solar cells. (May, 2011)., *Nanotechnology*, 22, 285201

Ghicov, A. & Schmuki, P. (2009). Self-ordering electrochemistry: a review on growth and functionality of TiO_2 NTs and other self aligned MO_x structures. *Chem. Commun.* (April, 2009), 2791-2808

Han, L.; Koide, N., Chiba, Y., Islam, A., Komiya, R., Fuke, N., Fukui, A. & Yamanaka, R. (2005). Improvement of efficiency of dye-sensitized solar cells by reduction of internal resistance. *Appl. Phys. Letts.* (May, 2005), 86, 213501

He, H.; Dubey, M., Zhong, Y., Shrestha, M. & Sykes, A.G. (2011). 2-(1-Acetyl-2-oxopropyl)-5,10,15,20-tetraphenyl Porphyrin and its transition metal complexes. (July, 2011). *Eur. J. Inorg. Chem.* 25, 3731-3738

He, H.; Sykes, A.G., Dubey, M., Yan, X., Galipeau, D. & Ropp, M. (2008). *Proc. 33rd IEEE Photovolt. Spec. Conf.*

In, S-I.; Hou, Y., Abrams, B.L., Vesborg, P.C.K. & Chorkendorff, I. (2010). Controlled directional growth of TiO_2 NTs. (March, 2010). *Journal of The Electrochemical Society*, 157, (5), E69-E74

Jennings, J. R.; Ghicov, A., Peter, L.M., Schmuki, P. & Walker, A.B. (2008). Dye-sensitized solar cells based on oriented TiO NT arrays: transport, trapping and transfer of electrons. (September, 2008). *J. Am. Chem. Soc.*, 130, 40, 13364-13372

Kang, S.H.; Kim, H.S., Kim, J-Y. & Sung, Y-E. (2009). An investigation on electron behavior employing vertically-aligned TiO_2 NT electrodes for dye-sensitized solar cells. (August, 2009), *Nanotechnology*, 20, 355307

Kang, S.H.; Kim, J-Y., Kim, H-S. & Sung, Y-E. (2008). Formation and mechanistic study of self-ordered TiO_2 NTs on Ti substrate. (June, 2007). *Journal of Industrial and Engineering Chemistry*, 14, 52-59

Karthikeyan, C.S. & Thelakkat, M (2008). Key aspects of individual layers in solid-state dye-sensitized solar cells and novel concepts to improve their performance (April, 2007). *Inorganica Chimica Acta*, 361, 635-655

Kontos, A.G.; Kontos, A.I., Tsoukleirs, D.S., Likodimos, V., Kunze, J., Schmuki, P. & Falaras, P. (2009). Photo-induced effects on self organized TiO_2 NT arrays: the influence of surface morphology. (December, 2008). *Nanotechnology*, 20, 045603

Lee, K-M.; Suryanarayanan, V. & Ho, K-C (2006). The influence of surface morphology of TiO_2 coating on the performance of dye-sensitized solar cells (May, 2006). *Solar Energy materials & Solar Cells*, 90, 2398-2404

Li, S.; Zhang, G, Guo, D, Yu, L & Zhang, W. (2009). Anodization fabrication of highly ordered TiO_2 nanotubes. (May, 2009), J. Phys. Chem. C, 113, 12759-12765

Liberator, M.; Burtone, L., Brown, T.M.., Reale, A., Carlo, A.D., Decker, F., Caramori, S. & Bignozzi, C.A. (2009). On the effect of Al_2O_3 blocking layer on the performance of dye solar cells with cobalt based electrolytes. (April, 2009). *Appl. Phys. Lett.* 94, 173113

Lin, C.J.; Yu, W-Y. & Chien, S-H. (2010). Transparent-electrodes of ordered opened-ended TiO_2-NT arrays for highly efficient dye-sensitized solar cells. (December, 2009). *J. Mater. Chem.*, 20, 1073-1077

Macak, J.M.; Hildebrand , H., Marten-Jhans, U. & Schmuki, P. (2008). Mechanistic aspects and growth of large diameter self-organized TiO_2 NTs. (January, 2008). *Journal of Electroanalytical Chemistry.*, 621, 254-266

Mohammadpour, A. & Shankar, K. (2010) Anodic TiO_2 NT arrays with optical wavelength-sized apertures. *J. Mater. Chem.* (September, 2010), 20, 8474-8477

Mor, G. K.; Varghese, O.K., Paulose, M. & Grimes, C.A. (2005). Transparent highly ordered TiO_2 NT arrays via anodization of titanium thin films. *Adv. Funct. Mater.*, 15, 1291-1296

Nazeeruddin, M. K.; Angelis, F. D., Fantacci, S., Selloni, A., Viscardi, G., Liska, P., Ito, S., Takeru, B. & Grätzel. M. (2005). Combined experimental and DFT-TDDFT computational study of photoelectrochemical cell ruthenium sensitizers. (November, 2005). *J. Am. Chem. Soc.*, 127, 48, 16835-16847

Nusbaumer, H.; Zakeeruddin, S.M., Moser, J-E. & Grätzel, M (2003). An alternative redox couple for the dye-sensitized solar cell system. *Chem. Eur. J.* 2003, 9, 3756-3763

O'Regan, B. & Grätzel, M (1991). A low-cost, high-efficiency solar cell based on dye-sensitized colloidal TiO_2 films. (October, 1991). *Nature*, 353, 737-740

Ofir, A.; Grinis, L. & Zaban, A. (2008). Direct measurement of the recombination losses via the transparent conductive substrate in dye sensitized solar cells. (January, 2008). *J. Phys. Chem. C.*, 112, 2279-2783

Oshaki, Y.; Masaki, N., Kitamura, T., Wada, Y., Okamoto, T., Sekino, T., Niihara, K. & Yanagida, S. (2005). Dye-sensitized TiO_2 NT solar cells: fabrication and electronic characterization. (October, 2005). *Phys. Chem. Chem. Phys.*, 7, 4157-4163

Park, J. H.; Lee, T-W. & Kang, M.G. (2008). Growth, detachment and transfer of highly-ordered TiO_2 NT arrays: use in dye-sensitized solar cells. (May, 2008). *Chem. Commun.*, 2867-2869

Prakasam, H. E.; Shankar, K., Paulose, M., Varghese, O.K. & Grimes, C.A. (2007). A new benchmark for TiO_2 NT array growth by anodization. (April, 2007). *J. Phys. Chem. C.* , 111, 20, 7235-7241

Robertson, N. (2006). Optimizing dyes for dye-sensitized solar cells (2006). *Angew. Chem. Int. Ed.* 45, 2338-2345

Roy, P.; Kim, D., Lee, K., Spiecker, E. & Schmuki, P. (2010). TiO_2 NTs and their application in dye-sensitized solar cells. *Nanoscale*. (December, 2009), 2, 45-49

Ruan, C.; Paulose, M., Varghese, O. K., Mor, G. K. & Grimes, C.A. (2005). Fabrication of highly ordered TiO_2 NT arrays using an organic electrolyte. (July, 2005)., *J. Phys. Chem. B.*, 109, 33, 15754-15759

Santiago, F.; Bisquert, J., Belmonte, G., Boschloo, G. & Hagfeldt, A. (2005). Influence of electrolyte in transport and recombination in dye-sensitized solar cells studied by impedance spectroscopy (November, 2004). *Solar Energy Material & Solar Cells*, 87, 117-131

Santiago, F.F.; Barea, E.M., Bisquert, J., Mor, G.K., Shankar, K. & Grimes, C.A. (2008). High carrier density and capacitance in TiO NT arrays induced by electrochemical doping. (August, 2008). *J. Am. Chem. Soc.*, 130, 34, 11312-11316

Santiago, F.F.; Belmonte, G.G., Bisquert, J., Zaban, A. & Salvador, P. (2002). Decoupling of transport, charge storage, and interfacial charge transfer in the nanocrystalline TiO_2/electrolyte system by impedance methods. (December, 2001). *J. Phys. Chem. B.*, 106, 334-339

Santiago, F.F.; Bisquert, J., Belmonte, G.G., Boschloo, G. & Hagfeldt, A. (2005). Influence of electrolyte in transport and recombination in dye-sensitized solar cells studied by impedance spectroscopy (November, 2004). *Solar Energy Materials & Solar Cells*, 87, 117-131

Sero, I. M.; Dittrich, T., Belaidi, A., Belmonte, G.G. & Bisquert, J. (2005). Observation of diffusion and tunneling recombination of dye-photoinjected electrons in ultrathin TiO_2 layers by surface photovoltage transients. (May, 2005). *J. Phys. Chem. B.* 109, 14932-14938

Sero, I.M. & Bisquert, J. (2003). Fermi level of surface states in TiO_2 NPs. (June, 2003). *Nano Letts.* 3, 7, 945-949

Sero, I.M.; Dittrich, T., Belmonte, G.G. & Bisquert, J. (2006). Determination of spatial charge separation of diffusing electrons by transient photovoltage measurements. (August, 2006). *J. App. Phys.* 100, 1

Shiga, A.; Tsujiko, A., Ide, T., Yae, S. & Nakato, Y (1998). Nature of electrical junction at the TiO_2/substrate interface for particulate TiO_2 film electrodes in aqueous electrolyte (May, 1998). *J. Phys. Chem. B.*, 102, 6049-6055

Shin, Y. & Lee, S. (2008). Self-organized regular arrays of anodic TiO_2 nanotubes. (September, 2008), Nano Letts. 8, 10, 3171-3173

Sun, L.; Zhang, S., Sun, X. & He, X. (2010). Effect of the geometry of the anodized titania NT array on the performance of dye-sensitized solar cells. *J. Nanosci. Nanotechnol.* , 10, 1-10

Tschirch, J.; Bahnemann, D., Wark, M. & Rathousky, J. (2008). A comparative study into the photocatalytic properties of thin mesoporous layers of TiO_2 with controlled mesoporosity (August, 2007). *Journal of Photochemistry and Photobiology A: Chemistry*, 194, 181-188

Vanmaekelbergh, D. & de Jongh, P.E (1999). Driving force for electron transport in porous nanostructured photoelectrode (January, 1999). *J. Phys. Chem. B.*, 103, 5, 747-750

Varghese, O.K.; Paulose, M. & Grimes, C.A. (2010). Long vertically aligned titania NTs on transparent conducting oxide for highly efficient solar cells. (August, 2009). *N. Nano*.2009.226

Wang, J. & Lin. Z. (2008). Freestanding NT arrays with ultrahigh aspect ratio via electrochemical anodization. (November, 2007). *Chem. Mater.* 20, 1257-1261

Wei, M.; Konishi, Y., Zhou, H., Yanagida, M., Sugihara, H. & Arakawa, H. (2006). Highly efficient dye-sensitized solar cells composed of mesoporous titanium dioxide (January, 2006). *J. Mater. Chem.*, 16, 1287-1293

Xu, T.; He, H., Wang, Q., Dubey, M., Galipeau, D. & Ropp, M. (2008). *Proc. 33rd IEEE Photovolt. Spec. Conf.*

Yuan, L.; Xurui, X., Dongshe, Z., Puhui, X. & Baowen, Z (2003). Light scattering characteristic of TiO_2 nanocrystalline porous films (May, 2003). *Chinese Science Bulletin*, 48, 9

Zhang, L. & Han, Y. (2010). Effect of nanostructured titanium on anodization growth of self-organized TiO_2 nanotubes. (December, 2009), Nanotechnology., 21, 055602

Zhu, J.; Hsu, C-M., Yu, Z., Fan, S. & Cui, Y. (2010). Nanodome solar cells with efficient light management and self-cleaning. (November, 2009). *Nano Lett.* , 10, 1979-1984

Zhu, K.; Neale, N.R., Miedaner, A. & Frank, A. J. (2007). Enhanced charge collection efficiencies and light scattering in dye-sensitized solar cells using oriented TiO_2 NT arrays. (December, 2006), *Nano Letts.*, 7, 1, 69-74

Zhu, K.; Vinzant, T.B., Neale, N.R. & Frank, A.J. (2007). Removing structural disorder from oriented TiO NT arrays: Reducing the dimensionality of transport and recombination in dye-sensitized solar cells. (November, 2007). *Nano Lett.*, 7, 12, 3739-3746

28

Synthesis and Characterisation of Silica/Polyamide-Imide Composite Film for Enamel Wire

Xiaokun Ma and Sun-Jae Kim*
Institute/Faculty of Nanotechnology and Adv. Materials Engin., Sejong University #98 Gunja-dong, Gwangjin-gu, Seoul, South Korea

1. Introduction

In the past decade, the demand for polyamide-imide (PAI) and other high-temperature resistant polymeric materials has grown steadily because of their outstanding mechanical properties and excellent thermal and oxidative stability (Zhong, 2002; Sun, 2006; Yanagishita, 2001; Babooram, 2008). PAI is well-known for its low thermal expansion coefficient and dielectric constant. In microelectronics, PAI has been widely used as an inter-dielectric material, and in the large-scale integrated circuit industry, as an electrical insulation for conventional appliances (Kawakami, 1996, 1998, 2003; Rupnowski, 2006; Wu, 2005). Compared with pure polyimide and polyamide, PAI exhibits better process ability and heat-resistant properties. The application of PAI as a wire-coating material with thermal-resistant properties has attracted increasing interest (Chen, 1997; Ranade, 2002; Ma, 2007). However, with the introduction of higher-surge voltage devices, an increasing number of insulation electric breakdown cases have been reported. Insulation electric breakdown must be prevented because it may lead to electrical component failure or may endanger the people handling the component. Thus, the development of an organic/inorganic composite insulating material is essential in designing insulation for continuous use (Alexandre, 2000; Hossein, 2007; David, 1995; Yang, 2006). Polymer composites have received much attention, as various properties of the original matrix polymer can be considerably improved by adding a limited percentage of inorganic filler (Jiao, 1989; Rangsunvigit, 2008; Xu, 2007; Hwang, 2008; Kim, 2007; Rankin, 1998).

Silica has been commonly used as an inorganic component because it is effective in enhancing the mechanical and thermal properties of polymers. Various studies on the preparation of polymer/silica composite films have been conducted (Butterworth, 1995; Mosher, 2006; Kim, 2006; Ahn, 2006; Stathatos, 2004).

The properties of hybrid composites are affected by many factors, such as particle size, size distribution, and filler content. In addition, the inorganic particle shape, surface structure,

---

* Corresponding Author

and mechanical properties of a filler (stiffness and strength, among others) play important roles in inorganic/organic composite material synthesis. In particular, the bond strength between the inorganic particles and the polymer matrix, influenced by the dispersion aid type or coupling agent used, should be improved (Kusakabe, 1996; Fuchigami, 2008; Castellano, 2005; Alexandre, 2000; Wu, 2006; Zheng, 2007; Ohki, 2005).

Silica nanoparticles, as important inorganic materials, have emerged as an area of intense interest because of their special physical and chemical properties, such as their small size, strong surface energy, high scattered performance, and thermal resistance (Ouabbas, 2009; Lee, 2006; Bhagat, 2008; Oh, 2009; Xue, 2009). However, the applications of silica nanoparticles are largely limited because of their highly energetic hydrophilic surface, which causes the silica nanoparticles to easily agglomerate. Surface modification methods using different surfactant agents may resolve this limitation. Thus, the strong interface adhesion between the organic matrix and the silica nanoparticles is the key to the application of silica nanoparticles as fillers.

Jadav et al. successfully synthesised a silica/polyamide nanocomposite film via interfacial polymerisation using two types of silica nanoparticles of 16 and 3 nm in size (Jadav, 2009). The nanocomposite films exhibited superior thermal stability to the pure polyamide membranes. In the current work, silica nanoparticle loading significantly modified the polyamide network structure, pore structure, and transport properties. The excellent membrane performance in terms of separation efficiency and productivity flux was also discussed. Zhang et al. prepared a novel isometric polyimide/silica hybrid material via sol-gel technique (Zhang, 2007). Initially, 3-[(4-phenylethynyl) phthalimide] propyl triethoxysilane was synthesised to modify the nanosilica precursor. Then, the isomeric polyimide/silica hybrid material was produced using isomeric polyimide resin solution and the modified nanosilica precursor after heat treatment. The isomeric polyimide/silica composite has much better thermal properties and nano-indenter properties than those of the isomeric polyimide.

In the current work, the commercial silica nanoparticles and self-synthesised spherical silica particles were successfully dispersed in the PAI polymer matrix after the surface modification process. The cationic surfactant cetyltrimethyl ammonium bromide (CTAB) was chosen to modify the silica nanoparticles. The amount of CTAB added in modifying the silica nanoparticles was increased from 0 to 3 wt%. After the surface modification process, the CTAB-modified silica nanoparticles showed better compatibility with the PAI polymer matrix. The results indicate that CTAB plays an important role in the preparation of silica/PAI composite film. The thermal stability improved and the decomposition temperature increased with increasing amounts of silica particles. The thermal expansion coefficient of the composite film was lower than that of the PAI polymer matrix, which is helpful in extending the life of the enameled wire.

2. Synthesis and characterisation of the spherical silica/PAI composite film

Spherical silica/PAI composite films have been successfully prepared via simple ultrasonic blending. In the current study, the spherical silica particles were prepared according to the Stőber procedure, and the size was controlled to approximately 300 nm at room temperature. After the surface modification process, the commercial silica nanoparticles and the self-synthesised spherical silica particles were dispersed in separate PAI polymer matrices. The correlation of the silica particle size with the amount of silica dispersed in the PAI under optimal experimental conditions was discussed.

2.1 Preparation of the silica/PAI composite film

2.1.1 Synthesis of the spherical silica submicron particles

The spherical silica particles were prepared according to the Stőber procedure, which allows the preparation of monodispersed silica particles with particle sizes in the nanometer to submicron range. A 500 ml three-necked flask equipped with a mechanical stirrer was filled with 45 ml ethanol, 2 ml NH_3 H_2O, and 1 ml deionised H_2O. The spherical silica synthesis was initiated by the rapid addition of 2 ml tetra-ethoxy-silane (TEOS) to a stirred solution. After the mixture was vigorously stirred for 2 h, spherical silica submicron particles of approximately 300 nm in size, were obtained at room temperature. Then, the silica particles were centrifuged at 10000 rpm for 30 min. The resultant silica particles were washed with ethanol and distilled H_2O, and then modified with CTAB.

2.1.2 Surface modification of the silica particles

CTAB can be directly added into the submicron silica solutions, and the CTAB amount can be increased from 0 to 3wt % at the optimal temperature of 60 °C. After the surface modification process, the modified silica particles were collected via centrifugation and then dried at 60 °C.

In the experiment, 100 ml deionised H_2O and 1 g commercial silica nanoparticles were added to a flask and the solution was adjusted to pH 8 by the addition of 0.1 M NaOH. The silica nanoparticles were modified at 65 °C with the addition of CTAB under constant stirring. The dispersal state of the silica nanoparticles in the PAI matrix was improved by increasing the amount of CTAB from 0 to 3.0 wt%. After the surface modification process, the modified nanosilica particles were collected via suction filtration and then dried at 90 °C for 6 h.

2.1.3 Preparation of the silica/PAI composite film

The nanosilica/PAI composite films were prepared via simple ultrasonic blending. Two grams of PAI powder were dissolved in 3 ml N,N-dimethyl form amide (DMF). The silica nanoparticles were added into the solution, and the amount of silica was increased from 2 to 10 wt%. The mixture was put under ultrasonic dispersion for approximately 3 h at room temperature. The mixture solution was then cast on a square glass plate (5 cm × 5 cm) using a bar coater. Bar coaters are primarily used in applying a variety of coatings or emulsions to a multitude of substrates. The bar coater used in the current experiment had a 0.2 mm diameter and was made of stainless steel wire, which resulted in more uniform silica/PAI composite films. The films were initially heated to remove the solvent in the vacuum oven. The temperature was controlled as follows: at 60 °C for 4 h, at 80 °C for 2 h, increased to 120 °C for 1 h, and finally, kept at 160 °C for 1 h. The experimental details of the silica/PAI composite film preparation are shown in Scheme 1.

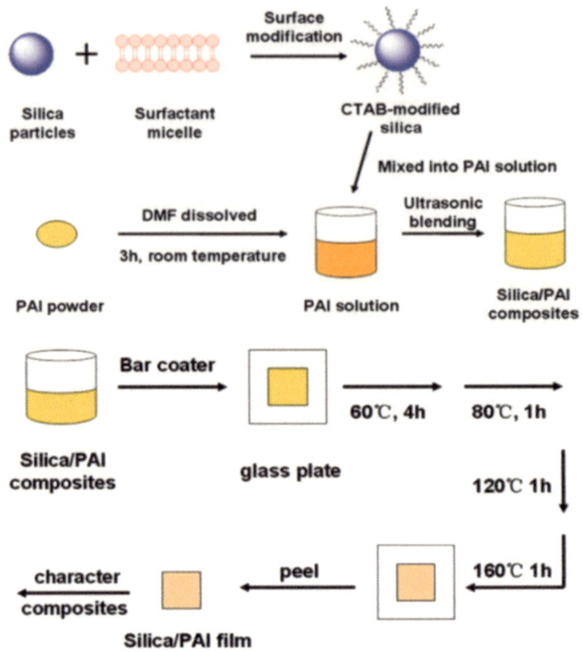

Scheme 1. Preparation of the silica/PAI nanocomposite films.

The spherical silica/PAI composite films were obtained under the same method. Spherical submicron silica particles do not easily agglomerate, so the amount of submicron silica added to PAI in the system was increased to 25%. The reactants of the submicron silica/PAI composites are listed in Table 1.

Reactant Sample	Silica Particles (g)	PAI (g)	DMF (ml)	Theoretical Weight Percent (wt%)
Composite 1	0.04	2.0	3	1.96
Composite 2	0.08	2.0	3	3.85
Composite 3	0.12	2.0	3	5.66
Composite 4	0.16	2.0	3.5	7.41
Composite 5	0.20	2.0	3.5	9.10
Composite 6	0.30	2.0	3.5	13.0
Composite 7	0.40	2.0	4	16.7
Composite 8	0.50	2.0	4	20.0

Table 1. Reactants of the different submicron silica/PAI composite samples.

2.2 Characterisation of the silica/PAI composite film

The fracture surfaces of the composite films were studied using a scanning electron microscope (SEM Hitachi S-4700, Hitachi Co.). Prior to SEM imaging, the samples were sputtered with thin layers of Pt-Pd. The silica/PAI nanocomposite films were characterised by an FT-IR (OMNIC NICOLET 380) spectrometer. The spectra were measured in the range 4000–650 cm^{-1}. A Scinco STA S-1500 simultaneous thermal analyser was then used to analyse the thermal stability of the nanosilica/PAI composite films. The samples were heated from 30 to 800 °C at 10 °C/min under air atmosphere. The coefficients of thermal expansion (CTE) of the silica/PAI composites films were evaluated using a Q 400 EM (U.S.A) thermomechanical analyser (5 °C/min from 25 to 300 °C, 50 mN). All the samples were 3 mm × 16 mm, cut from the original films using a razor blade.

2.2.1 CTAB effect on the synthesis of silica/PAI composite film

The FT-IR spectra of the silica/PAI composite films are shown in Fig. 1. The effect of the surfactant on the composite films was evaluated by increasing the CTAB dosage from 0 to 3 wt%. The amount of silica nanoparticles added to the PAI was 6 wt%. The character vibrations of the Si-O were observed at 1086, 945, and 796 cm^{-1}, as shown in Fig. 1 (e). After the surface modification process, the typical stretching vibrations of the C-H were found at 2855 and 2928 cm^{-1}, which resulted from the $-CH_2$ and $-CH_3$ in the CTAB. Figure 1 shows the typical characteristic bands of the PAI polymer matrix that were found, such as the N-H stretching band at 3317 cm^{-1}, the amide C=O region at around 1710 cm^{-1}, and the bands at 1771 and 1710 cm^{-1} associated with the imide carbonyl band. The bands are similar to one another because the same amount of silica was added into the composites. The characteristic stretching vibration of Si-O at 1086 cm^{-1} became wider when the silica nanoparticles were modified by CTAB. This peak broadening may be explained by the organic side-chain of CTAB grafted on the surface of the silica nanoparticles, which improved the interaction between the silica nanoparticles and the PAI polymer matrix.

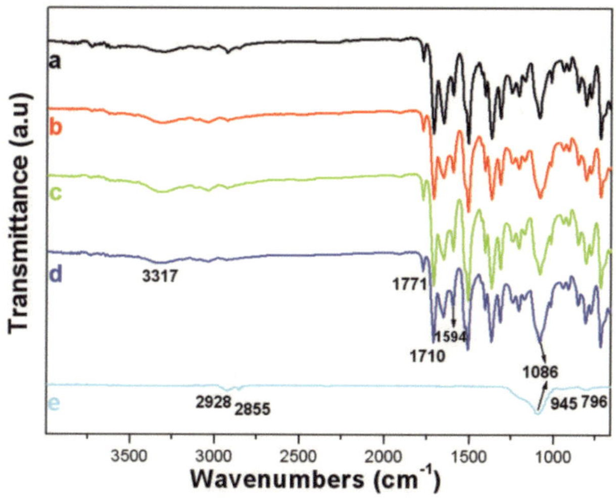

Fig. 1. FT-IR spectra of (a) unmodified-silica/PAI nanocomposites, (b) 1% CTAB-silica/PAI nanocomposites, (c) 2% CTAB-silica/PAI nanocomposites, (d) 3% CTAB-silica/PAI nanocomposites, and (e) CTAB-modified silica nanoparticles.

The fracture surface micrographs of the silica/PAI nanocomposite films are shown in Fig. 2. In Fig. 2 (a), some silica agglomerations are found in the micrograph when the unmodified-silica nanoparticles were added into the PAI matrix. The silica nanoparticles easily agglomerate because of their large surface-to-volume ratios and high surface tension. However, the CTAB-modified silica nanoparticles dispersed well in the PAI matrix with the increase in CTAB. The dispersal state of the silica nanoparticles improved when the silica nanoparticles were modified with 1 wt% CTAB, as shown in Fig. 2 (b). In Fig. 2 (c), the silica nanoparticles are almost monodispersed, and little agglomeration is observed when 2 wt% CTAB-modified silica nanoparticles were added into the PAI polymer. After the silica were modified with 3 wt% CTAB, the silica nanoparticles became monodispersed without any agglomerations, although the amount of silica nanoparticles added to the PAI was increased to 6 wt%, as shown in Fig. 2 (d). CTAB improves the dispersal state of the silica nanoparticles in a PAI polymer matrix, with an optimal dosage of 3 wt%.

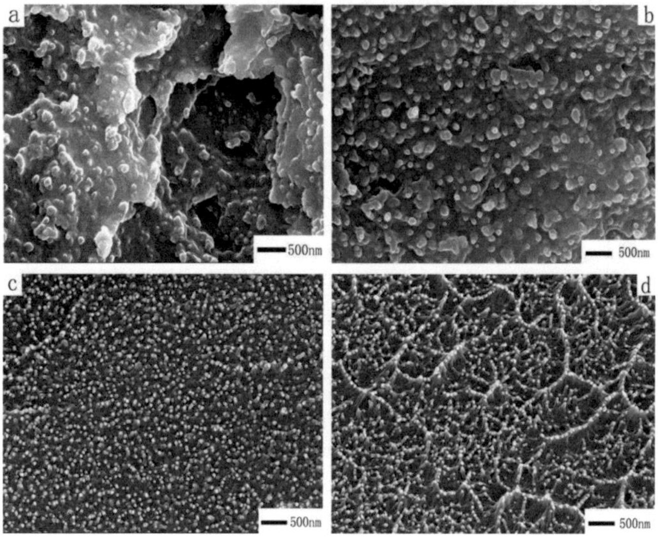

Fig. 2. Fracture surface micrographs of (a) unmodified-silica/PAI nanocomposites, (b) 1% CTAB-silica/PAI nanocomposites, (c) 2% CTAB-silica/PAI nanocomposites, and (d) 3% CTAB-silica/PAI nanocomposites.

Fig. 3 shows the thermogravimetric analysis (TGA) plots of the different CTAB modified-silica/PAI nanocomposites films when 6 wt% nanosilica was added into the PAI matrix. Some differences are found in the weight loss curves shown in Fig. 3. When the temperature was increased to 475 °C, the silica/PAI composites began to decompose. Compared with the unmodified-silica/PAI composites, the decomposition temperature increased after the silica nanoparticles were modified by CTAB. The results indicate that the CTAB-modified silica particles improve the thermal stability of the PAI polymer matrix. In addition, the decomposition temperature of composite films increased with increasing CTAB dosage. When the amount of CTAB added to the silica was increased from 0 to 3 wt%, the decomposition temperature of the composite films increased from 646 to 658, 671, and 682 °C, respectively. A thermal decomposition process occurs when the temperature approaches 595 °C, as shown in Figs. 3 (b)-(d). Therefore, the interaction between the silica nanoparticles and PAI polymer matrix is enhanced after the silica are modified by CTAB.

CTAB improves not only the dispersal state of silica in the PAI matrix, but also the thermal stability of the composite film because of the better interaction between the nanosilica and

PAI matrix. The amount of silica in the composites was calculated based on the plots, and the result was about 5.6 wt%, which is in accordance with the theoretical data.

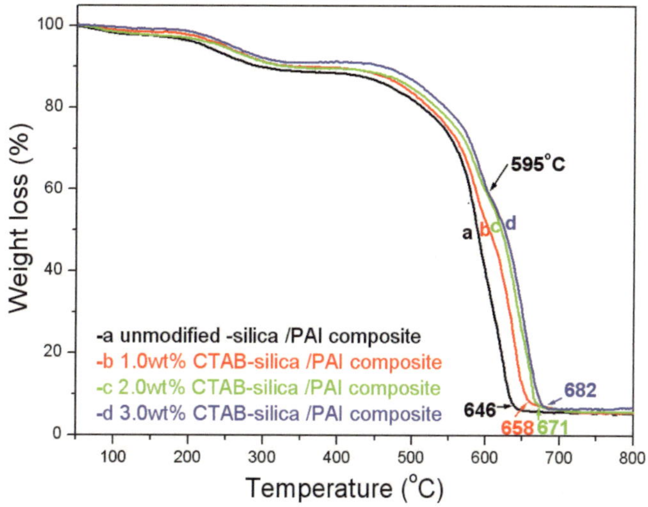

Fig. 3. TGA plots of (a) unmodified-silica/PAI nanocomposites, (b) 1% CTAB-silica/PAI nanocomposites, (c) 2% CTAB-silica/PAI nanocomposites, and (d) 3% CTAB-silica/PAI nanocomposites.

Thus, CTAB, with an optimal dosage of 3 wt%, was chosen to modify the nanosilica. The fracture surface micrographs of the composites show that the silica nanoparticles were well-dispersed in the PAI matrix after the surface modification process. In the TGA plots of the silica/PAI nanocomposites, the thermal stability and the decomposition temperature increased with increasing CTAB. Therefore, CTAB is important in the preparation of the silica/PAI nanocomposite film.

2.2.2 The effect of silica nanoparticle amount on the properties of silica/PAI composite film

The FT-IR spectra of the pure PAI and some silica/PAI nanocomposite films are shown in Fig. 4 The amount of silica nanoparticles added to PAI was changed from 2 to 10 wt%, and the dosage of CTAB added to the silica was 3 wt%. In the spectra shown in Fig. 4, the bands at 1771 and 1710 cm^{-1} are associated with the imide carbonyl band. Both bands are insensitive to the presence of the silica nanoparticles. The bands in the region from 945 to 650 cm^{-1} increased with the increase in silica content, caused by the presence of a broad band associated with the vibration of the Si-O bond. The N-H stretching band at 3317 cm^{-1} was slightly intensified with the increase in silica content, indicating the hydrogen-bonded N-H groups in the PAI polymer and the Si-O-Si or Si-O-H groups of the silica nanoparticles. The characteristic band at 1594 cm^{-1} comes from the benzene-ring stretch and a contribution from the O-H bond in monomeric H_2O, which also has a band at 1663 cm^{-1}. All the characteristic peaks in the composites indicate that the interaction between the silica nanoparticles and PAI polymer matrix is sensitive to the amount of silica in the composites.

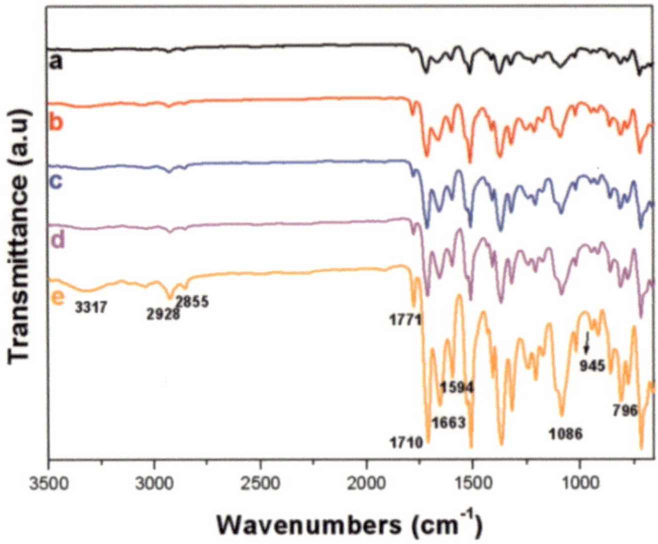

Fig. 4. FT-IR spectra of the silica/PAI composites with the amount of silica nanoparticles added to PAI at (a) 2 wt%, (b) 4 wt%, (c) 6 wt%, (d) 8 wt%, and (e) 10 wt%.

The fractured surface micrographs of the pure PAI and several composite films are shown in Fig. 5. The fracture surfaces of the pure PAI film are uniform, and the continuous polymer phase is shown in Fig. 5 (a). Figs. 5 (b)-(f) show the fracture surfaces of the different silica/PAI composites. The amount of silica added to the PAI was increased from 2 to 10 wt%. The larger the amount of silica nanoparticles added to the PAI, the greater their amount found in the fracture surface micrographs. In Fig. 5 (f), when 10 wt% silica nanoparticles was added into the PAI, the silica nanoparticles remained monodispersed without any agglomerations. The results indicate that the CTAB-modified silica nanoparticles have better dispersal state in the PAI polymer matrix. In addition, the CTAB-modified silica nanoparticles increased the silica nanoparticle content in the composites. Thus, the surface modification process is an effective method of preparing silica/PAI nanocomposites.

The thermal stability of the silica/PAI composite films was evaluated via TGA. The TGA plots of the PAI and the composites with the different amounts of silica nanoparticles are shown in Fig. 6. The amount of silica nanoparticles added to the PAI polymer matrix was increased from 0 to 10 wt%. The plots are shown in Figs. 6 (a)-(f). A weight loss is observed above 170 °C on all the TGA plots, which corresponds to water and solvent losses. When the temperature was increased to 450 °C, the PAI matrix began to decompose. The decomposition temperature increased when the silica nanoparticles were added into the PAI polymer matrix. At the same temperature, all the curves of the composites indicated that the composite weight loss was less than that of the pure PAI matrix. The silica/PAI composites have higher decomposition temperature when the PAI polymer matrix loses the same weight. The thermal stability of PAI was enhanced with the increase in the silica content. The amount of silica nanoparticles in the composites were calculated accurately in the TGA plots. The silica content in the composites based on Figs. 6 (b)-(f) are 1.9, 3.6, 5.8, 7.3, and 10.2 wt%, respectively.

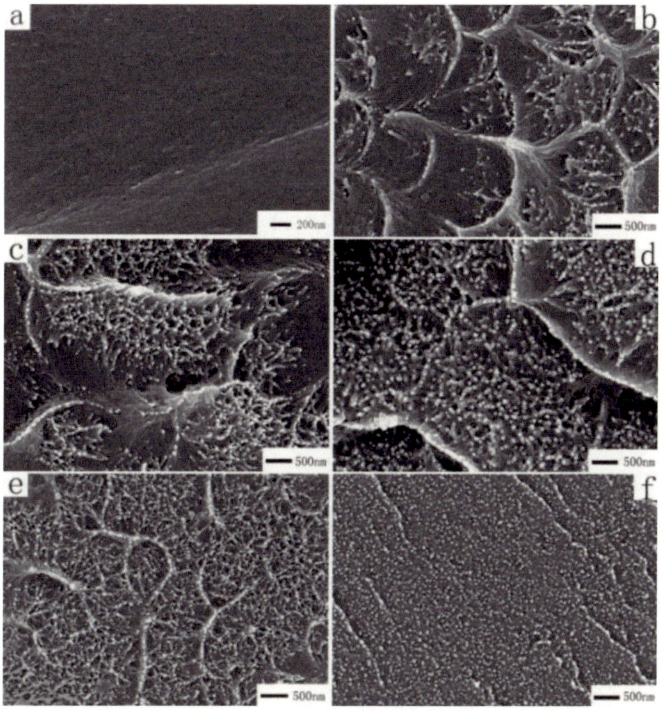

Fig. 5. Fracture surface micrographs of (a) pure PAI and silica/PAI nanocomposite, with the amount of silica added to PAI at (b) 2 wt%, (c) 4 wt%, (d) 6 wt%, (e) 8 wt%, and (f) 10 wt%.

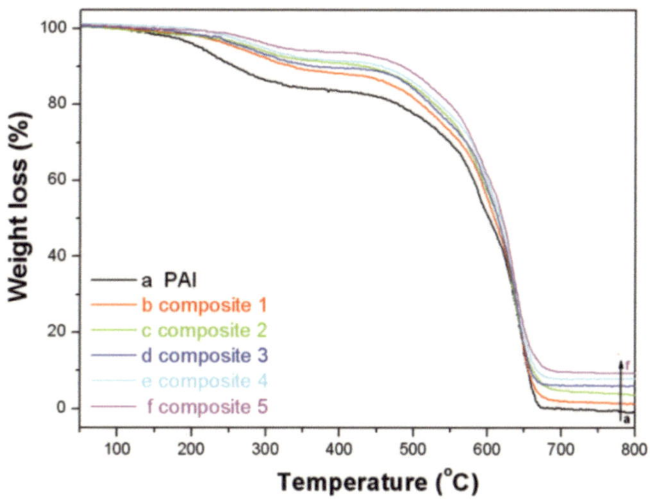

Fig. 6. TGA plots of the (a) PAI matrix, and the silica/PAI nanocomposite films with the amount of silica added to PAI at (b) 2 wt%, (c) 4 wt%, (d) 6 wt%, (e) 8 wt%, and (f) 10 wt%.

CTE is an important parameter in evaluating the properties of enamel wire. A low CTE can reduce thermal stress build-up and prevent device failure through peeling and cracking at the interface between the polymer film and the copper. The CTE curves of the pure PAI film and some composite films are shown in Fig. 7. The CTE value of the PAI films was 3.87×10^{-5} m/m/°C, whereas that of the silica/PAI composite film decreased to 3.69×10^{-5} and 3.51×10^{-5}m/m/°C when the amount of silica added to PAI was 4 and 6 wt%, respectively. The CTE values continuously decreased with increasing amount of the silica particles. In particular, the CTE value decreased to 3.35×10^{-5} m/m/°C when the amount of silica added to PAI was 10 wt%, as shown in Fig. 7 (d). Compared with the PAI polymer film, the silica/PAI composite films had lower CTE, which may be attributed to the rigidity and stiffness of the silica nanoparticles and the interaction between the silica and PAI polymer matrix. The rigidity and stiffness of the silica nanoparticles limit the polymer chain movement, resulting in the decrease of the PAI matrix thermal expansion.

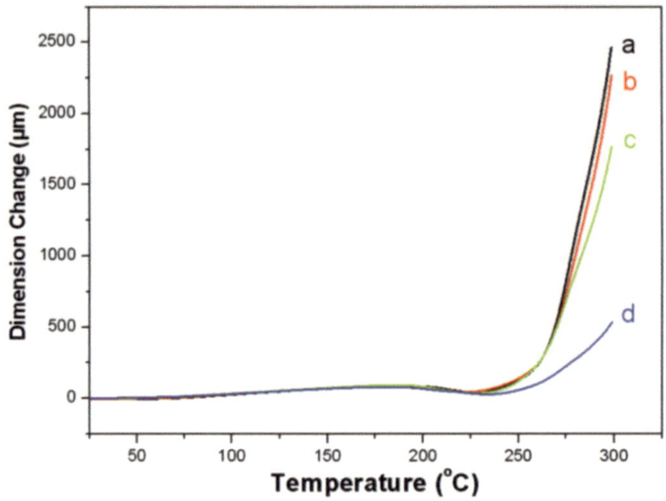

Fig. 7. CTE curves of the (a) pure PAI film and the composite films with silica content at (b) 4 wt%, (c) 6 wt%, and (d) 10 wt%.

The thermal stability of the silica/PAI nanocomposites improved and the decomposition temperature increased when the amount of silica nanoparticles was increased. The lower CTE of the composite films can reduce the peeling and cracking at the interface between the polymer film and the copper. In the current system, the high thermal stability and low CTE show that the silica/PAI nanocomposite films can be widely used in the enamel wire industry.

2.2.3 Effect of silica diameter on the properties of the silica/PAI composite film

In the past years, innovative inorganic/organic composite technology has been used to develop wires with better inverter-surge-resistance and mechanical properties than those of conventional enamelled wires. Kikuchi et al. (Kikuchi, 2002) emphasised that the inorganic/organic composite film can decrease erosion rate by increasing creeping distance and decreasing collision energy via reflection or scattering, as shown in Scheme 2.

Submicron spherical silica particles were added into the PAI polymer matrix during the synthesis of the silica/PAI composite films to evaluate the effects of silica diameter on their properties. The characterisation results of the spherical silica/PAI composites are discussed as follows.

Fig. 8 shows the SEM micrographs of (a) the spherical silica submicron particles and (b) the fracture surface of the pure PAI and several spherical silica/PAI composite films. In Fig. 8 (a), the mean diameter of the spherical silica nanoparticles is about 300 nm, and most are well-dispersed after the surface modification process. The fracture surfaces of the pure PAI film are uniform, and the continuous polymer phase is shown in Fig. 8 (b). Figs. 8 (c)–(f) show the fracture surfaces of the different silica/PAI composites. The amount of silica added to the PAI was increased from 2 to 8 wt%. When the submicron spherical silica was added into the PAI matrix, some prominent features were observed on the fracture surfaces of the composite films, as shown in Fig. 8 (c). The spherical silica particles were embedded in the PAI matrix, and the continuous PAI organic phase appeared when the amount of silica added to the PAI was 2 and 4 wt%, as shown in Figs. 8 (c) and (d). When the amount of submicron silica increased, more spherical silica particles were observed on the fracture surfaces of the composite films. The continuous organic PAI phase separated, as shown in Fig. 8 (e). Some alveolate pores were observed in Fig. 8(f), when 8 wt% spherical silica was added into the PAI matrix. These pores are caused by the removal of the submicron silica particles from the PAI matrix when the composite films were broken.

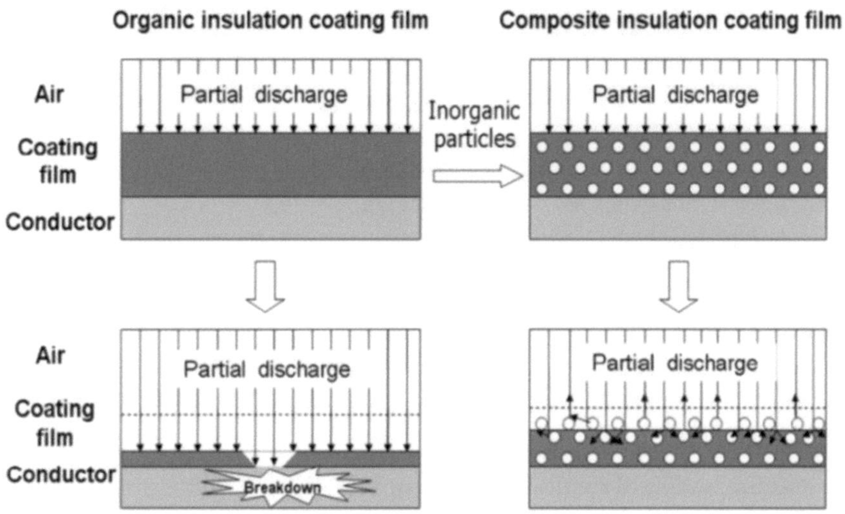

Scheme 2. Mechanism of erosion suppression.

More submicron spherical silica particles were subsequently added into the PAI matrix. The SEM micrographs of the composite film fracture surfaces are shown in Fig. 9. The micrographs show that the diameters of the submicron silica particles are uniform. In addition, the spherical silica particles are orderly arranged in the PAI matrix. An increase in the silica particles added into the films may result in a more compact framework. Partial discharge is a prime factor causing enamel wire breakdown, so the composite films caused a decrease in the erosion rate via reflection and scattering when the spherical silica particles were added into the PAI matrix. That is, the charged particles were reflected and scattered around the submicron silica, which slowed down the corrosion process. Therefore, the particle discharge resistance is improved with the increase in silica content. When the amount of silica added to the PAI was increased to 25 wt%, as shown in Fig. 9 (d), the fracture surfaces of the composite films remained well-integrated without the obvious phase separation. The results indicate that such a simple method effectively increases the amount of inorganic silica particles in the PAI matrix after the surface modification process.

Fig. 8. SEM micrographs of (a) spherical silica submicron particles and fracture surfaces of (b) pure PAI film, (c) composite 1 with 2 wt% submicron silica, (d) composite 2 with 4 wt% submicron silica, (e) composite 3 with 6 wt% submicron silica, and (f) composite 4 with 8 wt% submicron silica.

The FT-IR spectra of CTAB-modified silica, pure PAI, and some submicron silica/PAI composite films are shown in Fig. 10. As shown in Fig. 3 (a), typical bands of silica were observed at 1086, 950, and 809 cm^{-1}, indicating the stretching vibrations of Si-O. The stretching vibration peaks of C-H were found at 2861 and 2915cm^{-1}, which came from the -CH_2 and -CH_3 in CTAB. The FT-IR spectra of PAI show the presence of the imide carbonyl band at 1776 and 1715 cm^{-1}, and the peaks at 3320 and 1715 cm^{-1} come from the N-H stretching and C=O region, respectively. Other characteristic bands include the absorption at 1600 cm^{-1} caused by the benzene ring stretch and a contribution from the O-H bond in monomeric H_2O, which also has a band at 1663 cm^{-1}. The bands at 950–650 and 3320 cm^{-1} increased with the increase in silica content. Furthermore, the adsorption peak at 1086 cm^{-1} was intensified in the spectra of the composites. All the characteristic peaks in the composites were insensitive to the presence of the silica component, indicating good interaction between the spherical silica and the PAI polymer matrix.

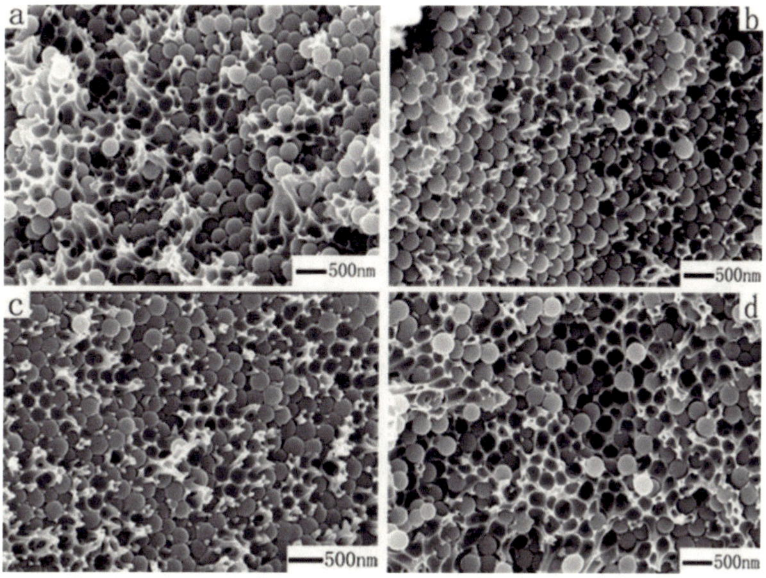

Fig. 9. SEM micrographs of (a) composite 5 with 10 wt% submicron silica, (b) composite 6 with 15 wt% submicron silica, (c) composite 7 with 20 wt% submicron silica, and (d) composite 8 with 25 wt% submicron silica.

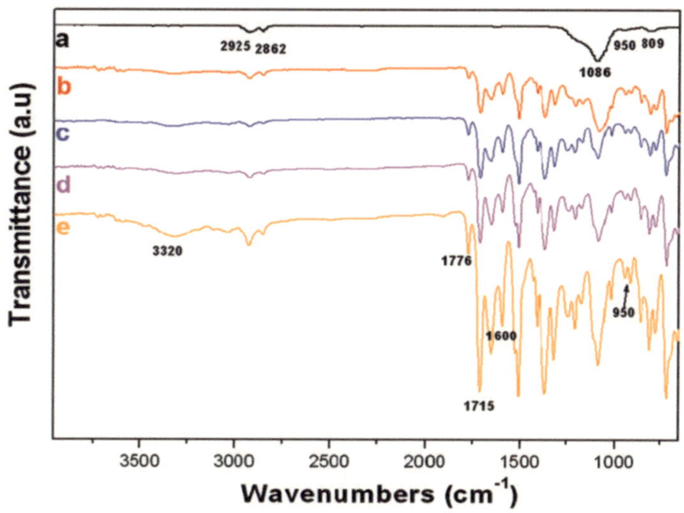

Fig. 10. FT-IR spectra of (a) CTAB-modified submicron silica particles, (b) pure PAI, (c) composite 1 with 2 wt% silica, (d) composite 2 with 4 wt% silica, and (e) composite 4 with 8 wt% silica.

The TGA plots of the PAI and the composites with the different amounts of submicron silica particles are shown in Fig. 11. Obvious weight loss is found in these plots. A weight loss is observed below 180 °C in all the TGA plots, corresponding to water and solvent losses. When the temperature was increased to 450 °C, the PAI matrix began to decompose with increasing decomposition temperature. The decomposition temperature increased from

640 to 678 °C after the silica particles were added to the PAI polymer matrix. The thermal stability of the PAI was increased by the incorporation of the submicron silica particles, as clearly shown in Fig. 11. The amount of silica particles in the composites were accurately calculated from the TGA plots. The silica content in the composites calculated based on the plots in Figs.11 (b)-(e) are 1.68, 3.80, 5.80, and 7.24 wt%. The calculated data are close to the theoretical values listed in Table 1.

Inorganic silica particles improve the thermal stabilities of the PAI polymer matrix and reduce the cost of enameled wire. To obtain the least expensive and most stable composite films, the amount of silica particles was increased from 10 to 25 wt%. In Fig. 12, the TGA curves are similar to the composite films shown in Fig. 11. The amount of silica particles was calculated based on the plots. The realistic data were 9.3, 13.7, 16.5, and 20.4 wt%.

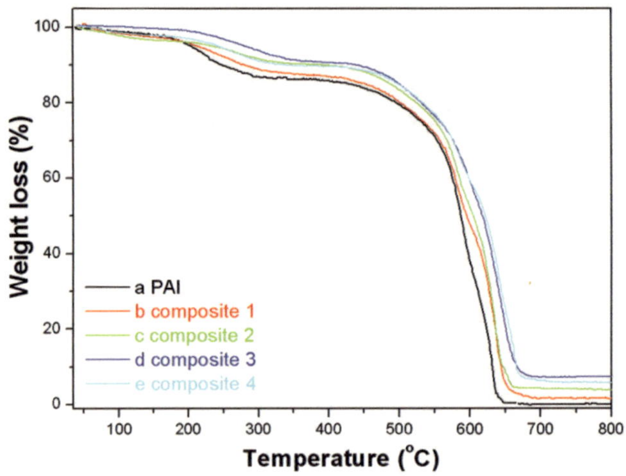

Fig. 11. TGA plots of (a) pure PAI, (b) composite 1 with 2 wt% silica, (c) composite 2 with 4 wt% silica, (d) composite 3 with 6 wt% silica, and (e) composite 4 with 8 wt% silica.

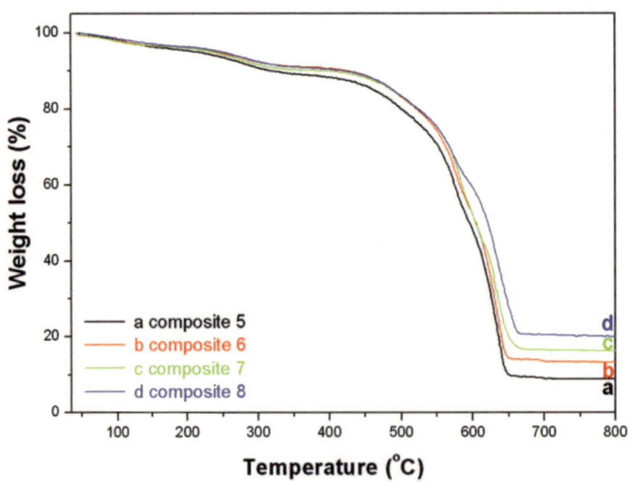

Fig. 12. TGA plots of (a) composite 5 with 10 wt% silica, (b) composite 6 with 15 wt% silica, (c) composite 7 with 20 wt% silica, and (d) composite 8 with 25 wt% silica.

Scheme 3. Interactions between the silica submicron particles and the PAI polymer matrix.

Second, the thermal stability of the composite film was further improved by increasing the silica particles. However, the decomposition temperature was increased from 650 to 668 °C, the results of which did not exceed that of the composites shown in Fig. 11. These results indicate that the interaction between the silica and the PAI matrix changed when more silica submicron particles were added into the composite films. Initially, the thermal stability of PAI improved after the silica submicron particles were added into the PAI matrix. Given the thermal motion of molecules, the cohesion between the adjacent polymer chains and the partial resistance and friction from the incorporation of silica in the PAI matrix had to be overcome. When more silica submicron particles were added into the PAI polymer, the interaction between the silica and the PAI matrix weakened because of the more compact and orderly arrangement of silica in the PAI matrix, as shown in Scheme 3. In addition, the silica submicron particles were easier to break away from the fracture surfaces, as confirmed by the SEM micrographs shown in Fig. 9. As a result, the decomposition temperature of the composite film did not increase with increase in the amount of silica particles, when the amount of silica added to PAI was more than 10 wt%. However, the thermal stability and decomposition temperature of the composite film were obviously more prominent than those of the pure PAI polymer matrix. Considering the similar thermal stability and lower price of the enameled wire, the amount of silica submicron particles can be modulated from 4 to 25 wt%.

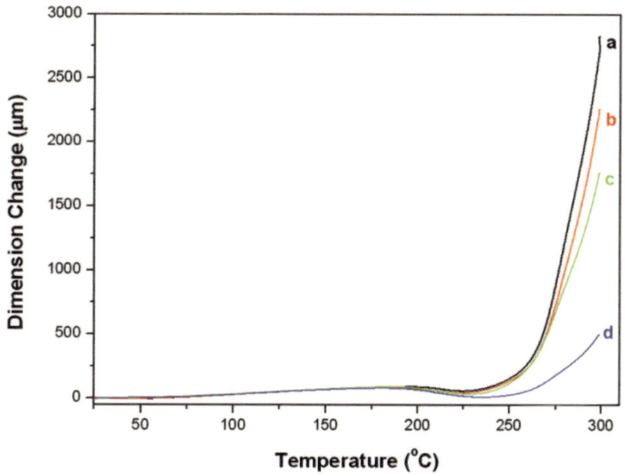

Fig. 13. CTE curves of the (a) pure PAI film, (b) composite 2 with 4 wt% silica, (c) composite 3 with 6 wt% silica, and (d) composite 6 with 15 wt% silica.

Controlling the CTE value of the dielectrical PAI is important because copper is the typical choice for defining the circuit lines. The selection of composites with the CTE close to that of copper is very critical, and can prevent damage to the copper interconnection; thus improving the reliability of the integrated circuits.

The CTE curves of the pure PAI film and some composite films are shown in Fig. 13. The CTE value of the pure PAI film is 3.87×10^{-5}m/m/°C, whereas that of the composite films in Figs. 13 (b)–(d) are 3.76×10^{-5}, 3.57×10^{-5}, and 3.25×10^{-5}m/m/°C, respectively. The CTE values continuously decreased with the increasing amount of the silica particles. The CTE value decreased by 16% when the amount of silica added to the PAI was 15 wt%. Such a variation in CTE is explained by the rigidity and stiffness of the silica submicron particles, which limit the movement of the polymer chain, resulting in the decrease of the thermal expansion of the PAI matrix. When the temperature increased to 220 °C, it approached the glass transition temperature of PAI. The dimension change significantly decreased with increased silica content. Therefore, the silica submicron particles effectively decrease the CTE value of the PAI polymer matrix, which in turn increases the thermal stress build-up, resulting in device failure through peeling and cracking at the interface between the PAI polymer film and the copper.

The thermal stability and the CTE value of the spherical silica/PAI composite films significantly improved when the submicron silica particles were added into the PAI polymer matrix. The submicron silica particles were obtained through the sol-gel method. Most previous studies used a sol-gel process because of the diameter of the silica particles. In the present study, the diameter effect was easily controlled using the Stőber procedure. Compared with the silica nanoparticles, the submicron silica particles have better dispersal state and do not easily agglomerate. Thus, more submicron silica particles are well-dispersed in the PAI polymer matrix after the surface modification process. Considering the higher thermal stability and lower CTE value, especially the lower cost, submicron silica/PAI composite films can be widely used in the enamel wire industry.

3. Conclusion

In the current work, silica particles with two different diameters were successfully added into the PAI polymer matrix in the synthesis of silica/PAI composite films via simple ultrasonic blending. First, the effect of CTAB on the synthesis of the silica/PAI composite film was investigated. The optimal dosage of CTAB is 3 wt%. The fracture surface micrographs of the composites show that the silica nanoparticles are well-dispersed in the PAI matrix after the surface modification process. In the TGA plots of the silica/PAI nanocomposites, the thermal stability and the decomposition temperature obviously increased with increasing CTAB dosage. Therefore, CTAB improves not only the dispersal state of silica in the PAI matrix, but also the thermal stability of the composite film because of better interaction between the nanosilica and the PAI matrix.

When the amount of silica nanoparticles added to the PAI was increased from 2 to 10 wt%, the thermal stability of the silica/PAI nanocomposites improved. The decomposition temperature increased with the increase in the amount of silica nanoparticles. In particular, the CTE value decreased when the silica particles were added into the PAI matrix. The CTE value further decreased with the increase in the amount of silica nanoparticles. When the submicron silica particles were added into the PAI polymer matrix, a similar conclusion was reached. However, more submicron silica particles are well-dispersed in the PAI polymer matrix after the surface modification process because the submicron silica particles have better dispersal state. Considering the higher thermal stability, lower CTE value, and the lower cost, silica/PAI composite films can be widely used in the enamel wire industry.

4. Acknowledgement

This research was supported by Basic Science Research Program through the National Research Foundation of Korea (NRF) Funded by the Ministry of Education, Science and Technology. (No. 2011-0016699). Also, it was supported by Sanhak Fellowship program funded of Korea Sanhak foundation.

5. References

Ahn, B. Y.; Seok, S. I.; Hong, S.; Oh, J.; Jung, H.; Chung, W. J. (2006). Optical properties of organic/inorganic nanocomposite sol-gel films containing $LaPO_4$:Er,Yb nanocrystals. *Optical Materials*, 28, 4, pp. 374-379

Alexandre, M.; Dubois, P. (2000). Polymer-layered silicate nanocomposites: preparation, properties and uses of a new class of materials. *Materials Science and Engineering*, 28, 1-2, pp. 1-63

Babooram, K.; Francis, B.; Bissessur, R.; Narain R. (2008). Synthesis and characterization of novel (amide–imide)-silica composites by the sol–gel process. *Composites Science and Technology*, 68, 3-4, pp. 617–624

Bhagat, S. D.; Kim, Y.; Suh, K.; Ahn, Y.; Yeo, J.; Han, J. (2008). Superhydrophobic silica aerogel powders with simultaneous surface modification, solvent exchange and sodium ion removal from hydrogels. *Microporous and Mesoporous Materials*, 112, 1-3, pp. 504-509

Butterworth, M. D.; Corradi, R.; Johal, J.; Lascelles, S. F.; Maeda, S.; Armes, S. P. (1995). Zeta Potential Measurements on Conducting Polymer-Inorganic Oxide Nanocomposite Particles. *Journal of Colloid and Interface Science*, 174, 2, pp. 510-517

Castellano, M.; Conzatti, L.; Costa, G.; Falqui, L.; Turturro, A.; Valenti, B.; Negroni, F. (2005). Surface modification of silica: 1. Thermodynamic aspects and effect on elastomer reinforcement. *Polymer*, 46, 3, pp. 695-703

Chen, L. W.; Ho, K. S. (1997). Synthesis of Polyamide-imide by Blocked-Methylene Diisocyanates. *Journal of Polymer Science Part A: Polymer Chemistry*, 35, 9, pp. 1711-1717

David, I. A.; Scherer, G. W. (1995). An Organic/Inorganic Single-Phase Composite. *Chemistry of Materials*, 7, pp. 1957-1967

Fuchigami, K.; Taguchi, Y.; Tanaka, M. (2008) Synthesis of spherical silica particles by sol-gel method and application. *Polymer for Advanced Technologies*, 19, pp. 977–983

Hossein, S. S.; Lia, Y.; Chunga, T.; Liu, Y. (2007). Enhanced gas separation performance of nanocomposite membranes using MgO nanoparticles. *Journal of Membrane Science*, 302, 1-2, pp. 207–217

Hwang, J.; Lee, B. I.; Klep, V.; Luzinov, I. (2008). Transparent hydrophobic organic-inorganic nanocomposite films. *Materials Research Bulletin*, 43, 10, pp. 2652-2657

Jadav, G. L.; Singh, P. S. (2009). Synthesis of novel silica-polyamide nanocomposite membrane with enhanced properties. *Journal of Membrane Science*, 328, 1-2, pp. 257-267

Jiao, W. M.; Vidal, A.; Papirer, E.; Donnet, J.B. (1989). Modification of Silica Surfaces by Grafting of Alkyl Chains Part III. Particle/Particle Interactions: Rheology of Silica Suspensions in Low Molecular Weight Analogs of Elastomers. *Colloids and Surfaces*, 40, pp. 279-291

Kawakami, H.; Mikawa, M.; Nagaoka, S. (1996). Gas permeability and selectivity through asymmetric polyimide membranes. *Journal of Appled Polymer Science*, 62, 7, pp. 965–971

Kawakami, H.; Mikawa, M.; Nagaoka, S. (1998). Gas transport properties of asymmetric polyimide membrane with an ultrathin surface skin layer. *Macromolecules*, 31, 19, pp. 6636–6638

Kawakami, H.; Nakajima, K.; Shimizu, H.; Nagaoka S. (2003). Gas permeation stability of asymmetric polyimide membrane with thin skin layer: effect of polyimide structure. *Journal of Membrane Science*, 212, 1-2, pp. 195–203

Kikuchi, H.; Yukimon, Y.; Itonaga, S. (2002). Inverter-surge-resistant enameled wire based on nano-composite insulating material. *Hitachi Cable Review*, 21, pp. 55-62

Kim, J. Y.; Mulmi, S.; Lee, C. H.; Park, H. B.; Chung, Y. S.; Lee, Y. M. (2006). Preparation of organic–inorganic nanocomposite membrane using a reactive polymeric dispersant and compatibilizer: Proton and methanol transport with respect to nano-phase separated structure. *Journal of Membrane Science*, 283, 1-2, pp. 172-181

Kim, T. K.; Kang, M.; Choi, Y. S .; Kim, H. K.; Lee, W.; Chang, H.; Seung, D. (2007). Preparation of Nafion-sulfonated clay nanocomposite membrane for direct menthol fuel cells via a film coating process. *Journal of Power Sources*, 165, 1, pp. 1-8

Kusakabe, K.; Ichiki, K.; Hayashi, J.; Maeda, H.; Morooka, S. (1996). Preparation and characterization of silica-polyimide composite membranes coated on porous tubes for CO_2 separation. *Journal of Membrane Science*, 115, 1, pp. 65-75

Lee, Y. L.; Du, Z. C.; Lin, W. X.; Yang, Y. M. (2006). Monolayer behavior of silica particles at air/water interface:A comparison between chemical and physical modifications of surface. *Journal of Colloid Interface Science*, 296, 1, pp. 233-241

Ma, J.; Yang, Z.; Wang, X.; Qu, X.; Liu, J.; Lu, Y.; Hu, Z.; Yang, Z. (2007). Flexible bi-continuous mesostructured inorganic/polymer composite membranes. *Polymer*, 48, 15, pp. 4305-4310

Mosher, B. P.; Wu, C.; Sun, T.; Zeng, T. (2006). Particle-reinforced water-based organic-inorgaic aocomposite coatings for tailored applications. *Journal of Non-Crystalline Solids*, 352, 30-31, pp. 3295-3301

Oh, C.; Lee, Y. G.; Jon, C. U.; Oh, S. G. (2009). Synthesis and characterization of hollow silica microspheres functionalized with magnetic particles using W/O emulsion method. *Colloids and Surfaces A: Physicochemical and Engineering Aspects*, 337, 1-3, pp. 208-212

Ohki, Y. (2005). Study on dielectric properties of LDPE-based nanocomposites by J-power systems. *IEEE Electrical Insulation Magazine*, 21, 3, pp. 55-56

Ouabbas, Y.; Chamayou, A.; Galet, L.; Baron, M.; Thomas, G.; Grosseau, G.; Guilhot, B. (2009). Surface modification of silica particles by dry coating: Characterization and powder aging. *Powder Technology*, 190, 1-2, pp. 200-209

Ranade, A.; Souza, N.; Gnade, B. (2002). Exfoliated and intercalated polyamide-imide nanocomposites with montmorllonite. *Polymer*, 43, 13, pp. 3759-3766

Rangsunvigit, P.; Imsawatgul, P.; Na-ranong, N.; O'Haver, J. H.; Chavadej, S. (2008) Mixed surfactants for silica surface modification by admicellar polymerization using a continuous stirred tank reactor. *Chemical Engineering Journal*, 136, 2-3, pp. 288-294

Rankin, S. E.; Macosko, C. W.; McCormick, A. V. (1998). Sol-Gel Polycondensation Kinetic Modeling: Methylethoxysilanes. *AIChE Journal*, 44, 5, pp. 1141-1156

Rupnowski, P.; Gentz, M.; Kumosa, M. (2006). Mechanical response of a unidirectional graphite fiber /polyimide composite as a function of temperature. *Composites Science and Technology*, 66, 7-8, pp. 1045–1055

Stathatos, E.; Lianos, P.; Tsakiroglou, C. (2004). Highly efficient nanocrystalline titania films made from organic/inorganic nanocomposite gels. *Microporous and Mesoporous Materials*, 75, 3, pp. 255-260

Sun, S.; Li, C.; Zhang L.; Du, H.L.; Burnell-Gray J.S. (2006). Effects of surface modification of fumed silica on interfacial structures and mechanical properties of poly(vinyl chloride) composites. *European Polymer Journal*, 42, 7, pp. 1643-1652

Wu, J.; Yang, S.; Gao, S.; Hu, A.; Liu, J.; Fan, L. (2005). Preparation, morphology and properties of nano-sized Al_2O_3/polyimide hybrid films. *European Polymer Journal*, 41, 1, pp. 73-81

Wu, T.; Ke, Y. (2006). Preparation of silica–PS composite particles and their application in PET. *European Polymer Journal*, 42, 2, pp. 274-285

Xu, J.; Wong, C.P. (2007). Characterization and properties of an organic–inorganic dielectric nanocomposite for embedded decoupling capacitor applications. *Composites Part A: Applied Science and Manufacturing*, 38, 1, pp. 13-19

Xue, L.; Li, J.; Fu, J.; Han, Y. (2009). Super-hydrophobicity of silica nanoparticles modified with vinyl groups. *Colloids and Surfaces A: Physicochemical and Engineering Aspects*, 338, 1, pp. 15-19

Yang, Y.; Wang, P. (2006). Preparation and characterizations of a new PS/TiO_2 hybrid membrane by sol–gel process. *Polymer*, 47, 8, pp. 2683–2688

Yanagishita, H.; Kitamoto, D.; Haraya, K.; Nakane, T.; Okada, T.; Matsuda, H.; Idemoto, Y.; Koura N. (2001). Separation performance of polyimide composite membrane prepared by dip coating process. *Journal of Membrane Science*, 188, 2, pp.165–172

Zhang, C.; Zhang, M.; Cao, H.; Zhang, Z.; Wang, Z.; Gao, L.; Ding, M. (2007). Synthesis and properties of a novel isomeric polyimide/SiO_2 hybrid material. *Composites Science and Technology*, 67, 3-4, pp. 380-389

Zheng, P.; Kong L. X.; Li, S.; Yin, C.; Huang M. F. (2007). Self-assembled natural rubber/silica nanocomposites: Its preparation and characterization. *Composites Science and Technology*, 67, 15-16, pp. 3130-3139

Zhong, S.; Li, C .; Xiao, X. (2002). Preparation and characterization of polyimide–silica hybrid membranes on kieselguhr–mullite supports. *Journal of Membrane Science*, 199, 1-3, pp. 53–58

29

Scanning Electron Microscope for Characterising of Micro- and Nanostructured Titanium Surfaces

Areeya Aeimbhu
Department of Physics, Faculty of Science, Srinakharinwirot University, Bangkok, Thailand

1. Introduction

Titanium and its alloys have been used broadly and successfully for numerous applications such as sport equipment [1], aerospace industry [2], marine application [3], medical applications [4] because of its optimum mechanical properties, outstanding corrosion resistance and bio-inert due to the presence of a thin layer of titanium oxide which a naturally formed onto the titanium surface [5-6]. This layer mainly consists of titanium dioxide or titania. Properties of oxide films covering titanium implant surfaces are a key role for a successful osseointegration [7-8], wear and corrosion resistance [9]. Moreover, titania has a large number of potentials in water photoelectrolysis and photocatalysis [10], sensors [11], wastewater remediation [12], automotive industry [13], industrial applications [14] and micro-optoelectronic applications [15]. However, there are some disadvantages of the native titanium oxide which has poor mechanical properties and easily fractured under small scale of fretting and wears conditions [16]. Therefore, many surface modification treatments for examples anodisation (anodic oxidation), cathodic electrodeposition and sol-gel reactions [17-22] have been studied in order to improve the performance of titanium. Moreover, surface properties such as topography, chemical composition and hydrophilicity have an effect on the mechanical stability of the implant-tissue interface. Various surface modification methods of titanium have been shown to improve interfacial interactions at the bone-implant interface and their clinical performance. Moreover, the biological performance of implantable titanium depends crucially on their surface topography in the micrometre (structures larger than 1 micron) and nanometre (structures smaller than 1 micron) range. Surface micro- and nano-topography can reduced inflammatory and guide direct osteoblast responses by altering adhesion, recruitment, movement, morphology, apoptosis and gene expression, and subsequently protein production [23-25]. An anodisation is a simple and an inexpensive technique to prepare thin film titania on titanium surface in different conditions and electrolytes such as acidic, basic, neutral, organic and inorganic which affect surface architecture and chemical composition [26-27]. Anodised oxide layer has thickness in the range of 20 to 180 nm which is thicker than a naturally formed oxide [28]. Moreover, this technique is presently used to achieve micro- and nano-topography surfaces. An anodisation is an electrolytic etching for coating the surface of a metal with an oxide layer which changes the microscopic texture of the surface and the crystal structure of the metal near the surface. An anodisation process accelerates the formation of an oxide coating under controlled conditions to provide the desired result. Since the coating is biocompatible as well as nontoxic, the process lends itself to achieve drastic improvement in implant performance. By adjusting the anodisation condition such as electrolyte, pH, voltage and time, micro- and nano-scale properties could be controlled.

In this article, the morphological of surface was studied by means of scanning electron microscope (SEM) after surface modification in order to evaluate qualitatively the effect of the anodisation conditions. For this purpose, the SEM uses to examine morphological of development of anodised film of titanium that were prepared under different controlled conditions.

2. Materials and methods

2.1 Preparation of surface

Prior to anodisation, commercially pure titanium grade 2 were provided by Prolog Titanium Co., Ltd and cut into 1 cm × 1 cm squares. Titanium sheets were abraded mechanically using silicon carbide (SiC) abrasive papers (Buehler) number 120 to 2000 and rinsed with distilled water. The surface of sheets were sequentially polished to a mirror finished with aqueous alumina (Al_2O_3) 5, 1, 0.3 and 0.05 micron (Buehler, Alpha Micropolish II). Afterward clean with acetone in ultrasonic bath.

2.2 Remove oxide

Titanium surfaces were immersed in an acid mixture (2 ml 48% HF (48% in water) which purchased from Panreac + 3 ml 70% HNO_3 (70% in water) which purchased from Fluka + 100 ml DI water) about 90 seconds to remove the naturally formed oxide layer and then immediately treated by anodisation. All electrolytes were prepared with reagent grade chemicals.

2.3 Anodisation

The electrochemical cell consists of a two electrodes system which graphite acting as the counter electrode (cathode) and titanium sheet acting as the working electrode (anode). The anode and cathode were connected by copper wires and were linked to a positive and negative port of a DC power supply (KMB 3002), respectively. During processing, the anode and cathode were kept parallel with a separation distance of about 1 cm, and were immersed into an electrolyte. Anodisation was performed under potentiostatic conditions at 20 volts. All experiments were carried out at room temperature. Test conditions applied vary in terms of the concentration of acetic acid, the concentration of hydrofluoric acid and an anodisation time. Different series of experiments have been performed by changing the associated experimental parameters as shown in Table 1. After the electrochemical process, the titanium sheets were immediately rinsed by deionised water and heated at 120°C for 30 minutes.

2.4 Surface characterisation

The surface morphology of the samples was observed by using Scanning electron microscope (SEM: JEOL 6300) with Dispersive X-Ray Spectroscopy (EDS) technique.

	Acetic concentration (M)	Hydrofluoric acid (wt%)	Anidising time (minutes)
Series 1	0.001, 0.01, 0.1, 1 and 10M	-	120
Series 2	1 M	-	30, 60, 120, 240, 360 and 480
Series 3	1 M	0.075	30, 60, 360, 480, 720, 1440, 2160 and 2880
Series 4	1 M	0.5, 1.5 and 2.0	60

Table 1. Parameters used for anodisation

3. Result and discussion

In order to find out the effect of anodisation electrolyte and anodisation time on the forming of the oxide at constant voltage in an ambient temperature; the structure of morphology were investigated. Experimental with different anodisation conditions: the concentration of acetic acid, the concentration of hydrofluoric acid and an anodisation time were lead to get an overall view of the formation process of titania nanotubes.

3.1 Effect of concentration of acetic acid on the surface topography

Figure 1 displayed scanning electron microscope (SEM) photographs of an untreated and the anodised titanium surface at 20 volts in different electrolytes concentration. An untreated sample has no visible scratches or wrinkles (figure 1a). The effect of concentration of electrolytes on the surface morphology of the anodic oxide film formed on titanium was monitored. Based on the experimental observation, it was showed that the passive film formed on titanium surface in low concentrations of acetic acid range which from 0.001 to 0.1 M (figure 1(b-d)). The flower-like structures [29] were visible on the substrates which anodised in 1 M acetic acid. Moreover, upon increase concentration of acetic acid the surface morphologies shown the concentration of the flower was developed and increased as shown in figure 1e - 1f. The average diameter of the flowers formed in an acetic acid for 120 minutes was around 300 nanometres.

3.2 Effect of anodisation time on the surface topography

The effect of anodisation time on the morphology of an oxide film in 1M acetic acid was investigated. Figure 2(b-g) summarised the effect of anodisation time on the development of an anodised film in 1M acetic acid at 20 volts. Within 30 minutes (figure 2b), SEM image revealed the formation of a continuous of oxide film over the surface. Flower-like structures were developed at 60 minutes (figure 2c). The density of the flower oxides was observed with the increasing of anodisation time. The average diameter of the flowers increases from a few hundred nanometres to 1200 nanometres.

The effect of hydrofluoric acid and anodisation time on the formation of titania nanotubes in 1M acetic acid with 0.075 wt% HF is studied. SEM image of the surface morphologies obtained in 1 M acetic acid with 0.075 %wt HF at different anodisation time are shown in figure 3(b-i).

The results demonstrated at the initial stage of anodisation (Figure 3(b-c)), the tube-like features appeared over the titanium surface. For longer anodisation time, it can be seen from figure 3(e-f) that the architecture transformation from tube-like nanostructure to a sponge-like titania film with nanoholes. Further anodisation time, the surface of nanoholes is covered with the oxide film (figure 3(g-h)). When the anodisation time is prolonged to 2880 minutes, the discontinuous oxide film covered the titanium surface. The results confirm that fluorine containing electrolyte is a capable electrolyte for anodic formation of titania nanotubes.

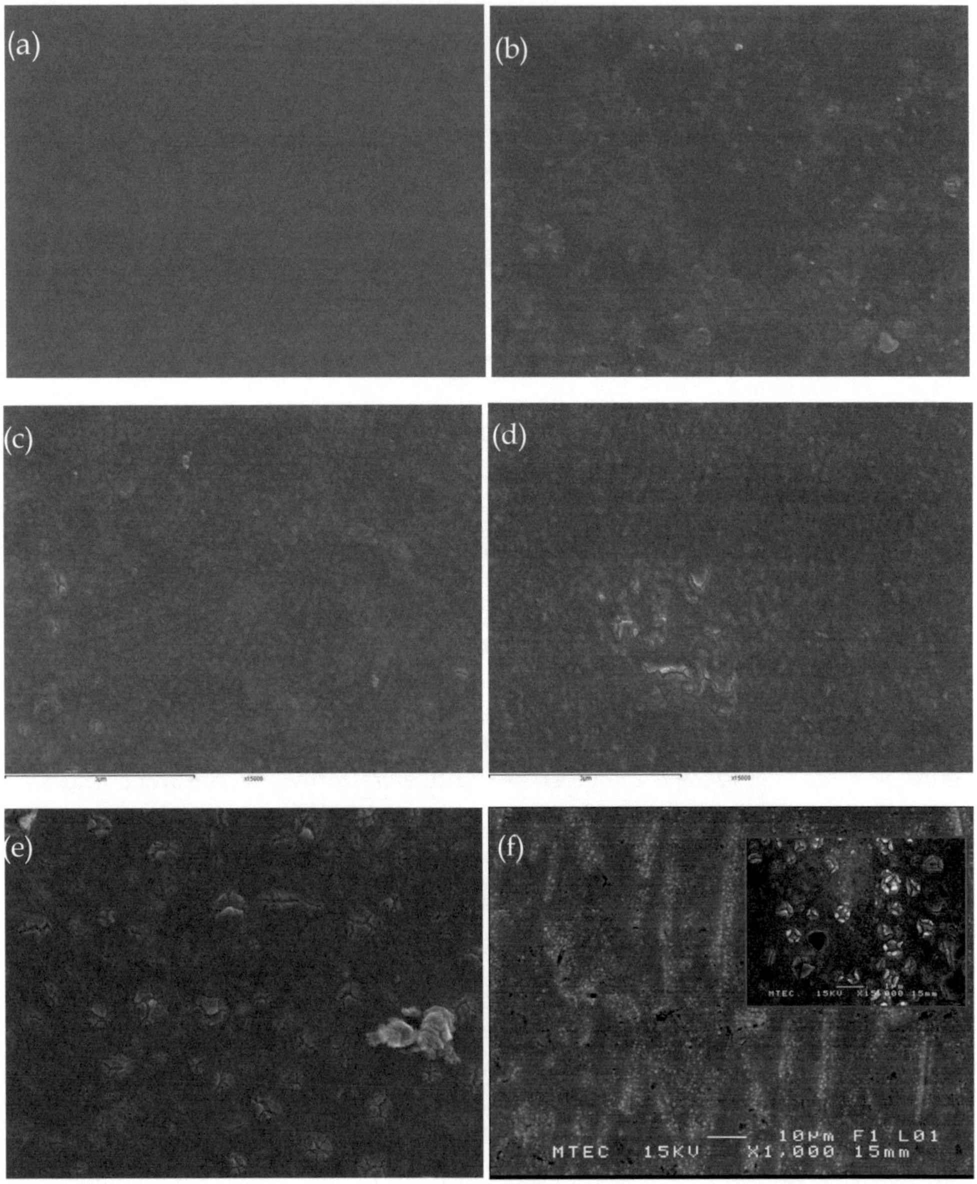

Fig. 1. SEM images of titanium sheets (a) untreated surface and anodised in (b) 0.001 (c) 0.01 (d) 0.1 (e) 1 and (f) 10M acetic acid at 20 volts for 120 minutes at room temperature.

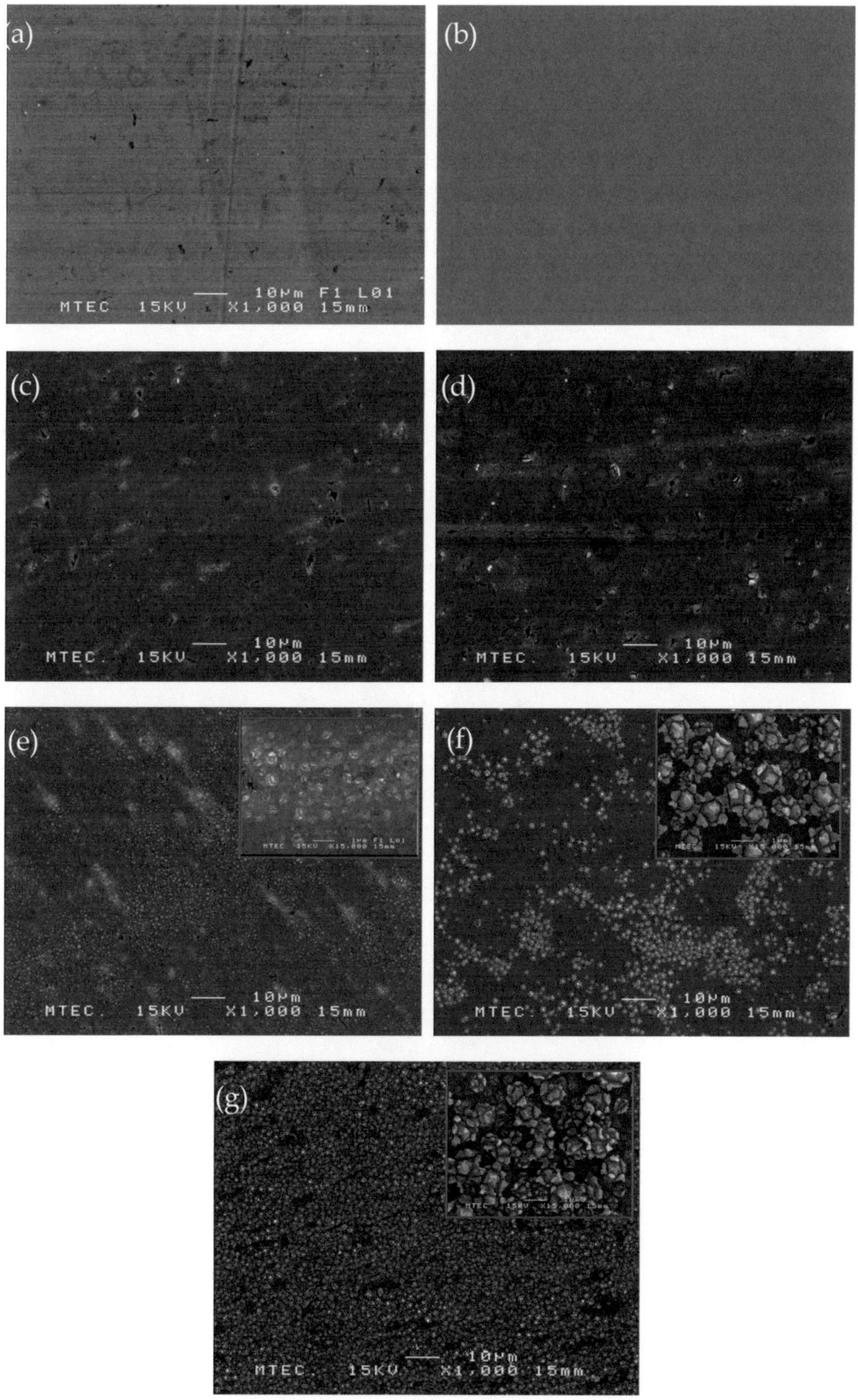

Fig. 2. SEM images of titanium sheets (a) untreated surface and anodised in 1M acetic acid at 20 volts for (b) 30 minutes (c) 60 minutes (d) 120 minutes (e) 240 minutes (f) 360 minutes and (g) 480 minutes at room temperature.

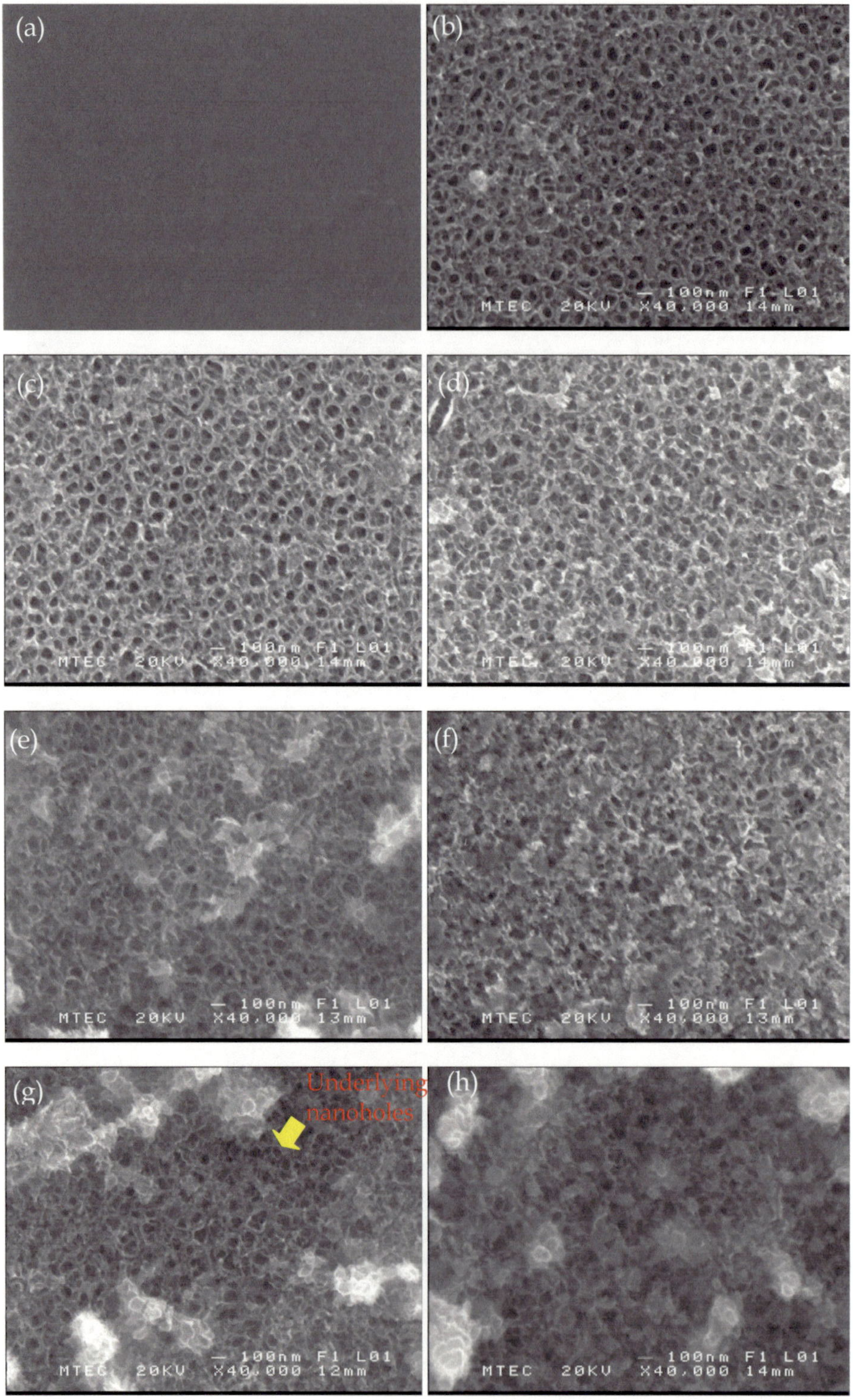
(a)
(b)
100nm F1 L01
MTEC 20KV X40,000 14mm
(c)
100nm F1 L01
MTEC 20KV X40,000 14mm
(d)
100nm F1 L01
MTEC 20KV X40,000 14mm
(e)
100nm F1 L01
MTEC 20KV X40,000 13mm
(f)
100nm F1 L01
MTEC 20KV X40,000 13mm
(g)
Underlying nanoholes
100nm F1 L01
MTEC 20KV X40,000 12mm
(h)
100nm F1 L01
MTEC 20KV X40,000 14mm

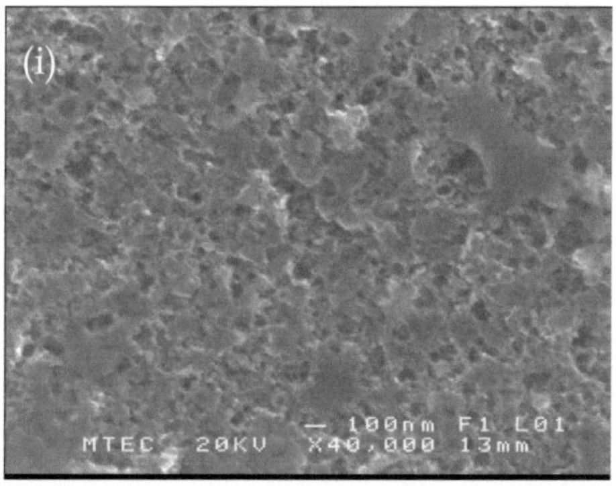

Fig. 3. SEM images of titanium sheets (a) untreated surface and anodised in 1M acetic acid with 0.075 wt% HF at 20 volts for (b) 30 minutes (c) 60 minutes (d) 360 minutes (e) 480 minutes (f) 720 minutes (g) 1440 minutes (h) 2160 minutes and (i) 2880 minutes at room temperature.

3.3 Effect of concentration of hydrofluoric acid on the surface topography

The surface topologies of titanium sheets anodised in electrolyte containing 0.1 M acetic acid with different concentration of HF: 0.075, 0.5, 1.5 and 2.0 wt% are shown in Figure 4. The anodisation was carried out at 20 volts. A network structure appears on the anodised titanium surfaces with concentration of 0.075 wt% HF (Figure 4b). Anodisation in 0.5 wt% and 1.5 wt% HF containing occur a highly ordered and uniform titanium oxide nanotube arrays (Figure 4 (d-e)). The average nanotube inner diameter is approximately 74.5 and 76.5 nanometre, respectively. As the concentration of hydrofluoric concentration was further increased to 2.0 wt%, the surface architecture developed sponge-like (Figure 4f). It points out that the concentration of hydrofluoric acid affect the morphology of titanium surface.

The anodic growth of compact oxides on titanium substrate and the formation of nanotubes in fluoride-containing electrolytes is the result of key processes [30] which are (1) Field-assisted oxidation of the titanium metal that leads to oxide growth at the surface due to interaction of titanium with O^{2-} or OH^- ions. An initial oxide layer formed on the substrate, these anions travel through the oxide layer reaching the titanium/oxide interface. (2) Titanium metal ion (Ti^{4+}) migrate from the substrate at the titanium/oxide interface; Ti^{4+} cations will ejected from the titanium/oxide interface under application of an electric field that move towards the oxide/electrolyte interface. (3) Field-assisted dissolution of titanium metal ion at the oxide/electrolyte interface into the electrolyte. (4) Chemical dissolution of titanium metal ion and TiO_2 due to etching away by fluoride ions. The reactions are [31]:

at anode:

1. oxidation of titanium metal which releases titanium metal ions (Ti^{4+}) and electron

$$2Ti \rightarrow 2Ti^{4+} + 8e^- \qquad (1)$$

2. interaction of titanium metal ions with O^{2-} or OH^- ions

$$Ti^{4+} + 4OH^- \rightarrow Ti(OH)_4 \qquad (2)$$

$$Ti^{4+} + 2O^{2-} \rightarrow TiO_2 \quad (3)$$

Equation (2) and (3) elucidate the hydrated anodic layer and the oxide layer. Further oxide is produced when the hydrated anodic layer releases water by a condensation reaction as the following equation:

$$Ti(OH)_4 \rightarrow TiO_2 + 2\,H_2O \quad (4)$$

at cathode:

$$8H^+ + 8e^- \rightarrow 4H_2 \quad (5)$$

The overall process for anodic oxidation of titanium can be represented as:

$$Ti + 2H_2O \rightarrow TiO_2 + 2H_2 \quad (6)$$

In the presence of fluoride ions electrolyte, fluoride ions enters the $Ti(OH)_4$ or anodic titanium oxide as the following equation:

$$TiO_2 + 6F^- + 4H^+ \rightarrow TiF_6^- + H_2O \quad (7)$$

$$Ti(OH)_4 + 6F^- \rightarrow TiF_6^- + 4OH^- \quad (8)$$

$$Ti^{4+} + 6F^- \rightarrow TiF_6 \quad (9)$$

Equation (7-8) is the mechanism of the pit formation due to the localised dissolution of the oxide and hydrated anodic layer. Then these pits transfer to bigger pores and the pore density increases subsequent to uniformly pores over the titanium surface, with the TiO_2 pores growing more and more deeply into the titanium metal. As described above, the formation of nanotubes govern by a competition between anodic oxidation and chemical dissolution of the oxide as soluble fluoride complexes.

The concentration of the HF added to the electrolyte was varied from 0.075 to 2.0 wt%. As a result, a large change in surface architecture was observed. An average diameter of the nanotubes increased as the fluoride content in electrolyte. As mentioned above, the formation of titania nanotubes determined by the oxide growth rate and the dissolution rate. A result shows that with 0.5 and 1.5 wt% HF, the dissolution rate was slow and resulted in small pore size. With increasing the concentration of HF, the dissolution rate increased and resulted in big pore. It is obvious that the dissolution rate was extremely high with 2 wt% HF because the morphology is not uniform.

The result of the chemical analysis by Energy Dispersive X-ray spectroscopy (EDS) indicated that anodised titanium in 1 M acetic acid with 0.075 wt% HF for 60 minutes at room temperature titania film are shown in figure 5b. The EDS measurements present an oxygen and titanium proportion of 34.914% and 65.086 %, respectively. The result reviewed that the chemical composition of anodic film was nonstoichiometric and the atomic ration of Ti/O is approximately 1.86. The presented of nonstoichiometric structure on the substrate layer implied that some defect exist in TiO_2 nanotubes due to oxygen deficiencies which can cause the formation of crystallographic shear planes and active Ti-sites for the adsorption and chemisorptions of OH groups or other contaminants [32].

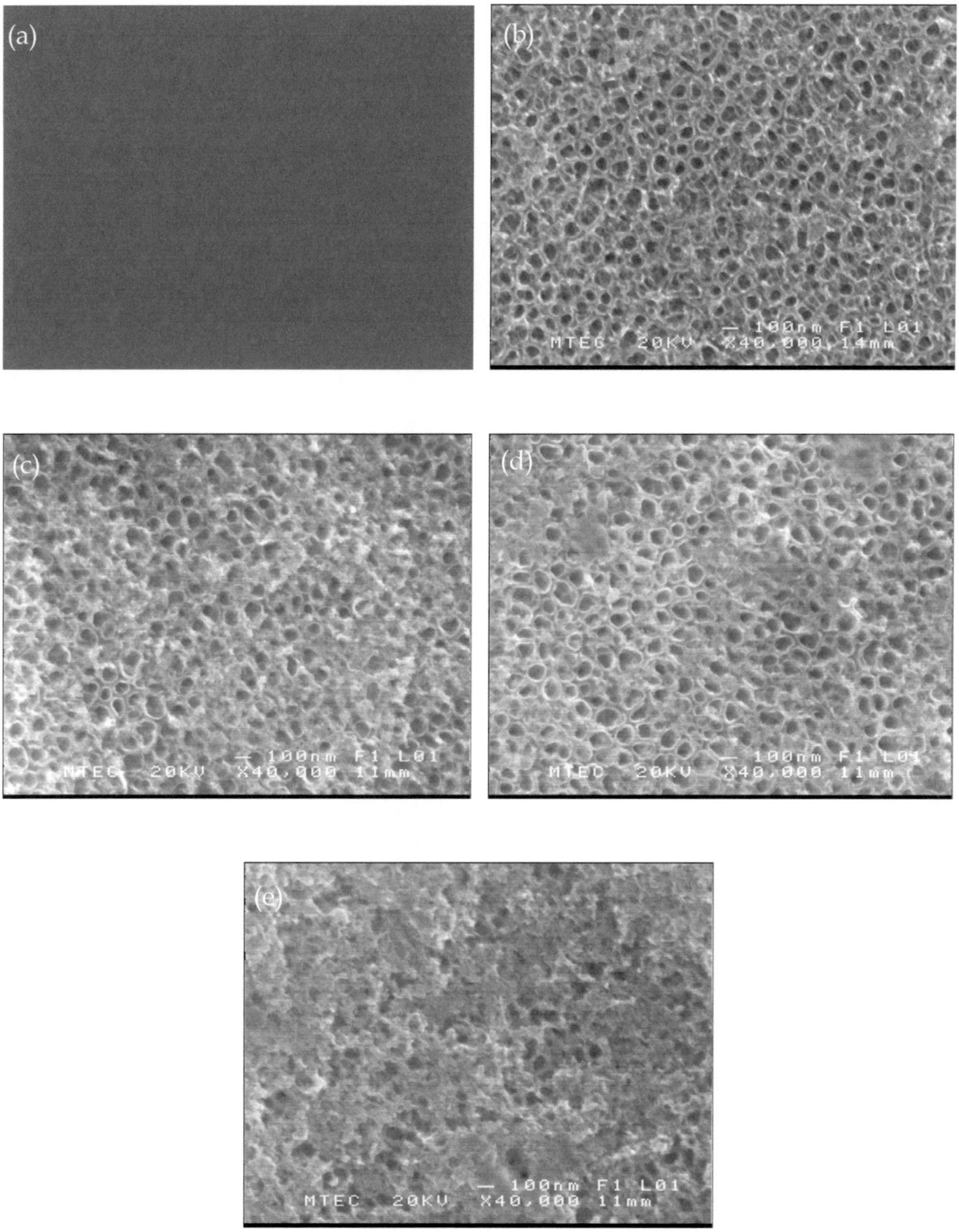

Fig. 4. SEM images of titanium sheets (a) untreated surface and anodised in 1 M acetic acid with (b) 0.075 wt% HF (c) 0.5 wt% HF (d) 1.5 wt% HF (e) 2.0 wt% HF for 60 minutes at room temperature.

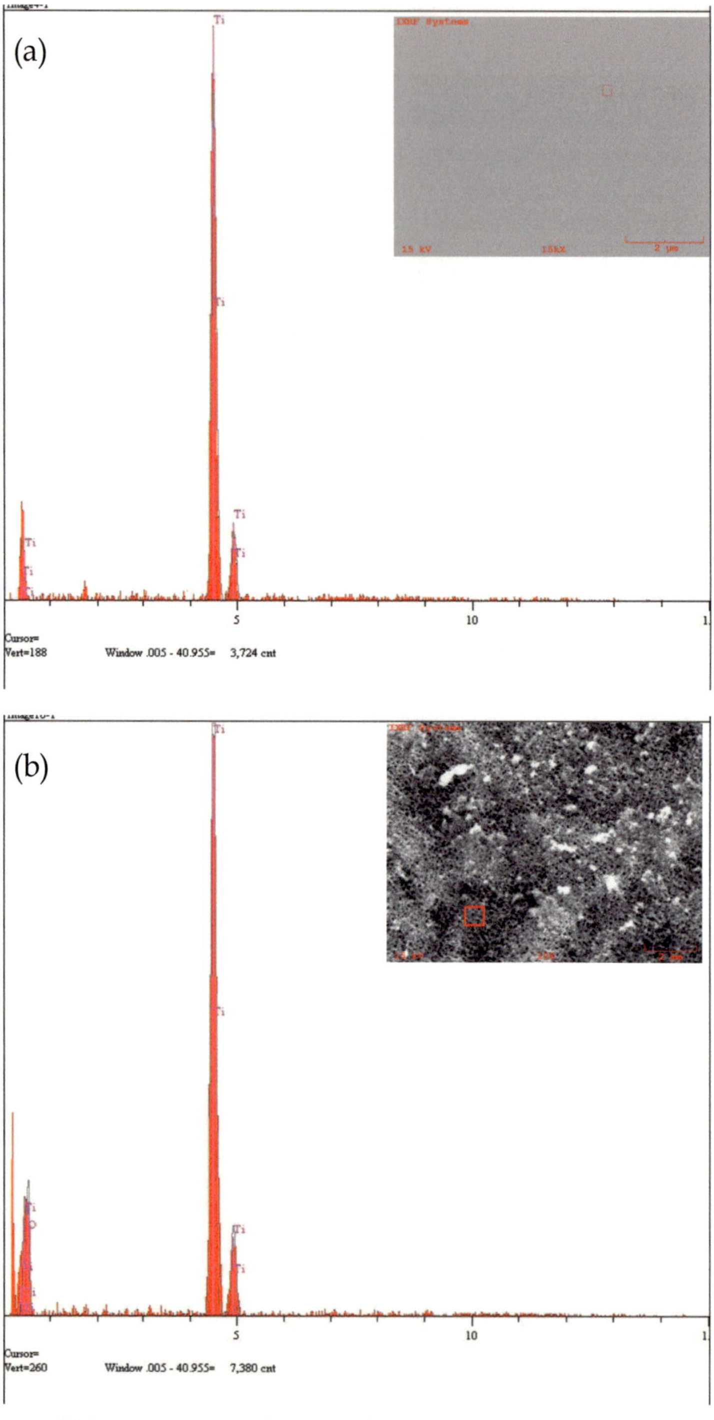

Fig. 5. SEM image of (a) untreated surface and (b) anodised in 1 M acetic acid with 0.075 wt% HF for 60 minutes at room temperature then EDS analysis of the surfaces for energy dispersive analysis was 15 kV, Takeoff Angle 35.0° and Elapsed Livetime 10.0.

4. Conclusion

SEM analysis can clearly showed that anodisation is a simple and economical method to synthesise various surface patterns and textures on the surface of a metallic titanium surface. Moreover, surface morphology is strongly affected by anodisation condition. SEM revealed that the anodisation condition caused micro- and nanomorphological alterations of titanium surface, whereas prolonged exposure to electrolyte resulted in micromorphological changes of the titanium surface. The above result clearly point out that the hydrofluoric acid play important role in controlling the formation of titania nanotubes. Moreover, by adding fluoride ions into the electrolyte, nanotubes can be fabricated under suitable conditions

5. References

[1] M Yamada, *Mats Sci & Eng A.*, 213 (1996) 8.
[2] R R Boyer, *Mats Sci & Eng* A., 213 (1996) 103.
[3] I V Gorynin, *Mats Sci & Eng A.*, 263 (1999) 112.
[4] M Balazic and J Kopac, *Int. J. Nano and Biomaterials.*, 1 (2007) 3.
[5] M Cortada, L L Giner, S Costa, F J Gil, D Rodríguez, J A Planell, *J of Mats Sci: Mats in Med.*, 11 (2000) 287.
[6] A W E Hodgson, Y Mueller, D Forster, S Virtanen, *Electrochemical Acta.*, 47 (2002) 1913.
[7] T Albrektsson and B Albrektsson, *Acta Orthopaedica.*, 58 (1987) 567.
[8] D Leonardis, A K Garg, G E Pecora, *Int. J. of Oral & Maxillofacial Implants.*, 14 (1999) 1.
[9] Y X Leng, J Y Chen, P Yang, H Sun, N Huang, *Surf & Coating Technol.*, 166 (2003) 176.
[10] D M Blake, P C Maness, Z Huang, E J Wolfrum, J Huang, *Separation and Purification Methods.*, 28 (1999) 1.
[11] J A Byrne, J W J Hamilton, T A McMurray, P S M Dunlop, V J A Donaldson, J Rankin, G Dale, D Al Rousan. *Clean Technol.*, (2007) 242.
[12] A S Stasinakis, *Global NEST J.*, 10 (2008) 376.
[13] Y Yamashita, I Takayama, H Fujii, T Yamazaki, *Nippon Steel Technical Report.*, 85 (2002) 11.
[14] G F New, *Current Science.*, 3 (1973) 133.
[15] R Vogel, P Meredith, I Kartini, M Harvey, J D Riches, A Bishop, N Heckenberg, M Trau, H Rubinsztein-Dunlop, *ChemPhysChem.*, 4 (2003) 595.
[16] S A Brown and J E Lemons, *ASTM STP 1272, ASTM.* (Philadelphia, 1996).
[17] X Liu, P K Chu, C Ding, *Mats Sci & Eng R.*, 47 (2004) 49.
[18] J M Lee, Y S Kim, C W Kim, K S Jang, Y J Lim, *J Korean Acad Prosthodont.*, 42 (2004) 352.
[19] D P Dowling, P V Kola, K Donnelly, T C Kelly, K Brumitt, L Lloyd, R Eloy, M Therin, N Weill, *Diamond and Related Mats.*, 6 (1997) 390.
[20] C C Chen, J H Chen, C G Chao, *J of Mats Sci.*, 40 (2005) 4053.
[21] A M PeirÓ, E Brillas, J Peral, X Domènech, J A AyllÓn, *J of Mats Chem.*, 12 (2002) 2769.
[22] Y Li, J Hagen, W Schaffrath, P Otschi, D Haarer, *Solar Energy Material and Solar Cells.*, 56 (1999) 167.
[23] S Lavanus, G Louarn, P Layrolle, *Int. J. of Biomaterials.*, 2010 (2010) 1-9.
[24] F Variola, F Vetrone, L Richert, P Jedrzejowski, J.-H Yi, S Zalzal, S Clair, A Sarkissian, D.F Perepichka, J.D Wuest, F Rosei, A Nanci, *Small.*, 5 (2009) 996-1006.
[25] C Toth, G Szabó, L Kovács, K Vargha, J Barabás, Z Németh, *Smart Mater. Struct.*, 11 (2002) 813-818.

[26] T Shibata and Y C Zhu, *Corrosion Science.*, 37 (1995) 253.
[27] K H Kim and N Ramaswamy, *Dental Mats J.*, 28 (2009) 20.
[28] Metalast Technical Bulletin. (NV: Metalast International Inc, 2000).
[29] L Bartlett, *Optics & Laser Technol.*, 38 (2006) 440.
[30] K G Mor, O K Varghese, M Paulose, K Shankar, C A Grimes, *Solar Energy Materials & Solar Cells.*, 90 (2006) 2011.
[31] A Jaroenworaluck, D Regonini, C R Bowen, R Stevens, D Allsopp, J Mater Sci., 42 (2007) 6729-6734.
[32] M K Kyung, S J Doo, S H Cheol, *Nanotechnology.*, 22 (2011) 1-17.

Part 5

Thin Films, Membranes, Ceramic

30

Application of Scanning Electron Microscopy for the Morphological Study of Biofilm in Medical Devices

R. M. Abd El-Baky
Microbiology Department, Faculty of Pharmacy, Minia University, Minia, Egypt

1. Introduction

The widespread use of medical devices has caused a great advance in the management of many diseases. Indwelling medical devices are being increasingly used for the treatment of functional deficits in numerous medical fields. Urinary tract infections (UTIs) represent the most commonly acquired bacterial infection. The risk of developing a urinary tract infection increases significantly with the use of indwelling devices such as catheters and urethral stents/sphincters. Although these catheters are valuable, they also have complications, the major complications are: encrustation, stone formation and biofilm formation. Microbial biofilms may pose a public health problem for persons requiring catheterization as the microorganisms in biofilms are difficult or impossible to be treated by antimicrobial agents.

Several approaches have been studied to prevent the formation of biofilms or to eradicate biofilm associated microorganisms. Some of these depends on coating medical devices with silver, antiseptics or by producing radio-opacity in catheters by the silicone material and some depends on the use of antimicrobial agents or non antimicrobial agents.

One approach to overcome the antimicrobial resistance of biofilm bacteria would be to enhance the penetration of agents through the biofilm matrix. Many trials were done to increase the efficacy of antimicrobial agents using some agents such as protamine sulfate (anticoagulant), EDTA, sodium citrate and penicillamine.

This work was done to determine whether N- acetylcysteine could aid ciprofloxacin in penetrating biofilm formed by some microorganisms on ureteral stents.

The use of biomaterials in the urinary tract dates back to ancient times, when the Egyptians described using lead and papyrus to construct urinary catheters (Bitschay and Brodny, 1956). Today, the majority of biomaterials used in urology are made from synthetic polymeric compounds, which were originally developed in the plastic injury. In the process of endourological development, a great variety of foreign bodies have been invented besides urethral catheters like ureter, prostatic stents, percutaneous nephrostomy, penile, testicular implants, and artificial urinary sphincters (AUS). Although the tendency of patients being predisposed to infections due to foreign bodies has been recognized since the fourteenth century, the mechanisms of device-related infections are still not completely understood (Tenke *et al.*, 2006).

Bacterial adherence and the growth of bacteria on solid surfaces as biofilm are both naturally occurring phenomena. Biofilm formation affects many aspects of our life and also plays an important role in medicine involving the field of urology. It is able to build up under natural circumstances, for instance on the urothelium or prostate stones and in the presence of temporarily or permanently implanted foreign bodies. The frequently used urethral catheters, double J stents and transrenal drains provide just as perfect surfaces to bacteria to adhere. Biofilms can have a positive impact as well, namely lining the healthy intestine and the female genito-urinary tract. Biofilms have significant implications for clinical pharmacology, particularly related to antibiotic resistance, drug adsorption onto and pealing off devices, and minimum inhibitory concentrations of drugs required for effective therapy (Mardis and Kroeger, 1988).

2. What is a biofilm

2.1 Definition

Biofilm is defined as structured communities of microbial species embedded in a biopolymer matrix on either biotic (living tissues) or a biotic (inert non living material) substrata. The general theory of biofilm predominance states that the majority of bacteria grow in matrix-enclosed biofilms adherent to surfaces in all nutrient-sufficient aquatic ecosystems and that these sessile bacterial cells differ greatly from their planktonic counterparts (Costerton *et al.*, 1978). The reason for this ubiquity is that the protective layer and distinct metabolic states of bacteria within biofilms provide them with special resistance to host defences and antimicrobials, including natural antibiotics.

2.2 Estructure

The basic structure unit of the biofilm is the microcolony. A mature biofilm is composed of cells (10-15% by volume) and of glycocalyx (85-90%). The cells embedded in glycocalyx form gross structure resembling towers and mushrooms, sometimes as high as a few millimeters. Open channels are interspersed between the microcolonies resembling a primitive circulatory system. Water and nutrients enter these channels and contribute to nutrition and formation of mature biofilms. Waste products are also removed through this system.

The cells composing a biofilm can be of single species or more commonly are heterogenous species of bacteria and fungi. In the latter case the metabolic by-products of one organism might serve to support the growth of another, and the adhesion of one species might provide ligands allowing the attachment of others. A mature biofilms contains thousands of bacteria (Dunne, 2002).

The glycocalyx is mainly composed of bacterial exopolysaccharides. Other components are nucleic acids, minerals and proteins. When fully hydrated, the glycocalyx is predominantly water, with an anionic charge that creates a scavenging system for trapping and concentrating essential nutrients from the environment. Glycocalyx also provides a protective layer against biocides and this is the most prominent morphologic feature addressed in this chapter.

2.3 Development

2.3.1 Reversible attachment

Once at the surface, different physical, chemical and biological processes take place during this initial interaction between bacteria and the surface. On the abiotic surface, primary attachment between bacteria and the surface is mediated by non-specific interactions such as electrostatic, hydrophobic, or vander waals forces. On biotic surfaces such as tissues, Primary attachment is through specific molecular adhesion by lectin or adhesin. The surface appendages and structures that bear adhesins include fimbriae, flagella, capsule, outer membrane and other appendages. Bacteria bearing adhesins can approach receptors at a distance, form complexes, and ultimately settle onto substratum (Ofek and Doyle, 1994).

In this stage, The organism must be brought into close approximation of the surface either randomly by a stream of fluid flowing over a surface or by a directed fashion via chemotaxsis and motility. Once the organisms reach critical proximity of the surface (usually < 1nm) the final determination of adhesion depends on:

The net sum of attractive or repulsive forces generated between the two surfaces. These forces include electrostatic and hydrophobic interactions, steric hindrance, vander waals forces, temperature and hydrodynamic forces. Electrostatic interactions tend to favor repulsion, because most bacteria and inert surfaces are negatively charged. Hydrophobic interactions probably have greater influence on the outcome of primary adhesion depending on the molecules in the conditioned film. Repulsive forces can be over come by specific molecular interactions mediated by adhesins located on structures extending from the cell surface such as pili (Carpentier and Cerf, 1993).

It was observed that flagellar and twitching motility are necessary for *pseudomonas aeruginosa* biofilm development. As when biofilm formation to a biotic surface of polyvinyl chloride plates (PVC) by *P. aeruginosa* and its 2 mutants is compared, one mutant is defective in flagellar mediated motility and the other is defective in biogenesis of polar type of iv pili. The biofilm of wild type is followed using phase contrast microscopy. First, the strain formed a monolayer of cells on the a biotic surface (PVC) followed by the appearance of microcolonies that were dispersed through the monolayer of cells. Then using time lapse microscopy, it showed that microcolonies were formed by aggregation of cells present in the monolayer. As observed in the wild type, the strain defective in type iv pili formed a monolayer of cells on the PVC but cells failed to develop microcolonies suggesting that these structures play an important role in microcolonies formation, while very few cells of flagellar defective non-motile mutant are attached to PVC surface even after 8 hours incubation showing that the role of flagella and/ or motility in initial cell-surface adhesion is very important (Arora *et al.*, 1998).

2.3.2 Irreversible attachment

After binding to the surface through exopolymeric matrix, bacterial cells start the process of irreversible adhesion, proliferation and accumulation as multilayered cell clusters. These extracellular matrices, composed of a mixture of materials such as polysaccharides, proteins, nucleic acids and other substances are considered to be essential in cementing bacterial cells together in the biofilm structure, in helping to trap and retain nutrients for biofilm growth and in protecting cells from dehydration and the effects of antimicrobial agents (Davis and Geesey, 1995).

2.3.3 Maturation of biofilm formation

Once having irreversibly attached to a surface, bacterial cells undergo phenotypic changes, and the process of biofilm maturation begins. Bacteria start to form microcolonies either by aggregation of already attached cells, clonal growth (cell division) or cell recruitment of planktonic cells or cell flocs from the bulk liquid. The attached cells generate a large amount of extracellular components which interact with organic and inorganic molecules in the immediate environment to create the glycocalyx (Jiang and Pace, 2006).

Mature biofilms consist of differentiated mushroom and pillar like structures of cells embedded in copious amounts of extracellular polymer matrix or glycocalyx, which are separated by water-filled channels and voids to allow convective flows that transport nutrients and oxygen from the interface to the interior parts of the biofilm and remove metabolic wastes (Stoodly *et al.*, 2002).

There are many environment within a biofilm, each varying because of difference in local conditions such as nutrient availability, PH, oxidizing potential (redox) and so on. Cells near the surface of the biofilm are exposed to high concentrations of oxygen, while near the center oxygen is rapidly depleted to near anaerobic levels (Lewandowski, 1994). The steep oxygen gradients are paralleled by gradients for other nutrients or metabolites from the biofilm (de Beer *et al.*, 1994). Apparently, biofilms display both structural and metabolic heterogenecity which provide this community the capability to resist stresses, whether from host defense systems or antimicrobial agents (Kumar and Anand, 1998).

2.3.4 Detachment

At some point the biofilm reaches a critical mass and the the outermost layer begins to generate planktonic organisms that may escap from the biofilm and colonize other surfaces. Dispersion of planktonic cells can be facilitated by digestion of glycocalyx by enzymes and quorum-sensing might be required for this phenomenon (Dagostino et al., 1991).

2.4 Factors affect the adherence of microorganisms to a device surface

2.4.1 Device related factors

Certain materials used in the design of Indwelling medical devices (IMDs) are more conductive to microbial adherence/biofilm formation than others. In vitro studies performed by many laboratories have determined that microbial adherence to biomaterials occurs in the following order: latex> silicone > PVC > Teflon > Polyurethane > stainless steel > titanium (Darouiche, 2001).

Surface characteristics determining the adherence properties of specific materials include: (a) surface texture, (b) surface charge, and (c) hydrophobicity.

a. Surface texture:
   Materials with irregular or rough surfaces tend to have enhanced microbial adherence compared to smooth surfaces. It is documented that surface irregularities in central

venous catheters (CVC) varied with different polymer materials so that bacteria preferentially adhered to surface defects within minutes after infusing the catheters with contaminated buffer solution (Locci *et al.,* 1981). Another study examined the surface of five commercially available polyurethane CVCs by scanning electron microscopy and found that the catheters with the most surface irregularities had significantly more adherent bacteria compared to catheters with smoother surfaces (Tebbs *et al.*, 1994).

b. Surface charge:
Biomaterial surface charge greatly influences adherence of microorganisms. Most microorganisms exhibit a negative surface charge in an aqueous environment. Therefore, a negatively charged biomaterial surface should lead to decreased adherence of microorganisms due to a repulsion effect between both negatively charged surfaces (Jansen *et al.*, 1988).

c. Hydrophobicity:
Bacterial cells, which tend to have hydrophobic cell surfaces, are attracted to the hydrophobic surfaces of many biomaterials currently used in IMDs (Schierholz and Beuth, 2001). This hydrophobic interaction between the microorganisms and the biomaterial leads to increased adherence and subsequent biofilm formation. An increase in the surface hydrophilicity of the polymers leads to weakened hydrophobic interactions and decreased adherence (Jansen *et al.*, 1988).

2.4.2 Host factors

The biomaterials used in IMDs result in the activation of the host immune response leading to local tissue damage and the development of an immuno-incompetent, fibro-inflammatory zone that increases the susceptibility of the IMD to infection (Schierholz and Beuth, 2001). The deposition of proteinaceous layer (including fibronectin, fibrinogen, fibrin, albumin, collagen, laminin) on the surface of the device forming conditioning film leads to the alteration of surface properties of the biomaterial and the increase of microbial adherence (Pascual, 2002).

2.4.3 Microbial factors

The cell surface of a bacterium possesses many structures and properties that contribute to bacterial adhesion including fimbriae (pili), the cell wall (teichoic acid in gram-positive bacteria) and outer cell membrane (lipopolysaccharides in gram negative pathogens). These characteristics influence the surface charge and hydrophobicity of the bacterial cell, thereby directly affecting adherence (Bonner *et al.,* 1997). The physico-chemical characters of microbial cell surface, i.e., hydrophobicity and charge will influence adherence to biomaterial surfaces since the process is strongly governed by hydrophobic and electrostatic interactions (Martinez-Martinez *et al.*, 1991).

2.4.4 The suspending medium

The absorption of components from the suspending fluid can affect the adhesive properties of microorganisms. The ionic strength, osmolarity, and pH all influence the initial attachment of bacteria (Denstedt *et al.*, 1998). In the process of adherence of microorganisms to an implanted device, one or both entities will be exposed to a biological secretion or body fluid of host origin. The subsequent conditioning of microbial cell or/and biomaterial surface will modify the nature of both surfaces, thereby determining the outcome of the adherence process. It is observed that prior colonization of endotracheal tubes, microorganisms preferentially adhere to a biological film of human origin rather than to the constituent biomaterial itself (Poisson *et al.*, 1991). Adherence of *E. coli* and *E. faecalis*, grown in Mueller- Hinton broth, was shown to increase after the biomaterial was exposed to human urine (Bonner *et al.*, 1997).

2.4.5 Bacteria- biomaterial interaction

The adhesion of microorganisms to biomaterial surfaces has been shown to require both non-specific reversible interactions and highly specific irreversible interactions. First, reversible adhesion of microorganisms to biomaterial surface is dependent upon the physical characteristics of the microorganisms, biomaterial and the surrounding environment (Gristina, 1987). Microorganisms randomly reach the surface of the biomaterial by several mechanisms: direct contamination, contiguous spread, hematogenous spread. Once near the surface, initial adherence of the microorganism depends upon microorganism-biomaterial interactions including van der waals forces and hydrophobic interactions (pascual, 2002). The common charges of the microorganisms and the IMD surfaces will repel each other, however the effect of van der waals forces overcome this repulsion beginning about 10 nm from the IMD surface keeping the microorganisms near the biomaterial surface (Gristina, 1987).

It has shown that hydrophobic forces are 10 to 100 times stronger than van der waals forces at 10 nm from the biomaterial surface. The hydrophobic forces easily overcome electrostatic repulsion and position the organisms 1-2 nm from IMD surface then allows irreversible adhesion to occur (Pashley *et al.*, 1985). Second, irreversible adhesion occurs with the binding of specific microorganism adhesins to receptors expressed by the conditioning film . i.e., *S. aureus* and *S. epidermidis* which are the most common microorganisms causing IMD-related infections relies on specific cell surface proteins called "microbial surface component recognizing adhesive matrix molecules" (MSCRAMM) which bind to specific host ligands that are found in the conditioning films. The most important MSCRAMMS are the fibronectin-binding proteins (FnBPs), the fibrinogen-binding proteins (clumping factors, CIf) and the collagen (Darouiche *et al.*, 1997).

Cell surface proteins also play an important role in *S. epidermidis* adhesion to IMDs. Proteinaceaous autolysin and polysaccharide adhesin (PSA) are two surface proteins that play an early role in the irreversible adhesion of *S. epidermidis* to IMD surfaces. Once adherent to the biomaterial surface, cell accumulation and early biofilm formation are dependent upon the polysaccharide intercellular adhesin (PIA), which promotes intercellular adhesion (Rupp *et al.*, 1999).

2.5 Defense mechanisms

The use of antibiotics is currently one of the possibilities for the prevention of biofilm formation. However, even in the presence of antibiotics bacteria can adhere, colonize and survive on implanted medical devices as has been shown for urinary catheters and ureteral stent surfaces in-vitro and in-vivo (Caldwell, 1995). In addition, resistance to antimicrobial agents and other chemicals is one of the greatest problems in the age of widely used medical devices. The problem in conventional clinical microbiology is how to treat patients in the best way when choosing antibiotics is based on bacterial cultures derived from planktonic bacterial cells which differ very from bacteria in the biofilm mode. This can stand behind the clinical failure rate of treating chronic bacterial infection (Choong and Whitfield, 2000).

The failure of antimicrobial agents to treat biofilms has been associated with a variety of mechanisms (Brown *et al.*, 1990):

1. The glycocalyx restricts access and diffusion of antimicrobial agents to the deeper lying bacteria (extrinsic resistance). In situ studies have shown that the surface film influences the transport of nutrients and interferes with the transport of antimicrobials (Nivens *et al.*, 1993).
2. The growth rates of bacteria within a biofilm vary widely. Slow-growing bacteria are particularly resistant to antimicrobial agents (Brown, 1997). The limitation of diffusion of nutrients in a biofilm results in spatial gradients of growth rate leading to a plethora of phenotypes within the biofilm. In general, the faster-growing, more susceptible bacteria lie superficially but the slow-growing, less susceptible bacteria being placed more deeply. The failure of antimicrobial agents to eradicate these slow-growing bacteria may exert selection pressures on the least susceptible genotype to select for a resistant population. Furthermore, antimicrobial binding proteins are poorly expressed in the slow-growing bacteria, rendering the antimicrobial agents ineffective (Cozens *et al.*, 1986). Commonly, the entire biofilm is coated with a complex of a hydrophilic polymer, the glycocalyx that is typically anionic in nature where the antimicrobial agents reacts chemically with exopolymer or is adsorbed to it, then the net effect is that of having the appearance of a penetration barrier. There will be a similar effect if antimicrobials adsorb onto cells, perhaps dead ones, in the outer parts of the biofilm (Sutherland, 2001).
3. Bacteria within a biofilm are phenotypically so different from their planktonic counterparts that antimicrobial agents developed against the latter often fail to eradicate organisms in the biofilms. Bacteria within a biofilm activate many genes which alter the cell envelope and molecular targets, and alter the susceptibility to antimicrobial agents (intrinsic resistance). Current opinion is that phenotypic changes brought on by a genetic switch, when 65-80 proteins change, play a much more important role in the protection from antimicrobial agents than the external resistance provided by the exopolysaccharide slime (Anonymous, 1999).
4. Bacteria within a biofilm can sense the external environment, communicate with each other and transfer genetic information and plasmids within biofilms (Trieu-Cuot *et al.*, 1987).
5. Bacteria in a biofilm can usually survive the presence of antimicrobial agents at concentrations 1000-1500 times higher than the concentrations that kill planktonic cells of the same species (Costerton, 1999).

Resistance may be due to:

- production of inactivating enzymes as it is found that a relatively large amounts of antibiotic-inactivating enzymes such as ß- lactamase which accumulate within the glycocalyx produce concentration gradients can protect underlying cells (Bagge *et al.*, 2000).

- Efflux pumps were also believed to play a role in the resistance of biofilms; however expression of the pumps decreases within the biofilm bacteria dependent on time and location of the cells within the community, as compared to planktonic bacterial form (De Kievit *et al.*, 2001).
- Recent work has highlighted the contribution of oxygen deprivation and anaerobic growth to antibiotic resistance, it is indicated that oxygen penetrate approximately to 25% of the depth of the biofilm, when challenged with antimicrobials, 4h old colony biofilms growing in the presence of air were susceptible, however similar aged biofilms grown anaerobically were much less susceptible. The authors calculated that oxygen limited could explain 70% or more of the protection of old biofilm cells (Borriello *et al.*, 2004).

2.6 Treatment and prevention of biofilms

Strategies for prevention of these infections include: (i) minimizing tissue destruction and removal of all extraneous biomaterials and devitalized tissues during surgery. (ii) development of biomaterials that resist the initial adherence of bacteria by surface characteristic or by promoting bactericidal, bacteriostatic or phagocytic activity at their surfaces. (iii) further study of the microstructure and chemical nature of the adherence mechanism and development of analogs and enzymes that might block the initial adherence by modification of receptors and ligands (Khardori and Yassien, 1995) .

Several approaches have been studied to prevent the formation of biofilms and to eradicate biofilms associated bacteria. Some of that depends on the use of antimicrobial agents or non antimicrobial agents.

a. Antimicrobial agents:

In the case of the use of antimicrobial agents, it was found that some antibiotic at sub-MIC inhibit the initial adherence. Dicloxacillin is the antibiotic that found to prevent the adherence to the greatest extent when it is used alone at 1/2 of the MIC (Cerca *et al.*, 2005). Also clindamycin at subinhibitory concentrations inhibits the adherence of *Pseudomonas aeruginosa*, *Staphylococcus aureus*, Bacteroids spp., *Escherichia coli* to bone surfaces (Lambe *et al.*, 1987).

Norfloxacin, ciprofloxacin, ofloxacin and azithromycin at sub inhibitory concentrations reduced the glycocalyx production and inhibited the adherence of *Staphylococcus epidermidis* and *Pseudomonas aeruginosa* (Pézer-Giraldo *et al.*, 1989 ; Yassien *et al.*, 1995). Ciprofloxacin was reported to eradicate the performed biofilms of *P. aeruginosa* (Reid *et al.*, 1994). It was reported also that 1/2 MIC of ciprofloxacin, 1/4 MIC of ofloxacin and 1/32 of levofloxacin caused significant inhibition of adherence of some uropathogenic strains of *E. coli* to periuretheral epithelial cells (Baskin *et al.*, 2002).

Macrolides are generally bacteriostatic in-vitro and in-vivo, and have useful activity versus gram-positive bacteria. Macrolides have been evaluated to affect the adherence of gram-negative bacteria at sub-MIC concentrations by 50 to 70% as it is found to affect the production of virulence determinants such as secreted virulence factor, motility, quorum sensing and biofilm production (Vranes, 2000). It is discovered that sub-MIC level of clarithromycin inhibits the twitching motility of *P. aeruginosa*, they do not affect the production of pili but inhibit their assembly on the surface of bacteria that should affect some steps in biofilm formation (Wozniak and keyser, 2004).

It is investigated that the antibiofilm effects by incubating ciprofloxacin with *P. aeruginosa* or in combination with macrolides. At twice the minimum bactericidal concentration of ciprofloxacin, 85% of the population of *P. aeruginosa* within the biofilm survived. In contrast, the killing effect of ciprofloxacin was greatly enhanced when combined with clarithromycin, erythromycin and azithromycin, but not with the 16-membered ring macrolides. It is speculated that the 14-membered and 15- membered ring macrolides posses an ability to increase the permeability of biofilms, there by facilitating the penetration of quinolone antibiotics. Tigecycline was observed to inhibit the growth of *S. epidermidis* which indicates that tigecycline is able to diffuse through the biofilm and act normally against its cellular target (Labthavikul *et al.*, 2003).

b. Non-antimicrobial agents:

It is observed that some drugs other than antimicrobial agents such as anti-inflammatory or antiseptic compounds reduce the adherence of bacteria. Coating the catheter with acetylsalicylic acid or sodium salicylate reduces or inhibits microbial adherence, Bandazac lysine (non steroidal anti-inflammatory) was found to prevented the adherence of bacteria to contact lenses (Arciola *et al.*, 1998).

Some mucolytics substances such as EDTA, sodium citrate and penicillamine may disperse the biofilms formed by *P. aeruginosa* (Gordon *et al.*, 1991). It is observed also that N-acetylcysteine (NAC) (a non antibiotic drug that has antibacterial properties (bacteriostatic) and a mucolytic agent that disrupts disulphide bonds in mucus and reduces the viscosity of secretions) decreases biofilm formation and therefore may be an effective alternative for preventing infections by *S. epidermidis* and other coagulase negative staphylococci (Pérez-Giraldo *et al.*, 1997).

It is observed that NAC not only reduced the adhesion but in fact also detached adhered cells from a steel surface. This has some importance since the initial adhesion often develops into a stronger interaction with time (bond ageing) (Meinders *et al.*, 1995). The reduction in the amount of exopolysaccharides (EPS) in the presence of NAC may have many explanations. The direct effects of NAC include a possible reaction of its sulfhydryl group with disulfide bonds of enzymes involved in EPS production or excretion, which renders these molecules less active, or competitive inhibition of cysteine utilization. Also, the possibility of interference of NAC with control or signaling systems that direct the EPS production at translation or at the enzymatic level cannot be excluded. The fact that NAC is an anti-oxidant may have indirect effects on cell metabolism and EPS production. NAC increases the wettability of surfaces. Moreover, NAC detached bacteria that were adhering to steel surfaces. Growth of various bacteria, as monocultures or in multi-species community, was inhibited at different concentrations of NAC. It is also found that there was no detectable degradation of EPS by NAC, indicating that NAC reduced the production of EPS in most bacteria tested, even at concentrations at which growth was not affected (Olofsson *et al.*, 2003).

Aspirin (acetylsalicylic acid) has a short half life in circulating blood (about 20 min) and is rapidly deacetylated to form salicylic acid in-vivo. Sodium salicylate and related compounds such as aspirin are known to have a variety of effects on microorganisms. Growth of certain bacteria in the presence of salicylate can induce multiple resistance to antibiotic. Paradoxically, it can also reduce resistance to some antibiotics (Price *et al.*, 2000). *E. coli,* for example exhibits increased resistance to chloramphenicol, ampicillin, naldixic acid and tetracycline after such treatment. On the other hand *E. coli* cells grown in the presence of salicylate are more sensitive to aminoglycosides (Aumercier *et al.*, 1990).

Sodium salicylate inhibits biofilm formation by *P. aeruginosa* and *S. epidermidis* on contact lenses and medical polymers such as polyethylene and polystyrene. It also decreases bacterial adhesion in a dose-dependent manner. Some strains of *S. epidermidis* secrete mucoid extracellular polymers (polysaccharides, proteins and teichoic acid) that promote biofilm formation and become important components of the biofilm matrix. Salicylate can inhibit the production of some of these components by as much as 95%. It has been suggested that the use of salicylate into contact lens solutions might decrease the incidence of some device- related infections (Farber and Wolff, 1992).

Chlorohexidine gluconate and silver sulfadiazine coated vascular catheter has been shown to be highly effective in decreasing catheter related infections (George *et al.*, 1997).

Protamine sulfate (a surface active, basic polypeptide presently used to reverse the anticoagulant effects of heparin) could aid antibiotics in penetrating a *P. aeruginosa* biofilm (Richards, 1976). Parsons and coworkers have shown that protamine sulfate penetrates and disrupts the protective glycosaminoglycan layer. There is a significant, synergistic effect observed between protamine sulfate and ciprofloxacin as protamine sulfate may have denatured the complex extracellular polymeric structure of the *P. aeruginosa* biofilm enhancing penetration of ciprofloxacin through the biofilm (Soboh *et al.*, 1995).

Gendine solution (a novel antiseptic solution) formed of gention violet and chlorohexidine has the ability to coat various polymers and devices. It has also a broad spectrum antiadherence activity and antimicrobial activity which decreases the risk of device colonization, which may in turn decreases the rates of nosocomial infection and their associated morbidity and mortality (Chaiban *et al.*, 2005).

3. Techniques for the study of biofilm

This work was done to detect biofilm formed on ureteral stents and to determine whether N-acetylcysteine could aid ciprofloxacin in penetrating biofilm formed by some microorganisms on ureteral stents. Several techniques were used in this study first Stents were removed by physicians and collected in sterile screw capped tubes, then cut into segments to be examined by Scanning electron microscope (SEM) and to be cultured on different media.

Catheter segment were fixed in 2.5% (vol/vol) glutaraldehyde in Dulbecco PBS (pH 7.2) for 1.5h, rinsed with Phosphate buffer saline (PBS), and then dehydrated through an ethanol series. Samples were critical point dried and gold-palladium coated. SEM examinations were made on a JSM-840 SEM (JEOL Ltd., Tokyo, Japan).

Urine samples were collected and streaked onto the culture media and incubated at 37°C for 24 hours (Benson, 2002). The resultant colonies were streaked and examined morphologically, microscopically and biochemically.

Catheter samples: Each catheter was placed in 10 ml of tryptic soy broth (TSB), sonicated for 1 min and then vortexed for 15 s. 0.1 ml of the sonicated broth were surface plated by using a wire loop on trypticase soy agar with 5% sheep blood and MacConkey agar. Organisms were then identified by routine microbiological techniques (Sherertz *et al.*, 1990).

Antibiotic susceptibility and MICs were determined for the isolated microorganisms by the agar dilution method, according to clinical laboratory standards institute (CLSI) (2007)

The isolated microorganisms were tested for their ability to form biofilm by tissue culture plate method (TCP). Effect of different concentrations of ciprofloxacin, N-acetylcysteine each alone and in combination on the bacterial adherence to plastic surfaces were determined by tissue culture plate assay (Christensen *et al*, 1985). The Effect of different concentrations of ciprofloxacin, N-acetylcysteine each alone and in combination on the bacterial adherence to the surface of ureteral catheter were determined by Static adhesion assay (Reid *et al.*, 1994) and their effects were determined also using scanning electron microscope.

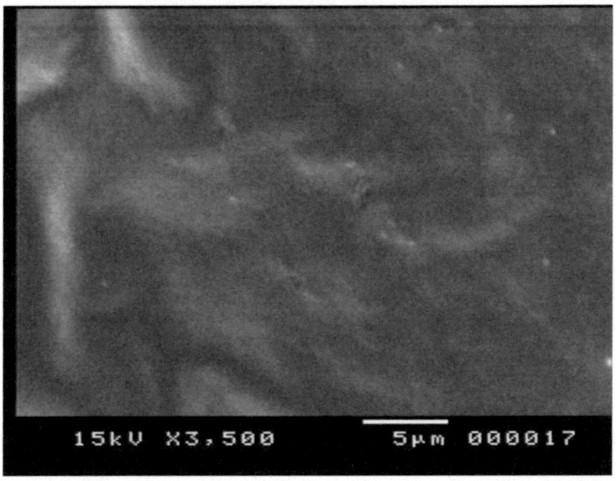

Fig. 1. Scanning electron micrograph of an empty ureteral stents incubated in saline for 24 h (control) (× 3500).

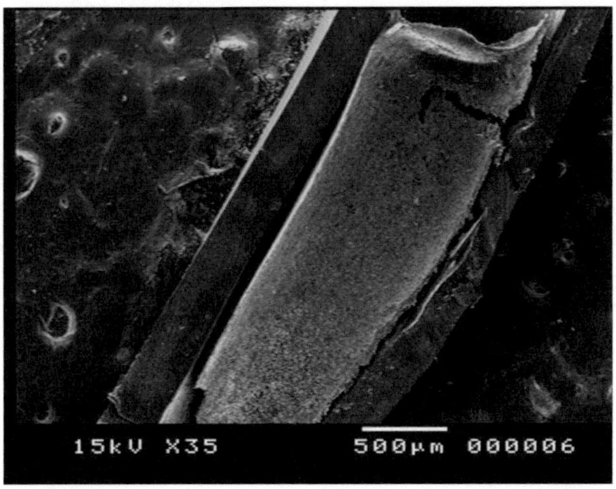

Fig. 2. Scanning electron micrograph showed the lumen of the ureteral stent (× 35) blocked with a dense mass of biofilm containing bacteria.

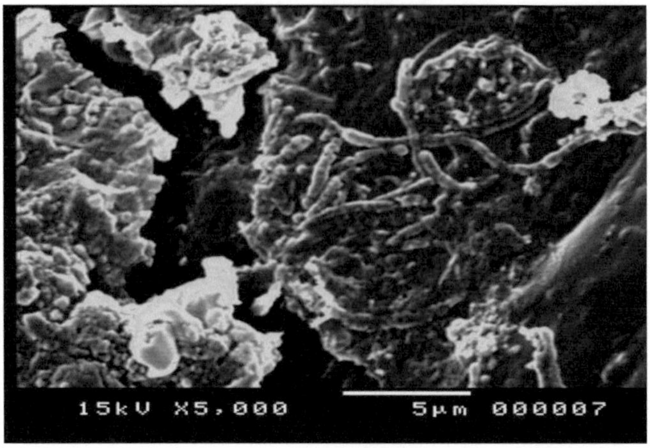

Fig. 3. Scanning electron micrograph showed the lumen of the ureteral stent covered with a densed mass of biofilm containing bacteria (*S. aureus* and *P. rettgeri*) and crystalline patches (× 5000).

In the present work, 292 strains were isolated and identified from 284 samples. As out of 100 urine samples (before catheterization), 76 (76%) were positive for bacterial growth. Out of 92 urine samples (after stent removal), 80 (86.95%) were positive for bacterial growth and out of 92 stent samples, 84 (91.3%) were positive for bacterial growth. Stents collected from patients were examined for biofilm using SEM and it was found that all stents positive for microbial growth were showing biofilm upon their examination.

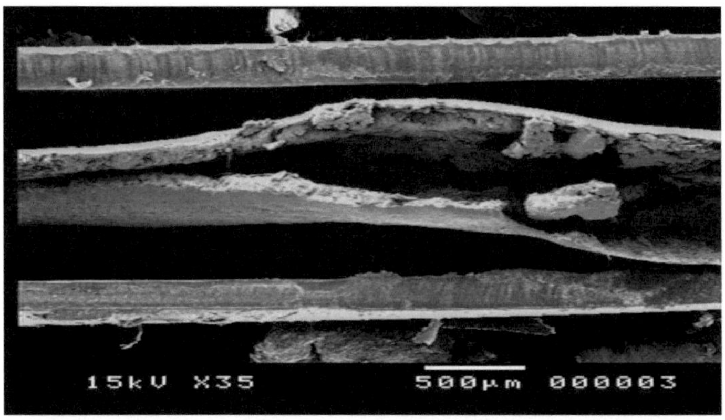

Fig. 4. Scanning electron micrograph showed the lumen of a ureteral stent obtained from patients treated with cefotaxime (× 35). It showed a dense mass of biofilm and a high level of encrustation.

Klebseilla spp. was the most prevalent (21.9%) microorganism followed by Pseudomonas spp. (18.8%), Staphylococci spp. (18.2%), *E. coli* (17.8%), Proteus spp. (11.3%), *Providencia rettgeri* (4.8%) *Citrobacter freundii* (4.8%) and *Serratia marcescens* (2.8%). Mixed infection represented 22.9%. All *S. aureus* and coagulase negative staphylococci isolates were polymicrobial with Klebseilla spp., Pseudomonas spp., *Providencia rettgeri* and *S. marcescens*.

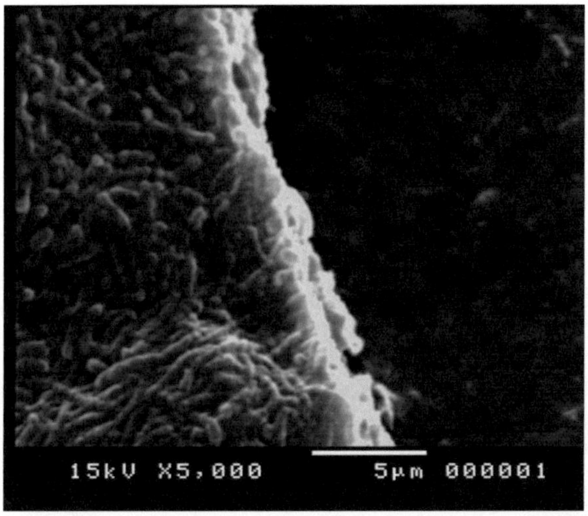

Fig. 5. Scanning electron micrograph showed the lumen of the ureteral stent (×5000). It showed a dense mass of biofilm (rods and cocci bacteria).

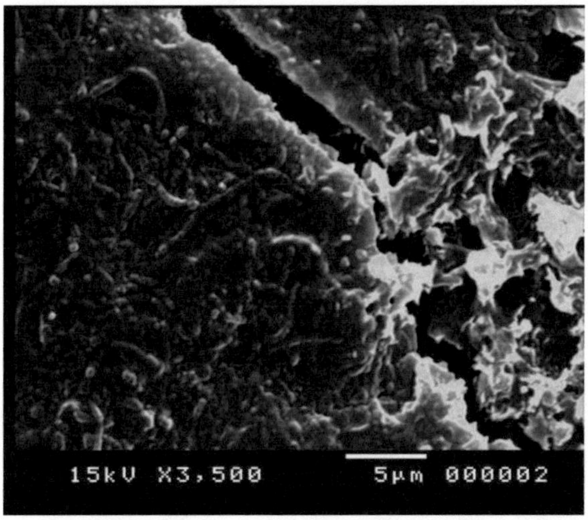

Fig. 6. Scanning electron micrograph showed the surface of the ureteral stent (× 3500). It showed a dense mass of biofilm containing microorganisms and a high level of encrustation.

The resistance pattern to cefotaxime, augmentin, ciprofloxacin, levofloxacin and ofloxacin revealed that the highest incidence of resistance to cefotaxime was shown by *K. oxytocae* (54.2%). Also the highest incidence of resistance to augmentin and levofloxacin was shown by Pseudomonas spp. (80 and 72.7%, respectively), while the highest resistance to ciprofloxacin and ofloxacin was shown by *C. freundii* (78.6% each).

Biofilm production was found in 84.6% of the isolates using TCP. Pseudomonas spp. were the highest biofilm producing microorganism. A dose related decrease in biofilm formation was observed by both ciprofloxacin and N-acetylcysteine. This was detected by a decrease in the optical density of the biofilm layer on microtiter plates and the number of viable cells attached to the catheter surfaces in comparison to controls. It was found also that CIP/NAC combinations have the highest inhibitory effect on the initial adherence (84-100% of the controls) and the highest disruptive effect to mature biofilms (87-100% of the controls).

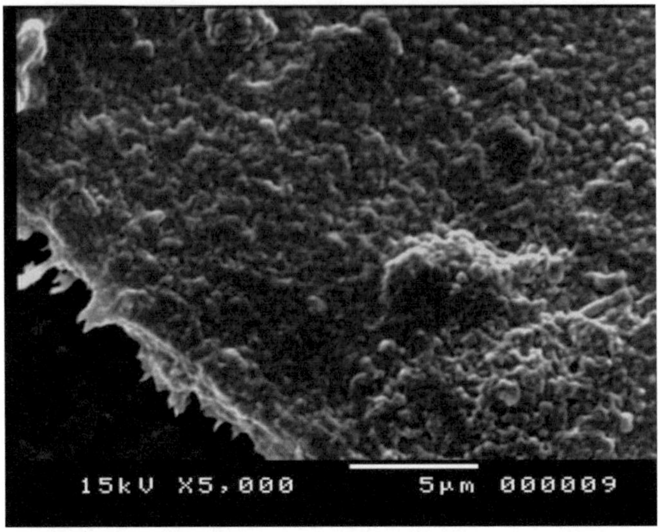

Fig. 7. Scanning electron micrograph showed the surface of a ureteral stent covered with high densed crystalline biofilm (× 5000).

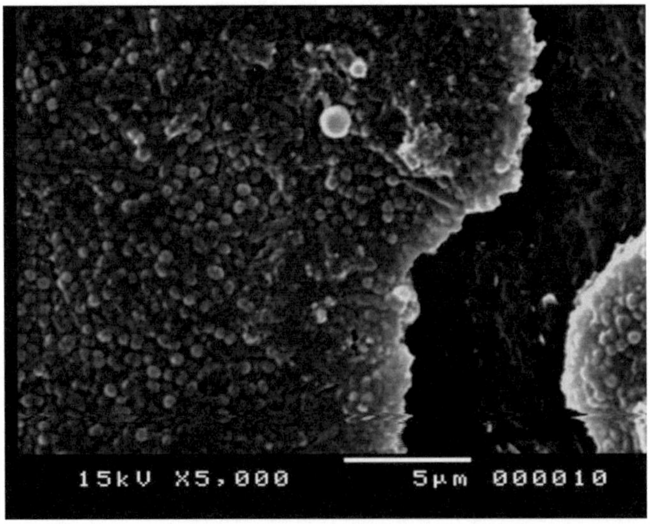

Fig. 8. Scanning electron micrograph showed the lumen of a ureteral stent covered with a big mass of biofilm containing bacteria (rods and cocci) (*K. pneumoniae* and *S. aureus*) (× 5000).

The inhibitory effects of the tested agents were also verified by (SEM). Scanning electron micrographs showed the morphological response of the tested organisms to ciprofloxacin and N-acetylcysteine. They showed also the decrease in the extent of biofilm formation in the presence of the tested agents.

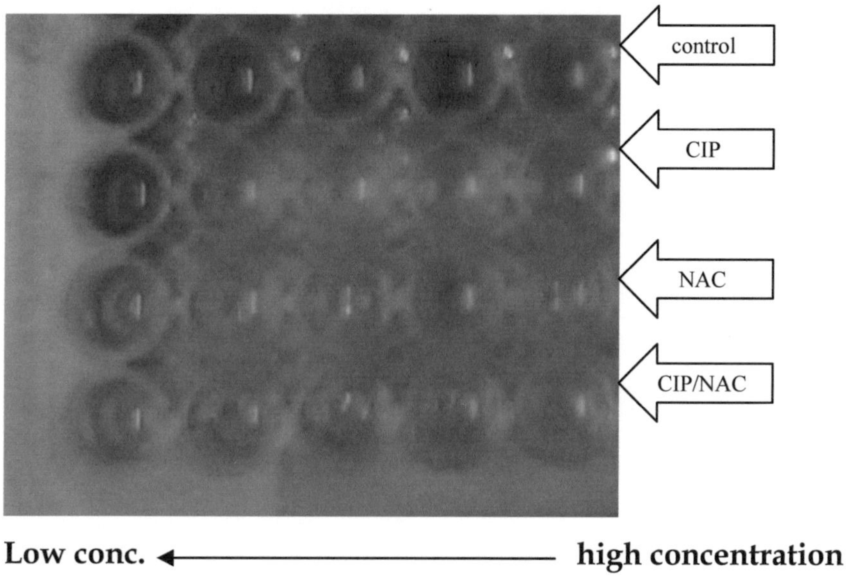

Fig. 9.

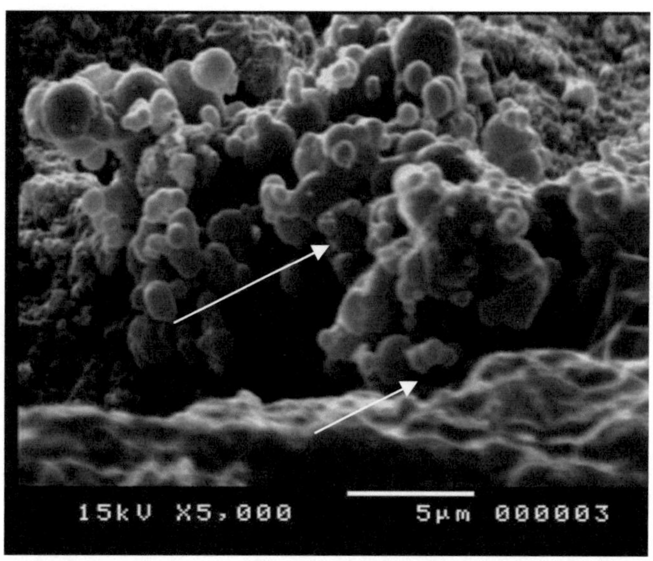

Fig. 10. a. Scanning electron micrograph of *S. aureus* biofilm on the surface. (a uretral stent incubated with *S. aureus* suspension for 24h as a control) (× 5000).

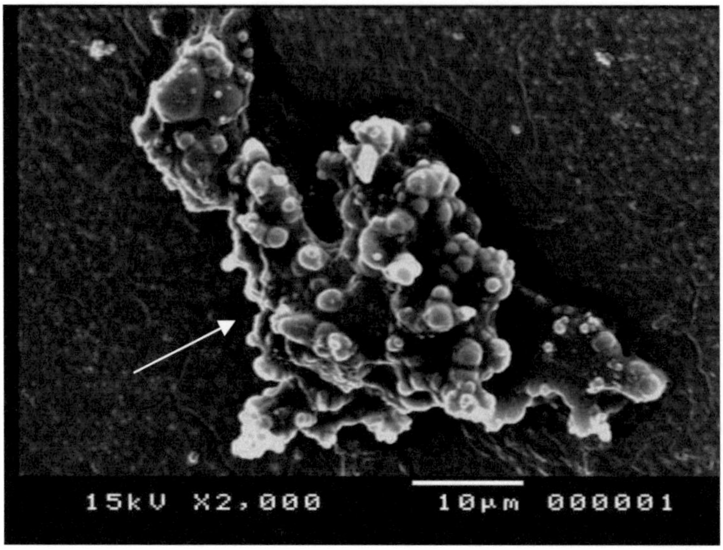

Fig. 10. b. Scanning electron micrograph showed the morphological response of *S. aureus* performed biofilm on the surface of a uretral stent exposed to sub-MIC concentration (CIP 4 µg/ml). there was a decrease in the amount of biofilm mass adhered to stent surface.

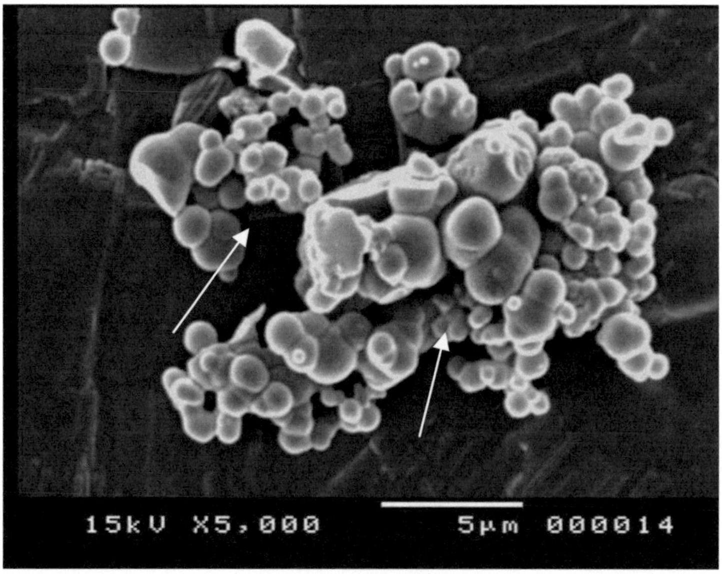

Fig. 10. c. Scanning electron micrograph showed the effect of N-acetylcysteine on a performed *S. aureus* biofilm. Cotton like mass disappeared and cells appeared swollen with disrupted cell wall (× 5000).

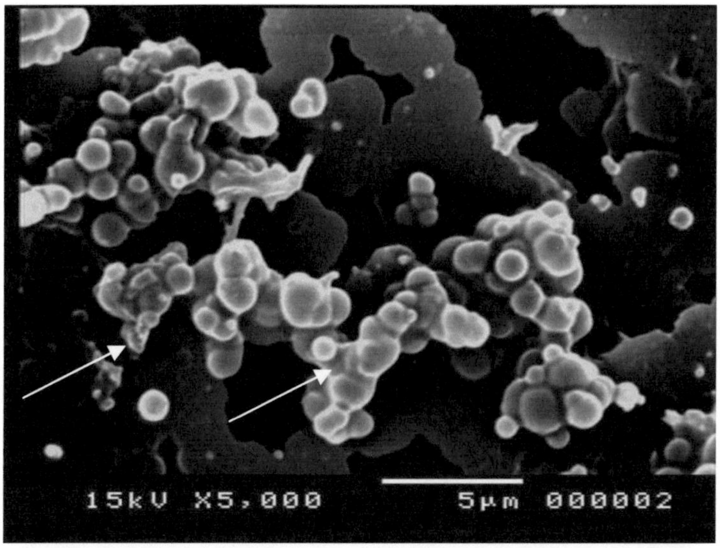

Fig. 10. d. Scanning electron micrograph showed the effect of ciprofloxacin-N-acetylcysteine combination on a performed *S. aureus* biofilm. Cell appeared swollen, disrupted and scattered (× 5000).

Scanning electron micrographs showed the effect of Ciprofloxacin, N-acetylcysteine each alone and in combination on a performed *S. aureus* biofilm developed *in-vitro* on stent surface.

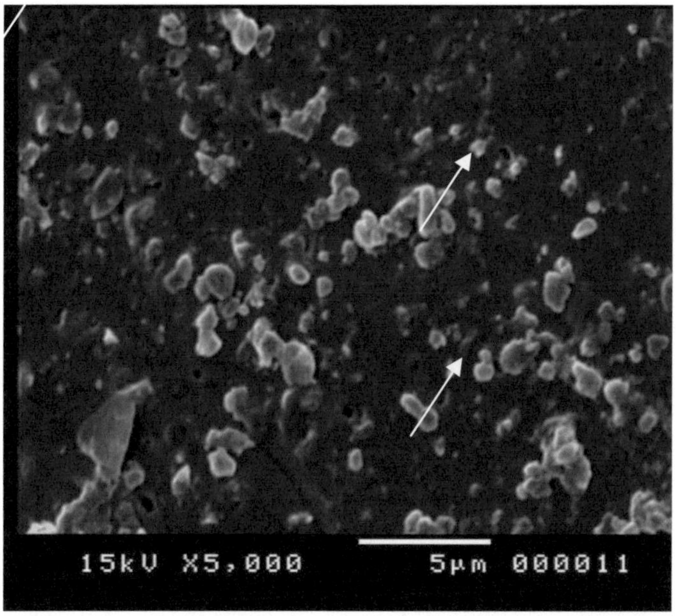

Fig. 11. a. Scanning electron micrograph showing the morphological responses of Pseudomonas spp. and *S. epidermidis* grown in the presence of sub-MIC concentration of ciprofloxacin. Cells appeared swolled and scattered with no biofilm mass.

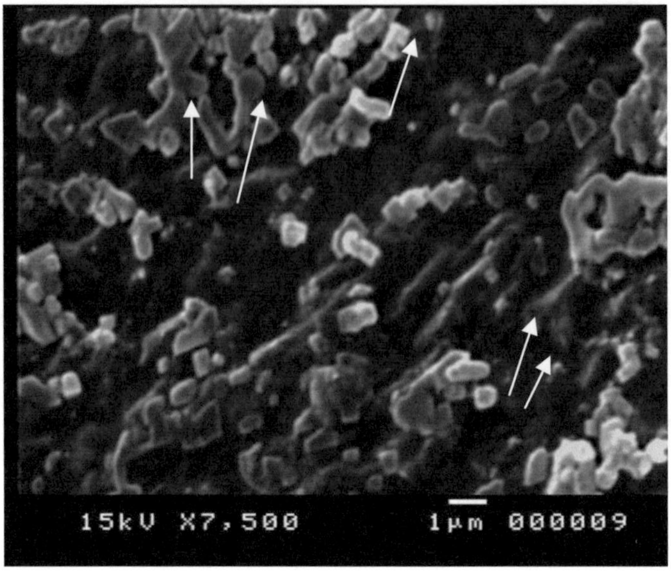

Fig 11. b. Scanning electron micrograph showing the effect of N-acetylcysteine on the biofilm formed by *S. epidermidis* and pseudomonas spp.. Cells showed membrane disorganization, appeared swolled and with disrupted outer membrane.

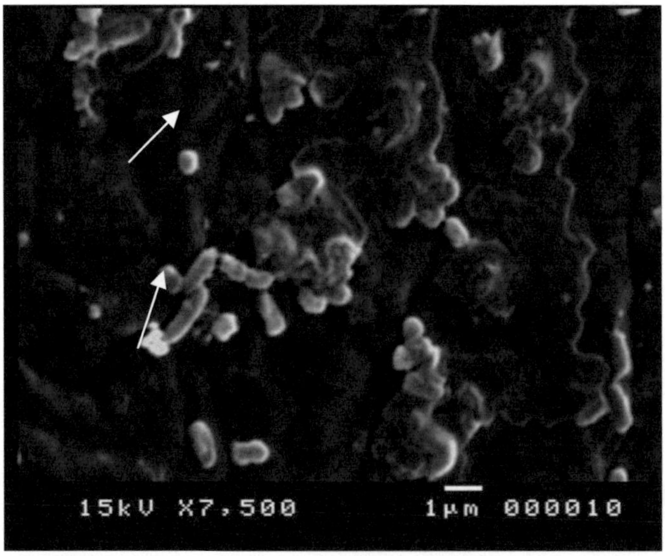

Fig 11. c. Scanning electron micrograph showed the effect of CIP/NAC (MIC/$4_{mg/ml}$) on the ability of *S. epidermidis* and Pseudomonas spp. to form biofilm. Cells appeared scattered, elongated, swollen, with disorganized (irregular) membrane and with no cotton like mass (biofilm) around cells.

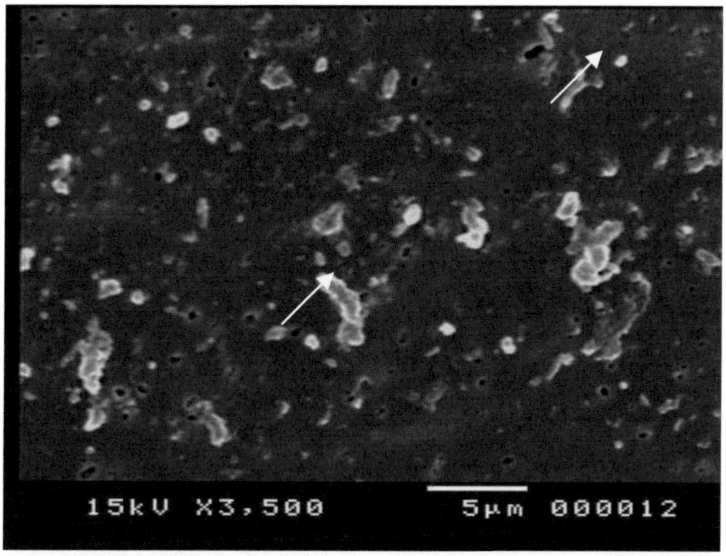

Fig 11. D. Scanning electron micrograph showed the effect of CIP/NAC combination (2 MIC/ 8 $_{mg/ml}$) on the ability of *S. epidermidis* and pseudomonas spp. To form biofilm. A high decrease in the nimber of adherent cells observed. Cells appeared large, swollen and with disrupted cell wall.

Scanning electron micrographs showed the morphological response and the ability of *S. epidermidis* and Pseudomonas spp. grown in the presence of Ciprofloxacin, N-acetylcysteine and their combinations to form biofilm on stent surfaces.

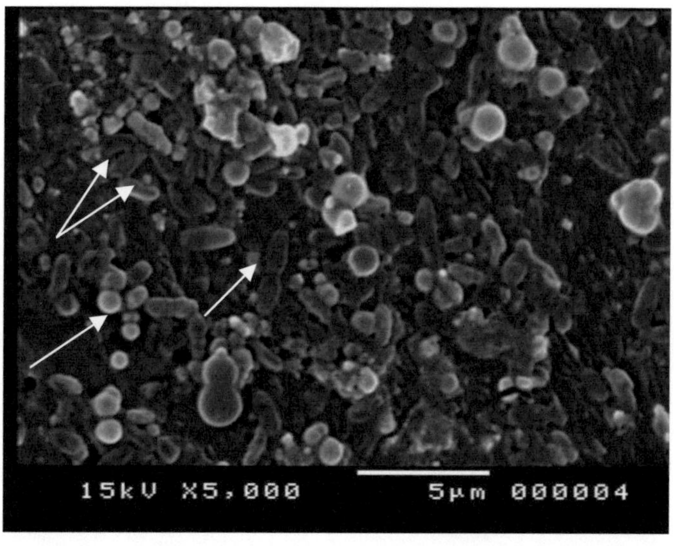

Fig. 12. a. Scanning electron micrograph showed the morphological response of *S. aureus* and pseudomonas spp. cells grown in the presence of ciprofloxacin at sub-MIC concentration. Cells appeared swollen, enlarged, with irregular cell wall, some showed v-shaped cells and small amount of biofilm mass observed.

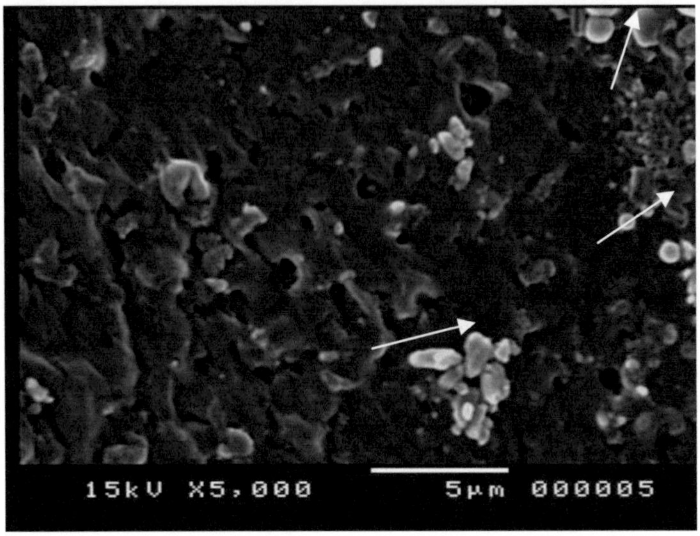

Fig. 12. b. Scanning electron micrograph showed the effect of N-acetylcysteine (4 mg/ml) on biofilm formation by *S. aureus* and Pseudomonas spp.. Cells appeared swollen, irregular in shape and small microcolonies observed scattered. A decrease in the numer of adherent cells was observed.

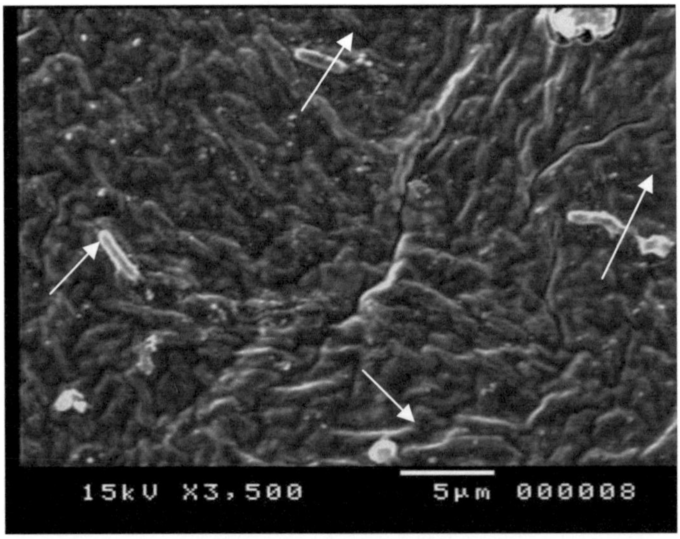

Fig. 12. c. Scanning electron micrograph showed the effect of CIP/NAC combination of ($MIC/4_{mg/ml}$) on *S. aureus* and pseudomonas spp. ability to form biofilm. Cells appeared elongated, enlarged and scattered with no biofilm mass observed on the surface.

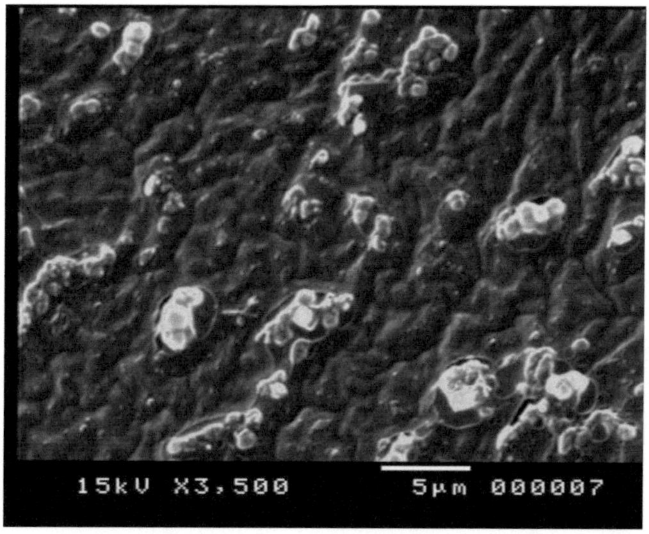

Fig. 12. d. Scanning electron micrograph showed the effect of CIP/NAC combination of (2 MIC/8 $_{mg/ml}$) on *S. aureus* and pseudomonas spp. ability to form biofilm. No biofilm observed on the surface of stent.

Scanning electron micrographs showed the morphological response and the ability of *S. aureus* and Pseudomonas spp. grown in the presence of Ciprofloxacin (sub-MIC), N-acetylcysteine and their combinations to form biofilm on stent surfaces.

4. Conclusion

The presence of non antimicrobial agent such as N-acetylcysteine (NAC), caused significant decrease in biofilm formation by a variety of bacteria and reduces the production of extracellular polysaccharide matrix while promoting the disruption of mature biofilms. It was found that the inhibitory effect of both ciprofloxacin and N-acetylcysteine was concentration dependent.

CIP/NAC combinations were found to show the highest effect on bacterial adherence inhibition and on the disruption of the already formed biofilms. As N-acetylcysteine increase the therapeutic activity of ciprofloxacin when used in combination by degrading the extracellular polysaccharide matrix of biofilm.

In the chapter, Scanning Electron Microscope is used for the evaluation of medical implants, detection of biofilm and studying the effect of different biofilm inhibitory agents. This technique provides excellent visualization of glycocalyx, which is one of the most prominent features of biofilms and a crucial research subject in the searching for alternative antimicrobial and anti adherent agents treatments.

5. Acknowledgment

Thanks for my professor doctors: Mohamed Ali Mohamed El-Feky, Mostafa Said Khalil El-Rehewy, Mona Amin Hassan(Department of microbiology), Faculty of medicine, Assuit university, professor doctor Hassan Abd El-latif Abolella, (Department of urology) Faculty of medicine, Assuit university and professor doctor: Gamal Fadl Mahmoud Gad, Department of microbiology, Faculty of pharmacy, Minia university for their scientific revision, their help and their technical support.

6. References

Anonymous, N. (1999): Panel discussion on biofilms in urinary tract infection. Int. J. Antimicrobial Agents, 11: 237-9.

Arciola, C.R.; Montanaro, L.; Caramazza, R.; Sassoli, V.; and Cavedagna, D. (1998): Inhibition of bacterial adherence to high water content polymer by a water soluble, non-steroidal, anti-inflammatory drug. J. Biomed. Maten. Res., 42: 1-5.

Arora, S. K.; Ritchings, B.W.; Almira, E.C.; Lary, S. and Ramphal, R. (1998): The *pseudomonas aeruginosa* flagellar cap proteins responsible for mucin adhesion. Infect. Immun., 66: 1000-7.

Aumercier, M.; Murray, D.M. and Rosner, J.L. (1990): Potentiation of susceptibility to aminoglycoside by salicylate on *Escherichia coli*. Antimicrob. Agents Chemother., 34: 786-791.

Bagge, N., Ciofu, O., Skovgaard, L.T. and Hoiby, N. (2000): Rapid development of in-vitro and in-vivo of resistance to ceftazidime in biofilm growing *pseudomonas aeruginosa* due to chromosomal ß-lactamase, APMIS., 108: 589-600.

Baskin, H.; Dogan, Y.; Bahar, N. and Yulug, N. (2002): Effect of sub-inhibitory concentration of three fluoroquinolones on adherence of uropathogenic strains of *Escherichia coli*. Inter. J. Antimicrob. Agents, 1979-82.

Benson, H.C. (2002): Microbiological Application: Laboratory manual in general microbiology, 11thed., McGram-Hill Higher Education, Sanfrancisco. pp.168.

Bitschay, J. and Brodny, M.L. (1956): A history of urology in Egypt. Riverside Press, New York, p. 76.

Bonner, M.C.; Tunney, M.M.; Jones, D.S. and Gorman, S.P. (1997): Factors affecting in-vitro adherence of ureteral stent biofilm isolates to polyurethane. Int. J. Pharmaceutics. 151: 201-207.

Borriello, G.; Warner, E.; Roe, F.; Kim, A.M.; Ehrlich, G.D. and Stewart, P.S. (2004): Oxygen limitation contributes to antibiotic tolerance of *Pseudomonas aeruginosa* biofilms. Antimicrob. Agents Chemother., 48: 2659-2664.

Brown, M.R.W. (1997): The role of the envelope in resistance. In: Brown, M.R.W. ed. Resistance of *Pseudomonas aeruginosa*. London: Wiley. 71-107.

Brown, M.R.W.; Collier, P.J. and Gilbert, P. (1990): Influence of growth rate on susceptibility to antimicrobial agents modification of the cell envelope, batch and continuous culture studies. *Antimicrob. Agents Chemother*. 34: 1623-1628.

Caldwell, D.E. (1995): Cultivation and study of biofilm communities. In: Lippincott, H.M. and Costerton, J.W. (eds). Microbial biofilms. Vol. 1195. Cambridge university press, Cambridge, pp.64-79.

Carpentier, B and cerf, O. (1993): Biofilms and their consequences, with particular references to hygiene in food industry. J. Appl. Bacteriol. 75: 499-511.

Cerca, N.; Martins, S.; Pier, B.G.; Oliveira, R. and Azeredo, J. (2005): The relationship between inhibition of bacterial adhesion to a solid surface by sub-MICs of antibiotics and subsequent development of a biofilm. Res. Microbiol., 156:650-655.

Chaiban, G.; Hanna, H.; Dvorak, T. and Raad, I. (2005): A rapid method of impregnated endotracheal tubes and urinary catheters with gendine: a novel antiseptic agent. J. Antimicrob. Agents Chemother., 55: 51-56.

Choong, S. and Whitfield, H. (2000): Biofilms and their role in infections in urology. B.J.U. international., 86: 935-941.

Christensen, J.H.; Simpson, W.A.; Younger, J.J.; Baddour, L.M.; Barrett, F.F.; Melton, D.M. and Beachey, E.H. (1985): Adherence of coagulase negative staphylococci to plastic tissue culture plates: A qualitative model for the adherence of staphylococci to medical devices. J. Clin. Microbiol., 22: 996-1006.

Clinical and laboratory standards institutes: Performance standards for antimicrobial susceptibility testing. Seventeenth informational supplement M100-S17. Wayne, PA: CLSI: 2007.

Costerton, J. W., Geesy, G.G., Cheng, G.K. (1978): How bacteria stick. Sci Am., 238: 86-95.

Costerton, J.W. (1999): Introduction to biofilm. Int. J. Antimicrobiol Agents., 11: 217-21.

Cozens, R.M.; Tuomanen, E.; Tosh, W.; *et al.* (1986): Evaluation of the bactericidal activity of ß-lactam antibiotics upon slowly growing bacteria cultured in the chemostat. Antimicrob. Agents Chemother., 29: 797-802.

Dagostino L., Goodman AE., Marshall KC., (1991): physiological responses induced in bacteria adhering to surfaces. Biofouling, 4: 113-119.

Darouiche, R.O. (2001): Device-associated infections: a macroproblem that starts with microadherence. Clin. Infect. Dis., 33:1567–1572.

Darouiche, R.O.; Landon, G.C.; Patti, J.M.; Nguyen, L.L.; Fernau, R.C. and McDevitt, D. (1997): Role of *Staphylococcus aureus* adhesions in orthopaedic device infections: are results model-dependent?. J. Med. Microbiol., 46: 75-79.

Davis, D.G. and Geessey, G.G. (1995): Regulation of the alginate biosynthesis gene *algC* in *pseudomonas aeruginosa* during biofilm development in continuous culture. Appl. Environ. Microbiol., 61: 860-867.

de Beer, D.; Stoodley, P.; Roe, F. and Lewandowski, Z. (1994): Effects of biofilm structure on oxygen distribution and mass transport. Biotechnol. Bioeng., 43: 1131-1138.

De Kievit, T.R.; Parkins, M.D.; Gillis, R.J.; Srikumar, R.; Ceri, H.; *et al.* (2001): Multidrug efflux pumps: Expression patterns and contribution to antibiotic resistance in *pseudomonas aeruginosa* biofilms. Antimicrob. Agents Chemother., 45: 1761-1770.

Denstedt, J.D.; Wollin, T.A. and Reid, G. (1998): Biomaterials used in urology: current issues of biocompatibility, infection and encrustation. J. Endourol., 12: 109-112.

Dunne, W. M. (2002): Bacterial adhesion: seen any good biofilm lately?. Clin. Microbiol. Rev., 15: 155-166.

Farber, B.F. and Wolff, A.G. (1992): The use of nonsteroidal anti-inflammatory drugs to prevent adherence of *Staphylococcus epidermidis* to medical polymers. J. Infect. Dis., 166: 861-865.

George, S.J.; vuddamalay, P. and Boscoe, M.J. (1997): Antiseptic-impregnated central venous catheters reduce the incidence of bacterial colonization and associated infection in immunocompromised transplant patients. Europ. J. Anesth., 14: 428-31.

Gristina, A.G. (1987): Biomaterial-centered infection: microbial adhesion versus tissue integration. Science., 237: 1588-1595.

Gordon, C.A.; Hodges, N.A. and Marriott, C. (1991): Use of slime dispersants to promote antibiotic penetration through the extracellular polysaccharide of mucoid *pseudomonas aeruginosa*. Antimicrob. Agents and chemother., 35: 1258-60.

Jansen, B.; Peters, G. and Pulverer, G. (1988): Mechanisms and clinical relevance of bacterial adhesion to polymers. J. Biomatr. Appl. 2: 520-543.

Jiang, X. and Pace, J. (2006): Microbial biofilms in: Pace, J., Rupp, M., Finch, R. eds., Biofilms, Infection and Antimicrobial Therapy. USA, 3-19.

Khardori, N. and Yassien, M. (1995): Biofims in device related infections. J. Ind. Microbiol. 15: 141-7.

Kumar, C.G. and Anand, S.K. (1998): Significance of microbial biofilms in the food industry: a review. Int. Food Microbiol., 42: 9-27.

Labthavikul, P.; Petersen, P. and Bradford, P. (2003): In-vitro activity of tigecycline against *Staphylococcus epidermidis* growing in adherent-cell biofilm model. Antimicrob. Agents and Chemother. 47: 3967-3969.

Lambe, D.W.; Mayberry-Carson, K.J.; Mayberry, W.R.; Tober-Meyer, B.K. and Costerton, J.W. (1987): The effect of sub-inhibitory concentrations of clindamycin on the adherence and glycocalyx of *Staphylococcus aureus* and Bacteroides species in-vitro and in-vivo, p.35-49. In: Szentivanly, A., Friedman, H. and Gillissen, G. (eds) Antibiosis and host immunity. Plenum publishing, New York.

Lewandowski, Z. (1994): Dissolved oxygen gradients near microbically colonized surfaces. In: Geesey, G.G., Lewandowski, Z., and Flemming, H.C., eds, Biofouling and biocorrosion in industrial water systems. Florida: Lewis 175-188.

Locci, R., Peters, G., and Pulverer, G. (1981): Microbial colonization of prosthetic devices. Microtopographical characteristics of intravenous catheters as detected by scanning electron microscopy. Zentralbl. Bakteriol. Mikrobiol. Hyg. 173: 285-292.

Mardis, H.K. and Kroeger, R.M. (1988): Ureteral stents. Urol. Clin. North Am., 15: 471–479.

Martinez-Martinez, L.; Pascual, A.; and Perea, E.J. (1991): Kinetics of adherence of mucoid and non-mucoid *pseudomonas aeruginosa* to plastic catheters. J. Med. Microbiol., 34: 7-12.

Meinders, H.; Vander Mei, H.C; and Busscher, H.J. (1995): Deposition efficiency and reversibility of bacterial adhesion under flow. J. Colloid Interface Sci., 176: 329-341.

Nivens, D.E.; Chambers, J.Q.; Anderson, T.R.; *et al.* (1993): Monitoring microbial adhesion and biofilm formation by attenuated total reflection? Fourier transform infrared spectroscopy. J. Microbiol. Methods., 17: 199-213.

Ofek, I. and Doyole, R.J. (1994): Animal cell membranes as substrata for bacterial adherence, p. 41-53. In Bacterial adhesion to cells and tissues, Chapmann and Hall, New york and London.

Olofsson, A. C.; Hermansson, M. and Elwing, H. (2003): N-acetyl-L-cysteine affects growth, extracellular polysaccharide production, and bacterial biofilm formation on solid surfaces. Appl. Environ. Microbiol. 69: 4814–4822.

Pascual, A. (2002): Pathogenesis of catheter related infections: lessons for new designs. Clin. Microbiol. Infect., 8: 256-264.

Pashley, R.M.; McGuiggan, P.M.; Ninham, B.W. and Evanes, D.F. (1985): Attractive forces between uncharged hydrophobic surfaces: direct measurement in aqueous solution. Science., 229: 1088-1089.

Pézer-Giraldo, C.; Rodriguez-Benito, A.; Maron, F.J.; Hurtado, C.; Blanco, M.T. and Gomez-Garcia, A.C. (1989): In-vitro slime production by *Staphylococcus epidermidis* in presence of subinhibitory concentrations of ciprofloxacin, ofloxacin and sparfloxacin. J. Antimicrob. Chemother., 33: 845-848.

Poisson, D.M.; Arbeille, B. and Laugier, J. (1991): Electron microscope studies of endotracheal tubes used in neonates: do microbes adhere to the polymer?. Res. Microbiol., 142: 1019-1027.

Price, C.T.; Lee, I.R. and Gustafson, J.E. (2000): The effects of salicylate on bacteria. Int. J. Biochem., 32: 1029-1043.

Reid G.; Sharma, S.; Advikolanu, K.; Tieszer, C.; Martin, R. A. and Bruce, A.W. (1994): Effects of ciprofloxacin, norfloxacin, and ofloxacin on In Vitro adhesion and survival of *Pseudomonas aeruginosa* AK1 on urinary catheters. Antimicrob. Agents and Chemother., 38: 1490-1495.

Richards, G.K. (1976): Resistance to infection p. 65-77. In Freedman, S.O. and Gol, P. (eds) clinical immunology. Harper and Row. Newyork

Rupp, M.E.; Ulphani, J.S.; Fey, P.D.; Bartscht, K. and Mack, D. (1999): Characterization of the importance of polysaccharide intercellular adhesion/hemagglutinin of *Staphylococcus epidermidis* in the pathogensis of biomaterial-based infection in a mouse foreign body infection model. Infect. Immun., 67: 2627-2632.

Schierholz, J.M and Beuth, J. (2001): Implant infections: a haven of opportunistic bacteria. J. Hosp. Infect., 49: 87-93.

Sheretz, R.J.; Raad, I.L.; Balani, A. (1990): Three-year experience with sonicated vascular catheter cultures in a clinical microbiology laboratory. J. Clin. Microbiol. 28: 76-82.

Soboh, F.; Khoury, A. E.; Zamboni, A.C.; Davidson, D. and Mittelman, M. W. (1995): Effects of ciprofloxacin and protamine sulfate combinations against catheter-associated *Pseudomonas aeruginosa* biofilms. Antimicrob. Agents Chemother., 39: 1281-1286.

Stoodly, P.; Saur, K.; Davis, D.G. and Costerton, J.W. (2002): Biofilms as complex differentiated communities. Annu. Rev. Microbiol., 56: 187-209.

Sutherland, I.W. (2001): The biofilm matrix-an immobilized but dynamic environment. Trends Microbiol., 9: 222-227.

Tebbs, S.E.; Sawyer, A. and Elliott, T.S. (1994): influence of surface morphology on in-vitro bacterial adherence to central venous catheters. Br. J. Anaesth., 72: 587-591.

Tenke, P.; Riedl, C.R.; Jones, G.L.I.; Williams, G.R.; Stickler, D. and Nagy, E. (2004): Bacterial biofilm formation on urologic devices and heparin-coating as preventive strategy. Int. J. Antimicrob. Agents., 23: 67-74.

Trieu-Cuot, P.; Carlier, C.; Martin, P. and Courvalin, P. (1987): Plasmid transfer by conjugation from Escherichia coli to gram-positive bacteria. FEMS Microbiol. Lett., 48: 289-94.

Vranes, J. (2000): Effect of sub minimal inhibitory concentrations of azithromycin on adherence of *pseudomonas aeruginosa* to polystyrene. J. Chemother., 12: 280-285.

Wozniak, D. and Keyser, R. (2004): Effects of subinhibitory concentrations of macrolide antibiotics on *pseudomonas aeruginosa*. Chest., 125: 62-69.

Yassien, M.A.; Khardori, N.; Ahmedy, A. and Toama, M. (1995): Modulation of biofilms *pseudomonas aeruginosa* by quinolones. Antimicrob. Agents Chemother. 39: 2262-2268.

31

Interrelated Analysis of Performance and Fouling Behaviors in Forward Osmosis by Ex-Situ Membrane Characterizations

Coskun Aydiner*, Semra Topcu, Caner Tortop, Ferihan Kuvvet, Didem Ekinci, Nadir Dizge and Bulent Keskinler
Gebze Institute of Technology, Faculty of Engineering, Department of Environmental Engineering, Gebze, Kocaeli, Turkey

1. Introduction

In membrane processes, flux decline takes place as an inherent result of membrane fouling that varies with specificity in their implementations. The membrane fouling having uncontrollable or unexplainable complexity in many cases leads to somewhat loss of process efficiency which results mainly in costly pretreatment, higher operating pressure requirement, limited recoveries, feed water loss, frequent chemical cleaning and short lifetimes of membranes as the factors increasing the water and energy costs (Aydiner, 2010, as cited in Tu et al., 2005; Hoek et al., 2008; Van der Bruggen et al., 2008). In recent years, more economical operation of membrane processes is to be taken into account based on lower energy and membrane costs in practice. At this point, understanding the reasons lying under the fouling phenomena as related with a membrane's performance is to be rather valuable task in terms of scientific and technological developments of these processes (Danis & Aydiner, 2009). However, non-generalization course of the fouling during a membrane filtration necessitates the use of either modeling tools or specific analyses for clarifying meaningful performance-fouling relationships in each specific application. The modeling solutions are widely utilized not only to expose these relations but to put forward performance dynamics intended for a main aim of systematic representation and reasoning. A specific modeling study for lab-scale researches mostly results in simulation deficiencies or key limitations in attainment of a definitive solution when compared to that for real-world implementations. As a matter of fact, the development of simple, accurate and effective models needs to produce a solution relying on a "solution-directed focus" approach which includes full-scale consideration of theoretical and practical issues of the events. But, it is explicit that successive synchronization of model assemblies with real-time could not be entirely accomplished by the community of membrane scientists and technologists at this time. In that sense, it can be said that specific membrane analyses based on either in-situ or ex-situ characterizations could be foreseen as a progressive tool on the purpose of obviating interruption or non-coordination of transition among small and large

* Corresponding Author

scale operations, especially for emerging membrane technologies such as forward osmosis and membrane distillation.

The applications on the membrane fouling characterization falls into two categories: *(i)* laboratory researches involving in-situ monitoring, and *(ii)* field-level studies employing on-line ex-situ scaling observation. In-situ monitoring techniques are to be used for analysing the membrane fouling comprising concentration polarization and cake formation, and are evaluated as an annotation tool in understanding the fouling characteristics (Huang et al., 2010). The most widely used techniques for in situ monitoring of concentration polarization are light deflection techniques (shadowgraphy and refractometry), magnetic resonance imaging, radio isotope labeling, electron diode array microscope, and direct pressure measurements. Whereas, the fouling analyses based on particle deposition or cake layer formation can be carried out by the techniques such as laser triangulometry, optical laser sensor, ultrasonic time-domain reflectometry, electrical impedance spectroscopy, and small-angle neutron scattering (Chen et al., 2004). The main advantage of monitoring-based techniques over traditional lab-scale systems is the ability to visually observe what really happens on the membrane surface simultaneously in real-time (Huang et al., 2010). Ex-situ fouling and scaling detectors are utilized as another important means for understanding of fouling-performance relationships. By the studies under this type of characterization, various observation detectors or fouling simulators can be developed with the intention of controlling the membrane fouling in real-time applications (Uchymiak et al., 2007). Also, various ex-situ membrane investigations based on different analytical techniques such as spectroscopic ellipsometry, x-ray photoelectron spectroscopy (XPS), scanning electron microscope (SEM), atomic force microscope (AFM), Fourier transform infrared (FTIR) spectroscopy, and contact angle etc. can be effectively used for associating the fouling dynamics with the performance (Darton et al., 2004; P. Xu et al., 2010). At the end of a general evaluation of literature on characterizing the membrane fouling, it can be stated that the interrelation of ex-situ characterizations of the membrane fouling concurrently with both process performance and various model response parameters would be a viable simulative tool oriented to removing the transition problems from lab-scale toward real-world. Already, the presence of many techniques and theoretical models developed during the past two decades could makes a sense to better comprehend the interactions between foulants and membrane, develop more viable membranes and employ the process more effectively.

Forward osmosis (FO) is an osmotically–driven membrane process that works spontaneously by osmosis across a semi–permeable membrane. The process possesses a water flow through the membrane from the solution having low concentration (feed solution) toward the solution having high concentration (draw solution) due to the osmotic pressure difference between the solutions (Cath et al., 2006). Along last decade, FO process can be favorably utilized in many applications such as electricity production (Aaberg, 2003; Gormly et al., 2011), power generation (Loeb, 2007; McGinnis et al., 2007), water or wastewater reclamation (Holloway et al., 2007; Cornelissen et al., 2008), seawater desalination or brine concentration (McCutcheon et al., 2006; Low, 2009), concentration of liquid foods (Petrotos & Lazarides, 2001; Dova et al., 2007), protein enrichment and concentration (Yang et al., 2009; Wang et al., 2011), and water purification and reuse in space (Cath et al., 2005a, 2005b). FO process as one of the foremost among processes which have been recently increasingly explored in separation science and technology pursues its development depending on desalination needs. It has remarkably lower cost due to no hydraulic pressure operation, nearly complete rejection of many contaminants, potentially low membrane fouling tendency (Holloway et al., 2007; Cornelissen et al., 2008; Y. Xu et al., 2010; Wang et al., 2010; Chung et al., 2011). But, there are still some constrains in front of becoming widespread of FO's industrial applications. Major drawback is the lack of a membrane having a relatively high flux compared to commercial reverse osmosis (RO) membranes, as an inherent result of the membrane fouling or internal concentration polarization significantly limiting flux efficiency (McCutcheon et al., 2006; Tang, et al., 2010; Wang et al., 2010). The other is the requirement to provide high osmotic pressure difference under continuously operating conditions in which draw solution needs to be concentrated for producing clean water by a complementary process such as reverse osmosis (RO), membrane distillation, membrane osmotic distillation, decomposition with heating, and magnetic separation (Cath et al., 2005a, 2005b; McCutcheon et al., 2005; Martinetti et al., 2009; Ling et al., 2010).

In spite of its development potential in membrane science and technology, very few publications on the membrane fouling in FO systems were presented in the literature when compared to those in pressure-driven membrane systems. Hence, FO process was especially preferred in this study to interrelate the performance and internal fouling with membrane surface characteristics. Besides, cheese whey was selected as the feed solution due to its high pollutant capacity with rich nutrient content. In order to render the operation of the system as independent of membrane type and operation mode, the process was employed at normal and reverse operation modes using one each of FO and RO membranes. First, the FO system performance was investigated, and internal membrane fouling was estimated by modeling of the performance data. Thereafter, internal fouling was associated with the results of ex–situ membrane surface characterizations. Solute resistivity which is defined as the internal fouling was estimated depending on salt permeability coefficient at the end of the modeling of the performance data. Ex-situ characterizations of the studied membranes were carried out by SEM, AFM, contact angle, and FTIR measurements. Afterwards, the relation equations among external fouling and each one of performance response parameter which comprises of the water flux, specific water fluxes, salt flux, and net and effective osmotic pressure differences were individually evolved by using contact angle determined as illustrative parameter for external fouling from the relation equations among external fouling and internal fouling in ex-situ characterizations. By this way, in light of significant perspectives obtained by joint interpretations of the results, whether or not the membrane fouling can be correlated with the whole performance was put forward. As concluding remarks, the individual mathematical representations of the relationships of internal and external foulings with the performance were elucidated as regards each dynamics of the whole performance. The prospects of more effective treatment of a membrane process were straightforwardly presented in special to osmotically–driven system as independent of membrane type and operating mode.

2. Materials and methods

2.1 Materials

In the experiments, two different membrane materials in a form of flat sheet were used, one being cellulose triacetate (CTA) FO membrane (Hydration Technologies Inc., OR) and the other being composite polyamide (CPA-3) RO membrane (Hydranautics Inc., CA). FO and RO membranes have salt rejection rates of about 95 and 99.6%, respectively. Their pure water permeabilities, *A* were determined to be approximately 8.1×10^{-3} $m^3/m^2\cdot h\cdot bar$ and 3.305×10^{-3} $m^3/m^2\cdot h\cdot bar$, respectively, by means of pure water permeation experiments in a pressure–driven cross–flow membrane system operated at transmembrane pressures of 5, 10, 15, 20 and 25 bar with a constant temperature of 25 °C. FO draw solution used was prepared by dissolving 3 M NaCl (Prolabo, >99%) into the distilled water in order to obtain a high net osmotic pressure difference in the system. Cheese whey was obtained from industrial facilities of Cayirova Milk&Milk Products Inc., located at Kocaeli, Turkey. The characteristics of the raw and FO concentrated cheese whey were presented in Table 1, together with the average water quality values measured in draw solutions after the processing.

parameter	unit	raw cheese whey	FO concentrated cheese whey	FO draw solution
pH		4.97±0.09	4.7±0.1	5.7±1.0
conductivity	mS/cm	6.73±0.08	9.8±3.1	179.8±17.5
Cl^-	mg/L	950±28	1,896±1,103	78,776±10,345
COD	mg/L	58,220±12,509	82,043±30,774	727±905
TOC	mg/L	39,261±2,611	49,013±13,649	30±35
NH_4-N	mg/L	142±8	167±24	2.0±1.1
NO_2-N	mg/L	0.04±0.02	0.2±0.2	0
NO_3-N	mg/L	254±15	280±34	0
TKN	mg/L	1,353±139	1,659±343	6.1±5.6
Org-N	mg/L	1,211±138	1,492±321	4.1±6.0
TN	mg/L	1,607±146	1,939±371	6.1±5.6
PO_4-P	mg/L	370±10	464±116	7.3±13.3
TP	mg/L	470±98	569±154	12.6±23.0
total protein	%	2.38±0.18	3.23±1.16	–
fat	%	0.37±0.07	0.50±0.12	–
SNF (fat–free dry matter)	%	6.39±0.22	8.82±3.04	–
total solid content	%	6.76±0.29	9.32±3.14	–
lactose	%	3.05±0.27	4.48±1.61	–
minerals	%	0.99±0.06	1.39±0.42	–

Table 1. Characteristics of the raw and FO concentrated cheese whey and the average water quality observed in draw solutions

It should be noted in Table 1 that COD and TOC parameters were measured as soluble COD and soluble TOC in cheese whey samples, and dash means that the relevant parameters were not measured in draw solution samples.

2.2 Methods

2.2.1 Experimental procedure

The experiments were carried out by a lab–scale FO system shown in Fig. 1. The process was operated in both normal and reverse orientation mode with a closed loop, which means that the active or selective layer of the membrane was faced on draw solution and whey, respectively. Cross–flow membrane module was a custom made cell with equivalent flow channel at both sides of the membrane. The membrane module which was made from Delrin acetal resin material (DuPont, Wilmington, Delaware) has an effective membrane area of 140 cm^2. The system was employed with 3 L volumes for both the feed (whey) and draw sides. Hydrodynamic flow at the channels was co–currently run to reduce strain on the suspended membrane. Two speed controllable peristaltic pumps (EW 77111–67, Cole Parmer, IL) were used to pump whey liquor and draw solution. Cross–flow velocities on both faces of the membrane were kept constant with 5 L/min (0.5 m/s). The setup was also equipped with a constant temperature water bath (462–7028, VWR Scientific, IL) to maintain the same temperature (25±0.5 °C) at both solutions during FO tests. Each experiment was conducted with 360 min duration time.

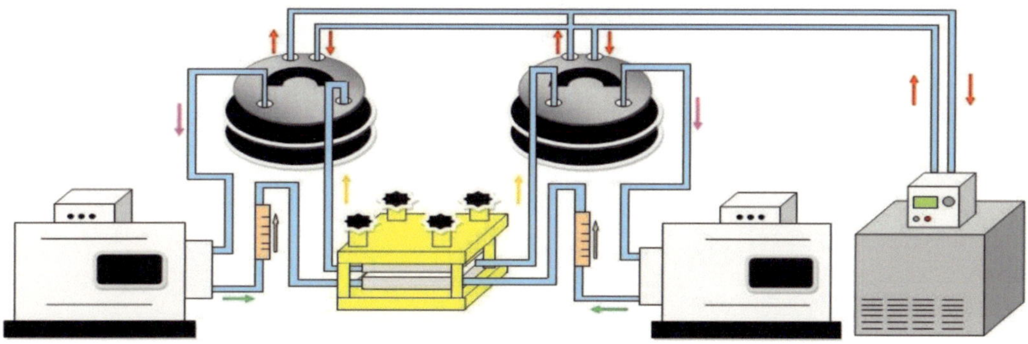

Fig. 1. Experimental setup of lab-scale FO system

2.2.2 Analytical procedure

Total protein, fat, fat–free dry matter (SNF), total solids, lactose and minerals contents of cheese whey samples were measured by Lactostar instrument equipped with thermal and optic sensors (Funke Gerber Company, Germany). The pH, conductivity, and temperature measurements were done by using WTW Multi 340i Meter (WTW, Weilheim, Germany). For density measurements in the samples, DA–130N density meter (KEM Co., Ltd., Kyoto, Japan) was used. Osmolalities of the samples were determined in duplicate for each data point by Advanced Osmometer instrument (Model 3250–Advanced Instruments Inc., USA) in accordance with freezing point depression method after completing the entire experiment. The analyses of water qualities in whey and draw solution were carried out in accordance with standard methods (American Public Health Association, 2005). Besides, nitrite, nitrate, and TOC, TN parameters were analyzed using GBC UV-visible Cintra 20 spectrometer (Cintra, Australia), and HACH IL 550 TOC-TN (Hach Lange Ltd., Germany) instruments, respectively.

2.2.3 FO performance calculations

The permeated water volume, *V* was determined from the osmolality differences of both solutions. First, the osmolalities of solutions were individually measured at definite time intervals along the experiments, and thereafter *V* was determined from the differences of sequential time points by calculations on osmolarity–based mass balance. These results were also made valid by their verification from time–dependent variations of total solid content (*TSC*) of cheese whey in the feed. The water flux, J_w was determined from the volume increase in the draw solution using Eq. (1).

$$J_w = \frac{1}{A_m} \cdot \frac{\Delta V_t}{\Delta t} \qquad (1)$$

where A_m is the membrane area, t the time, and V_t the volumetric water permeation at any time. The salt flux, J_s flowed in reverse direction from the water flux between both solutions was calculated using the following equation (Cornelissen et al., 2008).

$$J_S = \frac{1}{A_m} \cdot \frac{\Delta (C_t \cdot V_t)}{\Delta t} \qquad (2)$$

where C_t is the salt concentration at any time. In addition to J_w and J_s, the specific water fluxes as the water flux per net osmotic pressure difference (J_w'), and the water flux per net osmotic pressure difference normalized with respect to the osmotic pressure of the feed solution (J_w^*) were also calculated from Eqs. (3), and (4), respectively.

$$J_w' = \frac{J_w}{\Delta \pi_{net}} \qquad (3)$$

$$J_w^* = \frac{J_w \cdot \pi_{f,b}}{\Delta \pi_{net}} \qquad (4)$$

Time–dependent variations of two osmotically–driven forces comprising net osmotic pressure difference ($\Delta\pi_{net}$) and effective osmotic pressure difference ($\Delta\pi_{eff}$) were examined for various operating conditions. The net osmotic pressure refers to the osmotic pressure difference between the draw and feed solutions in the system. It was determined from difference of osmotic pressures of both solutions after analytically measurement of each solution osmolality by (Cornelissen et al., 2008; Tan & Ng, 2008)

$$\Delta \pi_{net} = \pi_{d,b} - \pi_{f,b} \qquad (5)$$

where $\pi_{d,b}$ and $\pi_{f,b}$ are the osmotic pressures of the draw and feed solutions, respectively. The osmotic pressures in the solutions were calculated in accordance with the van't Hoff equation.

$$\pi = R \cdot T \cdot [\ m \cdot d\ ] = R \cdot T \cdot [\ i \cdot C\ ] \qquad (6)$$

where R is the ideal gas constant (8.314 J/K·mol), T the absolute temperature (K), m the osmolality (mosm/kg), d the solution density (kg/L), i the van't Hoff factor, and C the molar concentration. It should be noted that the multiplication term in square bracket at left–hand side is referred to as the osmolarity (mosm/L) representing total solute concentration in the solution. The equivalence on the left–hand side of the equation was used to determine the osmotic pressure of the draw solution during the experiments. Whereas, the equivalence on the right–hand side was utilized for obtaining theoretical salt permeability coefficient in frame of the model calculations (see the section 2.2.4).

The effective osmotic pressure is the principal driven force that essentially governs osmotic water permeation into the draw solution. It takes place as an inherent event of internal concentration polarization which forms the most significant portion of the whole membrane fouling. It is theoretically predicted from the difference of osmotic pressures at the draw and feed sides of membrane's active layer using Eq. (7), as independent of the membrane operation mode (Tan & Ng, 2008).

$$\Delta\pi_{eff} = \pi_{d,w} - \pi_{f,w} \tag{7}$$

2.2.4 Modeling framework

The water flux, J_w is represented as a function of the effective osmotic pressure difference according to the osmotic–pressure model given as (Cornelissen et al., 2008; Tan & Ng, 2008)

$$J_w = A \cdot (\pi_{d,w} - \pi_{f,w}) \tag{8}$$

where A is the pure water permeability of the membrane. In the process, another flux term needs to be taken into account due to salt transport taking place simultaneously with water transport. Dissolved salt ions transport from the draw toward the feed, as being in reverse direction to the water flow. It should be designated that mutually transport dynamics goes on across the selective layer of the membrane until system reaches equilibrium. This second flux parameter known to be the salt flux, J_s is determined by Eq. (9) (Cornelissen et al., 2008; Tan & Ng, 2008).

$$J_s = B \cdot (C_{d,w} - C_{f,w}) \tag{9}$$

The modeling of FO process is widely carried out based on water flux from the osmotic pressure difference among the draw and feed solutions which is usually theoretically determined using the data obtained from the experiments. In an osmotically–driven membrane process, water and solute transport in a membrane matrix takes place across a selective interface inside the membrane by convection and diffusion. The fouling phenomena in the membrane become together simultaneously with the transport events, in which two distinctive behaviors known as internal and external concentration polarizations occurs inside and outside the matrix, respectively. The restrictiveness of the fouling on the process efficiency comes from its weakening effect on the effective osmotic pressure of which internal fouling portion plays dominant role. Hence, the main subject in assessing the process performance by the modeling is to be the molar salt concentrations at both sides of the membrane wall which are admitted as a key factor for driven force (Cath et al., 2006; McCutcheon & Elimelech 2006). But, different operation modes (normal and reverse modes) of FO process lead to quite differentiation of the effect of internal concentration polarization depending on the placement of the membrane's active layer toward the draw or feed solution. In a FO process being operated at normal mode in which membrane's selective layer is faced on the draw solution, concentrative internal concentration polarization is valid. Whereas, at reverse mode which means that the selective layer faced on the feed solution, dilutive concentration polarization governs the process. As the indicator of internal membrane fouling, solute resistivity resulted from soluble concomitant ions inside the active layer can be estimated by Eqs. (10), and (11), respectively (Loeb et al. 1997; Cath et al., 2006; McCutcheon & Elimelech 2006).

$$K=\left(\frac{1}{J_w}\right)\cdot\left[\ln\left(\frac{B+A\cdot\pi_{d,w}-J_w}{B+A\cdot\pi_{f,b}}\right)\right] \tag{10}$$

$$K=\left(\frac{1}{J_w}\right)\cdot\left[\ln\left(\frac{B+A\cdot\pi_{d,b}}{B+J_w+A\cdot\pi_{f,w}}\right)\right] \tag{11}$$

In accordance all the equations given above for performance and fouling, performance data related to the water and salt fluxes were simultaneously modeled as a result of which actual internal membrane fouling was defined as solute resistivity at the scope of computational methodology briefly outlined in Fig. 2. The modeling was based upon a framework rendering time–dependent estimation of FO data that was individually fulfilled for each selected time point of each time–dependent data set obtained experimentally. In the computations, theoretical salt permeability coefficient, $B_{\text{theoretical}}$ was determined using Eq. (12) which is obtained by the equalization of Eqs. (8), and (9), together with the use of the formula on right–hand side of Eq. (6).

$$B=\frac{-J_s\cdot i\cdot A\cdot R\cdot T}{J_w} \tag{12}$$

Osmotic pressures at the membrane's wall were predicted using π–C relationship given in Eq. (13) which was determined experimentally for NaCl solution up to 4 M concentration.

$$[\pi]=(5.8062\times C^2)+(40.091\times C)+0.7289\quad (r^2=1.000) \tag{13}$$

2.2.5 Ex-situ membrane characterizations

Ex-situ membrane investigations were carried out with the surface characterizations of active and support layers of the studied membranes by SEM, AFM, FTIR, and contact angle. Prior to analytical measurements, representative samples were washed twice with pure water and dried at room temperature. The views of membrane's active and support layers were observed by AFM (NanoScope IV AFM) and SEM (Philips XL30 SFEG). The membrane surface roughness was measured in contact mode by AFM. The mean surface roughness (R_A), the root mean square error (R_{RMSE}) of the average height of surface peaks, and the mean difference in height among the five highest peaks and the five lowest valleys (R_Z) were established to compare the roughnesses of FO and RO membranes depending on the membrane fouling. After Au coating of the samples, SEM images were taken at 5 kV to view the surface fouling. The contact angle was measured using a goniometry instrument (KSV Instruments, CAM 101) as an indicator for the hydrophilicity or wettability of the membrane surfaces. In the analysis, 2 µL of pure water in a tight syringe was dropped on the surface according to the sessile–drop technique. The results were obtained as the average values of contact angles at both sides of each drop fall on four arbitrary places of the samples. FTIR technique was employed to examine the interactions between whey components with the active and support layers of the membranes studied, respectively. The infrared spectra were recorded in a wave number range of 4000–650 cm^{-1} on a Bio Rad FTS 175C spectrophotometer.

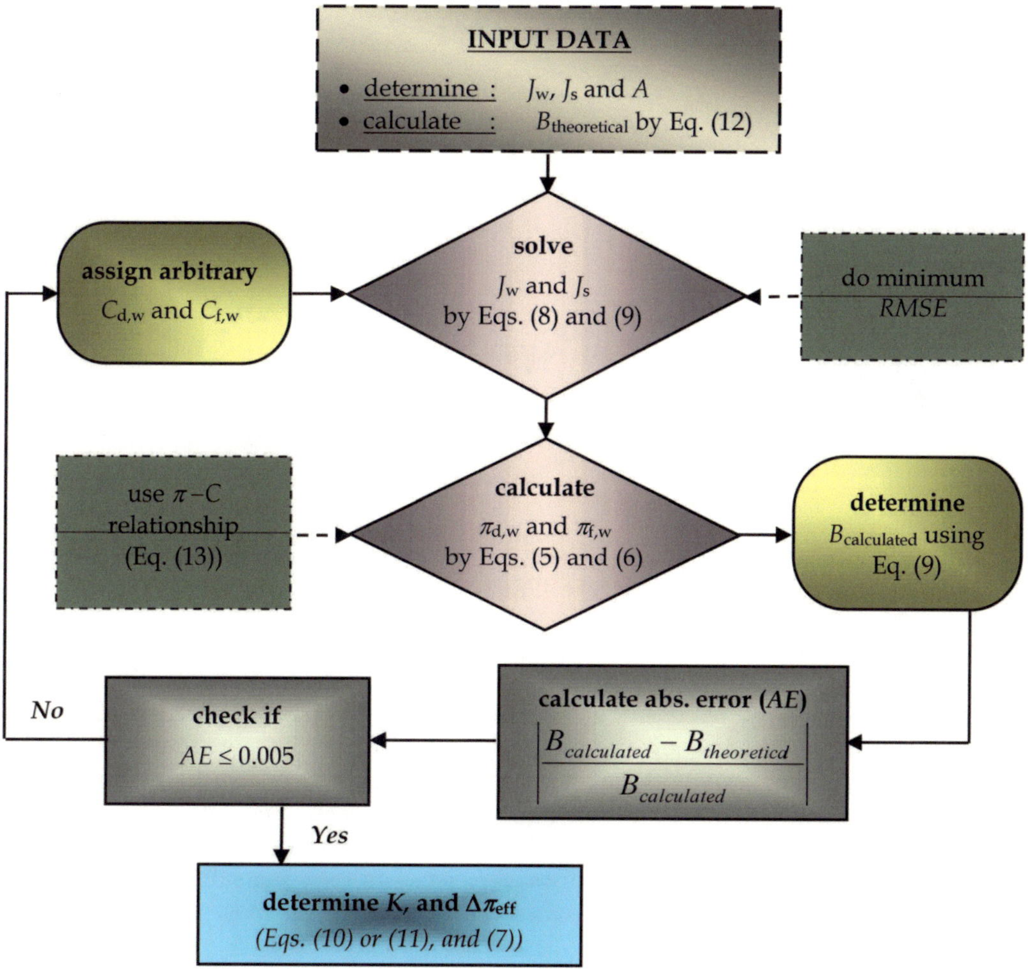

Fig. 2. Computational methodology used in the modeling of experimental data

3. Results and discussion

3.1 Analysis of FO Performance

3.1.1 Water permeation and fluxes

Time-dependent variations of water permeation and both fluxes depending on two different operation modes of both membranes were depicted in Fig. 3. In the figure, dotted lines are the results estimated from the modeling of experimental water and salt fluxes, whereas solid lines are the non-linear fitting curves for water permeation.

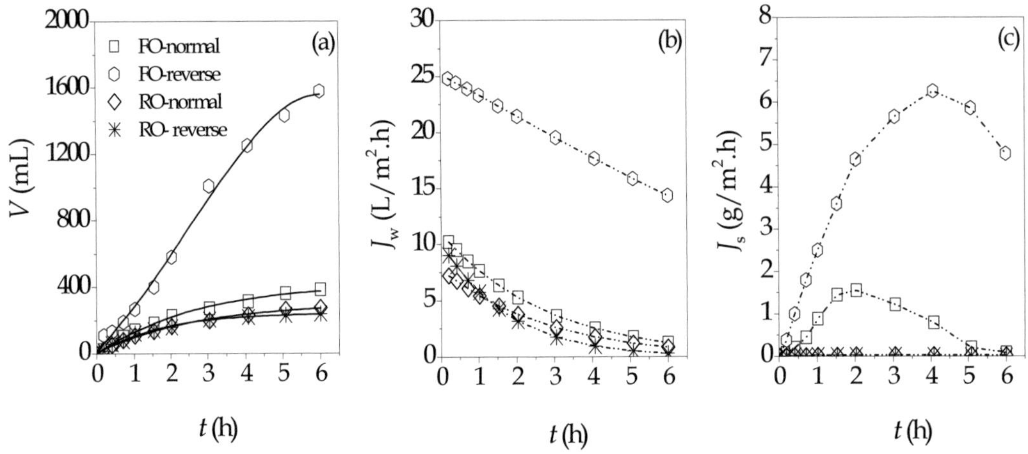

Fig. 3. FO performance at normal and reverse modes for both membranes ((a) volumetric water permeation, (b) water flux, and (c) salt flux)

The process instituted rather different performances at the conditions operated with FO and RO membranes. The system reached a steady state at the end of 360 min except for FO-reverse condition. A conspicuous superiority for reverse operation mode of FO membrane in whey concentration was established in which water more than half of initial water volume of whey was withdrawn as being approximately five times higher than other three varieties. But, FO membrane performed lower salt rejections due to relatively higher salt passages into the whey. It can be accordingly said that novel FO membranes need to be developed to remove the partial deficiency in rejecting salt ions as well as to allow higher water flux performance when also compared to those of RO membrane systems.

3.1.2 Driven forces, *B* and *K* variations

The effective osmotic pressure difference, salt permeability coefficient and solute resistivity were estimated by means of the model calculations based on computational methodology of experimental data presented in Fig. 2. Time evolutions of the estimated parameters via net driven force determined experimentally were presented in Fig. 4. In the figure, solid lines are the non-linear fitting curves, while dotted lines and symbols are the calculated and theoretical values of *B*, respectively.

In the process, net driven force decreased in company with the amount of water permeated into the draw solution. Quite different time variations were observed for the model parameters. FO membrane exhibited relatively higher salt permeability even at normal mode depending mainly on higher salt flux compared to that of RO membrane. Figs. 3-(a) and 4-(d) clearly indicated that the water flux performance of the process occurred as an inherent consequence of concentrative or dilutive concentration polarization influence of the membrane's active layer in different operation modes. Despite higher salt passages, FO membrane's inner was fouled less than RO membrane's by organic and inorganic solutes. These results gave some hints on the development of novel FO membranes. So, a FO membrane should preferably be as much thinly as possible designed by enabling high rejection rates especially for monovalent solutes and low molecular weight dissolved organics. In that circumstance, more effective operation of the FO system can be anticipated in terms of the operability with a longer period of time at high performance levels under the influence of lower internal fouling.

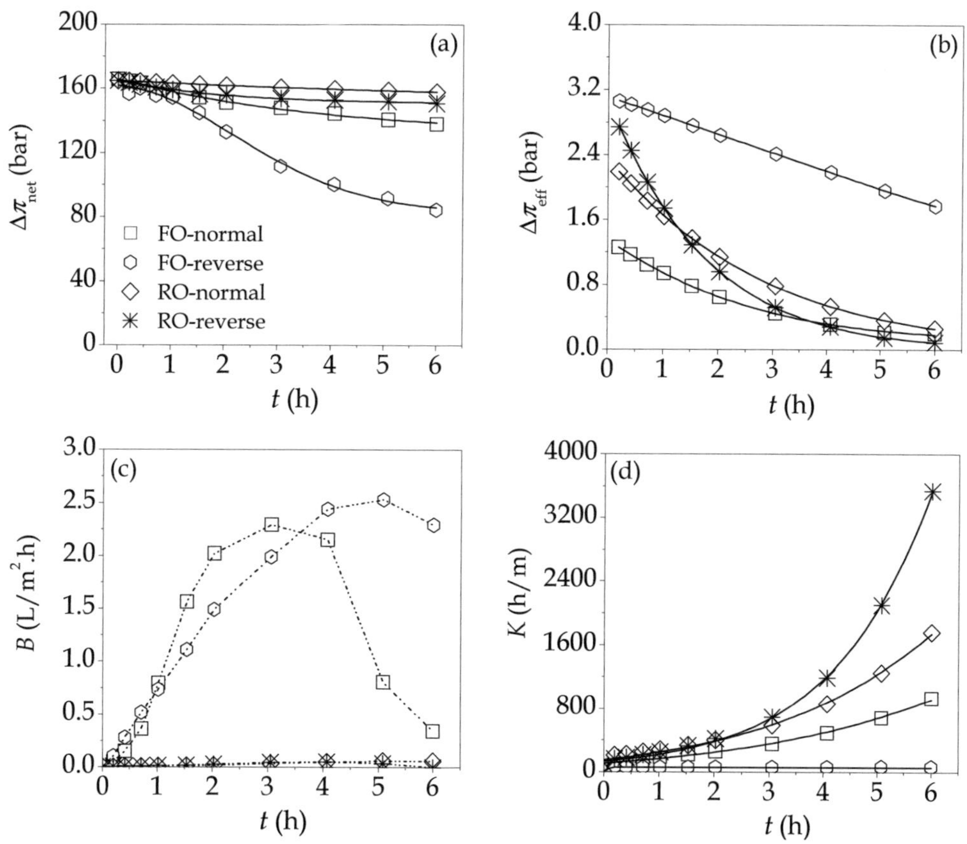

Fig. 4. Results for three modeling parameters via net driven force ((a) net osmotic pressure difference, (b) effective osmotic pressure difference, (c) salt permeability coefficient, and (d) solute resistivity)

3.2 Evaluation of ex-situ fouling

3.2.1 AFM, SEM and contact angle

Visual inspection of a membrane are employed by micro- or nano-scale membrane autopsy in which some properties such as microstructure and morphology can be observed to find any physical damage or changing on the surface. By means of analyses based on these observations, significant information are obtained with the intents of determining the membrane characteristics, analyzing the intensity or course of the membrane fouling, and improving the membrane properties for obtaining better performance. In that sense, top view AFM and SEM images of the surfaces of the studied membranes were shown in Figs. 5 and 6 for the active and support layers, respectively. In addition, contact angle and roughness results measured on the membrane surfaces were presented in Table 2.

membran–mode	active layer				support layer			
	$\theta°$	R_A (nm)	R_Z (nm)	R_{RMSE} (nm)	$\theta°$	R_A (nm)	R_Z (nm)	R_{RMSE} (nm)
FO clean	67.9±10.7	2.2	3.4	12.5	93.5±4.0	2.8	3.6	12.8
RO clean	62.5±2.9	30.1	50.2	191.6	43.0±8.1	74.5	93.4	368.2
FO–normal	43.8±12.4	19.8	30.6	103.1	61.5±3.4	30.1	35.2	115.4
FO–reverse	50.0±2.3	58.4	76.0	258.7	64.5±14.2	23.9	32.3	125.8
RO–normal	43.5±10.8	96.1	116.8	425.8	68.0±5.8	9.7	14.5	42.0
RO–reverse	41.3±17.8	76.2	96.6	326.9	72.8±4.9	36.8	44.8	181.7

Table 2. Contact angle ($\theta°$) and roughness (R, nm) of the active and support surfaces of FO and RO membranes

Table 2 indicates the general fouling trends in the FO system including organic and inorganic foulants on each opposite side as independent of orientation mode of the membranes. As can be seen from Figs. 5 and 6 and Table 2, RO membrane was fouled more intensively, and wettability and roughness on the active surface of both membranes increased. However, higher increasing rates in external fouling of FO membrane were observed as more than those of RO membrane. Wettability and roughness increased on the support layer of FO membrane, whereas decreased on the support layer of RO membrane in spite of increased fouling. AFM and SEM images and contact angle results pointed out that the surface foulings formed on the layers of FO and RO membranes resulted from different fouling dynamics depending on solute-membrane and solute-solute interactions.

3.2.2 FTIR Spectra

FTIR technique is widely used to characterize functional groups on a membrane surface via attenuated total reflection (ATR). As is known, organic and inorganic compounds absorb the infrared radiation energy corresponding to the vibrational energy of atomic bonds. By this unique property in the absorption spectrum, the functional groups of a specific compound can be identified by means of FTIR spectra. Hence, the variations of functional chemistry on both surfaces of FO and RO membranes were investigated by FTIR spectroscopy, and thereby general behaviors in the membrane fouling were comparatively evaluated.

(a)

(b)

(c)

(d)

(e)

(f)

Fig. 5. Top view AFM images and SEM microphotographs of active layer of clean and fouled membranes ((a) FO clean, (b) FO–normal, (c) FO–reverse, (d) RO clean, (e) RO–normal, and (f) RO–reverse)

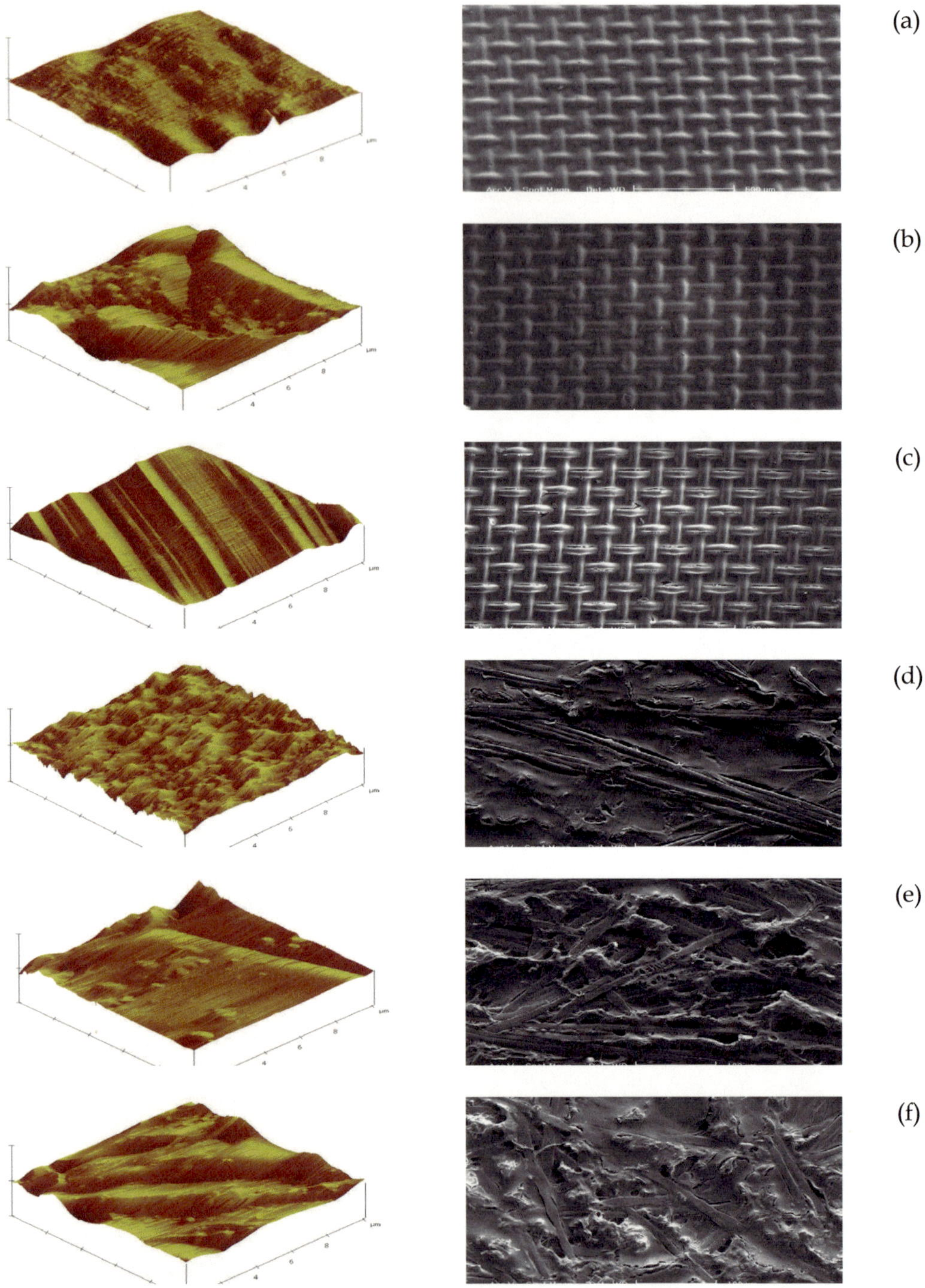

Fig. 6. Top view AFM images and SEM microphotographs of support layer of clean and fouled membranes ((a) FO clean, (b) FO–normal, (c) FO–reverse, (d) RO clean, (e) RO–normal, and (f) RO–reverse)

To determine surface groups responsible for external fouling, the results obtained by FTIR analyses were depicted in Figs. 7 and 8 for active and support layers of the studied membranes, respectively. In FTIR spectroscopy, whey containing rather different chemical groups absorbs infrared radiation energy at different band levels. Absorption bands at 1600–1700 and 1520–1565 cm^{-1} are responsible for amide I and amide II groups of protein structures, respectively. Besides, the bands at 1725–1745 and 1400–800 cm^{-1} are assigned to C–O stretching of fats, and coupled stretching and bending of carbohydrates, respectively. The vibration bands of clean FO membrane at 1730–1745, 1370–1390, 1235–1245, 1040–1100, and 880–995 cm^{-1} were possibly indicative for C=O stretching of carboxylic acid, symmetric CH_3 bending, C–O stretching, C–O stretching of carboxylic acid or C–N stretching, and asymmetric =C–H or =CH_2 stretching, respectively. Whereas, those of clean RO membrane at 1630-1680, 1500-1600, 1485, and 1040-1100 cm^{-1} were attributed to the stretchings of C=C, C=C ring, CH_2, and C–O, respectively. The bands at around 3445 and 3272 cm^{-1} corresponded to asymmetric and symmetric O–H stretching of water molecules on the surface due to deficient drying of the samples. As independent of membrane operation mode, lower transmission intensities on the surfaces of RO membrane indicated more severe fouling than those of FO membrane. It was also ascertained that similar transmission peaks corresponded to comparable foulings on the surfaces of both membranes. All of the peaks extended to a range of 700-1750 cm^{-1} proved that the surfaces of both membranes were fouled by severe adsorption of proteins, lactose and fats in whey.

3.3 Interrelated analysis of performance and fouling

3.3.1 Modeling-based approach

This approach, namely "*modeling-based performance analysis*" was focused on the membrane fouling results obtained by the modeling of the performance results. Herein, internal fouling-performance relationships can be explicitly derived with mathematical equations as related to two different groups of performance parameters, one being the measurable parameters ($\Delta\pi_{net}$, J_w and J_s), and the other being the prevailing estimated parameters ($\Delta\pi_{eff}$, J_w^* and B). At the basis of the membrane solute resistivity, non-linear relationships were evolved according to the statistical power law analysis. The equations belonging to data set at the end of the process were shown in Eqs. (14), and (15), while those belonging to the whole time–dependent data set were given in Eqs. (16) and (17). In these equations, internal fouling was represented separately for the measurable and the prevailing estimated responses. Standard error of estimate (S) and correlation (r^2) were also presented.

$$K_m = (0.00091)\cdot[\Delta\pi_{net}]^{2.780}\cdot[J_w]^{-0.368}\cdot[J_s]^{-0.083} \quad (r^2=1.000) \qquad (14)$$

$$K_{pe} = (192.4)\cdot\left[\Delta\pi_{eff}\right]^{0.018}\cdot\left[J_w^{*}\right]^{-0.810}\cdot[B]^{0.028} \quad (r^2=1.000) \qquad (15)$$

$$K_m = (0.127)\cdot[\Delta\pi_{net}]^{1.836}\cdot[J_w]^{-0.919}\cdot[J_s]^{0.015} \quad (r^2=0.993;\ S=60.3) \qquad (16)$$

$$K_{pe} = (132.7)\cdot\left[\Delta\pi_{eff}\right]^{-0.121}\cdot\left[J_w^{*}\right]^{-0.779}\cdot[B]^{0.017} \quad (r^2=0.989;\ S=73.0) \qquad (17)$$

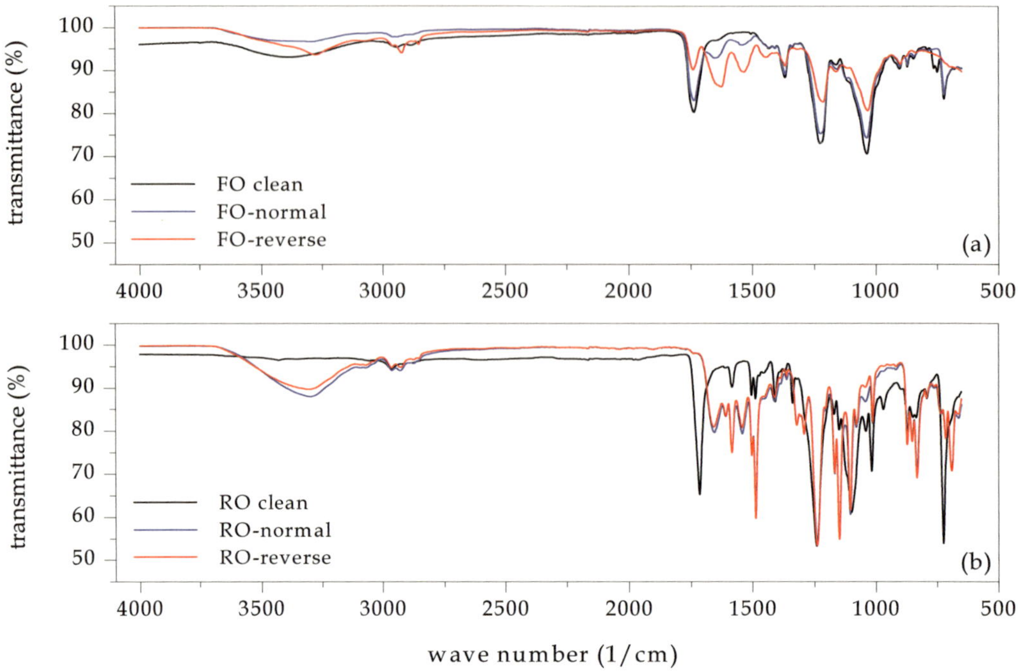

Fig. 7. FTIR spectra of active layers of FO (a) and RO (b) membranes

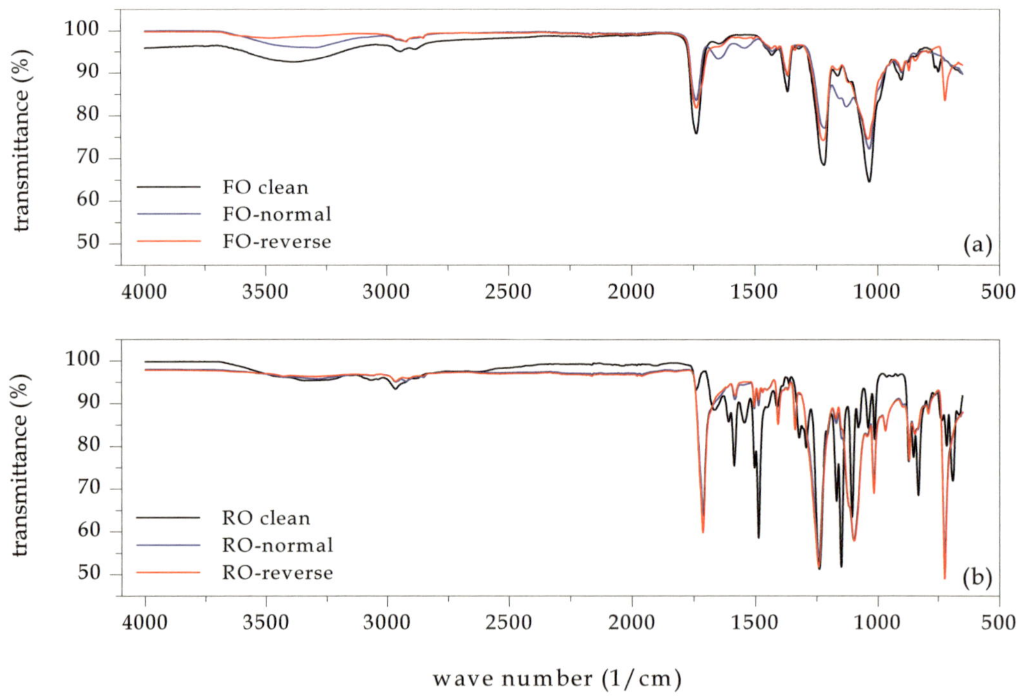

Fig. 8. FTIR spectra of support layers of FO (a) and RO (b) membranes

Eqs. (14)-(17) proved that internal fouling was remarkably interrelated one by one with both groups of the performance variables. The internal fouling estimates were considerably reliable for simulating the FO performance due to very good agreements. With respect to final performances, the internal fouling increased by increasing of driven forces and salt permeability, while decreased by increasing of water and salt fluxes. The major restricting factor for the performance distinctly seemed to be net pressure difference. Water and salt fluxes were determined to be the second and the third factors in the system efficiency, respectively. According to these results, a FO process has to be employed with a draw solution to be yielded high net driven force. It should be however emphasized that this choice would not be alone sufficient for the expected efficacy in practice. Furthermore, the modeling-based analysis indisputably revealed that the membranes to be specifically designed with the intents of higher driven forces and lower salt permeability would be able to be operated with enough high performances. In other words, a FO membrane having a high salt rejection (>99%) under relatively high effective driven force would be effectively utilized by higher clean water production rates even under a reasonable net driven force.

3.3.2 Ex-situ-based approach

Ex-situ-based approach can be described as "*ex-situ membrane characterization-based performance analysis*" of which the methodology is principally relied on an illustrative surface parameter to be representative for the internal membrane fouling. The approach follows a solution-directed strategy that includes a focal point toward predicting each performance component as associated individually with representative surface parameter. Accordingly, firstly roughness and contact angle of the membrane surfaces were associated with the internal fouling by appropriate one of linear and non-linear curve-fitting techniques to be given the highest correlation for the relation equations. Thereafter, each performance parameter was related to contact angle determined as representative for the internal fouling.

The best relationships obtained in the first step of ex-situ-based performance analysis were shown in Fig. 9. The surface morphologies of the fouled membranes did not give meaningful correlations for which belong to the membrane surface coated by foulants not to the membrane itself. Unlike this case, wettability of both surfaces of FO and RO membranes were relevant to the internal fouling. However, a high correlation of r^2=0.985 could only be obtained for the wettability of the active surface. This meant that the internal fouling was directly related with the hydrophilicity of the active layer, but not with that of the support layer. Fig. 9-(b) also showed that wettability of the active surface increased as internal fouling increased. Accordingly, internal and external foulings were interrelated with each other by the relation equations. More importantly, the increase in external fouling concurrently led to the increasing of internal fouling. Thereupon, it can be said that a novel FO membrane should be designed in a form of enabling less external fouling on its active surface without compromising the compatibility with net driven force.

At the second step of the approach, contact angles of the surfaces were interrelated to the performance responses comprising osmotically–driven forces, salt and water passages, and the results were depicted in Figs. 10, 11, and 12, respectively. Equating procedure of wettability of the support layer was determined as practically rather complex of which the relations performed distribution-type variations. Thus, performance simulations were relied on the wettability of the active layer possessed good agreements with the experimental data.

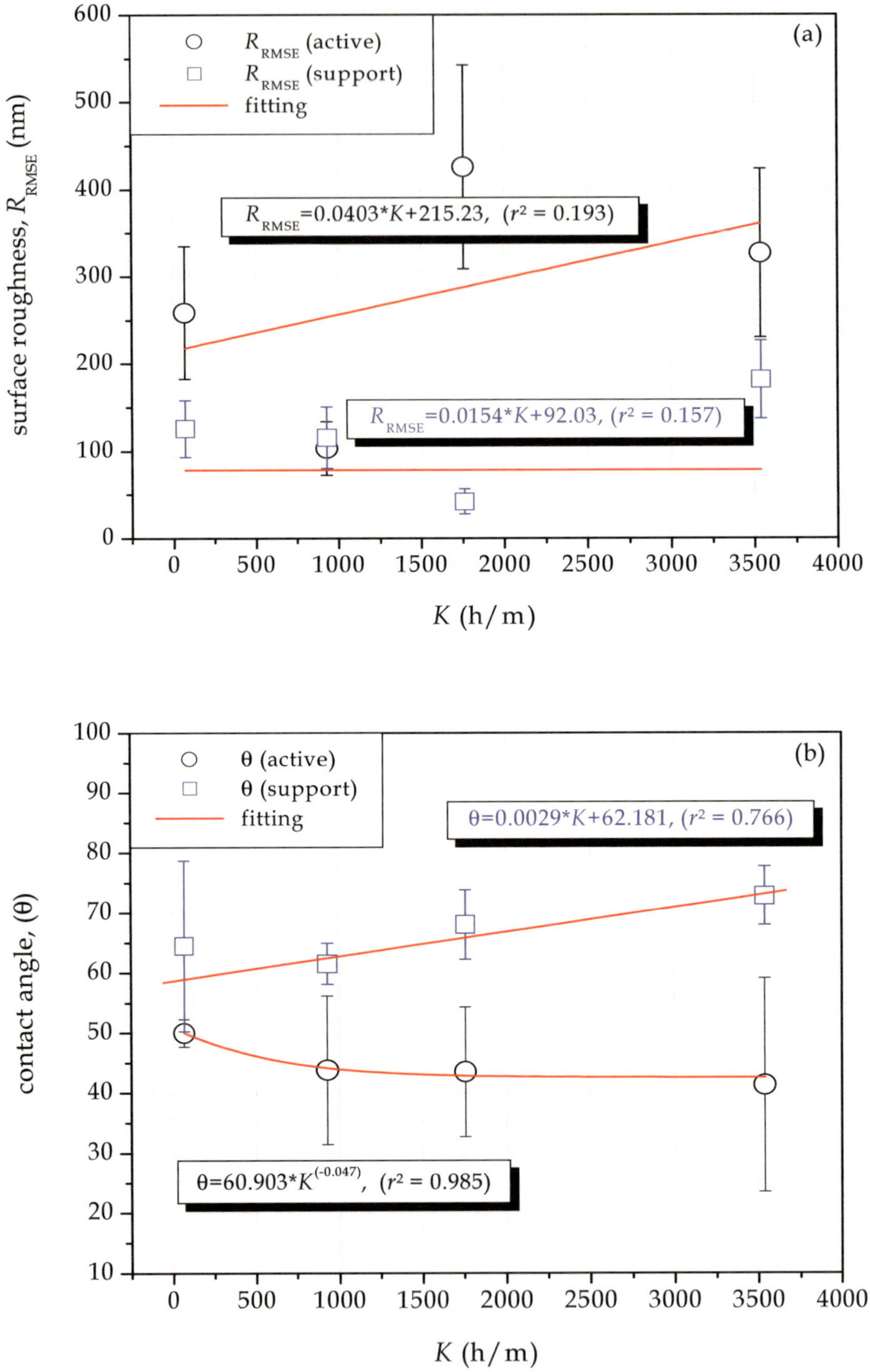

Fig. 9. Relationships of surface roughness (a) and wettability (b) to internal fouling

As can be seen from the graphs, the hydrophilicity of the active layer was the main representative for the whole system performance in second order polynomial structure. Fig. 10 proved that the performances based on net and effective driven forces realized as being under the predominant influences of external and internal foulings, respectively. The increase in the hydrophilicity yielded by the net force increase was very likely owing to the fact that soluble hydrophilic foulants increased on the surface by ionic binding or adsorption. As to the effective force increase, the hydrophilicity decrease pointed out an external fouling with the fact that soluble or insoluble hydrophobic organics were predominantly bound or attached on the surface as also supported by FTIR analyses. In FO-reverse mode, weakening in interactions among organic and inorganic solutes brought about a performance result of higher net force and excessive salt passage, whereby soluble whey organics did not notably penetrated inside the membrane's active layer. Hence, it can be said that targeting a too high rejection rate for organics, especially for ones with low-molecular weight should be seriously taken into consideration, if a FO membran devoted to the objective implementation of the process to water or wastewater containing organic substances will be improved.

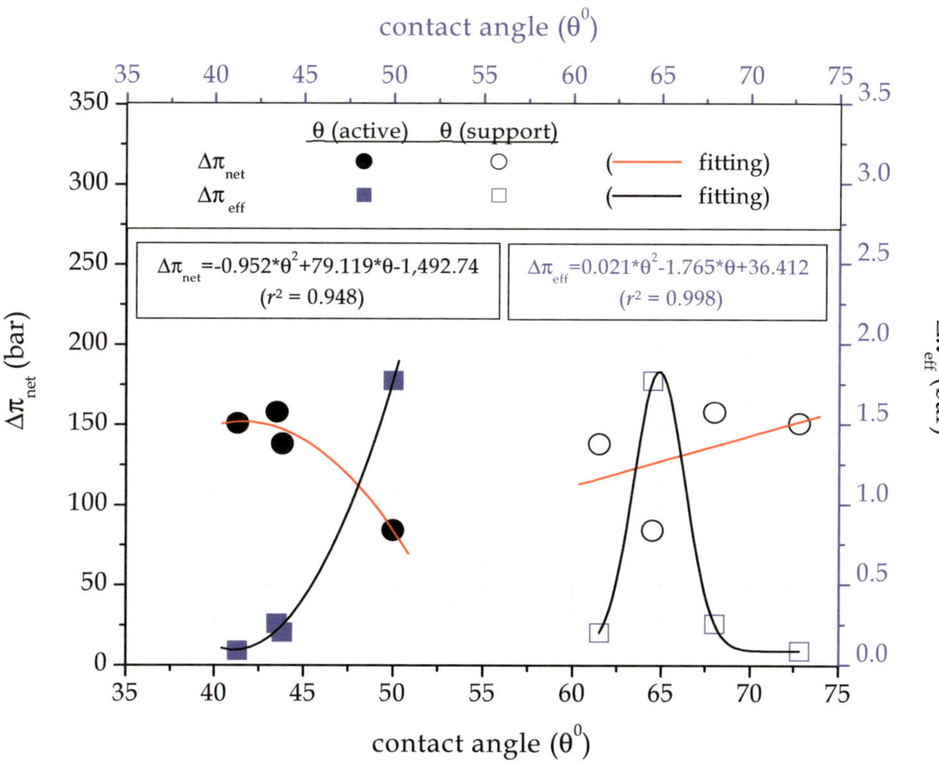

Fig. 10. Interrelationships of wettability of active and support layers with osmotically–driven forces

According to Fig. 11, the active surface became more hydrophobic as both salt flux and membrane's salt permeability increased. This concluded that the fouling on the active surface by hydrophobic foulants in whey brought increasing of internal fouling as also supported by Fig. 9-(b). That's why, for more effective operation of a FO system involving organic-inorganic binary, the membrane has to be devised by not only achievement of high organic rejection rate but also acquirement of low hydrophobic fouling on the selective surface. To this end, it can be foreseen that the process would become more economically viable with a higher performance by means of the arrangements to be made on the active surface.

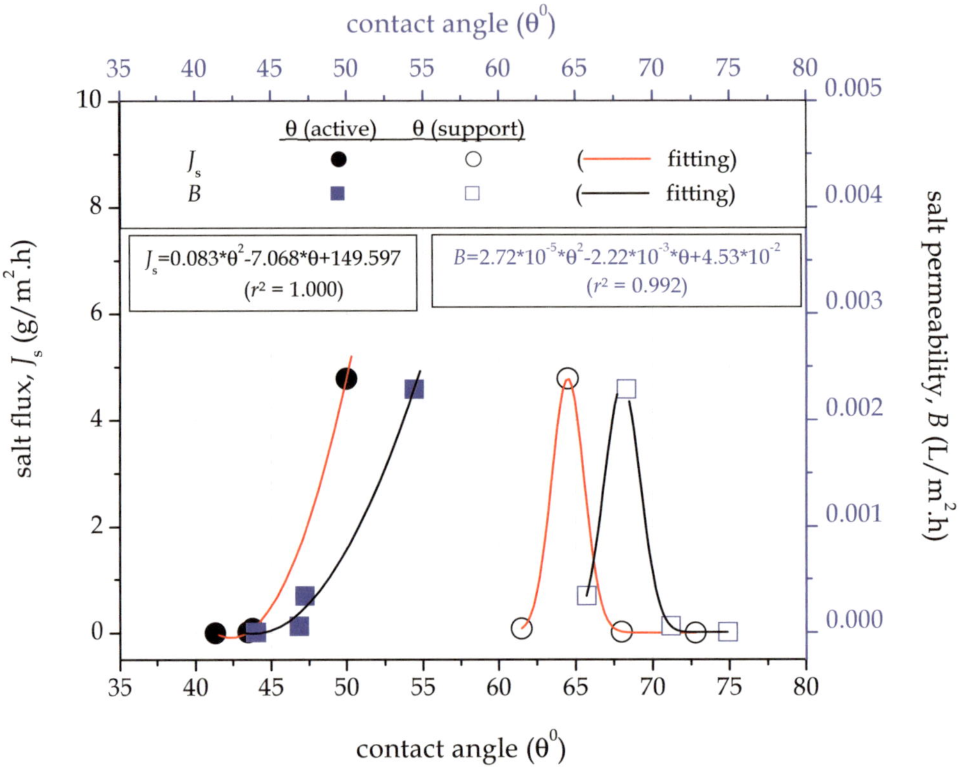

Fig. 11. Interrelationships of wettability of active and support layers with salt passage

Fig. 12 presented that each one of parameters belonging to water passage was well fitted with the active contact angles as in that of driven forces and salt passage. The decrease in the surface wettability accounted for the increase in water permeation performance of the membrane. This meant that, a FO membrane for organic-inorganic binary system have to be an optimum solution for the wettability on the active layer so that the surface should have a hydrophobic domain enough to enable as much as possible of water passage. In light of the results founded on the development of novel FO membranes, matters of fact or opinions deemed to require a thinner membrane than CTA FO membrane of about 50 μm in order to increase the corresponding effective osmotic pressure difference. It can be such that increasing membrane thickness (averagely 200-250 μm for a RO membrane) may be the major triggering factor for lower FO performance by simultaneous contribution of the increase of internal fouling and the decrease of effect of net driven force.

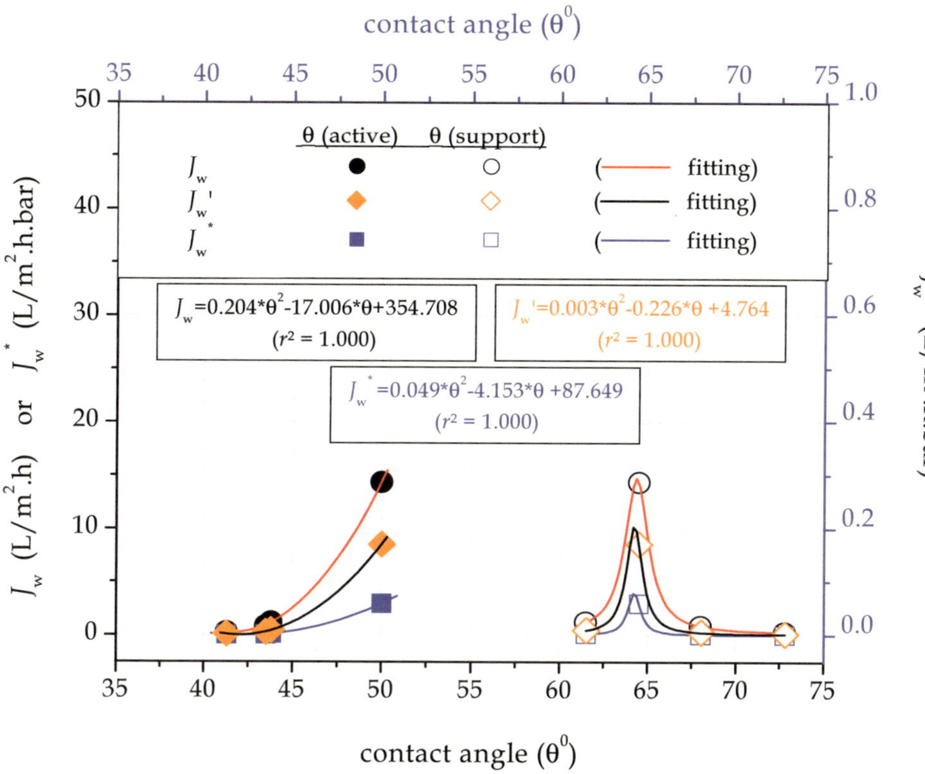

Fig. 12. Interrelationships of wettability of active and support layers with water passage

3.3.3 Performance simulation by integration of the approaches

Final performances were individually predicted for each response parameter using the second order polynomial relations obtained from the results of ex-situ-based approach (Figs. 10-12). By means of the predicted performance responses, analogical solute resistivities, K_m and K_{pe} were separately calculated for the measurable and the prevailing estimated parameters by Eqs. (14) and (15), respectively, to compare their consistency with internal fouling (actual solute resistivity, K). The consistency graph was shown in Fig. 13 in which dotted lines were the variation ranges of ±10 and ±20%. In the calculations, deviations from actual values of the responses involving driven forces, water and salt passages were determined by normalized root mean square error (Fig 14). As in modeling-based approach, the use of either the measurable or the prevailing estimated parameters in ex-situ-based approach sufficed for meaningfully describing the relationships of the fouling with the performance. The consistencies of actual and analogical solute resistivities were very close to each other. In order to estimate the performance, analogical internal foulings from the joint solution of the approaches were separately conformed by actual fouling in an acceptable variation range of almost ±10 and ±20% for the prevailing estimated and the measured parameters, respectively. In that sense, Fig. 14 presented that ex-situ-modeling-performance triple with the combination of both approaches was determined as having a capability of 10% in terms of *NRMSE* values, except for the membrane salt permeability.

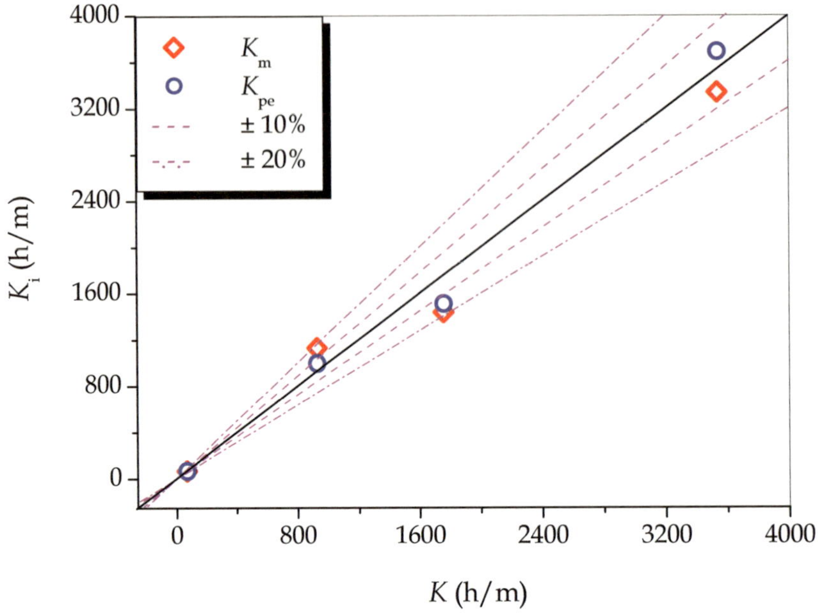

Fig. 13. Consistencies for solutions based on *"ex-situ-modeling-performance"* integration

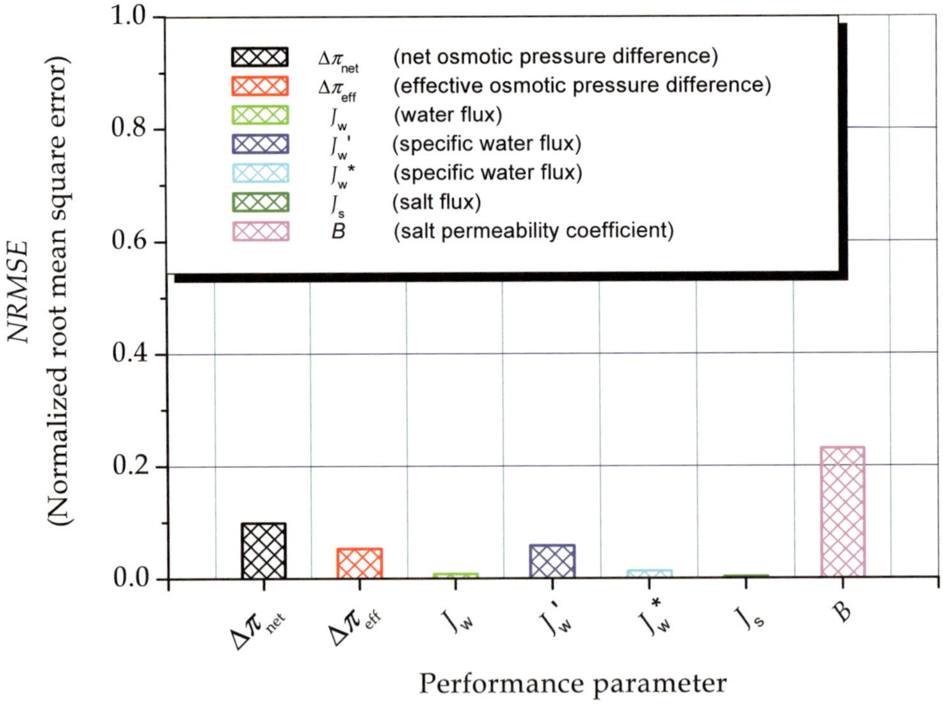

Fig. 14. Deviations from actual values of each performance response parameter with regard to the consistencies of the integrated approaches

4. Conclusion

In this study, two different approaches grounded on modeling- and ex-situ membrane characterization-based performance analysis were applied for the evaluation of performance-fouling relationships in a FO system operated as independent of the kind and orientation mode of FO and RO membranes. The prominent findings from normative interpretations together of the results related to the performance, modeling and ex-situ membrane characterizations were summarized below:

i. FO membrane operated at reverse mode exhibited better performance due to lower internal fouling compared to RO membrane. But its high salt permeability and higher increasing rate in external fouling were the main drawbacks to its more effective use.
ii. The modeling-based approach was admirably applied in interrelating internal fouling and process performance by statistical power law analyses of time-dependent and time-independent data. The approach proved that the performance restricting factors were, in decreasing order, the net pressure difference and water and salt fluxes, respectively.
iii. The ex-situ-based approach was successfully implemented in interrelating both internal-external foulings by non-linear curve-fitting and external fouling-process performance by second order polynomial curve fitting for time-independent data using the contact angle of active membrane surface. It was comprehended by the approach that the increase in external fouling increased the internal fouling. The performance of net and effective driven forces was determined to be in association with external and internal foulings, respectively.
iv. Some useful knowledge was obtained on the improvement of FO membranes by thorough evaluation of the results belonging to both the process performance analyses and the individual application of the approaches. According to the performance results, there need to be developed a novel FO membrane to be able to be employed with better salt rejection, higher water flux, and thinner membrane thickness. The modeling-based approach justified the inferences of the stand-alone performance analysis. Whereas, knowledge obtained by the ex-situ-based approach was to some extent greater. As exclusive of others two, without compromising the compatibility with net driven force, a novel FO membrane should have less external foulings on both surfaces for both organic and inorganic foulants to provide the continuity of lower internal fouling.

The integrated implementation of both approaches was also successfully carried out intended for the process simulation. The integrated approach was ascertained as a novel and progressive tool in both elucidating the performance-fouling relationships and simulating the whole performance and its components. But, the proposed methodology should amply be examined and confirmed by its experiencing in the performance predictions of full-scale membrane systems. In that sense, for the future works, testing the integrated approach for applications with real-time based on the fouling control may be offered to membrane specialists to be oriented to ex-situ-based performance simulations.

At the end of this chapter, it can be essentially said that, despite low fouling tendencies of FO membranes being easily removed by membrane cleaning, development of superior membranes having distinctive properties for various dewatering/purification applications would gain more prominence world-wide in the near future when technical and commercial inadequacies in their diversity are started to be considered more widely. By this means, common implementations of forward osmosis as an energy efficient and environmentally-friendly desalination technology would be much more possible.

5. Acknowledgment

This study was financially supported with a national project (No: 109Y300) by the TUBITAK, the Scientific and Technological Research Council of Turkey. Authors would like to thank to the Cayirova Milk&Milk Products Inc., especially to Mr. Seyhmuz Aslan and Ms. Neslihan Genal, for the cheese whey supplement. Authors would also like to thank to Hydration Technologies Inc., and Hydranautics Inc., due to the membrane supplements.

6. References

Aaberg, R.J. (2003). Osmotic Power: A New and Powerful Renewable Energy Source, ReFocus, Vol.4, No.6, (November-December 2003), pp. 48–50, ISSN 1471-0846

American Public Health Association (AWWA), Standard Methods for the Examination of Water and Wastewater, 21th ed., Washington DC., 2005.

Aydiner, C. (2010). A Novel Approach Based on Distinction of Actual and Pseudo Resistances in Membrane Fouling: "*Pseudo Resistance*" Concept and Its Implementation in Nanofiltration of Single Solutions. Journal of Membrane Science, Vol.361, No.1-2, (September 2010), pp. 96-112, ISSN 0376-7388

Cath, T.Y.; Gormly, S.; Beaudry, E.G.; Flynn, M.T.; Adams, V.D. & Childress, A.E. (2005a). Membrane Contactor Processes for Wastewater Reclamation in Space. I. Direct Osmotic Concentration as Pretreatment for Reverse Osmosis, Journal of Membrane Science, Vol.257, No.1-2, (July 2005), pp. 85–98, ISSN 0376-7388

Cath, T.Y.; Adams, V.D. & Childress, A.E. (2005b). Membrane Contactor Processes for Wastewater Reclamation in Space. II. Combined Direct Osmosis, Osmotic Distillation, and Membrane Distillation for Treatment of Metabolic Wastewater, Journal of Membrane Science, Vol.257, No.1-2, (July 2005), pp. 111–119, ISSN 0376-7388

Cath, T.Y.; Childress, A.E. & Elimelech, M. (2006). Forward Osmosis: Principles, Applications and Recent Developments, Journal of Membrane Science, Vol.281, No.1-2, (September 2006), pp. 70–87, ISSN 0376-7388

Chen, J.C.; Li, Q. & Elimelech, M. (2004). In Situ Monitoring Techniques for Concentration Polarization and Fouling Phenomena in Membrane Filtration, Advances in Colloid and Interface Science, Vol.107, No.2-3, (March 2004), pp. 83–108, ISSN: 0001-8686

Chung, T.S.; Zhang, S.; Wang, K.Y.; Su, J. & Ling, M.M. (2011). Forward Osmosis Processes: Yesterday, Today and Tomorrow, Desalination (2011), ISSN 0011-9164, (article in press)

Cornelissen, E.R.; Harmsen, D.; De Korte, K.F.; Ruiken, C.J.; Qin, J. J.; Oo, H. & Wessels, L.P. (2008). Membrane Fouling and Process Performance of Forward Osmosis Membranes on Activated Sludge, Journal of Membrane Science, Vol.319, No.1-2, (July 2008), pp. 158–168, ISSN 0376-7388

Danis, U. & Aydiner, C. (2009). Investigation of Process Performance and Fouling Mechanisms in Micellar-Enhanced Ultrafiltration of Nickel-Contaminated Waters. Journal of Hazardous Materials, Vol.162, No.2-3, (March 2009), pp. 577-587, ISSN 0304-3894

Darton, T.; Annunziata, U.; Pisano, F.V. & Gallego, S. (2004). Membrane Autopsy Helps to Provide Solutions to Operational Problems, Desalination, vol.167, No.1-3, (March 2004), pp. 239–245, ISSN 0011-9164

Dova, M.I.; Petrotos, K.B. & Lazarides, H.N. (2007). On the Direct Osmotic Concentration of Liquid Foods. Part I. Impact of Process Parameters on Process Performance, Journal of Food Engineering, Vol.78, No.2, (January 2007), pp. 422–430, ISSN 0260-8774

Gormly, S.; Herron, J.; Flynn, M.; Hammoudeh, M. & Shaw, H. (2011). Forward Osmosis for Applications in Sustainable Energy Development, Desalination and Water Treatment, vol.27, No.1-3, (March 2011), pp. 327–333, ISSN 1944-3994

Hoek, E.M.V.; Allred, J.; Knoell, T. & Jeong, B.H. (2008). Modeling the Effects of Fouling on Full-Scale Reverse Osmosis Processes, Journal of Membrane Science, vol. 314, No.1-2, (April 2008), pp. 33–49, ISSN 0376-7388

Holloway, R.W.; Childress, A.E.; Dennett, K.E. & Cath, T.Y. (2007). Forward Osmosis for Concentration of Anaerobic Digester Centrate, Water Research, Vol.41, No.17, (September 2007), pp. 4005–4014, ISSN 0043-1354

Huang, X.; Guillen, G.R. & Hoek, E.M.V. (2010). A New High-Pressure Optical Membrane Module for Direct Observation of Seawater RO Membrane Fouling and Cleaning, Journal of Membrane Science, Vol.364, No.1-2, (August 2010), pp. 149–156, ISSN 0376-7388

Ling, M.M.; Wang, K.Y. & Chung, T.S. (2010). Highly Water-Soluble Magnetic Nanoparticles as Novel Draw Solutes in Forward Osmosis for Water Reuse, Industrial Engineering Chemistry Research, Vol.49, No.12, (May 2010), pp. 5869–5876, ISSN 0888-5885

Loeb,S.; Titelman, L.; Korngold, E. & Freiman, J. (1997). Effect of Porous Support Fabric on Osmosis Through a Loeb–Sourirajan–Type Asymmetric Membrane, Journal of Membrane Science, Vol.129, No.2, (July 1997), pp. 243–249, ISSN 0376-7388

Loeb, S. (2001). One Hundred and Thirty Benign and Renewable Megawatts from Great Salt Lake? The Possibilities of Hydroelectric Power by Pressure Retarded Osmosis, Desalination, Vol.141 No.1, (December 2001), pp. 85–91, ISSN 0011-9164

Low, S.C. (2009). Preliminary Studies of Seawater Desalination using Forward Osmosis, Desalination and Water Treatment, vol.7, No.1-3, (2009), pp. 41–46, ISSN 1944-3994

Martinetti, C.R.; Childress, A.E. & Cath, T.Y. (2009). High Recovery of Concentrated RO Brines using Forward Osmosis and Membrane Distillation, Journal of Membrane Science, vol. 331, No.1-2, (April 2009), pp. 31–39, ISSN 0376-7388

Mccutcheon, J.R.; Mcginnis, R.L. & Elimelech, M. (2005). A Novel Ammonia–Carbon Dioxide Forward (Direct) Osmosis Desalination Process, Desalination vol.174, No.1, (April 2005), pp. 1–11, ISSN 0011-9164

Mccutcheon, J.R.; Mcginnis, R.L. & Elimelech, M. (2006). Desalination by Ammonia–Carbon Dioxide Forward Osmosis: Influence of Draw and Feed Solution Concentrations on Process Performance, Journal of Membrane Science,vol.278, No.1-2, (July 2006), pp. 114–123, ISSN 0376-7388

McCutcheon, J.R. & Elimelech, M. (2006). Influence of Concentrative and Dilutive Internal Concentration Polarization on Flux Behavior in Forward Osmosis, Journal of Membrane Science, Vol.284, No.1-2, (November 2006), pp. 237–247, ISSN 0376-7388

McGinnis, R.L.; McCutcheon, J.R. & Elimelech, M. (2007). A Novel Ammonia–Carbon Dioxide Osmotic Heat Engine for Power Generation, Journal of Membrane Science, Vol.305, No.1-2, (November 2007), pp. 13–19, ISSN 0376-7388

Petrotos, K.B. & Lazarides, H.N. (2001). Osmotic Processing of Liquid Foods, Journal of Food Engineering, Osmotic Processing of Liquid Foods, Vol.49, No.2-3, (2001), pp. 201–206, ISSN 0260-8774

Tan,C.H. & Ng, H.Y. (2008). Modified Models to Predict Flux Behavior in Forward Osmosis in Consideration of External and Internal Concentration Polarizations, Journal of Membrane Science, Vol.324, No.1-2, (October 2008), pp. 209–219, ISSN 0376-7388

Tang, C.Y.; She, Q.; Lay, W.C.L.; Wang, R. & Fane, A.G. (2010). Coupled Effects of Internal Concentration Polarization and Fouling on Flux Behavior of Forward Osmosis Membranes During Humic Acid Filtration, Journal of Membrane Science, Vol.354, No.1-2, (May 2010), pp. 123–133, ISSN 0376-7388

Tu, S.C.; Ravindran, V. & Pirbazari, M. (2005). A Pore Diffusion Transport Model for Forecasting the Performance of Membrane Processes, Journal of Membrane Science, vol.265, No.1-2, (November 2005), pp. 29–50, ISSN 0376-7388

Uchymiak, M.; Rahardianto, A.; Lyster, E.; Glater, J. & Cohen, Y. (2007). A Novel RO Ex situ Scale Observation Detector (EXSOD) for Mineral Scale Characterization and Early Detection, Journal of Membrane Science, vol.291, No.1-2, (January 2007), pp. 86–95, ISSN 0376-7388

Van der Bruggen, B.; Manttari, M. & Nystrom, M. (2008). Drawbacks of Applying Nanofiltration and How to Avoid Them: A Review, Separation and Purification Technology, vol.63, No.2, (October 2008), pp. 251–263, ISSN 1383-5866

Wang, R.; Shi, L.; Tang, C.Y.; Chou, S.; Qiu, C. & Fane, A.G. (2010) Characterization of Novel Forward Osmosis Hollow Fiber Membranes, Journal of Membrane Science, Vol.355, No.1-2, (June 2010), pp. 158–167, ISSN 0376-7388

Wang, K.Y.; Teoh, M.; Nugroho, A. & Chung, T.S. (2011). Integrated Forward Osmosis–Membrane Distillation (FO-MD) Hybrid System for the Concentration of Protein Solutions, Chemical Engineering Science, Vol.66, No.11, (June 2011), pp. 2421-2430, ISSN 0009-2509

Xu, P.; Bellona, C. & Drewes, J.E. (2010). Fouling of Nanofiltration and Reverse Osmosis Membranes during Municipal Wastewater Reclamation: Membrane Autopsy Results from Pilot-scale Investigations, Journal of Membrane Science, Vol.353, No.1-2, (February 2010), pp. 111–121, ISSN 0376-7388

Xu, Y.; Peng, X.; Tang, C.Y.; Fu, Q.S. & Nie, S. (2010). Effect of Draw Solution Concentration and Operating Conditions on Forward Osmosis and Pressure Retarded Osmosis Performance in a Spiral Wound Module, Journal of Membrane Science, Vol. 348, No.1-2, (February 2010), pp. 298–309, ISSN 0376-7388

Yang, Q.; Wang, K.Y. & Chung, T.S. (2009). A Novel Dual-Layer Forward Osmosis Membrane for Protein Enrichment and Concentration, Separation and Purification Technology, Vol.69, No.3, (October 2009), pp. 269–274, ISSN 1383-5866

32

Biodegradation of Pre-Aged Modified Polyethylene Films

Bożena Nowak[1], Jolanta Pająk[1] and Jagna Karcz[2]
[1]*Department of Biochemistry,*
[2]*Laboratory of Scanning Electron Microscopy,*
University of Silesia,
Poland

1. Introduction

Synthetic polymers, which are ubiquitous in modern industrial society, contribute to improving comfort and quality of our life. Currently more than 260 million tonnes of plastics are being produced each year (O'Brine & Thompson, 2010). Among them polyolefins constitute the majority of consumed thermoplastics. Polyolefin materials, such as low-density polyethylene (LDPE), due to the exceptional mechanical and thermal properties, ease of fabrication and low cost, find diverse applications in many fields. Polyethylenes represent 64% of materials used for various applications such as containers, bottles, tubing, plastic bags, greenhouses, mulching films, which are usually discarded after only brief use (Peacock, 2000). These mostly one-trip applications lead to a large quantity of plastic waste accumulating, at the rate of 25 million tons per year (Meenakshi et al., 2002) in landfill and in natural habitats (Thompson et al., 2009). Thus, plastics - the most visible form of trash - have become ubiquitous in our environment, leading to long-term environment, economic and waste management problems (Koutny et al., 2006). Since plastic waste are often soiled by biological substances, physical recycling of these materials turned out impractical and generally undesirable (El-Naggar & Farag, 2010). Incineration of plastics, in turn, has various environmental and social constraints. It seems that the use of plastics, which can re-enter the biological life cycle through biodegradation will be the best choice (Sivan, 2011; Soni et al., 2009).

The term "biodegradation" indicates the predominance of biological activity in this phenomenon. Until recently, biodegradation was perceived as a decomposition of substances solely by the action of microorganisms. At present, when the complexity of biodegradation of many substances, especially polymeric materials, is better understood, a new definition of this process divide it into several steps and the process can stop at each stage (Lucas et al., 2008). During the first, extracellular step, which is called biodeterioration, the action of microorganisms in combination with other decomposer organisms or/and abiotic factors fragment polymeric materials into small fractions. To go across the plasmic membrane these small fractions of materials must be depolymerised to low molecular weight products. Due to the mixed action of abiotic factors and microbial communities, which secrete enzymes and free radicals, long polymeric chains are cleaved and small oligomers are generated. After being transported into the cytoplasm the small molecules integrate the metabolism pathways. This so called assimilation is the essential step to produce microbial energy, biomass and primary and secondary metabolites. As mineralisation takes place CO_2, N_2, CH_4, H_2O and different salts from completely oxidised metabolites are released in the extracellular environment (Lucas et al., 2008).

According to the new definition, biodegradation of plastics results from combination of biotic and abiotic factors which act synergistically to decompose organic matter. Such interaction of environmental factors is not only beneficial but, in some cases, required to degrade particularly stable compounds.

It is also noteworthy that a frequent source of misunderstanding concerning the term 'biodegradability of polymer' between the polymer scientists and microbiologists originates from the fact that for polymer scientists, degradation means the loss of physical properties, whereas microbiologists are interested in the complete mineralisation of the material (Koutny et al., 2006).

Polyethylene, like many conventional plastics, is resistant to degradation. Its recalcitrance to degradation is traced back to physical and chemical properties that limit chemical reactivity in general (Koutny et al., 2006). In addition, stabilizers contained in industrial polyethylene prevent degradation during processing and usage (Reddy et al., 2008).

Ohtake et al., estimated that it takes about 300 years to degrade LDPE films with thickness of 60μm (Ohtake et al., 1998). However, according to many published reports this estimation implies unrealistic constant rate of LDPE biodegradation. Conventional supposition assumes, that it takes rather thousands than hundreds years for the LDPE to be completely degraded (Kyrikou & Briassoulis, 2007). Formerly, some authors claimed that the poor biodegradability of synthetic substances is a consequence of their short time of presence the environment, so that the enzymes capable of degrading e.g. plastics are not available (Müeller, 2006). However, the research on the microbial metabolism of xenobiotics showed that the evolution of specifity of oxidoreductases secreted by microorganisms is relatively fast (Koutny et al., 2006; Wojcieszyńska et al., 2011). These enzymes involved in the transformation of the molecular edifices increase the polarity of the molecule (Lucas et al., 2008). Nowadays, it also seems that even hydrophobic nature of polyethylene resulted from carbon-only backbone is not a hindrance during biodegradation, since fungi, due to their ability to form hydrophobic proteins, can easily attach to the polymer surface (Sahebnazar et al., 2010). Contrary to previously described features of polyethylene, the high molecular weight itself represents a serious problem because, as a molecule of this size cannot cross a cell wall and a cytoplasmic membrane, it is inaccessible to intracellular metabolic pathways (Koutny et al., 2006). It is therefore necessary to reduce its molecular weight drastically by some predegradation method prior to biological attack. It has been shown that linear paraffin molecules having a molecular weight below 500, or n-alkanes up to C_{44}, can be utilized as a carbon source by microorganisms (Jakubowicz et al., 2006).

Many sources clearly indicate that biodegradation of LDPE could be accelerated by polymer pretreatments, such as photo-oxidation, thermo-oxidation and chemical-oxidation (Sahebnazar et al., 2010).

Photo-oxidative degradation is the process of decomposition of the material by the action of light. It is well known that radiation in the wavelength region of 290–400 nm, which have sufficient energy to cleave C-C bond (Mark et al., 1986) is not absorbed by pure LDPE. Its photo-oxidation can only be induced by some impurities (Briassoulis et al., 2004).

Thermo-oxidative degradation is an exposition of polymer to high temperatures. The sensitivity of polyolefins towards thermal oxidation is largely due to the presence of impurities, hydroperoxides and carbonyl groups (Briassoulis et al., 2004).

Though both types of abiotic oxidation produce functional macromolecules susceptible to random cleavage with the formation of low molecular weight oxygenated products containing carbonyl residues (Chiellini et al., 2007), the main difference between photo- and thermal oxidation is that photochemical reactions occur only on the surface of the polymer sample, whereas thermal reactions occur throughout the bulk (Briassoulis et al., 2004).

Hydrolysis is another way by which polymers can undergo degradation (Sahebnazar et al., 2010). However, mechanism of the process strongly depend on polymer structure. Some polymer materials are hydrolysed via both bulk degradation and surface erosion, others hydrolyse through only one of the mechanisms.

Initial abiotic oxidation and/or hydrolysis of polyethylene is an important stage as it determines the rate of the further biodegradation process. At a second stage of environmental degradation polymer with increased bioavailability and biodegradability should be consumed by microorganisms in soil or during composting (Abrusci et al., 2011).

Another of the possible ways to accelerate biodegradation rate of polyethylene in the environment is copolymerisation, blending or grafting with functional polymers and compounds (Corti et al., 2010; Huang et al., 2005).

Few additives having hydrophilic groups make plastic less hydrophobic and susceptible for photo-, chemical and microbial degradation. The most desirable effect of this approach is microbial assimilation of the filler, serving as initial point of microbial attack, resulted in the increase of the surface area of the synthetic bulk material rendering it more susceptible not only to biotic but also abiotic oxidation (Chiellini et al., 2003; Singh & Sharma, 2008). As a result the remaining inert components should disintegrate and disappear (Chandra & Rustgi, 1998).

Among biodegradable plastics which can serve as degradable fillers are products of microbial fermentation (e.g., polyesters), modified natural products (e.g., starch, cellulose) and plastics based on chemical synthesis (e.g. polylactic acid, polycaprolactone) (El-Naggar & Farag, 2010). Despite advantages, such as good biodegradability, the commercial production of these polymers is 2.5–10 times more expensive than conventional polymers. Moreover, their physical or chemical properties often restrict their use (Ojeda et al., 2009). Therefore combination of their biodegradability with the excellent properties of conventional materials, such as low-priced LDPE, seems to be a promising alternative (Rosa et al., 2007).

Poly(butylene succinate) (PBS) is, next to poly(butylene succinate-co-butylene adipate) (PBSA) and poly(ethylene succinate) (PESu), a member of group of biodegradable aliphatic polyesters trademarked 'Bionolle' invented by Showa Denko (Japan) in 1990, and produced through a polycondensation reaction of glycols with aliphatic dicarboxylic acids and their derivatives (Tserki et al., 2006a). Among these three polyesters, PBS is the only one which is available commercially. However some studies revealed that the biodegradation of PBS is much slower than that of PBSA (H-S. Kim et al., 2006).

So a few years ago, we started testing the biodegradability PBSA. We are the only team which, in order to accelerate the biodegradation of synthetic polymers, applied modification of LDPE with poly(butylene succinate-co-butylene adipate) (PBSA). Since then we are investigating mechanisms of biodegradation of LDPE/PBSA compositions under different environmental conditions.

It is well known, that even though the mechanism of oxidation of LDPE is more or less understood, the knowledge of the behaviour of this polymer in blends with other materials is not sufficient. It was confirmed that the lack of additivity of component properties in blends is a main reason for difficulties in predicting their life-time (Ołdak & Kaczmarek, 2005).

The purpose of this study was to examine the synergistic or antagonistic effects on the oxidation, hydrolysis and biodegradation of a commercial PELD films filled with PBSA copolyester. Investigations were conducted by first exposing polymeric films to the abiotic oxidation (action of photo- and/or thermal degradation), followed by abiotic hydrolysis under mild conditions, and subsequently to microbial biodegradation.

Several techniques were employed to elucidate the chemical and physical polymers structure. Changes in chemical structure of polymeric films caused by various types of degradation were interpreted on the basis of IR spectra analysis. The scanning electron microscope (SEM) was used to examine these polymers morphologically (surface topography); the excellent resolution provided by the SEM makes it one of the best tools for this purpose.

2. Experimentals

2.1 Film preparation

Low-density polyethylene (LDPE, type "FGNX23-D022", MFR of 2,2g/10min) was purchased from POLICHEM in Kędzierzyn-Koźle. Bionolle® (type #3001, MFR of 1,5g/10min.) was obtained from Showa Denko (Europe) GmbH. The LDPE and Bionolle® were homogenised in a Co-Knetter Buss high-speed mixer at 170°C. The homogenised material was further processed on a PLV 151 type Plasti-Corder extruder for the production of thin films. The compositions 85/15, 70/30, 40/60 LDPE/ Bionolle® were prepared with ratio of 33 rpm and 220, 230, 230, and 235°C set temperatures. Polyethylene film without any additives 100/0 was used as a control material. The films were extruded at the Institute for Engineering Polymer Materials and Dies in Gliwice (Poland). Each film was cut into 40 mm *x* 40 mm squares.

2.2 Abiotic treatment

2.2.1 Hydrolytic aging

Hydrolysis of films was carried out in phosphate buffer (pH 7.4) with sodium azide to prevent growth of microorganisms for 84 days.

2.2.2 Oxidative aging

The photo- and/or thermal degradation of polymers was achieved by placing the samples in an adapted oven for 16 days. Every 24 hours, the location of the samples was changed in a clockwise direction. After four days, when a complete change of the position of samples within the chamber took place, they were inverted.

2.2.2.1 Photodegradation procedure

Films were positioned 15 cm from the lamp then UV-irradiated using a low-pressure mercury vapor lamp generating energy between 280 nm and 370 nm (TUV 6W, Philips, Holland) in air atmosphere at room temperature.

2.2.2.2 Thermodegradation procedure

The samples were subjected to dark heated exposure at 50°C in air atmosphere.

2.2.2.3 Photothermodegradation procedure

Photothermal degradation of polymers proceeded with simultaneous action of UV radiation and temperature under the same conditions as described for the individual processes.

2.3 Biodegradation experiments

2.3.1 Strains of fungi

Filamentous fungi *Aspergillus niger, Aspergillus terreus, Aureobasidium pullulans, Paecilomyces varioti, Penicillium funiculosum, Penicillium ochrochloron, Scopulariopsis brevicaulis, Trichoderma viride* and their mixture were employed for the biodegradation. *Aspergillus niger* and *Penicillium funiculosum* were isolated from dump in Sosnowiec and their identification was carried out by Institute for Ecology of Industrial Areas in Katowice, Poland. The others were purchased from Institute of Fermentation Technology and Microbiology in Łódź, Poland.

Fungi were maintained in test tubes containing Czapek-Doxa medium ($NaNO_3$, 2g; KH_2PO_4, 0,7g; K_2HPO_4, 0,3g; KCl, 0,5g; $MgSO_4x7H_2O$, 0,5g; $FeSO_4x7H_2O$, 0,01g; sucrose, 30g; Bacto Agar (Difco), 20g; distilled water, 1000ml; pH 6.0). Cultures were incubated at 28°C. After completion of sporulation, spores of fungi were separated from hyphae by centrifugation at 4000 rpm and resuspended in SDS solution. The spore suspension at a concentration of 10^6 spores ml^{-1} were either used for biodegradation tests or transferred to glycerol solution (50% v/v) before storage at -20°C.

2.3.2 Biodegradation procedure

Squares of each film (5 replicate samples) unexposed and pre-exposed to abiotic oxidation or/and hydrolysis were sterilised in 70% ethyl alcohol, rinsed with sterile distilled water and aseptically placed in Petri dishes containing modified sucrose-free Czapek-Doxa medium. Each film was covered with 0,1 ml spore suspension. Biodegradation was carried out at 28°C and relative humidity of > 90% for 84 days. Loss of water during incubation was supplemented with sterile distilled water.

After the incubation period, polymer samples were delicately removed from the soils and sterilised by immersion in 1% mercuric chloride for 5 minutes, rinsed in water and dried in a desiccator until a constant weight was obtained.

2.4 Assessment of degradation

Sample weight loss was determined gravimetrically on an analytical balance (Mettler Toledo, AB 240-S).

The tensile strength (R_m) and elongation at brake (ε_r) tests were carried out in accordance with EN ISO 527-3: 1995; EN ISO 527-1: 1996; EN ISO 527-2: 1996. Tests were performed on a tensile tester (INSTRON 4466). Results of mechanical strength evaluation were averaged over 5 replicate specimen.

Infrared spectra of the films were recorded on an FTS 40A spectrophotometer (BIO-RAD) over a range of 3700-700 cm^{-1} at a resolution of 2 cm^{-1} and over 32 scans. Samples were dissolved in a mixture of decahydronaphthalene and dimethylformamide at 70°C and analysed as thin films on the NaCl cell surfaces after evaporation of the solvent. Carbonyl index (CI) and terminal double bond index, were used as a parameters to monitor the degree of degradation of films. Carbonyl index is the ratio between the absorbance of the carbonyl peak (1712 cm^{-1}) and the absorbance of the CH_2 groups at 1465 cm^{-1}. Terminal double bond index is the ratio between the absorbance of the terminal double-bond peak (908 cm^{-1}) and the absorbance of the CH_2 groups at 1465 cm^{-1} (Gilan (Orr) et al., 2004).

The pieces of control and treated polyethylene films were cut with a sharp blade to obtain a small cube (5 mm). The cube samples were mounted on an aluminium stubs with double-sided adhesive carbon tape, and sputter-coated in Pelco SC-6 sputter coater (25 mA and 0,8 hPa) for 30 seconds with a thin film of gold to improve the electrically conducting properties of the sample surface. After sputtering the samples were imaged by the Tesla BS 340 scanning electron microscope (SEM) in a high-vacuum mode operating at 20 kV with secondary electron detector (ESD), and working distance (WD) of 10 mm. Collected images were compared with those recorded for the original untreated samples.

3. Results and discussion

The term degradation with respect to decomposition of polymeric materials has not been explicitly specified. The main problem is to determine the susceptibility of the polymer material to degradation in the environment and the length of time during which process will last. Several methods can be used to estimate polymer deterioration. Frequently used methods rely on gravimetric, spectroscopic and microscopic techniques, mostly in combination with each other (Sudhakar et al., 2008).

A simple and quick way to measure the degradation of polymers is by determining the weight variations. However, this measurement itself cannot be a reliable indicative of material degradability, since both an increase in weight and a weight loss of polymer sample, not directly related to the breakdown of polymer chain, may occur. A good example is an increase in weight due to accumulation of microorganism, whereas loss of weight can be due to the vanishing of volatile and soluble impurities (Lucas et al., 2008).

Deterioration of polymers can be also evaluated by change in their rheological properties. Contrary to the weight measurement, these properties directly depend on molecular weight of polymers, their crystallinity and the presence of branches and crosslinkings effects (Briassoulis et al., 2004).

Among others, Fourier transform infrared (FTIR) spectroscopy is most widely used in determining the structural changes in macromolecules. Since it is known that degradation of polymers can proceed *via* both hydrolysis and oxidation, with this tool it is possible to estimate the extend of modification of polymer main chain due to the action of abiotic or biotic factors. It is assumed, that the mechanism of polymer degradation can be determined by measuring the levels of ketone carbonyl, ester carbonyl and internal double bond absorbance peaks (Gilan (Orr) et al., 2004; Jakubowicz et al., 2006; Sudhakar et al., 2008).

Scanning electron microscopy (SEM) is a useful imaging approach for the visualization of different polymers, because it provides a consistent picture of the polymer morphology as a non-uniform structure characterised by variable thickness and variable polymer density. This technique allows to illustrate the surface topography of polymers with high resolution. Due to its high lateral resolution, its great depth of focus and its facility for x-ray microanalysis (SEM/EDX), SEM is often used in material science - including polymer sciences to elucidate the microscopic structure of polymers. In SEM, the surface of non-conductive samples must be coated with a thin layer of gold or platinum. Sometimes, a surface pretreatment (ion sputtering or chemical etching) is carried out to reveal structural details. Moreover, brittle fracture of samples (in liquid nitrogen in cryo-SEM) can give information about the internal morphology of bulk specimens. SEM micrographs indicate that polymers are characterised by different surface features and heterogenous local density of chemical components. They also show surface defects such as cracks, etching residues, differential swelling, depressions and perforations.

Currently, a number of different SEM techniques and sample preparation methods have been employed for study of polymers structure, including ultra-high resolution field emission SEM (UHR FE-SEM), scanning transmission SEM (STEM), low-vacuum SEM (LVSEM/cryo SEM) and environmental SEM (ESEM). In LVSEM mode, the delicate polymer samples are observed in frozen state, whereas in an ESEM mode the specimens can contain liquids. SEM equipped with an energy dispersive X-ray spectrometry profiling (SEM/EDX) is widely used to characterize the variation of chemical composition of polymers interface. STEM is used to analyze lamellar arrangements in polymers, their dimensions and crystallography. Especially the recent development of ultra-high resolution field emission scanning electron microscopy has opened new opportunities in polymer study at the molecular scale. These SEM techniques provide complementary data to transmission electron microscopy (TEM) and scanning probe microscopy (SPM).

In polymer studies the following applications of SEM have been made: study of surface microstructure of polymer films, fibres and powders (amorphous and crystalline); investigation of liquid crystals; control of the polymerization process; study of the structure of copolymers, polymer blends and networks (investigation of miscibility and adhesion of components); observation of structural defects and sample roughness; changes in the structure of polymers during stretching and upon loading (formation of crazes and cracks; fracture surface morphology; chemical agent transport processes through membranes (porosity of membranes) and after biotic (microbiological) treatment (Bonhomme et al., 2003; Borghei et al., 2010; González et al., 2006; Guise et al., 2011; Šašek et al., 2006; Vezie et al., 1995).

3.1 Biodegradation

The percentage weight loss of LDPE and LDPE/Bionolle® compositions after biodegradation with different filamentous fungi is shown in Table 1 (Łabużek et al., 2006a; Nowak et al., 2010).

Filamentous fungi	LDPE/Bionolle® film compositions			
	100/0	85/15	70/30	40/60
	Weight loss, %			
Aspergillus niger	0,12	0,16	1,29	24,0
Aspergillus terreus	0,08	0,08	0,17	0,35
Aureobasidium pullulans	0,04	0,05	0,28	0,25
Paecilomyces varioti	0,07	0,06	0,22	0,29
Penicillium funiculosum	0,15	0,22	1,24	60,0
Penicillium ochrochloron	0,06	0,04	0,29	0,27
Scopulariopsis brevicaulis	0,06	0,10	0,47	1,17
Trichoderma viride	0,02	0,04	0,18	0,34
Mixed fungal population	0,11	0,15	1,22	58,0

Table 1. Weight loss of films after biodegradation

It was found that pure LDPE and the blends 85/15 and 70/30 showed little loss of mass, which probably reflected the inertness of LDPE towards biological degradation. Unfortunately, even 30% Bionolle® wasn't enough to observe sufficient weight loss of film. Probably, the polyethylene matrix prevented microbes from accessing the polyester in the depth of the film. Similar relationships were observed for the polyolefins modified with 6-15% starch content, where only the surface of the material was susceptible to biodegradation (Nakamura et al., 2005). In a separate study, Rosa et al. found that the pure LDPE and the blends 25PHB/75LDPE and 50PHB/50LDPE showed little or no loss of mass during aging in simulated soil (Rosa et al., 2007). Also Lee et al. discovered that polystyrene (PS) in the P(3HB-co-3HV)/PS (95/5 by wt%) blend acts as a retardant of enzymatic attack to the surface of the blend film (Lee et al., 2003).

This phenomenon is likely due to the high-molecular weight hydrophobic chains of the synthetic polymer preventing enzymes from accessing the biodegradable polymers contained within the material (Nowak et al., 2011).

Among polymeric compositions, film containing 60% Bionolle® had the most obvious reduction in weight. Film lost 24%, 60% and 58% of its initial mass after incubation with *Aspergillus niger, Penicillium funiculosum* and mixed population of fungi, respectively. Other fungi caused slight film weight loss ranging from 0,25% to 1,17%. Taken together with our previous studies, the present findings show, that only fungi which were able to decompose Bionolle® (80-100% for 90 days), were also able to degrade 40/60 composition (Nowak et al., 2010).

Mechanical properties of the investigated films before degradation are shown in Table 2.

LDPE/Bionolle® film compositions	Mechanical properties	
	Rm, MPa	ε_r,%
100/0	16,40	516,0
85/15	14,02	518,4
70/30	10,80	336,8
40/60	20,50	485,0

Table 2. Mechanical properties of control films

The changes in tensile strength and elongation at break of films after biodegradation with selected fungi are shown in Figure 1. It has been reported that the changes in tensile strength and elongation loss are excellent indicators of degradation (Reddy et al., 2008).

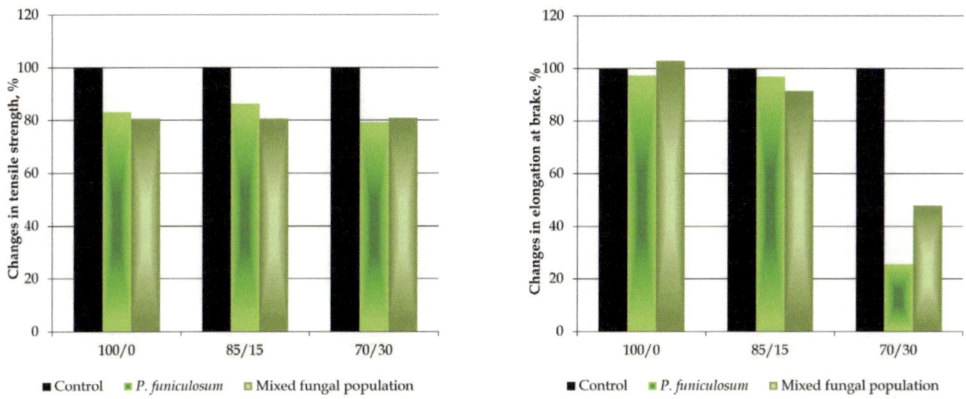

Fig. 1. Mechanical properties of films after biodegradation

Although little weight loss of polyethylene was observed after 84 days of biodegradation, a marked reduction in tensile strength imply that both *Penicillium funiculosum* and mixed fungi population excrete enzymes able to cleave macromolecules of LDPE. Slight increase in the elongation at break in samples incubated with mixed population of fungi, could be tentatively attributed to a sort of plasticisation effect exerted by low molecular weight fractions produced in the first stage of the biodegradation of the polymer matrix (Chiellini et al., 2003). After 84 days of biodegradation, the tensile strength of film containing 30% Bionolle® decreased by about 20%. At this time, elongation at break changed by about 74% and 52% after incubation with *Penicillium funiculosum* and mixed fungi population, respectively. Far-advanced decomposition of 40/60 blend, prevented determination of its mechanical properties.

Data obtained from FTIR spectra (spectra are presented elsewhere (Łabużek et al.) of examined films showed an increase in carbonyl and double bond indices, except for LDPE (decrease in carbonyl index by 57%), incubated with mixed population of fungi (Łabużek et al., 2006a). Increase in the internal double bond index is in accordance with the biodegradation mechanism suggested by Albertsson et al. who reported on the formation of terminal double bonds as a result of exposure to biotic environment (Albertsson et al., 1987). This can be attributed to biotic dehydrogenation (Chiellini et al., 2003). However, in contrast with study of Albertsson et al., we found an increase in the amount of carbonyl residues (up to 525%) in LDPE after 84 days of incubation with *Penicillium funiculosum* (Albertsson et al., 1987). Carbonyl residues have been reported as major products formed in the presence oxidoreductases (Sudhakar et al., 2008).

Important aspect during the biodegradation of a material is the sustained growth of microorganisms during the entire process (Abrusci et al., 2011). The changes on the surface of polymers as a result of biodegradation are no less important (Nowak et al., 2011).

Figure 2 shows the micrographs of control films.

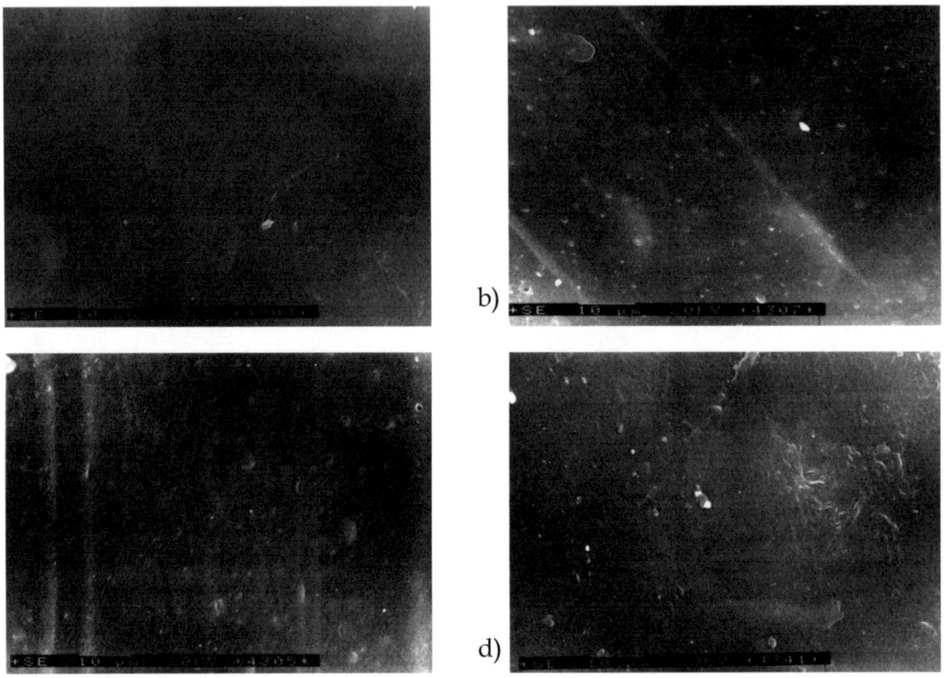

Fig. 2. SEM micrographs showing surface of control films a) LDPE b) composition 85/15 c) composition 70/30 d) composition 40/60

The photomicrograph shows that the surface of non-degraded material is smooth, without cracks and holes.

Neat polymer samples after biodegradation with filamentous fungi are presented in Figures 3-6.

Paecilomyces varioti (Figure 3c) and *Penicillium funiculosum* (Figure 3d) expanded their colonies over the entire surface of neat LDPE. Apart from hyphae and conidiophores, samples were covered with characteristic spores. *Aspergillus niger* (Figure 3a), *Aspergillus terreus, Aureobasidium pullulans* (Figure 3b), *Penicillium ochrochloron, Scopulariopsis brevicaulis, Trichoderma viride* and mixed fungal population grew less rapidly and primarily on the edges of the sample. Considerable change in mechanical properties (Figure 1) and FTIR spectrum of LDPE proved that growth of fungi cannot be considered only as a result of the surface moistness (Sahebnazar et al., 2010). Moreover, knowing that the biodegradation experiments were conducted in minimal solid medium, it is obvious that solid surface of LDPE, at least for some fungi, served as the source of carbon and energy.

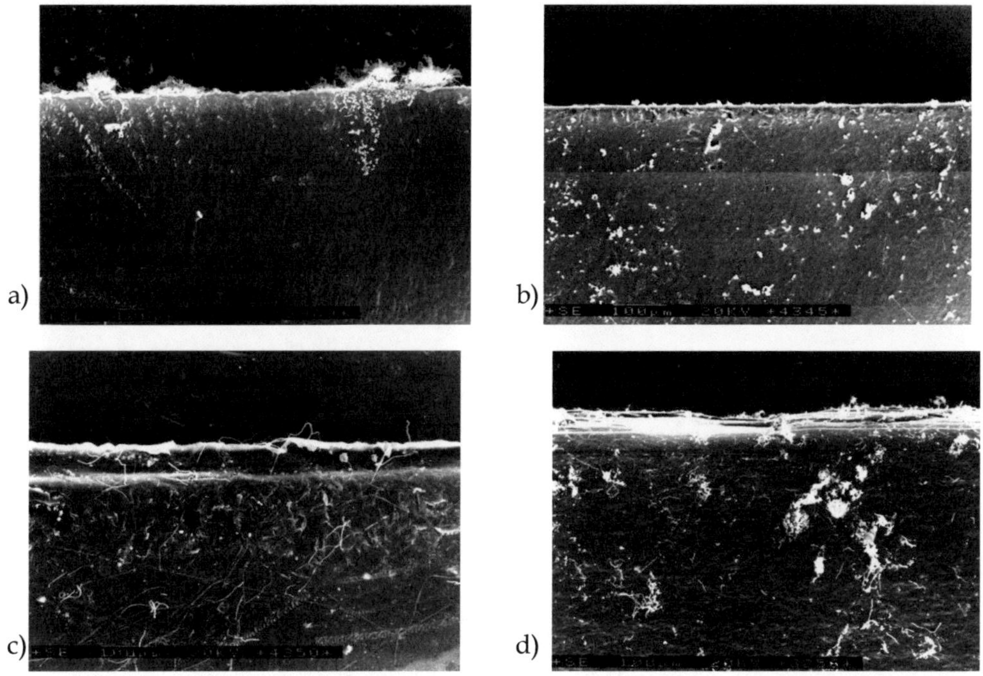

Fig. 3. SEM micrographs of neat LDPE film after biodegradation with a) *Aspergillus niger* b) *Aureobasidium pullulans* c) *Paecilomyces varioti* d) *Penicillium funiculosum*

As a result of fungal degradation, peeling and cracking in texture of film containing 15% Bionolle® were visible (Figure 4). The entire surface of film was densely covered with spores belonging to *Aspergillus niger* (Figure 4a) and *Aureobasidium pullulans* (Figure 4b). Scarce hyphae of *Paecilomyces varioti* (Figure 4c), *Penicillium funiculosum* (Figure 4d) and mixed population of fungi (Figure 4f) colonised both the edges and the surface of film. Agglomerations of *Trichoderma viride* conidiophores (Figure 4e) inhabited primarily the edges of the sample.

The destructive process of biodegradation was very prominent in film containing 30% Bionolle®. Small holes and cracks, surface irregularities, peeling and exfoliation appeared. *Aspergillus niger* (Figure 5a), *Paecilomyces varioti* (Figure 5c), *Penicillium funiculosum* (Figure 5d) and mixed population of fungi (Figure 5h) produced a well-developed mycelium over the entire surface of the film. Hyphae and conidiophores of *Penicillium ochrochloron* (Figure 5e), *Scopulariopsis brevicaulis* (Figure 5f) and *Trichoderma viride* (Figure 5g) was scattered on the film surface. *Aspergillus terreus* (Figure 5b), almost unable to colonise 85/15 composition, introduced deep cracks and holes, suggesting that the fungi penetrate into the sample matrix during degradation.

After 84-day incubation with filamentous fungi the most intense changes were found on the surface of 40/60 composition. Hyphae and conidiophores of *Aspergillus niger* (Figure 6a), *Aspergillus terreus* (Figure 6b), *Aureobasidium pullulans* (Figure 6c) and *Trichoderma viride*

(Figure 6g) grew out directly from the polymer sample. However, changes induced by the action of these microorganisms deep and distinct. Dense network of fractures was particularly visible after incubation with *Aspergillus terreus*. *Aspergillus niger* and *Aureobasidium pullulans* caused massive exfoliation of the plastic edges. Although *Penicillium ochrochloron* (Figure 6e) and *Scopulariopsis brevicaulis* (Figure 6f) created mycelium, a considerable part of the film surface was found to be unaffected. Loss of integrity of film resulted into fragile surface entirely covered by dense mycelia of *Penicillium funiculosum* (Figure 6d) and mixed population of fungi (Figure 6h).

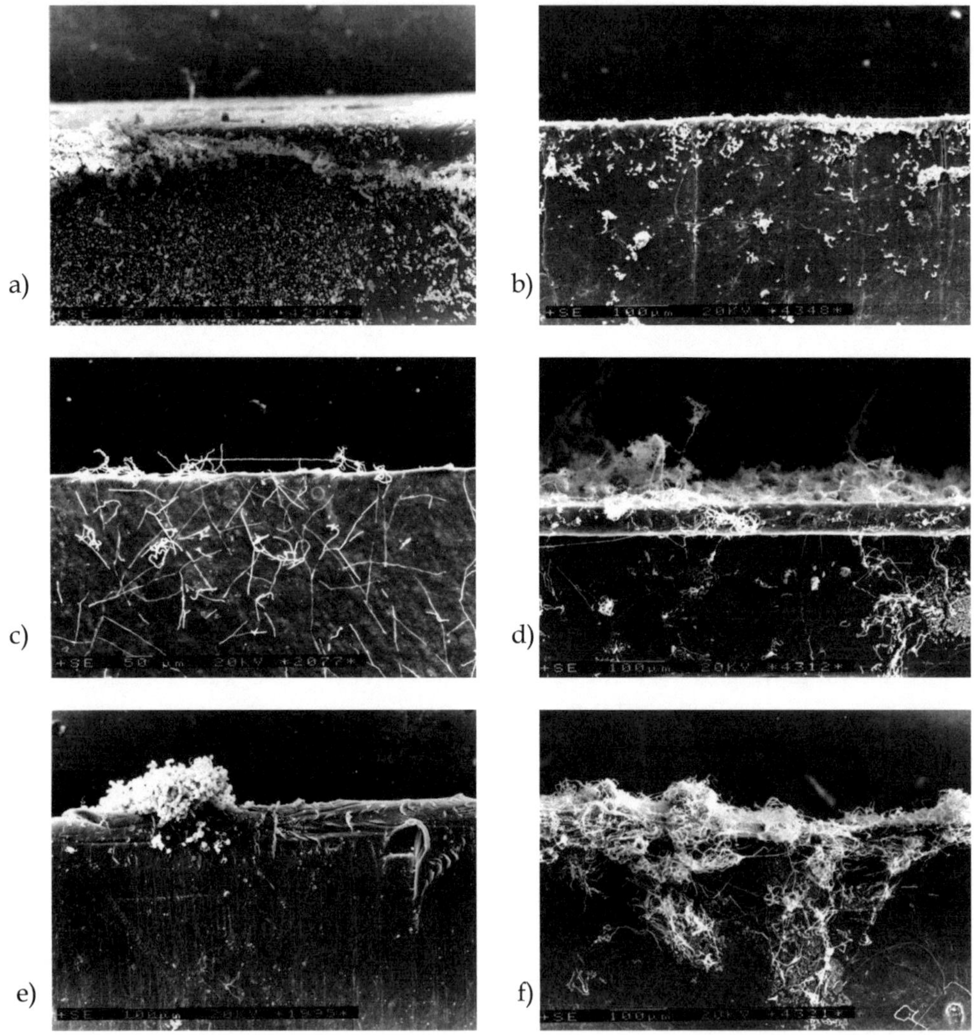

Fig. 4. SEM micrographs of neat 85/15 composition after biodegradation with a) *Aspergillus niger* b) *Aureobasidium pullulans* c) *Paecilomyces varioti* d) *Penicillium funiculosum* e) *Trichoderma viride* f) mixed fungal population

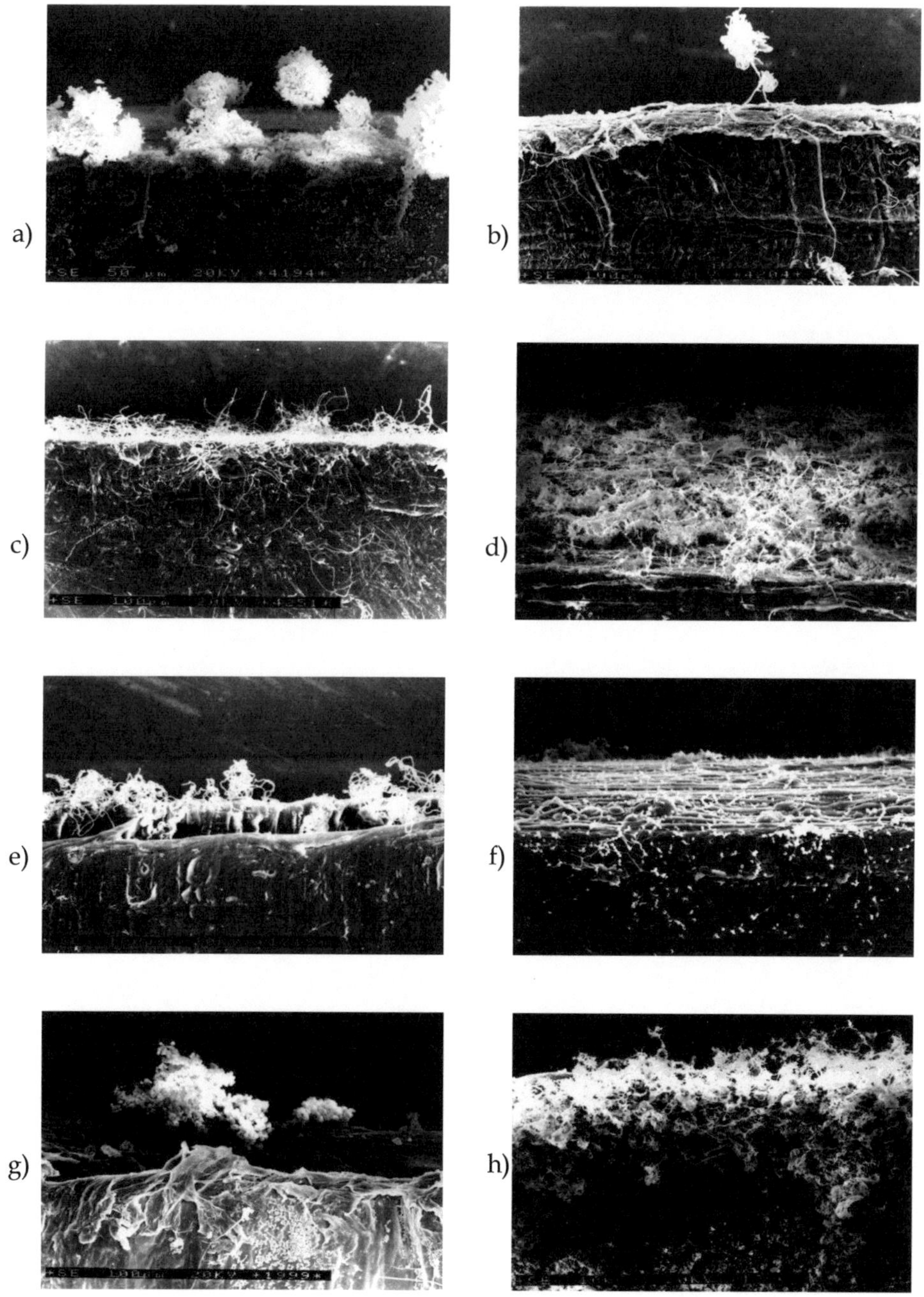

Fig. 5. SEM micrographs of neat 70/30 composition after biodegradation with a) *Aspergillus niger* b) *Aspergillus terreus* c) *Paecilomyces varioti* d) *Penicillium funiculosum* e) *Penicillium ochrochloron* f) *Scopulariopsis brevicaulis* g) *Trichoderma viride* h) mixed fungal population

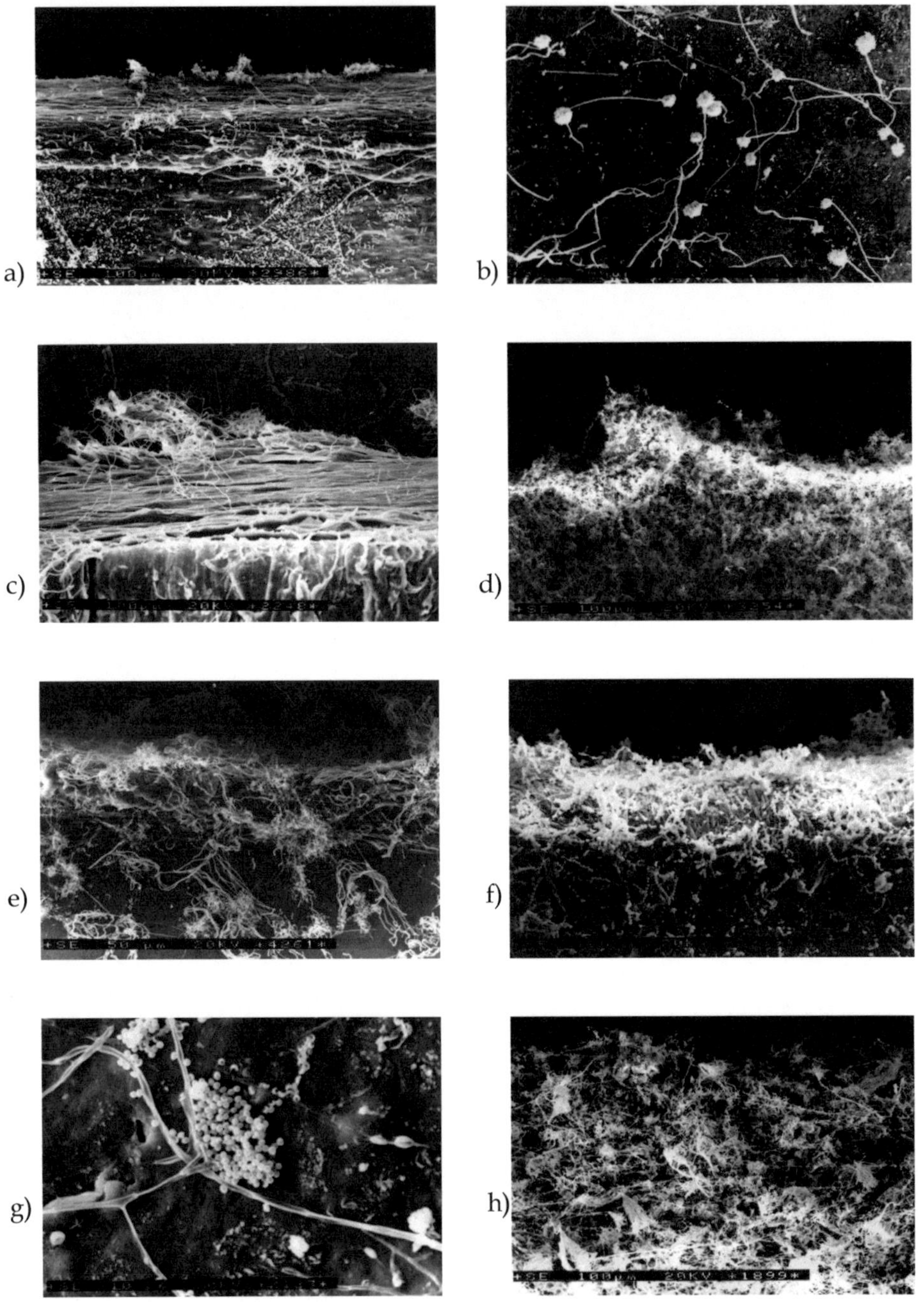

Fig. 6. SEM micrographs of neat 40/60 composition after biodegradation with a) *Aspergillus niger* b) *Aspergillus terreus* c) *Aureobasidium pullulans* d) *Penicillium funiculosum* e) *Penicillium ochrochloron* f) *Scopulariopsis brevicaulis* g) *Trichoderma viride* h) mixed fungal population

3.2 Abiotic degradation

Weight variations of LDPE film and LDPE/Bionolle® compositions recorded after exposing to abiotic treatment are shown in Table 3.

Abiotic degradation processes	LDPE/Bionolle® film compositions			
	100/0	85/15	70/30	40/60
	Weight loss, %			
Photodegradation	0,06	0,05	0,07	0,06
Thermodegradation	0,02	0,03	0,13	0,1
Photo- and thermodegradation	0,03	0,04	0,09	0,26
Hydrolysis	0	0,04	0,07	0,38
Photodegradation and hydrolysis	0,02	0,05	0,07	0,27
Thermodegradation and hydrolysis	0,02	0,03	0,08	0,34
Photo- thermodegradation and hydrolysis	0	0,04	0,09	0,31

Table 3. Weight loss of films after abiotic aging

Regardless of the type and combination of abiotic aging factors, there were no significant differences in weight loss observed for LDPE and its compositions containing up to 30% Bionolle®. More significant decrease of the weight in film containing 60% polyester was recorded not till then it was subjected to hydrolysis or when more than one aging factor was used during degradation experiments. Oxidative aging of films did not accelerate their loss of mass, most likely due to the presence of crosslinks evolved under the action of radiation and/or heat (Ojeda et al., 2011). It is often reported that, in LDPE films, crosslinking competes with the chain scission mechanism depending on the oxygen concentration at the reaction site (Feuilloley et al., 2005).

Changes in some mechanical properties of films after abiotic degradation are presented in Figure 7.

Values obtained by measuring elongation at break of polyethylene after photodegradation, thermodegradation and photothermodegradation slightly increased. As it was reported above, due to the formation of cross-linking bonds between the polyethylene chains. However, reduction in the mechanical properties after hydrolysis was observed. This decrease can be attributed to the chain scission of the polymer which, in this case, most intensively proceeded in films exposed earlier to both UV radiation and heat. This macromolecular chain scission is the cause of embrittlement of films (Abrusci et al., 2011). Amongst modified films, sensitivity toward abiotic treatment in terms of loss in the mechanical properties can be arranged as follows: 40/60>85/15>70/30. It seems that the exposure of the films containing 15% and 30% polyester to abiotic oxidation has accelerated their subsequent hydrolysis. On the contrary, in film consisted in most part of polyester, the most significant decrease in tensile strength by about 81% and elongation at break reduced by 99% was observed in samples subjected only to hydrolysis. From the above results it can be concluded that the samples with low content of Bionolle® behaved more like polyethylene while 40/60 composition more like polyester. It is known, that in hydrolytic degradation of biodegradable polyesters, elongation at break is the most sensitive property among the tensile properties (Tsuji et al., 2006).

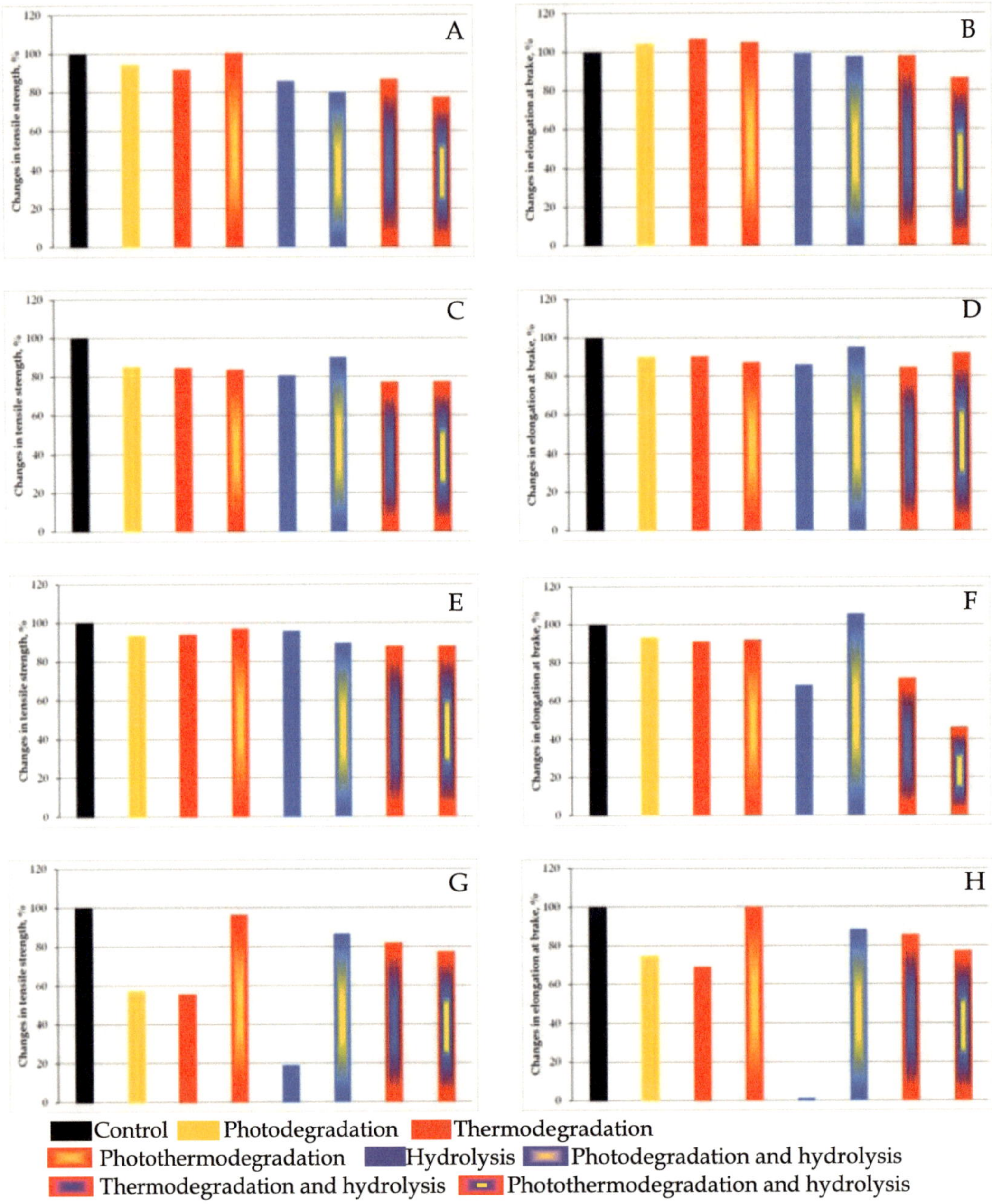

Fig. 7. Changes in some mechanical properties of films after abiotic degradation A) and B) LDPE C) and D) 85/15 blend E) and F) 70/30 blend G) and H) 40/60 blend

Figure 8 shows FTIR spectra of samples before and after exposing to selected abiotic and biotic factors.

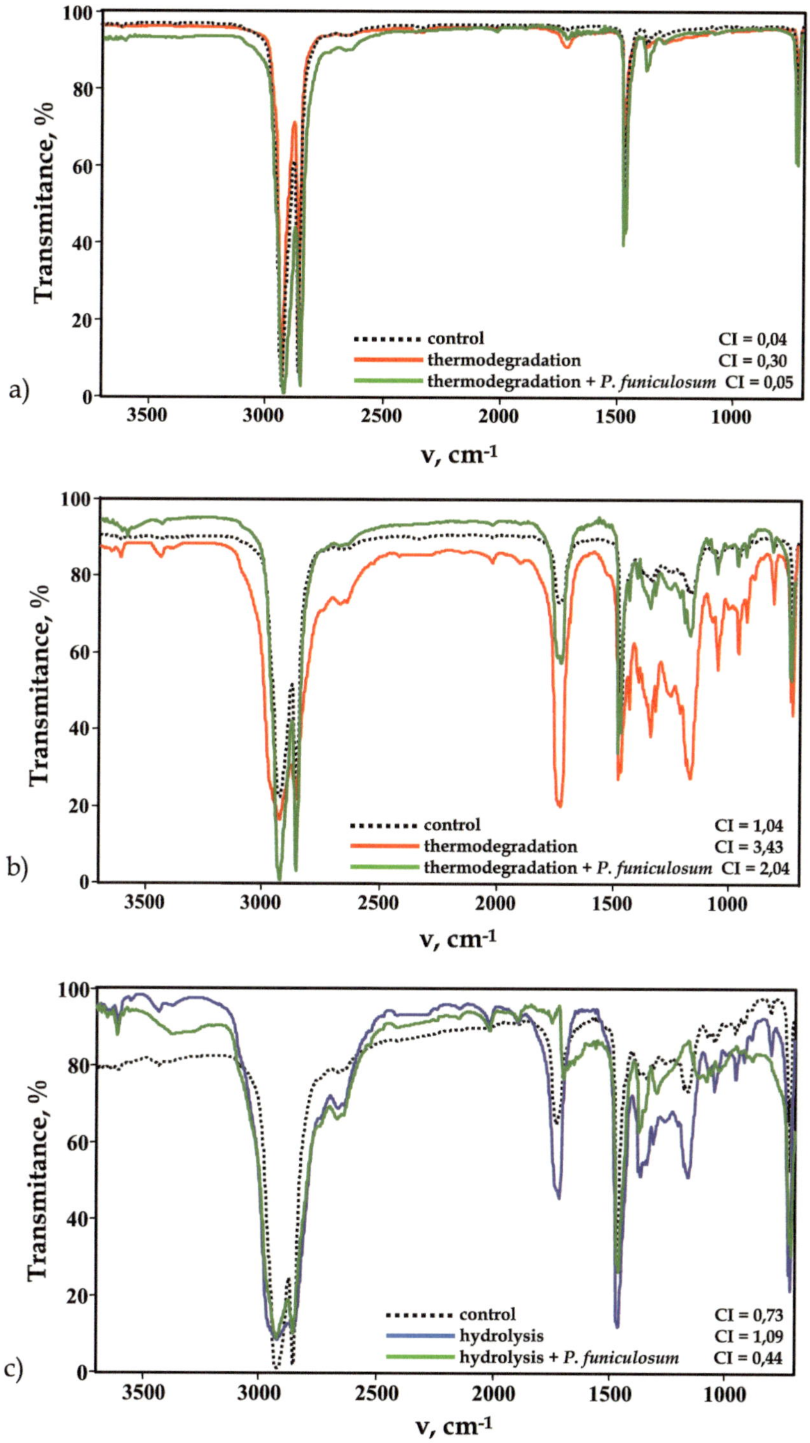

Fig. 8. FTIR spectra of films before and after degradation with selected abiotic and biotic factors a) LDPE b) 70/30 composition c) 40/60 composition

The data showed that the area corresponding to the carbonyl region has grown after abiotic degradation of all films indicating the formation of low molecular weight carbonyl compounds as a result of oxidation and/or hydrolysis. Carbonyl index of LDPE was increased by 25%(photothermodegradation)-650%(thermodegradation). As for polyethylene, the greatest impact on the oxidation of 70/30 composition had thermodegradation (230%) while the smallest - photothermodegradation (196%). Similar correlations were also observed for internal double bond index, which after thermodegradation of polyethylene and 70/30 composition increased by 167% and 316%, respectively. For film containing 60% Bionolle®, the largest increase in carbonyl index, amounting 50%, was found after hydrolysis. Such low degree of oxidation resulted probably from the fact that low-molecular fractions of polymer diffused out of the polymer matrix during the hydrolysis of film (Göpferich, 1996). The above findings suggest that single processes, especially thermo-oxidation, are more efficient in polymer degradation than simultaneous action of UV radiation and heat. These findings are in agreement with (Ram et al., 1980) who claims that the presence of oxygen in conjunction with high temperatures, plays more significant role in the increase of carbonyl concentration than when combined with UV exposure. Moreover, the air temperature as a critical factor increases the rate of various chemical reactions associated with degradation (Briassoulis et al., 2004). Additionally, as mentioned earlier, thermodegradation occurs throughout the bulk of polymer, not only on its surface.

Micrographs of films exposed to selected abiotic factors are presented in Figures 9 and 10.

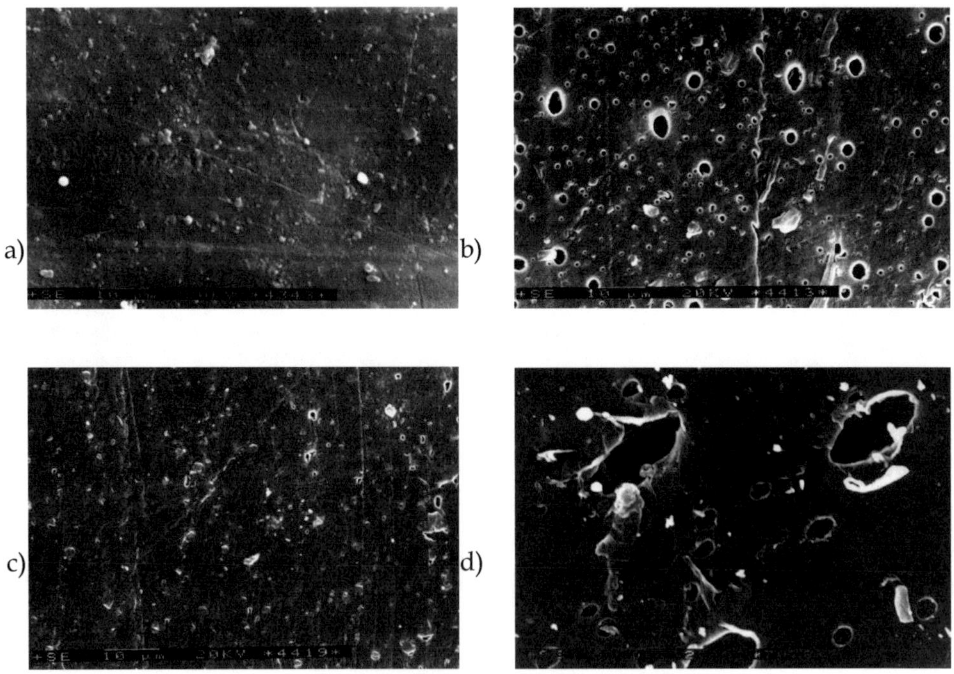

Fig. 9. SEM micrographs of 70/30 composition after a) photodegradation b) thermodegradation c) photothermodegradation d) hydrolysis

Compared to the smooth control film (Figure 2), only some irregularities were visible on the surface of 70/30 composition exposed to UV radiation (Figure 9a). After thermodegradation numerous and well distributed oval cavities with a diameter 1-8 mm were seen all over the surface (Figure 9b). Observations of changes resulting from the individual oxidative processes (photo- and thermodegradation) were helpful in interpreting the data obtained after simultaneous action of UV radiation and heat (Figure 9c). It is evident that both processes act antagonistically. This observation confirms the results obtained by other methods. The smallest weight loss and negligible increase in carbonyl index, probably resulted from the fact that sites of potential chain oxidation (macro radicals) were involved in the crosslinks formation (H-S. Kim & H-J Kim, 2008). Hence, observed reinforcement of plastics after photothermodegradation. However, the most pronounced changes were observed after hydrolysis of the film (Figure 9d). Holes on the surface were less in number but bigger in size than cavities observed after thermodegradation. Studies conducted by other researchers suggest that hydrolysis, in contrast to the photodegradation, occured mainly in the amorphous region of polymer (Bikiaris et al., 2006; Tsuji et al., 2006). Judging by the size, shape and distribution of holes, identical to that observed after thermodegradation, it can be concluded, that during the heat-treatment, the amorphous phase breaks down in the first place. This is due to the fact that, under the impact of warmth, the mobility of macromolecules within amorphous region increases more significantly, therefore they become more prone to degradation.

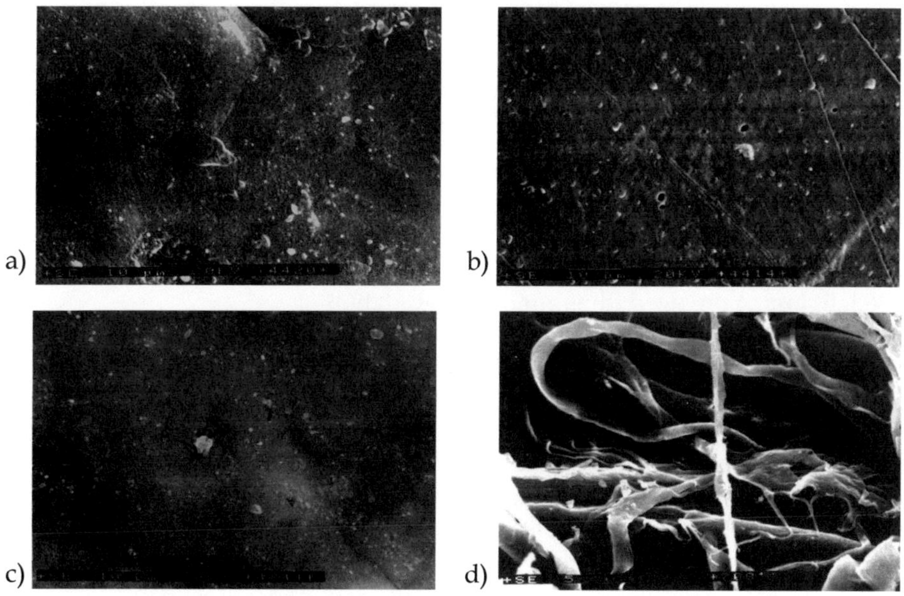

Fig. 10. SEM micrographs of 40/60 composition after a) photodegradation b) thermodegradation c) photothermodegradation d) hydrolysis

There were no visible changes of film texture after photo- and photothermodegradation (Figure 10a and c). Again, after another analysis of 40/60 composition, it could be stated that the behaviour of the composition under the influence of different factors is affected mainly by a large quantity of polyester in polyethylene. These findings are in agreement with our previous study on photodegradation of PBSA (Łabużek et al., 2006b). Thermodegradation of film containing 60% Bionolle® resulted in the formation of regular holes in small quantities (Figure 10b). The most significant surface fragmentation was observed after hydrolysis of the film (Figure 10d). As is evident from SEM, amorphous regions are preferably hydrolysed, revealing the crystalline ones which in turn affects the mechanical properties. It is postulated that mechanical properties, especially elongation at break (see Figure 7G and H), depend more on changes occurring in the amorphous phase (Briassoulis et al., 2004).

3.3 Influence of abiotic degradation on the rate of biodegradation

Penicillium funiculosum and mixed population of fungi were used in studies on biodegradation of abiotically aged films, since, in previous stage of experiments, they exhibited the greatest ability to degrade examined plastics.

Table 4 shows the percentage weight loss of pre-aged LDPE film and LDPE/Bionolle® compositions after biodegradation with *Penicillium funiculosum* and mixed fungal population.

Process	Fungi	LDPE/Bionolle® film compositions			
		100/0	85/15	70/30	40/60
		Weight loss, %			
Photodegradation and biodegradation	*P. funiculosum*	0,13	0,26	1,43	69,19
	Mixed fungal population	0,07	0,18	1,38	65,61
Thermodegradation and biodegradation	*P. funiculosum*	0,28	0,28	1,48	71,61
	Mixed fungal population	0,04	0,23	1,56	74,14
Photo- and thermodegradation and biodegradation	*P. funiculosum*	0,08	0,14	1,45	70,16
	Mixed fungal population	0,12	0,15	1,27	60,38
Hydrolysis and biodegradation	*P. funiculosum*	0,17	0,45	2,18	74,62
	Mixed fungal population	0,13	0,32	1,86	59,32

Table 4. Weight loss of pre-aged films after biodegradation

Based on material weight loss it cannot be clearly determined what impact have abiotic processes had on subsequent biodegradation of the material. Contrary to LDPE, photodegradation, thermodegradation and photothermodegradation of modified polyethylene films accelerated their biodegradation. Percentage weight loss of LDPE films modified with 30% Bionolle® after biodegradation with *Penicillium funiculosum* and mixed population of fungi increased by 15% and 13%; 19% and 28%; 17% and 4% in comparison to weight loss of neat films, respectively. The biodegradative propensity of abiotically degraded 40/60 composition was also markedly affected. The pre-treated film after being exposed to *Penicillium funiculosum* and mixed population of fungi exhibit a mass loss of 15-19% and 4-28% higher than observed for control film. However, the biggest mass loss was observed after hydrolytic aging and biodegradation with *Penicillium funiculosum*. Biodegradation of films containing 15%, 30% and 60% polyester after prior hydrolysis increased by 104.5%, 75.8% and 24%, respectively. Considering percentage weight loss after hydrolysis and biodegradation of 40/60 composition, we have to remark, that any amount greater than 60% meant the degradation of polyethylene.

Changes in some mechanical properties of film samples after the abiotic degradation and biodegradation are shown in Figure 11.

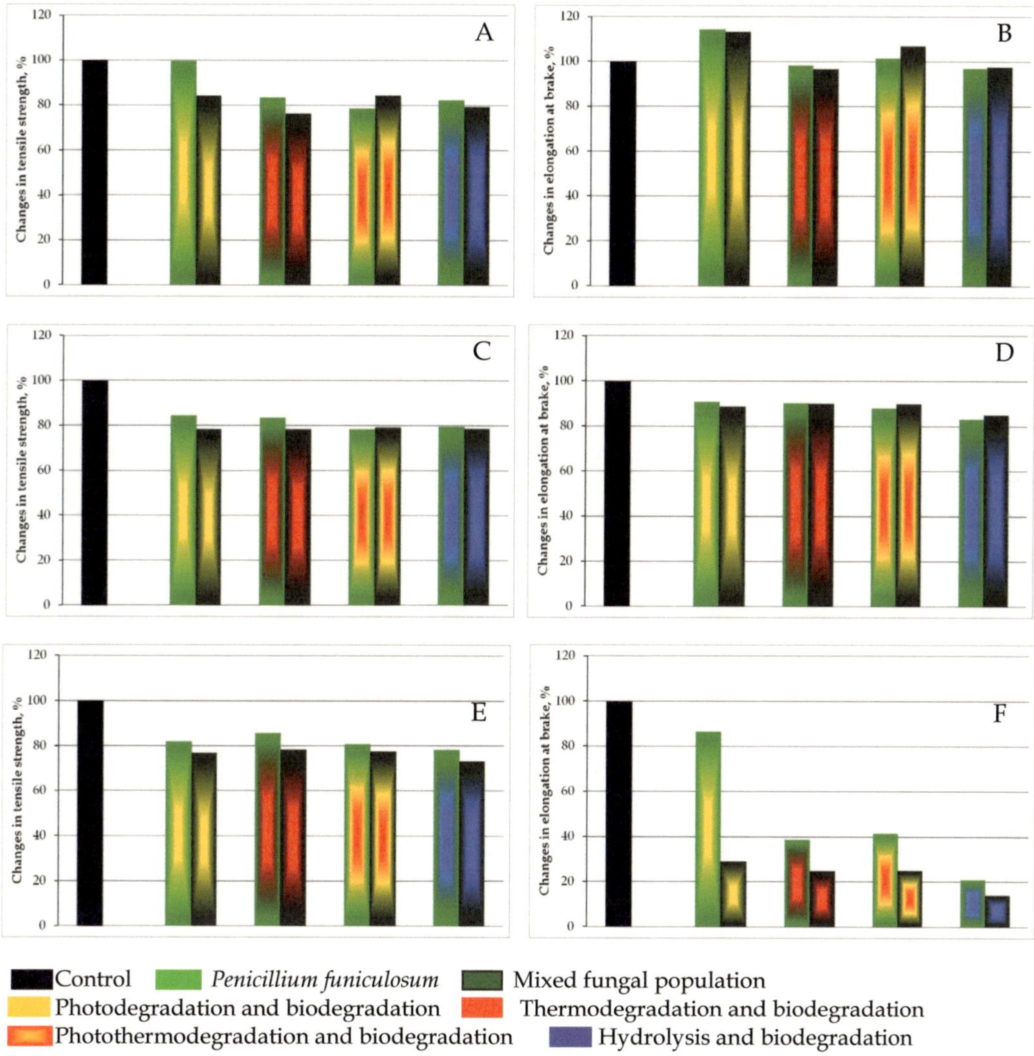

Fig. 11. Changes in some mechanical properties of films after abiotic and biotic degradation A) and B) LDPE C) and D) 85/15 blend E) and F) 70/30 blend.

A significant increase in elongation of LDPE after thermo-, photo-thermo- and biodegradation with both *Penicillium funiculosum* and mixed populations of fungi was noted. These values were significantly higher than observed after biodegradation of neat polyethylene. The highest impact of abiotic pretreatment on the subsequent biodegradability was observed for 85/15 composition which elongation decreased by 9,26%-16,82% as compared to 3,09%-8,49% for the film subjected only to biodegradation. As seen from Figure 11 there is a difference concerning the influence of abiotic aging to biodegradation of modified films between mixed fungi population and *Penicillium funiculosum*. Contrary to mixed fungal population, assimilation of products of polymer oxidation during incubation with *Penicillium funiculosum* occurred less efficiently than depolymerisation of long LDPE chains. Far-advanced decomposition of 40/60 blend, prevented determination of its mechanical properties.

The amount of carbonyl groups, resulting from thermodegradation, decreased by 83% as a consequence of the assimilation of the degradation products by *Penicillium funiculosum*. Similar mechanism of biodegradation of abiotically aged polymers is repeatedly reported. Also Albertsson et al. (1987) found a synergistic effect between photooxidation and the biodegradation of polyethylene. Carbonyl residues completely disappeared after the incubation of the polymer samples in the presence of *Arthrobacter paraffineus*. Decrease of carbonyl index by 30-35% with respect to the starting materials was described for LDPE film samples containing pro-oxidant additives exposed to thermal- and biological degradation (Chiellini et al., 2007). The spectroscopic investigations led by Roy et al. revealed that the bacterial consortium consisted of *Bacillus pumilus*, *Bacillus halodenitrificans* and *Bacillus cereus* preferentially consumed the oxygenated products, thus leading to a decrease in the carbonyl index from 1,29 to 0,31 (Roy et al., 2008). In our study, internal double bonds also proved to be equally easily digested, since their amount after thermo- and biodegradation decreased by 92%. Interestingly, our study also shown that the number of carbonyl residues was about 80% lower than that observed during the biodegradation alone. The amount of carbonyl groups in the film with 30% content of polyester after thermo- and biodegradation compared to control film increased by 96%, while in relation to the film after thermodegradation decreased by 41%. As shown on Figure 8 carbonyl index of aged 40/60 composition after hydrolysis and subsequent biodegradation was lower by 40%, 93% and 60% than the value obtained for the control, after biodegradation and after hydrolysis, respectively.

These results implies that the filamentous fungi having at their disposal oxidised degradation products of polyethylene become less effective in degradation of macromolecules. Decrease in both indices (carbonyl and internal double bond) is a clear evidence that microorganisms use other set of enzymes during biodegradation of aged films than when they grow on high-molecular neat LDPE and polyester. Another conclusion is that the main enzymes involved in degradation of polymers are not constitutive proteins, expressed and secreted by microorganism independently of the substrate. On the contrary, the difference in mechanism of biodegradation clearly indicates the participation of inducible enzymes expressed only under specific conditions.

Micrographs of films exposed to selected abiotic factors and subsequent biodegradation are presented in Figure 12.

Compared to the film exposed only to biodegradation, observations of LDPE after thermo- and biodegradation revealed filamentous fungi growing over the entire surface of the film (Figure 12a). Rough, peeling surface of the material was seen at higher magnification. It is likely that the fungi inhabiting film, used the degradation products of LDPE as a carbon and energy source. This surveillance was supported by a decrease of carbonyl index (Figure 8) and elongation at break (Figure 11) indicating fungal assimilation of low-molecular weight fractions.

SEM micrographs of 85/15 composition revealed deep cracks and holes with diameter 5-50 mm. The cavities on the surface suggested that microorganisms penetrated the polymer matrix during the degradation process.

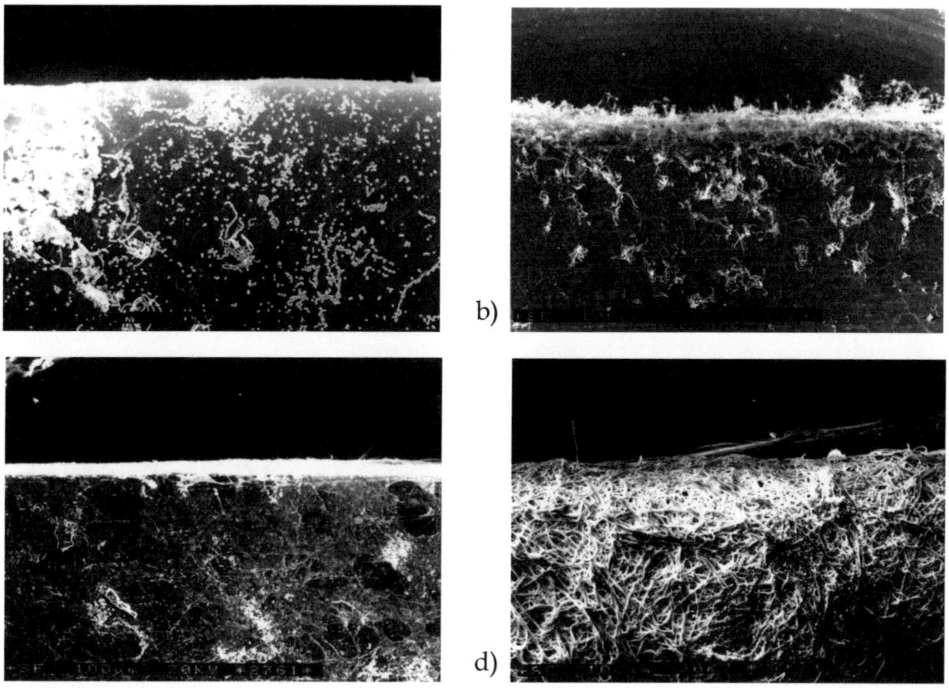

Fig. 12. SEM micrographs of films after exposing to different abiotic factors and subsequent biodegradation with mixed fungal population a) LDPE after thermo- and biodegradation b) 85/15 composition after photo- and biodegradation c) 70/30 composition after photo-thermo- and biodegradation d) 40/60 composition after hydrolysis and biodegradation

Massive erosion (holes with diameter about 200 mm) of film 70/30 and dense network of fungal hyphae indicated that, after abiotic degradation, surface of film become at least as available to microorganism as unaged 40/60 blend.

It was impossible to separate fungal hyphae from the residual 40/60 composition after 84 days of hydrolysis and subsequent biodegradation. As revealed earlier (Table 4), weight loss of film exposed to both factors increased slightly compared to film subjected only to biodegradation. In order to describe the possible mechanism of degradation of this material, it is essential that a few facts should be given. Firstly, hydrolytic degradation (abiotic or biotic) of the polymer occurs predominantly in the amorphous regions (H-S. Kim & H-J Kim, 2008) Secondly, hydrolysis of the crystalline material is slow, because of the limited water diffusion rates into the crystalline domains and stereochemical limitations (Bikiaris et al., 2006). In our study, after inoculation of the composition with fungi, developing hyphae at first assimilated low-molecular products of polymer hydrolysis then attacked its crystalline region (Tserki et al., 2006)). Moreover, by taking into account the weight loss of the composition (59-75%), it can be assumed that fungi (long before the end of the experiment) completely assimilated polyester. Remaining LDPE fibres are clearly visible in the micrographs. Hence, decrease in the rate of biodegradation could be the result of (i) slow biodegradation of crystalline phase of polyester (ii) complex biodegradation of LDPE and (iii) penetration of fungi into the depths of the polymer matrix.

4. Conclusion

In the environment plastics decompose under the influence of different abiotic and biotic factors. The abiotic factors such as radiation, temperature, humidity, chemical pollution and wind can act synergistically or antagonistically causing various types of structural and chemical changes in the polymer. Microorganisms, especially bacteria or fungi, play a crucial role in biological degradation of polymers.

Scanning electron microscopy (SEM) is an invaluable tool for polymer analysis, since it is extensively used to study changes in the texture and composition of biodegradable polymer materials exposed to various environmental factors. It allows for the exploration of large surfaces with excellent resolution of topographic features.

Especially the new generation of SEM technology, known as ultra-high resolution field emission SEM (UHR FE-SEM) presents a promising technique for polymer morphological characterization, and provides complementary data to other microscopic methods.

One of the possible ways to accelerate the degradation of so-called "stable polymers" in the environment is their blending with polymers containing ester, ether, carboxyl and hydroxyl groups that are susceptible to hydrolytic attack of microorganisms. The chapter describes studies on biodegradation of LDPE/Bionolle® blends. It was shown that the addition of polyester significantly accelerated biodegradation of material.

Another way to sensitize the polymer is to expose it to the abiotic aging (action of abiotic factors) followed by the action of microbes. Abiotically modified surface of plastics promotes growth of microorganisms thus accelerates biodegradation.

Indeed, examined films underwent rapid biodegradation, but the course of the process was significantly different from that seen earlier. Further investigation on the mechanism of biodegradation revealed that fungi secreted different sets of enzymes depending on molecular weight of substrate. Unlike some other researchers, we have shown that filamentous fungi used in our study, were capable of efficient oxidation and degradation of the high-molecular weight LDPE. However, when pre-degraded, low-molecular weight chains of polyethylene served as a source of carbon and energy, microorganisms assimilated only short oligomers resulted from prior abiotic degradation. They did not show strong oxidising activity.

In our opinion, LDPE/Bionolle® blends can be used in the production of environmentally degradable packagings. Particularly noteworthy is composition containing 60% Bionolle®, which decompose several dozen percent within 84 days.

5. References

Abrusci, C.; Pablos, J.L.; Corrales, T.; López-Marín, J.; Marín, I. & Catalina, F. (2011). Biodegradation of photo-degraded mulching films based on polyethylenes and stearates of calcium and iron as pro-oxidant additives. *International Biodeterioration & Biodegradation*, Vol.65, No.3, (June 2011), pp. 451-459, ISSN 0964-8305

Albertsson, A.C.; Andersson, S.O. & Karlsson, S. (1987). The mechanism of biodegradation of polyethylene. *Polymer Degradation and Stability*, Vol.18, No.1, pp.73-87, ISSN 0141-3910

Bikiaris, D.N.; Papageorgiou, G.Z. & Achilias, D.S. (2006). Synthesis and comparative biodegradability studies of three poly(alkylene succinate)s. *Polymer Degradation and Stability*, Vol.91, No.1, (January 2006), pp.31-43, ISSN 0141-3910

Bonhomme, S.; Cuer, A.; Delort, A-M.; Lemaire, J.; Sancelme, M. & Scott, G. (2003). Environmental biodegradation of polyethylene. *Polymer Degradation and Stability*, Vol.81, No.3, pp.441-452, ISSN 0141-3910

Borghei, M.; Karbassi, A.; Khoramnejadian, S.; Oromiehie, A. & Javid, A.H. (2010). Microbial biodegradable potato starch based low density polyethylene. *African Journal of Biotechnology*, Vol.9, No.26, (June 2010), pp.4075-4080, ISSN 1684-5315

Briassoulis, D.; Aristopoulou, A.; Bonora, M. & Verlodt, I. (2004). Degradation Characterisation of Agricultural Low-density Polyethylene Films. *Biosystems Engineering*, Vol.88, No.2, (June 2004), pp. 131-143, ISSN 1537-5110

Chandra, R. & Rustgi, R. (1998). Biodegradable polymers. Progress in Polymer Science, Vol.23, No.7, (November 1998), pp. 1273-13335, ISSN 0079-6700

Chiellini, E.; Corti, A. & D'Antone, S. (2007). Oxo-biodegradable full carbon backbone polymers - biodegradation behaviour of thermally oxidized polyethylene in an aqueous medium. *Polymer Degradation and Stability*, Vol.92, No.7, (July 2007), pp. 1378-1383, ISSN 0141-3910

Chiellini, E.; Corti, A. & Swift, G. (2003). Biodegradation of thermally-oxidized, fragmented low-density polyethylenes. *Polymer Degradation and Stability*, Vol.81, No.2, pp.341-351, ISSN 0141-3910

Corti, A.; Muniyasamy, S.; Vitali, M.; Imam, S.H. & Chiellini, E. (2010). Oxidation and biodegradation of polyethylene films containing pro-oxidant additives: Synergistic effects of sunlight exposure, thermal aging and fungal biodegradation. *Polymer Degradation and Stability*, Vol.95, No.6, (June 2010), pp. 1106-1114, ISSN 0141-3910

El-Naggar, M.M.A. & Farag, M.Gh. (2010). Physical and biological treatments of polyethylene–rice starch plastic films. *Journal of Hazardous Materials*, Vol.176, No.1-3, (April 15 2010), pp. 878-883, ISSN 0304-3894

Feuilloley, P.; César, G.; Benguigui, L.; Grohens, Y.; Pillin, I.; Bewa, H.; Lefaux, S. & Jamal, M. (2005). Degradation of Polyethylene Designed for Agricultural Purposes. *Joural of Polymers and the Environment*, Vol.13, No.4, (October 2005), pp. 349-355, ISSN 1566-2543

Gilan (Orr), I.; Hadar, Y. & Sivan, A. (2004). Colonization, biofilm formation and biodegradation of polyethylene by a strain of *Rhodococcus ruber*. *Applied Microbiology and Biotechnology*, Vol.65, No.1, (July 2004), pp. 97-104, ISSN 0175-7598

González, G.P.; Hernando, P.F. & Alegría, J.S.D. (2006). A morphological study of molecularly imprinted polymers using the scanning electron microscope. *Analytica Chimica Acta*, Vol.557, No.1-2, (January 31 2006), pp.179-183, ISSN 0003-2670

Göpferich, A. (1996). Mechanisms of Polymer Degradation and Erosion. *Biomaterials*, Vol.17, No.2, pp.103-114, ISSN 0142-9612

Guise, O.; Strom, C. & Preschilla, N. (2011). STEM-in-SEM method for morphology analysis of polymer systems. *Polymer*, Vol.52, No.5, (March 1 2011), pp.1278-1285, ISSN 0032-3861

Huang, C-Y.; Roan, M-L.; Kuo, M-C. & Lu W-L. (2005). Effect of compatibiliser on the biodegradation and mechanical properties of high-content starch/low-density polyethylene blends. *Polymer Degradation and Stability*, Vol.90, No.1, (October 2005), pp.95-105, ISSN 0141-3910

Jakubowicz, I.; Yarahmadi, N. & Petersen, H. (2006). Evaluation of the rate of abiotic degradation of biodegradable polyethylene in various environments. *Polymer Degradation and Stability*, Vol.91, No.7, (July 2006), pp. 1556-1562, ISSN 0141-3910

Kim, H-S. & Kim, H-J. (2008). Enhanced hydrolysis resistance of biodegradable polymers and bio-composites. *Polymer Degradation and Stability*, Vol.93, No.8, (August 2008), pp. 1544-1553, ISSN 0141-3910

Kim, H-S.; Kim, H-J.; Lee, J-W. & Choi, I-G. (2006). Biodegradability of bio-flour filled biodegradable poly(butylene succinate) bio-composites in natural and compost soil. *Polymer Degradation and Stability*, Vol.91, No.5, (May 2006), pp.1117-1127, ISSN 0141-3910

Koutny, M.; Lemaire, J. & Delort A-M. (2006). Biodegradation of polyethylene films with prooxidant additives. *Chemosphere*, Vol.64, No.8, (August 2006), pp. 1243-1252, ISSN 0045-6535

Kyrikou, I. & Briassoulis, D. (2007). Biodegradation of Agricultural Plastic Films: A Critical Review. *Journal of Polymers and the Environment*, Vol.15, No.3, (July 2007), pp. 125-150, ISSN 1566-2543

Łabużek, S.; Nowak, B. & Pająk J. (2006). Biodegradation of Aged Composition of Polyethylene with Synthetic Polyester. *Polimery*, Vol.51, No.1, pp.27-32, ISSN 0032-2725

Łabużek, S.; Pająk, J. & Nowak, B. (2006). Characteristic of Polyethylene Films Modified with Polyester Bionolle® after Biodegradation by Filamentous Fungi. *Corrosion Protection*, No.9s/A, pp. 85-88, ISSN 0473-7733

Lee, W-K.; Ryou, J-H. & Ha, C-S. (2003). Retardation of enzymatic degradation of microbial polyesters using surface chemistry: effect of addition of non-degradable polymers. *Surface Science*, Vol.542, No.3, (September 20 2003), pp. 235-243, ISSN 0039-6028

Lucas, N.; Bienaime, C.; Belloy, C.; Queneudec, M.; Silvestre, F. & Nava-Saucedo J-E. (2008). Polymer biodegradation: Mechanisms and estimation techniques. *Chemosphere*, Vol.73, No.4, (September 2008), pp. 429-442, ISSN 0045-6535

Mark, H.F.; Bikales, N.M.; Overberger, C.G. & Menges, G. (1986). *Encyclopedia of polymer science and engineering*, Wiley Interscience Publication, ISBN 978-047-1440-26-0, New York, USA

Meenakshi, P.; Noorjahan, S.E; Rajini, R.; Venkateswarlu, U.; Rose, C. & Sastry, T.P. (2002). Mechanical and microstructure studies on the modification of CA film by blending with PS. *Bulletin of Materials Science*, Vol.25, No.1, (February 2002), pp. 25-29, ISSN 0022-2461

Müeller, R.J. (2006). Biological degradation of synthetic polyesters - enzymes as potential catalysts for polyester recycling. Process Biochemistry, Vol.41, No.10, (October 2006), pp. 2124-2128, ISSN 1359-5113

Nakamura, E.M.; Cordi, L.; Almeida, G.S.G.; Duran, N. & Mei, L.H.I. (2005). Study and development of LDPE/starch partially biodegradable compounds. *Journal of Materials Processing Technology*, Vol.162/163, Complete, (May 15 2005), pp. 236-241, ISSN 0924-0136

Nowak, B.; Pająk, J. & Łabużek, S. (2010). Biodegradation of Compositions of Poly(ethylene terephthalate) with Bionolle® or Starch. *Composites*, Vol.10, No.3, pp. 224-228, ISSN 1641-8611

Nowak, B.; Pająk, K.; Drozd-Bratkowicz, M. & Rymarz, G. (2011) Microorganisms participating in the biodegradation of modified polyethylene films in different soils under laboratory conditions. *International Biodeterioration & Biodegradation*, Vol.65, No.6, (September 2011), pp. 757-767, ISSN 0964-8305

O'Brine, T. & Thompson, R.C. (2010). Degradation of plastic carrier bags in the marine environment. *Marine Pollution Bulletin*, Vol.60, No.12, (December 2010), pp. 2279-2283, ISSN 0025-326X

Ohtake, Y.; Kobayashi, T.; Asabe H. & Murakami, N. (1998). Studies on Biodegradation of LDPE - Observation of LDPE Films Scattered in Agricultural Fields or in Garden Soil. *Polymer Degradation and Stability*, Vol.60, No.1, (April 1998), pp. 79-84, ISSN 0141-3910

Ojeda, T.; Freitas, A.; Birck, K.; Dalmolin, E.; Jacques, R.; Bento, F. & Camargo, F. (2011). Degradability of linear polyolefins under natural weathering. *Polymer Degradation and Stability*, Vol.96, No.4, (April 2011), pp. 703-707, ISSN 0141-3910

Ojeda, T.F.M.; Dalmolin, E.; Forte, M.M.C.; Jacques, R.J.S.; Bento, F.M. & Camargo, F.A.O. (2009). Abiotic and biotic degradation of oxo-biodegradable polyethylenes. *Polymer Degradation and Stability*, Vol.94, No.6, (June 2009), pp. 965-970, ISSN 0141-3910

Ołdak, D. & Kaczmarek, H. (2005). Photo- and bio-degradation processes in polyethylene, cellulose and their blends studied by ATR-FTIR and Raman spectroscopies. *Journal of Materials Science*, Vol.40, No.16, (August 2005), pp. 4189-4198, ISSN 0922-2461

Peacock, A.J. (2000). *Handbook of polyethylene e structures, properties and applications*, Marcel Dekker, ISBN 978-082-4795-46-7, New York, USA

Ram, A.; Meir, T. & Miltz, J. (1980). Durability of polyethylene films. *International Journal of Polymeric Materials*, Vol.8, No.4, pp.323-336, ISSN 0091-4037

Reddy, M.M.; Gupta, R.K.; Gupta, R.K.; Bhattacharya, S.N. & Parthasarathy, R. (2008). Abiotic Oxidation Studies of Oxo-biodegradable Polyethylene. *Journal of Polymers and the Environment*, Vol.16, No.1, (January 2008), pp. 27-34, ISSN 1566-2543

Rosa, D.S.; Gaboardi, F.; Guedes, C.G.F. & Calil, M.R. (2007). Influence of oxidized polyethylene wax (OPW) on the mechanical, thermal, morphological and biodegradation properties of PHB/LDPE blends. *Journal of Materials Science*, Vol.42, No.19, (October 2007), pp. 8093-8100, ISSN 0022-2461

Roy, P.K.; Titus, S.; Surekha, P.; Tulsi, E.; Deshmukh, C. & Rajagopal, C. (2008). Degradation of abiotically aged LDPE films containing pro-oxidant by bacterial consortium. *Polymer Degradation and Stability*, Vol.93, No.10, (October 2008), pp. 1917-1922, ISSN 0141-3910

Sahebnazar, Z.; Abbas, S.S.; Mahsa, M-T. & Mohsen, N. (2010). Biodegradation of low-density polyethylene (LDPE) by isolated fungi in solid waste medium. *Waste Management*, Vol.30, No.3, (March 2010), pp. 396-401, ISSN 0956-053X

Šašek, V.; Vitásek, J.; Chromcová, D.; Prokopová, I.; Brožek, J. & Náhlík, J. (2006). Biodegradation of Synthetic Polymers by Composting and Fungal Treatment. *Folia Microbiologica*, Vol.51, No.5, (September 2006), pp. 425-430, ISSN 0015-3632

Singh, B. & Sharma, N. (2008). Mechanistic implications of plastic degradation. *Polymer Degradation and Stability*, Vol.93, No.3, (March 2008), pp. 561-584, ISSN 0141-3910

Sivan, A. (2011). New perspectives in plastic biodegradation. *Current Opinion in Biotechnology*, Vol.22, No.3, (June 2011), pp. 422-426, ISSN 0958-1669

Soni, R.; Kapri, A.; Zaidi, M.G.H. & Goel, R. (2009). Comparative Biodegradation Studies of Non-poronized and Poronized LDPE Using Indigenous Microbial Consortium. *Journal of Polymers and the Environment*, Vol.17, No.4, (December 2009), pp. 233-239, ISSN 1566-2543

Sudhakar, M.; Doble, M.; Murthy, P.S. & Venkatesan, R. (2008). Marine microbe-mediated biodegradation of low- and high-density polyethylenes. *International Biodeterioration & Biodegradation*, Vol.61, No.3, (April 2008), pp. 203-213, ISSN 0964-8305

Thompson, R.C.; Swan, S.H.; Moore, C.J. & vom Saal, F. (2009). Our plastic age. *Philosophical Transactions of the Royal Society Biological Sciences*, Vol.364, No.1526, (July 2009), pp. 1973-1976, ISSN 1471-2970

Tserki, V.; Matzinos, P.; Pavlidou, E.; Vachliotis, D. & Panayiotou, C. (2006). Biodegradable aliphatic polyesters. Part I. Properties and biodegradation of poly(butylene succinate-*co*-butylene adipate). *Polymer Degradation and Stability*, Vol.91, No.2, (February 2006), pp.367-376, ISSN 0141-3910

Tsuji, H.; Echizen, Y. & Nishimura, Y. (2006). Photodegradation of biodegradable polyesters: A comprehensive study on poly(L-lactide) and poly(ε-caprolactone). *Polymer Degradation and Stability*, Vol.91, No.5, (May 2006), pp.1128-1137, ISSN 0141-3910

Viezie, D.L.; Thomas, E.L. & Adams, W.W. (1995). Low-voltage, high-resolution scanning electron microscopy: a new characterization technique for polymer morphology. *Polymer*, Vol.36, No.9, (April 1995), pp.1761-1779, ISSN 0032-3861

Wojcieszyńska, D.; Hupert-Kocurek, K.; Greń, I. & Guzik, U. (2011). High activity catechol 2,3-dioxygenase from the cresols - Degrading *Stenotrophomonas maltophilia* strain KB2. *International Biodeterioration & Biodegradation*, Vol.65, No.6, (September 2011), pp. 853-858, ISSN 0964-8305

33

Surface Analysis Studies on Polymer Electrolyte Membranes Using Scanning Electron Microscope and Atomic Force Microscope

M. Ulaganathan[1,2,*], R. Nithya[3] and S. Rajendran[1,*]
[1]*Energy Research Institute @ NTU, Nanyang Technological University, Singapore*
[2]*School of Physics, Alagappa University, Karaikudi, Tamil Nadu,*
[3]*Dept. of Physics, Shanmuganathan Engineering College,*
Arasampatti, Pudukkottai, Tamil Nadu,
[1]*Singapore*
[2,3]*India*

1. Introduction

A battery generally consists of three important parts namely, anode, cathode and electrolyte. The batteries further classified into primary and secondary batteries. Among the different kinds of batteries, Li-ion batteries are plays a very important role in the development of modern technologies especially in the portable electronic device industries and in heavy electrical vehicles because of its advantages such as high theoretical capacity, improved safety, lower material costs, ease of fabrication into flexible geometries, and the absence of electrolyte leakage. In the battery system different kinds of electrolytes were used for promoting the ions from anode to cathode (during charge) and cathode to anode (during discharge). For this purpose, liquid electrolyte is identified as suitable electrolytes which facilitate higher ionic conductivity (10^{-2} Scm^{-1}) than other electrolyte systems. However, it has several disadvantages namely gas formation during the operation, leakage, difficult to utilise for portable applications and etc. To overcome these difficulties, many attempts were made on solid polymer electrolyte systems. The main objective of the researchers is to improve the ambient temperature ionic conductivity, mechanical stability, thermal and interfacial stability of the electrolytes. However, it is difficult task for the researchers, inorder to improve these basic requirements of the electrolytes simultaneously because the ionic conductivity and mechanical strength of a polymer electrolyte are disparate to each other, i.e., mechanical strength of the electrolyte decreases as conductivity increases.

In recent years, Polymer electrolytes have been attracted scientific and technological importance because of their potential applications in many areas such as Li-ion polymer batteries, super capacitor, electro chromic devices and etc. The idea of preparation of

---

* Corresponding Authors

polymer electrolytes was first proposed by Wright and Fenton et.al in 1973 [wright et al., 1973] but their technological significances are fulfilled and appreciated by Armand et.al few years later [Armand et al., 1998]. Poly (ethylene oxide) was the first solvating polymer to be proposed and studied in solid polymer electrolyte (SPE) Li-rechargeable batteries. Most of the solid polymer electrolytes (SPEs) are prepared by dissolving lithium salts in a solvating polymer using common solvent or by diffusion in the solid (or) molten state. An effort is also made to fix or immobilize the anion on the polymer matrix by covalent bonding (or) another chemical or physical process.

In general, Polymer electrolytes are plastic materials that can be modified and processed by conventional techniques. If the polymer chains are helped for a charge transport of the ionic type, often called "polymer electrolyte". Solid polymer electrolytes (SPEs) afford two important roles in Li-ion battery. 1) It is used as a separator in the battery system because of it rigid structure, at the same time to avoid the electrical contact between the anode and the cathode; 2) it is the medium in which the ions are transported between the anode and cathode during the cell operations. As a result, the polymer electrolyte should act as good electrical insulator but at the same time it should has high ionic conductivity.

1.1 Classification of polymer electrolyte systems

The polymer electrolyte systems could be classified into three categories, namely, i) Polyelectrolyte, ii) Solvent swollen polymer electrolyte and iii) solvent free polymer electrolytes.

1.2.1 Polyelectrolyte's

Polyelectrolytes are polymers which have their own ion-generating groups chemically bound to the macromolecular chain and the presence of a counter-ion maintains the electroneutrality of the salt. This class of materials either positively or negatively charged ions covalently attached to the polymer backbone and therefore only the unattached counter ion has long range mobility. The conductivity of these polymers is very low (10^{-10}-10^{-15} Scm^{-1}) in dry conditions but hydrated polyelectrolytes achieve high conductivity in the presence of high dielectric constant solvent (e.g. Water). In hydrated polyelectrolytes, ionic transport takes place through the aqueous medium in which the polymer is dispersed. Slade et.al. [Slade et al., 1983] reported high ambient temperature conductivity of 10^{-2} Scm^{-1} in hydrated Nafion.

1.2.2 Solvent swollen polymer electrolytes

In solvent swollen polymers, solvents (aqueous/non-aqueous) swell the basic polymer host [like poly(vinyl alcohol) or poly(vinyl pyrrolidone)]. The dopant ionic solutes like H_3PO_4 are accommodated in the swollen lattice which permits the ionic motion in solvent rich swollen region of the polymer host. These materials are, in general, unstable and their conductivity depends on the concentration of the solvent in the swollen region. The properties of such polymers depend on the pre-treatment, structure of the sample, temperature, relative humidity, etc.

Examples: Nafian, complexes of poly(vinyl alcohol) (PVA) with H_3PO_4 [Polak et al., 1986] poly (silamine) with H_3PO_4[Rekukawa et al.,, 1996], PEI with H_3SO_4 [Daniel et al., 1988].

1.2.3 Solvent free polymer electrolytes

The polymer –salt complexes are formed by complexes between salts of alkali metals and polymer containing solvating heteroatoms such as O, N, S, etc. The most common examples are complexation between poly (ethylene oxide) (PEO) and alkali metal salts. The polymer salt complexes are further classified into: i) Solid polymer electrolytes ii) Blend polymer electrolytes iii) Gel polymer electrolytes iv) Composite polymer electrolytes. Among the various polymer electrolytes which are used in Li-ion batteries, solvent free polymer electrolytes are the most favourable for device fabrications. The solvent free polymer salt complexes are further classified into: i) Solid polymer electrolytes ii) Gel polymer electrolytes iii) Composite polymer electrolytes.

a) Solid polymer electrolytes

Solid polymer electrolytes (SPEs) have an ionic conductivity when modified by dissolving alkali salts in suitable polymer matrix. SPEs are typically thin films, which have a wide range of electrochemical applications such as batteries and electrochromic devices. They have several advantages when used in a battery and can be formed into thin films of large surface area giving high power levels. The flexibility of the films allows space-efficient batteries to be constructed [Quartarone et al., 1998].

b) Gel polymer electrolytes

Plasticizers incorporated polymer- salt complex is called gel polymer electrolytes. The addition of plasticizers into the polymer matrix softens the polymers and they increase free volumes which are used for ion migration. Addition of plasticizer also increases the chain flexibility, reduces crystallinity, decreases the glass transition temperature and hence increases the ionic conductivity. The conductivity of PEO: $LiBF_4$ is of the order of $x10^{-6}$ Scm^{-1} which has been increased to the order of $x10^{-4}Scm^{-1}$ when the complex is plasticized at 25°C [Chiodelli et al., 1988] this is mainly due to the specific nature of the plasticizers and the prepared gels has both the cohesive properties of solids and the diffusive property of liquids. Even though the gel polymer electrolyte exhibits high ionic conductivity, its thermal and mechanical stability are poor and it has higher reactivity towards the electrode. Gel electrolytes may undergo solvent exudation upon long storage, especially under open atmosphere conditions. This phenomenon is known as 'Synerisis effect', and has been encountered in many systems such as PAN: EC: PC: $LiClO_4$, PAN: EC: PC: $LiAsF_6$ [Groce et al., 1994 and Slane & Salomon et al., 1995].

c) Composite polymer electrolytes

This is another approach in which both the ionic conductivity and the mechanical stability of the electrolytes were considerably enhanced simultaneously. Composite polymer electrolytes are prepared by the addition of high surface inorganic fillers such as Al_2O_3, SiO_2, MgO, $LiAlO_2$, TiO_2, $BaTiO_3$ and Zeolite powders. The mechanical strength and stiffness of the complex systems were improved appreciably when the fillers are incorporated into the polymer matrix. However the main advantages of the composite electrolyte is the enhancement of room temperature ionic conductivity and an improved stability at the electrode electrolyte interface. The inert fillers due to its large surface area prevent the local chain reorganization with the result of locking in at ambient temperature, a high degree of disorder characteristic of the amorphous phase, which is more favour for the high ionic transport [G. Nagasubramanian and S. Di Stefano, 1990, Peter P Chu,P.P. Reddy, M.J., 2003]. The nano sized $BaTiO_3$ incorporated PEO composite electrolytes exhibits ionic conductivity of the order of x10^{-3} Scm^{-1} and good electrochemical stability (4.0V).

1.3 Blend polymer electrolytes

Blend is a mixture of two or more polymers. Mixing of two polymers is a well established strategy for the purpose of obtaining materials with combined superior properties or avoiding the need to synthesize novel structures constitutes an attractive research area. As many emerging applications are of limited volume and require specific property profiles not suitable for broad application utility, polymer blend technology is often the only viable method to deliver the desired material. These polymer blends have some unique properties that are different from the basic polymers from which these have been produced. To improve the processing behaviour for end use, one polymer blending with another polymer is a common practice. The exploitation of certain unique set of properties of individual polymer for the benefit of the overall properties of a multi component system forms the basis of polymer blending. Hence blending of polymers has resulted in the development of polymeric materials with desirable combination of properties. Polymer blend electrolytes are developed in such a way that they remain structurally stable during manufacturing, cell assembling, storage and usages as well as to prevent leakage from the cell container or without the cell.

1.4 Some commonly available polymer electrolytes for lithium ion batteries

1.4.1 PEO based electrolytes

PEO is a crystalline polymer. The oxygen in PEO acts as a donor for the cation and the anion generally of large dimension stabilizes the PEO alkali salt complex.

The polymer electrolytes composed of a blend of poly (ethylene oxide) (PEO) and poly (vinylidenefluoride-hexa fluoropropylene) as a host polymer, mixture of EC and PC as plasticizer and $LiClO_4$ as a salt were prepared by Fan et al. [Fan et al., 2002]. The ionic conductivity of various compositions of blend polymers was found to be in the order of x10^{-4} Scm^{-1} at 30°C. On increasing the PEO content in the matrix, the conductivity decreased due to its high crystalline nature. The mechanical strength of the polymer electrolytes was measured from stress-strain tests. The electrolytes were also characterized by SEM, XRD and thermal analysis techniques. Xi et al. [Xi., 2006] aimed to improve ionic conductivity with a novel approach using PEO and PVdF as host polymers by phase inversion technique. The room temperature conductivity was measured as a function of PEO content. As the weight ratio of PEO was increased from 40 to 50%, the ionic conductivity increased more than one magnitude from 0.15 to 1.96 x$10^{-1}$$Scm^{-1}$ which is mainly due to the increasing of pore connectivity. This is very important for the transport of charge carriers in microporous polymer membrane. The plasticizer effect on PEO-salt complex was studied by Fan et al [Fan et al., 2008] using succinitrile (SN) as a plasticizer, $LiClO_4$, $LiPF_6$ and $LiCF_3SO_3$ as lithium salts. They found that the addition of plasticizer was responsible for high ionic conductivity which could be attributed to the high polarity and diffusivity of succinitrile. This, in turn, decreased the crystallinity of PEO polymer. Activation energy of the electrolytes was also estimated from Arrhenius plot. Itoh et al. [Itoh et al., 2003] prepared the composite polymer electrolytes using poly(ethylene oxide)/ poly(triethylene glycol) benzoate, $BaTiO_3$ and lithium imides. They estimated the ionic conductivity value as $1.6x10^{-3}$ Scm^{-1} at 80°C and the electrochemical stability window as 4.0V. The membrane was also characterized by TG/DTA and it is thermally stable upto 307°C. Novel effect of organic acids such as malonic, maleic and carboxylic acids on $PEO/LiClO_4/Al_2O_3$ complexes was studied by Park et al. [Park et al., 2006]. It was noted that the ionic conductivity of the film consisting of $PEO/LiClO_4$: Citric acid (99.95:0.05 wt%) was $3.25x10^{-4}Scm^{-1}$ at 30°C and it was further improved to $3.1x10^{-3}$ Scm^{-1} at 20°C by adding 20wt% of Al_2O_3 filler. The prepared membranes were also characterized by Brewster Angle Microscopy (BAM), thermal analysis and cyclic voltametry.

1.4.2 PVC based electrolytes

PVdF/PVC blend composite polymer electrolyte was prepared by Aravindhan et al. [Aravindhan et al., 2007] incorporating lithium bis(oxalate) borate and ZrO_2. All the prepared membranes were subjected to SEM, XRD and ac impedance studies. The maximum ionic conductivity ($1.53x10^{-3}$ Scm^{-1}) was obtained for 2.5wt% of ZrO_2 at 343K. The ionic conductivity and FTIR studies on plasticized polymer electrolyte based on PVC and PMMA as host polymer were studied by Manuel Stephan et al. It was found that $LiBF_4$ based PVC/PMMA/EC/PC complexes exhibited higher ionic conductivity compared to that with $LiClO_4$. The thermal stability of the films was also ascertained using TG/DTA analysis.

1.4.3 PAN based polymer electrolytes

Kim et.al [Kim,D.W. Sun,Y.K. 2001] prepared highly porous polymer electrolyte employing P(VdF-co-HFP) and PAN with a view to attain high ionic conductivity and good mechanical strength. Lithium-ion polymer battery using these gel polymer electrolytes was assembled, and its charge-discharge characteristics were also reported. Panero et al. [Panero et al., 2002] studied the characteristics and the properties of a polymer electrolyte formed by trapping $LiPF_6$-PC solution in a poly(acrylonitrile) matrix with the addition of Al_2O_3. They reported the ionic conductivity value as $0.8x10^{-2}Scm^{-1}$ at 25°C. The performances of the electrolyte were found to be promising in terms of cycle life and basic energy density content. Very recently, Moreno et.al [Moreno et al., 2010] reported a series of composite electrolytes basically constituted by poly(acrylonitrile), Clay and montmorillonite as filler. The structural and complex formations of the CPEs were also studied. However, the composite based on PAN system showed poor ionic conductivity of the order of $x10^{-6}$ Scm^{-1}.

Tsutsumi and Kitagawa [Tsutsumi and Kitagawa et al., 2010] synthesized a new type polymer electrolyte films based on poly(acrylonitrile), and Cyanoethylated poly(vinyl alcohol) (CN-PVA) and its conductivity behaviour was also investigated. They found the ionic conductivity value as $14.6x10^{-3}Scm^{-1}$ at 30°C for PAN (10)-CN-PVA (10) - $LiClO_4$ (8)-PC (4) complex system. The interactions of Li^+-ion and nitrile groups of PAN in the matrix were confirmed by FTIR analysis. The ionic conductivity and FTIR studies were carried out on PAN based gel electrolytes with EC: PC and EC: DMC mixtures as plasticizers, $LiClO_4$ or $LiBF_4$ as the salt by Amaral et al [Amaral et al., 2007]. The high ionic conductivity ($1.47x10^{-3}$ Scm^{-1}) was estimated for 20:28:45:7 molar ratio of PAN-PVA: EC: DMC: $LiBF_4$ system.

Charge/discharge performance of the maximum ionic conductivity complex was also studied. The practical performance and thermal stability of Li-ion polymer batteries with $LiNi_{0.8}Co_{0.2}O_2$, mesocarbon microbead-based graphite, and poly (acrylonitrile) (PAN) based gel electrolytes were reported by Akashi et al. [Akashi et al., 2002].

1.4.4 PVdF based polymer electrolytes

PVdF is a semicrystalline polymer and the electrolytes based on PVdF are expected to have high anodic stabilities due to strong electron withdrawing functional groups (-C-F). It also has high permittivity, relatively low dissipation factor and high dielectric constant (ε=8.4) which assist in high ionization of lithium salts, providing a high concentration of charge carriers. Choe et al. [Cheo et al., 1995] reported that the PVdF based electrolytes plasticized with a solution of $LiN(SO_2CF_3)_2$ in PC had a conductivity of 1.74×10^{-3} Scm^{-1} at 30°C and has a oxidatively stable potential limits between 3.9 and 4.3V vs Li^+/Li. Nicotera et al. [Nicotera et al., 2006] measured the ionic conductivity and the lithium salt diffusion coefficient of PMMA/PVdF based blend electrolytes with EC/PC as plasticizers and lithium perchlorate as salt by the PFG-NMR method, which revealed maximum lithium mobility for the composition PMMA 60%-PVdF 40%. Raman spectroscopic study confirmed the change of interaction between the lithium cations and the plasticizer molecules for different PMMA/PVdF ratios. Wang et al. [Wang et al., 2007] prepared the nanocomposite polymer electrolytes comprising of poly (vinylidene fluoride) (PVdF) as a host polymer, lithium perchlorate ($LiClO_4$) as salt and TiO_2 used as a filler by solvent-casting technique. The prepared films were characterized by XRD, DSC and SEM. The conductivity value was found to be of the order of 10^{-3} Scm^{-1} for the sample with 10% TiO_2.

1.4.5 PMMA based electrolytes

Ali et al. [Ali et al., 2007] reported the electrical properties of polymer electrolytes comprising PMMA, PC, EC as plasticizer and different lithium salts $LiCF_3SO_3$ and $LiN(CF_3SO_2)_2$. The polymer electrolytes exhibited high ionic conductivity at room temperature in the range of 10^{-6} to 10^{-4} Scm^{-1}. The temperature dependence studies confirmed that the conduction in electrolyte is only by ions and seemed to obey the VTF rule. FTIR spectroscopy studies confirmed the polymer-salt interactions. FTIR spectroscopic investigations coupled with ionic conductivity and viscosity measurements on lithium imide $LiN(CF_3SO_2)_2$–propylene carbonate (PC)–poly(methyl methacrylate) (PMMA) based liquid and gel electrolytes over a wide range of salt (0.025–3 M) and polymer (5–25 wt.%) concentration were reported by Deepa et al. [Deepa et al., 2004] and found that the high ionic conductivity occurs at salt concentrations ≥1.25 M.

1.4.6 PVdF-co-HFP based polymer electrolytes

In recent years, the studies on PVdF-co-HFP based systems are electrochemically stable and indispensable for the electrode properties. The PVdF-co-HFP based electrolyte system shows high electrochemical stability in the range 4V. Fan et al. [Fan et al., 2002] studied the thermal, electrical and mechanical properties of $EC/PC/LiClO_4$ based PEO/P(VdF-co-HFP) blends. They concluded that the polymers have good compatibility and PVdF-HFP hinders the crystallinity of PEO. Saika and Kumar [Saika and Kumar, 2004] made systematic studies on the ionic conductivity and transport properties of polymer electrolytes comprising of copolymer PVdF-co-HFP/ PC/ DEC/ $LiClO_4$ and PVdF/ PC/ DEC/ $LiClO_4$ separately. The co-polymer complex showed higher ionic conductivity and transport number compared to PVdF system. The higher conductivity of the polymer electrolyte based on copolymer was attributed to its higher amorphousity. Wang et.al [Wang et al, 2004] reported that the polymer electrolyte composed of poly(methyl methacrylate-co-acrylonitrile-co-lithium methacrylate) (PMAML) and poly(vinylidene fluoride-co-hexafluoropropylene) (PVDF-co-HFP) exhibited high ionic conductivity and good electrochemical stability. The ionic conductivity was about 2.6×10^{-3} Scm^{-1} at ambient temperature and the electrochemical window of the polymer electrolyte was about 4.6V. Charge -discharge test results revealed that lithium ion batteries with these gel polymer electrolytes have good electrochemical performance.

Manuel Stephan et.al [Manuel Stephan et al., 2006] prepared the composite polymer electrolyte membranes, comprising Poly(vinylidene fluoride–hexafluoropropylene), Aluminum Oxyhydroxide of two different sizes 7nm/14 nm and $LiN(C_2F_5SO_2)_2$ as lithium salt and they found that the incorporation of the inert filler not only reduces the crystallinity of the polymer host but also acts as 'solid plasticizer' capable of enhancing the transport properties and also provides a better interfacial property towards lithium metal anode. Nam-Soon Choi et.al [Choi et al., 2001] reported that the interfacial stability between the polymer electrolyte and the lithium electrode was enhanced by blending PVAc with P(VdF-co-HFP)(Kynar 2801). The ionic conductivity of the polymer electrolyte based on the Kynar 2801: PVAc (7:3, w:w) blend was 2.3×10^{-3} S cm^{-1} at 25°C. Kim et.al [Kim et al., 2005] prepared and characterized gel polymer electrolytes consisting 25wt% PVdF-co-HFP/65 (EC+PC)/10 wt% $LiN(CF_3SO_2)_2$.They reported the ionic conductivity value as $1.2x10^{-3}Scm^{-1}$. The electrochemical stability window of the membrane was obtained at around 4.8V Vs Li/Li^+ using linear sweep voltametry technique. The Charge - discharge behaviour of the membrane was also studied and they estimated the specific discharge capacity as 140 $mAhg^{-1}$ for upto 20 cycles at different current densities.

1.4.7 PVAc based polymer electrolytes

Though variety of polymer electrolytes were characterized for the fast four decades, a limited number of studies were made on PVAc based polymer electrolytes. PVAc polymer has a large dipole moment and high relaxation time. Baskaran et al [Baskaran et al., 2006] prepared the polymer electrolyte comprising of PVAc-PMMA and reported the conductivity value as $1.76x10^{-3}Scm^{-1}$ at 303K. The DSC thermograms of the blend electrolytes showed two Tg's and they decreased with an increase of $LiClO_4$ concentration. The structural and complex formations of the electrolytes were confirmed by X-ray diffraction analysis. They established that the optimized blend ratio of PVAc: PMMA: $LiClO_4$ is suitable for lithium battery applications. Structural, thermal and transport properties of PVAc-$LiClO_4$ base complexes were studied by Baskaran et al [Baskaran et al., 2007]. The bulk conductivity of PVAc: $LiClO_4$ system was found to vary between $7.6x10^{-7}$Scm and $6.2x10^{-5}$ Scm^{-1} at 303K with an increase in the salt concentration. The amorphous nature of the polymer complexes was confirmed by XRD and SEM analysis.

Surface morphology and ionic conductivity of the membrane based on P(EO)/PVAc /$LiClO_4$ were studied by Animitsa et al [Animitsa et al., 1998]. They reported the conductivity value as 10^{-5} Scm^{-1} for lower concentration of PVAc at 25°C. Baskaran et al.

[Baskaran et al., 2006] studied the ac impedance and dielectric properties of PVdF/PVAc blend electrolytes. They reported that the blend ratio (75:25) of PVAc/PVdF exhibited a maximum ionic conductivity value of the order of $6.4x10^{-4}Scm^{-1}$ at 343K. The ionic transference number of mobile ions was also estimated by Wagner's polarization method. The complex formation and thermal behaviour of the electrolytes were also studied by FTIR and DSC analysis respectively.

2. Basic requirements of polymers and the salt for polymer electrolytes

Atoms or groups of atoms with sufficient electron donor power to form coordinate bonds with cations.

- Low barriers to bond rotation so that segmental motion of the polymer chain can take place readily.
- A suitable distance between coordinating centres for the formation of multiple intrapolymers ion bonds appears to be important
- Low glass transition temperature to increase the segmental motion.
- The polymer should have amorphous phase which lowers the barrier for ionic movement and yields high ionic conductivity.
- The lattice energy of the salt should be low.
- High electrochemical reduction potential

Electron pair donicity (DN)

DN measures the ability of the solvent to donate electrons to solvate the cations considered as a Lewis acid. So the polymer host should have high DN number.

Acceptor number (AN)

The acceptor number quantifies the possibility of anion salvation. It should be less for an inorganic salt so that cationic salvation is high compared to anionic salvation.

Entropy term

The entropy term depends on the optimal spatial disposition of the solvating units which should be high for the polymer host.

2.1 Importance of Li^+ cation

Last four decades, many alkali salts consisting Li^+, Na^+, K^+, Ag^+, Mg^+, NH_4^+ cations were mixed with the polymers (PEO), (PPO) etc in the preparation of polymer electrolytes. Among the various cations in the periodic table, Li^+ is the most electropositive. Lithium easily gives up electrons to form a positive Li^+ which has small ionic radii (0.6Å). Lithium is promising candidate for high energy density batteries because of its high specific capacity of 3860 Ah/Kg, its light weight and high electrochemical reduction potential [Scrosati et al., 1994; Abraham et al., 1993; Dell, 2000].

2.2 Preparation of polymer blend electrolytes

Polymer blend electrolytes have been prepared using various approaches namely, Phase inversion method, Hot pressed method, Solvent casting method and etc. Each method has own advantages and disadvantages. Among the various method of preparation solvent costing is the cost effective one and easier to control the moistures during the preparation of electrolytes as well as in the cell assembling. However, when assembling the cells, these membranes show poor cyclic behaviour. As a result many research groups have been mainly focused on phase inversion method for preparing the suitable electrolytes. Preparation of polymer electrolyte system is explained using the following flow chart (Fig.1.).

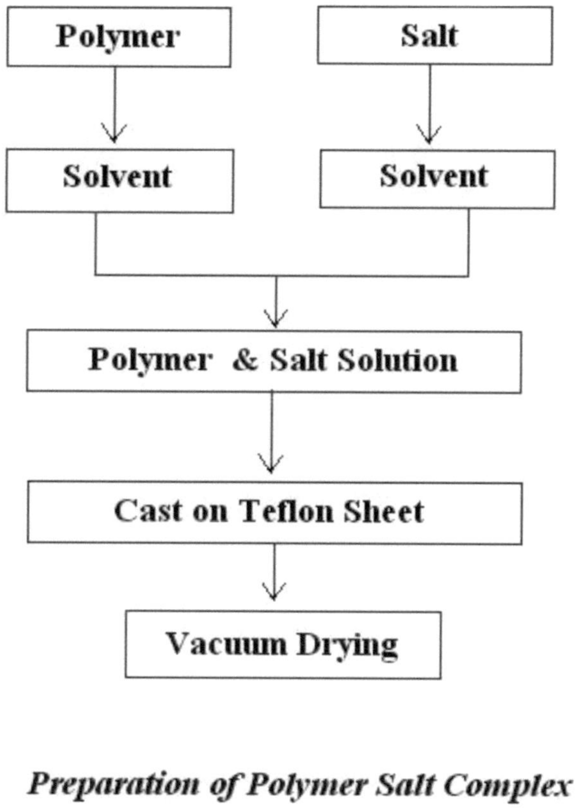

Fig. 1. Preparation of Polymer electrolyte

In the present study, all the electrolytes were prepared using solvent casting technique. The polymers PVAc, PVdF-co-HFP, and the salt $LiClO_4$ were dissolved in a common solvent (tetrahydrofuran) separately. All three solutions were mixed together and starrier continuously using magnetic stirrer until got a homogeneous mixture. The low molecular weight plasticizer and the inorganic fillers were added into the matrix inorder to get the gel and composite polymer electrolytes systems. In the present study, ethylene carbonate(EC) and barium titanate ($BaTiO_3$) were used a plasticizer and fillers respectively. Thus the obtained homogenous slurry was degassed to remove air bubbles for about five minutes and the slurry was poured on a well cleaned glass plate. The casted slurry was allowed to evaporate the solution at room temperature about 5h followed by the electrolyte membranes were heated using hot air oven at a temperature of 60 °C for 6h in order to removing the residual solvent present in the electrolyte films. Finally, the harvested electrolyte films were stored in highly evacuated desiccators to avoid the moistures absorption.

3. Characterization of polymer blend electrolytes

Ac impedance analysis was carried out with the help of stainless steel blocking electrodes by using a computer controlled micro auto lab type III Potentiostat/Galvanostat of frequency range 1 Hz–300 KHz in the temperature range 303–373 K. The XRD equipment used in this study was X'pert PRO PANlytical X-ray diffractometer. FTIR spectroscopy studies were carried out using SPECTRA RXI, Perkin Elmer spectro-photometer in the range 400–4000 cm^{-1}. FTIR spectroscopy studies were carried out for confirming the complex formation using SPECTRA RXI, Perkin-Elmer spectro-photometer in the range 400–4000 cm^{-1}. TG/DTA thermal analysis of the film having maximum ionic conductivity was studied using PYRIS DIAMOND under air atmosphere with the scan rate of 10 ◦C min−1. The electrolyte film having maximum ionic conductivity was subjected to atomic force microscopy [model Veeco-diCP-II]. The pore size and the root mean square (rms) roughness of the film were measured from the topography image. Secondary electron images of the sample were examined by using JEOL, JSM-840A scanning electron microscope

3.1 FTIR analysis

Infrared spectral (IR) analysis is a powerful tool for identifying the nature of bonding and different functional groups present in a sample by monitoring the vibrational energy levels of the molecules, which are essentially the fingerprint of different molecules [Nagatomo et al., 1987]. Fig.2. depicts the FTIR transmittance spectra in the range 400-4000 cm^{-1} for polymers, the $LiClO_4$ salt, the blend electrolyte with the incorporation of plasticizer ethylene carbonate and the filler $BaTiO_3$.

The vibrational bands observed at 2933 cm^{-1} and 2465 cm^{-1} are ascribed to $-CH_3$ asymmetric and symmetric stretching vibrations of PVAc respectively. The strong absorbance at 1734 cm^{-1} represents the C=O stretching vibration mode of PVAc polymer. The existence of C-O band has been confirmed by the strong absorbance band at around 1033 cm^{-1}. The strong band at 1373 cm^{-1} is ascribed to $-CH_3$ symmetric bending vibration of pure PVAc. The band at 1243 cm^{-1} is assigned to C-O-C symmetric stretching mode of vibration. The peak at 947 cm^{-1} is ascribed to CH bending vibration and the peak at 609 cm^{-1} is assumed to be linked with CH_3 (C-O) group. The C-H wagging mode of vibration has been confirmed by the presence of a band at 799 cm^{-1} [Baskaran et al., 2004]. The vibrational peaks at 502 and 416 cm^{-1} are assigned to bending and wagging vibrations of $-CF_2$ of PVdF-co-HFP polymer respectively. Crystalline phase of the PVdF-co-HFP polymer is identified by the vibrational bands at 985, 763, and 608 cm^{-1} and the amorphous phase of the co-polymer is confirmed by the presence of vibrational band at 872 cm^{-1} [Rajendran et al., 2002].

The strong absorption peak appeared at 1173 cm^{-1} is assigned to the symmetrical stretching of $-CF_2$ group. The peak appeared at 1390 cm^{-1} is assigned to the CH_2 groups [Rajendran et al., 2002; Singh Missan et al., 2006]. Table.1. shows the comparison of band spectra of pure polymers and their blends with different blend ratios. From the table, it is clear that the band assignments of FTIR spectra of blend samples are shifted from their pure spectra. For all the blends, some peaks are found above 3000 cm^{-1}, which correspond to the C-H stretching vibration modes of blend electrolytes.

In addition, some new peaks are present and some of them are absent in the blend electrolytes. Thus the spectral analysis confirms the complexation of these two polymers and lithium salt.

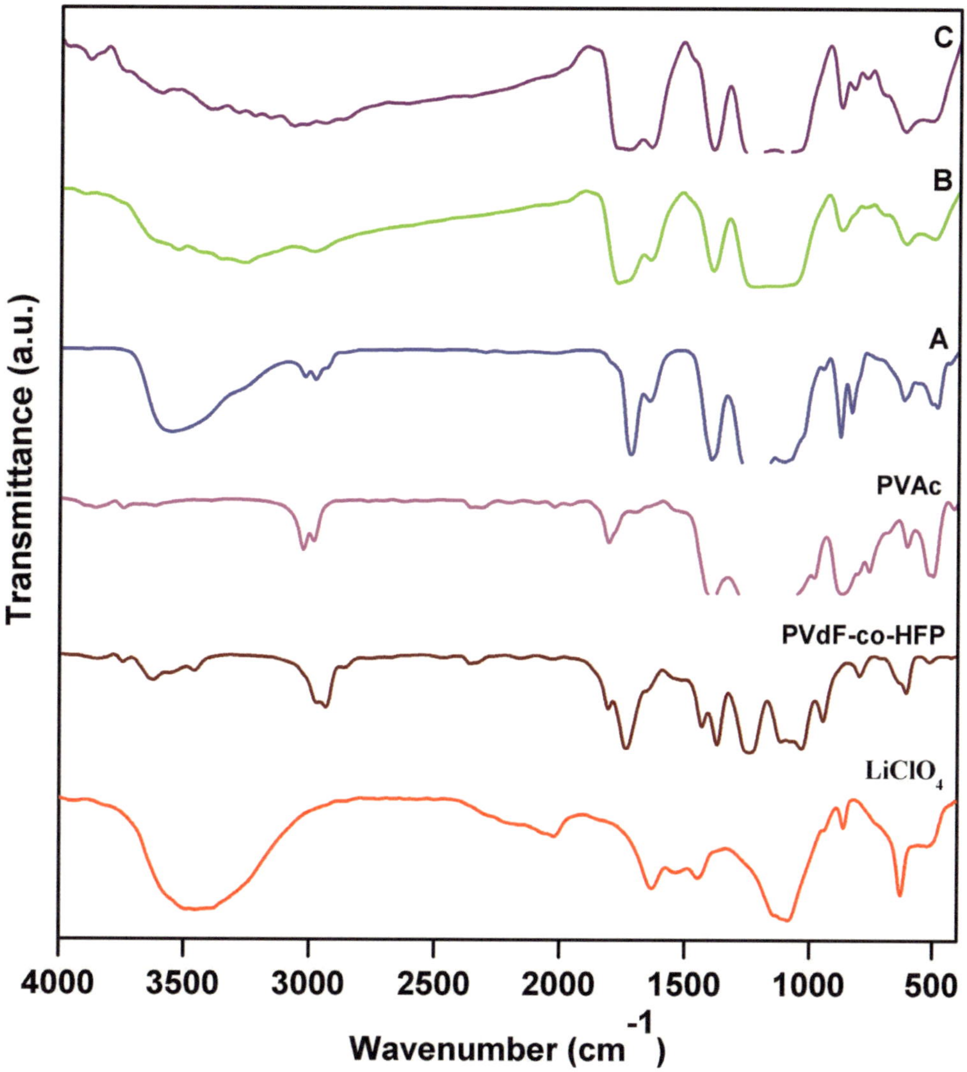

Fig. 2. FTIR analysis of $LiClO_4$, PVAc, PVdF-co-HFP and their complexes

3.2 X-ray diffraction analysis

XRD patterns of $LiClO_4$, PVdF-co-HFP, PVAc and their complexes A, B, C are shown in Fig. 3. The presence of characteristic peaks corresponding to the lithium salt reveals the high crystalline nature of the salt. Three peaks found at 2θ = 17.3, 18.59 and 38.78° for PVdF-co-

HFP confirms the partial crystallization of PVdF units present in the copolymer, to give an overall semi-crystalline morphology for PVdF-co-HFP [Saika, Kumar, 2004]. Presence of broad humps in the XRD pattern of PVAc confirms the complete amorphous nature of the polymer. It is observed that the characteristic peaks corresponding to the lithium salt in their respective electrolyte systems (A, B and C) were absent and it confirms the complete dissolution of the lithium salts in the complex matrix which implies that the salt do not have any separate phase in the electrolytes. The addition of plasticizer in the blend complex enhances the amorphous region thus permitting the free flow of ions from one site to another site; hence the overall ionic conductivity of the electrolyte has been significantly improved. According to Hodge et al. [Hodge et al., 1996] the ionic conduction in the polymer electrolytes occurs mostly in the amorphous region and it has been achieved by the addition of low molecular weight plasticizer. Further addition of inorganic filler into the polymer salt matrix would increase the dissolution of the charge carriers in the matrix; hence, the ionic conductivity was improved. The XRD pattern of the sample contains the filler $BaTiO_3$ shows a broad hump confirms the further enhancement of the amorphous region in the polymer electrolyte complex systems.

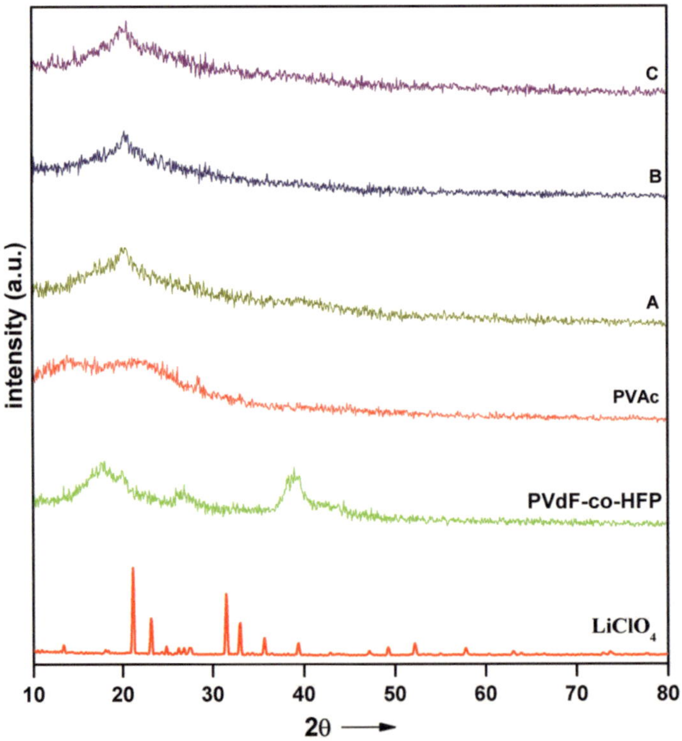

Fig. 3. X-ray diffraction analysis of $LiClO_4$, PVdF-co-HFP, PVAc and their complexes.

3.3 Ac impedance analysis

Ac studies are similar to the DC techniques in that the ratio of voltage to current is measured. For DC, this ratio provides the value of the resistance, R, measured in ohms.

For AC the ratio gives an analogous quantity, the impedance, Z, also measured in ohms. The impedance contains four main contributions; these are from resistance, capacitance, constant phase elements, and inductance. The induction is not an important factor for the polymer electrolytes although it can play a role in other electrochemical applications of polymers.

Measurement of the impedance as a function of frequency is called impedance spectroscopy. In general, impedance is complex quantity, in which the real and the imaginary parts are labelled Z′ and Z″ respectively. In the complex impedance plot, the real quantity Z′ (X-axis) is plotted against Z″ (Y-axis) which displayed the polymer electrolytes characteristics as an arc followed by the linear spike is straight line inclined to the real axis. From the plotted graph, we can easily read the bulk resistance of the electrolyte system.

The complex impedance plot of the PVAc/PVdF-co-HFP/$LiClO_4$ electrolyte is shown in Fig.4a. Figure shows the semicircular portion which is mainly due to the parallel combination of the geometrical capacitance, C_g and the bulk resistance, R. When adding the plasticizer and the filler (Fig.4b.) into the electrolyte matrix, the impedance spectra shows only a linear spike which corresponds to the lower frequency region. It confirms the idea that the current carriers are ions and the majority of the conduction only by the ions not by the electrons. And the disappearance of the semicircular portion is may be due to the fact the corresponding characteristic frequency is higher than the frequency 300kHz and it is mainly depends on the instrument limit. From the obtained bulk resistance value, we can estimate the ionic conductivity value of the electrolyte system using the relation $\sigma = l/R_b A$, where, R_b is the bulk resistance of the electrolyte film, A is the area of the electrode surface and is the thickness of the electrolyte medium and σ is the ionic conductivity.

It is noted from the spectra that the addition of plasticizer (ethylene carbonate) in to the polymer salt matrix greatly being reduced the bulk resistance of the system; this is because of the high dielectric nature of the low molecular weight plasticizer. The addition of plasticizer would considerably enhance the amorphous phase of the polymer electrolyte which will improve the ionic conductivity of the system. However, the gain in conductivity is adversely associated with a loss of the mechanical properties and by a loss of the compatibility with the lithium electrode, both effects resulting in serious problems since they affect the battery cycle life and increase the safety hazard. So it is necessary to identify the solid additives which would not affect the mechanical stability and interfacial stability of the electrolyte, at the same time will enhances the ionic conductivity. The addition of solid additives should improve the amorphicity of the electrolyte at room temperature. The addition of solid additives in the present study, such as nano filler $BaTiO_3$ highly being enhanced the amourphicity of the electrolyte medium, hence the room temperature ionic conductivity and the interfacial stability of the electrode-electrolyte interface is increased. The ceramic dispersed electrolyte shows good thermal stability. The $BaTiO_3$ incorporated sample is thermally stable up to 320 °C. The temperature dependence of the conductivity is given by the Vogel–Tamman–Fulcher (VTF) equation $\sigma = \sigma_0 \exp(-B/T\text{-}T_0)$, where σ_0 is the pre-exponential factor, B should not be confused with an activation energy in the Arrhenius expression and T0 is related to the so called thermodynamic Tg. Plots of logs vs 1/T are curved because of the reduced temperature (T-T0).

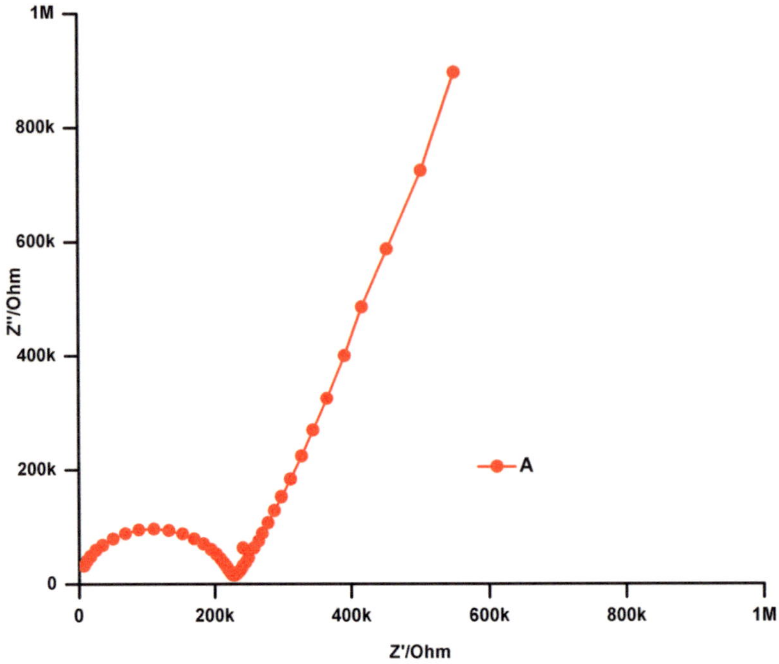

Fig. 4a. Room temperature complex ac impedance spectrum of PVAc/PVdF-co-HFP/$LiClO_4$electrolyte

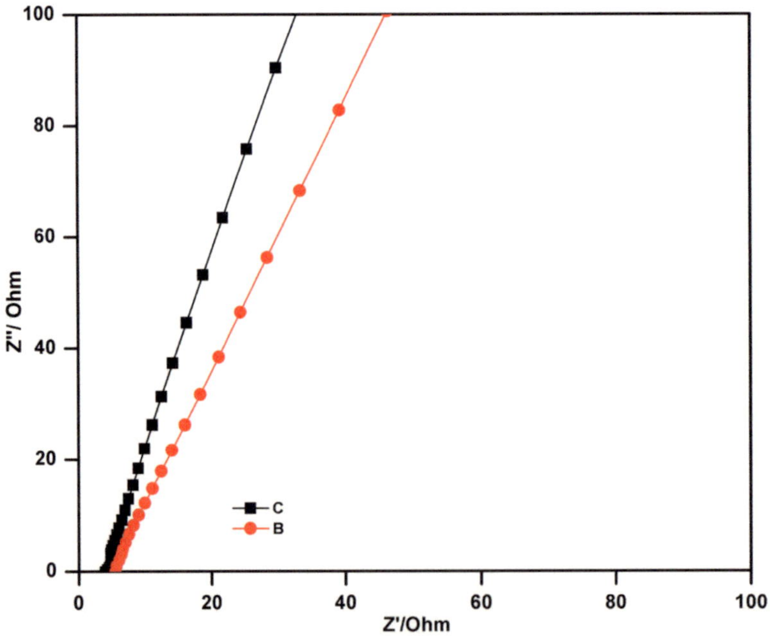

Fig. 4b. Room temperature complex ac impedance spectra of PVAc/PVdF-co-HFP/$LiClO_4$/EC and PVAc/PVdF-co-HFP/$LiClO_4$/EC+PC/$BaTiO_3$ electrolyte

3.4 TG/DTA analysis

Thermo gravimetric analysis / differential thermal analysis have been used widely to study all physical processes involving the weight changes. It is also used to investigate the thermal degradation, phase transitions and crystallization of polymers. Fig.5(a-c) shows the TG/DTA curves of PVAc/PVdF-co-HFP/$LiClO_4$, PVAc/PVdF-co-HFP/$LiClO_4$/EC and PVAc/PVdF-co-HFP/$LiClO_4$/EC+PC/$BaTiO_3$ polymer electrolytes. From the thermogram, it is observed that the sample A is thermally stable up to 238°C. The sample starts to decompose at 238°C, beyond which, there is a gradual weight loss of 20% in the temperature range 240-280°C. DTA curve of the sample shows an exothermic peak at 265°C, which is well correlated with the weight loss of the sample observed in TG curve. It is also observed that the complete decomposition of the sample takes place between 490-510°C with the corresponding weight loss of about 80-90%. After 520°C, there is no appreciable weight loss (Fig.5a). The remaining residue around 15% may be due to the formation of impure crystalline metal oxide and lithium fluoride. It is also observed from DTA curves that the exothermic peaks at 90, 225, and 445 °C are concurrent with the weight losses observed in the TG trace. The sample (Fig.5b.) exhibit gradual weight loss of about 10-15%, which is due to the removal of the residual solvent and the moisture from the electrolyte sample in the temperature range 90-115 °C. From the TG curve of the sample PVAc/PVdF-co-HFP/$LiClO_4$/EC, it is observed that the decomposition occurred at 229°C with the weight loss of about 20%. After the second decomposition, there is sudden weight loss of 40-45% in the temperature range 446-460 °C for the electrolyte. The thermogram of the sample having $BaTiO_3$ inert filler is shown Fig.5c. From the themogram, it is observed that the sample is thermally stable up to 320°C. The sample exhibits gradual weight loss of about 8% in the temperature range 100-110 °C, which is due to the removal of the residual solvent and the moisture. DTA curve of the sample shows an exothermic peak at 320°C, which is well correlated with the weight loss of the sample observed in the TG curve. The remaining residue around 30% may be due to the presence of $BaTiO_3$. It is noted from the above analysis that the additions of plasticizer into the polymer blend-salt matrix slightly influence the thermal stability of the electrolyte medium; however it has enhanced the ionic conductivity of the electrolyte. But, the addition of nano composite in to the matrix greatly being increased the thermal stability and the room temperature conductivity simultaneously. It is concluded that the incorporation/dispersion of inorganic filler in the electrolyte significantly increased the thermal stability of the electrolyte membrane.

3.5 SEM analysis

The scanning electron microscope (SEM) is one of the most versatile instruments available for the examination and analysis of the microstructure morphology of the conducting surfaces. Scanning electron microscope (SEM) image of PVAc-$LiClO_4$, PVdF-co-HFP-$LiClO_4$, PVAc/PVdF-co-HFP/$LiClO_4$, PVAc/PVdF-co-HFP/$LiClO_4$/EC and PVAc/PVdF-co-HFP/$LiClO_4$ /EC+PC/$BaTiO_3$ electrolyte films are shown in Fig .6(a-e). Fig.6a. clearly shows smooth and uniform surface morphology of the PVAc- $LiClO_4$. This smooth morphology confirms the complete amorphous nature of PVAc polymer and complete dissolution of the lithium salt, which also coincides with the XRD result. Fig.6b shows the photograph of PVdF-co-HFP-$LiClO_4$ salt complex with maximum number of pores giving rise to high ionic conductivity. Presence of the spherical grains in the microstructure image

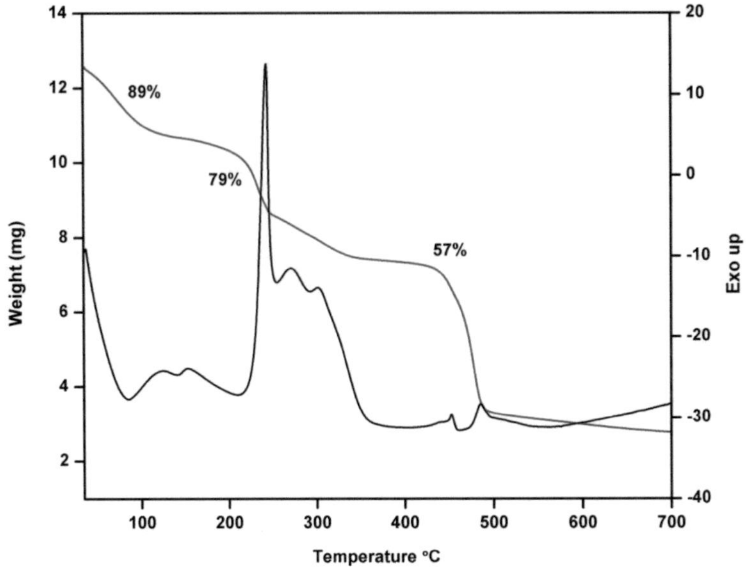

(a)

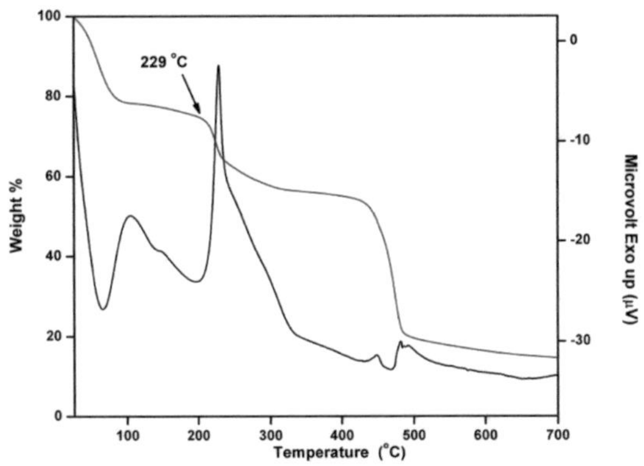

(b)

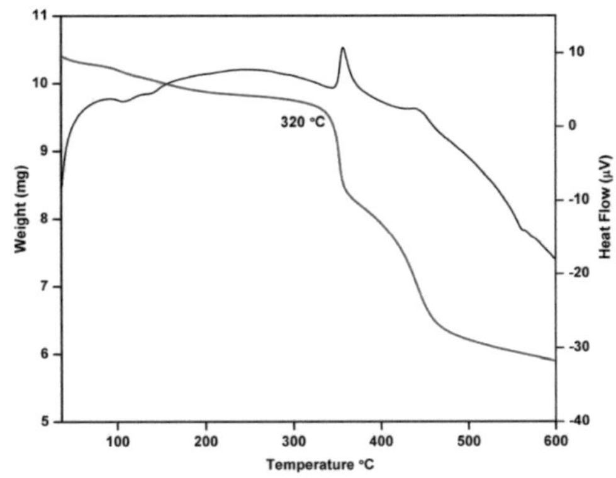

(c)

Fig. 5. TG/DTA Thermal analysis of a) PVAc/PVdF-co-HFP/$LiClO_4$; b) PVAc/PVdF-co-HFP/$LiClO_4$/EC; c) PVAc/PVdF-co-HFP/$LiClO_4$/EC+PC/$BaTiO_3$

of the samples A and B (Fig.6b and d) are belongs to the co-polymer and it means that the copolymer do not dissolve completely in the matrix which results, the membrane gets brittle nature. The appearance of number of uniform tracks of few micrometer sizes is responsible for the appreciable ionic conductivity of the electrolyte (Fig.6c). The maximum ionic conductivity of the polymer blend electrolyte also depends on the segmental motion of the PVAc and PVdF-co-HFP. The better miscibility of these two polymers can be depicted from the microstructural photograph. Fig.6d shows the scanning electron micrographs of PVAc/PVdF–co-HFP/LiClO4/EC-based electrolyte system. The micrograph shows the spherical grains, and they are uniformly distributed in the electrolyte system. It is observed that the numerous pores (dark region) with the size of 1–10µm are responsible for the high conductivity of the sample, i.e., the membrane shows highly porous structure. This increased number of porosity leads to entrapment of large volumes of the liquid in the pores accounting for the increased conductivity. The interconnected microspores in the membrane helped in absorbing liquid electrolytes and hence the ionic conductivity of the membrane is enhanced. The presence of pores in the microstructure is mainly due to the solvent removal and increased amorphous region and solvent retention ability in the electrolyte system. Surface images of the samples $BaTiO_3$ is shown in Fig.6e. The pores in the complexes are responsible for entrapping the large volume of the solution (plasticizer +salt) in the cavities accounting for the enhanced ionic conductivity. It is observed from image that the membrane has numerous number of randomly distributed spherical grains and shows maximum number of pores with very small size of the order of 50-100nm. The smooth surface of the sample reveals that the polymers and salt used in this study have a good compatible nature and the light gray region indicates the presence of plasticizer rich medium which assists for ionic motion. It is also studied that the content $BaTiO_3$ increases beyond certain percentage the film surface becomes rough above the optimum level the grain size increases, with a reduction in the number of grain aggregates, that tend to restrict the ionic movement. Finally, the SEM photograph of the polymer electrolyte indicates good compatibility of these two polymers and the other constituents which are used in the electrolyte preparation. The enhancement of the amorphous region in the matrix has also been confirmed from the images. The miscibility of these two polymers has also been confirmed from FTIR analysis.

3.6 AFM analysis

An AFM is a mechanical imaging instrument, which is used to obtain the three dimensional topography images of the samples. In the present study, the scanning probe spectroscopic method was used to measure the pore size of the prepared sample as well as the roughnees factor of the sample. The two dimensional and three dimensional topography images of PVAc/PVdF-co-HFP/$LiClO_4$ complex are shown in Fig.7a. This image clearly shows the presence of pores within the scanned area of 3x3μm and the measured pore size of the sample is approximately 600nm. The size of the chain segment is also obtained and it is in the order of 688 nm. The root mean square (rms) roughness of the topography image over the scanned area is found to be 122 nm. The topography image of the sample PVAc/PVdF-co-HFP/$LiClO_4$/EC is shown in Fig.7b. The two dimensional image shows smooth surface. The modified surface image of the sample is mainly due to the addition of plasticizer which increases the amorphous phase of the matrix and hence the ionic conductivity. In addition, small pores are also observed in the surface entrapping the liquid solution, which are responsible for easy ionic movement. From the topography image we have determined the pore size of the order of 100nm which is in close agreement with the value obtained from SEM photograph. In addition, the rms roughness of the sample over the scanned area 1.4x1.4μm has been estimated and it is of the order of 53nm, it is quiet low when compare with sample without plasticizers. The micropores, amorphous phase and the chain segments of the plasticized polymer electrolytes are responsible for the enhancement of ionic conductivity. Two and three dimensional topographic images of the sample having $BaTiO_3$ are shown in Fig.7c. The image shows the dispersion of the fillers and it also contains small pore with a size of 100nm entrapping the ionic liquid which assists for fast ionic motion. In addition, the rms roughness of the sample over the scanned area 1x1μm has been obtained and it is of the order of 4nm. It is noted that the sample contains $BaTiO_3$ showed lower roughness value than the other two samples which means that the incorporation of the fillers and the plasticizers are significantly improve d the amorphous phase in the matrix is helpful for the ionic movement.

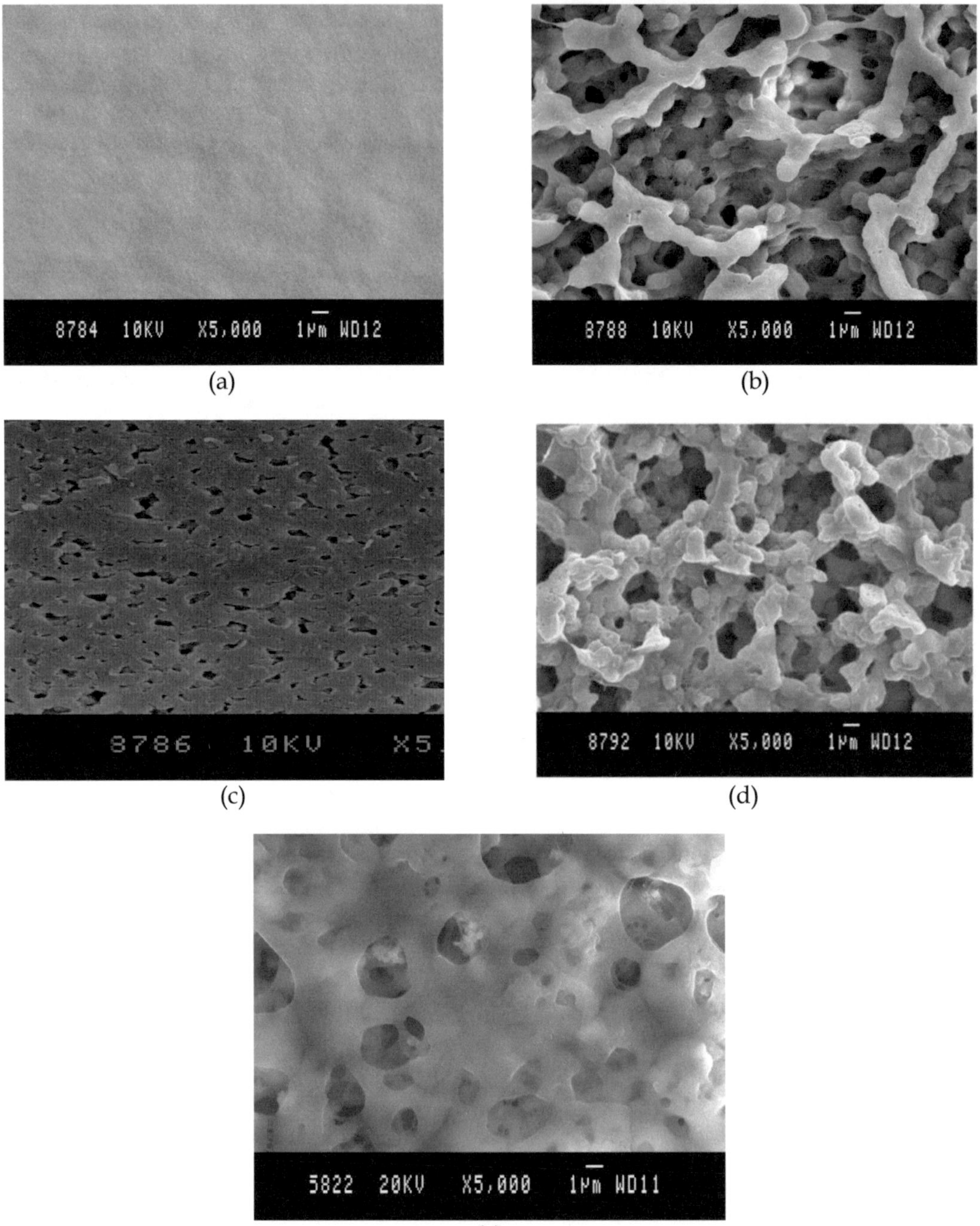

Fig. 6. SEM images of a) PVAc/$LiClO_4$; b) PVdF-co-HFP-$LiClO_4$; c) PVAc/PVdF-co-HFP/$LiClO_4$;d) PVAc/PVdF-co-HFP/$LiClO_4$/EC; e) PVAc/PVdF-co-HFP/$LiClO_4$/EC+PC

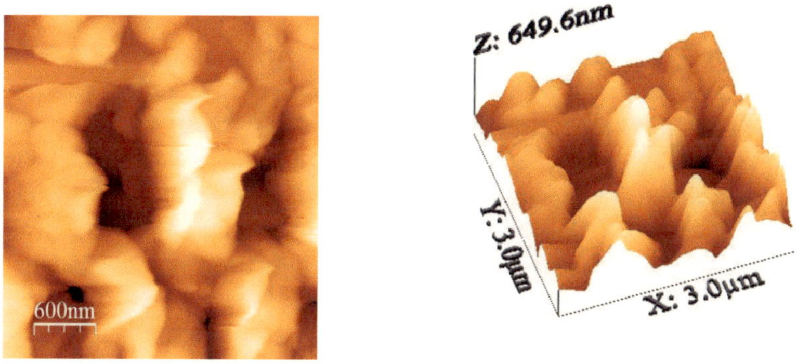

(a)

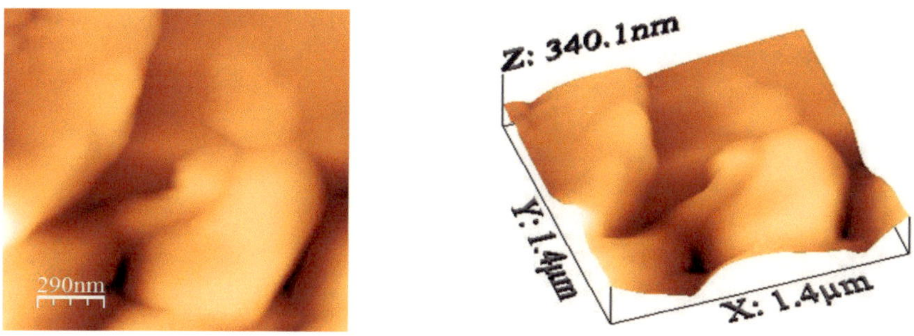

(b)

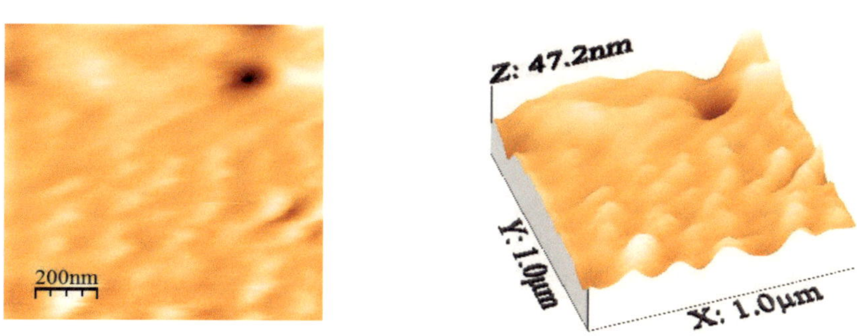

(c)

Fig. 7. AFM images of a) PVAc/$LiClO_4$; b) PVdF-co-HFP-$LiClO_4$; c) PVAc/PVdF-co-HFP/$LiClO_4$; d) PVAc/PVdF-co-HFP/$LiClO_4$/EC; e) PVAc/PVdF-co-HFP/$LiClO_4$/EC+PC

4. Conclusion

All the polymer electrolytes were prepared using solvent costing technique. The specific interactions of the constituents were confirmed using FTIR analysis. The enhanced amorphous region of the polymer electrolyte has been identified from X-ray diffraction analysis. The porous natures of the samples were identified using scanning electron microscope. The Atomic force microscope study was used to estimate the roughness factors of the scanned area. The thermal stability of the electrolyte samples were estimated using TG/DTA analysis. It is concluded that the addition of plasticizer (Ethylene Carbonate) and the dispersion of inorganic filler into the PVAc/PVdF-co-HFP/$LiClO_4$ electrolyte system significantly improve the amourphicity of the medium, which will helped for easy ionic motion. These enhanced regions have been confirmed from the impedance and the surface image studies. The change in bulk resistance of the electrolytes mainly due to the interactions of the basic constituents which cause produce more amorphous phase in the matrix. It is no doubt about that the addition of plasticizers and the fillers are greatly enhanced the amorphous phase of the electrolyte, hence the ionic conductivity is improved. To conclude, polymer electrolyte for possible applications as in high energy density batteries has been identified in terms of parameters such as conductivity, thermal stability.

5. References

Abraham, K.M. Brummer, S.B. "Secondary Lithium Cells": "Lithium Batteries"Editor, Gabano, J.P. Academic Press, London (1983).

Akashi, H. Shibuya, M. Orui, K. Shibamoto, G. Sekai, K. Practical performances of Li-ion polymer batteries with $LiNi_{0.8}Co_{0.2}O_2$, MCMB, and PAN-based gel electrolyte, *J.Power Sources*, Vol.112, No.2, (November 2002), pp.577-582, ISSN 0378-7753

Ali, A.M.M. Yahya, M.Z.A. Bahron, H. Subban, R.H.Y. Harun, M.K. Atan.I. Impedance studies on plasticized PMMA-LiX [X: $CF_3SO_3^-$, $N(CF_3SO_2)_2^-$] polymer electrolytes, *Mater.Lett.* Vol.61, No.10, (April 2007), pp.2026-2029, ISSN 0167-577X

Amaral, F.A. Dalmolin, C. Canobre, S.C.Nerilso Bocchi, Rocha-Filho, R.C. Biaggio, S.R. Electrochemical and physical properties of poly(acrylonitrile)/poly(vinyl acetate)-based gel electrolytes for lithium ion batteries, *J.Power Sources*, Vol.164, No.1, (January 2007), pp.379-385, ISSN 0378-7753

Animitsa, I.E. Kruglyashov, A.L. Bushkova, O.V. Zhukovsky, V.M. Morphology and ionic conductivity of poly(ethylene oxide)-poly(vinyl acetate)-$LiClO_4$ polymer electrolytes, *Solid State Ionics*, Vol.106, No.3-4, (February 1998), pp.321-327 , ISSN 0167-2738

Aravindan, V. Vickraman, P. Prem Kumar, T. ZrO_2 nanofiller incorporated PVC/PVdF blend-based composite polymer electrolytes (CPE) complexed with LiBOB, *J.Membr.Sci.*, Vol.305, No.1-2, (November2007), pp.146-151, ISSN 0376-7388

Armand, M.B, Chabango, J.M, Duclot, M (1998), Second International Meeting on Solid Electrolytes, Standee's, Scotland, Extended Absts.

B.Scrosati, "Insertion Compounds for Lithium Rocking Chair Batteries-the Electrochemistry of Novel materials" Editors: Lipkowski and Ross.VCH, Newyork (1994).

Baskaran, R. Selvasekarapandian, S. Kuwata, N. Kawamura, J. Hattori, T. Conductivity and thermal studies of blend polymer electrolytes based on PVAc–PMMA, *Solid State Ionics*, Vol.177, No.26-32, (October 2006), pp.2679-2682 , ISSN 0167-2738

Baskaran, R. Selvasekarapandian, S. Kuwata, N. Kawamura, J. Hattori, T. Structure, thermal and transport properties of PVAc-$LiClO_4$ solid polymer electrolytes, *J.Phys.Chem.Solids*, Vol.68, No.3, (March 2007), pp.40-412, ISSN 0022-3697

Chiodelli, G. Ferloni, P. Magistris, A. Sanesi, M, Ionic conduction and thermal properties of poly (ethylene oxide)-lithium tetrafluoroborate films, *Solid State Ionics*, Vol.28-30, No. 2, (September 1988), pp.1009-1013, ISSN 0167-2738

Choi, N-S. Lee, Y-G. Park, J-K. Ko, J.M. Preparation and electrochemical characteristics of plasticized polymer electrolytes based upon a P(VdF-co-HFP)/PVAc blend, *Electrochimi.Acta*, Vol.46, No.10-11, (March 2001), pp. 1581-1586, ISSN 0013-4686

D.E. Fenton, J.M. Parker, P.V. Wright Complexes of alkali metal ions with poly (ethylene oxide). *Polymer*, Vol.14, No. 11 (November 1973), pp-589. ISSN 0032-3861

Daniel, M.F. Desbat, B. Lassegues, J.C. Solid state protonic conductors: complexation of poly(ethylene oxide) or poly(acrylic acid) with NH_4HSO_4 , *Solid State Ionics*, Vol. 28-30, No. 1, (September 1988), pp.632-636, ISSN 0167-2738

Deepa, M. Agnihotry, S.A. Gupta, D. Chandra, R. Ion-pairing effects and ion–solvent–polymer interactions in $LiN(CF_3SO_2)_2$–PC–PMMA electrolytes: a FTIR study, *Electrochimi. Acta*, Vol.49, No.3, (January 2004), pp.373-383, ISSN 0013-4686

Dell.R.M. Batteries: fifty years of materials development, *Solid State Ionics*, Vol.134, No.(1-2), (October 2000), pp. 139-158, ISSN 0167-2738

Fan, L. Dang, Z. Nan, C-W, Li, M.Thermal, electrical and mechanical properties of plasticized polymer electrolytes based on PEO/P(VDF-HFP) blends, *Electrochimi.Acta*, Vol.48, No. 2, 21 (November 2002), pp.205-209, ISSN 0013-4686

Fan, L-Z. Wang, X-L. Long, F. Wang, X. Enhanced ionic conductivities in composite polymer electrolytes by using succinonitrile as a plasticizer, *Solid State Ionics*, Vol.179, No.(27-32), (September 2008), pp.1772-1775, ISSN 0167-2738

G. Nagasubramanian and S. Di Stefano 12-Crown-4 Ether-Assisted Enhancement of Ionic Conductivity and Interfacial Kinetics in Polyethylene Oxide Electrolytes, *J. Electrochem. Soc.*, Vol.137 (December 1990), pp.3830-3835, ISSN 1945-7111

Groce, F. Gerace, F. Dautzemberg, G. Passerini, S. Appetecchi, G.B. Scrosati, B. Synthesis and characterization of highly conducting gel electrolytes, *Electrochimi. Acta*, Vol.39, No.14, (October 1994), pp.2187-2194, ISSN 0013-4686

H.S. Choe, J. Giaccai, M. Alamgir, K.M. Abraham, Preparation and characterization of poly(vinyl sulfone)- and poly(vinylidene fluoride)-based electrolytes, *Electrochimi. Acta*, Vol.40, No.13-14, (October 1995), pp.2289-2293, ISSN 0013-4686

Hodge, R.M. Edward, G.H. Simon, G.P. Water absorption and states of water in semicrystalline poly(vinyl alcohol) films, *Polymer*, Vol.37, No. (1996) 1371, ISSN 0032-3861

Itoh, T. Miyamura, Y. Ichikawa, Y. Uno, T. Kubo, M.Yamamoto, O.Composite polymer electrolytes of poly(ethylene oxide)/$BaTiO_3$/Li salt with hyperbranched polymer, *J.Power Sources*, 119-121, (June 2003), pp. 403-408, ISSN 0378-7753

Kim, D.W. Sun, Y.K. Electrochemical characterization of gel polymer electrolytes prepared with porous membranes, *J.Power Sources*, Vol.102, No.1-2, (December 2001), pp.41-45, ISSN 0378-7753

Kim, H-S. Periasamy, P. Moon, S.I. Electrochemical properties of the Li-ion polymer batteries with P(VdF-co-HFP)-based gel polymer electrolyte, *J.Power Sources*, Vol.141, No.2, (March 2005), pp.293-297, ISSN 0378-7753

Li, Z. Su, G. Wang, X. Gao, D. Micro-porous P(VDF-HFP)-based polymer electrolyte filled with Al_2O_3 nanoparticles, *Solid State Ionics*, Vol.176, NO.(23-24), (July 2005), pp. 1903-1908, ISSN 0167-2738

Lizhen Fan, Zhimin Dang, Ce-Wen Nan, Ming Li Thermal, electrical and mechanical properties of plasticized polymer electrolytes based on PEO/P(VDF-HFP) blends, *Electrochimi.Acta*, Vol.48, No.2, (November 2002), pp.205-209, ISSN 0013-4686

Manuel Stephan, A. Nahm, K.S. Prem Kumar, T. Anbu Kulandainathan, M Ravi, G. Wilson, J. Nanofiller incorporated poly(vinylidene fluoride–hexafluoropropylene) (PVdF–HFP) composite electrolytes for lithium batteries, *J.Power Sources*, Vol.159, No.2, (September 2006), pp.1316-1321, ISSN 0378-7753

Moreno, M. Santa Ana, M.A. Gonzalez, G. Benavente, E.Poly (acrylonitrile)-montmorillonite nanocomposites: Effects of the intercalation of the filler on the conductivity of composite polymer electrolytes, *Electrochimi.Acta*, Vol.55, No.4, (January 2010), pp. 1323-1327, ISSN 0013-4686

Nagatomo, T. Ichikawa, C. Omoto, O.All-Plastic Batteries with Polyacetylene Electrodes, *J. Electrochem. Soc.* Vol.134, (February 1987), pp.305-308, ISSN 1945-7111

Nicotera, I. Coppola, L. Oliviero, C. Castriota, M. Cazzanelli, E. Investigation of ionic conduction and mechanical properties of PMMA–PVdF blend-based polymer electrolytes, *Solid State Ionics*, Vol.177, No.5-6, (February 2006), pp.581-588, ISSN 0167-2738

Panero, S. Satolli, D. D'Epifano, A. and Scrosati, B. High Voltage Lithium Polymer Cells Using a PAN-Based Composite Electrolyte, J. Electrochem. Soc., 149 (February 2002), pp.A414-A417,. ISSN 1945-7111.

Park, J.W. Jeong, E.D. Won, M-S. Shim, Y.B. Effect of organic acids and nano-sized ceramic doping on PEO-based olid polymer electrolytes, *J.Power sources*, Vol.160, No.1, (September 2006), pp. 674-680, ISSN 0378-7753

Peter P Chu, P.P. Reddy, M.J. Sm_2O_3 composite PEO solid polymer electrolyte, *J.Power Sources*, Vol.115, No.2, (April 2003), pp.288-294, ISSN 0378-7753

Polak, A.J. Petty-weeks, S. Beuhler, A.J. Applications of novel proton-conducting polymers to hydrogen sensing, *Sensors and Actuators*, Vol.9, No.1, (February 1986), pp. 1-7, ISSN 0924-4247

Quartarone, E. Mustarelli, P. Magistris, A. PEO-based composite polymer electrolytes, *Solid State Ionics*, Vol.110, No.1-2, (July 1998), pp.1-14, ISSN 0167-2738

R Baskaran, S Selvasekarapandian, G Hirankumar, M.S Bhuvaneswari, Vibrational, ac impedance and dielectric spectroscopic studies of poly(vinylacetate)-N, N-dimethylformamide-$LiClO_4$ polymer gel electrolytes, *J.Power Sources*, Vol.134, No.2(August 2004), pp.235-240, ISSN 0378-7753

R. Baskaran, S. Selvasekarapandian, N. Kuwata, J. Kawamura, T. Hattori ac impedance, DSC and FT-IR investigations on (x)PVAc-(1 − x)PVdF blends with $LiClO_4$, *Mater.Chem.Phys*, Vol.98, No.1, (July 2006), pp. 55-61, ISSN 0254-0584

Rajendran, S. Mahendran, O. Kannan, R. Lithium ion conduction in plasticized PMMA-PVdF polymer blend electrolytesm, Mater.Chem.Phys, Vol.74, No.1, (February 2002), pp.52-57, ISSN 0254-0584

Rajendran, S. Mahendran, O. Mahalingam.T. Thermal and ionic conductivity studies of plasticized PMMA/PVdF blend polymer electrolytes, *Euro.Polym.J.* Vol.38, No.1, (January 2002), pp. 49-55, ISSN 0014-3057

Rekukawa, M. Kawahara, M. Sanui, K. Ogata, N. in extended abstracts, 12'th Intl.Conf.on Solid State Ionics (Greece, 1996b), pp-777.

Saikia, D. Kumar, A. Ionic conduction in P(VDF-HFP)/PVDF-(PC + DEC)-$LiClO_4$ polymer gel electrolytes, *Electrochimi.Acta*, Vol.49, No.16, (July 2004), pp. 2581-2589, ISSN 0013-4686

Saikia, D. Kumar, A. Ionic conduction in P(VDF-HFP)/PVDF-(PC + DEC)-$LiClO_4$ polymer gel electrolytes, *Electrochimi.Acta*, Vol.49, No.16, (July 2004), pp.2581-2589, ISSN 0013-4686

Singh Missan, H.P. Chu, P.P. Sekhon, S.S. Ion conduction mechanism in non-aqueous polymer electrolytes based on oxalic acid: Effect of plasticizer and polymer, *J.Power Sources*, Vol.158, No.2, (August 2006), pp. 1472-1479, ISSN 0378-7753

Slade, R.C.T. Hardwick, A. Dickens, P.G. Investigation of H^+ motion in NAFION film by pulsed ^1H NMR and A.C. conductivity measurements, *Solid State Ionics*, Vol.9-10, No.2, (December 1983), pp. 1093-1098.

Slane, S.Salomon, M. Composite gel electrolyte for rechargeable lithium batteries, *J.Power Sources Vol.*55, No.1, (May 1995), pp.7-10, ISSN 0378-7753

Tsutsumi, H. Kitagawa, T. High ionic conductive behavior of cyanoethylated polyvinylalcohol- and polyacrylonitrile-based electrolytes, *Solid State Ionics*, Vol.177. No. 26-32, (October 2006), pp. 2683-2686, ISSN 0167-2738

Wang, Y-J. Dukjoon Kim, D.Crystallinity, morphology, mechanical properties and conductivity study of in situ formed PVdF/$LiClO_4$/TiO_2 nano composite polymer electrolytes, *Electrochimi.Acta*, Vol.52, No. 9, (February 2007), pp.3181-3189, ISSN 0013-4686

Wang, Z-L Tang, Z-Y. A novel polymer electrolyte based on PMAML/PVDF-HFP blend, *Electrochimi.Acta*, Vol.49, No.7, (March 2004), pp.1063-1068, ISSN 0013-4686

Xi, J. Qiu, X. Li, J. Tang, X. Zhu, W. Chen, L. PVDF–PEO blends based microporous polymer electrolyte: Effect of PEO on pore configurations and ionic conductivity, *J.Power Sources*, Vol.157, No.1, (June 2006), pp.501-506, ISSN 0378-7753

34

Characterization of Ceramic Materials Synthesized by Mechanosynthesis for Energy Applications

Claudia A. Cortés-Escobedo[1], Félix Sánchez-De Jesús[2,*], Gabriel Torres-Villaseñor[3], Juan Muñoz-Saldaña[4] and Ana M. Bolarín-Miró[2]

[1]Centro de Investigación e Innovación Tecnológica del IPN,
[2]Universidad Autónoma del Estado de Hidalgo-AACTyM,
[3]Instituto de Investigaciones en Materiales-UNAM,
[4]Centro de Investigación y Estudios Avanzados del IPN, Unidad Querétaro,
México

1. Introduction

The close relationship between processing, structure and properties of materials is well known. Some of the most useful tools to elucidate the best choice in processing for a given application are scanning electron microscopy (SEM), transmission electron microscopy (TEM), scanning transmission electron microscopy (STEM), selected area electron diffraction (SAED) and x-ray diffraction (XRD). In this chapter we will focus on the application of these techniques to the characterization of ceramic materials processed by mechanosynthesis, evaluating the effect of the milling process on their physical properties.

The ceramics that are the focus of this chapter, lanthanum manganites, have a Perovskite-structure (ABO_3), with the general chemical formula $R_{1-x}A_xMnO_3$ (where R^{3+} = La, and A^{2+} = Ca and Sr). Perovskite structures have been extensively studied for almost 50 years (Coey & Viret, 1999). Since the initial discovery of their electrical and magnetic properties, the interest in these compounds has remained high, and they have been the focus of significant scientific activity throughout the past decade. This kind of ceramic material has a wide variety of applications due to its ionic conduction, magnetic, thermal and mechanical properties, etc. Furthermore, it is known that the physical properties of Perovskite manganites depend on many factors, such as external pressure (Hwang & Palstra, 1995; Neumeier et al., 1995), magnetic field (Asamitsu et al., 1995; Kuwahara et al., 1995), structure (Tokura et al., 1994) and chemical composition (Hwang et al., 1995; Mahesh et al., 1995; Schiffer et al., 1995). For example, the ionic conduction of lanthanum manganates has led to their use in oxygen sensors and solid oxide fuel cells (Shu et al., 2009). Their ionic conduction is due to punctual defects, in the form of oxygen vacancies, in the crystalline structure of the Perovskite lattice. The generation of punctual defects is activated by the addition of doping atoms, as well as changing synthesis precursors and by applying mechanical energy, all of which are typical processes going on during the high-energy ball-milling process. These effects are described in detail in this chapter.

* Corresponding Author

There are several methods to synthesize lanthanum manganites. The traditional method is via solid-state reaction of the components (mixed oxides route). Controlling the temperature during the solid state reaction is the major challenge to obtaining homogeneity in the stoichiometry, grain size, porosity and purity. Alternatively, chemical methods such as sol-gel (Zhou et al., 2010), solution combustion (Shinde et al., 2010), co-precipitation (Uskokovic & Drofenik, 2007), and others (Jafari et al., 2010) have been used, resulting in lanthanum manganites with a wide variety of physical properties.

Another technique, high-energy ball-milling, used to promote mechanosynthesis of nanostructured manganites, such as $La_{1-x}Ca_xMnO_3$, by mechanical activation of chloride and oxide compounds, has shown excellent results (Muroi et al., 2000; Bolarín et al., 2006, 2007). Zhang et al. (Zhang & Saito, 2000, Zhang et al., 2000) synthesized $LaMnO_3$ Perovskites at room temperature by milling a mixture of Mn_2O_3 and La_2O_3 powders using a planetary ball mill. Additionally, K. Sato et al. (Sato et al., 2006) used an alternative mechanical synthesis route to produce fine $LaMnO_3$ powder - compression and shear stress were repeatedly applied to a mixture of La_2O_3 and Mn_3O_4 using an attrition type milling apparatus.

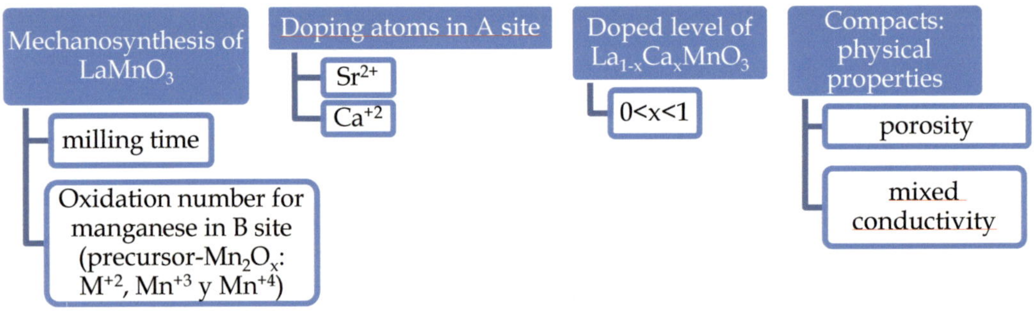

Fig. 1. Variables used to maximize punctual defects, and therefore ionic conduction in doped lanthanum manganites.

The chapter is divided in four main parts. In the first, details of the mechanosynthesis of lanthanum manganites, showing the process parameters that are necessary to carry out the synthesis of this ceramic material without a posterior annealing treatment are presented. The effect of these process parameters on the crystalline structure of the Perovskite-type manganites and mixed ionic-electronic conductivity will be presented and discussed (Figure 1). Specifically, the synthesis of lanthanum manganese oxide ($LaMnO_3$) by solid-state reaction using high-energy reactive ball milling from manganese oxides (MnO, Mn_2O_3, MnO_2) mixed with lanthanum oxide in stoichiometric ratios, for different milling times is reported. This is a useful route used to maximize punctual defects while preserving the initial structure by using manganese oxides with different oxidation numbers for manganese (Mn^{2+}, Mn^{3+} and Mn^{4+}). The mechanical energy of the mechanosynthesis processs induces a higher level of intrinsic defects than can be created by other synthesis methods (Cortés-Escobedo et al., 2008). In addition, a comparative study of mechanosynthesis with the solid-state reaction at high temperature is presented. In this section, we demonstrate the usefulness of electronic microscopy and x-ray diffraction to provide a structural description of the synthesized material.

In the second part of the chapter, the effects of doping on the structure and physical properties of lanthanum manganite is emphasized, using Ca^{+2} and Sr^{+2} as doping atoms in

A-sites of the Perovskite structure. Results of crystal structure analyses of calcium-doped lanthanum manganites, $La_{1-x}Ca_xMnO_3$ have been previously reported. The calcium-to-lanthanum ratio x was varied from 0 to 1 in increments of 0.1, allowing the study of changes in crystal structure with different degrees of calcium substitution, from $LaMnO_3$ (x=0) to $CaMnO_3$ (x=1) (Lira-Hernández et al., 2010). Here we present a study to increase the calcium content while maintaining the orthorhombic in order to obtain the best ion-electronic conduction. The repercussions of the doping on the microstructural characteristics were studied by means of XRD and TEM /SAED.

The third part of the chapter is dedicated to the process of consolidating manganite powder in order to see how the compaction and sintering process affects the physical properties of the ceramic. A study of the relationship between the crystalline structure, the mixed ionic-electronic conductivity and calcium content in calcium-doped lanthanum manganites, $La_{1-x}Ca_xMnO_3$, is presented.

Finally, in the fourth part of the chapter, the interaction between lanthanum manganites and yttria-stabilized zirconia, which are used as cathode and electrolyte, respectively, in both solid oxide fuel cells and sensors is discussed. A reduction in ionic conductivity is observed at certain temperatures, due to the formation of a very stable pyrochlore phase (lanthanum zirconate). Diffusion from manganite to zirconia to form zirconate has been well demonstrated by backscattered electron microscopy, showing the microstructure of cubic zirconia, in contrast with tetragonal zirconia, lanthanum manganite and lanthanum zirconate.

2. Theoretical aspects of mechanochemical processing (MCP)

Mechanochemical processing (MCP), or mechanosynthesis, is a synthesis method that uses mechanical energy (e.g., ball milling) to activate chemical reactions and structural changes in powder mixtures. In some cases, milling is followed by a low-temperature heat treatment to complete the reaction. In this study, lanthanum manganites were produced from oxide precursors La_2O_3, Mn_2O_x and CaO or SrO by means of high-energy ball millling, with a subsequent heat treatment.

Lanthanum manganites, $La_{1-x}M_xMnO_3$ ($0 < x < 1$), where M was Ca^{+2} or Sr^{+2}, were prepared using a SPEX 8000 D high-energy ball mill. Powders of Mn_2O_x and La_2O_3 were mixed in stoichiometric proportion for obtaining different level of doped manganite. MCP was carried out at room temperature in air atmosphere, in hardened-steel vials with steel balls as milling elements; ball-to-powder weight ratio was 10:1 at different milling times. X-ray diffraction (XRD) was used to evaluate phase transformations as function of milling time. Morphology and particle size of the manganites were characterized by scanning electron microscopy (SEM). Particle size distribution was measured by a zeta size analyzer. Selected area electron diffraction (SAED) patterns, obtained using a transmission electron microscope, were indexed to identify the resultant crystalline phase. Rietveld refinements of the XRD patterns were performed with the purpose of identifying and quantifying the phases and to identify the structural changes in the starting oxides until the formation of manganites.

For applications as sensors, as well as cathodes in solid oxide fuel cells, ionic conductivity of lanthanum manganites is desired. However, ionic conductivity is favored by punctual defects in the structure of the Perovskite ceramic. For magnetic applications, it has been discussed that punctual defects have an effect on the colossal magnetoresistance of the material. For these reasons, punctual defects were induced, firstly by changing the oxidation number of manganese in precursor manganese oxide in Mn^{+2}, Mn^{+3} and Mn^{+4}, it is to say, by inducing intrinsic defects and secondly by inserting calcium or strontium atoms which will suit in A sites (in place of lanthanum), inducing extrinsic defects in the atomic structure.

3. Defects in ceramic materials

In general, properties of solid ionics and in particular ceramic materials depend on crystalline structure and chemical bonding, and sometimes these properties can be improved by means of the processing or synthesis route employed. One of the ways to improve mechanical, electrical, magnetic, and structural properties is by inducing defects in crystalline structure. Defects are deviations from ideality in crystals and can be volumetric (3-dimensional), surface (2-dimensional), or punctual (1-dimensional). Punctual defects in ionic solids can be intrinsic, this is, due to disorder in atomic arrangement. Cations may be shifted to interstitial sites (Frenkel defect) or there may be a lack of an anion-cation pair (Schottky defect), both cases leading to a vacancy. Punctual defects can also be extrinsic, due to a substitution of an atom with another that has a different size and shell configuration, resulting in vacancy generation to maintain electroneutrality. The presence of intrinsic or extrinsic defects in ceramics promotes mobility of ions driven by local electrical charges present in vacancies, improving ionic conduction properties.

3.1 Intrinsic defects in ceramic materials

Intrinsic defects can be promoted by different synthesis or processing routes. One way to promote intrinsic defects is synthesizing by mechanical milling. This process generates nanometric-sized particles and disorder at the atomic level. It is important to consider that vacancies in the crystalline structure can also be induced by using precursors composed of cations with non-stoichiometric oxidation numbers.

With the intention of promoting intrinsic defects, a stoichiometric mixture of MnO and La_2O_3 were milled to follow the reaction:

$$La_2O_3 + 2MnO \xrightarrow{MCP} 2LaMnO_3 + 2Mn'_{Mn} + 2V_O^{\cdot} \quad (1)$$

After 630 min of milling, the x-ray diffraction pattern (Figure 2) reveals the amorphization of the precursor mixture. But after 600 min of heat treatment at 1000°C, orthorhombic $LaMnO_3$ is formed. This can be represented by the following reaction:

$$0.5La_2O_3 + MnO + 0.5O_2 \overset{MCP/630\,min}{\rightarrow} (0.5La_2O_3 + MnO)_{fine-grained} + 0.5O_2 \overset{1000°C/600\,min}{\rightarrow} LaMnO_3 \quad (2)$$

A stoichiometric mixture of precursors Mn_2O_3 and La_2O_3 were milled from 0 to 540 min, and x-ray diffraction patterns are shown in Figure 3. Formation of lanthanum manganite can be detected after 60 min, and the precursors were completely consumed after 210 min of milling, following the reaction:

$$
\begin{aligned}
0.5La_2O_3 + 0.5Mn_2O_3 &\xrightarrow{MCP/30\,\min} (La_2O_3 + Mn_2O_3)_{fine\ grained} \\
&\xrightarrow{MCP/60\,\min} LaMnO_3 + (La_2O_3 + Mn_2O_3)^{remnant}_{fine\ grained} \\
LaMnO_3 + (La_2O_3 + Mn_2O_3)^{remnant}_{fine\ grained} &\xrightarrow{MCP/210\,\min} LaMnO_3
\end{aligned} \tag{3}
$$

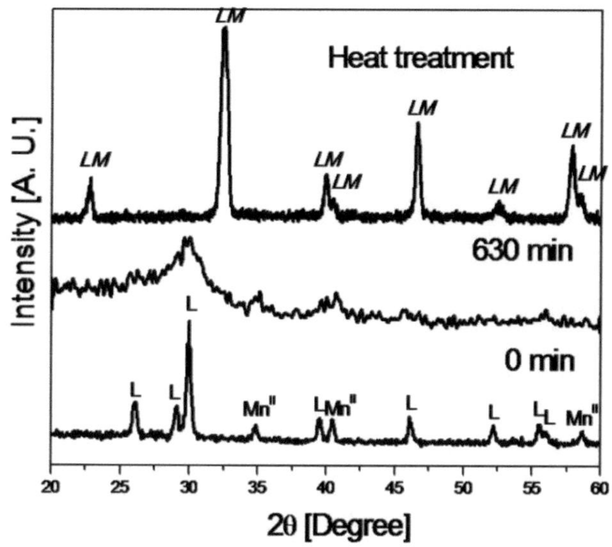

Fig. 2. X-ray diffraction patterns obtained from La_2O_3 + MnO powder mixtures before milling, milled for 630 min, and heat treated at 1000°C for 600 min (L: La_2O_3, JCPDS No. 05-0602; M^{II}: MnO, JCPDS No. 07-0230; LM: $La_{1-x}Mn_{1-z}O_3$). The pattern for Mn^{II}+L milled for 630 min was rescaled for easier comparison (Cortés-Escobedo et al., 2007).

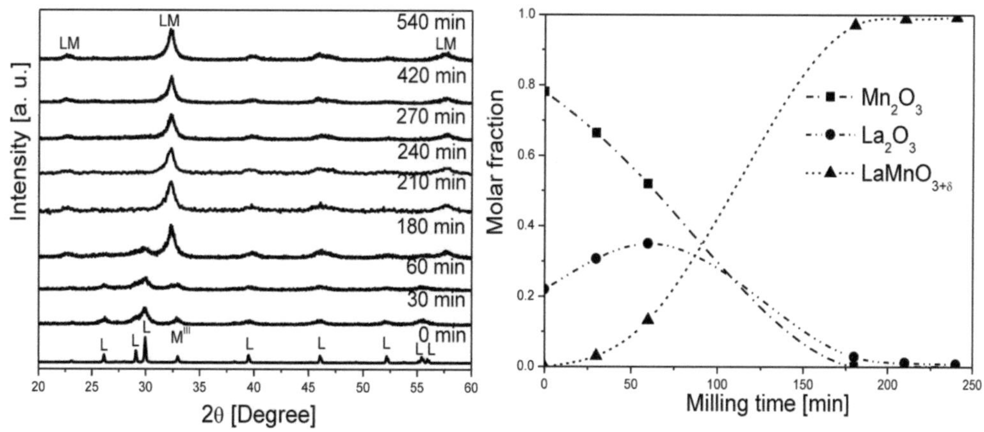

Fig. 3. Left: X-ray diffraction patterns obtained from a mixture of La_2O_3 and Mn_2O_3 powder for various milling times (M^{III}: Mn_2O_3, JCPDS No. 24-0508). Right: mol fraction of phases as analyzed using Rietveld refinement of the x-ray diffraction patterns. LM corresponds to the lanthanum manganite phases irrespective of the structure (Cortés-Escobedo et al., 2007).

Figure 4 shows the electron diffraction patterns for lanthanum manganite obtained from its stoichiometric precursors, and the rings correspond to reflections from the orthorhombic and rhombohedral structure of lanthanum manganite.

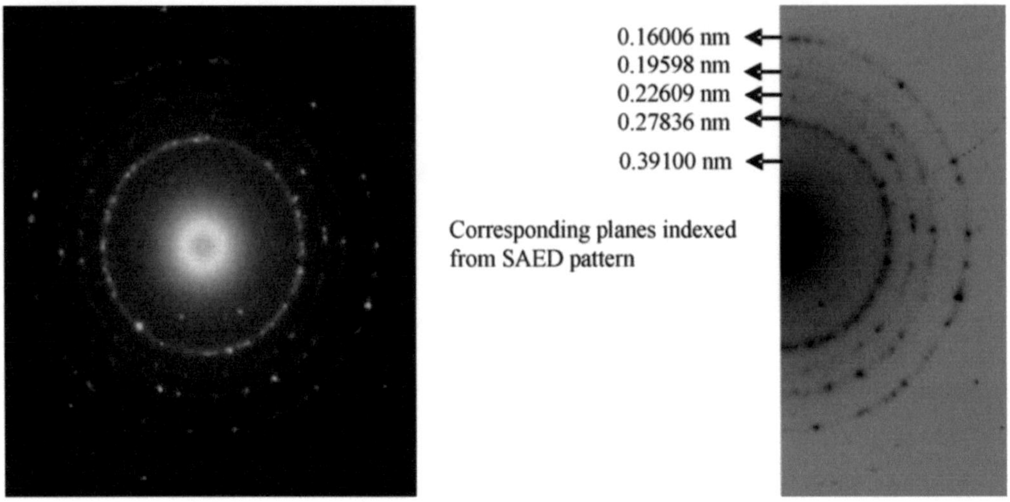

Fig. 4. Electron diffraction patterns of $LaMnO_3$ milled for 7 h, with the measured interplanar spacings (d) noted (Bolarín et al., 2007).

Finally, milling was performed using a mixture according to the following reaction:

$$La_2O_3 + 2MnO_2 \xrightarrow{MCP} 2LaMnO_3 + \frac{1}{2}O_2 + 2Mn_{Mn}^{\cdot} + V_O^x \tag{4}$$

After 270 of milling time $LaMnO_3$ is completely formed following the steps:

$$\begin{aligned} &0.5La_2O_3 + MnO_2 \xrightarrow{MCP/90\,min} (La_2O_3 + MnO_2)_{fine\ grained} \xrightarrow{MCP/120\,min} (La_2O_3 + MnO_2)_{fine\ grained} \\ &+(LaMnO_3)_{fine\ grained} \xrightarrow{MCP/270\,min} LaMnO_3 \end{aligned} \tag{5}$$

The corresponding x-ray diffraction patterns are shown in Figure 5. After Rietveld refinement, three different structures of lanthanum manganite were identified: orthorhombic, rhombohedral and cubic, in varying proportions.

A discussion of the structures can be read in (Cortés-Escobedo et al., 2008). The particle size and morphology of the milled powders as a function of time and precursor is shown in Figure 6. The decrease in particle size as a function of time is consistent with the diminishing crystallite size observed with x-ray diffraction (Figures 3-5).

Figure 7 shows the particle size as measured by laser diffraction and image analysis of the SEM photomicrographs. In both cases, after 100 min of milling time particles of 500 nm are predominant.

In Figure 8 photomicrographs of the crystallite size for the La_2O_3+Mn_2O_3 mixture milled for 90 min show crystallites from 5 to 20 nm in size and the deformation of the particles.

Thermodynamic calculations for ΔG and ΔH were made using HSC Chemistry 5.11[2] software (Figure 9) in order to know the feasibility of the formation of $LaMnO_3$ in the three cases, and the reaction with lowest ΔG was $MnO_2 + La_2O_3 = LaMnO_3$, opposite to what was observed for the complete formation of the manganite in the milling time experiments. This difference can be attributed to structural and redox transformations required to obtain the final Perovskite structure.

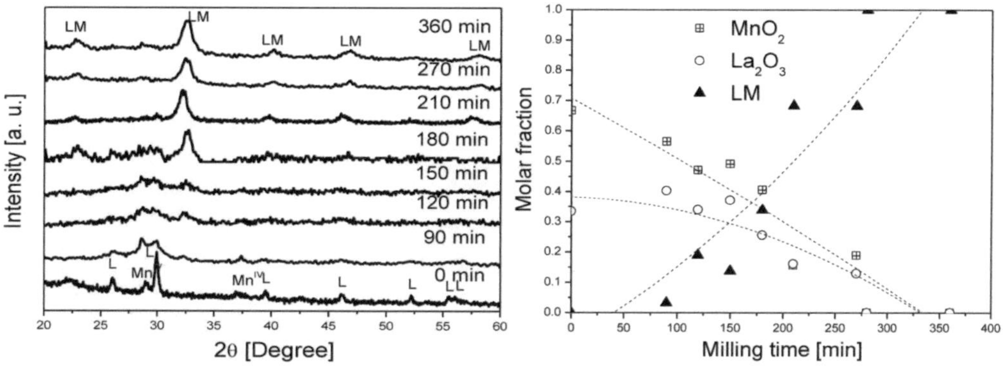

Fig. 5. Left: X-ray diffraction patterns obtained from a La_2O_3+MnO_2 powder mixture at different milling times (M^{IV}: MnO_2, JCPDS No.72-1984). Right: mol fraction of phases as analysed by the Rietveld x-ray diffraction pattern refinement, LM corresponds to the lanthanum manganite phases irrespective to the structure. (Cortés-Escobedo et al., 2007)

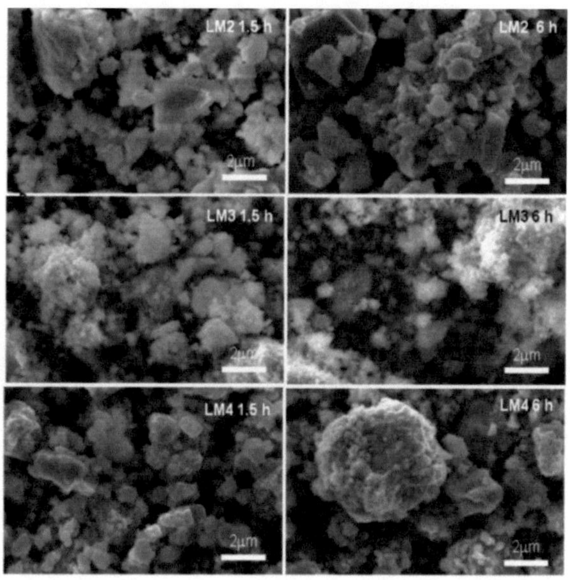

Fig. 6. Typical SEM micrographs of powder mixtures from different Mn precursors and milling times: LM2 1.5 h: La_2O_3+MnO for 90 min; LM2 6 h: La_2O_3+MnO for 360 min; LM3 1.5 h: La_2O_3+Mn_2O_3 at 90 min; LM3 6 h: La_2O_3+Mn_2O_3 at 360 min; LM4 1.5 h: La_2O_3+MnO_2 at 90 min; LM4 6 h: La_2O_3+MnO_2 at 360 min. (Cortés-Escobedo et al., 2007).

[2] HSC Chemistry 5.11, Copyright (C) Outokumpu Research Oy, Pori, Finland, A. Roine.

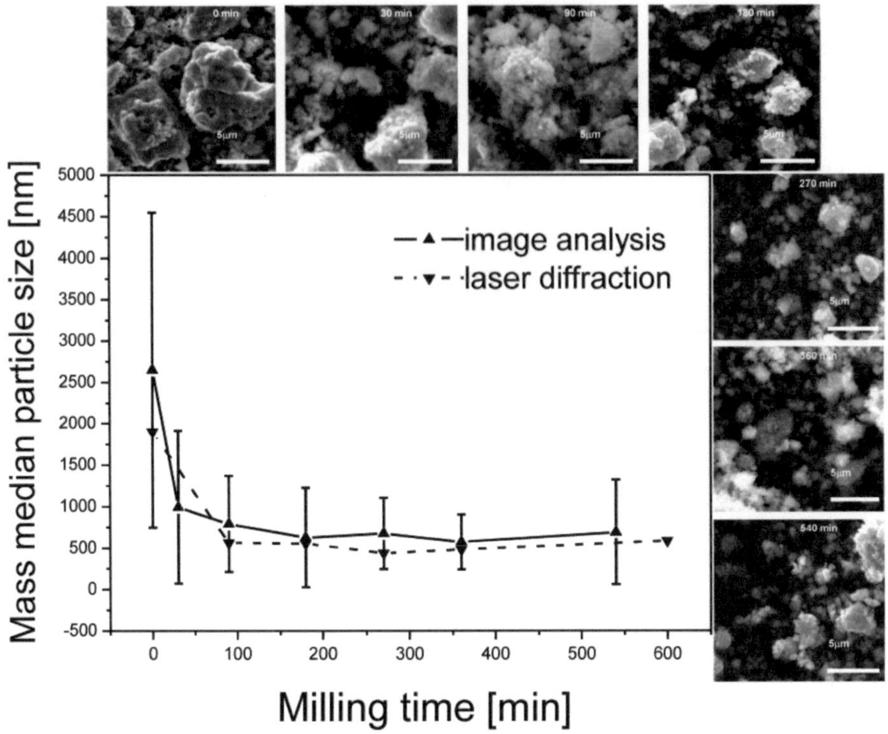

Fig. 7. Particle size results from Zetasizer measurements and image analysis of La_2O_3+Mn_2O_3 milled powders at different milling times (Cortés-Escobedo et al., 2008).

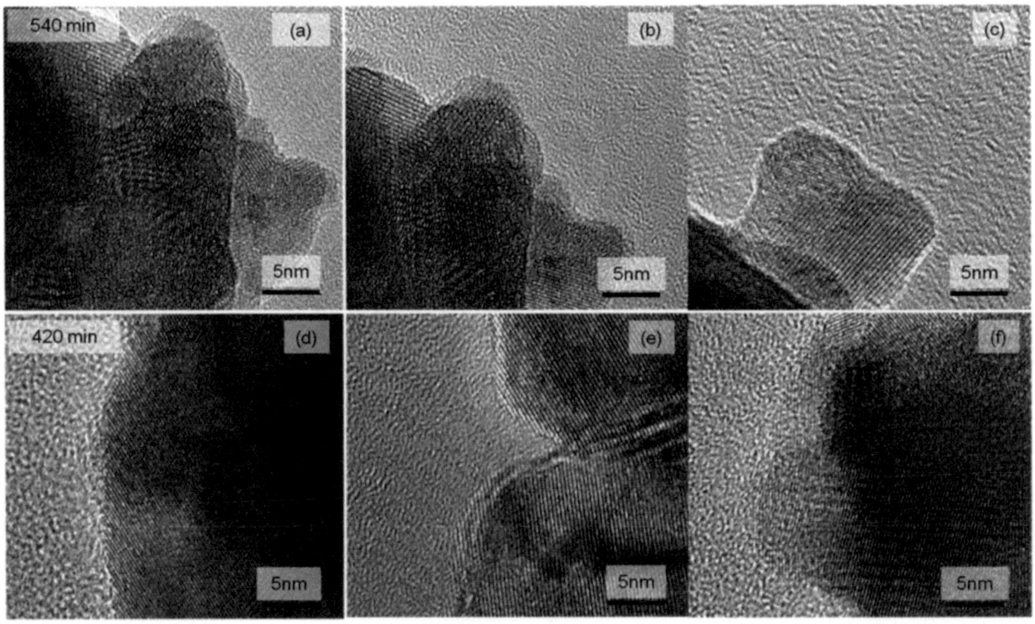

Fig. 8. HRTEM photomicrographs of the La_2O_3+Mn_2O_3 mixture milled for 540 and 420 min.

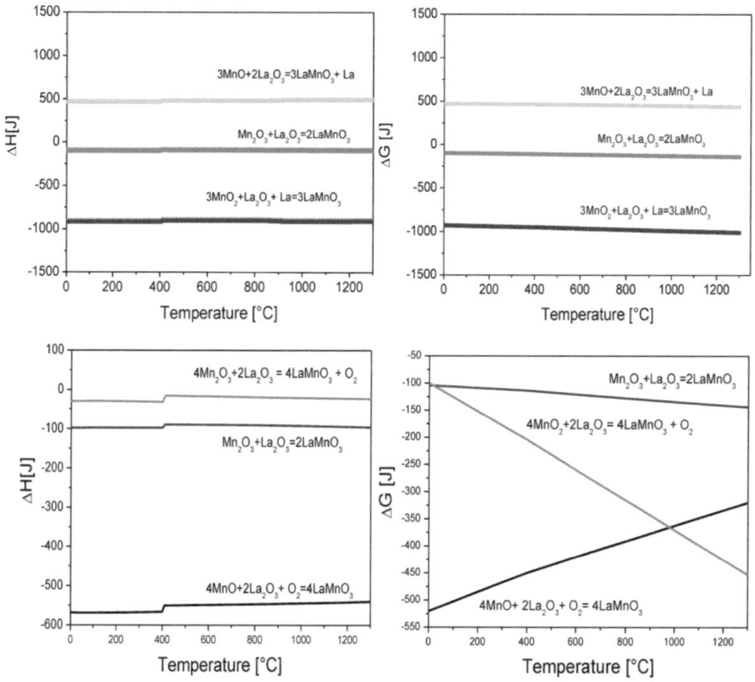

Fig. 9. Thermodynamic calculations for the reaction of formation of $LaMnO_3$ with different precursor oxides.

3.2 Extrinsic defects: Doped effect

The other way used to promote ionic conductivity was to introduce a different cation in the place of lanthanum. The cations chosen for their ionic radiuses were Ca^{2+} and Sr^{2+} (see Table 1). In the case of Ca^{2+}, the reaction expected was:

$$2La_2O_3 + 2Mn_2O_3 \xrightarrow{CaO} 2LaMnO_3 + 2Ca'_{La} + V_O^{xx} \tag{6}$$

Ion	Ionic radius (Å)
La+	1.39
La3+	1.06
Sr2+	1.12
Ca2+	0.99

Table 1. Ionic radius for different cations

Figure 10 shows x-ray diffraction patterns for $CaO+Mn_2O_3+La_2O_3$ powder mixtures milled for 7 h, with different doping levels ($0 \leq x \leq 1$). The milled precursors react according to the following expression:

$$\frac{(1-x)}{2}La_2O_3 + \frac{1}{2}Mn_2O_3 + xCaO - \frac{x}{4}O_2 \xrightarrow{MCP} La_{1-x}Ca_xMnO_3 \tag{7}$$

After 7 h of milling, there is a shift to the right of the mean peak of the lanthanum manganite as the amount of doping is increased, indicating a decrease in the interplanar distance, corresponding to deformation of the network as proposed in Figure 11.

In Figures 12- 14, morphology of the oxide precursors is shown. In the micrographs particle sizes of up to 20 μm before milling can be observed.

After 4.5 h of milling, particle sizes have decreased to around 1 μm with aggregates. Figures 15 and 16 show the aggregates and particles after 4.5 and 7 h of milling, respectively, for the stoichiometric mixture used to form $LaMnO_3$.

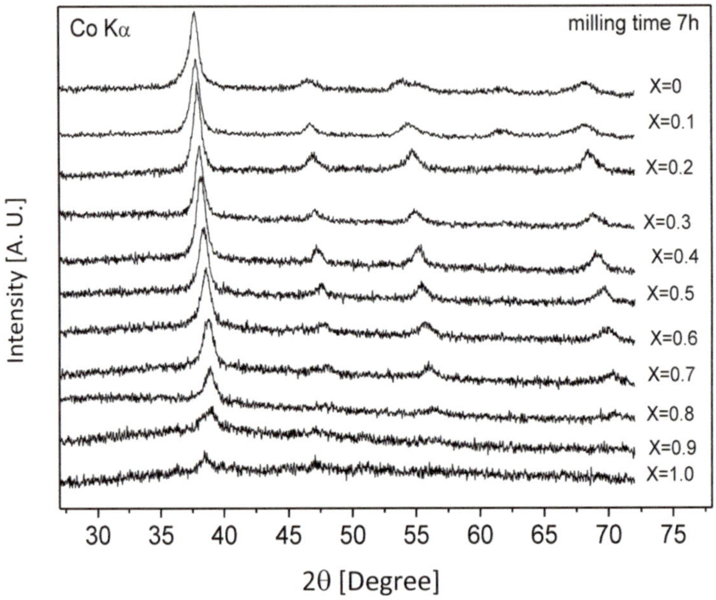

Fig. 10. X-ray powder diffraction patterns of different mixtures milled at 7 h, modifying the level of doping from x = 0 to x = 1 (Lira-Hernández et al., 2010).

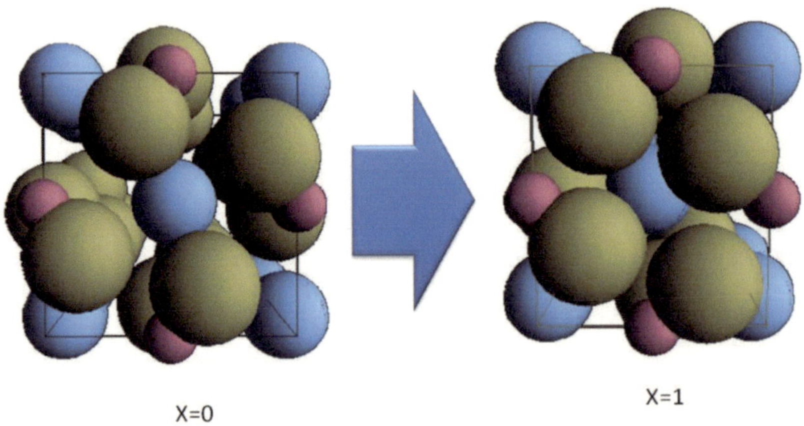

Fig. 11. Distortion of the $La_{(1-x)}Ca_xMnO_3$ unit cell.

Figures 17-19 show the morphology of the nanoparticles aggregated after 4.5 and 7 h of milling of the (Mn_2O_3+CaO+La_2O_3) mixtures. The morphology is the same for compositions x = 0.3 and 0.4, but for x = 0.8 after 7 h of milling a different morphology is observed - larger and continuous particles with softer edges.

Figure 20 shows the morphology for the extreme case of x = 1, the Mn_2O_3+CaO mixture after 4.5 h of milling, and the absence of large agglomerates is noticeable; the particle size is on the order of nanometers.

Fig. 12. Calcium oxide precursor for the reaction of formation of $La_{(1-x)}Ca_xMnO_3$.

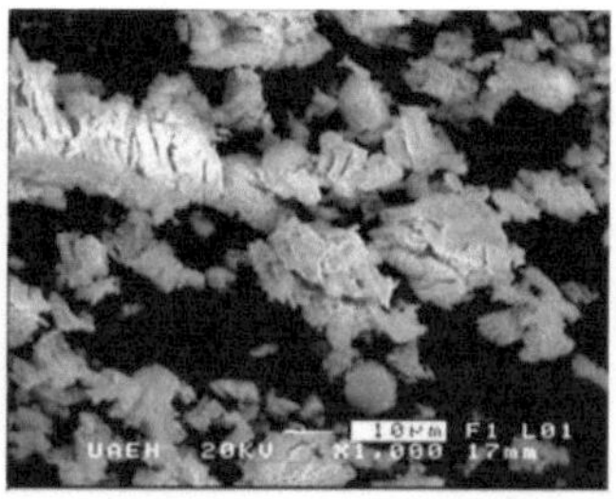

Fig. 13. Lanthanum oxide precursor for the reaction of formation of $La_{(1-x)}Ca_xMnO_3$.

Fig. 14. Manganesum oxide precursor (Mn_2O_3) for the reaction of formation of $La_{(1-x)}Ca_xMnO_3$.

Fig. 15. (Mn_2O_3+La_2O_3) mixture after 4.5 h of milling.

Fig. 16. (Mn_2O_3+La_2O_3) mixture after 7 h of milling.

Fig. 17. (Mn_2O_3+CaO+La_2O_3) mixture after 4.5 h of milling (x = 0.3).

Fig. 18. (Mn_2O_3+CaO+La_2O_3) mixture after 7 h of milling (x = 0.4).

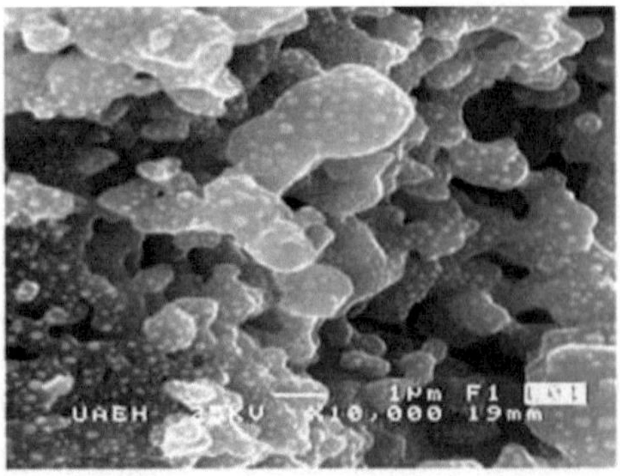

Fig. 19. (Mn2O3+CaO+La2O3) mixture after 7 h of milling (x = 0.8).

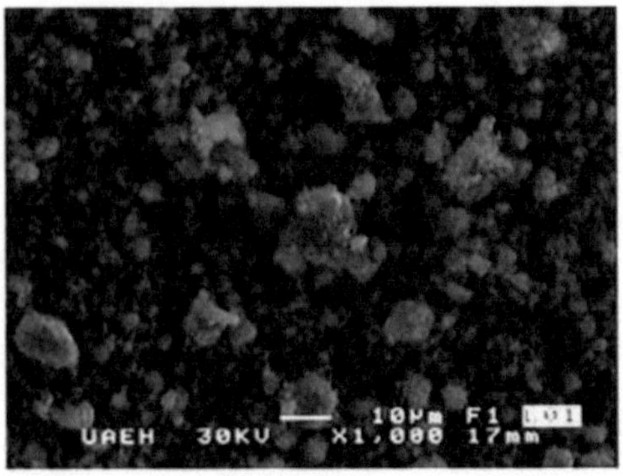

Fig. 20. (Mn2O3+CaO) mixture after 4.5 h of milling.

When substituting strontium (Sr^{2+}) for lanthanum, the expected reaction was:

$$2La_2O_3 + 2Mn_2O_3 \xrightarrow{SrO} 2LaMnO_3 + 2Sr'_{La} + V_O^{xx} \tag{8}$$

Strontium was substituted in the amounts of x = 0.15 and 0.2 according to the following reactions:

$$\frac{(1-x)}{2}La_2O_3 + MnO + xSrO + \frac{1}{4}(1-x)O_2 \xrightarrow{MCP} La_{1-x}Sr_xMnO_3 \ldots \text{LSMx2} \tag{9}$$

$$\frac{(1-x)}{2}La_2O_3 + \frac{1}{2}Mn_2O_3 + xSrO \xrightarrow{MCP} La_{1-x}Sr_xMnO_3 + \frac{(-x)}{4}O_2 \ldots \text{LSMx3} \tag{10}$$

$$\frac{(1-x)}{2}La_2O_3 + MnO_2 + xSrO \xrightarrow{MCP} La_{1-x}Sr_xMnO_3 + \frac{(1-x)}{4}O_2 \ldots \text{LSMx4} \tag{11}$$

LSMx2 is shorthand for $LaMnO_3$ produced using MnO as the precursor and with *x* percent of Sr. LSMx3 is shorthand when Mn_2O_3 is the precursor, and LSMx4 is for when MnO_2 is the precursor. Thus, lanthanum manganite produced from MnO and doped with 20 at% Sr ($La_{0.8}Sr_{0.2}MnO_3$) is represented by LSM202, and lanthanum manganite produced from MnO_2 and doped with 15 at% Sr ($La_{0.85}Sr_{0.15}MnO_3$) is represented by LSM153.

The x-ray diffraction patterns for this collection of milled samples are shown in Figure 21. It is important to note that lanthanum manganites from MnO (LM2, LSM152, and LSM202) were not obtained only by MCP, but subsequently it was necessary to heat treat (1000°C for 600 min) the mixtures to produce lanthanum manganite. On the other hand, mixtures with precursors MnO_2 and Mn_2O_3 developed lanthanum manganites without heat treatment.

4. Consolidation of ceramic powders

Compaction of lanthanum manganites was carried out by using two agents, corn starch and etilen-bis-estearamide (EBS, a lubricant commonly used in powder metallurgy industry) at 200 MPa. With corn starch porosity in range from 10 to 30% was observed, and LSM204 had the largest surface area of all samples (Figure 22). Differences in morphology with different strontium contents and manganese oxide precursors were also observed. Samples of consolidated manganites obtained from MnO had a grain size of 300-800 μm, larger than those obtained from MnO_2 (100-250 μm). Grain growth can be associated with cohesion between particles due to internal repulsi on forces which prevent recrystallization or grain growth during the sintering process.

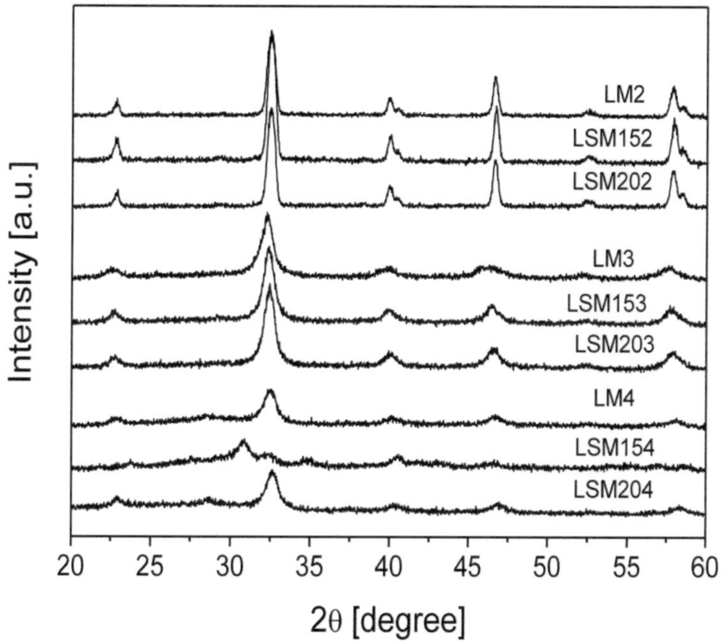

Fig. 21. X-ray diffraction patterns for the mechanosynthesis and heat treatment process (1000°C for 600 min) to obtain lanthanum manganites with intrinsic and extrinsic defects.

The microstructure is similar for pellets with 10, 15 and 20 wt% EBS (Figures 23-25). Sintering of the ceramics is indicated by necks formed after sintering at 1100°C. Heat treating at 1300°C for 3 h results in a decrease in porosity as well as particle growth.

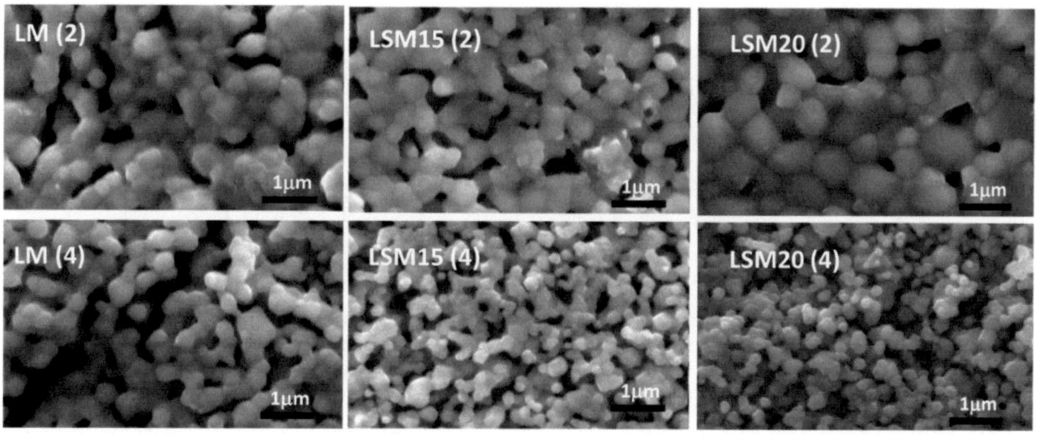

Fig. 22. SEM micrographs for lanthanum manganites consolidated with corn starch.

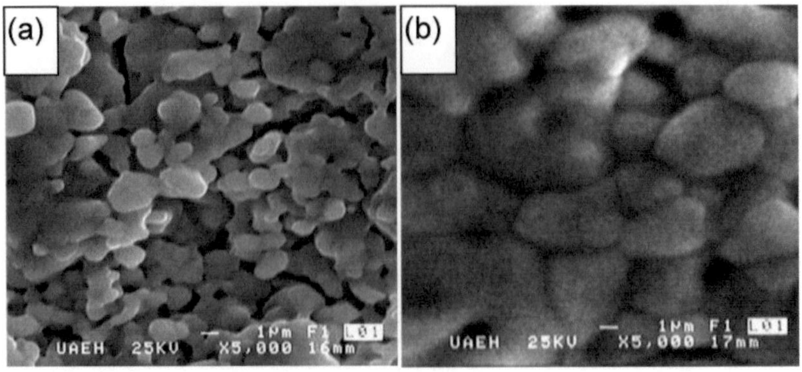

Fig. 23. SEM images of $La_{1-x}Ca_xMnO_3$ consolidated with 10 wt% EBS at: a) 1100°C for 1 h and b) 1300°C for 3 h.

Fig. 24. SEM images of $La_{1-x}Ca_xMnO_3$ consolidated with 15 wt% EBS at: a) 1100°C for 1 h and b) 1300°C for 3 h.

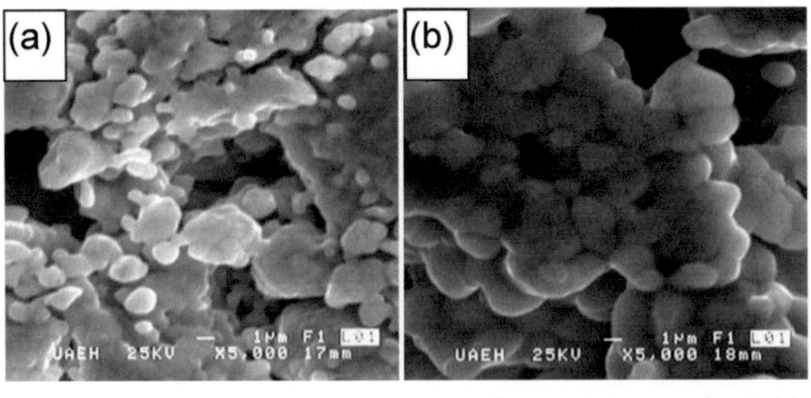

Fig. 25. SEM images of $La_{1-x}Ca_xMnO_3$ consolidated with 20 wt% EBS at: a) 1100°C for 1 h and b) 1300°C for 3 h.

5. YSZ-LSM interactions

As was mentioned previously, for applications as sensors and as cathodes in solid oxide fuel cells, the desired functionality of the material is based on the ionic conduction of oxygen through the interface between yttrium-doped zirconia (YSZ) and lanthanum manganite (LSM). This functionality can be potentiated by increasing the triple phase boundaries (TPB) formed between oxygen, YSZ and LSM (Figure 26).

But in some conditions, an undesirable reaction occurs in the TPBs, giving rise to lanthanum zirconate formation through the diffusion of lanthanum atoms from manganite to zirconia (Figure 27).

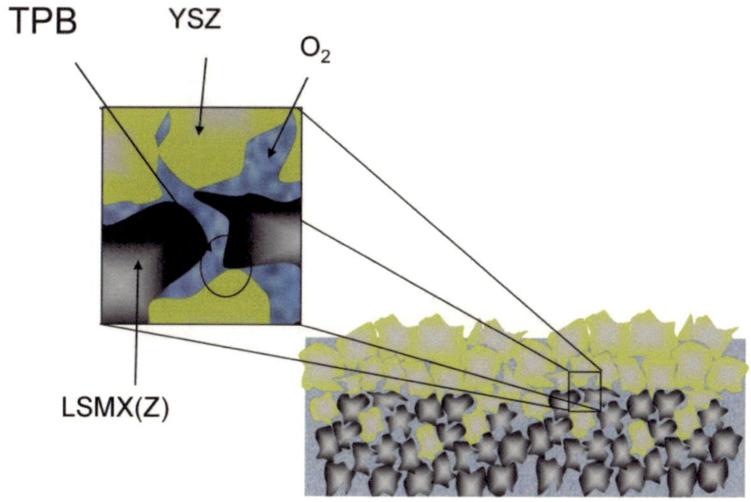

Fig. 26. Schematic of the triple phase boundaries in LSM-YSZ interfaces.

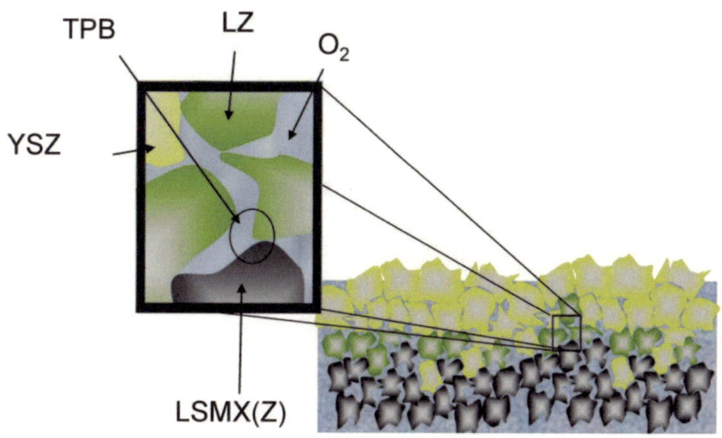

Fig. 27. Schematic of lanthanum zirconate formation in TPBs.

For this reason, it is important to determine the optimal conditions required to avoid the formation of lanthanum zirconate, which diminishes the ionic conduction through TPBs.

Heat treatment (1000 - 1300°C) and spark plasma sintering were used to obtain the conditions in which the formation of lanthanum zirconates is avoided.

Figure 28 shows the x-ray diffraction patterns of mixtures of yttria partially stabilized tetragonal zirconia (YPSZ) with lanthanum manganites synthesized from different manganese oxide precursors (LMz) without heat-treatment (Figures 28a, 28e, and 28i) and heat-treated at 1300°C (Figures 28b-28d, 28f-28h and 28j-28l).

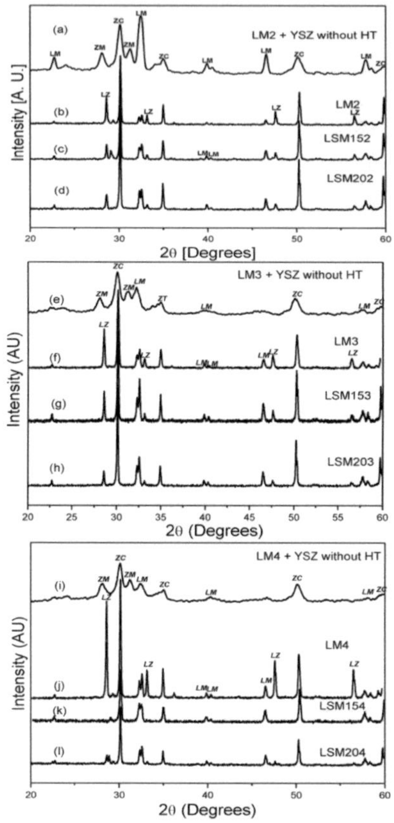

ZM = monoclinic zirconia, *ZC-T* = cubic + tetragonal zirconia, *ZT* = tetragonal zirconia, *LZ* = lanthanum zirconate, *LM* = lanthanum manganite (Cortés-Escobedo et al., 2008).

Fig. 28. X-ray diffraction patterns of mixtures of YPSZ with lanthanum manganites prepared from: (a)-(d) MnO, La_2O_3 and SrO by mechanosynthesis and heat treatment in air; (e)-(h) Mn_2O_3, La_2O_3 and SrO by mechanosynthesis; (i)-(l) MnO_2, La_2O_3 and SrO by mechanosynthesis. Manganites in mixtures (a), (b), (e), (f), (i) and (j) are undoped, while manganites in mixtures (c), (g) and (k) are doped with 15 at% Sr in La sites and (d), (h) and (l) have 20 at% Sr in La sites. All mixtures were heat-treated at 1300°C except (a), (e) and (i).

Figures 28a-b, 28e-f and 28i-j correspond to mixtures of YPSZ with undoped manganites prepared from Mn^{II}, Mn^{III} and Mn^{IV}, respectively. In these figures, increasing heat-treatment temperature is shown to result in increased intensity and a widening of the peaks (Cortés-Escobedo et al., 2008).

In all cases, grain growth is reflected in the decrease of the peak width and increase in the intensity of the peaks. The relative intensity of the main lanthanum zirconate peak is greatest for lanthanum manganite obtained from MnO_2+La_2O_3 without strontium and is smallest when doped with 15 at% strontium. In addition, there is a phase change for zirconia in the presence of lanthanum zirconate, tending to the cubic phase instead of its usual tetragonal structure (without lanthanum zirconate).

The last point can be also observed in the backscattered electron SEM micrograph (Figure 29), in which the cubic zirconia morphology (shown with small dots and in dark gray) is clearly distinguished from the lanthanum zirconate (bright gray) and the lanthanum manganite (medium gray). From this image we can also deduce the direction of the diffusion of the atoms. That is to say, lanthanum zirconate grains tend to invade zirconate grains from lanthanum manganite grains.

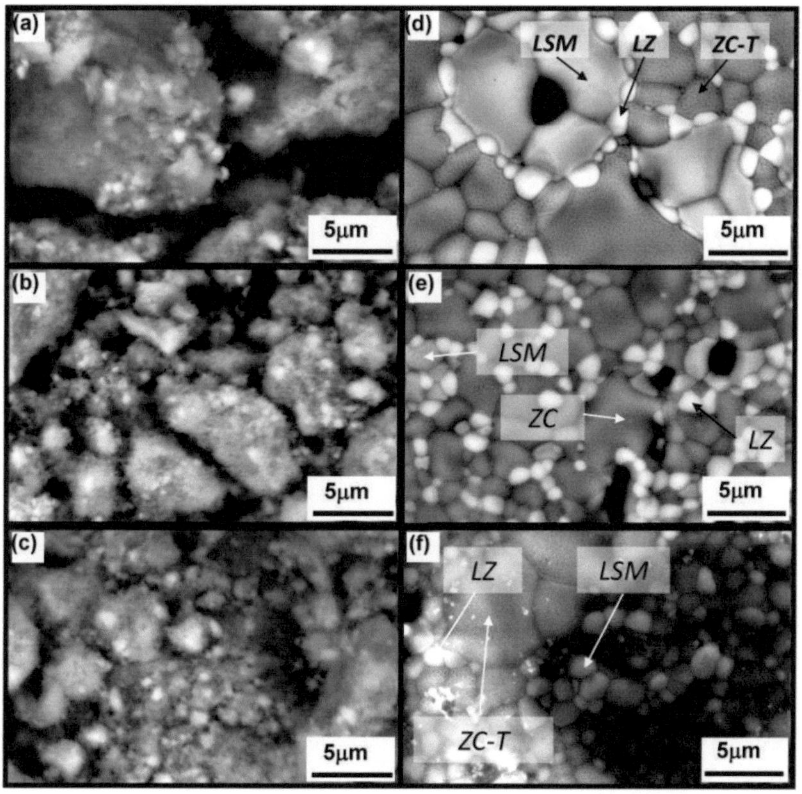

Fig. 29. SEM micrographs of the mixture of YPSZ with lanthanum manganite obtained from undoped Mn^{II} (a), (b), (c) before and (d), (e), (f) after sintering at 1300°C (Cortés-Escobedo et al., 2008). LZ: Lanthanum zirconate; LSM: Strontium doped lanthanum manganite; ZC: cubic zirconia; ZC-T: tetragonal zirconia.

In another experiment, using spark plasma sintering, for samples heat treated at 1000°C, lanthanum zirconate is observed only for sintered $LaMnO_3$ mechanosinthesized from MnO_2+La_2O_3 (LM4). But by treating at 1300°C, lanthanum zirconate is detected in all the samples (Figures 30-31).

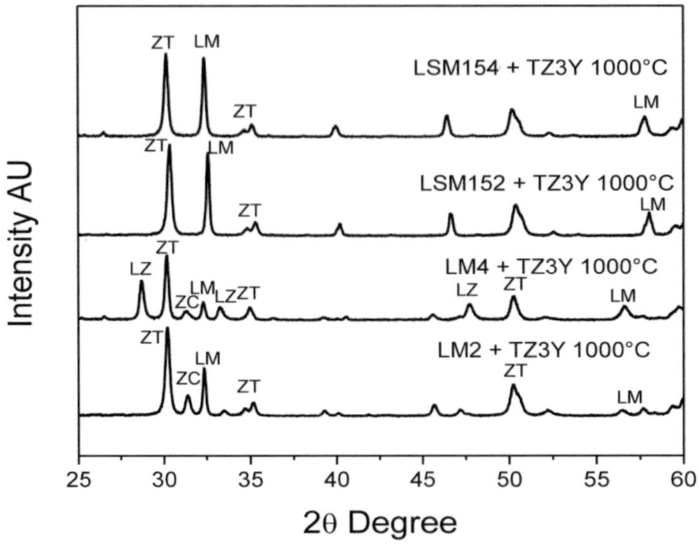

Fig. 30. X-ray diffraction of lanthanum manganite – yttria-stabilized zirconia mixtures, which have been spark plasma sintered at 1000°C for 10 min. LM2: Lanthanum manganite from $MnO+La_2O_3$; LM4: Lanthanum manganite from $MnO_2+La_2O_3$; LSM152: 15 at% Sr-doped lanthanum manganite from $MnO+La_2O_3+ SrO$; LSM154: 15 at% Sr-doped lanthanum manganite from $MnO_2+La_2O_3+SrO$.

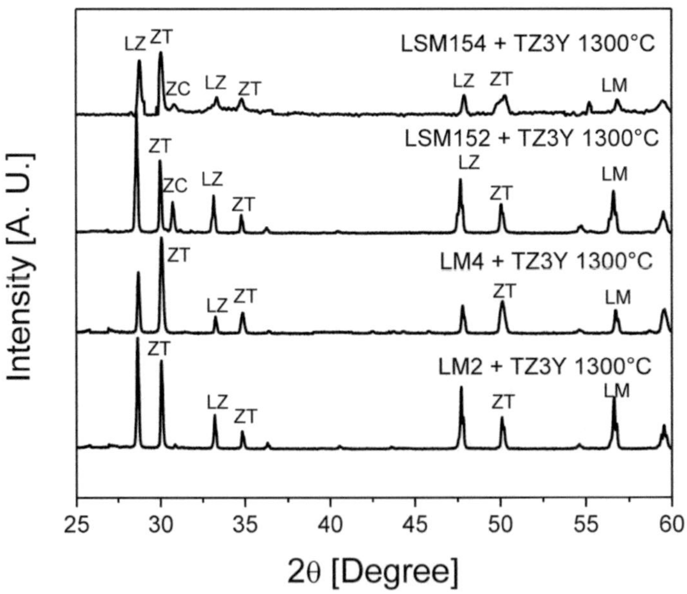

Fig. 31. X-ray diffraction of lanthanum manganite – yttria-stabilized zirconia mixtures, which have been spark plasma sintered at 1300°C for 10 min. LM2: Lanthanum manganite from $MnO+La_2O_3$; LM4: Lanthanum manganite from $MnO_2+La_2O_3$; LSM152: 15 at% Sr-doped lanthanum manganite from $MnO+La_2O_3+SrO$; LSM154: 15 at% Sr-doped lanthanum manganite from $MnO_2+La_2O_3+SrO$.

In Figure 32 SEM images show the microstructure of spark plasma sintered $LaMnO_3$ mechanosinthesized from $MnO+La_2O_3$ (LM2) and $MnO_2+La_2O_3$ (LM4) and $La_{0.85}Sr_{0.15}MnO_3$ mechanosinthesized from $MnO+SrO+La_2O_3$ (LSM152) and $MnO_2+SrO+La_2O_3$ (LSM154), treated at 1000°C and 1300°C. In the case of treatment at 1000°C, pores are present in all samples, but the differentiation between grains is diffuse. Samples treated at 1300°C show continuity in the microstructure without differentiation in grains.

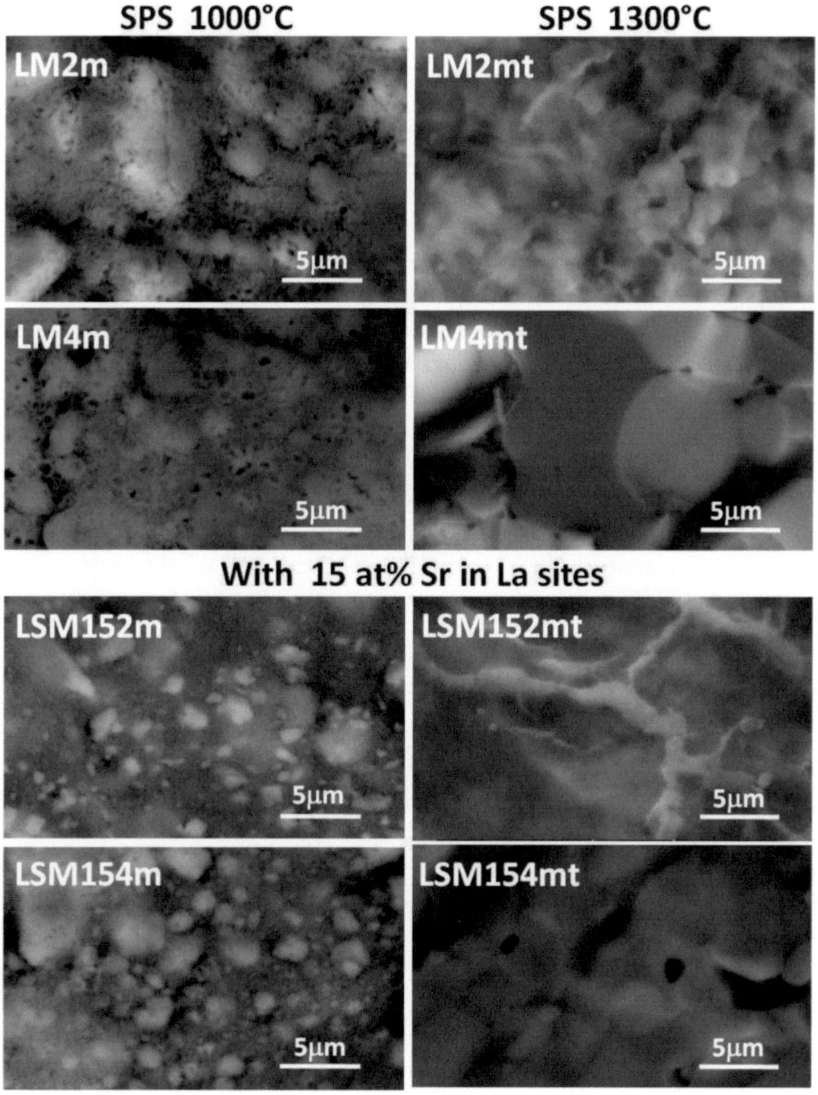

Fig. 32. SEM images of the microstructure of spark plasma sintered $LaMnO_3$ mechanosinthesized from $MnO+La_2O_3$ (LM2) and $MnO_2+La_2O_3$ (LM4) and $La_{0.85}Sr_{0.15}MnO_3$ mechanosinthesized from $MnO+SrO+La_2O_3$ (LSM152) and $MnO_2+SrO+La_2O_3$ (LSM154). On the left are samples spark plasma sintered at 1000°C, and on right are samples treated at 1300°C.

6. Acknowledgment

The results presented in this chapter were financially assisted by the Consejo Nacional de Ciencia y Tecnología, CONACyT under grants no. 157925, 129910 and 130413, and by the National Polytechnic Institute under grant no. MULT-1338 MOD-20113549. The authors thank Arturo Ponce and Adriana Tejeda Cruz, of the University of Texas at San Antonio and the Institute of Materials of UNAM, respectively, for their support in HRTEM and XRD measurements. Also, authors thank M. Boldrick (Peace Corps) for his many useful suggestions.

7. References

Asamitsu, A., Moritomo, Y., Tomioka, Y., Arima, T. & Tokura, Y. (1995). A Structural Phase Transition Induced by an External Magnetic Field. *Nature*. Vol. 373, (February 1995), pp. (407–409), ISSN 0028-0836.

Bolarín, A. M., Sánchez, F., Palomares, S., Aguilar, J.A. & Torres, G. (2007). Synthesis of calcium doped lanthanum manganite by mechanosynthesis. *Journal of Alloys and Compounds*. Vol. 436, (June 2007), pp. (335-340), ISSN 0925-8388.

Bolarín, A. M., Sánchez, F., Ponce, A. & Martinez, E. E. (2007). Mechanosynthesis of lanthanum manganite. *Materials Science and Engineering A*. Vol. 454-455, (April 2007), pp. (69-74), ISSN 0921-5093.

Coey, J.M.D. & Viret, M. (1999). Mixed Valence Manganites. *Advances in Physics*. Vol. 48 (December 1997), pp. (167-293), ISSN 1460-6976.

Cortés-Escobedo, C. A., Muñoz-Saldaña, J., Bolarín-Miró, A. M., Sánchez de Jesús, F. (2008). Determination of strontium and lanthanum zirconates in YPSZ-LSM mixtures for SOFC. *Journal of Power Sources*. Vol. 180, (May 2008), pp. (209-214), ISSN 0378-7753.

Cortés-Escobedo, C. A., Sánchez de Jesús, F., Bolarín Miró, A. M., Muñoz-Saldaña, J. (2007). Mechanosynthesis of $LaMnO_3$ from different manganese oxides. *Physica Status Solidi C*. Vol. 4, (November 2007), pp. (4054-4063), ISSN 1610-1642.

Hwang, H.Y. & Palstra, T.T.M. (1995). Pressure Effects on the Magnetoresistance in Doped Manganese Perovskites. *Physical Review B: Condensed Matter and Materials Physics*. Vol. 52, (December 1995), pp. (15046–15049), ISSN 1550-235x.

Hwang, H.Y., Cheong, S.W., Radaelli, P.G., Marezio, M., Batlogg, B. (1995). Lattice Effects on the Magnetoresistance in Doped $LaMnO_3$. *Physical Review Letters*. Vol. 75, (October 1995), pp. (914-917), ISSN 0031-9007.

Jafari-Nejad, Sh., Abolghasemi, H., Moosavian, M.A., Golzary, A. & Maragheh, M. G. (2010). Fractional factorial design for the optimization of hydrothermal synthesis of lanthanum oxide nanoparticles under supercritical water condition. *Journal of Supercritical Fluids*. Vol. 52, (April 2010), pp. (292-297), ISSN 0896-8446.

Kuwahara, H., Tomioka Y. & Asamitsu A., Moritomo, Y. & Tokura, Y. (1995). A First Order Phase Transition Induced by a Magnetic Field. *Science*. Vol. 270, (November 1995), pp. (961–963), ISSN 1095-9203.

Lira-Hernández, I. A., Sánchez-De Jesús, F., Cortés-Escobedo, C.A. & Bolarín-Miró, A.M. (2010). Crystal structure analysis of calcium doped lanthanum manganites prepared by mechanosynthesis. *Journal of the American Ceramic Society*. Vol. 93, (October 2010), pp. (3474-3477), ISSN 0002-7820.

Mahesh, R., Mahendiran, R., Raychaudhuri, A.K. & Rao, C.N.R. (1995). Effect of the Internal Pressure Due to the A-Site Cations on the Giant Magnetoresistance and Related Properties of Doped Rare-Earth Manganates $Ln_{1-x}A_xMnO_3$ (Ln = La, Nd, Gd, Y; A = Ca, Sr, Ba, Pb). *Journal of Solid State Chemistry*. Vol. 120, (November 1995), pp. (204–207), ISSN 0022-4596.

Muroi, M., Street, R. & McCormick, P.G. (2000). Structural and Magnetic Properties of Ultrafine $La_{0.7}Ca_{0.3}MnO_z$ Powders Prepared by Mechanical Alloying. *Journal of Solid State Chemistry*. Vol. 152, (July 2000), pp. (503–510), ISSN 0022-4596.

Neumeier, J.J., Wundley, M.F., Thomson, J.D. & Heffner R.H. (1995). Substantial Pressure Effects on the Electrical Resistivity and Ferromagnetic Transition Temperature of $La_{1-x}Ca_xMnO_3$. *Physical Review B: Condensed Matter and Materials Physics*. Vol. 52, (December 1995), pp. (R7006–R7009), ISSN 1550-235x.

Sato, K., Chaichanawong, J., Abe, H., & Naito, M. (2006). Mechanochemical synthesis of $LaMnO_{3+\delta}$ fine powder assisted with water vapor. *Materials Letters*. Vol. 60, (May 2006), pp. (1399-1402), ISSN 0167-577x.

Schiffer, P., Ramirez, A.P., Bao, W. & Cheong, S-W. (1995). Low Temperature Magnetoresistance and the Magnetic Phase Diagram of $La_{1-x}A_xMnO_3$. *Physical Review Letters*. Vol. 75, (October 1995), pp. (3336–3339), ISSN 0031-9007.

Shinde, K. P., Deshpande, N.G., Eom, T., Lee, Y.P., & Pawar, S.H. (2010). Solution-combustion synthesis of $La_{0.65}Sr_{0.35}MnO_3$ and the magnetocaloric properties. *Materials Science and Engineering B*. Vol. 167, (March 2010), pp. (202-205), ISSN 0921-5107.

Shu, Q., Zhang, J., Yan, B. & Liu, J. (2009). Phase formation mechanism and kinetics in solid-state synthesis of undoped and calcium-doped lanthanum manganite. *Materials Research Bulletin*. Vol. 44, (March 2009), pp. (649-653), ISSN 0025-5408.

Tokura, Y., Urushibara, A., Moritomo, Y., Arima, T., Asamitsu, A., Kido, G. & Furukawa, N. (1994). Giant Magnetotransport Phenomena in Filling-Controlled Kondo Lattice System $La_{1-x}Sr_xMnO_3$. *Journal of the Physical Society of Japan*. Vol. 63, (September 1994), pp. (3931–3935), ISSN 1347-4073.

Uskokovic, V. & Drofenik, M. (2007). Four novel co-precipitation procedures for the synthesis of lanthanum-strontium manganites. *Materials & Design*. Vol. 18, (august 2005), pp. (667-672), ISSN 0261-3069.

Zhang, Q. & Saito, F. (2000). Mechanochemical synthesis of $LaMnO_3$ from La_2O_3 and Mn_2O_3 powders (2000). *Journal of Alloys and Compounds*. Vol. 297, (February 2000), pp. (99-103), ISSN 0925-8388.

Zhang, Q., Nakagawa, T. & Saito, F. (2000). Mechanochemical synthesis of $La_{0.7}Sr_{0.3}MnO_3$ by grinding constituent oxides. *Journal of Alloys and Compounds*. Vol. 308, (August 2000), pp. (121-125), ISSN 0925-8388.

Zhou, S. M., Zhao, S. Y., He, L. F., Guo, Y.Q. & Shi, L. (2010). Facile synthesis of Ca-doped manganite nanoparticles by a nonaqueous sol–gel method and their magnetic properties. *Materials Chemistry and Physics*. Vol. 120, (March 2010), pp. (75-78), ISSN 0254-0584.

35

Scanning Electron Microscopy (SEM) and Environmental SEM: Suitable Tools for Study of Adhesion Stage and Biofilm Formation

Soumya El Abed[1,2], Saad Koraichi Ibnsouda[1,2],
Hassan Latrache[3] and Fatima Hamadi[3]
[1]*Laboratory of Microbial Biotechnology, Faculty of Science and Technics, Fez,*
[2]*Regional University Center of Interface, University Sidi Mohamed Ben Abdellah, Fez,*
[3]*Laboratory of Valorization and Security Food Products,*
Faculty of Science and Technics, Beni Mellal,
Morocco

1. Introduction

For most of the history of microbiology, microorganisms have primarily been characterized as planktonic, freely suspended cells and described on the basis of their growth characteristics in nutritionally rich culture media. The discovery of microorganisms, 1684, is usually ascribed to Antoni van Leeuwenhoek, who was the first person to publish microscopic observations of bacteria. The direct quantitative recovery techniques showed unequivocally that more than 99.9% of the bacteria grow in biofilms on a wide variety of surfaces. Although the most common mode of growth for microorganisms on earth is in surface associated communities (Stoodley et al., 2002; Sutherland, 2001), the first reported findings of microorganisms "attached in layers" were not made until the 1940s. During the 1960s and 70s the research on "microbial slimes" accelerated but the term "biofilm" was not unanimous formulated until 1984 (Bryers, 2000). Biofilm has three-dimensional (3D) structured, heterogeneous community of microbial cells enclosed in an exopolysaccharide matrix (also called glycocalyx) that are irreversibly attached to an inert or living surface. As establish, biofilm formation has a serious implications in public health and medicine. In the case of human health, a number of microbial infections are associated with surface colonization not only on live surfaces (sinusitis, pulmonary infection in cystic fibrosis patients, periodontitis, etc. (Hall-Stoodley et al., 2004) but also on medical implants (contact lenses, dental implants, intravascular catheters, urinary stents) etc. (Donlan, 2001; Hall-Stoodley et al., 2004). Biofilms affect heat exchangers, filters, etc. because they induce biocorrosion and biofouling, producing damages on metallic surfaces and the efficiency loss in industrial set-up (Dunne, 2002; Garret et al., 2008). However,biofilms have also useful applications in bioremediation (Vidali, 2001) of different environments (microorganisms degrade and convert pollutants into less toxic forms) and biolixiviation (bacteria can efficiently dissolve minerals used in industry, to obtain copper and gold).

In order that we may gain a greater insight into the ecology of the microorganisms that exist in biofilm, it is necessary not only to be able to isolate them by traditional culture methods but also to have some understanding of the way in which these individual microorganisms interact in situ in their environment. Different microscopic techniques for biofilm monitoring including Scanning Electron microscopy (SEM) have been proved to be suitable tools in order to follow the study of adhesion stage and biofilm formation. Scanning electron microscopy as a specialized field of science that employs the electron microscope as a tool and uses a beam of electrons to form an image of a specimen allowing imaging and quantification of surface topographic features.

The scope of this chapter is to illustrate the importance of scanning electron microscopy and environnemental scanning electron microscopy in biofilm examination and control. Furthermore, although we are conscious about the vast variety of biofilms in natural, clinical and industrial environments, this chapter will mainly concentrate on imaging application of SEM and ESEM biofilms.

2. Step of biofilm formation

Planktonic cells are able to attach on the surfaces and form biofilm through a process that include several steps:

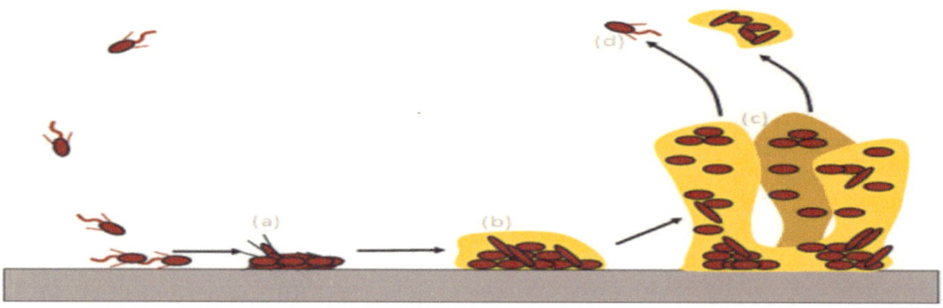

Fig. 1. Schematic illustrations of biofilm formation and development. (Filloux & Vallet, 2003).

2.1 Attachment/colonization

The primary adhesion stage constitutes the beneficial contact between a conditioned surface and planktonic microorganisms. During the process of attachment, the organism must be brought into close proximity of the surface, propelled either randomly or in a directed fashion via chemotaxis and mobility (Prakash et al., 2003). This step is reversible and it is characterized by a number of physicochemical variables that defines the interaction between the microbial cell surface and the conditioned surface of interest (An et al., 2000; Liu et al., 2004; Singh et al., 2002).

2.2 Irreversible adhesion

The second step is the irreversible adhesion during which bacteria start to express adhesion protein such as curli or fimbriae to adhere to the surface. Microorganisms starts to produce intercellular connections (intercellular curli for example) and a polymeric matrix, usually called extracellular polymeric substances (EPS). This matrix is a complex hydrogel embedding the bacteria community and building up in three dimensions. The backbone of this gel is mainly composed of polysaccharides produced by bacteria (such as colanic acid, chitosan, alginate), other components such as enzymes, DNA, RNA, nutrients, proteins, surfactants (Flemming et al., 2007). The exact role of the matrix is not yet completely elucidated but it has been demonstrated that the matrix acts as a protective layer (Fux et al., 2005) and is microenvironment-conservative (Beech, 2004).

After the adherence of microorganism to the inert surface, the association becomes stable for micro-colonies formation (Bechmann & Eduvean, 2006; O'Toole et al., 2000). The microorganism begin to multiply while sending out chemical signals that intercommunicate among the bacterial cells. In this way, the bacteria multiply within the embedded exopolysaccharide matrix, thus giving rise to formation of a micro-colonies (Prakash et al., 2003).

2.3 Maturation of biofilm

Once bacteria have irreversibly attached to a surface, the process of biofilm maturation begins. The overall density and complexity of the biofilm increase as surface-bound organisms begin to actively replicate and extracellular components generated by attached bacteria interact with organic and inorganic molecules in the immediate environment to create the glycocalyx (Carpentier & Cerf, 1993). The maturation of biofilm generate many process already having taken place, such as quorum sensing (Nadell et al., 2008), gene transfer (Molin, 2003), persister development (Lewis, 2005) etc. All of these processes contribute to the community life of the biofilm and play an important role in biofilm survival and biofilm spreading, since they allow also detachment of biofilm parts and release of free bacteria, which is the most common way for biofilm to spread (Kaplan et al., 2003).

2.4 Detachment and dispersal of biofilm cells

As the biofilm gets older, cells detach, disperse and colonize a new niche. This detachment can be due to various factors including, fluid dynamics and shear effects of the bulk fluid (Brugnoni et al., 2007). At some point of biofilms may partially dissolve releasing cells that more away to other where a new cycle begins (Prakash et al., 2003; Singh et al., 2002).

3. Imaging application

SEM is a well-established basic method to observe the morphology of bacteria adhered on a material surfaces, the morphology of the material surface, and the relationships between them (Peters et al., 1982). SEM has been used for enumeration of adhered bacteria or tissue large number of samples. It is as a key technique that provides also information about the morphology of biofilm, presence of EPS and the nature of corrosion products (crystalline or amorphous).

3.1 SEM applied of adhesion stage

Microbial adhesion is the first step of the formation of biofilm and an extremely complicated process that is affected by many factors. In this regard, detailed investigation of microbial adhesion involved in the developmental process from single sessile bacteria to multicellular biofilm is crucial to elaborate strategies to control biofilm development. Moreover, submicrometer-scale cell surface polymers and appendages, such as curli, flagella, and exocellular polymers, have been shown to play essential roles during cell adhesion and biofilm formation (Busscher et al., 2008; Dufrêne, 2008; Rodrigues & Elimelech, 2009). A SEM image of such a curli is depicted in Figure 2.

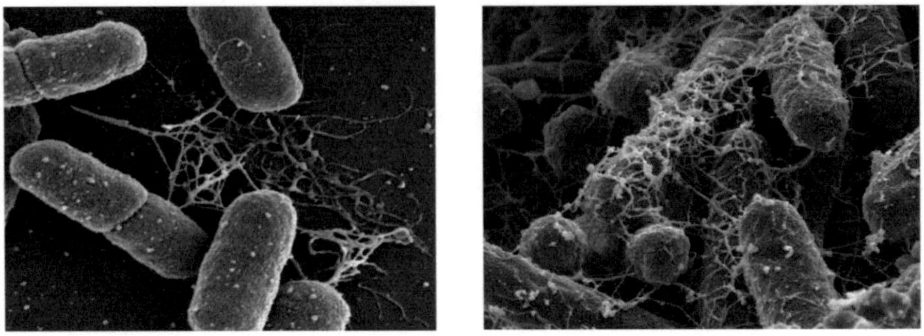

Fig. 2. SEM images of *E.coli* K-12 MG 1655 ompR234 producing curli (Olsen et al., 1989)

Adhesion phenomena has been evaluated as function of substratum, liquid medium, carbone source, pH and hydrodynamics parameters including flow rate. Many of the conclusions about biofilm development, composition, distribution, and relationship to substratum have been derived from scanning electron microscopy (Bragadeewaran et al., 2010; Herald & Zottola, 1988; Pinna et al., 2000). We report here several investigations made in our laboratory used scanning electron microscopy to study adhesion phenomena. Hamadi et al., (2005) have investigated the adhesion of *Staphylococcus aureus* ATCC 25923 to glass at different pH values using scanning electron microscopy and image analysis with the Mathlab® program is shown in Figure 3.

The surface topography has been widely discussed as a parameter influencing microbial adhesion. In this regard, experiments made by Kouider et al., (2010) using SEM to determine the effect of stainless steel surface roughness on *Staphylococcus aureus* adhesion shown that adhesion level was found to largely depend on the substrate roughness with maximum at Ra = 0.025µm and minimum at Ra= 0.8µm. Mallouki et al., (2007) have studied the anti-adhesive effect of fucans by SEM and a MATLAB program to determine the number and characteristics of adhered cells.

3.2 SEM applied of biofilm formation

Scanning electron microscopy (SEM) is a useful technique for the investigation of surface structure of biological samples (Duckett & Ligrone, 1995; Minoura et al., 1995; Motta et al., 1994). For instance, much of the current knowledge about biofilms is due to the advances in imaging studies, especially the SEM. Early microscopic techniques used in biofilm monitoring, mainly applied during the 1980s, include scanning electron microscopy. SEM has been previously used to show a clear visualization of bacteria within a biofilm and is capable of demonstrating even a single bacterium and the relation of the biofilm to the underlying surface.

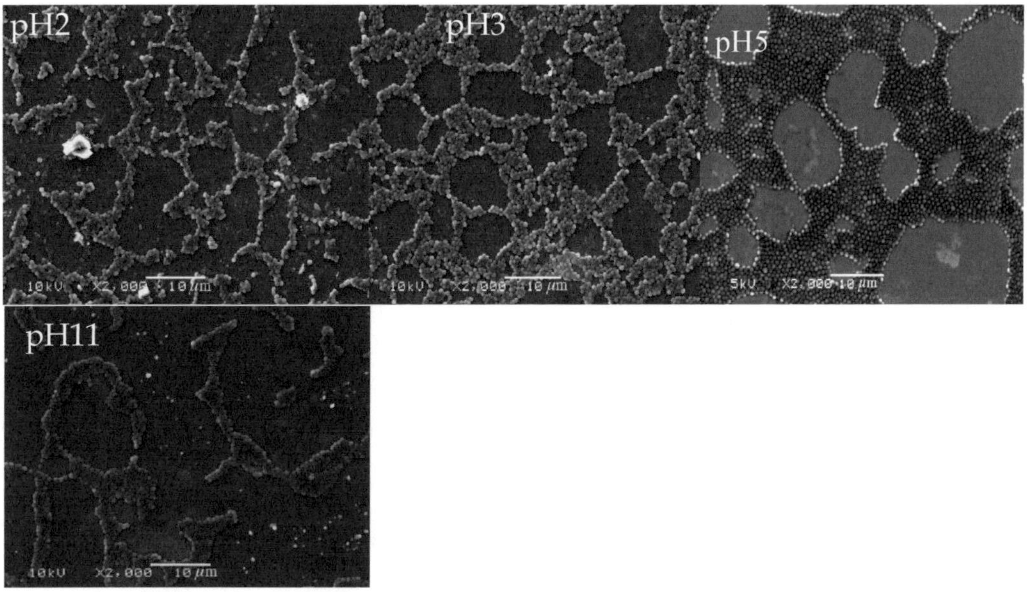

Fig. 3. SEM images of *S. aureus* adhered to glass as a function of pH (Hamadi et al., 2005)

Biofilm morphology and mass are important characteristics that control the kinetics of substrate removal by biofilms. SEM is a powerful technique for revealing the fine structure of living systems and has been applied to biofilms (Eighmy et al., 1983; Richards and Turner, 1984; Weber et al., 1978). It has also been of special importance in elucidating biofilm structure for understanding the physiology and ecology of these microbial systems (Blenkinsopp & Costerton 1991). For example, electron-microscopic studies proved that the biofilm is composed of bacterial cells "wrapped" in a dense "glycocalyx", i.e. exopolysaccharide matrix (Blenkinsopp & Costerton, 1991; Eighmy et al. 1983). In medical applications, for example, Storti et al., (2005) used scanning electron microscopy and reported that the extracellular biofilm matrix appears as an amorphous material on the catheter surface. In the same context, scanning electron microscopy (SEM) images of matrix-enclosed microbial assemblages on leaf surfaces (Surico, 1993) have led some authors to suggest that biofilms occur in the phyllosphere (Beattie and Lindow, 1995). Morris et al., (1997) have been to observe microbial biofilms directly on leaf surfaces. Bacterial aggregates in the phyllosphere have been observed previously with SEM (Surico,1993), but most have been very small (less than 20 mm long) or have lacked an obvious exopolymeric matrix (Surico,1993). Previous studies have claimed to demonstrate the presence of biofilms in situ on plant aerial surfaces using SEM (Gras et al., 1994).

Biofilm thickness is also especially important for calculation of heat exchange or diffusion rates of antimicrobials or nutrients through a biofilm and for evaluation of the mechanical properties of a biofilm (Korstgens et al., 2001). As reported elsewhere, SEM sample (freeze-dried cross-section of Foley bladder catheter) revealed the thickness of biofilm and also the layers of embedded of slime by different strains and species of bacterial cells (Ganderton et al., 1992).

In general, other application of SEM techniques may be mentioned. Akernan et al., (1993) used scanning electron microscopy of nanobacteria - novel biofilm producing organisms in blood. Indeed, nanoscale characterization of *Escherichia coli* Biofilm formed in the glass surface using scanning electron microscopy has been reported by Lim et al., (2008). He showed reticular structures on the surface of biofilms. The reticular structures consist of nanopores having diameter ranging from 14 nm to 100 nm.

Scanning electron microscopy (SEM) is one of the many methods available for the visual the effect of antibacterial or antifungal on biofilm development (Camargo et al., 2005; McDowell et al., 2004; Sasidharan et al., 2010; Sevinç & Hanley, 2010; Zameer & Gopal, 2010; Zeraik & Nitschke, 2010). Sasidharan et al., (2010) used SEM for studied The effects of potential antifungal extracts from natural sources in *Candida albicans* biofilm (Figure.4).

Fig. 4. Scanning electron micrograph reduction in *Candida albicans* biofilm after 36 h treatment. (a) Control and (b) Cassia spectabilis extract treated *C. albicans* cells.

3.3 Advantages and disadvantages of SEM

In part, it is true that Scanning electron microscopy (SEM) present a many advantages, the more important are: (i) higher resolution of visualization microbial biofilms (Walker et al., 2001) than other imaging techniques, typically 3.5 nm, (ii) able to measure and quantify data in three dimensions. However, this technique utilizes graded solvents (alcohol, acetone, and xylene) to gradually dehydrate the specimen prior to examination, since water of hydration is not compatible with the vacuum used with the electron beam. While any pretreatment can alter specimen morphology, drying appears to significantly alter biofilms due to EPS polymers collapsing (Fassel & Edmiston, 1999; Little et al., 1991). The dehydration process results in significant sample distortion and artifacts; the extracellular polymeric substances, which are approximately 95% water and the liquid loss led them to appear more like fibers surrounding the cells than like a gelatinous matrix (Characklis & Marshall, 1990). Several ultrastructural studies have used conventional scanning electron microscopy (SEM) to investigate the glycocalyx, but these studies (Costerton et al., 1981; Fassel et al., 1991; Marshall et al., 1971) were hampered by low resolution and also by the inability to use low voltages (<5 keV), which yield increased information from small topographical features (Pawley & Erlandsen, 1989).

Typically, SEM imaging requires a high vacuum, $\leq 10^{-8}$ Torr (reviewed in Stewart, 1985), having first been chemically fixed, dehydrated, and coated with a conductive material (e.g. gold) to prevent charge buildup from the electron beam. Few biological specimens tolerate these conditions without rapid collapse (Heslop-Harrison, 1970) and fewer still survive (Read & Lord, 1991). Uncoated non-conductors build up local concentration of electron, referred to as-charging- that prevent the formation of usable images. Energy X-ray Spectroscopy (EDS) can be used to determine the elemental composition of surface films in the SEM, but EDS analyses must be completed prior to deposition of the thin metal coating EDS data are typically collected from an area, the specimen must be removed from the specimen chamber and coating with a conductive layer, and returned to the SEM.

To allow observations under the high vacuum conditions of SEM, many preparations of biological samples have been developed, e.g., glutaraldehyde fixation, negative staining, the Sputter–Cryo technique, and coating with gold or osmium (Allan-Wojtas et al., 2008; Hassan et al., 2003; Lamed et al., 1987). Moreover, these preparations have some positive effects on the biological sample; for instance, they enhance contrast, reduce damage, and are uncharged up by the electron beam.

4. Biofilm formation: Environmental Scanning Electron Microscopy (ESEM)

A new SEM technique is now available which allows overcoming these obstacles. a modified, low-vacuum scanning electron microscopy technique for biofilm monitoring that enables imaging of hydrated specimens, termed environmental scanning electron microscopy (ESEM) also called variable pressure SEM (VP-SEM),was introduced in the mid-1990s (Little et al., 1991). The environmental SEM (reviewed in Stokes & Donald, 2000) uses a series of pressure limiting apertures (Muscariello et al., 2005) while preventing gas leakage from the specimen chamber, which can be maintained at 1–20 Torr. The ESEM is based upon the gaseous detection device (GDD). The main feature distinguishing ESEM from conventional SEM is the presence of a gas in the specimen chamber. Gases may include nitrous oxide, helium, argon and other, but water vapour is the most efficient amplifying gas found and the most common gas used in ESEM. The ionization GDD uses the ionization of the gas for the detection of secondary electrons from the specimen surface. It is a conical electrode about 1 cm in diameter that is positioned with the apex downward and concentric with the beam at the bottom of the pole piece. Secondary electrons emitted from the sample collide with water molecules in the chamber producing additional electrons and positive ions. The positive ions are attracted to the sample surface and eliminate the charging artifacts. A proportional cascade amplification of the original secondary electron signal results. With the GDD both secondary and backscattered electron images can be produced. Detailed technical explanations about this device can be found elsewhere (Danilatos, 1990).

The balance of gas flows into and out of the ESEM sample chamber determines its pressure.

The multiple apertures are situated below the objective lens and separate the sample chamber from the column. This feature allows the column to remain at high vacuum while the specimen chamber may sustain pressures as high as 50 Torr. The temperature and humidity of the sample can also be manually controlled to provide a suitable environment for maintaining the biological samples in their natural state.

The relative humidity in an ESEM specimen chamber can be controlled (Stokes & Donald, 2000), so ESEM is particularly useful for hydrated materials (Muscariello et al., 2005; Stokes & Donald, 2000; Stokes, 2001). A gaseous secondary electron detector (GSED) exploits the gas in the specimen chamber for signal amplification. BSED operation produces positive ions that have the added benefit of limiting charging of non-conductive specimens (Stokes & Donald, 2000). It does not require prior fixing and staining of the biofilm, minimizes biofilm dehydration and thus preserves native morphologies including surface structures (Walker et al., 2001) and native morphologies of bacteria and biofilms (e.g. Priester et al., 2007) and is able to achieve high magnifications, comparable with SEM. Shrinkage is prevented and artefact formation is reduced.

Additional advantages of ESEM include minimal processing of samples. It results in shorter time scales and lower costs while reducing the possibility of introducing artefacts. Samples can be preserved in saline in a common refrigerator (in fresh) if examination is to be deferred a few hours (Ramírez-Camacho et al., 2008). ESEM provides spatial resolutions of 10 nm or less. Compared to SEM, ESEM produces different, perhaps complementary, information for biological specimens (Doucet et al., 2005; Surman et al., 1996). Cell structures are visible with SEM, but external polymers around cells are more apparent in ESEM (Callow et al., 2003; Doucet et al., 2005; S. Douglas & D.D. Douglas, 2001).

4.1 ESEM applied of biofilm formation

Sutton et al., (1994) used this technique to study the structure of a *Streptococcus crista* CR3 biofilm. Gilpin & Sigee (1995) showed that biological samples can be imaged in the ESEM in wet or partially hydrated states with a minimum of sample damage and changes in specimen morphology. This gave the possibility to the visualization of biofilm surfaces in their natural wet anaerobic state (Darkin et al., 2001). Recently, Schwartz et al., (2009) used ESEM imaging to obtain information about the bacterial composition, matrix composition, and spatial biofilm structures of natural biofilms grown on filter materials at waterworks.

Scanning electron microscopes are frequently equipped with an energy dispersive x-ray analyser. This equipment permits elemental analysis with a high horizontal resolution of the inspected specimens. In this same context, mineral structures formed by bacterial and microalgal biofilms growing on the archaeological surface in Maltese hypogea were studied using Energy Dispersive X-Ray Spectroscopy (EDS) coupled to Environmental Scanning Electron Microscopy (ESEM), are reported by Zammit et al., (2011). These techniques have shown that mineral structures having different morphologies and chemical composition were associated with the microorganisms in the subaerophytic biofilm (Figure.5).

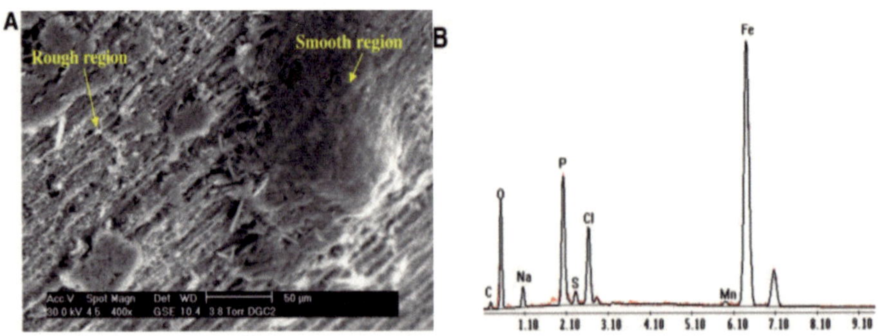

Fig. 5. ESEM and EDS analysis for the system under SRB-biofilm influence. (A) SEM Image of carbon steel exposed to sterile artificial seawater (supplemented with nutrients) and with SRB, (B) EDS analysis corresponding to the ESEM smooth region.

Interestingly, Shen et al., (2011) have been proposed a novel method for measuring an adhesion force of single yeast cell based on a nanorobotic manipulation system inside an environmental scanning electron microscope (ESEM) and Dubey & Ben-Yehuda (2011) report the identification of analogous nanotubular channels formed among bacterial cells grown on solid surface. They demonstrate that nanotubes connect bacteria of the same and different species, thereby providing an effective conduit for exchange of intracellular content.

5. Conclusion

Scanning electron microscopy is a key tool to study the effect of physicochemical properties on adhesion phenomena (pH, roughness, topography, temperature, etc). SEM plays also a paramount role for assessing the microbial populations, three-dimensional structure, physiology, thickness, etc.

SEM proved to be an invaluable method for ultra-structural investigation, allowing imaging of the overall appearance and/or specific features of biofilms formed in different environments , e.g. microbial colonies and individual cells, the glycocalyx, and the presence of inorganic products within the biofilm.

Surely, Scanning Electron Microscope (SEM) is a powerful research tool, but since it requires high vacuum conditions, the wet materials and biological samples must undergo a complex preparation that limits the application of SEM on this kind of specimen and often causes the introduction of artifacts. The introduction of Environmental Scanning Electron Microscope (ESEM), working in gaseous atmosphere, represented a new perspective in biofilm monitoring with high resolution without prior fixing and staining.

ESEM could be useful as a complementary technique to help in the characterization of the structure and architecture of biofilms. In fact, ESEM could reveal the exact topography of intact, live and fully hydrated biofilms, with a higher magnification than the other microscopy techniques. In general, a combination of several techniques is to be recommended when investigating biofilms as the different techniques offer distinctly valuable information about different aspects of biofilm development.

6. References

Akernan, K.K. Kuronen,Ilpo. Olavi Kajander, E. (1993). Scanning electron microscopy of nanobacteria - Novel biofilm producing organisms in blood. *Scanning,* Vol.15, Supplement III.

Allan-Wojtas, P. Hansen, L.T. & Paulson, A.T. (2008). Microstructural studies of probiotic bacteria loaded alginate microcapsules using standard electron microscopy techniques and anhydrous fixation. *LWT-Food Science and Technology,* Vol.41, No.1,(January 2008), pp.101-108.

An, Y. H, Dickinson, R. B. & Doyle, R. J. (2000). Mechanisms of bacterial adhesion and pathogenesis of implant and tissue infections. pp. 1-27. In An, Y. H. & Friedman, R. J. (ed.), Handbook of bacterial adhesion: principles, methods, and applications. Humana Press, Totowa, N.J.

Bacteria to Polystyrene Surfaces: Effect of Temperature and hydrophobicity. *Current of Microbiology,*Vol.61, (December 2010), pp.554-559.

Beattie, G.A. Lindow, S. E. (1995). The secret life of foliar bacterial pathogens on leaves. *Annual Reviews of Phytopathology,* Vol.33, (September 1995), pp.145–17.

Bechmann, R.T. & Eduvean, R.G.C. (2006). AFM Study of the colonization of stainless steel by *Aquabecterium commune. International Biodeterioration & Biodegradation,*Vol.58, No.3-4, (October-December 2006), pp.112-118.

Beech, I. (2004). Biocorrosion: towards understanding interactions between biofilms and metals. *Current Opinion in Biotechnology,* Vol.15, No. 3, (Jun 2004), pp.181-186.

Blenkinsopp, AS. & Costerton, JW. (1991). Understanding bacterial biofilms. *Trends in Biotechnology,* Vol.9, No.1, (January 1991), pp. 138–143.

Bragadeeswaran, S. Balasubramanian, ST. Raffi, SM. & Rani, Sophia S. (2010). Scanning electron microscopy elemental studies of primary film. *World Applied Sciences Journal.* Vol.10, No.2, pp.169-172.

Brugnoni, L-I. Lozano, J-E. & Cubitto, M-A. (2007). Potential of yeast isolated from apple juice to adhere to stainless steel surfaces in the apple juice processing industry. *Food Research International,* Vol.40, No.3, (April 2007), pp. 332–340.

Bryers, JD, (2000). Biofilms: an introduction, in Biofilms II: process analysis and applications, In: Bryers, JD, (Ed.), 3-11, Wiley-Liss, New York.

Busscher, H. J. van de Belt-Gritter, B. Dijkstra, R. J. B. Norde, W. & van der Mei, H. C. (2008). *Streptococcus mutans* and *Streptococcus intermedius* adhesion to fibronectin films are oppositely influenced by ionic strength. *Langmuir,* Vol24, N .19,(August 2008),pp.10968–10973.

Callow, J.A. Osborne, M.P. Callow, M.E. Baker, F. & Donald, A.M. (2003). Use of environmental scanning electron microscopy to image the spore adhesive of the marine alga Enteromorpha in its natural hydrated state. *Colloids Surface B: Biointerfaces,* Vol.27, No.4, (Jun 2003), pp.315–321.

Camargo, G.M.P.A. Pizzolitto, A.C. & Pizzolitto, E.L. (2005). Biofilm formation on catheters used after cesarean section as observed by scanning electron microscopy. *International Journal of Gynecology and Obstetrics,* Vol.90, (August 2005), pp.148—149.

Carpentier, B. & Cerf, O. (1993) Biofilms and their consequences, with particular reference to hygiene in the food industry. *Journal of Applied Bacteriology,* Vol.75, No. 6, (March 1993),pp.499-511.

Characklis, W. G. & Marshall, K.C. (1990). Biofilms: a basis for an interdisciplinary approach, pp. 3–15. In: Characklis, W.G. & Marshall, K.C. (ed.), Biofilms. John Wiley & Sons, New York, N.Y.

Costerton, J.W. Irvin, R.T. Cheng, K-J. (1981). The bacterial glycocalyx in nature and disease. *Annual Reviews of Microbiolology,* Vol.35, (October 1981), pp.299–324

Costerton, JW, Stewart, PS. & Greenberg, EP. (1999). Bacterial biofilms: a common cause of persistent infections. *Science,* Vol.284, No.5418, (May 1999), pp. 1318-1322.

Danilatos, G.D. (1990). Theory of the gaseous detector device in the environmental scanning electron microscope. *Advances in Electronics and Electron Physics,* Vol.78, pp.1-102.

Darkin, M.G. Gilpin, C. Williams, J.B. & Sangha, C.M. (2001). Direct wet surface imaging of an anaerobic biofilm by environmental scanning electron microscopy: application to landfill clay liner barriers, *Scanning,* Vol.23, No .5,pp. 346–350.

Donlan, RM. (2001) Biofilms and Device-Associated Infections. *Emerging Infectious Diseases,* vol.7, No .2, (March-April 2001),pp.277-281.

Doucet, F.J. Lead, J.R. Maguire, L. Achterberg, E.P. & Millward, G.E. 2005. Visualisation of natural aquatic colloids and particles - a comparison of conventional high vacuum and environmental scanning electron microscopy. *Journal of Environmental Monitoring,* Vol.7, No.2, (January 2005),pp.115–121.

Douglas, S. & Douglas, D.D. (2001). Structural and geomicrobiological characteristics of a microbial community from a cold sulfide spring. *Geomicrobiology Journal,* Vol.18, No.4, (November 2001),pp.401–422.

Dubey, G.P. and Ben-Yehuda, S. (2011). Intercellular nanotubes mediate bacterial communication. *Cell,* Vol.144, No. 4, (February 2011),pp. 590–600.

Duckett, J.G. & Ligrone, R. (1995). The formation of catenate foliar gemmae and the origin of oil bodies in the liverwort Odontoschisma denudatum (Mart.) dum (Jungermanniales): a light and electron microscope study. *Annals of Botany,* Vol.76, (October 1995),pp.405–419.

Dufrêne, Y.F. (2008).Towards nanomicrobiology using atomic force microscopy. *Nature Reviews Microbiology,* Vol.6, N .9,(September 2008),pp.674-680.

Dunne, WM. (2002). Bacterial Adhesion: Seen Any Good Biofilms Lately? *Clinical Microbiology Reviews,* Vol.15, No.2, (April 2002), pp. 155-166.

Eighmy, T.T. Maratea, D. & Bishop P.L. (1983). Electron microscopic examination of wastewater biofilm formation and structural components. *Applied and Environmental Microbiology,* Vol.45, No.6, pp.1921-1931.

Fassel, T.A. & Edmiston, C.E. (1999). Bacterial biofilms: strategies for preparing glycocalyx for electron microscopy. *Methods in Enzymology,* Vol.310, pp.194–203.

Fassel, T.A. Van Over, J.E. Hauser, C.C. Edmiston, C.E. Sanger, J.R. (1991). Adhesion of staphylococci to breast prosthesis biomaterials: an electron microscopic evaluation. *Cells Materials,* Vol.1,pp.199–208.

Filloux, A. & Vallet, I. (2003). Biofilm: set-up and organization of a bacterial community. *Medical Science, Vol.*19, No.1, (January 2003), pp. 77-83.

Flemming, H. Neu, T.R. & Wozniak, D.J. (2007). The EPS Matrix: The "House of Biofilm Cells". *Journal of Bacteriology,* Vol.189, No. 22, (November 2007),pp.7945-7947.

Fux, C. Costerton, J. Stewart, P. & Stoodley P. (2005). Survival strategies of infectious biofilms. *Trends in Microbiology,* Vol.13, No. 1, (January 2005), pp.34-40.

Ganderton, L. Chawla, J. Winters, C. Wimpenny, J. & Stickler, D. (1992). Scanning electron microscopy of bacterial biofilms on indwelling bladder catheters. *European Journal Of clinical Microbiololgy Infection Diseases,* Vol.11, No.9, (September 1992), pp. 789–796.

Garrett, TR. Bhakoo, M. & Zhang, Z. (2008). Bacterial adhesion and biofilms on surfaces. *Progress in Natural Science,* Vol.18, No.9, (September 2008), pp. 1049-1056.

Gilpin, C.J. & Sigee, D.C. (1995). X-ray microanalysis of wet biological specimens in the environmental scanning electron microscope. 1. Reduction of specimen distance under different atmospheric conditions. *Journal of Microscopy,* Vol. 179,No .1,(July 1995),pp.22–28.

Gras, M. H. Druetmichaud, C. & Cerf, O. (1994). La flore bactérienne des feuilles de salade fraiche. *Sciences des Aliments.* Vol.14, No. 2,pp.173–188.

Hall-Stoodley, L. Costerton, J.W. & Stoodley, P. (2004). Bacterial Biofilms: from the natural environment to infectious diseases. *Nature Reviews,* Vol.2, No .2, (February 2004),pp.95-108.

Hamadi, F. Latrache, H. Mabrrouki, M. Elghmari, A. Outzourhit,A. Ellouali, M. & Chtaini, A. (2005). Effect of pH on distribution and adhesion of *Staphylococcus aureus* to glass. *Journal of Adhesion Science and Technology,* Vol.19, No.1, (November 2004), pp. 73-85.

Hassan, A.N. Frank, J.F. &Elsoda, M. (2003). Observation of bacterial exopolysaccharide in dairy products using cryo-scanning electron microscopy. *International Dairy Journal,* Vol.13, No.9,(July 2003), pp.755–762.

Herald, PJ. & Zottola, EA. (1988). Scanning electron microscopic examination of *Yersinia enterocolitica* attached to stainless steel at selected temperatures and pH values. *Journal of Food Protection*, Vol.51, No.6, (Jun 1988), pp. 445–448

Heslop-Harrison, Y.(1970). Scanning electron microscopy of fresh leaves of *Pinguicula. Science*, Vol.167, No .3815, (January 1970),pp.172–174.

Kaplan, J.B. Meyenhofer, M.F. & Fine, D.H. (2003). Biofilm Growth and Detachment of *Actinobacillus actinomycetemcomitans. Journal of Bacteriology*, Vol.185, No.4, (February 2003), pp.1399-1404.

Korstgens, V, Flemming, HC. Wingender, J. & Borchard, W. (2001). Influence of calcium ions on the mechanical properties of a model biofilm of mucoid *Pseudomonas aeruginosa. Water Science. Technology*, Vol.13, No.6, pp.49-57.

Kouider, N. Hamadi, F. Mallouki, B. Bengoram, J. Mabrouki, M. Zekraoui, M. Ellouali, M. & Latrache, H. (2010). Effect of stainless steel surface roughness on Staphylococcus aureus adhesion. *International Journal of Pure and Applied Science*, Vol.4, No.1, (August 2009),pp. 1-7.

Lamed, R. Naimark, J. Morgenstern, E. & Bayer, E.A (1987). Scanning electron microscopic delineation of bacterial surface topology using cationized ferritin. *Journal Microbiological. Methods*, Vol.7, No.4-5,(December 1987), pp.233–240.

Lewis, K. (2005). Persister cells and the riddle of biofilm survival. *Biochemistry (Moscow)*, Vol.70, No. 2, (February 2005), pp.267-274.

Lim,J. (2008). Nanoscale characterization of *Escherichia coli* biofilm formed under laminar flow using atomic force microscopy (AFM) and scanning electron microscopy. *Bulletin of the Korean Chemical Society*,Vol.29,No .11,pp.2114-2118.

Little, B. Wagner, P. Ray, R. Pope, R. & Scheetz, R. (1991). Biofilms: an ESEM evaluation of artifacts introduced during SEM preparation. *Journal of Industrial Microbiology*, Vol. 8, No.4, pp. 213-222.

Liu, Y-Q. Liu, Y. & Tay, J-H. (2004). The effects of extracellular polymeric substances on the formation and stability of biogranules. *Applied Microbiology and Biotechnology*, Vol.65, No. 2, (Jun 2004),pp.143-148.

Mallouki, B. Latrache, H. Mabrouki, M. Outzourhit, A. Hamadi, F. Muller, D. & Ellouali, M. (2007).The inhibitory effect of fucans on adhesion and production of slime of *Staphylococcus aureus. Microbiologie Hygiène Alimentaire*, Vol.19, No.55, (July 2007), pp. 64-71.

Marshall,K.C. Stout, R. Mitchell, R. (1971). Mechanism of the initial events in the sorption of marine bacteria to surfaces. *Journal of Genetic of Microbiology*, Vol.68, No.3, (November 1971), pp.337–348

McDowell, J.W. Daryl, B.S. Paulson, S. & Mitchell, J.A. (2004). A simulated-use evaluation of a strategy for preventing biofilm formation in dental unit waterlines. *The Journal of the American Dental Association*, Vol 135, No. 6, (Jun 2004),pp.799-805.

Minoura, N. Aiba, S.I. Higuchi, M. Gotoh, Y. Tsukada, M. & Imai, Y. (1995). Attachment and growth of fibroblast cells on silk fibroin. *Biochemical and Biophysical Research Communications*, Vol.208, No.2, (March 1995),pp.511–516.

Molin, S. (2003). Gene transfer occurs with enhanced efficiency in biofilms and induces enhanced stabilisation of the biofilm structure. *Current Opinion in Biotechnology*, Vol.14, No. 3, (Jun 2003),pp.255-261.

Morris, C. E, Monier, J.-M. & Jacques, M.-A. (1997). Methods for observing microbial biofilms directly on leaf surfaces and recovering them for isolation of culturable microorganism. *Applied Environmental of Microbiology*,Vol.63, No. 4, (April 1997),pp.1570-1576.

Motta, P.M. Makabe, S. Naguro, T. & Correr, S. (1994). Oocyte follicle cells association during development of human ovarian follicle. A study by high resolution scanning and transmission electron microscopy. *Archives of Histology and Cytology,* Vol.57, No.4, (October 1994),pp.369–394.

Muscariello, L. Rosso, F. Marino, G. Giordano, A. Barbarisi, M. Cafiero, G. & Barbarisi, A. (2005). A critical overview of ESEM applications in the biological field. *Journal of Cellular Physiology,* Vol.205, (Jun 2005),pp.328–334.

Nadell, C.D. Xavier, J.B. Levin, S.A. & Foster, K.R. (2008). The Evolution of Quorum Sensing in Bacterial Biofilms, *Plos Biology,*Vol.6, No. 1, (January 2008), e14.

O'Toole. G, Kaplan, H.B. & Kolter, R. (2000). Biofilm formation as microbial development. *Annual Reviews of Microbiology,* Vol.54, pp.49-79.

Pawley, J.B. Erlandsen, S.L. (1989). The case for low voltage high resolution scanning electron microscopy of biological samples. *Scanning Microscopy,* 3(suppl),pp.16–173.

Peters, G. Locci, R. & Pulverer, G. (1982). Adherence and growth of coagulase-negative *staphylococci* on surfaces of intravenous catheters, *Journal of Infectious Diseases,* Vol.146, No.4, pp.479–482.

Pinna, A. Sechi, LA. Zanetti, S. Delogu, D. & Carta, F. (2000). Adherence of Ocular Isolates of *Staphylococcus Epidermidis* to ACRYSOF Intraocular Lenses. *Ophthalmology,* Vol.107, No.12, (October 1982), pp. 2162- 2166.

Prakash, B. Veeregowda, B.M. & Krishnappa, G. (2003). Biofilms: A survival strategy of bacteria. *Current Science,* Vol. 85, No. 9, (November 2003),pp.9-10.

Priester, J.H. Horst, A.M. Van De Werfhorst, L.C. Saleta, J.L. Mertes, L.A.K. & Holden, P.A. (2007). Enhanced visualization of microbial biofilms by staining and environmental scanning electron microscopy. *Journal of Microbiological Methods,* Vol.68, No .2, (March 2007), pp.577–587.

Ramírez-Camacho, R. González-Tallón, A.I. Gómez, D. Trinidad, A. Ibáñez, A. García-Berrocal, J.R. Verdaguer, J.M. González-García, J.A. & San Román, J. (2008). Environmental scanning electron microscopy for biofilm detection in tonsils. *Acta Otorrinolaringol Esp,* Vol.59, No.1, (January 2008),pp.16-20.

Read, N.D. & Lord, K.M. (1991). Examination of living fungal spores by scanning electron microscopy. *Experimental mycology,* Vol.15, No .2,pp.132–139.

Richards, S.R. & Turner, R.J. (1984). A comparative study of techniques for the examination of biofilms by scanning electron microscopy. *Water Research,* Vol.18, No.6, pp.767-773.

Rodrigues, D.F. & Elimelech, M. (2009). Role of Type 1 Fimbriae and Mannose in the Development of *E. coli* K12 Biofilm: From Initial Cell Adhesion to Biofilm Formation. *Biofouling,* Vol.25, No .5,(July 2009), pp.401-411.

Sasidharan, S. Yoga Latha, L. & Angeline, T. (2010). Imaging In vitro Anti-biofilm Activity to Visualize the Ultrastructural Changes. Microscopy: Science, Technology, Applications and Education A. Méndez-Vilas & J. Díaz (Eds.) Formatex, 2010,pp. 622-626.

Schwartz, T. Jungfer, C. Heißler, S. Friedrich, F. Faubel, W. & Obst, U. (2009). Combined use of molecular biology taxonomy, Raman spectrometry, and ESEM imaging to study natural biofilms grown on filter materials at waterworks. *Chemosphere,*Vol.77, No .2, (September 2009),pp.249–257.

Sevinç, B.A. & Hanley, L. (2010). Antibacterial activity of dental composites containing zinc oxide nanoparticles. *Journal of Biomedical Materials Research Part B: Applied Biomaterials,* Vol.94, No.1, (July 2010), pp.22–31.

Shen, Y. Nakajima, M. Ahmad, M.R. Kojima, S. Hommac, M. & Fukuda, T. (2011). Effect of ambient humidity on the strength of the adhesion force of single yeast cell inside environmental-SEM. *Ultramicroscopy,* Vol.111,No .8, pp:1176-1183.

Singh, P. K. Parsek, M. R. Greenberg, E. P. & Welsh, M. J. (2002). A component of innate immunity prevents bacterial biofilm development. *Nature,* Vol.417, No. 6888, (May 2002), pp.552-555.

Stewart, A.D.G. (1985). The origins and development of scanning electron microscopy. *Journal of Microscopy,* Vol.139, No .2, (August 1985), pp.121–127.

Stokes, D.J. & Donald, A.M. (2000). In situ mechanical testing of dry and hydrated breadcrumb in the environmental scanning electron microscope (ESEM). *Journal of Materials Science,* Vol.35, No .3, (December 2000),pp.599–607.

Stokes, D.J. & Donald, A.M. (2000). In situ mechanical testing of dry and hydrated breadcrumb in the environmental scanning electron microscope (ESEM). *Journal of Materials Science,* Vol.35, No .3, (December 2000),pp.599–607.

Stokes, D.J. (2001). Characterization if soft condensed matter and delicate materials using environmental scanning electron microscopy (ESEM). *Advanced Engineering Materials,* Vol.3, No.3, pp.126–130.

Stoodley, P, Sauer K, Davies, DG. & Costerton, JW. (2002). Biofilms as complex differentiated communities. *Annual Review of* Microbiology, Vol.56, (January 2002), pp. 187-209.

Storti, A., Pizzolitto, CA. & Pizzolitto, LE. (2005) Detection of mixed microbial biofilms on central venous catheters removed from intensive care unit patients. *Brazilian Journal of Microbiology,* Vol.36, pp.275-80.

Surico, G. (1993). Scanning electron microscopy of olive and oleander leaves colonized by *Pseudomonas syringae* subsp. *savastanoi. Journal of Phytopathology,*Vol.138, No. 1, (May 1993), pp.31–40.

Surman, S.B. Walker, J.T. Goddard, D.T. Morton, L.H.G. Keevil, C.W. Weaver, W. Skinner, A. Hanson, K. & Caldwell, D. (1996). Comparison of microscope techniques for the examination of biofilms. *Journal of Microbiological Methods,* Vol.25, No.1, (March 1996) pp.57–70.

Sutherland, IW. (2001). The biofilm matrix - an immobilized but dynamic microbial environment. *Trends of Microbiology,* Vol.9, No.5, (May 2001), pp. 222-227.

Sutton, N.A. Hughes, N. & Handley, PS. (1994). A comparison of conventional SEM techniques, low temperature SEM and the electroscan wet scanning electron microscope to study the structure of a biofilm of *Streptococcus crista* CR3. *Journal of Applied Bacteriology,*Vol.76, No .5,(May 1994),pp.448-454.

Vidali, M. (2001). Bioremediation. A overview. *Pure and Applied Chemistry,* Vol.73, No.7, pp. 1163-1172.

Walker, JT. Verran, J. Boyd, RD. & Percival, S. (2001). Microscopy methods to investigate structure of potable water biofilms. *Methods in Enzymology,* Vol.337, No.2001, (July 2004), pp.243–255.

Weber, W.J.J. Pirbazari, M. & Melson, G.L. (1978). Biological growth on activated carbon: an investigation by scanning electron microscopy. *Environmental Science & Technology,*Vol.12, No. 7, (July 1978),pp. 817-819.

Zameer, F. & Gopal, S. (2010). Evaluation of antibiotic susceptibility in mixed culture biofilms. *International Journal of Biotechnology and Biochemistry,*Vol.6, No.1, pp. 93–99.

Zammit, G. Sánchez-Moral, S. & Albertano P. (2011). Bacterially mediated mineralisation processes lead to biodeterioration of artworks in Maltese catacombs. *Science of the Total Environment,* Vol.409, No. 14, (Jun 2011),pp.2773-2783.

Zeraik, A.E. & Nitschke, M.(2010). Biosurfactants as agents to reduce adhesion of pathogenic

36

Scanning Electron Microscopy Study of Fiber Reinforced Polymeric Nanocomposites

Mohammad Kamal Hossain
Department of Mechanical Engineering, Tuskegee University, USA

1. Introduction

The nanoscale materials offer the opportunity to explore new behavior beyond those established in conventional materials. It has been established that mechanical and thermal properties, moisture barrier, and flame resistance of polymeric composites can be improved by adding a small amount of nanoparticles as filler particles without compromising the density, toughness, storage life, weight or processability of the composite [Sandler et al., 2002; Bruzaud & Bourmaud, 2007]. The higher surface area is one of the most promising characteristics of nanoparticles due to their ability in creating a good interface in a composite. The dispersion of nanoparticles in the matrix is one of the most important parameters in fabricating nanophased composites. It depends on processing techniques such as solution blending, shear mixing, in-situ polymerization, ultrasonic cavitation, and high pressure mixing [Giannelis, 1998; Yasmin et al., 2003; Vaia et al., 1996]. Nanomaterials have enhanced various characteristics in a given polymer. However, these enhancements have limitations at higher loadings due to increased agglomeration causing premature failure.

During service life, composite structures might encounter high stresses resulting in crack propagation through fiber matrix interfaces. Therefore, stronger adhesion between fiber and matrix, higher strength, and higher toughened matrix are desired. Improvement of flexural strength by addition of nanofillers into a matrix is expected to be observed for several reasons. Young's modulus of the second phase dispersed particles is higher than that of the matrix and thus stress transfer from the matrix to the particles will take place. As a result, the strength of the composites is increased. Strong interfacial bonding between the fiber and matrix also contributes to higher flexural strength. Dispersed filler particles act as a mechanical interlocking between the fiber and matrix which creates a high friction coefficient. Finally, a mixed mode of fracture (flexural and shear) occurs under bending-load conditions. After an initial failure of fibers at the tensile side of the specimen, cracks are deflected parallel to the fibers and also to the applied load direction. The stress-strain curve shows a sharp increment with increasing load before reaching the maximum stress and then irregularities and staggered decrease in stress were observed for both conventional and nanophased composites [M.K. Hossain et al., 2011]. However, the initial load and the crack arrest area are higher in nanophased composites which lead to high energy absorbing mechanisms [Hussain et al., 1996].

The morphological study of the nanoparticle dispersion in the CNF-loaded polyester nanocomposites using various processing techniques, including mechanical mixing, magnetic stirring, and sonication is evaluated through SEM studies in this article. Nanoparticles facilitate proper wetting out fibers with resin, void reduction, enhanced crosslinking, and increased friction co-efficient. Sometimes it also works as a nucleating agent in a fiber reinforced polymeric composite. Morphology of the glass/polyester-CNF composites manufactured by the VARTM process has also been studied using the SEM for various applications, including civil infrastructure, automotive, aerospace, sporting, and marine industries. Quantitative enhancement in properties is characterized through the flexural test in this article. Qualitative and visual analyses are important to support the resultant experimental quantitative data and fracture morphology evaluation of different tested specimens using SEM is inevitable in this matter. Therefore, the fracture morphology of all types of tested specimens has been evaluated in this article.

2. Research methodology

2.1 Neat and CNF-loaded polyester sample fabrication

B-440 premium polyester resin, styrene, and heat treated PR-24 CNF were used as matrix, thinner, and nanoparticle, respectively. Polyester resin contains two parts: part-A (polyester resin) and part-B (MEKP- methyl ethyl ketone peroxide) as crosslinking agent. Sonication was performed in a glass beaker using a high intensity ultrasonic irradiation (Ti-horn, 20 kHz Sonics Vibra Cell, Sonics Mandmaterials, Inc, USA) for 60, 90, and 120 minutes, respectively, adding 0.1-0.4 wt.% CNFs to polyester resin while adding 10 wt.% styrene. The mixing process was carried out in a pulse mode of 30 sec. on and 15 sec. off at an amplitude of 50%. To study other types of mixing methods, CNFs were mixed using mechanical mixing and magnetic stirring methods. The mechanical mixer was run for 90 minutes at 300 rpm at room temperature. The magnetic stirring was carried out for 5 hours at 500 rpm at room temperature. To lessen the void formation, vacuum was applied using Brand Tech Vacuum system for about 90-120 minutes and 0.7 wt.% catalyst was then added to the mixer using a high-speed mechanical stirrer for about 2-3 minutes and vacuum was again applied for about 6-8 minutes to degasify the bubbles produced during the catalyst mixing. The as-prepared resin was poured into the mold and kept at room temperature for 12-15 hours. Controlled polyester samples were fabricated to compare with the nanophased samples. All samples were kept in a mechanical convection oven at 110 °C for 3 hours for post curing [M.E. Hossain et al., 2011].

2.2 Conventional and nanophased fiber reinforced composite sample cabrication

Both conventional and nanophased E-glass/polyester-CNF composites were manufactured by the VARTM process. Vacuum was maintained until the end of cure to remove any volatiles generated during the polymerization process. The panels were cured for about 12-15 hours at room temperature and then thermally post cured at 110 °C for 3 hours in a mechanical convection oven. The fiber volume fraction for the nanophased glass reinforced polyester composites fabricated by VARTM was found to be around 56%. The void content (3-4%) was also within a reasonable limit in these composites [M.K. Hossain et al., 2011]. The overall sample fabrication procedure is presented in Figure 1.

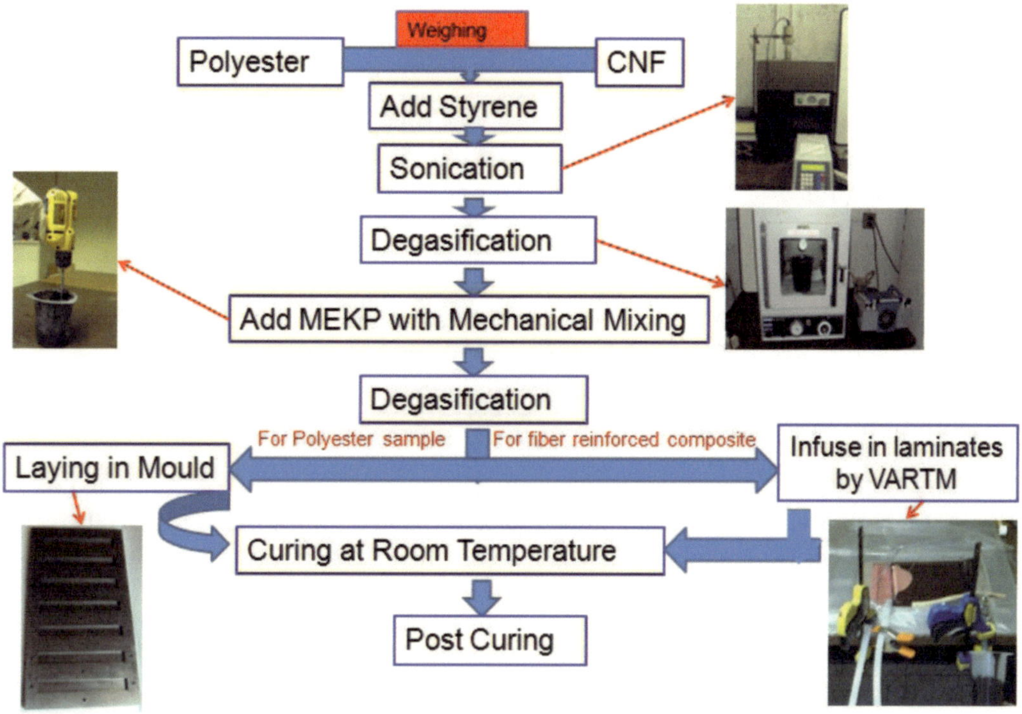

Fig. 1. Flow chart of sample fabrication

2.3 Flexural test

Flexural tests under three-point bend configuration were performed using a Zwick Roell testing unit according to the ASTM D790-02 standard to evaluate flexural modulus and strength of each of the material systems of the polymer nanocomposites and its laminates [M.E. Hossain et al., 2011; M.K. Hossain et al., 2011]. The machines were run under displacement control mode at a crosshead speed of 2.0 mm/min and tests were performed at room temperature. The span to depth ratio was maintained at 16:1. The maximum stress at failure on the tension side of a flexural specimen was considered as the flexural strength of the material. Flexural modulus was calculated from the slope of the stress-strain plot. Five samples of each type were tested. The average values and standard deviation of flexural strength and modulus were determined.

2.4 Scanning Electron Microscopy (SEM)

SEM studies were carried out to examine change in the microstructure due to the addition of CNFs using a JEOL JSM 5800 microscope. SEM also facilitates to monitor the failure approach at micro level. Failed samples from the three point flexure test were examined to distinguish the changes in the failure mode. The samples were cut through the cross-section of the failed region. The samples were positioned on a sample holder with a silver paint and coated with gold to prevent charge build-up by the electron absorbed by the specimen. A 15 kilovolt accelerating voltage was applied to achieve desired magnification [M.E. Hossain et al., 2011; M.K. Hossain et al., 2011].

2.5 SEM sample preparation

SEM samples must have an appropriate size to fit in the specimen chamber and is generally mounted rigidly on a specimen holder. Specimens must be electrically conductive, specially the surface, for imaging, and electrically grounded to prevent the accumulation of electrostatic charge at the surface of the specimen during electron irradiation [Suzuki, 2002]. Metal objects require little special preparation for SEM except for cleaning and mounting on the specimen holder. Nonconductive specimens tend to charge when scanned by the electron beam, especially in the secondary electron imaging mode. This causes scanning faults and other image artifacts. They are therefore usually coated with an ultrathin electrically-conducting material, commonly gold, deposited on the sample either using a low vacuum sputtering machine or a high vacuum evaporation unit.

3. Results and discussion

3.1 Flexural test results

Typical stress–strain curves of neat and nanophased polyester samples as well as their fiber reinforced laminates generated from flexural tests are illustrated in Figures 2 and 3. Flexural strength, modulus, and the strain at maximum strength for all CNF-loaded samples were larger than those of the neat samples. CNF has high aspect ratio which can prevent crack generation and propagation in the polyester matrix (Figure 2). In all cases, the samples failed rapidly after experiencing the maximum load showing induced brittle nature of failure due to the addition of CNFs. The 0.2 wt.% CNFs loading and 90 minutes sonication time were observed to be the optimal condition for this nanocomposite system. The 0.2 wt.% CNF-loaded samples enhanced the flexural strength and modulus by about 88% and 16%, respectively, compared to the neat ones. The failure strain also increased significantly with the addition of CNFs into the system. Flexural properties were slightly decreased at higher CNF content. It might be due to the creation of micro aggregates of CNFs in various regions of the polymer matrix, which act as areas of weakness [M.E. Hossain et al., 2011].

Typical stress-strain curves of conventional and nanophased glass/polyester composites presented in Figure 3 demonstrated significant improvement in the mechanical properties up to the 0.2 wt.% of CNFs loading, beyond which there was a decreasing trend. These curves showed considerable nonlinear deformation before reaching the maximum stress. This was attributed to the random fiber breakage during loading. However, more or less ductility was observed in each type of laminate sample and cracking noise was heard while the individual fiber broke or the inter-layer delaminated. No obvious yield point was found. From the resultant data, it was concluded that the 0.2 wt.% CNF was the optimum amount for this material system to achieve the maximum flexural modulus and strength. These specimens showed approximately 49% and 31% increase in the flexural strength and modulus, respectively. There are several reasons for the better mechanical properties observed in the CNF-infused glass fabric reinforced polyester laminates. First, CNFs increase the strength and modulus of the polyester matrix, which was observed in the CNF-loaded polyester in this study. Second, the presence of CNFs increases the crack propagation resistance and prevents crack generation by bridging effect at the interface region of the long glass fiber, CNF, and polyester matrix. Moreover, CNF has high aspect ratio, which improves the strength and modulus [M.K. Hossain et al., 2011].

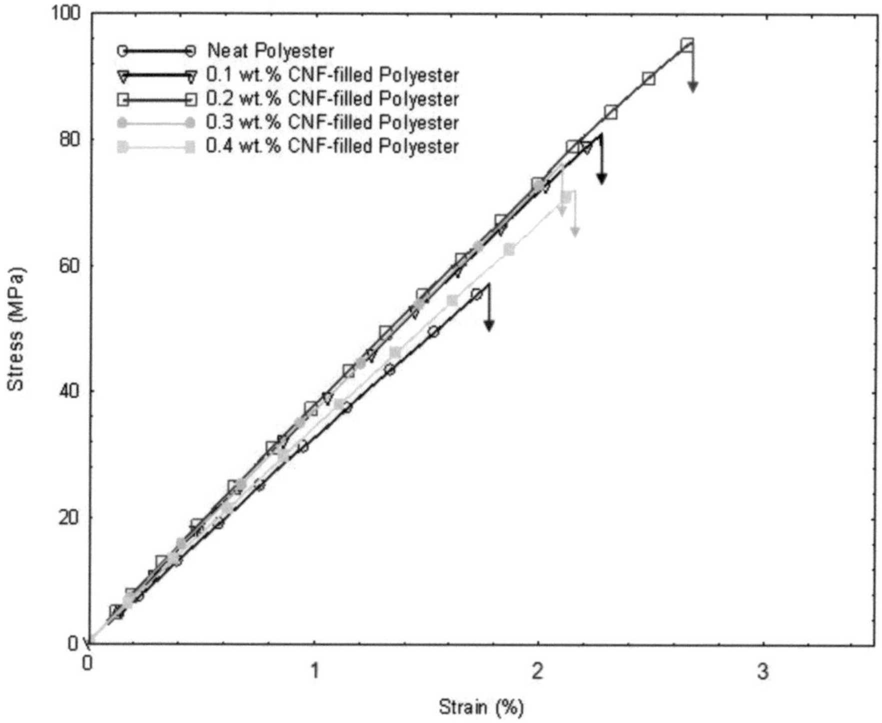

Fig. 2. Flexural stress-strain plot of polyester samples with different wt.% of CNFs.

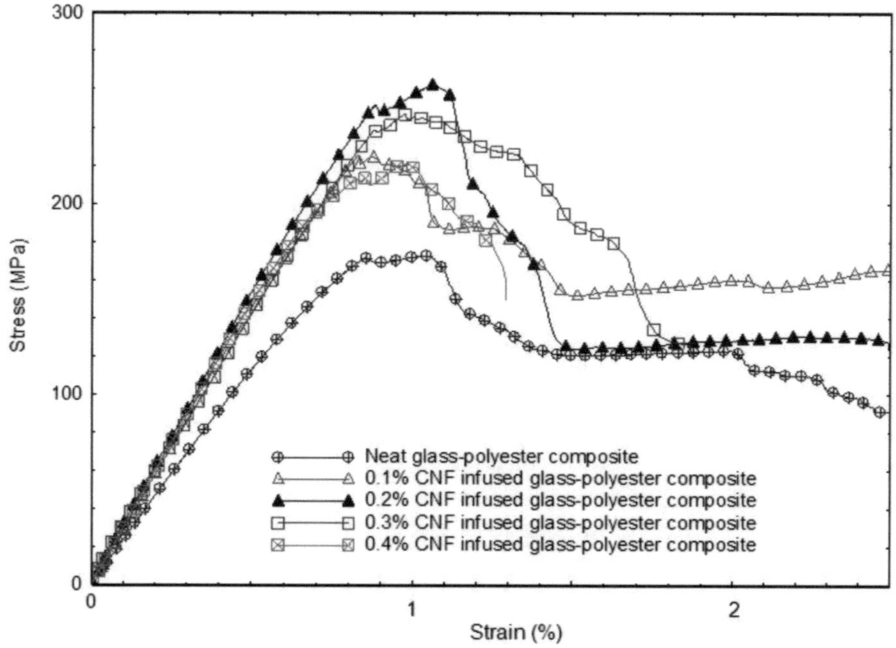

Fig. 3. Flexural stress- strain plot of GRPC laminates with different wt.% of CNF.

3.2 Microstructure and morphological analyses

The SEM micrographs of as-received PR-24 CNF and the neat polyester matrix are shown in Figures 4 (a) and 4 (b), respectively. To investigate the dispersion properties of CNFs in polyester, drops of concentrated HNO_3 acid were added on the cleavage surfaces to partly unveil the CNFs formerly covered by the polyester. From the micrograph of 0.2 wt.% CNF-filled polyester, excellent dispersion of CNFs was found (Figure 5). Only broken ends of CNFs were observed near the surface. Some CNFs broke in a brittle manner and some were pulled out.

(a) (b)

Fig. 4. SEM micrographs of (a) as-received PR-24 CNF and (b) neat polyester matrix.

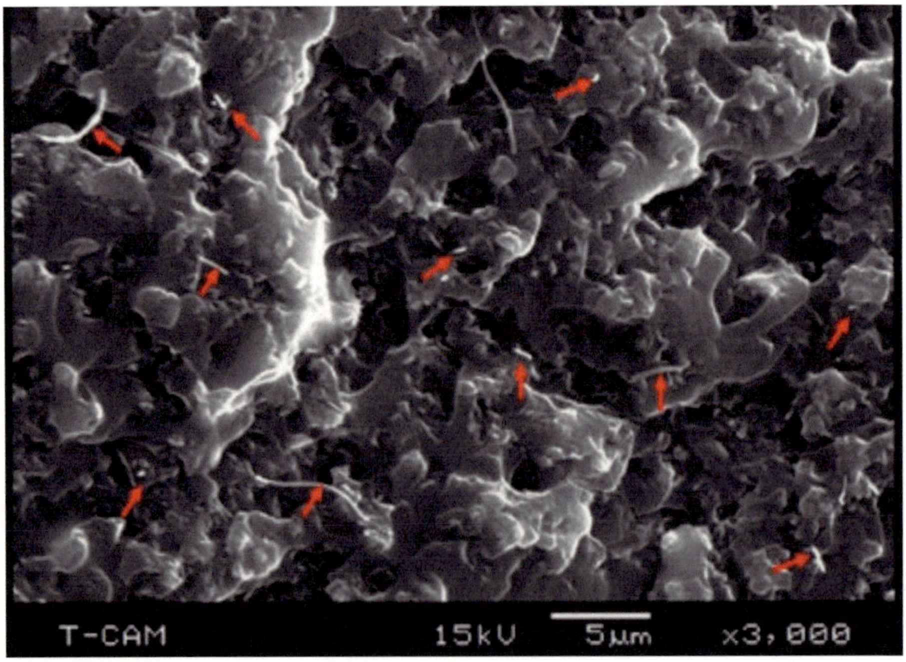

Fig. 5. SEM micrograph of acid-etched 0.2 wt.% CNF-loaded polyester at 3000X.

Agglomerations in the polyester matrix were observed from the micrographs (Figure 6) of 0.2 wt.% infused polyester samples prepared through mechanical mixing and magnetic stirring methods, respectively. These agglomerates area create stress concentration zones which might act as a crack initiator.

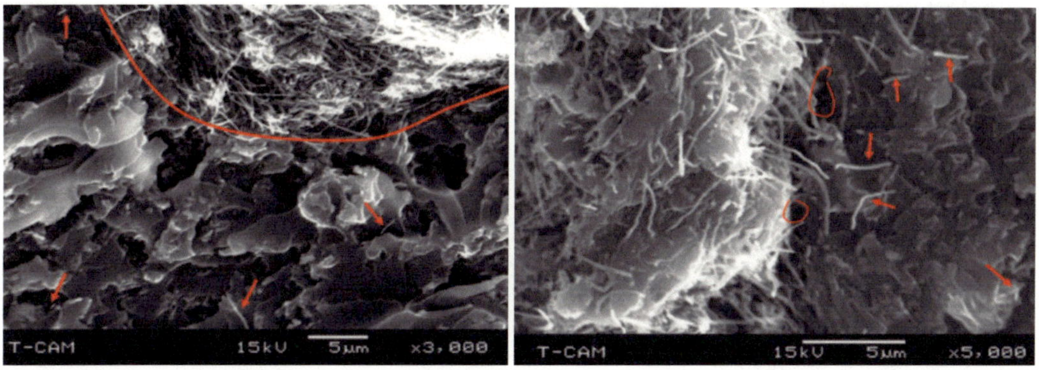

(a) (b)

Fig. 6. Micrographs of acid-etched 0.2 wt.% CNF-loaded polyester: (a) Mechanical stirring and (b) Magnetic stirring

Strong attractive fiber van der Waals forces cause CNFs to agglomerate, which reduces the strength of the nanocomposite by stress concentration effect. Agglomerates of CNFs, called nanoropes, are difficult to separate and infiltrate with matrix. They entangle and form nest-like structures due to their curvature and high aspect ratios. Both disagglomeration and dispersion in resins depend on the relative van der Waals forces, curvature, and on the relative surface energy of CNFs versus that of the resin. To overcome attractive forces, researchers have been extensively using mechanical energy, intense ultrasonication, and high speed shearing. Some rebundling of the aggregates is possible even after discontinuation of the external force [Yoonessi et al., 2008]. However, optimal loading and uniform dispersion of CNFs in matrix are the key parameters to promote better nanofiber-matrix interface properties to reach an efficient load transfer between two constituents of the nanocomposite [Kozey et al., 1995; Ma et al., 2003].

Uniform dispersion of 0.2 wt.% CNFs into the polyester resin was achieved using the sonication mixing method for 90 minutes. High magnification SEM micrograph in Figure 7 clearly exhibits that CNFs are well separated and uniformly embedded in the 0.2% polyester resin system. It can also be easily observed that the interfacial bonding between the CNF and matrix was very compact which would allow CNFs to be anchored in the embedding matrix. In essence, these CNFs are likely to interlock and entangle with the polymer chains in the matrix [Li et al., 2008]. Thus, addition of CNFs enhanced the crosslinking between polymer chains and provided better interfacial bonding.

Figures 8 (a) and 8 (b) show the woven glass reinforced polyester laminates with 0.2 wt.% CNFs. It was found that the resin was distributed uniformly over the fabric, and the interfacial bonding between matrix and fiber was very good. Resin flow and impregnation of the glass fibers can be observed in the SEM micrographs. Clear resin matrix adhesion is present in these micrographs, and glass fibers are observed to be embedded within the matrix. Good matrix-fiber wetting was achieved and resin is also visible in between the glass fiber filaments. It appears that better interfacial bonding between the nanophased polymer matrix and glass fiber is present due to the presence of CNFs [Green et al., 2009]. The fiber volume fraction as determined from matrix digestion method for the nanophased glass reinforced polyester composites fabricated by the VARTM process was found to be around 56%.

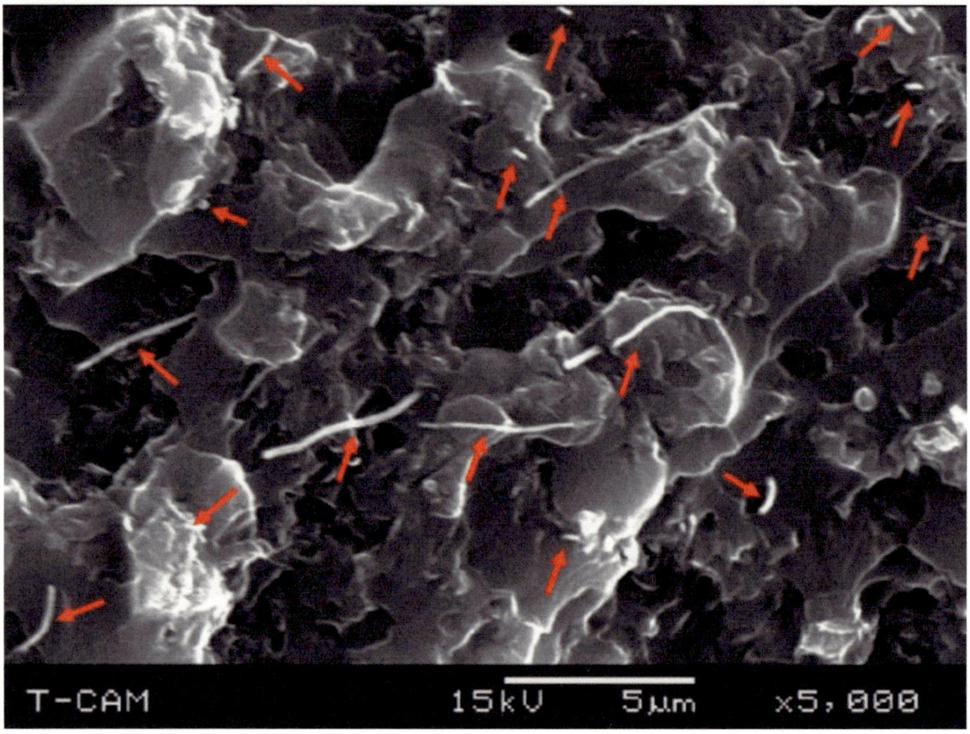

Fig. 7. SEM micrograph of acid-etched 0.2 wt.% CNF-loaded polyester at 5000X

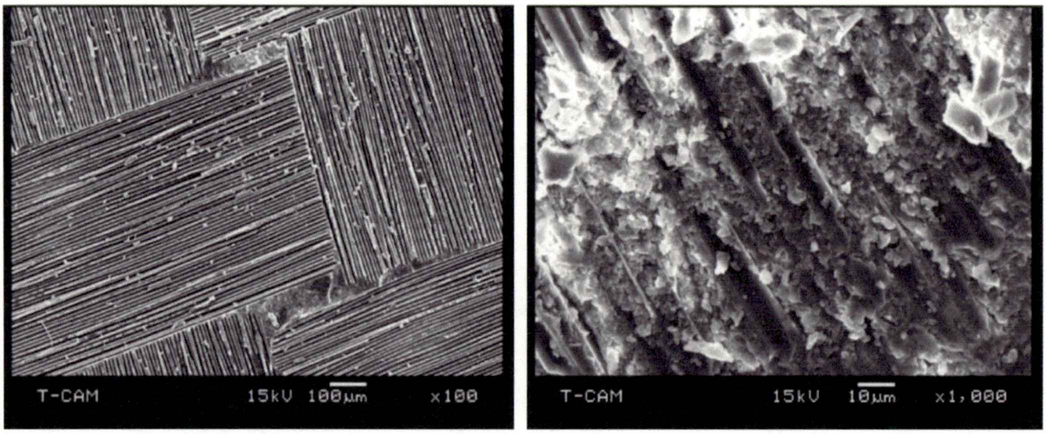

(a) (b)

Fig. 8. 0.2 wt.% CNF-loaded GRPC laminates (a and b).

3.3 Fracture morphology analysis

Results from the SEM study substantiate the quantitative results obtained through flexural test [M.E. Hossain et al., 2011; M.K. Hossain et al., 2011]. SEM performed on the fractured samples of flexure tests revealed rough and smooth fracture surfaces in 0.2 CNF-loaded sample and neat sample, respectively (Figure 9). The bonding between polyester and CNF was seen to be strong and attributed to cause deviation in the path of crack front as it propagated, thus requiring more energy to fracture the samples. This has resulted in increased strength and stiffness of samples. The effect was most pronounced at 0.2 wt.% loading of CNFs.

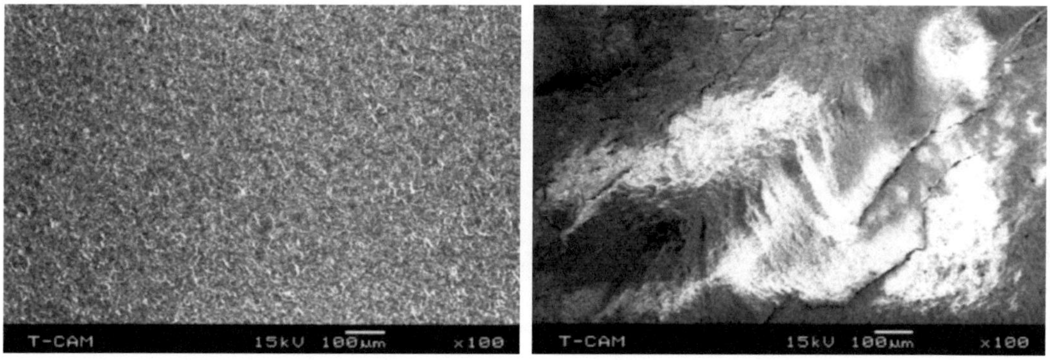

(a) Controlled polyester samples (b) 0.2 CNF loaded polyester samples

Fig. 9. SEM micrographs of fracture surface (a & b) after flexural test

From the SEM micrograph taken at higher magnification as shown in Figure 10 (a), excellent bridging effect in the interfacial region of the long glass fiber, CNF, and matrix was observed. CNF has high aspect ratio which can prevent crack propagation and crack generation resulting in improved performance. Some resin was stacked on the fractured glass fiber as shown in Figure 10 (b), which represents better adhesion due to the addition of CNFs. The presence of polyester adhering to the fiber surface also suggests that interfacial adhesion is stronger than matrix strength in nanophased composites [Hussain et al., 1996]. Thus, it is evident from these micrographs that CNFs are anchored with both resin and fiber tightly that promotes a better interfacial bonding between the matrix and fiber. Better fiber-matrix interfacial bonding, and CNFs' crack generation and propagation resistance result in higher strength in nanocomposites. On the other hand, the addition of the CNFs led to an improvement in the modulus of elasticity of the nanophased composites. This is attributed to the stiffening of the matrix of these composites (Figure 2). The interfacial area between the resin matrix and CNFs was increased because of the high aspect ratio of the CNFs, which in turn led to better mechanical properties [Tsantzalis et al., 2007]. The nanoparticles also act as reinforcing element and bear the load in the composite material system [Jawahar et al., 2006]. Again, both CNFs and fibers are stronger than matrix. Thus, when load is applied to the composite structures, matrix starts to crack first and stress is then transferred from the lower modulus matrix to the CNFs to the long fiber by bridging effect and ultimately, the composites' properties enhance.

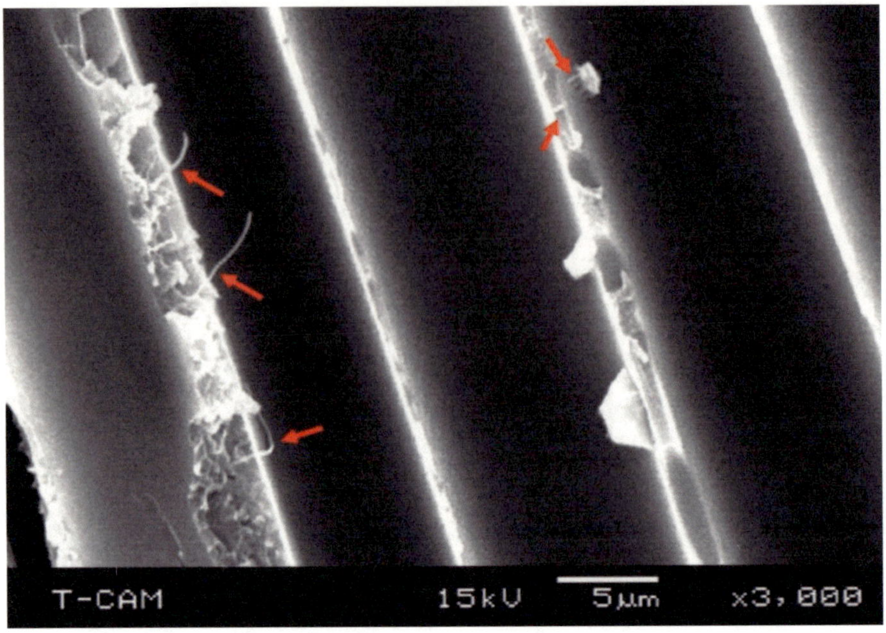

(a)

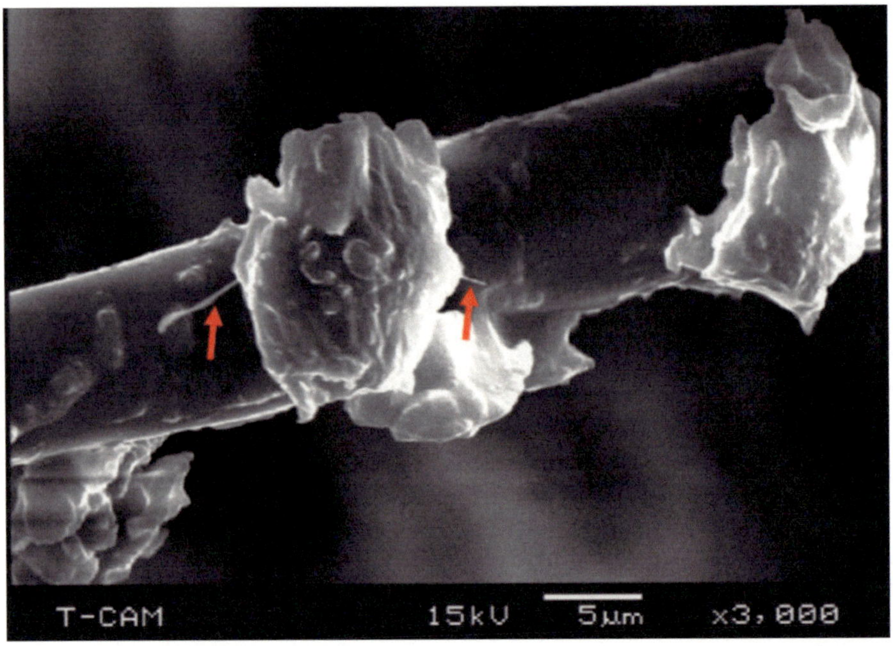

(b)

Fig. 10. (a) Bridging effect at the interface region of the long glass fiber, CNF and the resin and (b) 0.2 wt.% CNF-loaded polyester matrix stacked with glass fabric after fractured laminate.

For a better understanding of the fracture process, fracture morphology of samples was studied using higher magnification SEM micrographs. The SEM micrographs of the fractured surfaces of the conventional and 0.2 wt.% CNF-loaded GRPC are illustrated in Figure 11. For conventional composite shown in Figure 11 (a), the surface of the fiber was clean, and no matrix adhered to the fiber. The fracture surface of the matrix was flat, and some cracks were seen in the matrix side near the fiber-matrix interface. Resin appears not to protrude from the surface of fibers. These results indicate that the interfacial bonding between the fiber and matrix was weak. The fracture surface of the nanophased composite (Figure 11 (b)) shows that the surface of the matrix was rougher than that of neat composite. CNFs were observed to be randomly but uniformly distributed in the matrix. The resin appears to cling to fibers well. The strengthened matrix held the glass fabrics together. The protrusion of the resin from the surface of the fibers accounts for the increase in fracture toughness of the samples. Moreover, the resin appears to be sticking to the fiber surface giving rise to a significant plastic deformation [Xu & Hoa, 2008]. The plastic deformation enhances mechanical properties significantly in the nanophased composites (Figure 3) [M.K. Hossain et al., 2011].

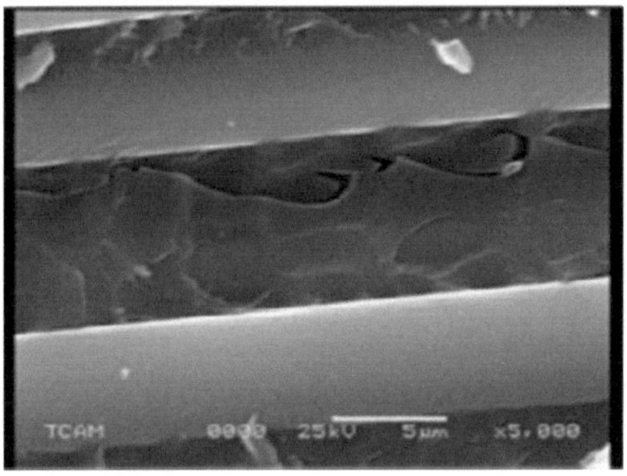

(a) (b)

Fig. 11. Fracture morphology of (a) conventional, and (b) 0.2 wt.% CNF-loaded GRPC.

4. Conclusion

Sonication, mechanical mixing, and magnetic stirring were performed to infuse 0.1-0.4 wt.% carbon nanofibers (CNFs) into the polyester resin. CNFs were used as nanoparticle fillers in woven glass fiber-reinforced polyester composites. Better dispersion of CNFs was observed in the 0.2 wt.% CNF-loaded polyester resin while CNFs were mixed using the sonication. The fiber volume fraction for the nanophased GRPC fabricated by the VARTM process was found around 56%. The void content was also within a reasonable limit in these composites. CNFs infusion even at quite low concentrations enhanced the mechanical properties of the system. This SEM investigation visually demonstrated that CNFs can be used without difficulty to modify the conventional fiber reinforced composite materials. Thus, SEM micrographs confirm that uniform dispersion and optimal loading of nanoparticles improve the mechanical properties of composites with the following outcomes:

- SEM micrographs revealed better dispersion of CNFs in the 0.2 wt.% CNF-loaded polyester prepared by sonication mixing and agglomeration in samples prepared by mechanical and magnetic mixing methods.
- Uniform resin flow and proper impregnation of the glass fibers were observed in the SEM micrographs due to the presence of CNFs.
- Uniform resin flow and proper impregnation of the glass fibers appeared to result in a better interaction between the fiber and matrix that aided to an efficient stress transfer from the continuous polymer matrix to the dispersed fiber reinforcement through the mechanical interlocking of the CNFs with the fibers.
- Excellent bridging effect in the interfacial region of the long glass fiber, CNF, and matrix was observed in SEM micrographs of nanophased GRPC.
- SEM micrographs exhibited rougher fracture surface in the CNF-loaded polyester sample compared to the neat sample due to the presence of well dispersed and well separated CNFs.
- SEM micrographs revealed the flat and clean fracture surface of the matrix with some cracks in the matrix side of the fiber-matrix interface in the conventional GRPC whereas nanophased GRPC showed rougher fracture of the matrix with randomly but uniformly distributed CNFs throughout the matrix that appeared to cling to fibers well.

5. Aknowledgements

The author is truly grateful to his hardworking and dedicated graduate students: Mr. Muhammad Enayet Hossain, Mr. Mohammad Washim Dewan, Mr. Kazi Al Imran, and Mr. Chinedu Okoro for their direct and indirect contribution in this article. The author also likes to express his gratitude towards Dr. Mahesh Hosur and Dr. Vijay Rangari for their cooperation and advice. The author also appreciates the help from the staff members of Tuskegee Center for Advanced Materials (T-CAM).

The author's special thanks are extended to Dr. Shaik Jeelani, Vice President, Research and Sponsored Programs at Tuskegee University for his invaluable support and guidance throughout this work both professionally and personally.

The author acknowledges the financial support of NSF-EPSCoR Grant No. EPS-0814103 and NSF-RISE Grant No. HRD-0833158 for this research work.

Finally, special thanks go to his wife, Dr. Shamim Ara Begum, and son, Ahnaf Hossain, for their love, patience, and mental support. Above all, the author thanks to God, Almighty for His love and mercy.

6. References

Bruzaud, S., & Bourmaud, A. (2007). Thermal degradation and (nano) mechanical behavior of layered silicate reinforced poly(3-hydroxybutyrate-co-3-hydroxyvalerate) nanocomposites. *Polym Test*, 26, (2007), pp. (652-659).

Giannelis, E.P. (1998). Polymer-layered silicate nanocomposites: synthesis, properties and applications. *Appl Organomet Chem*, 12, (1998), pp. (675-680).

Green, K.J., Dean, D.R., Vaidya, U.K., & Nyairo, E. (2009). Multiscale fiber reinforced composites based on a carbon nanofiber/epoxy nanophased polymer matrix: Synthesis, mechanical and thermomechanical behavior. *Comp: Part A.*, 40, 9, (2009), pp. (1470-1475).

Hossain, M.E., Hossain, M.K., Hosur, M.V., & Jeelani, S. (2011). Study of Mechanical Responses and Thermal Expansion of CNF-modified Polyester Nanocomposites Processed by Different Mixing Systems. *Cambridge Journal Online 2011*, DOI:10.1557/opl.2011.111.

Hossain, M.K., Hossain, M.E., Hosur, M.V., & Jeelani, S. (2011). Flexural and Compression Response of Woven E-Glass/Polyester–CNF Nanophased Composites. *Comp: Part A.*, 42, 11, (2011), pp. (1774-1782).

Hussain, M., Nakahira, A., & Niihara, K. (1996). Mechanical property improvement of carbon fiber reinforced epoxy composites by Al_2O_3 filler dispersion. *Mater Lett.*, 26, 3, (1996), pp. (185-191).

Jawahar, P., Gnanamoorthy, R., & Balasubramanian, M. (2006). Tribological behavior of clay-thermoset polyester nanocomposites. *Wear*, 261, (2006), pp. (835-840).

Kozey, V.V., Jiang, H., Mehta, V.R., & Kumar, S. (1995). Compressive behavior of materials, Part II: High performance fibers. *J Mater Res.*, 10, 4, (1995), pp. (1044-1061).

Li, X.F., Lau, K.T., & Yin, Y.S. (2008). Mechanical properties of epoxy-based composites using coiled carbon naotubes. *Comp Sci Technol.*, 68, (2008), pp. (2876-2881).

Ma, H., Zeng, J., Realff, M.L., Kumar, S., & Schiraldi, D.A. (2003). Processing, structure, and properties of fibers from polyester/carbon naofiber composites. *Comp Sci Technol.*, 63, (2003), pp. (1617-1628).

Sandler, J., Werner, P., Shaffer, M.S.P., Denchuk, V., Altstadt, V., & Windle, A.H. (2002). Carbon nanofibers reinforced poly (ether ether ketone) composites. *Comp Part A*, 33, (2002), pp. (1033-1039).

Suzuki, E. (2002). High-Resolution Scanning Electron Microscopy of Immunogold-Labelled Cells by the Use of Thin Plasma Coating of Osmium. *Journal of Microscopy* 208, 3, (2002), pp. (153–157).

Tsantzalis, S., Karapappas, P., Vavouliotis, A., Tsotra, P., Paipetis, A., Kostopoulos, V., & Friedrich, K. (2007). Enhancement of the mechanical performance of an epoxy resin and fiber reinforced epoxy resin composites by the introduction of CNF and PZT particles at the microscale. *Comp: Part A*, 38, 4, (2007), pp. (1076-1081).

Vaia, R.A., Jandt, K.D., Kramer, E.J., & Giannelis E.P. (1996). Microstructural evolution of melt intercalated polymer-organically modified layered silicates nanocomposites. *Chem Mater*, 8, (1996), pp. (2628- 2635).

Xu, Y. & Hoa, S.V. (2008). Mechanical properties of carbon fiber reinforced epoxy/clay nanocomposites. *Comp Sci Technol.*, 68, 3-4, (2008), pp. (854-861).

Yasmin, A., Abot, J.L., & Daniel, I.M. (2003). Processing of clay/epoxy nanocomposites by Shear mixing. *Scripta Materia*, 49, 1, (2003), pp. (81–86).

Yoonessi, M., Toghiani, H., Wheeler, R., Porcar, L., Kline, S., & Pittman, C. (2008). Neutron scattering, electron microscopy and dynamic mechanical studies of carbon nanofiber/phenolic resin composites. *Carbon*, 46, (2008), pp. (577-588).

Preparation and Characterization of Dielectric Thin Films by RF Magnetron-Sputtering with $(Ba_{0.3}Sr_{0.7})(Zn_{1/3}Nb_{2/3})O_3$ Ceramic Target

Feng Shi
College of Physics & Electronics, Shandong Normal University, P.R. China

1. Introduction

Barium-strontium-zinc niobate (BSZN) [$(Ba_xSr_{1-x})(Zn_{1/3}Nb_{2/3})O_3$, where x is the mole fraction and $0 \le x \le 1$] is one of the $A\left(B'_{1/3}B''_{2/3}\right)O_3$ type microwave dielectric ceramics which have many advantage properties at very high microwave frequency, such as extremely low dielectric loss and near zero temperature coefficient of resonance frequency with low cost (Xu et al., 2006; Yu et al., 2006; Ianculescu et al., 2007; Huang et al., 2006; Varma et al., 2006). Due to these properties, BSZN has several potential applications in the fields of satellite communication and radar and mobile communication systems. As promising materials used in microelectronic and microwave integration circuit, microwave dielectric ceramic thin films will attract great attention in the near future (Huang et al., 2006).

Radio frequency (RF) magnetron sputtering is a dominant technique to grow thin films because a large quantity of thin films can be prepared at relatively high purity and low cost. The growth of thin films using $(Ba_xSr_{1-x})(Zn_{1/3}Nb_{2/3})O_3$ microwave dielectric ceramic as target materials has not been reported all over the world except my group (Cui et al., 2010; Shi et al., 2010) and there is very little information for direct reference. In this article, the thin films were initially prepared by radio frequency magnetron sputtering system using sintered $(Ba_xSr_{1-x})(Zn_{1/3}Nb_{2/3})O_3$ microwave dielectric ceramic as target. Then the deposited samples were annealed in oxygen ambience at different powers of 150 W, 200 W, and 250 W, different pressures of 0.1 Pa, 0.25 Pa, 0.5 Pa, 0.7 Pa, and 1.0 Pa, different annealing temperatures of 850 ℃, 1000 ℃, and 1150 ℃ and different annealing times of 15 min, 30 min, 45 min, and 60 min. The microstructure, components and surface morphology properties are investigated in detail, and the effect of experimental conditions on the growth of the thin films is studied in particular.

2. Experimental procedure

Ceramic thin films were deposited on SiO_2 (110) substrates by adopting sintered $(Ba_xSr_{1-x})(Zn_{1/3}Nb_{2/3})O_3$ microwave dielectric ceramics as sputtering target with a size of 62- × 3 mm in a JGP-450 radio frequency magnetron sputtering system. The SiO_2 (110) substrate was ultrasonically cleaned in an acetone and followed by rinsing in de-ionized water several times. The chamber pressure was maintained at 1.0×10^{-3} Pa, and the argon gas (99.999%) was then introduced into the chamber. The distance between the target and the substrates was 11 cm, the substrate was heated to 610 °C and the sputtering time is 3 hrs per time. A post-deposition annealing was needed so that the deposited films have a well-crystallized structure. When the conventional tube furnace was heated up to a certain temperature, the quartz boat with the samples was placed into the constant temperature region. Then the flowing O_2 (99.999%) was introduced into the tube and the samples were annealed at O2 atmosphere with a flow rate of 500 ml/min. After being annealed, the samples were taken out for characterization.

A Rigaku D/max-rB X-ray diffraction (XRD) meter with Cu K_α-line, X-ray photoelectron spectroscope (XPS), a Hitachi S-450 scanning electron microscope (SEM), a Park Autoprobe CP atomic force microscope (AFM) and a Hitachi H-8010 transmission electron microscopy (TEM) were applied to characterize the microstructure, components and surface morphology properties of the thin films and to study their crystallinity. The morphology of cross-section of the thin film was examined by SEM (JEOL JSM-6390).

3. Results and discussion

3.1 Microstructure and components analysis

Figure 1 shows the X-ray diffraction patterns of the as-cast thin films at different powers (the sputtering pressure is 0.25 Pa).

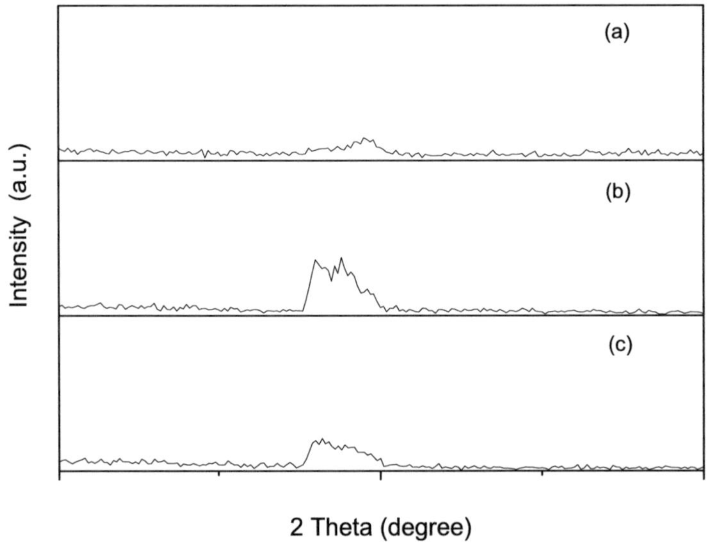

Fig. 1. X-ray diffraction patterns of the as-cast thin films at different powers.
(a) 150 W, (b) 200 W, (c) 250 W.

Figure 1 shows no diffraction peak exists, only three weak diffraction packets, which indicates that the thin films do not crystallize, but only exist in an amorphous state. The intensities and shapes are different obviously for these samples sputtered at different powers. The intensity is the strongest when sputtering power is 200 W.

Figure 2 shows the XRD spectra obtained from the thin films deposited at the pressure of 0.25 Pa, annealing at 1150 °C for 30 min and different powers of (a) 150 W, (b) 200 W, (c) 250 W.

In Figure 2 (a), there are a small number of diffraction peaks with low intensity in comparison to those of the other two samples. This phenomenon is due to the fact that the low kinetic energies of the sputtered-ejected ions are not sufficient for the arrangement and crystallization of particles on the substrates when sputtered at lower RF power (Wang et al., 2004). On the other hand, at 200 W (as shown in Figure 1 (b)), the amorphous thin films begin to crystallize, and more diffraction peaks appear in the sample with the strongest intensity. In Figure 2 (b), major peaks are identified to be $Ba_{0.5}Sr_{0.5}Nb_2O_6$ and $Ba_{0.27}Sr_{0.75}Nb_2O_{5.78}$ as compared with JCPDS files No. 39-0265 and No. 31-0166 (International Center for Diffraction Data, 2002), which is different from the target components.

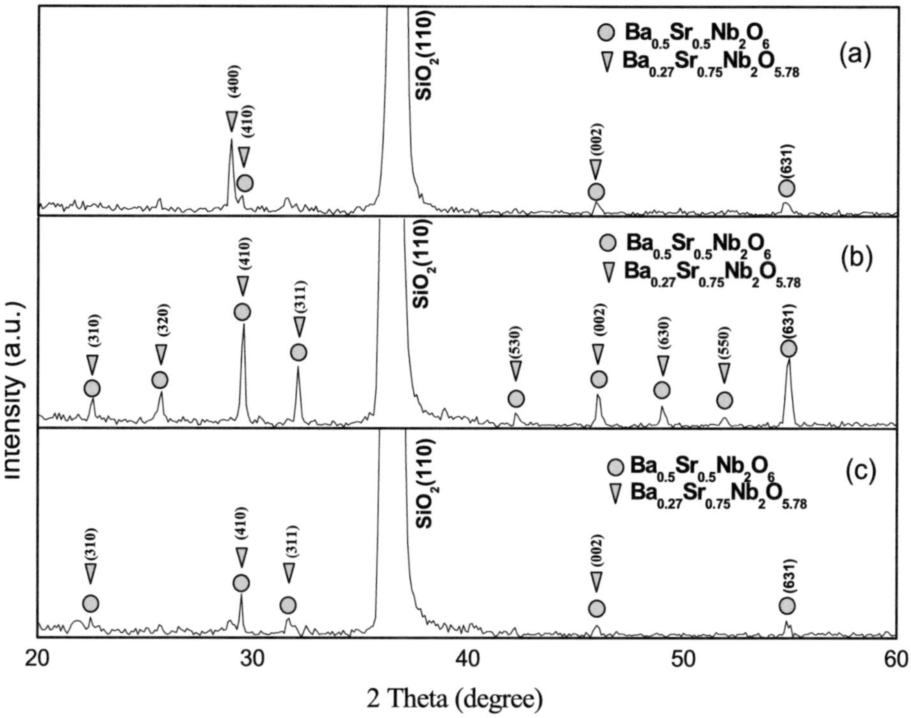

Fig. 2. X-ray diffraction patterns of the thin films deposited at 0.25 Pa and different powers: (a) 150 W, (b) 200 W, (c) 250 W.

These peaks are caused by the volatilization of ZnO during the process of sputtering and annealing; this is the same as that stated in reference (Huang et al., 2006). There is no preferential orientation for the thin films, i.e., they are randomly orientated. The diffraction peaks in every crystal plane of the films are almost complete, which indicates the grains have excellent crystal quality. When sputtered at 250 W (as shown in Figure 2 (c)), both the quantity and intensity of the thin film diffraction peaks are lower and weaker than that of the sample sputtered at 200 W. Both the crystal quality of the thin films and the intensity of the reflection peaks are at their best when the RF power is of 200 W, because the increase in kinetic energies of the sputter-ejected species accelerates the arrangement and crystallization of particles on the substrate surface (Hsi et al., 2003). However, when the power is of 250 W, the kinetic energy of the sputter -ejected species is very high. As a result, the particles excited from the ceramic target were thrown away and not deposited on the substrate.

On the other hand, the greater the sputtering powers, the higher the kinetic energies, and the quicker the deposition rate. Deposition rate is lower at 150 W, and is higher at 250 W, i.e., appropriate deposition rate attributes to the film growth; therefore, 200 W is the optimum condition for the growth of the thin films.

In order to further investigate the growth of the thin films, the full-width at half-maximum (FWHM) of the (410) peaks of the $Ba_{0.5}Sr_{0.5}Nb_2O_6$ and $Ba_{0.27}Sr_{0.75}Nb_2O_{5.78}$ phases is analyzed. According to Scherrer`s formula (Klung et al., 1974), the grain size in (410) orientation can be estimated as follows:

$$L_{(410)} = \frac{k\lambda}{\beta_0 \cos\theta},$$

where k is a constant with a value of about 0.89 for Cu target, λ is the X-ray wavelength, β_0 is the FWHM of (410) peak, and θ is the diffraction angle. The estimated results are listed in Table 1.

Sample	(hkl)	$2\theta/^0$	FWHM/rad	L/nm
B_1(150 w)	(410)	29.51	0.0061	24.1
B_2(200 w)	(410)	29.57	0.0043	36.3
B_3(250 w)	(410)	29.51	0.0050	30.6

Table 1. The grain size of $Ba_{0.5}Sr_{0.5}Nb_2O_6$ and $Ba_{0.27}Sr_{0.75}Nb_2O_{5.78}$ phases in the thin films estimated by Scherrer`s formula.

Therefore, with an increase in the sputtering power, the grains sizes of the thin films reach a maximum of 36.3 nm at 200 W and then decrease to 30.6 nm at 250 W.

Figure 3 shows X-ray diffraction patterns of the as-cast thin films and the samples annealing at different temperatures for 30 min (sputtered at 200 W and 0.25 Pa).

Figure 3 (a) indicates there is only a single crystal SiO_2 peak instead of diffraction peaks of other phases. It is well known that the formation of crystalline phase in thin films is affected by two important factors: one is the temperature in the sputtering process and the other is the annealing temperature. Therefore, during sputtering, the temperature of SiO_2 substrate, about 610 °C, is insufficient to initiate the formation of crystallization, thus the annealing treatment after sputtering is crucial. When the annealing temperature is 850 °C, as shown in Figure 3 (b), one weak peak appears, which is too weak to separate clearly and shows the thin film is in an amorphous state. But when the annealing temperature increases to 1000 °C, more and stronger peaks occur in Figure 3 (c). The main phases are also $Ba_{0.5}Sr_{0.5}Nb_2O_6$ and $Ba_{0.27}Sr_{0.75}Nb_2O_{5.78}$, but there is no peak of $(Ba_{0.3}Sr_{0.7})(Zn_{1/3}Nb_{2/3})O_3$. The difference between the components of the films and target attributes to the volatilization of ZnO during the process of sputtering and annealing, which is the same as that stated in Figure 2. The intensity of diffraction peak in Figure 3 (c) is stronger than that of the sample shown in

Figure 3 (b), because this thin film has transited from the amorphous state to the crystalline state at higher annealing temperature. When the annealing temperature increases to 1150 °C, as shown in Figure 3 (d), which is the same as Figure 2 (b), all peaks of crystal planes have appeared, which reveals that the thin film crystallizes completely, but there is no preferential growth. Moreover, the intensity of diffraction peak in Figure 3 (d) enhances greatly due to good crystallization at the highest annealing temperature. Figure 3 states there is steady growth for the strength and quantity of the thin film diffraction peaks with the increase in annealing temperature.

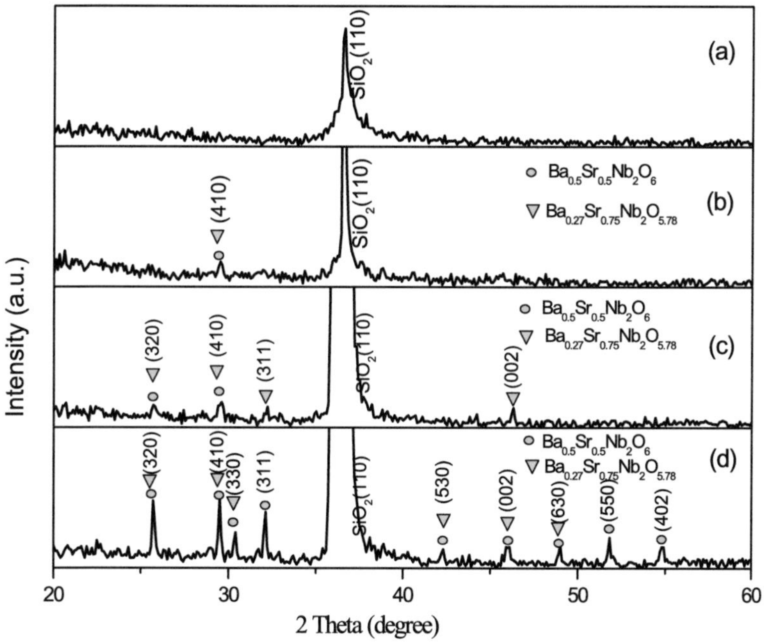

Fig. 3. X-ray diffraction patterns of the as-cast thin films and the thin films annealing at different temperatures for 30 min,(a) as-cast, (b) 850 ℃, (c) 1000 ℃, (d) 1150 ℃.

According to the (320), (410), and (330) peaks, through the formula (Wang et al., 2004) $d_{hkl} = \left[(h/a)^2 + (k/a)^2 + (l/c)^2 \right]^{-\frac{1}{2}}$, where d_{hkl} is interplanar crystal spacing, and h, k and l are crystal indices, we can calculate the lattice constants of $Ba_{0.5}Sr_{0.5}Nb_2O_6$ and $Ba_{0.27}Sr_{0.75}Nb_2O_{5.78}$ phases: a=b=12.456 Å and c=3.952 Å of $Ba_{0.5}Sr_{0.5}Nb_2O_6$ phase and a=b=12.430 Å and c=3.913 Å of $Ba_{0.27}Sr_{0.75}Nb_2O_{5.78}$ phase. A higher annealing temperature accelerates the material migration and diminishes the crystallographic defects and dislocations, thus promots crystallization of the thin films, and therefore, the excellent crystal structures are obtained accordingly.

The intensity is related with the crystallinity, the thickness of the thin films and the consistency, whereas the full width at half maximum (FWHM) value only reveals the crystallinity of the thin films. Due to different volatilization at different annealing temperatures, the thickness and consistency of peaks are different, therefore, the intensities of diffraction peaks are also different. Generally speaking, the annealing treatment can promote atomic mobility and enhance the ability of atoms so that they can find the most energetically favored points; therefore, with the increase in the annealing temperature, the crystallographic defects including dislocations and vacancies in the thin films decrease rapidly, i.e., the appropriate annealing temperature is the key factor for a better crystal structure. The Scherrer formula (Schroeder et al., 1968) tells us that the smaller of the FWHM value, the better of the C-axis-preferred orientation, and the larger of the grain size and the better of the crystalline quality of the thin films. That is, the thin film which has the smallest FWHM value possesses the best crystalline quality and the largest grain size. The relation between FWHM and the annealing temperature is listed in Table 2.

Sample	phase	(hkl)	$2\theta/^0$	FWHM/^{0}
B_1(850 °C)	$Ba_{0.5}Sr_{0.5}Nb_2O_6$ $Ba_{0.27}Sr_{0.75}Nb_2O_{5.78}$	(410)	29.6085	0.291
B_2(1000 °C)	$Ba_{0.5}Sr_{0.5}Nb_2O_6$ $Ba_{0.27}Sr_{0.75}Nb_2O_{5.78}$	(410)	29.6085	0.215
B_3(1150 °C)	$Ba_{0.5}Sr_{0.5}Nb_2O_6$ $Ba_{0.27}Sr_{0.75}Nb_2O_{5.78}$	(410)	29.6085	0.125

Table 2. The relation between FWHM and annealing temperatures of the samples.

As shown in Table 2, the value of FWHM decreases with the increase in annealing temperature. The sample annealed at 1150 °C has the smallest value of 0.125, which shows the best crystalline quality (Zheng et al., 1993), because the energy supplied from the outside environment during the annealing process is sufficient to make the films crystallize completely at higher temperature.

Figure 4 shows X-ray diffraction patterns of the thin films annealing at 1150 °C for different times (sputtered at 200 W and 0.25 Pa).

In Figure 4 (a), there is only one diffraction peak with low intensity, and the phase is $Ba_{0.27}Sr_{0.75}Nb_2O_{5.78}$. This thin film just began to crystallize at 1150 °C for 15 min, because there was no enough time for the atoms migration on the substrate surface (Zannetti et al., 1969; Yang et al., 1998). Figure 4 (b) is the same as shown in Figure 3 (d): there is no other peak, which indicates that the thin film has high purity. Figure 4 (c) shows the intensity and number of the diffraction peaks decrease greatly in comparison to that shown in Figure 4 (b) because of excess volatilization of ZnO and recrystallization of the grains at longer annealing time, and the trend is clear as seen in Figure 4 (d). A new phase has appeared, named $Ba_{5.75}Nb_{2.25}O_{11.375}$ in Figure 4 (c) and Figure 4 (d).

Figure 5 is the relation between the FWHM values and the grain sizes with the different annealing times.

Annealing for 15 min, the samples cannot crystallize because of deficiency of the heat energy for material migration, thus the FWHM value is large; whereas after annealing for longer time, such as 45 min and 60 min, more heat energy will lead to excess volatilization of ZnO and recrystallization of the grains, thus lessening the FWHM value. As for the sample annealing for 30 min, the FWHM value is at its minimum and the grain size is the largest with the best crystallinity, as compared with the other samples annealing for 15 min, 45 min, and 60 min, because of the sufficient heat energy supplied by the outside environment improve crystallization of the thin films, but the heat energy is not excessive. That is, it is appropriate for the thin films being annealed for 30 min.

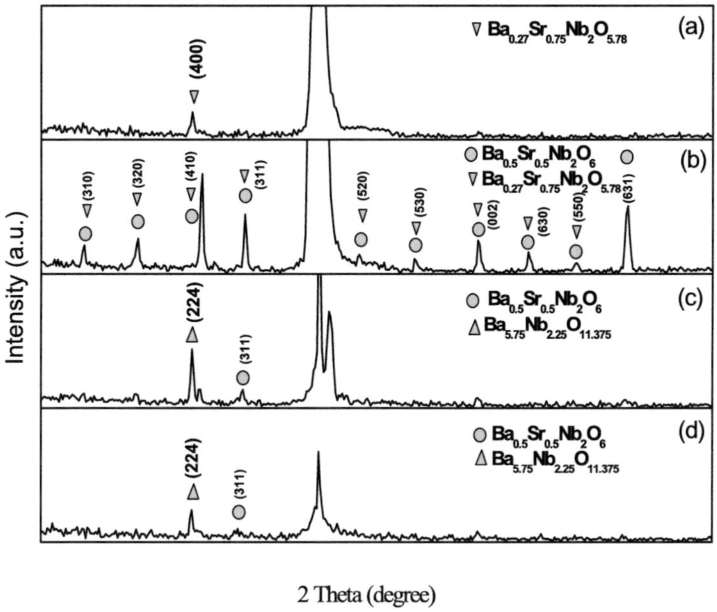

Fig. 4. X-ray diffraction patterns of the thin films annealing at 1150 °C for different times (sputtered at 200 w and 0.25 Pa).(a) 15 min, (b) 30 min, (c) 45 min, (d) 60 min.

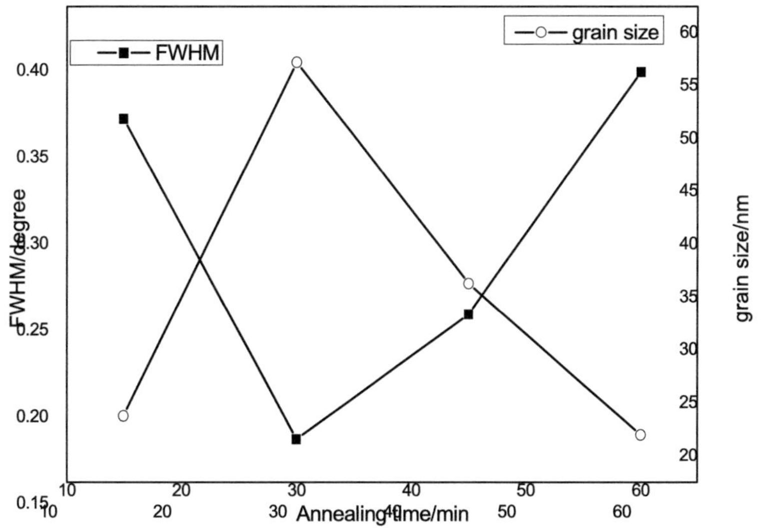

Fig. 5. The relation between the FWHM values and the grain size with the different annealing times.

Figure 6 shows X-ray diffraction patterns of the thin films at different pressures (sputtered at 200 W, annealing at 1150 °C for 30 min).

As seen from Figure 6, the trend is the same as that stated above in Figure 2 - Figure 4, i.e., the higher of the pressure, the lower of the intensity and the less of the number of the diffraction peaks. There is greater atomic density in vacuum chamber at higher gaseous pressure, i.e., the chances of atomic collision increase, and the energy loss of atoms is enhanced during material migration, thus lessening the quantity of the atoms reaching the substrate surface. The nucleation density decreases subsequently due to less particles; therefore, the growth of the thin films is weakened. The imbalance between Ba and Sr ions is closely related with saturated gaseous pressure, for instance, the stoichiometric ratio imbalance of Ba and Sr elements becomes inconspicuous at lower pressure as compared to the phases in Figure 6 (a), (b) and (c).

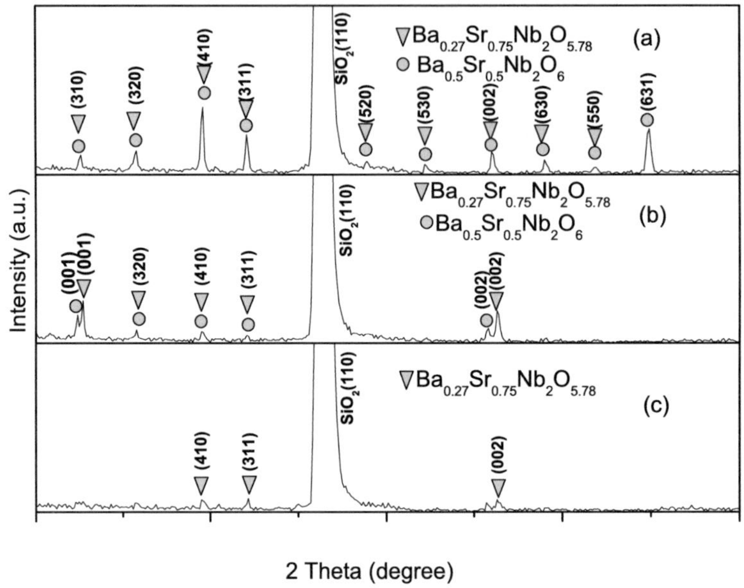

Fig. 6. X-ray diffraction patterns of the thin films at different pressures (sputtered at 200 W, annealed at 1150 °C for 30 min). (a) 0.25 Pa, (b) 0.50 Pa, (c) 0.70 Pa.

In Figure 6 (a), the grain growth is relatively complete, and therefore, more and stronger diffraction peaks appear. When the pressure increases to 0.5 Pa, the intensity and number of peaks decrease because of more collision among particles caused by higher pressure, thus material migration is inhibited and it's the same trend as the sample shown in Figure 6 (c).

According to the results of XRD, we pay much attention to the samples deposited at 150 W, 200 W, and 250 W (the sputtering pressure are of 0.25 Pa and annealed at 1150 °C for 30 min) and study their components; Figure 7 shows the XPS spectra of these samples. In Figure 7, the general scan of binding energy ranging from 0 to 1100 eV, the peaks of Sr3d, $Nb3p_{3/2}$, Ba3d, C1s, Nb3d, Zn3d and O1s are observed.

As seen in Figure 7, the main elements in the thin films are Ba, Sr, Nb, Zn, and O element. Since the samples selected for XPS analysis are exposed to the air, the elements of O and C arise from the thin film solution. O1s and C1s core level peaks are always detected, and O1s peak centers at 529.8 eV. Amanullah et al., (Amanullah et al., 1998) have reported that O1s peak is usually observed in binding energy area of 529-535 eV; the peaks in 529-530 eV are ascribed to the lattice oxygen, the peaks for binding energy ranging from 530.0 eV to 530.9 eV correspond to chemisorbed oxygen, and therefore, the O1s peak corresponds to lattice oxygen. The peaks of C1s attribute to the standard sample during XPS testing. There is a weak peak corresponding to the Zn element in Figure 7, indicating that a part of Zn element exists in the thin film although the other part of ZnO is evaporated during the sputtering and annealing process, which is not consistent with the results shown in XRD because of the systematic error of the XRD.

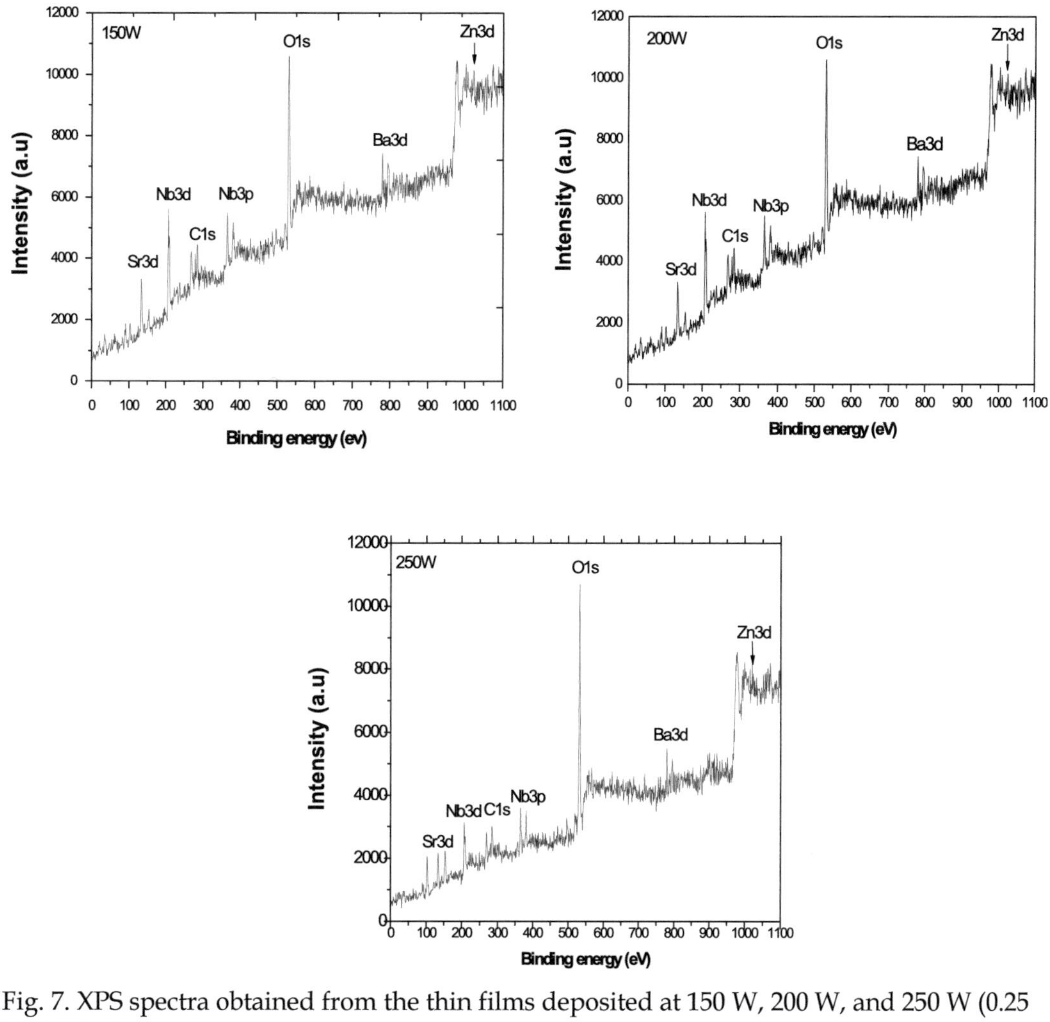

Fig. 7. XPS spectra obtained from the thin films deposited at 150 W, 200 W, and 250 W (0.25 Pa, 1150 °C, 30 min).

3.2 Morphology analysis

Figure 8 shows the typical SEM images of the samples sputtered at different powers and annealing at 1150 °C for 30 min (sputtering pressure: 0.25 Pa).

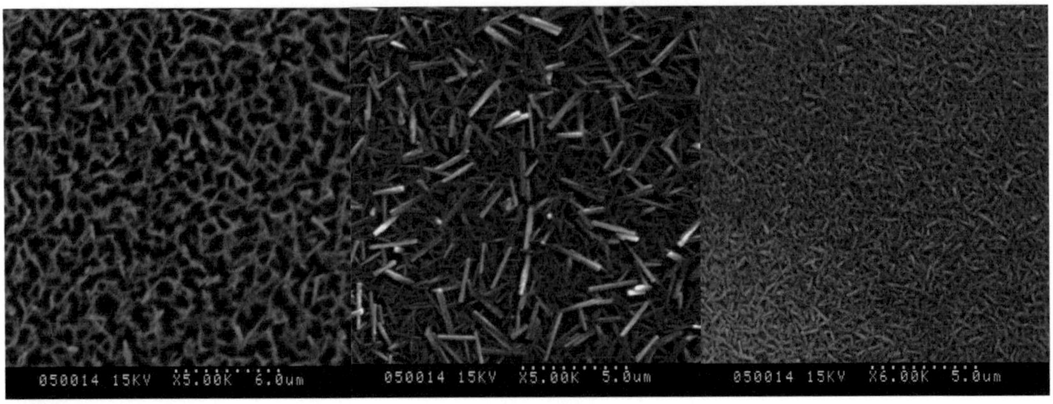

(a) (b) (c)

Fig. 8. Typical SEM images of the thin films sputtered at different powers. (a) 150 W, (b) 200 W, (c) 250 W.

When deposited at 150 W, the grains of the thin films have been formed with uniform and loose surface, and there are many holds on the thin film surface. During the coalescence stage of the sputtering process, fine particles do not have enough kinetic energy to grow into larger grains. After the annealing process, the thin films have changed from amorphous phase to crystal phase partially. Therefore, the intensities of diffraction peaks in XRD spectra are at their lowest, probably due to small deposition rate at 150 W. In Figure 8 (b), it is evident that the grains have well-defined boundaries around them and homogenously distribute on the substrates surface with uniform grain size. The grains are rod-like structures which form during the annealing process because the greater grains can combine with the smaller ones around them to generate rod-like grains. With higher energy gas plasma caused by higher RF power, more fine particles were sputtered from the target and these high-energy particles can migrate and combine together with each other on the substrate surface and form high quality crystalline films, which complies with the result of the XRD analysis (as shown in Figure 2 (b)). The sample in Figure 8 (c) is uniform and dense with strip shape grains, because adequate energy supplied during material migration promotes the formation of the thin films and crystallization of grains. Therefore, the thin films sputtered at higher powers possess perfect surface morphology.

Figure 9 shows the typical SEM images of the as-cast sample and the samples sputtered at 200 W, 0.25 Pa after annealing at different temperatures for 30 min.

As compared with the sample annealing at 1150 °C in Figure 8 (b), the as-cast sample (Figure 9 (a)) has a clean surface, but no grain; whereas when annealing at 850 °C, less grains appear; when the annealing temperature is 1000 °C, the amount of grains increases with fine particles covered on the surface. That is, crystallization is not complete at lower annealing temperature.

Figure 10 shows the SEM images of the samples sputtered at 200 W and 250 W, after annealing at 1150 °C for 45 min (sputtering pressure: 0.25 Pa).

In comparison to Figure 8 (b), there are more holes in Figure 10 (a) and the thin films are composed of block structures; this situation is similar to that shown in Figure 10 (b) and

Figure 8 (c). As seen in Figure 10 (b), the film is cracked and many gaps and holes exist among each fragmental structure, which attributes to recrystallization of grains due to longer material migration time.

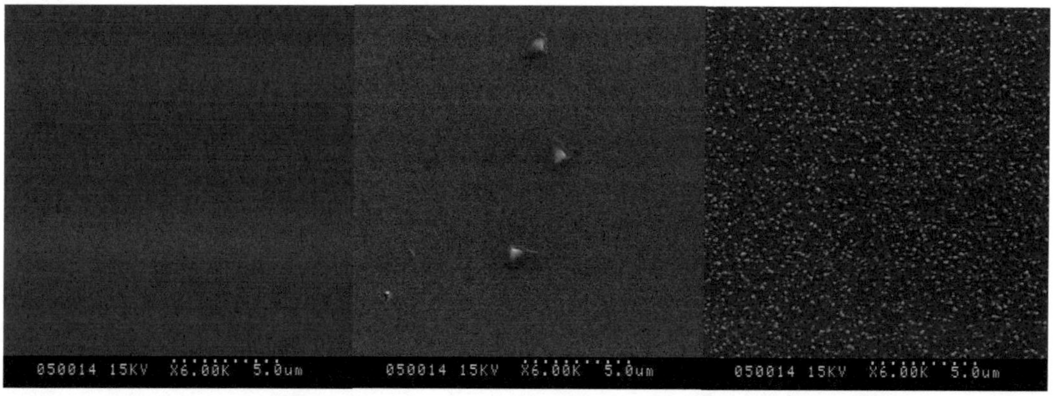

(a) (b) (c)

Fig. 9. Typical SEM images of the as-cast thin film and the thin films sputtered at 200 W, 0.25 Pa after annealing at different temperatures. (a) as-cast, (b) 850 °C, (c) 1000 °C.

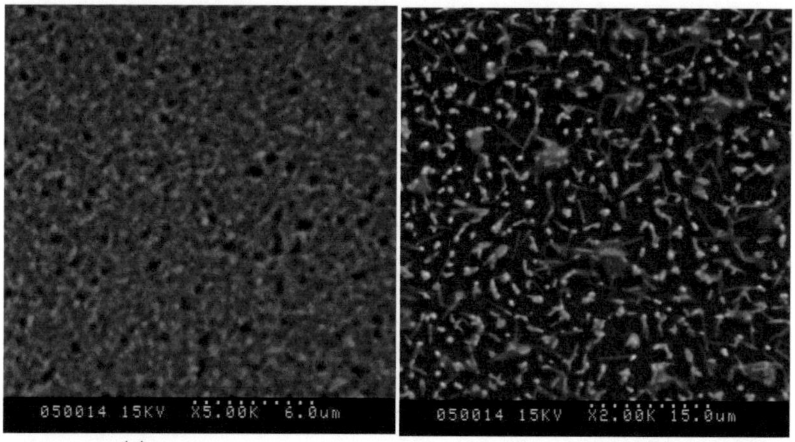

(a) (b)

Fig. 10. SEM images of the thin films sputtered at 200 W and 250 W, respectively, after annealing at 1150 °C for 45 min. (a)200 W, (b)250 W.

Figure 11 shows the SEM images of the samples sputtered at 200 W and different pressures of 0.10 Pa, 0.50 Pa, and 0.70 Pa, after annealing at 1150 °C for 30 min.

As seen from Figure 11, surface morphology differs with the sputtering gaseous pressures. Morphology shown in Figure 11 (a) is similar to that shown in Figure 8 (b), that is, the sample in Figure 11 (a) has perfect crystalline and surface morphology, which indicates that an appropriate low pressure can contribute to the film growth. With the increase of sputtering pressures, more holes and gaps form on the surface accompanied by the decrease in grain size, which are shown in Figs.11 (b) and (c). During sputtering, the variation of sputtering gaseous pressure leads to the change of transmission capacity of Ba, Sr, Zn and Nb elements. The higher the sputtering gaseous pressure, the smaller the mean free path. When sputtered at lower pressures, the mean free path and transmission capacity of Ba, Sr, Zn and Nb elements increase, and particles sputtered to the substrate surface multiply rapidly (Wang et al., 2002; Liu et al., 2005). Therefore, the content of elements deposited on the surface increases at the lower pressure, and there are sufficient materials on the substrate surface which can promote the film growth and crystallization of grains. This conclusion can also be proved by Figure 12 as compared with Figure 8(c).

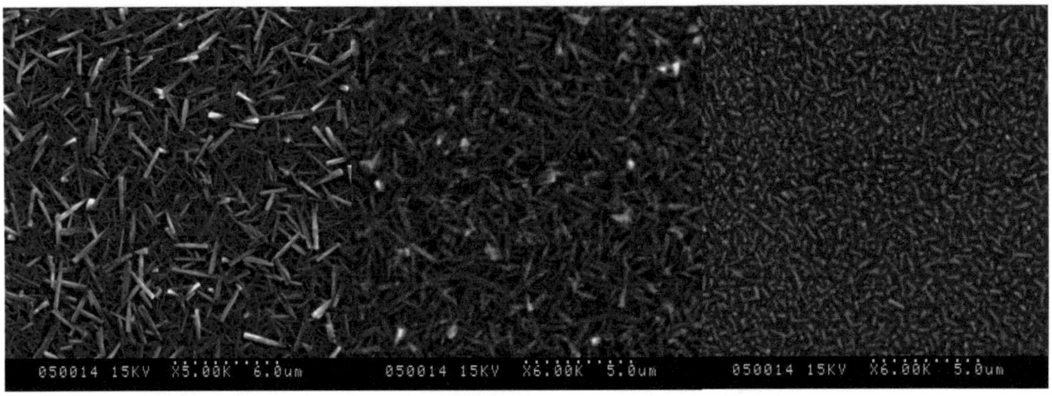

(a) (b) (c)

Fig. 11. SEM images of the thin films sputtered at 200 W and different pressures after annealing at 1150 °C for 30 min. (a) 0.10 Pa, (b) 0.50 Pa, (c) 0.70 Pa.

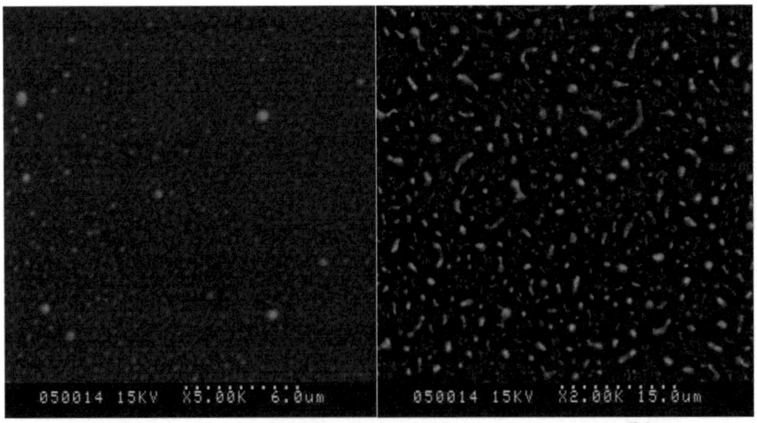

(a) (b)

Fig. 12. SEM images of the thin films sputtered at 250 W and different pressures annealing at 150 °C for 30 min, (a) 0.50 Pa, (b) 0.70 Pa.

Figure 13 shows the SEM micrograph of the cross section of the thin films prepared at 200 W, 0.25 Pa, with annealing 3 times (3 hrs per schedule) at 1150 °C for 30 min.

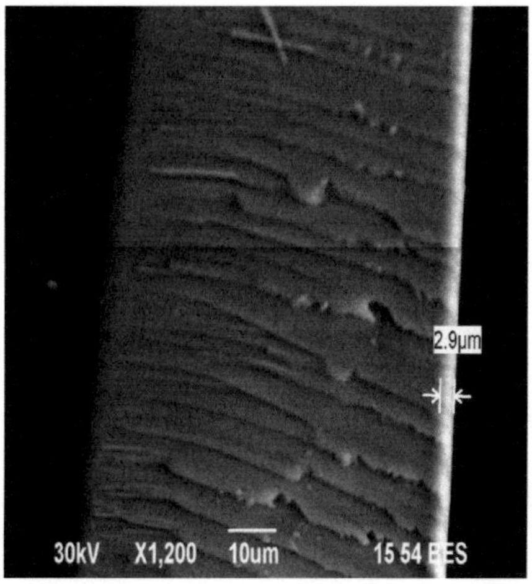

Fig. 13. The SEM micrograph of cross-section of the thin films deposited at 200 W for 3 times (3 hrs per schedule).

It can be seen that the thin film is dense and crack-free, with the thickness of 2.8 µm. That is, the thickness is about 0.9 um after sputtering for 3 hrs.

Figure 14 shows the AFM images of the as-cast thin film and the thin films deposited at 200 W, 0.25 Pa and annealing at different temperatures for 30 min (longitudinal cross-section).

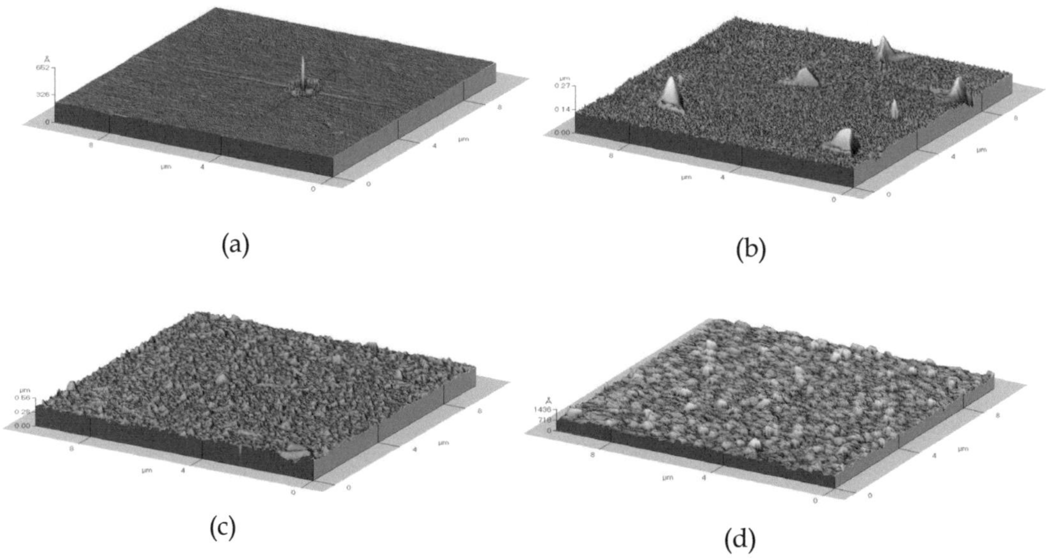

Fig. 14. AFM images of the as-cas thin film and the thin films deposited at 200 W and annealed at different temperatures for 30 min (longitudinal cross-section), (a)as-cast, (b) 850 °C, (c) 1000 °C, (d) 1150 °C

As shown in Figure 14, the grains on the surface of the samples become greater gradually with the increase of annealing temperature, which indicates that surface morphology and crystallization of the grains depend on annealing temperature significantly – the higher the annealing temperature, the better the morphology and crystalline quality. Besides, the thin film surface becomes rougher with the increase in annealing temperature. Figure 14 also tells us the fact that the growth mechanism complies with the nucleus growth mode (Tang et al., 2003). At the beginning of the thin film growth, some atoms or molecules are deposited on the substrate surface and form the so-called nucleation phases. These atoms or molecules form some homogenous and fine atomic groups which can move and are called "islands". These islands can accept new atoms continuously and merge with other islands, and nascent islands could appear at the empty area at the same time, therefore, the amount of islands saturates quickly. The progress of formation and consolidation of islands was ongoing till these isolated islands combine into flakes, with less holes and gaps, which are filled with new atoms. Therefore, the continuous films form. The consolidation of islands ends when the thickness of the thin films is about several tens of nanometers (Tang et al., 2003).

Figure 15 shows the AFM images (horizontal cross-section) of the thin films deposited at different powers (0.25 Pa of sputtering pressure and annealing at 1150 °C for 30 min).

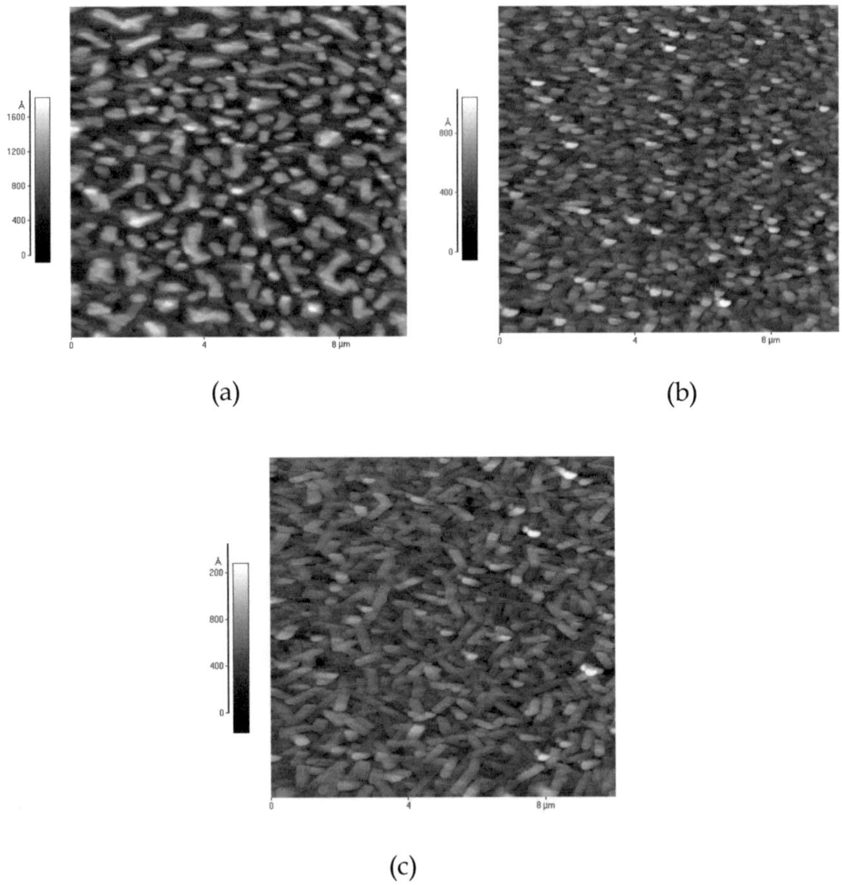

Fig. 15. AFM images (horizontal cross-section,10 µm×10 µm) of the thin films deposited at different RF powers: (a)150 W, (b) 200 W, (c) 250 W.

It is evident that different surface morphologies rely on sputtering powers significantly. Figure 15 (a) shows the surface morphology of the thin film at 150 W and the surface is smooth and compact with the smallest roughness. In Figure 15 (b), the film surface is obviously rough with high roughness value, and it can be seen that a few spikes grow along the c-axis that might have been caused by the agglomeration of the particles during crystallization of grains. Figure 15 shows that the films with high-quality crystalline form and the grain sizes of the thin films are at the greatest when the sputtering power is 200 W. The crystal quality is better for the sample deposited at 250 W than that of the sample deposited at 150 W, but worse than that of the sample deposited at 200 w. As seen from Figure 15 (c), the surface roughness is greater than that of the sample in Figure 15 (a), but less than that of the sample in Figure 15(b). As seen from Figure 15 (b) and (c), it can be inferred that the thin films consist of tightly packed particles with an average size of 92.8 nm and 58.5 nm, respectively. The grain size is considerably larger than that obtained by Scherrer`s formula from XRD. Obviously, each particle contains many single-crystal grains. In addition, the overall observation of the thin films indicate a superior microstructure for the thin films sputtered at 200 W, 0.25 Pa, after annealing at 1150 °C for 30 min.

Figure 16 shows the AFM images of the as-cast thin film and the films deposited at 200 W, 0.25 Pa and annealing at 1150 °C for different times (longitudinal cross-section).

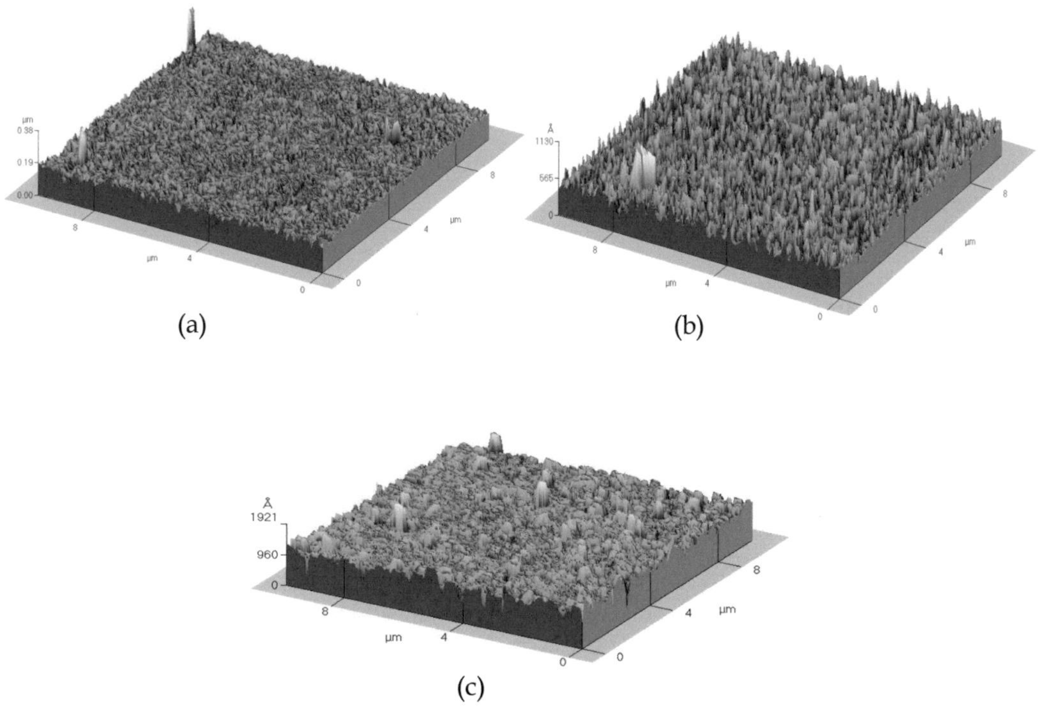

Fig. 16. AFM images of the as-cast thin film and the thin films deposited at 200 W and annealing at 1150 °C for different times (longitudinal cross-section).(a)15 min, (b) 45 min, (c) 60 min.

As compared with Figure 14 (d), Figure 16 reveals that the morphologies of the thin films are closely related with the annealing time and the roughness of the thin films changes greatly with the variation in the annealing times. Annealing after 15 min, surface irregularity appears but root-mean-square surface roughness (RMS, calculated from the AFM data using software) is lower than that of the sample in Figure 14 (d), because the time for material migration is not sufficient. While annealing after 45 min and 60 min, surface irregularities enhance and the roughness is larger than that of the sample in Figure 16 (a), due to availability of more time for material migration, but lower than that of the sample in Figure 14 (d), because of excess volatilization of ZnO and recrystallization of grains. That is, surface irregularities increase from 15 min to 30 min, and then decrease gradually from 30 min to 60 min, which is the same as that shown in Figure 17, the relation between RMS and annealing time.

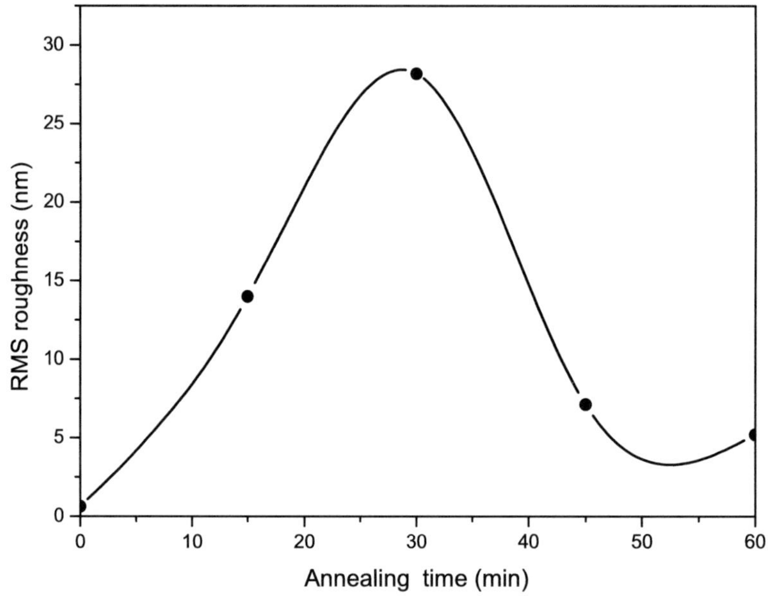

Fig. 17. Relation between RMS and annealing time.

Figure 17 shows that the roughness of the samples annealing after different times are 0.633 nm (as-cast), 14 nm (15 min), 28.2 nm (30 min), 7.12 nm (45 min), and 5.2 nm (60 min). Therefore, annealing times can affect RMS deeply, as the results show in Figure 5.

Transmission electron microscopy (TEM) is carried out for further investigating the crystal property of the sample deposited at 200 W, 0.25 Pa and annealing at 1150 ^{o}C for 30min, as shown in Figure 18. In Figure 18 (a), the grains are relatively straight, and the shape of most grains is rod-like structure, as indicated in Figure 8 (b). Figure 18 (b) shows the selected area electron diffraction (SAED) spectrum of the sample likes a ring, which indicates that the thin films are polycrystalline, and there are many flakes like dancing butterflies in Figure 18 (c), which indicates that the shape of the sample spurtered at 250 W is different from that shown in Figure 18 (b). The grain sizes in Figure 18 (b) are greater than that of the samples sputtered at 200 W.

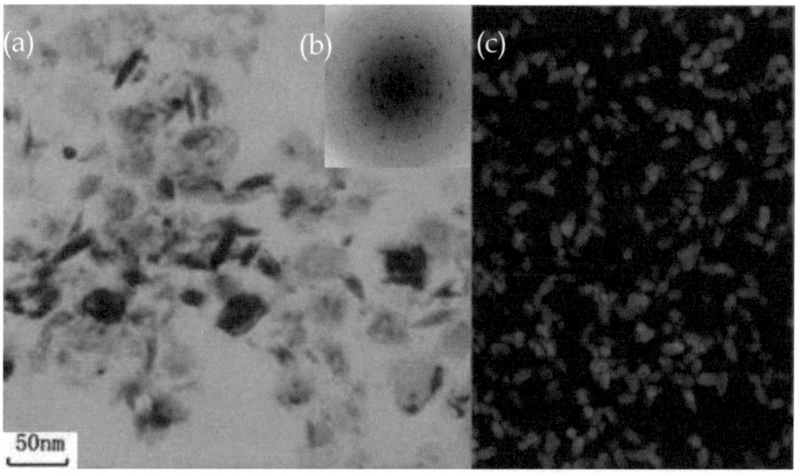

Fig. 18. (a) TEM image (bright field image) and (b) SAED spectrum of the thin film deposited at 200 W, 0.25 Pa and annealing at 1150 °C for 30 min. (c) TEM image (dark field image) of the thin film deposited at 250 W, 0.25 Pa and annealed at 1150 °C for 30 min.

4. Conclusion

The thin films were successfully fabricated on SiO_2(110) substrates through radio frequency (RF) magnetron sputtering system using sintered $(Ba_{0.3}Sr_{0.7})(Zn_{1/3}Nb_{2/3})O_3$ microwave dielectric ceramics as sputtering target, and next annealing in O_2 ambient. The thin films cannot crystallize without annealing and the main phases after annealing at 1150 °C for 30 min are $Ba_{0.5}Sr_{0.5}Nb_2O_6$ and $Ba_{0.27}Sr_{0.75}Nb_2O_{5.78}$ with less $Ba_{5.75}Nb_{2.25}O_{11.375}$, and the difference between the ceramic target and the thin films arises from the Zn loss through volitilization during sputtering and annealing process. The experimental conditions of sputtering powers, annealing temperatures, annealing times and sputtering pressures have deep influence on the microstructures and morphologies of the thin films. The samples deposited at 200 W, 0.25 Pa, and annealed at 1150 °C for 30 min have the highest crystalline quality. This thin film is polycrystalline with a dense and rod-like structure. The growth mechanism complies with the nucleus growth mode.

5. Using Zn - enriched $(Ba_{0.3}Sr_{0.7})(Zn_{1/3}Nb_{2/3})O_3$ as Target

5.1 Introduction

To obtain a dense thin film with fewer oxygen vacancies and less Zn loss, in this article, we adjusted the target component to fabricate ceramic thin films using magnetron sputtering deposition, which differs from those targets reported previously (Cui et al., 2010; Shi et al., 2010), i.e., 1 mol excess ZnO was incorporate in this stoichiometric $(Ba_{0.3}Sr_{0.7})(Zn_{1/3}Nb_{2/3})O_3$ target so as to compensate the ZnO volatilization during the sputtering and annealing process.

5.2 Experimental

Ceramic thin films were deposited on SiO_2(110) substrates by RF magnetron sputtering, using a Zn-enriched target comprised of a homogeneous mixture of 1mol $(Ba_{0.3}Sr_{0.7})(Zn_{1/3}Nb_{2/3})O_3$

and 1 mol ZnO, which were synthesized by a conventional solid-state sintering technique. The monocrystal $SiO_2(110)$ substrates were placed on the substrate holder, which can rotate around a central axis to improve the homogeneity of the thin films. The thin films were deposited using an Ar-O_2 gas mixture in a JGP450 RF magnetron sputtering systerm with the process parameters: 200 W of sputtering power, 0.25 Pa of sputtering pressure, and 610 °C of the substrate temperature. The Ar gas flow rate was fixed at 16 ml/min and the oxygen flow rate was fixed at 3.2 ml/min, i.e., the O_2/Ar flow ratio of 0.2:1 was used as working atmosphere. The distance between the substrates and the targets was 11cm, the sputtering time was 180 min, and the base pressure of the chamber was 1.0×10^{-3} Pa. As-deposited films were annealed in flowing oxygen (99.999%) at 1150 °C for 30 min in a tube furnace.

5.3 Results and discussion

Figure 19 is the X-ray diffraction patterns of the thin films, exhibiting the presence of the well-crystallized thin films. No preferential orientation for the thin film is observed.The thin films contain a mixture of $Ba_{0.5}Sr_{0.5}Nb_2O_6$ (as compared with the JCPDS card of No.39-0265, International Center for Diffraction Data, 2002), $Ba_{0.67}Sr_{0.33}Nb_2O_6$ (No.73-0126), $Sr_{0.744}Ba0.247Nb_2O_6$ (No.70-3747), and $SrNb_2O_6$ (No.72-2088), which can be written as $Ba_xSr_{1-x}Nb_2O_6$. And ZnO is also observed, as compared with the characteristic peak of ZnO. As we all know, the various sputtering yields of Ba, Sr, Nb, Zn, and O elements are a knotty problem in the deposition of $(Ba_{0.3}Sr_{0.7})(Zn_{1/3}Nb_{2/3})O_3$ – based thin films. The discrepancy between thin film and target probably results from the volatilization of ZnO during the process of sputtering and annealing, because Zn has a significant larger vapor pressure than the other metal elements (Ba, Sr, Nb) (Tang et al., 2009). Therefore, ZnO may either not stick to or be re-evaporated from the growing surface (Tang et al., 2009), worsening the composition deviation with the target.

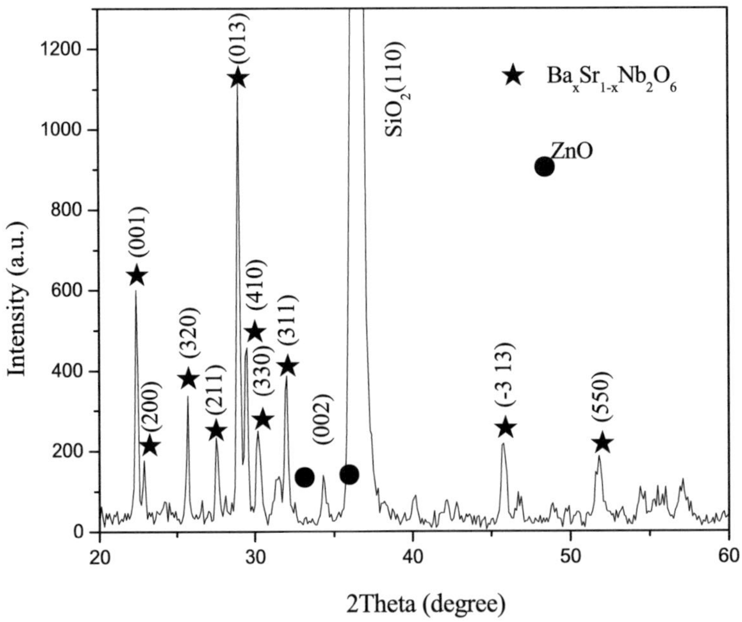

Fig. 19. XRD spectrum of the thin film.

According to the Scherrer`s formula (Ji et al., 2011), the grain size in the (013) plane can be estimated by the following expression:

$$L_{(013)} = \frac{\kappa\lambda}{\beta_0 \cos\theta},$$

where κ is a constant with a value of about 0.89 for the Cu target, λ is the X-ray wavelength with a value of about 1.54718 Å, β_0 is the FWHM of the (013) peak, and θ is the diffraction angle. In the Scherrer`s formula, β_0 = 0.003 rad and θ =29.065°; therefore, the grain size in the (013) plane of the thin film is about 52.76 nm.

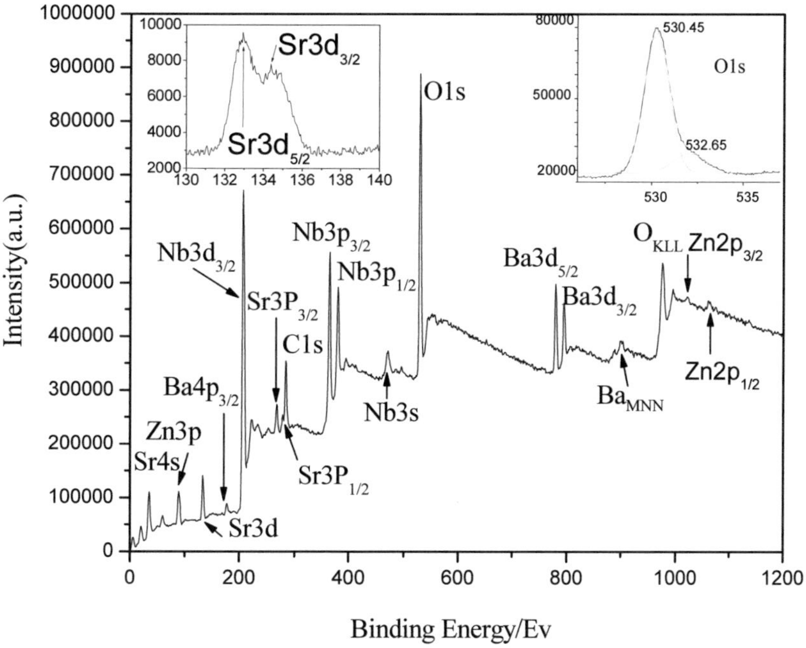

Fig. 20. XPS spectrum of the thin film.

Figure 20 shows the XPS spectrum of the thin film in the binding energy range from 0 eV to 1200 eV, and the binding energies at various peaks were calibrated using the C1s (284.6 eV) as a standard sample. The inset images in Figure 20 shows the XPS spectrum of O1s peak. All the XPS spectra of Ba3d, Nb3d, Sr3d, Sr3p, and Zn2p of the thin films consist of two peaks corresponding to their angular momentum of electron. Only one spin-orbit doublet is observed for the individual element, i.e., $Ba3d_{5/2}$ and $Ba3d_{3/2}$ peaks at 780.67 and 795.95 eV, $Nb3d_{5/2}$ and $Nb3d_{3/2}$ peaks at 206.57 and 209.47 eV, $Sr3d_{5/2}$ and $Sr3d_{3/2}$ peaks at 133.07 and 135.02 eV, $Sr3p_{3/2}$ and $Sr3p_{1/2}$ peaks at 267.84 and 278.37 eV, $Zn2p_{3/2}$ and $Zn2p_{1/2}$ peaks at 1023.27 and 1046.51 eV, indicating that only one chemical state exists in the thin films for each element of Ba, Nb, Sr, Zn, i.e., chemical state of Ba^{2+}, Sr^{2+}, Nb^{5+}, and Zn^{2+}. A doublet structure is observed in the XPS spectrum of O1s peak. Its component peak in the spectrum is fitted to a Gaussian-type distribution with the lower binding energy of 530.45 eV and the higher binding energy of 532.65 eV, as shown in the inset spectrum, corresponding to the lattice oxygen and the adsorbed oxygen, respectively. In general, the peak of adsorptional oxygen is much weaker than that of lattice oxygen, which would be favorable for the dielectric property of the thin films, because the oxygen vacancies result in the dielectric loss. The quantitative analysis using Ba3d, Sr3d, Zn2p, Nb3d, and O1s peaks reveals that the Ba:Sr:Zn:Nb:O atomic ratio is 3.01: 5.33: 0.96: 27.90: 62.80.

Figure 21. shows the SEM image of the thin film, which shows that fine spherical particles are distributed on the sample's surface. The surface morphology of the thin film is of dense structure with less small holes. It is evident that the grains are of uniform grain size, and are distributed homogenously on the surface of the thin film.

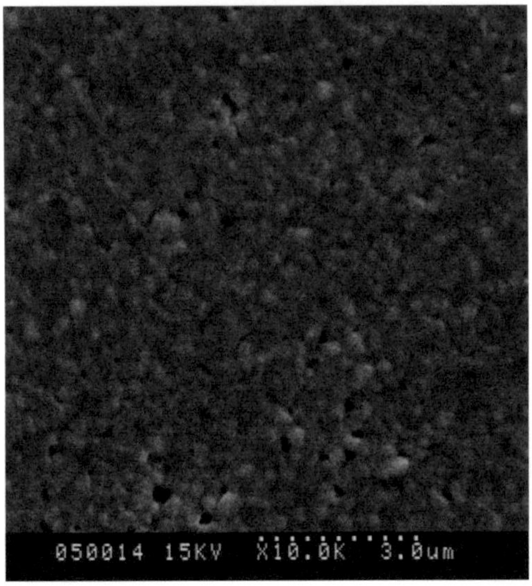

Fig. 21. SEM image of the thin film.

Fig. 22. Cross-sectional SEM images of the thin film.

Figure 22. shows the thickness of the thin film is about 1.38 µm with dense structure and homogeneous grains and well-defined grain boundaries, which are consistent with the results shown in Figure 21.

Figure 23. shows the AFM images of the sample, with a scan area of 10 × 10 µm. As seen in Figure 23(a), the sample is regular in surface with fine and uniform grains, and a few spherical particles are observed growing along the c axis. The root-mean-square surface roughness (RMS, calculated from the AFM data using the PSI ProScan Image Processing software package version 1.0, Park Scientific Instruments, Sunnyvale, CA) of the thin films is about 27.4 nm. Figure 23(b) shows the formation of the ceramic thin film with high-quality crystalline.

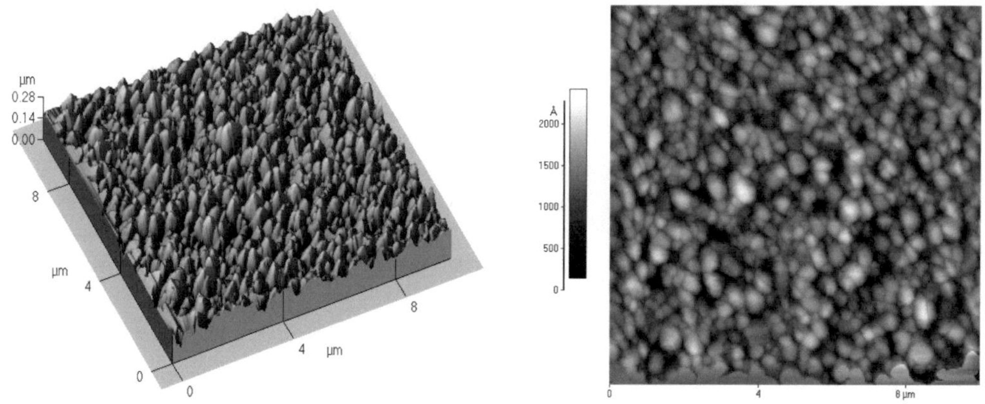

(a) (b)

Fig. 23. AFM images of the thin film. (a) longitudinal cross-section; (b) horizontal cross-section.

The TEM image and the selected area electron diffraction (SAED) spectrum are shown in Figure 24. The spherical particles with the well-defined boundaries are observed in Figure 24(a). The spherical shape of the grains is consistent with the results acquired by the SEM and AFM. The SAED spectrum indicats that the thin films are polycrystalline, as shown in Figure 24(b). About the (013) grain size, there is considerable discrepancy between the measured grain size (approximate 1600 nm by TEM) and calculation size (53 nm obtains from the Scherrer`s formula), which may be because the grain size obtained through the Scherrer`s formula is the average size of the orderly arrangement grains, however, the grain size measured by the TEM is the size of the clusters consisting of many small grains, which have not special periodicity (Ji et al., 2011).

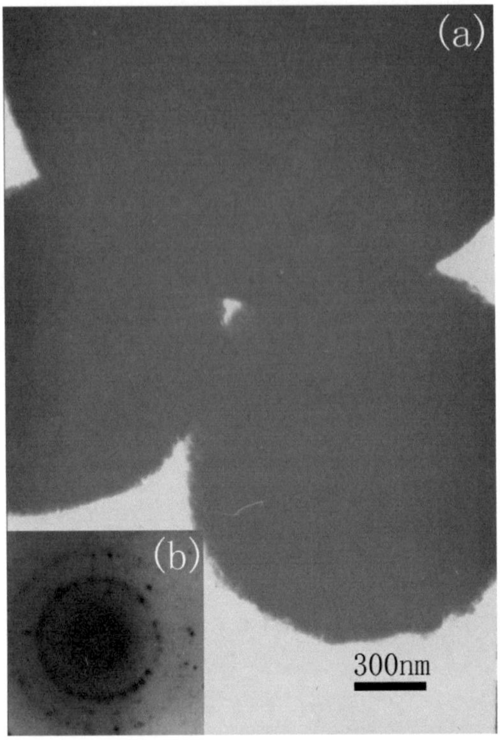

Fig. 24. TEM image and the SAED spectrum of the thin film. (a) TEM; (b) SAED.

5.4 Conclusion

The main phase of the ceramic thin film is $Ba_xSr_{1-x}Nb_2O_6$. The thin film presences a high crystalline quality, with few adsorbed oxygen. The surface morphology indicates that the thin film is of dense structure. The grains are polycrystalline and uniform in size with the spherical shape and well-defined grain boundaries.

6. Acknowledgements

The authors would like to thank the National Natural Science Foundation of China (No.51042001). The authors are also grateful to Professor Chengshan Xue for his help in discussion our results.

7. References

Amanullah, F.M.; Pratap, K.I.; & Hari, V.B. (1998). Compositional Analysis and Depth Profile Studies on Undoped and Doped Tin Oxide Films Prepared by Spray Technique, *Materials Science and Engineering B*, Vol. 52, No.2-3. (April 1998). pp.93-98. ISBN 0921-5107

Cui, C.W.; Shi, F.; Li, Y.G.; & Wang, S.Y. (2010). Orthogonal Analysis for Perovskite Structure Microwave Dielectric Ceramic Thin Films Fabricated by the RF

Magnetron-Sputtering Method, *Journal of Materials Science: Materials in Electronics*, Vol. 21, No. 4, (April 2010), pp. 349-354, ISBN 0957-4522

HSi, C.S.; Chen, C.Y.; Wu, N.C.; & Wang, M.C. (2003). Dielectric Properties of $Ba(Zr_xTi_{1-x})O_3$ Thin Films Prepared Using Radio Frequency Magnetron Sputtering, *Journal of Applied Physics*, Vol. 94, No. 1, (January 2003), pp. 598-604, ISBN 0021-8979

Huang, H.H., Wang, M.C., Chen, C.Y., Wu, N.C., and Lin, H.J. (2006). Effect of Deposition Parameters on the Growth Rate and Dielectric Properties of the $Ba(Sn_xTi_{1-x})O_3$ Thin Films Prepared by Radio Frequency Magnetron Sputtering, *Journal of the European Ceramics Society*, Vol. 26, No. 15, (June 2006), pp. 3211-3219, ISBN 0955-2219

Ianculescu, A.; Despax, B.; Bley, V.; & Lebey, T. (2007). Structure-Properties Correlations for Barium Titanate Thin Films Obtained by RF-Sputtering, *Journal of the European Ceramics Society*, Vol. 27, No. 2-3, (June 2007), pp. 1129-1135, ISBN 0955-2219

Ji, F.F.; Shi, F. (2011). Fabrication and Charicaterization of Ceramic Thin Films by RF Magnetron Sputtering Using Zn - Enriched $(Ba_{0.3}Sr_{0.7})(Zn_{1/3}Nb_{2/3})O_3$ as Target, *Journal of Alloys and Compounds*, Vol.509, No. 5, (Febrary 2011), ppL95-L98, ISBN 0925-8388

Klung, P.; Alexander, L.E. *X-ray diffraction procedures*, John Wiley & Sons, New York,1974.

Liu, L.Y.; Zhu M.K.; Hou, Y.D.; Shao, M.M., & Yan, H. (2005), Effect of Sputtering Pressure on Microstructure of Ferroelectric $Ba_{2-x}TiSi_{2+y}O_8$ Thin Films, *Journal of Wuhan University of Technology*, Vol. 27, No.7, (July 2005). pp.1-3. ISBN 1671-6477

Ma, J.; Feng, J.; Yang, C.S.; & Ding, G.F. (2006). Key Problems in the Preparation of $(Ba,Sr)TiO_3$ Thin Films by RF Magnetron Reactive Sputter Deposition, *Piezoelectrics & Acoustooptics*, Vol. 28, No. 2, (March 2006), pp. 5-8, ISBN 1004-2474

Schroeder, J. B.; Klimasewski, R.G. (1968). Scatter from X-ray Reflecting Surfaces, *Applied Optics*, Vol. 7, No. 10, (October 1968). pp. 1921-1927. ISBN 1559-128X

Shi, F.; Cui, C.W. (2010). Microstructure of Microwave Dielectric Thin Films by RF Magnetron Sputtering, *Applied Surface Science*, Vol. 256, No. 8, (February 2010), pp. 2626-2629, ISBN 0169-4332

Tang, W.Z. (2003), *Theory, Technology and Application for the Fabrication of Thin Film Materials*, Metallurgy Industry Press, Beijing, 2003.

Tang, Z.Z.; Liu, S.J.; Singh, R.K.; Bandyopadhyay, S.; Sus, I.; Kotani, T.; Schilfgaarde, M.; & Newman, N. (2009). Growth and characterization of epitaxial $Ba(Zn_{1/3}Ta_{2/3})O_3$ (1 0 0) thin films, *Acta Materialia*. Vol. 57, No.2, (January 2009). pp.432-440. ISBN 1359-6454

Varma, M.R.; Biju, S.; & Sebastian, M.T. (2006). Preparation of Phase Pure $Ba(Zn_{1/3}Ta_{2/3})O_3$ Nanopowders for Microwave Dielectric Resonator Applications, *Journal of the European Ceramics Society*, Vol. 26, No. 15, pp. 1903 -1907, ISBN 0955-2219

Wang, B.; Wang, M.; Zhang, D.X.; Huang, A.P.; Song, X.M.; Zou, Y.J.; & Yan, H. (2002). Dependence of e-BN Film Structure on Working Pressure and Growth Time, *Functinal Materials*, Vol.33, No. 1, (January 2002). pp. 82-83. ISBN 1001-9731

Wang, C.F. *Solid State Physics Tutorial*, Shandong University Press, Shan Dong, 2004.

Wang, M.C.; HSi, C.S.; & Wu,N.C. (2004). Characterization and Leakage Current Density of Radio Frequency Magnetron Sputtered Nanocrstalline $SrTiO_3$ Thin Films, *Journal of Crystal Growth*, Vol. 264, No. 1-3, (March 2004), pp. 271-277, ISBN 0022-0248

Yang, H.N.; Tweet, D. J.; Ma, Y.J.; & Tue, N.Y. (1998). Deposition of Highly Crosslinked Fluorinated Amorphous Carbon Film and Structural Evolution During Thermal

Annealing, *Applied Physics Letter*, Vol. 73, No.11, (September 1998). pp.1514-1517. ISBN 1882-0778

Yu, L.H.; Fu, X.H. (2006). The Influences of Annealed Temperatures on The Properties and Microstructure of $Ba_{0.5}Sr_{0.5}TiO_3$ Thin Film, *Advanced Ceramics*, Vol. 27, No. 3, (March 2006), pp. 16-19. ISBN 1005-1198

Zannetti, R.; Celotti, G.; Fichera A.; & Francesconi, R. (1969). The Structural Effects of Annealing Time and Temperature on the Paracrystal-Crystal Transition in Isotactic Polypropylene, Die Makromolekulare Chemie, Vol. 128, No.1, (October 1969). pp.137-142. ISBN 0003-3146

Zheng, J.P.; Kwok, H.S. (1993). Preparation of indium tin oxide films at room temperature by pulsed laser deposition, *Thin Solid films*, Vol. 232, No. 1, (September 1993). pp.99-104. ISBN 0040-6090

Part 6

Geoscience, Mineralogy

38

Microstructural and Mineralogical Characterization of Clay Stabilized Using Calcium-Based Stabilizers

Pranshoo Solanki[1] and Musharraf Zaman[2]
[1]*Illinois State University*
[2]*University of Oklahoma*
USA

1. Introduction

The properties of clays can be significantly improved by treating with calcium-based stabilizers (or additives) such as hydrated lime (or lime), Portland cement, cement kiln dust (CKD), and class C fly ash (CFA). In the presence of water, the calcium ions released from these stabilizers reduce the thickness of double diffused layer through cation-exchange and flocculation-agglomeration reactions. This is primarily responsible for improvement in workability through reduction of adsorbed water and decrease in plasticity index. In long-term, pozzolanic reactions occur between the calcium ions of the stabilizer and the silica and alumina of the clay minerals resulting in the formation of cementitious products such as calcium-silicate-hydrates (C-S-H), calcium-aluminate-hydrates (C-A-H), and calcium-aluminum-silicate-hydrates (C-A-S-H). The reaction may be written as:

$$Ca(OH)_2 \text{ (ionization of lime)} \rightarrow Ca^{2+} + 2(OH)^- \quad (1)$$

$$Ca^{2+} + OH^- + SiO_2 \text{ (soluble clay silica)} \rightarrow \text{calcium-silicate-hydrate} \quad (2)$$

$$Ca^{2+} + OH^- + Al_2O_3 \text{ (soluble clay alumina)} \rightarrow \text{calcium-aluminate-hydrate} \quad (3)$$

The hyphens indicate that the composition is indefinite. The specific composition is defined by pH level, solubility of silica and alumina, clay mineralogy, and curing conditions among other reasons. C-S-H, formed by the hydration of C_3S, is also commonly known as tobermorite gel. The tobermorite gel is poorly crystalline, with only a few broad, weak bands in its X-ray diffraction pattern (Mohamed, 2002). C-A-H is formed as platelets with hexagonal symmetry. The morphology of both C-A-H and C-A-S-H resembles that of tobermorite gel. In the presence of calcium sulfate, an additional product of hydration, known as ettringite ($Ca_6[Al(OH)_6]_2.(SO_4)_3.26H_2O$) is formed (Eq. 4). Ettringite mineral consists of a prismatic crystal and a hexagonal cross section.

$$6Ca^{2+} + 2Al(OH)_4^- + 4OH^- + 3SO_4^{2-} + 26H_2O \rightarrow Ca_6[Al(OH)_6]_2.(SO_4)_3.26H_2O \quad (4)$$

The formation of aforementioned cementitious products in the soil-stabilizer matrix are responsible for increase in the internal friction and shear strength of the stabilized soil.

However, the efficacy of stabilization depends on the soil mineralogy, type and amount of stabilizer, and curing conditions (e.g., time, temperature, moisture). Although several researchers studied the improvement in engineering properties of stabilized soil at macro level (e.g, shear strength, unconfined compressive strength, swell behavior), very few studies discussed the changes in soil-stabilizer matrix at micro level. Consequently, the primary objective of this study is to examine the changes in microstructure and mineralogy of soil due to stabilization with calcium-based stabilizer. Additionally, at macro level the modulus of elasticity values are evaluated for both raw and stabilized specimens and correlated with microstructural and mineralogical characteristics.

2. Literature review

Several investigations were carried out to study the changes in the microstructural and mineralogical characteristics of stabilized soils. Rajasekaran et al. (1995) studied the influence of sodium hydroxide on the fabric of lime treated marine clays using scanning electron microscopy (SEM) technique. It was found that lime stabilization is very effective for marine clays. Adding sodium hydroxide additive resulted in better formation of pozzolanic compounds. Formation of cementitious compounds such as C-A-H and C-S-H due to the soil-lime reactions were observed in all lime treated soils which were further confirmed by using X-Ray Diffraction (XRD) technique. The study conducted by using SEM indicated that there is an overall improvement in the structure of the soil system resulting in a porous system and aggregate formation. However, this study was limited to only lime-stabilization.

In another study, Lav and Lav (2000) investigated the effects of cement- and lime-stabilization on class F fly ash in terms of change in chemical composition, crystalline structures, and hydration products. The unconfined compressive strength (UCS) of samples was also evaluated over time to observe the effect of stabilization. Cement- and lime-stabilized fly ash produced similar hydration products. None of these produced was recognizable by XRD except weak calcium hydroxide (CH) and calcium carbonate due to carbonation. The improvement in microstructure was found initiating from fly ash particles serving as nucleation centers for hydration or pozzolanic reaction products. The increasing strength gain was attributed to growth of hydrates in the voids between the particles. However, no attempt was made to study the influence of aforementioned additives on soil.

Ghosh and Subbarao (2001) studied the physicochemical and microstructural developments of fly ash-lime- and fly ash-lime-gypsum-stabilized materials. Different analytical techniques namely, XRD, differential thermal analysis, SEM, and Energy Dispersive Spectroscopy (EDS) were used for studying the microstructure. The SEM micrographs revealed evidence of the development of a compact matrix after three months of curing time and a densified compact network of pozzolanic reaction products of fly ash-lime-gypsum with the increase in the curing period to ten months. The XRD analysis results indicated appearance of new peaks of low intensity in the modified fly ash specimens. Some of these peaks were not recognized as part of any new crystalline phases. Similar to previous study, this study was limited to fly ash-lime mixtures and no attempt was made to study the influence of these additives on soil.

In a laboratory study, Al-Rawas (2002) investigated the microfabric and mineralogical aspects of the expansive soil using cement by-pass dust (CBPD), copper slag, slag-cement, and ground granulated blast furnace slag (GGBFS). The highest swell potential of untreated soil was explained by the presence of the highest percentage of sodium smectite clay mineral along with palygorskite and illite in soil. The fabric of the untreated soil was found composed of dense clay matrices with no appearance of aggregations and increasing amount of pore spaces. However, stabilization resulted in the formation of aggregations and few connectors. It was found that higher amount of sodium ions and lower amount of calcium ions promotes swelling and vice versa. Further, the XRD results showed a general reduction in all the clay minerals' peak intensities particularly in the case of CBPD treated samples. This study addressed most of the properties that will be evaluated in the present study. However, it was carried out on predominantly silty soil stabilized with non-traditional additives. It is also important to note that the mineralogical and textural characteristics of fat clay used in the present study are different than silty soil. Also, non-traditional additives are not commonly available and are expensive. This makes it necessary to investigate locally available additives such as lime, fly ash, and CKD.

Stutzman (2004) studied the bulk and surface phase composition of hydraulic cement using SEM in conjunction with XRD. Direct imaging of hydraulic cements obtained through SEM yielded complete picture of both bulk and surface phase compositions. Mass percent and volume percent were calculated by analyzing resulting composite image from SEM. A good agreement was found between mass percentages obtained by SEM imaging and percentages based upon quantitative XRD. The finer grained phases (e.g., gypsum, tricalcium aluminate, and ferrite) showed much higher surface areas per unit mass than the coarser-grained phases such as alite and belite. But no attempt was made to compare the microstructures of stabilized soils.

In another laboratory study, Kolias et al. (2005) investigated the effectiveness of using high calcium fly ash and cement in stabilizing lean and fat clays. Strength tests in compression, indirect tension, and flexure modes were conducted on stabilized samples. Additionally, thermogravimetric-single differential thermal analysis (TG-SDTA) and XRD tests were conducted on selected samples to study the hydraulic compounds. The study showed the potential benefit of stabilizing clays with high calcium fly ash. However, it was found that the effectiveness of stabilization is dependent on the type of soil, the amount of stabilizer and the curing time. The study of formation of hydraulic products showed that a significant amount of tobermorite gel is formed due to stabilization leading to a denser and more stable structure of the samples. The mechanical properties such as strength and modulus of elasticity showed considerable enhancement due to stabilization. This study, however, did not examine the microstructure of lime- and CKD-stabilized soils.

Horpibulsuk et al. (2010) analyzed the strength development in cement-stabilized silty clay based on microstructural considerations. A qualitative and quantitative study was conducted on the microstructure using a SEM, mercury intrusion pore size distribution measurements, and thermal gravity analysis. A total of three zones of improvement namely, active, inert, and deterioration zones, were observed. The active zone was found to be most effective for stabilization where the cementitious products increased with cement content and filled the pore space. In the inert zone, both pore size distribution and cementitious products change insignificantly with increasing cement. In deterioration zone, the water is not adequate for hydration because of the excess of cement input. It was found that in short stabilization period, the volume of large pores ($> 0.1\ \mu m$) increases because of input of coarser particles while the volume of small pores (< 0.1 μm) decreases because of the solidification of the hydrated cement. With time, the large pores are filled with the cementitious products; thus, the small pore volume increases, and the total pore volume decreases resulting in the development of strength with time. This study, however, was limited to only cement and no attempt was made to compare the microstructure of soil stabilized with other additives.

In a recent study, Chaunsali and Peethamparan (2011) characterized a nontraditional binding material containing cement kiln dust (CKD) and ground GGBFS. The CKD used in this study contained low free lime and high sulfate and alkali content, and proved effective in accelerating the hydration of GGBFS. The strength rate development was found to be dependent on the curing conditions but eventually all the samples achieved similar compressive strengths independent of the curing conditions. The microstructural and mineralogical examinations showed that the strength development was mainly due to the formation of C-S-H. Additionally, aluminum and magnesium incorporated C-S-H phases were also identified in CKD-GGBFS blends. The formation of ettringite appeared as a contributing factor in the development of early age strength in CKD-GGBFS binder.

It is clear from the aforementioned literature review that there is a lack of detailed comparative studies on microstructure and mineralogy of expansive soils stabilized with various stabilizers. Therefore, there was a need to undertake a detailed investigation to fully characterize the microstructure and mineralogy of soil stabilized with locally available additives.

3. Materials and sources

In the present study, one fat clay and three cementitious additives are used. This section describes the fundamental properties including grain size distribution, index properties and chemical compositions of the soils and additives.

3.1 Native soils

The soil used in this study is Carnasaw series soil. According to the Unified Soil Classification System (USCS), Carnasaw series soil is classified as fat clay (CH) with an average liquid limit of approximately 58% and a plasticity index (PI) of 29 in accordance with ASTM D 4318 test method. The gradation tests revealed percent passing No. 200 sieve (< 0.075 mm) and clay content (< 0.002 mm) as 94% and 48%, respectively. A summary of the physical and chemical properties of the selected soil is presented in Table 1.

3.2 Cementitious additives

As noted earlier, three different cementitious additives, namely, hydrated lime, CFA, and CKD were used. Hydrated lime was supplied by the Texas Lime Company, Cleburne, Texas. It is a dry powder manufactured by treating quicklime (calcium oxide) with sufficient water to satisfy its chemical affinity with water, thereby converting the oxides to hydroxides. CFA from Lafarge North America (Tulsa, Oklahoma) was brought in well-sealed plastic buckets. It was produced in a coal-fired electric utility plant, American Electric Power (AEP), located in Muskogee, Oklahoma. CKD used was provided by Lafarge North America located in Tulsa, Oklahoma. It is an industrial waste collected during the production of Portland cement. The physical and chemical properties of the stabilizing agents are presented in Table 1. The X-Ray Fluorescence (XRF) analysis was conducted using a Panalytical 2403 spectrometer on specimens obtained by using fused bead preparation method. The fused bead preparation technique consists of dissolving the specimen in a solvent called a flux at high temperature (>1000°C) in a platinum crucible and to cast it in a casting-dish. It is evident from Table 1 that the calcium oxide content in hydrated lime is 68.6%. This can be explained using the stoichiometry of the chemical reaction taking place during the specimen preparation for XRD.

$$Ca(OH)_2 \rightarrow CaO + H_2O \quad (5)$$

$$(74)=40x1+16x2+1x2 \qquad (56)=40x1+16x1 \qquad (18)=1x2+16x1$$

Using above chemical equation, it can be shown that 95.9% of $Ca(OH)_2$ (reactant) will produce approximately 72% of CaO (product). Further, the free lime content (i.e., any lime not bound up in glassy phase compounds such as tricalcium silicate and tricalcium aluminate) was determined in accordance with ASTM C 114 (Alternate Test Method B, ammonium acetate titration). Although CFA is having a very low lime content (0.2%), specimens stabilized with CFA showed enhancement in strength and modulus values as will be discussed later in this chapter. It is speculated that during the reaction process some lime is liberated from the bound state which takes part in the cementitious reactions and thus, producing increase in strength and modulus values.

Chemical compound/Property	Percentage by weight, (%)			
	Lime	CFA	CKD	Soil
Silica (SiO_2)[a]	0.6	37.7	14.1	63.4
Alumina (Al_2O_3)[a]	0.4	17.3	3.1	21.5
Ferric oxide (Fe_2O_3)[a]	0.3	5.8	1.4	9.1
Calcium oxide (CaO)[a]	68.6	24.4	47	0.1
Calcium hydroxide ($Ca(OH)_2$)[a]	95.9**	...	...	...
Magnesium oxide (MgO)[a]	0.7	5.1	1.7	1.2
Sulfur trioxide (SO_3)[a]	0.1	1.2	4.4	0
Alkali content ($Na_2O + K_2O$)[a]	0.1	2.2	1.7	3.0
Loss on ignition[b]	31.8*	1.2	27	...
Free lime[b]	46.1	0.2	6.7	...
Percentage passing No. 325[c]	98.4	85.8	94.2	87.2
pH (pure material)[d]	12.58	11.83	12.55	4.17
Specific surface area (m^2/gm)[e]	17.0	6.0	12.0	118.5
28-day UCS (kPa)	...	708	17	207

[a]X-ray Fluorescence analysis; [b]ASTM C 114; [c]ASTM C 430; [d]ASTM D 6276; [e]Ethylene glycol monoethyl ether method (Cerato and Lutenegger 2001); UCS: Unconfined compressive strength;
*Ca(OH)2 decomposes at 512oC; **Before ignition

Table 1. Chemical and physical properties of soil and additives

4. Factors affecting cementitious stabilization

The effectiveness of cementitious stabilization depends on properties of both soil and additive (AFJMAN, 1994, Al-Rawas et al., 2002, Parsons et al., 2004, Evangelos, 2006). A description of the pertinent factors intrinsic to the soils and additives which influence the efficiency of cementitious stabilization is presented herein.

4.1 Soil properties

4.1.1 Gradation and plasticity index

Several researchers (e.g., Diamond and Kinter, 1964; Haston and Wohlgemuth, 1985; Prusinski and Bhattacharja, 1999; Little, 2000; Qubain et al., 2000; Kim and Siddiki, 2004; Mallela et al., 2004; Puppala et al., 2006; Consoli et al., 2009) recommended use of lime with fine-grained soils. However, CFA (see e.g., McManis and Arman, 1989; Chang, 1995; Misra, 1998; Zia and Fox, 2000; Puppala et al., 2003; Bin-Shafique et al., 2004; Phanikumar and Sharma, 2004; Nalbantoglu, 2004; Camargo et al., 2009; Li et al., 2009) and CKD (e.g., McCoy and Kriner, 1971; Baghdadi and Rahman, 1990; Zaman et al., 1992; Sayah, 1993; Miller and Azad, 2000; Miller and Zaman, 2000; Parsons and Kneebone, 2004; Sreekrishnavilasam et al., 2007; Peethamparan et al., 2008; Gomez, 2009) is used successfully with both fine- and coarse-grained soils. Lower effectiveness of lime with coarse-grained soil can be attributed to scarcity of pozzolana (silicious and aluminacious material) in coarse-grained soils which is required for pozzolanic (or cementitious) reactions. Little (2000) and Mallela et al. (2004) recommend a soil with a minimum clay content (< 0.002 mm) of 10% and a plasticity index of 10 for lime-stabilization. In this study, Carnasaw series soil fulfils this requirement with a clay content of 48%. Also, mineralogical analyses conducted using XRF revealed that the soil used in this study is having high (85%) amount of pozzolana, as presented in Table 1.

4.1.2 Cation exchange capacity

Cation Exchange Capacity (CEC) is the quantity of exchangeable cations required to balance the charge deficiency on the surface of the clay particles (Mitchell, 1993). During ion-exchange reaction of soil with cementitious additive, cation of soil (e.g., Na^+, K^+) is replaced by cation of additive (Ca^{2+}) and the thickness of double diffused layer is reduced. Hence, the replacement of cations results in an increase in workability and strength of soil-additive mixture. The rate of exchange depends on clay type, solution concentrations and temperature (Gomez, 2009). In soil stabilization studies, CEC values have been used to a limited extent to explain the effectiveness of soil stabilization (Nalbantoglu and Tuncer, 2001; Al-Rawas et al., 2002; Nalbantoglu, 2004; Gomez, 2009).

In this study, CEC was measured by sodium acetate method in accordance with the EPA 9081 test method (Chapman, 1965). As evident from Table 1, Carnasaw soil showed CEC value of 5.2 meq/100 gm.

4.1.3 Sulfate content

Primary "sulfate-induced heaving" problems arise when natural sulfate rich soils are stabilized with calcium-based additives (Puppala et al., 2004), also known as "sulfate attack." This heave is known to severely affect the performance of pavements, and other geotechnical structures built on sulfate rich soils stabilized with calcium-based additive (Hunter, 1988; Mitchell and Dermatas, 1990; Petry and Little, 1992; Rajendran and Lytton, 1997; Rollings et al., 1999; Puppala et al., 2004). According to current understanding, "low to moderate" and high sulfate soils are those with sulfate less than 2,000 ppm and more than 2,000 ppm, respectively (Kota et al., 1996; Mitchell and Dermatas, 1990; Puppala et al., 2002; Rao and Shivananda, 2005). In this study, soluble sulfate content in the soil was measured using the Oklahoma Department of Transportation procedure for determining soluble sulfate content: OHD L-49 (ODOT, 2006). No detectable sulfate content (< 200 ppm) was found in the soil used in this study.

4.1.4 Specific surface area

Surface phenomena have an important influence on the behavior of fine-grained soils; they affect many physical and chemical properties (Cerato and Lutenegger, 2002). The specific surface area (SSA), refers to the area per unit mass of soil, may be a dominant factor in controlling the fundamental behavior of many fine-grained soils (Gomez, 2009). The mineralogy of fine-grained soils is the dominant factor in determining the effect of SSA. For this study, only total SSA measurement was conducted using the polar liquid Ethylene Glycol Monoethyl Ether (EGME) method (Cerato and Lutenegger, 2002) and results are presented in Table 1. The Carnasaw series soil showed a SSA value of 118.5 m^2/gm.

4.2 Additive properties

4.2.1 Free-lime content

In calcium-based stabilizers (e.g., Portland cement, CFA, CKD) most of the lime (CaO) is bound up in compounds such as tricalcium silicate and tricalcium aluminate. The unreacted lime that is not combined in any of these compounds is called free-lime, which is expected to play a major role in stabilization (Collins and Emery, 1983; Misra, 1998; Zaman et al., 1998; Ferguson and Levorson, 1999; Miller and Azad, 2000; Miller and Zaman, 2000; Sezer et al., 2006; Khoury and Zaman, 2007; Peethamparan and Olek, 2008). Free-lime content was determined by conducting titration in accordance with ASTM C 114 alternative test method B and results are presented in Table 1. It is clear that lime is having the highest free-lime content of 46.7% followed by 6.7% for CKD and 0.2% for CFA.

4.2.2 Specific surface area

The specific surface area (SSA) of additives, as measured by using the ethylene glycol monoethyl ether (EGME) method (Cerato and Lutenegger, 2002), were 17.0, 6.0, and 12.0 m^2/gm, respectively, for lime, CFA and CKD. It can be seen that lime and CFA had the highest and the lowest SSA values, respectively. A higher SSA indicates more reactivity of additive (Nalbantoglu and Tuncer, 2001; Sreekrishnavilasam et al., 2007).

4.2.3 Loss on ignition

A higher loss on ignition (LOI) value indicates high carbonates for CFA/CKD and high hydroxides for lime. Some researchers reported that high LOI indicates low free-lime content for CKDs, making CKDs less reactive, and therefore lower improvements (Bhatty et al., 1996; Miller and Azad, 2000). In the laboratory, LOI was evaluated by igniting additive inside a muffle furnace at a temperature of 950°C (1742°F) in accordance with ASTM C 114 test method for hydraulic cements. As evident from results presented in Table 1, lime and CFA produced highest and lowest LOI values of 31.8% and 1.2%, respectively. On the other hand, approximately 27% of CKD is lost on ignition.

4.2.4 Percent passing No. 325 sieve

Several researchers noticed increased reactivity of additive with increase in amount of additive passing No. 325 (45 μm) sieve (NCHRP, 1976; Bhatty et al., 1996; Zaman et al., 1998; Zheng and Qin, 2003; Khoury, 2005). The percentage of passing No. 325 sieve for lime, CFA and CKD determined in accordance with ASTM C 430 test method are 98.4, 85.8 and 94.2, respectively. It is clear that lime is finest among all the additives used in this study.

4.2.5 pH and pH response

The elevated pH level of soil-lime mixture is important because it provides an adequate alkaline environment for ion-exchange reactions (Little, 2000). In the laboratory, pH is determined using the method recommended by ASTM D 6276 for lime-stabilization, which involves mixing the solids with de-ionized (DI) water, periodically shaking samples, and then testing with a pH meter after 1 h. The procedure specifies that enough lime must be added to a soil-water system to maintain a pH of 12.4 after 1 h. This ensures that adequate lime is provided to sustain the saturation during the 1-h period (Prusinski and Bhattacharja, 1999).

Additive Content (%)	Lime		Additive Content (%)	CFA		Additive Content (%)	CKD	
	pH value	% Increase[a]		pH value	% Increase[a]		pH value	% Increase[a]
0	4.17	---	0	4.17	---	0	4.17	---
1	9.22	121.1	2.5	5.19	24.5	2.5	7.05	69.1
3	12.23	193.3	5	5.93	42.2	5	8.8	111.0
5	**12.54**	**200.7**	7.5	6.55	57.1	7.5	10.11	142.4
6	12.55	201.0	10	7.79	86.8	10	10.88	160.9
7	12.55	201.0	12.5	8.32	99.5	12.5	11.28	170.5
9	12.57	201.4	15	8.86	112.5	15	11.62	178.7
100	12.58	201.7	17.5	9.47	127.1	17.5	11.98	187.3
			100	4.17	---	100	4.17	---

[a]Increase in pH w.r.t. pH value of raw soil; Bold values represent minimum additive content providing asymptotic behavior (< 1% increase)

Table 2. Variation of pH values with soil and additive type

Several researchers (e.g., Haston and Wohlgemuth, 1985; Prusinski and Bhattacharja, 1999; IRC, 2000; Little, 2000; Qubain et al., 2000; Mallela et al., 2004; Puppala et al., 2006; Consoli et al., 2009) used pH values on soil-lime mixture as an indicator of reactivity of lime. However, only limited studies (see e.g., Miller and Azad, 2000; Parsons et al., 2004; Peethamparan and Olek, 2008; Gomez, 2009) evaluated pH response of soil-CFA or soil-CKD mixtures. Hence, the pH values of soil-additive mixtures were determined to investigate whether pH would reflect the effectiveness of soil stabilization with lime, CFA or CKD.

The pH results of raw soil, raw additive and soil-additive mixtures are presented in Table 2 and are used as the primary guide for determining the amount of additive required to stabilize each soil. It is clear that Carnasaw soil is acidic with a pH value approximately 4.17. Also, it was found that raw lime, CFA and CKD had a pH value of 12.58, 11.83 and 12.55, respectively. The pH values of raw CFA and CKD are consistent with the results reported by other researchers (e.g., Miller and Azad, 2000; Sear, 2001; Parsons et al., 2004; Peethamparan and Olek, 2008; Gomez, 2009). The pH trend of raw additives is similar to the trend of available free-lime content in additive, as shown in Table 1.

For all the soil-additive mixtures, pH values increase with the increase in the percentage of additive and show an asymptotic behavior after a certain percentage. In the current study, an increase of less than 1% in pH with respect to raw soil is assumed as starting point of the asymptotic behavior. As evident from Table 2, pH values started showing an asymptotic behavior with 5% lime. Additionally, soil never attained an asymptotic behavior with CFA and CKD contents up to 17.5%. This can be attributed to the acidic behavior of Carnasaw soil which requires higher amounts of moderately basic CFA and CKD for neutralization. Based on the aforementioned observations, it was decided to select 9% of lime and 15% of CFA and CKD for laboratory performance evaluation.

5. Experimental methodology

5.1 Specimen preparation

In this study, a total of 16 specimens were prepared for evaluating modulus of elasticity. The procedure consists of adding a specific amount of additive to the raw soil desired. The amount of additive (9% for lime and 15% for CFA and CKD) was added based on the dry weight of the soil. The additive and soil were mixed manually for uniformity. After the blending process, a desired amount of water was added based on the optimum moisture content (OMC). The mixture was then compacted in a mold having a diameter of 101.6 mm and a height of 203.2 mm to reach a dry density of between 95%-100% of the maximum dry unit weight (MUW). After compaction, specimens were cured at a temperature of 23.0 ± 1.7°C and a relative humidity of approximately 96% for 28 days. A total of four replicates were prepared for each additive type and tested for modulus of elasticity in accordance with ASTM D 1633 test method.

5.2 Modulus of elasticity

As noted earlier, modulus of elasticity test was conducted in accordance with the ASTM D 1633 test method. Specimens were loaded in a MTS frame at a constant strain rate of 0.63% (of sample height) per minute, which is equivalent to 1.27 mm (0.05 in.) per minute for the specimen configuration used here. Deformation values were recorded during the test using LVDTs fixed to opposite sides of and equidistant from piston rod with a maximum stroke length of ±12.7 mm (±0.5 in). The load values were obtained from a load cell having a capacity of 22.7 kN (5,000 lb). Each specimen was subjected to two unloading-reloading cycles. Straight lines were drawn through the first two unloading-reloading curves (secant modulus) and the average slope of these lines is the modulus of elasticity of the stabilized clay specimen.

5.3 Mineralogical studies

To facilitate the macro-behavior comparison and explanation, the mineralogical study techniques namely SEM and EDS were employed to qualitatively identify the micro-structural developments in the matrix of the stabilized soil specimens. The SEM technique was employed using a JEOL JSM 880 microscope to qualitatively identify the micro-structural developments in the matrix of the stabilized soil specimens. After the modulus of elasticity test, specimens were broken and mix was air-dried for approximately two days. Three representative tiny pieces were mounted on stubs (1 cm, i.e., 0.4 in. wide discs having a pin-mount on the base of the disc). The samples were not electrically conductive; therefore, they were initially coated by Iridium to maintain conductivity. The quality of images was not satisfactory, so it was decided to use gold-palladium alloy for the process instead of Iridium coating. Hence, pieces were coated with a thin layer (≈ 5 nm) of an alloy of gold-palladium by sputter coating technique to provide surface conductivity. A JEOL JSM 880 scanning electron microscope operating at 15 kV was used to visually observe the coated specimens. The JEOL JSM 880 was fitted with an energy-dispersive X-ray spectrometer (EDS). The EDS was used to analyze chemical compositions of the specimen. In this technique, electrons are bombarded in the area of desired elemental composition; the elements present will emit characteristic X-rays, which are then recorded on a detector. The micrographs were taken using EDS2000 software. It must be noted that SEM study allows only a tiny area of raw and stabilized specimen to be examined (unlike engineering laboratory specimens). However, it is believed to be representative of the reaction process of stabilized specimens. Figure 1 (a) shows a photographic view of JEOL JSM 880 setup along with Hummer VI triode sputter coater.

(a) (b)

Fig. 1. (a) JEOL JSM 880 setup for SEM, Hummer VI triode sputter coater with sample (corner picture), and (b) Rigaku D/Max X-ray diffractometer

The XRD tests were performed on raw soil and stabilized specimens. Two-day air dried mix was pulverized with a mortar and pestle, sieved through a U.S. standard No. 325 sieve (45 μm). Then, the powder finer than 45 μm was collected, mixed with methanol, and placed on a specimen holder. The holder containing specimen was oven dried for approximately 15 minutes prior to testing. This holder was then mounted on a Rigaku D/Max X-ray diffractometer for analysis. This diffractometer is equipped with bragg-brentano parafocusing geometry, a diffracted beam monochromator, and a conventional copper target X-ray tube set to 40 kV and 30 mA. The measurements were performed from 4° to 70° (2θ range), with 0.05^o step size and 5 seconds count (dwell time) at each step. Data obtained by the diffractometer were analyzed with Jade 3.1, an X-ray powder diffraction analytical software, developed by Materials Data, Inc. (Jade, 1999). Generated diffractograms (using the peaks versus 2θ and d-spacing) were used to determine the presence of minerals. Figure 1 (b) shows a photographic view of Rigaku D/Max X-ray diffractometer.

6. Presentation and discussion of modulus of elasticity test results

The variation of modulus of elasticity values with the additive content is shown in Figure 2. It is clear from Figure 2 that the modulus of elasticity of stabilized specimens is influenced by the type of additive. For example, an increase of approximately 99%, 94% and 193% in modulus of elasticity values was observed for 9% lime-, 15% CFA- and 15% CKD-stabilized specimens, respectively. Overall, 15% CKD-stabilized specimens showed the highest improvement.

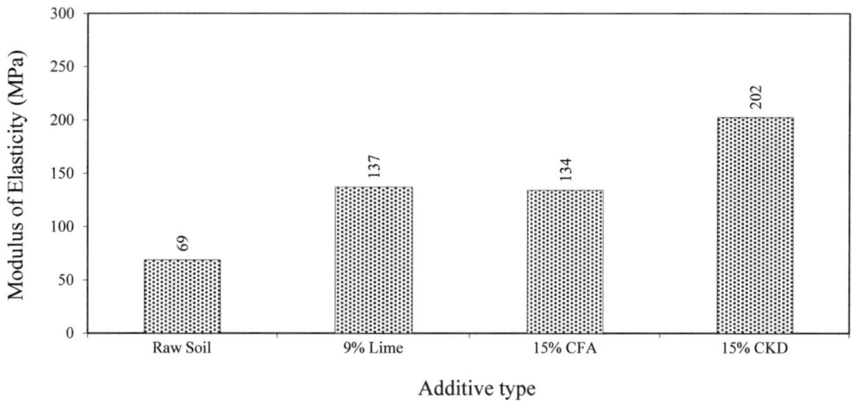

Fig. 2. Variation of modulus of elasticity values with additive type

Attempts were made to observe the effect of additive properties, namely, free-lime content, alkali content, loss on ignition, specific surface area (SSA), pH, and passing No. 325 sieve, on the modulus of elasticity. The effect of these additive properties on normalized modulus of elasticity (modulus of elasticity value/percent additive) is depicted in Figure 3. Here, it is clear that the normalized modulus of elasticity value increases with the free-lime content. The Carnasaw soil specimens exhibited an increase of approximately 13 to 15 as the free-lime content increased from 6.7% (CKD) to 46.1% (lime). A decrease in normalized modulus of elasticity values with alkali content can be observed; however, increase in normalized modulus of elasticity values with loss on ignition was observed. This trend is contrary to the behavior reported by other researchers for different type of CKDs (e.g., Bhatty et al., 1996; Miller and Azad, 2000; Peethamparan and Olek, 2008). For example, Bhatty et al. (1996) reported that CKDs containing less than 6% alkalis and low LOI values are reactive and produces higher strength. This difference in behavior could be attributed to other factors such as free-lime content that might have influenced in enhancing the effectiveness of the additives. Although CFA had higher alkali content and lower LOI than lime, it also had lower free-lime content (0.2% for CFA versus 46.1% for lime).

Further, it is clear from Figure 3 that percent passing No. 325 sieve influences the Mr values. An increase in percent passing No. 325 sieve from 85.8% (CFA) to 98.4% (lime) increased the normalized modulus of elasticity values from 9 to 15 for P-soil. This can be attributed to increase in fine contents in the soil and thus increased surface area for pozzolanic reactivity. Normalized modulus of elasticity values versus SSA of additive are shown in Figure 3. It is clear that normalized modulus of elasticity increases with increase in SSA. The fact that the additive particles have a larger surface to interact with the soil can explain this behavior. Larger SSA values imply more available surface for soil-additive interaction resulting in more cementitious products and thus higher gain in modulus values. The pH value of additive also plays an important role in enhancing the modulus values, as evident from Figure 3. An increase in normalized modulus values with pH can be observed from Figure 3. Lime-stabilized specimens having highest pH value of 12.58 produced the highest modulus value followed by CKD- (pH = 12.55) and CFA- (pH = 11.83) stabilized specimens. As discussed earlier, high pH value causes silica from the clay minerals to dissolve and, in combination with Ca^{2+} form calcium silicate and calcium aluminate hydrate (Eades, 1962; Diamond and Kinter, 1964).

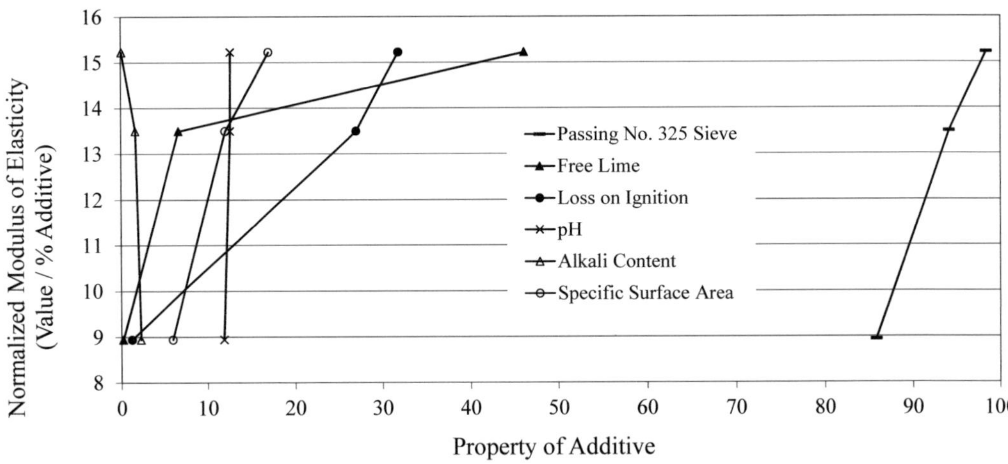

Fig. 3. Variation of normalized modulus of elasticity values with different properties of additives

7. Microstructure and mineralogical characteristics

As noted earlier, mineralogical studies namely SEM and EDS were conducted on all the raw soils, raw additives powder, raw additives paste and 28-day cured stabilized Carnasaw soil specimens to study the influence of stabilization on microstructure and mineralogical characteristics.

7.1 Raw soil and additives powder

Figure 4 (a) shows the SEM micrographs of raw Carnasaw soil sample at high magnification (10,000 times). It is clear that the raw soil has a discontinuous structure, where the voids are more visible because of the absence of hydration products. The EDS results showed majority of silicon (Si) and aluminium (Al) minerals and trace amounts of potassium (K), iron (Fe) and sodium (Na) minerals in the raw soil. The raw additives used in this study were also studied using the SEM/EDS methods. Figures 4 (b), (c) and (d) show SEM/EDS of raw lime, CFA and CKD powder, respectively. As evident from Figure 4 (b), raw lime is an amorphous powder consisting mainly of calcium (Ca) compounds. This is in agreement with the XRF results reported in Table 1. On the other hand, CFA and CKD are more complex compounds. The EDS results indicated presence of Ca, Al, Si, Fe, sulphur (S), phosphorous (P), titanium (Ti), and magnesium (Mg) minerals in CFA. Whereas EDS results of CKD indicated presence of Ca, Si, Mg, S, and K minerals.

The SEM micrographs of raw CFA showed that CFA is composed of different size spherical particles (or cenosphere); however, CKD micrographs showed particles with poorly defined shapes. The gold (Au) and palladium (Pd) peaks that appeared in all EDS spectra is due to the gold-palladium sputter coating used on SEM samples for making them electrically conductive.

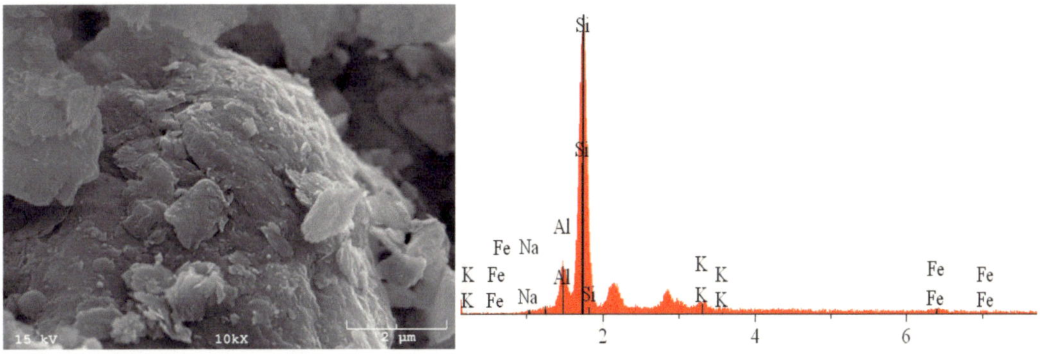

(a)

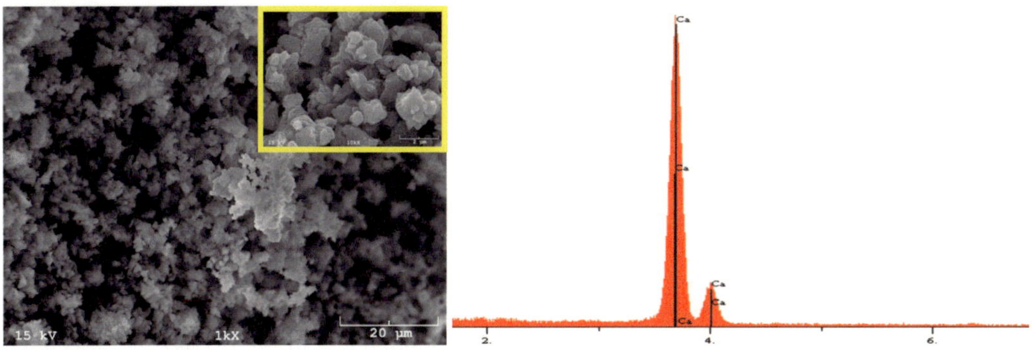

(b)

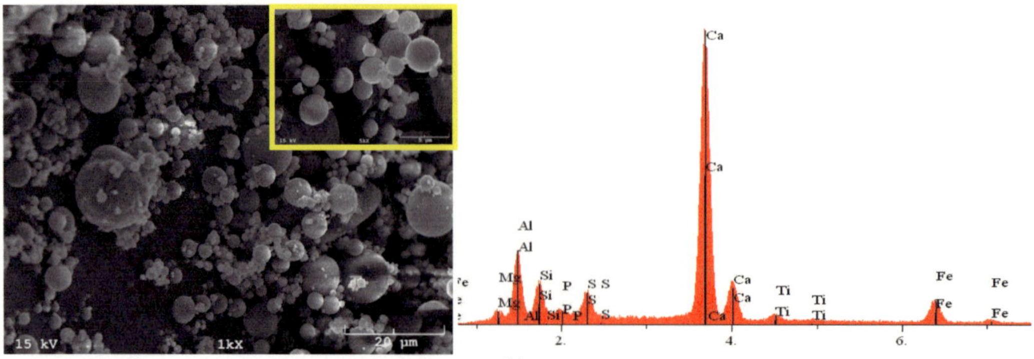

(c)

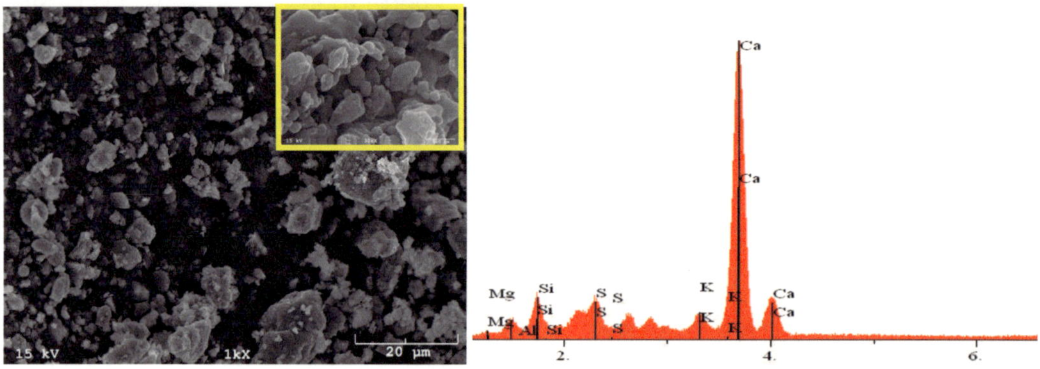

(d)

Fig. 4. SEM/EDS of (a) Raw Soil (b) Lime, (c) CFA, and (d) CKD Powder

7.2 Raw additives paste

The representative SEM micrographs of raw lime, CFA and CKD paste samples that had been subjected to 28 day curing and air dried for approximately two days are presented in Figures 5 (a) through (c). Figures 4 (b) and 5 (a) show very similar microstructure and EDS results, as expected in lime due to its negligible self-cementing properties. At a higher magnification (x 30,000 times), a flower-like structure of calcium hydroxide crystals is evident. Figure 5 (b) shows the SEM micrographs of raw CFA paste at a magnification level of 5,000. Overall, the microstructure of CFA paste had a relatively finer matrix with cenospheres covered with cementitious products. Based on EDS at different locations, the microstructure consisted of C-A-S-H like crystals with variable amounts of Al, Si, and Mg incorporated phases and traces of Fe and Ti as impurities. The Ca/Si ratio was observed to be approximately 2. The microstructure of CKD paste (Figure 5 (c)) is clearly denser and compact as compared to the microstructure of raw CKD powder (Figure 4 (d)). At a higher magnification (x 20,000 times), C-S-H gel is evident. The microstructure in combination with EDS spectrum gave an indication of the formation of C-S-H phases with Ca/Si ratio of less than 1. Please note that the EDS spectrum presented in this study were collected from a fractured surface of specimen and not from the polished smooth surface of specimen. Hence, the heights of the peaks were used as a qualitative measure rather than a quantitative measure of different crystal phases.

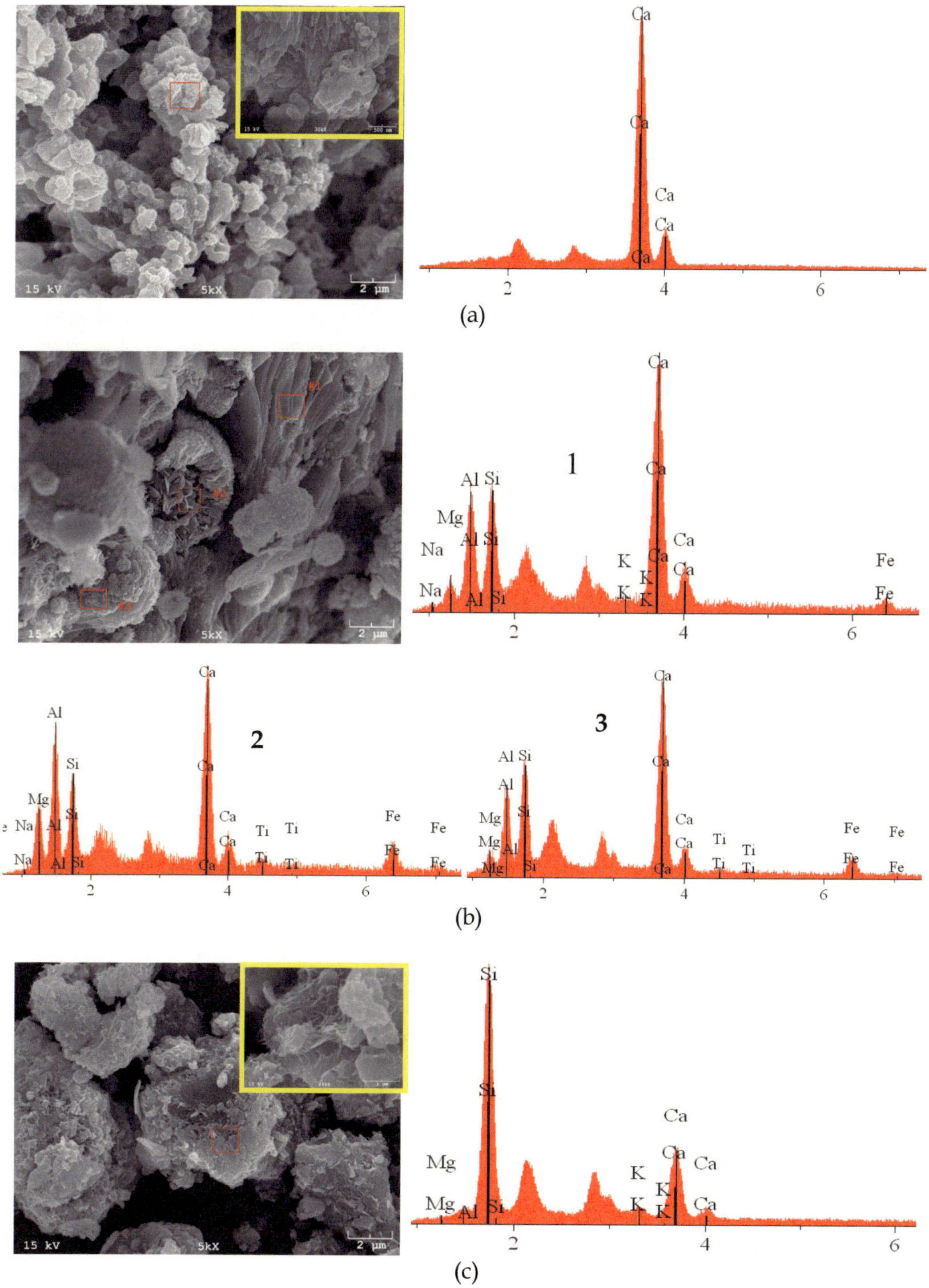

Fig. 5. SEM/EDS of (a) Lime, (b) CFA, and (c) CKD Pastes

7.3 Carnasaw soil with 9% lime

To study the microstructure of 9% lime-stabilized Carnasaw soil specimens, 28-day modulus of elasticity tested specimens were examined using SEM micrographs. Figure 6 (a) shows the microstructure at a magnification level of 10,000, which when compared with the raw soil micrograph of Figure 4 (a) shows marked change in morphology. From Figure 6 (a), it is clear that the raw soil structure has transformed from a particle based form to a more integrated composition due to cementitious reactions. At a higher magnification (x 25,000 times), the cementing phases could clearly be seen. Further, EDS pattern was used as a basis to monitor the changes occurring in the chemical composition at selected locations within the Carnasaw soil after stabilization with 9% lime. As evident from Figure 6 (a), analysis on the cementing phases showed presence of Ca and Si with high Ca/Si ratio (>7), which is an indication of the presence of C-S-H ($xCaO.ySiO_2.zH_2O$).The other two peaks not marked in Figure 6 (a) belong to Au-Pd coating The cementing phases, due to gradual crystallization of the new secondary minerals, caused an increase in the modulus of elasticity of the stabilized soil, as discussed in Section 6. Similar observations were reported by other researchers (see e.g., Locat et al., 1996; Ghosh and Subbarao, 2001; Nalbantoglu, 2006; Kavak and Akyarh, 2007). Figure 6 (b) shows micrograph of same specimen with EDS spectra collected at three different locations. Three locations from which the EDS was secured are marked as 1, 2 and 3 in the micrographs. All three EDS spectra indicated presence of C-A-S-H with different Ca/Si ratios. For example, the Ca/Si ratio is higher at location 1 (0.8) as compared to location 2 (0.3). This gives an indication that C-A-S-H is at different levels of development at different locations, as expected.

Figure 6 (c) shows EDS spectra at a magnification level of 5,000 taken from a different location. A flower-like structure of calcium hydroxide crystals is evident, which indicates presence of un-reacted hydrated lime in the stabilized specimen. This is in agreement with the micrograph presented in Figure 5 (a) for the lime paste.

7.4 Carnasaw soil with 15% CFA

The SEM micrographs of Carnasaw soil stabilized with 15% CFA are presented in Figures 7 (a) through (d). Figure 7 (a) reveals the formation of cementing products, with lamellar form, adjacent to the fly ash particles. The EDS analysis showed presence of Ca and Si indicating presence of C-S-H, the main cementing product responsible for strength gain (Choquette et al., 1987; Lav and Lav, 2000). The Ca/Si ratio of C-S-H phases identified in the CFA-stabilized soil was qualitatively determined to be approximately 3. Also, two additional peaks of Au and Pd appeared because specimens were sputter coated with alloy of gold-palladium. In viewing these samples, one would notice that the spherical particles of fly ash are joined strongly to the clay particles in its surrounding (Chang, 1995). It was also apparent that the fly ash particles served as nucleation sites for the growth of the hydration products (or coatings), as shown in Figure 7 (b). Formation of ettringite, $Ca_6[Al(OH)_6]_2.(SO_4)_3.26H_2O$, was also observed in the form of heaps of rod-like crystals (Figure 7 c). This observation was further confirmed by conducting EDS analysis, which suggested presence of Ca, Al and S with traces of Si and Ti as impurities. No areas were found showing normal ettringite spectra without traces of Si and Ti. Similar structure, as shown in Figure 2.25 (c), was reported as ettringite by other researchers (e.g., Mitchell and Dermatas, 1992; Intharasombat, 2003).

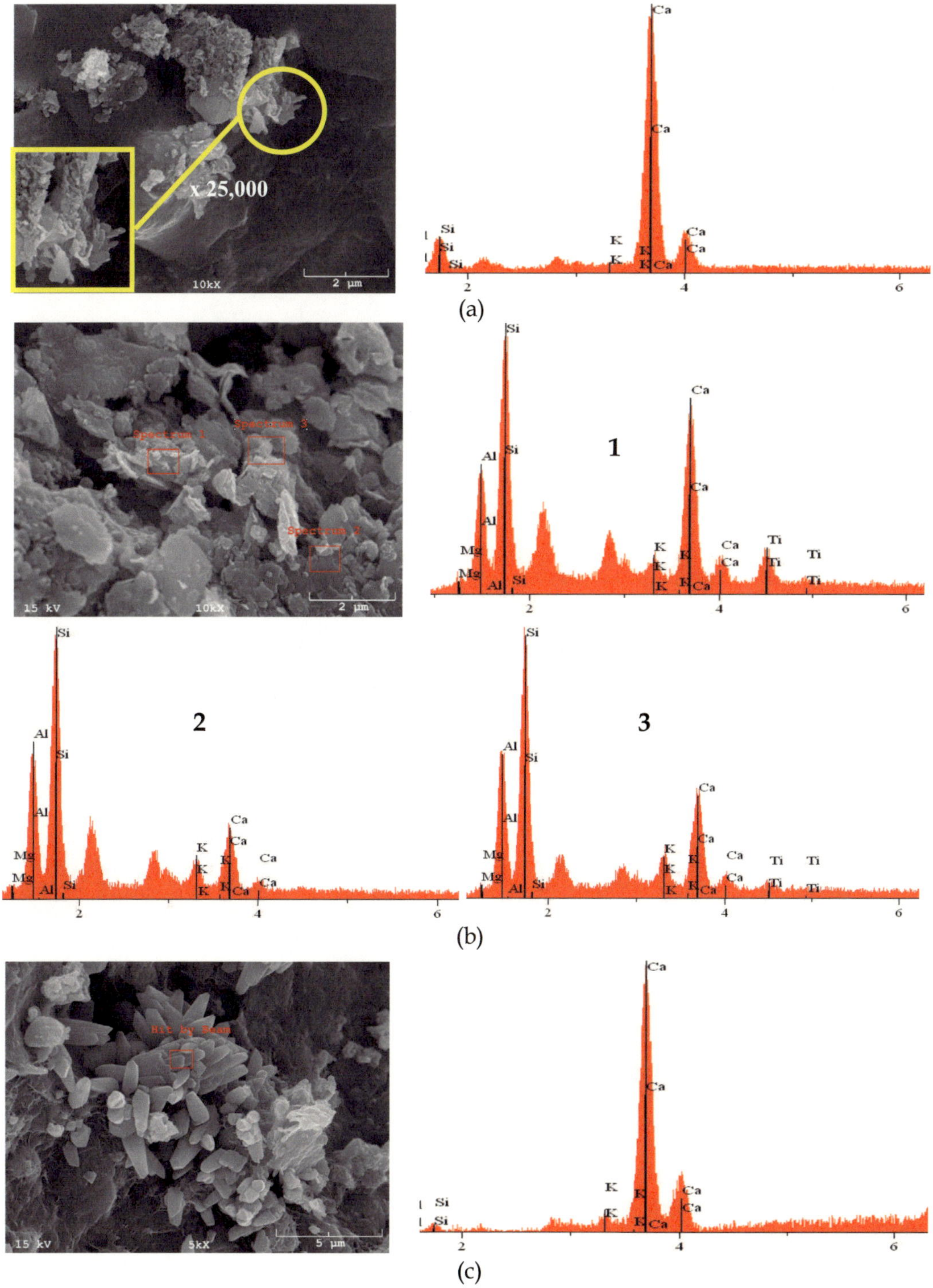

Fig. 6. SEM/EDS of 28-Day Cured 9% Lime-Stabilized Caranasaw Soil Specimens (a) C-S-H, (b) C-A-S-H, and (c) CH

Further, the SEM micrographs revealed that most of the fly ash particles were covered with a reaction shell as seen in Figure 7 (d). The approximate chemical composition of the outer shell was determined at location 1 and 3 by the EDS analysis and a typical composition is presented in pattern marked as point and 1 and 3. The composition of the shell was slightly different from that of the un-reacted inner fly ash surface which is shown in spectrum 2. The higher Ca peak in 1 and 3 compared to spectrum 2 suggests the initiation of reaction products (e.g., C-A-S-H) formation on the surface of fly ash particle. It should be noted that the exact quantitative composition cannot be obtained using the EDS analysis of the stabilized specimens.

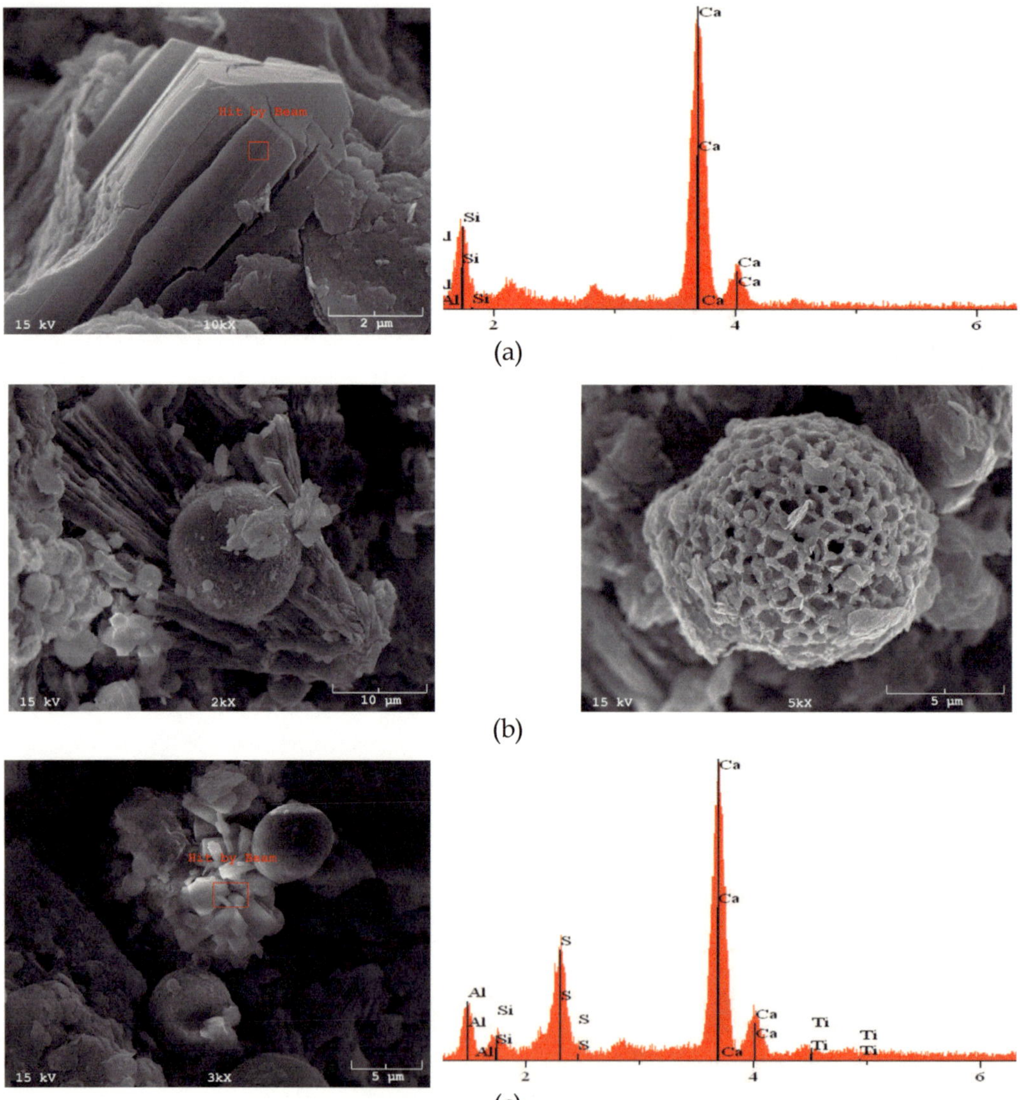

Fig. 7. SEM/EDS of 28-Day Cured 15% CFA-Stabilized Caranasaw Soil Specimens (a) C-S-H, (b) hydration coatings, and (c) ettringite crystal

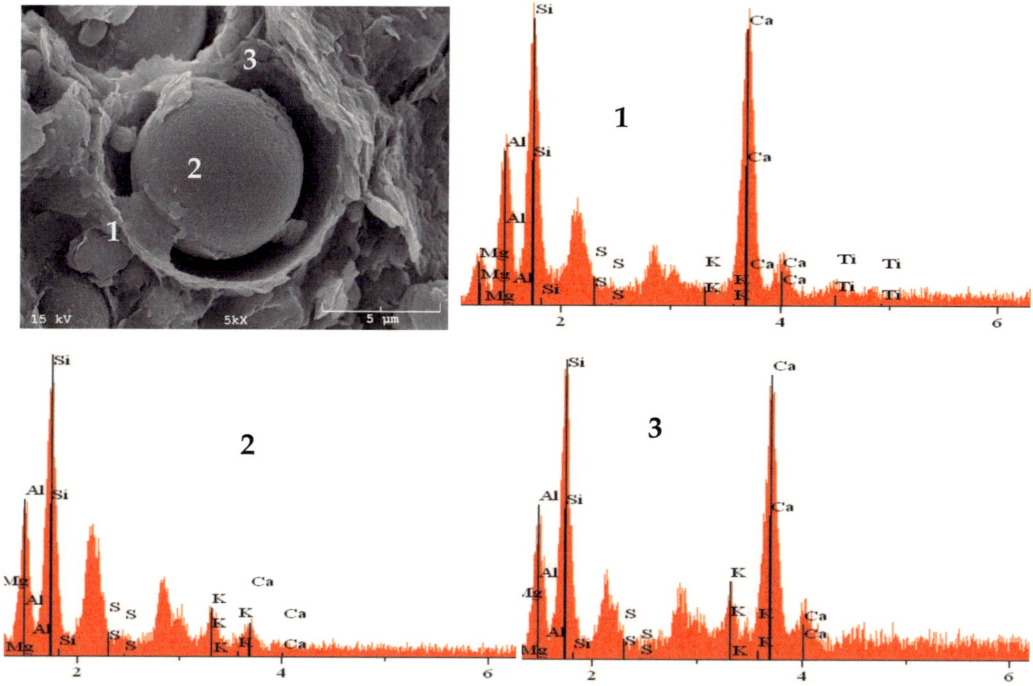

(d)

Fig. 7. (Cont'd). SEM/EDS of 28-Day Cured 15% CFA-Stabilized Caranasaw Soil Specimens (d) reaction shell of CFA particle

8.5 Carnasaw soil with 15% CKD

The SEM micrographs, as illustrated in Figures 8 (a) through (e) show significant changes in the microstructure of raw soil when mixed with CKD and cured for 28 days. It could be observed that flat clay structure surface observed in Figure 4 (a) is covered with cementitious reaction products, as shown in Figure 8 (a). Figure 8 (a) shows the C-A-S-H ($xCaO.yAl_2O_3.zSiO_2.wH_2O$) phase development which contains distinct peaks of Ca, Si and Al elements based on the EDS analysis, consistent with observation reported by Chaunsali and Peethamparan (2010). The Ca/Si ratio of approximately 3 is also evident from Figure 8 (a). The SEM micrograph at a different location revealed presence of C-S-H phase with Ca/Si ratio less than 1 (Figure 8 (b)). Additionally, C-S-H phases with very high Ca/Si ratio (>10) are also evident from the SEM micrograph and EDS spectra taken at different locations (Figure 8 (c)).

Figure 8 (d) shows micrographs of rose-shaped and web-shaped hydration coatings and bonds developed in 15% CKD-stabilized Carnasaw soil. Another prominent feature of the microstructure of 15% CKD-stabilized soil was the presence of needle-shaped ettringite crystals (Figure 8 (e)). The presence of ettringite crystals in CKD-stabilized soil is consistent with the observations reported by Peethamparan et al. (2008), Moon et al. (2009), and Chaunsali and Peethamparan (2011). Hence, improved modulus of elasticity exhibited by CKD-stabilized soil specimens after curing could be attributed to aforementioned reaction products.

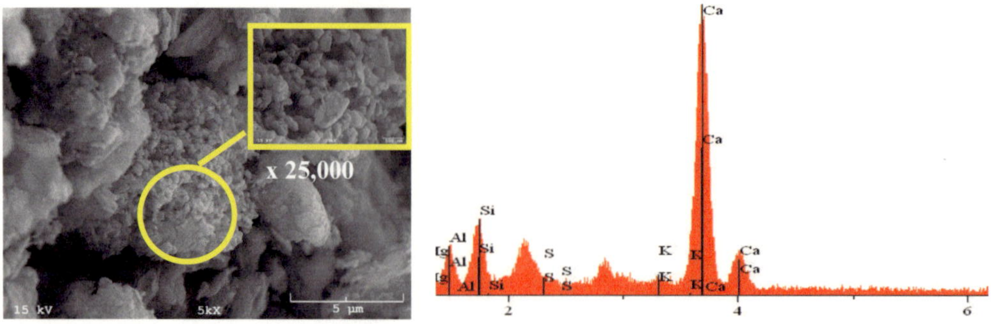

(a)

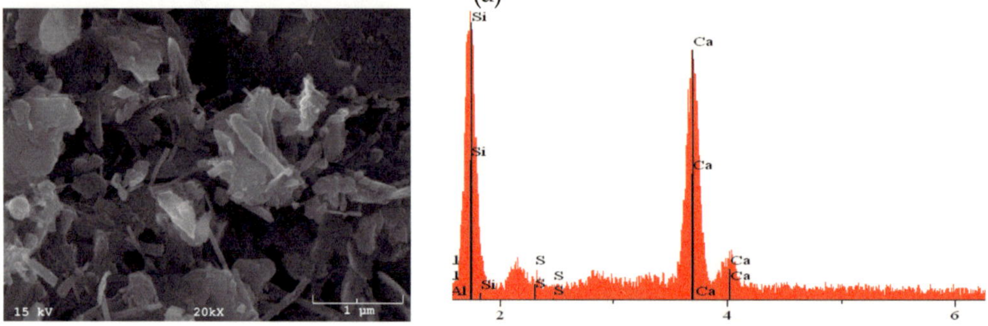

(b)

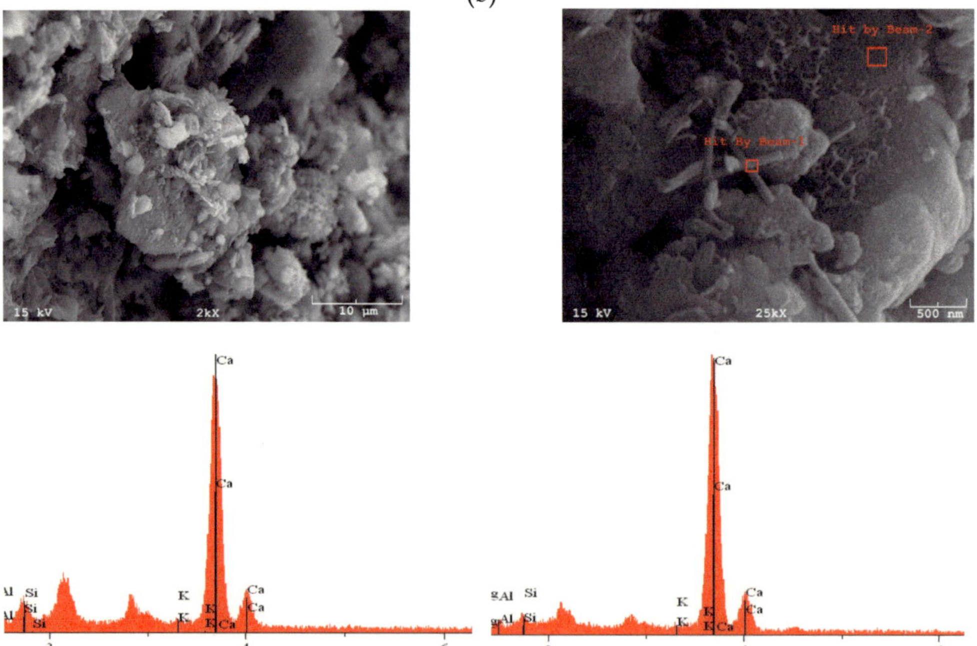

(c)

Fig. 8. SEM/EDS of 28-Day Cured 15% CKD-Stabilized Caranasaw Soil Specimens (a) C-A-S-H, (b) C-S-H, and (c) C-S-H

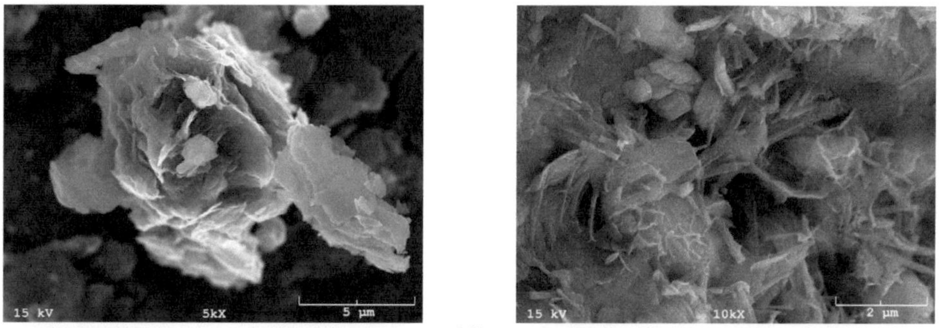

(d)

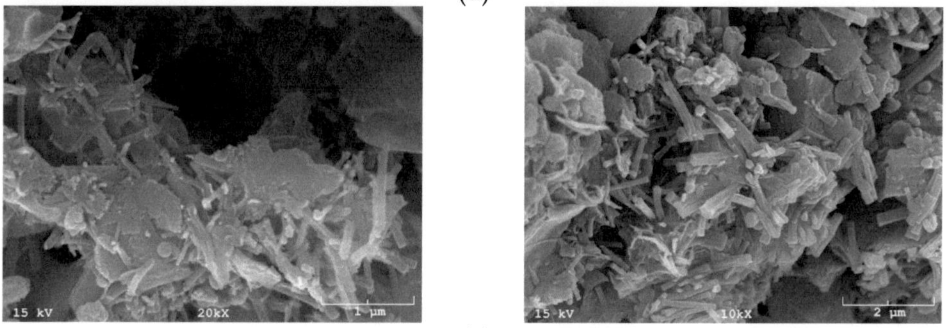

(e)

Fig. 8. (Cont'd). SEM/EDS of 28-Day Cured 15% CKD-Stabilized Caranasaw Soil Specimens (d) hydration coatings, and (e) ettringite crystals

9. XRD results

The XRD patterns of the raw soil and additives powder is presented in Figure 9. The raw soil showed presence of clay minerals namely, illite, $(K,H_3O)(Al,Mg,Fe)_2(Si,Al)_4O_{10}[(OH)_2,(H_2O)])$ and kaolinite $(Al_2Si_2O_5(OH)_4)$ which are expected in a fat clay with a PI value of 29. The presence of elements (Si, Al, Na, Fe, K) in aforementioned minerals of raw soil is also in agreement with elements observed from, EDS spectrum (Figure 4 (a)). The XRD pattern of raw lime powder indicates only presence of calcite ($CaCO_3$) and calcium hydroxide ($Ca(OH)_2$). The diffractogram of CFA powder revealed presence of three minerals namely, quartz (SiO_2), mullite ($Al_6Si_2O_{13}$), and merwinite ($Ca_3Mg(SiO_4)_2$). This is in agreement with XRF results presented in Table 1. Similar minerals were reported by other researchers for class C fly ash (e.g., McCarthy, 2000; Chaunsali and Peethamparan, 2010). The CKD diffractogram showed presence of calcite, quick lime (CaO), quartz, and anhydrite ($CaSO_4$). Similar observations were reported by Chaunsali and Peethamparan (2010).

The XRD patterns of raw and stabilized specimens are presented in one figure for comparison purpose (Figure 10). In general, there was a reduction in the peak intensity of most of the stabilized specimens, as can be seen by the reduction in peak heights of some of the peaks, particularly for the samples stabilized with 9% lime. This could be attributed to cementitious reactions between additive and clay minerals resulting in the reduction of the clay mineral intensities. Similar behavior was reported by Al-Rawas (2002). None of the peaks of the raw soil disappeared due to stabilization. It was noted that with the addition of 15% CFA, the peaks shifted away to the right from their original positions while maintaining the same patterns. This could be due to instrument distortion during the XRD scanning process or conditions of the sample. The XRD pattern of CKD-stabilized specimens revealed additional peaks of calcite formed due to cementitious reactions.

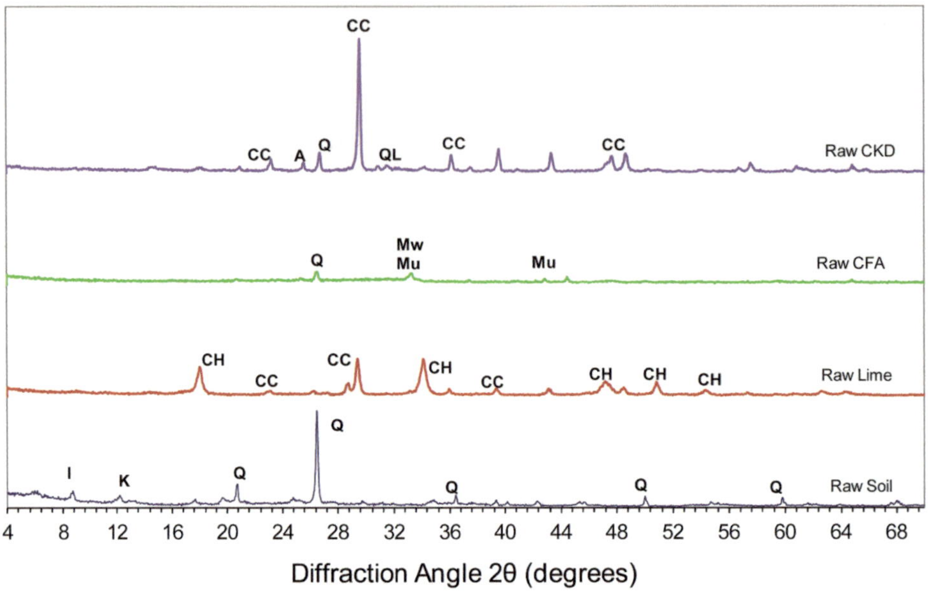

Fig. 9. XRD pattern for raw soil, lime, CFA and CKD powders [CC-calcite ($CaCO_3$), A-anhydrite ($CaSO_4$), QL-quicklime (CaO), Q-quartz (SiO_2), Mu-mullite ($Al_6Si_2O_{13}$), Mw-merwinite ($Ca_3Mg(SiO_4)_2$), CH-calcium hydroxide ($Ca(OH)_2$), I-illite, K-kaolinite)

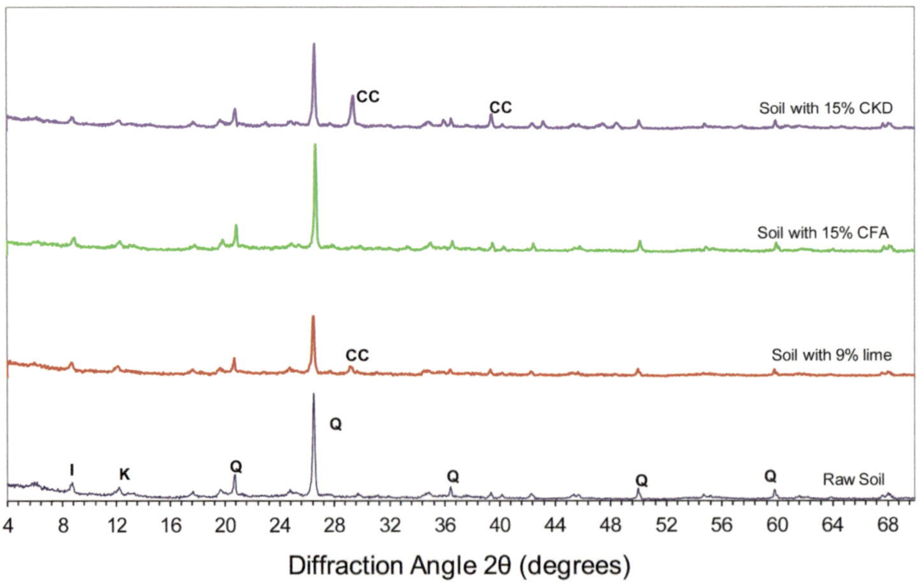

Fig. 10. XRD pattern for 9% lime-, 15% CFA- and 15% CKD-stabilized specimens [CC-calcite ($CaCO_3$), Q-quartz (SiO_2), I-illite, K-kaolinite)

10. Conclusions

Based on the study presented, in this chapter the following conclusions can be derived:

1. The results from pH tests showed that 5% lime provide an asymptotic behavior (less than 1% increase in pH w.r.t raw soil pH) in Carnasaw soil-lime mixtures. No such asymptotic behavior was observed for Carnasaw soil stabilized with CFA and CKD.
2. All three additives improved the modulus of elasticity values of Carnasaw soil (fat clay) specimens; however, degree of improvement varied with the type of additive.
3. The normalized modulus of elasticity values is better correlated with additive properties - free-lime content, alkali content, loss on ignition, percent passing No. 325 sieve, specific surface area, and pH.
4. In general, microscopic analyzes confirm that the addition of lime or CFA or CKD to soil induces beneficial reactions and significant improvements in stiffness. Also, it could be concluded that the formation of reaction products such as C-S-H, C-A-S-H and ettringite contributed to strength development of stabilized soil.
5. The lime-stabilized specimens indicated presence of both C-S-H and C-A-S-H phases in addition to unreacted calcium hydroxide. Additionally, C-A-S-H at different levels of development at different locations was identified.
6. The SEM and EDS results showed presence of C-S-H, C-A-S-H and ettringite crystals in both CFA-and CKD-stabilized specimens. The fly ash particles served as nucleation sites for the growth of the hydration products with reaction products initiating from the surface.
7. The Ca/Si ratio in C-S-H and C-A-S-H phases was found vary between 0.3 - 12 at different locations of the stabilized specimen. The presence of higher Ca/Si ratio phases in the microstructure of the stabilized specimen could be due to possible secondary cementitious reactions between soil and Ca^{2+} ion of additive.
8. The XRD results showed a general reduction in all clay minerals' peak intensities particularly in the case of lime-stabilized samples. However, none of the peaks of the raw soil disappeared due to stabilization.

11. Acknowledgment

The authors are thankful to the Oklahoma Department of Transportation (ODOT) and Oklahoma Transportation Center (OTC) for providing funds for this project. Technical assistance from Dr. Preston Larson (University of Oklahoma) and Tim Rawlsky (Lafarge North America) is gratefully acknowledged.

12. References

Air Force Manual (AFJMAN) (1994). "Soil Stabilization for Pavements," *Technical Manual No. 5-822-14*, Departments of the Army and Air Force, Washington, DC.

Al-Rawas, A. A. (2002). Microfabric and mineralogical studies on the stabilization of an expansive soil using cement by-pass dust and some types of slags. *Canadian Geotechnical Journal*, Vol. 39, Issue 5, pp. 1150-1167

Al-Rawas, A. A., Taha, R., Nelson, J. D., Al-Shab, T. B., and Al-Siyabi, H. (2002). "A Comparative Evaluation of Various Additives Used in the Stabilization of Expansive Soils," *ASTM Geotechnical Testing Journal*, Vol. 25, No. 2, pp. 199 - 209.

Baghdadi, Z. A., and Rahman, M. A. (1990). "The Potential of Cement Kiln Dust for the Stabilization of Dune Sand in Highway Construction," *Building and Environment*, Vol. 25, No. 4, pp. 285 - 289.

Bhatty, J. I., Bhattacharja, S., and Todres, H. A. (1996). "Use of Cement Kiln Dust in Stabilizing Clay Soils," Portland Cement Association, Skokie, IL.

Bin-Shafique, S., Edil, T., and Benson, C. (2004). "Incorporating a Fly Ash Stabilized Layer into Pavement Design Case Study," *Geotechnical Engineering*, Vol. 157, No. 4, pp. 239 - 249.

Camargo, F. F., Edil, T. B., and Benson, C. H. (2009). "Strength and Stiffness of Recycled Base Materials Blended with Fly Ash," *Transportation Research Board 88th Annual Meeting*, CD-ROM Publication, Transportation Research Board, Washington, DC.

Cerato, A. B., and Lutenegger, A. J. (2002). "Determination of surface are of fine-grained soils by the ethylene glycol monoethyl ether (EGME) method," *ASTM Geotechnical Testing Journal*, Vol. 25, No. 3, pp. 1 - 7.

Chang, D.T. (1995). "Resilient Properties and Microstructure of Modified Fly Ash-stabilized Fine Grained Soils." *Transportation Research Record: Journal of the Transportation Research Board*, Vol. 1486, pp. 88 - 96.

Chapman, H. D. (1965). "Cation Exchange Capacity," *Methods of Soil Analysis*, American Society of Soil Agronomy, C. A. Black et al., Eds., Madison, WI, pp. 891 - 901.

Chaunsali, P., and Peethamparan, S. (2010). "Microstructural and Mineralogical Characterization of Cement Kiln Dust Activated Fly Ash Binder," *Transportation Research Board 89th Annual Meeting*, CD-ROM Publication, National Research Council, Washington, DC.

Chaunsali, P., and Peethamparan, S. (2011). Evloution of Strength, Microstructure and Mineralogical Composition of a CKD-GGBFS Binder. *Cement and Concrete Research*. Vol. 41, pp. 197-208.

Choquette, M., Berube, M. –A., and Locat, J. (1987). "Mineralogical and Microtextural Changes Associated with Lime Stabilization of Marine Clays from Eastern Canada," *Applied Clay Science*, Vol. 2, pp. 215 - 232.

Collins, R. J., and Emery, J. J. (1983). "Kiln Dust/Fly Ash Systems for Highway Bases and Subbases," U.S. Department of Transportation - Department of Energy Report, FHWA/RD-82/167.

Consoli, N. C., Lopes, L. S., and Heineck, K. S. (2009). "Key Parameters for the Strength Control of Lime Stabilized Soils," *ASCE Journal of Materials in Civil Engineering*, Vol. 21, No. 5, pp. 210 - 216.

Diamond, S., and Kinter, E. B. (1964). "Mechanisms of Soil-Lime Stabilization," *Highway Research Record*, Vol. 92, pp. 83 - 96.

Eades, J. (1962). "Reactions of $Ca(OH)_2$ with Clay Minerals in Soil Stabilization," PhD Thesis, University of Illinois, Urbana, IL.

Evangelos, S. (2006). "A Solution to the Problem of Predicting the Suitability of Silty-Clayey Materials for Cement-Stabilization," *Geotechnical and Geological Engineering*, Vol. 24, No. 2, pp. 379 - 398.

Ferguson, G., and Levorson, S. M. (1999). "Soil and Pavement Base Stabilization with Self-Cementing Coal Fly Ash," Final Report for American Coal Ash Association, Alexandria, VA.

Ghosh, A., and Subbarao, C. (2001). "Microstructural Development in Fly Ash Modified with Lime and Gypsum," *ASCE Journal of Materials in Civil Engineering*, Vol. 13, No. 1, pp. 65 – 70.

Ghosh, A., and Subbarao, C. (2001). "Microstructural Development in Fly Ash Modified with Lime and Gypsum," *Journal of Materials in Civil Engineering*, Vol. 13, No. 1, pp. 65-70.

Gomez, J. D. P. (2009). "Influence of Curing Time on the Resilient Modulus of Chemically Stabilized Soils," MS Thesis, University of Oklahoma, Norman, OK.

Haston, J.S., and Wohlgemuth, S.K. (1985). "Experiences in the Selection of the Optimum Lime Content for Soil Stabilization." *Texas Civil Engineer*, November 1985, pp.17 – 20.

Horpibulsuk, S., Rachan, R., Chinkulkijniwat, A., Raksachon, Y., and Suddeepong, A. (2010).Analysis of Strength Development in Cement-stabilized Silty Clay from Microstructural Considerations, *Construction and Building Materials*, Vol. 24, pp. 2011-2021.

Hunter, D. (1988). "Lime-induced Heave in Sulfate-bearing Clay Soils," *Journal of Geotechnical Engineering*, Vol.114, No.2, pp.150 – 167.

Indian Road Congress (IRC) (2000). "State of the Art: Lime-Soil Stabilization," *Special Report*, IRC Highway Research Board, New Delhi, India.

Intharasombat, N. (2003). "Ettringite Formation in Lime Treated Sulfate Soils: Verification by Mineralogical and Swell Testing," MS Thesis, University of Texas, Arlington, TX

Jade (1999). *Materials Data Manual*, MDI, Livermore, CA, 94550.

Kavak, A., and Akyarh, A. (2007). "A field Application for Lime Stabilization," *Environmental Geology*, Vol. 51, pp. 987 – 997.

Khoury, N. N. (2005). "Durability of Cementitiously Stabilized Aggregate Bases for Pavement Application," PhD Thesis, University of Oklahoma, Norman, OK.

Khoury, N., and Zaman, M. M. (2007). "Durability of Stabilized Base Courses Subjected to Wet-Dry Cycles," *International Journal of Pavement Engineering*, Vol. 8, No. 4, pp. 265 – 276.

Kim, D., and Siddiki, N. (2004). "Lime Kiln Dust and Lime – A Comparative Study in Indiana," *Transportation Research Board 83rd Annual Meeting*, CD-ROM Publication, Transportation Research Board, Washington, DC.

Kolias, S., Kasselouri-Rigopoulou, V., and Karahalios, A. (2005). *Cement & Concrete Composites*, Vol. 27, pp. 301-313.

Kota, P. B. V. S., Hazlett, D., and Perrin, L. (1996). "Sulfate-Bearing Soils: Problems with Calcium-based Additives," *Transportation Research Record: Journal of the Transportation Research Board*, Vol. 1546, pp. 62 – 69.

Lav, A. H., and Lav, M. A. (2000). "Microstructural Development of Stabilized Fly Ash as Pavement Base Material," *ASCE Journal of Materials in Civil Engineering*, Vol. 12, No. 2, pp. 157 – 163.

Li, L., Edil, T. B., and Benson, C. H. (2009). "Properties of Pavement Geomaterials Stabilized with Fly Ash," *Proceedings of 2009 World of Coal Ash Conference*, Lexington, KY.

Little, D. N. (2000). "Evaluation of Structural Properties of Lime Stabilized Soils and Aggregates," *Mixture Design and Testing Protocol for Lime Stabilized Soils*, Vol. 3, National Lime Association report, http://www.lime.org/SOIL3.PDF, Last Accessed: Jan., 2007.

Locat, J., Tremblay, H., and Leroueil, S. (1996). "Mechanical and Hydraulic Behavior of a Soft Inorganic Clay Treated with Lime," *Canadian Geotechnical Journal*, Vol. 33, pp. 654 - 669.

Mallela, J., Quintus, H. V., and Smith, K. L. (2004). "Consideration of Lime-Stabilized Layers in Mechanistic-Empirical Pavement Design," F*inal Report submitted to the National Lime Association*, Arlington, VA.

McCarthy, G. J. (2000). "Energy: By-Products of Coal Combustion in Power Plants," Chapter 23, *Industrial Applications of X-Ray Diffraction*, editors: Chung, F. H. and Smith, D. K., Marcel Dekker Inc., New York, NY.

McCoy, W. J., and Kriner, R. W. (1971). "Use of Waste Kiln Dust for Soil Consolidation," Lehigh Portland Cement Co., Allentown, PA.

McManis, K. L., and Arman, A. (1989). "Class C Fly Ash as a Full or Partial Replacement for Portland Cement or Lime," *Transportation Research Record: Journal of the Transportation Research Board*, Vol. 1219, pp. 68 - 61.

Miller, G.A. and Azad, S. (2000). "Influence of Soil Type on Stabilization with Cement Kiln Dust." *Construction and Building Materials*, Vol. 14, pp. 89 - 97.

Miller, G. A. and Zaman, M. (2000). "Field and Laboratory Evaluation of Cement Kiln Dust as a Soil Stabilizer," *Transportation Research Record: Journal of the Transportation Research Board*, Vol. 1714, pp. 25 - 32.

Misra, A. (1998). "Stabilization Characteristics of Clays Using Class C Fly Ash," *Transportation Research Record: Journal of the Transportation Research Board*, Vol. 1611, pp. 46 - 54.

Mitchell, J. K., and Dermatas, D. (1990). "Clay Soil Heave Caused by Lime-Sulfate Reactions," *ASTM Special Technical Publication*, Vol. 1135, pp. 41 - 64.

Mitchell, J. K., and Dermatas, D. (1992). "Clay Soil Heave Caused by Lime-Sulfate Reactions," *ASTM STP 1135: Innovations and Uses for Lime*, Philadelphia, PA.

Mitchell, J.K. (1993). *Fundamentals of Soil Behavior*. 2nd Edition. John Wiley & Sons, NY.

Mohamed, A. M. O. (2002). "Hydro-mechanical Evaluation of Soil Stabilized with Cement-Kiln Dust in Arid Lands," *Engineering Geology*, Vol. 42, No. 8, pp. 901 - 921.

Moon, D. H., Grubb, D. G., and Reilly, T. L. (2009). "Stabilization/Solidification of Selenium-Impacted Soils Using Portland Cement and Cement Kiln Dust," *Journal of Hazardous Materials*, Vol. 168, pp. 944 - 951.

Nalbantoglu, Z. (2004). "Effectiveness of Class C Fly Ash as an Expansive Soil Stabilizer," *Construction and Building Materials*, Vol. 18, pp. 377 - 381.

Nalbantoglu, Z. (2006). "Lime Stabilization of Expansive Clay," *Expansive Soils*, Part 7, Vol. 1, pp. 341 - 348.

Nalbantoglu, Z. and Tuncer, E. R. (2001). "Compressibility and Hydraulic Conductivity of a Chemically Treated Expansive Clay," *Canadian Geotechnical Journal*, Vol. 38, pp. 154 - 160.

National Cooperative Highway Research Report (NCHRP) (1976). "Lime-Fly Ash-Stabilized Bases and Subbases," Transportation Research Board, National Council, Washington, DC.

ODOT (Oklahoma Department of Transportation) (2006). OHD L-49: Method of Test for Determining Soluble Sulfate Content in Soil, *Oklahoma Highway Department Test Method*, Oklahoma City, Oklahoma.

Parsons, R.L. and E. Kneebone. (2004). "Use of Cement Kiln Dust for the Stabilization of Soils," *Proceedings of Geo-Trans 2004*, Los Angeles, California, No. 1, pp. 1124 - 1131.

Parsons, R.L., Kneebone, E. and Milburn, J.P. (2004). "Use of Cement Kiln Dust for Subgrade Stabilization." *Final Report No. KS-04-03*, Kansas Department of Transportation, Topeka, KS.

Peethamparan, S. and Olek, J. (2008). "Study of the Effectiveness of Cement Kiln Dusts in Stabilizing Na-montmorillonite Clays," *ASCE Journal of Materials in Civil Engineering*,Vol. 20 No.2, pp.137 - 146.

Peethamparan, S., Olek, J. and Diamond, S. (2008). "Physicochemical Behavior of Cement Kiln Dust-Treated Kaolinite Clay," *Transportation Research Record: Journal of the Transportation Research Board*, Vol. 2059, pp. 80 - 88.

Petry, T. M., and Little, D. N. (1992). "Update on Sulfate Induced Heave in Treated Clays: Problematic Sulfate Levels," *Transportation Research Record: Journal of the Transportation Research Board*, Vol. 1362, pp. 51 - 55.

Phanikumar, B. R., and Sharma, R. S. (2004). "Effect of Fly Ash on Engineering Properties of Expansive Soils," *ASCE Journal of Geotechnical and Geoenvironmental Engineering*, Vol. 130, No. 7, pp. 764 - 767.

Prusinski, J. R., and Bhattacharia, S. (1999). "Effectiveness of Portland cement and lime in stabilizing clay soils," *Transportation Research Record: Journal of the Transportation Research Board*, Vol. 1632, pp. 215 - 227.

Puppala, A. J., E. Wattanasanticharoen, and A. Porbaha. (2006). "Expansive soils: Recent advances in characterization and treatment, A. Combined lime and polypropylene fiber stabilization for modification of expansive soils," Chapter 24, Taylor and Francis, NY.

Puppala, A. J., Griffin, J. A., Hoyos, L. R. and Chomtid, S. (2004). "Studies of Sulfate-Resistant Cement Stabilization Methods to Address Sulfate-Induced Soil Heave," *ASCE Journal of Geotechnical and GeoEnvironmental Engineering*, Vol.130, No.4, pp.391 - 402.

Puppala, A. J., Viyanant, C., Kruzic, A., and Perrin, L. (2002). "Evaluation of a modified sulfate determination method for cohesive soils," *Geotechnical Testing Journal*, Vol. 25, No. 1, pp. 85 - 94.

Puppala, A. J., Wattanasanticharoen, E., Intharasombat, N., and Hoyos, L. R. (2003). "Studies to Understand Soil Compositional and Environmental Variables on Sulfate Heave Problems," *Proceedings of Soil Rock America: 12th Pan American Conference on Soil Mechanics and Geotechnical Engineering*, Boston, MA.

Qubain, B. S., Seksinsky, E. J., and Li, J. (2000). "Incorporating Subgrade Lime Stabilization into Pavement Design," *Transportation Research Record: Journal of the Transportation Research Board*, Vol. 1721, pp. 3 - 8.

Rajasekaran, G., Murali, K., & Srinivasaraghavan, R. (1995). Fabric and mineralogical studies on lime treated marine clays. *Ocean Engng*, 24(3), 227-234.

Rajendran, D., and Lytton, R. L. (1997). "Reduction of Sulfate Swell in Expansive clay Subgrades in the Dallas district," *Texas Transportation Institute Report No. TX-98/3929-1*, Bryan, TX

Rao, S. M., and Shivananda, P. (2005). "Impact of Sulfate Contamination on Swelling Behavior of Lime-Stabilized Clays," *Journal of ASTM International*, Vol. 2, No. 6, pp. 1-10

Rollings, R. S., Burkes, J. P., and Rollings, M. P. (1999). "Sulfate Attack on Cement-stabilized Sand," *ASCE Journal of Geotechnical and Geoenvironmental Engineering*, Vol. 125, No. 5, pp. 364–372.

Sayah, A. I. (1993). "Stabilization of a Highly Expansive Clay Using Cement Kiln Dust," MS Thesis, University of Oklahoma, Norman, OK.

Sear, L.K.A. (2001). *Properties and Use of Coal Fly Ash – A Valuable Industrial By-Product*, Thomas Telford, Reston, VA.

Sezer, A., Inan, G., Yilmaz, R. and Ramyar, K. (2006). "Utilization of a Very High Lime Fly Ash for Improvement of Lzmir Clay," *Building and Environment*, Vol. 41, No. 2, pp. 150 - 155.

Sreekrishnavilasam, A., Rahardja, S., Kmetz, R. and Santagata, M. (2007). "Soil Treatment Using Fresh and Landfilled Cement Kiln Dust," *Construction and Building Materials*, Vol. 21, pp. 318 - 327.

Stutzman, P. (2004). "Scanning Electron Microscopy Imaging of Hydraulic Cement Microstructure," *Cement and Concrete Research*, Vol. 26, pp. 957-966.

Zaman, M., Laguros, J. G., and Sayah, A. I. (1992). "Soil Stabilization using Cement Kiln Dust," *Proceeding of the 7th International Conference on Expansive Soils*, Dallas, TX, pp. 1 - 5.

Zaman, M., Laguros, J., Tian, P., Zhu, J, and Pandey, K. (1998). "Resilient Moduli of Raw and Stabilized Aggregate Bases and Evaluations of Layer Coefficients for AASHTO Flexible Pavement Design." *ORA 125-4262*, Item 2199, Department of Transportation, Oklahoma City, OK.

Zheng, J. and Qin, W. (2003). "Performance Characteristics of Soil-Cement from Industry Waste Binder," *ASCE Journal of Materials in Civil Engineering*, Vol. 15, No. 6, pp. 616 - 618.

Zia, N., and Fox, P. J. (2000). "Engineering Properties of Loess-Fly Ash Mixtures for Roadbase Construction," *Transportation Research Record: Journal of the Transportation Research Board*, Vol. 1714, pp. 49 - 56.

39

The Use of ESEM in Geobiology

Magnus Ivarsson[1,2] and Sara Holmström[2]
[1]*Swedish Museum of Natural History, Department of Palaeozoology,*
[2]*Nordic Center for Earth Evolution (NordCEE),*
[3]*Stockholm University, Department of Geological Sciences,*
[1,3]*Sweden*
[2]*Denmark*

1. Introduction

Geobiology is an interdisciplinary field of research that explores the interaction between the biosphere and the geosphere and/or the atmosphere. It involves researchers from numerous fields such as paleontology, microbiology, mineralogy, geochemistry, biochemistry, sedimentology and genetics. Geobiological research cover a wide range of areas like, for example, the origin and evolution of life, environmental microbiology, microbe-mineral interactions, molecular ecology and detection of biomarkers. It is responsible for at least two major subdisciplines: geomicrobiology (the study of microbe-mineral interactions) and astrobiology (a discipline focused on the conditions for life in the universe, including the search for life on other planets).

A major part of most geobiological research is focused on interactions between microorganisms and minerals or other substrates. This includes everything from microbes in soil and sediments, via fossilized microorganisms in rock and minerals, to extremophiles at hydrothermal vents. There are numerous methods, protocols and instruments that are used for this kind of research but one of the most basic methods and commonly used is the Scanning Electron Microscope (SEM). SEM is easy operated and can give high resolved images down to micro meter size, which is a requirement when analysing microorganisms. Coupled with, for example, an energy dispersive X-ray spectroscopy (EDS or EDX) detector it becomes a valuable tool for elemental analysis which is a critical part of geobiological research. Microbe-mineral interactions most commonly result in micro-sized biomineralizations or amorphous precipitates that may contain important information about the metabolism and life-cycles of the microbes, redox chemistry in the microbial habitat, and paleoenvironmental conditions.

Geobiological samples usually involve living species of microorganisms or fragile materials like fossilized microorganisms, organic matter or hydrated minerals like clays that collapse in conventional SEM. Environmental Scanning Electron Microscopy (ESEM) is a modification of conventional SEM originally developed for the study of biological samples but has also become more frequently used within geobiological related research. With its gaseous environment in the specimen chamber as well as other technical applications ESEM makes it possible to study wet and uncoated specimens in their natural state. This is a substantial advantage opposed to conventional SEM.

The main difference between ESEM and SEM is that the first has a specimen chamber where the specimen can be imaged while gas is present. The specimen chamber is designed to maintain water in its liquid phase and for that a minimum water vapour pressure of 609 Pa (6.09 mbar or 4.579 Torr) is required at O° C. This creates new possibilities: A) Hydrated specimens can be examined in contrast to SEM where specimens are desiccated by the vacuum. Thus biological specimens can be maintained fresh and live. B) Specimens do not require preparation techniques used in SEM such as the deposition of a thin gold or carbon coating. Such techniques sometimes require vacuum and can disturb the samples. Biological samples also need to be dehydrated before coating which is a time consuming process. The gas in ESEM is electrically conductive due to the ionization, which prevents that negative charge accumulates and this is the reason why specimens do not need to be coated prior to examination. Thus, with ESEM specimens can be examined faster and more easily, without complex and time consuming preparation methods and without modifications or in worst case damage to the sample surface by preparation work and exposure to vacuum. The aim with this paper is to give a brief background to geobiology and ESEM, and to show the advantages of ESEM over conventional SEM in the study of geobiology.

2. Geobiology: The link between geology and biology

Geobiology as an independent discipline is relatively new and has attracted a lot of attention during the last decade with the result of an increased number of active researchers, foundation of international scientific journals as well as centers and institutions worldwide devoted to geobiological research. However, the link between life and geological processes can be traced back as far as the foundational text in modern geology by James Hutton (1788). He documented quite ordinary observations that anyone could have done about erosion of land into the oceans by rivers and the presence of fossilized shells in sedimentary rocks in the mountains of Scotland. What Hutton managed to do was to put these observations in a context where he questioned the surface of Earth as a sustainable habitat for life. In the early part of the 20th century Vernadsky (1926) further explored the connection between life and geological processes, and Baas Becking (1934) coined the term geobiology and outlined many geobiological processes much as we see them today.

The last decades have involved a growing awareness of the close connection between the physical world and life sciences. Earth as a system is complex and not as black and white as previously thought. The traditional way when studying the Earth system in dividing it into separated disciplines like geology, biology, chemistry or physics are not always the most practical approach. An interdisciplinary perspective and awareness when looking at the Earth system is almost a requirement to understand it and move forward in Earth sciences. Geobiology is a result of interdisciplinary thinking within geology and biology and their subdisciplines, and geobiology as a science has shown that there is no distinct boundary between the both. They are tightly connected and interact with each other on many levels, both at the present but also throughout Earth´s history (Knoll, 2003). The evolution of life has been intimately connected with the mineral evolution (Hazen et al., 2008), the rise of continents (Rosing et al., 2006), emergence of the aerobic biosphere (Melezhik et al., 2005), formation of fossil fuel and ore formations (Southam & Saunders, 2005). Ever since their emergence on Earth microbes have played an important role as geological agents involved in mineral growth and dissolution, rock and mineral weathering and alteration, mobilization of metals, cycling of elements in the ocean and the lithosphere, metabolism of hydrocarbons and transformation of organic carbon, fractionation of stable isotopes etc (e.g. Lindsay and Brasier, 2002; Tice and Lowe, 2004; Lowe and Tice, 2007; Furnes et al., 2008). But it is not only biological activity that influences geological processes. It works both ways and geological processes control and influence the microbial ecology as well, something that is explicit in, for example, extreme environments (Huber et al., 2007). The discovery of the subsurface biosphere has deepen the knowledge of life´s distribution, adaptability and variety (Pedersen, 1993). Earths surface is no longer the limit for habitability. The subsurface is just as colonized and may contain as much as one third of the Earths biomass (Gold, 1992; Staudigel et al., 2004). Life in extreme environments and in the subsurface further show that Earth itself may not be the limit for life but that it may extend beyond. Astrobiology has shown that the conditions for life exist on other planets as well, and that life could have originated or been transported there, and possibly be able to sustain (Farmer and Des Marais, 1999).

3. Environmental Electron Scanning Microscope (ESEM)

3.1 History

Experimental approaches to examine specimens in chambers filled with water or atmospheric gas with conventional and scanning transmission types of electron microscopes were reported of as early as the 1940s (Ardenne and Beischer, 1940; Abrams and McBain, 1944; Swift and Brown, 1970; Parsons et al., 1974). Such experiments used different kinds of "environmental cells" where gas was introduced temporarily during the examination, however, neither of these experiments succeeded in creating a stable environmental cell for routine analysis. In 1970 the first images of wet specimens in an SEM were published by Lane (1970), who injected a jet of water vapor over the point of observation at the specimen surface. The gas diffused in the chamber without any damage to the instrument. The need for differentially pumped chambers to allow for the transfer of the electron beam from the high vacuums in the gun area to the high pressures in the specimen chamber forced developments during the 1970s. Shah and Beckett (1977) reported of differentially pumped chambers to maintain botanical specimens conductive for signal detection, and Robinson (1974) reported of improvements by combining a backscattered electron detector with differential vacuum pumping and introduction of water vapour around 600 Pa at the freezing point of temperature. In 1978, Gerasimos Danilatos, a Greek-Australian physicist started to work with Robinson at the University of New South Wales in Sydney, Australia, and designed the original ESEM that were operable at room temperature and high pressures up to 7000 Pa (Danilatos and Robinson, 1979). During the 1980s and early 1990s Danilatos developed and optimized the design of the ESEM (Danilatos, 1981, 1985, 1988, 1990a, b; Danilatos and Postle, 1983). He reported the construction of an ESEM capable of working at any pressure from vacuum up to one atmosphere, optimization of the use of differential pumping systems combined with electron backscatter detectors, the idea of the environmental gas itself as detection medium, and the invention of the gaseous detection device (GDD).

3.2 Description of ESEM

The functioning of an ESEM is in many ways identical to a conventional SEM and it is assumed that the reader is familiar with the operation of a SEM. Basically, an ESEM is a

SEM that can operate at the low pressure of a usual SEM through to, at least, the pressure required to observe liquid distilled water. An ESEM, just like a SEM, employs a scanned electron beam and electromagnetic lenses to focus and direct the beam on the specimen surface. A very small focused electron spot is scanned in a raster form over a small specimen area and the beam electrons interact with the specimen surface layer and produce various signals. These signals are collected with appropriate detectors and can be monitored in the form of images, graphs, digital recordings etc. Beyond these common principles, the ESEM deviates substantially from a SEM in several respects briefly outlined below.

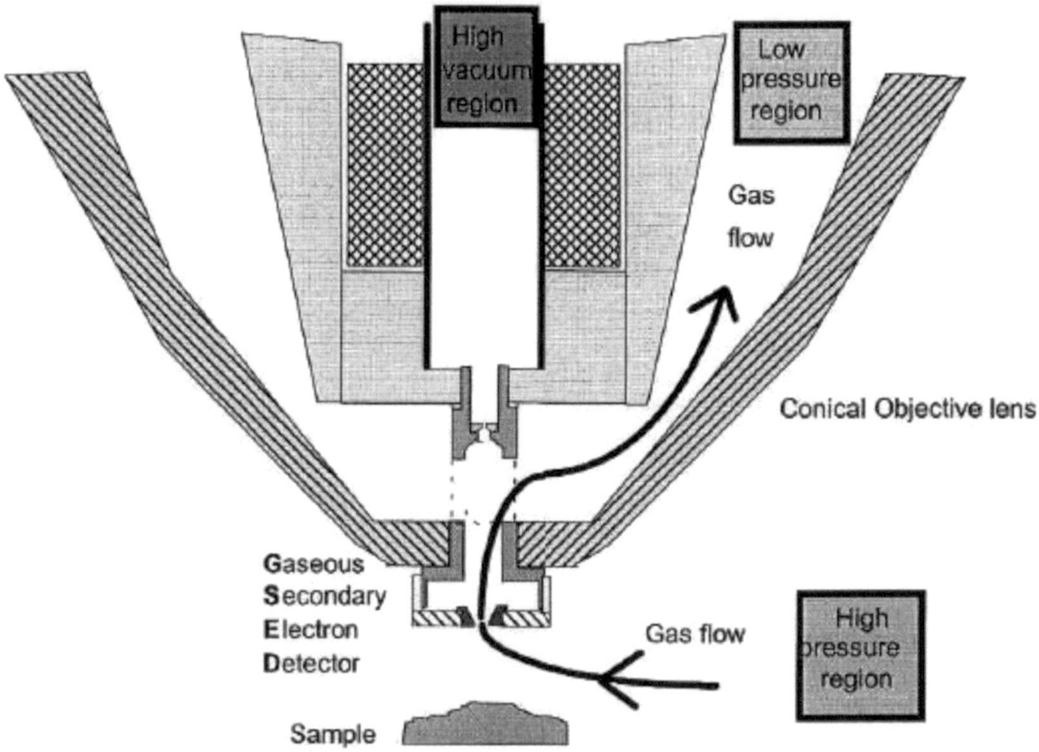

Fig. 1. Schematic illustration of an ESEM lens. Cross section showing the high vacuum region and low pressure region as well as the direction of the gas flow. From the XL30 Options Manual, FEI and Philips (1998).

3.2.1 Separated regions

The ESEM must have, just like a conventional SEM, a high vacuum region (usually with pressure less than 10^{-2} Pa) for the generation and focusing of the electron beam. In a SEM the high vacuum region and the region of the specimen is the same but in an ESEM these two regions must be separated (Fig. 1). The specimen chamber in an ESEM is designed to maintain water in its liquid phase and for that a minimum water vapour pressure of 609 Pa (6.09 mbar or 4.579 Torr) is required at O° C.

3.2.2 Differential pumping

The basic principle of an ESEM is, thus, to have a high pressure and gaseous specimen chamber, and a high vacuum electron optics area separated from each other but still connected to allow for the transfer of the electron beam. The two regions are separated by at least two small pressure limiting apertures, one aperture that separates the high vacuum region of the electron gun and an intermediate cavity. The second aperture separates the intermediate cavity and the high pressure specimen cavity. Gas leaking from the specimen chamber through the aperture to the intermediate cavity is instantly removed by a pump system. This is called differential pumping. Gas that escapes further into the high vacuum area of the electron optics is similarly removed by a pump to maintain required vacuum. Additional pumping stages may be added to achieve an even higher vacuum in the electron optics area.

3.2.3 Electron beam transfer

An electron beam generated in the vacuum of the upper column will on its way through the intermediate cavity and in the specimen chamber come in contact with an increasing amount of gas molecules. This will result in a gradual loss of electrons due to electron scattering by the gas molecules and eventually total loss of the beam. However, the electrons are scattered over a broad skirt-like area around the focused spot and since the skirt width is orders of magnitude greater than the spot width, the skirt only contribute background noise and the amount of electrons in the original focused spot is enough for imaging of the specimen. The remaining electron beam is, however, only a fraction of what it was in the upper column and merely enough for imaging. The particular conditions of pressure, distance and beam voltage is crucial for signal detection and the operation of an ESEM is centered on refining the instrument for optimum performance and achieving precision for the instrument to operate close to its physical limit (Danilatos, 2009). By doing this it is possible to use an ESEM in much the same way as a SEM. Secondary and backscattered electrons, X-rays and cathodoluminescence is generated as in a SEM and can be detected with slight modifications to the detectors. The main difference regarding detectors is that the conventional secondary electron detector of SEM, the Everhart-Thornley detector, can not be used in the presence of gas and thus the gaseous detection device (GDD) has been developed. The principle of the GDD is that the environmental gas itself is used for beam transmission and as a detector of the electrons (Danilatos, 1997, 1990a), compared to the Everhart-Thornley detector where light guide the transmitted electrons. In a GDD the signals emanating from the beam specimen-interaction interact with the surrounding gas in the form of gaseous ionization and excitation. The ionized gas is then collected by electrodes and the signal is amplified for its purpose.

3.2.4 Specimen charging

In conventional SEM negative charge is accumulated as the electron beam impinge on the surface of the specimen. This tends to deflect the electron beam from the scanned point with the result of charging artefacts on the image which greatly disturbs the imaging and analyses. This is normally eliminated in conventional SEM by coating the specimen prior to examination by a thin layer of usually gold or carbon. The gas in ESEM is electrically conductive due to the ionization, which prevents that negative charge accumulates and this is the reason why specimens do not need to be coated prior to examination.

3.2.5 Disadvantages

Even though ESEM involve several substantial advantages over conventional SEM there are a few disadvantages. Some of these disadvantages can be limited by instrument design and the disadvantages differ between various instrument manufacturers. The main disadvantage is the distance in the specimen chamber over which the electron beam remains usable in the gaseous environment. The useful distance is a function of accelerating voltage, beam current, nature and pressure of the gas, and of the aperture diameter used. The distance varies from ~10mm to less than 1mm depending on the gas pressure. Another result of the limitation of useful specimen distance is the limitation of magnification. At very high pressure the distance becomes so small that the field of view is limited by the aperture width. The vacuum in a conventional SEM result in a superior magnification range compared to an ESEM.

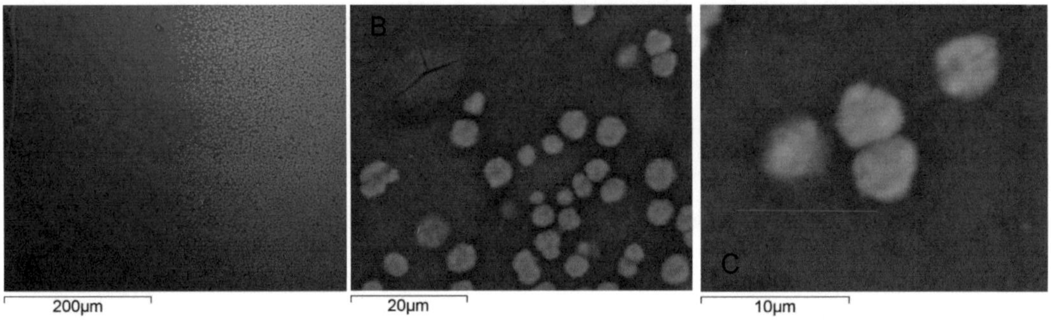

Fig. 2. ESEM images of plant-growth-promoting rhizobacteria. A) Image showing an overview of the bacteria grown on an agar plate. B) A close up showing the coccoidal morphology of the bacteria. C) Image showing mitosis.

The presence of gas may also generate various disturbances in certain applications, like, for instance, the resolution of the image. This issue can, however, be limited by altering chamber pressure and accelerating voltage. It is needed for each instrument to find the most useable combination and correlation between the parameters.

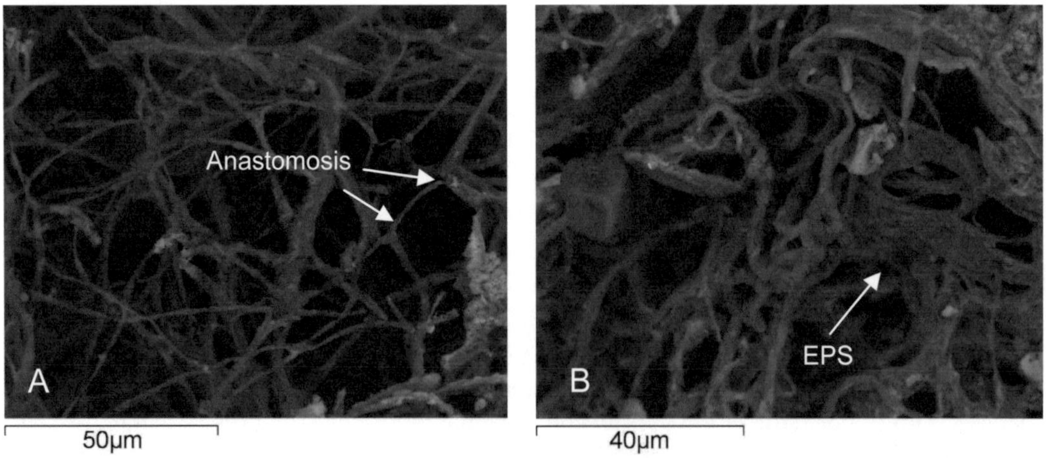

Fig. 3. ESEM images of an ectomychorrizal fungus. Note the frequent branching and anstomoses between branches. In B EPS has been precipitated in the fungal mycelium.

4. The use of ESEM in geobiology

As mentioned in sections 1 and 2 geobiological research is performed at the intersection where geology and biology meet. One could ask why the use of ESEM in geobiology differs from conventional geological or biological use of ESEM, and of course, they sometimes overlap. Microorganisms are studied with ESEM within microbiology (e.g. Bergmans et al., 2005; Ahmad, 2010), and minerals, sediments and substrates is studied with ESEM within traditional geology (e.g. Donald, 2003; Reed, 2005; Huiming et al., 2011). However, the main difference is that within geobiology the interaction of these two fields is examined. Geobiology represent the point at which life starts to interact with the physical world and the outcome of this is usually very fragile such as living, encrusted or fossilised microorganisms, amorphous, hydrated minerals or substrates like clays, and combinations of these like biomineralisations or mineral trapping biofilms and EPS (extra cellular polymeric substances). ESEM has been used in various geobiological studies (e.g. Little et al., 1991; Douglas and Douglas, 2000; Nealson et al., 2002; Hallberg and Ferris, 2004; Waters, 2008), and in the following sections we will try to illustrate some of its applications. In example I live microorganisms (bacteria and fungus) from soils collected at various locations in Sweden will be studied. In example II drilled rock samples from the oceanic seafloor in the Pacific Ocean will be used to illustrate the exploration of the deep subseafloor biosphere and how the interaction between microorganisms and mineral substrates can be studied, and also how sensitive hydrated minerals can be if they are not treated in a proper way.

4.1 Instrument

In this study an XL30 environmental scanning electron microscope with a field emission gun (XL30 ESEM-FEG) was used. The ESEM was equipped with an Oxford x-act energy dispersive spectrometer (EDS), backscatter electron detector (BSE) and a secondary electron detector (SE). The acceleration voltage was 20, 15 or 10 kV depending on the nature of the sample and the instrument was calibrated with a cobolt and a carbon standard. Peak and element analyses were made using INCA Suite 4.11 software.

The high vacuum mode was in some tests used as an equivalent to the conditions of a conventional SEM.

4.2 Example I: Live microorganisms

Microorganisms used in geobiological related research is usually collected from natural environments, isolated and grown in laboratory for further studies or experiments. This is a time consuming and expensive procedure but beyond the scope of this paper and, thus, will not be described in further detail.

First of all, we have used a plant-growth-promoting rhizobacteria. These bacteria have been isolated from soil samples where they existed in symbiosis with fungi and then grown on an agar plate (Fig. 2). With the aim of ESEM we are able to view the bacteria at micrometer scale and study their morphology. The bacteria are coccoidal and their main diameter range between ~1 to 3 µm. It is also possible to observe such feature as mitosis (cell division) (Fig. 2C).

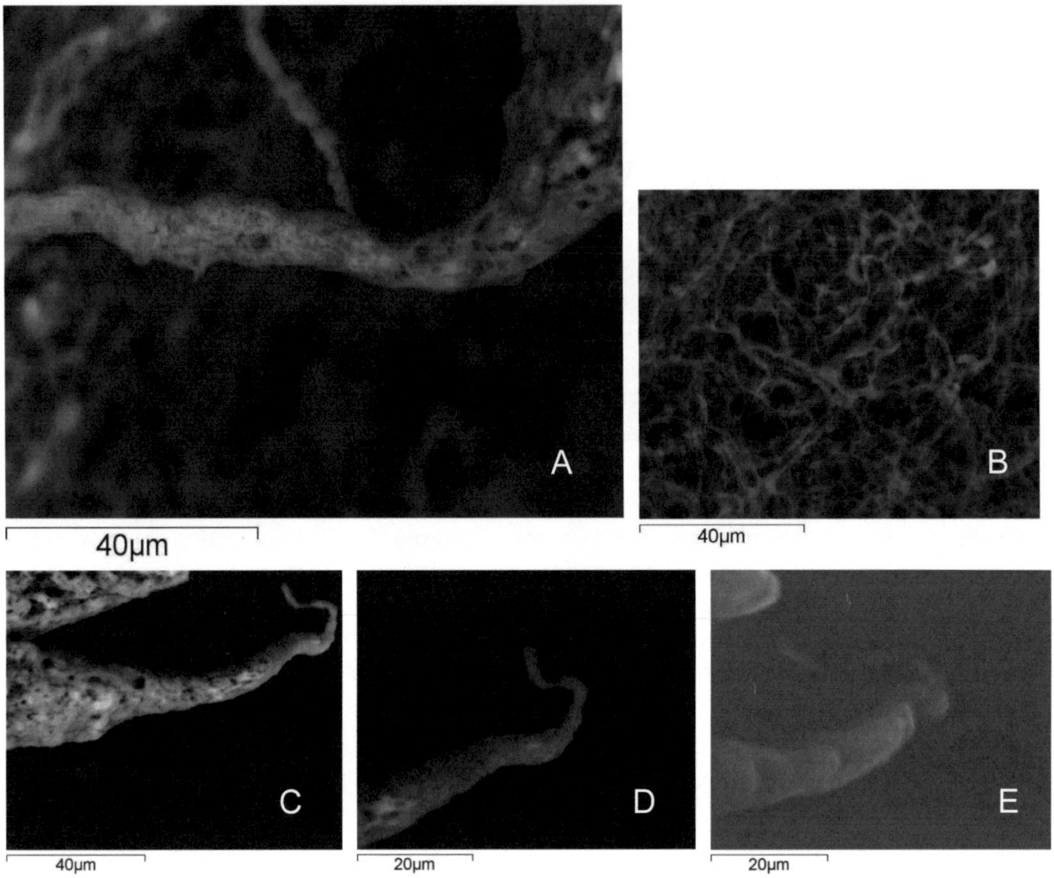

Fig. 4. ESEM images of *Suillus Granulatus* grown in a liquid media. A) Close up of a hypha. B) An overview image showing the mycelium. C-E) Images showing the same hyphae at various conditions: C) low vacuum (2mbar), D) 1mbar, and E) wet mode.

Secondly, an ectomychorrizal fungus grown on an agar plate has been used (Fig. 3). This fungal mycelium show characteristic fungal morphologies as frequent branching hyphae and anastomosis between branches (Fig. 3A). The diameter of the hyphae varies between a few micro meter to ~10 µm. It is also possible to see production of EPS on the fungal mycelium (Fig. 3B).

Thirdly, a fungus, *Suillus Granulatus*, grown in liquid media has been used. It is a fungus with traditional fungal morphology (Fig. 4). It is characterized by long, curvi-linear hyphae, 3-10 µm in diameter and several hundred µm in length, creating a complex mycelium. Biomineralizations are frequently occurring in this mycelium and ESEM mode makes it possible to analyse them with EDS without coating, which may disturb the analysis (Fig C in Table 1).

Figures 4C-D show one of the issues that may occur with ESEM. The presence of gas in the specimen chamber results in a blurry and unfocused image at high magnifications. To achieve the best quality of the images we experimented by altering the pressure in the specimen chamber. Figures 4C-E illustrate how the quality of the images change with varied pressure. Wet mode resulted in poor contrast, high vac resulted in charging artifacts, but the best image quality was achieved with low vacuum at 2mbar. At that pressure and in that mode a high quality image could be produced as well as an element spectrum that is of great advantage. We tried to document the microorganisms in an ordinary SEM as an illustrative example but failed to produce images due to charging artefacts. However, conventional SEM is not a realistic option to study these samples. The microorganisms would collapse and to prevent that a dehydration process and coating would be required, which is time consuming and would destroy the samples and a lot of information that they contain.

4.2.1 EDS analysis

Element analysis with EDS is also possible to perform in ESEM mode on the live microorganisms and associated biomineralizations without gold or carbon coating (Table 1). In live material and in biomineralizations analyses of the carbon content is sometimes of highest priority and would be impossible to do with carbon coating.

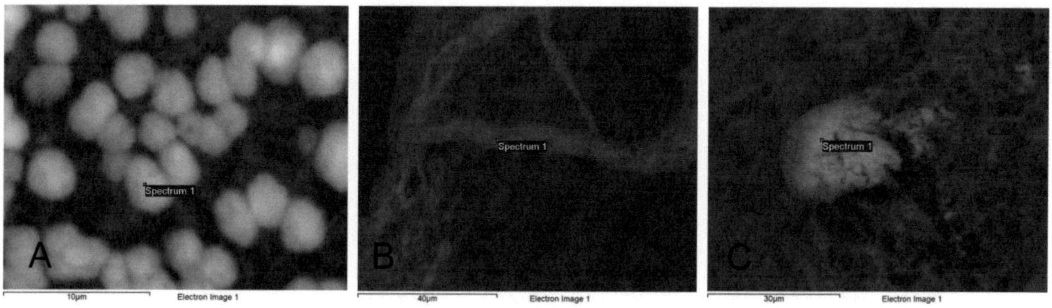

	A	B	C
C	24.72		
O	21.71	57.87	46.63
Na	14.18	1.81	
Mg		1.95	
P	2.79	20.38	22.08
S	1.65	1.64	
K	8.96	6.12	3.36
Cl	25.98	0.91	
Ca		4.45	0.99
Fe		4.87	26.94
Total	100.00	100.00	100.00

Table 1. EDS data in wt %. A) A plant-growth-promoting rhizobacteria, B) *Suillus Granulatus*, C) biomineralization in the hyphal network of *Suillus Granulatus*. It is difficult to identify the mineral but it appears to be a Fe and P-rich oxide. Normally a mineral phase with such high Fe content needs to be coated before analysed by EDS.

4.3 Example II: Hydrated minerals and fossilized microorganisms

Samples of subseafloor basalts from the Emperor Seamounts in the Pacific Ocean, drilled and collected during Ocean Drilling Program (ODP) Leg 197, are used to present the advantages of ESEM in the study of hydrated minerals, fossilized microorganisms and interactions between the both. For a detailed description of the sampling sites, mineralogy and biogenicity of the microfossils see Ivarsson et al. (2008a, b) and Ivarsson and Holm (2008). Briefly, the samples consist of veins and vesicles in basalts. These veins and vesicles are partly filled by hydrothermally formed secondary mineralisations of calcite, zeolites and clays (Fig. 5). The vein walls are usually coated with a montmorillonite phase ($(Na,Ca)_{0.33}(Al,Mg)_2(Si_4O_{10})(OH)_2 \cdot nH_2O$) and successively with carbonates ($CaCO_3$) or zeolites species such as phillipsite ($K_2(Ca_{0.5}, Na)_4(Al_6Si_{10}O_{32}) \cdot 12H_2O$), chabazite ($Ca_2(Al_4Si_8O_{24}) \cdot 12H_2O$), and tetranatrolite ($Na_{16}(Al_{16}Si_{24}O_{80}) \cdot 16H_2O$). Zeolites and calcite seldom occur in the same void probably due to local differences in the composition of the hydrothermal fluids but they do exist in the same system with interconnected veins and vesicles. In addition, several of these veins and vesicles contain complex networks of fossilized filamentous microorganisms (Figure 6A-C). These microfossils are composed of a similar montmorillonite phase as is found on the vein walls.

Fig. 5. Optical micrograph. A vein in basalt from the Emperor Seamounts in the Pacific Ocean, drilled and collected during Ocean Drilling Program (ODP) Leg 197. The vein contain secondary mineralisations of calcite, zeolites and montmorillonite (clay), and fossilized microorganisms on the vein walls.

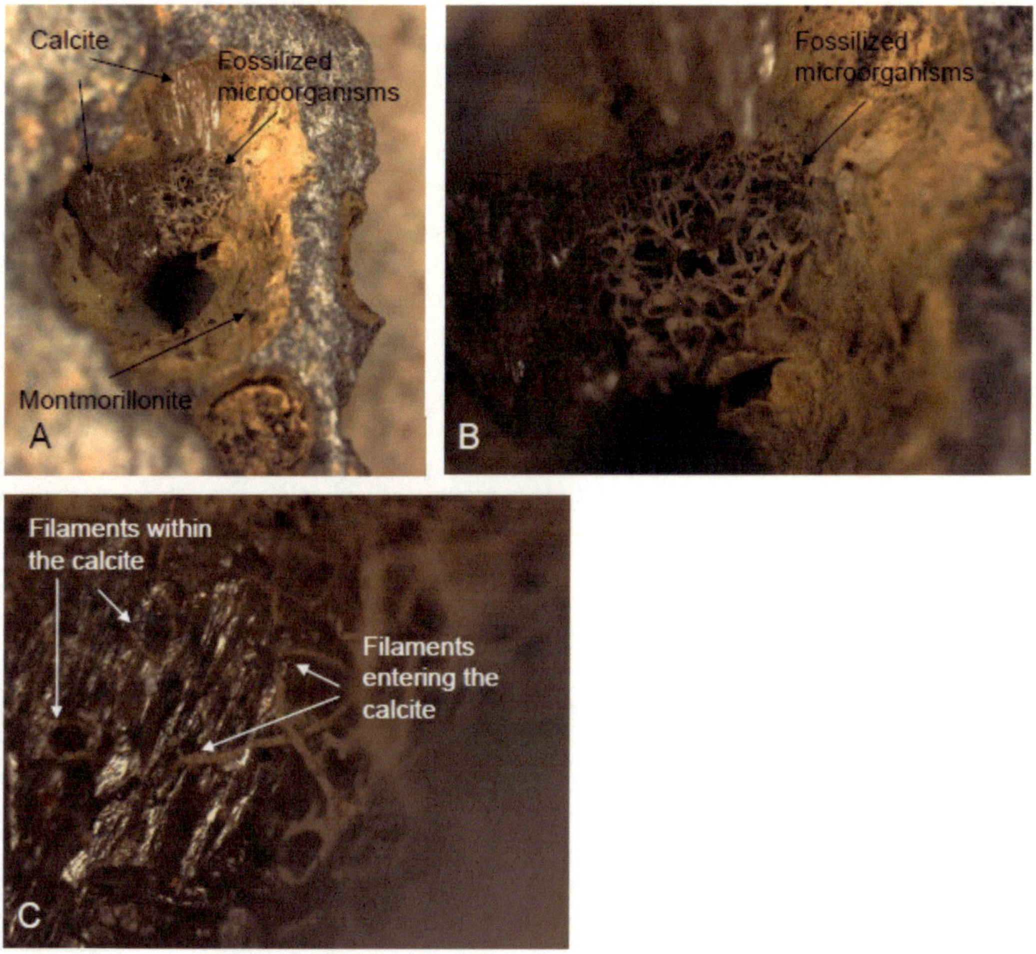

Fig. 6. Optical micrographs. A) Showing a vesicle in basalt from ODP Leg 197 with calcite crystals and montmorillonite grown on the vein walls and an assemblage of fossilized microorganisms. B) Close up of the network of fossilized microorganisms seen in A. C) Image that shows how the fossilized filaments have penetrated into the mineral substrate.

4.3.1 Preparation

Due to the nature of the samples with partly filled veins or vesicles, and the fragile nature of the microfossils the samples are not prepared as thin sections but the veins and the vesicles are sawed to small cubes (~1x1 cm in diameter) from the original drill cores. Attempts were made to expose as much as possible of the content of the vesicles but differences in height are difficult to avoid and the samples are far from being as horizontal as the surface of thin sections. Coating these samples is also not an option due to their nature as sawed cubes and due to the fragile nature of the fossilized microorganisms.

The zeolites and the clays are both hydrated minerals and contain crystalline H_2O. Analysing them in SEM would result in desiccation and eventually crystal collapse.

However, by studying the minerals with a gaseous atmosphere in ESEM they maintain their mineral structure and collapse is avoided. This is a substantial advantage over conventional SEM. Another advantage is that the samples do not need to be coated, which in this case is impossible due to the sensitivity of the samples.

4.3.2 Fossilized microorganisms

The ESEM analysis of the fossilized microorganisms gives us a much more detailed description of the morphology, occurrence, preservation and composition of these structures than optical microscopy do (Fig 7). The filaments are up to several hundred μm in length and 5- 20 μm in width depending on where in the network they occur, smaller diameter closer to the attachment in the minerals and wider diameters further from the attachment. They branch frequently and in some cases anastomoses between branches occur. They consist of an inner part and an outer part. Both the inner and the outer part are usually 5 to 10 μm in width depending on the total diameter of the filament. They mainly consist of a clay phase that compositionally corresponds to montmorillonite. It is possible to see that the filamentous networks are attached directly onto the vein walls but also penetrating calcite or zeolite crystals. The morphology and occurrence of these fossilized microorganisms resolved by ESEM correspond to fungal morphology rather than filamentous prokaryotes, thus, with the aim of ESEM it is possible to characterize the microfossils and determine what type of microorganisms they once were.

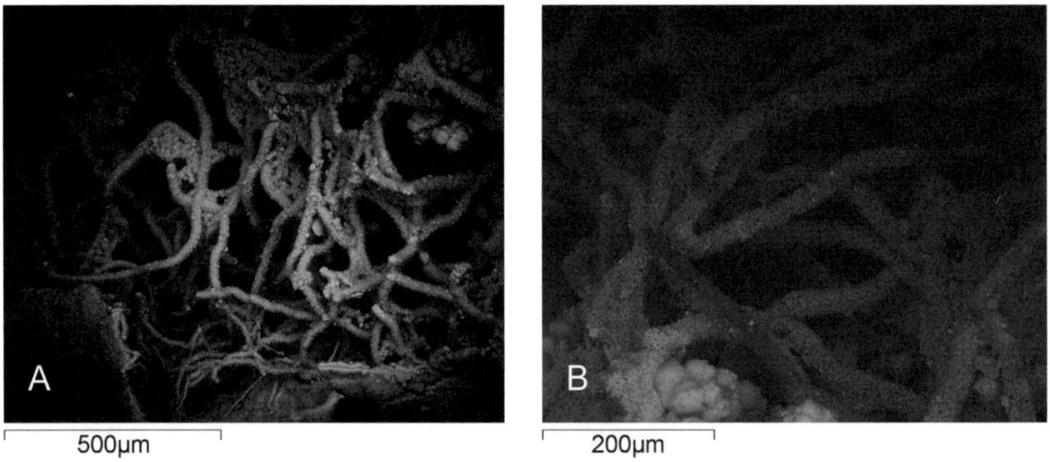

Fig. 7. ESEM images. Images showing how the fossilized microorganisms occur in the veins.

4.3.3 Hydrated minerals

The minerals in the investigated veins and vesicles consist of clays and zeolites, two groups of minerals that contain crystalline H_2O in their crystal structure. The ESEM images show how the montmorillonite looks on a micro meter scale and it allow relative good EDS analyses despite the differences in focal depth (Fig. 8, Table 2). Montmorillonite form in aqueous environments thus the fact that the microfossils consist of montmorillonite indicate that the microorganisms lived while the vesicles were circulated by hydrothermal fluids. This makes it possible to constrain a time window when the microorganisms existed in the system – while the hydrothermal system was active and the volcanism still was active. This is a strong argument for interpreting the microfossils as syngenetic with the rock, the secondary mineralizations and the hydrothermal activity, and not being a modern contaminant.

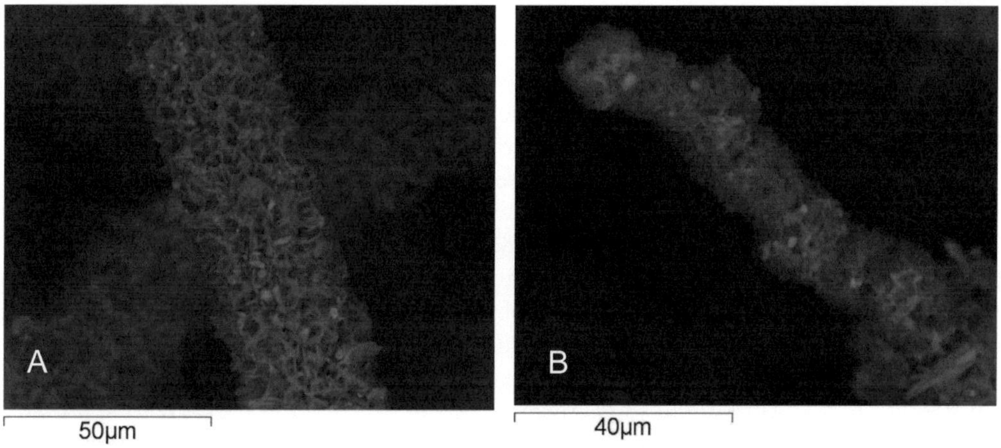

Fig. 8. ESEM images of fossilized microorganisms. A) ESEM image showing the well crystalline clay phase the microfossils consist of. B) ESEM image after the microfossil have been subject to high vacuum. Note how the clay phase has been desiccated and appear collapsed.

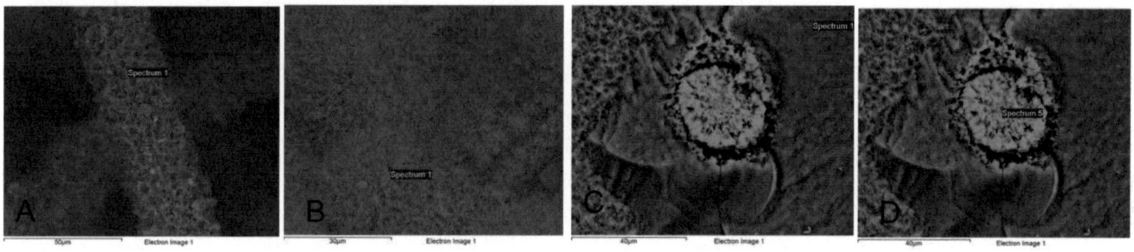

	A	B	C	D
Mg	7.96	3.65		11.40
Na			3.61	0.84
Al		2.99	12.27	8.43
Si	18.95	9.65	30.13	22.62
K			1.15	0.44
Ca		1.07	4.37	1.12
Fe	32.36	46.26		9.48
O	40.73	36.37	48.47	45.67
Total	100.00	100.00	100.00	100.00

Table 2. EDS data in wt%. A) Close up of a fossilized microfossil showing the structure of the clay phase. B) The clay phase of the vein walls. C) Zeolite. D) Cross section of a fossilized microorganism that dissolved the zeolite producing a tunnel structure while it was alive.

Tests were done in high vacuum to analyse these mineral phases but failed due to charging artifacts. However, ESEM mode made it possible to produce high resolved images and EDS analyses. Figure 8 illustrate how a fossilized filament consisting of montmorillonite looks like in ESEM and how a fossilized filament look like after being subject to high vacuum mode which is equivalent to SEM mode.

4.3.4 Microbe-mineral interactions

In optical microscopy it was possible to observe that the microorganisms had penetrated the zeolite and to some extent the calcite crystals during their existence with the result of long micro sized tunnel structures. Microbially produced cavities or tunnel-like structures in minerals are either produced by mechanical force or by chemical dissolution (McLoughlin et al., 2010). It is not possible to determine in optical microscopy how the tunnel-structures in our samples have been produced, however, in ESEM images it is possible to see that the minerals are clearly dissolved at the margins of these structures (Fig. 9). It is also possible to view dissolved patches on the mineral surface where the microfossils have been attached but for some reason been removed (Fig. 9B). Several types of microorganisms are known to produce long, curvi-linear structures in minerals due to directed dissolution of the minerals by the production of acids and siderophores. The cause of mineral dissolution could be just to expand the microbial habitat but there could also be other causes like searching for elements or compounds within the minerals that the microorganisms could use for their metabolism. In subseafloor environments iron and manganese are elements with redox potential that microorganisms commonly use (Edwards et al., 2005). Iron oxidising autotrophic bacteria as well as manganese oxidising heterotrophic bacteria are common at Seamounts in the Pacific Ocean i.e. the Loihi Seamount which is the active seamount of the Emperor-Hawaiian chain today (Emerson and Moyer, 2002; Templeton et al., 2005). The production of tunnel structures in volcanic glass from subseafloor pillow lavas have also been interpreted to be caused by the oxidation of iron since volcanic glass contain high amounts of reduced iron easy to be oxidised by microbes. However, zeolites do not contain iron or any other element that could be used by microorganisms. Thus the question remains why the microorganisms once dissolved the zeolites to such an extent. One explanation could be that they actually did obtain elements or compounds that they could use in their metabolism by boring through the minerals. Zeolites are well known for their capacity to adsorb various elements and compounds like metals, hydrocarbons and molecular hydrogen within their crystal framework of molecular-sized channels (Sheta et al., 2003; Langmi et al., 2003). Zeolites are frequently used in industrial processes as ion exchangers, catalysts and molecular sieves, and it is most likely that zeolites in subseafloor settings adsorb compounds like Fe, CH_4 or H_2 from hydrothermal fluids which microorganisms can scavenge when dissolving the minerals. The question is whether zeolites can adsorb enough compounds and elements with redox potential to make microbial mining worth the effort or not.

In conclusion, it was possible by the aim of ESEM to characterize the fossilized microorganisms, perform element analyses of the hydrated minerals and study the interaction between the microorganisms and the mineral phases which gives us information about the living conditions of the microorganisms and perhaps even their metabolism, which is interesting in a geobiological context.

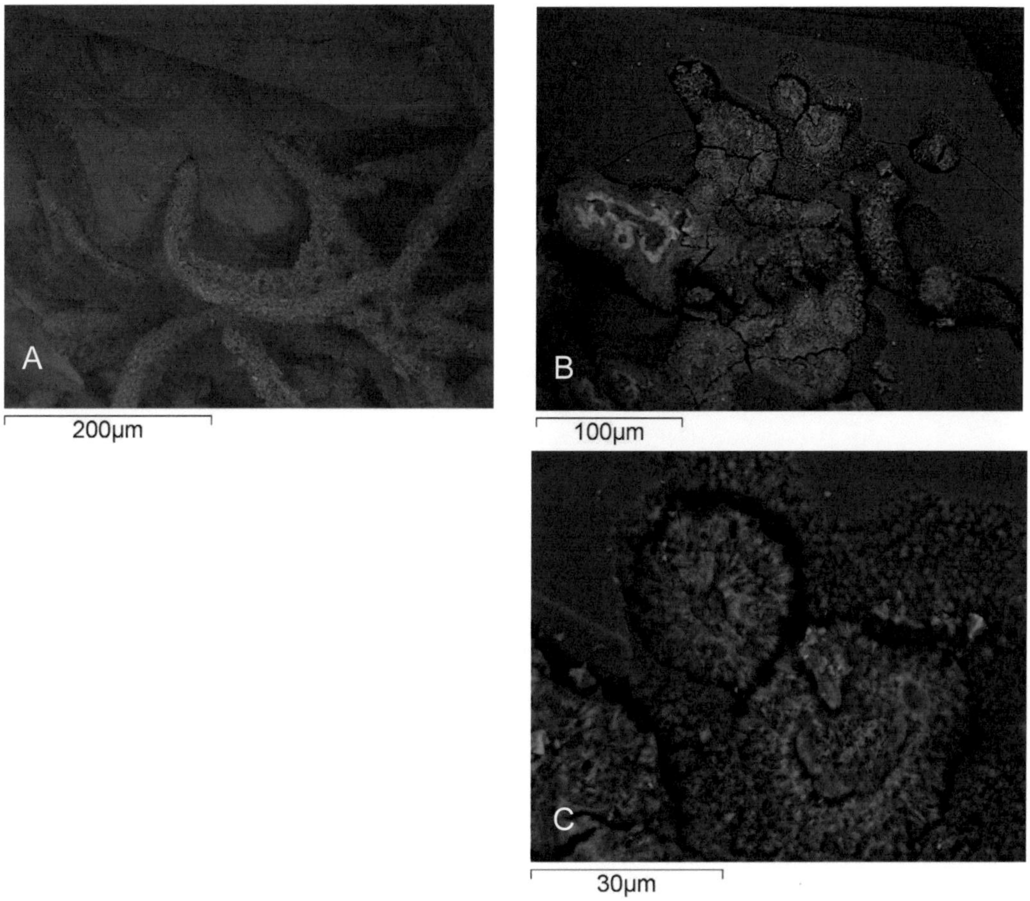

Fig. 9. ESEM images. Images showing the interaction between the mineral substrate and the fossilized microorganisms. In B and C it is possible to see that the microorganisms produced the tunnels in the mineral by dissolution. In B dissolved patches are viewable where fossilized microorganisms have been attached but later removed.

5. Concluding remarks

The ambition with this chapter was to show the advantage of ESEM in geobiological related research over conventional SEM. We have choosen samples that represent that specific intersection in nature where biology meets geology. We have seen how live microorganisms collected and isolated from soil samples can be viewed and analysed in ESEM. We have seen how the deep subseafloor biosphere can be explored by the study of fossilized microorganisms in rock samples. We have further seen how microorganisms interact with their close environment by, for instance, dissolving minerals to scavenge elements for their metabolism, but we have also seen how microorganisms form minerals by precipitation. All this have been made possible by the use of ESEM. The samples used in this study are extremely sensitive and fragile and procedures for ordinary SEM analysis such as desiccation and coating would have substantially damaged the samples. We have in this chapter used modern or quite young samples (even though the samples from the Emperor

Seamount are 48Ma) but a large part of geobiological research is carried out on samples much older and, thus, much more fragile than this. The oldest traces of life on Earth are about 3.5 Ga and need to be handled with extreme care (Schopf, 1993). The same is for extraterrestrial samples, such as putative microfossils in meteorites (McKay et al., 1996) or near future sample return from Mars (MEPAG, 2008). Thus, ESEM is an instrument with great future potential and we hope that we with this chapter have shown its capability within geobiological research.

6. Acknowledgment

We would like to thank Marianne Ahlbom at the Department of Geological Sciences, Stockholm University, for assistance, technical guidance and discussions during the ESEM sessions.

7. References

Abrams, I.M. & McBain, J.W. (1944). A closed cell for electron microscopy. *Journal of Applied Physics,* Vol.15, No.8, (August 1944), pp. 607-609, ISSN 1089-7550.

Ahmad, M.R.; Nakajima, M.; Kojima, S.; Homma, M. & Fukuda, T. (2010) Single cell analysis inside environmental scanning electron microscope (ESEM)-nanomanipulator system, In: *Cutting Edge Nanotechnology*, D. Vasileska (Ed.), 413-438, InTech, ISBN 978-953-7619-93-0, Vukovar, Croatia.

Ardenne, M. & Beischer, D. (1940). Untersuchung von metalloxyd-rauchen mit dem universal-elektronenmikroskop. *Zeitsschrift für Elektrochemie und Angewandte Physikalische Chemie,* Vol. 46, No. 4, (April 1940), pp. 270-277, ISSN 0005-9021.

Baas Becking, L.G.M. (1934). *Geobiologie: of Inleiding tot de Milieukunde,* Van Stockum and Zoon, The hague, The Netherlands.

Bergmans, L.; Moisiadis, P.; Van Meerbeek, B.; Quirynen, M. & Lambrechts, P. (2005) Microscopic observation of bacteria: review highlighting the use of environmental SEM. *International Endodontic Journal,* Vol. 38, No. 11 (November 2005), pp. 775-788, ISSN 1365-2591.

Danilatos, G.D. (1981). Design and construction of an atmospheric or environmental SEM (part 1). *Scanning,* Vol. 4, No. 1, pp. 9-20, ISSN 1932-8745.

Danilatos, G.D. (1985). Design and construction of an atmospheric or environmental SEM (part 3). *Scanning,* Vol. 7, No. 1, pp. 26-42, ISSN 1932-8745.

Danilatos, G.D. (1988). Foundations of Environmental Scanning Electron Microscopy. *Advances in Electronics and Electron Physics,* Vol. 71, 109-250, Academic Press, ISBN 9780120146710.

Danilatos, G.D. (1990a). Theory of the Gaseous Detector Device in the ESEM. *Advances in Electronics and Electron Physics,* Vol. 78, 1-102, Academic Press, ISBN 9780120146789.

Danilatos, G.D. (1990b). Design and construction of an atmospheric or environmental SEM (part 4). *Scanning,* Vol.12, No. 1, (August 2011), pp. 23-27, ISSN 1932-8745.

Danilatos, G.D, (1997). Environmental scanning electron microscopy. In: *In Situ Microscopy in Materials Research,* P.L. Gai, (Ed.), 14-44, Kluwer Academic Publishers, ISBN 978-0-7923-9989-6, Dordrecht, Germany.

Danilatos, G.D. (2009). Optimum beam transfer in the environmental scanning electron microscope. *Journal of Microscopy*, Vol. 234, No. 1, (April 2009), pp. 26-37, ISSN 0022-2720.

Danilatos, G.D. & Robinson, V.N.E. (1979). Principles of scanning electron microscopy at high specimen pressures. *Scanning*, Vol 2, No. 1, pp. 72-82, ISSN 1932-8745.

Danilatos, G.D. & Postle, R. (1983). Design and construction of an atmospheric or environmental SEM (part 2). *Micron*, Vol. 14, No. 1 (December 2002), pp. 41-52, ISSN 0968-4328.

Donald, A.M. (2003) The use of environmental scanning electron microscopy for imaging wet and insulating materials. *Nature materials*, Vol. 2, No. 8 (August 2003), pp. 511-516, ISSN 1476-1122.

Douglas, S. & Douglas, D.D. (2000) Environmental scanning electron microscopy studies of colloidal sulphur deposition in a natural microbial community from a cold sulfide spring near Ancaster, Ontario, Canada. *Geomicrobiology Journal*, Vol. 17, No. 4, pp. 275-289, ISSN 0149-0451.

Edwards, K.J.; Bach, W. & McCollom, T.M. (2005). Geomicrobiology in oceanography: microbe-mineral interactions at and below the seafloor. *TRENDS in Microbiology*, Vol. 13, No. 9, (September 2005), pp. 449-456, ISSN 0966-842X.

Emerson, D. & Moyer, C.L. (2002). Neutrophilic Fe-oxidising bacteria are abundant at the Loihi Seamount hydrothermal vents and play a major role in Fe oxide deposition. *Applied and Environmental Microbiology*, Vol. 68, No. 6 (June 2002), pp. 3085-3093, ISSN 1098-5336.

Farmer, J.D. & Des Marais, D.J. (1999). Exploring for a record of ancient Martian life. *Journal of Geophysical Research*, Vol. 104, No. E11, (November 1999), pp. 26977-26995, ISSN 0148-0227.

Furnes, H.; McLoughlin, N.; Muehlenbachs, K.; Banerjee, N.; Staudigel, H.; Dilek, Y.; de Wit M.; Van Kranendonk, M. & Schiffman, P. (2008) Oceanic pillow lavas and hyaloclastites as habitats for microbial life through time-A review, In: *Links between Geological Processes, Microbial Activities and Evolution of Life*, Y. Dilek, H. Furnes, & K. Muehlenbachs, (Eds.), 1-68, Springer, ISBN 978-1-4020-8305-1, Berlin, Germany.

Gold, T. (1992). The deep, hot biosphere. *Proceedings of the National Academy of Sciences of the United States of America*, Vol. 89, No. 13, (July 1992), pp. 6045-6049, ISSN 0027-8424.

Hallberg, R. & Ferris, F.G. (2004) Biomineralization by *Gallionella*. *Geomicrobiology Journal*, Vol. 21, No. 5 (July 2004) , pp. 325-330. ISSN 0149-0451.

Hazen, R.M.; Papineau, D.; Bleeker, W.; Downs, R.T.; Ferry, J.M.; McCoy, T.J.; Sverjensky, D.A. & Yang, H. (2008). Mineral evolution. *American Mineralogist*, Vol. 93, No. 11-12, (Nov-Dec 2008), pp.1693-1720, ISSN 0003-004X.

Huber, J.A.; Mark Welch, D.B.; Morrison, H.G.; Huse, S.M.; Neal, P.R.; Butterfield, D.A. & Sogin, M.L. (2007). Microbial population structures in the deep marine biosphere. *Science*, Vol. 318, No, 5847, (October 2007), pp. 97-100, ISSN 0036-8075.

Huiming, Z.; Hongwei, F. & Minghong, C. (2011) Floc architecture of bioflocculation sediment by ESEM and CLSM. *Scanning*, Vol. 33, No. 1 (January/February 2011) , pp. 1-9, ISSN 1932-8745.

Hutton, J. (1788). Theory of the Earth; or an investigation of the laws observable in the composition, dissolution, and restoration of land upon the globe. *Transactions of the Royal Society of Edinburgh*, Vol. 1, Part II, pp. 209-304.

Ivarsson, M.; Lindblom, S.; Broman, C. & Holm, N.G. (2008a). Fossilized microorganisms associated with zeolite-carbonate interfaces in sub-seafloor hydrothermal environments. *Geobiology*, Vol. 6, No. 2, (March 2008), pp. 155-170, ISSN 1472-4677.

Ivarsson, M.; Lausmaa, J.; Lindblom, S.; Broman, C. & Holm, N.G. (2008b). Fossilized microorganisms from the Emperor Seamounts: implications for the search for a subsurface fossil record on Earth and Mars. *Astrobiology*, Vol. 8, No. 6, (December 2008), pp.1139-1157, ISSN 1531-1074.

Ivarsson, M. & Holm, N.G. (2008). Microbial colonization of various habitable niches during alteration of oceanic crust, In: *Links between geological processes, microbial activities and evolution of life*, Y. Dilek, H. Furnes, & K. Muehlenbachs, (Eds.), 69-111, Springer, ISBN 978-1-4020-8305-1, Berlin, Germany.

Knoll, A.H. (2003). The geological consequences of evolution. *Geobiology*, Vol. 1, No. 1, (July 2003), pp. 3-14, ISSN 1472-4677.

Lane, W.C. (1970). The environmental control stage. In: *Scanning Electron Microscopy*, 41-48, IIT Research Institute, Chicago, USA.

Langmi, H.W.; Walton, A.; Al-Mamouri, M.M.; Johnson, S.R.; Book, D.; Speight, J.D.; Edwards, P.P.; Gameson, I.; Anderson, P.A. & Harris, I.R. (2003). Hydrogen adsorption in zeolites A, X, Y and RHO. *Journal of Alloys and Compounds*, Vol. 356-357, (August 2003), pp. 710-715, ISSN 0925-8388.

Lindsay, J.F. & Brasier, M.D. (2002). Did global tectonics drive early biological evolution? Carbon isotopic record from 2.6 to 1.9 Ga carbonates of Western Australian basins. *Precambrian Research*, Vol. 114, No. 1-2, (February 2002), pp. 1-34, ISSN 0301-9268.

Little, B.; Wagner, P.; Ray, R.; Pope, R. & Scheetz, R. (1991) Biofilms: an ESEM evaluation of artifacts introduced during SEM preparation. *Journal of Industrial Microbiology*, Vol. 8, No. 4 (November 1991), pp. 213-222, ISSN 1476-5535.

Lowe, D.R. & Tice, M.M. (2007). Tectonic controls on atmospheric, climatic, and biological evolution 3.5-2.4 Ga. *Precambrian Research*, Vol. 158, No. 3-4, (October 2007), pp. 177-197, ISSN 0301-9268.

McKay, D.S.; Gibson Jr, E.K.; Thomas-Keprta, K.L.; Vali, H.; Romanek, C.S.; Clemett, S.J.; Chillier, X.D.F.; Maechling, C.R. & Zare, R.N. (1996) Search for past life on Mars: possible relic biogenic activity in Martian meteorite ALH84001. *Science*, Vol. 273, No. 5277, (August 1996), pp. 924-930, ISSN 0036-8075.

McLoughlin, N.; Staudigel, H.; Furnes, H.; Eickmann, B. & Ivarsson, M. (2010). Mechanisms of microtunneling in rock substrates: distinguishing endolithic biosignatures from abiotic microtunnels. *Geobiology*, Vol. 8, No. 4, (September 2010), pp. 245-255, ISSN 1472-4669.

Melezhik, V.A.; Fallick, A.; Hanski, E.J.; Kump, L.R.; Lepland, A.; Prave, A.R. & Strauss, H. (2005). Emergence of the aerobic biosphere during the Archean-Proterozoic transition: challenges of future research. *GSA Today*, Vol. 15, No. 11, (November 2005), pp. 4-11, ISSN 1052-5173.

The MEPAG Next Decade Science Analysis Group (2008) Science priorities for Mars sample return. *Astrobiology*, Vol. 8, No. 3, (June 2008), pp. 489-535, ISSN 1531-1074.

Nealson, K.H.; Belz, A. & McKee, B. (2002) Breathing metals as a way of life: geobiology in action. *Antonie van Leeuwenhoek*, Vol. 81, No. 1-4, pp. 215-222, ISSN 1572-9699.

Parsons, D.F.; Matricardi, V.R.; Moretz, R.C. & Turner, J.N. (1974). Electron microscopy and diffraction of wet unstained and unfixed biological objects. *Advances in Biological and Medical Physics*, Vol. 15, (June 1974), pp. 161-270, ISSN 0065-2245.

Pedersen, K. (1993). The deep subterranean biosphere. *Earth-Science Review*, Vol. 34, No. 4, (August 1993), pp. 243–260, ISSN 0012-8252.

Reed, S.J.B. (2005) *Electron microprobe analysis and scanning electron microscopy in geology*, Cambridge University Press, ISBN 13 978-0-521-84875-6, Cambridge, United Kingdom.

Robinson, V.N.E, (1974) A wet stage modification to a scanning electron microscope. *8th International Congress of Electron Microscopy*, Australian Academy of Science, Vol. II, pp. 50–51.

Rosing, M.T.; Bird, D.K.; Sleep, N.H.; Glassley, W. & Albarede, F. (2006). The rise of continents - an essay on the geologic consequences of photosyntesis. *Palaeogeography, Palaeoclimatology, Palaeoecology*, Vol. 232, No. 2-4, (March 2006), pp. 99-113, ISSN 0031-0182.

Schopf, J.W. (1993) Microfossils of the early Archean Apex chert: new evidence of the antiquity of life. *Science*, Vol. 260, No. 5108, (April 1993), pp. 640-646, ISSN 0036-8075.

Shah, J.S. & Beckett, A. (1979). A preliminary evaluation of moist environment ambient temperature scanning electron microscopy. *Micron*, Vol. 10, No. 1, (November 2002), pp.13-23, ISSN 0968-4328.

Sheta, A.S.; Falatah, A.M.; Al-Sewailem, M.S.; Khaled, E.M. & Sallam, A.S.H. (2003). Sorption characteristics of zinc and iron by natural zeolite and bentonite. *Microporous and Mesoporous Materials*, Vol. 61, No. 1-3, (July 2003), pp. 127-136, ISSN 1387-1811.

Southam, G. & Saunders, J.A. (2005) The geomicrobiology of ore deposits. *Economic Geology*, Vol. 100, No. 6, (September-October), pp. 1067-1084, ISSN 0013-0109.

Staudigel, H.; Tebo, B.; Yayanos, A.; Furnes, H.; Kelley, K.; Plank, T. & Muehlenbachs, K. (2004). The oceanic crust as a bioreactor, In: *The Subseafloor Biosphere at Mid-Ocean Ridges*. W.S.D. Wilcock, E.F. DeLong, D.S. Kelley, J.A. Baross, & S.C. Cary, (Eds.), 325-341, Geophysical Monograph Series 144, American Geophysical Union, ISBN 0-87590-409-2, Washington DC.

Swift, J.A, and Brown, A.C., (1970) An environmental cell for the examination of wet biological specimens at atmospheric pressure by transmission scanning electron microscopy. *Journal of Physics E: Scientific Instruments*, Vol. 3, No. 11, (November 1970), pp. 924-926, ISSN 0022-3735.

Templeton, A.; Staudigel, H. & Tebo, B.M. (2005). Diverse Mn(II)-oxidizing bacteria isolated from submarine basalts at Loihi Seamount. *Geomicrobiology Journal*, Vol. 22, No. 3-4, (February 2007), pp. 127-139, ISSN 0149-0451.

Tice, M.M. & Lowe, D.R. (2004). Photosynthetic microbial mats in the 3416-Myr-old ocean. *Nature*, Vol. 431, No. 7008, (September 2004), pp. 549-552, ISSN 0028-0836.

Vernadsky, V. (1926). The Biosphere. Abridged English edition published in 1988 by Springer-Verlag, ISBN 0-387-98268-X, New York, USA.

Waters, M.S.; Sturm, C.A.; El-Naggar, M.Y.; Luttge, A.; Udwadia, F.E.; Cvitkovitch, D.G.; Goodman, S.D. & Nealson, K.H. (2008) In search for the microbe/mineral interface: quantitative analysis og bacteria on metal surfaces using vertical scanning interferometry. *Geobiology*, Vol. 6, No. 3 (June 2008), pp. 254-262, ISSN 1472-4677.

XL30 Options Manual (1998). *XL30 ESEM Series, Scanning Electron Microscope*, No 2.6, (December 1998), FEI Company and Philips.

40

How Log Interpreter Uses SEM Data for Clay Volume Calculation

Mohammadhossein Mohammadlou and Mai Britt Mørk
Norwegian University of Science and Technology, NTNU, Trondheim, Norway

1. Introduction

Guggenheim and Martin (1995), define the clay as a naturally occurring aluminum silicate composed dominantly of fine-grained minerals. Several other definitions and classifications based on the grain size, pore size, sedimentation, lattice, and other properties can be found in the literature. The attention here is, however, not related to these definitions instead it is about "how to calculate reservoir clay volume using Scanning Electron Microscope (SEM) data".

The rock that predominantly contains clay minerals is by definition called Shale. Shales are mostly fine-grained clastic sedimentary rocks composed of clays and fragments of other minerals i.e., carbonates and siliciclastics. Most of the clay minerals contain variable amounts of water trapped in the mineral structure (Wiki definition). Clay volume is one of the key parameters used to correct porosity and water saturation for the effects of clay bound water in petrophysical evaluation (Crain, 2000).

A problematic aspect of evaluating hydrocarbon bearing reservoirs is to accurately model the effects of clay and shale components on reservoir properties. Clay minerals and other sedimentary materials as detrital shale components in the form of shale lamina, structural clasts and dispersed shale matrix can be deposited in the sedimentary rocks. Clay minerals are also commonly present as diagenetic clays, including pore-filling clays, pore-lining clays and pore-bridging clays (Wilson and Pittman 1977).

Accurate quantification of the clay content, distribution and clay type in the reservoir rock using core and wireline log data is rather complicated by a number of factors including: rock heterogeneity, mud filtrate invasion and fluid contamination, alteration of clay microstructure and wettability by mud invasion. On the other hand, a number of geological and petrophysical methods have been developed to run the clay volume calculation.

Traditionally, a log interpreter i.e., petrophysicist, uses wireline log data including clay indicator logs of gamma rays and spectral gamma rays in combination with other porosity and resistivity logs to estimate reservoir clay volume. Integrated analyses of these logs together with geological information are generally used for clay volume calculation. Clay volume in homogeneous reservoir rocks can be estimated by conventional methods however, a mixture of clay minerals, quartz, and calcite notably complicates the interpretation of wireline log data. Rock heterogeneity influences the resistivity logs and thereby water saturation as well. The focus here, however, is to discuss the use of SEM data in combination with wireline logs to calculate reservoir clay volume.

2. Conventional shale volume calculation

Generally, porosity logging tools (neutron, density and sonic) display higher porosity measurements in clay-rich rocks, whereas much of the porosity is neither effective for hydrocarbon accumulation nor for hydrocarbons transmission through the rock. The shale porosity has to be removed from the total porosity to obtain effective reservoir porosity. The clay correction is therefore an essential step to carry out before estimating reservoir porosity and saturation. Figure 1 displays the gamma-ray, spectral gamma-ray, and calculated clay volume from these logs for a reservoir in the offshore Norwegian Barents Sea. Presence of siliciclastic fines and diagenetic minerals (e.g. dolomite) within carbonate breccias has resulted in a heterogeneous carbonate reservoir in the prospect area. Interactive Petrophysics (IP) software was used for wireline log interpretation and petrophysical analysis of the well.

Carbonate rocks, depending on the diagenesis and sedimentary environment, are generally uranium-rich compared to siliciclastic rocks (Luczaj and Goldstein, 2000). A large amount of uranium may precipitate during and after carbonate rock deposition. The process of uranium precipitation is more abundant in the dolomitic sediments. The easy solution to distinguish the uranium-rich from non-uranium rocks is to compare the two most well known logs of natural gamma ray spectrometry (NGS) against natural gamma ray log (GR). Natural gamma-ray (NGS) spectrometry allows estimation of the elemental concentrations of potassium (K), uranium (U), and thorium (Th), which can be used to interpret sediment composition, clay volume, and diagenesis.

Comparison of these logs in Figure 1 shows uranium-rich carbonate rocks in zones 2 and 3 compared to the more silica-rich intervals of zone 5. The higher uranium content originates from uranium-rich organic matter in the reservoir rock. To calculate the shale volume of the reservoir, the spectral gamma-ray log is used. It is also used to verify the calculated shale volume from the gamma-ray log. In Figure 2, track numbers 4, 5, and 6 show the results of the clay volumes for all reservoir sections in this well. The results are shown for potassium, thorium and corrected gamma-ray logs respectively. The shale and clean (sand) lines are drawn on the potassium, thorium, and corrected gamma-ray logs to estimate the bounding values for zero and 100% shale volume. The bounding values to the corrected gamma-ray response are assumed to be 25 API for zero shale volume and 85 API for 100% shale volume. (Crain Petrophysical Handbook). According to the results in figure 2, discrepancy of the results is considerable. The higher uncertainty in clay volume calculation impacts more effectively the reservoir porosity and saturation estimation afterward.

3. NMR log application for shale volume calculation

The NMR log T2 distribution and its characteristics are dependent on the fluid content and pore properties of the formation, and independent of mineral composition. It is by default used for pore size distribution, capillary bound water and free fluid estimation. Basically, modern NMR tools determine the fluid content in the pore space near the wellbore in the order of few inches into the formation. They measure the total rock porosity regardless of the pore fluid type (Dunn et al., 1998). The two T2 distributions, one from the fully polarized echo train and one from the partially polarized echo train, are merged together to build a T2 distribution from 0.3 ms to over 3000 ms for estimating the total porosity. In other words, the sum of the volume fractions occupied by water, oil, and gas will be the total NMR porosity. Traditional NMR tools miss the very fast T2 decay times; however, in modern tools the T2 spectra include all types and sizes of pores and cracks (Coates et al., 1999).

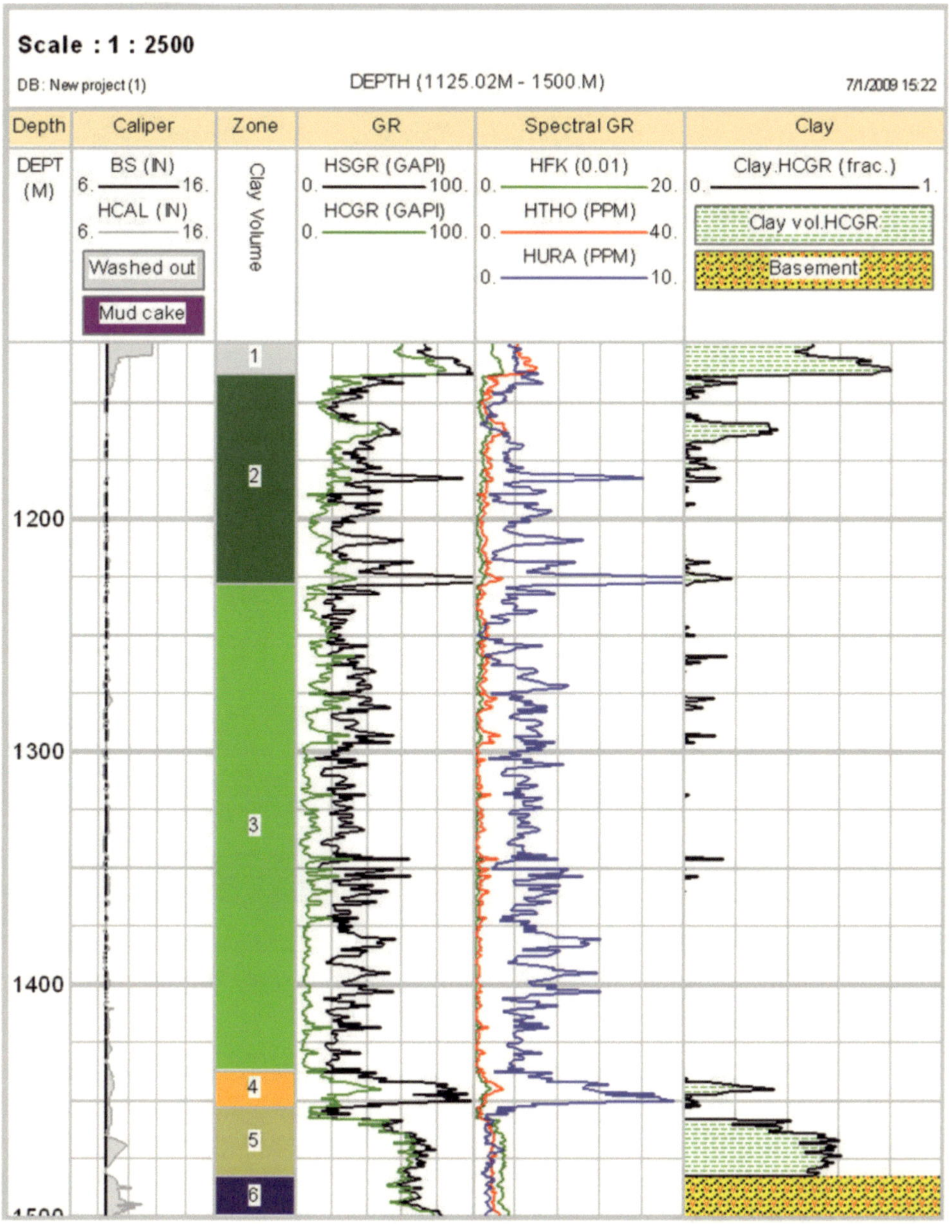

Fig. 1. Well log data, reservoir zonation and lithology.

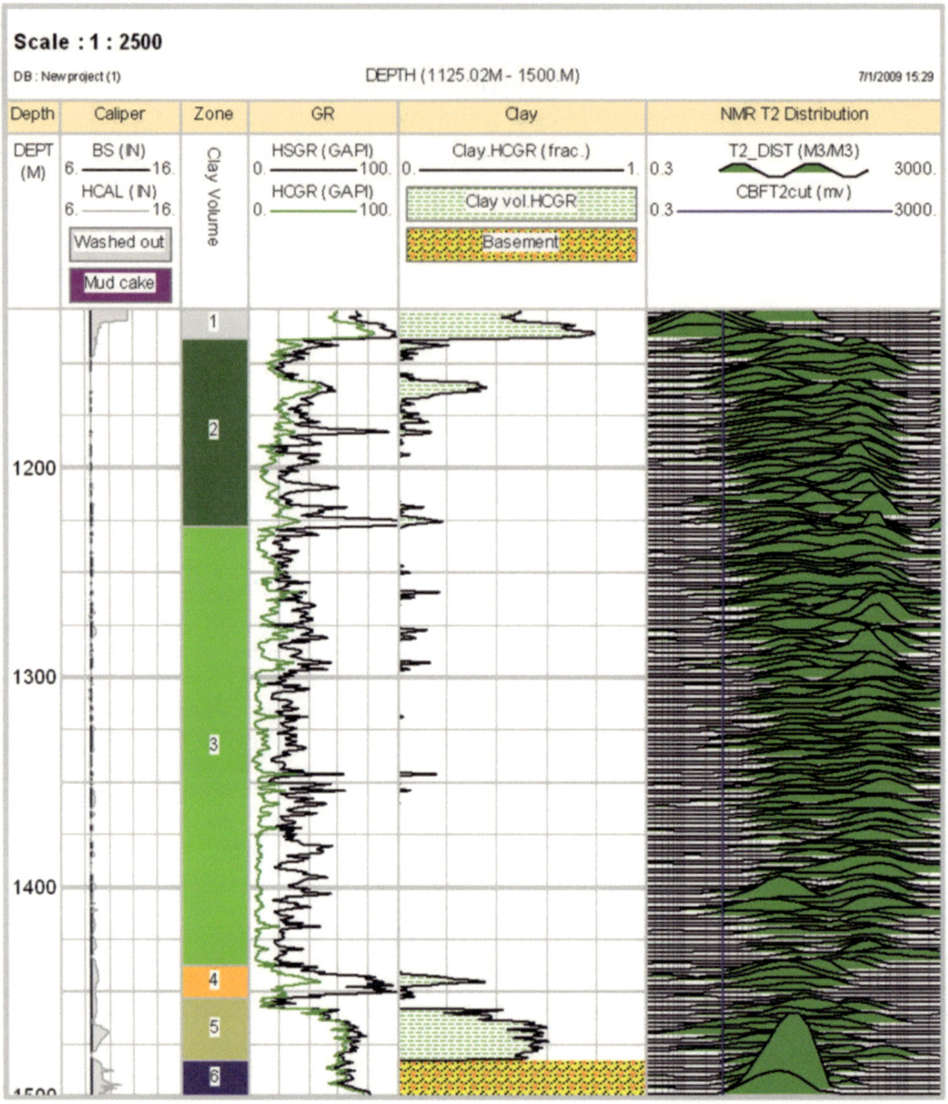

Fig. 2. Shale volume calculation from different methods: track-4 shale volume from potassium log; track-5 shale volume from thorium log; track-6 shale volume from uranium-corrected gamma-ray; track-7 shale volume from NMR log.

The NMR T2 distribution displays a distinct peak at small relaxation times, proving the existence of clay minerals in the logged intervals. To categorize the contribution to porosity from clay, a T2 cut-off value for clay-bound water is applied to the T2 distribution over the reservoir section. Morriss et al. (1997) proposed a T2 clay-bound water cut-off of 3 ms by studying 45 siliciclastic reservoirs. In a supplementary study by Prammer et al. (1996) different T2 cut-off values were found for particular clay types at shorter decay times. Knowing the clay type and implementing an appropriate T2 cut-off value, more accurate determination of clay-bound water volume is possible.The clay-bound water T2 cut-off and clay type/shale volume estimation are important parameters in reservoir characterization. The shale volume estimation is more complicated when the rock is composed of a mixture of clay minerals and tiny fragments of other minerals. Greater amounts of fragments of non-clay minerals increase the uncertainty of obtaining an accurate clay bound T2 cut-off value for clay-bound water (Chitale et al., 1999). Although estimation of the T2 cut-off has been discussed by many authors, there is still uncertainty in the optimum choice for the value. With reference to the work of Matteson et al. (2000), the standard relaxation time of 3.3 ms is used, which is appropriate where the clay type is not kaolinite.

The calculated shale volumes from gamma-ray, spectral gamma-ray (thorium and potassium only), and NMR logs (Figure 7, track 4,5,6 and 7)) show that the calculated volumes are not identical from these methods. Depending upon the chosen gamma-ray and spectral gamma-ray values for the clean and shaly rock samples, the results for the shale volume can differ significantly with different methods. The results from the spectral gamma-ray log in tracks 4 and 5 are in good agreement within the whole reservoir section except in zone 5. The uranium-corrected gamma-ray log, however, shows shale-free reservoir except in zone 5 and parts of zones 2 and 4. To estimate shale volume from the NMR log, representative clayey intervals (100% clay volume) and the amount of clay-bound water in those intervals estimated from the NMR log are classified as references for particular clay types in the well. The amount of clay-bound water estimated from the NMR log response at any other point in the well is compared to that in the selected reference clay in order to estimate the shale volume at that point. There may be errors in this estimate if the selected reference clay was inappropriate. The estimated shale volume from the NMR log is significantly less than that from the gamma-ray log indicators (track 6, Figure 2), presumably due to the errors in conversion of clay-bound water to the shale volume.

A clear response to the presence of clay in the T2 distribution is seen right at the top of Figure 2 in zone 1, the top seal of marine claystones. However, assuming that the top shale seal is totally composed of clay minerals, the NMR estimated shale volume is also not accurate within this interval. Selection of too low a T2 cut-off for clay-bound water or too rapid a decay of the clay-bound water could explain the errors. In addition to the clays, viscose hydrocarbons (tar) relax at much shorter relaxation times than the clay-bound water T2 cut-offs. The NMR log interpretation can be more complicated where both clay and tar-mats are present in the reservoir rock. Retrieved fluid samples from the wireline formation tester and recovered hydrocarbons from core material obtained from the studied well have not shown any tar in this reservoir. Overall, zones 4 and 5 of the reservoir show uncorrelated results compared to the rest of the well. Assuming a formation rich in potassium minerals coupled with high gamma-ray response, the rock type could be either shale, a mixture of shale and other potassium-bearing rocks, or other rock types rich in radioactive minerals.

To identify the clay minerals and potassium sources of zones 1 and 5, the thorium-potassium reference cross-plot of Schlumberger (1985) and corresponding cross-plot of the well data are shown in Figure 3. According to these plots, potassium sources in zone 5 are glauconitic or feldspathic sandstone together with illite and mica, whereas the clay minerals of zone 1 are mixed-layer clay and/or illite. The result gives valuable information about the source of the clays but it could not provide a volumetric response for the shale volume of the rock. Further calibration of the logs and petrographical analysis of the rock samples are necessary to quantify the reservoir shale volume.

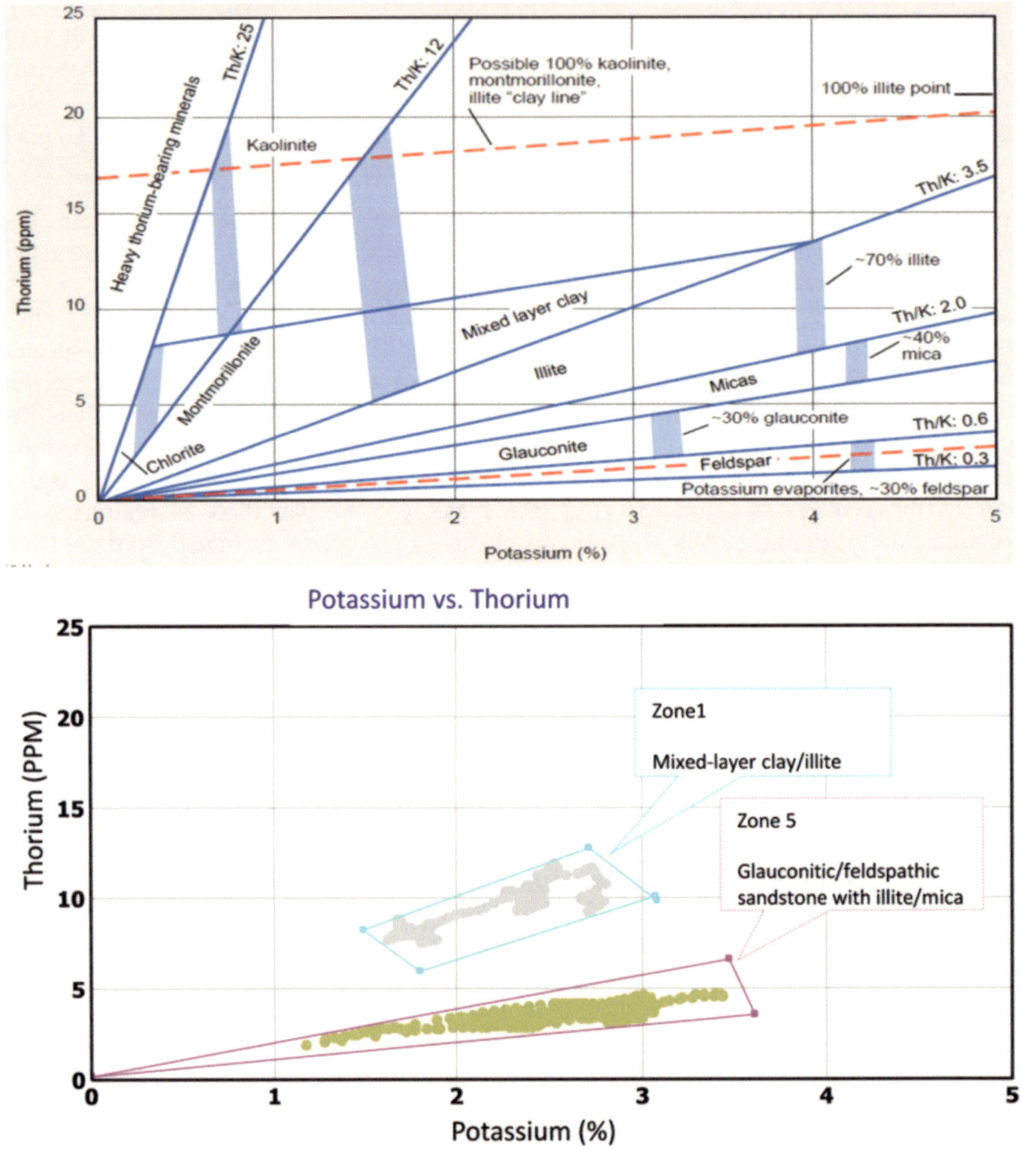

Fig. 3. (a) Schlumberger reference crossplot of thorium/potassium; and (b) corresponding crossplots for zones 1 and 5 to identify the clay types.

4. Scanning electron microscope (SEM) analyses

To identify the rock mineralogy and quantify the mineral volume fractions, SEM analysis was carried out on the selected reservoir samples. Because of inconsistencies in the amount of calculated shale volume in zone 5 (Figure 7), most of the samples are chosen from this zone. Backscattered electron imaging of polished thin sections is used to map mineral distributions in a grey scale with intensity related to the average atomic number of the minerals (Reed, 2005). Mineral identification was supplemented by energy dispersive spectroscopy analysis, and X-ray mapping of element distributions was done to further identify and quantify the mineral distributions. A backscattered electron image of a sample of sandy carbonate rock in zone 5 is shown in Figure 4a. To estimate the proportion of one particular mineral, the image is processed and one component, e.g., potassium feldspar, is characterized (Figure 4b). The energy dispersive spectrum of potassium feldspars (black in Figure 4b) is shown in Figure 4c. The proportion of elements (K, Si, Al, and O) in this figure verify that the selected grains are potassium feldspar. Energy dispersive spectroscopy is used to recognize every selected mineral in a thin section sample by its elemental composition, but X-ray mapping of the section provides a map of each element.

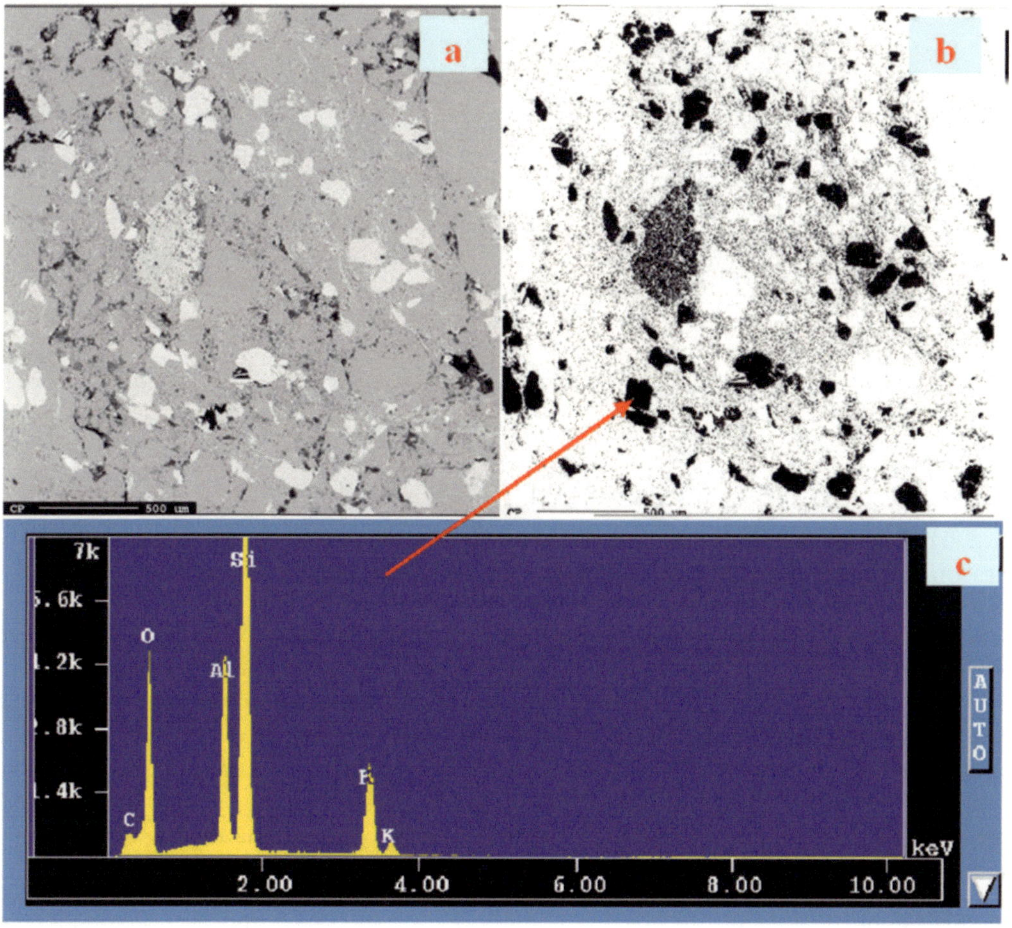

Fig. 4. Details of scanning electron microscope imaging of one sample from zone 5. (a) Backscattered electron image. (b) Processed image for the characterization of potassium feldspar, showing up as black. (c) Energy dispersive spectrum of a selected clay mineral to identify it.

The X-ray elemental mapping of the selected thin section in Figure 4 also confirms the abundance of potassium feldspar in this section, but not clays. Figure 5 shows the elemental map of Na, O, K, C, S, Si, Mg, Ca, Ti, Al, and Fe for the selected sample in Figure 4. The volume fraction of potassium feldspar from this analysis is calculated as 16–23 %. The map of feldspar content in other thin sections in zone 5 shows that the potassium volume increases toward the bottom of the formation. A lesser amount of illite/mica is estimated in comparison with the estimated volume from the conventional method. The clay types and volume estimations from SEM analysis are also consistent with the Schlumberger crossplot. The results show that the clay minerals of the sample from zone 5 are illite and mica with a volume fraction of 15–21 % (Figure 6).

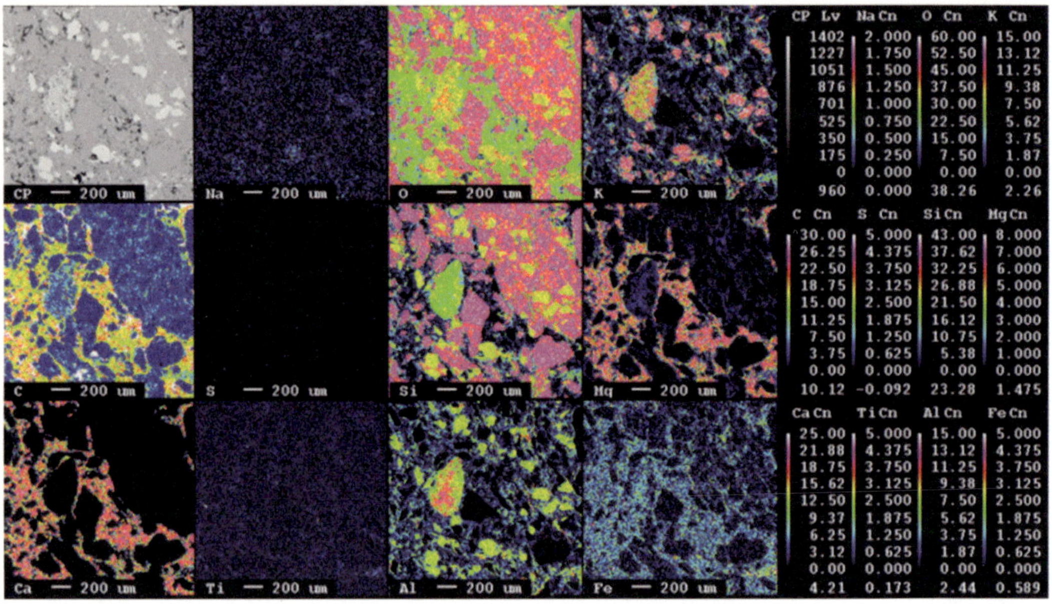

Fig. 5. The backscattered electron image of the thin sample section in Figure 4 with details of elemental mapping for Na, O, K, C, S, Si, Mg, Ca, Ti, Al, and Fe.

A few thin sections cannot be representative of the whole reservoir section because they have been selected from a heterogeneous interval. However, SEM analysis together with the NMR and spectral gamma-ray logs facilitates shale volume calculation. SEM data provide shale volume and clay types at the selected depths which can be correlated with the spectral gamma-ray log to extract the shale and potassium feldspar volume at those depths. Comparison of these results with the NMR-estimated shale volumes shows that the NMR log underestimates the shale volume. In contrast, the spectral gamma-ray log overestimates the shale volume.

The potassium readings on the spectral gamma-ray log in zone 5 and part of zone 2 come from a mixture of clay minerals and potassium feldspar. As discussed by Ellis and Singer (2008) with reference to the gamma-ray response in shaly rocks, the correlation between clay minerals and thorium is largest, because of the potassium association with other components of the shale, such as feldspars. They also established an empirical equation for the total gamma-ray response based on the three radioactive components of the spectral gamma-ray signal, however potassium-rich shaly sand could be inaccurately interpreted as shale by this equation. Therefore, the contribution of potassium feldspar has to be removed from the spectral gamma-ray log to estimate an accurate shale volume in the selected samples. Using Ellis and Singer's (2008) approximation and estimated potassium values from the SEM samples, the corrected gamma-ray log values were calibrated for selected samples and subsequent shale volumes. The new potassium-corrected log was then used with the thorium log to estimate a revised shale volume in zone 5. Figure 7 shows the calculated shale volumes from all the above-mentioned methods. The calculated clay volumes using potassium, thorium, and uranium-corrected gamma-ray (CGR) and the integration of SEM, CGR and NMR logs are shown in Figure 7 from tracks 4 to 7, respectively. In track 7, the final corrected shale volume has been plotted for the whole interval.

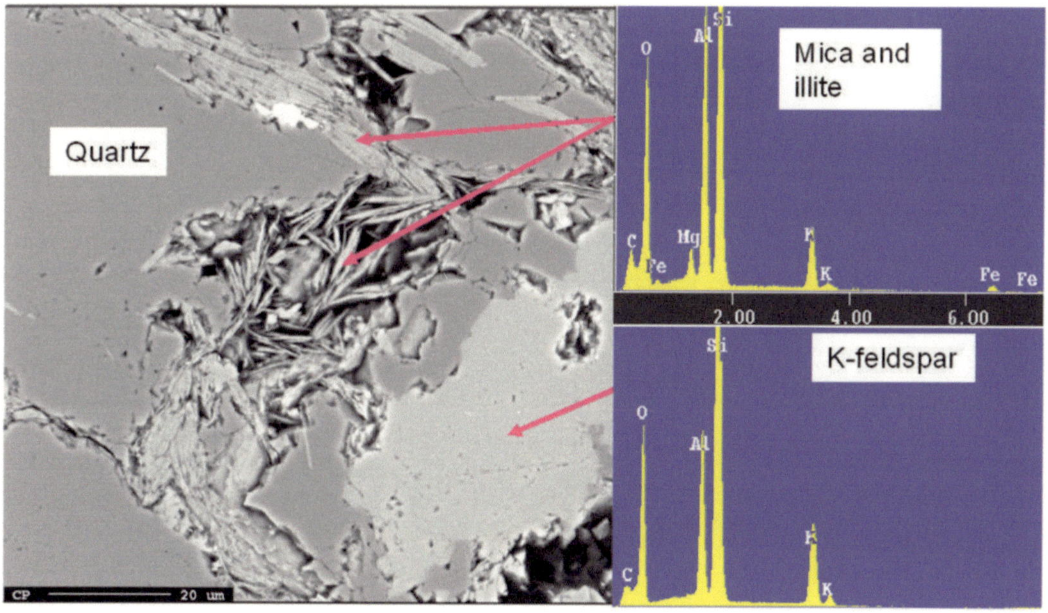

Fig. 6. Backscattered electron image of the sample section in zone 5 showing illite, mica, and feldspar.

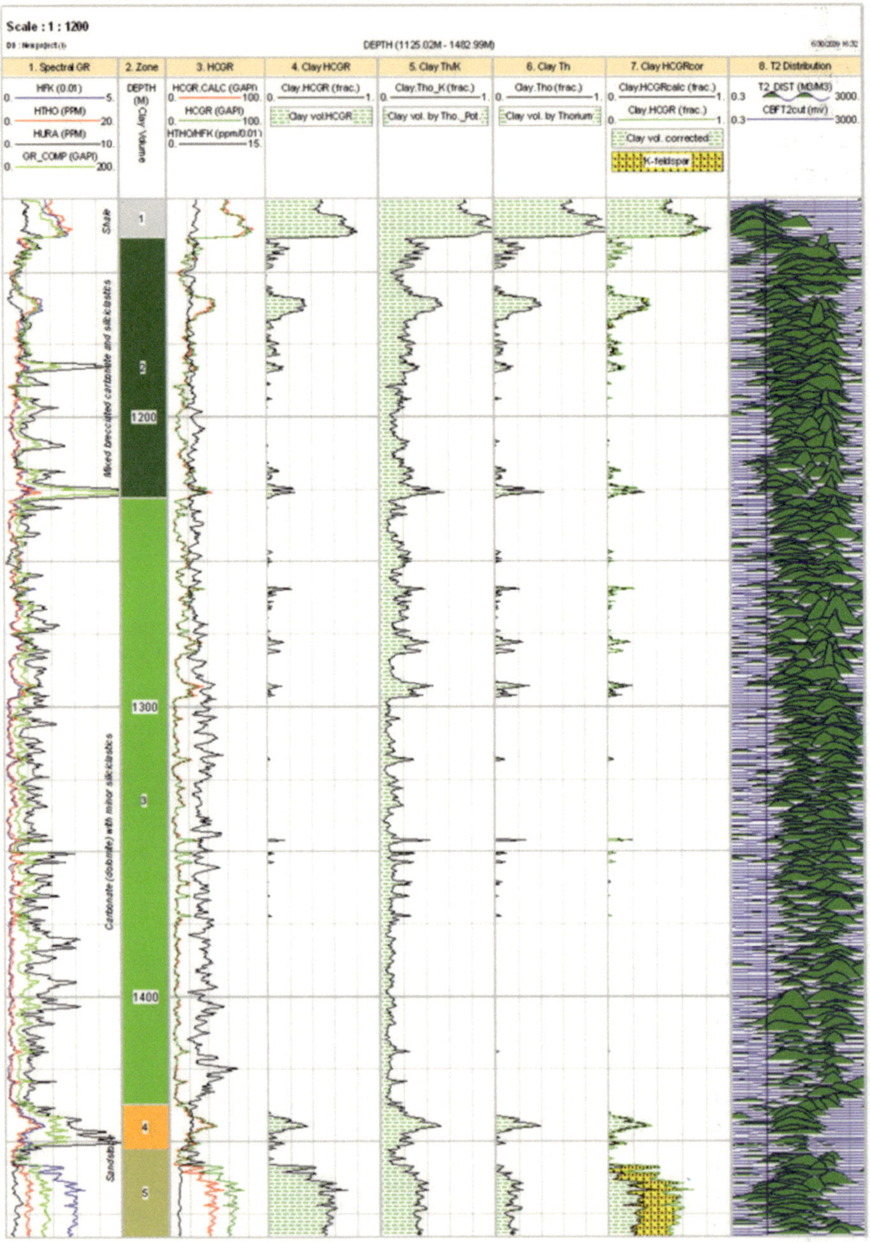

Fig. 7. Shale volume calculation by using HCGR, thorium-potassium ratio, thorium, and integration of SEM, thorium, and potassium.

5. Summary

Although the gamma-ray log has traditionally been used for the analysis of shaly formations, the shale volume estimation from this measurement is, to a greater or lesser extent, inaccurate. The GR log responds to the natural-gamma radiation from the formation.

The spectral gamma-ray log works on the same principle, except that the gamma rays are assigned to three different energy bins, showing the concentrations of K, U, and Th in the formation. Generally, carbonate rocks show higher uranium measurements on the spectral gamma-ray log because of the presence of organic matter. Uranium is, therefore, removed prior to the use in clay volume calculation.The sources of potassium also have to be distinguished because potassium is present in shaly and non-shaly rocks. In this study, the NMR log is used as an easy tool to check the reliability of the shale volume calculation from gamma-ray and spectral gamma-ray logs. Inconsistency between the shale volumes estimated from these methods is significant in the lowermost reservoir section. To solve the problem, SEM analysis was done to identify the mineralogy and mineral volume fractions. Backscattered electron images and X-ray mapping of selected samples show a noticeable contribution of potassium feldspar in zone 5. The potassium feldspar content of the rock influences both the gamma-ray and spectral gamma-ray logs and thus the shale volume calculation. The spectral gamma-ray log of potassium was therefore corrected on the basis of SEM information. Subsequently, the SEM information was used as a reference point for the calibration of the spectral gamma-ray log to estimate the shale volume. The uniformity of the spectral gamma-ray logs within zone 5 used for calibration over the entire zone 5. The NMR log was also used to verify the calculation to some extent. The volume of potassium feldspar was then removed from the total potassium reading in the spectral gamma-ray log to obtain the actual shale volume of the rock. The results show a considerable reduction in the estimated shale volume of the reservoir rock. Overestimation of shale volume has a direct impact on the reservoir evaluation by causing underestimation of the effective porosity and, consequently, hydrocarbon volume.

6. Acknowledgements

Authors thank Norwegian University of Science and Technology and Statoil for support and access to petrophysical data, and Stephen Lippard for his advice on the manuscript.

7. Nomenclature

Name	Description (unit)
BS	Bit size (inch)
CBW	Clay-bound water (decimal fraction)
Depth	Depth from rotary table (m)
Dt	Sonic transit time (μs/ft-1)
HCAL	Caliper log (inch)
HCGR	Compensated gamma-ray (API)
HCGR_SHALE	Shale volume from HCGR log (decimal fraction)
HFK	Spectral gamma-ray log potassium (%)
HTHO	Spectral gamma-ray log thorium (ppm)
HTHO_SHALE	Shale volume from thorium log (decimal fraction)
HURA	Spectral gamma-ray log uranium (ppm)
NMR_SHALE	Shale volume from NMR log (decimal fraction)
nmrT2cutoff	T2 cut-off (ms)
NPHIC	Neutron log (corrected) (decimal fraction)

PHIE	Effective porosity (decimal fraction)
PHIT	Total porosity (decimal fraction)
RHOC	Density log (corrected) (g cm-3)
Vanhydrite	Anhydrite volume (decimal fraction)
VDol	Dolomite volume (decimal fraction)
VLime	Limestone volume (decimal fraction)
VSand	Sandstone volume (decimal fraction)
VWL	Wey clay volume (decimal fraction)

8. References

Chitale, D.V. Day, P. I. and Coates, G. R. [1999] Petrophysical implications of laboratory NMR and petrographical investigation on a shaly sand core. SPE Annual Technical Conference and Exhibition, Houston, Texas.

Coates, G.R. Xiao, L. and Prammer, M.G. [1999] NMR logging principles and applications. Halliburton Energy Services Publication H02308, USA, 253 pp.

Darling, T. [2005] Well logging and formation evaluation. Gulf professional publishing, Oxford OX2 8DP, UK, 302 pp.

Dunn, K. J. Bergman, D. J. LaTorraca, G. A. Stonard, S. W. and Crowe, M. B. [1998] A method for inverting NMR data sets with different signal to noise ratios. 39th Annual Logging Symposium of SPWLA, Keystone, USA.

Ellis, D, V. and Singer, J. M. [2008] Well logging for earth scientists. Springer, 2nd edition Netherlands, 698 pp.

Guggenheim, S. and Martin RT [1995] Definition of clay minerals. Joint report of the AIPEA nomenclature and CMS nomenclature committees. Clays Clay Miner 43:255-256.

Luczaj, J. A. and Goldstein R. H. [2000] Diagenesis of the lower Permian Krider Member, southwest Kansas, U.S.A.: Fluid-inclusion, U-Pb, and fission-track evidence for reflux dolomitization during latest Permian time. Journal of Sedimentary Research, 70, 762-773.

Matteson, A. Tomanic, J.P. Herron, M.M. Allen, D.F. and Kenyon, W.E. [2000] NMR relaxation of clay/brine mixtures. SPE Reservoir Eval. & Eng. 3 (5), 408-413.

Mohammadlou, M., Mork M. B., Langeland, H., [2010] Quantification of shale volume from borehole logs calibrated by SEM analysis: a case study. First break, 28, 21-29.

Reed, S. J. B. [2005] Electron microscope analysis and scanning electron microscopy in geology. Cambridge University Press, UK, 2nd Edition, 232 pp.

Ruppel, S. C. [1992] Styles of deposition and diagenesis in Leonardian carbonate reservoirs in West Texas. Annual exhibition and technical conference, SPE 24691.

Schlumberger [1985] Log interpretation charts. Schlumberger, New York, USA, 207 pp.

Schlumberger [1989] Schlumberger log principles and applications. Schlumberger, Wireline & Testing, Texas, USA, 230 pp.

Straley, C. et al. [1994] Core Analysis by Low Field NMR. SCA-9404 presented at the 1994 Intl. Symposium of the Soc. of Core Analysts, Stavanger, Norway.

Prammer, M.G. et al. [1996] Measurements of clay-bound water and total porosity by magnetic resonance. The Log Analyst. 37(6), 61.

Tiab, D. and Donaldson E. C. [2004] Petrophysics, theory and practice of measuring reservoir rock and fluid transport properties. Gulf professional publishing, 2nd edition, Oxford OX2 8DP, UK, 926 pp.